PROPERTIES OF ABSOLUTE VALUE

$|a| \geq 0$

$|a| = |-a|$

$|a - b| = |b - a|$

$|a^2| = |a|^2 = a^2$

MULTIPLICATION PATTERNS

$(a + b)^2 = a^2 + 2ab + b^2$

$(a - b)^2 = a^2 - 2ab + b^2$

$(a + b)(a - b) = a^2 - b^2$

$(a + b)^3 = a^3 + 3a^2b + 3ab^2 + b^3$

$(a - b)^3 = a^3 - 3a^2b + 3ab^2 - b^3$

$(a + b)^n = \binom{n}{0}a^n + \binom{n}{1}a^{n-1}b + \binom{n}{2}a^{n-2}b^2 + \cdots + \binom{n}{n}b^n$

PROPERTIES OF EXPONENTS AND RADICALS

$b^n \cdot b^m = b^{n+m}$

$(b^n)^m = b^{mn}$

$(ab)^n = a^nb^n$

$\left(\frac{a}{b}\right)^n = \frac{a^n}{b^n}$

$\frac{b^n}{b^m} = b^{n-m}$

$\sqrt[n]{ab} = \sqrt[n]{a}\sqrt[n]{b}$

$\sqrt[n]{\frac{a}{b}} = \frac{\sqrt[n]{a}}{\sqrt[n]{b}}$

EQUATIONS DETERMINING FUNCTIONS

Linear function:	$f(x) = ax + b$
Quadratic function:	$f(x) = ax^2 + bx + c$
Polynomial function:	$f(x) = a_nx^n + a_{n-1}x^{n-1} + \cdots + a_1x + a_0$
Rational function:	$f(x) = \frac{g(x)}{h(x)}$, where g and h are polynomial functions
Exponential function:	$f(x) = b^x$, where $b > 0$ and $b \neq 1$
Logarithmic function:	$f(x) = \log_b x$, where $b > 0$ and $b \neq 1$

INTERVAL NOTATION / SET NOTATION

INTERVAL NOTATION	SET NOTATION
(a, ∞)	$\{x \mid x > a\}$
$(-\infty, b)$	$\{x \mid x < b\}$
(a, b)	$\{x \mid a < x < b\}$
$[a, \infty)$	$\{x \mid x \geq a\}$
$(-\infty, b]$	$\{x \mid x \leq b\}$
$(a, \mathrm{b}]$	$\{x \mid a < x \leq b\}$
$[a, b)$	$\{x \mid a \leq x < b\}$
$[\mathrm{a}, b]$	$\{x \mid a \leq x \leq b\}$

PROPERTIES OF LOGARITHMS

$\log_b b = 1$

$\log_b 1 = 0$

$\log_b rs = \log_b r + \log_b s$

$\log_b\left(\frac{r}{s}\right) = \log_b r - \log_b s$

$\log_b r^p = p(\log_b r)$

FACTORING PATTERNS

$a^2 - b^2 = (a + b)(a - b)$

$a^3 - b^3 = (a - b)(a^2 + ab + b^2)$

$a^3 + b^3 = (a + b)(a^2 - ab + b^2)$

College Algebra and Trigonometry

THE PRINDLE, WEBER & SCHMIDT SERIES IN DEVELOPMENTAL, BUSINESS, AND TECHNICAL MATHEMATICS

Baley/Holstege, *Algebra: A First Course, Third Edition*
Cass/O'Connor, *Beginning Algebra*
Cass/O'Connor, *Beginning Algebra with Fundamentals*
Cass/O'Connor, *Fundamentals with Elements of Algebra, Second Edition*
Ewen/Nelson, *Elementary Technical Mathematics, Fifth Edition*
Geltner/Peterson, *Geometry for College Students, Second Edition*
Hyde, *College Algebra and Trigonometry Explorations for the TI-81 and TI-82 Graphics Calculators*
Johnston/Willis/Hughes, *Developmental Mathematics, Third Edition*
Johnston/Willis/Lazaris, *Essential Algebra, Sixth Edition*
Johnston/Willis/Lazaris, *Essential Arithmetic, Sixth Edition*
Johnston/Willis/Lazaris, *Intermediate Algebra, Fifth Edition*
Jordan/Palow, *Integrated Arithmetic and Algebra*
Kennedy/Green, *Prealgebra for College Students*
Lee, *Self-Paced Business Mathematics, Fourth Edition*
McCready, *Business Mathematics, Sixth Edition*
McKeague, *Basic Mathematics, Third Edition*
McKeague, *Introductory Mathematics*
McKeague, *Prealgebra, Second Edition*
Mangan, *Arithmetic for Self-Study, Second Edition*
Perez/Weltman/Byrum/Polito, *Beginning Algebra: A Worktext*
Perez/Weltman/Byrum/Polito, *Intermediate Algebra: A Worktext*
Pierce/Tebeaux, *Operational Mathematics for Business, Second Edition*
Proga, *Arithmetic and Algebra, Third Edition*
Proga, *Basic Mathematics, Third Edition*
Rogers/Haney/Laird, *Fundamentals of Business Mathematics*
Szymanski, *Algebra Facts*
Szymanski, *Math Facts: Survival Guide to Basic Mathematics*
Wood/Capell/Hall, *Developmental Mathematics, Fourth Edition*

THE PRINDLE, WEBER & SCHMIDT SERIES IN PRECALCULUS, LIBERAL ARTS MATHEMATICS, AND TEACHER TRAINING

Barnett, *Analytic Trigonometry with Applications, Fifth Edition*
Bean/Sharp/Sharp, *Precalculus*
Boye/Kavanaugh/Williams, *Elementary Algebra*
Boye/Kavanaugh/Williams, *Intermediate Algebra*
Davis/Moran/Murphy, *Precalculus in Context: Functioning in the Real World*
Drooyan/Franklin, *Intermediate Algebra, Seventh Edition*
Franklin/Drooyan, *Modeling, Functions and Graphs*
Fraser, *Elementary Algebra*
Fraser, *Intermediate Algebra: An Early Functions Approach*
Gantner/Gantner, *Trigonometry*
Gobran, *Beginning Algebra, Fifth Edition*
Grady/Drooyan/Beckenbach, *College Algebra, Eighth Edition*
Hall, *Beginning Algebra*
Hall, *Intermediate Algebra*
Hall, *Algebra for College Students, Second Edition*
Hall, *College Algebra with Applications, Third Edition*
Holder, *A Primer For Calculus, Sixth Edition*
Huff/Peterson, *College Algebra Activities for the TI-81 Graphics Calculator*
Johnson/Mowry, *Mathematics: A Practical Odyssey*
Kaufmann, *Elementary Algebra for College Students, Fourth Edition*
Kaufmann, *Intermediate Algebra for College Students, Fourth Edition*
Kaufmann, *Elementary and Intermediate Algebra: A Combined Approach*
Kaufmann, *Algebra for College Students, Fourth Edition*
Kaufmann, *Algebra with Trigonometry for College Students, Third Edition*
Kaufmann, *College Algebra, Third Edition*
Kaufmann, *Trigonometry, Second Edition*
Kaufmann, *College Algebra and Trigonometry, Third Edition*
Kaufmann, *Precalculus, Second Edition*
Laughbaum, *Intermediate Algebra; A Graphical Approached, Revised Preliminary Edition*
Lavoie, *Discovering Mathematics*
McCown/Sequeira, *Patterns in Mathematics: Problem Solving from Counting to Chaos*
Rice/Strange, *Plane Trigonometry, Sixth Edition*
Riddle, *Analytic Geometry, Fifth Edition*
Ruud/Shell, *Prelude to Calculus, Second Edition*
Sgroi/Sgroi, *Mathematics for Elementary School Teachers*
Swokowski/Cole, *Fundamentals of College Algebra, Eighth Edition*
Swokowski/Cole, *Fundamentals of Algebra and Trigonometry, Eighth Edition*
Swokowski/Cole, *Fundamentals of Trigonometry, Eighth Edition*
Swokowski/Cole, *Algebra and Trigonometry with Analytic Geometry*, Eighth Edition
Swokowski/Cole, *Precalculus: Functions and Graphs, Seventh Edition*

Weltman/Perez, *Beginning Algebra, Second Edition*
Weltman/Perez, *Intermediate Algebra, Third Edition*

THE PRINDLE, WEBER & SCHMIDT SERIES IN CALCULUS AND UPPER-DIVISION MATHEMATICS

Althoen/Bumcrot, *Introduction to Discrete Mathematics*
Andrilli/Hecker, *Linear Algebra*
Burden/Faires, *Numerical Analysis, Fifth Edition*
Crooke/Ratcliffe, *A Guidebook to Calculus with Mathematica*
Cullen, *An Introduction to Numerical Linear Algebra*
Cullen, *Linear Algebra and Differential Equations, Second Edition*
Denton/Nasby, *Finite Mathematics, Preliminary Edition*
Dick/Patton, *Calculus*
Dick/Patton, *Single Variable Calculus*
Dick/Patton, *Technology in Calculus: A Sourcebook of Activities*
Edgar, *A First Course in Number Theory*
Eves, *In Mathematical Circles*
Eves, *Mathematical Circles Revisited*
Eves, *Mathematical Circles Squared*
Eves, *Return to Mathematical Circles*
Faires/Burden, *Numerical Methods*
Finizio/Ladas, *Introduction to Differential Equations*
Finizio/Ladas, *Ordinary Differential Equations with Modern Applications, Third Edition*
Fletcher/Hoyle/Patty, *Foundations of Discrete Mathematics*
Fletcher/Patty, *Foundations of Higher Mathematics, Second Edition*
Gilbert/Gilbert, *Elements of Modern Algebra, Third Edition*
Gordon, *Calculus and the Computer*
Hartfiel/Hobbs, *Elementary Linear Algebra*
Hartig, *Guidebook to Linear Algebra for Theorist*
Hill/Ellis/Lodi, *Calculus Illustrated*
Hillman/Alexanderson, *Abstract Algebra: A First Undergraduate Course, Fifth Edition*
Humi/Miller, *Boundary-Value Problems and Partial Differential Equations*
Laufer, *Discrete Mathematics and Applied Modern Algebra*
Leinbach, *Calculus Laboratories Using Derive*
Maron/Lopez, *Numerical Analysis, Third Edition*
Miech, *Calculus with Mathcad*
Mizrahi/Sullivan, *Calculus with Analytic Geometry, Third Edition*
Molluzzo/Buckley, *A First Course in Discrete Mathematics*
Nicholson, *Elementary Linear Algebra with Applications, Third Edition*
Nicholson, *Introduction to Abstract Algebra*
O'Neil, *Advanced Engineering Mathematics, Third Edition*
Pence, *Calculus Activities for Graphic Calculators*
Pence, *Calculus Activities for the TI Graphic Calculator, Second Edition*
Plybon, *An Introduction to Applied Numerical Analysis*
Powers, *Elementary Differential Equations, Brief Edition*
Powers, *Elementary Differential Equations with Boundary-Value Problems*
Prescience Corporation, *The Student Edition of Theorist*
Riddle, *Calculus and Analytic Geometry, Fourth Edition*
Schelin/Bange, *Mathematical Analysis for Business and Economics, Second Edition*
Sentilles, *Applying Calculus in Economics and Life Science*
Swokowski/Olinick/Pence, *Calculus, Sixth Edition*
Swokowski/Olinick/Pence, *Calculus of a Single Variable, Second Edition*
Swokowski, *Calculus, Fifth Edition (Late Trigonometry Version)*
Swokowski, *Elements of Calculus with Analytic Geometry: High School Edition*
Tan, *Applied Finite Mathematics, Fourth Edition*
Tan, *Calculus for the Managerial, Life, and Social Sciences, Third Edition*
Tan, *Applied Calculus, Third Edition*
Tan, *College Mathematics, Third Edition*
Trim, *Applied Partial Differential Equations*
Venit/Bishop, *Elementary Linear Algebra, Third Edition*
Venit/Bishop, *Elementary Linear Algebra, Alternate Second Edition*
Wattenberg, *Calculus in a Real and Complex World*
Wiggins, *Problem Solver for Finite Mathematics and Calculus*
Zill, *Calculus, Third Edition*
Zill, *A First Course in Differential Equations, Fifth Edition*
Zill/Cullen, *Differential Equations with Boundary-Value Problems, Third Edition*
Zill/Cullen, *Advanced Engineering Mathematics*

THE PRINDLE, WEBER & SCHMIDT SERIES IN ADVANCED MATHEMATICS

Ehrlich, *Fundamental Concepts of Abstract Algebra*
Judson, *Abstract Algebra: Theory and Applications*
Kirkwood, *An Introduction to Real Analysis*
Patty, *Foundations of Topology*
Ruckle, *Modern Analysis: Measure Theory and Functional Analysis with Applications*
Sieradski, *An Introduction to Topology and Homotopy*
Steinberger, *Algebra*
Strayer, *Elementary Number Theory*
Troutman, *Boundary-Value Problems of Applied Mathematics*

INSTRUCTOR'S EDITION

COLLEGE ALGEBRA AND TRIGONOMETRY

THIRD EDITION

Jerome E. Kaufmann

PWS PUBLISHING COMPANY
BOSTON

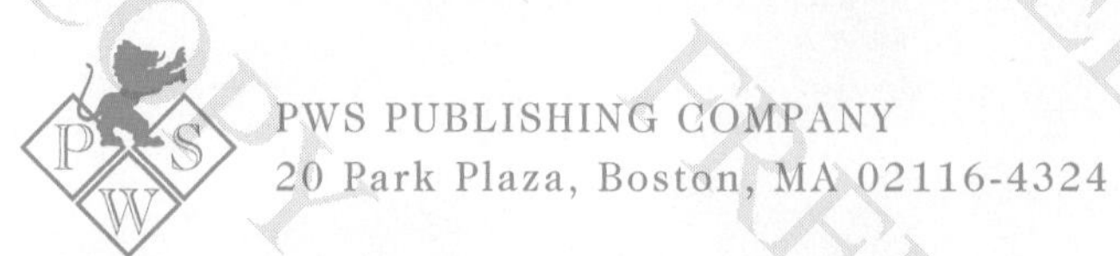

PWS PUBLISHING COMPANY
20 Park Plaza, Boston, MA 02116-4324

PWS Publishing Company is a division of Wadsworth, Inc.

International Thomson Publishing
The trademark ITP is used under license.

Library of Congress Cataloging-in-Publication Data
Kaufmann, Jerome E.
College algebra and trigonometry / Jerome E. Kaufmann.–3rd ed.
p. cm.
Includes index.
ISBN 0-534-93525-7
1. Algebra. 2. Trigonometry. I. Title.
QA154.2.K367 1993 93-6078
512′.13–dc20 CIP

Chapter Opening Photograph Credits
Chapter 1–Hank deLespinasse/The Image Bank (p. 1); *Chapter 2*–Flip Chalfant/The Image Bank (p. 81); *Chapter 3*–The Image Bank (p. 153); *Chapter 4*–Santi Visalli/The Image Bank (p. 217); *Chapter 5*–Louis H. Jawitz/The Image Bank (p. 293); *Chapter 6*–Robert Kristofik/The Image Bank (p. 347); *Chapter 7*–Michael Salas/The Image Bank (p. 413); *Chapter 8*–© 1990 Steve Krongard/The Image Bank (p. 481); *Chapter 9*–Joseph Van Os/The Image Bank (p. 537); *Chapter 10*–© 1988 Frank Whitney/The Image Bank (p. 593); *Chapter 11*–Harald Sund/The Image Bank (p. 645); *Chapter 12*–Joe McNally/The Image Bank (p. 683); *Chapter 13*–Murray Alcosser/The Image Bank (p. 725); *Chapter 14*–Bob Daemmrich/Stock Boston (p. 759).

Sponsoring Editor: Timothy L. Anderson
Production Coordinator: Robine Andrau
Manufacturing Coordinator: Lisa Flanagan
Assistant Editor: Kelle Flannery
Editorial Assistant: Clark Benbow
Marketing Manager: Marianne C. P. Rutter
Production: Susan Graham
Interior Designer: Julia Gecha
Interior Illustrator: Network Graphics
Cover Designer: Elise S. Kaiser
Cover Photo: J. Carmichael, Jr./The Image Bank
Cover Printer: John Pow Company
Typesetter: Jonathan Peck Typographers
Printer and Binder: Arcata Graphics/Hawkins

Printed and bound in the United States of America
93 94 95 96 97 98 – 10 9 8 7 6 5 4 3 2 1

CONTENTS

3 Coordinate Geometry and Graphing Techniques 153

4 Functions 217

5 Exponential and Logarithmic Functions 293

6 Polynomial and Rational Functions 347

PREFACE

College Algebra and Trigonometry, Third Edition is written for students who need a college algebra and trigonometry course to serve as a prerequisite for the standard calculus sequence or to satisfy a liberal arts requirement.

Four major ideas serve as this text's unifying themes–namely, solving equations and inequalities, solving problems, developing graphing techniques, and understanding the concept of a function.

College Algebra amd Trigonometry, Third Edition presents basic concepts of algebra in a simple, straightforward way. A wealth of examples motivates students and reinforces algebraic concepts by the application of these examples to situations with which students can identify. The examples also guide students in organizing their work logically and using meaningful shortcuts whenever appropriate.

In the preparation of the third edition, special effort was made to incorporate improvements suggested by reviewers and by users of the earlier editions, while at the same time preserving the book's many successful features.

New in This Edition

- *Graphics Calculator Examples* incorporated throughout the text enable students to experience the power of this new technology in demonstrating how to visualize solutions to problems.
- *Graphics Calculator Activities*, a new category of problems, were added to give students practice with graphics calculators. About half of the problem sets contain these problems, a total of about 500 new exercises, which are flagged with a graphing utility icon. These exercises were designed to reinforce concepts already presented and lay the groundwork for concepts about to be discussed. They also help students to predict shapes and locations of graphs based on earlier graphing experience and to solve problems that are best expressed graphically with the aid of the graphics calculator. Through working these problems, students become more familiar with the capabilities and limitations of the graphics calculator. See, for example, Problem Sets 3.2, 4.2, 4.3, 5.1, 6.3, 8.1, 8.3, and 9.1.

- *Thoughts into Words*, another new category of problems, are designed to encourage students to express in written form their thoughts about various mathematical ideas. For example, see Problem Sets 1.2, 1.4, 2.1, 2.7, 6.1, 8.1, and 10.5.
- A new section in Chapter 4 introduces transformations (translations, reflections, stretchings, and shrinkings) of some basic curves. These ideas are then used throughout the text when various functions are graphed.
- A more up-to-date coverage of exponents and logarithms in Chapter 5 underscores the important role of the calculator; tables were placed in the appendixes.
- Problem sets were very carefully revised to achieve an even better balance in the level of difficulty.
- Many examples were enhanced or changed to better illustrate the concept under discussion.
- Chapter summaries were revised for maximum use by the student.

Other Special Features

- *Student "Self-Review"* in Chapter 1, a review of intermediate algebra concepts, was written so that students can work through this material with a minimum of assistance from the instructor.
- *Miscellaneous Problems*, an optional set of exercises in many of the problem sets, encompass a variety of ideas: Some are proofs, some exhibit different approaches to topics covered in the text, some bring in supplementary topics and relationships, and some are more challenging problems. Note that, although these problems add variety and flexibility to the exercise sets, they can be omitted entirely without disrupting the continuity of the text.
- *Problem Sets* were constructed on an even/odd basis; that is, all variations of skill-development exercises are contained in the even- and odd-numbered problems.
- *Review Problem Sets* appear at the end of each chapter and are designed to help students pull together all of the ideas presented in the chapter.
- *Cumulative Review Problem Sets* appear at the ends of Chapters 3, 6, and 9.

Additional Comments About Some of the Chapters

- With problem solving as its focus, Chapter 2 pulls together and expands on a variety of approaches to the process of solving equations and inequalities. Polya's four-phase plan is used as a basis for developing different problem-solving strategies.
- Chapter 3 has as its premise the need for more work with concepts of coordinate geometry–specifically graphing techniques–*before* the notion

of a function is introduced. In this chapter, varying the coefficients of the equation $Ax^2 + By^2 = F$ results in various ellipses and hyperbolas (centers at the origin). In Chapter 12 the standard approach of developing basic forms from the definitions is used.

- The concept of a function is introduced in Chapter 4, unclouded by the need to jump back and forth between functions and relations that are not functions.
- Chapter 5 underscores the importance of understanding the concept of an exponent and a logarithm, along with relevant applications.
- The trigonometry material (Chapters 7, 8, and 9) is organized around three central themes: problem solving (in Chapter 7), graphing (in Chapter 8), and solving equations (in Chapter 9). After using angles to introduce the basic trigonometric functions in Chapter 7, we use functions to solve a variety of interesting problems involving right triangles and oblique triangles. The short introduction to vectors in Section 7.7 also focuses on solving problems. Chapter 8 presents variations of all six basic trigonometric curves in a carefully organized manner consistent with the graphing discussions in previous chapters. Section 8.5 introduces the polar coordinate system and establishes some groundwork for graphing polar equations in the next section. Section 8.6 is devoted to the graphing of polar equations and includes some work with symmetry tests. Chapter 9 is centered around solving trigonometric equations and verifying trigonometric identities.
- Chapters 10 and 11 provide the instructor with some flexibility as to choice of topics. Chapter 10 is devoted entirely to solving systems of linear equations, including the use of matrices and determinants. In Chapter 11 the algebra of matrices is the focal point.
- Problem solving is the unifying theme of Chapters 13 and 14. In contrast to many college algebra and trigonometry books, Chapter 14 contains a significant amount of material on probability.

Supplements for Instructors

The following supplements are available to adopters of this text:

- An *Instructor's Edition* of the textbook provides answers to all of the exercises.
- An *Instructor's Solutions Manual* offers detailed solutions to most of the text's exercises.
- *Test Bank with Chapter Tests* may be photocopied by instructors and used by them to test their students. Answers to test questions are provided at the back of the book for the instructor only. There are two multiple-choice and one short-answer test for each text chapter.
- *EXPTest*, a computerized test bank for IBM PCs and compatibles, contains hundreds of problems. Questions are multiple choice, true/false, and open-

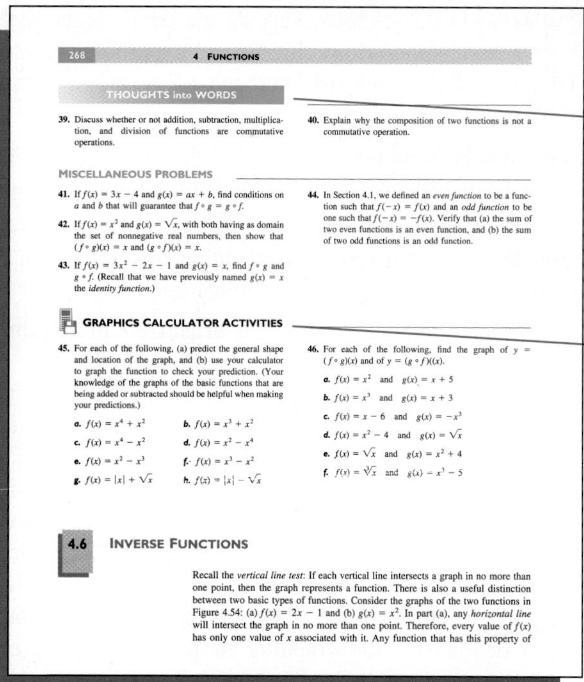

268 4 FUNCTIONS

THOUGHTS into WORDS

39. Discuss whether or not addition, subtraction, multiplication, and division of functions are commutative operations.

40. Explain why the composition of two functions is not a commutative operation.

MISCELLANEOUS PROBLEMS

41. If $f(x) = 3x - 4$ and $g(x) = ax + b$, find conditions on a and b that will guarantee that $f \circ g = g \circ f$.

42. If $f(x) = x^2$ and $g(x) = \sqrt{x}$, with both having as domain the set of nonnegative real numbers, then show that $(f \circ g)(x) = x$ and $(g \circ f)(x) = x$.

43. If $f(x) = 3x^2 - 2x - 1$ and $g(x) = x$, find $f \circ g$ and $g \circ f$. (Recall that we have previously named $g(x) = x$ the *identity function*.)

44. In Section 4.1, we defined an *even function* to be a function such that $f(-x) = f(x)$ and an *odd function* to be one such that $f(-x) = -f(x)$. Verify that (a) the sum of two even functions is an even function, and (b) the sum of two odd functions is an odd function.

GRAPHICS CALCULATOR ACTIVITIES

45. For each of the following, (a) predict the general shape and location of the graph, and (b) use your calculator to graph the function to check your prediction. (Your knowledge of the graphs of the basic functions that are being added or subtracted should be helpful when making your predictions.)

a. $f(x) = x^4 + x^2$ b. $f(x) = x^3 + x^2$
c. $f(x) = x^4 - x^2$ d. $f(x) = x^2 - x^4$
e. $f(x) = x^2 - x^3$ f. $f(x) = x^3 - x^2$
g. $f(x) = |x| + \sqrt{x}$ h. $f(x) = |x| - \sqrt{x}$

46. For each of the following, find the graph of $y = (f \circ g)(x)$ and of $y = (g \circ f)((x)$.

a. $f(x) = x^2$ and $g(x) = x + 5$
b. $f(x) = x^3$ and $g(x) = x + 3$
c. $f(x) = x - 6$ and $g(x) = -x^3$
d. $f(x) = x^2 - 4$ and $g(x) = \sqrt{x}$
e. $f(x) = \sqrt{x}$ and $g(x) = x^2 + 4$
f. $f(x) = \sqrt[3]{x}$ and $g(x) - x^3 - 5$

4.6 INVERSE FUNCTIONS

Recall the *vertical line test*: If each vertical line intersects a graph in no more than one point, then the graph represents a function. There is also a useful distinction between two basic types of functions. Consider the graphs of the two functions in Figure 4.54: (a) $f(x) = 2x - 1$ and (b) $g(x) = x^2$. In part (a), any *horizontal line* will intersect the graph in no more than one point. Therefore, every value of $f(x)$ has only one value of x associated with it. Any function that has this property of

Thoughts into Words encourage students to express in written form their thoughts about various mathematical ideas.

Graphics Calculator Activities enable students with a graphics calculator to predict shapes and locations of graphs and solve problems that would be difficult or impossible to graph manually.

A new section in Chapter 4 now introduces transformations of some basic curves, so that these concepts can be developed later in the text.

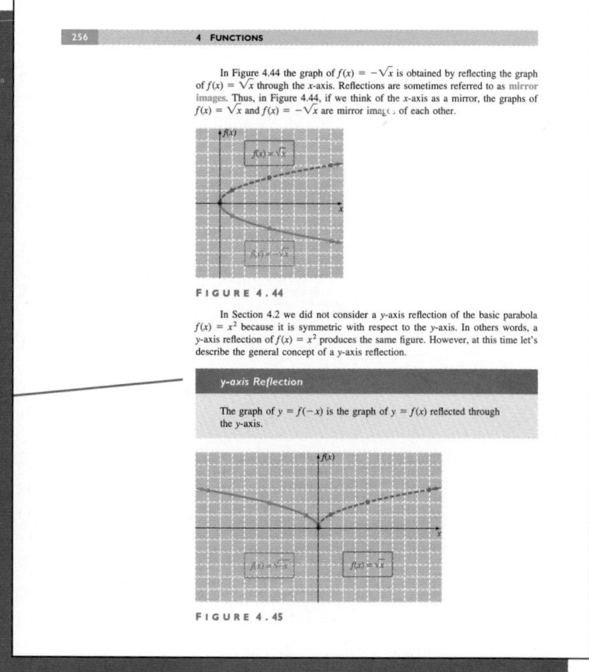

256 4 FUNCTIONS

In Figure 4.44 the graph of $f(x) = -\sqrt{x}$ is obtained by reflecting the graph of $f(x) = \sqrt{x}$ through the x-axis. Reflections are sometimes referred to as mirror images. Thus, in Figure 4.44, if we think of the x-axis as a mirror, the graphs of $f(x) = \sqrt{x}$ and $f(x) = -\sqrt{x}$ are mirror images of each other.

FIGURE 4.44

In Section 4.2 we did not consider a y-axis reflection of the basic parabola $f(x) = x^2$ because it is symmetric with respect to the y-axis. In others words, a y-axis reflection of $f(x) = x^2$ produces the same figure. However, at this time let's describe the general concept of a y-axis reflection.

y-axis Reflection

The graph of $y = f(-x)$ is the graph of $y = f(x)$ reflected through the y-axis.

FIGURE 4.45

y-axis Reflection

Miscellaneous Problems add variety to the exercise sets but can be omitted without disrupting the continuity of the text.

4.3 Quadratic Functions and Problem Solving 249

40. A motel advertises that they will provide dinner, dancing, and drinks for \$50 per couple for a New Year's Eve party. They must have a guarantee of 30 couples. Furthermore, they will agree that for each couple in excess of 30, they will reduce the price per couple for all attending by \$.50. How many couples will it take to maximize the motel's revenue?

41. A cable TV company has 1000 subscribers who each pay \$15 per month. Based on a survey, they feel that for each decrease of \$.25 on the monthly rate, they could obtain 20 additional subscribers. At what rate will maximum revenue be obtained and how many subscribers will it take at that rate?

42. A manufacturer finds that for the first 500 units of his product that are produced and sold, the profit is \$50 per unit. The profit on each of the units beyond 500 is decreased by \$.10 times the number of additional units sold. What level of output will maximize profit?

MISCELLANEOUS PROBLEMS

43. Suppose that an arch is shaped like a parabola. It is 20 feet wide at the base and 100 feet high. How wide is the arch 50 feet above the ground? (See Figure 4.37.)

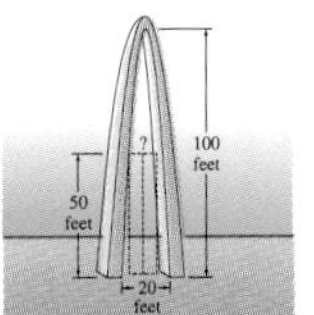

FIGURE 4.37

44. A parabolic arch 27 feet high spans a parkway. If the center section of the parkway is 50 feet wide, how wide is the arch if it has a minimum clearance of 15 feet above the center section?

45. A parabolic arch spans a stream 200 feet wide. How high must the arch be above the stream to give a minimum clearance of 40 feet over a 120-foot-wide channel in the center?

GRAPHICS CALCULATOR ACTIVITIES

46. Suppose that the viewing window on your graphics calculator is set so that $-15 \le x \le 15$ and $-10 \le y \le 10$. Now try to graph the function $f(x) = x^2 - 8x + 28$. Since nothing appears on the screen, the parabola must be outside the viewing window. We could arbitrarily expand the window until the parabola appears. However, let's be a little more systematic and use $\left(-\frac{b}{2a}, f\left(\frac{-b}{2a}\right)\right)$ to find the vertex; thus we find the vertex is at (4, 12). So let's change the y-values of the window so that $0 \le y \le 25$. Now we get a good picture of the parabola.

Graph each of the following parabolas, and keep in mind that you may need to change the dimensions of the viewing window to obtain a good picture.

a. $f(x) = x^2 - 2x + 12$

b. $f(x) = -x^2 - 4x - 16$

c. $f(x) = x^2 + 12x + 44$

4.4 Transformations of Some Basic Curves 257

Now suppose that we want to do a y-axis reflection of $f(x) = \sqrt{x}$. Since $f(x) = \sqrt{x}$ is defined for $x \ge 0$, the y-axis reflection $f(x) = \sqrt{-x}$ is defined for $-x \ge 0$, which is equivalent to $x \le 0$. Figure 4.45 shows the y-axis reflection of $f(x) = \sqrt{x}$.

Vertical Stretching and Shrinking

Translations and reflections are called rigid transformations because the basic shape of the curve being transformed is not changed. In other words, only the positions of the graphs are changed. Now we want to consider some transformations that distort the shape of the original figure somewhat.

In Section 4.2 we graphed the equation $y = 2x^2$ by doubling the y-coordinates of the ordered pairs that satisfy the equation $y = x^2$. We obtained a parabola with its vertex at the origin, symmetric to the y-axis, but *narrower* than the basic parabola. Likewise, we graphed the equation $y = \frac{1}{2}x^2$ by halving the y-coordinates of the ordered pairs that satisfy $y = x^2$. In this case, we obtained a parabola with its vertex at the origin, symmetric to the y-axis, but *wider* than the basic parabola.

The concepts of *narrower* and *wider* can be used to describe parabolas but cannot be used to accurately describe some other curves. Instead, we use the more general concepts of vertical stretching and shrinking.

Vertical Stretching and Shrinking

The graph of $y = cf(x)$ is obtained from the graph of $y = f(x)$ by mulitplying the y-coordinates of $y = f(x)$ by c. If $c > 1$, the graph is said to be *stretched* by a factor of c, and if $0 < c < 1$, the graph is said to be *shrunk* by a factor of c.

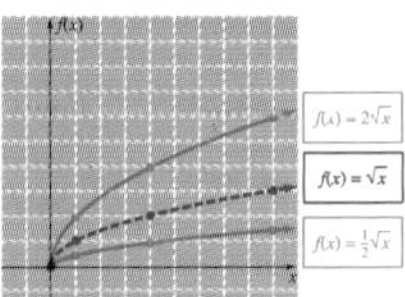

FIGURE 4.46

Vertical Stretching and Shrinking

ended. Intructors can interact with the program by adding to existing questions and producing individual tests. A demonstration disk is available.

- *ExamBuilder*, a computerized test bank for the Macintosh, has features and questions similar to those of EXPTest. A demonstration disk is available.
- *Transparencies*: 33 full-color acetates provide enlarged versions of illustrations similar to those found in the text.
- *A Quick Reference Card*: Packaged with this edition of the text is a new problem-solving tool–a formula card. This perforated card, found in the back of the book, will help students to master key formulas, equations, and graphs in the course. By serving as a quick reference and minimizing the need for page turning, the formula card reduces the time spent on tedious tasks so that students can focus on the central concepts and principles of the course.

Supplements for Students

- A *Students' Solutions Manual* provides detailed solutions to about one-fourth of the text's exercises.
- *Investigate*, tutorial software for the Macintosh and IBM PCs and compatibles, helps students review college algebra as needed. Questions are presented to the students with full mathematical notation and graphs. When students give incorrect answers, they are stepped through explanations of the problems to give immediate feedback and to correct misunderstandings. The program comes with a pop-up calculator, which instructors may disable at their discretion. *Investigate* may be set up to record individual students' grades and is fully network compatible. Operation of both the MS-DOS and Macintosh versions is identical, allowing easy training and use in labs having both types of computers. Demonstration disks are available for instructors.
- *College Algebra and Trigonometry Videotapes* are available to qualified adopters. Through the departmental or college audiovisual library, students can check out these videos and use them to review material when they need additional help.
- *Precalculus in Context: Functioning in the Real World* by Davis, Moran, and Murphy is a lab manual consisting of twelve projects that encourages students to explore precalculus concepts. Graphics calculators or computer graphing software are required to solve each experiment and its corresponding exercises.
- *Student Edition of Theorist* is software for the Macintosh that combines powerful algebra and graphics capabilities with an intuitive, user-friendly interface. Once they purchase it, students will be able to make use of this software for this course as well as for future mathematics courses.

- *College Algebra and Trigonometry Explorations for the TI-81 and TI-82 Graphics Calculators* by Nancy Hyde, Broward Community College, is designed to complement the Kaufmann precalculus series of texts. The purpose of these explorations is to guide the student into discovering or reinforcing important mathematical concepts through visualization, computation, and programming. This book provides (1) an introduction to the TI-81 and TI-82 graphics calculators; (2) examples to be used in the classroom, in small groups, or by the student working independently; and (3) corresponding exercises. Each college algebra or trigonometry activity in the book includes a brief explanation, relevant graphics calculator techniques, and examples that show the graphical interpretation of the concept. The author demonstrates how these concepts can be explored graphically or how the graphical representation reinforces a concept.

Acknowledgments

I would like to take this opportunity to thank the following people who served as reviewers for this text:

Charles A. Demetriou
State Technical Institute at Memphis

Sheldon M. Eisenberg
University of Hartford

William L. Hoard
Front Range Community College–Larimer Campus

J. Chris Neve
Front Range Community College

Patsy Norwood Newman
Richard Bland College

Robert A. Powers
Front Range Community College–Larimer Campus

Eric Schulz
Walla Walla Community College

Richard D. Semmler
Northern Virginia Community College

Debbye Stapleton
Georgia Southern University

Eleanor Storey
Front Range Community College

I am very grateful to the staff of PWS, especially Timothy Anderson, Kelle Karshick, and Clark Benbow for their continuous cooperation and assistance throughout this project. I would also like to express my sincere gratitude to Robine Andrau and Susan Graham. They continue to make my life as an author so much easier by carrying out the details of production in a dedicated and caring way. My thanks go out to Joan and Stuart Thomas for all of their hard work on the creation and programming of questions for the computerized test banks and chapter tests.

Again, very special thanks are due to my wife, Arlene, who spends numerous hours typing and proofreading manuscripts, answer keys, and solutions manuals.

Jerome E. Kaufmann
Marble Falls, Texas

Some Basic Concepts of Algebra

Algebra is often described as generalized arithmetic. That description may not tell the whole story, but it does indicate an important idea, namely, that a good understanding of arithmetic provides a sound basis for the study of algebra. Furthermore, a good understanding of some basic algebra concepts provides an even better basis for the study of more advanced algebraic ideas. Be sure that you can work effectively with the algebraic concepts reviewed in this first chapter.

1

The general nature of algebra makes it applicable to a large variety of occupations.

1.1 SOME BASIC IDEAS

Let's begin by pulling together some basic ideas that are needed in the study of algebra. In arithmetic, symbols such as 6, $\frac{2}{3}$, 0.27, and π are used to represent numbers. The operations of addition, subtraction, multiplication, and division are commonly indicated by the symbols $+$, $-$, $\times$, and $\div$, respectively. These symbols enable us to form specific **numerical expressions**. For example, the indicated sum of six and eight can be written $6 + 8$.

In algebra, the concept of a variable provides the basis for generalizing arithmetic ideas. For example, by using x and y to represent *any* two numbers, the expression $x + y$ can be used to represent the indicated sum of *any* two numbers. The x and y in such an expression are called **variables** and the phrase $x + y$ is called an **algebraic expression**.

Many of the notational agreements made in arithmetic are extended to algebra, with a few modifications. The following chart summarizes those notational agreements regarding the four basic operations.

OPERATION	ARITHMETIC	ALGEBRA	VOCABULARY
Addition	$4 + 6$	$x + y$	The sum of x and y
Subtraction	$14 - 10$	$a - b$	The difference of a and b
Multiplication	7×5 or $7 \cdot 5$	$a \cdot b$, $a(b)$, $(a)b$, $(a)(b)$, or ab	The product of a and b
Division	$8 \div 4$, $\frac{8}{4}$, $8/4$, or $4\overline{)8}$	$x \div y$, x/y, $\frac{x}{y}$, or $y\overline{)x}$ $(y \neq 0)$	The quotient of x divided by y

Note the different ways of indicating a product, including the use of parentheses. The ab form is the simplest and probably the most widely used form. Expressions such as abc, $6xy$, and $14xyz$ all indicate multiplication. Notice the various forms used to indicate division. In algebra, the fractional forms $\frac{x}{y}$ or x/y are usually used, although the other forms do serve a purpose at times.

The Use of Sets

Some of the vocabulary and symbolism associated with the concept of sets can be effectively used in the study of algebra. A **set** is a collection of objects; the objects are called **elements** or **members of the set**. The use of capital letters to name sets and the use of set braces, { }, to enclose the elements or a description of the elements provides a convenient way to communicate about sets. For example, a set A, which consists of the vowels of the alphabet, can be represented as follows.

	$A = \{\text{vowels of the alphabet}\}$	Word description
or	$A = \{a, e, i, o, u\}$	List or roster description
or	$A = \{x \mid x \text{ is a vowel}\}$	Set builder notation

A set consisting of no elements is called the **null** or **empty set** and is written $\varnothing$.

The **set-builder notation** combines the use of braces and the concept of a variable. For example, $\{x \mid x \text{ is a vowel}\}$ is read *The set of all x such that x is a vowel.* Note that the vertical line is read *such that.*

Two sets are said to be **equal** if they contain exactly the same elements. For example, $\{1, 2, 3\} = \{2, 1, 3\}$ because both sets contain exactly the same elements; the order in which the elements are listed does not matter. A slash mark through an equality symbol denotes *not equal to.* Thus if $A = \{1, 2, 3\}$ and $B = \{3, 6\}$, we can write $A \neq B$, which is read *Set A is not equal to set B.*

Real Numbers

The following terminology is commonly used to classify different types of numbers.

$\{1, 2, 3, 4, \ldots\}$	Natural numbers, counting numbers, positive integers
$\{0, 1, 2, 3, \ldots\}$	Whole numbers, nonnegative integers
$\{\ldots, -3, -2, -1\}$	Negative integers
$\{\ldots, -3, -2, -1, 0\}$	Nonpositive integers
$\{\ldots, -2, -1, 0, 1, 2, \ldots\}$	Integers

A **rational number** is defined as any number that can be expressed in the form a/b, where a and b are integers and b is not zero. The following are examples of rational numbers.

$\frac{2}{3}$ $\quad -\frac{3}{4}$ $\qquad 6\frac{1}{2}$ because $6\frac{1}{2} = \frac{13}{2}$

6 because $6 = \frac{6}{1}$ $\qquad -4$ because $-4 = \frac{-4}{1} = \frac{4}{-1}$

0 because $0 = \frac{0}{1} = \frac{0}{2} = \frac{0}{3}$, etc. $\qquad 0.3$ because $0.3 = \frac{3}{10}$

A rational number can also be defined in terms of a decimal representation. Before doing so, let's briefly review the different possibilities for decimal representations. Decimals can be classified as **terminating**, **repeating**, or **nonrepeating**. Here are some examples of each.

$$\begin{pmatrix} 0.3 \\ 0.46 \\ 0.789 \\ 0.2143 \end{pmatrix} \qquad \text{Terminating decimals}$$

$$\left.\begin{array}{l} 0.3333\ldots \\ 0.141414\ldots \\ 0.712712712\ldots \\ 0.24171717\ldots \\ 0.9675283283283\ldots \end{array}\right\} \quad \text{Repeating decimals}$$

$$\left.\begin{array}{l} 0.472195631\ldots \\ 0.21411711191111\ldots \\ 0.752389433215333\ldots \end{array}\right\} \quad \text{Nonrepeating decimals}$$

A **repeating decimal** has a block of digits that repeats indefinitely. This repeating block of digits may be of any size and may or may not begin immediately after the decimal point. A small horizontal bar is commonly used to indicate the repeating block. Thus, $0.3333\ldots$ can be expressed as $0.\overline{3}$ and $0.24171717\ldots$ as $0.24\overline{17}$.

In terms of decimals, a rational number is defined as a number with either a terminating or a repeating decimal representation. The following examples illustrate some rational numbers written in $\frac{a}{b}$ form and in the equivalent decimal form.

$$\frac{3}{4} = 0.75 \qquad \frac{3}{11} = 0.\overline{27} \qquad \frac{1}{8} = 0.125 \qquad \frac{1}{7} = 0.\overline{142857} \qquad \frac{1}{3} = 0.\overline{3}$$

An **irrational number** is defined as a number that cannot be expressed in $\frac{a}{b}$ form, where a and b are integers and b is not zero. Furthermore, an irrational number has a nonrepeating decimal representation. The following are some examples of irrational numbers and a partial decimal representation for each.

$$\sqrt{2} = 1.414213562373095\ldots$$
$$\sqrt{3} = 1.73205080756887\ldots$$
$$\pi = 3.14159265358979\ldots$$

The entire set of **real numbers** is composed of the rational numbers along with the irrationals. The following tree diagram can be used to summarize the various classifications of the real number system.

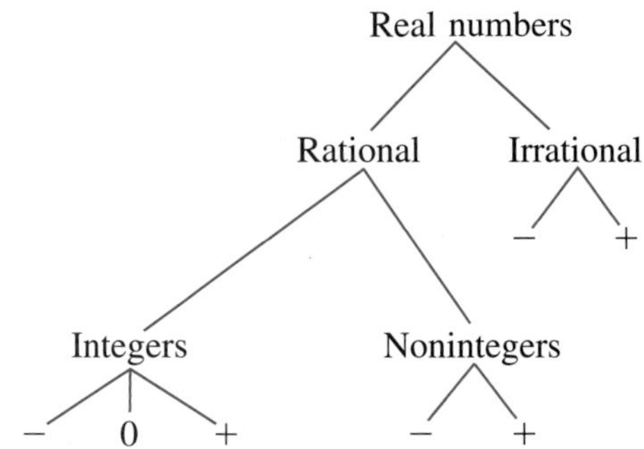

Any real number can be traced down through the tree. For example,

7 is real, rational, an integer, and positive;

$-\frac{2}{3}$ is real, rational, a noninteger, and negative;

$\sqrt{7}$ is real, irrational, and positive;

0.59 is real, rational, a noninteger, and positive.

The concept of subset is convenient to use at this time. A set A is a **subset** of another set B if and only if every element of A is also an element of B. For example, if $A = \{1, 2\}$ and $B = \{1, 2, 3\}$, then A is a subset of B. This is written $A \subseteq B$ and is read *A is a subset of B*. The slash mark can also be used here to denote negation. If $A = \{1, 2, 4, 6\}$ and $B = \{2, 3, 7\}$, we can say *A is not a subset of B* by writing $A \nsubseteq B$. The following statements use the subset vocabulary and symbolism; they are represented in Figure 1.1.

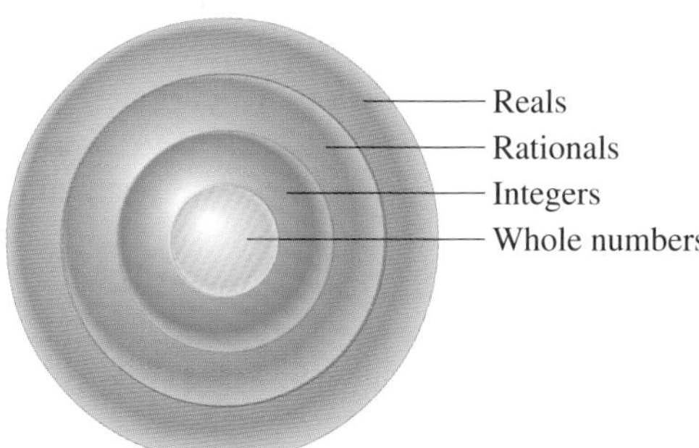

FIGURE 1.1

1. The set of whole numbers is a subset of the set of integers.

$$\{0, 1, 2, 3, \ldots\} \subseteq \{\ldots, -2, -1, 0, 1, 2, \ldots\}.$$

2. The set of integers is a subset of the set of rational numbers.

$$\{\ldots, -2, -1, 0, 1, 2, \ldots\} \subseteq \{x \mid x \text{ is a rational number}\}.$$

3. The set of rational numbers is a subset of the set of real numbers.

$$\{x \mid x \text{ is a rational number}\} \subseteq \{y \mid y \text{ is a real number}\}.$$

Real Number Line and Absolute Value

It is often helpful to have a geometric representation of the set of real numbers in front of us, as indicated in Figure 1.2. Such a representation, called the **real number**

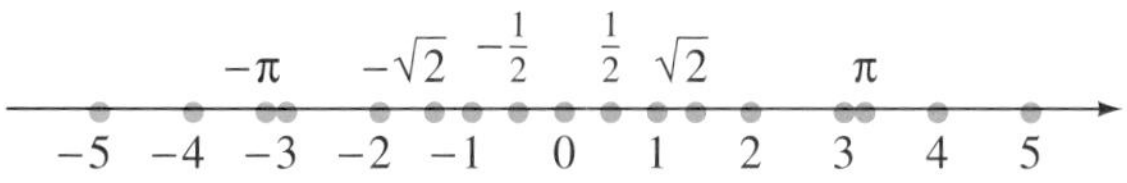

$-\pi$ $-\sqrt{2}$ $-\frac{1}{2}$ $\frac{1}{2}$ $\sqrt{2}$ π

-5 -4 -3 -2 -1 0 1 2 3 4 5

FIGURE 1.2

line, indicates a one-to-one correspondence between the set of real numbers and the points on a line. In other words, to each real number there corresponds one and only one point on the line, and to each point on the line there corresponds one and only one real number. The number that corresponds to a particular point on the line is called the **coordinate** of that point.

Many operations, relations, properties, and concepts pertaining to real numbers can be given a geometric interpretation on the number line. For example, the addition problem $(-1) + (-2)$ can be interpreted on the number line as in Figure 1.3.

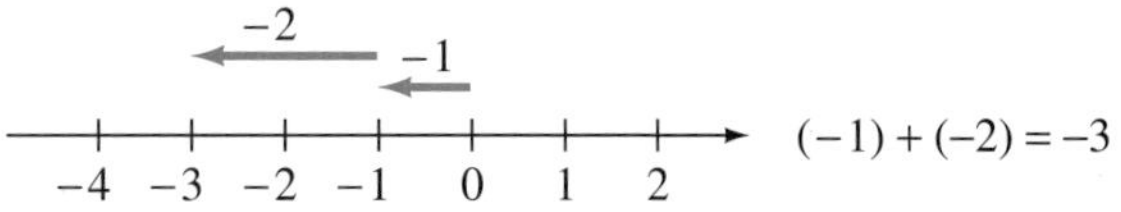

-2 -1

-4 -3 -2 -1 0 1 2 $(-1) + (-2) = -3$

FIGURE 1.3

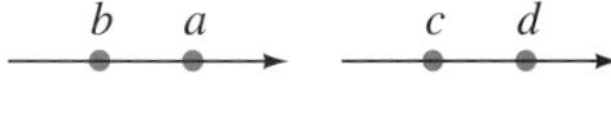

FIGURE 1.4

The inequality relations also have a geometric interpretation. The statement $a > b$ (read *a is greater than b*) means that a is to the right of b, and the statement $c < d$ (read *c is less than d*) means that c is to the left of d (see Figure 1.4).

The property $-(-x) = x$ can be pictured on the number line in a sequence of steps. See Figure 1.5.

1. Choose a point having a coordinate of x.
2. Locate its opposite (written as $-x$) on the other side of zero.
3. Locate the opposite of $-x$ (written as $-(-x)$) on the other side of zero.

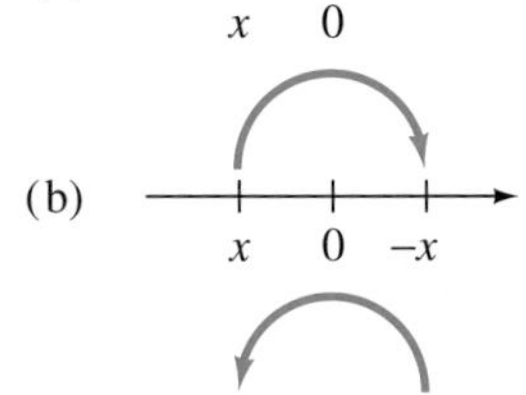

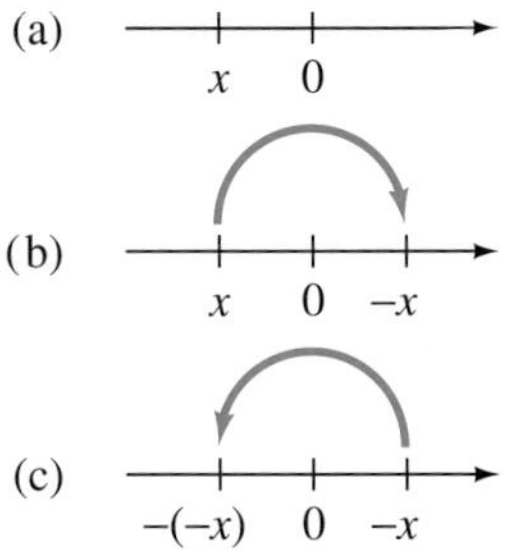

FIGURE 1.5

Therefore, we conclude that **the opposite of the opposite of any real number is the number itself**, and we symbolically express this by $-(-x) = x$.

REMARK The symbol -1 can be read *negative one*, the *negative of one*, the *opposite of one*, or the *additive inverse of one*. The opposite-of and additive-inverse-of terminology is especially meaningful when working with variables. For example, the symbol $-x$, read *the opposite of x* or *the additive inverse of x* emphasizes an important issue. Since x can be any real number, $-x$ (opposite of x) can be zero, positive, or negative. If x is positive, then $-x$ is negative. If x is negative, then $-x$ is positive. If x is zero, then $-x$ is zero.

The concept of absolute value can be interpreted on the number line. Geometrically, the **absolute value** of any real number is the distance between that number and zero on the number line. For example, the absolute value of 2 is 2, the absolute value of -3 is 3, and the absolute value of zero is zero (see Figure 1.6). Symbolically, absolute value is denoted with vertical bars. Thus, we write $|2| = 2$, $|-3| = 3$, and $|0| = 0$. More formally, the concept of absolute value is defined as follows.

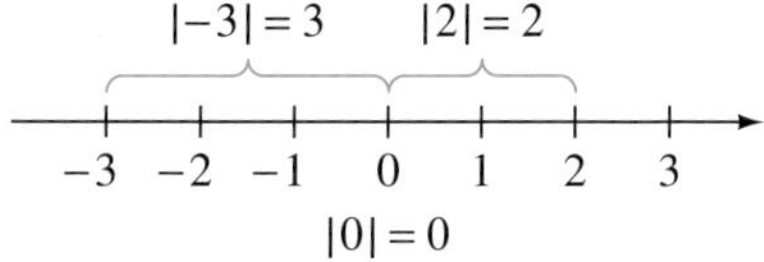

FIGURE 1.6

DEFINITION 1.1

For all real numbers a,

1. If $a \geq 0$, then $|a| = a$;
2. If $a < 0$, then $|a| = -a$.

According to Definition 1.1, we obtain

$|6| = 6$	by applying part 1;
$|0| = 0$	by applying part 1;
$|-7| = -(-7) = 7$	by applying part 2.

Notice that the absolute value of a positive number is the number itself, but the absolute value of a negative number is its opposite. Thus, the absolute value of any number except zero is positive, and the absolute value of zero is zero. Together, these facts indicate that the absolute value of any real number is equal to the absolute value of its opposite. All of these ideas are summarized in the following properties.

Properties of Absolute Value

The variables a and b represent any real number.

1. $|a| \geq 0$
2. $|a| = |-a|$
3. $|a - b| = |b - a|$ — $a - b$ and $b - a$ are opposites of each other.

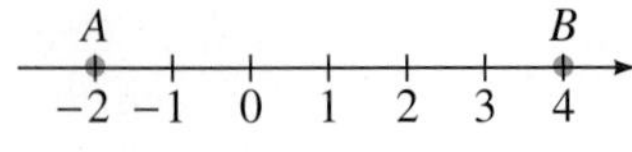

FIGURE 1.7

In Figure 1.7 we located points A and B at -2 and 4, respectively. The distance between A and B is 6 units and can be calculated by using either $|-2 - 4|$ or $|4 - (-2)|$. In general, if two points have coordinates x_1 and x_2, the distance between the two points is determined by using either $|x_2 - x_1|$ or $|x_1 - x_2|$, since they are the same quantity by the third property.

Properties of Real Numbers

As you work with the set of real numbers, the basic operations, and the relations of equality and inequality, the following properties will guide your study. Be sure that you understand these properties, for not only do they facilitate manipulations with real numbers, they also serve as a basis for many algebraic computations. The variables a, b, and c represent real numbers.

Properties of Real Numbers

Closure properties	$a + b$ is a unique real number. ab is a unique real number.
Commutative properties	$a + b = b + a$ $ab = ba$
Associative properties	$(a + b) + c = a + (b + c)$ $(ab)c = a(bc)$
Identity properties	There exists a real number 0 such that $a + 0 = 0 + a = a$. There exists a real number 1 such that $a(1) = 1(a) = a$.
Inverse properties	For every real number a, there exists a unique real number $-a$ such that $a + (-a) = (-a) + a = 0$. For every nonzero real number a, there exists a unique real number $\frac{1}{a}$ such that $a\left(\frac{1}{a}\right) = \frac{1}{a}(a) = 1.$
Multiplication property of zero	$a(0) = (0)(a) = 0$
Multiplication property of negative one	$a(-1) = -1(a) = -a$
Distributive property	$a(b + c) = ab + ac$

Let's make a few comments about the previous list of properties. The set of real numbers is said to be **closed** with respect to addition and multiplication. That is to say, the sum of two real numbers is a real number and the product of two real numbers is a real number. **Closure** plays an important role when proving additional properties pertaining to real numbers.

Addition and multiplication are said to be **commutative operations**. This means that the order in which you add or multiply two real numbers does not affect the result. For example, $6 + (-8) = -8 + 6$ and $(-4)(-3) = (-3)(-4)$. It is important to realize that subtraction and division are *not* commutative operations; order does make a difference. For example, $3 - 4 = -1$, but $4 - 3 = 1$. Likewise, $2 \div 1 = 2$, but $1 \div 2 = \frac{1}{2}$.

Addition and multiplication are **associative operations**. The associative properties are grouping properties. For example, $(-8 + 9) + 6 = -8 + (9 + 6)$; changing the grouping of the numbers does not affect the final sum. Likewise, for multiplication, $[(-4)(-3)](2) = (-4)[(-3)(2)]$. Subtraction and division are *not* associative operations. For example, $(8 - 6) - 10 = -8$, but $8 - (6 - 10) = 12$. An example showing that division is not associative is $(8 \div 4) \div 2 = 1$, but $8 \div (4 \div 2) = 4$.

Zero is the **identity element for addition**. This means that the sum of any real number and zero is identically the same real number. For example, $-87 + 0 = 0 + (-87) = -87$. One is the **identity element for multiplication**. The product of any real number and 1 is identically the same real number. For example, $(-119)(1) = (1)(-119) = -119$.

The real number $-a$ is called the **additive inverse of** a or the **opposite of** a. The sum of a number and its additive inverse is the identity element for addition. For example, 16 and -16 are additive inverses and their sum is zero. The additive inverse of zero is zero.

The real number $1/a$ is called the **multiplicative inverse** or **reciprocal of** a. The product of a number and its multiplicative inverse is the identity element for multiplication. For example, the reciprocal of 2 is $\frac{1}{2}$ and $2\left(\frac{1}{2}\right) = \frac{1}{2}(2) = 1$.

The product of any real number and zero is zero. For example, $(-17)(0) = (0)(-17) = 0$. The product of any real number and -1 is the opposite of the real number. For example, $(-1)(52) = (52)(-1) = -52$.

The distributive property ties together the operations of addition and multiplication. We say that *multiplication distributes over addition.* For example, $7(3 + 8) = 7(3) + 7(8)$. Furthermore, because $b - c = b + (-c)$, it follows that *multiplication also distributes over subtraction.* This can be symbolically expressed as $a(b - c) = ab - ac$. For example, $6(8 - 10) = 6(8) - 6(10)$.

Algebraic Expressions

Algebraic expressions such as

$$2x, \quad 8xy, \quad -3xy, \quad -4abc, \quad \text{and} \quad z$$

are called **terms**. A term is an indicated product and may have any number of factors. The variables of a term are called **literal factors** and the numerical factor is called the **numerical coefficient**. Thus in $8xy$, the x and y are literal factors and 8 is the numerical coefficient. Since $1(z) = z$, the numerical coefficient of the term z is understood to be 1. Terms having the same literal factors are called **similar** or **like terms**. The distributive property in the form $ba + ca = (b + c)a$ provides the basis for simplifying algebraic expressions by *combining similar terms*, as illustrated in the following examples.

$$\begin{aligned} 3x + 5x &= (3 + 5)x \\ &= 8x \end{aligned} \qquad \begin{aligned} -6xy + 4xy &= (-6 + 4)xy \\ &= -2xy \end{aligned} \qquad \begin{aligned} 4x - x &= 4x - 1x \\ &= (4 - 1)x \\ &= 3x \end{aligned}$$

Sometimes an algebraic expression can be simplified by applying the distributive property to remove parentheses and combine similar terms, as the next examples illustrate.

$$\begin{aligned} 4(x + 2) + 3(x + 6) &= 4(x) + 4(2) + 3(x) + 3(6) \\ &= 4x + 8 + 3x + 18 \\ &= 7x + 26 \end{aligned}$$

$$\begin{aligned} -5(y + 3) - 2(y - 8) &= -5(y) - 5(3) - 2(y) - 2(-8) \\ &= -5y - 15 - 2y + 16 \\ &= -7y + 1 \end{aligned}$$

An algebraic expression takes on a numerical value whenever each variable in the expression is replaced by a real number. For example, if x is replaced by 5 and y by 9, the algebraic expression $x + y$ becomes the numerical expression $5 + 9$, which is equal to 14. We say that $x + y$ has a value of 14 when $x = 5$ and $y = 9$.

Consider the following examples, which illustrate the process of finding a value of an algebraic expression. The process is commonly referred to as **evaluating an algebraic expression**.

EXAMPLE 1

Find the value of $3xy - 4z$ when $x = 2$, $y = -4$, and $z = -5$.

Solution

$$\begin{aligned} 3xy - 4z &= 3(2)(-4) - 4(-5) \qquad \text{when } x = 2,\ y = -4, \text{ and } z = -5 \\ &= -24 + 20 \\ &= -4 \end{aligned}$$

EXAMPLE 2

Find the value of $a - [4b - (2c + 1)]$ when $a = -8$, $b = -7$, and $c = 14$.

Solution

$$a - [4b - (2c + 1)] = -8 - [4(-7) - (2(14) + 1)]$$

$$= -8 - [-28 - 29]$$
$$= -8 - [-57]$$
$$= 49$$

EXAMPLE 3

Evaluate $\dfrac{a - 2b}{3c + 5d}$ when $a = 14$, $b = -12$, $c = -3$, and $d = -2$.

Solution

$$\frac{a - 2b}{3c + 5d} = \frac{14 - 2(-12)}{3(-3) + 5(-2)}$$
$$= \frac{14 + 24}{-9 - 10}$$
$$= \frac{38}{-19} = -2$$

Look back at the previous examples and notice that the following **order of operations** was followed when simplifying numerical expressions.

1. Perform the operations inside the symbols of inclusion (parentheses, brackets, and braces) and above and below each fraction bar. Start with the innermost inclusion symbol.
2. Perform all multiplications and divisions in the order in which they appear, from left to right.
3. Perform all additions and subtractions in the order in which they appear, from left to right.

You should also realize that first simplifying by combining similar terms can sometimes aid in the process of evaluating algebraic expressions. The last example of this section illustrates this idea.

EXAMPLE 4

Evaluate $2(3x + 1) - 3(4x - 3)$ when $x = -5$.

Solution

$$2(3x + 1) - 3(4x - 3) = 2(3x) + 2(1) - 3(4x) - 3(-3)$$
$$= 6x + 2 - 12x + 9$$
$$= -6x + 11$$

Now substituting -5 for x, we obtain

$$-6x + 11 = -6(-5) + 11$$
$$= 30 + 11$$
$$= 41.$$

PROBLEM SET 1.1

Identify each of the following as *true* or *false*.

1. Every rational number is a real number.
2. Every irrational number is a real number.
3. Every real number is a rational number.
4. If a number is real, then it is irrational.
5. Some irrational numbers are also rational numbers.
6. All integers are rational numbers.
7. The number zero is a rational number.
8. Zero is a positive integer.
9. Zero is a negative number.
10. All whole numbers are integers.

In the set of numbers $\left[0, \sqrt{5}, -\sqrt{2}, \frac{7}{8}, -\frac{10}{13}, 7\frac{1}{8}, 0.279, 0.4\overline{67}, -\pi, -14, 46, 6.75\right]$, list those elements that belong to each of the following sets. (Problems 11–18)

11. The natural numbers
12. The whole numbers
13. The integers
14. The rational numbers
15. The irrational numbers
16. The nonnegative integers
17. The nonpositive integers
18. The real numbers

For Problems 19–32, use the following set designations.

$N = \{x \mid x \text{ is a natural number}\}$
$W = \{x \mid x \text{ is a whole number}\}$
$I = \{x \mid x \text{ is an integer}\}$
$Q = \{x \mid x \text{ is a rational number}\}$
$H = \{x \mid x \text{ is an irrational number}\}$
$R = \{x \mid x \text{ is a real number}\}$

Place $\subseteq$ or $\nsubseteq$ in each blank to make a true statement.

19. N ______ R
20. R ______ N
21. N ______ I
22. I ______ Q
23. H ______ Q
24. Q ______ H
25. W ______ I
26. N ______ W
27. I ______ W
28. I ______ N
29. $\{0, 2, 4, \ldots\}$ ______ W
30. $\{1, 3, 5, 7, \ldots\}$ ______ I
31. $\{-2, -1, 0, 1, 2\}$ ______ W
32. $\{0, 3, 6, 9, \ldots\}$ ______ N

For Problems 33–42, list the elements of each set. For example, the elements of $\{x \mid x \text{ is a natural number less than } 4\}$ can be listed $\{1, 2, 3\}$.

33. $\{x \mid x \text{ is a natural number less than } 2\}$
34. $\{x \mid x \text{ is a natural number greater than } 5\}$
35. $\{n \mid n \text{ is a whole number less than } 4\}$
36. $\{y \mid y \text{ is an integer greater than } -3\}$
37. $\{y \mid y \text{ is an integer less than } 2\}$
38. $\{n \mid n \text{ is a positive integer greater than } -4\}$
39. $\{x \mid x \text{ is a whole number less than } 0\}$
40. $\{x \mid x \text{ is a negative integer greater than } -5\}$
41. $\{n \mid n \text{ is a nonnegative integer less than } 3\}$
42. $\{n \mid n \text{ is a nonpositive integer greater than } 1\}$
43. Find the distance on the real number line between two points whose coordinates are as follows.

    (a) 17 and 35 **(b)** -14 and 12

    (c) 18 and -21 **(d)** -17 and -42

    (e) -56 and -21 **(f)** 0 and -37

44. Evaluate each of the following if x is a nonzero real number.

(a) $\frac{|x|}{x}$ (b) $\frac{x}{|x|}$

(c) $\frac{|-x|}{-x}$ (d) $|x| - |-x|$

In Problems 45–58, state the property that justifies each of the statements. For example, $3 + (-4) = (-4) + 3$ because of the commutative property of addition.

45. $x(2) = 2(x)$

46. $(7 + 4) + 6 = 7 + (4 + 6)$

47. $1(x) = x$

48. $43 + (-18) = (-18) + 43$

49. $(-1)(93) = -93$

50. $109 + (-109) = 0$

51. $5(4 + 7) = 5(4) + 5(7)$

52. $-1(x + y) = -(x + y)$

53. $7yx = 7xy$

54. $(x + 2) + (-2) = x + [2 + (-2)]$

55. $6(4) + 7(4) = (6 + 7)(4)$

56. $\left(\frac{2}{3}\right)\left(\frac{3}{2}\right) = 1$

57. $4(5x) = (4 \cdot 5)x$

58. $[(17)(8)](25) = (17)[(8)(25)]$

For Problems 59–79, evaluate each of the algebraic expressions for the given values of the variables.

59. $5x + 3y$; $x = -2$ and $y = -4$

60. $7x - 4y$; $x = -1$ and $y = 6$

61. $-3ab - 2c$; $a = -4$, $b = 7$, and $c = -8$

62. $x - (2y + 3z)$; $x = -3$, $y = -4$, and $z = 9$

63. $(a - 2b) + (3c - 4)$; $a = 6$, $b = -5$, and $c = -11$

64. $3a - [2b - (4c + 1)]$; $a = 4$, $b = 6$, and $c = -8$

65. $\frac{-2x + 7y}{x - y}$; $x = -3$ and $y = -2$

66. $\frac{x - 3y + 2z}{2x - y}$; $x = 4$, $y = 9$, and $z = -12$

67. $(5x - 2y)(-3x + 4y)$; $x = -3$ and $y = -7$

68. $(2a - 7b)(4a + 3b)$; $a = 6$ and $b = -3$

69. $5x + 4y - 9y - 2y$; $x = 2$ and $y = -8$

70. $5a + 7b - 9a - 6b$; $a = -7$ and $b = 8$

71. $-5x + 8y + 7y + 8x$; $x = 5$ and $y = -6$

72. $|x - y| - |x + y|$; $x = -4$ and $y = -7$

73. $|3x + y| + |2x - 4y|$; $x = 5$ and $y = -3$

74. $\left|\frac{x - y}{y - x}\right|$; $x = -6$ and $y = 13$

75. $\left|\frac{2a - 3b}{3b - 2a}\right|$; $a = -4$ and $b = -8$

76. $5(x - 1) + 7(x + 4)$; $x = 3$

77. $2(3x + 4) - 3(2x - 1)$; $x = -2$

78. $-4(2x - 1) - 5(3x + 7)$; $x = -1$

79. $5(a - 3) - 4(2a + 1) - 2(a - 4)$; $a = -3$

80. You should be able to do calculations like those in Problems 59–79 with and without a calculator. Different types of calculators handle the priority of operations issue in different ways. Be sure you can do Problems 59–79 with your calculator.

1.2 EXPONENTS

Positive integers are used as **exponents** to indicate repeated multiplication. For example, $4 \cdot 4 \cdot 4$ can be written 4^3, where the raised 3 indicates that 4 is to be used as a factor three times. The following general definition is helpful.

DEFINITION 1.2

If n is a positive integer and b is any real number, then

$$b^n = \underbrace{bbb \cdots b}_{n \text{ factors of } b}.$$

The number b is referred to as the **base** and n is called the **exponent**. The expression b^n can be read *b to the nth power*. The terms **squared** and **cubed** are commonly associated with exponents of 2 and 3, respectively. For example, b^2 is read *b squared* and b^3 as *b cubed*. An exponent of 1 is usually not written, so b^1 is simply written b. The following examples illustrate Definition 1.2.

$$2^3 = 2 \cdot 2 \cdot 2 = 8 \qquad \left(\frac{1}{2}\right)^5 = \frac{1}{2} \cdot \frac{1}{2} \cdot \frac{1}{2} \cdot \frac{1}{2} \cdot \frac{1}{2} = \frac{1}{32}$$

$$3^4 = 3 \cdot 3 \cdot 3 \cdot 3 = 81 \qquad (0.7)^2 = (0.7)(0.7) = 0.49$$

$$(-5)^2 = (-5)(-5) = 25 \qquad -5^2 = -(5 \cdot 5) = -25$$

We especially want to call your attention to the last two examples. Note that $(-5)^2$ means that -5 is the base used as a factor twice. However, -5^2 means that 5 is the base and after it is squared, we take the opposite of that result.

Properties of Exponents

In a previous algebra course, you have seen some properties pertaining to the use of positive integers as exponents. Those properties can be summarized as follows.

PROPERTY 1.1 ***Properties of Exponents***

If a and b are real numbers and m and n are positive integers, then

1. $b^n \cdot b^m = b^{n+m}$;
2. $(b^n)^m = b^{mn}$;
3. $(ab)^n = a^n b^n$;
4. $\left(\frac{a}{b}\right)^n = \frac{a^n}{b^n}, \qquad b \neq 0$;
5. $\frac{b^n}{b^m} = b^{n-m}$ when $n > m, \qquad b \neq 0$;

$$\frac{b^n}{b^m} = 1 \quad \text{when } n = m, \qquad b \neq 0;$$

$$\frac{b^n}{b^m} = \frac{1}{b^{m-n}} \quad \text{when } n < m, \qquad b \neq 0.$$

Each part of Property 1.1 can be justified by using Definition 1.2. For example, to justify part 1 we can reason as follows;

$$b^n \cdot b^m = \underbrace{(bbb \cdots b)}_{\substack{n \text{ factors} \\ \text{of } b}} \cdot \underbrace{(bbb \cdots b)}_{\substack{m \text{ factors} \\ \text{of } b}}$$

$$= \underbrace{bbb \cdots b}_{\substack{n + m \text{ factors} \\ \text{of } b}}$$

$$= b^{n+m}.$$

Similar reasoning can be used to verify the other parts of Property 1.1. The following examples illustrate the use of Property 1.1 along with the commutative and associative properties of the real numbers. The steps enclosed in the dashed boxes can be performed mentally.

EXAMPLE 1

$$(3x^2y)(4x^3y^2) = \boxed{3 \cdot 4 \cdot x^2 \cdot x^3 \cdot y \cdot y^2}$$

$$= \boxed{12x^{2+3}y^{1+2}} \qquad b^n \cdot b^m = b^{n+m}$$

$$= 12x^5y^3$$

EXAMPLE 2

$$\boxed{(-2y^3)^5 = (-2)^5(y^3)^5} \qquad (ab)^n = a^nb^n$$

$$= -32y^{15} \qquad (b^n)^m = b^{mn}$$

EXAMPLE 3

$$\left(\frac{a^2}{b^4}\right)^7 = \boxed{\frac{(a^2)^7}{(b^4)^7}} \qquad \left(\frac{a}{b}\right)^n = \frac{a^n}{b^n}$$

$$= \frac{a^{14}}{b^{28}} \qquad (b^n)^m = b^{mn}$$

EXAMPLE 4

$$\frac{-56x^9}{7x^4} = \boxed{-8x^{9-4}} \qquad \frac{b^n}{b^m} = b^{n-m} \quad \text{when } n > m$$

$$= -8x^5$$

Zero and Negative Integers as Exponents

Now we can extend the concept of an exponent to include the use of zero and negative integers. First let's consider the use of zero as an exponent. We want to use zero in a way that Property 1.1 will continue to hold. For example, if $b^n \cdot b^m = b^{n+m}$ is to hold, then $x^4 \cdot x^0$ should equal x^{4+0}, which equals x^4. In other words, x^0 *acts like* 1 because $x^4 \cdot x^0 = x^4$. Look at the following definition.

DEFINITION 1.3

If b is a nonzero real number, then

$$b^0 = 1.$$

Therefore, according to Definition 1.3, the following statements are all true.

$$5^0 = 1 \qquad (-413)^0 = 1$$

$$\left(\frac{3}{11}\right)^0 = 1 \qquad (x^3y^4)^0 = 1 \quad \text{if } x \neq 0 \text{ and } y \neq 0.$$

A similar line of reasoning can be used to motivate a definition for the use of negative integers as exponents. Consider the example $x^4 \cdot x^{-4}$. If $b^n \cdot b^m = b^{n+m}$ is to hold, then $x^4 \cdot x^{-4}$ should equal $x^{4+(-4)}$, which equals $x^0 = 1$. Therefore x^{-4} must be the reciprocal of x^4, since their product is 1. That is to say, $x^{-4} = 1/x^4$. This suggests the following definition.

DEFINITION 1.4

If n is a positive integer and b is a nonzero real number, then

$$b^{-n} = \frac{1}{b^n}.$$

According to Definition 1.4, the following statements are true.

$$x^{-5} = \frac{1}{x^5} \qquad 2^{-4} = \frac{1}{2^4} = \frac{1}{16}$$

$$\left(\frac{3}{4}\right)^{-2} = \frac{1}{\left(\frac{3}{4}\right)^2} = \frac{1}{\frac{9}{16}} = \frac{16}{9} \qquad \frac{2}{x^{-3}} = \frac{2}{\frac{1}{x^3}} = 2x^3$$

The first four parts of Property 1.1 hold true *for all integers*. Furthermore, we do not need all three equations in part 5 of Property 1.1. The first equation,

$$\frac{b^n}{b^m} = b^{n-m},$$

can be used for *all integral exponents*. Let's restate Property 1.1 as it holds for all integers. We will include name tags for easy reference.

PROPERTY 1.2

If m and n are integers and a and b are real numbers, with $b \neq 0$ whenever it appears in a denominator, then

1. $b^n \cdot b^m = b^{n+m}$ — Product of two powers
2. $(b^n)^m = b^{mn}$ — Power of a power
3. $(ab)^n = a^n b^n$ — Power of a product
4. $\left(\frac{a}{b}\right)^n = \frac{a^n}{b^n}$ — Power of a quotient
5. $\frac{b^n}{b^m} = b^{n-m}$. — Quotient of two powers

Having the use of all integers as exponents allows us to work with a large variety of numerical and algebraic expressions. Let's consider some examples that illustrate the various parts of Property 1.2.

EXAMPLE 5

Evaluate each of the following numerical expressions.

a. $(2^{-1} \cdot 3^2)^{-1}$ **b.** $\left(\frac{2^{-3}}{3^{-2}}\right)^{-2}$

Solutions

a.
$$\begin{aligned} (2^{-1} \cdot 3^2)^{-1} &= (2^{-1})^{-1}(3^2)^{-1} && \text{Power of a product} \\ &= (2^1)(3^{-2}) && \text{Power of a power} \\ &= (2)\left(\frac{1}{3^2}\right) \\ &= 2\left(\frac{1}{9}\right) = \frac{2}{9} \end{aligned}$$

b.
$$\begin{aligned} \left(\frac{2^{-3}}{3^{-2}}\right)^{-2} &= \frac{(2^{-3})^{-2}}{(3^{-2})^{-2}} && \text{Power of a quotient} \\ &= \frac{2^6}{3^4} && \text{Power of a power} \\ &= \frac{64}{81} \end{aligned}$$

EXAMPLE 6

Find the indicated products and quotients and express the final results with positive integral exponents only.

a. $(3x^2y^{-4})(4x^{-3}y)$ **b.** $\dfrac{12a^3b^2}{-3a^{-1}b^5}$ **c.** $\left(\dfrac{15x^{-1}y^2}{5xy^{-4}}\right)^{-1}$

Solutions

a. $(3x^2y^{-4})(4x^{-3}y) = 12x^{2+(-3)}y^{-4+1}$ Product of powers

$= 12x^{-1}y^{-3}$

$= \dfrac{12}{xy^3}$

b. $\dfrac{12a^3b^2}{-3a^{-1}b^5} = -4a^{3-(-1)}b^{2-5}$ Quotient of powers

$= -4a^4b^{-3}$

$= -\dfrac{4a^4}{b^3}$

c. $\left(\dfrac{15x^{-1}y^2}{5xy^{-4}}\right)^{-1} = (3x^{-1-1}y^{2-(-4)})^{-1}$ First simplify inside parentheses.

$= (3x^{-2}y^6)^{-1}$ Power of a product

$= 3^{-1}x^2y^{-6}$

$= \dfrac{x^2}{3y^6}$

The next three examples illustrate the simplification of numerical and algebraic expressions involving sums and differences. In such cases, Definition 1.4 can be used to change from negative to positive exponents so that we can proceed in the usual ways.

EXAMPLE 7

Simplify $2^{-3} + 3^{-1}$.

Solution

$$2^{-3} + 3^{-1} = \frac{1}{2^3} + \frac{1}{3^1}$$

$$= \frac{1}{8} + \frac{1}{3}$$

$$= \frac{3}{24} + \frac{8}{24}$$

$$= \frac{11}{24}$$

EXAMPLE 8

Simplify $(4^{-1} - 3^{-2})^{-1}$.

Solution

$$\begin{aligned}(4^{-1} - 3^{-2})^{-1} &= \left(\frac{1}{4^1} - \frac{1}{3^2}\right)^{-1} \\ &= \left(\frac{1}{4} - \frac{1}{9}\right)^{-1} \\ &= \left(\frac{9}{36} - \frac{4}{36}\right)^{-1} \\ &= \left(\frac{5}{36}\right)^{-1} \\ &= \frac{1}{\left(\frac{5}{36}\right)^1} = \frac{36}{5}\end{aligned}$$

EXAMPLE 9

Express $a^{-1} + b^{-2}$ as a single fraction involving positive exponents only.

Solution

$$\begin{aligned}a^{-1} + b^{-2} &= \frac{1}{a^1} + \frac{1}{b^2} \\ &= \left(\frac{1}{a}\right)\left(\frac{b^2}{b^2}\right) + \left(\frac{1}{b^2}\right)\left(\frac{a}{a}\right) \\ &= \frac{b^2}{ab^2} + \frac{a}{ab^2} \\ &= \frac{b^2 + a}{ab^2}\end{aligned}$$

Scientific Notation

The expression $(n)(10)^k$ (where n is a number greater than or equal to 1 and less than 10, written in decimal form, and k is any integer) is commonly called **scientific notation** or the **scientific form** of a number. The following are examples of numbers expressed in scientific form.

$$(4.23)(10)^4 \qquad (8.176)(10)^{12} \qquad (5.02)(10)^{-3} \qquad (1)(10)^{-5}$$

Very large and very small numbers can be conveniently expressed in scientific notation. For example, a light year (the distance that a ray of light travels in one year) is approximately 5,900,000,000,000 miles and this can be written as $(5.9)(10)^{12}$. The weight of an oxygen molecule is approximately 0.000000000000000000000053 of a gram and this can be expressed as $(5.3)(10)^{-23}$.

To change from ordinary decimal notation to scientific notation, the following procedure can be used.

> Write the given number as the product of a number greater than or equal to 1 and less than 10, and a power of 10. The exponent of 10 is determined by counting the number of places that the decimal point was moved when going from the original number to the number greater than or equal to 1 and less than 10. This exponent is (a) negative if the original number is less than 1, (b) positive if the original number is greater than 10, and (c) 0 if the original number itself is between 1 and 10.

Thus, we can write

$$0.00092 = (9.2)(10)^{-4};$$
$$872{,}000{,}000 = (8.72)(10)^{8};$$
$$5.1217 = (5.1217)(10)^{0}.$$

To change from scientific notation to ordinary decimal notation, the following procedure can be used.

> Move the decimal point the number of places indicated by the exponent of 10. We move the decimal point to the right if the exponent is positive and to the left if it is negative.

Thus, we can write

$$(3.14)(10)^{7} = 31{,}400{,}000;$$
$$(7.8)(10)^{-6} = 0.0000078.$$

Scientific notation can be used to simplify numerical calculations. We merely change the numbers to scientific notation and use the appropriate properties of exponents. Consider the following examples.

EXAMPLE 10

Perform the indicated operations.

a. $\dfrac{(0.00063)(960{,}000)}{(3200)(0.0000021)}$ **b.** $\sqrt{90{,}000}$

Solution

a.
$$\begin{aligned}\frac{(0.00063)(960{,}000)}{(3200)(0.0000021)} &= \frac{(6.3)(10)^{-4}(9.6)(10)^{5}}{(3.2)(10)^{3}(2.1)(10)^{-6}}\\ &= \frac{(6.3)(9.6)(10)^{1}}{(3.2)(2.1)(10^{-3})}\\ &= (9)(10)^{4}\\ &= 90{,}000\end{aligned}$$

b.
$$\begin{aligned}\sqrt{90{,}000} &= \sqrt{(9)(10)^{4}}\\ &= \sqrt{9}\,\sqrt{10^{4}}\\ &= (3)(10)^{2}\\ &= 3(100)\\ &= 300\end{aligned}$$

Many calculators are equipped to display numbers in scientific notation. The display panel shows the number between 1 and 10 and the appropriate exponent of 10. For example, evaluating $(3{,}800{,}000)^2$ yields

| 1.444 13 |.

Thus, $(3{,}800{,}000)^2 = (1.444)(10)^{13} = 14{,}440{,}000{,}000{,}000$. Similarly, the answer for $(0.000168)^2$ is displayed as

| 2.8224 −08 |.

Thus, $(0.000168)^2 = (2.8224)(10)^{-8} = 0.000000028224$.

Calculators vary as to the number of digits displayed in the number between 1 and 10 when using scientific notation. For example, we used two different calculators to estimate $(6729)^6$ and obtained the following results.

| 9.2833 22 |

| 9.283316768 22 |

Obviously, you need to know the capabilities of your calculator when working with problems in scientific notation.

Many calculators also allow the entry of a number in scientific notation. Such calculators are equipped with an enter-the-exponent key (often labeled as | EE | or | E EX |). Thus, a number such as $(3.14)(10)^8$ might be entered as follows.

ENTER	PRESS	DISPLAY
3.14	EE	3.14 00
8		3.14 08

However, your calculator may perform the switch from ordinary notation to scientific notation with the following routine.

ENTER	PRESS	DISPLAY
4721	EE	4721 00
	=	4.721 03

Be sure that you know how to accomplish this switch on your calculator. It should be evident that even when using a calculator, you need to have an understanding of scientific notation.

PROBLEM SET 1.2

Evaluate each of the following numerical expressions.

1. 2^{-3}
2. 3^{-2}
3. -10^{-3}
4. 10^{-4}
5. $\frac{1}{3^{-3}}$
6. $\frac{1}{2^{-5}}$
7. $\left(\frac{1}{2}\right)^{-2}$
8. $-\left(\frac{1}{3}\right)^{-2}$
9. $\left(-\frac{2}{3}\right)^{-3}$
10. $\left(\frac{5}{6}\right)^{-2}$
11. $\left(-\frac{1}{5}\right)^{0}$
12. $\frac{1}{\left(\frac{3}{5}\right)^{-2}}$
13. $\frac{1}{\left(\frac{4}{5}\right)^{-2}}$
14. $\left(\frac{4}{5}\right)^{0}$
15. $2^5 \cdot 2^{-3}$
16. $3^{-2} \cdot 3^5$
17. $10^{-6} \cdot 10^4$
18. $10^6 \cdot 10^{-9}$
19. $10^{-2} \cdot 10^{-3}$
20. $10^{-1} \cdot 10^{-5}$
21. $(3^{-2})^{-2}$
22. $((-2)^{-1})^{-3}$
23. $(4^2)^{-1}$
24. $(3^{-1})^3$
25. $(3^{-1} \cdot 2^2)^{-1}$
26. $(2^3 \cdot 3^{-2})^{-2}$
27. $(4^2 \cdot 5^{-1})^2$
28. $(2^{-2} \cdot 4^{-1})^3$
29. $\left(\frac{2^{-2}}{5^{-1}}\right)^{-2}$
30. $\left(\frac{3^{-1}}{2^{-3}}\right)^{-2}$
31. $\left(\frac{3^{-2}}{8^{-1}}\right)^{2}$
32. $\left(\frac{4^2}{5^{-1}}\right)^{-1}$
33. $\frac{2^3}{2^{-3}}$
34. $\frac{2^{-3}}{2^3}$
35. $\frac{10^{-1}}{10^4}$
36. $\frac{10^{-3}}{10^{-7}}$
37. $3^{-2} + 2^{-3}$
38. $2^{-3} + 5^{-1}$
39. $\left(\frac{2}{3}\right)^{-1} - \left(\frac{3}{4}\right)^{-1}$
40. $3^{-2} - 2^3$
41. $(2^{-4} + 3^{-1})^{-1}$
42. $(3^{-2} - 5^{-1})^{-1}$

Simplify each of the following; express final results without using zero or negative integers as exponents.

43. $x^3 \cdot x^{-7}$
44. $x^{-2} \cdot x^{-3}$
45. $a^2 \cdot a^{-3} \cdot a^{-1}$
46. $b^{-3} \cdot b^5 \cdot b^{-4}$
47. $(a^{-3})^2$
48. $(b^5)^{-2}$
49. $(x^3y^{-4})^{-1}$
50. $(x^4y^{-2})^{-2}$
51. $(ab^2c^{-1})^{-3}$
52. $(a^2b^{-1}c^{-2})^{-4}$
53. $(2x^2y^{-1})^{-2}$
54. $(3x^4y^{-2})^{-1}$
55. $\left(\frac{x^{-2}}{y^{-3}}\right)^{-2}$
56. $\left(\frac{y^4}{x^{-1}}\right)^{-3}$
57. $\left(\frac{2a^{-1}}{3b^{-2}}\right)^{-2}$
58. $\left(\frac{3x^2y}{4a^{-1}b^{-3}}\right)^{-1}$
59. $\frac{x^{-5}}{x^{-2}}$
60. $\frac{a^{-3}}{a^5}$
61. $\frac{a^2b^{-3}}{a^{-1}b^{-2}}$
62. $\frac{x^{-1}y^{-2}}{x^3y^{-1}}$

Find the indicated products, quotients, and powers; express answers without using zero or negative integers as exponents.

63. $(4x^3y^2)(-5xy^3)$
64. $(-6xy)(3x^2y^4)$
65. $(-3xy^3)^3$
66. $(-2x^2y^4)^4$
67. $\left(\frac{2x^2}{3y^3}\right)^3$
68. $\left(\frac{4x}{5y^2}\right)^3$
69. $\frac{72x^8}{-9x^2}$
70. $\frac{108x^6}{-12x^2}$

Find the indicated products and quotients; express results using positive integral exponents only.

71. $(2x^{-1}y^2)(3x^{-2}y^{-3})$
72. $(4x^{-2}y^3)(-5x^3y^{-4})$
73. $(-6a^5y^{-4})(-a^{-7}y)$
74. $(-8a^{-4}b^{-5})(-6a^{-1}b^8)$

75. $\dfrac{24x^{-1}y^{-2}}{6x^{-4}y^{3}}$

76. $\dfrac{56xy^{-3}}{8x^{2}y^{2}}$

77. $\dfrac{-35a^{3}b^{-2}}{7a^{5}b^{-1}}$

78. $\dfrac{27a^{-4}b^{-5}}{-3a^{-2}b^{-4}}$

79. $\left(\dfrac{14x^{-2}y^{-4}}{7x^{-3}y^{-6}}\right)^{-2}$

80. $\left(\dfrac{24x^{5}y^{-3}}{-8x^{6}y^{-1}}\right)^{-3}$

Express each of the following as a single fraction involving positive exponents only.

81. $x^{-1} + x^{-2}$

82. $x^{-2} + x^{-4}$

83. $x^{-2} - y^{-1}$

84. $2x^{-1} - 3y^{-3}$

85. $3a^{-2} + 2b^{-3}$

86. $a^{-2} + a^{-1}b^{-2}$

87. $x^{-1}y - xy^{-1}$

88. $x^{2}y^{-1} - x^{-3}y^{2}$

Find each of the following products and quotients. Assume that all variables appearing as exponents represent integers. For example,

$$(x^{2b})(x^{-b+1}) = x^{2b+(-b+1)} = x^{b+1}.$$

89. $(3x^{a})(4x^{2a+1})$

90. $(5x^{-a})(-6x^{3a-1})$

91. $(x^{a})(x^{-a})$

92. $(-2y^{3b})(-4y^{b+1})$

93. $\dfrac{x^{3a}}{x^{a}}$

94. $\dfrac{4x^{2a+1}}{2x^{a-2}}$

95. $\dfrac{-24y^{5b+1}}{6y^{-b-1}}$

96. $(x^{a})^{2b}(x^{b})^{a}$

97. $\dfrac{(xy)^{b}}{y^{b}}$

98. $\dfrac{(2x^{2b})(-4x^{b+1})}{8x^{-bp2}}$

For Problems 99–102, express each number in scientific notation.

99. 62,400,000

100. 17,000,000,000

101. 0.000412

102. 0.000000078

For Problems 103–106, change each of the scientific notations to ordinary decimal form.

103. $(1.8)(10)^{5}$

104. $(5.41)(10)^{7}$

105. $(2.3)(10)^{-6}$

106. $(4.13)(10)^{-9}$

For Problems 107–112, use scientific notation and the properties of exponents to help perform the indicated operations.

107. $\dfrac{0.00052}{0.013}$

108. $\dfrac{(0.000075)(4{,}800{,}000)}{(15{,}000)(0.0012)}$

109. $\sqrt{900{,}000{,}000}$

110. $\sqrt{0.000004}$

111. $\sqrt{0.0009}$

112. $\dfrac{(0.00069)(0.0034)}{(0.0000017)(0.023)}$

THOUGHTS into WORDS

113. Explain how you would simplify $(3^{-1} \cdot 2^{-2})^{-1}$ and also how you would simplify $(3^{-1} + 2^{-2})^{-1}$.

114. How would you explain to someone why the product of x^2 and x^4 is x^6 and not x^8?

1.3 Polynomials

Recall that algebraic expressions such as $5x$, $-6y^2$, $2x^{-1}y^{-2}$, $14a^2b$, $5x^{-4}$, and $-17ab^2c^3$ are called **terms**. Terms containing variables with only nonnegative integers as exponents are called **monomials**. Of the previously listed terms, $5x$, $-6y^2$, $14a^2b$, and $-17ab^2c^3$ are monomials. The **degree** of a monomial is the sum of the exponents of the literal factors. For example, $7xy$ is of degree 2, while $14a^2b$ is of degree 3, and $-17ab^2c^3$ is of degree 6. If the monomial contains only one variable, then the exponent of that variable is the degree of the monomial. For example, $5x^3$ is of degree 3 and $-8y^4$ is of degree 4. Any nonzero constant term, such as 8, is of degree zero.

A **polynomial** is a monomial or a finite sum of monomials. Thus

$$4x^2, \quad 3x^2 - 2x - 4, \quad 7x^4 - 6x^3 + 5x^2 - 2x - 1,$$
$$3x^2y + 2y, \quad \frac{1}{5}a^2 - \frac{2}{3}b^2, \quad \text{and} \quad 14$$

are examples of polynomials. In addition to calling a polynomial with one term a monomial, we also classify polynomials with two terms as **binomials** and those with three terms, **trinomials**. The **degree of a polynomial** is the degree of the term with the highest degree in the polynomial. The following examples illustrate some of this terminology.

The polynomial $4x^3y^4$ is a monomial in two variables of degree 7.

The polynomial $4x^2y - 2xy$ is a binomial in two variables of degree 3.

The polynomial $9x^2 - 7x - 1$ is a trinomial in one variable of degree 2.

Addition and Subtraction of Polynomials

Both adding and subtracting polynomials rely on basically the same ideas. The commutative, associative, and distributive properties provide the basis for rearranging, regrouping, and combining similar terms. Consider the following addition problems.

$$\begin{aligned}(4x^2 + 5x + 1) + (7x^2 - 9x + 4) &= (4x^2 + 7x^2) + (5x - 9x) + (1 + 4)\\ &= 11x^2 - 4x + 5;\end{aligned}$$

$$\begin{aligned}(5x - 3) + (3x + 2) + (8x + 6) &= (5x + 3x + 8x) + (-3 + 2 + 6)\\ &= 16x + 5.\end{aligned}$$

The definition of subtraction as *adding the opposite* $(a - b = a + (-b))$ extends to polynomials in general. The opposite of a polynomial can be formed by taking the opposite of each term. For example, the opposite of $3x^2 - 7x + 1$ is $-3x^2 + 7x - 1$. Symbolically this is expressed

$$-(3x^2 - 7x + 1) = -3x^2 + 7x - 1.$$

You can also think in terms of the property $-x = -1(x)$ and the distributive property. Therefore,

$$-(3x^2 - 7x + 1) = -1(3x^2 - 7x + 1) = -3x^2 + 7x - 1.$$

Now consider the following subtraction problems.

$$\begin{aligned}(7x^2 - 2x - 4) - (3x^2 + 7x - 1) &= (7x^2 - 2x - 4) + (-3x^2 - 7x + 1)\\ &= (7x^2 - 3x^2) + (-2x - 7x) + (-4 + 1)\\ &= 4x^2 - 9x - 3\end{aligned}$$

$$\begin{aligned}(4y^2 + 7) - (-3y^2 + y - 2) &= (4y^2 + 7) + (3y^2 - y + 2)\\ &= (4y^2 + 3y^2) + (-y) + (7 + 2)\\ &= 7y^2 - y + 9\end{aligned}$$

Multiplying Polynomials

The distributive property is usually stated as $a(b + c) = ab + ac$, but it can be extended as follows.

$$a(b + c + d) = ab + ac + ad$$

$$a(b + c + d + e) = ab + ac + ad + ae \qquad \text{etc.}$$

The commutative and associative properties, the properties of exponents, and the distributive property work together to form a basis for finding the product of a monomial and a polynomial. The following example illustrates this idea.

$$\begin{aligned}3x^2(2x^2 + 5x + 3) &= 3x^2(2x^2) + 3x^2(5x) + 3x^2(3)\\ &= 6x^4 + 15x^3 + 9x^2\end{aligned}$$

Extending the method of finding the product of a monomial and a polynomial to finding the product of two polynomials is again based on the distributive property.

$$\begin{aligned}(x + 2)(y + 5) &= x(y + 5) + 2(y + 5)\\ &= x(y) + x(5) + 2(y) + 2(5)\\ &= xy + 5x + 2y + 10\end{aligned}$$

Notice that each term of the first polynomial multiplies each term of the second polynomial.

$$\begin{aligned}(x - 3)(y + z + 3) &= x(y + z + 3) - 3(y + z + 3)\\ &= xy + xz + 3x - 3y - 3z - 9\end{aligned}$$

Frequently, multiplying polynomials will produce similar terms that can be combined, which simplifies the resulting polynomial.

$$\begin{aligned}(x + 5)(x + 7) &= x(x + 7) + 5(5x + 7)\\ &= x^2 + 7x + 5x + 35\\ &= x^2 + 12x + 35\end{aligned}$$

$$\begin{aligned}(x - 2)(x^2 - 3x + 4) &= x(x^2 - 3x + 4) - 2(x^2 - 3x + 4)\\ &= x^3 - 3x^2 + 4x - 2x^2 + 6x - 8\\ &= x^3 - 5x^2 + 10x - 8\end{aligned}$$

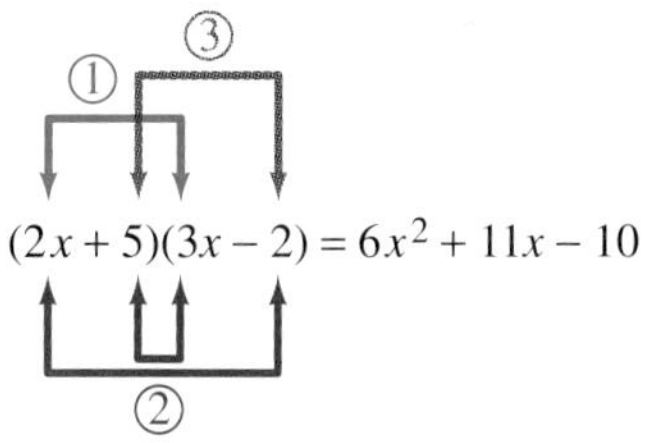

In a previous algebra course, you may have developed a shortcut for multiplying binomials, as illustrated by Figure 1.8.

① ③

$(2x + 5)(3x - 2) = 6x^2 + 11x - 10$

②

FIGURE 1.8

STEP ① Multiply $(2x)(3x)$.

STEP ② Multiply $(5)(3x)$ and $(2x)(-2)$ and combine.

STEP ③ Multiply $(5)(-2)$.

REMARK Shortcuts can be very helpful for certain manipulations in mathematics. But a word of caution: Do not lose the understanding of what you are doing. Make sure that you are able to do the manipulation without the shortcut.

Exponents can also be used to indicate repeated multiplication of polynomials. For example, $(3x - 4y)^2$ means $(3x - 4y)(3x - 4y)$, and $(x + 4)^3$ means $(x + 4)(x + 4)(x + 4)$. Therefore, raising a polynomial to a power is merely another multiplication problem.

$$\begin{aligned}(3x - 4y)^2 &= (3x - 4y)(3x - 4y)\\ &= 9x^2 - 24xy + 16y^2\end{aligned}$$

[*Hint:* When squaring a binomial, be careful not to forget the middle term. That is to say, $(x + 5)^2 \neq x^2 + 25$; instead, $(x + 5)^2 = x^2 + 10x + 25$.]

$$\begin{aligned}(x + 4)^3 &= (x + 4)(x + 4)(x + 4)\\ &= (x + 4)(x^2 + 8x + 16)\\ &= x(x^2 + 8x + 16) + 4(x^2 + 8x + 16)\\ &= x^3 + 8x^2 + 16x + 4x^2 + 32x + 64\\ &= x^3 + 12x^2 + 48x + 64\end{aligned}$$

Special Patterns

When multiplying binomials, some special patterns occur that you should learn to recognize. These patterns can be used to find products, and some of them are very helpful later when factoring polynomials.

$$\begin{aligned}(a + b)^2 &= a^2 + 2ab + b^2\\ (a - b)^2 &= a^2 - 2ab + b^2\\ (a + b)(a - b) &= a^2 - b^2\\ (a + b)^3 &= a^3 + 3a^2b + 3ab^2 + b^3\\ (a - b)^3 &= a^3 - 3a^2b + 3ab^2 - b^3\end{aligned}$$

The three following examples illustrate the first three patterns, respectively.

$$\begin{aligned}(2x + 3)^2 &= (2x)^2 + 2(2x)(3) + (3)^2\\ &= 4x^2 + 12x + 9\end{aligned}$$

$$\begin{aligned}(5x - 2)^2 &= (5x)^2 - 2(5x)(2) + (2)^2\\ &= 25x^2 - 20x + 4\end{aligned}$$

$$(3x + 2y)(3x - 2y) = (3x)^2 - (2y)^2 = 9x^2 - 4y^2$$

In the first two examples, the resulting trinomial is called a **perfect-square trinomial**; it is the result of squaring a binomial. In the third example, the resulting binomial is called the **difference of two squares**. Later, we will use both of these patterns extensively when factoring polynomials.

The cubing-of-a-binomial patterns are helpful primarily when multiplying. These patterns can shorten your work when cubing a binomial, as the next two examples illustrate.

$$\begin{aligned}(3x + 2)^3 &= (3x)^3 + 3(3x)^2(2) + 3(3x)(2)^2 + (2)^3\\ &= 27x^3 + 54x^2 + 36x + 8\end{aligned}$$

$$\begin{aligned}(5x - 2y)^3 &= (5x)^3 - 3(5x)^2(2y) + 3(5x)(2y)^2 - (2y)^3\\ &= 125x^3 - 150x^2y + 60xy^2 - 8y^3\end{aligned}$$

Keep in mind that these multiplying patterns are useful shortcuts, but if you forget them, simply revert to applying the distributive property.

Binomial Expansion Pattern

It is possible to write the expansion of $(a + b)^n$, where n is *any* positive integer, without showing all of the intermediate steps of multiplying and combining similar terms. To do this, let's observe some patterns in the following examples; each one can be verified by direct multiplication.

$$(a + b)^1 = a + b$$

$$(a + b)^2 = a^2 + 2ab + b^2$$

$$(a + b)^3 = a^3 + 3a^2b + 3ab^2 + b^3$$

$$(a + b)^4 = a^4 + 4a^3b + 6a^2b^2 + 4ab^3 + b^4$$

$$(a + b)^5 = a^5 + 5a^4b + 10a^3b^2 + 10a^2b^3 + 5ab^4 + b^5$$

First, note the patterns of the exponents for a and b on a term-by-term basis. The exponents of a begin with the exponent of the binomial and decrease by 1, term-by-term, until the last term, which has $a^0 = 1$. The exponents of b begin with zero ($b^0 = 1$) and increase by 1, term-by-term, until the last term, which contains b to the power of the original binomial. In other words, the variables in the expansion of $(a + \mathrm{b})^n$ have the pattern

$$a^n, \qquad a^{n-1}b, \qquad a^{n-2}b^2, \qquad \ldots, \qquad ab^{n-1}, \qquad b^n,$$

where for each term, the *sum* of the exponents of a and b is n.

Next, let's arrange the *coefficients* in a triangular formation; this yields an easy-to-remember pattern.

```
        1   1
      1   2   1
    1   3   3   1
  1   4   6   4   1
1   5  10  10   5   1
```

Row number n in the formation contains the coefficients of the expansion of $(a + b)^n$. For example, the fifth row contains 1 5 10 10 5 1, and these numbers are the coefficients of the terms in the expansion of $(a + b)^5$. Furthermore, each row can be formed from the previous row as follows.

1. Start and end each row with 1.
2. All other entries result from adding the two numbers in the row immediately above, one number to the left and one number to the right.

Thus, from row 5 we can form row 6.

```
Row 5:   1     5     10     10     5     1
          \   / \   /  \   /  \   / \   /
          Add  Add    Add    Add    Add
Row 6:   1     6     15     20     15    6    1
```

Now we can use these 7 coefficients and our discussion about the exponents to write out the expansion for $(a + b)^6$.

$$(a + b)^6 = a^6 + 6a^5b + 15a^4b^2 + 20a^3b^3 + 15a^2b^4 + 6ab^5 + b^6.$$

REMARK The triangular formation of numbers that we have been discussing is often referred to as *Pascal's triangle*. This is in honor of Blaise Pascal, a seventeenth century mathematician, to whom the discovery of this pattern is attributed.

Let's consider two more examples using Pascal's triangle and the exponent relationships.

EXAMPLE 1

Expand $(a - b)^4$.

Solution

We can treat $a - b$ as $a + (-b)$ and use the fourth row of Pascal's triangle to obtain the coefficients.

$$\begin{aligned}[a + (-b)]^4 &= a^4 + 4a^3(-b) + 6a^2(-b)^2 + 4a(-b)^3 + (-b)^4 \\ &= a^4 - 4a^3b + 6a^2b^2 - 4ab^3 + b^4\end{aligned}$$

EXAMPLE 2

Expand $(2x + 3y)^5$.

Solution

Let $2x = a$ and $3y = b$. The coefficients come from the fifth row of Pascal's triangle.

$$\begin{aligned}(2x + 3y)^5 &= (2x)^5 + 5(2x)^4(3y) + 10(2x)^3(3y)^2 + 10(2x)^2(3y)^3 + 5(2x)(3y)^4 + (3y)^5 \\ &= 32x^5 + 240x^4y + 720x^3y^2 + 1080x^2y^3 + 810xy^4 + 243y^5\end{aligned}$$

Dividing Polynomials by Monomials

In Section 1.5 we will review the addition and subtraction of rational expressions using the properties

$$\frac{a}{b} + \frac{c}{b} = \frac{a + c}{b} \quad \text{and} \quad \frac{a}{b} - \frac{c}{b} = \frac{a - c}{b}.$$

These properties can also be viewed as

$$\frac{a + c}{b} = \frac{a}{b} + \frac{c}{b} \quad \text{and} \quad \frac{a - c}{b} = \frac{a}{b} - \frac{c}{b}.$$

Together with our knowledge of dividing monomials, these properties provide the basis for dividing polynomials by monomials. Consider the following examples.

$$\frac{18x^3 + 24x^2}{6x} = \frac{18x^3}{6x} + \frac{24x^2}{6x} = 3x^2 + 4x$$

$$\frac{35x^2y^3 - 55x^3y^4}{5xy^2} = \frac{35x^2y^3}{5xy^2} - \frac{55x^3y^4}{5xy^2} = 7xy - 11x^2y^2$$

Therefore, to divide a polynomial by a monomial, we divide each term of the polynomial by the monomial. As with many skills, once you feel comfortable with the process you may then choose to perform some of the steps mentally. Your work could take the following format.

$$\frac{40x^4y^5 + 72x^5y^7}{8x^2y} = 5x^2y^4 + 9x^3y^6$$

$$\frac{36a^3b^4 - 48a^3b^3 + 64a^2b^5}{-4a^2b^2} = -9ab^2 + 12ab - 16b^3$$

PROBLEM SET 1.3

In Problems 1–10, perform the indicated operations.

1. $(5x^2 - 7x - 2) + (9x^2 + 8x - 4)$

2. $(-9x^2 + 8x + 4) + (7x^2 - 5x - 3)$

3. $(14x^2 - x - 1) - (15x^2 + 3x + 8)$

4. $(-3x^2 + 2x + 4) - (4x^2 + 6x - 5)$

5. $(3x - 4) - (6x + 3) + (9x - 4)$

6. $(7a - 2) - (8a - 1) - (10a - 2)$

7. $(8x^2 - 6x - 2) + (x^2 - x - 1) - (3x^2 - 2x + 4)$

8. $(12x^2 + 7x - 2) - (3x^2 + 4x + 5) + (-4x^2 - 7x - 2)$

9. $5(x - 2) - 4(x + 3) - 2(x + 6)$

10. $3(2x - 1) - 2(3x + 4) - 4(5x - 1)$

In Problems 11–54, find the indicated products. Remember the special patterns that we discussed in this section.

11. $3xy(4x^2y + 5xy^2)$

12. $-2ab^2(3a^2b - 4ab^3)$

13. $6a^3b^2(5ab - 4a^2b + 3ab^2)$

14. $-xy^4(5x^2y - 4xy^2 + 3x^2y^2)$

15. $(x + 8)(x + 12)$

16. $(x - 9)(x + 6)$

17. $(n - 4)(n - 12)$

18. $(n + 6)(n - 10)$

19. $(s - t)(x + y)$

20. $(a + b)(c + d)$

21. $(3x - 1)(2x + 3)$

22. $(5x + 2)(3x + 4)$

23. $(4x - 3)(3x - 7)$

24. $(4n + 3)(6n - 1)$

25. $(x + 4)^2$

26. $(x - 6)^2$

27. $(2n + 3)^2$

28. $(3n - 5)^2$

29. $(x + 2)(x - 4)(x + 3)$

30. $(x - 1)(x + 6)(x - 5)$

31. $(x- 1)(2x + 3)(3x - 2)$

32. $(2x + 5)(x - 4)(3x + 1)$

33. $(x -1)(x^2 + 3x - 4)$

34. $(t + 1)(t^2 - 2t - 4)$

35. $(t - 1)(t^2 + t + 1)$

36. $(2x - 1)(x^2 + 4x + 3)$

37. $(3x + 2)(2x^2 - x - 1)$

38. $(3x - 2)(2x^2 + 3x + 4)$

39. $(x^2 + 2x - 1)(x^2 + 6x + 4)$

40. $(x^2 - x + 4)(2x^2 - 3x - 1)$

41. $(5x - 2)(5x + 2)$

42. $(3x - 4)(3x + 4)$

43. $(x^2 - 5x - 2)^2$

44. $(-x^2 + x - 1)^2$

45. $(2x + 3y)(2x - 3y)$

46. $(9x + y)(9x - y)$

47. $(x + 5)^3$

48. $(x - 6)^3$

49. $(2x + 1)^3$

50. $(3x + 4)^3$

51. $(4x - 3)^3$

52. $(2x - 5)^3$

53. $(5x - 2y)^3$

54. $(x + 3y)^3$

For Problems 55–66, use Pascal's triangle to help expand each of the following.

55. $(a + b)^7$

56. $(a + b)^8$

57. $(x - y)^5$

58. $(x - y)^6$

59. $(x + 2y)^4$

60. $(2x + y)^5$

61. $(2a - b)^6$

62. $(3a - b)^4$

63. $(x^2 + y)^7$

64. $(x + 2y^2)^7$

65. $(2a - 3b)^5$

66. $(4a - 3b)^3$

In Problems 67–72, perform the indicated divisions.

67. $\dfrac{15x^4 - 25x^3}{5x^2}$

68. $\dfrac{-48x^8 - 72x^6}{-8x^4}$

69. $\dfrac{30a^5 - 24a^3 + 54a^2}{-6a}$

70. $\dfrac{18x^3y^2 + 27x^2y^3}{3xy}$

71. $\dfrac{-20a^3b^2 - 44a^4b^5}{-4a^2b}$

72. $\dfrac{21x^5y^6 + 28x^4y^3 - 35x^5y^4}{7x^2y^3}$

In Problems 73–82, find the indicated products. Assume all variables that appear as exponents represent integers.

73. $(x^a + y^b)(x^a - y^b)$

74. $(x^{2a} + 1)(x^{2a} - 3)$

75. $(x^b + 4)(x^b - 7)$

76. $(3x^a - 2(x^a + 5)$

77. $(2x^b - 1)(3x^b + 2)$

78. $(2x^a - 3)(2x^a + 3)$

79. $(x^{2a} - 1)^2$

80. $(x^{3b} + 2)^2$

81. $(x^a - 2)^3$

82. $(x^b + 3)^3$

THOUGHTS into WORDS

83. Describe how to multiply two binomials.

84. Describe how to multiply a binomial and a trinomial.

1.4 FACTORING POLYNOMIALS

If a polynomial is equal to the product of other polynomials, then each polynomial in the product is called a **factor** of the original polynomial. For example, since $x^2 - 4$ can be expressed as $(x + 2)(x - 2)$, we say that $x + 2$ and $x - 2$ are factors of $x^2 - 4$. The process of expressing a polynomial as a product of polynomials is called **factoring**. In this section we will consider methods of factoring polynomials with integer coefficients.

In general, factoring is the reverse of multiplication, so we can use our knowledge of multiplication to help develop factoring techniques. For example, we previously used the distributive property to find the product of a monomial and a polynomial, as the next examples illustrate.

$$3(x + 2) = 3(x) + 3(2) = 3x + 6$$

$$3x(x + 4) = 3x(x) + 3x(4) = 3x^2 + 12x$$

For factoring purposes, the distributive property (now in the form $ab + ac = a(b + c)$) can be used to reverse the process. (The steps indicated in the dashed boxes can be done mentally.)

$$3x + 6 = 3(x) + 3(2) = 3(x + 2)$$

$$3x^2 + 12x = 3x(x) + 3x(4) = 3x(x + 4)$$

Polynomials can be factored in a variety of ways. Consider some factorizations of $3x^2 + 12x$.

$$3x^2 + 12x = 3x(x + 4) \quad \text{or} \quad 3x^2 + 12x = 3(x^2 + 4x) \quad \text{or}$$

$$3x^2 + 12x = x(3x + 12) \quad \text{or} \quad 3x^2 + 12x = \frac{1}{2}(6x^2 + 24x)$$

We are, however, primarily interested in the first of these factorization forms; we shall refer to it as the **completely factored form**. A polynomial with integral coefficients is in completely factored form if:

1. it is expressed as a product of polynomials with *integral coefficients*, and
2. no polynomial, other than a monomial, within the factored form can be further factored into polynomials with integral coefficients.

Do you see why only the first of the factored forms of $3x^2 + 12x$ is said to be in completely factored form? In each of the other three forms, the polynomial inside the parentheses can be further factored. Furthermore, in the last form, $\frac{1}{2}(6x^2 + 24x)$, the condition of using only integers is violated.

This application of the distributive property is often referred to as **factoring out the highest common monomial factor**. The following examples further illustrate the process.

$$12x^3 + 16x^2 = 4x^2(3x + 4)$$

$$8ab - 18b = 2b(4a - 9)$$

$$6x^2y^3 + 27xy^4 = 3xy^3(2x + 9y)$$

$$30x^3 + 42x^4 - 24x^5 = 6x^3(5 + 7x - 4x^2)$$

Sometimes there may be a common *binomial* factor rather than a common monomial factor. For example, each of the two terms in the expression $x(y + 2) + z(y + 2)$ has a binomial factor of $y + 2$. Thus we can factor $y + 2$ from each term and obtain the following result.

$$x(y + 2) + z(y + 2) = (y + 2)(x + z)$$

Consider a few more examples involving a common binomial factor.

$$a^2(b + 1) + 2(b + 1) = (b + 1)(a^2 + 2)$$

$$x(2y - 1) - y(2y - 1) = (2y - 1)(x - y)$$

$$x(x + 2) + 3(x + 2) = (x + 2)(x + 3)$$

It may seem as if a given polynomial exhibits no apparent common monomial or binomial factor. Such is the case with $ab + 3c + bc + 3a$. However, by using the commutative property to rearrange the terms it can be factored as follows.

$ab + 3c + bc + 3a = ab + 3a + bc + 3c$

$= a(b + 3) + c(b + 3)$ — Factor a from the first two terms and c from the last two terms.

$= (b + 3)(a + c)$ — Factor $b + 3$ from both terms.

This factoring process is referred to as **factoring by grouping**. Let's consider another example of this type.

$ab^2 - 4b^2 + 3a - 12 = b^2(a - 4) + 3(a - 4)$ — Factor b^2 from first two terms, 3 from last two.

$= (a - 4)(b^2 + 3).$ — Factor common binomial from both terms.

Difference of Two Squares

In Section 1.3 we called your attention to some special multiplication patterns. One of these patterns was the following.

$$(a + b)(a - b) = a^2 - b^2$$

This same pattern, viewed as a factoring pattern,

$$a^2 - b^2 = (a + b)(a - b)$$

is referred to as the **difference of two squares**. Applying the pattern is a fairly simple process, as these next examples illustrate. Again, the steps we have included in dashed boxes are usually performed mentally.

$$x^2 - 16 = \boxed{(x)^2 - (4)^2} = (x + 4)(x - 4)$$

$$4x^2 - 25 = \boxed{(2x^2) - (5)^2} = (2x + 5)(2x - 5)$$

Since multiplication is commutative, the order of writing the factors is not important. For example, $(x + 4)(x - 4)$ can also be written $(x - 4)(x + 4)$.

You must be careful not to assume an analogous factoring pattern for the *sum* of two squares; *it does not exist*. For example, $x^2 + 4 \neq (x + 2)(x + 2)$ because $(x + 2)(x + 2) = x^2 + 4x + 4$. We say that a polynomial such as $x^2 + 4$ is **not factorable using integers**.

Sometimes the difference-of-two-squares pattern can be applied more than once, as the next example illustrates.

$$16x^4 - 81y^4 = (4x^2 + 9y^2)(4x^2 - 9y^2) = (4x^2 + 9y^2)(2x + 3y)(2x - 3y)$$

It may also happen that the squares are not just simple monomial squares. These next three examples illustrate such polynomials.

$$(x + 3)^2 - y^2 = [(x + 3) + y][(x + 3) - y] = (x + 3 + y)(x + 3 - y)$$

$$\begin{aligned} 4x^2 - (2y + 1)^2 &= [2x + (2y + 1)][2x - (2y + 1)] \\ &= (2x + 2y + 1)(2x - 2y - 1) \end{aligned}$$

$$\begin{aligned} (x - 1)^2 - (x + 4)^2 &= [(x - 1) + (x + 4)][(x - 1) - (x + 4)] \\ &= (x - 1 + x + 4)(x - 1 - x - 4) \\ &= (2x + 3)(-5) \end{aligned}$$

It is possible that both the technique of factoring out a common monomial factor and the pattern of the difference of two squares can be applied to the same problem. *In general, it is best to look first for a common monomial factor.* Consider the following examples.

$$\begin{aligned} 2x^2 - 50 &= 2(x^2 - 25) \\ &= 2(x + 5)(x - 5) \end{aligned}$$

$$48y^3 - 27y = 3y(16y^2 - 9)$$
$$= 3y(4y + 3)(4y - 3)$$
$$9x^2 - 36 = 9(x^2 - 4)$$
$$= 9(x + 2)(x - 2)$$

Factoring Trinomials

Expressing a trinomial as the product of two binomials is one of the most common factoring techniques used in algebra. Like before, to develop a factoring technique we first look at some multiplication ideas. Let's consider the product $(x + a) \cdot (x + b)$, using the distributive property to show how each term of the resulting trinomial is formed.

$$(x + a)(x + b) = x(x + b) + a(x + b)$$
$$= x(x) + x(b) + a(x) + a(b)$$
$$= x^2 + (a + b)x + ab$$

Notice that the coefficient of the middle term is the *sum* of a and b and that the last term is the *product* of a and b. These two relationships can be used to factor trinomials. Let's consider some examples.

EXAMPLE 1

Factor $x^2 + 12x + 20$.

Solution

We need two integers whose sum is 12 and whose product is 20. The numbers are 2 and 10, and we can complete the factoring as follows.

$$x^2 + 12x + 20 = (x + 2)(x + 10)$$

EXAMPLE 2

Factor $x^2 - 3x - 54$.

Solution

We need two integers whose sum is -3 and whose product is -54. The integers are -9 and 6, and we can factor as follows.

$$x^2 - 3x - 54 = (x - 9)(x + 6)$$

EXAMPLE 3

Factor $x^2 + 7x + 16$.

Solution

We need two integers whose sum is 7 and whose product is 16. The only possible pairs of factors of 16 are $1 \cdot 16$, $2 \cdot 8$, and $4 \cdot 4$. Since a sum of 7 is not produced by any of these pairs, the polynomial $x^2 + 7x + 16$ is *not factorable using integers*.

Trinomials of the Form $ax^2 + bx + c$

Now let's consider factoring trinomials where the coefficient of the squared term is not one. First, let's illustrate an informal trial-and-error technique that works quite well for certain types of trinomials. This technique is based on our knowledge of multiplication of binomials.

EXAMPLE 4

Factor $3x^2 + 5x + 2$.

Solution

By looking at the first term, $3x^2$, and the positive signs of the other two terms, we know that the binomials are of the form

$$(x + \_\_)(3x + \_\_).$$

Since the factors of the last term, 2, are 1 and 2, we have only the following two possibilities to try.

$$(x + 2)(3x + 1) \qquad \text{or} \qquad (x + 1)(3x + 2)$$

By checking the middle term formed in each of these products, we find that the second possibility yields the desired middle term of $5x$. Therefore,

$$3x^2 + 5x + 2 = (x + 1)(3x + 2).$$

EXAMPLE 5

Factor $8x^2 - 30x + 7$.

Solution

First, observe that the first term, $8x^2$, can be written as $2x \cdot 4x$ or $x \cdot 8x$. Secondly, since the middle term is negative and the last term is positive, we know that the binomials are of the form

$$(2x - \_\_)(4x - \_\_) \qquad \text{or} \qquad (x - \_\_)(8x - \_\_)$$

Thirdly, since the factors of the last term, 7, are 1 and 7, the following possibilities exist.

$$(2x - 1)(4x - 7) \qquad (2x - 7)(4x - 1)$$
$$(x - 1)(8x - 7) \qquad (x - 7)(8x - 1)$$

By checking the middle term formed in each of these products, we find that $(2x - 7)(4x - 1)$ produces the desired middle term of $-30x$. Therefore,

$$8x^2 - 30x + 7 = (2x - 7)(4x - 1).$$

EXAMPLE 6

Factor $5x^2 - 18x - 8$.

Solution

The first term, $5x^2$, can be written as $x \cdot 5x$. The last term, -8, can be written as

$(-2)(4)$, $(2)(-4)$, $(-1)(8)$, or $(1)(-8)$. Therefore, we have the following possibilities to try.

$$\begin{array}{ll} (x-2)(5x+4) & (x+4)(5x-2) \\ (x+2)(5x-4) & (x-4)(5x+2) \\ (x-1)(5x+8) & (x+8)(5x-1) \\ (x+1)(5x-8) & (x-8)(5x+1) \end{array}$$

By checking the middle terms, we find that $(x - 4)(5x + 2)$ yields the desired middle term of $-18x$. Thus,

$$5x^2 - 18x - 8 = (x - 4)(5x + 2).$$

EXAMPLE 7

Factor $4x^2 + 6x + 9$.

Solution

The first term, $4x^2$, and the positive signs of the middle and last terms indicate that the binomials are of the form

$$(x + \_\_)(4x + \_\_) \quad \text{or} \quad (2x + \_\_)(2x + \_\_).$$

Since the factors of the last term, 9, are 1 and 9 or 3 and 3, we have the following possibilities to try.

$$\begin{array}{ll} (x+1)(4x+9) & (x+9)(4x+1) \\ (x+3)(4x+3) & (2x+1)(2x+9) \\ (2x+3)(2x+3) & \end{array}$$

None of these possibilities yields a middle term of $6x$. Therefore, $4x^2 + 6x + 9$ is *not factorable using integers.*

Certainly as the number of possibilities increases, this trial-and-error technique for factoring becomes more tedious. The key idea is to organize your work so that all possibilities are considered. We have suggested one possible format in the previous examples. However, as you practice such problems, you may devise a format that works better for you. Whatever works best for you is the right approach.

There is another more systematic technique that you may wish to use with some trinomials. It is an extension of the technique we used earlier with trinomials where the coefficient of the squared term was one. To see the basis of this technique, consider the following general product.

$$\begin{aligned} (px + r)(qx + s) &= px(qx) + px(s) + r(qx) + r(s) \\ &= (pq)x^2 + ps(x) + rq(x) + rs \\ &= (pq)x^2 + (ps + rq)x + rs \end{aligned}$$

Notice that the product of the coefficient of x^2 and the constant term is $pqrs$. Likewise, the product of the two coefficients of x (ps and rq) is also $pqrs$. Therefore, the coefficient of x must be a sum of the form $ps + rq$, such that the product of the coefficient of x^2 and the constant term is $pqrs$. Now let's see how this works in some specific examples.

EXAMPLE 8

Factor $6x^2 + 17x + 5$.

Solution

$6x^2 + 17x + 5$ Sum of 17

Product of $6 \cdot 5 = 30$

We need two integers whose sum is 17 and whose product is 30. The integers 2 and 15 satisfy these conditions. Therefore, the middle term, $17x$, of the given trinomial can be expressed as $2x + 15x$ and we can proceed as follows.

$$\begin{aligned} 6x^2 + 17x + 5 &= 6x^2 + 2x + 15x + 5 \\ &= 2x(3x + 1) + 5(3x + 1) \\ &= (3x + 1)(2x + 5) \end{aligned}$$

EXAMPLE 9

Factor $5x^2 - 18x - 8$.

Solution

$5x^2 - 18x - 8$ Sum of -18

Product of $5(-8) = -40$

We need two integers whose sum is -18 and whose products is -40. The integers -20 and 2 satisfy these conditions. Therefore the middle term, $-18x$, of the trionomial can be written $-20x + 2x$ and we can factor as follows.

$$\begin{aligned} 5x^2 - 18x - 8 &= 5x^2 - 20x + 2x - 8 \\ &= 5x(x - 4) + 2(x - 4) \\ &= (x - 4)(5x + 2) \end{aligned}$$

EXAMPLE 10

Factor $24x^2 + 2x - 15$.

Solution

$24x^2 + 2x - 15$ Sum of 2

Product of $24(-15) = -360$

We need two integers whose sum is 2 and whose product is -360. To help find these integers, let's factor 360 into primes.

$$360 = 2 \cdot 2 \cdot 2 \cdot 3 \cdot 3 \cdot 5$$

Now by grouping these factors in various ways, we find that $2 \cdot 2 \cdot 5 = 20$ and $2 \cdot 3 \cdot 3 = 18$, so we can use the integers 20 and -18 to produce a sum of 2 and a product of -360. Therefore the middle term, $2x$, of the trinomial can be expressed as $20x - 18x$ and we can proceed as follows.

$$\begin{aligned} 24x^2 + 2x - 15 &= 24x^2 + 20x - 18x - 15 \\ &= 4x(6x + 5) - 3(6x + 5) \\ &= (6x + 5)(4x - 3) \end{aligned}$$

Sum and Difference of Two Cubes

Earlier in this section we discussed the difference-of-squares factoring pattern. We pointed out that no analogous sum-of-squares pattern exists; that is to say, a polynomial such as $x^2 + 9$ is not factorable using integers. However, there do exist patterns for both the *sum* and the *difference of two cubes*. These patterns come from the following special products.

$$\begin{aligned} (x + y)(x^2 - xy + y^2) &= x(x^2 - xy + y^2) + y(x^2 - xy + y^2) \\ &= x^3 - x^2y + xy^2 + x^2y - xy^2 + y^3 \\ &= x^3 + y^3 \end{aligned}$$

$$\begin{aligned} (x - y)(x^2 + xy + y^2) &= x(x^2 + xy + y^2) - y(x^2 + xy + y^2) \\ &= x^3 + x^2y + xy^2 - x^2y - xy^2 - y^3 \\ &= x^3 - y^3 \end{aligned}$$

Thus, we can state the following factoring patterns.

$$x^3 + y^3 = (x + y)(x^2 - xy + y^2)$$
$$x^3 - y^3 = (x - y)(x^2 + xy + y^2)$$

Note how these patterns are used in the next three examples.

$$x^3 + 8 = x^3 + 2^3 = (x + 2)(x^2 - 2x + 4)$$

$$8x^3 - 27y^3 = (2x)^3 - (3y)^3 = (2x - 3y)(4x^2 + 6xy + 9y^2)$$

$$8a^6 + 125b^3 = (2a^2)^3 + (5b)^3 = (2a^2 + 5b)(4a^4 - 10a^2b + 25b^2)$$

We do want to leave you with one final word of caution: **Be sure to factor completely.** Sometimes more than one technique needs to be applied, or perhaps the same technique can be applied more than once. Study the following examples very carefully.

$$2x^2 - 8 = 2(x^2 - 4) = 2(x + 2)(x - 2)$$

$$3x^2 + 18x + 24 = 3(x^2 + 6x + 8) = 3(x + 4)(x + 2)$$

$$3x^3 - 3y^3 = 3(x^3 - y^3) = 3(x - y)(x^2 + xy + y^2)$$

$$a^4 - b^4 = (a^2 + b^2((a^2 - b^2) = (a^2 + b^2)(a + b)(a - b)$$

$$x^4 - 6x^2 - 27 = (x^2 - 9)(x^2 + 3) = (x + 3)(x - 3)(x^2 + 3)$$

$$\begin{aligned} 3x^4y + 9x^2y - 84y &= 3y(x^4 + 3x^2 - 28) \\ &= 3y(x^2 + 7)(x^2 - 4) \\ &= 3y(x^2 + 7)(x + 2)(x - 2) \end{aligned}$$

$$\begin{aligned} x^2 - y^2 + 8y - 16 &= x^2 - (y^2 - 8y + 16) \\ &= x^2 - (y - 4)^2 \\ &= (x - (y - 4))(x + (y - 4)) \\ &= (x - y + 4)(x + y - 4) \end{aligned}$$

PROBLEM SET 1.4

Factor each of the following completely. Indicate any that are not factorable using integers.

1. $6xy - 8xy^2$

2. $4a^2b^2 + 12ab^3$

3. $x(z + 3) + y(z + 3)$

4. $5(x + y) + a(x + y)$

5. $3x + 3y + ax + ay$

6. $ac + bc + a + b$

7. $ax - ay - bx + by$

8. $2a^2 - 3bc - 2ab + 3ac$

9. $9x^2 - 25$

10. $4x^2 + 9$

11. $1 - 81n^2$

12. $9x^2y^2 - 64$

13. $(x + 4)^2 - y^2$

14. $x^2 - (y - 1)^2$

15. $9s^2 - (2t - 1)^2$

16. $4a^2 - (3b + 1)^2$

17. $x^2 - 5x - 14$

18. $a^2 + 5a - 24$

19. $15 - 2x - x^2$

20. $40 - 6x - x^2$

21. $x^2 + 7x - 36$

22. $x^2 - 4xy - 5y^2$

23. $3x^2 - 11x + 10$

24. $2x^2 - 7x - 30$

25. $10x^2 - 33x - 7$

26. $8y^2 + 22y - 21$

27. $x^3 - 8$

28. $x^3 + 64$

29. $64x^3 + 27y^3$

30. $27x^3 - 8y^3$

31. $4x^2 + 16$

32. $n^3 - 49n$

33. $x^3 - 9x$

34. $12n^2 + 59n + 72$

35. $9a^2 - 42a + 49$

36. $1 - 16x^4$

37. $2n^3 + 6n^2 + 10n$

38. $x^2 - (y - 7)^2$

39. $10x^2 + 39x - 27$

40. $3x^2 + x - 5$

41. $36a^2 - 12a + 1$

42. $18n^3 + 39n^2 - 15n$

43. $8x^2 + 2xy - y^2$

44. $12x^2 + 7xy - 10y^2$

45. $2n^2 - n - 5$

46. $25t^2 - 100$

47. $2n^3 + 14n^2 - 20n$

48. $25n^2 + 64$

49. $4x^3 + 32$

50. $2x^3 - 54$

51. $x^4 - 4x^2 - 45$

52. $x^4 - x^2 - 12$

53. $2x^4y - 26x^2y - 96y$

54. $3x^4y - 15x^2y - 108y$

55. $(a + b)^2 - (c + d)^2$

56. $(a - b)^2 - (c - d)^2$

57. $x^2 + 8x + 16 - y^2$

58. $4x^2 + 12x + 9 - y^2$

59. $x^2 - y^2 - 10y - 25$

60. $y^2 - x^2 + 16x - 64$

61. $60x^2 - 32x - 15$

62. $40x^2 + 37x - 63$

63. $84x^3 + 57x^2 - 60x$

64. $210x^3 - 102x^2 - 180x$

For Problems 65–74, factor each of the following, and assume that all variables appearing as exponents represent integers.

65. $x^{2a} - 16$

66. $x^{4n} - 9$

67. $x^{3n} - y^{3n}$

68. $x^{3a} + y^{6a}$

69. $x^{2a} - 3x^a - 28$

70. $x^{2a} + 10x^a + 21$

71. $2x^{2n} + 7x^n - 30$

72. $3x^{2n} - 16x^n - 12$

73. $x^{4n} - y^{4n}$

74. $16x^{2a} + 24x^a + 9$

75. Suppose that we want to factor $x^2 + 34x + 288$. We need to complete the following with two numbers whose sum is 34 and whose product is 288.

$$x^2 + 34x + 288 = (x + \_\_)(x + \_\_)$$

These numbers can be found as follows: Since we need a product of 288, let's consider the prime factorization of 288.

$$288 = 2^5 \cdot 3^2$$

Now we need to use five 2s and two 3s in the statement

$$(\quad) + (\quad) = 34$$

Since 34 is divisible by 2 but not by 4, four factors of 2 must be in one number and one factor of 2 in the other number. Also, since 34 is not divisible by 3, both factors of 3 must be in the same number. These facts aid us in determining that

$$(2 \cdot 2 \cdot 2 \cdot 2) + (2 \cdot 3 \cdot 3) = 34$$

or

$$16 \qquad + \qquad 18 \qquad = 34.$$

Thus, we can complete the original factoring problem.

$$x^2 + 34x + 288 + (x + 16)(x + 18)$$

Use this approach to factor each of the following expressions.

a. $x^2 + 35x + 96$

b. $x^2 + 27x + 176$

c. $x^2 - 45x + 504$

d. $x^2 - 26x + 168$

e. $x^2 + 60x + 896$

f. $x^2 - 84x + 1728$

THOUGHTS into WORDS

76. Describe, in words, the pattern for factoring the sum of two cubes.

77. What does it mean to say that the polynomial $x^2 + 5x + 7$ is not factorable using integers?

78. What role does the distributive property play when factoring polynomials?

1.5 RATIONAL EXPRESSIONS

Indicated quotients of algebraic expressions are called **algebraic fractions**, or **fractional expressions**. The indicated quotient of two polynomials is called a **rational expression**. (This is analogous to defining a rational number as the indicated quotient of two integers.) The following are examples of rational expressions.

$$\frac{3x^2}{5} \qquad \frac{x-2}{x+3} \qquad \frac{x^2+5x-1}{x^2-9} \qquad \frac{xy^2+x^2y}{xy} \qquad \frac{a^3-3a^2-5a-1}{a^4+a^3+6}$$

Because division by zero must be avoided, no values can be assigned to variables that will create a denominator of zero. Thus, the rational expression $\frac{x-2}{x+3}$ is meaningful for all real number values of x except $x = -3$. Rather than making restrictions for each individual expression, we will merely assume that **all denominators represent nonzero real numbers**.

The basic properties of the real numbers can be used for working with rational expressions. For example, the property

$$\frac{a \cdot k}{b \cdot k} = \frac{a}{b},$$

which is used to reduce rational numbers, is also used to *simplify* rational expressions. Consider the following examples.

$$\frac{15xy}{25y} = \frac{3 \cdot \cancel{5} \cdot x \cdot \cancel{y}}{\cancel{5} \cdot 5 \cdot \cancel{y}} = \frac{3x}{5}$$

$$\frac{-9}{18x^2y} = -\frac{\overset{1}{\cancel{9}}}{\underset{2}{\cancel{18}}x^2y} = -\frac{1}{2x^2y}$$

Notice that slightly different formats were used in these two examples. In the first one we factored the coefficients into primes and then proceeded to simplify; however, in the second problem we simply divided a common factor of 9 out of both the numerator and the denominator. This is basically a format issue and depends upon your personal preference. Also notice that in the second example, we applied the property $\frac{-a}{b} = -\frac{a}{b}$. This is part of the general property that states

$$\frac{-a}{b} = \frac{a}{-b} = -\frac{a}{b}.$$

The factoring techniques discussed in the previous section can be used to factor numerators and denominators so that the property $(a \cdot k)/(b \cdot k) = a/b$ can be applied. Consider the following examples.

$$\frac{x^2+4x}{x^2-16} = \frac{x\cancel{(x+4)}}{(x-4)\cancel{(x+4)}} = \frac{x}{x-4}$$

$$\frac{5n^2+6n-8}{10n^2-3n-4} = \frac{\cancel{(5n-4)}(n+2)}{\cancel{(5n-4)}(2n+1)} = \frac{n+2}{2n+1}$$

$$\frac{x^3+y^3}{x^2+xy+2x+2y} = \frac{(x+y)(x^2-xy+y^2)}{x(x+y)+2(x+y)}$$

$$= \frac{\cancel{(x+y)}(x^2-xy+y^2)}{\cancel{(x+y)}(x+2)} = \frac{x^2-xy+y^2}{x+2}$$

$$\frac{6x^3y - 6xy}{x^3 + 5x^2 + 4x} = \frac{6xy(x^2 - 1)}{x(x^2 + 5x + 4)} = \frac{6\cancel{x}y\cancel{(x + 1)}(x - 1)}{\cancel{x}\cancel{(x + 1)}(x + 4)} = \frac{6y(x - 1)}{x + 4}$$

Note that in the last example we left the numerator of the final fraction in factored form. This is often done if expressions other than monomials are involved. Either

$$\frac{6y(x - 1)}{x + 4} \quad \text{or} \quad \frac{6xy - 6y}{x + 4}$$

is an acceptable answer.

Remember that the quotient of any nonzero real number and its opposite is -1. For example, $6/-6 = -1$ and $-8/8 = -1$. Likewise, the indicated quotient of any polynomial and its opposite is equal to -1. For example,

$\frac{a}{-a} = -1$ because a and $-a$ are opposites,

$\frac{a - b}{b - a} = -1$ because $a - b$ and $b - a$ are opposites,

$\frac{x^2 - 4}{4 - x^2} = -1$ because $x^2 - 4$ and $4 - x^2$ are opposites.

The next example illustrates the use of this idea when simplifying rational expressions.

$$\begin{aligned} \frac{4 - x^2}{x^2 + x - 6} &= \frac{(2 + x)}{(x + 3)}\boxed{\frac{(2 - x)}{(x - 2)}} \\ &= (-1)\left(\frac{x + 2}{x + 3}\right) \qquad \frac{2 - x}{x - 2} = -1 \\ &= -\frac{x + 2}{x + 3} \quad \text{or} \quad \frac{-x - 2}{x + 3} \end{aligned}$$

Multiplying and Dividing Rational Expressions

Multiplication of rational expressions is based on the following property.

$$\frac{a}{b} \cdot \frac{c}{d} = \frac{ac}{bd}$$

In other words, we multiply numerators and we multiply denominators and express the final product in simplified form. Study the following examples carefully and pay special attention to the formats used to organize the computational work.

$$\frac{3x}{4y} \cdot \frac{8y^2}{9x} = \frac{\cancel{3} \cdot \overset{2}{\cancel{8}} \cdot \cancel{x} \cdot \overset{y}{\cancel{y^2}}}{\cancel{4} \cdot \underset{3}{\cancel{9}} \cdot \cancel{x} \cdot \cancel{y}} = \frac{2y}{3}$$

$$\frac{12x^2y}{-18xy} \cdot \frac{-24xy^2}{56y^3} = \frac{\overset{2}{\cancel{12}} \cdot \overset{8}{\cancel{24}} \cdot \overset{x^2}{\cancel{x^3}} \cdot \cancel{y^3}}{\underset{3}{\cancel{18}} \cdot \underset{7}{\cancel{56}} \cdot \cancel{x} \cdot \underset{y}{\cancel{y^4}}} = \frac{2x^2}{7y}$$

$\frac{12x^2y}{-18xy} = -\frac{12x^2y}{18xy}$ and $\frac{-24xy^2}{56y^3} = -\frac{24xy^2}{56y^3}$

so the product is positive.

$$\frac{y}{x^2-4} \cdot \frac{x+2}{y^2} = \frac{\cancel{y(x+2)}}{\underset{y}{\cancel{y^2}}\cancel{(x+2)}(x-2)} = \frac{1}{y(x-2)}$$

$$\frac{x^2-x}{x+5} \cdot \frac{x^2+5x+4}{x^4-x^2} = \frac{\cancel{x(x-1)(x+1)}(x+4)}{(x+5)\underset{x}{\cancel{(x^2)}}\cancel{(x+1)(x-1)}} = \frac{x+4}{x(x+5)}$$

To divide rational expressions we merely apply the following property.

$$\frac{a}{b} \div \frac{c}{d} = \frac{a}{b} \cdot \frac{d}{c} = \frac{ad}{bc}$$

That is to say, the quotient of two rational expressions is the product of the first expression times the reciprocal of the second. Consider the following examples.

$$\frac{16x^2y}{24xy^3} \div \frac{9xy}{8x^2y^2} = \frac{16x^2y}{24xy^3} \cdot \frac{8x^2y^2}{9xy} = \frac{16 \cdot \cancel{8} \cdot \overset{x^2}{\cancel{x^4}} \cdot \cancel{y^3}}{\underset{3}{\cancel{24}} \cdot 9 \cdot \cancel{x^2} \cdot \underset{y}{\cancel{y^4}}} = \frac{16x^2}{27y}$$

$$\begin{aligned}\frac{3a^2+12}{3a^2-15a} \div \frac{a^4-16}{a^2-3a-10} &= \frac{3a^2+12}{3a^2-15a} \cdot \frac{a^2-3a-10}{a^4-16} \\ &= \frac{\cancel{3(a^2+4)(a-5)(a+2)}}{\cancel{3}a\cancel{(a-5)(a^2+4)(a+2)}(a-2)} \\ &= \frac{1}{a(a-2)}\end{aligned}$$

Adding and Subtracting Rational Expressions

The following two properties provide the basis for adding and subtracting rational expressions.

$$\frac{a}{b} + \frac{c}{b} = \frac{a+c}{b}$$

$$\frac{a}{b} - \frac{c}{b} = \frac{a-c}{b}$$

These properties state that rational expressions with a common denominator can be added (or subtracted) by adding (or subtracting) the numerators and placing the result over the common denominator. Let's illustrate this idea.

$$\frac{8}{x-2} + \frac{3}{x-2} = \frac{8+3}{x-2} = \frac{11}{x-2}$$

$$\frac{9}{4y} - \frac{7}{4y} = \frac{9-7}{4y} = \frac{2}{4y} = \frac{1}{2y}$$

Don't forget to simplify the final result.

$$\frac{n^2}{n-1} - \frac{1}{n-1} = \frac{n^2-1}{n-1} = \frac{(n+1)\cancel{(n-1)}}{\cancel{n-1}} = n+1$$

If we need to add or subtract rational expressions that do not have a common denominator, then we apply the property $a/b = (a \cdot k)/(b \cdot k)$ to obtain equivalent fractions with a common denominator. Study the next examples and again pay special attention to the format we used to organize our work.

> **REMARK** Remember that the **least common multiple** of a set of whole numbers is the smallest nonzero whole number divisible by each of the numbers in the set. When we add or subtract rational numbers, the least common multiple of the denominators of those numbers is the **least common denominator (LCD)**. This concept of a least common denominator can be extended to include polynomials.

EXAMPLE 1

Add $\frac{x+2}{4} + \frac{3x+1}{3}$.

Solution

By inspection we see that the LCD is 12.

$$\begin{aligned}\frac{x+2}{4} + \frac{3x+1}{3} &= \left(\frac{x+2}{4}\right)\left(\frac{3}{3}\right) + \left(\frac{3x+1}{3}\right)\left(\frac{4}{4}\right)\\ &= \frac{3(x+2)}{12} + \frac{4(3x+1)}{12}\\ &= \frac{3x+6+12x+4}{12}\\ &= \frac{15x+10}{12}\end{aligned}$$

EXAMPLE 2

Perform the indicated operations.

$$\frac{x+3}{10} + \frac{2x+1}{15} - \frac{x-2}{18}$$

Solution

If you cannot determine the LCD by inspection, then use the prime-factored forms of the denominators.

$$10 = 2 \cdot 5 \qquad 15 = 3 \cdot 5 \qquad 18 = 2 \cdot 3 \cdot 3$$

The LCD must contain one factor of 2, two factors of 3, and one factor of 5. Thus the LCD is $2 \cdot 3 \cdot 3 \cdot 5 = 90$.

$$\begin{aligned}\frac{x+3}{10} + \frac{2x+1}{15} - \frac{x-2}{18} &= \left(\frac{x+3}{10}\right)\left(\frac{9}{9}\right) + \left(\frac{2x+1}{15}\right)\left(\frac{6}{6}\right) - \left(\frac{x-2}{18}\right)\left(\frac{5}{5}\right)\\ &= \frac{9(x+3)}{90} + \frac{6(2x+1)}{90} - \frac{5(x-2)}{90}\\ &= \frac{9x + 27 + 12x + 6 - 5x + 10}{90}\\ &= \frac{16x+43}{90}\end{aligned}$$

The presence of variables in the denominators does not create any serious difficulty; our approach remains the same. Study the following examples very carefully. For each problem notice the same basic procedure: (1) find the LCD; (2) change each fraction to an equivalent fraction having the LCD as its denominator; (3) add or subtract numerators and place this result over the LCD; and (4) look for possibilities to simplify the resulting fraction.

EXAMPLE 3

Add $\frac{3}{2x} + \frac{5}{3y}$.

Solution

Using an LCD of $6xy$, we can proceed as follows.

$$\begin{aligned}\frac{3}{2x} + \frac{5}{3y} &= \left(\frac{3}{2x}\right)\left(\frac{3y}{3y}\right) + \left(\frac{5}{3y}\right)\left(\frac{2x}{2x}\right)\\ &= \frac{9y}{6xy} + \frac{10x}{6xy} = \frac{9y+10x}{6xy}\end{aligned}$$

EXAMPLE 4

Subtract $\frac{7}{12ab} - \frac{11}{15a^2}$.

Solution

We can factor the numerical coefficients of the denominators into primes to help find the LCD.

$$\left.\begin{aligned}12ab &= 2 \cdot 2 \cdot 3 \cdot a \cdot b\\ 15a^2 &= 3 \cdot 5 \cdot a^2\end{aligned}\right\} \quad \text{LCD} = 2 \cdot 2 \cdot 3 \cdot 5 \cdot a^2 \cdot b = 60a^2b$$

$$\begin{aligned}\frac{7}{12ab} - \frac{11}{15a^2} &= \left(\frac{7}{12ab}\right)\left(\frac{5a}{5a}\right) - \left(\frac{11}{15a^2}\right)\left(\frac{4b}{4b}\right)\\ &= \frac{35a}{60a^2b} - \frac{44b}{60a^2b} = \frac{35a - 44b}{60a^2b}\end{aligned}$$

EXAMPLE 5 Add $\dfrac{8}{x^2 - 4x} + \dfrac{2}{x}$.

Solution

$$\left.\begin{aligned} x^2 - 4x &= x(x-4) \\ x &= x \end{aligned}\right\} \quad \text{LCD} = x(x-4)$$

$$\begin{aligned}
\frac{8}{x(x-4)} + \frac{2}{x} &= \frac{8}{x(x-4)} + \left(\frac{2}{x}\right)\left(\frac{x-4}{x-4}\right) \\
&= \frac{8}{x(x-4)} + \frac{2(x-4)}{x(x-4)} \\
&= \frac{8 + 2x - 8}{x(x-4)} \\
&= \frac{2\not{x}}{\not{x}(x-4)} \\
&= \frac{2}{x-4}
\end{aligned}$$

EXAMPLE 6 Add $\dfrac{3n}{n^2 + 6n + 5} + \dfrac{4}{n^2 - 7n - 8}$.

Solution

$$\left.\begin{aligned} n^2 + 6n + 5 &= (n + 5(n + 1) \\ n^2 - 7n - 8 &= (n - 8)(n + 1) \end{aligned}\right\} \quad \text{LCD} = (n+1)(n+5)(n-8)$$

$$\begin{aligned}
\frac{3n}{n^2 + 6n + 5} + \frac{4}{n^2 - 7n - 8} &= \left[\frac{3n}{(n+5)(n+1)}\right]\left(\frac{n-8}{n-8}\right) + \left[\frac{4}{(n-8)(n+1)}\right]\left(\frac{n+5}{n+5}\right) \\
&= \frac{3n(n-8)}{(n+5)(n+1)(n-8)} + \frac{4(n+5)}{(n+5)(n+1)(n-8)} \\
&= \frac{3n^2 - 24n + 4n + 20}{(n+5)(n+1)(n-8)} \\
&= \frac{3n^2 - 20n + 20}{(n+5)(n+1)(n-8)}
\end{aligned}$$

Simplifying Complex Fractions

Fractional forms that contain rational expressions in the numerator and/or denominator are called **complex fractions**. The following examples illustrate some approaches to simplifying complex fractions.

EXAMPLE 7

Simplify $\dfrac{\dfrac{3}{x}+\dfrac{2}{y}}{\dfrac{5}{x}-\dfrac{6}{y^2}}$.

Solution A

Treating the numerator as the sum of two rational expressions and the denominator as the difference of two rational expressions, we can proceed as follows.

$$\frac{\frac{3}{x}+\frac{2}{y}}{\frac{5}{x}-\frac{6}{y^2}}=\frac{\left(\frac{3}{x}\right)\left(\frac{y}{y}\right)+\left(\frac{2}{y}\right)\left(\frac{x}{x}\right)}{\left(\frac{5}{x}\right)\left(\frac{y^2}{y^2}\right)-\left(\frac{6}{y^2}\right)\left(\frac{x}{x}\right)}$$

$$=\frac{\frac{3y}{xy}+\frac{2x}{xy}}{\frac{5y^2}{xy^2}-\frac{6x}{xy^2}}=\frac{\frac{3y+2x}{xy}}{\frac{5y^2-6x}{xy^2}}$$

$$=\frac{3y+2x}{\cancel{xy}}\cdot\frac{\overset{y}{\cancel{xy^2}}}{5y^2-6x}$$

$$=\frac{y(3y+2x)}{5y^2-6x}$$

Solution B

The LCD of all four denominators (x, y, x, and y^2) is xy^2. Let's multiply the entire complex fraction by a form of 1, namely, $(xy^2)/(xy^2)$.

$$\frac{\frac{3}{x}+\frac{2}{y}}{\frac{5}{x}-\frac{6}{y^2}}=\left(\frac{\frac{3}{x}+\frac{2}{y}}{\frac{5}{x}-\frac{6}{y^2}}\right)\left(\frac{xy^2}{xy^2}\right)$$

$$=\frac{(xy^2)\left(\frac{3}{x}\right)+(xy^2)\left(\frac{2}{y}\right)}{(xy^2)\left(\frac{5}{x}\right)-(xy^2)\left(\frac{6}{y^2}\right)}$$

$$=\frac{3y^2+2xy}{5y^2-6x}\quad\text{or}\quad\frac{y(3y+2x)}{5y^2-6x}$$

Certainly either approach (Solution A or Solution B) will work with a problem such as Example 7. We suggest that you study Solution B very carefully. This approach works effectively with complex fractions when the LCD of all the denominators is easy to find. Let's look at a type of complex fraction used in certain calculus problems.

EXAMPLE 8

Simplify $\dfrac{\dfrac{1}{x+h} - \dfrac{1}{x}}{h}$.

Solution

$$\frac{\dfrac{1}{x+h} - \dfrac{1}{x}}{\dfrac{h}{1}} = \left[\frac{x(x+h)}{x(x+h)}\right]\left[\frac{\dfrac{1}{x+h} - \dfrac{1}{x}}{\dfrac{h}{1}}\right]$$

$$= \frac{x(x+h)\left(\dfrac{1}{x+h}\right) - x(x+h)\left(\dfrac{1}{x}\right)}{x(x+h)(h)}$$

$$= \frac{x-(x+h)}{hx(x+h)} = \frac{x-x-h}{hx(x+h)}$$

$$= \frac{-h}{hx(x+h)} = -\frac{1}{x(x+h)}$$

Example 9 gives us another way to simplify complex fractions.

EXAMPLE 9

Simplify $1 - \dfrac{n}{1 - \dfrac{1}{n}}$.

Solution

We first simplify the complex fraction by multiplying by n/n.

$$\left(\frac{n}{1 - \dfrac{1}{n}}\right)\left(\frac{n}{n}\right) = \frac{n^2}{n-1}$$

Now we can perform the subtraction.

$$1 - \frac{n^2}{n-1} = \left(\frac{n-1}{n-1}\right)\left(\frac{1}{1}\right) - \frac{n^2}{n-1} = \frac{n-1}{n-1} - \frac{n^2}{n-1}$$

$$= \frac{n-1-n^2}{n-1} \quad \text{or} \quad \frac{-n^2+n-1}{n-1}$$

Finally, we need to recognize that complex fractions are sometimes the result of applying the definition $b^{-n} = \dfrac{1}{b^n}$. Our final example illustrates this idea.

EXAMPLE 10

Simplify $\dfrac{2x^{-1} + y^{-1}}{x - 3y^{-2}}$.

Solution

First, let's apply $b^{-n} = \dfrac{1}{b^n}$.

$$\frac{2x^{-1} + y^{-1}}{x - 3y^{-2}} = \frac{\dfrac{2}{x} + \dfrac{1}{y}}{x - \dfrac{3}{y^2}}$$

Now we can proceed as in the previous examples.

$$\left(\frac{\dfrac{2}{x} + \dfrac{1}{y}}{x - \dfrac{3}{y^2}}\right)\left(\frac{xy^2}{xy^2}\right) = \frac{\dfrac{2}{x}(xy^2) + \dfrac{1}{y}(xy^2)}{x(xy^2) - \dfrac{3}{y^2}(xy^2)}$$

$$= \frac{2y^2 + xy}{x^2y^2 - 3x}$$

PROBLEM SET 1.5

Simplify each of the following rational expressions.

1. $\dfrac{14x^2y}{21xy}$

2. $\dfrac{-26xy^2}{65y}$

3. $\dfrac{-63xy^4}{-81x^2y}$

4. $\dfrac{x^2 - y^2}{x^2 + xy}$

5. $\dfrac{a^2 + 7a + 12}{a^2 - 6a - 27}$

6. $\dfrac{6x^2 + x - 15}{8x^2 - 10x - 3}$

7. $\dfrac{2x^3 + 3x^2 - 14x}{x^2y + 7xy - 18y}$

8. $\dfrac{3x - x^2}{x^2 - 9}$

9. $\dfrac{x^3 - y^3}{x^2 + xy - 2y^2}$

10. $\dfrac{ax - 3x + 2ay - 6y}{2ax - 6x + ay - 3y}$

11. $\dfrac{2y - 2xy}{x^2y - y}$

12. $\dfrac{16x^3y + 24x^2y^2 - 16xy^3}{24x^2y + 12xy^2 - 12y^3}$

Perform the following indicated operations involving rational expressions. Express final answers in simplest form.

13. $\dfrac{4x^2}{5y^2} \cdot \dfrac{15xy}{24x^2y^2}$

14. $\dfrac{5xy}{8y^2} \cdot \dfrac{18x^2y}{15}$

15. $\dfrac{-14xy^4}{18y^2} \cdot \dfrac{24x^2y^3}{35y^2}$

16. $\dfrac{6xy}{9y^4} \cdot \dfrac{30x^3y}{-48x}$

17. $\dfrac{7a^2b}{9ab^3} \div \dfrac{3a^4}{2a^2b^2}$

18. $\dfrac{9a^2c}{12bc^2} \div \dfrac{21ab}{14c^3}$

19. $\dfrac{5xy}{x + 6} \cdot \dfrac{x^2 - 36}{x^2 - 6x}$

20. $\dfrac{2a^2 + 6}{a^2 - a} \cdot \dfrac{a^3 - a^2}{8a - 4}$

21. $\dfrac{5a^2 + 20a}{a^3 - 2a^2} \cdot \dfrac{a^2 - a - 12}{a^2 - 16}$

22. $\dfrac{t^4 - 81}{t^2 - 6t + 9} \cdot \dfrac{6t^2 - 11t - 21}{5t^2 + 8t - 21}$

23. $\dfrac{x^2 + 5xy - 6y^2}{xy^2 - y^3} \cdot \dfrac{2x^2 + 15xy + 18y^2}{xy + 4y^2}$

24. $\dfrac{10n^2 + 21n - 10}{5n^2 + 33n - 14} \cdot \dfrac{2n^2 + 6n - 56}{2n^2 - 3n - 20}$

25. $\dfrac{9y^2}{x^2 + 12x + 36} \div \dfrac{12y}{x^2 + 6x}$

26. $\dfrac{x^2 - 4xy + 4y^2}{7xy^2} \div \dfrac{4x^2 - 3xy - 10y^2}{20x^2y + 25xy^2}$

27. $\dfrac{2x^2 + 3x}{2x^3 - 10x^2} \cdot \dfrac{x^2 - 8x + 15}{3x^3 - 27x} \div \dfrac{14x + 21}{x^2 - 6x - 27}$

28. $\dfrac{a^2 - 4ab + 4b^2}{6a^2 - 4ab} \cdot \dfrac{3a^2 + 5ab - 2b^2}{6a^2 + ab - b^2} \div \dfrac{a^2 - 4b^2}{8a + 4b}$

29. $\dfrac{x + 4}{6} + \dfrac{2x - 1}{4}$

30. $\dfrac{3n - 1}{9} - \dfrac{n + 2}{12}$

31. $\dfrac{x + 1}{4} + \dfrac{x - 3}{6} - \dfrac{x - 2}{8}$

32. $\dfrac{x - 2}{5} - \dfrac{x + 3}{6} + \dfrac{x + 1}{15}$

33. $\dfrac{7}{16a^2b} + \dfrac{3a}{20b^2}$

34. $\dfrac{5b}{24a^2} - \dfrac{11a}{32b}$

35. $\dfrac{1}{n^2} + \dfrac{3}{4n} - \dfrac{5}{6}$

36. $\dfrac{3}{n^2} - \dfrac{2}{5n} + \dfrac{4}{3}$

37. $\dfrac{3}{4x} + \dfrac{2}{3y} - 1$

38. $\dfrac{5}{6x} - \dfrac{3}{4y} + 2$

39. $\dfrac{3}{2x + 1} + \dfrac{2}{3x + 4}$

40. $\dfrac{5}{x - 1} - \dfrac{3}{2x - 3}$

41. $\dfrac{4x}{x^2 + 7x} + \dfrac{3}{x}$

42. $\dfrac{6}{x^2 + 8x} - \dfrac{3}{x}$

43. $\dfrac{4a - 4}{a^2 - 4} - \dfrac{3}{a + 2}$

44. $\dfrac{6a + 4}{a^2 - 1} - \dfrac{5}{a - 1}$

45. $\dfrac{3}{x + 1} + \dfrac{x + 5}{x^2 - 1} - \dfrac{3}{x - 1}$

46. $\dfrac{5}{x} - \dfrac{5x - 30}{x^2 + 6x} + \dfrac{x}{x + 6}$

47. $\dfrac{5}{x^2 + 10x + 21} + \dfrac{4}{x^2 + 12x + 27}$

48. $\dfrac{8}{a^2 - 3a - 18} - \dfrac{10}{a^2 - 7a - 30}$

49. $\dfrac{5}{x^2 - 1} - \dfrac{2}{x^2 + 6x - 16}$

50. $\dfrac{4}{x^2 + 2} - \dfrac{7}{x^2 + x - 12}$

51. $x - \dfrac{x^2}{x - 1} + \dfrac{1}{x^2 - 1}$

52. $x - \dfrac{x^2}{x + 7} - \dfrac{x}{x^2 - 49}$

53. $\dfrac{2n^2}{n^4 - 16} - \dfrac{n}{n^2 - 4} + \dfrac{1}{n + 2}$

54. $\dfrac{n}{n^2 + 1} + \dfrac{n^2 + 3n}{n^4 - 1} - \dfrac{1}{n - 1}$

55. $\dfrac{2x + 1}{x^2 - 3x - 4} + \dfrac{3x - 2}{x^2 + 3x - 28}$

56. $\dfrac{3x - 4}{2x^2 - 9x - 5} - \dfrac{2x - 1}{3x^2 - 11x - 20}$

57. Consider the addition problem $\dfrac{8}{x - 2} + \dfrac{5}{2 - x}$. Note that the denominators are opposites of each other. If the property $\dfrac{a}{-b} = -\dfrac{a}{b}$ is applied to the second fraction, we obtain $\dfrac{5}{2 - x} = -\dfrac{5}{x - 2}$. Thus we can proceed as follows.

$$\dfrac{8}{x - 2} + \dfrac{5}{2 - x} = \dfrac{8}{x - 2} - \dfrac{5}{x - 2}$$
$$= \dfrac{8 - 5}{x - 2} = \dfrac{3}{x - 2}$$

Use this approach to do the following problems.

a. $\dfrac{7}{x - 1} + \dfrac{2}{1 - x}$

b. $\dfrac{5}{2x - 1} + \dfrac{8}{1 - 2x}$

c. $\dfrac{4}{a - 3} - \dfrac{1}{3 - a}$

d. $\dfrac{10}{a - 9} - \dfrac{5}{9 - a}$

e. $\dfrac{x^2}{x - 1} - \dfrac{2x - 3}{1 - x}$

f. $\dfrac{x^2}{x - 4} - \dfrac{3x - 28}{4 - x}$

Simplify each of the following complex fractions.

58. $\dfrac{\dfrac{2}{x} + \dfrac{7}{y}}{\dfrac{3}{x} - \dfrac{10}{y}}$

59. $\dfrac{\dfrac{5}{x^2} - \dfrac{3}{x}}{\dfrac{1}{y} + \dfrac{2}{y^2}}$

60. $\dfrac{\dfrac{1}{x} + 3}{\dfrac{2}{y} + 4}$

61. $\dfrac{1 + \dfrac{1}{x}}{1 - \dfrac{1}{x}}$

62. $\dfrac{3 - \dfrac{2}{n - 4}}{5 + \dfrac{4}{n - 4}}$

63. $\dfrac{1 - \dfrac{1}{n + 1}}{1 + \dfrac{1}{n - 1}}$

64. $\dfrac{\dfrac{2}{x-3}-\dfrac{3}{x+3}}{\dfrac{5}{x^2-9}-\dfrac{2}{x-3}}$

65. $\dfrac{\dfrac{-2}{x}-\dfrac{4}{x+2}}{\dfrac{3}{x^2+2x}+\dfrac{3}{x}}$

66. $\dfrac{\dfrac{-1}{y-2}+\dfrac{5}{x}}{\dfrac{3}{x}-\dfrac{4}{xy-2x}}$

67. $1+\dfrac{x}{1+\dfrac{1}{x}}$

68. $2-\dfrac{x}{3-\dfrac{2}{x}}$

69. $\dfrac{a}{\dfrac{1}{a}+4}+1$

70. $\dfrac{3a}{2-\dfrac{1}{a}}-1$

71. $\dfrac{\dfrac{1}{(x+h)^2}-\dfrac{1}{x^2}}{h}$

72. $\dfrac{\dfrac{1}{(x+h)^3}-\dfrac{1}{x^3}}{h}$

73. $\dfrac{\dfrac{1}{x+h+1}-\dfrac{1}{x+1}}{h}$

74. $\dfrac{\dfrac{3}{x+h}-\dfrac{3}{x}}{h}$

75. $\dfrac{\dfrac{2}{2x+2h-1}-\dfrac{2}{2x-1}}{h}$

76. $\dfrac{\dfrac{3}{4x+4h+5}-\dfrac{3}{4x+5}}{h}$

77. $\dfrac{x^{-1}+y^{-1}}{x-y}$

78. $\dfrac{x+y}{x^{-1}+y^{-1}}$

79. $\dfrac{x+x^{-1}y^{-2}}{x^{-1}-y^{-2}}$

80. $\dfrac{x^{-2}-2y^{-1}}{3x^{-1}+y^{-2}}$

THOUGHTS into WORDS

81. What role does factoring play in the simplifying of rational expressions?

82. Explain in your own words how to multiply two rational expressions.

83. Give a step-by-step description of how to add $\dfrac{2x-1}{4}+\dfrac{3x+5}{14}$.

84. Look back at the two approaches shown in Example 7. Which approach would you use to simplify $\dfrac{\dfrac{1}{4}+\dfrac{1}{6}}{\dfrac{1}{2}-\dfrac{3}{4}}$? Which approach would you use to simplify $\dfrac{\dfrac{5}{8}+\dfrac{4}{9}}{\dfrac{5}{14}-\dfrac{2}{21}}$? Explain the reason for your choice of approach for each problem.

1.6 RADICALS

Recall from our work with exponents that to **square a number** means to raise it to the second power, that is, to use the number as a factor twice. For example, $4^2 = 4 \cdot 4 = 16$ and $(-4)^2 = (-4)(-4) = 16$. A **square root of a number** is one of its two equal factors. Thus, 4 and -4 are both square roots of 16. In general, a is a square root of b if $a^2 = b$. The following statements generalize these ideas.

1. Every positive real number has two square roots; one is positive and the other is negative. They are opposites of each other.
2. Negative real numbers have no real number square roots because the square of any nonzero real number is positive.

3. The square root of zero is zero.

The symbol $\sqrt{\ }$, called a **radical sign**, is used to designate the *nonnegative* square root, which is called the **principal square root**. The number under the radical sign is called the **radicand** and the entire expression, such as $\sqrt{16}$, is referred to as a **radical**.

The following examples demonstrate the use of the square root notation.

$\sqrt{16} = 4$	$\sqrt{16}$ indicates the *nonnegative* or *principal square root* of 16.
$-\sqrt{16} = -4$	$-\sqrt{16}$ indicates the negative square root of 16.
$\sqrt{0} = 0$	Zero has only one square root. Technically, we could also write $-\sqrt{0} = -0 = 0$.
$\sqrt{-4}$	Not a real number
$-\sqrt{-4}$	Not a real number

To **cube a number** means to raise it to the third power, that is, to use the number as a factor three times. For example, $2^3 = 2 \cdot 2 \cdot 2 = 8$ and $(-2)^3 = (-2)(-2)(-2) = -8$. A **cube root of a number** is one of its three equal factors. Thus 2 is a cube root of 8 and, as we will discuss later, it is the only real number that is a cube root of 8. Furthermore, -2 is the only real number that is a cube root of -8. In general, a is a cube root of b if $a^3 = b$. The following statements generalize these ideas.

1. Every positive real number has one positive real number cube root.

2. Every negative real number has one negative real number cube root.

3. The cube root of zero is zero.

REMARK Every nonzero real number has three cube roots, but only one of them is a real number. The other two roots are complex numbers, which we will discuss in Section 1.8.

The symbol $\sqrt[3]{\ }$ is used to designate the cube root of a number. Thus we can write

$$\sqrt[3]{8} = 2, \qquad \sqrt[3]{-8} = -2, \qquad \sqrt[3]{\frac{1}{27}} = \frac{1}{3} \qquad \text{and} \qquad \sqrt[3]{-\frac{1}{27}} = -\frac{1}{3}.$$

The concept of root can be extended to fourth roots, fifth roots, sixth roots, and, in general, nth roots. If n is an *even positive integer*, then the following statements are true.

1. Every positive real number has exactly two real nth roots, one positive and one negative. For example, the real fourth roots of 16 are 2 and -2.

2. Negative real numbers do not have real nth roots. For example, there are no real fourth roots of -16.

If n is an *odd positive integer* greater than 1, then the following statements are true.

1. Every real number has exactly one real nth root.
2. The real nth root of a positive number is positive. For example, the fifth root of 32 is 2.
3. The real nth root of a negative number is negative. For example, the fifth root of -32 is -2.

In general, the following definition is useful.

DEFINITION 1.5

$\sqrt[n]{b} = a$ if and only if $a^n = b$.

In Definition 1.5, if n is an even positive integer, then a and b are both nonnegative. If n is an odd positive integer greater than 1, then a and b are both nonnegative or both negative. The symbol $\sqrt[n]{}$ designates the principal root.

The following examples are applications of Definition 1.5.

$$\sqrt[4]{81} = 3 \quad \text{because } 3^4 = 81.$$
$$\sqrt[5]{32} = 2 \quad \text{because } 2^5 = 32.$$
$$\sqrt[5]{-32} = -2 \quad \text{because } (-2)^5 = -32.$$

To complete our terminology, the n in the radical $\sqrt[n]{b}$ is called the **index** of the radical. If $n = 2$, we commonly write $\sqrt{b}$ instead of $\sqrt[2]{b}$. In this text, when we use symbols such as $\sqrt[n]{b}$, $\sqrt[m]{y}$, and $\sqrt[r]{x}$, we will assume the previous agreements relative to the existence of real roots, without listing the various restrictions, unless a special restriction is needed.

From Definition 1.5 we see that if n is any positive integer greater than 1 and $\sqrt[n]{b}$ exists, then

$$\left(\sqrt[n]{b}\right)^n = b.$$

For example, $\left(\sqrt{4}\right)^2 = 4$, $\left(\sqrt[3]{-8}\right)^3 = -8$, and $\left(\sqrt[4]{81}\right)^4 = 81$. Furthermore, if $b \geq 0$ and n is any positive integer greater than 1 or if $b < 0$ and n is an odd positive integer greater than 1, then

$$\sqrt[n]{b^n} = b.$$

For example, $\sqrt{4^2} = 4$, $\sqrt[3]{(-2)^3} = -2$, and $\sqrt[5]{6^5} = 6$. But we must be careful, because

$$\sqrt{(-2)^2} \neq -2 \qquad \text{and} \qquad \sqrt[4]{(-2)^4} \neq -2.$$

Simplest Radical Form

Let's use some examples to motivate another useful property of radicals.

$$\sqrt{16 \cdot 25} = \sqrt{400} = 20 \quad \text{and} \quad \sqrt{16} \cdot \sqrt{25} = 4 \cdot 5 = 20$$
$$\sqrt[3]{8 \cdot 27} = \sqrt[3]{216} = 6 \quad \text{and} \quad \sqrt[3]{8} \cdot \sqrt[3]{27} = 2 \cdot 3 = 6$$
$$\sqrt[3]{-8 \cdot 64} = \sqrt[3]{-512} = -8 \quad \text{and} \quad \sqrt[3]{-8} \cdot \sqrt[3]{64} = -2 \cdot 4 = -8$$

In general, the following property can be stated.

PROPERTY 1.3

$\sqrt[n]{bc} = \sqrt[n]{b}\sqrt[n]{c}$ if $\sqrt[n]{b}$ and $\sqrt[n]{c}$ are real numbers.

Property 1.3 states that **the nth root of a product is equal to the product of the nth roots**.

The definition of nth root, along with Property 1.3, provides the basis for changing radicals to simplest radical form. The concept of **simplest radical form** takes on additional meaning as we encounter more complicated expressions, but for now it simply means that the radicand does not contain any perfect powers of the index. Consider the following examples of reductions to simplest radical form.

$$\sqrt{45} = \sqrt{9 \cdot 5} = \sqrt{9}\sqrt{5} = 3\sqrt{5}$$
$$\sqrt{52} = \sqrt{4 \cdot 13} = \sqrt{4}\sqrt{13} = 2\sqrt{13}$$
$$\sqrt[3]{24} = \sqrt[3]{8 \cdot 3} = \sqrt[3]{8}\sqrt[3]{3} = 2\sqrt[3]{3}$$

A variation of the technique for changing radicals with index n to simplest form is to factor the radicand into primes and then to look for the perfect nth powers in exponential form, as in the following examples.

$$\sqrt{80} = \sqrt{2^4 \cdot 5} = \sqrt{2^4}\sqrt{5} = 2^2\sqrt{5} = 4\sqrt{5}$$
$$\sqrt[3]{108} = \sqrt[3]{2^2 \cdot 3^3} = \sqrt[3]{3^3}\sqrt[3]{2^2} = 3\sqrt[3]{4}$$

The distributive property can be used to combine radicals that have the same index and the same radicand.

$$3\sqrt{2} + 5\sqrt{2} = (3 + 5)\sqrt{2} = 8\sqrt{2}$$
$$7\sqrt[3]{5} - 3\sqrt[3]{5} = (7 - 3)\sqrt[3]{5} = 4\sqrt[3]{5}$$

Sometimes it is necessary to simplify the radicals first and then to combine them by applying the distributive property.

$$\begin{aligned} 3\sqrt{8} + 2\sqrt{18} - 4\sqrt{2} &= 3\sqrt{4}\sqrt{2} + 2\sqrt{9}\sqrt{2} - 4\sqrt{2} \\ &= 6\sqrt{2} + 6\sqrt{2} - 4\sqrt{2} \\ &= (6 + 6 - 4)\sqrt{2} \\ &= 8\sqrt{2} \end{aligned}$$

Property 1.3 can also be viewed as $\sqrt[n]{b}\sqrt[n]{c} = \sqrt[n]{bc}$. Then, along with the commutative and associative properties of the real numbers, it provides the basis for multiplying radicals that have the same index. Consider the following two examples.

$$\begin{aligned}(7\sqrt{6})(3\sqrt{8}) &= 7 \cdot 3 \cdot \sqrt{6} \cdot \sqrt{8}\\ &= 21\sqrt{48}\\ &= 21\sqrt{16}\sqrt{3}\\ &= 21 \cdot 4 \cdot \sqrt{3}\\ &= 84\sqrt{3}\end{aligned}$$

$$\begin{aligned}(2\sqrt[3]{6})(5\sqrt[3]{4}) &= 2 \cdot 5 \cdot \sqrt[3]{6} \cdot \sqrt[3]{4}\\ &= 10\sqrt[3]{24}\\ &= 10\sqrt[3]{8}\sqrt[3]{3}\\ &= 10 \cdot 2 \cdot \sqrt[3]{3}\\ &= 20\sqrt[3]{3}\end{aligned}$$

The distributive property, along with Property 1.3, provides a way of handling special products involving radicals, as the next examples illustrate.

$$\begin{aligned}2\sqrt{2}(4\sqrt{3} - 5\sqrt{6}) &= (2\sqrt{2})(4\sqrt{3}) - (2\sqrt{2})(5\sqrt{6})\\ &= 8\sqrt{6} - 10\sqrt{12}\\ &= 8\sqrt{6} - 10\sqrt{4}\sqrt{3}\\ &= 8\sqrt{6} - 20\sqrt{3}\end{aligned}$$

$$\begin{aligned}(2\sqrt{2} - \sqrt{7})(3\sqrt{2} + 5\sqrt{7}) &= 2\sqrt{2}(3\sqrt{2} + 5\sqrt{7}) - \sqrt{7}(3\sqrt{2} + 5\sqrt{7})\\ &= (2\sqrt{2})(3\sqrt{2}) + (2\sqrt{2})(5\sqrt{7}) - (\sqrt{7})(3\sqrt{2}) - (\sqrt{7})(5\sqrt{7})\\ &= 6 \cdot 2 + 10\sqrt{14} - 3\sqrt{14} - 5 \cdot 7\\ &= -23 + 7\sqrt{14}\end{aligned}$$

$$\begin{aligned}(\sqrt{5} + \sqrt{2})(\sqrt{5} - \sqrt{2}) &= \sqrt{5}(\sqrt{5} - \sqrt{2}) + \sqrt{2}(\sqrt{5} - \sqrt{2})\\ &= (\sqrt{5})(\sqrt{5}) - (\sqrt{5})(\sqrt{2}) + (\sqrt{2})(\sqrt{5}) - (\sqrt{2}))(\sqrt{2})\\ &= 5 - \sqrt{10} + \sqrt{10} - 2\\ &= 3\end{aligned}$$

Pay special attention to the last example. It fits the special product pattern $(a + b)(a - b) = a^2 - b^2$. We will use that idea in a moment.

More About Simplest Radical Form

Another property of nth roots is motivated by the following examples.

$$\sqrt{\frac{36}{9}} = \sqrt{4} = 2 \qquad \text{and} \qquad \frac{\sqrt{36}}{\sqrt{9}} = \frac{6}{3} = 2$$

$$\sqrt[3]{\frac{64}{8}} = \sqrt[3]{8} = 2 \quad \text{and} \quad \frac{\sqrt[3]{64}}{\sqrt[3]{8}} = \frac{4}{2} = 2$$

In general, the following property can be stated.

PROPERTY 1.4

$\sqrt[n]{\frac{b}{c}} = \frac{\sqrt[n]{b}}{\sqrt[n]{c}}$ if $\sqrt[n]{b}$ and $\sqrt[n]{c}$ are real numbers and $c \neq 0$.

Property 1.4 states that **the nth root of a quotient is equal to the quotient of the nth roots.**

To evaluate radicals such as $\sqrt{\frac{4}{25}}$ and $\sqrt[3]{\frac{27}{8}}$, where the numerator and denominator of the fractional radicands are perfect nth powers, we can either use Property 1.4 or rely on the definition of nth root:

$$\sqrt{\frac{4}{25}} = \frac{\sqrt{4}}{\sqrt{25}} = \frac{2}{5} \quad \text{or} \quad \sqrt{\frac{4}{25}} = \frac{2}{5} \quad \text{because } \frac{2}{5} \cdot \frac{2}{5} = \frac{4}{25};$$

$$\sqrt[3]{\frac{27}{8}} = \frac{\sqrt[3]{27}}{\sqrt[3]{8}} = \frac{3}{2} \quad \text{or} \quad \sqrt[3]{\frac{27}{8}} = \frac{3}{2} \quad \text{because } \frac{3}{2} \cdot \frac{3}{2} \cdot \frac{3}{2} = \frac{27}{8}$$

Radicals such as $\sqrt{\frac{28}{9}}$ and $\sqrt[3]{\frac{24}{27}}$, where only the denominators of the radicand are perfect nth powers, can be simplified as follows:

$$\sqrt{\frac{28}{9}} = \frac{\sqrt{28}}{\sqrt{9}} = \frac{\sqrt{4}\sqrt{7}}{3} = \frac{2\sqrt{7}}{3};$$

$$\sqrt[3]{\frac{24}{27}} = \frac{\sqrt[3]{24}}{\sqrt[3]{27}} = \frac{\sqrt[3]{8}\sqrt[3]{3}}{3} = \frac{2\sqrt[3]{3}}{3}.$$

Before we consider more examples, let's summarize some ideas about simplifying radicals. A radical is said to be in **simplest radical form** if the following conditions are satisfied.

1. No fraction appears within a radical sign. — Thus $\sqrt{\frac{3}{4}}$ violates this condition.
2. No radical appears in the denominator. — So $\frac{\sqrt{2}}{\sqrt{3}}$ violates this condition.
3. No radicand contains a perfect power of the index. — Therefore $\sqrt{7^2 \cdot 5}$ violates this condition.

Now let's consider an example in which neither the numerator nor the denominator of the radicand is a perfect nth power.

$$\sqrt{\frac{2}{3}} = \frac{\sqrt{2}}{\sqrt{3}} = \frac{\sqrt{2}}{\sqrt{3}} \cdot \frac{\sqrt{3}}{\sqrt{3}} = \frac{\sqrt{6}}{3}$$

Form of 1

The process used to simplify the radical in the previous example is referred to as **rationalizing the denominator**. There is more than one way to rationalize the denominator, as illustrated by the next example.

EXAMPLE 1

Simplify $\frac{\sqrt{5}}{\sqrt{8}}$.

Solution A

$$\frac{\sqrt{5}}{\sqrt{8}} = \frac{\sqrt{5}}{\sqrt{8}} \cdot \frac{\sqrt{8}}{\sqrt{8}} = \frac{\sqrt{40}}{8} = \frac{\sqrt{4}\sqrt{10}}{8} = \frac{2\sqrt{10}}{8} = \frac{\sqrt{10}}{4}$$

Solution B

$$\frac{\sqrt{5}}{\sqrt{8}} = \frac{\sqrt{5}}{\sqrt{8}} \cdot \frac{\sqrt{2}}{\sqrt{2}} = \frac{\sqrt{10}}{\sqrt{16}} = \frac{\sqrt{10}}{4}$$

Solution C

$$\frac{\sqrt{5}}{\sqrt{8}} = \frac{\sqrt{5}}{\sqrt{4}\sqrt{2}} = \frac{\sqrt{5}}{2\sqrt{2}} = \frac{\sqrt{5}}{2\sqrt{2}} \cdot \frac{\sqrt{2}}{\sqrt{2}} = \frac{\sqrt{10}}{4}$$

The three approaches to Example 1 again illustrate the need to think first and then to push the pencil. You may find one approach easier than another.

EXAMPLE 2

Simplify $\frac{\sqrt{6}}{\sqrt{8}}$.

Solution

$$\frac{\sqrt{6}}{\sqrt{8}} = \sqrt{\frac{6}{8}}$$ Remember that $\frac{\sqrt{a}}{\sqrt{b}} = \sqrt{\frac{a}{b}}$.

$$= \sqrt{\frac{3}{4}}$$ Reduce the fraction.

$$= \frac{\sqrt{3}}{\sqrt{4}}$$

$$= \frac{\sqrt{3}}{2}$$

EXAMPLE 3

Simplify $\frac{\sqrt[3]{5}}{\sqrt[3]{9}}$.

Solution

$$\frac{\sqrt[3]{5}}{\sqrt[3]{9}} = \frac{\sqrt[3]{5}}{\sqrt[3]{9}} \cdot \frac{\sqrt[3]{3}}{\sqrt[3]{3}}$$
$$= \frac{\sqrt[3]{15}}{\sqrt[3]{27}}$$
$$= \frac{\sqrt[3]{15}}{3}$$

Now let's consider an example in which the denominator is of binomial form.

EXAMPLE 4

Simplify $\frac{4}{\sqrt{5} + \sqrt{2}}$ by rationalizing the denominator.

Solution

Remember a moment ago we found that $(\sqrt{5} + \sqrt{2})(\sqrt{5} - \sqrt{2}) = 3$. Let's use that idea here.

$$\frac{4}{\sqrt{5} + \sqrt{2}} = \left(\frac{4}{\sqrt{5} + \sqrt{2}}\right)\left(\frac{\sqrt{5} - \sqrt{2}}{\sqrt{5} - \sqrt{2}}\right)$$
$$= \frac{4(\sqrt{5} - \sqrt{2})}{(\sqrt{5} + \sqrt{2})(\sqrt{5} - \sqrt{2})} = \frac{4(\sqrt{5} - \sqrt{2})}{3}$$

The process of rationalizing the denominator does agree with the previously listed conditions. However, for certain problems in calculus it is necessary to **rationalize the numerator**. Again, the fact that $(\sqrt{a} + \sqrt{b})(\sqrt{a} - \sqrt{b}) = a - b$ can be used.

EXAMPLE 5

Change the form of $\frac{\sqrt{x + h} - \sqrt{x}}{h}$ by rationalizing the *numerator*.

Solution

$$\frac{\sqrt{x + h} - \sqrt{x}}{h} = \left(\frac{\sqrt{x + h} - \sqrt{x}}{h}\right)\left(\frac{\sqrt{x + h} + \sqrt{x}}{\sqrt{x + h} + \sqrt{x}}\right)$$
$$= \frac{(x + h) - x}{h(\sqrt{x + h} + \sqrt{x})}$$
$$= \frac{\cancel{h}}{\cancel{h}(\sqrt{x + h} + \sqrt{x})}$$
$$= \frac{1}{\sqrt{x + h} + \sqrt{x}}$$

Radicals Containing Variables

Before we illustrate how to simplify radicals that contain variables, there is one important point we should call to your attention. Let's look at some examples to illustrate the idea.

Consider the radical $\sqrt{x^2}$ for different values of x.

Let $x = 3$; then $\sqrt{x^2} = \sqrt{3^2} = \sqrt{9} = 3$.

Let $x = -3$; then $\sqrt{x^2} = \sqrt{(-3)^2} = \sqrt{9} = 3$.

Thus if $x \geq 0$ then $\sqrt{x^2} = x$, but if $x < 0$ then $\sqrt{x^2} = -x$. Using the concept of absolute value, we can state that **for all real numbers, $\sqrt{x^2} = |x|$**

Now consider the radical $\sqrt{x^3}$. Since x^3 is negative when x is negative, we need to restrict x to the nonnegative real numbers when working with $\sqrt{x^3}$. Thus we can write,

$$\text{if } x \geq 0, \quad \text{then } \sqrt{x^3} = \sqrt{x^2}\sqrt{x} = x\sqrt{x},$$

and no absolute value sign is needed.

Finally, let's consider the radical $\sqrt[3]{x^3}$.

Let $x = 2$; then $\sqrt[3]{x^3} = \sqrt[3]{2^3} = \sqrt[3]{8} = 2$.

Let $x = -2$; then $\sqrt[3]{x^3} = \sqrt[3]{(-2)^3} = \sqrt[3]{-8} = -2$.

Thus it is correct to write,

$$\sqrt[3]{x^3} = x \quad \text{for all real numbers,}$$

and again, no absolute value sign is needed.

The previous discussion indicates that, technically, every radical expression with variables in the radicand needs to be analyzed individually to determine the necessary restrictions on the variables. However, to avoid having to do this on a problem-by-problem basis, we shall merely **assume that all variables represent positive real numbers**.

Let's conclude this section by simplifying some radical expressions that contain variables.

$$\sqrt{72x^3y^7} = \sqrt{36x^2y^6}\sqrt{2xy} = 6xy^3\sqrt{2xy}$$

$$\sqrt[3]{40x^4y^8} = \sqrt[3]{8x^3y^6}\sqrt[3]{5xy^2} = 2xy^2\sqrt[3]{5xy^2}$$

$$\frac{\sqrt{5}}{\sqrt{12a^3}} = \frac{\sqrt{5}}{\sqrt{12a^3}} \cdot \frac{\sqrt{3a}}{\sqrt{3a}} = \frac{\sqrt{15a}}{\sqrt{36a^4}} = \frac{\sqrt{15a}}{6a^2}$$

$$\frac{3}{\sqrt[3]{4x}} = \frac{3}{\sqrt[3]{4x}} \cdot \frac{\sqrt[3]{2x^2}}{\sqrt[3]{2x^2}} = \frac{3\sqrt[3]{2x^2}}{\sqrt[3]{8x^3}} = \frac{3\sqrt[3]{2x^2}}{2x}$$

PROBLEM SET 1.6

Evaluate each of the following.

1. $\sqrt{81}$
2. $-\sqrt{49}$
3. $\sqrt[3]{125}$
4. $\sqrt[4]{81}$
5. $\sqrt{\frac{36}{49}}$
6. $\sqrt{\frac{256}{64}}$
7. $\sqrt[3]{-\frac{27}{8}}$
8. $\sqrt[3]{\frac{64}{27}}$

Express each of the following in simplest radical form. All variables represent positive real numbers.

9. $\sqrt{24}$
10. $\sqrt{54}$
11. $\sqrt{112}$
12. $6\sqrt{28}$
13. $-3\sqrt{44}$
14. $-5\sqrt{68}$
15. $\frac{3}{4}\sqrt{20}$
16. $\frac{3}{8}\sqrt{72}$
17. $\sqrt{12x^2}$
18. $\sqrt{45xy^2}$
19. $\sqrt{64x^4y^7}$
20. $3\sqrt{32a^3}$
21. $\frac{3}{7}\sqrt{45xy^6}$
22. $\sqrt[3]{32}$
23. $\sqrt[3]{128}$
24. $\sqrt[3]{54x^3}$
25. $\sqrt[3]{16x^4}$
26. $\sqrt[3]{81x^5y^6}$
27. $\sqrt[4]{48x^5}$
28. $\sqrt[4]{162x^6y^7}$
29. $\sqrt{\frac{12}{25}}$
30. $\sqrt{\frac{75}{81}}$
31. $\sqrt{\frac{7}{8}}$
32. $\frac{\sqrt{35}}{\sqrt{7}}$
33. $\frac{4\sqrt{3}}{\sqrt{5}}$
34. $\frac{\sqrt{27}}{\sqrt{18}}$
35. $\frac{6\sqrt{3}}{7\sqrt{6}}$
36. $\sqrt{\frac{3x}{2y}}$
37. $\frac{\sqrt{5}}{\sqrt{12x^4}}$
38. $\frac{\sqrt{5y}}{\sqrt{18x^3}}$
39. $\frac{\sqrt{12a^2b}}{\sqrt{5a^3b^3}}$
40. $\frac{5}{\sqrt[3]{3}}$
41. $\frac{\sqrt[3]{27}}{\sqrt[3]{4}}$
42. $\sqrt[3]{\frac{5}{2x}}$
43. $\frac{\sqrt[3]{2y}}{\sqrt[3]{3x}}$
44. $\frac{\sqrt[3]{12xy}}{\sqrt[3]{3x^2y^5}}$

Use the distributive property to help simplify each of the following. For example,

$$\begin{aligned} 3\sqrt{8} + 5\sqrt{2} &= 3\sqrt{4}\sqrt{2} + 5\sqrt{2} \\ &= 6\sqrt{2} + 5\sqrt{2} \\ &= (6 + 5)\sqrt{2} \\ &= 11\sqrt{2}. \end{aligned}$$

45. $5\sqrt{12} + 2\sqrt{3}$
46. $4\sqrt{50} - 9\sqrt{32}$
47. $2\sqrt{28} - 3\sqrt{63} + 8\sqrt{7}$
48. $4\sqrt[3]{2} + 2\sqrt[3]{16} - \sqrt[3]{54}$
49. $\frac{5}{6}\sqrt{48} - \frac{3}{4}\sqrt{12}$
50. $\frac{2}{5}\sqrt{40} + \frac{1}{6}\sqrt{90}$
51. $\frac{2\sqrt{8}}{3} - \frac{3\sqrt{18}}{5} - \frac{\sqrt{50}}{2}$
52. $\frac{3\sqrt[3]{54}}{2} + \frac{5\sqrt[3]{16}}{3}$

Multiply and express the results in simplest radical form. All variables represent nonnegative real numbers.

53. $(4\sqrt{3})(6\sqrt{8})$
54. $(5\sqrt{8})(3\sqrt{7})$
55. $2\sqrt{3}(5\sqrt{2} + 4\sqrt{10})$
56. $3\sqrt{6}(2\sqrt{8} - 3\sqrt{12})$
57. $3\sqrt{x}(\sqrt{6xy} - \sqrt{8y})$
58. $\sqrt{6y}(\sqrt{8x} + \sqrt{10y^2})$
59. $(\sqrt{3} + 2)(\sqrt{3} + 5)$
60. $(\sqrt{2} - 3)(\sqrt{2} + 4)$
61. $(4\sqrt{2} + \sqrt{3})(3\sqrt{2} + 2\sqrt{3})$
62. $(2\sqrt{6} + 3\sqrt{5})(3\sqrt{6} + 4\sqrt{5})$
63. $(6 + 2\sqrt{5})(6 - 2\sqrt{5})$
64. $(7 - 3\sqrt{2})(7 + 3\sqrt{2})$
65. $(\sqrt{x} + \sqrt{y})^2$
66. $(2\sqrt{x} - 3\sqrt{y})^2$
67. $(\sqrt{a} + \sqrt{b})(\sqrt{a} - \sqrt{b})$
68. $(3\sqrt{x} + 5\sqrt{y})(3\sqrt{x} - 5\sqrt{y})$

For each of the following, rationalize the denominator and simplify. All variables represent positive real numbers.

69. $\frac{3}{\sqrt{5} + 2}$
70. $\frac{7}{\sqrt{10} - 3}$
71. $\frac{4}{\sqrt{7} - \sqrt{3}}$
72. $\frac{2}{\sqrt{5} + \sqrt{3}}$
73. $\frac{\sqrt{2}}{2\sqrt{5} + 3\sqrt{7}}$
74. $\frac{5}{5\sqrt{2} - 3\sqrt{5}}$
75. $\frac{\sqrt{x}}{\sqrt{x} - 1}$
76. $\frac{\sqrt{x}}{\sqrt{x} + 2}$

77. $\dfrac{\sqrt{x}}{\sqrt{x}+\sqrt{y}}$

78. $\dfrac{2\sqrt{x}}{\sqrt{x}-\sqrt{y}}$

79. $\dfrac{2\sqrt{x}+\sqrt{y}}{3\sqrt{x}-2\sqrt{y}}$

80. $\dfrac{3\sqrt{x}-2\sqrt{y}}{2\sqrt{x}+5\sqrt{y}}$

For each of the following, *rationalize the numerator.* All variables represent positive real numbers.

81. $\dfrac{\sqrt{2x+2h}-\sqrt{2x}}{h}$

82. $\dfrac{\sqrt{x+h+1}-\sqrt{x+1}}{h}$

83. $\dfrac{\sqrt{x+h-3}-\sqrt{x-3}}{h}$

84. $\dfrac{2\sqrt{x+h}-2\sqrt{x}}{h}$

THOUGHTS into WORDS

85. Is the equation $\sqrt{x^2y} = x\sqrt{y}$ true for all real number values for x and y? Defend your answer.

86. Is the equation $\sqrt{x^2y^2} = xy$ true for all real number values for x and y? Defend your answer.

87. Give a step-by-step description of how you would change $\sqrt{252}$ to simplest radical form.

1.7 Relationship Between Exponents and Roots

Recall that we used the basic properties of positive integral exponents to motivate a definition of negative integers as exponents. In this section, we shall use the properties of integral exponents to motivate definitions for rational numbers as exponents. These definitions will tie together the concepts of *exponent* and *root.* Let's consider the following comparisons.

From our study of radicals we know that	If $(b^m)^n = b^{mn}$ is to hold when m is a rational number of the form $1/p$, where p is a positive integer greater than 1 and $n = p$, then
$(\sqrt{5})^2 = 5;$	$(5^{1/2})^2 = 5^{2(1/2)} = 5^1 = 5;$
$(\sqrt[3]{8})^3 = 8;$	$(8^{1/3})^3 = 8^{3(1/3)} = 8^1 = 8;$
$(\sqrt[4]{21})^4 = 21.$	$(21^{1/4})^4 = 21^{4(1/4)} = 21^1 = 21.$

Such examples motivate the following definition.

DEFINITION 1.6

If b is a real number, n is a positive integer greater than 1, and $\sqrt[n]{b}$ exists, then

$$b^{1/n} = \sqrt[n]{b}.$$

Definition 1.6 states that $b^{1/n}$ means the nth root of b. We shall assume that b and n are chosen so that $\sqrt[n]{b}$ exists in the real number system. For example, $(-25)^{1/2}$ is not meaningful at this time because $\sqrt{-25}$ is not a real number. The following examples illustrate the use of Definition 1.6.

$$25^{1/2} = \sqrt{25} = 5 \qquad 16^{1/4} = \sqrt[4]{16} = 2$$

$$8^{1/3} = \sqrt[3]{8} = 2 \qquad (-27)^{1/3} = \sqrt[3]{-27} = -3$$

Now the following definition provides the basis for the use of *all* rational numbers as exponents.

DEFINITION 1.7

If m/n is a rational number expressed in lowest terms, where n is a positive integer greater than one, and m is any integer, and if b is a real number such that $\sqrt[n]{b}$ exists, then

$$b^{m/n} = \sqrt[n]{b^m} = \left(\sqrt[n]{b}\right)^m.$$

In Definition 1.7, whether we use the form $\sqrt[n]{b^m}$ or $\left(\sqrt[n]{b}\right)^m$ for computational purposes depends somewhat on the magnitude of the problem. Let's use both forms on the two following problems.

$$8^{2/3} = \sqrt[3]{8^2} = \sqrt[3]{64} = 4 \qquad \text{or} \qquad 8^{2/3} = \left(\sqrt[3]{8}\right)^2 = (2)^2 = 4$$

$$27^{2/3} = \sqrt[3]{27^2} = \sqrt[3]{729} = 9 \qquad \text{or} \qquad 27^{2/3} = \left(\sqrt[3]{27}\right)^2 = (3)^2 = 9$$

To compute $8^{2/3}$, both forms work equally well. However, to compute $27^{2/3}$, the form $\left(\sqrt[3]{27}\right)^2$ is much easier to handle. The following examples further illustrate Definition 1.7.

$$25^{3/2} = \left(\sqrt{25}\right)^3 = 5^3 = 125$$

$$(32)^{-2/5} = \frac{1}{(32)^{2/5}} = \frac{1}{\left(\sqrt[5]{32}\right)^2} = \frac{1}{2^2} = \frac{1}{4}$$

$$(-64)^{2/3} = \left(\sqrt[3]{-64}\right)^2 = (-4)^2 = 16$$
$$-8^{4/3} = -\left(\sqrt[3]{8}\right)^4 = -(2)^4 = -16$$

It can be shown that all of the results pertaining to integral exponents listed in Property 1.2 (on page 17) also hold for all rational exponents. Let's consider some examples to illustrate each of those results.

$$x^{1/2} \cdot x^{2/3} = x^{1/2+2/3} = x^{3/6+4/6} = x^{7/6} \qquad b^n \cdot b^m = b^{n+m}$$

$$(a^{2/3})^{3/2} = a^{(3/2)(2/3)} = a^1 = a \qquad (b^n)^m = b^{nm}$$

$$(16y^{2/3})^{1/2} = (16)^{1/2}(y^{2/3})^{1/2} = 4y^{1/3} \qquad (ab)^n = a^n b^n$$

$$\frac{y^{3/4}}{y^{1/2}} = y^{3/4-1/2} = y^{3/4-2/4} = y^{1/4} \qquad \frac{b^n}{b^m} = b^{n-m}$$

$$\left(\frac{x^{1/2}}{y^{1/3}}\right)^6 = \frac{(x^{1/2})^6}{(y^{1/3})^6} = \frac{x^3}{y^2} \qquad \left(\frac{a}{b}\right)^n = \frac{a^n}{b^n}$$

The link between exponents and roots provides a basis for multiplying and dividing some radicals even if they have a different index. The general procedure is one of changing from radical to exponential form, applying the properties of exponents, and then changing back to radical form. Let's apply these procedures in the next three examples.

$$\sqrt{2}\sqrt[3]{2} = 2^{1/2} \cdot 2^{1/3} = 2^{1/2+1/3} = 2^{5/6} = \sqrt[6]{2^5} = \sqrt[6]{32}$$

$$\begin{aligned}\sqrt{xy}\sqrt[5]{x^2y} &= (xy)^{1/2}(x^2y)^{1/5}\\ &= x^{1/2}y^{1/2}x^{2/5}y^{1/5}\\ &= x^{1/2+2/5}y^{1/2+1/5}\\ &= x^{9/10}y^{7/10}\\ &= (x^9y^7)^{1/10}\\ &= \sqrt[10]{x^9y^7}\end{aligned}$$

$$\frac{\sqrt{5}}{\sqrt[3]{5}} = \frac{5^{1/2}}{5^{1/3}} = 5^{1/2-1/3} = 5^{1/6} = \sqrt[6]{5}$$

Earlier we agreed that a radical such as $\sqrt[3]{x^4}$ is not in simplest form because the radicand contains a perfect power of the index. Thus we simplified $\sqrt[3]{x^4}$ by expressing it as $\sqrt[3]{x^3}\sqrt[3]{x}$, which in turn can be written $x\sqrt[3]{x}$. Such simplification can also be done in exponential form, as follows.

$$\sqrt[3]{x^4} = x^{4/3} = x^{3/3} \cdot x^{1/3} = x \cdot x^{1/3} = x\sqrt[3]{x}$$

Note the use of this type of simplification in the following examples.

EXAMPLE 1

Perform the indicated operations and express the answers in simplest radical form.

a. $\sqrt[3]{x^2}\sqrt[4]{x^3}$ **b.** $\sqrt{2}\sqrt[3]{4}$ **c.** $\dfrac{\sqrt{27}}{\sqrt[3]{3}}$

Solutions

a. $\sqrt[3]{x^2}\sqrt[4]{x^3} = x^{2/3} \cdot x^{3/4} = x^{2/3+3/4} = x^{17/12} = x^{12/12} \cdot x^{5/12} = x\sqrt[12]{x^5}$

b. $\sqrt{2}\sqrt[3]{4} = 2^{1/2} \cdot 4^{1/3} = 2^{1/2}(2^2)^{1/3} = 2^{1/2} \cdot 2^{2/3}$

$= 2^{1/2+2/3} = 2^{7/6} = 2^{6/6} \cdot 2^{1/6} = 2\sqrt[6]{2}$

c. $\dfrac{\sqrt{27}}{\sqrt[3]{3}} = \dfrac{27^{1/2}}{3^{1/3}} = \dfrac{(3^3)^{1/2}}{3^{1/3}} = \dfrac{3^{3/2}}{3^{1/3}} = 3^{3/2-1/3} = 3^{7/6}$

$= 3^{6/6} \cdot 3^{1/6} = 3\sqrt[6]{3}$

The process of rationalizing the denominator can sometimes be handled more easily in exponential form. Consider the following examples, which illustrate this procedure.

EXAMPLE 2

Rationalize the denominator and express the answer in simplest radical form.

a. $\dfrac{2}{\sqrt[3]{x}}$ **b.** $\dfrac{\sqrt[3]{x}}{\sqrt{y}}$

Solutions

a. $\dfrac{2}{\sqrt[3]{x}} = \dfrac{2}{x^{1/3}} = \dfrac{2}{x^{1/3}} \cdot \dfrac{x^{2/3}}{x^{2/3}} = \dfrac{2x^{2/3}}{x} = \dfrac{2\sqrt[3]{x^2}}{x}$

b. $\dfrac{\sqrt[3]{x}}{\sqrt{y}} = \dfrac{x^{1/3}}{y^{1/2}} = \dfrac{x^{1/3}}{y^{1/2}} \cdot \dfrac{y^{1/2}}{y^{1/2}} = \dfrac{x^{1/3} \cdot y^{1/2}}{y} = \dfrac{x^{2/6} \cdot y^{3/6}}{y} = \dfrac{\sqrt[6]{x^2y^3}}{y}$

Note in part (b) that if we had changed back to radical form at the step $\dfrac{x^{1/3}y^{1/2}}{y}$, we would have obtained the product of two radicals, $\sqrt[3]{x}\sqrt{y}$, in the numerator. Instead we used the exponential form to find this product and express the final result with a single radical in the numerator. Finally, let's consider an example involving *the root of a root* situation.

EXAMPLE 3 Simplify $\sqrt[3]{\sqrt{2}}$.

Solution

$$\sqrt[3]{\sqrt{2}} = (2^{1/2})^{1/3} = 2^{1/6} = \sqrt[6]{2}$$

PROBLEM SET 1.7

Evaluate each of the following.

1. $49^{1/2}$ **2.** $64^{1/3}$ **3.** $32^{3/5}$

4. $(-8)^{1/3}$ **5.** $-8^{2/3}$ **6.** $64^{-1/2}$

7. $\left(\frac{1}{4}\right)^{-1/2}$ **8.** $\left(-\frac{27}{8}\right)^{-1/3}$ **9.** $16^{3/2}$

10. $(0.008)^{1/3}$ **11.** $(0.01)^{3/2}$ **12.** $\left(\frac{1}{27}\right)^{-2/3}$

13. $64^{-5/6}$ **14.** $-16^{5/4}$ **15.** $\left(\frac{1}{8}\right)^{-1/3}$

16. $\left(-\frac{1}{8}\right)^{2/3}$

Perform the indicated operations and simplify. Express final answers using positive exponents only.

17. $(3x^{1/4})(5x^{1/3})$ **18.** $(2x^{2/5})(6x^{1/4})$

19. $(y^{2/3})(y^{-1/4})$ **20.** $(2x^{1/3})(x^{-1/2})$

21. $(4x^{1/4}y^{1/2})^3$ **22.** $(5x^{1/2}y)^2$

23. $\frac{24x^{3/5}}{6x^{1/3}}$ **24.** $\frac{18x^{1/2}}{9x^{1/3}}$

25. $\frac{56a^{1/6}}{8a^{1/4}}$ **26.** $\frac{48b^{1/3}}{12b^{3/4}}$

27. $\left(\frac{2x^{1/3}}{3y^{1/4}}\right)^4$ **28.** $\left(\frac{6x^{2/5}}{7y^{2/3}}\right)^2$

29. $\left(\frac{x^2}{y^3}\right)^{-1/2}$ **30.** $\left(\frac{a^3}{b^{-2}}\right)^{-1/3}$

31. $\left(\frac{4a^2x}{2a^{1/2}x^{1/3}}\right)^3$ **32.** $\left(\frac{3ax^{-1}}{a^{1/2}x^{-2}}\right)^2$

Perform the indicated operations and express the answers in simplest radical form.

33. $\sqrt{2}\sqrt[4]{2}$ **34.** $\sqrt[3]{3}\sqrt{3}$

35. $\sqrt[3]{x}\sqrt[4]{x}$ **36.** $\sqrt[3]{x^2}\sqrt[5]{x^3}$

37. $\sqrt{xy}\sqrt[4]{x^3y^5}$ **38.** $\sqrt[3]{x^2y^4}\sqrt[4]{x^3y}$

39. $\sqrt[3]{a^2b^2}\sqrt[4]{a^3b}$ **40.** $\sqrt{ab}\sqrt[3]{a^4b^5}$

41. $\sqrt[3]{4}\sqrt{8}$ **42.** $\sqrt[3]{9}\sqrt{27}$

43. $\frac{\sqrt{2}}{\sqrt[3]{2}}$ **44.** $\frac{\sqrt{9}}{\sqrt[3]{3}}$

45. $\frac{\sqrt[3]{8}}{\sqrt[4]{4}}$ **46.** $\frac{\sqrt[3]{16}}{\sqrt[6]{4}}$

47. $\frac{\sqrt[4]{x^9}}{\sqrt[3]{x^2}}$ **48.** $\frac{\sqrt[5]{x^7}}{\sqrt[3]{x}}$

Rationalize the denominators and express the final answers in simplest radical form.

49. $\frac{5}{\sqrt[3]{x}}$ **50.** $\frac{3}{\sqrt[3]{x^2}}$ **51.** $\frac{\sqrt{x}}{\sqrt[3]{y}}$

52. $\frac{\sqrt[4]{x}}{\sqrt{y}}$ **53.** $\frac{\sqrt[4]{x^3}}{\sqrt[5]{y^3}}$ **54.** $\frac{2\sqrt{x}}{3\sqrt[3]{y}}$

55. $\frac{5\sqrt[3]{y^2}}{4\sqrt[4]{x}}$ **56.** $\frac{\sqrt{xy}}{\sqrt[3]{a^2}}$

57. Simplify each of the following, expressing the final result as one radical. For example,

$$\sqrt{\sqrt{3}} = (3^{1/2})^{1/2} = 3^{1/4} = \sqrt[4]{3}.$$

a. $\sqrt{\sqrt[3]{2}}$ **b.** $\sqrt[3]{\sqrt[4]{3}}$

c. $\sqrt[3]{\sqrt{x^3}}$ **d.** $\sqrt{\sqrt[3]{x^4}}$

MISCELLANEOUS PROBLEMS

58. If your calculator has $\boxed{y^x}$ and $\boxed{1/x}$ keys, then they can be used to evaluate cube roots, fourth roots, fifth roots, and so on. For example, $\sqrt[3]{4913}$ can be evaluated as follows.

$$4913 \boxed{y^x} 3 \boxed{1/x} = 17.$$

Use your calculator to evaluate each of the following.

a. $\sqrt[3]{1728}$ **b.** $\sqrt[3]{5832}$

c. $\sqrt[4]{2401}$ **d.** $\sqrt[4]{65536}$

e. $\sqrt[5]{161051}$ **f.** $\sqrt[5]{6436343}$

59. In Definition 1.7 we stated that $b^{m/n} = \sqrt[n]{b^m} = \left(\sqrt[n]{b}\right)^m$. Use your calculator to verify each of the following.

a. $\sqrt[3]{27^2} = \left(\sqrt[3]{27}\right)^2$ **b.** $\sqrt[3]{8^5} = \left(\sqrt[3]{8}\right)^5$

c. $\sqrt[4]{16^3} = \left(\sqrt[4]{16}\right)^3$ **d.** $\sqrt[5]{16^2} = \left(\sqrt[5]{16}\right)^2$

e. $\sqrt[5]{9^4} = \left(\sqrt[5]{9}\right)^4$ **f.** $\sqrt[3]{12^4} = \left(\sqrt[3]{12}\right)^4$

60. Use your calculator to evaluate each of the following.

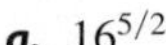

a. $16^{5/2}$ **b.** $25^{7/2}$ **c.** $16^{9/4}$

d. $27^{5/3}$ **e.** $343^{2/3}$ **f.** $512^{4/3}$

61. Use your calculator to estimate each of the following to the nearest thousandth.

a. $7^{4/3}$ **b.** $10^{4/5}$ **c.** $12^{2/5}$

d. $19^{2/5}$ **e.** $7^{3/4}$ **f.** $10^{5/4}$

Sometimes we meet the following type of simplification problem in calculus.

$$\frac{(x-1)^{1/2} - x(x-1)^{-(1/2)}}{[(x-1)^{1/2}]^2}$$

$$= \left(\frac{(x-1)^{1/2} - x(x-1)^{-(1/2)}}{(x-1)^{2/2}}\right) \cdot \left(\frac{(x-1)^{1/2}}{(x-1)^{1/2}}\right)$$

$$= \frac{x - 1 - x(x-1)^0}{(x-1)^{3/2}}$$

$$= \frac{x - 1 - x}{(x-1)^{3/2}} \qquad (x-1)^0 = 1$$

$$= \frac{-1}{(x-1)^{3/2}} \quad \text{or} \quad -\frac{1}{(x-1)^{3/2}}$$

For Problems 62–67, simplify each expression as we did in the previous example.

62. $\dfrac{2(x+1)^{1/2} - x(x+1)^{-(1/2)}}{[(x+1)^{1/2}]^2}$

63. $\dfrac{2(2x-1)^{1/2} - 2x(2x-1)^{-(1/2)}}{[(2x-1)^{1/2}]^2}$

64. $\dfrac{2x(4x+1)^{1/2} - 2x^2(4x+1)^{-(1/2)}}{[(4x+1)^{1/2}]^2}$

65. $\dfrac{(x^2+2x)^{1/2} - x(x+1)(x^2+2x)^{-(1/2)}}{[(x^2+2x)^{1/2}]^2}$

66. $\dfrac{(3x)^{1/3} - x(3x)^{-(2/3)}}{[(3x)^{1/3}]^2}$

67. $\dfrac{3(2x)^{1/3} - 2x(2x)^{-(2/3)}}{[(2x)^{1/3}]^2}$

1.8 COMPLEX NUMBERS

So far we have dealt only with real numbers. However, as we get ready to solve equations in the next chapter, there is a need for *more numbers*. There are some very simple equations that do not have solutions with the set of real numbers. For example, the equation $x^2 + 1 = 0$ has no solutions among the real numbers. To solve such equations we need to extend the real number system. In this section we will introduce a set of numbers that contains some numbers whose squares are negative real numbers. Then, in the next chapter and in Chapter 6, we will see that this set of numbers, called the **complex numbers**, provides solutions not only for equations such as $x^2 + 1 = 0$, but also for *any* polynomial equation in general.

Let's begin by defining a number i such that

$$i^2 = -1.$$

The number i is not a real number and is often called the **imaginary unit**, but the number i^2 is the real number -1. The imaginary unit i is used to define a complex number as follows.

DEFINITION 1.8

A **complex number** is any number that can be expressed in the form

$$a + bi,$$

where a and b are real numbers.

The form $a + bi$ is called the **standard form** of a complex number. The real number a is called the **real part** of the complex number and b is called the **imaginary part**. (Note that b is a real number even though it is called the imaginary part.) Each of the following represents a complex number.

$6 + 2i$ is already expressed in the form $a + bi$. Traditionally, complex numbers for which $a \neq 0$ and $b \neq 0$ have been called imaginary numbers.

$5 - 3i$ can be written $5 + (-3i)$ even though the form $5 - 3i$ is often used.

$-8 + i\sqrt{2}$ can be written $-8 + \sqrt{2}i$. It is easy to mistake $\sqrt{2}i$ for $\sqrt{2i}$. Thus, we commonly write $i\sqrt{2}$ instead of $\sqrt{2}i$ to avoid any difficulties with the radical sign.

$-9i$ can be written $0 + (-9i)$. Complex numbers such as $-9i$, for which $a = 0$ and $b \neq 0$, traditionally have been called **pure imaginary numbers**.

5 can be written $5 + 0i$.

The set of real numbers is a subset of the set of complex numbers. The following diagram indicates the organizational format of the complex number system.

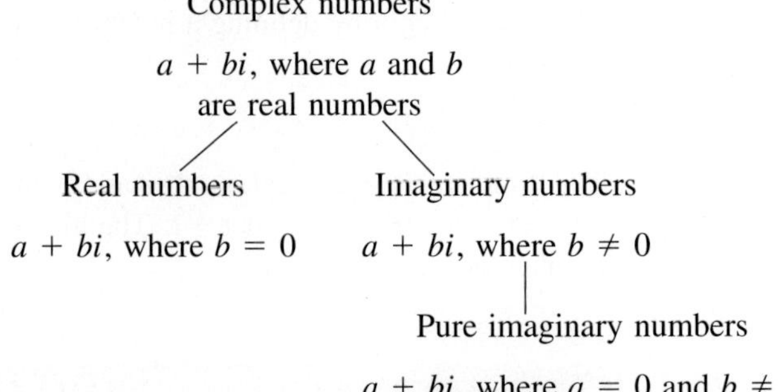

Two complex numbers $a + bi$ and $c + di$ are said to be *equal* if and only if $a = c$ and $b = d$. In other words, two complex numbers are equal if and only if their real parts are equal and their imaginary parts are equal.

Adding and Subtracting Complex Numbers

The following definition provides the basis for adding complex numbers.

$$(a + bi) + (c + di) = (a + c) + (b + d)i$$

We can use this definition to find the sum of two complex numbers.

$$(4 + 3i) + (5 + 9i) = (4 + 5) + (3 + 9)i = 9 + 12i$$

$$(-6 + 4i) + (8 - 7i) = (-6 + 8) + (4 - 7)i = 2 - 3i$$

$$\left(\frac{1}{2} + \frac{3}{4}i\right) + \left(\frac{2}{3} + \frac{1}{5}i\right) = \left(\frac{1}{2} + \frac{2}{3}\right) + \left(\frac{3}{4} + \frac{1}{5}\right)i$$

$$= \left(\frac{3}{6} + \frac{4}{6}\right) + \left(\frac{15}{20} + \frac{4}{20}\right)i = \frac{7}{6} + \frac{19}{20}i$$

$$\left(3 + i\sqrt{2}\right) + \left(-4 + i\sqrt{2}\right) = [3 + (-4)] + \left(\sqrt{2} + \sqrt{2}\right)i = -1 + 2i\sqrt{2}$$

Note the form for writing $2\sqrt{2}i$.

The set of complex numbers is **closed with respect to addition**; that is, the sum of two complex numbers is a complex number. Furthermore, the commutative and associative properties of addition hold for all complex numbers. The additive identity element is $0 + 0i$, or simply the real number 0. The additive inverse of $a + bi$ is $-a - bi$ because

$$(a + bi) + (-a - bi) = [a + (-a)] + [b + (-b)]i = 0.$$

Therefore, to *subtract* $c + di$ from $a + bi$, we add the additive inverse of $c + di$.

$$(a + bi) - (c + di) = (a + bi) + (-c - di)$$
$$= (a - c) + (b - d)i$$

The following examples illustrate the subtraction of complex numbers.

$$(9 + 8i) - (5 + 3i) = (9 - 5) + (8 - 3)i = 4 + 5i$$

$$(3 - 2i) - (4 - 10i) = (3 - 4) + [-2 - (-10)]i = -1 + 8i$$

$$\left(-\frac{1}{2} + \frac{1}{3}i\right) - \left(\frac{3}{4} + \frac{1}{2}i\right) = \left(-\frac{1}{2} - \frac{3}{4}\right) + \left(\frac{1}{3} - \frac{1}{2}\right)i = -\frac{5}{4} - \frac{1}{6}i$$

Multiplying and Dividing Complex Numbers

Since $i^2 = -1$, the number i is a square root of -1; so we write $i = \sqrt{-1}$. It should also be evident that $-i$ is a square root of -1 because

$$(-i)^2 = (-i)(-i) = i^2 = -1.$$

Therefore, in the set of complex numbers, -1 has two square roots, namely, i and $-i$. This is symbolically expressed

$$i = \sqrt{-1} \qquad \text{and} \qquad -i = -\sqrt{-1}.$$

Let's extend the definition so that in the set of complex numbers, every negative real number has two square roots. For any positive real number b,

$$\left(i\sqrt{b}\right)^2 = i^2(b) = -1(b) = -b.$$

Therefore, let's denote the **principal square root of $-b$** by $\sqrt{-b}$ and define it to be

$$\sqrt{-b} = i\sqrt{b},$$

where b is any positive real number. In other words, the principal square root of any negative real number can be represented as the product of a real number and the imaginary unit i. Consider the following examples.

$$\sqrt{-4} = i\sqrt{4} = 2i$$

$$\sqrt{-17} = i\sqrt{17}$$

$$\sqrt{-24} = i\sqrt{24} = i\sqrt{4}\sqrt{6} = 2i\sqrt{6}$$

Note that we simplified the radical $\sqrt{24}$ to $2\sqrt{6}$.

We should also observe that $-\sqrt{-b}$, where $b > 0$, is a square root of $-b$ because

$$\left(-\sqrt{-b}\right)^2 = \left(-i\sqrt{b}\right)^2 = i^2(b) = (-1)b = -b.$$

Thus, in the set of complex numbers, $-b$ (where $b > 0$) has two square roots, $i\sqrt{b}$ and $-i\sqrt{b}$. These are expressed as

$$\sqrt{-b} = i\sqrt{b} \qquad \text{and} \qquad -\sqrt{-b} = -i\sqrt{b}.$$

We must be careful with the use of the symbol $\sqrt{-b}$, where $b > 0$. Some properties that are true in the set of real numbers involving the square root symbol do not hold if the square root symbol does not represent a real number. For example, $\sqrt{a}\sqrt{b} = \sqrt{ab}$ *does not hold* if a and b are both negative numbers.

Correct $\quad \sqrt{-4}\sqrt{-9} = (2i)(3i) = 6i^2 = 6(-1) = -6$

Incorrect $\quad \sqrt{-4}\sqrt{-9} = \sqrt{(-4)(-9)} = \sqrt{36} = 6$

To avoid difficulty with this idea, you should rewrite all expressions of the form $\sqrt{-b}$, where $b > 0$ in the form $i\sqrt{b}$ *before* doing any computations. The following examples further illustrate this point.

$$\sqrt{-5}\sqrt{-7} = \left(i\sqrt{5}\right)\left(i\sqrt{7}\right) = i^2\sqrt{35} = (-1)\sqrt{35} = -\sqrt{35};$$

$$\sqrt{-2}\sqrt{-8} = \left(i\sqrt{2}\right)\left(i\sqrt{8}\right) = i^2\sqrt{16} = (-1)(4) = -4;$$

$$\sqrt{-2}\sqrt{8} = \left(i\sqrt{2}\right)\left(\sqrt{8}\right) = i\sqrt{16} = 4i;$$

$$\sqrt{-6}\sqrt{-8} = \left(i\sqrt{6}\right)\left(i\sqrt{8}\right) = i^2\sqrt{48} = i^2\sqrt{16}\sqrt{3} = 4i^2\sqrt{3} = -4\sqrt{3};$$

$$\frac{\sqrt{-2}}{\sqrt{3}} = \frac{i\sqrt{2}}{\sqrt{3}} = \frac{i\sqrt{2}}{\sqrt{3}} \cdot \frac{\sqrt{3}}{\sqrt{3}} = \frac{i\sqrt{6}}{3};$$

$$\frac{\sqrt{-48}}{\sqrt{12}} = \frac{i\sqrt{48}}{\sqrt{12}} = i\sqrt{\frac{48}{12}} = i\sqrt{4} = 2i.$$

Since complex numbers have a *binomial form*, we can find the product of two complex numbers in the same way that we find the product of two binomials. Then, by replacing i^2 with -1, we can simplify and express the final product in the standard form of a complex number. Consider the following examples.

$$\begin{aligned}(2 + 3i)(4 + 5i) &= 2(4 + 5i) + 3i(4 + 5i)\\ &= 8 + 10i + 12i + 15i^2\\ &= 8 + 22i + 15(-1)\\ &= 8 + 22i - 15\\ &= -7 + 22i\end{aligned}$$

$$\begin{aligned}(1 - 7i)^2 &= (1 - 7i)(1 - 7i)\\ &= 1(1 - 7i) - 7i(1 - 7i)\\ &= 1 - 7i - 7i + 49i^2\\ &= 1 - 14i + 49(-1)\\ &= 1 - 14i - 49\\ &= -48 - 14i\end{aligned}$$

$$\begin{aligned}(2 + 3i)(2 - 3i) &= 2(2 - 3i) + 3i(2 - 3i)\\ &= 4 - 6i + 6i - 9i^2\\ &= 4 - 9(-1)\\ &= 4 + 9\\ &= 13\end{aligned}$$

REMARK Don't forget that the multiplication patterns

$$(a + b)^2 = a^2 + 2ab + b^2$$
$$(a - b)^2 = a^2 - 2ab + b^2$$
$$(a + b)(a - b) = a^2 - b^2$$

can also be used when multiplying complex numbers.

The last example illustrates an important idea. The complex numbers $2 + 3i$ and $2 - 3i$ are called conjugates of each other. In general, the two complex numbers $a + bi$ and $a - bi$ are called **conjugates** of each other and **the product of a complex number and its conjugate is a real number**. This can be shown as follows.

$$\begin{aligned}(a + bi)(a - bi) &= a(a - bi) + bi(a - bi)\\ &= a^2 - abi + abi - b^2i^2\\ &= a^2 - b^2(-1)\\ &= a^2 + b^2\end{aligned}$$

Conjugates are used to simplify an expression such as $3i/(5 + 2i)$, which *indicates the quotient of two complex numbers.* To eliminate i in the denominator and to change the indicated quotient to the standard form of a complex number, we can multiply both the numerator and denominator by the conjugate of the denominator.

$$\begin{aligned}\frac{3i}{5 + 2i} &= \frac{3i}{5 + 2i} \cdot \frac{5 - 2i}{5 - 2i}\\ &= \frac{3i(5 - 2i)}{(5 + 2i)(5 - 2i)}\\ &= \frac{15i - 6i^2}{25 - 4i^2}\\ &= \frac{15i - 6(-1)}{25 - 4(-1)}\\ &= \frac{6 + 15i}{29}\\ &= \frac{6}{29} + \frac{15}{29}i\end{aligned}$$

The following examples further illustrate the process of dividing complex numbers.

$$\begin{aligned}\frac{2 - 3i}{4 - 7i} &= \frac{2 - 3i}{4 - 7i} \cdot \frac{4 + 7i}{4 + 7i}\\ &= \frac{(2 - 3i)(4 + 7i)}{(4 - 7i)(4 + 7i)}\\ &= \frac{8 + 14i - 12i - 21i^2}{16 - 49i^2}\end{aligned}$$

$$= \frac{8 + 2i - 21(-1)}{16 - 49(-1)}$$

$$= \frac{29 + 2i}{65} = \frac{29}{65} + \frac{2}{65}i$$

$$\frac{4 - 5i}{2i} = \frac{4 - 5i}{2i} \cdot \frac{-2i}{-2i}$$

$$= \frac{(4 - 5i)(-2i)}{(2i)(-2i)}$$

$$= \frac{-8i + 10i^2}{-4i^2}$$

$$= \frac{-8i + 10(-1)}{-4(-1)}$$

$$= \frac{-10 - 8i}{4} = -\frac{5}{2} - 2i$$

For a problem such as the last one, in which the denominator is a pure imaginary number, we can change to standard form by choosing a multiplier other than the conjugate of the denominator. Consider the following alternate approach.

$$\frac{4 - 5i}{2i} = \frac{4 - 5i}{2i} \cdot \frac{i}{i}$$

$$= \frac{(4 - 5i)(i)}{(2i)(i)}$$

$$= \frac{4i - 5i^2}{2i^2}$$

$$= \frac{4i - 5(-1)}{2(-1)}$$

$$= \frac{5 + 4i}{-2}$$

$$= -\frac{5}{2} - 2i$$

PROBLEM SET 1.8

Add or subtract as indicated.

1. $(5 + 2i) + (8 + 6i)$

2. $(-9 + 3i) + (4 + 5i)$

3. $(8 + 6i) - (5 + 2i)$

4. $(-6 + 4i) - (4 + 6i)$

5. $(-7 - 3i) + (-4 + 4i)$

6. $(6 - 7i) - (7 - 6i)$

7. $(-2 - 3i) - (-1 - i)$

8. $\left(\frac{1}{3} + \frac{2}{5}i\right) + \left(\frac{1}{2} + \frac{1}{4}i\right)$

9. $\left(-\frac{3}{4} - \frac{1}{4}i\right) + \left(\frac{3}{5} + \frac{2}{3}i\right)$

10. $\left(\frac{5}{8} + \frac{1}{2}i\right) - \left(\frac{7}{8} + \frac{1}{5}i\right)$

11. $\left(\frac{3}{10} - \frac{3}{4}i\right) - \left(-\frac{2}{5} + \frac{1}{6}i\right)$

12. $\left(4 + i\sqrt{3}\right) + \left(-6 - 2i\sqrt{3}\right)$

13. $(5 + 3i) + (7 - 2i) + (-8 - i)$

14. $(5 - 7i) - (6 - 2i) - (-1 - 2i)$

Write each of the following in terms of i and simplify. For example,

$$\sqrt{-20} = i\sqrt{20} = i\sqrt{4}\sqrt{5} = 2i\sqrt{5}.$$

15. $\sqrt{-9}$ 16. $\sqrt{-49}$ 17. $\sqrt{-19}$

18. $\sqrt{-31}$ 19. $\sqrt{-\frac{4}{9}}$ 20. $\sqrt{-\frac{25}{36}}$

21. $\sqrt{-8}$ 22. $\sqrt{-18}$ 23. $\sqrt{-27}$

24. $\sqrt{-32}$ 25. $\sqrt{-54}$ 26. $\sqrt{-40}$

27. $3\sqrt{-36}$ 28. $5\sqrt{-64}$ 29. $4\sqrt{-18}$

30. $6\sqrt{-8}$

Write each of the following in terms of i, perform the indicated operations, and simplify. For example,

$$\sqrt{-9}\sqrt{-16} = \left(i\sqrt{9}\right)\left(i\sqrt{16}\right) = (3i)(4i)$$
$$= 12i^2 = 12(-1) = -12.$$

31. $\sqrt{-4}\sqrt{-16}$ 32. $\sqrt{-25}\sqrt{-9}$

33. $\sqrt{-2}\sqrt{-3}$ 34. $\sqrt{-3}\sqrt{-7}$

35. $\sqrt{-5}\sqrt{-4}$ 36. $\sqrt{-7}\sqrt{-9}$

37. $\sqrt{-6}\sqrt{-10}$ 38. $\sqrt{-2}\sqrt{-12}$

39. $\sqrt{-8}\sqrt{-7}$ 40. $\sqrt{-12}\sqrt{-5}$

41. $\dfrac{\sqrt{-36}}{\sqrt{-4}}$ 42. $\dfrac{\sqrt{-64}}{\sqrt{-16}}$

43. $\dfrac{\sqrt{-54}}{\sqrt{-9}}$ 44. $\dfrac{\sqrt{-18}}{\sqrt{-3}}$

Find each of the following products and express the answers in standard form.

45. $(3i)(7i)$ 46. $(-5i)(8i)$

47. $(4i)(3 - 2i)$ 48. $(5i)(2 + 6i)$

49. $(3 + 2i)(4 + 6i)$ 50. $(7 + 3i)(8 + 4i)$

51. $(4 + 5i)(2 - 9i)$ 52. $(1 + i)(2 - i)$

53. $(-2 - 3i)(4 + 6i)$ 54. $(-3 - 7i)(2 + 10i)$

55. $(6 - 4i)(-1 - 2i)$ 56. $(7 - 3i)(-2 - 8i)$

57. $(3 + 4i)^2$ 58. $(4 - 2i)^2$

59. $(-1 - 2i)^2$ 60. $(-2 + 5i)^2$

61. $(8 - 7i)(8 + 7i)$ 62. $(5 + 3i)(5 - 3i)$

63. $(-2 + 3i)(-2 - 3i)$ 64. $(-6 - 7i)(-6 + 7i)$

Find each of the following quotients expressing answers in standard form.

65. $\dfrac{4i}{3 - 2i}$ 66. $\dfrac{3i}{6 + 2i}$ 67. $\dfrac{2 + 3i}{3i}$

68. $\dfrac{3 - 5i}{4i}$ 69. $\dfrac{3}{2i}$ 70. $\dfrac{7}{4i}$

71. $\dfrac{3 + 2i}{4 + 5i}$ 72. $\dfrac{2 + 5i}{3 + 7i}$ 73. $\dfrac{4 + 7i}{2 - 3i}$

74. $\dfrac{3 + 9i}{4 - i}$ 75. $\dfrac{3 - 7i}{-2 + 4i}$ 76. $\dfrac{4 - 10i}{-3 + 7i}$

77. $\dfrac{-1 - i}{-2 - 3i}$ 78. $\dfrac{-4 + 9i}{-3 - 6i}$

79. Using $a + bi$ and $c + di$ to represent two complex numbers, verify the following properties.

   a. The conjugate of the sum of two complex numbers is equal to the sum of the conjugates of the two numbers.

   b. The conjugate of the product of two complex numbers is equal to the product of the conjugates of the numbers.

THOUGHTS into WORDS

80. Is every real number also a complex number? Explain your answer.

81. Can the product of two nonreal complex numbers be a real number? Explain your answer.

MISCELLANEOUS PROBLEMS

82. Observe the following powers of i.

$$i = \sqrt{-1}, \quad i^2 = -1, \quad i^3 = i^2 \cdot i = -1(i) = -i, \quad i^4 = i^2 \cdot i^2 = (-1)(-1) = 1$$

Any power of i greater than 4 can be simplified to i, -1, $-i$, or 1 as follows.

$$i^9 = (i^4)^2(i) = (1)(i) = i$$

$$i^{14} = (i^4)^3(i^2) = (1)(-1) = -1$$

$$i^{19} = (i^4)^4(i^3) = (1)(-i) = -i$$

$$i^{28} = (i^4)^7 = (1)^7 = 1$$

Express each of the following as i, -1, $-i$, or 1.

a. i^5 **b.** i^6 **c.** i^{11} **d.** i^{12}

e. i^{16} **f.** i^{22} **g.** i^{33} **h.** i^{63}

83. We can use the information from Problem 82 and the binomial expansion patterns to find powers of complex numbers as follows.

$$\begin{aligned}(3 + 2i)^3 &= (3)^3 + 3(3)^2(2i) + 3(3)(2i)^2 + (2i)^3 \\ &= 27 + 54i + 36i^2 + 8i^3 \\ &= 27 + 54i + 36(-1) + 8(-i) = -9 + 46i\end{aligned}$$

Find the indicated power of each of the following.

a. $(2 + i)^3$ **b.** $(1 - i)^3$ **c.** $(1 - 2i)^3$

d. $(1 + i)^4$ **e.** $(2 - i)^4$ **f.** $(-1 + i)^5$

CHAPTER I SUMMARY

Be sure of the following key concepts from this chapter: set, null set, equal sets, subset, natural numbers, whole numbers, integers, rational numbers, irrational numbers, real numbers, complex numbers, absolute value, similar terms, exponent, monomial, binomial, polynomial, degree of a polynomial, perfect-square trinomial, factoring polynomials, rational expression, least common denominator, radical, simplest radical form, root, and conjugate of a complex number.

The following properties of the real numbers provide a basis for arithmetic and algebraic computation: closure for addition and multiplication, commutativity for addition and multiplication, associativity for addition and multiplication, identity properties for addition and multiplication, inverse properties for addition and multiplication, multiplication property of zero, multiplication property of -1, and distributive property.

The following properties of absolute value are useful.

1. $|a| \geq 0$
2. $|a| = |-a|$ *a* and *b* are real numbers
3. $|a - b| = |b - a|$

The following properties of exponents provide the basis for much of our computational work with polynomials.

1. $b^n \cdot b^m = b^{n+m}$

2. $(b^n)^m = b^{mn}$

3. $(ab)^n = a^n b^n$

4. $\left(\dfrac{a}{b}\right)^n = \dfrac{a^n}{b^n}$

5. $\dfrac{b^n}{b^m} = b^{n-m}$

m and *n* are rational numbers and *a* and *b* are real numbers, except $b \neq 0$ whenever it appears in the denominator

The following product patterns are helpful to recognize when multiplying polynomials.

1. $(a + b)^2 = a^2 + 2ab + b^2$

2. $(a - b)^2 = a^2 - 2ab + b^2$

3. $(a + b)(a - b) = a^2 - b^2$

4. $(a + b)^3 = a^3 + 3a^2b + 3ab^2 + b^3$

5. $(a - b)^3 = a^3 - 3a^2b + 3ab^2 - b^3$

Be sure you know how to do the following techniques.

1. Factor out the highest common monomial factor;

2. Factor by grouping;

3. Factor a trinomial into the product of two binomials;

4. Recognize some basic factoring patterns, namely,

$$a^2 + 2ab + b^2 = (a + b)^2;$$

$$a^2 - 2ab + b^2 = (a - b)^2;$$

$$a^2 - b^2 = (a + b)(a - b);$$

$$a^3 + b^3 = (a + b)(a^2 - ab + b^2);$$

$$a^3 - b^3 = (a - b)(a^2 + ab + b^2).$$

Be sure that you can simplify, add, subtract, multiply, and divide rational expressions using the following properties and definitions.

1. $\dfrac{a \cdot k}{b \cdot k} = \dfrac{a}{b}$

2. $\dfrac{-a}{b} = \dfrac{a}{-b} = -\dfrac{a}{b}$

3. $\frac{a}{b} \cdot \frac{c}{d} = \frac{ac}{bd}$

4. $\frac{a}{b} \div \frac{c}{d} = \frac{a}{b} \cdot \frac{d}{c} = \frac{ad}{bc}$

5. $\frac{a}{c} + \frac{b}{c} = \frac{a + b}{c}$

6. $\frac{a}{c} - \frac{b}{c} = \frac{a - b}{c}$

Be sure that you can simplify, add, subtract, multiply, and divide radicals using the following definitions and properties.

1. $\sqrt[n]{b} = a$ if and only if $a^n = b$
2. $\sqrt[n]{bc} = \sqrt[n]{b}\sqrt[n]{c}$
3. $\sqrt[n]{\frac{b}{c}} = \frac{\sqrt[n]{b}}{\sqrt[n]{c}}$

The following definition provides the link between exponents and roots.

$$b^{m/n} = \sqrt[n]{b^m} = \left(\sqrt[n]{b}\right)^m$$

This link, along with the properties of exponents, allows us (1) to multiply and divide some radicals with different indices, (2) to change to simplest radical form while in exponential form, and (3) to simplify expressions that are roots of roots.

A complex number is any number that can be expressed in the form $a + bi$, where a and b are real numbers and i is the imaginary unit such that $i^2 = -1$.

Addition and subraction of complex numbers are defined as follows.

$$(a + bi) + (c + di) = (a + c) + (b + d)i$$

$$(a + bi) - (c + di) = (a - c) + (b - d)i$$

Since complex numbers have a binomial form, we can multiply two complex numbers in the same way that we multiply two binomials. Thus, i^2 can be replaced with -1 and the final result can be expressed in the standard form of a complex number. For example,

$$\begin{aligned}(3 + 2i)(4 - 3i) &= 12 - i - 6i^2 \\ &= 12 - i - 6(-1) \\ &= 18 - i.\end{aligned}$$

The two complex numbers $a + bi$ and $a - bi$ are called **conjugates** of each other. The product $(a - bi)(a + bi)$ equals the real number $a^2 + b^2$ and this property is used to help with dividing complex numbers.

CHAPTER 1 REVIEW PROBLEM SET

Evaluate each of the following.

1. 5^{-3}

2. -3^{-4}

3. $\left(\frac{3}{4}\right)^{-2}$

4. $\dfrac{1}{\left(\frac{1}{3}\right)^{-2}}$

5. $-\sqrt{64}$

6. $\sqrt[3]{\frac{27}{8}}$

7. $\sqrt[5]{-\frac{1}{32}}$

8. $36^{-1/2}$

9. $\left(\frac{1}{8}\right)^{-2/3}$

10. $-32^{3/5}$

Perform the indicated operations and simplify. Express the final answers using positive exponents only.

11. $(3x^{-2}y^{-1})(4x^4y^2)$

12. $(5x^{2/3})(-6x^{1/2})$

13. $(-8a^{-1/2})(-6a^{1/3})$

14. $(3x^{-2/3}y^{1/5})^3$

15. $\dfrac{64x^{-2}y^3}{16x^3y^{-2}}$

16. $\dfrac{56x^{-1/3}y^{2/5}}{7x^{1/4}y^{-3/5}}$

17. $\left(\dfrac{-8x^2y^{-1}}{2x^{-1}y^2}\right)^2$

18. $\left(\dfrac{36a^{-1}b^4}{-12a^2b^5}\right)^{-1}$

Perform the indicated operations.

19. $(-7x - 3) + (5x - 2) + (6x + 4)$

20. $(12x + 5) - (7x - 4) - (8x + 1)$

21. $3(a - 2) - 2(3a + 5) + 3(5a - 1)$

22. $(4x - 7)(5x + 6)$

23. $(-3x + 2)(4x - 3)$

24. $(7x - 3)(-5x + 1)$

25. $(x + 4)(x^2 - 3x - 7)$

26. $(2x + 1)(3x^2 - 2x + 6)$

27. $(5x - 3)^2$

28. $(3x + 7)^2$

29. $(2x - 1)^3$

30. $(3x + 5)^3$

31. $(x^2 - 2x - 3)(x^2 + 4x + 5)$

32. $(2x^2 - x - 2)(x^2 + 6x - 4)$

33. $\dfrac{24x^3y^4 - 48x^2y^3}{-6xy}$

34. $\dfrac{-56x^2y + 72x^3y^2}{8x^2}$

Factor each of the following polynomials completely. Indicate any that are not factorable using integers.

35. $9x^2 - 4y^2$

36. $3x^3 - 9x^2 - 120x$

37. $4x^2 + 20x + 25$

38. $(x - y)^2 - 9$

39. $x^2 - 2x - xy + 2y$

40. $64x^3 - 27y^3$

41. $15x^2 - 14x - 8$

42. $3x^3 + 36$

43. $2x^2 - x - 8$

44. $3x^3 + 24$

45. $x^4 - 13x^2 + 36$

46. $4x^2 - 4x + 1 - y^2$

Perform the following indicated operations involving rational expressions. Express final answers in simplest form.

47. $\dfrac{8xy}{18x^2y} \cdot \dfrac{24xy^2}{16y^3}$

48. $\dfrac{-14a^2b^2}{6b^3} \div \dfrac{21a}{15ab}$

49. $\dfrac{x^2 + 3x - 4}{x^2 - 1} \cdot \dfrac{3x^2 + 8x + 5}{x^2 + 4x}$

50. $\dfrac{9x^2 - 6x + 1}{2x^2 + 8} \cdot \dfrac{8x + 20}{6x^2 + 13x - 5}$

51. $\dfrac{3x - 2}{4} + \dfrac{5x - 1}{3}$

52. $\dfrac{2x - 6}{5} - \dfrac{x + 4}{3}$

53. $\dfrac{3}{n^2} + \dfrac{4}{5n} - \dfrac{2}{n}$

54. $\dfrac{5}{x^2 + 7x} - \dfrac{3}{x}$

55. $\dfrac{3x}{x^2 - 6x - 40} + \dfrac{4}{x^2 - 16}$

56. $\dfrac{2}{x - 2} - \dfrac{2}{x + 2} - \dfrac{4}{x^3 - 4x}$

Simplify each of the following complex fractions.

57. $\dfrac{\dfrac{3}{x} - \dfrac{2}{y}}{\dfrac{5}{x^2} + \dfrac{7}{y}}$

58. $\dfrac{3 - \dfrac{2}{x}}{4 + \dfrac{3}{x}}$

59. $\dfrac{\dfrac{3}{(x + h)^2} - \dfrac{3}{x^2}}{h}$

60. Simplify the expression

$$\frac{6(x^2+2)^{1/2}-6x^2(x^2+2)^{-1/2}}{[(x^2+2)^{1/2}]^2}.$$

Express each of the following in simplest radical form. All variables represent positive real numbers.

61. $5\sqrt{48}$ **62.** $3\sqrt{24x^3}$ **63.** $\sqrt[3]{32x^4y^5}$

64. $\frac{3\sqrt{8}}{2\sqrt{6}}$ **65.** $\sqrt{\frac{5x}{2y^2}}$ **66.** $\frac{3}{\sqrt{2}+5}$

67. $\frac{4\sqrt{2}}{3\sqrt{2}+\sqrt{3}}$ **68.** $\frac{3\sqrt{x}}{\sqrt{x}-2\sqrt{y}}$

Perform the indicated operations and express the answers in simplest radical form.

69. $\sqrt{5}\sqrt[3]{5}$ **70.** $\sqrt[3]{x^2}\sqrt[4]{x}$ **71.** $\sqrt{x^3}\sqrt[3]{x^4}$

72. $\sqrt{xy}\sqrt[5]{x^3y^2}$ **73.** $\frac{\sqrt{5}}{\sqrt[3]{5}}$ **74.** $\frac{\sqrt[3]{x^2}}{\sqrt[4]{x^3}}$

Perform the indicated operations and express the resulting complex number in standard form.

75. $(-7+3i)+(-4-9i)$

76. $(2-10i)-(3-8i)$

77. $(-1+4i)-(-2+6i)$

78. $(3i)(-7i)$

79. $(2-5i)(3+4i)$ **80.** $(-3-i)(6-7i)$

81. $(4+2i)(-4-i)$ **82.** $(5-2i)(5+2i)$

83. $\frac{5}{3i}$ **84.** $\frac{2+3i}{3-4i}$

85. $\frac{-1-2i}{-2+i}$ **86.** $\frac{-6i}{5+2i}$

Write each of the following in terms of i and simplify.

87. $\sqrt{-100}$ **88.** $\sqrt{-40}$

89. $4\sqrt{-80}$ **90.** $(\sqrt{-9})(\sqrt{-16})$

91. $(\sqrt{-6})(\sqrt{-8})$ **92.** $\frac{\sqrt{-24}}{\sqrt{-3}}$

Use scientific notation and the properties of exponents to help with the following computations.

93. $\frac{(0.0064)(420000)}{(0.00014)(0.032)}$ **94.** $\frac{(8600)(0.0000064)}{(0.0016)(0.000043)}$

EQUATIONS, INEQUALITIES, AND PROBLEM SOLVING

A common thread throughout precalculus algebra courses is one of developing algebraic skills, then using the skills to solve equations and inequalities, and then using equations and inequalities to solve applied problems. In this chapter we shall review and extend a variety of concepts pertaining to that thread.

2

We can use the equation $0.2(8) - 0.2x + x = 0.9(8)$ to determine the change of the mixture in a radiator from 20% antifreeze to 90% antifreeze. We would drain 7 liters of coolant from a radiator that holds 8 liters and replace them with pure antifreeze, which will change the protection against temperature from 12° F to –20° F.

2.1 LINEAR EQUATIONS AND PROBLEM SOLVING

An algebraic equation such as $5x + 2 = 12$ is neither true nor false as it stands; it is sometimes referred to as an **open sentence**. Each time that a number is substituted for x, the algebraic equation $5x + 2 = 12$ becomes a **numerical statement** that is either true or false. For example, if $x = 5$ then $5x + 2 = 12$ becomes $5(5) + 2 = 12$, which is a false statement. If $x = 2$ then $5x + 2 = 12$ becomes $5(2) + 2 = 12$, which is a true statement. **Solving an equation** refers to the process of finding the number (or numbers) that makes an algebraic equation a true numerical statement. Such numbers are called the **solutions** or **roots** of the equation and are said to **satisfy the equation**. The set of all solutions of an equation is called its **solution set**. Thus, $\{2\}$ is the solution set of $5x + 2 = 12$.

An equation that is satisfied by all numbers that can meaningfully replace the variable is called an **identity**. For example,

$$3(x + 2) = 3x + 6, \qquad x^2 - 4 = (x + 2)(x - 2), \qquad \text{and} \qquad \frac{1}{x} + \frac{1}{2} = \frac{2 + x}{2x}$$

are all identities. In the last identity x cannot equal zero; thus the statement

$$\frac{1}{x} + \frac{1}{2} = \frac{2 + x}{2x}$$

is true for all real numbers except zero. An equation that is true for some but not all permissible values of the variable is called a **conditional equation**. Thus the equation $5x + 2 = 12$ is a conditional equation.

Equivalent equations are equations that have the same solution set. For example,

$$7x - 1 = 20, \qquad 7x = 21, \qquad \text{and} \qquad x = 3$$

are all equivalent equations because $\{3\}$ is the solution set of each. The general procedure for solving an equation is to continue replacing the given equation with equivalent but simpler equations until an equation of the form *variable* = *constant* or *constant* = *variable* is obtained. Thus, in the example above, $7x - 1 = 20$ was simplified to $7x = 21$, which was further simplified to $x = 3$, which gives us the solution set, $\{3\}$.

Techniques for solving equations are centered around properties of equality. The following list summarizes some basic properties of equality.

PROPERTY 2.1 *Properties of Equality*

For all real numbers, a, b, and c,

1. $a = a$; Reflexive property

2. If $a = b$ then $b = a$; Symmetric property
3. If $a = b$ and $b = c$, then $a = c$; Transitive property
4. If $a = b$, then a may be replaced by b, or b may be replaced by a, in any statement without changing the meaning of the statement; Substitution property
5. $a = b$ if and only if $a + c = b + c$; Addition property
6. $a = b$ if and only if $ac = bc$, where $c \neq 0$. Multiplication property

The addition property of equality states that any number can be added to both sides of an equation to produce an equivalent equation. The multiplication property of equality states that an equivalent equation is produced whenever both sides of an equation are multiplied by the same nonzero real number.

Now let's consider how these properties of equality can be used to solve a variety of linear equations. A **linear equation** in the variable x is one that can be written in the form

$$ax + b = 0,$$

where a and b are real numbers and $a \neq 0$.

EXAMPLE 1

Solve the equation $2x - 3 = 9$.

Solution

$$2x - 3 = 9$$

$$2x - 3 + 3 = 9 + 3 \quad \text{Add 3 to both sides.}$$

$$2x = 12$$

$$\frac{1}{2}(2x) = \frac{1}{2}(12) \quad \text{Multiply both sides by } \frac{1}{2}.$$

$$x = 6$$

Check To check an apparent solution we can substitute it into the original equation to see if we obtained a true numerical statement.

$$2x - 3 = 9$$

$$2(6) - 3 \stackrel{?}{=} 9$$

$$12 - 3 \stackrel{?}{=} 9$$

$$9 = 9$$

Now we know that the solution set is $\{6\}$.

EXAMPLE 2

Solve the equation $-4x - 3 = 2x + 9$.

Solution

$$\begin{aligned} -4x - 3 &= 2x + 9 \\ -4x - 3 + (-2x) &= 2x + 9 + (-2x) && \text{Add } -2x \text{ to both sides.} \\ -6x - 3 &= 9 \\ -6x - 3 + 3 &= 9 + 3 && \text{Add 3 to both sides.} \\ -6x &= 12 \\ -\frac{1}{6}(-6x) &= -\frac{1}{6}(12) && \text{Multiply both sides by } -\frac{1}{6}. \\ x &= -2. \end{aligned}$$

Check

$$\begin{aligned} -4x - 3 &= 2x + 9 \\ -4(-2) - 3 &\stackrel{?}{=} 2(-2) + 9 \\ 8 - 3 &\stackrel{?}{=} -4 + 9 \\ 5 &= 5. \end{aligned}$$

Now we know that the solution set is $\{-2\}$.

EXAMPLE 3

Solve $4(n - 2) - 3(n - 1) = 2(n + 6)$.

Solution

First let's use the distributive property to remove parentheses and combine similar terms.

$$\begin{aligned} 4(n - 2) - 3(n - 1) &= 2(n + 6) \\ 4n - 8 - 3n + 3 &= 2n + 12 \\ n - 5 &= 2n + 12 \end{aligned}$$

Now we can apply the addition property of equality.

$$\begin{aligned} n - 5 + (-n) &= 2n + 12 + (-n) \\ -5 &= n + 12 \\ -5 + (-12) &= n + 12 + (-12) \\ -17 &= n \end{aligned}$$

Check

$$\begin{aligned} 4(n - 2) - 3(n - 1) &= 2(n + 6) \\ 4(-17 - 2) - 3(-17 - 1) &\stackrel{?}{=} 2(-17 + 6) \\ 4(-19) - 3(-18) &\stackrel{?}{=} 2(-11) \\ -76 + 54 &\stackrel{?}{=} -22 \\ -22 &= -22 \end{aligned}$$

The solution set is $\{-17\}$.

As you study these examples, pay special attention to the steps shown in the solutions. Certainly, there are no rules as to which steps should be performed mentally; this is an individual decision. We would suggest that you show enough steps so that the flow of the process is understood and so that the chances of making careless computational errors are minimized. We shall discontinue showing the check for each problem, but remember that checking an answer is the only way of being sure of your result.

EXAMPLE 4

Solve $\frac{1}{4}x - \frac{2}{3}x = \frac{5}{6}$.

Solution

$$\frac{1}{4}x - \frac{2}{3}x = \frac{5}{6}$$

$$12\left(\frac{1}{4}x - \frac{2}{3}x\right) = 12\left(\frac{5}{6}\right) \qquad \text{Multiply both sides by the LCD.}$$

$$12\left(\frac{1}{4}x\right) - 12\left(\frac{2}{3}x\right) = 12\left(\frac{5}{6}\right)$$

$$3x - 8x = 10$$

$$-5x = 10$$

$$x = -2$$

The solution set is $\{-2\}$.

EXAMPLE 5

Solve $\frac{2y - 3}{3} + \frac{y + 1}{2} = 3$.

Solution

$$\frac{2y - 3}{3} + \frac{y + 1}{2} = 3$$

$$6\left(\frac{2y - 3}{3} + \frac{y + 1}{2}\right) = 6(3) \qquad \text{Multiply both sides by the LCD.}$$

$$6\left(\frac{2y - 3}{3}\right) + 6\left(\frac{y + 1}{2}\right) = 6(3) \qquad \text{Apply the distributive property on the left side.}$$

$$2(2y - 3) + 3(y + 1) = 18$$

$$4y - 6 + 3y + 3 = 18$$

$$7y - 3 = 18$$

$$7y = 21$$

$$y = 3$$

The solution set is $\{3\}$. (Check it!)

EXAMPLE 6

Solve $\frac{4x-1}{10} - \frac{5x+2}{4} = -3$.

Solution

$$\begin{aligned}
\frac{4x-1}{10} - \frac{5x+2}{4} &= -3 \\
20\left(\frac{4x-1}{10} - \frac{5x+2}{4}\right) &= 20(-3) \\
20\left(\frac{4x-1}{10}\right) - 20\left(\frac{5x+2}{4}\right) &= -60 \\
2(4x-1) - 5(5x+2) &= -60 \\
8x - 2 - 25x - 10 &= -60 \\
-17x - 12 &= -60 \\
-17x &= -48 \\
x &= \frac{48}{17}
\end{aligned}$$

The solution set is $\left\{\frac{48}{17}\right\}$.

Problem Solving

The ability to use the tools of algebra to solve problems requires that we be able to translate the English language into the language of algebra. More specifically, at this time we need to translate *English sentences* into *algebraic equations* so that we can use our equation solving skills. Let's work through an example and then comment on some of the problem solving aspects of it.

PROBLEM 1

If 2 is subtracted from five times a certain number, the result is 28. Find the number.

Solution

Let n represent the number to be found. The sentence, *If 2 is subtracted from five times a certain number, the result is 28* translates into the equation $5n - 2 = 28$. Solving this equation, we obtain

$$\begin{aligned}
5n - 2 &= 28 \\
5n &= 30 \\
n &= 6.
\end{aligned}$$

The number to be found is 6.

Now let's make a few comments about our approach to Problem 1. Making a statement such as *Let n represent the number to be found* is often referred to as **declaring the variable**. It amounts to choosing a letter to use as a variable and indicating what the variable represents for a specific problem. This may seem like an insignificant idea, but as the problems become more complex, the process of declaring the variable becomes more important. It is also a good idea to choose a *meaningful* variable. For example, if the problem involves finding the width of a rectangle, then a choice of w for the variable is reasonable. Furthermore, it is true that some people can solve a problem such as Problem 1 without setting up an algebraic equation. However, as problems increase in difficulty, the translation from English to algebra becomes a key issue. Therefore, even with these relatively easy problems, we suggest that you concentrate on the translation process.

To check our answer for Problem 1, we must determine whether it satisfies the conditions stated in the original problem. Because 2 subtracted from 5(6) equals 28, we know that our answer of 6 is correct. Remember, when you are checking a potential answer for a word problem, it is *not* sufficient to check the result in the equation used to solve the problem, because the equation itself may be in error.

Sometimes it is necessary to not only declare the variable but also to represent other unknown quantities in terms of that variable. Let's consider a problem that illustrates this idea.

PROBLEM 2

Find three consecutive integers whose sum is -45.

Solution

Let n represent the smallest integer; then $n + 1$ is the next integer and $n + 2$ is the largest of the three integers. Since the sum of the three consecutive integers is to be -45, we have the following equation.

$$\begin{aligned} n + (n + 1) + (n + 2) &= -45 \\ 3n + 3 &= -45 \\ 3n &= -48 \\ n &= -16 \end{aligned}$$

If $n = -16$, then $n + 1$ is -15 and $n + 2$ is -14. Thus, the three consecutive integers are -16, -15, and -14.

PROBLEM 3

Tina is paid time-and-a-half for each hour worked over 40 hours in a week. Last week she worked 45 hours and earned \$380. What is her normal hourly rate?

Solution

Let r represent Tina's normal hourly rate. Then $\frac{3}{2}r$ represents $1\frac{1}{2}$ times her normal hourly rate (time-and-a-half). The following guideline can be used to help set up the equation.

Regular wages for first 40 hours	+	Wages for 5 hours of overtime	=	Total wages
↓		↓		↓
$40r$	+	$5\left(\frac{3}{2}r\right)$	=	\$380.

Solving this equation, we obtain

$$2\left[40r + 5\left(\frac{3}{2}r\right)\right] = 2(380)$$
$$2(40r) + 2\left[5\left(\frac{3}{2}r\right)\right] = 760$$
$$80r + 15r = 760$$
$$95r = 760$$
$$r = 8.$$

Her normal hourly rate is thus \$8 per hour. (Check the answer back in the statement of the problem!)

PROBLEM 4

There are 51 students in a certain class. The number of females is 5 less than three times the number of males. Find the number of females and the number of males in the class.

Solution

Let m represent the number of males; then $3m - 5$ reprsents the number of females. Since the total number of students is 51, we can set up and solve the following equation.

$$m + (3m - 5) = 51$$
$$4m - 5 = 51$$
$$4m = 56$$
$$m = 14$$

Therefore there are 14 males and $3(14) - 5 = 37$ females.

PROBLEM SET 2.1

Solve each of the following equations.

1. $9x - 3 = -21$ **2.** $-5x + 4 = -11$

3. $13 - 2x = 14$ **4.** $17 = 6a + 5$

5. $3n - 2 = 2n + 5$ **6.** $4n + 3 = 5n - 9$

7. $-5a + 3 = -3a + 6$

8. $4x - 3 + 2x = 8x - 3 - x$

9. $-3(x + 1) = 7$ **10.** $5(2x - 1) = 13$

11. $4(2x - 1) = 3(3x + 2)$ **12.** $5x - 4(x - 6) = -11$

13. $4(n - 2) - 3(n - 1) = 2(n + 6)$

14. $-3(2t - 5) = 2(4t + 7)$

15. $3(2t - 1) - 2(5t + 1) = 4(3t + 4)$

16. $-(3x - 1) + (2x + 3) = -4 + 3(x - 1)$

17. $-2(y - 4) - (3y - 1) = -2 + 5(y + 1)$

18. $\frac{-3x}{4} = \frac{9}{2}$ **19.** $-\frac{6x}{7} = 12$

20. $\frac{n}{2} - \frac{1}{3} = \frac{13}{6}$ **21.** $\frac{3}{4}n - \frac{1}{12}n = 6$

22. $\frac{2}{3}x - \frac{1}{5}x = 7$

23. $\frac{h}{2} + \frac{h}{5} = 1$

24. $\frac{4y}{5} - 7 = \frac{y}{10}$

25. $\frac{y}{5} - 2 = \frac{y}{2} + 1$

26. $\frac{x+2}{3} + \frac{x-1}{4} = \frac{9}{2}$

27. $\frac{c+5}{7} + \frac{c-3}{4} = \frac{5}{14}$

28. $\frac{2x-5}{6} - \frac{3x-4}{8} = 0$

29. $\frac{n-3}{2} - \frac{4n-1}{6} = \frac{2}{3}$

30. $\frac{3x-1}{2} + \frac{x-3}{4} = \frac{1}{2}$

31. $\frac{2t+3}{6} - \frac{t-9}{4} = 5$

32. $\frac{2x+7}{9} - 4 = \frac{x-7}{12}$

33. $\frac{3n-1}{8} - 2 = \frac{2n+5}{7}$

34. $\frac{x+2}{3} + \frac{3x+1}{4} + \frac{2x-1}{6} = 2$

35. $\frac{2t-3}{6} + \frac{3t-2}{4} + \frac{5t+6}{12} = 4$

36. $\frac{3y-1}{8} + y - 2 = \frac{y+4}{4}$

37. $\frac{2x+1}{14} - \frac{3x+4}{7} = \frac{x-1}{2}$

38. $n + \frac{2n-3}{9} - 2 = \frac{2n+1}{3}$

39. $(x-3)(x-1) - x(x+2) = 7$

40. $(3n+4)(n-2) - 3n(n+3) = 3$

41. $(2y+1)(3y-2) - (6y-1)(y+4) = -20y$

42. $(4t-3)(t+2) - (2t+3)^2 = -1$

Solve each of the following problems by setting up and solving an algebraic equation.

43. Three is subtracted from a certain number and then that result is multiplied by 4 to produce 152. Find the number.

44. One is subtracted from twice a certain number and then that result is multiplied by 3 to produce -75. Find the number.

45. The sum of two numbers is 30. The larger number is 5 less than four times the smaller number. Find the two numbers.

46. One number is 5 less than another number. Find the numbers if five times the smaller number is 11 less than four times the larger number.

47. The sum of three consecutive integers is 21 larger than twice the smallest integer. Find the integers.

48. Find three consecutive even integers such that if the largest integer is subtracted from four times the smallest, the result is 6 more than twice the middle integer.

49. Find three consecutive odd integers such that three times the largest is 23 less than twice the sum of the two smallest integers.

50. Find two consecutive integers such that the difference of their squares is 37.

51. Find three consecutive integers such that the product of the two largest is 20 more than the square of the smallest integer.

52. Find four consecutive integers such that the product of the two largest is 46 more than the product of the two smallest integers.

53. The average of the salaries of Kelly, Renee, and Nina is \$20,000 a year. If Kelly earns \$4000 less than Renee, and Nina's salary is two-thirds of Renee's salary, find the salary of each person.

54. Barry is paid double-time for each hour worked over 40 hours in a week. Last week he worked 47 hours and earned \$378. What is his normal hourly rate?

55. Greg had 80 coins consisting of pennies, nickels, and dimes. The number of nickels was 5 more than one-third the number of pennies and the number of dimes was 1 less than one-fourth the number of pennies. How many coins of each kind did he have?

56. Rita has a collection of 105 coins consisting of nickels, dimes, and quarters. The number of dimes is 5 more than one-third the number of nickels, and the number of quarters is twice the number of dimes. How many coins of each kind does she have?

57. In a class of 43 students, the number of males is 8 less than twice the number of females. How many females and how many males are there in the class?

58. A precinct reported that 316 people had voted in an election. The number of Republican voters was 6 more than two-thirds the number of Democrats. How many Republicans and how many Democrats voted in that precinct?

59. Two years ago Janie was half as old as she will be 9 years from now. How old is she now?

60. The sum of the present ages of Eric and his father is 58 years. In 10 years his father will be twice as old as Eric will be at that time. Find their present ages.

61. Brad is presently 6 years older than Pedro. Five years ago Pedro's age was three-fourths of Brad's age at that time. Find the present ages of Brad and Pedro.

62. Tina is 4 years older than Sherry. In 5 years the sum of their ages will be 48. Find their present ages.

THOUGHTS into WORDS

63. What is an algebraic equation?

64. Illustrate how the symmetric property of equality (if $a = b$, then $b = a$) can be used when solving equations.

65. How do you defend the statement that the equation $x + 1 = x + 2$ has no real number solutions?

MISCELLANEOUS PROBLEMS

66. Verify that for any three consecutive integers, the sum of the smallest and largest is equal to twice the middle integer.

67. Verify that no four consecutive integers can be found such that the product of the smallest and largest is equal to the product of the other two integers.

2.2 MORE EQUATIONS AND APPLICATIONS

In the previous section we considered linear equations, such as

$$\frac{x-1}{3} + \frac{x+2}{4} = \frac{1}{6},$$

that have fractional coefficients with constants as denominators. Now let's consider equations that contain the variable in one or more of the denominators. Our approach to solving such equations remains essentially the same except **we must avoid any values of the variable that make a denominator zero**. Consider the following examples.

EXAMPLE 1

Solve $\dfrac{5}{3x} - \dfrac{1}{9} = \dfrac{1}{x}$.

Solution

First we need to realize that *x cannot equal zero*. Let's indicate this restriction so that it is not forgotten; then we can proceed as follows.

$$\frac{5}{3x} - \frac{1}{9} = \frac{1}{x}, \qquad x \neq 0$$

$$9x\left(\frac{5}{3x} - \frac{1}{9}\right) = 9x\left(\frac{1}{x}\right) \qquad \text{Multiply both sides by the LCD.}$$

$$9x\left(\frac{5}{3x}\right) - 9x\left(\frac{1}{9}\right) = 9x\left(\frac{1}{x}\right)$$

$$15 - x = 9$$

$$-x = -6$$

$$x = 6$$

The solution set is $\{6\}$. (Check it!)

EXAMPLE 2

Solve $\dfrac{65 - n}{n} = 4 + \dfrac{5}{n}$.

Solution

$$\frac{65 - n}{n} = 4 + \frac{5}{n}, \qquad n \neq 0$$

$$n\left(\frac{65 - n}{n}\right) = n\left(4 + \frac{5}{n}\right)$$

$$65 - n = 4n + 5$$

$$60 = 5n$$

$$12 = n$$

The solution set is $\{12\}$.

EXAMPLE 3

Solve $\dfrac{a}{a - 2} + \dfrac{2}{3} = \dfrac{2}{a - 2}$.

Solution

$$\frac{a}{a - 2} + \frac{2}{3} = \frac{2}{a - 2}, \qquad a \neq 2$$

$$3(a - 2)\left(\frac{a}{a - 2} + \frac{2}{3}\right) = 3(a - 2)\left(\frac{2}{a - 2}\right)$$

$$3a + 2(a - 2) = 6$$

$$3a + 2a - 4 = 6$$

$$5a = 10$$

$$a = 2$$

Because our initial restriction was $a \neq 2$, we conclude that this equation *has no solution*. The solution set is $\varnothing$.

Example 3 illustrates the importance of recognizing the restrictions that must be made to exclude division by zero.

Ratio and Proportion

A **ratio** is the comparison of two numbers by division. The fractional form is frequently used to express ratios. For example, the ratio of a to b can be written a/b. A statement of equality between two ratios is called a **proportion**. Thus, if a/b and c/d are equal ratios, the proportion $a/b = c/d$ ($b \neq 0$ and $d \neq 0$) can be formed. There is a useful property of proportions.

$$\text{If } \frac{a}{b} = \frac{c}{d} \quad \text{then } ad = bc.$$

This property can be deduced as follows.

$$\frac{a}{b} = \frac{c}{d}, \qquad b \neq 0 \text{ and } d \neq 0$$

$$bd\left(\frac{a}{b}\right) = bd\left(\frac{c}{d}\right) \qquad \text{Multiply both sides by } bd.$$

$$ad = bc$$

This is sometimes referred to as the **cross-multiplication property of proportions**

Some equations can be treated as proportions and solved by using the cross-multiplication idea, as the next example illustrates.

EXAMPLE 4

Solve $\dfrac{3}{3x - 2} = \dfrac{4}{2x + 1}$.

Solution

$$\frac{3}{3x - 2} = \frac{4}{2x + 1}, \qquad x \neq \frac{2}{3}, x \neq -\frac{1}{2}$$

$$3(2x + 1) = 4(3x - 2) \qquad \text{Apply the cross-multiplication property.}$$

$$6x + 3 = 12x - 8$$

$$11 = 6x$$

$$\frac{11}{6} = x$$

The solution set is $\left\{\dfrac{11}{6}\right\}$.

Linear Equations Involving Decimals

To solve an equation such as $x + 2.4 = 0.36$ we can add -2.4 to both sides. However, as equations containing decimals become more complex, it is often easier to begin by *clearing the equation of all decimals*, which we accomplish by multiplying both sides by an appropriate power of 10. Let's consider two examples.

EXAMPLE 5

Solve $0.12t - 2.1 = 0.07t - 0.2$.

Solution

$$\begin{aligned} 0.12t - 2.1 &= 0.07t - 0.2 \\ 100(0.12t - 2.1) &= 100(0.07t - 0.2) \qquad \text{Multiply both sides by 100.} \\ 12t - 210 &= 7t - 20 \\ 5t &= 190 \\ t &= 38 \end{aligned}$$

The solution set is $\{38\}$.

EXAMPLE 6

Solve $0.8x + 0.9(850 - x) = 715$.

Solution

$$\begin{aligned} 0.8x + 0.9(850 - x) &= 715 \\ 10[0.8x + 0.9(850 - x)] &= 10(715) \qquad \text{Multiply both sides by 10.} \\ 10(0.8x) + 10[0.9(850 - x)] &= 10(715) \\ 8x + 9(850 - x) &= 7150 \\ 8x + 7650 - 9x &= 7150 \\ -x &= -500 \\ x &= 500 \end{aligned}$$

The solution set is $\{500\}$.

Changing Forms of Formulas

Many practical applications of mathematics involve the use of formulas. For example, to find the distance traveled in 4 hours at a rate of 55 miles per hour, we multiply the rate times the time; thus the distance is $55(4) = 220$ miles. The rule *distance equals rate times time* is commonly stated as a formula: $d = rt$. When using a formula, it is sometimes convenient first to change its form. For example, multiplying both sides of $d = rt$ by $1/t$ produces the equivalent form $r = d/t$. Multiplying both sides of $d = rt$ by $1/r$ produces another equivalent form, $t = d/r$. The following two examples further illustrate the process of obtaining equivalent forms of certain formulas.

EXAMPLE 7

If P dollars are invested at a simple rate of r percent, then the amount, A, accumulated after t years is given by the formula $A = P + Prt$. Solve this formula for P.

Solution

$$\begin{aligned} A &= P + Prt \\ A &= P(1 + rt) \qquad \text{Apply the distributive property to the right side.} \end{aligned}$$

$$\frac{A}{1 + rt} = P \qquad \text{Multiply both sides by } \frac{1}{1 + rt}.$$

$$P = \frac{A}{1 + rt} \qquad \text{Apply the symmetric property of equality.}$$

EXAMPLE 8

The area (A) of a trapezoid (see Figure 2.1) is given by the formula $A = \frac{1}{2}h(b_1 + b_2)$. Solve this equation for b_1.

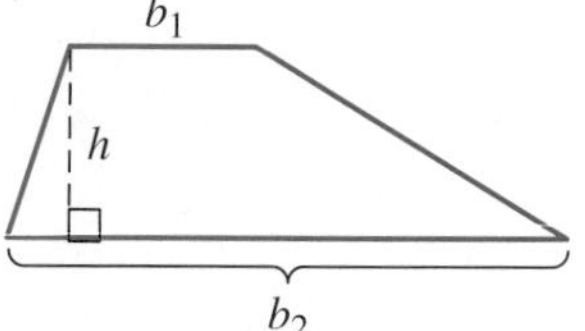

FIGURE 2.1

Solution

$$A = \frac{1}{2}h(b_1 + b_2)$$

$$2A = h(b_1 + b_2) \qquad \text{Multiply both sides by 2.}$$

$$2A = hb_1 + hb_2 \qquad \text{Apply the distributive property to the right side.}$$

$$2A - hb_2 = hb_1 \qquad \text{Add } -hb_2 \text{ to both sides.}$$

$$\frac{2A - hb_2}{h} = b_1 \qquad \text{Multiply both sides by } \frac{1}{h}.$$

In Example 7, notice that the distributive property is used to change from the form $P + Prt$ to $P(1 + rt)$. However, in Example 8 the distributive property was used to change $h(b_1 + b_2)$ to $hb_1 + hb_2$. In both examples the goal is to *isolate the term* containing the variable being solved for so that an appropriate application of the multiplication property will produce the desired result. Also note the use of *subscripts* to identify the two bases of the trapezoid. Subscripts allow us to use the same letter b to identify the bases, but b_1 represents one base and b_2 the other.

More on Problem Solving

Volumes have been written on the topic of problem solving, but certainly one of the best known sources is George Polya's book *How to Solve It.** In this book, Polya suggests the following four-phase plan for solving problems.

1. *Understand the problem.*
2. *Devise a plan* to solve the problem.
3. *Carry out the plan* to solve the problem.
4. *Look back* at the completed solution to review and discuss it.

We will comment briefly on each of the phases and offer some suggestions for using an algebraic approach to solve problems.

*George Polya, *How to Solve It* (Princeton: Princeton University Press), 1945.

Understand the Problem Read the problem carefully, making certain that you understand the meanings of all the words. Be especially alert for any technical terms used in the statement of the problem. Often it is helpful to sketch a figure, diagram, or chart to visualize and organize the conditions of the problem. Determine the known and unknown facts and if one of the previously mentioned pictorial devices is used, record these facts in the appropriate places of the diagram or chart.

Devise a Plan This is the key part of the four-phase plan. It is sometimes referred to as the *analysis* of the problem. There are numerous strategies and techniques used to solve problems. We shall discuss some of these strategies at various places throughout this text; however, at this time we offer the following general suggestions.

1. Choose a meaningful variable to represent an unknown quantity in the problem (perhaps t if time is an unknown quantity) and represent any other unknowns in terms of that variable.
2. Look for a *guideline* that can be used to set up an equation. A guideline might be a formula such as $A = P + Prt$ from Example 7, or a statement of a relationship such as *the sum of the two numbers is 28.* Sometimes a relationship suggested by a pictorial device can be used as a guideline for setting up the equation. Also, be alert to the possiblity that this *new* problem might really be an *old* problem in a new setting, perhaps even stated in different vocabulary.
3. Form an equation containing the variable so that the conditions of the guideline are translated from English into algebra.

Carry out the Plan This phase is sometimes referred to as the *synthesis* of the plan. If phase two has been successfully completed, then carrying out the plan may simply be a matter of solving the equation and doing any further computations to answer all of the questions in the problem. Confidence in your plan creates a better working atmosphere for carrying it out. It is also in this phase that the calculator may become a valuable tool. The type of data and the amount of complexity involved in the computations are two factors that can influence your decision to use one.

Look Back This is an important but often overlooked part of problem solving. The following list of questions suggests some things for you to consider in this phase.

1. Is your answer to the problem a *reasonable* answer?
2. Have you *checked* your answer by substituting it back into the conditions stated in the problem?
3. Looking back over your solution, do you now see another plan that could be used to solve the problem?
4. Do you see a way of generalizing your procedure for this problem that could be used to solve other problems of this type?

5. Do you now see that this problem is closely related to another problem that you have previously solved?
6. Have you tucked away for future reference the technique used to solve this problem?

Looking back over the solution of a newly solved problem can lay important groundwork for solving problems in the future.

Keep the previous suggestions in mind as we tackle some more word problems. Perhaps it would also be helpful for you to attempt to solve these problems on your own before looking at our approach.

PROBLEM 1

One number is 65 larger than another number. If the larger number is divided by the smaller, the quotient is 6 and the remainder is 5. Find the numbers.

Solution

Let n represent the smaller number. Then $n + 65$ represents the larger number. We can use the following relationship as a *guideline.*

$$\frac{\text{Dividend}}{\text{Divisor}} = \text{Quotient} + \frac{\text{Remainder}}{\text{Divisor}}$$

$$\downarrow \qquad\qquad \downarrow \qquad\qquad \downarrow$$

$$\frac{n + 65}{n} = 6 + \frac{5}{n}$$

We solve this equation by multiplying both sides by n.

$$n\left(\frac{n + 65}{n}\right) = n\left(6 + \frac{5}{n}\right), \qquad n \neq 0$$

$$n + 65 = 6n + 5$$

$$60 = 5n$$

$$12 = n$$

If $n = 12$ then $n + 65$ equals 77. The two numbers are 12 and 77.

Sometimes the concepts of ratio and proportion can be used to set up an equation and solve a problem as the next example illustrates.

EXAMPLE 2

The ratio of male students to female students at a certain university is 5 to 7. If there is a total of 16,200 students, find the number of male and the number of female students.

Solution

Let m represent the number of male students; then $16200 - m$ represents the number of female students. The following proportion can be set up and solved.

$$\frac{m}{16200 - m} = \frac{5}{7}$$
$$7m = 5(16200 - m)$$
$$7m = 81000 - 5m$$
$$12m = 81000$$
$$m = 6750$$

Therefore there are 6750 male students and $16200 - 6750 = 9450$ female students.

The next problem has a geometric setting. In such cases, the use of figures is very helpful.

PROBLEM 3

If two opposite sides of a square are each increased by 3 centimeters and the other two sides are each decreased by 2 centimeters, the area is increased by 8 square centimeters. Find the length of the side of the square.

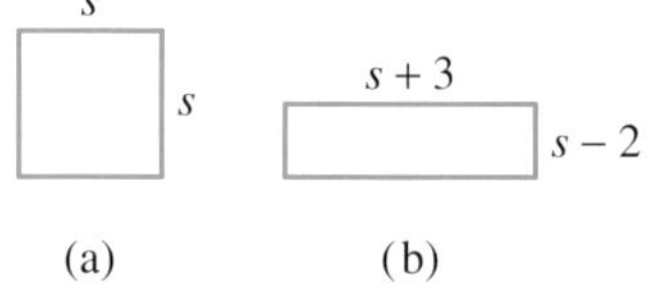

FIGURE 2.2

Solution

Let s represent the side of the square. Then Figures 2.2(a) and 2.2(b) represent the square and the rectangle formed by increasing two opposite sides of the square by 3 centimeters and decreasing the other two sides by 2 centimeters. Since the area of the rectangle is 8 square centimeters more than the area of the square, the following equation can be set up and solved.

$$(s + 3)(s - 2) = s^2 + 8$$
$$s^2 + s - 6 = s^2 + 8$$
$$s = 14$$

Thus the length of a side of the original square is 14 centimeters.

Many consumer problems can be solved by using an algebraic approach. For example, let's consider a discount sale problem involving the relationship, *original selling price minus discount equals discount sale price.*

PROBLEM 4

Jim bought a pair of slacks at a 30% discount sale for \$28. What was the original price of the slacks?

Solution

Let p represent the original price of the slacks.

Original price − Discount = Discount sale price

$$(100\%)(p) - (30\%)(p) = \$28$$

We switch this equation to decimal form to solve it.

$$p - 0.3p = 28$$
$$0.7p = 28$$
$$p = 40$$

The original price of the slacks was \$40.

Another basic relationship pertaining to consumer problems is *selling price equals cost plus profit*. Profit, also called markup, markon, and margin of profit, may be stated in different ways. It can be expressed as a percent of the cost, a percent of the selling price, or simply in terms of dollars and cents. Let's consider a problem where the profit is stated as a percent of the selling price.

PROBLEM 5

A retailer of sporting goods bought a putter for \$25. He wants to price the putter to make a profit of 20% of the selling price. What price should he mark on the putter?

Solution

Let s represent the selling price.

Selling price = Cost + Profit

$$s = \$25 + (20\%)(s)$$

Solving this equation involves using the methods we developed earlier for working with decimals.

$$s = 25 + (20\%)(s)$$
$$s = 25 + 0.2s$$
$$10s = 250 + 2s$$
$$8s = 250$$
$$s = 31.25$$

The selling price should be \$31.25.

Certain types of investment problems can be solved by using an algebraic approach. As our final example of this section, let's consider one such problem.

PROBLEM 6

Cindy invested a certain amount of money at 10% interest and \$1500 more than that amount at 11%. Her total yearly interest was \$795. How much did she invest at each rate?

Solution

Let d represent the amount invested at 10%; then $d + 1500$ represents the amount invested at 11%. The following guideline can be used to set up an equation.

Interest earned at 10% + Interest earned at 11% = Total interest

$$(10\%)(d) \quad + \quad (11\%)(d + 1500) \quad = \quad \$795$$

We can solve this equation by multiplying both sides by 100.

$$\begin{aligned} 0.1d + 0.11(d + 1500) &= 795 \\ 10d + 11(d + 1500) &= 79500 \\ 10d + 11d + 16500 &= 79500 \\ 21d &= 63000 \\ d &= 3000 \end{aligned}$$

Cindy invested \$3000 at 10% and \$3000 + \$1500 = \$4500 at 11%.

Don't forget phase four of Polya's problem solving plan. We have not taken the space to look back over and discuss each of our examples. However, it would be beneficial for you to do so, keeping in mind the questions posed earlier regarding this phase.

PROBLEM SET 2.2

Solve each of the following equations.

1. $\frac{x-2}{3} + \frac{x+1}{4} = \frac{1}{6}$
2. $\frac{5n-1}{4} - \frac{2n-3}{10} = \frac{3}{5}$
3. $\frac{5}{x} + \frac{1}{3} = \frac{8}{x}$
4. $\frac{5}{3n} - \frac{1}{9} = \frac{1}{n}$
5. $\frac{1}{3n} + \frac{1}{2n} = \frac{1}{4}$
6. $\frac{1}{x} - \frac{3}{2x} = \frac{1}{5}$
7. $\frac{35-x}{x} = 7 + \frac{3}{x}$
8. $\frac{n}{46-n} = 5 + \frac{4}{46-n}$
9. $\frac{n+67}{n} = 5 + \frac{11}{n}$
10. $\frac{n+52}{n} = 4 + \frac{1}{n}$
11. $\frac{5}{3x-2} = \frac{1}{x-4}$
12. $\frac{-2}{5x-3} = \frac{4}{4x-1}$
13. $\frac{4}{2y-3} - \frac{7}{3y-5} = 0$
14. $\frac{3}{2n+1} + \frac{5}{3n-4} = 0$
15. $\frac{n}{n+1} + 3 = \frac{4}{n+1}$
16. $\frac{a}{a+5} - 2 = \frac{3a}{a+5}$
17. $\frac{3x}{2x-1} - 4 = \frac{x}{2x-1}$
18. $\frac{x}{x-8} - 4 = \frac{8}{x-8}$
19. $\frac{3}{x+3} - \frac{1}{x-2} = \frac{5}{2x+6}$
20. $\frac{6}{x+3} + \frac{20}{x^2+x-6} = \frac{5}{x-2}$
21. $\frac{n}{n-3} - \frac{3}{2} = \frac{3}{n-3}$
22. $\frac{4}{x-2} + \frac{x}{x+1} = \frac{x^2-2}{x^2-x-2}$
23. $s = 9 + 0.25s$
24. $s = 1.95 + 0.35s$
25. $0.09x + 0.1(700 - x) = 67$
26. $0.08x + 0.09(950 - x) = 81$

27. $0.09x + 0.11(x + 125) = 68.75$

28. $0.08(x + 200) = 0.07x + 20$

29. $0.8(t - 2) = 0.5(9t + 10)$

30. $0.3(2n - 5) = 11 - 0.65n$

31. $0.92 + 0.9(x - 0.3) = 2x - 5.95$

32. $0.5(3x + 0.7) = 20.6$

Solve each of the following formulas for the indicated variable.

33. $P = 2l + 2w$ for w (Perimeter of a rectangle)

34. $V = \frac{1}{3}Bh$ for B (Volume of a pyramid)

35. $A = 2\pi r^2 + 2\pi rh$ for h (Surface area of a right circular cylinder)

36. $A = \frac{1}{2}h(b_1 + b_2)$ for h (Area of a trapezoid)

37. $C = \frac{5}{9}(F - 32)$ for F (Fahrenheit to Celsius)

38. $F = \frac{9}{5}C + 32$ for C (Celsius to Fahrenheit)

39. $V = C\left(1 - \frac{T}{N}\right)$ for T (Linear depreciation)

40. $V = C\left(1 - \frac{T}{N}\right)$ for N (Linear depreciation)

41. $I = kl(T - t)$ for T (Expansion allowance in highway construction)

42. $S = \frac{CRD}{12d}$ for d (Cutting speed of a circular saw)

43. $\frac{1}{R_n} = \frac{1}{R_1} + \frac{1}{R_2}$ for R_n (Resistance in parallel circuit design)

44. $f = \frac{1}{\frac{1}{a} + \frac{1}{b}}$ for b (Focal length of a camera lens)

Set up an equation and solve each of the following problems.

45. The sum of two numbers is 98. If the larger is divided by the smaller, the quotient is 4 and the remainder is 13. Find the numbers.

46. One number is 100 larger than another number. If the larger number is divided by the smaller, the quotient is 15 and the remainder is 2. Find the numbers.

47. What number must be added to both the numerator and denominator of $\frac{3}{5}$ to produce a rational number equivalent to $\frac{6}{7}$?

48. The denominator of a fraction is 4 more than its numerator. If 5 is added to the numerator and 11 is added to the denominator, a rational number equivlent to $\frac{1}{2}$ is produced. Find the original fraction.

49. A sum of $2250 is to be divided between two people in the ratio of 2 to 3. How much does each person receive?

50. One type of motor requires a mixture of oil and gasoline in a ratio of 1 to 15 (that is, 1 part of oil to 15 parts of gasoline). How many liters of each are contained in a 20-liter mixture?

51. The ratio of students to teaching faculty in a certain high school is 20 to 1. If the total number of students and faculty is 777, find the number of each.

52. The ratio of sodium to chlorine in common table salt is 5 to 3. Find the amount of each element in a salt compound weighing 200 pounds.

53. Gary bought a coat at a 20% discount sale for $52. What was the original price of the coat?

54. Roya bought a pair of slacks at a 30% discount sale for $33.60. What was the original price of the slacks?

55. After a 7% increase in salary, Laurie makes $1016.50 per month. How much did she earn per month before the increase?

56. Russ bought a car with 5% sales tax included for $11,025. What was the selling price of the car without the tax?

57. A retailer has some shoes that cost $28 per pair. At what price should they be sold to obtain a profit of 15% of the cost?

58. If a head of lettuce costs a retailer $.40, at what price should it be sold to make a profit of 45% of the cost?

59. Karla sold a bicycle for \$97.50. This selling price represented a 30% profit for her, based on what she had originally paid for the bike. Find Karla's original cost for the bicycle.

60. If a ring costs a jeweler \$250, at what price should it be sold to make a profit of 60% of the selling price?

61. A retailer has some skirts that cost \$18 each. She wants to sell them at a profit of 40% of the selling price. What price should she charge for the skirts?

62. Suppose that an item costs a retailer \$50. How much more profit could be gained by fixing a 50% profit based on selling price rather than a 50% profit based on cost?

63. Derek has some nickels and dimes worth \$3.60. The number of dimes is one more than twice the number of nickels. How many nickels and dimes does he have?

64. Robin has a collection of nickels, dimes, and quarters worth \$38.50. She has 10 more dimes than nickels and twice as many quarters as dimes. How many coins of each kind does she have?

65. A collection of 70 coins consisting of dimes, quarters, and half-dollars has a value of \$17.75. There are three times as many quarters as dimes. Find the number of each kind of coin.

66. A certain amount of money is invested at 8% per year and \$1500 more than that amount is invested at 9% per year. The annual interest from the 9% investment exceeds the annual interest from the 8% investment by \$160. How much is invested at each rate?

67. A total of \$5500 was invested, part of it at 9% per year and the remainder at 10% per year. If the total yearly interest amounted to \$530, how much was invested at each rate?

68. A sum of \$3500 is split between two investments, one paying 9% yearly interest and the other 11%. If the return on the 11% investment exceeds that on the 9% investment by \$85 the first year, how much is invested at each rate?

69. Celia has invested \$2500 at 11% yearly interest. How much must she invest at 12% so that the interest from both investments totals \$695 after a year?

70. The length of a rectangle is 2 inches less than three times its width. If the perimeter of the rectangle is 108 inches, find its length and width.

71. The length of a rectangle is 4 centimeters more than its width. If the width is increased by 2 centimeters and the length increased by 3 centimeters, a new rectangle is formed having an area of 44 square centimeters more than the area of the original rectangle. Find the dimensions of the original rectangle.

72. The length of a picture without its border is 7 inches less than twice its width. If the border is 1 inch wide and its area is 62 square inches, what are the dimensions of the picture alone?

THOUGHTS into WORDS

73. Give a step-by-step description of how you would solve the formula $F = \frac{9}{5}C + 32$, for C.

74. What does the phrase *declare a variable* mean when we solve a word problem?

75. Why must potential answers to word problems be checked back into the original statement of the problem?

2.3 QUADRATIC EQUATIONS

A **quadratic equation** in the variable x is defined as any equation that can be written in the form

$$ax^2 + bx + c = 0$$

where a, b, and c are real numbers and $a \neq 0$. The form $ax^2 + bx + c = 0$ is called the **standard form** of a quadratic equation. The choice of x for the variable is arbitrary. An equation such as $3t^2 + 5t - 4 = 0$ is a quadratic equation in the variable t.

Quadratic equations such as $x^2 + 2x - 15 = 0$, where the polynomial is factorable, can be solved by applying the following property: **$ab = 0$ if and only if $a = 0$ or $b = 0$.** Our work might take on the following format.

$$\begin{aligned} x^2 + 2x - 15 &= 0 \\ (x + 5)(x - 3) &= 0 \\ x + 5 = 0 \quad &\text{or} \quad x - 3 = 0 \\ x = -5 \quad &\text{or} \quad x = 3 \end{aligned}$$

The solution set for this equation is $\{-5, 3\}$.

Let's consider another example of this type.

EXAMPLE 1

Solve the equation $n = -6n^2 + 12$.

Solution

$$\begin{aligned} n &= -6n^2 + 12 \\ 6n^2 + n - 12 &= 0 \\ (3n - 4)(2n + 3) &= 0 \\ 3n - 4 = 0 \quad &\text{or} \quad 2n + 3 = 0 \\ 3n = 4 \quad &\text{or} \quad 2n = -3 \\ n = \frac{4}{3} \quad &\text{or} \quad n = -\frac{3}{2} \end{aligned}$$

The solution set is $\left\{-\frac{3}{2}, \frac{4}{3}\right\}$.

Now suppose that we want to solve $x^2 = k$, where k is any real number. We can proceed as follows.

$$\begin{aligned} x^2 &= k \\ x^2 - k &= 0 \\ \left(x + \sqrt{k}\right)\left(x - \sqrt{k}\right) &= 0 \\ x + \sqrt{k} = 0 \quad &\text{or} \quad x - \sqrt{k} = 0 \\ x = -\sqrt{k} \quad &\text{or} \quad x = \sqrt{k} \end{aligned}$$

Thus, we can state the following property for any real number k.

PROPERTY 2.2

The solution set of $x^2 = k$ is $\left\{-\sqrt{k}, \sqrt{k}\right\}$, which can also be written $\left\{\pm\sqrt{k}\right\}$.

Property 2.2, along with our knowledge of the square root, makes it very easy to solve quadratic equations of the form $x^2 = k$.

EXAMPLE 2

Solve each of the following.

a. $x^2 = 72$ **b.** $(3n - 1)^2 = 26$ **c.** $(y + 2)^2 = -24$

Solutions

a.
$$x^2 = 72$$
$$x = \pm\sqrt{72}$$
$$x = \pm 6\sqrt{2}$$

The solution set is $\{\pm 6\sqrt{2}\}$.

b.
$$(3n - 1)^2 = 26$$
$$3n - 1 = \pm\sqrt{26}.$$
$$3n - 1 = \sqrt{26} \quad \text{or} \quad 3n - 1 = -\sqrt{26}$$
$$3n = 1 + \sqrt{26} \quad \text{or} \quad 3n = 1 - \sqrt{26}$$
$$n = \frac{1 + \sqrt{26}}{3} \quad \text{or} \quad n = \frac{1 - \sqrt{26}}{3}$$

The solution set is $\left\{\frac{1 \pm \sqrt{26}}{3}\right\}$.

c.
$$(y + 2)^2 = -24$$
$$y + 2 = \pm\sqrt{-24}$$
$$y + 2 = \pm 2i\sqrt{6}.$$

Remember that $\sqrt{-24} = i\sqrt{24} = i\sqrt{4}\sqrt{6} = 2i\sqrt{6}$.

$$y + 2 = 2i\sqrt{6} \quad \text{or} \quad y + 2 = -2i\sqrt{6}$$
$$y = -2 + 2i\sqrt{6} \quad \text{or} \quad y = -2 - 2i\sqrt{6}$$

The solution set is $\{-2 \pm 2i\sqrt{6}\}$.

Completing the Square

A factoring technique we reviewed in Chapter 1 relied on recognizing *perfect-square trinomials*. In each of the following examples, the perfect-square trinomial on the right side of the identity is the result of squaring the binomial on the left side.

$$(x + 5)^2 = x^2 + 10x + 25 \qquad (x - 7)^2 = x^2 - 14x + 49$$
$$(x + 9)^2 = x^2 + 18x + 81 \qquad (x - 12)^2 = x^2 - 24x + 144$$

Notice that in each of the square trinomials, the constant term is equal to the square of one-half of the coefficient of the x-term. This relationship allows us to *form* a perfect-square trinomial by adding a proper constant term. For example, suppose that we want to form a perfect-square trinomial from $x^2 + 8x$. Since $\frac{1}{2}(8) = 4$ and

$4^2 = 16$, the perfect-square trinomial is $x^2 + 8x + 16$. Now let's use this idea to solve a quadratic equation.

EXAMPLE 3

Solve $x^2 + 8x - 2 = 0$

Solution

$$x^2 + 8x - 2 = 0$$
$$x^2 + 8x = 2$$
$$x^2 + 8x + 16 = 2 + 16$$

We added 16 to the left side to form a perfect-square trinomial. Thus, 16 has to be added to the right side.

$$(x + 4)^2 = 18$$
$$x + 4 = \pm\sqrt{18}$$
$$x + 4 = \pm 3\sqrt{2}$$
$$x + 4 = 3\sqrt{2} \quad \text{or} \quad x + 4 = -3\sqrt{2}$$
$$x = -4 + 3\sqrt{2} \quad \text{or} \quad x = -4 - 3\sqrt{2}$$

The solution set is $\{-4 \pm 3\sqrt{2}\}$.

We have been using a relationship for a perfect-square trinomial that states, *The constant term is equal to the square of one-half of the coefficient of the x-term.* This relationship holds only if the coefficient of x^2 is 1. Thus, a slight adjustment needs to be made when we are solving quadratic equations having a coefficient of x^2 other than 1. The next example shows how to make this adjustment.

EXAMPLE 4

Solve $2x^2 + 6x - 3 = 0$.

Solution

$$2x^2 + 6x - 3 = 0$$
$$2x^2 + 6x = 3$$
$$x^2 + 3x = \frac{3}{2}$$

Multiply both sides by $\frac{1}{2}$.

$$x^2 + 3x + \frac{9}{4} = \frac{3}{2} + \frac{9}{4}$$

Add $\frac{9}{4}$ to both sides.

$$\left(x + \frac{3}{2}\right)^2 = \frac{15}{4}$$
$$x + \frac{3}{2} = \pm\frac{\sqrt{15}}{2}.$$
$$x + \frac{3}{2} = \frac{\sqrt{15}}{2} \quad \text{or} \quad x + \frac{3}{2} = -\frac{\sqrt{15}}{2}$$
$$x = -\frac{3}{2} + \frac{\sqrt{15}}{2} \quad \text{or} \quad x = -\frac{3}{2} - \frac{\sqrt{15}}{2}$$

$$x = \frac{-3 + \sqrt{15}}{2} \quad \text{or} \quad x = \frac{-3 - \sqrt{15}}{2}$$

The solution set is $\left\{\frac{-3 \pm \sqrt{15}}{2}\right\}$.

Quadratic Formula

The process used in Examples 3 and 4 is called **completing the square**. It can be used to solve *any* quadratic equation. If we use this process of completing the square to solve the general quadratic equation $ax^2 + bx + c = 0$, we obtain a formula known as the **quadratic formula**. The details are as follows.

$$ax^2 + bx + c = 0, \qquad a \neq 0$$

$$ax^2 + bx = -c$$

$$x^2 + \frac{b}{a}x = -\frac{c}{a}$$ Multiply both sides by $\frac{1}{a}$.

$$x^2 + \frac{b}{a}x + \frac{b^2}{4a^2} = -\frac{c}{a} + \frac{b^2}{4a^2}$$ Complete the square by adding $\frac{b^2}{4a^2}$ to both sides.

$$\left(x + \frac{b}{2a}\right)^2 = \frac{b^2 - 4ac}{4a^2}$$ Combine the right side into a single fraction.

$$x + \frac{b}{2a} = \pm\sqrt{\frac{b^2 - 4ac}{4a^2}}$$

$$x + \frac{b}{2a} = \pm\frac{\sqrt{b^2 - 4ac}}{\sqrt{4a^2}}$$

$$x + \frac{b}{2a} = \pm\frac{\sqrt{b^2 - 4ac}}{2a}$$ $\sqrt{4a^2} = |2a|$ but $2a$ can be used because of the use of $\pm$.

$$x + \frac{b}{2a} = \frac{\sqrt{b^2 - 4ac}}{2a} \quad \text{or} \quad x + \frac{b}{2a} = -\frac{\sqrt{b^2 - 4ac}}{2a}$$

$$x = -\frac{b}{2a} + \frac{\sqrt{b^2 - 4ac}}{2a} \quad \text{or} \quad x = -\frac{b}{2a} - \frac{\sqrt{b^2 - 4ac}}{2a}$$

$$x = \frac{-b + \sqrt{b^2 - 4ac}}{2a} \quad \text{or} \quad x = \frac{-b - \sqrt{b^2 - 4ac}}{2a}$$

The quadratic formula can be stated as follows.

Quadratic Formula

If $a \neq 0$, then the solutions (roots) of the equation $ax^2 + bx + c = 0$ are given by

$$x = \frac{-b \pm \sqrt{b^2 - 4ac}}{2a}.$$

This formula can be used to solve any quadratic equation by expressing the equation in the standard form, $ax^2 + bx + c = 0$, and substituting the values for a, b, and c into the formula. Let's consider some examples.

EXAMPLE 5

Solve each of the following by using the quadratic formula.

a. $3x^2 - x - 5 = 0$ **b.** $25n^2 - 30n = -9$ **c.** $t^2 - 2t + 4 = 0$

Solutions

a. We need to think of $3x^2 - x - 5 = 0$ as $3x^2 + (-x) + (-5) = 0$; thus, $a = 3$, $b = -1$, and $c = -5$. We then substitute these values into the quadratic formula and simplify.

$$x = \frac{-b \pm \sqrt{b^2 - 4ac}}{2a}$$

$$x = \frac{-(-1) \pm \sqrt{(-1)^2 - 4(3)(-5)}}{2(3)}$$

$$x = \frac{1 \pm \sqrt{61}}{6}$$

The solution set is $\left\{\frac{1 \pm \sqrt{61}}{6}\right\}$.

b. The quadratic formula is usually stated in terms of the variable x, but again the choice of variable is arbitrary. The given equation, $25n^2 - 30n = -9$, needs to be changed to standard form: $25n^2 - 30n + 9 = 0$. From this we obtain $a = 25$, $b = -30$, and $c = 9$. Now we use the formula.

$$n = \frac{-(-30) \pm \sqrt{(-30)^2 - 4(25)(9)}}{2(25)}$$

$$n = \frac{30 \pm \sqrt{0}}{50}$$

$$n = \frac{3}{5}$$

The solution set is $\left\{\frac{3}{5}\right\}$.

c. We substitute $a = 1$, $b = -2$, and $c = 4$ into the quadratic formula.

$$t = \frac{-(-2) \pm \sqrt{(-2)^2 - 4(1)(4)}}{2(1)}$$

$$= \frac{2 \pm \sqrt{-12}}{2}$$

$$= \frac{2 \pm 2i\sqrt{3}}{2}$$

$$= \frac{\cancel{2}\left(1 \pm i\sqrt{3}\right)}{\cancel{2}}$$

The solution set is $\left\{1 \pm i\sqrt{3}\right\}$.

From Example 5 we see that different kinds of solutions are obtained depending upon the radicand $(b^2 - 4ac)$ inside the radical in the quadratic formula. For this reason, the number $b^2 - 4ac$ is called the **discriminant** of the quadratic equation. It can be used to determine the nature of the solutions as follows.

1. If $b^2 - 4ac > 0$, the equation has two unequal real solutions.

2. If $b^2 - 4ac = 0$, the equation has one real solution.

3. If $b^2 - 4ac < 0$, the equation has two complex but nonreal solutions.

The following examples illustrate each of these situations. (You may want to solve the equations completely to verify our conclusions.)

EQUATION	DISCRIMINANT	NATURE OF SOLUTIONS
$4x^2 - 7x - 1 = 0$	$b^2 - 4ac = (-7)^2 - 4(4)(-1)$ $= 49 + 16$ $= 65$	Two real solutions
$4x^2 + 12x + 9 = 0$	$b^2 - 4ac = (12)^2 - 4(4)(9)$ $= 144 - 144$ $= 0$	One real solution
$5x^2 + 2x + 1 = 0$	$b^2 - 4ac = (2)^2 - 4(5)(1)$ $= 4 - 20$ $= -16$	Two complex solutions

There is another useful relationship involving the solutions of a quadratic equation of the form $ax^2 + bx + c = 0$ and the numbers a, b, and c. Suppose that we let x_1 and x_2 be the two roots of the equation. (If $b^2 - 4ac = 0$, then $x_1 = x_2$ and the one-solution situation can be thought of as two equal solutions.) By the quadratic formula we have

$$x_1 = \frac{-b + \sqrt{b^2 - 4ac}}{2a} \quad \text{and} \quad x^2 = \frac{-b - \sqrt{b^2 - 4ac}}{2a}.$$

Now let's consider both the sum and the product of the two roots.

Sum $\quad x_1 + x_2 = \frac{-b + \sqrt{b^2 - 4ac}}{2a} + \frac{-b - \sqrt{b^2 - 4ac}}{2a}$

$$= \frac{-2b}{2a} = -\frac{b}{a}$$

Product $$(x_1)(x_2) = \left(\frac{-b + \sqrt{b^2 - 4ac}}{2a}\right)\left(\frac{-b - \sqrt{b^2 - 4ac}}{2a}\right)$$

$$= \frac{b^2 - (b^2 - 4ac)}{4a^2}$$

$$= \frac{b^2 - b^2 + 4ac}{4a^2}$$

$$= \frac{4ac}{4a^2} = \frac{c}{a}$$

These relationships provide another way of checking potential solutions when solving quadratic equations. We will illustrate this point in a moment.

Solving Quadratic Equations: Which Method?

Which method should be used to solve a particular quadratic equation? There is no definite answer to that question; it depends upon the type of equation and perhaps your personal preference. However, it is to your advantage to be able to use all three techniques and to know the strengths and weaknesses of each technique. In the next two examples we will give our reasons for choosing a specific technique.

EXAMPLE 6

Solve $x^2 - 4x - 192 = 0$.

Solution

The size of the constant term makes the factoring approach a little cumbersome for this problem. However, since the coefficient of the x^2-term is 1 and the coefficient of the x-term is even, the method of completing the square should work effectively.

$$x^2 - 4x - 192 = 0$$
$$x^2 - 4x = 192$$
$$x^2 - 4x + 4 = 192 + 4$$
$$(x - 2)^2 = 196$$
$$x - 2 = \pm\sqrt{196}$$
$$x - 2 = \pm 14.$$
$$x - 2 = 14 \quad \text{or} \quad x - 2 = -14$$
$$x = 16 \quad \text{or} \quad x = -12$$

Check

Sum of roots $16 + (-12) = 4$ and $-\frac{b}{a} = -\left(\frac{-4}{1}\right) = 4.$

Product of roots $(16)(-12) = -192$ and $\frac{c}{a} = \frac{-192}{1} = -192.$

The solution set is $\{-12, 16\}$.

EXAMPLE 7

Solve $2x^2 - x + 3 = 0$.

Solution

It would be reasonable first to try factoring the polynomial $2x^2 - x + 3$. Unfortunately, it is not factorable using integers; thus we must solve the equation by completing the square or by using the quadratic formula. Since the coefficient of the x^2-term is not 1, let's avoid completing the square and use the formula instead.

$$\begin{aligned} x &= \frac{-b \pm \sqrt{b^2}}{2a} \\ &= \frac{-(-1) \pm \sqrt{(-1)^2 - 4(2)(3)}}{2(2)} \\ &= \frac{1 \pm \sqrt{-23}}{4} \\ &= \frac{1 \pm i\sqrt{23}}{4} \end{aligned}$$

Check

Sum of roots $\frac{1 + i\sqrt{23}}{4} + \frac{1 - i\sqrt{23}}{4} = \frac{2}{4} = \frac{1}{2}$ and

$-\frac{b}{a} = -\frac{-1}{2} = \frac{1}{2}$

Product of roots $\left(\frac{1 + i\sqrt{23}}{4}\right)\left(\frac{1 - i\sqrt{23}}{4}\right) = \frac{1 - 23i^2}{16}$

$= \frac{1 + 23}{16} = \frac{24}{16} = \frac{3}{2}$ and $\frac{c}{a} = \frac{3}{2}.$

The solution set is $\left\{\frac{1 \pm i\sqrt{23}}{4}\right\}$.

The ability to solve quadratic equations enables us to solve more word problems. Some of these problems involve geometric formulas and relationships. We have included a brief summary of some basic geometric formulas in the back sheets of this text.

PROBLEM

One leg of a right triangle is 7 meters longer than the other leg. If the length of the hypotenuse is 17 meters, find the length of each leg.

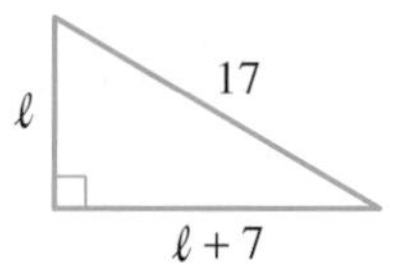

FIGURE 2.3

Solution

Look at Figure 2.3. Let l represent the length of one leg; then $l + 7$ represents the length of the other leg. Using the Pythagorean theorem as a guideline, we can set up and solve a quadratic equation.

$$l^2 + (l + 7)^2 = 17^2$$
$$l^2 + l^2 + 14l + 49 = 289$$
$$2l^2 + 14l - 240 = 0$$
$$l^2 + 7l - 120 = 0$$
$$(l + 15)(l - 8) = 0$$
$$l + 15 = 0 \quad \text{or} \quad l - 8 = 0$$
$$l = -15 \quad \text{or} \quad l = 8$$

The negative solution must be disregarded (since l is a length), so the length of one leg is 8 meters. The other leg, represented by $l + 7$, is $8 + 7 = 15$ meters long.

PROBLEM SET 2.3

Solve each of the following equations by factoring or by using the property, *If* $x^2 = k$ *then* $x = \pm\sqrt{k}$.

1. $x^2 - 3x - 28 = 0$
2. $x^2 - 4x - 12 = 0$
3. $3x^2 + 5x - 12 = 0$
4. $2x^2 - 13x + 6 = 0$
5. $2x^2 - 3x = 0$
6. $3n^2 = 3n$
7. $9y^2 = 12$
8. $(4n - 1)^2 = 16$
9. $(2n + 1)^2 = 20$
10. $3(4x - 1)^2 + 1 = 16$
11. $15n^2 + 19n - 10 = 0$
12. $6t^2 + 23t - 4 = 0$
13. $(x - 2)^2 = -4$
14. $24x^2 + 23x - 12 = 0$
15. $10y^2 + 33y - 7 = 0$
16. $(x - 3)^2 = -9$

Use the method of completing the square to solve each of the following equations. Check your solutions by using the sum-and-product-of-roots relationships.

17. $x^2 - 10x + 24 = 0$
18. $x^2 + x - 20 = 0$
19. $n^2 + 10n - 2 = 0$
20. $n^2 + 6n - 1 = 0$
21. $y^2 - 3y = -1$
22. $y^2 + 5y = -2$
23. $x^2 + 4x + 6 = 0$
24. $x^2 - 6x + 21 = 0$
25. $2t^2 + 12t - 5 = 0$
26. $3p^2 + 12p - 2 = 0$
27. $x^2 - 2x - 288 = 0$
28. $x^2 + 4x - 221 = 0$
29. $3n^2 + 5n - 1 = 0$
30. $2n^2 + n - 4 = 0$

Use the quadratic formula to solve each of the following equations. Check your solutions by using the sum-and-product-of-roots relationships.

31. $n^2 - 3n - 54 = 0$
32. $y^2 + 13y + 22 = 0$
33. $3x^2 + 16x = -5$
34. $10x^2 - 29x - 21 = 0$
35. $y^2 - 2y - 4 = 0$
36. $n^2 - 6n - 3 = 0$
37. $2a^2 - 6a + 1 = 0$
38. $2x^2 + 3x - 1 = 0$
39. $n^2 - 3n = -7$
40. $n^2 - 5n = -8$
41. $x^2 + 4 = 8x$
42. $x^2 + 31 = -14x$
43. $4x^2 - 4x + 1 = 0$
44. $x^2 + 24 = 0$

Solve each of the following quadratic equations by using the method that seems most appropriate to you.

45. $8x^2 + 10x - 3 = 0$
46. $18x^2 - 39x + 20 = 0$
47. $x^2 + 2x = 168$
48. $x^2 + 28x = -187$

49. $2t^2 - 3t + 7 = 0$

50. $3n^2 - 2n + 5 = 0$

51. $(3n - 1)^2 + 2 = 18$

52. $20y^2 + 17y - 10 = 0$

53. $4y^2 + 4y - 1 = 0$

54. $(5n + 2)^2 + 1 = -27$

55. $x^2 - 16x + 14 = 0$

56. $x^2 - 18x + 15 = 0$

57. $t^2 + 20t = 25$

58. $n^2 - 18n = 9$

59. $5x^2 - 2x - 1 = 0$

60. $7x^2 - 17x + 6 = 0$

61. Find the discriminant of each of the following quadratic equations and determine whether the equation has (1) two complex but nonreal solutions, (2) one real solution, or (3) two unequal real solutions.

a. $4x^2 + 20x + 25 = 0$

b. $x^2 + 4x + 7 = 0$

c. $x^2 - 18x + 81 = 0$

d. $36x^2 - 31x + 3 = 0$

e. $2x^2 + 5x + 7 = 0$

f. $16x^2 = 40x - 25$

g. $6x^2 - 4x - 7 = 0$

h. $5x^2 - 2x - 4 = 0$

Set up a quadratic equation and solve each of the following problems.

62. Find two consecutive even integers whose product is 528.

63. Find two consecutive whole numbers such that the sum of their squares is 265.

64. The sum of two integers is 4. The sum of the squares of the integers is 136. Find the integers.

65. Find two positive numbers having a sum of 22 and a product of 112.

66. One leg of a right triangle is 4 inches longer than the other leg. If the length of the hypotenuse is 20 inches, find the length of each leg.

67. The sum of the lengths of the two legs of a right triangle is 34 meters. If the length of the hypotenuse is 26 meters, find the length of each leg.

68. The lengths of the three sides of a right triangle are consecutive even integers. Find the length of each side.

69. The perimeter of a rectangle is 44 inches and its area is 112 square inches. Find the length and width of the rectangle.

70. A page of a magazine contains 70 square inches of type. The height of the page is twice the width. If the margin around the type is 2 inches uniformly, what are the dimensions of the page?

71. The length of a rectangle is 4 meters more than twice its width. If the area of the rectangle is 126 square meters, find its length and width.

72. The length of one side of a triangle is 3 centimeters less than twice the length of the altitude to that side. If the area of the triangle is 52 square centimeters, find the length of the side and the length of the altitude to that side.

73. A rectangular plot of ground measuring 12 meters by 20 meters is surrounded by a sidewalk of uniform width. The area of the sidewalk is 68 square meters. Find the width of the walk.

74. A piece of wire 60 inches long is cut into two pieces and then each piece is bent into the shape of a square. If the sum of the areas of the two squares is 117 square inches, find the length of each piece of wire.

75. A rectangular piece of cardboard is 4 inches longer than it is wide. From each of its corners a square piece 2 inches on a side is cut out. The flaps are then turned up to form an open box, which has a volume of 42 cubic inches. Find the length and width of the original piece of cardboard. See Figure 2.4.

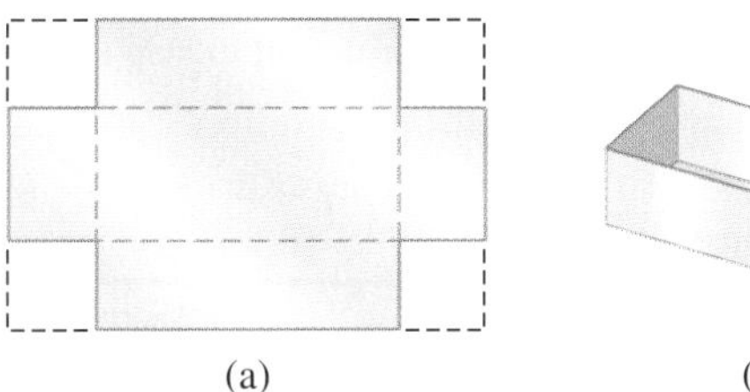

FIGURE 2.4

76. The area of a rectangular region is 52 square feet. If the length of the rectangle is increased by 4 feet and the width by 2 feet, the area is increased by 50 square feet. Find the length and width of the original rectangular region.

77. The area of a circular region is numerically equal to four times the circumference of the circle. Find the length of a radius of the circle.

THOUGHTS into WORDS

78. Explain how you would solve $(x - 3)(x + 4) = 0$ and also how you would solve $(x - 3)(x + 4) = 8$.

79. Explain the process of completing the square to solve a quadratic equation.

80. Explain how to use the quadratic formula to solve $3x = x^2 - 2$.

MISCELLANEOUS PROBLEMS

81. Solve each of the following equations for x.

a. $x^2 - 7kx = 0$ **b.** $x^2 = 25kx$

c. $x^2 - 3kx - 10k^2 = 0$ **d.** $6x^2 + kx - 2k^2 = 0$

e. $9x^2 - 6kx + k^2 = 0$ **f.** $k^2x^2 - kx - 6 = 0$

g. $x^2 + \sqrt{2}x - 3 = 0$ **h.** $x^2 - \sqrt{3}x + 5 = 0$

82. Solve each of the following for the indicated variable. (Assume that all letters represent positive numbers.)

a. $A = \pi r^2$ for r **b.** $E = c^2m - c^2m_0$ for c

c. $s = \frac{1}{2}gt^2$ for t **d.** $\frac{x^2}{a^2} + \frac{y^2}{b^2} = 1$ for x

e. $\frac{x^2}{a^2} - \frac{y^2}{b^2} = 1$ for y

f. $s = \frac{1}{2}gt^2 + V_0t$ for t

For Problems 83–85, use the discriminant to help solve each problem.

83. Determine k so that the solutions of $x^2 - 2x + k = 0$ are complex but nonreal.

84. Determine k so that $4x^2 - kx + 1 = 0$ has two equal real solutions.

85. Determine k so that $3x^2 - kx - 2 = 0$ has real solutions.

86. The solution set for $x^2 - 4x - 37 = 0$ is $\{2 \pm \sqrt{41}\}$. With a calculator, we found a rational approximation, to the nearest one-thousandth, for each of these solutions.

$$2 - \sqrt{41} = -4.403 \quad \text{and}$$

$$2 + \sqrt{41} = 8.403$$

Thus, the solution set is $\{-4.403, 8.403\}$, with answers rounded to the nearest one-thousandth.

Solve each of the following equations and express solutions to the nearest one-thousandth.

a. $x^2 - 6x - 10 = 0$ **b.** $x^2 - 16x - 24 = 0$

c. $x^2 +6x - 44 = 0$ **d.** $x^2 + 10x - 46 = 0$

e. $x^2 +8x + 2 = 0$ **f.** $x^2 + 9x + 3 = 0$

g. $4x^2 - 6x + 1 = 0$ **h.** $5x^2 - 9x + 1 + 0$

i. $2x^2 - 11x - 5 = 0$ **j.** $3x^2 - 12x - 10 = 0$

2.4 APPLICATIONS OF LINEAR AND QUADRATIC EQUATIONS

Let's begin this section by considering two fractional equations, one that is equivalent to a linear equation and one that is equivalent to a quadratic equation.

EXAMPLE 1

Solve $\frac{3}{2x - 8} - \frac{x - 5}{x^2 - 2x - 8} = \frac{7}{x + 2}$.

Solution

$$\frac{3}{2x - 8} - \frac{x - 5}{x^2 - 2x - 8} = \frac{7}{x + 2}$$

$$\frac{3}{2(x - 4)} - \frac{x - 5}{(x - 4)(x + 2)} = \frac{7}{x + 2}, \quad x \neq 4, x \neq -2$$

$$2(x - 4)(x + 2)\left(\frac{3}{2(x - 4)} - \frac{x - 5}{(x - 4)(x + 2)}\right) = 2(x - 4)(x + 2)\left(\frac{7}{x + 2}\right)$$

$$3(x + 2) - 2(x - 5) = 14(x - 4)$$

$$3x + 6 - 2x + 10 = 14x - 56$$

$$x + 16 = 14x - 56$$

$$72 = 13x$$

$$\frac{72}{13} = x$$

The solution set is $\left\{\frac{72}{13}\right\}$.

In Example 1, notice that we did not indicate the restrictions until the denominators were expressed in factored form. It is usually easier to determine the necessary restrictions at that step.

EXAMPLE 2

Solve $\frac{3n}{n^2 + n - 6} + \frac{2}{n^2 + 4n + 3} = \frac{n}{n^2 - n - 2}$.

Solution

$$\frac{3n}{n^2 + n - 6} + \frac{2}{n^2 + 4n + 3} = \frac{n}{n^2 - n - 2}$$

$$\frac{3n}{(n + 3)(n - 2)} + \frac{2}{(n + 3)(n + 1)} = \frac{n}{(n - 2)(n + 1)}, \quad n \neq -3, n \neq 2, n \neq -1$$

$$(n + 3)(n - 2)(n + 1)\left(\frac{3n}{(n + 3)(n - 2)} + \frac{2}{(n + 3)(n + 1)}\right) = (n + 3)(n - 2)(n + 1)\left(\frac{n}{(n - 2)(n + 1)}\right)$$

$$3n(n + 1) + 2(n - 2) = n(n + 3)$$

$$3n^2 + 3n + 2n - 4 = n^2 + 3n$$

$$3n^2 + 5n - 4 = n^2 + 3n$$

$$2n^2 + 2n - 4 = 0$$

$$n^2 + n - 2 = 0$$

$$(n + 2)(n - 1) = 0$$

$$n + 2 = 0 \quad \text{or} \quad n - 1 = 0$$

$$n = -2 \quad \text{or} \quad n = 1$$

The solution set is $\{-2, 1\}$.

More on Problem Solving

Before tackling a variety of applications of linear and quadratic equations, let's restate some suggestions made earlier in this chapter for solving word problems.

Suggestions for Solving Word Problems

1. Read the problem carefully, making certain that you understand the meanings of all the words. Be especially alert for any technical terms used in the statement of the problem.
2. Read the problem a second time (perhaps even a third time), to get an overview of the situation being described and to determine the known facts, as well as what is to be found.
3. Sketch any figure, diagram, or chart that might be helpful in analyzing the problem.
4. Choose a meaningful variable to represent an unknown quantity in the problem (for example, l for the length of a rectangle), and represent any other unknowns in terms of that variable.
5. Look for a *guideline* that can be used in setting up an equation. A guideline might be a formula, such as $A = lw$, or a relationship, such as *the fractional part of the job done by Bill plus the fractional part of the job done by Mary equals the total job.*
6. Form an equation containing the variable to translate the conditions of the guideline from English to algebra.
7. Solve the equation and use the solution to determine all the facts requested in the problem.
8. Check all answers back into the *original statement of the problem.*

Suggestion 5 is a key part of the analysis of a problem. A formula to be used as a guideline may or may not be explicitly stated in the problem. Likewise, a relationship to be used as a guideline may not be actually stated in the problem but must be determined from what is stated. Let's consider some examples.

PROBLEM 1

A room contains 120 chairs. The number of chairs per row is one less than twice the number of rows. Find the number of rows and the number of chairs per row.

Solution

Let r represent the number of rows. Then $2r - 1$ represents the number of chairs per row. The statement of the problem implies a formation of chairs such that the total number of chairs is equal to the number of rows times the number of chairs per row. This gives us an equation.

$$\begin{array}{ccccc} \text{Number of rows} & \times & \text{Number of chairs per row} & = & \text{Total number of chairs} \\ \downarrow & & \downarrow & & \downarrow \\ r & \times & (2r - 1) & = & 120 \end{array}$$

We solve this equation by the factorization method.

$$2r^2 - r = 120$$
$$2r^2 - r - 120 = 0$$
$$(2r + 15)(r - 8) = 0$$
$$2r + 15 = 0 \quad \text{or} \quad r - 8 = 0$$
$$2r = -15 \quad \text{or} \quad r = 8$$
$$r = -\frac{15}{2} \quad \text{or} \quad r = 8$$

The solution $-\frac{15}{2}$ must be disregarded, so there are 8 rows and $2(8) - 1 = 15$ chairs per row.

The basic relationship *distance equals rate times time* is used to help solve a variety of *uniform-motion problems*. This relationship may be expressed by any one of the following equations.

$$d = rt \qquad r = \frac{d}{t} \qquad t = \frac{d}{r}$$

PROBLEM 2

Domenica and Javier start from the same location at the same time and ride their bicycles in opposite directions for 4 hours, at which time they are 140 miles apart. If Domenica rides 3 miles per hour faster than Javier, find the rate of each rider.

Solution

Let r represent Javier's rate; then $r + 3$ represents Domenica's rate. A sketch as in Figure 2.5 may help in our analysis. The fact that the total distance is 140 miles can be used as a guideline. We use the $d = rt$ equation.

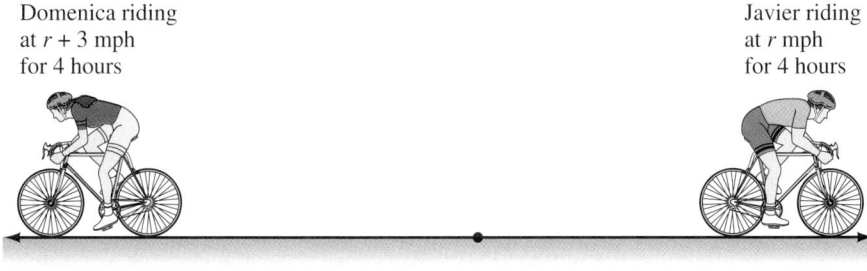

FIGURE 2.5

Distance Domenica rides + Distance Javier rides = 140

$$4(r + 3) \quad + \quad 4r \quad = 140$$

Solving this equation yields Javier's speed.

$$4r + 12 + 4r = 140$$
$$8r = 128$$
$$r = 16$$

Thus Javier rides at 16 miles per hour and Domenica at $16 + 3 = 19$ miles per hour.

Some people find that it is helpful to use a chart or a table to organize the known and unknown facts in uniform-motion problems. Let's illustrate this approach.

PROBLEM 3

Riding on a moped, Sue takes 2 hours less to travel 60 miles than Ann takes to travel 50 miles on a bicycle. Sue travels 10 miles per hour faster than Ann. Find the times and rates of both girls.

Solution

Let t represent Ann's time; then $t - 2$ represents Sue's time. We can record the information in a table as shown below. The fact that Sue travels 10 miles per hour faster than Ann can be used as a guideline.

	DISTANCE	TIME	$r = \frac{d}{t}$
Ann	50	t	$\frac{50}{t}$
Sue	60	$t - 2$	$\frac{60}{t - 2}$

Sue's rate = Ann's rate + 10

$$\frac{60}{t - 2} = \frac{50}{t} + 10$$

Solving this equation yields Ann's time.

$$t(t-2)\left(\frac{60}{t-2}\right) = t(t-2)\left(\frac{50}{t}+10\right), \qquad t \neq 0,\ t \neq 2$$

$$\begin{aligned} 60t &= 50(t-2) + 10t(t-2) \\ 60t &= 50t - 100 + 10t^2 - 20t \\ 0 &= 10t^2 - 30t - 100 \\ 0 &= t^2 - 3t - 10 \\ 0 &= (t-5)(t+2) \end{aligned}$$

$$t - 5 = 0 \quad \text{or} \quad t + 2 = 0$$

$$t = 5 \quad \text{or} \quad t = -2$$

The solution -2 must be disregarded, since we're solving for time. Therefore, Ann rides for 5 hours at $\frac{50}{5} = 10$ miles per hour and Sue rides for $5 - 2 = 3$ hours at $\frac{60}{3} = 20$ miles per hour.

There are various applications commonly classified as *mixture problems*. Even though these problems arise in many different areas, essentially the same mathematical approach can be used to solve them. A general suggestion for solving mixture-type problems is to *work in terms of a pure substance*. We will illustrate what we mean by that statement.

PROBLEM 4

How many milliliters of pure acid must be added to 50 milliliters of a 40% acid solution to obtain a 50% acid solution?

Solution

Let a represent the number of milliliters of pure acid to be added. Thinking in terms of pure acid, we know that *the amount of pure acid to start with plus the amount of pure acid added equals the amount of pure acid in the final solution*. Let's use that as a guideline and set up an equation.

Pure acid to start with + Pure acid added = Pure acid in final solution

↓ ↓ ↓

$$40\%(50) + a = 50\%(50 + a)$$

Solving this equation we obtain the amount of acid we must add.

$$\begin{aligned} 0.4(50) + a &= 0.5(50 + a) \\ 4(50) + 10a &= 5(50 + a) \\ 200 + 10a &= 250 + 5a \\ 5a &= 50 \\ a &= 10 \end{aligned}$$

We need to add 10 milliliters of pure acid.

There is another class of problems commonly referred to as *work problems*, or sometimes as *rate-time problems*. For example, if a certain machine produces 120 items in ten minutes, then we say that it is working at a rate of $\frac{120}{10} = 12$ items per minute. Likewise, if a person can do a certain job in five hours, then he is working at a rate of $\frac{1}{5}$ of the job per hour. In general, if Q is the quantity of something done in t units of time, then the rate, r, is given by $r = Q/t$. The rate is stated in terms of *so much quantity per unit of time*. The uniform-motion problems discussed earlier are a special kind of rate-time problems when the quantity is distance. Likewise, the use of tables to organize information (as we illustrated with the motion problems) is a convenient aid for rate-time problems in general. Let's consider some problems.

PROBLEM 5

Printing press A can produce 35 fliers per minute and press B can produce 50 fliers per minute. Suppose that 2225 fliers are produced by first using press A alone for 15 minutes and then using presses A and B together until the job is done. How long does press B need to be used?

Solution

Let m represent the number of minutes that press B is used. Then $m + 15$ represents the number of minutes press A is used. The information in the problem can be organized in a table as shown below. Since the total quantity (the total number of fliers) is 2225, we can set up and solve the following equation.

	QUANTITY	TIME	RATE
A	$35(m + 15)$	$m + 15$	35
B	$50m$	m	50

$$\begin{aligned} 35(m + 15) + 50m &= 2225 \\ 35m + 525 + 50m &= 2225 \\ 85m &= 1700 \\ m &= 20 \end{aligned}$$

Therefore press B must be used for 20 minutes.

PROBLEM 6

It takes Amy twice as long to deliver newspapers as it does Nancy. How long does it take each girl by herself if they can deliver the papers together in 40 minutes?

Solution

Let m represent the number of minutes that it takes Nancy by herself. Then $2m$ represents Amy's time by herself. Thus, the information can be organized as shown below. (Notice that the *quantity* is 1; there is one job to be done.) Since their combined rate is $\frac{1}{40}$, we can solve the following equation.

	QUANTITY	TIME	RATE
Nancy	1	m	$\frac{1}{m}$
Amy	1	$2m$	$\frac{1}{2m}$

$$\frac{1}{m} + \frac{1}{2m} = \frac{1}{40} \qquad m \neq 0$$
$$40m\left(\frac{1}{m} + \frac{1}{2m}\right) = 40m\left(\frac{1}{40}\right)$$
$$40 + 20 = m$$
$$60 = m$$

Therefore, Nancy can deliver the papers by herself in 60 minutes and Amy can deliver them by herself in $2(60) = 120$ minutes.

Our final examples of this section illustrate another approach that some people find works well for rate-time problems. The basic idea used in this approach involves representing the fractional parts of a job. For example, if a person can do a certain job in 7 hours, then at the end of 3 hours he has finished $\frac{3}{7}$ of the job. (Again, a constant rate of work is being assumed.) At the end of 5 hours he has finished $\frac{5}{7}$ of the job and, in general, at the end of h hours he has finished $\frac{h}{7}$ of the job.

PROBLEM 7

Carlos can mow a lawn in 45 minutes and Felipe can mow the same lawn in 30 minutes. How long would it take the two of them working together to mow the lawn?

Solution

(Before you read any further, *estimate* an answer for this problem. Remember that Felipe can mow the lawn by himself in 30 minutes.) Let m represent the number of minutes that it takes them working together. Then we can set up the following equation.

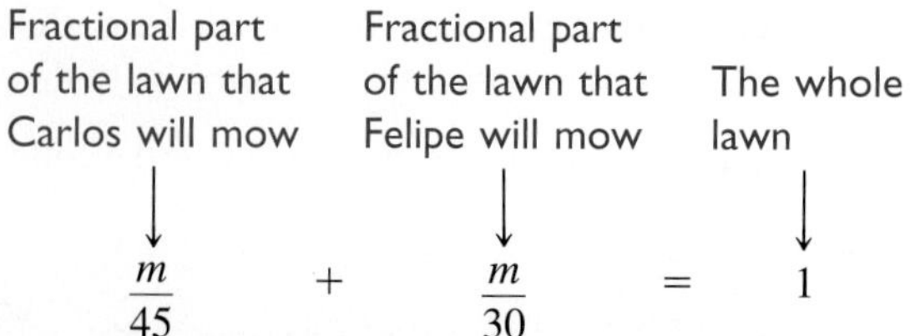

$$\frac{m}{45} + \frac{m}{30} = 1$$

Solving this equation yields the time that it will take when they work together.

$$90\left(\frac{m}{45} + \frac{m}{30}\right) = 90(1)$$

$$2m + 3m = 90$$

$$5m = 90$$

$$m = 18$$

It should take them 18 minutes to mow the lawn when they work together.

PROBLEM 8

Walt can mow a lawn in 50 minutes and his son, Mike, can mow the same lawn in 40 minutes. One day Mike started to mow the lawn by himself and worked for 10 minutes. Then Walt joined him with another mower and they finished the lawn. How long did it take them to finish mowing the lawn after Walt started to help?

Solution

Let m represent the number of minutes that it takes for them to finish the mowing after Walt starts to help. Since Mike has been mowing for 10 minutes, he has done $\frac{10}{40}$ or $\frac{1}{4}$ of the lawn when Walt starts. Thus there is $\frac{3}{4}$ of the lawn yet to mow. The following guideline can be used to set up an equation.

Fractional part of the remaining 3/4 of the lawn that Mike will mow in m minutes + Fractional part of the remaining 3/4 of the lawn that Walt will mow in m minutes $= \frac{3}{4}$

$$\frac{m}{40} + \frac{m}{50} = \frac{3}{4}$$

Solving this equation yields the time they mow the lawn together.

$$200\left(\frac{m}{40} + \frac{m}{50}\right) = 200\left(\frac{3}{4}\right)$$

$$5m + 4m = 150$$

$$9m = 150$$
$$m = \frac{150}{9} = \frac{50}{3}$$

They should finish the lawn in $16\frac{2}{3}$ minutes.

As you tackle word problems throughout this text, keep in mind that our primary objective is to expand your repertoire of problem solving techniques. In the examples, we are sharing some of our ideas for solving problems, but don't hesitate to use your own ingenuity. Furthermore, don't become discouraged–all of us have difficulty with some problems. Give it your best shot.

PROBLEM SET 2.4

Solve each of the following equations.

1. $\frac{x}{2x-8} + \frac{16}{x^2-16} = \frac{1}{2}$

2. $\frac{3}{n-5} - \frac{2}{2n+1} = \frac{n+3}{2n^2-9n-5}$

3. $\frac{5t}{2t+6} - \frac{4}{t^2-9} = \frac{5}{2}$

4. $\frac{x}{4x-4} + \frac{5}{x^2-1} = \frac{1}{4}$

5. $2 + \frac{4}{n-2} = \frac{8}{n^2-2n}$

6. $3 + \frac{6}{t-3} = \frac{6}{t^2-3t}$

7. $\frac{a}{a+2} + \frac{3}{a+4} = \frac{14}{a^2+6a+8}$

8. $\frac{3}{x+1} + \frac{2}{x+3} = 2$

9. $\frac{-2}{3x+2} + \frac{x-1}{9x^2-4} = \frac{3}{12x-8}$

10. $\frac{-1}{2x-5} + \frac{2x-4}{4x^2-25} = \frac{5}{6x+15}$

11. $\frac{n}{2n-3} + \frac{1}{n-3} = \frac{n^2-n-3}{2n^2-9n+9}$

12. $\frac{3y}{y^2+y-6} + \frac{2}{y^2+4y+3} = \frac{y}{y^2-y-2}$

13. $\frac{3y+1}{3y^2-4y-4} + \frac{9}{9y^2-4} = \frac{2y-2}{3y^2-8y+4}$

14. $\frac{4n+10}{2n^2-n-6} - \frac{3n+1}{2n^2-5n+2} = \frac{2}{4n^2+4n-3}$

15. $\frac{x+1}{2x^2+7x-4} - \frac{x}{2x^2-7x+3} = \frac{1}{x^2+x-12}$

16. $\frac{3}{x-2} + \frac{5}{x+3} = \frac{8x-1}{x^2+x-6}$

17. $\frac{7x+2}{12x^2+11x-15} - \frac{1}{3x+5} = \frac{2}{4x-3}$

18. $\frac{2n}{6n^2+7n-3} - \frac{n-3}{3n^2+11n-4} = \frac{5}{2n^2+11n+12}$

19. $\frac{3}{5x^2+18x-8} + \frac{x+1}{x^2-16} = \frac{5x}{5x^2-22x+8}$

20. $\frac{2}{4x^2+11x-3} - \frac{x+1}{3x^2+8x-3} = \frac{-4x}{12x^2-7x+1}$

Solve each of the following problems.

21. An apple orchard contains 126 trees. The number of trees in each row is 4 less than twice the number of rows. Find the number of rows and the number of trees per row.

22. The sum of a number and its reciprocal is $\frac{10}{3}$. Find the number.

23. Jill starts at city A and travels toward city B at 50 miles per hour. At the same time, Russ starts at city B and travels on the same highway toward city A at 52 miles per hour. How long will it take before they meet if the two cities are 459 miles apart?

24. Two cars, which are 510 miles apart and whose speeds differ by 6 miles per hour, are moving toward each other. If they meet in 5 hours, find the speed of each car.

25. Rita rode her bicycle out into the country at a speed of 20 miles per hour and returned along the same route at 15 miles per hour. If the round trip took 5 hours and 50 minutes, how far out did she ride?

26. A jogger who can run an 8-minute mile starts a half-mile ahead of a jogger who can run a 6-minute mile. How long will it take the faster jogger to catch the slower jogger?

27. It takes a freight train 2 hours more to travel 300 miles than it does an express train to travel 280 miles. The rate of the express train is 20 miles per hour greater than the rate of the freight train. Find the rates of both trains.

28. An airplane travels 2050 miles in the same time that a car travels 260 miles. If the rate of the plane is 358 miles per hour greater than the rate of the car, find the rate of the plane.

29. A container has 6 liters of a 40% alcohol solution in it. How much pure alcohol should be added to raise it to a 60% solution?

30. How many liters of a 60% acid solution must be added to 14 liters of a 10% acid solution to produce a 25% acid solution?

31. One solution contains 50% alcohol and another solution contains 80% alcohol. How many liters of each solution should be mixed to produce 10.5 liters of a 70% alcohol solution?

32. A contractor has a 24-pound mixture that is one-fourth cement and three-fourths sand. How much of a mixture that is half cement and half sand needs to be added to produce a mixture that is one-third cement?

33. A 10-quart radiator contains a 40% antifreeze solution. How much of the solution needs to be drained out and replaced with pure antifreeze in order to raise the solution to 70% antifreeze?

34. How much water must be evaporated from 20 gallons of a 10% salt solution in order to obtain a 20% salt solution?

35. One pipe can fill a tank in 4 hours and another pipe can fill the tank in 6 hours. How long will it take to fill the tank if both pipes are used?

36. Lolita and Doug working together can paint a shed in 3 hours and 20 minutes. If Doug can paint the shed by himself in 10 hours, how long would it take Lolita to paint the shed by herself?

37. An inlet pipe can fill a tank in 10 minutes. A drain can empty the tank in 12 minutes. If the tank is empty and both the pipe and drain are open, how long will it be before the tank overflows?

38. Mark can overhaul an engine in 20 hours and Phil can do the same job by himself in 30 hours. If they both work together for a time and then Mark finishes the job by himself in 5 hours, how long did they work together?

39. Pat and Mike working together can assemble a bookcase in 6 minutes. It takes Mike, working by himself, 9 minutes longer than it takes Pat working by himself to assemble the bookcase. How long does it take each, working alone, to do the job?

40. A computer company markets two card readers. Card reader B can read 8000 cards in 2 minutes less time than it takes card reader A to read 7500 cards. If the rate of card reader A is 250 cards per minute less than the rate of card reader B, find each rate.

41. Amelia can type 600 words in 5 minutes less time than it takes Paul to type 600 words. If Amelia types at a rate of 20 words per minute more than Paul types, find the rate of each.

42. A car that averages 16 miles per gallon of gasoline for city driving and 22 miles per gallon for highway driving uses 14 gallons in 296 miles of driving. How much of the driving was city driving?

43. Angie bought some golf balls for \$14. If each ball had cost \$.25 less, she could have purchased one more ball for the same amount of money. How many golf balls did Angie buy?

44. A new labor contract provides for a wage increase of \$1 per hour and a reduction of 5 hours in the workweek.

A worker who received \$320 per week under the old contract will receive \$315 per week under the new contract. How long was the workweek under the old contract?

45. Todd contracted to paint a house for \$480. It took him 4 hours longer than he had anticipated, so he earned \$.50 per hour less than he originally calculated. How long had he anticipated it would take him to paint the house?

THOUGHTS into WORDS

46. How would you solve the equation $x^2 - 4x = 252$? Explain your choice of the method that you would use.

47. Discuss any new ideas relative to problem solving that you have acquired in this course.

2.5 MISCELLANEOUS EQUATIONS

Our previous work with solving linear and quadratic equations provides us with a basis for solving a variety of other types of equations. For example, the technique of factoring and applying the property

$$ab = 0 \quad \text{if and only if } a = 0 \text{ or } b = 0$$

can sometimes be used for equations other than quadratic equations.

EXAMPLE 1

Solve $x^3 + 2x^2 - 9x - 18 = 0$.

Solution

$$\begin{aligned}
x^3 + 2x^2 - 9x - 18 &= 0\\
x^2(x + 2) - 9(x + 2) &= 0\\
(x + 2)(x^2 - 9) &= 0\\
(x + 2)(x + 3)(x - 3) &= 0
\end{aligned}$$

$$x + 2 = 0 \quad \text{or} \quad x + 3 = 0 \quad \text{or} \quad x - 3 = 0$$

$$x = -2 \quad \text{or} \quad x = -3 \quad \text{or} \quad x = 3.$$

The solution set is $\{-3, -2, 3\}$.

EXAMPLE 2

Solve $3x^5 + 5x^4 = 3x^3 + 5x^2$.

Solution

$$\begin{aligned}
3x^5 + 5x^4 &= 3x^3 + 5x^2\\
3x^5 + 5x^4 - 3x^3 - 5x^2 &= 0\\
x^4(3x + 5) - x^2(3x + 5) &= 0\\
(3x + 5)(x^4 - x^2) &= 0\\
(3x + 5)(x^2)(x^2 - 1) &= 0\\
(3x + 5)(x^2)(x + 1)(x - 1) &= 0.
\end{aligned}$$

$$3x + 5 = 0 \quad \text{or} \quad x^2 = 0 \quad \text{or} \quad x + 1 = 0 \quad \text{or} \quad x - 1 = 0$$
$$3x = -5$$
$$x = -\frac{5}{3} \quad \text{or} \quad x = 0 \quad \text{or} \quad x = -1 \quad \text{or} \quad x = 1.$$

The solution set is $\left\{-\frac{5}{3}, 0, -1, 1\right\}$.

Be careful with an equation like the one in Example 2. Don't be tempted to divide both sides of the equation by x^2. In so doing, the solution of zero will be lost. *In general, don't divide both sides of an equation by an expression containing the variable.*

Radical Equations

An equation such as

$$\sqrt{2x - 4} = x - 2,$$

which contains a radical with the variable in the radicand, is often referred to as a **radical equation**. To solve radical equations we need the following additional property of equalty.

PROPERTY 2.3

Let a and b be real numbers and n a positive integer.

If $a = b$ then $a^n = b^n$.

Property 2.3 states that *we can raise both sides of an equation to a positive integral power*. However, when applying Property 2.3 we must be very careful. Raising both sides of an equation to a positive integral power sometimes produces results that do not satisfy the original equation. Consider the following examples.

EXAMPLE 3

Solve $\sqrt{3x + 1} = 7$.

Solution

$$\sqrt{3x + 1} = 7$$
$$\left(\sqrt{3x + 1}\right)^2 = 7^2 \qquad \text{Square both sides.}$$
$$3x + 1 = 49$$
$$3x = 48$$
$$x = 16$$

Check

$$\begin{aligned} \sqrt{3x + 1} &= 7 \\ \sqrt{3(16) + 1} &\stackrel{?}{=} 7 \\ \sqrt{49} &\stackrel{?}{=} 7 \\ 7 &= 7 \end{aligned}$$

The solution set is $\{16\}$.

EXAMPLE 4

Solve $\sqrt{2x - 1} = -5$.

Solution

$$\begin{aligned} \sqrt{2x - 1} &= -5 \\ \left(\sqrt{2x - 1}\right)^2 &= (-5)^2 && \text{Square both sides.} \\ 2x - 1 &= 25 \\ 2x &= 26 \\ x &= 13 \end{aligned}$$

Check

$$\begin{aligned} \sqrt{2x - 1} &= -5 \\ \sqrt{2(13) - 1} &\stackrel{?}{=} -5 \\ \sqrt{25} &\stackrel{?}{=} -5 \\ 5 &\neq -5 \end{aligned}$$

Since 13 does not check, the equation has no solutions. The solution set is $\varnothing$.

REMARK It is true that the equation in Example 4 could be solved by inspection since the symbol $\sqrt{}$ refers to nonnegative numbers. However, we did want to demonstrate what happens if Property 2.3 is used.

EXAMPLE 5

Solve $\sqrt{x} + 6 = x$.

Solution

$$\begin{aligned} \sqrt{x} + 6 &= x \\ \sqrt{x} &= x - 6 \\ \left(\sqrt{x}\right)^2 &= (x - 6)^2 && \text{Square both sides.} \\ x &= x^2 - 12x + 36 \\ 0 &= x^2 - 13x + 36 \\ 0 &= (x - 4)(x - 9) \end{aligned}$$

$$x - 4 = 0 \quad \text{or} \quad x - 9 = 0$$

$$x = 4 \quad \text{or} \quad x = 9$$

Check

$$\begin{aligned} \sqrt{x} &= x - 6 & \qquad \sqrt{x} &= x - 6 \\ \sqrt{4} &\stackrel{?}{=} 4 - 6 & \qquad \sqrt{9} &\stackrel{?}{=} 9 - 6 \\ 2 &\neq -2 & \qquad 3 &= 3 \end{aligned}$$

The only solution is 9, so the solution set is $\{9\}$.

REMARK Notice what happens if we square both sides of the original equation. We obtain $x + 12\sqrt{x} + 36 = x^2$, an equation more complex than the original one and still containing a radical. Therefore it is important first to isolate the term containing the radical on one side of the equation and then to square both sides of the equation.

In general, raising both sides of an equation to a positive integral power produces an equation that has all of the solutions of the original equation, *but* it may also have some extra solutions that will not satisfy the original equation. Such extra solutions are called **extraneous solutions**. Therefore, when using Property 2.3, you *must* check each potential solution in the original equation.

EXAMPLE 6

Solve $\sqrt[3]{2x + 3} = -3$.

Solution

$$\begin{aligned} \sqrt[3]{2x + 3} &= -3 \\ \left(\sqrt[3]{2x + 3}\right)^3 &= (-3)^3 && \text{Cube both sides.} \\ 2x + 3 &= -27 \\ 2x &= -30 \\ x &= -15 \end{aligned}$$

Check

$$\begin{aligned} \sqrt[3]{2x + 3} &= -3 \\ \sqrt[3]{2(-15) + 3} &\stackrel{?}{=} -3 \\ \sqrt[3]{-27} &\stackrel{?}{=} -3 \\ -3 &= -3 \end{aligned}$$

The solution set is $\{-15\}$.

EXAMPLE 7

Solve $\sqrt{x + 4} = \sqrt{x - 1} + 1$.

Solution

$$\begin{aligned} \sqrt{x + 4} &= \sqrt{x - 1} + 1 \\ \left(\sqrt{x + 4}\right)^2 &= \left(\sqrt{x - 1} + 1\right)^2 && \text{Square both sides.} \\ x + 4 &= x - 1 + 2\sqrt{x - 1} + 1 && \text{Remember the middle term when squaring the binomial.} \\ 4 &= 2\sqrt{x - 1} \\ 2 &= \sqrt{x - 1} \\ 2^2 &= \left(\sqrt{x - 1}\right)^2 && \text{Square both sides.} \\ 4 &= x - 1 \\ 5 &= x \end{aligned}$$

Check

$$\begin{aligned} \sqrt{x + 4} &= \sqrt{x - 1} + 1 \\ \sqrt{5 + 4} &\stackrel{?}{=} \sqrt{5 - 1} + 1 \end{aligned}$$

$$\sqrt{9} \stackrel{?}{=} \sqrt{4} + 1$$
$$3 = 3$$

The solution set is $\{5\}$.

Equations of Quadratic Form

An equation such as $x^4 + 5x^2 - 36 = 0$ is not a quadratic equation. However, if we let $u = x^2$, then we get $u^2 = x^4$. Substituting u for x^2 and u^2 for x^4 in $x^4 + 5x^2 - 36 = 0$ produces

$$u^2 + 5u - 36 = 0$$

which is a quadratic equation. In general, an equation in the variable x is said to be of **quadratic form** if it can be written in the form

$$au^2 + bu + c = 0,$$

where $a \neq 0$ and u is some algebraic expression in x. We have two basic approaches to solving equations of quadratic form, as illustrated by the next two examples.

EXAMPLE 8

Solve $x^{2/3} + x^{1/3} - 6 = 0$.

Solution

Let $u = x^{1/3}$; then $u^2 = x^{2/3}$ and the given equation can be rewritten $u^2 + u - 6 = 0$. Solving this equation yields two solutions.

$$u^2 + u - 6 = 0$$
$$(u + 3)(u - 2) = 0.$$
$$u + 3 = 0 \quad \text{or} \quad u - 2 = 0$$
$$u = -3 \quad \text{or} \quad u = 2$$

Now, substituting $x^{1/3}$ for u, we have

$$x^{1/3} = -3 \quad \text{or} \quad x^{1/3} = -2,$$

from which we obtain

$$(x^{1/3})^3 = (-3)^3 \quad \text{or} \quad (x^{1/3})^3 = 2^3,$$
$$x = -27 \quad \text{or} \quad x = 8$$

Check

$$x^{2/3} + x^{1/3} - 6 = 0 \qquad x^{2/3} + x^{1/3} - 6 = 0$$
$$(-27)^{2/3} + (-27)^{1/3} - 6 \stackrel{?}{=} 0 \qquad (8)^{2/3} + (8)^{1/3} - 6 \stackrel{?}{=} 0$$
$$9 + (-3) - 6 \stackrel{?}{=} 0 \qquad 4 + 2 - 6 \stackrel{?}{=} 0$$
$$0 = 0 \qquad 0 = 0$$

The solution set is $\{-27, 8\}$.

EXAMPLE 9

Solve $x^4 + 5x^2 - 36 = 0$.

Solution

$$x^4 + 5x^2 - 36 = 0$$
$$(x^2 + 9)(x^2 - 4) = 0$$
$$x^2 + 9 = 0 \quad \text{or} \quad x^2 - 4 = 0$$
$$x^2 = -9 \quad \text{or} \quad x^2 = 4$$
$$x = \pm 3i \quad \text{or} \quad x = \pm 2$$

The solution set is $\{\pm 3i, \pm 2\}$.

Notice in Example 8 that we made a substitution (u for $x^{1/3}$) to change the original equation to a quadratic equation in terms of the variable u. Then, after solving for u, we substituted $x^{1/3}$ for u to obtain the solutions of the original equation. However, in Example 9 we factored the given polynomial and proceeded without changing to a quadratic equation. Which approach you use may depend upon the complexity of the given equation.

EXAMPLE 10

Solve $15x^{-2} - 11x^{-1} - 12 = 0$.

Solution

Let $u = x^{-1}$; then $u^2 = x^{-2}$ and the given equation can be written and solved as follows.

$$15u^2 - 11u - 12 = 0$$
$$(5u + 3)(3u - 4) = 0$$
$$5u + 3 = 0 \quad \text{or} \quad 3u - 4 = 0$$
$$5u = -3 \quad \text{or} \quad 3u = 4$$
$$u = -\frac{3}{5} \quad \text{or} \quad u = \frac{4}{3}$$

Now substituting x^{-1} back for u we have

$$x^{-1} = -\frac{3}{5} \quad \text{or} \quad x^{-1} = \frac{4}{3},$$

from which we obtain

$$\frac{1}{x} = \frac{-3}{5} \quad \text{or} \quad \frac{1}{x} = \frac{4}{3}$$
$$-3x = 5 \quad \text{or} \quad 4x = 3$$
$$x = -\frac{5}{3} \quad \text{or} \quad x = \frac{3}{4}$$

The solution set is $\left\{-\frac{5}{3}, \frac{3}{4}\right\}$.

PROBLEM SET 2.5

Solve each of the following equations. Don't forget that you *must* check potential solutions whenever Property 2.3 is applied.

1. $x^3 + x^2 - 4x - 4 = 0$
2. $x^3 - 5x^2 - x + 5 = 0$
3. $2x^3 - 3x^2 + 2x - 3 = 0$
4. $3x^3 + 5x^2 + 12x + 20 = 0$
5. $8x^5 + 10x^4 = 4x^3 + 5x^2$
6. $10x^5 + 15x^4 = 2x^3 + 3x^2$
7. $x^{3/2} = 4x$
8. $5x^4 = 6x^3$
9. $n^{-2} = n^{-3}$
10. $n^{4/3} = 4n$
11. $\sqrt{3x - 2} = 4$
12. $\sqrt{5x - 1} = -4$
13. $\sqrt{3x - 8} - \sqrt{x - 2} = 0$
14. $\sqrt{2x - 3} = 1$
15. $\sqrt{4x - 3} = -2$
16. $\sqrt{2x - 1} - \sqrt{x + 2} = 0$
17. $\sqrt[3]{2x + 3} + 3 = 0$
18. $\sqrt[3]{n^2 - 1} + 1 = 0$
19. $2\sqrt{n} + 3 = n$
20. $\sqrt{3t} - t = -6$
21. $\sqrt{3x - 2} = 3x - 2$
22. $5x - 4 = \sqrt{5x - 4}$
23. $\sqrt{2t - 1} + 2 = t$
24. $p = \sqrt{-4p + 17} + 3$
25. $\sqrt{x + 2} - 1 = \sqrt{x - 3}$
26. $\sqrt{x + 5} - 2 = \sqrt{x - 7}$
27. $\sqrt{7n + 23} - \sqrt{3n + 7} = 2$
28. $\sqrt{5t + 31} - \sqrt{t + 3} = 4$
29. $\sqrt{3x + 1} + \sqrt{2x + 4} = 3$
30. $\sqrt{2x - 1} - \sqrt{x + 3} = 1$
31. $\sqrt{x - 2} - \sqrt{2x - 11} = \sqrt{x - 5}$
32. $\sqrt{-2x - 7} + \sqrt{x + 9} = \sqrt{8 - x}$
33. $\sqrt{1 + 2\sqrt{x}} = \sqrt{x + 1}$
34. $\sqrt{7 + 3\sqrt{x}} = \sqrt{x} + 1$
35. $x^4 - 5x^2 + 4 = 0$
36. $x^4 - 25x^2 + 144 = 0$
37. $2n^4 - 9n^2 + 4 = 0$
38. $3n^4 - 4n^2 + 1 = 0$
39. $x^4 - 2x^2 - 35 = 0$
40. $2x^4 + 5x^2 - 12 = 0$
41. $x^4 - 4x^2 + 1 = 0$
42. $x^4 - 8x^2 + 11 = 0$
43. $x^{2/3} + 3x^{1/3} - 10 = 0$
44. $x^{2/3} + x^{1/3} - 2 = 0$
45. $6x^{2/3} - 5x^{1/3} - 6 = 0$
46. $3x^{2/3} - 11x^{1/3} - 4 = 0$
47. $x^{-2} + 4x^{-1} - 12 = 0$
48. $12t^{-2} - 17t^{-1} - 5 = 0$
49. $x - 11\sqrt{x} + 30 = 0$
50. $2x - 11\sqrt{x} + 12 = 0$
51. $x + 3\sqrt{x} - 10 = 0$
52. $6x - 19\sqrt{x} - 7 = 0$

MISCELLANEOUS PROBLEMS

53. Verify that $x = a$ and $x^2 = a^2$ are *not* equivalent equations.

54. Solve each of the following equations, and express solutions to the nearest hundredth.

a. $x^4 - 3x^2 + 1 = 0$ **b.** $x^4 - 5x^2 + 2 = 0$

c. $2x^4 - 7x^2 + 2 = 0$

d. $3x^4 - 9x^2 + 1 = 0$ **e.** $x^4 - 100x^2 + 2304 = 0$

f. $4x^4 - 373x^2 + 3969 = 0$

2.6 INEQUALITIES

Just as we use the symbol $=$ to represent *is equal to*, we also use the symbols $<$ and $>$ to represent *is less than* and *is greater than*, respectively. Thus, various **statements of inequality** can be made as follows.

$a < b$ means a is less than b.

$a \leq b$ means a is less than or equal to b.

$a > b$ means a is greater than b.

$a \geq b$ means a is greater than or equal to b.

The following are examples of **numerical statements of inequality**.

$$7 + 8 > 10 \qquad -4 + (-6) \geq -10$$

$$-4 > -6 \qquad 7 - 9 \leq -2$$

$$7 - 1 < 20 \qquad 3 + 4 > 12$$

$$8(-3) < 5(-3). \qquad 7 - 1 < 0.$$

Notice that only $3 + 4 > 12$ and $7 - 1 < 0$ are *false*; the other six are *true* numerical statements.

Algebraic inequalities contain one or more variables. The following are examples of algebraic inequalities.

$$x + 4 > 8 \qquad 3x + 2y \leq 4$$

$$(x - 2)(x + 4) \geq 0 \qquad x^2 + y^2 + z^2 \leq 16$$

An algebraic inequality such as $x + 4 > 8$ is neither true nor false as it stands and is called an **open sentence**. For each numerical value substituted for x, the algebraic inequality $x + 4 > 8$ becomes a numerical statement of inequality that is true or false. For example, if $x = -3$, then $x + 4 > 8$ becomes $-3 + 4 > 8$, which is false. If $x = 5$, then $x + 4 > 8$ becomes $5 + 4 > 8$, which is true. **Solving an algebraic inequality** refers to the process of finding the numbers that make it a true numerical statement. Such numbers are called the **solutions** of the inequality and are said to **satisfy** it.

The general process for solving inequalities closely parallels that for solving equations. We repeatedly replace the given inequality with equivalent but simpler inequalities until the solution set is obvious. The following property provides the basis for producing equivalent inequalities.

PROPERTY 2.4

1. For all real numbers a, b, and c,

$$a > b \quad \text{if and only if } a + c > b + c.$$

2. For all real numbers a, b, and c, **with $c > 0$**,

$$a > b \quad \text{if and only if } ac > bc.$$

3. For all real numbers a, b, and c, **with $c < 0$**,

$$a > b \quad \text{if and only if } ac < bc.$$

Similar properties exist if $>$ is replaced by $<$, $\leq$, or $\geq$. Part 1 of Property 2.4 is commonly called the **addition property of inequality**. Parts 2 and 3 together make up the **multiplication property of inequality**. Pay special attention to part 3. **If both sides of an inequality are multiplied by a negative number, the inequality symbol must be reversed.** For example, if both sides of $-3 < 5$ are multiplied by -2, the equivalent inequality $6 > -10$ is produced. Now let's consider using the addition and multiplication properties of inequality to help solve some inequalities.

EXAMPLE 1

Solve $3(2x - 1) < 8x - 7$.

Solution

$$3(2x - 1) < 8x - 7$$

$$6x - 3 < 8x - 7 \qquad \text{Apply distributive property to left side.}$$

$$-2x - 3 < -7 \qquad \text{Add } -8x \text{ to both sides.}$$

$$-2x < -4 \qquad \text{Add 3 to both sides.}$$

$$-\frac{1}{2}(-2x) > -\frac{1}{2}(-4) \qquad \text{Multiply both sides by } -\frac{1}{2}\text{, which reverses the inequality.}$$

$$x > 2$$

The solution set is $\{x \mid x > 2\}$.

A graph of the solution set $\{x \mid x > 2\}$ in Example 1 is shown in Figure 2.6. The parenthesis indicates that 2 does not belong to the solution set.

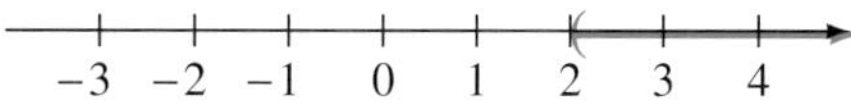

FIGURE 2.6

Checking the solutions of an inequality presents a problem. Obviously, we cannot check all of the infinitely many solutions for a particular inequality. However, by checking at least one solution, especially when the multiplication property has been used, we might catch a mistake of forgetting to change the type of inequality. In Example 1 we are claiming that all numbers greater than 2 will satisfy the original inequality. Let's check the number 3.

$$3(2x - 1) < 8x - 7$$
$$3[2(3) - 1] \overset{?}{<} 8(3) - 7$$
$$3(5) \overset{?}{<} 17$$
$$15 < 17 \qquad \text{It checks!}$$

It is also convenient to express solution sets of inequalities using **interval notation**. For example, the symbol $(2, \infty)$ refers to the interval of all real numbers greater than 2. As on the graph in Figure 2.6, the left-hand parenthesis indicates that 2 is not to be included. The infinity symbol, ∞, along with the right-hand parenthesis, indicates that there is no right-hand endpoint. Following is a list of interval notations along with the sets and graphs (Figure 2.7) that they represent. Notice the use of square brackets to *include* endpoints. Also recall that the notation $a < x < b$ is a compact way of expressing *x is greater than a and x is less than b*. From now on, we will express solution sets of inequalities using interval notation.

TYPE OF INTERVAL	SET	INTERVAL NOTATION	GRAPH (FIGURE 2.7)
open interval	$\{x \mid x > a\}$	(a, ∞)	(a)
	$\{x \mid a < x < b\}$	(a, b)	(b)
	$\{x \mid x < b\}$	$(-\infty, b)$	(c)
	$\{x \mid x \text{ is a real number}\}$	$(-\infty, \infty)$	(d)
half-open interval	$\{x \mid x \geq a\}$	$[a, \infty)$	(e)
	$\{x \mid a < x \leq b\}$	$(a, b]$	(f)
	$\{x \mid a \leq x < b\}$	$[a, b)$	(g)
	$\{x \mid x \leq b\}$	$(-\infty, b]$	(h)
closed interval	$\{x \mid a \leq x \leq b\}$	$[a, b]$	(i)

EXAMPLE 2

Solve $\dfrac{-3x + 1}{2} > 4$.

Solution

$$\frac{-3x+1}{2} > 4$$

$$2\left(\frac{-3x+1}{2}\right) > 2(4) \qquad \text{Multiply both sides by 2.}$$

$$-3x + 1 > 8$$

$$-3x > 7$$

$$-\frac{1}{3}(-3x) < -\frac{1}{3}(7) \qquad \text{Multiply both sides by } -\frac{1}{3}\text{, which reverses the inequality.}$$

$$x < -\frac{7}{3}$$

The solution set is $\left(-\infty, -\dfrac{7}{3}\right)$.

EXAMPLE 3

Solve $\dfrac{x-4}{6} - \dfrac{x-2}{9} \leq \dfrac{5}{18}$.

Solution

$$\frac{x-4}{6} - \frac{x-2}{9} \leq \frac{5}{18}$$

$$18\left(\frac{x-4}{6} - \frac{x-2}{9}\right) \leq 18\left(\frac{5}{18}\right) \qquad \text{Multiply both sides by the LCD.}$$

$$18\left(\frac{x-4}{6}\right) - 18\left(\frac{x-2}{9}\right) \leq 18\left(\frac{5}{18}\right)$$

$$3(x-4) - 2(x-2) \leq 5$$

$$3x - 12 - 2x + 4 \leq 5$$

$$x - 8 \leq 5$$

$$x \leq 13$$

The solution set is $(-\infty, 13]$.

The notation $a < x < b$ is a compact way of writing $x > a$ *and* $x < b$. It is a convenient form for solving **compound inequalities** involving the connecting word *and*, as the next example illustrates.

EXAMPLE 4

Solve $-2 < \dfrac{3x + 2}{2} < 7$.

Solution

$$-2 < \frac{3x+2}{2} < 7$$

$$2(-2) < 2\left(\frac{3x+2}{2}\right) < 2(7) \qquad \text{Multiply through by 2.}$$

$$-4 < 3x + 2 < 14$$

$$-6 < 3x < 12 \qquad \text{Add } -2 \text{ to all three quantities.}$$

$$-2 < x < 4 \qquad \text{Multiply through by } \frac{1}{3}.$$

The solution set is the interval $(-2, 4)$.

Quadratic Inequalities

The equation $ax^2 + bx + c = 0$ has been referred to as the standard form of a quadratic equation in one variable. Similarly, the form $ax^2 + bx + c < 0$ is used to represent a **quadratic inequality**. (The symbol $<$ can be replaced by $>$, $\leq$, or $\geq$ to produce other forms of quadratic inequalities.)

The number line becomes a very useful tool for helping to analyze quadratic inequalities. Let's consider some examples to illustrate the procedure.

EXAMPLE 5

Solve $x^2 + x - 6 < 0$.

Solution

First let's factor the polynomial.

$$x^2 + x - 6 < 0$$

$$(x + 3)(x - 2) < 0$$

Second, let's locate the values where the product $(x + 3)(x - 2)$ is equal to zero. The numbers -3 and 2 divide the number line into three intervals (see Figure 2.8):

the numbers less than -3,

the numbers between -3 and 2, and

the numbers greater than 2.

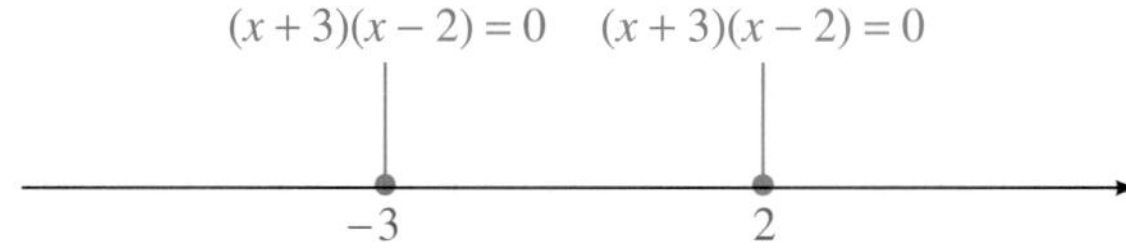

FIGURE 2.8

We can choose a **test number** from each of these intervals and see how it affects the signs of the factors $x + 3$ and $x - 2$ and, consequently, the sign of the product of these factors. For example, if $x < -3$ (try $x = -4$) then $x + 3$ is negative and $x - 2$ is negative; so their product is positive. If $-3 < x < 2$ (try $x = 0$) then $x + 3$ is positive and $x - 2$ is negative; so their product is negative. If $x > 2$ (try $x = 3$) then $x + 3$ is positive and $x - 2$ is positive; so their product is positive. This information can be conveniently arranged using a number line as in Figure 2.9.

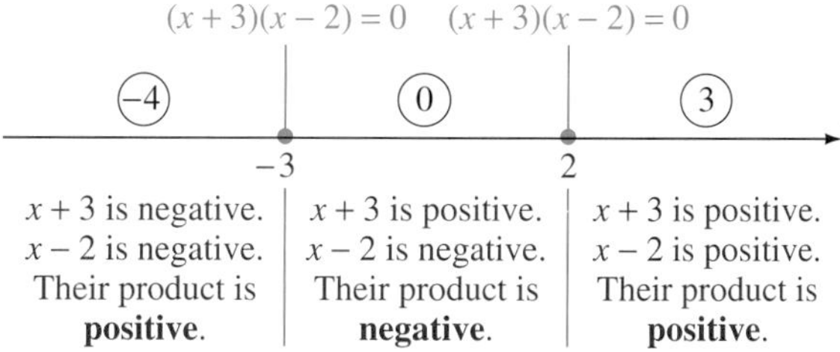
(x + 3)(x − 2) = 0
(x + 3)(x − 2) = 0
−4
0
3
−3
2
x + 3 is negative. x − 2 is negative. Their product is positive.
x + 3 is positive. x − 2 is negative. Their product is negative.
x + 3 is positive. x − 2 is positive. Their product is positive.

FIGURE 2.9

Therefore the given inequality, $x^2 + x - 6 < 0$, is satisfied by the numbers between -3 and 2. That is, the solution set is the open interval $(-3, 2)$.

Numbers such as -3 and 2 in the preceding example, where the given polynomial or algebraic expression equals zero or is undefined, are referred to as **critical numbers**. Let's consider some additional examples making use of critical numbers and test numbers.

EXAMPLE 6

Solve $6x^2 + 17x - 14 \geq 0$.

Solution

First, we factor the polynomial.

$$6x^2 + 17x - 14 \geq 0$$
$$(2x + 7)(3x - 2) \geq 0$$

Second, we locate the values where the product $(2x + 7)(3x - 2)$ equals zero. We suggest putting dots at $-\frac{7}{2}$ and $\frac{2}{3}$ (see Figure 2.10) to remind ourselves that these

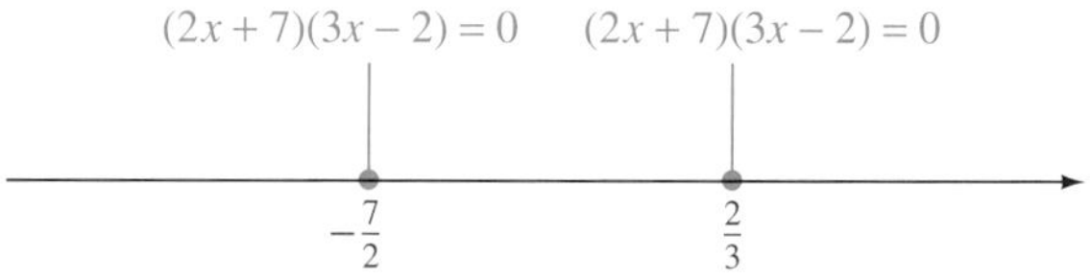

$(2x + 7)(3x - 2) = 0$ $(2x + 7)(3x - 2) = 0$

$-\frac{7}{2}$ $\frac{2}{3}$

FIGURE 2.10

two numbers must be included in the solution set, since the given statement includes equality. Now let's choose a test number from each of the three intervals and observe the sign behavior of the factors, as in Figure 2.11.

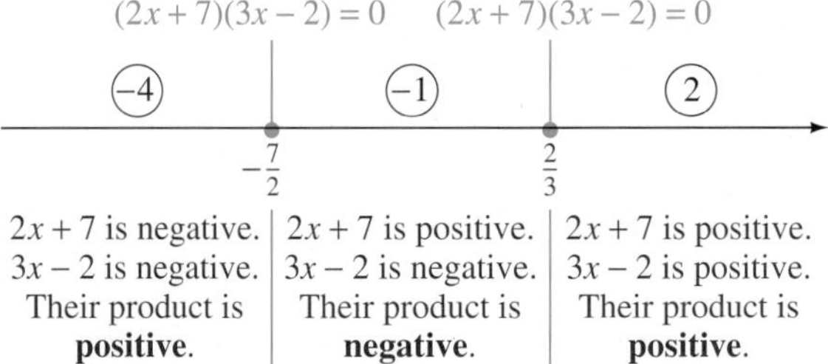

$2x + 7$ is negative.	$2x + 7$ is positive.	$2x + 7$ is positive.
$3x - 2$ is negative.	$3x - 2$ is negative.	$3x - 2$ is positive.
Their product is **positive**.	Their product is **negative**.	Their product is **positive**.

FIGURE 2.11

Using the concept of set union,* the solution set can be written $\left(-\infty, -\frac{7}{2}\right] \cup \left[\frac{2}{3}, \infty\right)$.

REMARK As you work with quadratic inequalities like those in Examples 5 and 6, you may be able to use a more abbreviated format than what we demonstrated. Basically, it is necessary to keep track of the sign of each factor in each of the intervals.

Let's conclude this section by considering a word problem that involves an inequality situation. All of the problem solving techniques offered earlier continue to apply except that now we look for a guideline that can be used to generate an inequality rather than an equation.

PROBLEM 1

Lance has \$500 to invest. If he invests \$300 at 9%, at what rate must he invest the remaining \$200 so that the total yearly interest from the two investments exceeds \$47?

Solution

Let r represent the unknown rate of interest. The following guideline can be used to set up an inequality.

Interest from 9% investment + Interest from r percent investment > \$47

$$(9\%)(\$300) \quad + \quad r(\$200) \quad > \$47$$

---

*The **union** of sets A and B, written $A \cup B$, is defined as $A \cup B = \{x \mid x \in A \text{ or } x \in B\}$.

We solve this inequality using methods we have already acquired.

$$(0.9)(300) + 200r > 47$$
$$27 + 200r > 47$$
$$200r > 20$$
$$r > 0.1$$

The other \$200 must be invested at a rate greater than 10%.

PROBLEM SET 2.6

Express each of the following in interval notation and graph each of the intervals.

1. $x \leq -2$

2. $x > -1$

3. $1 < x < 4$

4. $-1 < x \leq 2$

5. $2 > x > 0$

6. $-3 \geq x$

7. $-2 \leq x \leq -1$

8. $1 \leq x$

Solve each of the following inequalities. Express the solution sets in interval notation.

9. $-2x + 1 > 5$

10. $6 - 3x < 12$

11. $-3n + 5n - 2 \geq 8n - 7 - 9n$

12. $3n - 5 > 8n + 5$

13. $6(2t - 5) - 2(4t - 1) \geq 0$

14. $3(2x + 1) - 2(2x + 5) < 5(3x - 2)$

15. $\frac{2}{3}x - \frac{3}{4} \leq \frac{1}{4}x + \frac{2}{3}$

16. $\frac{3}{5} - \frac{x}{2} \geq \frac{1}{2} + \frac{x}{5}$

17. $\frac{n + 2}{4} + \frac{n - 3}{8} < 1$

18. $\frac{2n + 1}{6} + \frac{3n - 1}{5} > \frac{2}{15}$

19. $\frac{x}{2} - \frac{x - 1}{5} \geq \frac{x + 2}{10} - 4$

20. $\frac{4x - 3}{6} - \frac{2x - 1}{12} < -2$

21. $0.09x + 0.1(x + 200) > 77$

22. $0.06x + 0.08(250 - x) \geq 19$

23. $0 < \frac{5x - 1}{3} < 2$

24. $-3 \leq \frac{4x + 3}{2} \leq 1$

25. $3 \geq \frac{7 - x}{2} \geq 1$

26. $-2 \leq \frac{5 - 3x}{4} \leq \frac{1}{2}$

27. $x^2 + 3x - 4 < 0$

28. $x^2 - 4 < 0$

29. $x^2 - 2x - 15 > 0$

30. $x^2 - 12x + 32 \geq 0$

31. $n^2 - n \leq 2$

32. $n^2 + 5n \leq 6$

33. $3t^2 + 11t - 4 > 0$

34. $2t^2 - 9t - 5 > 0$

35. $15x^2 - 26x + 8 \leq 0$

36. $6x^2 + 25x + 14 \leq 0$

37. $4x^2 - 4x + 1 > 0$

38. $9x^2 + 6x + 1 \leq 0$

39. $(x + 1)(x - 3) > (x + 1)(2x - 1)$

40. $(x - 2)(2x + 5) > (x - 2)(x - 3)$

41. $(x + 1)(x - 2) \geq (x - 4)(x + 6)$

42. $(2x - 1)(x + 4) \geq (2x + 1)(x - 3)$

43. $(x - 1)(x - 2)(x + 4) > 0$

44. $(x + 1)(x - 3)(x + 7) \geq 0$

45. $(x + 2)(2x - 1)(x - 5) \leq 0$

46. $(x - 3)(3x + 2)(x + 4) < 0$

47. $x^3 - 2x^2 - 24x \geq 0$

48. $x^3 + 2x^2 - 3x > 0$

49. $(x - 2)^2(x + 3) > 0$

50. $(x + 4)^2(x + 5) > 0$

Solve each of the following problems.

51. Felix has \$1000 to invest. Suppose he invests \$500 at 8% interest. At what rate must he invest the other \$500 so that the two investments yield more than \$100 of yearly interest?

52. Suppose that Annette invests \$700 at 9%. How much must she invest at 11% so that the total yearly interest from the two investments exceeds \$162?

53. Rhonda had scores of 94, 84, 86, and 88 on her first four history exams of the semester. What score must she obtain on the fifth exam to have an average of 90 or better for the five exams?

54. The average height of the two forwards and the center of a basketball team is 6 feet 8 inches. What must the average height of the two guards be so that the team average is at least 6 feet 4 inches?

55. If the temperature for a 24-hour period ranged between 41°F and 59°F, inclusive, what was the range in Celsius degrees? $\left(F = \frac{9}{5}C + 32\right)$

56. If the temperature for a 24-hour period ranged between -20°C and -5°C, inclusive, what was the range in Fahrenheit degrees? $\left(C = \frac{5}{9}(F - 32)\right)$

57. A person's intelligence quotient (IQ) is found by dividing mental age (M), as indicated by standard tests, by the chronological age (C), and then multiplying this ratio by 100. The formula $IQ = 100M/C$ can be used. If the IQ range of a group of 11-year-olds is given by $80 \leq IQ \leq 140$, find the mental-age range of this group.

58. Repeat Problem 57 for an IQ range of 70 to 125, inclusive, for a group of 9-year-olds.

59. A car can be rented from agency A at \$75 per day plus \$.10 a mile or from agency B at \$50 a day plus \$.20 a mile. If the car is driven m miles, for what values of m does it cost less to rent from agency A?

60. Fired from ground level, the distance that a projectile is above the ground is given by $30t - 16t^2$, where t is measured in seconds. During what time interval is the projectile at least 14 feet above ground level?

THOUGHTS into WORDS

61. Give a step-by-step description of how you would solve the inequality $-4 < 2(x - 1) - 3(x + 2)$.

62. Explain how you would solve the inequality $(x - 1)^2(x + 2)^2 > 0$.

MISCELLANEOUS PROBLEMS

63. The product $(x - 2)(x + 3)$ is positive if both factors are negative *or* if both factors are positive. Therefore, we can solve $(x - 2)(x + 3) > 0$ as follows:

$$(x - 2 < 0 \text{ and } x + 3 < 0) \text{ or } (x - 2 > 0 \text{ and } x + 3 > 0)$$

$$(x < 2 \text{ and } x < -3) \quad \text{or} \quad (x > 2 \text{ and } x > -3)$$

$$x < -3 \quad \text{or} \quad x > 2$$

The solution set is $(-\infty, -3) \cup (2, \infty)$. Use this type of analysis to solve each of the following.

a. $(x - 1)(x + 5) > 0$

b. $(x + 2)(x - 4) \geq 0$

c. $(x + 4)(x - 3) < 0$

d. $(2x - 1)(x + 5) \leq 0$

e. $(x + 4)(x + 1)(x - 2) > 0$

f. $(x + 2)(x - 1)(x - 3) < 0$

64. If $a > b > 0$, verify that $1/a < 1/b$.

65. If $a > b$, is it always true that $1/a < 1/b$? Defend your answer.

2.7 Inequalities Involving Quotients and Absolute Value

The same type of number-line analysis that we did in the previous section can be used for indicated quotients as well as for indicated products. In other words, inequalities such as

$$\frac{x-2}{x+3} > 0$$

can be solved very efficiently using the same basic approach that we used with quadratic inequalities in the previous section. Let's illustrate this procedure.

EXAMPLE 1

Solve $\frac{x-2}{x+3} > 0$.

Solution

First we find that at $x = 2$ the quotient $\frac{x-2}{x+3}$ equals zero and at $x = -3$ the quotient is undefined. The critical numbers -3 and 2 divide the number line into three intervals. Then, using a test number from each interval (such as -4, 1, and 3), we can observe the sign behavior of the quotient, as in Figure 2.12.

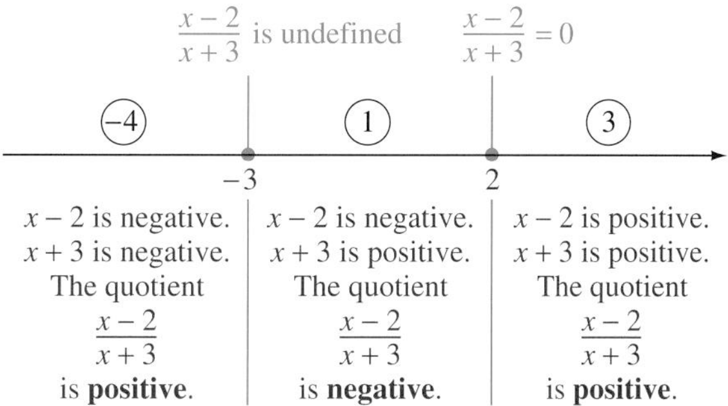

FIGURE 2.12

Therefore, the solution set for $\frac{x-2}{x+3} > 0$ is $(-\infty, -3) \cup (2, \infty)$.

EXAMPLE 2

Solve $\frac{x+2}{x+4} \leq 3$.

Solution

First let's change the form of the given inequality.

$$\frac{x+2}{x+4} \leq 3$$

$$\frac{x+2}{x+4} - 3 \leq 0$$

$$\frac{x+2-3(x+4)}{x+4} \leq 0$$

$$\frac{x+2-3x-12}{x+4} \leq 0$$

$$\frac{-2x-10}{x+4} \leq 0$$

Now we can proceed as before. If $x = -5$ then the quotient $\frac{-2x-10}{x+4}$ equals zero, and if $x = -4$, the quotient is undefined. Then, using test numbers such as -6, $-4\frac{1}{2}$, and -3, we can study the sign behavior of the quotient, as in Figure 2.13.

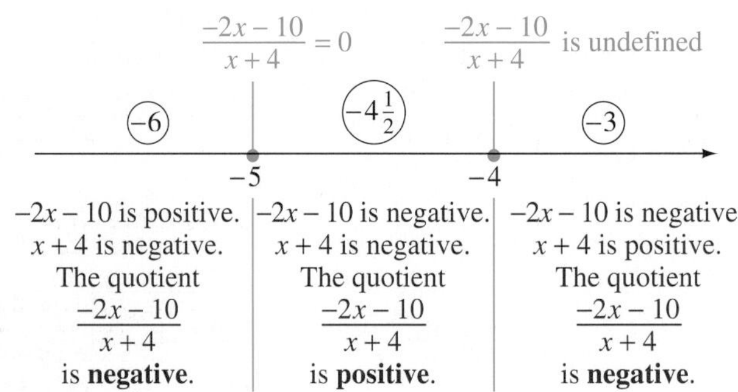

FIGURE 2.13

Therefore, the solution set for $\frac{x+2}{x+4} \leq 3$ is $(-\infty, -5] \cup (-4, \infty)$.

Absolute Value

In Section 1.1 we defined the **absolute value** of a real number by

$$|a| = \begin{cases} a, & \text{if } a \geq 0, \\ -a, & \text{if } a < 0. \end{cases}$$

We also interpreted the absolute value of any real number to be the distance between the number and zero on the real number line. For example, $|6| = 6$ because the distance between 6 and 0 is six units. Likewise, $|-8| = 8$ because the distance between -8 and 0 is eight units.

Both the definition and the number-line interpretation of absolute value provide ways of analyzing a variety of equations and inequalities involving absolute value. For example, suppose that we need to solve the equation $|x| = 4$. Thinking in terms of distance on the number line, the equation $|x| = 4$ means that we are looking for numbers that are four units from zero. Thus, x must be 4 or -4. From the definition viewpoint, we could proceed as follows.

If $x \geq 0$, then $|x| = x$ and the equation $|x| = 4$ becomes $x = 4$.

If $x < 0$, then $|x| = -x$ and the equation $|x| = 4$ becomes $-x = 4$, which is equivalent to $x = -4$.

Using either approach we see that the solution set for $|x| = 4$ is $\{-4, 4\}$.

The following property should seem reasonable from the distance interpretation and can be verified using the definition of absolute value.

PROPERTY 2.5

For any real number $k > 0$,

if $|x| = k$ then $x = k$ or $x = -k$.

To verify Property 2.5 using the definition of absolute value we can reason as follows.

If $x \geq 0$, then $|x| = x$ and the equation $|x| = k$ becomes $x = k$.

If $x < 0$, then $|x| = -x$ and the equation $|x| = k$ becomes $-x = k$, which is equivalent to $x = -k$.

Therefore, the equation $|x| = k$ is equivalent to $x = k$ or $x = -k$. Now let's use Property 2.5 to solve an equation of the form $|ax + b| = k$.

EXAMPLE 3

Solve $|3x - 2| = 7$.

Solution

$$|3x - 2| = 7$$

$$\begin{aligned} 3x - 2 &= 7 &\quad \text{or} \quad 3x - 2 &= -7 \\ 3x &= 9 &\quad \text{or} \quad 3x &= -5 \\ x &= 3 &\quad \text{or} \quad x &= -\frac{5}{3} \end{aligned}$$

The solution set is $\left\{-\frac{5}{3}, 3\right\}$.

The distance interpretation for absolute value also provides a good basis for solving some inequalities. For example, to solve $|x| < 4$, we know that the distance between x and 0 must be less than four units. In other words, x is to be less than four units away from zero. Thus, $|x| < 4$ is equivalent to $-4 < x < 4$ and the solution set is the interval $(-4, 4)$. We will have you use the definition of absolute value and verify the following general property in the next set of exercises.

PROPERTY 2.6

For any real number $k > 0$,

$$\text{if } |x| < k, \text{ then } -k < x < k.$$

Example 4 illustrates the use of Property 2.6.

EXAMPLE 4

Solve $|2x + 1| < 5$.

Solution

$$\begin{aligned} |2x + 1| &< 5 \\ -5 < 2x + 1 &< 5 \\ -6 < 2x &< 4 \\ -3 < x &< 2 \end{aligned}$$

The solution set is the interval $(-3, 2)$.

Property 2.6 can also be expanded to include the $\leq$ situation, that is, *if* $|x| \leq k$, *then* $-k \leq x \leq k$.

EXAMPLE 5

Solve $|-3x - 2| \leq 6$.

Solution

$$\begin{aligned} |-3x - 2| &\leq 6 \\ -6 \leq -3x - 2 &\leq 6 \\ -4 \leq -3x &\leq 8 \\ \frac{4}{3} \geq x &\geq -\frac{8}{3} \end{aligned}$$

Notice that multiplying through by $-\frac{1}{3}$ reverses the inequalities

The statement $\frac{4}{3} \geq x \geq -\frac{8}{3}$ is equivalent to $-\frac{8}{3} \leq x \leq \frac{4}{3}$. Therefore, the solution set is $\left[-\frac{8}{3}, \frac{4}{3}\right]$.

Now suppose that we want to solve $|x| > 4$. The distance between x and 0 must be more than four units or, in other words, x is to be more than four units away from zero. Therefore, $|x| > 4$ is equivalent to $x < -4$ or $x > 4$, and the solution set is $(-\infty, -4) \cup (4, \infty)$. The following general property can be verified by using the definition of absolute value.

PROPERTY 2.7

For any real number $k > 0$,

if $|x| > k$ then $x < -k$ or $x > k$.

EXAMPLE 6

Solve $|4x - 3| > 9$.

Solution

$$|4x - 3| > 9$$

$$4x - 3 < -9 \quad \text{or} \quad 4x - 3 > 9$$

$$4x < -6 \quad \text{or} \quad 4x > 12$$

$$x < -\frac{6}{4} \quad \text{or} \quad x > 3$$

$$x < -\frac{3}{2} \quad \text{or} \quad x > 3$$

The solution set is $\left(-\infty, -\frac{3}{2}\right) \cup (3, \infty)$.

Property 2.7 can also be expanded to include the $\geq$ situation, that is, *if* $|x| \geq k$*, then* $x \leq -k$ *or* $x \geq k$.

EXAMPLE 7

Solve $|-2 - x| \geq 9$.

Solution

$$|-2 - x| \geq 9$$

$$-2 - x \leq -9 \quad \text{or} \quad -2 - x \geq 9$$

$$-x \leq -7 \quad \text{or} \quad -x \geq 11$$

$$x \geq 7 \quad \text{or} \quad x \leq -11$$

The solution set is $(-\infty, -11] \cup [7, \infty)$.

Properties 2.5, 2.6, and 2.7 provide a sound basis for solving many equations and inequalities involving absolute value. However, if at any time you become doubtful as to which property applies, don't forget the definition and the distance interpretation for absolute value. Furthermore, there are some equations and inequalities where the properties do not apply. Let's consider one such example.

EXAMPLE 8

Solve the equation $|3x - 1| = |x + 4|$.

Solution

We could solve this equation by applying the definition of absolute value to both expressions; however, let's approach it in a less formal way. For the two numbers, $3x - 1$ and $x + 4$, to have the same absolute value, they must be equal or they must be opposites of each other. Therefore, the equation $|3x - 1| = |x + 4|$ is equivalent to $3x - 1 = x + 4$ or $3x - 1 = -(x + 4)$, which can be solved as follows.

$$
\begin{array}{rcl}
3x - 1 = x + 4 & \text{or} & 3x - 1 = -(x + 4) \\
2x = 5 & \text{or} & 3x - 1 = -x - 4 \\
x = \frac{5}{2} & \text{or} & 4x = -3 \\
x = \frac{5}{2} & \text{or} & x = -\frac{3}{4}
\end{array}
$$

The solution set is $\left\{-\frac{3}{4}, \frac{5}{2}\right\}$.

We should also note that in Properties 2.5 through 2.7, k is a positive number. This is not a serious restriction because problems where k is nonpositive are easily solved as follows.

$\lvert x - 2\rvert = 0$	The solution set is $\{2\}$ because $x - 2$ has to equal zero.
$\lvert 3x - 7\rvert = -4$	The solution set is $\varnothing$. For any real number, the absolute value of $3x - 7$ will always be nonnegative.
$\lvert 2x - 1\rvert < -3$	The solution set is $\varnothing$. For any real number, the absolute value of $2x - 1$ will always be nonnegative.
$\lvert 5x + 2\rvert > -4$	The solution set is $(-\infty, \infty)$. The absolute value of $5x + 2$, regardless which real number is substituted for x, will always be greater than -4.

The number-line approach used in Examples 1 and 2 of this section, along with Properties 2.6 and 2.7, provide a systematic way of solving absolute-value inequalities that have the variable in the denominator of a fraction. Let's analyze one such problem.

EXAMPLE 9

Solve $\left|\frac{x - 2}{x + 3}\right| < 4$.

Solution

By Property 2.6, $\left|\frac{x-2}{x+3}\right| < 4$ becomes $-4 < \frac{x-2}{x+3} < 4$, which can be written

$$\frac{x-2}{x+3} > -4 \quad \text{and} \quad \frac{x-2}{x+3} < 4.$$

Each part of this *and* statement can be solved as we handled Example 2 earlier.

(a)		**(b)**
$\frac{x-2}{x+3} > -4$	and	$\frac{x-2}{x+3} < 4$
$\frac{x-2}{x+3} + 4 > 0$	and	$\frac{x-2}{x+3} - 4 < 0$
$\frac{x-2+4(x+3)}{x+3} > 0$	and	$\frac{x-2-4(x+3)}{x+3} < 0$
$\frac{x-2+4x+12}{x+3} > 0$	and	$\frac{x-2-4x-12}{x+3} < 0$
$\frac{5x+10}{x+3} > 0$	and	$\frac{-3x-14}{x+3} < 0$
This solution set is shown in Figure 2.14(a) (see Example 2 of this section).		This solution set is shown in Figure 2.14(b) (see Example 2 of this section).

(a) −3 −2 (b) $-\frac{14}{3}$ −3

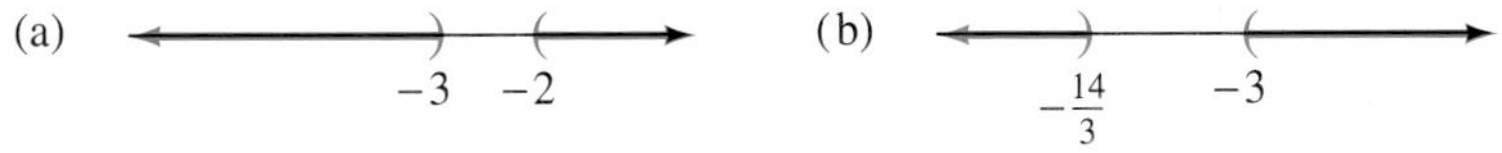

FIGURE 2.14

The intersection of the two pictured solution sets is the following set. See Figure 2.15.

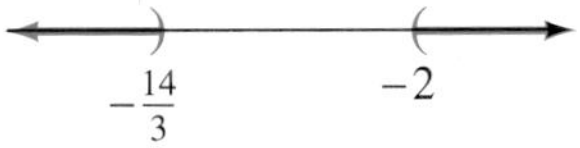

$-\frac{14}{3}$ −2

FIGURE 2.15

Therefore the solution set of $\left|\frac{x-2}{x+3}\right| < 4$ is $\left(-\infty, -\frac{14}{3}\right) \cup (-2, \infty)$.

Yes, Example 9 is a little messy, but it does illustrate the weaving together of previously used techniques to solve a more complicated problem. Don't be in a hurry when doing such problems. First, analyze the general approach to be taken and then carry out the details in a neatly organized format to minimize your chances of making careless errors.

PROBLEM SET 2.7

Solve each of the following inequalities and express the solution sets in interval notation.

1. $\frac{x+1}{x-5} > 0$

2. $\frac{x+2}{x+4} \le 0$

3. $\frac{2x-1}{x+2} < 0$

4. $\frac{3x+2}{x-1} > 0$

5. $\frac{-x+3}{3x-1} \ge 0$

6. $\frac{-n-2}{n+4} < 0$

7. $\frac{n}{n+2} \ge 3$

8. $\frac{x}{x-1} > 2$

9. $\frac{x-1}{x+2} < 2$

10. $\frac{t-1}{t-5} \le 2$

11. $\frac{t-3}{t+5} > 1$

12. $\frac{x+2}{x+7} < 1$

13. $\frac{1}{x-2} < \frac{1}{x+3}$

14. $\frac{2}{x+1} > \frac{3}{x-4}$

Solve each of the following equations.

15. $|x-2| = 6$

16. $|x+3| = 4$

17. $\left|x + \frac{1}{4}\right| = \frac{2}{5}$

18. $\left|x - \frac{2}{3}\right| = \frac{3}{4}$

19. $|2n-1| = 7$

20. $|2n+1| = 11$

21. $|3x+4| = 5$

22. $|5x-3| = 10$

23. $|7x-1| = -4$

24. $|-2x-1| = 6$

25. $|-3x-2| = 8$

26. $|5x-4| = -3$

27. $\left|\frac{3}{k-1}\right| = 4$

28. $\left|\frac{-2}{n+3}\right| = 5$

29. $|3x-1| = |2x+3|$

30. $|2x+1| = |4x-3|$

31. $|-2n+1| = |-3n-1|$

32. $|-4n+5| = |-3n-5|$

33. $|x-2| = |x+4|$

34. $|2x-3| = |2x+5|$

Solve each of the following inequalities and express the solution sets in interval notation.

35. $|x| < 6$

36. $|x| \ge 4$

37. $|x| > 8$

38. $|x| \le 1$

39. $|x| \ge -4$

40. $|x| < -5$

41. $|t-3| > 5$

42. $|n+2| < 1$

43. $|2x-1| \le 7$

44. $|2x+1| \ge 3$

45. $|3n+2| > 9$

46. $|5n-2| < 2$

47. $|4x-3| < -5$

48. $|2-x| > 1$

49. $|3-2x| < 4$

50. $|4x+5| > -3$

51. $|7x+2| \ge -2$

52. $|-1-x| \ge 8$

53. $|-2-x| \le 5$

54. $|x-1| + 2 < 4$

55. $|x+3| - 2 < 1$

56. $|x-5| + 4 \le 2$

57. $|x+4| - 1 > 1$

58. $|x-2| + 3 > 6$

59. $3|x-2| \ge 6$

60. $2|x+1| < 8$

61. $-2|x+1| > -10$

62. $-|x-4| \le -4$

63. $\left|\frac{x+1}{x-2}\right| < 3$

64. $\left|\frac{x-1}{x-4}\right| < 2$

65. $\left|\frac{x-1}{x+3}\right| > 1$

66. $\left|\frac{x+4}{x-5}\right| \ge 3$

67. $\left|\frac{n+2}{n}\right| \ge 4$

68. $\left|\frac{t+6}{t-2}\right| < 1$

69. $\left|\frac{k}{2k-1}\right| \le 2$

70. $\left|\frac{k}{k+2}\right| > 4$

THOUGHTS into WORDS

71. Explain how you would solve the inequality $\frac{x-2}{(x+1)^2} > 0$.

72. Explain how you would solve the inequality $|3x - 7| > -2$.

73. Why is $\left\{\frac{3}{2}\right\}$ the solution set for $|2x - 3| \leq 0$?

74. Consider the following approach for solving the inequality in Example 2 of this section.

$$\frac{x+2}{x+4} \leq 3$$

$$(x+4)\left(\frac{x+2}{x+4}\right) \leq 3(x+4) \qquad \text{Multiply both sides by } x + 4.$$

$$x + 2 \leq 3x + 12$$
$$-2x \leq 10$$
$$x \geq -5$$

Obviously, the solution set that we obtain using this approach differs from what we obtained in the text. What is wrong with this approach? Can we make any adjustments so that this basic approach works?

MISCELLANEOUS PROBLEMS

75. Use the definition of absolute value and prove Property 2.6.

76. Use the definition of absolute value and prove Property 2.7.

77. Solve each of the following inequalites by using the definition of absolute value. Do not use Properties 2.6 and 2.7.

a. $|x + 5| < 11$ **b.** $|x - 4| \leq 10$

c. $|2x - 1| > 7$ **d.** $|3x + 2| \geq 1$

e. $|2 - x| < 5$ **f.** $|3 - x| > 6$

CHAPTER 2 SUMMARY

Three large topics summarize this chapter: (1) solving equations, (2) solving inequalities, and (3) problem solving.

Solving Equations

The following properties are used extensively in the equation solving process.

1. $a = b$ if and only if $a + c = b + c$. Addition property of equality

2. $a = b$ if and only if $ac = bc$, $c \neq 0$. Multiplication property of equality

3. If $ab = 0$ then $a = 0$ or $b = 0$.

4. If $a = b$ then $a^n = b^n$, where n is a positive integer.

Remember that applying the fourth property may result in some extraneous solutions, so you *must check* all potential solutions.

The cross-multiplication property of proportions (if $a/b = c/d$ then $ad = bc$) can be used to solve some equations.

Quadratic equations can be solved by (1) factoring, (2) completing the square, or (3) using the quadratic formula, which can be stated as

$$x = \frac{-b \pm \sqrt{b^2 - 4ac}}{2a}.$$

The discriminant of a quadratic equation, $b^2 - 4ac$, indicates the nature of the solutions of the equation.

1. If $b^2 - 4ac > 0$, the equation has two unequal real solutions.
2. If $b^2 - 4ac = 0$, the equation has one real solution.
3. If $b^2 - 4ac < 0$, the equation has two complex but nonreal solutions.

If x_1 and x_2 are the solutions of a quadratic equation $ax^2 + bx + c = 0$, then (1) $x_1 + x_2 = -b/a$ and (2) $x_1x_2 = c/a$. These relationships can be used to check potential solutions.

The property *If* $|x| = k$, *then* $x = k$ *or* $x = -k (k > 0)$ is often helpful for solving equations involving absolute value.

Solving Inequalities

The following properties form a basis for solving inequalities.

1. If $a > b$ then $a + c > b + c$.
2. If $a > b$ and $c > 0$, then $ac > bc$.
3. If $a > b$ and $c < 0$, then $ac < bc$.

Quadratic inequalities such as $(x + 3)(x - 7) > 0$ can be solved by considering the *sign behavior* of the individual factors.

The following properties play an important role in solving inequalities involving absolute value.

1. If $|x| < k$, where $k > 0$, then $-k < x < k$.
2. If $|x| > k$, where $k > 0$, then $x > k$ or $x < -k$.

Problem Solving

It would be helpful for you to reread pages 86, 87, 94–96, 114. Some key problem solving suggestions are given on those pages.

CHAPTER 2 REVIEW PROBLEM SET

Solve each of the following equations.

1. $2(3x - 1) - 3(x - 2) = 2(x - 5)$

2. $\frac{n-1}{4} - \frac{2n+3}{5} = 2$

3. $\frac{2}{x+2} + \frac{5}{x-4} = \frac{7}{2x-8}$

4. $0.07x + 0.12(550 - x) = 56$

5. $(3x - 1)^2 = 16$

6. $4x^2 - 29x + 30 = 0$

7. $x^2 - 6x + 10 = 0$

8. $n^2 + 4n = 396$

9. $15x^3 + x^2 - 2x = 0$

10. $\frac{t+3}{t-1} - \frac{2t+3}{t-5} = \frac{3-t^2}{t^2-6t+5}$

11. $\frac{5-x}{2-x} - \frac{3-2x}{2x} = 1$

12. $x^4 + 4x^2 - 45 = 0$

13. $2n^{-4} - 11n^{-2} + 5 = 0$

14. $\left(x - \frac{2}{x}\right)^2 + 4\left(x - \frac{2}{x}\right) = 5$

15. $\sqrt{5 + 2x} = 1 + \sqrt{2x}$

16. $\sqrt{3 + 2n} + \sqrt{2 - 2n} = 3$

17. $\sqrt{3 - t} - \sqrt{3 + t} = \sqrt{t}$

18. $|5x - 1| = 7$

19. $|2x + 5| = |3x - 7|$

20. $\left|\frac{-3}{n-1}\right| = 4$

21. $x^3 + x^2 - 2x - 2 = 0$

22. $2x^{2/3} + 5x^{1/3} - 12 = 0$

Solve each of the following inequalities. Express the solution sets using interval notation.

23. $3(2 - x) + 2(x - 4) > -2(x + 5)$

24. $\frac{3}{5}x - \frac{1}{3} \le \frac{2}{3}x + \frac{3}{4}$

25. $\frac{n-1}{3} - \frac{2n+1}{4} > \frac{1}{6}$

26. $0.08x + 0.09(700 - x) \ge 59$

27. $-16 \le 7x - 2 \le 5$

28. $5 > \frac{3y+4}{2} > 1$

29. $x^2 - 3x - 18 < 0$

30. $n^2 - 5n \ge 14$

31. $(x - 1)(x - 4)(x + 2) < 0$

32. $\frac{x+4}{2x-3} \le 0$

33. $\frac{5n-1}{n-2} > 0$

34. $\frac{x-1}{x+3} \ge 2$

35. $\frac{t+5}{t-4} < 1$

36. $|4x - 3| > 5$

37. $|3x + 5| \le 14$

38. $|-3 - 2x| < 6$

39. $\left|\frac{x-1}{x}\right| > 2$

40. $\left|\frac{n+1}{n+2}\right| < 1$

Solve each of the following problems.

41. The sum of three consecutive odd integers is 31 less than four times the largest integer. Find the integers.

42. The sum of two numbers is 74. If the larger is divided by the smaller, the quotient is 7 and the remainder is 2. Find the numbers.

43. The perimeter of a rectangle is 38 centimeters and its area is 84 square centimeters. Find the dimensions of the rectangle.

44. A sum of money amounting to \$13.55 consists of nickels, dimes, and quarters. There are three times as many dimes as nickels and three less quarters than dimes. How many coins of each denomination are there?

45. A retailer has some shirts that cost him \$14 each. He wants to sell them to make a profit of 30% of the selling price. What price should he charge for the shirts?

46. How many gallons of a solution of glycerine and water containing 55% glycerine should be added to 15 gallons of a 20% solution to give a 40% solution?

47. The sum of the present ages of Rosie and her mother is 47 years. In 5 years, Rosie will be one-half as old as her mother at that time. Find the present ages of both Rosie and her mother.

48. Kelly invested \$800, part of it at 9% and the remainder at 12%. Her total yearly interest from the two investments was \$85.50. How much did she invest at each rate?

49. Regina has scores of 93, 88, 89, and 95 on her first four math exams. What score must she get on the fifth exam to have an average of 92 or better for the five exams?

50. At how many minutes after 2 p.m. will the minute hand of a clock overtake the hour hand?

51. Russ started to mow the lawn, a task that usually takes him 40 minutes. After he had been working for 15 minutes, his friend Jay came along with his mower and began to help Russ. Working together, they finished the lawn in 10 minutes. How long would it have taken Jay to mow the lawn by himself?

52. Barry bought a number of shares of stock for \$600. A week later the value of the stock increased \$3 per share and he sold all but 10 shares and regained his original investment of \$600. How many shares did he sell and at what price per share?

53. Larry drove 156 miles in one hour more than it took Mike to drive 108 miles. Mike drove at an average rate of 2 miles per hour faster than Larry. How fast did each one travel?

54. It takes Bill 2 hours longer to do a certain job than it takes Cindy. They worked together for 2 hours; then Cindy left and Bill finished the job in 1 hour. How long would it take each of them to do the job alone?

55. One leg of a right triangle is 5 centimeters longer than the other leg. If the hypotenuse is 25 centimeters long, find the length of each leg.

56. The area of a rectangle is 35 square inches. If both the length and width are increased by 3 inches, the area is increased by 45 square inches. Find the dimensions of the original rectangle.

Coordinate Geometry and Graphing Techniques

René Descartes, a French mathematician of the seventeenth century, was able to transform geometric problems into an algebraic setting so that he could use the tools of algebra to solve the problems. This merging of algebraic and geometric ideas is the foundation of a branch of mathematics called **analytic geometry**, today more commonly called **coordinate geometry**. Basically, there are two kinds of problems in coordinate geometry: Given an algebraic equation, find its geometric graph; and given a set of conditions pertaining to a geometric graph, find its algebraic equation. We will discuss problems of both types in this chapter.

3

Bridges are often constructed with semi-elliptical arches.

3.1 COORDINATE GEOMETRY

Recall that the real number line (Figure 3.1) exhibits a **one-to-one correspondence** between the set of real numbers and the points on a line. That is to say, to each real number there corresponds one and only one point on the line, and to each point on the line there corresponds one and only one real number. The number that corresponds to a particular point on the line is called the **coordinate** of that point.

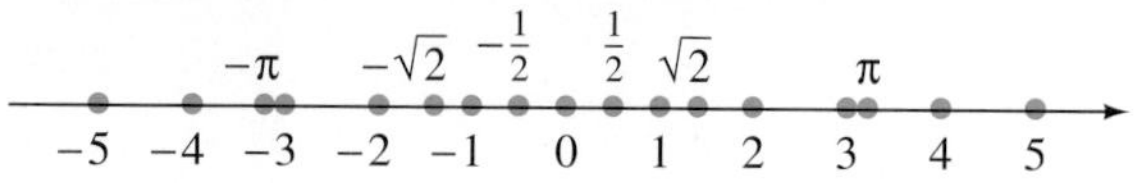

$-\pi$ $-\sqrt{2}$ $-\frac{1}{2}$ $\frac{1}{2}$ $\sqrt{2}$ π

-5 -4 -3 -2 -1 0 1 2 3 4 5

FIGURE 3.1

Suppose that on the number line we want to know the distance *from* -2 *to* 6. The *from-to* vocabulary implies a **directed distance**, which is $6 - (-2) = 8$ units. In other words, it is 8 units in a *positive direction* from -2 to 6. Likewise, the distance from 9 to -4 is $-4 - 9 = -13$, that is, 13 units in a *negative direction*. In general, if x_1 and x_2 are the coordinates of two points on the number line, then the distance **from x_1 to x_2** is given by $x_2 - x_1$, and the distance **from x_2 to x_1** is given by $x_1 - x_2$.

Now suppose that we want to find the distance *between* -2 and 6. The *between* vocabulary implies **distance without regard to direction**. Thus, the distance between -2 and 6 can be found by using either $|6 - (-2)| = 8$, or $|-2 - 6| = 8$. In general, if x_1 and x_2 are the coordinates of two points on the number line, the distance **between x_1 and x_2** can be found by using either $|x_2 - x_1|$ or $|x_1 - x_2|$.

Sometimes it is necessary to find the coordinate of a point located somewhere between the two given points. For example, in Figure 3.2 suppose that we want to find the coordinate (x) of the point located two-thirds of the distance *from 2 to 8*. Since the total distance from 2 to 8 is $8 - 2 = 6$ units, we can start at 2 and move $\frac{2}{3}(6) = 4$ units toward 8. Thus,

$$x = 2 + \frac{2}{3}(6) = 2 + 4 = 6.$$

2 x 8

FIGURE 3.2

The following examples further illustrate the process of finding the coordinate of a point somewhere between two given points (Figure 3.3).

Problem	Solution
a. Three-fourths of the distance from −2 to 10	$x = -2 + \frac{3}{4}[10 - (-2)]$ $= -2 + \frac{3}{4}(12)$ $= 7$

FIGURE 3.3a

Problem	Solution
b. Two-fifths of the distance from −1 to 7	$x = -1 + \frac{2}{5}[7 - (-1)]$ $= -1 + \frac{2}{5}(8)$ $= \frac{11}{5}$

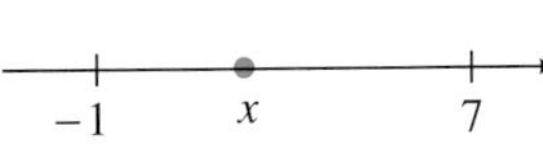

FIGURE 3.3b

Problem	Solution
c. One-third of the distance from 9 to 1	$x = 9 + \frac{1}{3}(1 - 9)$ $= 9 + \frac{1}{3}(-8)$ $= \frac{19}{3}$

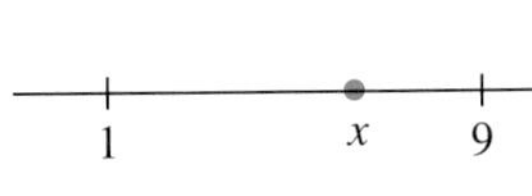

FIGURE 3.3c

Problem	Solution
d. a/b of the distance from x_1 to x_2	$x = x_1 + \frac{a}{b}(x_2 - x_1)$

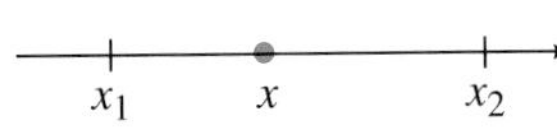

FIGURE 3.3d

Problem (d) indicates that a general formula can be developed for this type of problem. However, it may be easier to remember the basic approach than it is to memorize the formula.

As we saw in Chapter 2, the real number line provides a geometric model for graphing solutions of algebraic equations and inequalities involving *one variable*. For example, the solutions of "$x > 2$ or $x \leq -1$" are graphed in Figure 3.4.

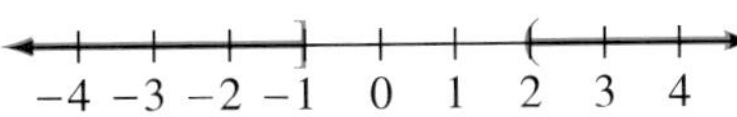

FIGURE 3.4

Rectangular Coordinate System

To expand our work with coordinate geometry, we now consider two number lines, one vertical and one horizontal, perpendicular to each other at the point associated with zero on both lines (Figure 3.5). We refer to these number lines as the **horizontal** and **vertical axes** and together as the **coordinate axes**. They partition the plane into four regions called **quadrants**. The quadrants are numbered counterclockwise from I through IV, as indicated in Figure 3.5. The point of intersection of the two axes is called the **origin**.

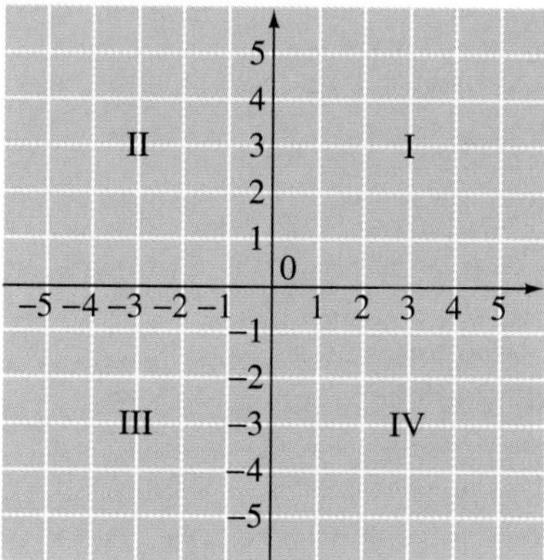

FIGURE 3.5

It is now possible to set up a one-to-one correspondence between *ordered pairs* of real numbers and the points in a plane. To each ordered pair of real numbers there corresponds a unique point in the plane and to each point in the plane there corresponds a unique ordered pair of real numbers. We have illustrated examples of this correspondence in Figure 3.6. The ordered pair (3, 2) means that the point

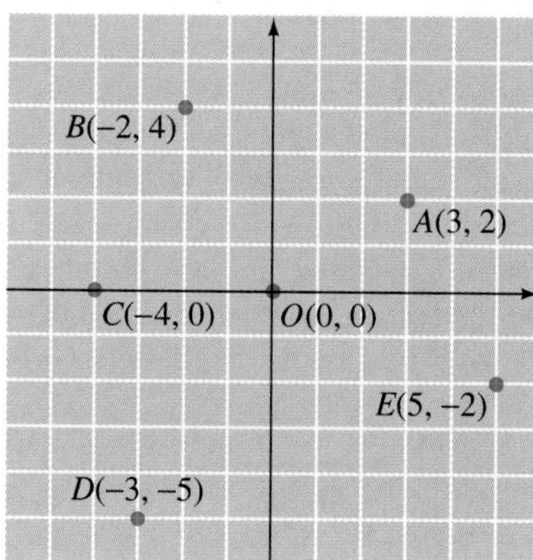

FIGURE 3.6

A is located three units to the right and two units up from the origin. The ordered pair $(-3, -5)$ means that point D is located three units to the left and five units down from the origin. The ordered pair $(0, 0)$ is associated with the origin.

REMARK The notation $(-2, 4)$ was used in Chapter 2 to indicate an interval of the real number line. Now we are using the same notation to indicate an ordered pair of real numbers. This double meaning should not be confusing since the context of the material will definitely indicate the meaning at a particular time. Throughout this chapter we will be using the ordered-pair interpretation.

In general, the real numbers a and b in the ordered pair (a, b) are associated with a point; they are referred to as the **coordinates of the point**. The first number, a, is called the **abscissa**; it is the directed distance of the point from the vertical axis, measured parallel to the horizontal axis. The second number, b,is called the **ordinate**; it is the directed distance from the horizontal axis, measured parallel to the vertical axis (Figure 3.7(a)). Thus, in the first quadrant, all points have a positive abscissa and a positive ordinate. In the second quadrant, all points have a negative abscissa and a positive ordinate. We have indicated the sign situations for all four quadrants in Figure 3.7(b). This system of associating points in a plane with pairs of real numbers is called the **rectangular coordinate system** or the **Cartesian coordinate system**.

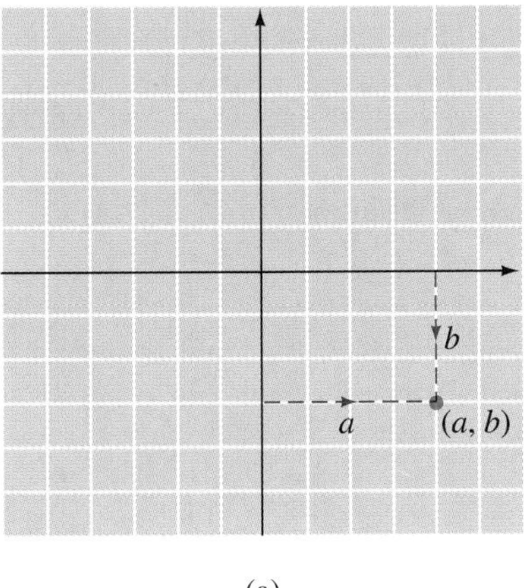

(a)

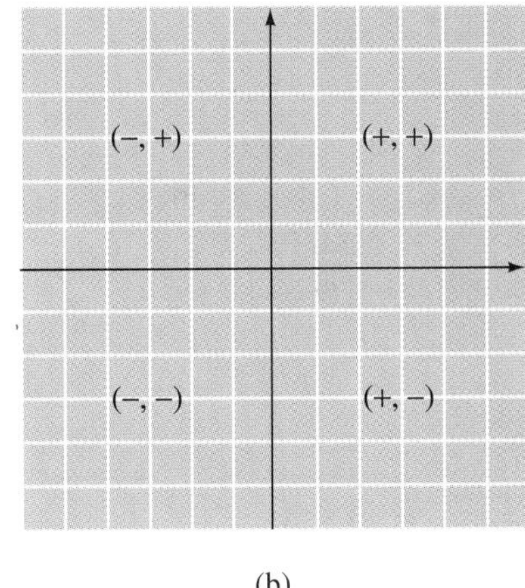

(b)

FIGURE 3.7

Distance Between Two Points

As we work with the rectangular coordinate system, it is sometimes necessary to express the length of certain line segments. In other words, we need to be able to find the *distance* between two points. Let's first consider two specific examples and then develop a general distance formula.

EXAMPLE 1

Find the distance between the points $A(2, 2)$ and $B(5, 2)$ and also between the points $C(-2, 5)$ and $D(-2, -4)$.

Solution

Let's plot the points and draw $\overline{AB}$ and $\overline{CD}$ as in Figure 3.8. (The symbol $\overline{AB}$ denotes the line segment with endpoints A and B.) Since $\overline{AB}$ is parallel to the horizontal axis, its length can be expressed as $|5 - 2|$ or $|2 - 5|$. Thus, the length of $\overline{AB}$ (we shall use the notation AB to represent the length of $\overline{AB}$) is $AB = 3$ units. Likewise, since $\overline{CD}$ is parallel to the vertical axis, we obtain $CD = |5 - (-4)| = 9$ units.

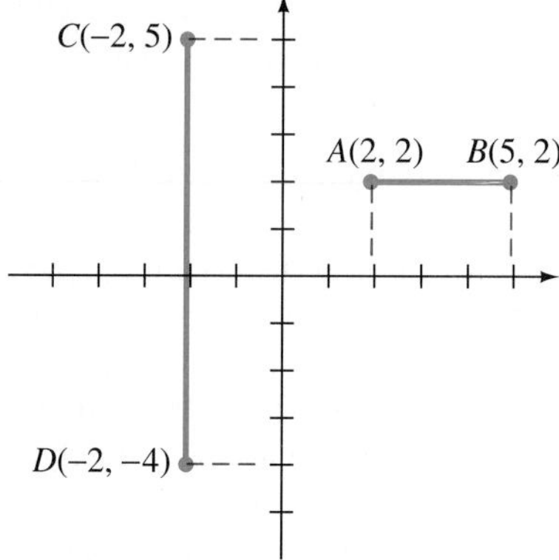

FIGURE 3.8

EXAMPLE 2

Find the distance between the points $A(2, 3)$ and $B(5, 7)$.

Solution

Let's plot the points and form a right triangle using point D, as indicated in Figure 3.9. Notice that the coordinates of point D are $(5, 3)$. Because $\overline{AD}$ is parallel

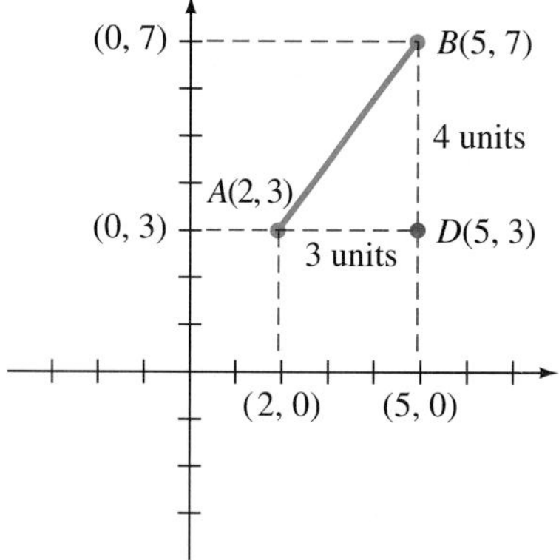

FIGURE 3.9

to the horizontal axis, as in Example 1, we have $AD = |5 - 2| = 3$ units. Likewise, $\overline{DB}$ is parallel to the vertical axis and therefore $DB = |7 - 3| = 4$ units. Applying the Pythagorean theorem, we obtain

$$\begin{aligned}(AB)^2 &= (AD)^2 + (DB)^2 \\ &= 3^2 + 4^2 \\ &= 9 + 16 \\ &= 25.\end{aligned}$$

Thus

$$AB = \sqrt{25} = 5 \text{ units.}$$

Before we use the approach in Example 2 to develop a general distance formula, let's make another notational agreement. For most problems in coordinate geometry it is customary to label the horizontal axis the **x-axis** and the vertical axis the **y-axis**. Then, ordered pairs representing points in the xy-plane are of the form (x, y); that is, x is the first coordinate and y is the second coordinate. Now let's develop a general distance formula.

Let $P_1(x_1, y_1)$ and $P_2(x_2, y_2)$ represent any two points in the xy-plane. Form a right triangle using point R, as indicated in Figure 3.10. The coordinates of the vertex of the right angle, point R, are (x_2, y_1). The length of $\overline{P_1R}$ is $|x_2 - x_1|$ and

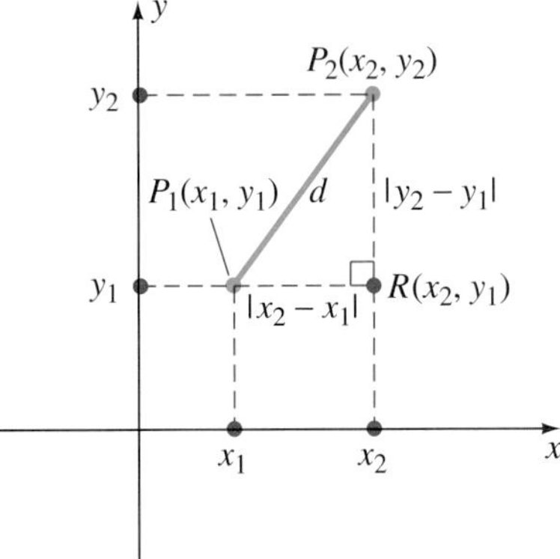

FIGURE 3.10

the length of $\overline{RP_2}$ is $|y_2 - y_1|$. Let d represent the length of $\overline{P_1P_2}$ and apply the Pythagorean theorem to obtain

$$d^2 = |x_2 - x_1|^2 + |y_2 - y_1|^2.$$

Since $|a|^2 = a^2$ for any real number a, the distance formula can be stated

$$d = \sqrt{(x_2 - x_1)^2 + (y_2 - y_1)^2}.$$

It makes no difference which point you call P_1 or P_2 when using the formula. Also, remember that if you forget the formula, there is no need to panic: Form a right triangle and apply the Pythagorean theorem as we did in Example 2.

Let's consider some examples that illustrate the use of the distance formula.

EXAMPLE 3

Find the distance between $(-2, 5)$ and $(1, -1)$.

Solution

Let $(-2, 5)$ be P_1 and $(1, -1)$ be P_2. Use the distance formula to obtain

$$\begin{aligned} d &= \sqrt{(x_2 - x_1)^2 + (y_2 - y_1)^2} \\ &= \sqrt{[1 - (-2)]^2 + (-1 - 5)^2} \\ &= \sqrt{3^2 + (-6)^2} \\ &= \sqrt{9 + 36} \\ &= \sqrt{45} = 3\sqrt{5}. \end{aligned}$$

The distance between the two points is $3\sqrt{5}$ units.

In Example 3, notice the simplicity of the approach when we use the distance formula. No diagram was needed; we merely plugged in the values and did the computation. However, many times a figure *is* helpful in the analysis of the problem, as we will see in the next example.

EXAMPLE 4

Verify that the points $(-3, 6)$, $(3, 4)$, and $(1, -2)$ are vertices of an isosceles triangle. (An isosceles triangle has two sides of the same length.)

Solution

Let's plot the points and draw the triangle (Figure 3.11). The lengths d_1, d_2, and d_3 can all be found by using the distance formula.

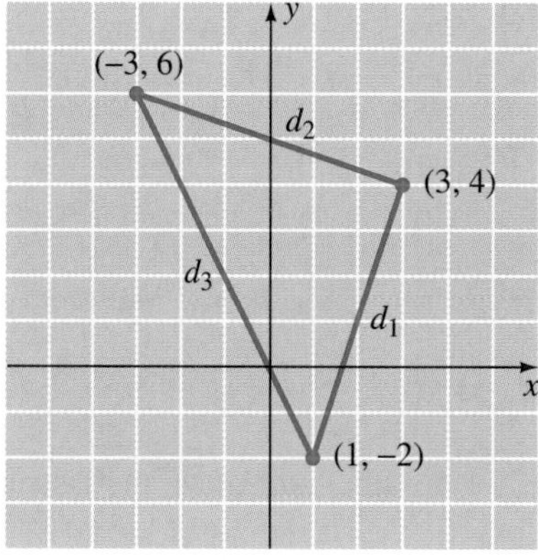

FIGURE 3.11

$$\begin{aligned} d_1 &= \sqrt{(3-1)^2 + [4-(-2)]^2} \\ &= \sqrt{4+36} \\ &= \sqrt{40} = 2\sqrt{10} \end{aligned}$$

$$\begin{aligned} d_2 &= \sqrt{(-3-3)^2 + (6-4)^2} \\ &= \sqrt{36+4} \\ &= \sqrt{40} = 2\sqrt{10} \end{aligned}$$

$$\begin{aligned} d_3 &= \sqrt{(-3-1)^2 + [6-(-2)]^2} \\ &= \sqrt{16+64} \\ &= \sqrt{80} = 4\sqrt{5} \end{aligned}$$

Since $d_1 = d_2$, it is an isosceles triangle.

Points of Division of a Line Segment

Earlier in this section we discussed the process of finding the coordinate of a point on a number line, given that it is located somewhere between two other points on the line. This same type of problem can occur in the *xy*-plane, and the approach we used earlier can be extended to handle it. Let's consider some examples.

EXAMPLE 5

Find the coordinates of the point P, which is two-thirds of the distance from $A(1, 2)$ to $B(7, 5)$.

Solution

In Figure 3.12 we plotted the given points A and B and completed a figure to help with the analysis of the problem. To find the coordinates of point P, we can proceed as follows. Point D is two-thirds of the distance from A to C because parallel lines

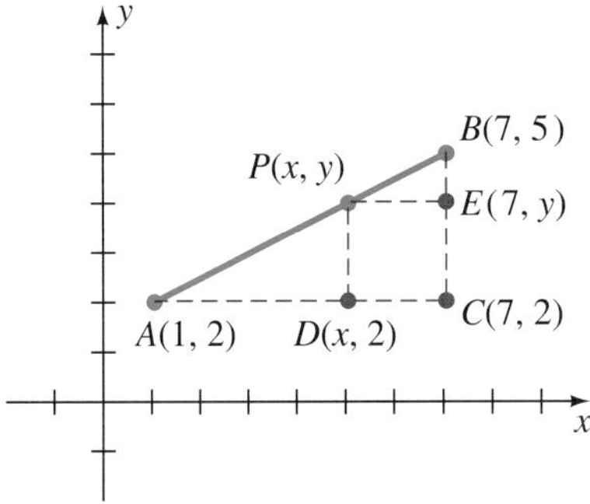

FIGURE 3.12

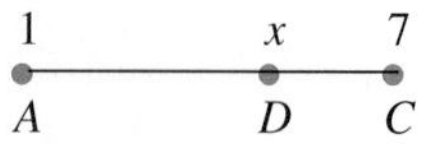

FIGURE 3.13

cut off proportional segments on every transversal that intersects the lines. Therefore, since $\overline{AC}$ is parallel to the x-axis, it can be treated as a segment of the number line (see Figure 3.13). Thus, we have

$$x = 1 + \frac{2}{3}(7 - 1) = 1 + \frac{2}{3}(6) = 5.$$

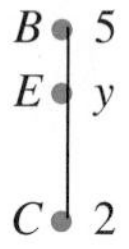

FIGURE 3.14

Similarly, $\overline{CB}$ is parallel to the y-axis, so it can also be treated as a segment of the number line (see Figure 3.14). Thus we obtain

$$\begin{aligned} y &= 2 + \frac{2}{3}(5 - 2) \\ &= 2 + \frac{2}{3}(3) = 4. \end{aligned}$$

The point P has coordinates (5, 4).

EXAMPLE 6

Find the coordinates of the midpoint of the line segment determined by the points $P_1(x_1, y_1)$ and $P_2(x_2, y_2)$.

Solution

Figure 3.15 helps with the analysis of the problem. The line segment $\overline{P_1R}$ is parallel to the x-axis and $S(x, y_1)$ is the midpoint of $\overline{P_1R}$ (see Figure 3.16). Thus we can determine the x-coordinate of S.

$$\begin{aligned} x &= x_1 + \frac{1}{2}(x_2 - x_1) \\ &= x_1 + \frac{1}{2}x_2 - \frac{1}{2}x_1 \\ &= \frac{1}{2}x_1 + \frac{1}{2}x_2 = \frac{x_1 + x_2}{2} \end{aligned}$$

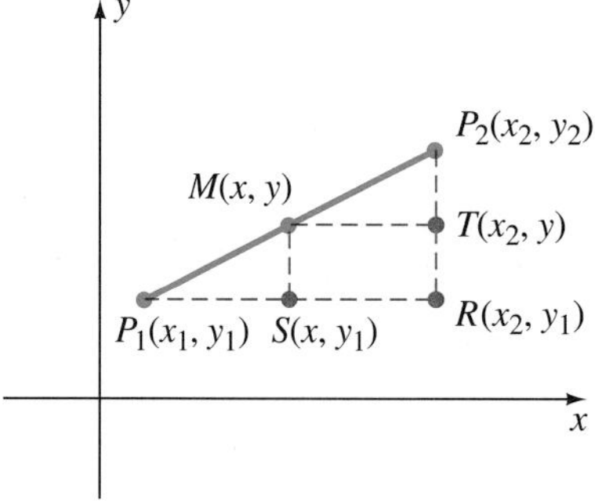

FIGURE 3.15

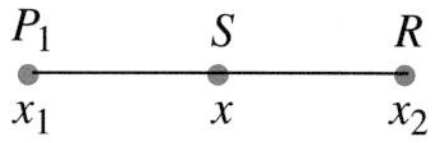

FIGURE 3.16

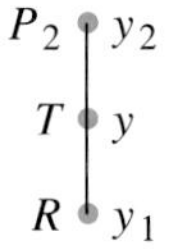

P_2 y_2
T y
R y_1

FIGURE 3.17

Similarly, $\overline{RP_2}$ is parallel to the y-axis and $T(x_2, y)$ is the midpoint of $\overline{RP_2}$ (see Figure 3.17). Therefore we can calculate the y-coordinate of T.

$$\begin{aligned} y &= y_1 + \frac{1}{2}(y_2 - y_1) \\ &= y_1 + \frac{1}{2}y_2 - \frac{1}{2}y_1 \\ &= \frac{1}{2}y_1 + \frac{1}{2}y_2 = \frac{y_1 + y_2}{2} \end{aligned}$$

Thus, the coordinates of the midpoint of a line segment determined by $P_1(x_1, y_1)$ and $P_2(x_2, y_2)$ are given by the following.

$$\left(\frac{x_1 + x_2}{2}, \frac{y_1 + y_2}{2}\right)$$

EXAMPLE 7

Find the coordinates of the midpoint of the line segment determined by the points $(-2, 4)$ and $(6, -1)$.

Solution

Using the midpoint formula, we obtain

$$\begin{aligned} \left(\frac{x_1 + x_2}{2}, \frac{y_1 + y_2}{2}\right) &= \left(\frac{-2 + 6}{2}, \frac{4 + (-1)}{2}\right) \\ &= \left(\frac{4}{2}, \frac{3}{2}\right) \\ &= \left(2, \frac{3}{2}\right). \end{aligned}$$

We want to emphasize a couple of ideas from Examples 5, 6, and 7. If we want to find a point of division of a line segment, we will use the same approach as in Example 5. However, for the special case of the midpoint, the formula developed in Example 6 is convenient to use.

PROBLEM SET 3.1

On a number line, find the following indicated distances.

1. From -4 to 6

2. From 5 to -14

3. From -6 to -11

4. From -7 to 10

5. Between -2 and 4

6. Between -4 and -12

7. Between 5 and -10

8. Between -2 and 13

For each of the following, find the coordinate of the indicated point on a number line.

9. Two-thirds of the distance from 1 to 10

10. Three-fourths of the distance from -2 to 14

11. One-third of the distance from -3 to 7

12. Two-fifths of the distance from -5 to 6

13. Three-fifths of the distance from -1 to -11

14. Five-sixths of the distance from 3 to -7

For each of the following, find the length of $\overline{AB}$ and the midpoint of $\overline{AB}$.

15. $A(2, 1)$, $B(10, 7)$

16. $A(-2, -1)$, $B(7, 11)$

17. $A(1, -1)$, $B(3, -4)$

18. $A(-5, 2)$, $B(-1, 6)$

19. $A(6, -4)$, $B(9, -7)$

20. $A(-3, 3)$, $B(0, -3)$

21. $A\left(\frac{1}{2}, \frac{1}{3}\right)$, $B\left(-\frac{1}{3}, \frac{3}{2}\right)$

22. $A\left(-\frac{3}{4}, 2\right)$, $B\left(-1, -\frac{5}{4}\right)$

For Problems 23–28, find the coordinates of the indicated point in the xy-plane.

23. One-third of the distance from (2, 3) to (5, 9)

24. Two-thirds of the distance from (1, 4) to (7, 13)

25. Two-fifths of the distance from $(-2, 1)$ to (8, 11)

26. Three-fifths of the distance from $(2, -3)$ to $(-3, 8)$

27. Five-eighths of the distance from $(-1, -2)$ to $(4, -10)$

28. Seven-eighths of the distance from $(-2, 3)$ to $(-1, -9)$

Solve each of the following problems.

29. Find the coordinates of the point that is one-fourth of the distance from (2, 4) to (10, 13) by (a) using the midpoint formula twice, and (b) using the same approach as for Problems 23–28.

30. If one endpoint of a line segment is $(-6, 4)$ and the midpoint of the segment is $(-2, 7)$, find the other endpoint.

31. Use the distance formula to verify that the points $(-2, 7)$, $(2, 1)$, and $(4, -2)$ lie on a straight line.

32. Use the distance formula to verify that the points $(-3, 8)$, $(7, 4)$, and $(5, -1)$ are vertices of a right triangle.

33. Verify that the points (0, 3), $(2, -3)$, and $(-4, -5)$ are vertices of an isosceles triangle.

34. Verify that the points (7, 12) and (11, 18) divide the line segment joining (3, 6) and (15, 24) into three segments of equal length.

35. Find the perimeter of the triangle whose vertices are $(-6, -4)$, $(0, 8)$, and $(6, 5)$.

36. Verify that $(-4, 9)$, $(8, 4)$, $(3, -8)$, and $(-9, -3)$ are vertices of a square.

37. Verify that the points $(4, -5)$, $(6, 7)$, and $(-8, -3)$ lie on a circle that has its center at $(-1, 2)$.

38. Suppose that $(-2, 5)$, $(6, 3)$, and $(-4, -1)$ are three vertices of a parallelogram. How many possibilities are there for the fourth vertex? Find the coordinates of each of these points. [*Hint:* The diagonals of a parallelogram bisect each other.]

39. Find x such that the line segment determined by $(x, -2)$ and $(-2, -14)$ is 13 units long.

40. Consider the triangle whose vertices are $(4, -6)$, $(2, 8)$, and $(-4, 2)$. Verify that the medians of this triangle intersect at a point that is two-thirds of the distance from a vertex to the midpoint of the opposite side. (A **median** of a triangle is the line segment determined by a vertex and the midpoint of the opposite side. Every triangle has three medians.)

41. Consider the line segment determined by $A(-1, 2)$ and $B(5, 11)$. Find the coordinates of a point P such that $AP/PB = 2/1$.

42. Verify that the midpoint of the hypotenuse of the right triangle formed by the points $A(4, 0)$, $B(0, 0)$, and $C(0, 6)$ is the same distance from all three vertices.

43. Consider the parallelogram determined by the points $A(1, 1)$, $B(5, 1)$, $C(6, 4)$, and $D(2, 4)$. Verify that the diagonals of this parallelogram bisect each other.

44. Consider the quadrilateral determined by the points $A(5, -3)$, $B(3, 4)$, $C(-2, 1)$, and $D(-1, -2)$. Verify that the line segments joining the midpoints of the opposite sides of this quadrilateral bisect each other.

THOUGHTS into WORDS

45. Consider the line segment determined by the two endpoints $A(2, 1)$ and $B(5, 10)$. Describe how you would find the coordinates of the point that is two-thirds of the distance from A to B. Then describe how you would find the point that is two-thirds of the distance from B to A.

MISCELLANEOUS PROBLEMS

46. The tools of coordinate geometry can be used to prove various geometric properties. For example, consider the following way of proving that the diagonals of a rectangle are equal in length.

First we draw a rectangle and coordinatize it by using a convenient position for the origin. Now we can use the distance formula to find the lengths of the diagonals $\overline{AC}$ and $\overline{BD}$. See Figure 3.18.

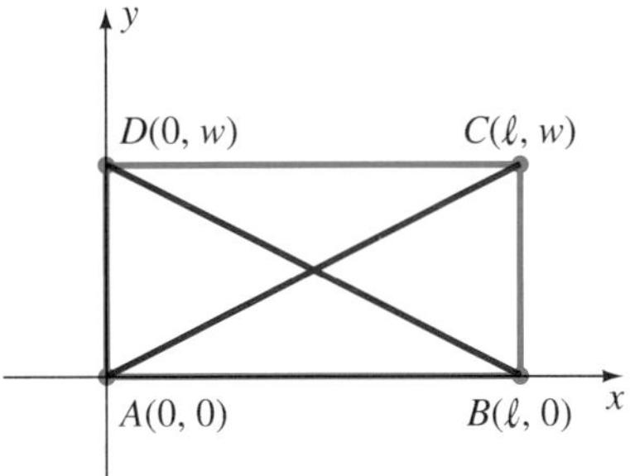

FIGURE 3.18

$$AC = \sqrt{(l - 0)^2 + (w - 0)^2} = \sqrt{l^2 + w^2}$$

$$BD = \sqrt{(0 - l)^2 + (w - 0)^2} = \sqrt{l^2 + w^2}$$

Thus, $AC = BD$, and we have proven that the diagonals are equal in length. Prove each of the following.

a. The diagonals of an isosceles trapezoid are equal in length.

b. The line segment joining the midpoints of two sides of a triangle is equal in length to one-half of the third side.

c. The midpoint of the hypotenuse of a right triangle is equally distant from all three vertices.

d. The diagonals of a parallelogram bisect each other.

e. The line segments joining the midpoints of the opposite sides of a quadrilateral bisect each other.

f. The medians of a triangle intersect at a point that is two-thirds of the distance from a vertex to the midpoint of the opposite side. (See Problem 40.)

3.2 GRAPHING TECHNIQUES: LINEAR EQUATIONS AND INEQUALITIES

As you continue to study mathematics, you will find that the ability to quickly sketch the graph of an equation is an important skill. Therefore, various curve sketching techniques are discussed throughout precalculus and calculus courses. We will use a good portion of this chapter to expand your repertoire of graphing techniques.

First, let's briefly review some basic ideas by considering the solutions for the equation $y = x + 2$. A **solution** of an equation in two variables is an ordered pair of real numbers that satisfy the equation. When using the variables x and y, the ordered pairs are of the form (x, y). We see that $(1, 3)$ is a solution for $y = x + 2$ because if x is replaced by 1 and y by 3, the true numerical statement $3 = 1 + 2$ is obtained. Likewise, $(-2, 0)$ is a solution because $0 = -2 + 2$ is a true statement. An infinite number of pairs of real numbers that satisfy $y = x + 2$ can be found by arbitrarily choosing values for x and, for each value of x chosen, determining a corresponding value for y. Let's use a table to record some of the solutions for $y = x + 2$.

CHOOSE x	DETERMINE y FROM y = x + 2	SOLUTIONS FOR y = x + 2
0	2	(0, 2)
1	3	(1, 3)
3	5	(3, 5)
5	7	(5, 7)
−2	0	(−2, 0)
−4	−2	(−4, −2)
−6	−4	(−6, −4)

Plotting the points associated with the ordered pairs from the table produces Figure 3.19(a). The straight line containing the points (Figure 3.19(b)) is called the **graph of the equation** $y = x + 2$.

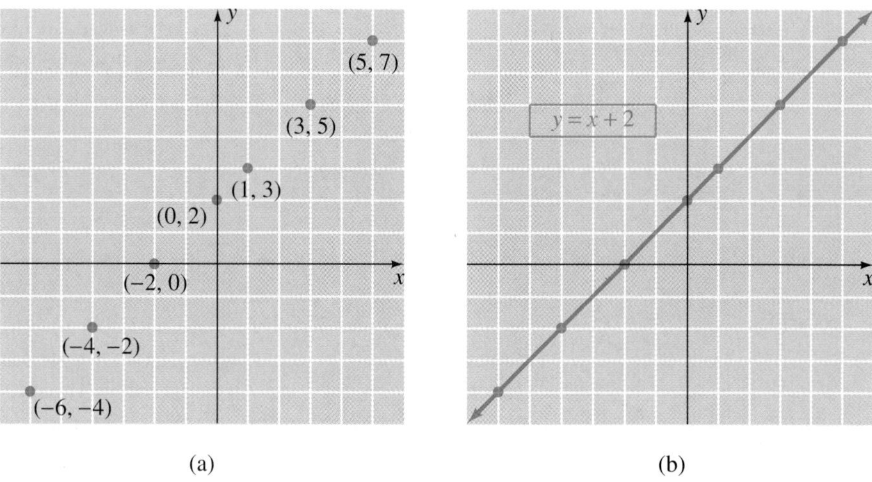

FIGURE 3.19

Graphing Linear Equations

Probably the most valuable graphing technique is the ability to recognize the kind of graph that is produced by a particular type of equation. For example, from previous mathematics courses you may remember that any equation of the form $Ax + By = C$, where A, B, and C are constants (A and B not both zero) and x and y are variables, is a **linear equation** and its graph is a **straight line**. Two comments about this description of a linear equation should be made. First, the choice of x and y as variables is arbitrary; any two letters can be used to represent the variables. For example, an equation such as $3r + 2s = 9$ is also a linear equation in two variables. In order to avoid constantly changing the labeling of the coordinate axes when graphing equations, we will use the same two variables, x and y, in all equations. Second, the statement *any equation of the form* $Ax + By = C$ technically means any equation of that form or equivalent to that form. For example, the equation $y = 2x - 1$ is equivalent to $-2x + y = -1$ and therefore is linear and produces a straight-line graph.

Before we graph some linear equations, let's define in general the **intercepts** of a graph.

The x-coordinates of the points that a graph has in common with the x-axis are called the **x-intercepts** of the graph. (To compute the x-intercepts, let $y = 0$ and solve for x.)

The y-coordinates of the points that a graph has in common with the y-axis are called the **y-intercepts** of the graph. (To compute the y-intercepts, let $x = 0$ and solve for y.)

Once we know that any equation of the form $Ax + By = C$ produces a straight-line graph, along with the fact that two points determine a straight line, graphing linear equations becomes a simple process. We can find two points on the graph and draw the line determined by those two points. Usually the two points involving the intercepts are easy to find, and generally it's a good idea to plot a third point to serve as a check.

EXAMPLE 1

Graph $3x - 2y = 6$.

Solution

First, let's find the intercepts. If $x = 0$, then

$$\begin{aligned} 3(0) - 2y &= 6 \\ -2y &= 6 \\ y &= -3. \end{aligned}$$

Therefore, the point $(0, -3)$ is on the line. If $y = 0$, then

$$\begin{aligned} 3x - 2(0) &= 6 \\ 3x &= 6 \\ x &= 2. \end{aligned}$$

Thus, the point $(2, 0)$ is also on the line. Now let's find a check point. If $x = -2$, then

$$\begin{aligned} 3(-2) - 2y &= 6 \\ -6 - 2y &= 6 \\ -2y &= 12 \\ y &= -6. \end{aligned}$$

So the point $(-2, -6)$ is also on the line. In Figure 3.20, the three points are plotted and the graph of $3x - 2y = 6$ is drawn.

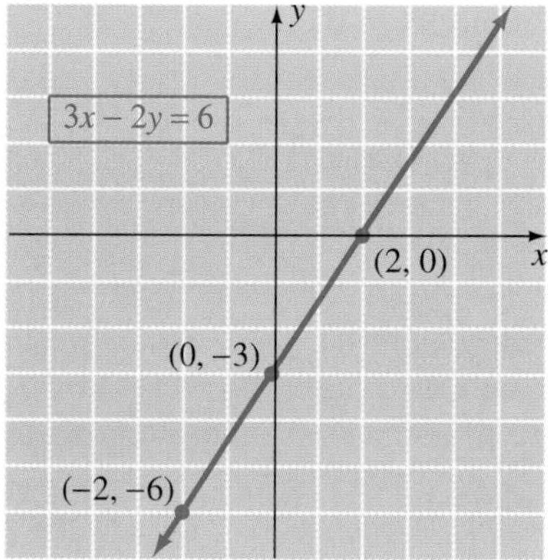

FIGURE 3.20

Notice in Example 1 that we did not solve the given equation for y in terms of x or for x in terms of y. Since we know it is a straight line, there is no need for an extensive table of values; thus, there is no need to change the form of the original equation. Furthermore, the point $(-2, -6)$ served as a check point. If it had not been on the line determined by the two intercepts, then we would have known that an error had been made in finding the intercepts.

EXAMPLE 2

Graph $y = -2x$.

Solution

If $x = 0$, then $y = -2(0) = 0$; so the origin $(0, 0)$ is on the line. Since both intercepts are determined by the point $(0, 0)$, another point is necessary to determine the line. Then a third point should be found as a check point. The graph of $y = -2x$ is shown in Figure 3.21.

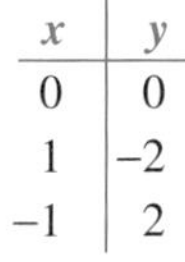

x	y
0	0
1	−2
−1	2

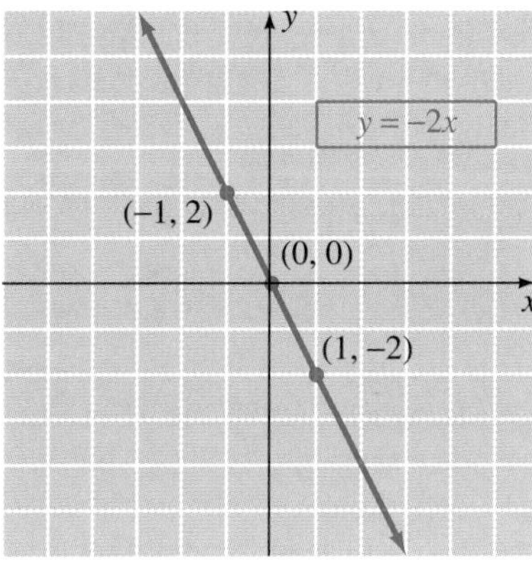

FIGURE 3.21

Example 2 illustrates the general concept that for the form $Ax + By = C$, if $C = 0$ then the line contains the origin. Stated another way, **the graph of any equation of the form $y = kx$**, where k is any real number, **is a straight line containing the origin**.

EXAMPLE 3

Graph $x = 2$.

Solution

Since we are considering linear equations *in two variables*, the equation $x = 2$ is equivalent to $x + 0(y) = 2$. Any value of y can be used, but the x-value must always be 2. Therefore, some of the solutions are (2, 0), (2, 1), (2, 2), (2, −1), and (2, −2). The graph of $x = 2$ is the vertical line shown in Figure 3.22.

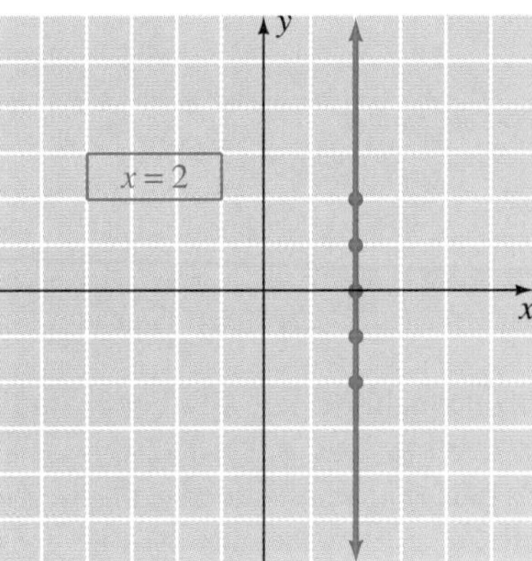

FIGURE 3.22

In general, the graph of any equation of the form $Ax + By = C$, where $A = 0$ or $B = 0$ (not both), is a line parallel to one of the axes. More specifically, **any equation of the form $x = a$**, where a is any nonzero real number, **is a line parallel to the y-axis** having an x-intercept of a. **Any equation of the form $y = b$**, where b is a nonzero real number, **is a line parallel to the x-axis** having a y-intercept of b.

Graphing Linear Inequalities

Linear inequalities in two variables are of the form $Ax + By > C$ or $Ax + By < C$, where A, B, and C are real numbers. (**Combined linear equality and inequality statements** are of the form $Ax + By \geq C$ or $Ax + By \leq C$.) Graphing linear inequalities is almost as easy as graphing linear equations. The following discussion will lead us to a simple, step-by-step process.

Let's consider the following equation and related inequalities.

$$x + y = 2, \qquad x + y > 2, \qquad x + y < 2$$

The straight line in Figure 3.23 is the graph of $x + y = 2$. The line divides the plane into two **half-planes**, one above the line and one below the line. For each point in the half-plane *above* the line, the ordered pair (x, y) associated with the

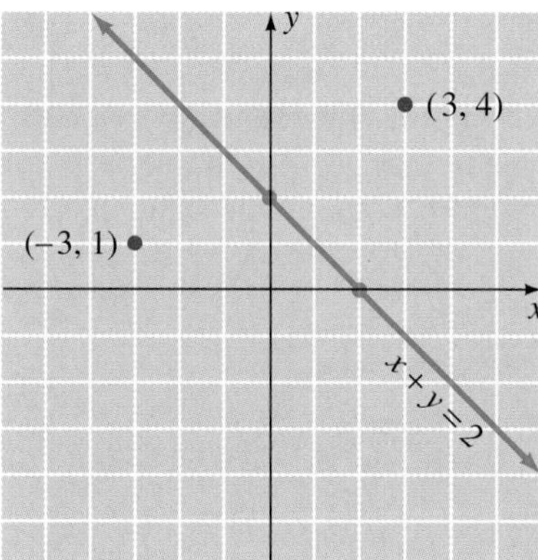

FIGURE 3.23

point satisfies the inequality $x + y > 2$. For example, the ordered pair (3, 4) produces the true statement $3 + 4 > 2$. Likewise, for each point in the half-plane *below* the line, the ordered pair (x, y) associated with the point satisfies the inequality $x + y < 2$. For example, $(-3, 1)$ produces the true statement $-3 + 1 < 2$.

Now let's use these ideas from the previous discussion to help graph some inequalities.

EXAMPLE 4

Graph $x - 2y > 4$.

Solution

First, graph $x - 2y = 4$ as a dashed line since equality is not included in $x - 2y > 4$ (Figure 3.24). Second, since *all* of the points in a specific half-plane satisfy either $x - 2y > 4$ or $x - 2y < 4$, let's try a *test point*. For example, try the origin.

$$x - 2y > 4 \quad \text{becomes } 0 - 2(0) > 4, \text{ which is a false statement.}$$

Because the ordered pairs in the half-plane containing the origin do not satisfy $x - 2y > 4$, the ordered pairs in the other half-plane must satisfy it. Therefore, the graph of $x - 2y > 4$ is the half-plane *below* the line, as indicated by the shaded portion in Figure 3.25.

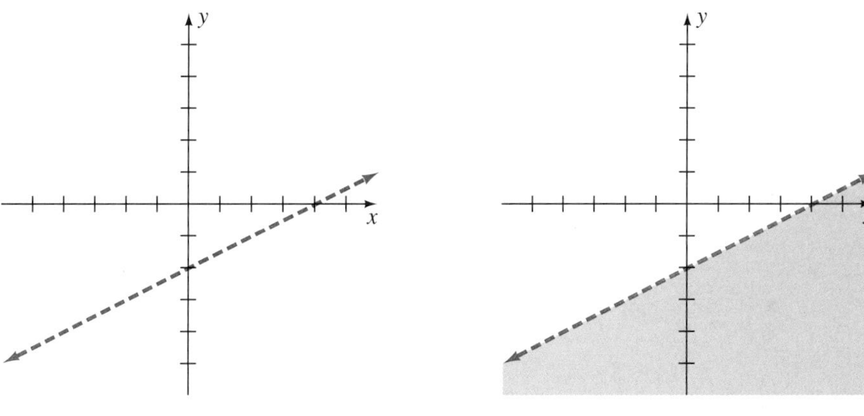

FIGURE 3.24 FIGURE 3.25

To graph a linear inequality, we suggest the following steps.

1. Graph the corresponding equality. Use a solid line if equality is included in the original statement and a dashed line if equality is not included.
2. Choose a *test point* not on the line and substitute its coordinates into the inequality. (The origin is a convenient point if it is not on the line.)
3. The graph of the original inequality is
   - **a.** the half-plane containing the test point if the inequality is satisfied by that point; or
   - **b.** the half-plane not containing the test point if the inequality is not satisfied by the point.

EXAMPLE 5

Graph $2x + 3y \geq -6$.

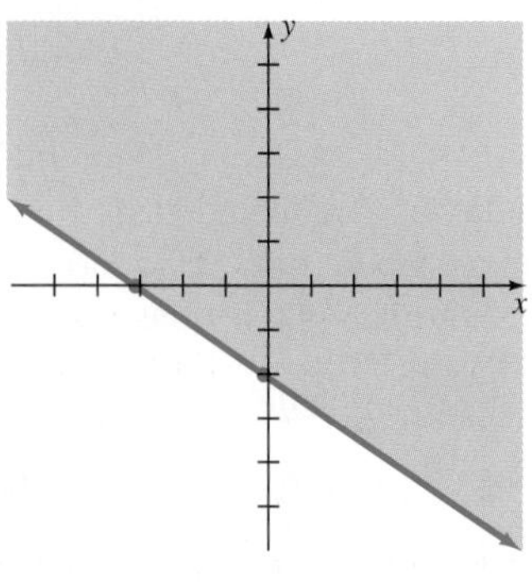

FIGURE 3.26

Solution

STEP 1 Graph $2x + 3y = -6$ as a solid line (Figure 3.26).

STEP 2 Choose the origin as a test point.

$$2x + 3y \geq -6 \quad \text{becomes } 2(0) + 3(0) \geq -6,$$

which is true.

STEP 3 Since the test point satisfies the given inequality, all points in the same half-plane as the test point satisfy it. The graph of $2x + 3y \geq -6$ is the line and the half-plane above the line (Figure 3.26).

Graphing Utilities

The term **graphing utility** is used in current literature to refer to either a graphics calculator or a computer with a graphics software package. (We will frequently use the phrase *use a graphics calculator to* . . . to mean a graphics calculator or a computer with the appropriate software.) These devices have a large range of capabilities that allows the user to not only obtain a quick sketch of a graph, but also to study various characteristics of it; for example, x-intercepts, y-intercepts, and turning points of a graph. We will introduce some of these features of graphing utilities as we need them in the text. Since there are so many different types of graphing utilities available, we will use mostly generic terminology and let you consult your user's manual for specific key punching instructions. We also suggest that you study the graphing utility examples in this text even if you do not have access to a graphics calculator or a computer. The examples were chosen to reinforce the concepts we are discussing.

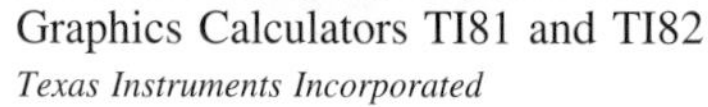
Graphics Calculators TI81 and TI82
Texas Instruments Incorporated

EXAMPLE 6

Use a graphing utility to obtain a graph of the line $2.1x + 5.3y = 7.9$.

Solution

First, we need to solve the equation for y in terms of x.

$$2.1x + 5.3y = 7.9$$

$$5.3y = 7.9 - 2.1x$$

$$y = \frac{7.9 - 2.1x}{5.3}$$

Now we can enter the expression $\frac{7.9 - 2.1x}{5.3}$ for Y_1 and obtain the graph as shown in Figure 3.27.

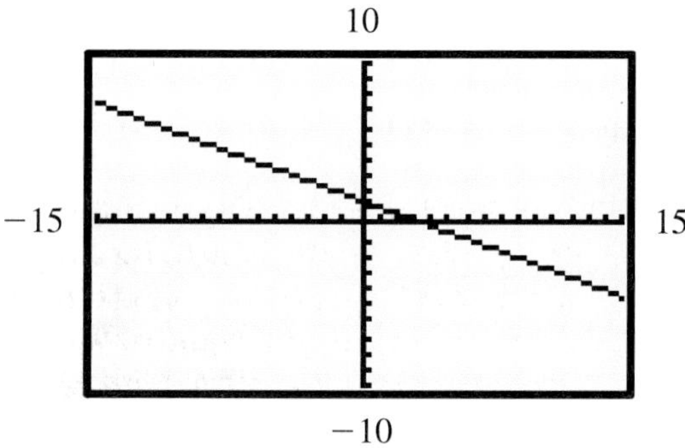

FIGURE 3.27

As indicated in Figure 3.27, the **viewing rectangle** of a graphing utility is a portion of the xy-plane shown on the display of the utility. In this display the boundaries were set so that $-15 \leq x \leq 15$ and $-10 \leq y \leq 10$. These boundaries were set automatically; however, the fact that boundaries can be assigned as necessary is an important feature of graphing utilities.

PROBLEM SET 3.2

Graph each of the following linear equations.

1. $x - 2y = 4$

2. $2x + y = -4$

3. $3x + 2y = 6$

4. $2x - 3y = 6$

5. $4x - 5y = 20$

6. $5x + 4y = 20$

7. $x - y = 3$

8. $-x + y = 4$

9. $y = 3x - 1$

10. $y = -2x + 3$

11. $y = -x$

12. $y = 4x$

13. $x = 0$

14. $y = -1$

15. $y = \frac{2}{3}x$

16. $y = -\frac{1}{2}x$

Graph each of the following linear inequalities.

17. $x + 2y > 4$

18. $2x - y < -4$

19. $3x - 2y < 6$

20. $2x + 3y < 6$

21. $2x + 5y \leq 10$

22. $4x + 5y \leq 20$

23. $y > -x - 1$

24. $y < 3x - 2$

25. $y \leq -x$

26. $y \geq x$

27. $x + 2y < 0$

28. $3x - y > 0$

29. $x > -1$

30. $y < 3$

THOUGHTS into WORDS

31. Explain how you would graph the inequality $-x + 2y > -4$.

32. What is the graph of $x = 0$ *or* $y = 0$? What is the graph of $x = 0$ *and* $y = 0$? Explain your answers.

MISCELLANEOUS PROBLEMS

From our work with absolute value, we know that $|x + y| = 4$ is equivalent to $x + y = 4$ or $x + y = -4$. Therefore, the graph of $|x + y| = 4$ is the two lines $x + y = 4$ and $x + y = -4$. Graph each of the following.

33. $|x - y| = 2$

34. $|2x + y| = 1$

35. $|x - 2y| \leq 4$

36. $|3x - 2y| \geq 6$

37. $|2x + 3y| > 6$

38. $|5x + 2y| < 10$

Using the definition of absolute value, the equation $y = |x| + 2$ becomes $y = x + 2$ for $x \geq 0$ and $y = -x + 2$ for $x < 0$. Therefore, the graph of $y = |x| + 2$ is shown in Figure 3.28. Graph each of the following.

39. $y = |x| - 1$

40. $y = |x - 2|$

41. $|y| = x$

42. $|y| = |x|$

43. $y = 2|x|$

44. $|x| + |y| = 4$

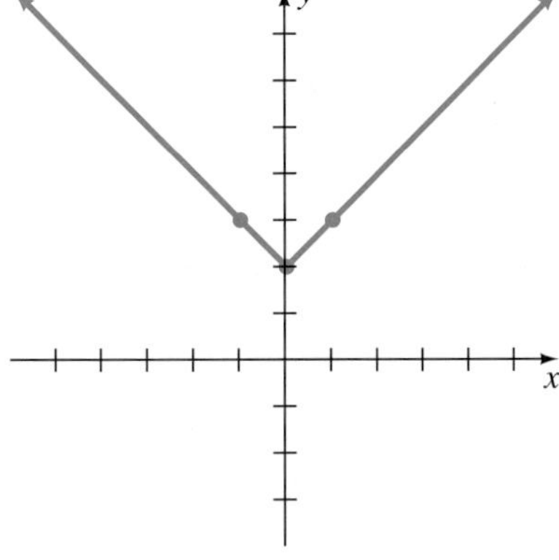

FIGURE 3.28

GRAPHICS CALCULATOR ACTIVITIES

This is the first of many appearances of a group of problems called Graphics Calculator Activities. These problems are specifically designed for those of you who have access to a graphics calculator or a computer with an appropriate software graphing package. Within the framework of these problems you will be given the opportunity to reinforce concepts discussed in the text, lay groundwork for concepts to be introduced later in the text, predict shapes and locations of graphs based on previous graphing experiences, solve problems that are unreasonable or perhaps impossible to solve without a graphing utility, and in general become familiar with the capabilities and limitations of your graphing utility.

The following problems are designed to get you started using your graphing utility and to lay some groundwork for concepts we present in the next section. Set your boundaries so that the distance between tic marks is the same on both axes.

45. a. Graph $y = 4x$, $y = 4x - 3$, $y = 4x + 2$, and $y = 4x + 5$ on the same set of axes. Do they appear to be parallel lines?

b. Graph $y = -2x + 1$, $y = -2x + 4$, $y = -2x - 2$, and $y = -2x - 5$ on the same set of axes. Do they appear to be parallel lines?

c. Graph $y = -\frac{1}{2}x + 3$, $y = -\frac{1}{2}x + 1$, $y = -\frac{1}{2}x - 1$, and $y = -\frac{1}{2}x - 4$ on the same set of axes. Do they appear to be parallel lines?

d. Graph $2x + 5y = 1$, $2x + 5y = -3$, $2x + 5y = 4$, and $2x + 5y = -5$ on the same set of axes. Do they appear to be parallel lines?

e. Graph $3x - 4y = 7$, $-3x + 4y = 8$, $3x - 4y = -2$, and $4x - 3y = 6$ on the same set of axes. Do they appear to be parallel lines?

f. Based on your results in parts (a) through (e), make a statement as to how we can recognize parallel lines from their equations.

46. a. Graph $y = 4x$ and $y = -\frac{1}{4}x$ on the same set of axes. Do they appear to be perpendicular lines?

b. Graph $y = 3x$ and $y = \frac{1}{3}x$ on the same set of axes. Do they appear to be perpendicular lines?

c. Graph $y = \frac{2}{5}x - 1$ and $y = -\frac{5}{2}x + 2$ on the same set of axes. Do they appear to be perpendicular lines?

d. Graph $y = \frac{3}{4}x - 3$, $y = \frac{4}{3}x + 2$, and $y = -\frac{4}{3}x + 2$ on the same set of axes. Does there appear to be a pair of perpendicular lines?

e. Based on your results in parts (a) through (d), make a statement as to how we can recognize perpendicular lines from their equations.

47. For each of the following pairs of equations, (1) predict if they represent parallel lines, perpendicular lines, or lines that intersect but are not perpendicular, and (2) graph each pair of lines to check your prediction.

a. $5.2x + 3.3y = 94.$ and $5.2x + 3.3y = 12.6$

b. $1.3x - 4.7y = 3.4$ and $1.3x - 4.7y = 11.6$

c. $2.7x + 3.9y = 1.4$ and $2.7x - 3.9y = 8.2$

d. $5x - 7y = 17$ and $7x + 5y = 19$

e. $9x + 2y = 14$ and $2x + 9y = 17$

f. $2.1x + 3.4y = 11.7$ and $3.4x - 2.1y = 17.3$

3.3 DETERMINING THE EQUATION OF A LINE

As we stated earlier, there are basically two types of problems in coordinate geometry: Given an algebraic equation, find its geometric graph; and given a set of conditions pertaining to a geometric figure, find its algebraic equation. In the previous section, we considered some type 1 problems; that is, we did some graphing. Now we want to consider some type 2 problems that deal specifically with straight lines; in other words, given certain facts about a line, we need to be able to determine its algebraic equation.

As we work with straight lines, it is often helpful to be able to refer to the *steepness* or *slant* of a particular line. The concept of *slope* is used as a measure of the slant of a line. The **slope** of a line is the ratio of the vertical change of distance compared to the horizontal change of distance as we move from one point on a line to another. Consider the line in Figure 3.29. From point A to point B there is a vertical change of two units and a horizontal change of three units; therefore, the slope of the line is $\frac{2}{3}$.

A precise definition for slope can be given by considering the coordinates of the points P_1, P_2, and R in Figure 3.30. The horizontal change of distance as we move from P_1 to P_2 is $x_2 - x_1$, and the vertical change is $y_2 - y_1$. Thus, we have the following definition.

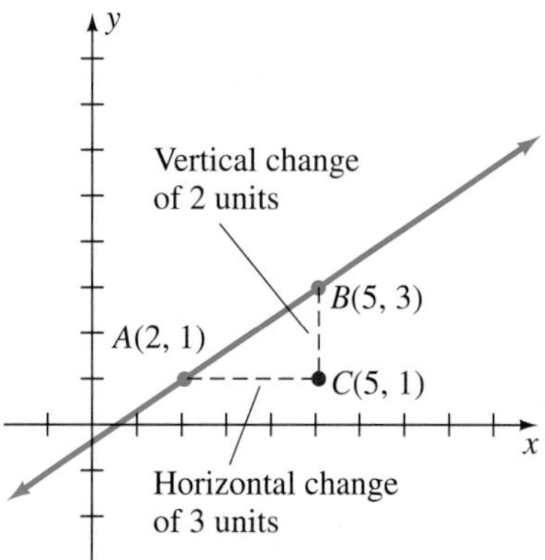

FIGURE 3.29

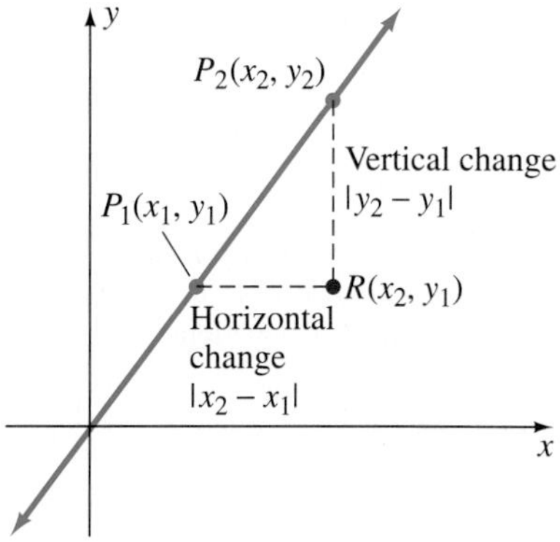

FIGURE 3.30

DEFINITION 3.1

If P_1 and P_2 are any two different points on a line, P_1 with coordinates (x_1, y_1) and P_2 with coordinates (x_2, y_2), then the **slope** of the line (denoted by m) is

$$m = \frac{y_2 - y_1}{x_2 - x_1}, \qquad x_2 \neq x_1.$$

Since

$$\frac{y_2 - y_1}{x_2 - x_1} = \frac{y_1 - y_2}{x_1 - x_2},$$

how we designate P_1 and P_2 is not important. Let's use Definition 3.1 to find the slopes of some lines.

EXAMPLE 1

Find the slope of the line determined by each of the following pairs of points and graph each line.

a. $(-1, 1)$ and $(3, 2)$ **b.** $(4, -2)$ and $(-1, 5)$ **c.** $(2, -3)$ and $(-3, -3)$

Solutions

a. Let $(-1, 1)$ be P_1 and $(3, 2)$ be P_2 (Figure 3.31).

$$m = \frac{y_2 - y_1}{x_2 - x_1} = \frac{2 - 1}{3 - (-1)} = \frac{1}{4}$$

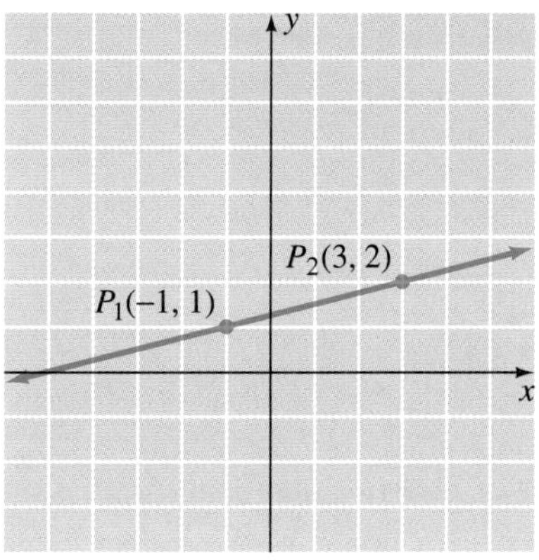

FIGURE 3.31

b. Let $(4, -2)$ be P_1 and $(-1, 5)$ be P_2 (Figure 3.32).

$$m = \frac{5 - (-2)}{-1 - 4} = \frac{7}{-5} = -\frac{7}{5}$$

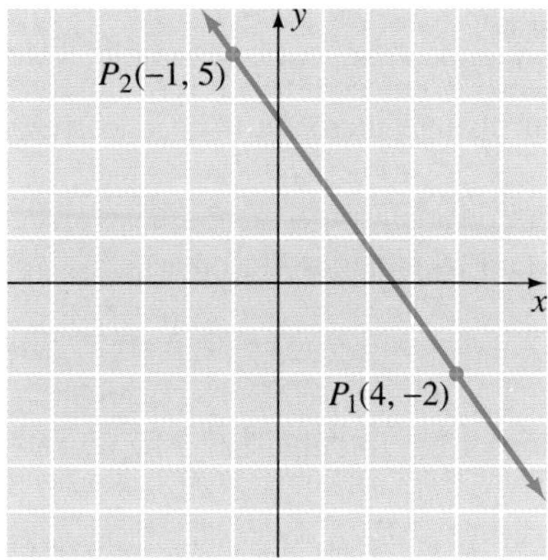

FIGURE 3.32

c. Let $(2, -3)$ be P_1 and $(-3, -3)$ be P_2 (Figure 3.33).

$$m = \frac{-3 - (-3)}{-3 - 2} = \frac{0}{-5} = 0$$

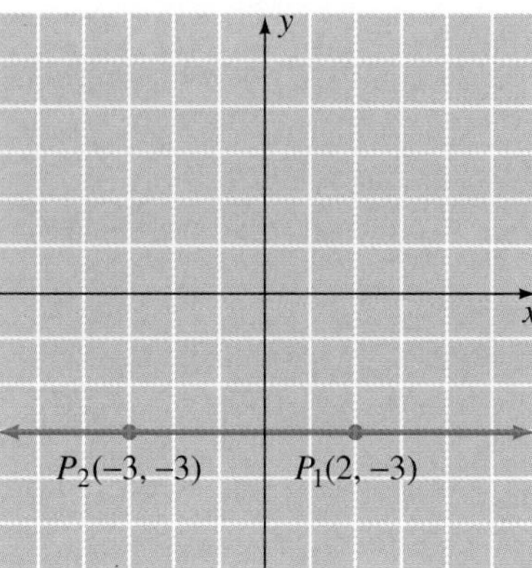

FIGURE 3.33

The three parts of Example 1 illustrate the three basic possibilities for slope; that is, the slope of a line can be positive, negative, or zero. A line that has a positive slope rises as we move from left to right, as in Figure 3.31. A line that has a negative slope falls as we move from left to right, as in Figure 3.32. A horizontal line, as in Figure 3.33, has a slope of zero. Finally, we need to realize that **the concept of slope is undefined for vertical lines**. This is due to the fact that for any vertical line, the horizontal change is zero as we move from one point on the line to another. Thus, the ratio $(y_2 - y_1)/(x_2 - x_1)$ will have a denominator of zero and be undefined. So, the restriction $x_2 \neq x_1$ is made in Definition 3.1.

Don't forget that **the slope of a line is a ratio**, the ratio of vertical change compared to horizontal change. For example, a slope of $\frac{2}{3}$ means that for every two units of vertical change there must be a corresponding three units of horizontal change.

Now let's consider some techniques for determining the equation of a line when given certain facts about the line.

EXAMPLE 2 Find the equation of the line that has a slope of $\frac{2}{5}$ and contains the point $(3, 1)$.

Solution

First, let's draw the line and record the given information (Figure 3.34).

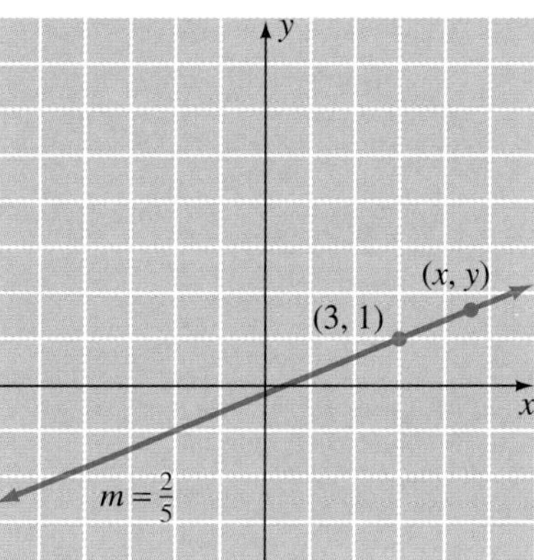

FIGURE 3.34

Then choose a point (x, y) that represents any point on the line other than the given point (3, 1). The slope determined by (3, 1) and (x, y) is to be $\frac{2}{5}$. Thus,

$$\frac{y-1}{x-3} = \frac{2}{5}$$
$$2(x-3) = 5(y-1)$$
$$2x - 6 = 5y - 5$$
$$2x - 5y = 1.$$

EXAMPLE 3

Find the equation of the line determined by (1, −2) and (−3, 4).

Solution

First, let's draw the line determined by the two given points (Figure 3.35). These two points determine the slope of the line.

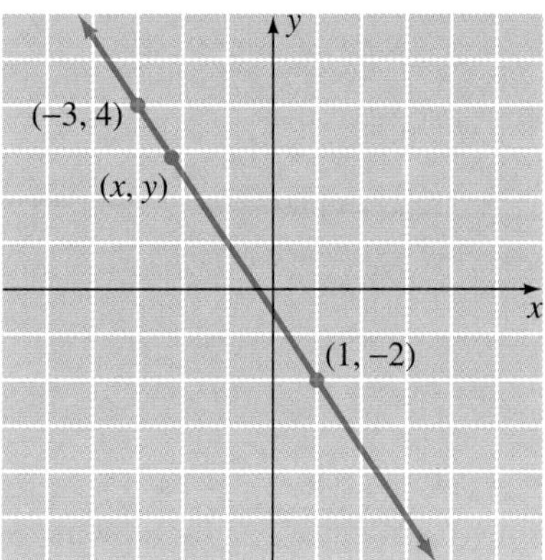

FIGURE 3.35

$$m = \frac{4-(-2)}{-3-1} = \frac{6}{-4} = -\frac{3}{2}$$

Now we can use the same approach as in Example 2. Form an equation using one of the two given points, a point (x, y), and the slope of $-\frac{3}{2}$.

$$\begin{aligned} \frac{y+2}{x-1} &= \frac{3}{-2} \\ 3(x-1) &= -2(y+2) \\ 3x-3 &= -2y-4 \\ 3x+2y &= -1 \end{aligned}$$

EXAMPLE 4

Find the equation of the line that has a slope of $\frac{1}{4}$ and a y-intercept of 2.

Solution

A y-intercept of 2 means that the point (0, 2) is on the line (Figure 3.36). Choosing a point (x, y), we can proceed as in the previous examples.

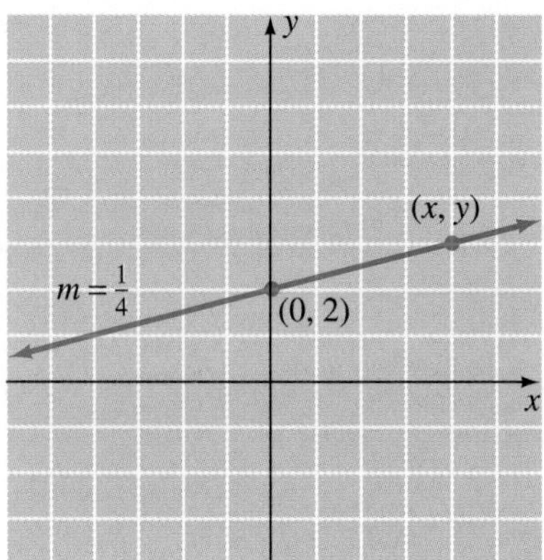

FIGURE 3.36

$$\begin{aligned} \frac{y-2}{x-0} &= \frac{1}{4} \\ 1(x-0) &= 4(y-2) \\ x &= 4y-8 \\ x-4y &= -8 \end{aligned}$$

At this point you might pause for a moment and look back over Examples 2, 3, and 4. Notice that the same basic approach is used in all three examples, that is, choosing a point (x, y) and using it to determine the equation that satisfies the conditions stated in the problem. This same approach will be used later with figures other than straight lines. Furthermore, you should realize that this approach can be used to develop some general forms of equations of straight lines.

Point-Slope Form

EXAMPLE 5

Find the equation of the line that has a slope of m and contains the point (x_1, y_1).

Solution

Choosing (x, y) to represent another point on the line (Figure 3.37), the slope of the line is given by

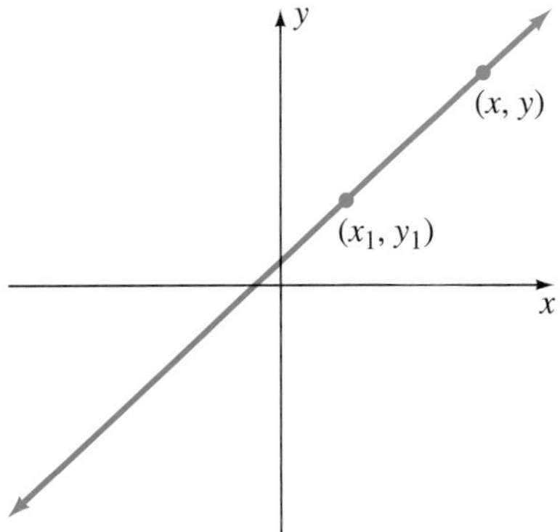

FIGURE 3.37

$$m = \frac{y - y_1}{x - x_1}, \qquad x \neq x_1,$$

from which we obtain

$$y - y_1 = m(x - x_1).$$

We refer to the equation

$$y - y_1 = m(x - x_1)$$

as the **point-slope form** of the equation of a straight line. Therefore, instead of using the approach of Example 2, we can substitute information into the point-slope form to write the equation of a line with a given slope that contains a given point. For example, the equation of the line that has a slope of $\frac{3}{5}$ and contains the point (2, 4) can be determined this way. We substitute (2, 4) for (x_1, y_1) and $\frac{3}{5}$ for m in the point-slope equation.

$$\begin{aligned} y - 4 &= \frac{3}{5}(x - 2) \\ 5(y - 4) &= 3(x - 2) \\ 5y - 20 &= 3x - 6 \\ -14 &= 3x - 5y \end{aligned}$$

Slope-Intercept Form

EXAMPLE 6

Find the equation of the line that has a slope of m and a y-intercept of b.

Solution

A y-intercept of b means that $(0, b)$ is on the line (Figure 3.38). Therefore, using the point-slope form with $(x_1, y_1) = (0, b)$, we obtain

$$\begin{aligned} y - y_1 &= m(x - x_1) \\ y - b &= m(x - 0) \\ y - b &= mx \\ y &= mx + b. \end{aligned}$$

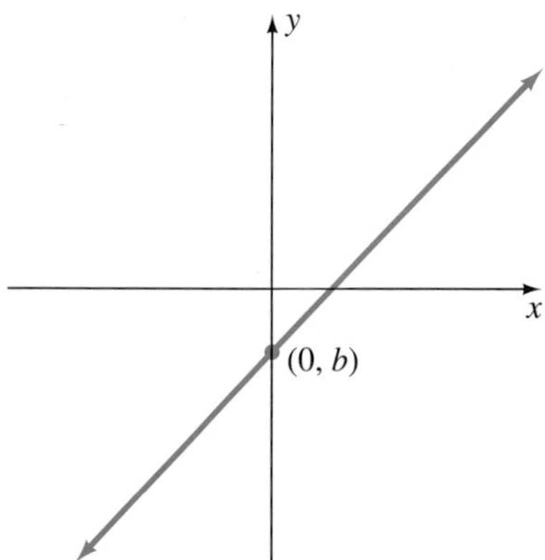

FIGURE 3.38

We refer to the equation

$$y = mx + b$$

as the **slope-intercept form** of the equation of a straight line. It can be used for two primary purposes, as the next two examples illustrate.

EXAMPLE 7

Find the equation of the line that has a slope of $\frac{1}{4}$ and a y-intercept of 2.

Solution

This is a restatement of Example 4, but this time we will use the slope-intercept form ($y = mx + b$) of the equation of a line to write its equation. Since $m = \frac{1}{4}$ and $b = 2$, we obtain

$$y = mx + b$$
$$y = \frac{1}{4}x + 2$$
$$4y = x + 8$$
$$-8 = x - 4y. \quad \text{Same result as in Example 4}$$

REMARK Sometimes we leave linear equations in slope-intercept form. We did not do so in Example 7 because we wanted to show that it was the same result as in Example 4.

EXAMPLE 8

Find the slope and y-intercept of the line that has an equation $2x - 3y = 7$.

Solution

We can solve the equation for y in terms of x and then compare the result to the general slope-intercept form.

$$2x - 3y = 7$$
$$-3y = -2x + 7$$
$$y = \frac{2}{3}x - \frac{7}{3} \qquad y = mx + b$$

The slope of the line is $\frac{2}{3}$ and the y-intercept is $-\frac{7}{3}$.

In general, **if the equation of a nonveritcal line is written in slope-intercept form, the coefficient of x is the slope of the line and the constant term is the y-intercept.**

Parallel and Perpendicular Lines

Since the concept of slope is used to indicate the slant of a line, it seems reasonable to expect slope to be related to the concepts of parallelism and perpendicularity. Such is the case, and the following two properties summarize this link.

PROPERTY 3.1

If two lines have slopes of m_1 and m_2, then

1. The two lines are parallel if and only if $m_1 = m_2$;
2. The two lines are perpendicular if and only if $m_1 m_2 = -1$.

We will test your ingenuity in devising proofs of these properties in the next problem set; however, for now we will illustrate their use.

EXAMPLE 9

a. Verify that the graphs of $3x + 2y = 9$ and $6x + 4y = 19$ are parallel lines.

b. Verify that the graphs of $5x - 3y = 12$ and $3x + 5y = 27$ are perpendicular lines.

Solution

a. Let's change each equation to slope-intercept form.

$$3x + 2y = 9 \longrightarrow 2y = -3x + 9$$
$$y = -\frac{3}{2}x + \frac{9}{2}$$
$$6x + 4y = 19 \longrightarrow 4y = -6x + 19$$
$$y = -\frac{6}{4}x + \frac{19}{4}$$
$$y = -\frac{3}{2}x + \frac{19}{4}$$

Both lines have the same slope but different y-intercepts. Therefore, the two lines are parallel.

b. Change each equation to slope-intercept form.

$$5x - 3y = 12 \longrightarrow -3y = -5x + 12$$
$$y = \frac{5}{3}x - 4$$
$$3x + 5y = 27 \longrightarrow 5y = -3x + 27$$
$$y = -\frac{3}{5}x + \frac{27}{5}$$

Because $\left(\frac{5}{3}\right)\left(-\frac{3}{5}\right) = -1$, the product of the two slopes is -1 and the lines are perpendicular.

REMARK The statement, *The product of two slopes is −1* is equivalent to saying that the two slopes are **negative reciprocals** of each other, that is, $m_1 = -1/m_2$.

EXAMPLE 10

Find the equation of the line that contains the point $(-1, 2)$ and is parallel to the line with equation $2x - y = 4$.

Solution

First, let's draw a figure to help in our analysis of the problem (Figure 3.39). Since the line through $(-1, 2)$ is to be parallel to the given line, it must have the same slope. So let's find the slope by changing $2x - y = 4$ to slope-intercept form.

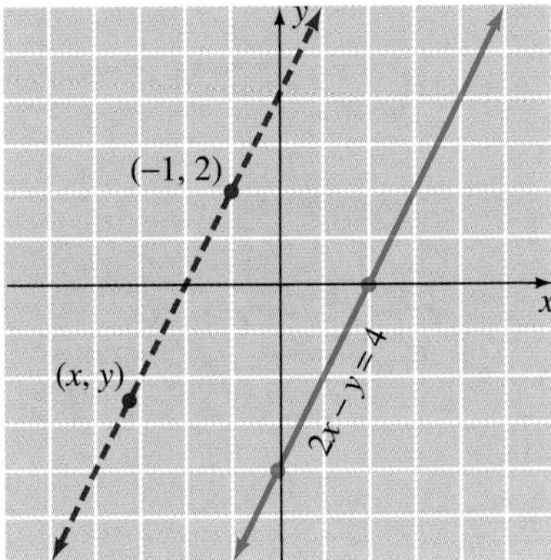

FIGURE 3.39

$$\begin{aligned} 2x - y &= 4 \\ -y &= -2x + 4 \\ y &= 2x - 4 \end{aligned}$$

The slope of both lines is 2. Now, using the point-slope form with $(x_1, y_1) = (-1, 2)$, we obtain the equation of the line.

$$\begin{aligned} y - y_1 &= m(x - x_1) \\ y - 2 &= 2[x - (-1)] \\ y - 2 &= 2(x + 1) \\ y - 2 &= 2x + 2 \\ -4 &= 2x - y \end{aligned}$$

EXAMPLE 11

Find the equation of the line that contains the point $(-1, -3)$ and is perpendicular to the line determined by $3x + 4y = 12$.

Solution

Again let's start by drawing a figure to help with our analysis (Figure 3.40). Because the line through $(-1, -3)$ is to be perpendicular to the given line, its slope must be the negative reciprocal of the slope of the line with equation $3x + 4y = 12$. So let's find the slope of $3x + 4y = 12$ by changing to slope-intercept form.

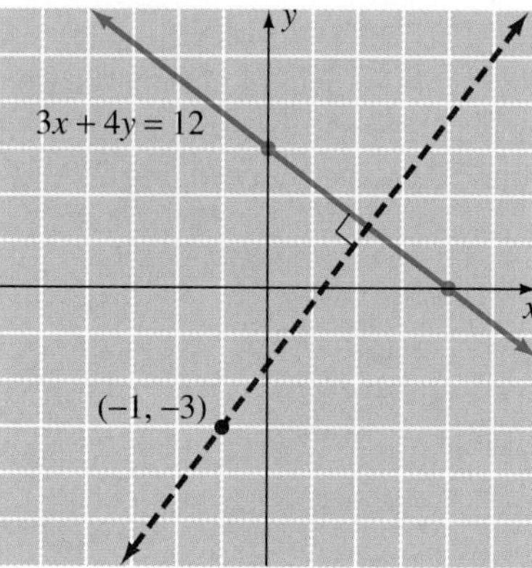

FIGURE 3.40

$$\begin{aligned} 3x + 4y &= 12 \\ 4y &= -3x + 12 \\ y &= -\frac{3}{4}x + 3 \end{aligned}$$

The slope of the desired line is $\frac{4}{3}\left(\text{the negative reciprocal of } -\frac{3}{4}\right)$ and we can proceed as before to obtain its equation.

$$\begin{aligned} y - y_1 &= m(x - x_1) \\ y - (-3) &= \frac{4}{3}[x - (-1)] \\ y + 3 &= \frac{4}{3}(x + 1) \\ 3y + 9 &= 4x + 4 \\ 5 &= 4x - 3y \end{aligned}$$

Two forms of equations of straight lines are used extensively. They are the *standard form* and the *slope-intercept form* and we can describe them as follows.

Standard Form $Ax + By = C$, where B and C are integers and A is a nonnegative integer (A and B are not both zero).

Slope-Intercept Form $y = mx + b$, where m is a real number representing the slope of the line and b is a real number representing the y-intercept.

PROBLEM SET 3.3

For Problems 1–8, find the slope of the line determined by each pair of points.

1. (3, 1) and (7, 4)
2. (−1, 2) and (5, −3)
3. (−2, −1) and (−1, −6)
4. (−2, −4) and (3, 7)
5. (−4, 2) and (−2, 2)
6. (4, −5) and (−1, −5)
7. $(a, 0)$ and $(0, b)$
8. (a, b) and (c, d)
9. Find x if the line through (−2, 4) and $(x, 6)$ has a slope of $\frac{2}{9}$.
10. Find y if the line through $(1, y)$ and (4, 2) has a slope of $\frac{5}{3}$.
11. Find x if the line through $(x, 4)$ and (2, −5) has a slope of $-\frac{9}{4}$.
12. Find y if the line through (5, 2) and $(-3, y)$ has a slope of $-\frac{7}{8}$.

For each of the following lines you are given one point and the slope of the line. Find the coordinates of three other points on the line.

13. (3, 2); $m = \frac{2}{3}$
14. (−4, 4); $m = \frac{5}{6}$
15. (−1, −4); $m = 4$
16. (−5, −3); $m = 2$
17. (2, −1); $m = -\frac{3}{5}$
18. (5, −1); $m = -\frac{2}{3}$

Write the equation of each of the following lines that has the indicated slope and contains the indicated point. Express final equations in standard form.

19. $m = \frac{1}{3}$; (2, 4)
20. $m = \frac{3}{5}$; (−1, 4)
21. $m = 2$; (−1, −2)
22. $m = -3$; (2, 5)
23. $m = -\frac{2}{3}$; (4, −3)
24. $m = -\frac{1}{5}$; (−3, 7)
25. $m = 0$; (5, −2)
26. $m = \frac{4}{3}$; (−4, −5)

Write the equation of each of the following lines that contains the indicated pair of points. Express final equations in standard form.

27. (2, 3) and (9, 8)
28. (1, −4) and (4, 4)
29. (−1, 7) and (5, 2)
30. (−3, 1) and (6, −2)
31. (4, 2) and (−1, 3)
32. (2, 7) and (2, 5)
33. (4, −3) and (−7, −3)
34. (−4, 2) and (2, −3)

Write the equation of each of the following lines that has the indicated slope (m) and y-intercept (b). Express final equations in slope-intercept form.

35. $m = \frac{1}{2}$, $b = 3$
36. $m = \frac{5}{3}$, $b = -1$
37. $m = -\frac{3}{7}$, $b = 2$
38. $m = -3$, $b = -4$
39. $m = 4$, $b = \frac{3}{2}$
40. $m = \frac{2}{3}$, $b = \frac{3}{5}$
41. $m = -\frac{5}{6}$, $b = \frac{1}{4}$
42. $m = -\frac{4}{5}$, $b = 0$

Write the equation of each of the following lines satisfying the given conditions. Express final equations in standard form.

43. The x-intercept is 4 and the y-intercept is −5.
44. Contains the point (3, −1) and is parallel to the x-axis
45. Contains the point (−4, 3) and is parallel to the y-axis
46. Contains the point (1, 2) and is parallel to the line $3x - y = 5$

47. Contains the point $(4, -3)$ and is parallel to the line $5x + 2y = 1$

48. Contains the origin and is parallel to the line $5x - 2y = 10$

49. Contains the point $(-2, 6)$ and is perpendicular to the line $x - 4y = 7$

50. Contains the point $(-3, -5)$ and is perpendicular to the line $3x + 7y = 4$

For each pair of lines in Problems 51–58, determine whether they are parallel, perpendicular, or intersecting lines that are not perpendicular.

51. $y = \frac{5}{6}x + 2$
$y = \frac{5}{6}x - 4$

52. $y = 5x - 1$
$y = -\frac{1}{5}x + \frac{2}{3}$

53. $5x - 7y = 14$
$7x + 5y = 12$

54. $2x - y = 4$
$4x - 2y = 17$

55. $4x + 9y = 13$
$-4x + y = 11$

56. $y = 5x$
$y = -5x$

57. $x + y = 0$
$x - y = 0$

58. $2x - y = 14$
$3x - y = 17$

For Problems 59–66, find the slope and the y-intercept of each of the given lines.

59. $2x - 3y = 4$

60. $3x + 4y = 7$

61. $x - 2y = 7$

62. $2x + y = 9$

63. $y = -3x$

64. $x - 5y = 0$

65. $7x - 5y = 12$

66. $-5x + 6y = 13$

67. The slope-intercept form of a line can also be used for graphing purposes. Suppose that we want to graph $y = \frac{2}{3}x + 1$. Since the y-intercept is 1, the point $(0, 1)$ is on the line. Furthermore, since the slope is $\frac{2}{3}$, another point can be found by moving two units *up* and three units to the *right*. Thus, the point $(3, 3)$ is also on the line. The two points $(0, 1)$ and $(3, 3)$ determine the line.

Use the slope-intercept form to help graph each of the following lines.

a. $y = \frac{3}{4}x + 2$

b. $y = \frac{1}{2}x - 4$

c. $y = -\frac{4}{5}x + 1$

d. $y = -\frac{2}{3}x - 6$

e. $y = -2x + \frac{5}{4}$

f. $y = x - \frac{3}{2}$

68. Use the concept of slope to verify that $(-4, 6)$, $(6, 10)$, $(10, 0)$, and $(0, -4)$ are the vertices of a square.

69. Use the concept of slope to verify that $(6, 6)$, $(2, -2)$, $(-8, -5)$, and $(-4, 3)$ are vertices of a parallelogram.

70. Use the concept of slope to verify that the triangle determined by $(4, 3)$, $(5, 1)$, and $(3, 0)$ is a right triangle.

71. Use the concept of slope to verify that the quadrilateral whose vertices are $(0, 7)$, $(-2, -1)$, $(2, -2)$, and $(4, 6)$ is a rectangle.

72. Use the concept of slope to verify that the points $(8, -3)$, $(2, 1)$, and $(-4, 5)$ lie on a straight line.

73. The midpoints of the sides of a triangle are $(-3, 4)$, $(1, -4)$, and $(7, 2)$. Find the equations of the lines that contain the sides of the triangle.

74. The vertices of a triangle are $(2, 6)$, $(5, 1)$, and $(1, -4)$. Find the equations of the lines that contain the three altitudes of the triangle. (An altitude of a triangle is the perpendicular line segment from a vertex to the opposite side.)

75. The vertices of a triangle are $(1, -6)$, $(3, 1)$ and $(-2, 2)$. Find the equations of the lines that contain the three medians of the triangle. (A median of a triangle is the line segment from a vertex to the midpoint of the opposite side.)

MISCELLANEOUS PROBLEMS

76. The concept of slope is used for highway construction. The grade of a highway, expressed as a percent, means the number of feet that the highway changes in elevation for each 100 feet of horizontal change.

a. A certain highway has a 2% grade. How many feet does it rise in a horizontal distance of 1 mile? (1 mile = 5280 feet)

b. The grade of a highway up a hill is 30%. How much change in horizontal distance is there if the vertical height of the hill is 75 feet?

77. Slope is often expressed as the ratio **rise-to-run** in the construction of stairs.

a. If the ratio rise-to-run is to be $\frac{3}{5}$ for some stairs and the rise is 19 centimeters, find the measure of the run to the nearest centimeter.

Application of slope
Larry Dale Gordon Studio/ The Image Bank

b. If the ratio rise-to-run is to be $\frac{2}{3}$ for some stairs and the run is 28 centimeters, find the rise to the nearest centimeter.

78. Suppose that a county ordinance requires a $2\frac{1}{4}\%$ fall for a sewage pipe from the house to the main pipe at the street. How much vertical drop must there be for a horizontal distance of 45 feet? Express the answer to the nearest tenth of a foot.

79. The form

$$\frac{y - y_1}{x - x_1} = \frac{y_2 - y_1}{x_2 - x_1}$$

is called the **two-point form** of the equation of a straight line. (1) Using points (x_1, y_1) and (x_2, y_2), develop the two-point form for the equation of a line. (2) Use the two-point form to write the equation of each of the following lines containing the indicated pair of points. Express the final equations in standard form.

a. (4, 3) and (5, 6) **b.** (−3, 5) and (2, −1)

c. (0, 0) and (−7, 2) **d.** (−3, −4) and (5, −1)

80. The form $(x/a) + (y/b) = 1$ is called the **intercept form** of the equation of a straight line. (1) Using a to represent the x-intercept and b the y-intercept, develop the intercept form. (2) Use the intercept form to write the equation of each of the following lines. Express the final equations in standard form.

a. $a = 2, b = 5$ **b.** $a = -3, b = 1$

c. $a = 6, b = -4$ **d.** $a = -1, b = -2$

81. Prove each of the following statements.

a. Two nonvertical parallel lines have the same slope.

b. Two lines with the same slope are parallel.

c. If two nonvertical lines are perpendicular, then their slopes are negative reciprocals of each other.

d. If the slopes of two lines are negative reciprocals of each other, then the lines are perpendicular.

82. Let $Ax + By = C$ and $A'x + B'y = C'$ represent two lines. Verify each of the following properties.

a. If $(A/A') = (B/B') \neq (C/C')$, then the lines are parallel.

b. If $AA' = -BB'$, then the lines are perpendicular.

83. The properties in Problem 82 give us another way to write the equation of a line parallel or perpendicular to a given line through a point not on the given line. For example, suppose that we want the equation of the line perpendicular to $3x + 4y = 6$ that contains the point (1, 2). The form $4x - 3y = k$, where k is a constant, represents a family of lines perpendicular to $3x + 4y = 6$ because we have satisfied the condition $AA' = -BB'$. Therefore, to find the specific line of the family containing (1, 2), we substitute 1 for x and 2 for y to determine k.

$$\begin{aligned} 4x - 3y &= k \\ 4(1) - 3(2) &= k \\ -2 &= k \end{aligned}$$

Thus, the equation of the desired line is $4x - 3y = -2$. Use the properties from Problem 82 to help write the equation of each of the following lines.

a. Contains (5, 6) and is parallel to the line $2x - y = 1$

b. Contains (−3, 4) and is parallel to the line $3x + 7y = 2$

c. Contains (2, −4) and is perpendicular to the line $2x - 5y = 9$

d. Contains (−3, −5) and is perpendicular to the line $4x + 6y = 7$

84. Some real world situations can be described by the use of linear equations in two variables. If two pairs of values are known, then the equation can be determined by using the approach we used in Example 2 of this section. For each of the following, assume that the relationship can be expressed as a linear equation in two variables, and use the given information to determine the equation. Express the equation in standard form.

a. A company produces 10 fiberglass shower stalls for \$2015 and 15 stalls for \$3015. Let y be the cost and x the number of stalls.

b. A company can produce 6 boxes of candy for \$8 and 10 boxes of candy for \$13. Let y represent the cost and x the number of boxes of candy.

c. Two banks on opposite corners of a town square have signs displaying the up-to-date temperature. One bank displays the temperature in Celsius degrees and the other in Fahrenheit. A temperature of 10°C was displayed at the same time as a temperature of 50°F. On another day, a temperature of −5°C was displayed at the same time as a temperature of 23°F. Let y represent the temperature in Fahrenheit and x the temperature in Celsius.

85. The relationships that tie slope to parallelism and perpendicularity provide fire power for constructing coordinate geometry proofs. (See Problem 81 on Page 189.) Prove each of the following using a coordinate geometry approach.

a. The diagonals of a square are perpendicular.

b. The line segment joining the midpoints of two sides of a triangle is parallel to the third side.

c. The line segments joining successive midpoints of the sides of a quadrilateral form a parallelogram.

d. The line segments joining successive midpoints of the sides of a rectangle form a rhombus. (A rhombus is a parallelogram with all sides of the same length.)

3.4 MORE ON GRAPHING

As we stated earlier, it is very helpful to recognize that a certain type of equation produces a particular kind of graph. In a later chapter we will pursue that idea in much more detail. However, we also need to develop some general graphing techniques to use with equations where we do not recognize the graph. Let's begin with the following suggestions and then add to the list throughout the remainder of the text. (Even though you may recognize some of the graphs in this section from previous graphing experiences, keep in mind that the primary objective at this time is the development of some additional graphing techniques.)

1. Find the intercepts.
2. Solve the equation for y in terms of x or for x in terms of y if it is not already in such a form.
3. Set up a table of ordered pairs that satisfy the equation.
4. Plot the points associated with the ordered pairs and connect them with a smooth curve.

EXAMPLE 1

Graph $y = x^2 - 4$.

Solution

First, let's find the intercepts. If $x = 0$, then

$$y = 0^2 - 4$$
$$y = -4.$$

This determines the point $(0, -4)$. If $y = 0$, then

$$0 = x^2 - 4$$
$$4 = x^2$$
$$\pm 2 = x.$$

Thus, the points $(2, 0)$ and $(-2, 0)$ are determined.

Second, since the given equation expresses y in terms of x, the form is convenient for setting up a table of ordered pairs. Plotting these points and connecting them with a smooth curve produces Figure 3.41.

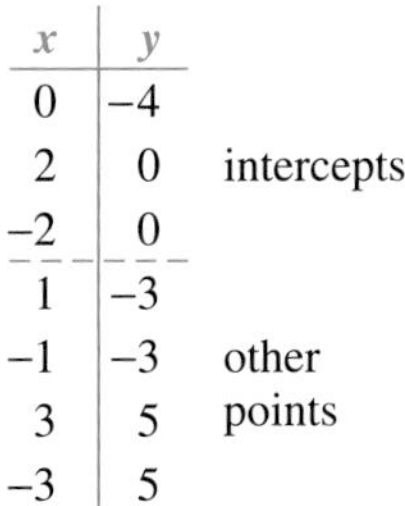

x	y	
0	−4	
2	0	intercepts
−2	0	
1	−3	
−1	−3	other
3	5	points
−3	5	

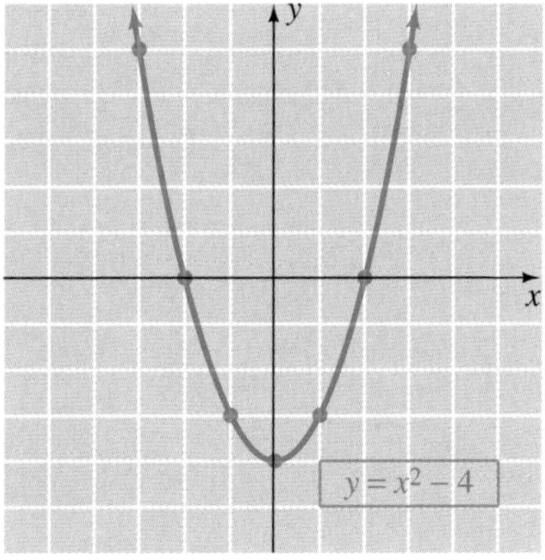

FIGURE 3.41

The curve in Figure 3.41 is said to be **symmetric with respect to the y-axis**. Stated another way, each half of the curve is a mirror image of the other half through the y-axis. Notice in the table of values that for each ordered pair (x, y), the ordered pair $(-x, y)$ is also a solution. Thus, a general test for y-axis symmetry can be stated as follows.

y-Axis Symmetry The graph of an equation is symmetric with respect to the y-axis if replacing x with $-x$ results in an equivalent equation.

So the equation $y = x^2 - 4$ exhibits y-axis symmetry because replacing x with $-x$ produces $y = (-x)^2 - 4 = x^2 - 4$. Likewise, the equations $y = x^2 + 6$, $y = x^4$, and $y = x^4 + 2x^2$ exhibit y-axis symmetry.

EXAMPLE 2

Graph $x - 1 = y^2$.

Solution

If $x = 0$, then

$$0 - 1 = y^2$$
$$-1 = y^2.$$

The equation $y^2 = -1$ has no real number solutions; therefore, this graph has no points on the y-axis. If $y = 0$, then

$$x - 1 = 0$$
$$x = 1.$$

So the point (1, 0) is determined.

Solving the original equation for x produces $x = y^2 + 1$, for which the table of values is easily determined. Plotting these points and connecting them with a smooth curve produces Figure 3.42.

x	y	
1	0	intercept
2	1	
2	−1	other
5	2	points
5	−2	

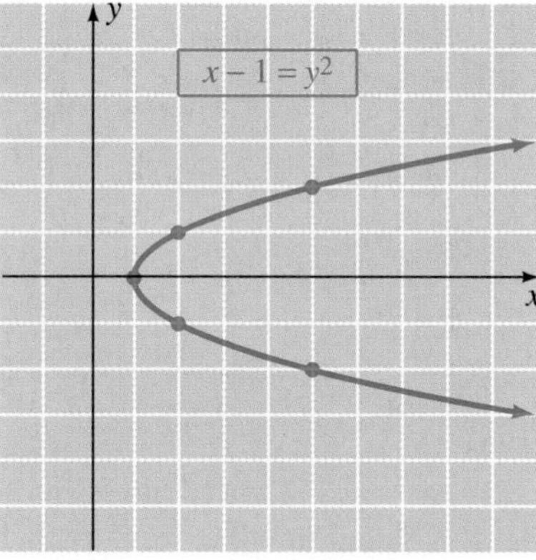

FIGURE 3.42

The curve in Figure 3.42 is said to be **symmetric with respect to the x-axis**. That is to say, each half of the curve is a mirror image of the other half through the x-axis. Notice in the table of values that for each ordered pair (x, y), the ordered pair $(x, -y)$ is also a solution. The following general test for x-axis symmetry can be stated.

x-Axis Symmetry The graph of an equation is symmetric with respect to the x-axis if replacing y with $-y$ results in an equivalent equation.

Thus, the equation $x - 1 = y^2$ exhibits x-axis symmetry because replacing y with $-y$ produces $x - 1 = (-y)^2 = y^2$. Likewise, the equations $x = y^2$, $x = y^4 + 2$, and $x^3 = y^2$ exhibit x-axis symmetry.

EXAMPLE 3

Graph $y = x^3$.

Solution

If $x = 0$, then

$$y = 0^3 = 0.$$

Thus, the origin (0, 0) is on the graph.

The table of values is easily determined from the equation. Plotting these points and connecting them with a smooth curve produces Figure 3.43.

x	y	
0	0	intercept
1	1	
2	8	other
–1	–1	points
–2	–8	

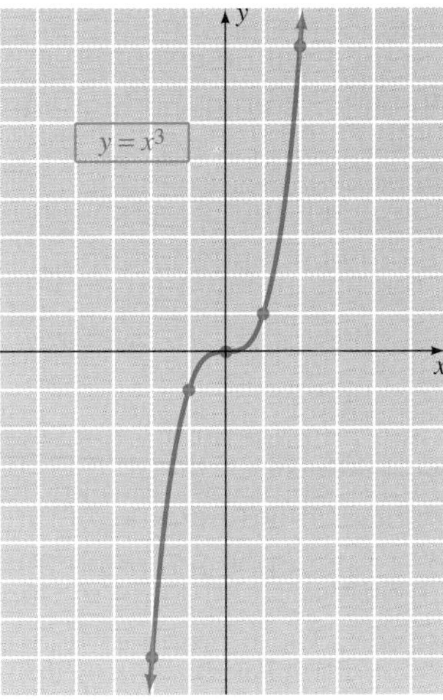

FIGURE 3.43

The curve in Figure 3.43 is said to be **symmetric with respect to the origin**. Each half of the curve is a mirror image of the other half through the origin. In the table of values we see that for each ordered pair (x, y), the ordered pair $(-x, -y)$ is also a solution. The following general test for origin symmetry can be stated.

Origin Symmetry The graph of an equation is symmetric with respect to the origin if replacing x with $-x$ and y with $-y$ results in an equivalent equation.

The equation $y = x^3$ exhibits origin symmetry because replacing x with $-x$ and y with $-y$ produces $-y = -x^3$, which is equivalent to $y = x^3$. (Multiplying both sides of $-y = -x^3$ by -1 produces $y = x^3$.) Likewise, the equations $xy = 4$, $x^2 + y^2 = 10$, and $4x^2 - y^2 = 12$ exhibit origin symmetry.

REMARK From the symmetry tests, we should observe that if a curve has both x-axis and y-axis symmetry, then it must have origin symmetry. However, it is possible for a curve to have origin symmetry and not be symmetrical to either axis. Figure 3.43 is an example of such a curve.

Another graphing consideration is that of **restricting a variable** to ensure real number solutions. The following example illustrates this point.

EXAMPLE 4

Graph $y = \sqrt{x - 1}$.

Solution

The radicand, $x - 1$, must be nonnegative. Therefore,

$$x - 1 \geq 0$$
$$x \geq 1.$$

The restriction $x \geq 1$ indicates that there is no y-intercept. The x-intercept can be found as follows: If $y = 0$, then

$$0 = \sqrt{x - 1}$$
$$0 = x - 1$$
$$1 = x.$$

The point $(1, 0)$ is on the graph.

Now, keeping the restriction in mind, the table of values can be determined. Plotting these points and connecting them with a smooth curve produces Figure 3.44.

x	y	
1	0	intercept
2	1	other points
5	2	
10	3	

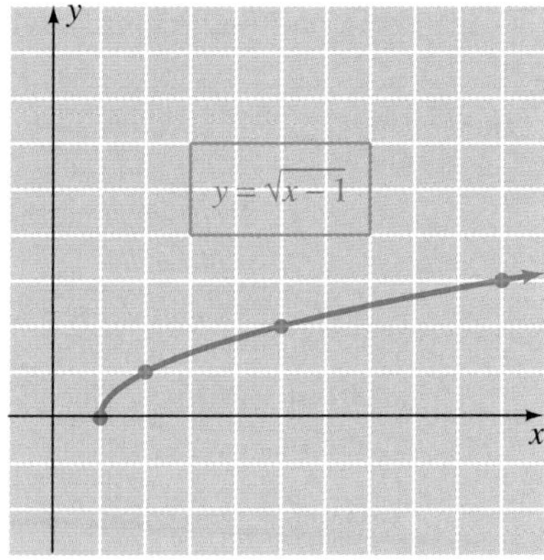

FIGURE 3.44

Now let's restate and add the concepts of symmetry and restrictions to the list of graphing suggestions. The order of the suggestions also indicates the order in which we usually attack a graphing problem if it is a new graph, that is, one that we do not recognize from its equation.

1. Determine the type of symmetry that the equation exhibits.
2. Find the intercepts.
3. Solve the equation for y in terms of x or for x in terms of y, if it is not already in such a form.
4. Determine the necessary restrictions so as to ensure real number solutions.
5. Set up a table of ordered pairs that satisfy the equation. The type of symmetry and the restrictions will affect your choice of values in the table.
6. Plot the points associated with the ordered pairs and connect them with a smooth curve. Then, if appropriate, reflect this curve according to the symmetry possessed by the graph.

The final two examples of this section should help you pull these ideas together and demonstrate the power of having these techniques at your fingertips.

EXAMPLE 5

Graph $x = -y^2 - 3$.

Solution

Symmetry The graph is symmetric with respect to the x-axis because replacing y with $-y$ produces $x = -(-y)^2 - 3$, which is equivalent to $x = -y^2 - 3$.

Intercepts If $x = 0$, then

$$\begin{aligned} 0 &= -y^2 - 3 \\ y^2 &= -3. \end{aligned}$$

Therefore, the graph contains no points on the y-axis. If $y = 0$, then

$$x = -0^2 - 3$$
$$x = -3.$$

Therefore, the point $(-3, 0)$ is on the graph.

Restrictions Since $x = -y^2 - 3$, y can take on any real number value and for every value of y, x will be less than or equal to -3.

Table of Values Because of the x-axis symmetry, let's choose only nonnegative values for y.

Plotting the Graph Plotting the points determined by the table and connecting them with a smooth curve produces Figure 3.45(a). Then reflecting that portion of the curve across the x-axis produces the complete curve in Figure 3.45(b).

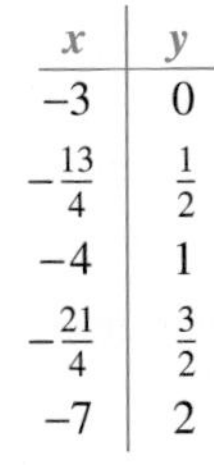

x	y
-3	0
$-\frac{13}{4}$	$\frac{1}{2}$
-4	1
$-\frac{21}{4}$	$\frac{3}{2}$
-7	2

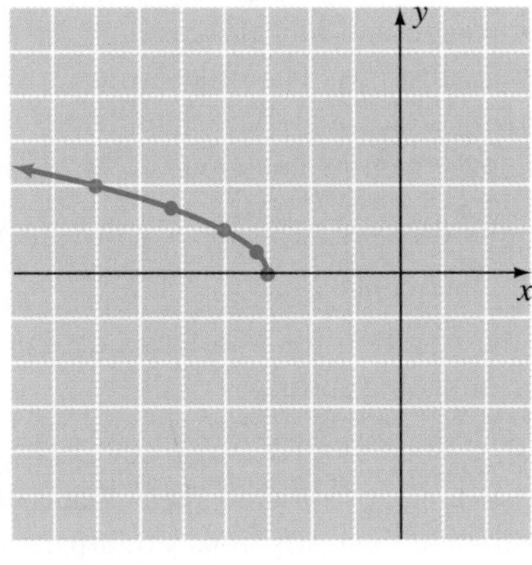

(a)

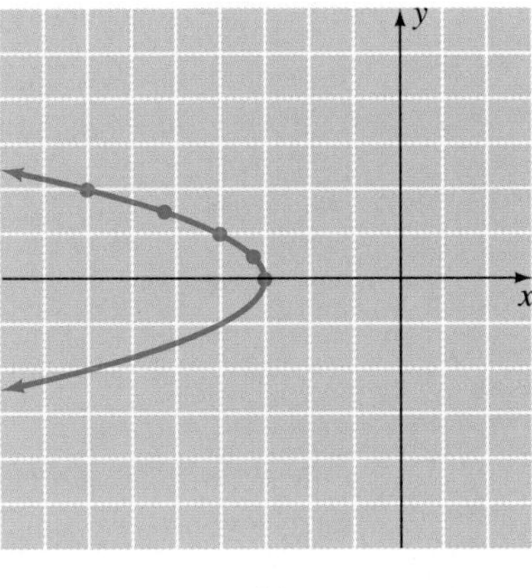

(b)

FIGURE 3.45

EXAMPLE 6

Graph $x^2 - y^2 = 4$.

Solution

Symmetry The graph is symmetric with respect to both axes and the origin because replacing x with $-x$ and y with $-y$ produces $(-x)^2 - (-y)^2 = 4$, which is equivalent to $x^2 - y^2 = 4$.

Intercepts If $x = 0$, then

$$0^2 - y^2 = 4$$
$$-y^2 = 4$$
$$y^2 = -4.$$

Therefore, the graph contains no points on the y-axis. If $y = 0$, then

$$\begin{aligned} x^2 - 0^2 &= 4 \\ x^2 &= 4 \\ x &= \pm 2. \end{aligned}$$

Therefore, the points $(2, 0)$ and $(-2, 0)$ are on the graph.

Restrictions Solving the given equation for y produces

$$\begin{aligned} x^2 - y^2 &= 4 \\ -y^2 &= 4 - x^2 \\ y^2 &= x^2 - 4 \\ y &= \pm\sqrt{x^2 - 4}. \end{aligned}$$

Therefore, $x^2 - 4 \geq 0$, which is equivalent to $x \geq 2$ or $x \leq -2$.

Table of Values Because of the restrictions and symmetries, we need only choose values corresponding to $x \geq 2$.

Plotting the Graph Plotting the points in the table of values and connecting them with a smooth curve produces Figure 3.46(a). Because of the symmetry with respect to both axes and the origin, the portion of the curve in Figure 3.46(a) can be reflected across both axes and through the origin to produce the complete curve in Figure 3.46(b).

x	y
2	0
3	$\sqrt{5} \approx 2.2$
4	$2\sqrt{3} \approx 3.5$
5	$\sqrt{21} \approx 4.6$
6	$4\sqrt{2} \approx 5.7$

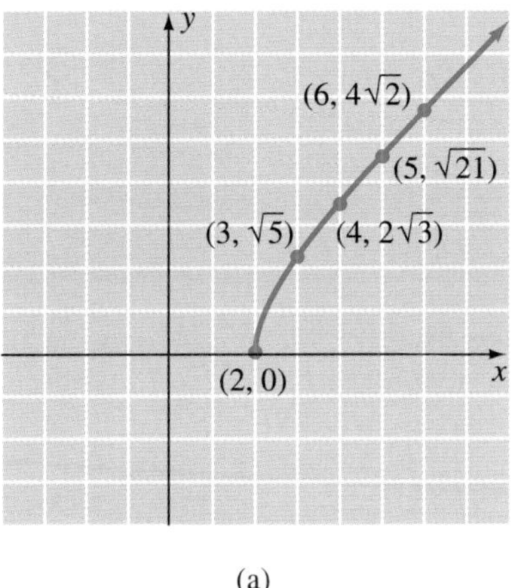

(a)

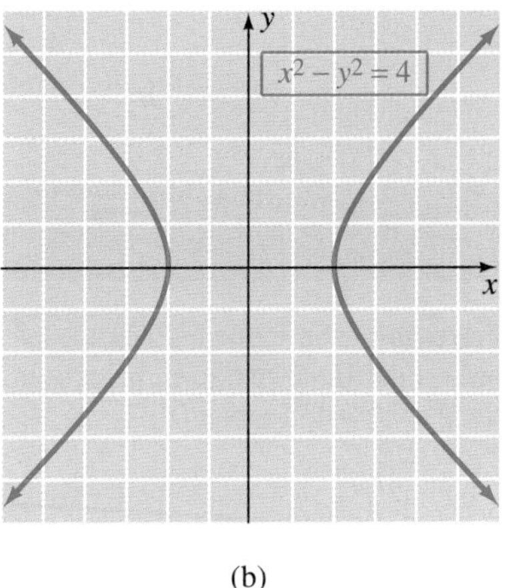

(b)

FIGURE 3.46

Even when using a graphing utility, it is often helpful to determine symmetry, intercepts, and restrictions before graphing the equation. This can serve as a partial check against using the utility incorrectly.

EXAMPLE 7

Use a graphing utility to obtain the graph of $y = \sqrt{x^2 - 49}$.

Solution

Symmetry The graph is symmetric with respect to the y-axis because replacing x with $-x$ produces the same equation.

Intercepts If $x = 0$, then $y = \sqrt{-49}$; so the graph has no points on the y-axis. If $y = 0$, then $x = \pm 7$; so the points (7, 0) and (−7, 0) are on the graph.

Restrictions Since $x^2 - 49$ has to be nonnegative, we know that $x \leq -7$ or $x \geq 7$. Now let's enter the expression $\sqrt{x^2 - 49}$ for Y_1 and obtain the graph in Figure 3.47. Note that the graph does exhibit the symmetry, intercepts, and restrictions that we determined earlier.

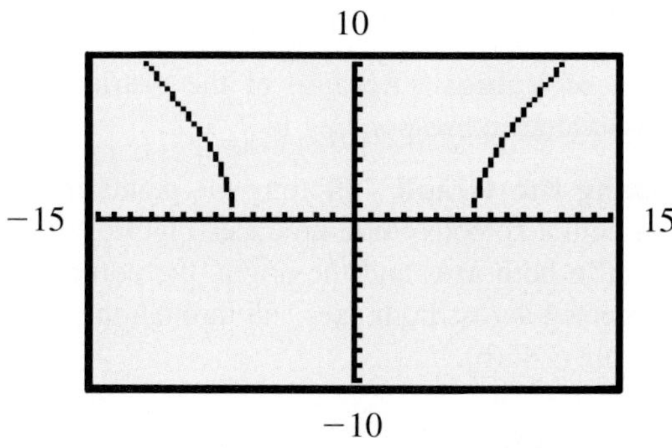

FIGURE 3.47

PROBLEM SET 3.4

For each of the following points, determine the points that are symmetric to the given point with respect to the x-axis, the y-axis, and the origin.

1. (4, 3) **2.** (−2, 5) **3.** (−6, −1)

4. (3, −7) **5.** (0, 4) **6.** (−5, 0)

Determine the type of symmetry (x-axis, y-axis, origin) possessed by each of the following graphs. Do not sketch the graph.

7. $y = x^2 - 6$ **8.** $x = y^2 + 1$

9. $x^3 = y^2$ **10.** $x^2y^2 = 4$

11. $x^2 + 2y^2 = 6$ **12.** $3x^2 - y^2 + 4x = 6$

13. $x^2 - 2x + y^2 - 3y - 4 = 0$

14. $xy = 4$ **15.** $y = x$

16. $2x - 3y = 15$ **17.** $y = x^3 + 2$

18. $y = x^4 + x^2$

19. $5x^2 - y^2 + 2y - 1 = 0$

20. $x^2 + y^2 - 2y - 4 = 0$

Use symmetry, intercepts, restrictions, and point plotting to help graph each of the following.

21. $y = x^2$ **22.** $y = -x^2$

23. $y = x^2 + 2$ **24.** $y = -x^2 - 1$

25. $xy = 4$

26. $xy = -2$

27. $y = -x^3$

28. $y = x^3 + 2$

29. $y^2 = x^3$

30. $y^3 = x^2$

31. $y^2 - x^2 = 4$

32. $x^2 - 2y^2 = 8$

33. $y = -\sqrt{x}$

34. $y = \sqrt{x + 1}$

35. $x^2y = 4$

36. $xy^2 = 4$

37. $x^2 + 2y^2 = 8$

38. $2x^2 + y^2 = 4$

39. $y = \dfrac{4}{x^2 + 1}$

40. $y = \dfrac{-2}{x^2 + 1}$

41. $y = \sqrt{x - 2}$

42. $y = 3 - x$

43. $-xy = 3$

44. $-x^2y = 4$

45. $x = y^2 + 2$

46. $x = -y^2 + 4$

47. $x = -y^2 - 1$

48. $x = y^2 - 3$

THOUGHTS into WORDS

49. How does the concept of symmetry help when graphing equations?

50. Explain how you would go about graphing $x^2y^2 = 4$.

GRAPHICS CALCULATOR ACTIVITIES

51. Graph $y = \dfrac{4}{x^2}$, $y = \dfrac{4}{(x - 2)^2}$, $y = \dfrac{4}{(x - 4)^2}$, and $y = \dfrac{4}{(x + 2)^2}$ on the same set of axes. Now predict the graph for $y = \dfrac{4}{(x - 6)^2}$. Check your prediction.

52. Graph $y = \sqrt{x}$, $y = \sqrt{x + 1}$, $y = \sqrt{x - 2}$, and $y = \sqrt{x - 4}$ on the same set of axes. Now predict the graph for $y = \sqrt{x + 3}$. Check your prediction.

53. Graph $y = \sqrt{x}$, $y = 2\sqrt{x}$, $y = 4\sqrt{x}$, and $y = 7\sqrt{x}$ on the same set of axes. How does the constant in front of the radical seem to affect the graph?

54. Graph $y = \dfrac{8}{x^2}$ and $y = -\dfrac{8}{x^2}$ on the same set of axes. How does the negative sign seem to affect the graph?

55. Graph $y = \sqrt{x}$ and $y = -\sqrt{x}$ on the same set of axes. How does the negative sign seem to affect the graph?

56. Graph $y = \sqrt{x}$, $y = \sqrt{x} + 2$, $y = \sqrt{x} + 4$, and $y = \sqrt{x} - 3$ on the same set of axes. How does the constant term seem to affect the graphs?

57. Graph $y = \sqrt{x}$, $y = \sqrt{x + 3}$, $y = \sqrt{x - 1}$, and $y = \sqrt{x - 5}$ on the same set of axes. How are the graphs related? Predict the location of $y = \sqrt{x + 5}$. Check your prediction.

58. To graph $x = y^2$ we need to first solve for y in terms of x. This produces $y = \pm\sqrt{x}$. Now we can let $Y_1 = \sqrt{x}$ and $Y_2 = -\sqrt{x}$ and graph the two equations on the same set of axes. Then graph $x = y^2 + 4$ on this same set of axes. How are the graphs related? Predict the location of the graph of $x = y^2 - 4$. Check your prediction.

59. To graph $x = y^2 + 2y$ we need to first solve for y in terms of x. Let's complete the square to do this.

$$y^2 + 2y = x$$
$$y^2 + 2y + 1 = x + 1$$
$$(y + 1)^2 = \left(\sqrt{x + 1}\right)^2$$
$$y + 1 = \sqrt{x + 1} \quad \text{or} \quad y + 1 = -\sqrt{x + 1}$$
$$y = -1 + \sqrt{x + 1} \quad \text{or} \quad y = -1 - \sqrt{x + 1}$$

So let's make the assignments $Y_1 = -1 + \sqrt{x + 1}$ and $Y_2 = -1 - \sqrt{x + 1}$ and graph them on the same set of axes to produce the graph of $x = y^2 + 2y$. Then graph $x = y^2 + 2y - 4$ on this same set of axes. Now predict the location of the graph of $x = y^2 + 2y + 4$. Check your prediction.

3.5 CIRCLES, ELLIPSES, AND HYPERBOLAS

When we apply the distance formula

$$d = \sqrt{(x_2 - x_1)^2 + (y_2 - y_1)^2},$$

(developed in Section 3.1) to the definition of a circle, we get what is known as the **standard form of the equation of a circle**. We start with a precise definition of a circle.

DEFINITION 3.2

A **circle** is the set of all points in a plane equidistant from a given fixed point called the **center**. A line segment determined by the center and any point on the circle is called a **radius**.

Now let's consider a circle that has a radius of length r and a center at (h, k) on a coordinate system (Figure 3.48). For any point P on the circle with coordinates (x, y), the length of a radius, denoted by r, can be expressed

$$r = \sqrt{(x - h)^2 + (y - k)^2}.$$

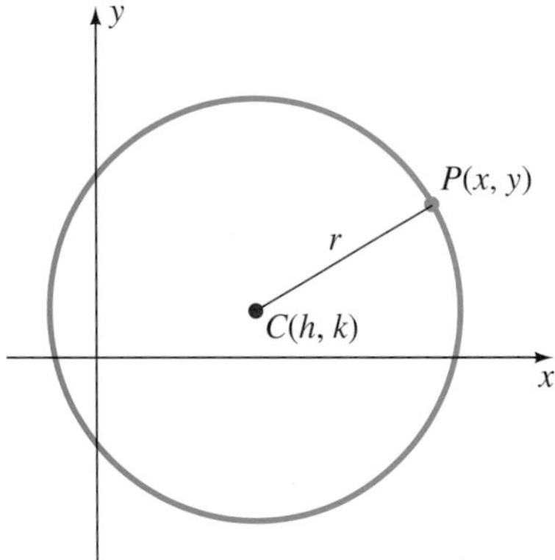

FIGURE 3.48

Squaring both sides of this equation, we obtain the standard form of the equation of a circle.

$$(x - h)^2 + (y - k)^2 = r^2$$

This form of the equation of a circle can be used to solve the two basic kinds of problems: (1) given the coordinates of the center of a circle and the length of a radius of a circle, find its equation; (2) given the equation of a circle, determine its graph. Let's illustrate each of these types of problems.

EXAMPLE 1

Find the equation of a circle that has its center at $(-3, 5)$ and has a radius of length four units.

Solution

Substitute -3 for h, 5 for k, and 4 for r in the standard equation and simplify to give us the equation of the circle.

$$\begin{aligned}
(x - h)^2 + (y - k)^2 &= r^2 \\
[x - (-3)]^2 + (y - 5)^2 &= 4^2 \\
(x + 3)^2 + (y - 5)^2 &= 4^2 \\
x^2 + 6x + 9 + y^2 - 10y + 25 &= 16 \\
x^2 + y^2 + 6x - 10y + 18 &= 0
\end{aligned}$$

Notice in Example 1 that we simplified the equation to the form $x^2 + y^2 + Dx + Ey + F = 0$, where D, E, and F are constants. This is another commonly used form when working with circles.

EXAMPLE 2

Graph $x^2 + y^2 - 6x + 4y + 9 = 0$.

Solution

We can change the given equation into the standard form for a circle by completing the square on x and on y.

$$\begin{aligned}
x^2 + y^2 - 6x + 4y + 9 &= 0 \\
(x^2 - 6x \quad\;) + (y^2 + 4y \quad\;) &= -9 \\
(x^2 - 6x + \underset{\uparrow}{9}) + (y^2 + 4y + \underset{\uparrow}{4}) &= -9 + \underset{\uparrow}{9} + \underset{\uparrow}{4}
\end{aligned}$$

Add 9 to complete the square on x.

Add 4 to complete the square on y.

Add 9 and 4 to compensate for the 4 and 9 added on the left side.

$$\begin{aligned}
(x - 3)^2 + (y + 2)^2 &= 2^2 \\
(x - \underset{\substack{\uparrow \\ h}}{3})^2 + [y - (\underset{\substack{\uparrow \\ k}}{-2})]^2 &= \underset{\substack{\uparrow \\ r}}{2}{}^2
\end{aligned}$$

The center is at (3, −2) and the length of a radius is two units. The circle is drawn in Figure 3.49.

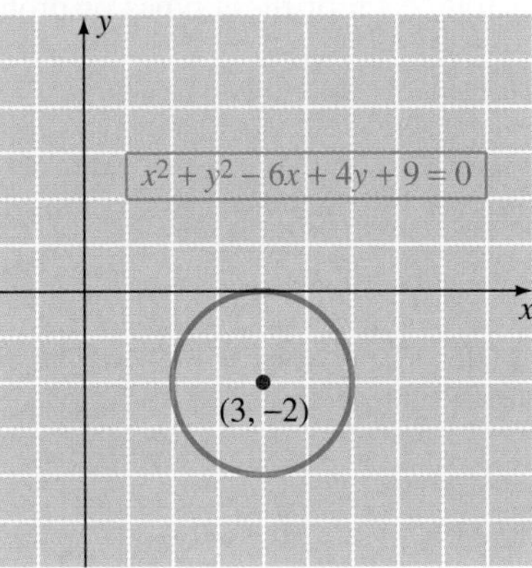

FIGURE 3.49

EXAMPLE 3

Find the center and length of a radius of the circle

$$4x^2 + 4x + 4y^2 - 12y - 26 = 0.$$

Solution

$$4x^2 + 4x + 4y^2 - 12y - 26 = 0$$

$$4(x^2 + x + \_) + 4(y^2 - 3y + \_) = 26$$

$$4\left(x^2 + x + \frac{1}{4}\right) + 4\left(y^2 - 3y + \frac{9}{4}\right) = 26 + 1 + 9$$

$$4\left(x + \frac{1}{2}\right)^2 + 4\left(y - \frac{3}{2}\right)^2 = 36$$

$$\left(x + \frac{1}{2}\right)^2 + \left(y - \frac{3}{2}\right)^2 = 9$$

$$\left[x - \left(-\frac{1}{2}\right)^2 + \left(y - \frac{3}{2}\right)^2\right] = 3^2$$

$$\uparrow h \qquad \uparrow k \qquad \uparrow r$$

Therefore, the center is at $\left(-\frac{1}{2}, \frac{3}{2}\right)$ and the length of a radius is 3 units.

Now suppose that we substitute 0 for h and 0 for k in the standard form of the equation of a circle.

$$\begin{aligned}(x - h)^2 + (y - k)^2 &= r^2\\(x - 0)^2 + (y - 0)^2 &= r^2\\x^2 + y^2 &= r^2\end{aligned}$$

The form $x^2 + y^2 = r^2$ is called the **standard form of the equation of a circle that has its center at the origin**. For example, by inspection we can recognize that $x^2 + y^2 = 9$ is a circle with its center at the origin and a radius of length three units. Likewise, the equation $5x^2 + 5y^2 = 10$ is equivalent to $x^2 + y^2 = 2$; therefore its graph is a circle with its center at the origin and a radius of length $\sqrt{2}$ units. Furthermore, we can easily determine that the equation of the circle with its center at the origin and a radius of 8 units is $x^2 + y^2 = 64$.

Ellipses

Generally, it is true that any equation of the form $Ax^2 + By^2 = F$ (where $A = B$ and A, B, and F are nonzero constants having the same sign) is a circle with its center at the origin. The general equation $Ax^2 + By^2 = F$ can be used to describe other geometric figures by changing the restrictions on A and B. For example, if A, B, and F are of the same sign but $A \neq B$, then the graph of the equation $Ax^2 + By^2 = F$ is an **ellipse**. Let's consider two examples.

EXAMPLE 4

Graph $4x^2 + 9y^2 = 36$.

Solution

Let's find the intercepts. If $x = 0$, then

$$\begin{aligned}4(0)^2 + 9y^2 &= 36\\9y^2 &= 36\\y^2 &= 4\\y &= \pm 2.\end{aligned}$$

Thus the points (0, 2) and (0, −2) are on the graph. If $y = 0$, then

$$\begin{aligned}4x^2 + 9(0)^2 &= 36\\4x^2 &= 36\\x^2 &= 9\\x &= \pm 3.\end{aligned}$$

The points (3, 0) and (−3, 0) are on the graph.

Plotting the four points that we have and knowing that it is an ellipse gives us a pretty good sketch of the figure (Figure 3.50).

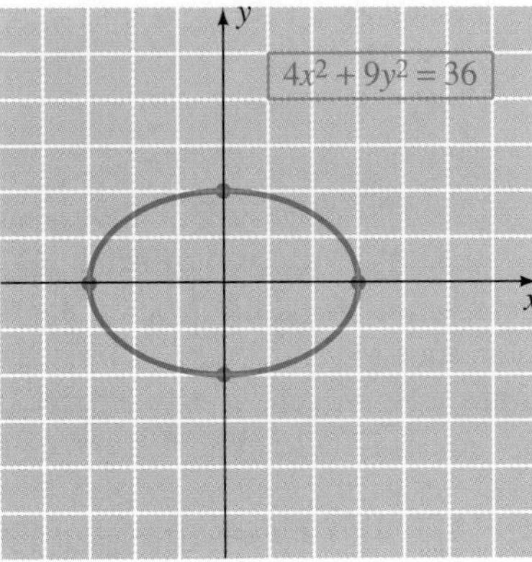

FIGURE 3.50

In Figure 3.50, the line segment with endpoints at $(-3, 0)$ and $(3, 0)$ is called the **major axis** of the ellipse. The shorter segment with endpoints at $(0, -2)$ and $(0, 2)$ is called the **minor axis**. Establishing the endpoints of the major and minor axes provides a basis for sketching an ellipse. Also notice that the equation $4x^2 + 9y^2 = 36$ exhibits symmetry with respect to both axes and the origin, as we see in Figure 3.50.

EXAMPLE 5

Graph $25x^2 + y^2 = 25$.

Solution

The endpoints of the major and minor axes can be found by finding the intercepts. If $x = 0$, then

$$\begin{aligned} 25(0)^2 + y^2 &= 25 \\ y^2 &= 25 \\ y &= \pm 5. \end{aligned}$$

The endpoints of the major axis are therefore at $(0, 5)$ and $(0, -5)$. If $y = 0$, then

$$\begin{aligned} 25x^2 + (0)^2 &= 25 \\ 25x^2 &= 25 \\ x^2 &= 1 \\ x &= \pm 1. \end{aligned}$$

The endpoints of the minor axis are at $(1, 0)$ and $(-1, 0)$. The ellipse is sketched in Figure 3.51.

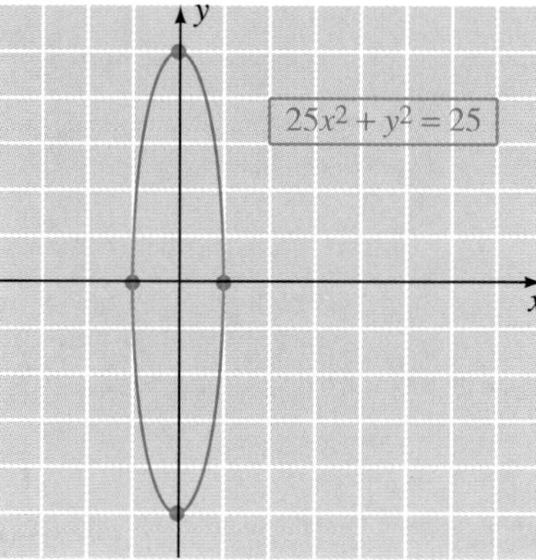

FIGURE 3.51

Hyperbolas

The graph of an equation of the form $Ax^2 + By^2 = F$, where A and B are of *unlike* signs, is a **hyperbola**. The next two examples illustrate the graphing of hyperbolas.

EXAMPLE 6

Graph $x^2 - 4y^2 = 4$.

Solution

If we let $y = 0$, then

$$\begin{aligned} x^2 - 4(0)^2 &= 4 \\ x^2 &= 4 \\ x &= \pm 2. \end{aligned}$$

Thus, the points (2, 0) and (−2, 0) are on the graph. If we let $x = 0$, then

$$\begin{aligned} 0^2 - 4y^2 &= 4 \\ -4y^2 &= 4 \\ y^2 &= -1. \end{aligned}$$

Since $y^2 = -1$ has no real number solutions, there are no points of the graph on the y-axis.

Notice that the equation $x^2 - 4y^2 = 4$ exhibits symmetry with respect to both axes and the origin. Now let's solve the given equation for y to get a more convenient form for finding other solutions.

$$\begin{aligned} x^2 - 4y^2 &= 4 \\ -4y^2 &= 4 - x^2 \end{aligned}$$

$$4y^2 = x^2 - 4$$
$$y^2 = \frac{x^2 - 4}{4}$$
$$y = \frac{\pm\sqrt{x^2 - 4}}{2}$$

Since the radicand, $x^2 - 4$, must be nonnegative, the values chosen for x must be such that $x \geq 2$ or $x \leq -2$. Symmetry and the points determined by the table provide the basis for sketching Figure 3.52.

x	y	
2	0	intercepts
−2	0	
3	$\pm\frac{\sqrt{5}}{2}$	other points
4	$\pm\sqrt{3}$	
5	$\pm\frac{\sqrt{21}}{2}$	

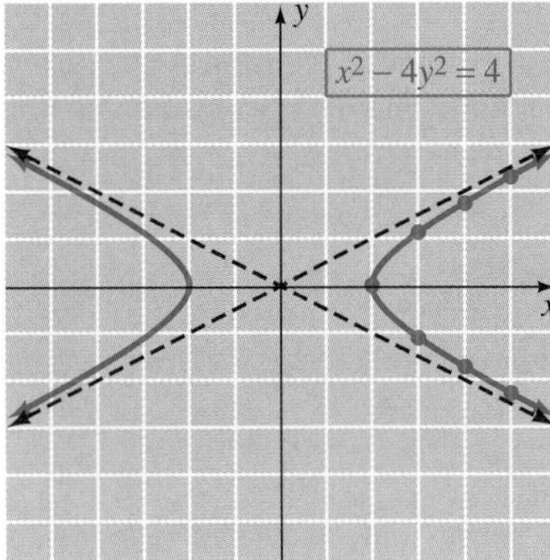

FIGURE 3.52

Notice the dashed lines in Figure 3.52; they are called **asymptotes**. Each **branch** of the hyperbola approaches one of these lines but does not intersect it. Therefore, being able to sketch the asymptotes of a hyperbola is very helpful for graphing purposes. Fortunately, the equations of the asymptotes are easy to determine. They can be found by replacing the constant term in the given equation of the hyperbola with zero and then solving for y. (The reason this works will be discussed in a later chapter.) Thus, for the hyperbola in Example 6 we obtain

$$x^2 - 4y^2 = 0$$
$$-4y^2 = -x^2$$

$$y^2 = \frac{1}{4}x^2$$

$$y = \pm\frac{1}{2}x.$$

So the lines $y = \frac{1}{2}x$ and $y = -\frac{1}{2}x$ are the asymptotes indicated by the dashed lines in Figure 3.52.

EXAMPLE 7

Graph $4y^2 - 9x^2 = 36$.

Solution

If $x = 0$, then

$$\begin{aligned} 4y^2 - 9(0)^2 &= 36 \\ 4y^2 &= 36 \\ y^2 &= 9 \\ y &= \pm 3. \end{aligned}$$

The points (0, 3) and (0, −3) are on the graph. If $y = 0$, then

$$\begin{aligned} 4(0)^2 - 9x^2 &= 36 \\ -9x^2 &= 36 \\ x^2 &= -4. \end{aligned}$$

Since $x^2 = -4$ has no real number solutions, we know that this hyperbola does not intersect the x-axis. Solving the equation for y yields

$$\begin{aligned} 4y^2 - 9x^2 &= 36 \\ 4y^2 &= 9x^2 + 36 \\ y^2 &= \frac{9x^2 + 36}{4} \\ y &= \frac{\pm\sqrt{9x^2}}{2} = \pm\frac{3\sqrt{x^2}}{2}. \end{aligned}$$

The table shows some additional solutions. The equations of the asymptotes are determined as follows.

$$\begin{aligned} 4y^2 - 9x^2 &= 0 \\ 4y^2 &= 9x^2 \\ y^2 &= \frac{9}{4}x^2 \\ y &= \pm\frac{3}{2}x \end{aligned}$$

Sketching the asymptotes, plotting the points from the table, and using symmetry determines the hyperbola in Figure 3.53.

x	y	
0	3	intercepts
0	-3	
1	$\pm\frac{3\sqrt{5}}{2}$	other points
2	$\pm 3\sqrt{2}$	
3	$\pm\frac{3\sqrt{13}}{2}$	

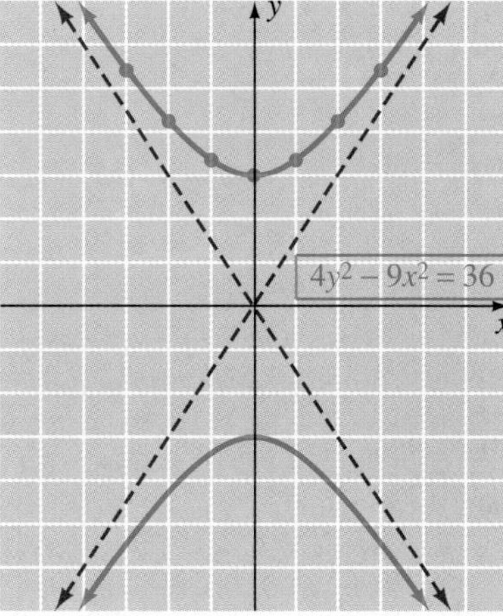

FIGURE 3.53

When using a graphing utility, it may be necessary to change the boundaries on x or y (or both) to obtain a complete graph. Consider the following example.

EXAMPLE 8

Use a graphing utility to graph $x^2 - 40x + y^2 + 351 = 0$.

Solution

First, we need to solve for y in terms of x.

$$x^2 - 40x + y^2 + 351 = 0$$
$$y^2 = -x^2 + 40x - 351$$
$$y = \pm\sqrt{-x^2 + 40x - 351}$$

Now we can make the following assignments.

$$Y_1 = \sqrt{-x^2 + 40x - 351}$$
$$Y_2 = -Y_1$$

(Note that we assigned Y_2 in terms of Y_1. By doing this we avoid repetitive key strokes and hopefully reduce the chance for errors. You may need to consult your user's manual for instructions on how to key stroke $-Y_1$.) Figure 3.54 shows the graph.

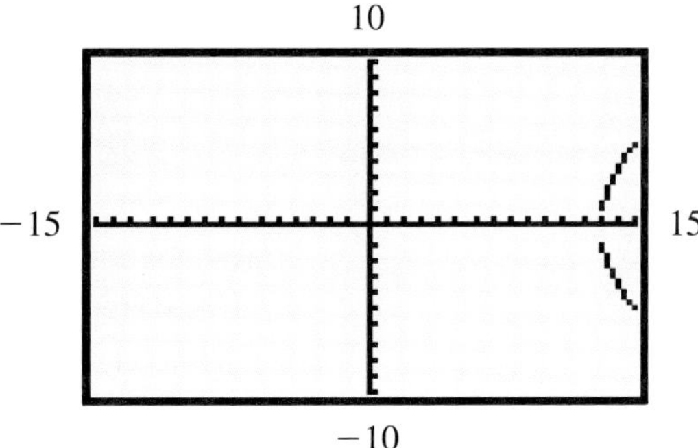

FIGURE 3.54

Since we know from the original equation that this graph should be a circle, we need to make some adjustments on the boundaries in order to get a complete graph. This can be done by completing the square on the original equation to change its form to $(x - 20)^2 + y^2 = 49$ or simply by a trial-and-error process. By changing the boundaries on x such that $-15 \leq x \leq 30$, we obtain Figure 3.55.

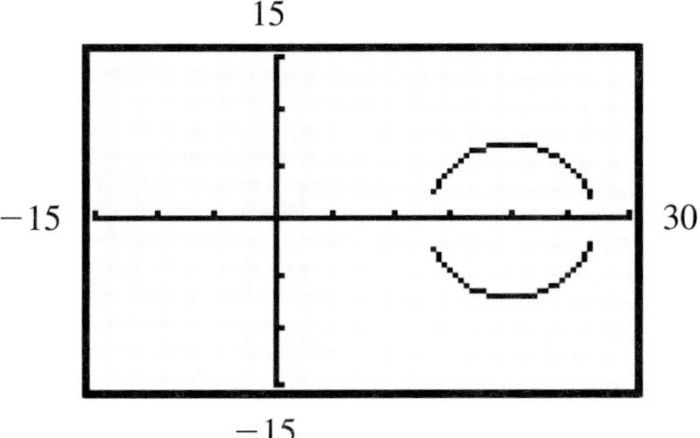

FIGURE 3.55

In summarizing this section, we do want you to be aware of the continuity pattern used. We started by using the definition of a circle to generate the standard form of the equation of a circle. Then ellipses and hyperbolas were discussed, not from a definition viewpoint, but by considering variations of the general equation ($Ax^2 + By^2 = F$, where A, B, and F are of the same sign and $A = B$) of a circle with its center at the origin. In the chapter titled "*Conic Sections*" (Chapter 9), we develop parabolas, ellipses, and hyperbolas from a definition viewpoint. In other words, we first define each of the concepts and then use those definitions to generate standard forms for their equations.

PROBLEM SET 3.5

Write the equation of each of the following circles. Express the final equations in the form $x^2 + y^2 + Dx + Ey + F = 0$.

1. Center at (2, 3) and $r = 5$.
2. Center at $(-3, 4)$ and $r = 2$.
3. Center at $(-1, -5)$ and $r = 3$.
4. Center at $(4, -2)$ and $r = 1$.
5. Center at (3, 0) and $r = 3$.
6. Center at $(0, -4)$ and $r = 6$.
7. Center at the origin and $r = 7$.
8. Center at the origin and $r = 1$.

For Problems 9–18, find the center and length of a radius of each of the following circles.

9. $x^2 + y^2 - 6x - 10y + 30 = 0$
10. $x^2 + y^2 + 8x - 12y + 43 = 0$
11. $x^2 + y^2 + 10x + 14y + 73 = 0$
12. $x^2 + y^2 + 6y - 7 = 0$
13. $x^2 + y^2 - 10x = 0$
14. $x^2 + y^2 - 4x + 2y = 0$
15. $x^2 + y^2 = 8$
16. $4x^2 + 4y^2 = 1$
17. $4x^2 + 4y^2 - 4x - 8y - 11 = 0$
18. $36x^2 + 36y^2 + 48x - 36y - 11 = 0$
19. Find the equation of the circle where the line segment determined by $(-4, 9)$ and $(10, -3)$ is a diameter.
20. Find the equation of the circle that passes through the origin and has its center at $(-3, -4)$
21. Find the equation of the circle that is tangent to both axes, has a radius of length seven units, and its center is in the fourth quadrant.
22. Find the equation of the circle that passes through the origin, has an x-intercept of -6, and a y-intercept of 12. (The perpendicular bisector of a chord contains the center of the circle.)
23. Find the equations of the circles that are tangent to the x-axis and have a radius of length five units. In each case the abscissa of the center is -3. (There is more than one circle that satisfies these conditions.)

Graph each of the following.

24. $4x^2 + 25y^2 = 100$
25. $9x^2 + 4y^2 = 36$
26. $x^2 - y^2 = 4$
27. $y^2 - x^2 = 9$
28. $x^2 + y^2 - 4x - 2y - 4 = 0$
29. $x^2 + y^2 - 4x = 0$
30. $4x^2 + y^2 = 4$
31. $x^2 + 9y^2 = 36$
32. $x^2 + y^2 + 2x - 6y - 6 = 0$
33. $y^2 - 3x^2 = 9$
34. $4x^2 - 9y^2 = 16$
35. $x^2 + y^2 + 4x + 6y - 12 = 0$
36. $2x^2 + 5y^2 = 50$
37. $4x^2 + 3y^2 = 12$
38. $x^2 + y^2 - 6x + 8y = 0$
39. $3x^2 - 2y^2 = 3$
40. $y^2 - 8x^2 = 9$

The graphs of equations of the form $xy = k$, where k is a nonzero constant, are also hyperbolas, sometimes referred to as **rectangular hyperbolas**. Graph each of the following.

41. $xy = 2$
42. $xy = 4$
43. $xy = -3$
44. $xy = -2$

45. What is the graph of $xy = 0$? Defend your answer.

46. We have graphed various equations of the form $Ax^2 + By^2 = F$, where F is a nonzero constant. Now describe the graph of each of the following.

a. $x^2 + y^2 = 0$ **b.** $2x^2 + 3y^2 = 0$

c. $x^2 - y^2 = 0$ **d.** $4x^2 - 9y^2 = 0$

MISCELLANEOUS PROBLEMS

47. By expanding $(x - h)^2 + (y - k)^2 = r^2$, we obtain $x^2 - 2hx + h^2 + y^2 - 2ky + k^2 - r^2 = 0$. Comparing this result to the form $x^2 + y^2 + Dx + Ey + F = 0$, we see that $D = -2h$, $E = -2k$, and $F = h^2 + k^2 - r^2$. Therefore, the center and the length of a radius of a circle can be found by using $h = D/-2$, $k = E/-2$, and $r = \sqrt{h^2 + k^2 - F}$. Use these relationships to find the center and the length of a radius of each of the following circles.

a. $x^2 + y^2 - 2x - 8y + 8 = 0$

b. $x^2 + y^2 + 4x - 14y + 49 = 0$

c. $x^2 + y^2 + 12x + 8y - 12 = 0$

d. $x^2 + y^2 - 16x + 20y + 115 = 0$

e. $x^2 + y^2 - 12y - 45 = 0$

f. $x^2 + y^2 + 14x = 0$

48. Use a coordinate geometry approach to prove that an angle inscribed in a semicircle is a right angle. (See Figure 3.56.)

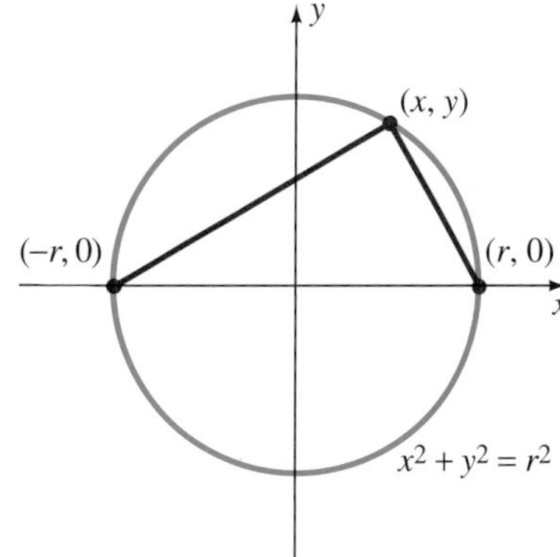

FIGURE 3.56

49. Use a coordinate geometry approach to prove that a line segment from the center of a circle bisecting a chord is perpendicular to the chord. [*Hint:* Let the ends of the chord be $(r, 0)$ and (a, b).]

GRAPHICS CALCULATOR ACTIVITIES

50. For each of the following equations, (a) predict the type and location of the graph, and (b) use your graphics calculator to check your prediction.

a. $x^2 + y^2 = 9$ **b.** $2x^2 + y^2 = 4$

c. $x^2 - y^2 = 9$ **d.** $4x^2 - y^2 = 16$

e. $x^2 + 2x + y^2 - 4 = 0$ **f.** $x^2 + y^2 - 4y - 2 = 0$

g. $(x - 2)^2 + (y + 1)^2 = 4$

h. $(x + 3)^2 - (y - 4)^2 = 9$

i. $9y^2 - 4x^2 = 36$ **j.** $9y^2 + 4x^2 = 36$

CHAPTER 3 SUMMARY

We emphasized throughout this chapter that coordinate geometry contains two basic kinds of problems:

1. Given an algebraic equation, determine its geometric graph;
2. Given a set of conditions pertaining to a geometric figure, determine its algebraic equation.

Let's review this chapter in terms of those two kinds of problems.

Graphing

The following graphing techniques were discussed in this chapter.

1. Recognize the type of graph that a certain kind of equation produces.
   - **a.** $Ax + By = C$ produces a straight line.
   - **b.** $x^2 + y^2 + Dx + Ey + F = 0$ produces a circle. The center and the length of a radius can be found by completing the square and comparing to the standard form of the equation of a circle:

     $$(x - h)^2 + (y - k)^2 = r^2.$$

   - **c.** $Ax^2 + By^2 = F$, where A, B, and F have the same sign and $A = B$, produces a circle with the center at the origin.
   - **d.** $Ax^2 + By^2 = F$, where A, B, and F are of the same sign but $A \neq B$, produces an ellipse.
   - **e.** $Ax^2 + By^2 = F$, where A and B are of unlike signs, produces a hyperbola.
2. Determine the symmetry that a graph possesses.
   - **a.** The graph of an equation is symmetric with respect to the y-axis if replacing x with $-x$ results in an equivalent equation.
   - **b.** The graph of an equation is symmetric with respect to the x-axis if replacing y with $-y$ results in an equivalent equation.
   - **c.** The graph of an equation is symmetric with respect to the origin if replacing x with $-x$ and y with $-y$ results in an equivalent equation.
3. Find the intercepts. The x-intercept is found by letting $y = 0$ and solving for x. The y-intercept is found by letting $x = 0$ and solving for y.
4. Determine the necessary restrictions to ensure real number solutions.
5. Set up a table of ordered pairs that satisfy the equation. The type of symmetry and the restrictions will affect your choice of values in the table. Furthermore, it may be convenient to change the form of the original equation by solving for y in terms of x or for x in terms of y.

6. Plot the points associated with the ordered pairs in the table and connect them with a smooth curve. Then, if appropriate, reflect the curve according to any symmetries possessed by the graph.

Determining Equations When Given Certain Conditions

You should review Examples 2, 3, and 4 of Section 3.3 to feel comfortable with the general approach of choosing a point (x, y) and using it to determine the equation that satisfies the conditions stated in the problem.

We developed some special forms that can be used to determine equations:

Point-slope form of a straight line: $y - y_1 = m(x - x_1)$

Slope-intercept form of a straight line: $y = mx + b$

Standard form of a circle: $(x - h)^2 + (y - k)^2 = r^2$

The following formulas were used in different parts of the chapter.

Distance formula: $d = \sqrt{(x_2 - x_1)^2 + (y_2 - y_1)^2}$

Midpoint formula: The coordinates of the midpoint of a line segment determined by (x_1, y_1) and (x_2, y_2) are

$$\left(\frac{x_1 + x_2}{2}, \frac{y_1 + y_2}{2}\right).$$

Slope formula: $m = \dfrac{y_2 - y_1}{x_2 - x_1}$

CHAPTER 3 REVIEW PROBLEM SET

1. On a number line, find the coordinate of the point located three-fifths of the distance from -4 to 11.

2. On a number line, find the coordinate of the point located four-ninths of the distance from 3 to -15.

3. On the xy-plane, find the coordinates of the point located five-sixths of the distance from $(-1, -3)$ to $(11, 1)$.

4. If one endpoint of a line segment is at $(8, 14)$ and the midpoint of the segment is $(3, 10)$, find the coordinates of the other endpoint.

5. Verify that the points $(2, 2)$, $(6, 4)$, and $(5, 6)$ are vertices of a right triangle.

6. Verify that the points $(-3, 1)$, $(1, 3)$, and $(9, 7)$ lie in a straight line.

For Problems 7–12, identify any symmetries (x-axis, y-axis, origin) that each of the equations exhibits.

7. $x = y^2 + 4$

8. $y = x^2 + 6x -.1$

9. $5x^2 - y^2 = 4$

10. $x^2 + y^2 - 2y - 4 = 0$

11. $y = -x$ **12.** $y = \dfrac{6}{x^2 + 4}$

Graph each of the following (Problems 13–22).

13. $x^2 + y^2 - 6x + 4y - 3 = 0$

14. $x^2 + 4y^2 = 16$ **15.** $x^2 - 4y^2 = 16$

16. $-2x + 3y = 6$ **17.** $2x - y < 4$

18. $x^2y^2 = 4$ **19.** $4y^2 - 3x^2 = 8$

20. $x^2 + y^2 + 10y = 0$ **21.** $9x^2 + 2y^2 = 36$

22. $y \leq -2x - 3$

23. Find the slope of the line determined by $(-3, -4)$ and $(-5, 6)$.

24. Find the slope of the line with equation $5x - 7y = 12$.

For Problems 25–28, write the equation of the lines satisfying the stated conditions. Express final equations in standard form $(Ax + By = C)$.

25. Contains the point $(7, 2)$ and has a slope of $-\dfrac{3}{4}$

26. Contains the points $(-3, -2)$ and $(1, 6)$

27. Contains the point $(2, -4)$ and is parallel to $4x + 3y = 17$

28. Contains the point $(-5, 4)$ and is perpendicular to $2x - y = 7$

For Problems 29–32, write the equation of each of the circles satisfying the stated conditions. Express final equations in the form $x^2 + y^2 + Dx + Ey + F = 0$.

29. Center at $(5, -6)$ and $r = 1$

30. The endpoints of a diameter are $(-2, 4)$ and $(6, 2)$

31. Center at $(-5, 12)$ and passes through the origin

32. Tangent to both axes, $r = 4$, and center is in the third quadrant

CUMULATIVE REVIEW PROBLEM SET - CHAPTERS 1–3

Evaluate each of the following.

1. 3^{-3} **2.** -4^{-2} **3.** $\left(\dfrac{2}{3}\right)^{-2}$

4. $-\sqrt[3]{\dfrac{8}{27}}$ **5.** $\left(\dfrac{1}{27}\right)^{-2/3}$ **6.** $\dfrac{1}{\left(\dfrac{3}{4}\right)^{-2}}$

Perform the indicated operations and simplify. Express final answers using positive exponents only.

7. $(5x^{-3}y^{-2})(4xy^{-1})$ **8.** $(-7a^{-3}b^2)(8a^4b^{-3})$

9. $\left(\dfrac{1}{2}x^{-2}y^{-1}\right)^{-2}$ **10.** $\dfrac{80x^{-3}y^{-4}}{16xy^{-6}}$

11. $\left(\dfrac{102x^{2/3}y^{3/4}}{6xy^{-1}}\right)^{-1}$ **12.** $\left(\dfrac{14a^3b^{-4}}{7a^{-1}b^3}\right)^2$

Express each of the following in simplest radical form. All variables represent positive real numbers.

13. $-5\sqrt{72}$ **14.** $2\sqrt{27x^3y^2}$ **15.** $\sqrt[3]{56x^4y^7}$

16. $\dfrac{3\sqrt{18}}{5\sqrt{12}}$ **17.** $\sqrt{\dfrac{3x}{7y}}$ **18.** $\dfrac{5}{\sqrt{2} - 3}$

19. $\dfrac{3\sqrt{7}}{2\sqrt{2} - \sqrt{6}}$ **20.** $\dfrac{4\sqrt{x}}{\sqrt{x} + 3\sqrt{y}}$

Perform the following indicated operations involving rational expressions. Express final answers in simplest form.

21. $\dfrac{12x^2y}{18x} \cdot \dfrac{9x^3y^3}{16xy^2}$ **22.** $\dfrac{-15ab^2}{14a^3b} \div \dfrac{20a}{7b^2}$

23. $\dfrac{3x^2 + 5x - 2}{x^2 - 4} \cdot \dfrac{5x^2 - 9x - 2}{3x^2 - x}$

24. $\dfrac{2x - 1}{4} + \dfrac{3x + 2}{6} - \dfrac{x - 1}{8}$

25. $\dfrac{5}{3n^2} - \dfrac{2}{n} + \dfrac{3}{2n}$ **26.** $\dfrac{5x}{x^2 + 6x - 27} + \dfrac{3}{x^2 - 9}$

Solve each of the following equations.

27. $3(-2x - 1) - 2(3x + 4) = -4(2x - 3)$

28. $(2x - 1)(3x + 4) = (x + 2)(6x - 5)$

29. $\frac{3x - 1}{4} - \frac{2x - 1}{5} = \frac{1}{10}$

30. $9x^2 - 4 = 0$

31. $5x^3 + 10x^2 - 40x = 0$

32. $7t^2 - 31t + 12 = 0$

33. $x^4 + 15x^2 - 16 = 0$

34. $|5x - 2| = 3$

35. $2x^2 - 3x - 1 = 0$

36. $(3x - 2)(x + 4) = (2x - 1)(x - 1)$

37. $\sqrt{5 - t} + 1 = \sqrt{7 + 2t}$

38. $(2x - 1)^2 + 4 = 0$

Solve each of the following inequalities. Express the solution sets using interval notation.

39. $-2(x - 1) + (3 - 2x) > 4(x + 1)$

40. $2n + 1 + \frac{3n - 1}{4} \geq \frac{n - 1}{2}$

41. $0.09x + 0.12(450 - x) \geq 46.5$

42. $n^2 + 5n > 24$

43. $6x^2 + 7x - 3 < 0$

44. $(2x - 1)(x + 3)(x - 4) > 0$

45. $\frac{3x - 2}{x + 1} \leq 0$

46. $\frac{x + 5}{x - 1} \geq 2$

47. $|3x - 1| > 5$

48. $|5x - 3| < 12$

Graph each of the following equations in Problems 49–54.

49. $x^2 + 4y^2 = 36$

50. $4x^2 - y^2 = 4$

51. $y = -x^3 - 1$

52. $y = -x + 3$

53. $y^2 - 5x^2 = 9$

54. $y = -\frac{3}{4}x - 1$

Solve each of the following problems.

55. Find the center and the length of a radius of the circle with equation $x^2 + y^2 + 14x - 8y + 56 = 0$.

56. Write the equation of the line that is parallel to $3x - 4y = 17$ and contains the point (2, 8).

57. Find the coordinates of the point located one-fifth of the distance from (−3, 4) to (2, 14).

58. Write the equation of the perpendicular bisector of the line segment determined by (−3, 4) and (5, 10).

For each of the following problems, set up an equation and solve the problem.

59. A retailer has some shirts that cost $22 per shirt. At what price should they be sold to obtain a profit of 30% of the cost? At what price should they be sold to obtain a profit of 30% of the selling price?

60. A total of $7500 was invested, part of it at 5% yearly interest and the remainder at 6%. If the total yearly interest was $420, how much was invested at each rate?

61. The length of a rectangle is 1 inch less than twice the width. If the area of the rectangular region is 36 square inches, find the length and width of the rectangle.

62. The length of one side of a triangle is 4 centimeters less than three times the length of the altitude to that side. If the area of the triangle is 80 square centimeters, find the length of the side and the length of the altitude to that side.

63. How many milliliters of pure acid must be added to 40 milliliters of a 30% acid solution to obtain a 50% acid solution?

64. Amanda rode her bicycle out into the country at a speed of 15 miles per hour and returned along the same route at 10 miles per hour. If the round trip took 5 hours, how far out did she ride?

65. If two inlet pipes are both open, they can fill a pool in 1 hour and 12 minutes. One of the pipes can fill the pool by itself in 2 hours. How long would it take the other pipe to fill the pool by itself?

FUNCTIONS

One of the fundamental concepts of mathematics is that of a function. Functions are used to unify different areas of mathematics and they also serve as a meaningful way of applying mathematics to many real world problems. They provide a means of studying quantities that vary with one another, that is, quantities such that a change in one produces a corresponding change in another. In this chapter we will (1) introduce the basic ideas pertaining to the function concept, (2) use the idea of a function to unify some concepts from Chapter 3, and (3) discuss some applications using functions.

4

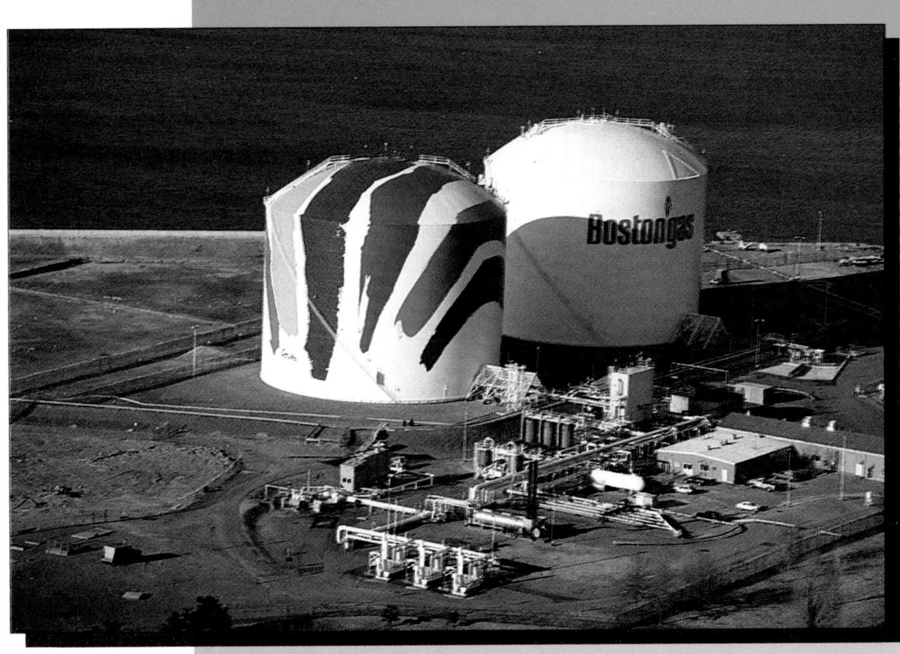

The volume of a right circular cylinder is a function of its height and the length of a radius of its base.

4.1 CONCEPT OF A FUNCTION

The notion of correspondence is used in everyday situations and is central to the concept of a function. Consider the following correspondences.

1. To each person in a class, there corresponds an assigned seat.
2. To each day of a year, there corresponds an assigned integer 1 that represents the average temperature for that day in a certain geographical location.
3. To each book in a library, there corresponds a whole number that represents the number of pages in the book.

Such correspondences can be depicted as in Figure 4.1. To each member in set A there corresponds *one and only one* member in set B. For example, in correspondence 1, set A would consist of the students in a class and set B would be the assigned seats. In the second example, set A would consist of the days of a year and set B would be a set of integers. Furthermore, the same integer might be

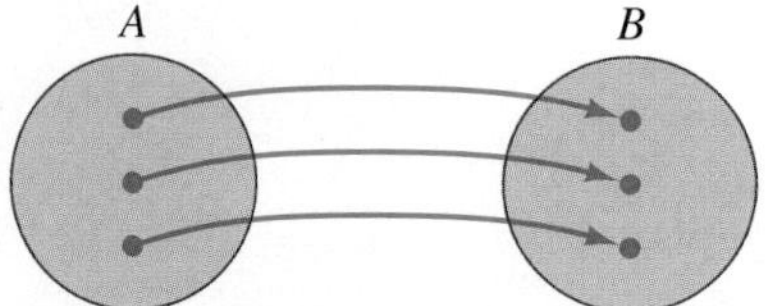

FIGURE 4.1

assigned to more than one day of the year. (Different days might have the same average temperature.) The key idea is that *one and only one* integer is assigned to *each* day of the year. Likewise, in the third example, more than one book may have the same number of pages, but to each book there is assigned one and only one number of pages.

Mathematically, the general concept of a function can be defined as follows.

DEFINITION 4.1

A **function** f is a correspondence between two sets X and Y that assigns to each element x of set X one and only one element y of set Y. The element y being assigned is called the **image** of x. The set X is called the **domain** of the function and the set of all images is called the **range** of the function.

In Definition 4.1, the image y is usually denoted by $f(x)$. Thus, the symbol $f(x)$ (read *f of x* or *the **value** of f at x*) represents the element in the range associated with the element x from the domain. Figure 4.2 depicts this situation. Again, we emphasize that each member of the domain has precisely one image in the range; however, different members in the domain, such as a and b in Figure 4.2, may have the same image.

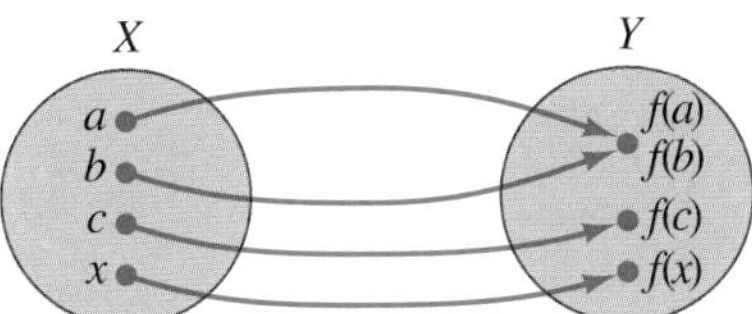

FIGURE 4.2

In Definition 4.1 we named the function f. It is common to name functions by means of a single letter and the letters f, g, and h are often used. We would suggest more meaningful choices when functions are used in real world situations. For example, if a problem involves a profit function, then naming the function p or even P would seem natural. Be careful not to confuse f and $f(x)$. Remember that f is used to name a function, whereas $f(x)$ is an element of the range, namely the element assigned to x by f.

The assignments made by a function are often expressed as ordered pairs. For example, referring back to Figure 4.2, the assignments could be expressed as $(a, f(a))$, $(b, f(b))$, $(c, f(c))$, and $(x, f(x))$, where the first components are from the domain and the second components are from the range. Thus, a function can also be thought of as **a set of ordered pairs where no two of the ordered pairs have the same first component**.

REMARK In some texts, the concept of a **relation** is introduced first and then functions are defined as special kinds of relations. A relation is defined as *a set of ordered pairs* and a function is defined as *a relation in which no two ordered pairs have the same first element.*

The ordered pairs representing a function can be generated by various means, such as a graph or a chart. However, one of the most common ways of generating ordered pairs is by use of equations. For example, the equation $f(x) = 2x + 3$ indicates that to each value of x in the domain, we assign $2x + 3$ from the range. For example,

$f(1) = 2(1) + 3 = 5$ produces the ordered pair $(1, 5)$;

$f(4) = 2(4) + 3 = 11$ produces the ordered pair $(4, 11)$;

$f(-2) = 2(-2) + 3 = -1$ produces the ordered pair $(-2, -1)$.

It may be helpful for you to picture mentally the concept of a function in terms of a *function machine*, as illustrated in Figure 4.3. Each time that a value of x is put into the machine, the equation $f(x) = 2x + 3$ is used to generate one and only one value for $f(x)$ to be ejected from the machine.

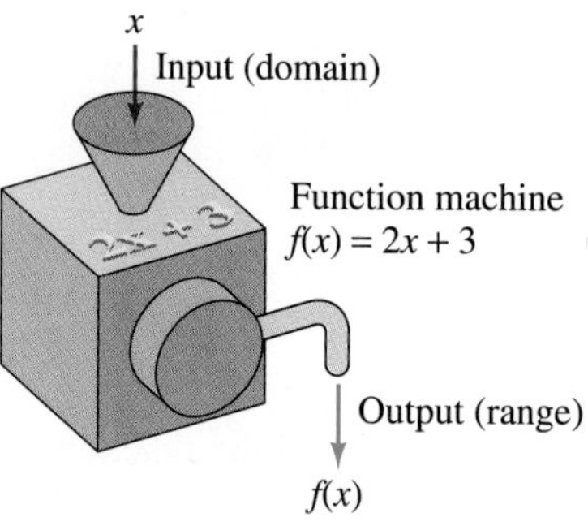

FIGURE 4.3

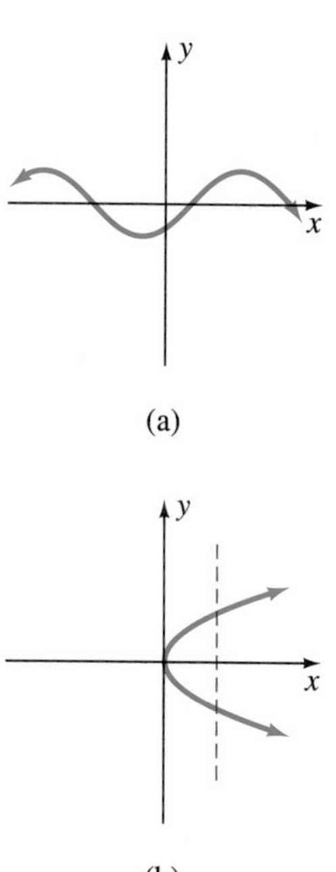

(a)

(b)

FIGURE 4.4

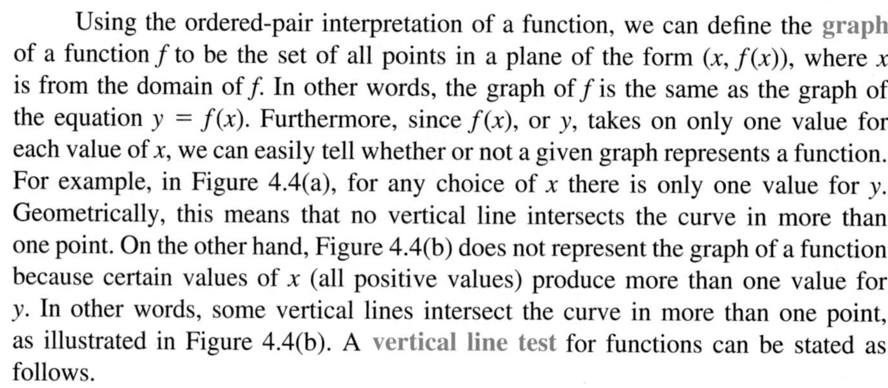

Using the ordered-pair interpretation of a function, we can define the **graph** of a function f to be the set of all points in a plane of the form $(x, f(x))$, where x is from the domain of f. In other words, the graph of f is the same as the graph of the equation $y = f(x)$. Furthermore, since $f(x)$, or y, takes on only one value for each value of x, we can easily tell whether or not a given graph represents a function. For example, in Figure 4.4(a), for any choice of x there is only one value for y. Geometrically, this means that no vertical line intersects the curve in more than one point. On the other hand, Figure 4.4(b) does not represent the graph of a function because certain values of x (all positive values) produce more than one value for y. In other words, some vertical lines intersect the curve in more than one point, as illustrated in Figure 4.4(b). A **vertical line test** for functions can be stated as follows.

Vertical Line Test If each vertical line intersects a graph in no more than one point, then the graph represents a function.

Let's consider some examples to help pull together some of these ideas about functions.

EXAMPLE 1

If $f(x) = x^2 - x + 4$ and $g(x) = x^3 - x^2$, find $f(3)$, $f(-2)$, $g(4)$, and $g(-3)$.

Solution

$$f(3) = 3^2 - 3 + 4 = 10 \qquad f(-2) = (-2)^2 - (-2) + 4 = 10$$

$$g(4) = 4^3 - 4^2 = 48 \qquad g(-3) = (-3)^3 - (-3)^2 = -36$$

In Example 1, notice that we were working with two different functions in the same problem. That is why we used two different names, f and g. Sometimes the rule of assignment for a function may consist of more than one part. We often refer to such functions as **piecewise-defined** functions. Let's consider an example of such a function.

EXAMPLE 2

If $f(x) = \begin{cases} 2x + 1 & \text{for } x \geq 0 \\ 3x - 1 & \text{for } x < 0 \end{cases}$, find $f(2)$, $f(4)$, $f(-1)$, and $f(-3)$.

Solution

For $x \geq 0$, we use the assignment $f(x) = 2x + 1$.

$$f(2) = 2(2) + 1 = 5$$

$$f(4) = 2(4) + 1 = 9$$

For $x < 0$, we use the assignment $f(x) = 3x - 1$.

$$f(-1) = 3(-1) - 1 = -4$$

$$f(-3) = 3(-3) - 1 = -10$$

The quotient $\dfrac{f(a + h) - f(a)}{h}$ is often called a **difference quotient** and is used extensively with functions when studying the limit concept in calculus. The next examples illustrate finding the difference quotient for specific functions.

EXAMPLE 3

Find $\dfrac{f(a + h) - f(a)}{h}$ for each of the following functions.

a. $f(x) = x^2 + 6$ **b.** $f(x) = 2x^2 + 3x - 4$ **c.** $f(x) = \dfrac{1}{x}$

Solutions

a.

$$f(a) = a^2 + 6$$

$$f(a + h) = (a + h)^2 + 6 = a^2 + 2ah + h^2 + 6$$

Therefore,

$$\begin{aligned} f(a + h) - f(a) &= (a^2 + 2ah + h^2 + 6) - (a^2 + 6) \\ &= a^2 + 2ah + h^2 + 6 - a^2 - 6 \\ &= 2ah + h^2 \end{aligned}$$

and

$$\frac{f(a+h)-f(a)}{h}=\frac{2ah+h^2}{h}=\frac{h(2a+h)}{h}=2a+h.$$

b.

$$\begin{aligned} f(a) &= 2a^2+3a-4 \\ f(a+h) &= 2(a+h)^2+3(a+h)-4 \\ &= 2(a^2+2ha+h^2)+3a+3h-4 \\ &= 2a^2+4ha+2h^2+3a+3h-4 \end{aligned}$$

Therefore

$$\begin{aligned} f(a+h)-f(a) &= (2a^2+4ha+2h^2+3a+3h-4)-(2a^2+3a-4) \\ &= 2a^2+4ha+2h^2+3a+3h-4-2a^2-3a+4 \\ &= 4ha+2h^2+3h \end{aligned}$$

and

$$\begin{aligned} \frac{f(a+h)-f(a)}{h} &= \frac{4ha+2h^2+3h}{h} \\ &= \frac{h(4a+2h+3)}{h} \\ &= 4a+2h+3. \end{aligned}$$

c.

$$\begin{aligned} f(a) &= \frac{1}{a} \\ f(a+h) &= \frac{1}{a+h} \end{aligned}$$

Therefore,

$$\begin{aligned} f(a+h)-f(a) &= \frac{1}{a+h}-\frac{1}{a} \\ &= \frac{a-(a+h)}{a(a+h)} \\ &= \frac{a-a-h}{a(a+h)} \\ &= \frac{-h}{a(a+h)} \quad \text{or} \quad -\frac{h}{a(a+h)} \end{aligned}$$

and

$$\begin{aligned} \frac{f(a+h)-f(a)}{h} &= -\frac{\frac{h}{a(a+h)}}{h} \\ &= -\frac{h}{a(a+h)}\cdot\frac{1}{h} \\ &= -\frac{1}{a(a+h)}. \end{aligned}$$

For our purposes in this text, if the domain of a function is not specifically indicated or determined by a real world application, then we will assume the domain to be *all real-number* replacements for the variable, provided they represent elements in the domain and produce *real-number* functional values.

EXAMPLE 4

For the function $f(x) = \sqrt{x - 1}$, (a) specify the domain, (b) determine the range, and (c) evaluate $f(5)$, $f(50)$, and $f(25)$.

Solutions

a. The radicand must be nonnegative, so $x - 1 \geq 0$ and thus $x \geq 1$. Therefore the domain (D) is

$$D = \{x | x \geq 1\}.$$

b. The symbol $\sqrt{}$ indicates the nonnegative square root; thus, the range (R) is

$$R = \{f(x) | f(x) \geq 0\}.$$

c.
$$f(5) = \sqrt{4} = 2$$
$$f(50) = \sqrt{49} = 7$$
$$f(25) = \sqrt{24} = 2\sqrt{6}$$

As we will see later, the range of a function is often easier to determine after having graphed the function. However, our equation- and inequality-solving processes are frequently sufficient to determine the domain of a function. Let's consider some examples.

EXAMPLE 5

Determine the domain for each of the following functions.

a. $f(x) = \dfrac{3}{2x - 5}$ **b.** $g(x) = \dfrac{1}{x^2 - 9}$ **c.** $f(x) = \sqrt{x^2 + 4x - 12}$

Solutions

a. We can replace x with any real number except $\frac{5}{2}$, because $\frac{5}{2}$ makes the denominator zero. Thus, the domain is

$$D = \left\{x | x \neq \frac{5}{2}\right\}.$$

b. We need to eliminate any values of x that will make the denominator zero. Therefore, let's solve the equation $x^2 - 9 = 0$.

$$x^2 - 9 = 0$$
$$x^2 = 9$$
$$x = \pm 3$$

The domain is thus the set

$$D = \{x | x \neq 3 \text{ and } x \neq -3\}.$$

c. The radicand, $x^2 + 4x - 12$, must be nonnegative. Therefore, let's use a number line approach as we did in Chapter 2, to solve the inequality $x^2 + 4x - 12 \geq 0$ (see Figure 4.5).

$$x^2 + 4x - 12 \geq 0$$
$$(x + 6)(x - 2) \geq 0$$

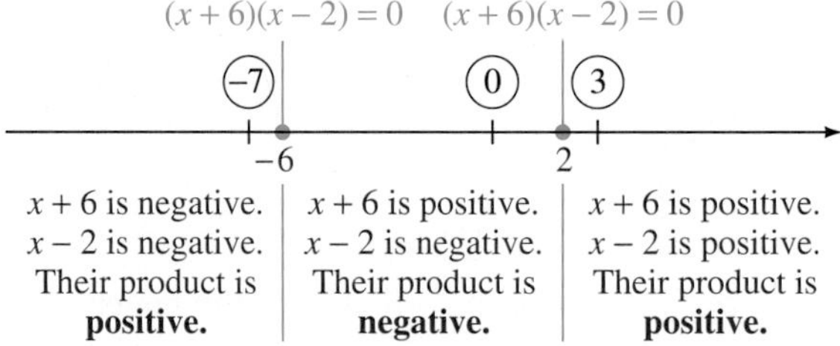

(x + 6)(x − 2) = 0
(x + 6)(x − 2) = 0
−7
0
3
−6
2
x + 6 is negative. x − 2 is negative. Their product is positive.
x + 6 is positive. x − 2 is negative. Their product is negative.
x + 6 is positive. x − 2 is positive. Their product is positive.

FIGURE 4.5

The product $(x + 6)(x - 2)$ is nonnegative if $x \leq -6$ or $x \geq 2$. Using interval notation, the domain can be expressed $(-\infty, -6] \cup [2, \infty)$.

Functions and function notation provide the basis for describing many real world relationships. The next example illustrates this point.

EXAMPLE 6

Suppose a factory determines that the overhead for producing a quantity of a certain item is \$500 and the cost for each item is \$25. Express the total expenses as a function of the number of items produced and compute the expenses for producing 12, 25, 50, 75, and 100 items.

Solution

Let n represent the number of items produced. Then $25n + 500$ represents the total expenses. Using E to represent the *expense function*, we have

$$E(n) = 25n + 500, \quad \text{where } n \text{ is a whole number.}$$

Therefore, we obtain

$$E(12) = 25(12) + 500 = 800;$$
$$E(25) = 25(25) + 500 = 1125;$$
$$E(50) = 25(50) + 500 = 1750;$$
$$E(75) = 25(75) + 500 = 2375;$$
$$E(100) = 25(100) + 500 = 3000.$$

So the total expenses for producing 12, 25, 50, 75, and 100 items are \$800, \$1125, \$1750, \$2375, and \$3000, respectively.

As we stated before, an equation such as $f(x) = 5x - 7$ that is used to determine a function can also be written $y = 5x - 7$. In either form, x is referred to as the **independent variable** and y (or $f(x)$) as the **dependent variable**. Many formulas in mathematics and other related areas also determine functions. For example, the area formula for a circular region, $A = \pi r^2$, assigns to each positive real value for r a unique value for A. This formula determines a function f, where $f(r) = \pi r^2$. The variable r is the independent variable and A (or $f(r)$) is the dependent variable.

Many functions that we will study throughout this text can be classified as even or odd functions. A function f having the property that $f(-x) = f(x)$ for every x in the domain of f is called an **even function**. A function f having the property that $f(-x) = -f(x)$ for every x in the domain of f is called an **odd function**.

EXAMPLE 7

For each of the following, classify the function as even, odd, or neither even nor odd.

a. $f(x) = 2x^3 - 4x$ **b.** $f(x) = x^4 - 7x^2$ **c.** $f(x) = x^2 + 2x - 3$

Solution

a. The function $f(x) = 2x^3 - 4x$ is an odd function because $f(-x) = 2(-x)^3 - 4(-x) = -2x^3 + 4x$, which equals $-f(x)$.

b. The function $f(x) = x^4 - 7x^2$ is an even function because $f(-x) = (-x)^4 - 7(-x)^2 = x^4 - 7x^2$, which equals $f(x)$.

c. The function $f(x) = x^2 + 2x - 3$ is neither even nor odd because $f(-x) = (-x)^2 + 2(-x) - 3 = x^2 - 2x - 3$, which does not equal $f(x)$ nor $-f(x)$.

PROBLEM SET 4.1

1. If $f(x) = -2x + 5$, find $f(3)$, $f(5)$, and $f(-2)$.

2. If $f(x) = x^2 - 3x - 4$, find $f(2)$, $f(4)$, and $f(-3)$.

3. If $g(x) = -2x^2 + x - 5$, find $g(3)$, $g(-1)$, and $g(-4)$.

4. If $g(x) = -x^2 - 4x + 6$, find $g(0)$, $g(5)$, and $g(-5)$.

5. If $h(x) = \frac{2}{3}x - \frac{3}{4}$, find $h(3)$, $h(4)$, and $h\left(-\frac{1}{2}\right)$.

6. If $h(x) = -\frac{1}{2}x + \frac{2}{3}$, find $h(-2)$, h (6), and $h\left(-\frac{2}{3}\right)$.

7. If $f(x) = \sqrt{2x - 1}$, find $f(5)$, $f\left(\frac{1}{2}\right)$, and $f(23)$.

8. If $f(x) = \sqrt{3x + 2}$, find $f\left(\frac{14}{3}\right)$, $f(10)$, and $f\left(-\frac{1}{3}\right)$.

9. If $f(x) = \begin{cases} x & \text{for } x \geq 0 \\ x^2 & \text{for } x < 0 \end{cases}$, find $f(4)$, $f(10)$, $f(-3)$, and $f(-5)$.

10. If $f(x) = \begin{cases} 3x + 2 & \text{for } x \geq 0 \\ 5x - 1 & \text{for } x < 0 \end{cases}$, find $f(2)$, $f(6)$, $f(-1)$, and $f(-4)$.

11. If $f(x) = \begin{cases} 2x & \text{for } x \geq 0 \\ -2x & \text{for } x < 0 \end{cases}$, find $f(3)$, $f(5)$, $f(-3)$, and $f(-5)$.

12. If $f(x) = \begin{cases} 2 & \text{for } x < 0 \\ x^2 + 1 & \text{for } 0 \le x \le 4, \\ -1 & \text{for } x > 4 \end{cases}$ find $f(3), f(6), f(0),$ and $f(-3)$.

13. If $f(x) = \begin{cases} 1 & \text{for } x > 0 \\ 0 & \text{for } -1 < x \le 0, \\ -1 & \text{for } x \le -1 \end{cases}$ find $f(2)$, $f(0)$, $f\left(-\frac{1}{2}\right)$, and $f(-4)$.

For each of the given functions in Problems 14–25, find $\dfrac{f(a+h) - f(a)}{h}$.

14. $f(x) = 4x + 5$

15. $f(x) = -7x - 2$

16. $f(x) = x^2 - 3x$

17. $f(x) = -x^2 + 4x - 2$

18. $f(x) = 2x^2 + 7x - 4$

19. $f(x) = 3x^2 - x - 4$

20. $f(x) = x^3$

21. $f(x) = x^3 - x^2 + 2x - 1$

22. $f(x) = \dfrac{1}{x+1}$

23. $f(x) = \dfrac{2}{x-1}$

24. $f(x) = \dfrac{x}{x+1}$

25. $f(x) = \dfrac{1}{x^2}$

For Problems 26–33 (Figures 4.6 through 4.13), determine whether or not the indicated graph represents a function of x.

26.

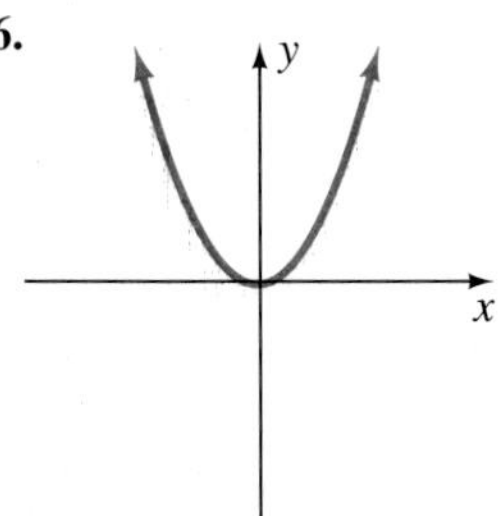

FIGURE 4.6

27.

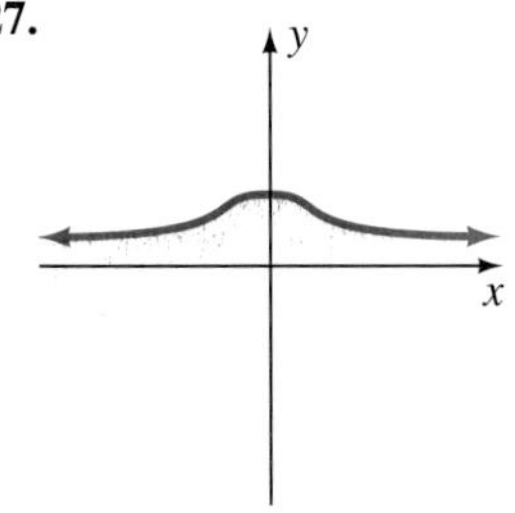

FIGURE 4.7

28.

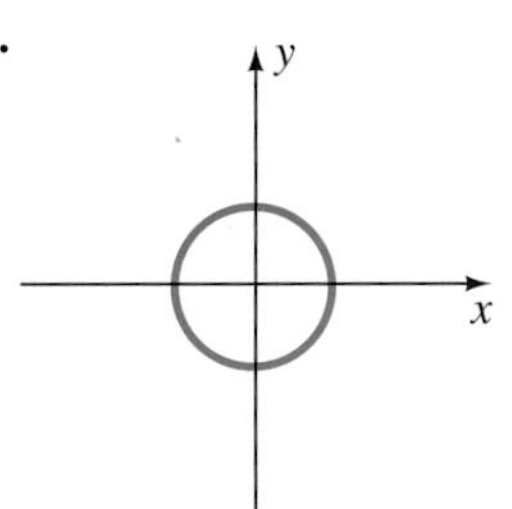

FIGURE 4.8

29.

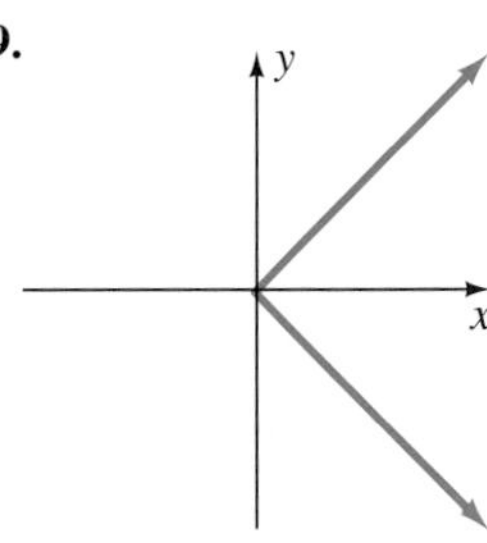

FIGURE 4.9

30.

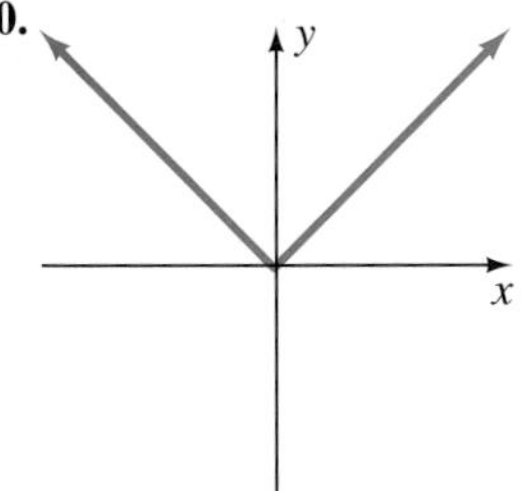

FIGURE 4.10

31.

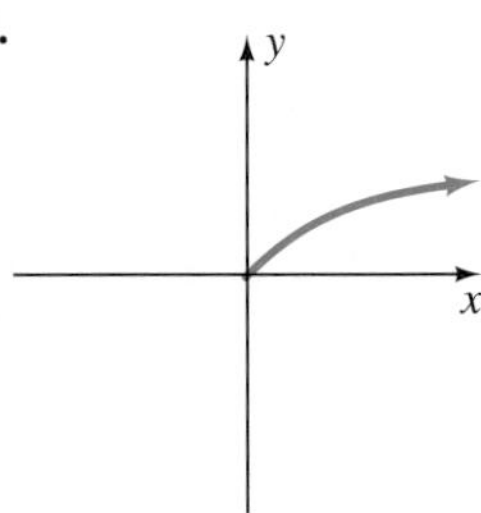

FIGURE 4.11

32.

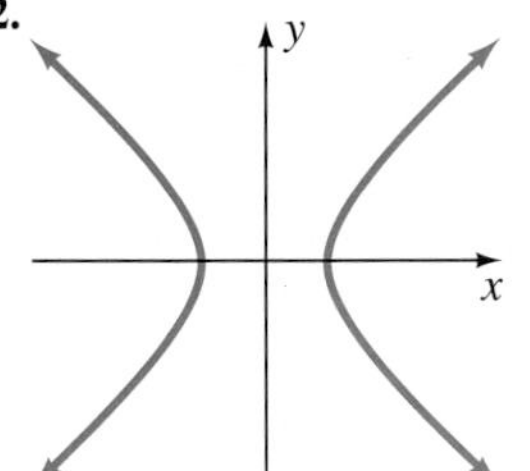

FIGURE 4.12

33.

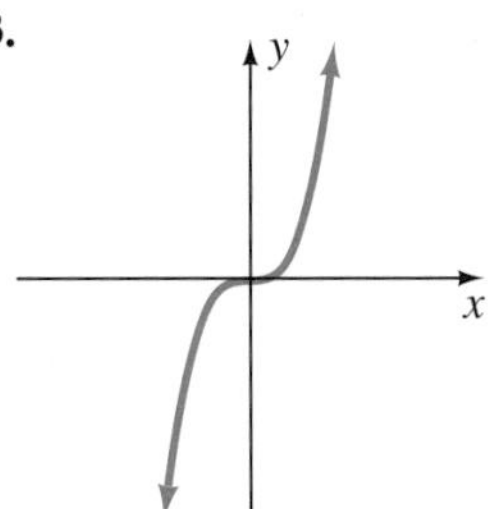

FIGURE 4.13

For Problems 34–41, determine the domain and the range of the given function.

34. $f(x) = \sqrt{x}$

35. $f(x) = \sqrt{3x - 4}$

36. $f(x) = x^2 + 1$

37. $f(x) = x^2 - 2$

38. $f(x) = x^3$

39. $f(x) = |x|$

40. $f(x) = x^4$

41. $f(x) = -\sqrt{x}$

For Problems 42–51, determine the domain of the given function.

42. $f(x) = \dfrac{3}{x-4}$

43. $f(x) = \dfrac{-4}{x+2}$

44. $f(x) = \frac{2x}{(x-2)(x+3)}$

45. $f(x) = \frac{5}{(2x-1)(x+4)}$

46. $f(x) = \sqrt{5x+1}$

47. $f(x) = \frac{1}{x^2-4}$

48. $g(x) = \frac{3}{x^2+5x+6}$

49. $f(x) = \frac{4x}{x^2-x-12}$

50. $g(x) = \frac{5}{x^2+4x}$

51. $g(x) = \frac{x}{6x^2+13x-5}$

For Problems 52–59, express the domain of the given function using interval notation.

52. $f(x) = \sqrt{x^2-1}$

53. $f(x) = \sqrt{x^2-16}$

54. $f(x) = \sqrt{x^2+4}$

55. $f(x) = \sqrt{x^2+1} - 4$

56. $f(x) = \sqrt{x^2-2x-24}$

57. $f(x) = \sqrt{x^2-3x-40}$

58. $f(x) = \sqrt{12x^2+x-6}$

59. $f(x) = -\sqrt{8x^2+6x-35}$

60. Suppose that the profit function for selling n items is given by

$$P(n) = -n^2 + 500n - 61500.$$

Evaluate $P(200)$, $P(230)$, $P(250)$, and $P(260)$.

61. The equation $A(r) = \pi r^2$ expresses the area of a circular region as a function of the length of a radius (r). Compute $A(2)$, $A(3)$, $A(12)$, and $A(17)$ and express your answers to the nearest hundredth.

62. In a physics experiment, it is found that the equation $V(t) = 1667t - 6940t^2$ expresses the velocity of an object as a function of time (t). Compute $V(0.1)$, $V(0.15)$, and $V(0.2)$.

63. The height of a projectile fired vertically into the air (neglecting air resistance) at an initial velocity of 64 feet per second is a function of the time (t) and is given by the equation $h(t) = 64t - 16t^2$. Compute $h(1)$, $h(2)$, $h(3)$, and $h(4)$.

64. A car rental agency charges \$50 per day plus \$.32 a mile. Therefore, the daily charge for renting a car is a function of the number of miles traveled (m) and can be expressed as $C(m) = 50 + 0.32\,m$. Compute $C(75)$, $C(150)$, $C(225)$, and $C(650)$.

65. The equation $I(r) = 500r$ expresses the amount of simple interest earned by an investment of \$500 for one year as a function of the rate of interest (r). Compute $I(0.11)$, $I(0.12)$, $I(0.135)$, and $I(0.15)$.

66. Suppose that the height of a semielliptical archway is given by the function $h(x) = \sqrt{64 - 4x^2}$, where x is the distance from the center line of the arch. Compute $h(0)$, $h(2)$, and $h(4)$.

67. The equation $A(r) = 2\pi r^2 + 16\pi r$ expresses the total surface area of a right circular cylinder of height 8 centimeters as a function of the length of a radius (r). Compute $A(2)$, $A(4)$, and $A(8)$ and express your answers to the nearest hundredth.

For Problems 68–77, determine whether f is even, odd, or neither even nor odd.

68. $f(x) = x^2$

69. $f(x) = x^3$

70. $f(x) = x^2 + 1$

71. $f(x) = 3x - 1$

72. $f(x) = x^2 + x$

73. $f(x) = x^3 + 1$

74. $f(x) = x^5$

75. $f(x) = x^4 + x^2 + 1$

76. $f(x) = -x^3$

77. $f(x) = x^5 + x^3 + x$

4.2 LINEAR AND QUADRATIC FUNCTIONS

As we use the function concept in our study of mathematics, it is helpful to classify certain types of functions and become familiar with their equations, characteristics, and graphs. In this section we will discuss two special types of functions: *linear* and *quadratic functions*. These functions are a natural extension of our earlier study of linear and quadratic equations.

Linear Functions

Any function that can be written in the form

$$f(x) = ax + b,$$

where a and b are real numbers, is called a **linear function**. The following are examples of linear functions.

$$f(x) = -2x + 4 \qquad f(x) = 7x - 9 \qquad f(x) = \frac{2}{3}x + \frac{5}{6}$$

The equation $f(x) = ax + b$ can also be written $y = ax + b$. From our work with the slope-intercept form in Chapter 3, we know that $y = ax + b$ is the equation of a straight line having a slope of a and a y-intercept of b. This information can be used to graph linear functions, as illustrated by the following example.

EXAMPLE 1

Graph $f(x) = -2x + 4$.

Solution

Since the y-intercept is 4, the point (0, 4) is on the line. Furthermore, because the slope is -2, we can move two units down and one unit to the right of (0, 4) to determine the point (1, 2). The line determined by (0, 4) and (1, 2) is drawn in Figure 4.14.

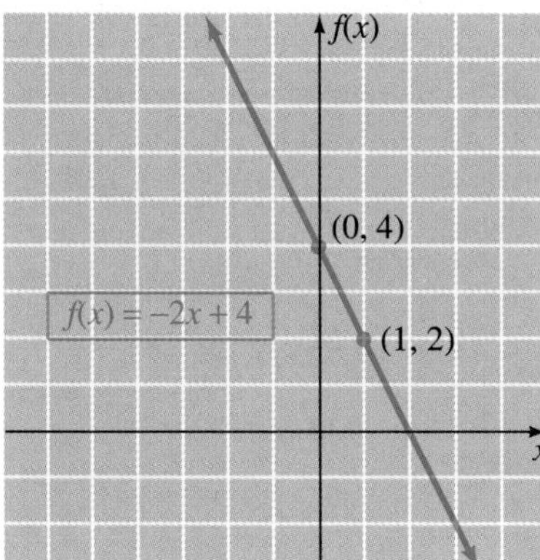

FIGURE 4.14

Note that in Figure 4.14 we labeled the vertical axis $f(x)$. It could also be labeled y, since $y = f(x)$. We will use the $f(x)$ labeling for most of our work with functions; however, we will continue to refer to y-axis symmetry instead of $f(x)$-axis symmetry.

Recall from Chapter 3 that we often graphed linear equations by finding the two intercepts. This same approach can be used with linear functions, as illustrated by the next example.

EXAMPLE 2

Graph $f(x) = 3x - 6$.

Solution

First, we see that $f(0) = -6$; thus, the point $(0, -6)$ is on the graph. Second, by setting $3x - 6$ equal to zero and solving for x, we obtain

$$\begin{aligned} 3x - 6 &= 0 \\ 3x &= 6 \\ x &= 2. \end{aligned}$$

Therefore, $f(2) = 3(2) - 6 = 0$ and the point $(2, 0)$ is on the graph. The line determined by $(0, -6)$ and $(2, 0)$ is drawn in Figure 4.15.

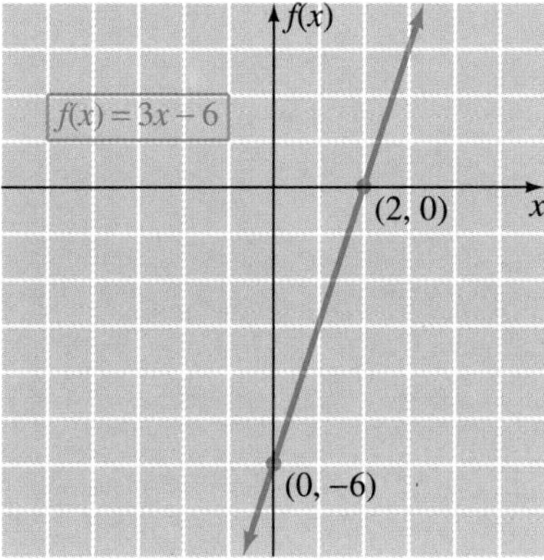

FIGURE 4.15

As you graph functions by using function notation, it is often helpful to think of the ordinate of every point on the graph as the value of the function at a specific value of x. Geometrically, **the functional value is the directed distance of the point from the x-axis**. We have illustrated this idea in Figure 4.16 for the function $f(x) = x$ and in Figure 4.17 for the function $f(x) = 2$.

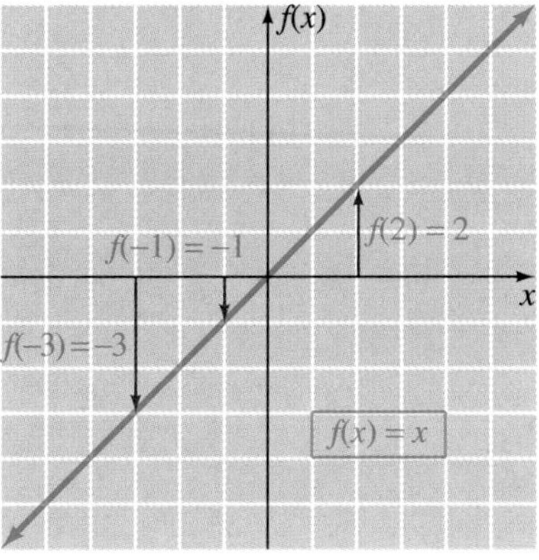

FIGURE 4.16

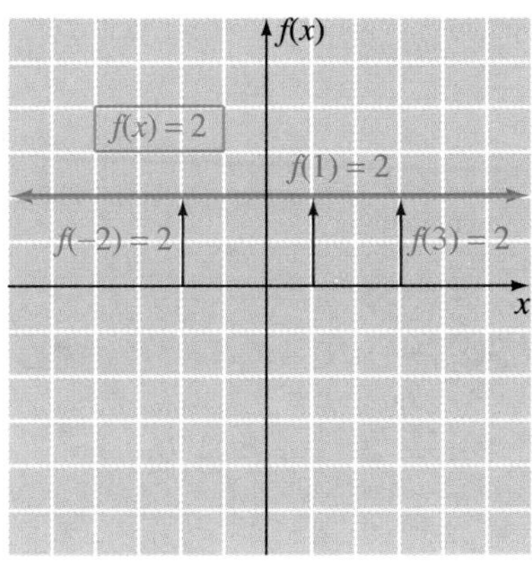

FIGURE 4.17

The linear function $f(x) = x$ is often called the **identity function**. Any linear function of the form $f(x) = ax + b$, where $a = 0$, is called a **constant function** and its graph is a horizontal line.

Quadratic Functions

Any function that can be written in the form

$$f(x) = ax^2 + bx + c,$$

where a, b, and c are real numbers and $a \neq 0$, is called a **quadratic function**. Furthermore, the graph of any quadratic function is a **parabola**. As we work with parabolas, the vocabulary indicated in Figure 4.18 will be used.

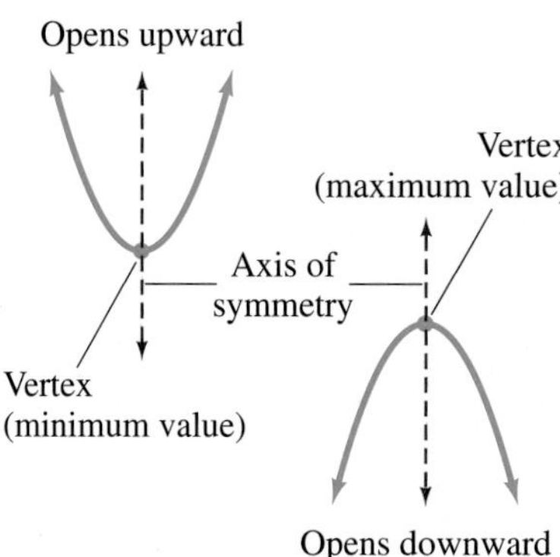

FIGURE 4.18

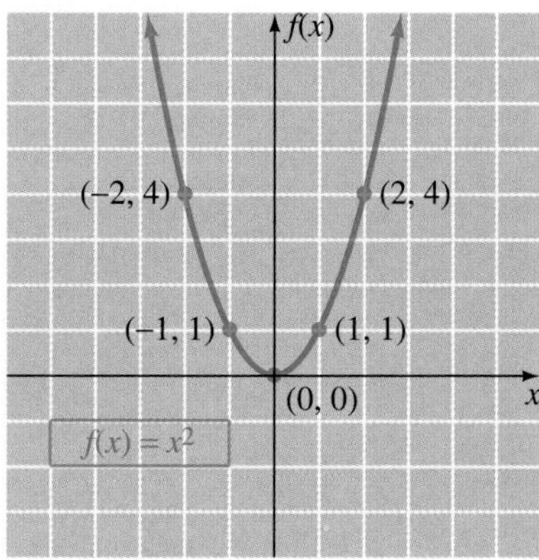

FIGURE 4.19

Graphing parabolas relies on finding the vertex, determining whether the parabola opens upward or downward, and locating two points on opposite sides of the axis of symmetry. It is also very helpful to compare the parabolas produced by various types of equations, such as $f(x) = x^2 + k$, $f(x) = ax^2$, $f(x) = (x - h)^2$, and $f(x) = a(x - h)^2 + k$. We are especially interested in how they compare to the **basic parabola** produced by the equation $f(x) = x^2$. The graph of $f(x) = x^2$ is shown in Figure 4.19. Notice that the graph of $f(x) = x^2$ is symmetric with respect to the y- or $f(x)$-axis. Remember that y-axis symmetry is exhibited by an equation if replacing x with $-x$ produces an equivalent equation. Therefore, since $f(-x) = (-x)^2 = x^2$, the equation $f(x) = x^2$ exhibits y-axis symmetry.

Now let's consider an equation of the form $f(x) = x^2 + k$, where k is a constant. (Keep in mind that all such equations exhibit y-axis symmetry.)

EXAMPLE 3

Graph $f(x) = x^2 - 2$.

Solution

It should be observed that functional values for $f(x) = x^2 - 2$ are 2 less than corresponding functional values for $f(x) = x^2$. For example, $f(1) = -1$ for $f(x) =$

$x^2 - 2$, but $f(1) = 1$ for $f(x) = x^2$. Thus, the graph of $f(x) = x^2 - 2$ is the same as the graph of $f(x) = x^2$ except *moved down 2 units* (Figure 4.20).

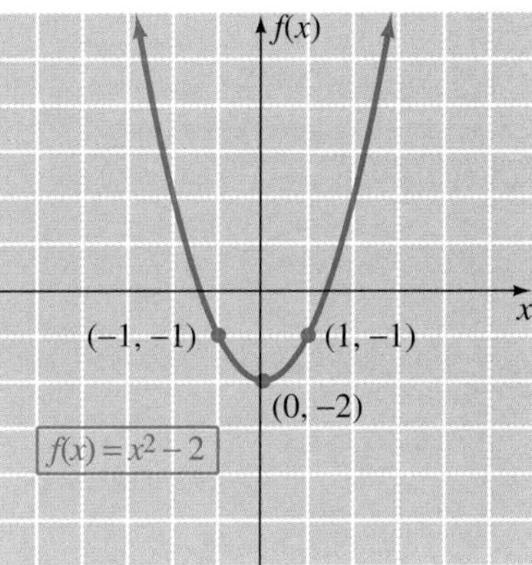

FIGURE 4.20

In general, the graph of a quadratic function of the form $f(x) = x^2 + k$ is the same as the graph of $f(x) = x^2$ except moved up or down k units, depending on whether k is positive or negative. We say that the graph of $f(x) = x^2 + k$ is a vertical translation of the graph of $f(x) = x^2$.

Now let's consider some quadratic functions of the form $f(x) = ax^2$, where a is a nonzero constant. (The graphs of these equations also have y-axis symmetry.)

EXAMPLE 4

Graph $f(x) = 2x^2$.

Solution

Let's set up a table to make some comparisons of functional values. Notice in the table that the functional values for $f(x) = 2x^2$ are *twice* the corresponding functional values for $f(x) = x^2$. Thus, the parabola associated with $f(x) = 2x^2$ has the same vertex (the origin) as the graph of $f(x) = x^2$, but it is *narrower*, as shown in Figure 4.21.

x	$f(x) = x^2$	$f(x) = 2x^2$
0	0	0
1	1	2
2	4	8
−1	1	2
−2	4	8

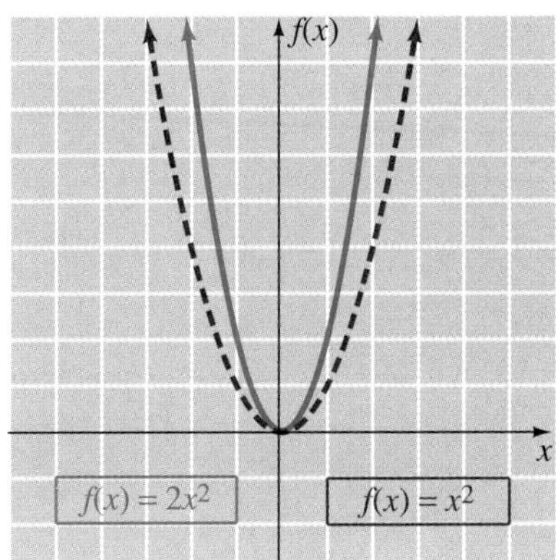

FIGURE 4.21

EXAMPLE 5

Graph $f(x) = \frac{1}{2}x^2$.

Solution

As we see from the table, the functional values for $f(x) = \frac{1}{2}x^2$ are *one-half* of the corresponding functional values for $f(x) = x^2$. Therefore, the parabola associated with $f(x) = \frac{1}{2}x^2$ is *wider* than the basic parabola, as shown in Figure 4.22.

x	$f(x) = x^2$	$f(x) = \frac{1}{2}x^2$
0	0	0
1	1	$\frac{1}{2}$
2	4	2
−1	1	$\frac{1}{2}$
−2	4	2

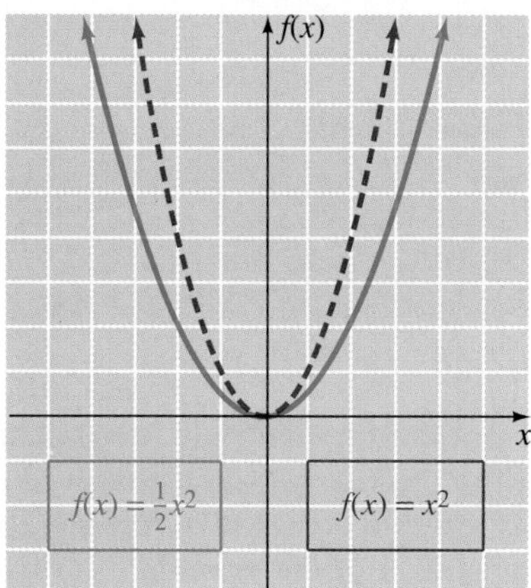

FIGURE 4.22

EXAMPLE 6

Graph $f(x) = -x^2$.

Solution

It should be evident that the functional values for $f(x) = -x^2$ are the *opposites* of the corresponding functional values for $f(x) = x^2$. Therefore, the graph of $f(x) = -x^2$ is a reflection across the x-axis of the basic parabola (Figure 4.23).

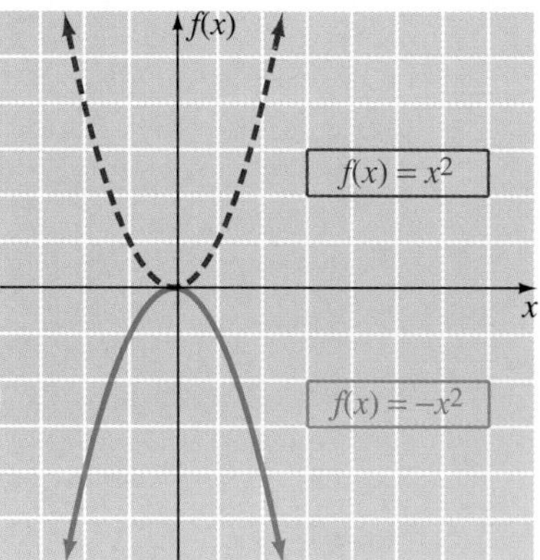

FIGURE 4.23

In general, the graph of a quadratic function of the form $f(x) = ax^2$ has its vertex at the origin and opens upward if a is positive and downward if a is negative. The parabola is narrower than the basic parabola if $|a| > 1$ and wider if $|a| < 1$.

Let's continue our investigation of quadratic functions by considering those of the form $f(x) = (x - h)^2$, where h is a nonzero constant.

EXAMPLE 7

Graph $f(x) = (x - 3)^2$.

Solution

A fairly extensive table of values illustrates a pattern. Notice that $f(x) = (x - 3)^2$ and $f(x) = x^2$ take on the same functional values, *but* for different values of x. More specifically, if $f(x) = x^2$ achieves a certain functional value at a specific value of x, then $f(x) = (x - 3)^2$ achieves that same functional value at x *plus three*. In other words, the graph of $f(x) = (x - 3)^2$ is the graph of $f(x) = x^2$ *moved three units to the right* (Figure 4.24).

x	$f(x) = x^2$	$f(x) = (x - 3)^2$
−1	1	16
0	0	9
1	1	4
2	4	1
3	9	0
4	16	1
5	25	4
6	36	9
7	49	16

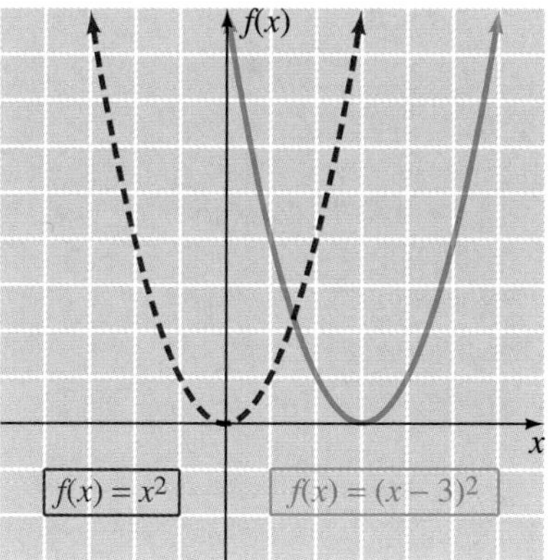

FIGURE 4.24

In general, the graph of a quadratic function of the form $f(x) = (x - h)^2$ is the same as the graph of $f(x) = x^2$ except moved to the right h units if h is positive or moved to the left h units if h is negative. We say that the graph of $f(x) = (x - h)^2$ is a horizontal translation of the graph of $f(x) = x^2$.

The following diagram summarizes our work thus far for graphing quadratic functions.

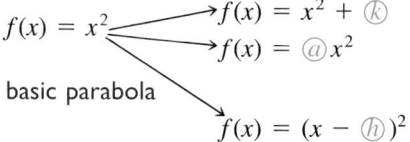

$f(x) = x^2$ (basic parabola)	$f(x) = x^2 + (k)$	Moves the parabola up or down
	$f(x) = (a)x^2$	Affects the width and the way the parabola opens
	$f(x) = (x - (h))^2$	Moves the parabola right or left

Now let's consider two examples that combine the previous ideas.

EXAMPLE 8

Graph $f(x) = 3(x - 2)^2 + 1$.

Solution

$$f(x) = 3(x - 2)^2 + 1$$

Narrows the parabola and opens it upward	Moves the parabola 2 units to the right	Moves the parabola 1 unit up

The vertex is (2, 1) and the line $x = 2$ is the axis of symmetry. If $x = 1$, then $f(1) = 3(1 - 2)^2 + 1 = 4$. Thus, the point (1, 4) is on the graph and so is its reflection, (3, 4), across the line of symmetry. The parabola is drawn in Figure 4.25.

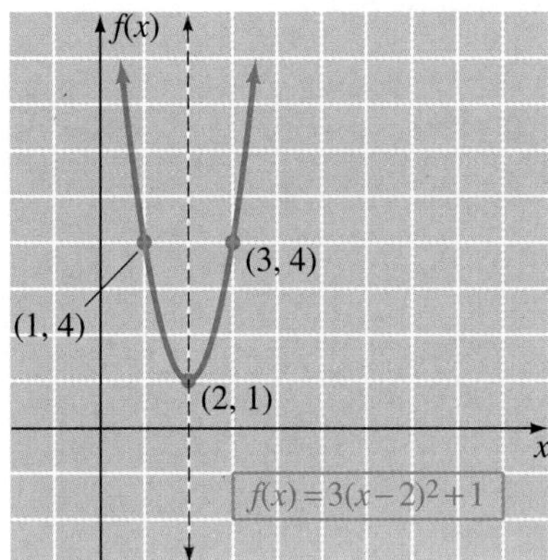

FIGURE 4.25

EXAMPLE 9

Graph $f(x) = -\frac{1}{2}(x + 1)^2 - 3$.

Solution

$$f(x) = -\frac{1}{2}[x - (-1)]^2 - 3$$

Widens the parabola and opens it downward	Moves the parabola 1 unit to the left	Moves the parabola 3 units down

The vertex is at $(-1, -3)$ and the line $x = -1$ is the axis of symmetry. If $x = 0$, then $f(0) = -\frac{1}{2}(0 + 1)^2 - 3 = -\frac{7}{2}$. So the point $\left(0, -\frac{7}{2}\right)$ is on the graph and so is its reflection, $\left(-2, -\frac{7}{2}\right)$, across the line of symmetry. The parabola is drawn in Figure 4.26.

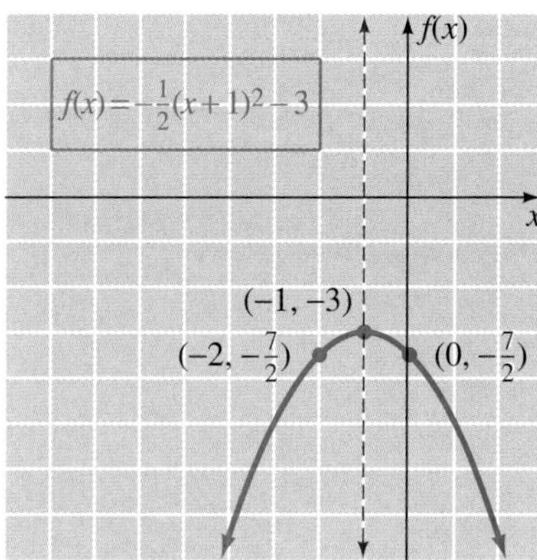

FIGURE 4.26

Quadratic Functions of the Form $f(x) = ax^2 + bx + c$

We are now ready to graph quadratic functions of the form $f(x) = ax^2 + bx + c$. The general approach is one of changing from the form $f(x) = ax^2 + bx + c$ to the form $f(x) = a(x - h)^2 + k$ and then proceeding as we did in Examples 8 and 9. The process of *completing the square* serves as the basis for making the change in form. Let's consider two examples to illustrate the details.

EXAMPLE 10

Graph $f(x) = x^2 - 4x + 3$.

Solution

$$\begin{aligned} f(x) &= x^2 - 4x + 3 \\ &= (x^2 - 4x \quad\;) + 3 \\ &= (x^2 - 4x + 4) + 3 - 4 \\ &= (x - 2)^2 - 1 \end{aligned}$$

Add 4, which is the square of one-half of the coefficient of x.

Subtract 4 to compensate for the 4 that was added.

The graph of $f(x) = (x - 2)^2 - 1$ is the basic parabola moved 2 units to the right and 1 unit down (Figure 4.27).

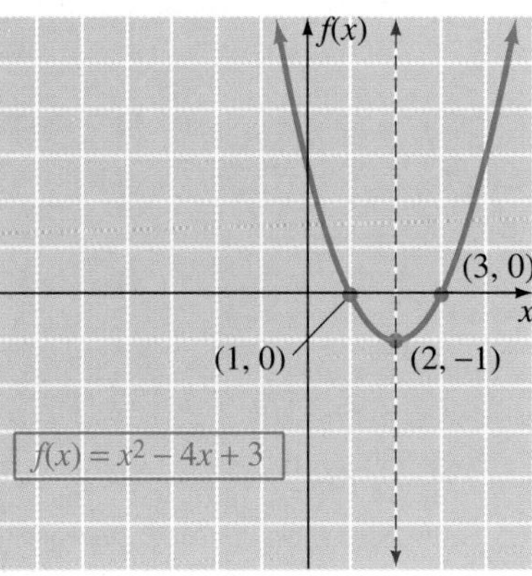

FIGURE 4.27

EXAMPLE 11

Graph $f(x) = -2x^2 - 4x + 1$.

Solution

$$\begin{aligned} f(x) &= -2x^2 - 4x + 1 \\ &= -2(x^2 + 2x \quad) + 1 && \text{Factor } -2 \text{ from the first two terms.} \\ &= -2(x^2 + 2x + 1) + 1 + 2 && \text{Add 1 inside the parentheses to complete the square.} \\ & && \text{Add 2 to compensate for the 1 inside the parentheses times the factor } -2. \\ &= -2(x + 1)^2 + 3 \end{aligned}$$

The graph of $f(x) = -2(x + 1)^2 + 3$ is drawn in Figure 4.28.

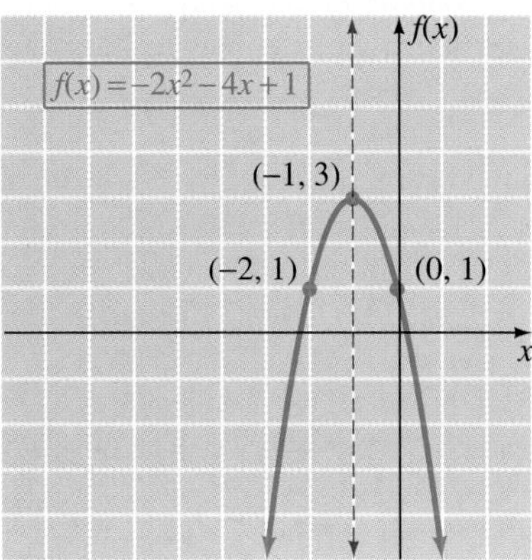

FIGURE 4.28

Now let's graph a piecewise-defined function involving both linear and quadratic rules of assignment.

EXAMPLE 12

Graph $f(x) = \begin{cases} 2x & \text{for } x \geq 0 \\ x^2 + 1 & \text{for } x < 0 \end{cases}$.

Solution

If $x \geq 0$, then $f(x) = 2x$. Thus, for nonnegative values of x we graph the linear function $f(x) = 2x$. If $x < 0$, then $f(x) = x^2 + 1$. Thus, for negative values of x we graph the quadratic function $f(x) = x^2 + 1$. The complete graph is shown in Figure 4.29.

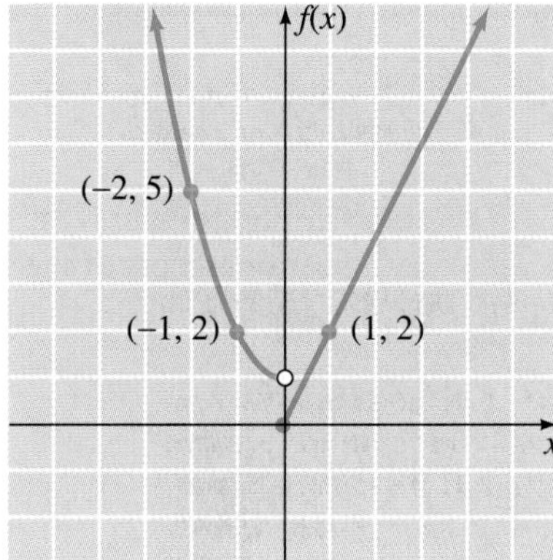

FIGURE 4.29

What we know about parabolas and the process of completing the square can be helpful when using a graphing utility to graph a quadratic function. Consider the following example.

EXAMPLE 13

Use a graphing utility to obtain the graph of the quadratic function

$$f(x) = -x^2 + 37x - 311.$$

Solution

First, we know that the parabola opens downward and its width is the same as that of the basic parabola $f(x) = x^2$. Then we can start the process of completing the square to determine an approximate location of the vertex.

$$\begin{aligned} f(x) &= -x^2 + 37x - 311 \\ &= -(x^2 - 37x \quad) - 311 \\ &= -\left(x^2 - 37x + \left(\frac{37}{2}\right)^2\right) - 311 + \left(\frac{37}{2}\right)^2 \\ &= -(x^2 - 37x + (18.5)^2) - 311 + 342.25 \end{aligned}$$

Thus, the vertex is near $x = 18$ and $y = 31$. Therefore, setting the boundaries of the viewing rectangle so that $-2 \leq x \leq 25$ and $-10 \leq y \leq 35$ we obtain the graph shown in Figure 4.30.

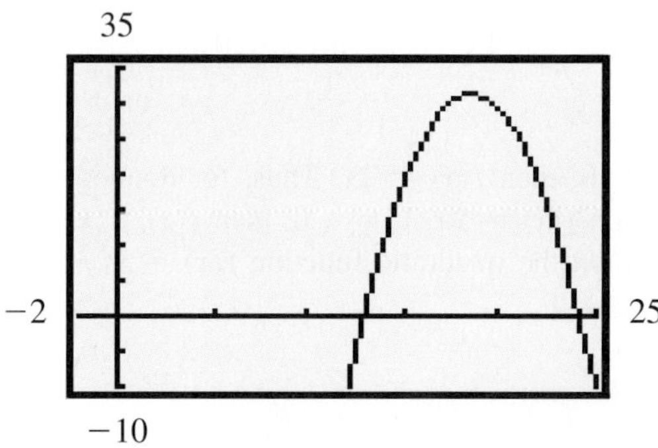

FIGURE 4.30

REMARK The graph in Figure 4.30 is sufficient for most purposes since it shows the vertex and the x-intercepts of the parabola. Certainly other boundaries could be used that also would give this information.

PROBLEM SET 4.2

Graph each of the following linear functions.

1. $f(x) = 2x - 4$
2. $f(x) = 3x + 6$
3. $f(x) = -x + 1$
4. $f(x) = -2x - 4$
5. $f(x) = -2x$
6. $f(x) = 3x$
7. $f(x) = \frac{1}{2}x - \frac{3}{4}$
8. $f(x) = -\frac{2}{3}x + \frac{1}{2}$
9. $f(x) = -1$
10. $f(x) = -3$

Graph each of the following quadratic functions.

11. $f(x) = x^2 + 1$
12. $f(x) = x^2 - 3$
13. $f(x) = 3x^2$
14. $f(x) = -2x^2$
15. $f(x) = -x^2 + 2$
16. $f(x) = -3x^2 - 1$
17. $f(x) = (x + 2)^2$
18. $f(x) = (x - 1)^2$
19. $f(x) = -2(x + 1)^2$
20. $f(x) = 3(x - 2)^2$
21. $f(x) = (x - 1)^2 + 2$
22. $f(x) = -(x + 2)^2 + 3$
23. $f(x) = \frac{1}{2}(x - 2)^2 - 3$
24. $f(x) = 2(x - 3)^2 - 1$
25. $f(x) = x^2 + 2x + 4$
26. $f(x) = x^2 - 4x + 2$
27. $f(x) = x^2 - 3x + 1$
28. $f(x) = x^2 + 5x + 5$
29. $f(x) = 2x^2 + 12x + 17$
30. $f(x) = 3x^2 - 6x$
31. $f(x) = -x^2 - 2x + 1$
32. $f(x) = -2x^2 + 12x - 16$
33. $f(x) = 2x^2 - 2x + 3$
34. $f(x) = 2x^2 + 3x - 1$
35. $f(x) = -2x^2 - 5x + 1$
36. $f(x) = -3x^2 + x - 2$

Graph each of the following functions.

37. $f(x) = \begin{cases} x & \text{for } x \geq 0 \\ 3x & \text{for } x < 0 \end{cases}$

38. $f(x) = \begin{cases} -x & \text{for } x \geq 0 \\ 4x & \text{for } x < 0 \end{cases}$

39. $f(x) = \begin{cases} 2x + 1 & \text{for } x \geq 0 \\ x^2 & \text{for } x < 0 \end{cases}$

40. $f(x) = \begin{cases} -x^2 & \text{for } x \geq 0 \\ 2x^2 & \text{for } x < 0 \end{cases}$

41. $f(x) = \begin{cases} 2 & \text{if } x \geq 0 \\ -1 & \text{if } x < 0 \end{cases}$

42. $f(x) = \begin{cases} 2 & \text{if } x > 2 \\ 1 & \text{if } 0 < x \leq 2 \\ -1 & \text{if } x \leq 0 \end{cases}$

43. $f(x) = \begin{cases} 1 & \text{if } 0 \leq x < 1 \\ 2 & \text{if } 1 \leq x < 2 \\ 3 & \text{if } 2 \leq x < 3 \\ 4 & \text{if } 3 \leq x < 4 \end{cases}$

44. $f(x) = \begin{cases} 2x + 3 & \text{if } x < 0 \\ x^2 & \text{if } 0 \leq x < 2 \\ 1 & \text{if } x \geq 2 \end{cases}$

45. The **greatest integer function** is defined by the equation $f(x) = [x]$, where $[x]$ refers to the largest integer less than or equal to x. For example, $[2.6] = 2$, $[\sqrt{2}] = 1$, $[4] = 4$, and $[-1.4] = -2$. Graph $f(x) = [x]$ for $-4 \leq x < 4$.

THOUGHTS into WORDS

46. Does the equation $x^2 + y^2 = 4$ define a function or a relation? Explain your answer.

47. Explain the concept of a piecewise-defined function.

48. Suppose that Julian walks at a constant rate of 3 miles per hour. Explain what it means to say that the distance Julian walks is a *function* of the time that he walks.

GRAPHICS CALCULATOR ACTIVITIES

49. This problem is designed to reinforce some graphing ideas presented in this section. For each part, first predict the shapes and locations of the parabolas, and then use your graphics calculator to graph them on the same set of axes.

a. $f(x) = x^2$, $f(x) = x^2 - 4$, $f(x) = x^2 + 1$, $f(x) = x^2 + 5$

b. $f(x) = x^2$, $f(x) = (x - 5)^2$, $f(x) = (x + 5)^2$, $f(x) = (x - 3)^2$

c. $f(x) = x^2$, $f(x) = 5x^2$, $f(x) = \frac{1}{3}x^2$, $f(x) = -2x^2$

d. $f(x) = x^2$, $f(x) = (x - 7)^2 - 3$, $f(x) = -(x + 8)^2 + 4$, $f(x) = -3x^2 - 4$

e. $f(x) = x^2 - 4x - 2$, $f(x) = -x^2 + 4x + 2$, $f(x) = -x^2 - 16x - 58$, $f(x) = x^2 + 16x + 58$

50. **a.** Graph both $f(x) = x^2 - 14x + 51$ and $f(x) = x^2 + 14x + 51$ on the same set of axes. What relationship seems to exist between the two graphs?

b. Graph both $f(x) = x^2 + 12x + 34$ and $f(x) = x^2 - 12x + 34$ on the same set of axes. What relationship seems to exist between the two graphs?

c. Graph both $f(x) = -x^2 + 8x - 20$ and $f(x) = -x^2 - 8x - 20$ on the same set of axes. What relationship seems to exist between the two graphs?

d. Make a statement that generalizes your findings in parts (a) through (c).

51. Use your graphics calculator to graph the piecewise-defined functions in Problems 37–44. You may need to consult your user's manual for instructions on graphing these functions. (This problem is designed to help you become more familiar with the capabilities and limitations of your graphics calculator.)

4.3 QUADRATIC FUNCTIONS AND PROBLEM SOLVING

In the previous section we used the process of completing the square to change a quadratic function such as $f(x) = x^2 - 4x + 3$ to the form $f(x) = (x - 2)^2 - 1$. From the form $f(x) = (x - 2)^2 - 1$, the vertex $(2, -1)$ and the axis of symmetry $x = 2$ of the parabola are easy to identify. In general, if we complete the square on

$$f(x) = ax^2 + bx + c,$$

we obtain

$$\begin{aligned} f(x) &= a\left(x^2 + \frac{b}{a}x\right) + c \\ &= a\left(x^2 + \frac{b}{a}x + \frac{b^2}{4a^2}\right) + c - \frac{b^2}{4a} \\ &= a\left(x + \frac{b}{2a}\right)^2 + \frac{4ac - b^2}{4a}. \end{aligned}$$

Therefore, the parabola associated with the function $f(x) = ax^2 + bx + c$ has its vertex at

$$\left(-\frac{b}{2a}, \frac{4ac - b^2}{4a}\right)$$

and the equation of its axis of symmetry is $x = -b/2a$. These facts are illustrated in Figure 4.31.

By using the information from Figure 4.31 we now have another way of graphing quadratic functions of the form $f(x) = ax^2 + bx + c$, as indicated by the following steps.

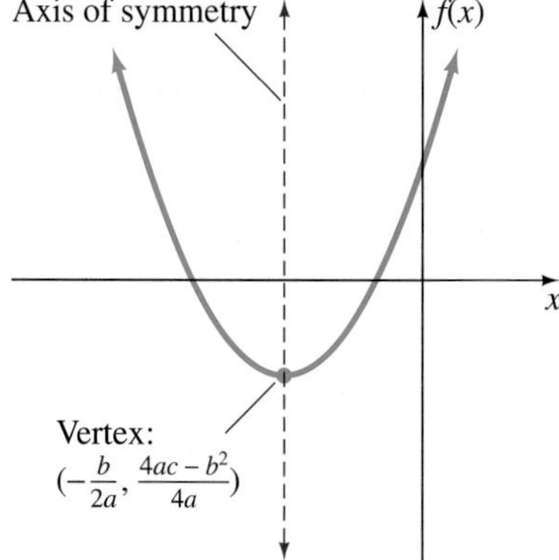

FIGURE 4.31

1. Determine whether the parabola opens upward (if $a > 0$) or downward (if $a < 0$).
2. Find $-b/2a$, which is the x-coordinate of the vertex.
3. Find $f(-b/2a)$, which is the y-coordinate of the vertex, or find the y-coordinate by evaluating

$$\frac{4ac - b^2}{4a}.$$

4. Locate another point on the parabola and also locate its image across the axis of symmetry, which is the line with equation $x = b/2a$.

The three points found in steps 2, 3, and 4 should determine the general shape of the parabola. Let's illustrate this procedure with two examples.

EXAMPLE 1

Graph $f(x) = 3x^2 - 6x + 5$.

Solution

STEP 1 Because $a > 0$, the parabola opens upward.

STEP 2 $-\frac{b}{2a} = -\frac{-6}{6} = 1.$

STEP 3 $f\left(-\frac{b}{2a}\right) = f(1) = 3 - 6 + 5 = 2$. Thus, the vertex is at (1, 2).

STEP 4 Letting $x = 2$, we obtain $f(2) = 12 - 12 + 5 = 5$. Thus, (2, 5) is on the graph and so is its reflection, (0, 5), across the line of symmetry, $x = 1$.

The three points (1, 2), (2, 5), and (0, 5) are used to graph the parabola in Figure 4.32.

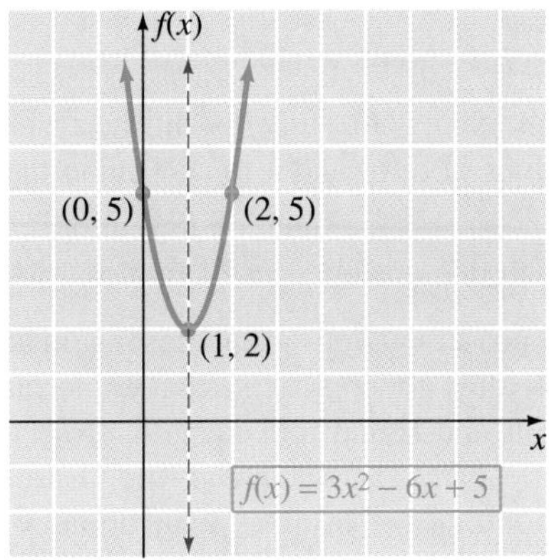

FIGURE 4.32

EXAMPLE 2

Graph $f(x) = -x^2 - 4x - 7$.

Solution

STEP 1 Since $a < 0$, the parabola opens downward.

STEP 2 $-\dfrac{b}{2a} = -\dfrac{-4}{-2} = -2$.

STEP 3 $f\left(-\dfrac{b}{2a}\right) = f(-2) = -(-2)^2 - 4(-2) - 7 = -3$. Thus, the vertex is at $(-2, -3)$.

STEP 4 Letting $x = 0$, we obtain $f(0) = -7$. Thus, $(0, -7)$ is on the graph and so is its reflection, $(-4, -7)$, across the line of symmetry, $x = -2$.

The three points $(-2, -3)$, $(0, -7)$, and $(-4, -7)$ are used to draw the parabola in Figure 4.33.

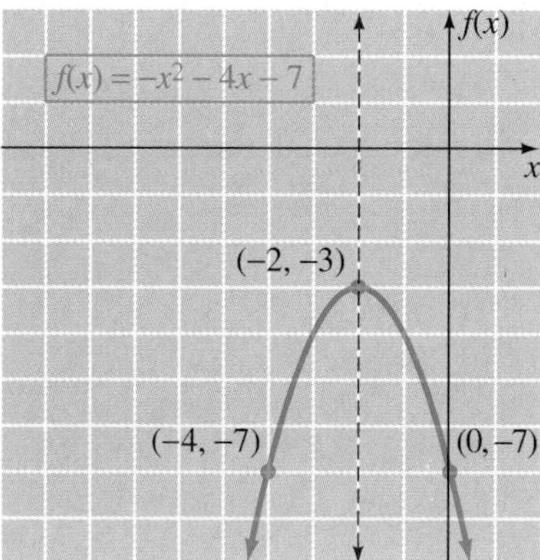

FIGURE 4.33

In summary, to graph a quadratic function we basically have two methods.

1. We can express the function in the form $f(x) = a(x - h)^2 + k$ and use the values of a, h, and k to determine the parabola.
2. We can express the function in the form $f(x) = ax^2 + bx + c$ and use the approach demonstrated in Examples 1 and 2.

Parabolas possess various properties that make them very useful. For example, if a parabola is rotated about its axis, a parabolic surface is formed and such surfaces are used for light and sound reflectors. A projectile fired into the air will follow the curvature of a parabola. The *trend line* of profit and cost functions sometimes follows a parabolic curve. In most applications of the parabola, we are primarily interested in the x-intercepts and the vertex. Let's consider some examples of finding the x-intercepts and the vertex.

EXAMPLE 3

Find the x-intercepts and the vertex for each of the following parabolas.

a. $f(x) = -x^2 + 11x - 18$ **b.** $f(x) = x^2 - 8x - 3$

c. $f(x) = 2x^2 - 12x + 23$

Solutions

a. To find the x-intercepts, let $y = 0$ and solve the resulting equation.

$$\begin{aligned} -x^2 + 11x - 18 &= 0 \\ x^2 - 11x + 18 &= 0 \\ (x - 2)(x - 9) &= 0 \end{aligned}$$

$$\begin{aligned} x - 2 &= 0 \quad \text{or} \quad & x - 9 &= 0 \\ x &= 2 \quad \text{or} \quad & x &= 9 \end{aligned}$$

Therefore, the x-intercepts are 2 and 9. To find the vertex, let's determine the point $\left(-\frac{b}{2a}, f\left(-\frac{b}{2a}\right)\right)$.

$$\begin{aligned} f(x) &= -x^2 + 11x - 18 \\ -\frac{b}{2a} &= -\frac{11}{2(-1)} = -\frac{11}{-2} = \frac{11}{2} \\ f\left(\frac{11}{2}\right) &= -\left(\frac{11}{2}\right)^2 + 11\left(\frac{11}{2}\right) - 18 \\ &= -\frac{121}{4} + \frac{121}{2} - 18 \\ &= \frac{-121 + 242 - 72}{4} \\ &= \frac{49}{4} \end{aligned}$$

Therefore, the vertex is at $\left(\frac{11}{2}, \frac{49}{4}\right)$.

b. To find the x-intercepts, let $y = 0$ and solve the resulting equation.

$$\begin{aligned} x^2 - 8x - 3 &= 0 \\ x &= \frac{-(-8) \pm \sqrt{(-8)^2 - 4(1)(-3)}}{2(1)} \\ &= \frac{8 \pm \sqrt{76}}{2} \\ &= \frac{8 \pm 2\sqrt{19}}{2} \\ &= 4 \pm \sqrt{19} \end{aligned}$$

Therefore, the x-intercepts are $4 + \sqrt{19}$ and $4 - \sqrt{19}$. This time, to find the vertex let's complete the square on x.

$$\begin{aligned} f(x) &= x^2 - 8x - 3 \\ &= x^2 - 8x + 16 - 3 - 16 \\ &= (x - 4)^2 - 19 \end{aligned}$$

Therefore the vertex is at $(4, -19)$.

c. To find the x-intercepts, let $y = 0$ and solve the resulting equation.

$$2x^2 - 12x + 23 = 0$$

$$\begin{aligned} x &= \frac{-(-12) \pm \sqrt{(-12)^2 - 4(2)(23)}}{2(2)} \\ &= \frac{12 \pm \sqrt{-40}}{4} \end{aligned}$$

Since these solutions are nonreal complex numbers, there are no x-intercepts. To find the vertex, let's determine the point $\left(-\frac{b}{2a}, f\left(-\frac{b}{2a}\right)\right)$.

$$f(x) = 2x^2 - 12x + 23$$

$$-\frac{b}{2a} = -\frac{-12}{2(2)} = 3$$

$$\begin{aligned} f(3) &= 2(3)^2 - 12(3) + 23 \\ &= 18 - 36 + 23 \\ &= 5 \end{aligned}$$

Therefore, the vertex is at $(3, 5)$.

REMARK Note that in parts (a) and (c) we used the general point $\left(-\frac{b}{2a}, f\left(-\frac{b}{2a}\right)\right)$ to find the vertices. However, in part (b) we completed the square and used that form to determine the vertex. Which approach you use is an individual preference. We chose to complete the square in part (b) because the algebra involved was quite easy.

Back to Problem Solving

As we have seen, the vertex of the graph of a quadratic function is either the lowest or the highest point on the graph. Thus, the vocabulary **minimum value** or **maximum value** of a function is often used in applications of the parabola. The x-value of the vertex indicates where the minimum or maximum occurs and $f(x)$ yields the minimum or maximum value of the function. Let's consider some examples that use these ideas.

EXAMPLE 4

A farmer has 120 rods of fencing and wants to enclose a rectangular plot of land that requires fencing on only three sides, since it is bounded by a river on one side. Find the length and width of the plot that will maximize the area.

Solution

Let x represent the width; then $120 - 2x$ represents the length, as indicated in Figure 4.34. The function $A(x) = x(120 - 2x)$ represents the area of the plot in terms of the width x. Since

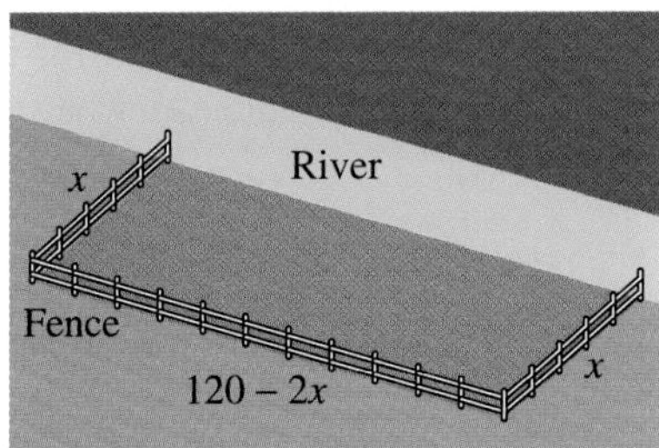

FIGURE 4.34

$$\begin{aligned} A(x) &= x(120 - 2x) \\ &= 120x - 2x^2 \\ &= -2x^2 + 120x, \end{aligned}$$

we have a quadratic function with $a = -2$, $b = 120$, and $c = 0$. Therefore, the *maximum* value ($a < 0$ so the parabola opens downward) of the function is obtained where the x-value is

$$-\frac{b}{2a} = -\frac{120}{2(-2)} = 30.$$

If $x = 30$, then $120 - 2x = 120 - 2(30) = 60$.

Thus, the farmer should make the plot 30 rods wide and 60 rods long to maximize the area at $(30)(60) = 1800$ square rods.

EXAMPLE 5

Find two numbers whose sum is 30, such that the sum of their squares is a minimum.

Solution

Let x represent one of the numbers; then $30 - x$ represents the other number. By expressing the sum of their squares as a function of x, we obtain

$$f(x) = x^2 + (30 - x)^2,$$

which can be simplified to

$$\begin{aligned} f(x) &= x^2 + 900 - 60x + x^2 \\ &= 2x^2 - 60x + 900. \end{aligned}$$

This is a quadratic function with $a = 2$, $b = -60$, and $c = 900$. Therefore, the x-value where the *minimum* occurs is

$$-\frac{b}{2a} = -\frac{-60}{4} = 15.$$

If $x = 15$, then $30 - x = 30 - 15 = 15$. Thus, the two numbers should both be 15.

EXAMPLE 6

A golf pro shop operator finds that she can sell 30 sets of golf clubs at \$500 per set in a year. Furthermore, she predicts that for each \$25 decrease in price, three extra sets of golf clubs could be sold. At what price should she sell the clubs to maximize gross income?

Solution

Sometimes in analyzing such a problem it helps to start by setting up a table, as follows.

	NUMBER OF SETS	PRICE PER SET	INCOME
	30	\$500	\$15,000
Three additional sets can be sold for → a \$25 decrease in price.	33	\$475	\$15,675
	36	\$450	\$16,200

Let x represent the number of \$25 decreases in price. Then the income can be expressed as a function of x.

$$f(x) = \underset{\substack{\uparrow \\ \text{Number of sets}}}{(30 + 3x)}\ \underset{\substack{\uparrow \\ \text{Price per set}}}{(500 - 25x)}.$$

Simplifying this, we obtain

$$\begin{aligned} f(x) &= 15{,}000 - 750x + 1500x - 75x^2 \\ &= -75x^2 + 750x + 15{,}000. \end{aligned}$$

We complete the square in order to analyze the parabola.

$$\begin{aligned} f(x) &= -75x^2 + 750x + 15{,}000 \\ &= -75(x^2 - 10x \quad) + 15{,}000 \\ &= -75(x^2 - 10x + 25) + 15{,}000 + 1875 \\ &= -75(x - 5)^2 + 16{,}875 \end{aligned}$$

From this form we know that the vertex of the parabola is at (5, 16875), and because $a = -75$, we know that a *maximum* occurs at the vertex. So five decreases of \$25, that is, a \$125 reduction in price, will give a maximum income of \$16,875. The golf clubs should be sold at \$375 per set.

We have determined that the vertex of a parabola associated with $f(x) = ax^2 + bx + c$ is located at $\left(-\frac{b}{2a}, f\left(-\frac{b}{2a}\right)\right)$, and the x-intercepts of the graph can be found by solving the quadratic equation $ax^2 + bx + c = 0$. Therefore, a graphing utility does not provide us with much extra firepower when working with quadratic functions. However, as functions become more complex, a graphing utility becomes more helpful. So let's build our confidence in the use of a graphing utility at this time when we have a way of checking our results.

EXAMPLE 7

Use a graphing utility to graph $f(x) = x^2 - 8x - 3$ and find the x-intercepts of the graph. (This is the parabola from part (b) of Example 3.)

Solution

A graph of the parabola is shown in Figure 4.35.

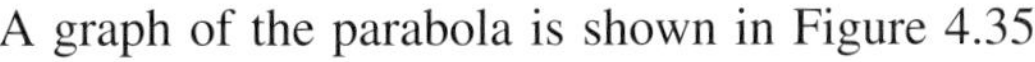

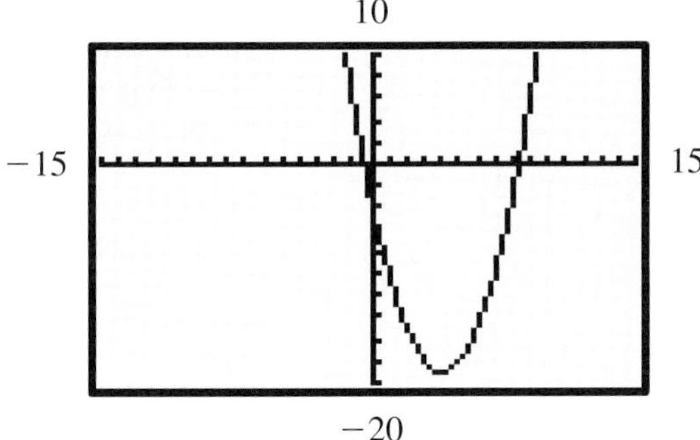

FIGURE 4.35

One x-intercept appears to be between 0 and −1 and the other between 8 and 9. Let's zoom in on the x-intercept between 8 and 9. This produces a graph like Figure 4.36.

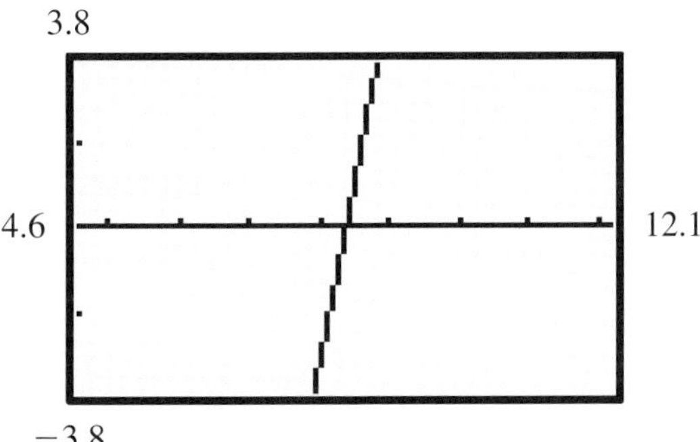

FIGURE 4.36

Now we can use the trace function to determine that this x-intercept is at approximately 8.4. (This agrees with the answer of $4 + \sqrt{19}$ obtained in Example 3.) In a similar fashion we can determine that the other x-intercept is at -0.4.

PROBLEM SET 4.3

Use the approach of Examples 1 and 2 of this section to graph each of the following quadratic functions.

1. $f(x) = x^2 - 8x + 15$

2. $f(x) = x^2 + 6x + 11$

3. $f(x) = 2x^2 + 20x + 52$

4. $f(x) = 3x^2 - 6x - 1$

5. $f(x) = -x^2 + 4x - 7$

6. $f(x) = -x^2 - 6x - 5$

7. $f(x) = -3x^2 + 6x - 5$

8. $f(x) = -2x^2 - 4x + 2$

9. $f(x) = x^2 + 3x - 1$

10. $f(x) = x^2 + 5x + 2$

11. $f(x) = -2x^2 + 5x + 1$

12. $f(x) = -3x^2 + 2x - 1$

For Problems 13–20, use the approach that you think is the most appropriate to graph each of the quadratic functions.

13. $f(x) = -x^2 + 3$

14. $f(x) = (x + 1)^2 + 1$

15. $f(x) = x^2 + x - 1$

16. $f(x) = -x^2 + 3x - 4$

17. $f(x) = -2x^2 + 4x + 1$

18. $f(x) = 4x^2 - 8x + 5$

19. $f(x) = -\left(x + \frac{5}{2}\right)^2 + \frac{3}{2}$

20. $f(x) = x^2 - 4x$

For Problems 21–32, find the x-intercepts and the vertex for each parabola.

21. $f(x) = x^2 - 8x + 15$

22. $f(x) = x^2 - 16x + 63$

23. $f(x) = 2x^2 - 28x + 96$

24. $f(x) = 3x^2 - 60x + 297$

25. $f(x) = -x^2 + 10x - 24$

26. $f(x) = -2x^2 + 36x - 160$

27. $f(x) = x^2 - 14x + 44$

28. $f(x) = x^2 - 18x + 68$

29. $f(x) = -x^2 + 9x - 21$

30. $f(x) = 2x^2 + 3x + 3$

31. $f(x) = -4x^2 + 4x + 4$

32. $f(x) = -2x^2 + 3x + 7$

33. Suppose that the equation $p(x) = -2x^2 + 280x - 1000$, where x represents the number of items sold, describes the profit function for a certain business. How many items should be sold to maximize the profit?

34. Suppose that the cost function for the production of a particular item is given by the equation $C(x) = 2x^2 - 320x + 12{,}920$, where x represents the number of items. How many items should be produced to minimize the cost?

35. The height of a projectile fired vertically into the air (neglecting air resistance) at an initial velocity of 96 feet per second is a function of time x and is given by the equation $f(x) = 96x - 16x^2$. Find the highest point reached by the projectile.

36. Find two numbers whose sum is 30, such that the sum of the square of one number plus ten times the other number is a minimum.

37. Find two numbers whose sum is 50 and whose product is a maximum.

38. Find two numbers whose difference is 40 and whose product is a minimum.

39. Two hundred and forty meters of fencing is available to enclose a rectangular playground. What should be the dimensions of the playground to maximize the area?

40. A motel advertises that they will provide dinner, dancing, and drinks for \$50 per couple for a New Year's Eve party. They must have a guarantee of 30 couples. Furthermore, they will agree that for each couple in excess of 30, they will reduce the price per couple for all attending by \$.50. How many couples will it take to maximize the motel's revenue?

41. A cable TV company has 1000 subscribers who each pay \$15 per month. Based on a survey, they feel that for each decrease of \$.25 on the monthly rate, they could obtain 20 additional subscribers. At what rate will maximum revenue be obtained and how many subscribers will it take at that rate?

42. A manufacturer finds that for the first 500 units of his product that are produced and sold, the profit is \$50 per unit. The profit on each of the units beyond 500 is decreased by \$.10 times the number of additional units sold. What level of output will maximize profit?

MISCELLANEOUS PROBLEMS

43. Suppose that an arch is shaped like a parabola. It is 20 feet wide at the base and 100 feet high. How wide is the arch 50 feet above the ground? (See Figure 4.37.)

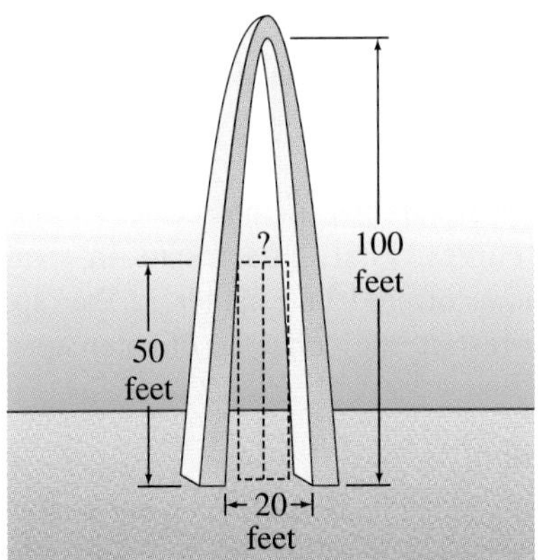

FIGURE 4.37

44. A parabolic arch 27 feet high spans a parkway. If the center section of the parkway is 50 feet wide, how wide is the arch if it has a minimum clearance of 15 feet above the center section?

45. A parabolic arch spans a stream 200 feet wide. How high must the arch be above the stream to give a minimum clearance of 40 feet over a 120-foot-wide channel in the center?

GRAPHICS CALCULATOR ACTIVITIES

46. Suppose that the viewing window on your graphics calculator is set so that $-15 \leq x \leq 15$ and $-10 \leq y \leq 10$. Now try to graph the function $f(x) = x^2 - 8x + 28$. Since nothing appears on the screen, the parabola must be outside the viewing window. We could arbitrarily expand the window until the parabola appears. However, let's be a little more systematic and use $\left(-\frac{b}{2a}, f\left(\frac{-b}{2a}\right)\right)$ to find the vertex; thus we find the vertex is at (4, 2). So let's change the y-values of the window so that $0 \leq y \leq 25$. Now we get a good picture of the parabola.

Graph each of the following parabolas, and keep in mind that you may need to change the dimensions of the viewing window to obtain a good picture.

a. $f(x) = x^2 - 2x + 12$

b. $f(x) = -x^2 - 4x - 16$

c. $f(x) = x^2 + 12x + 44$

d. $f(x) = x^2 - 30x + 229$ **e.** $f(x) = -2x^2 + 8x - 19$

47. For each of the following parabolas use a graphics calculator to graph the parabola, and use the trace function to help estimate the x-intercepts and the vertex. Then use the approach of Example 3 to find the x-intercepts and the vertex.

a. $y = x^2 - 6x + 3$

b. $y = x^2 - 18x + 66$

c. $y = -x^2 + 8x - 3$

d. $y = -x^2 + 24x - 129$

e. $y = 14x^2 - 7x + 1$

f. $y = -\frac{1}{2}x^2 + 5x - \frac{17}{2}$

4.4 TRANSFORMATIONS OF SOME BASIC CURVES

From our work in Section 4.2, we know that the graph of $f(x) = (x - 5)^2$ is the basic parabola, $f(x) = x^2$, translated five units to the right. Likewise, we know that the graph of $f(x) = -x^2 - 2$ is the basic parabola reflected across the x-axis and translated downward two units. Translations and reflections apply not only to parabolas but to curves in general. Therefore, if we know the shapes of a few basic curves, then numerous variations of these curves can be easily sketched using the concepts of translation and reflection.

Let's begin this section by establishing the graphs of four basic curves, and then apply some transformations to these curves. First, let's restate in terms of function vocabulary, the graphing suggestions offered in Chapter 3. Pay special attention to suggestions 2 and 3 where we restate the concepts of intercepts and symmetry using function notation.

1. Determine the domain of the function.
2. Find the y-intercept (we are labeling the y-axis with $f(x)$) by evaluating $f(0)$. Find the x-intercept by finding the value(s) of x such that $f(x) = 0$.
3. Determine any types of symmetry that the equation possesses. If $f(-x) = f(x)$, then the function exhibits y-axis symmetry. If $f(-x) = -f(x)$, then the function exhibits origin symmetry. (Note that the definition of a function rules out the possibility that the graph of a function has x-axis symmetry.)
4. Set up a table of ordered pairs that satisfy the equation. The type of symmetry and the domain will affect your choice of values of x in the table.
5. Plot the points associated with the ordered pairs and connect them with a smooth curve. Then, if appropriate, reflect this part of the curve according to any symmetries possessed by the graph.

EXAMPLE 1

Graph $f(x) = x^3$.

Solution

The domain is the set of real numbers. Since $f(0) = 0$, the origin is on the graph. Because $f(-x) = (-x)^3 = -x^3 = -f(x)$, the graph is symmetrical with respect to the origin. Therefore, we can concentrate our table on the positive values of x. By connecting the points associated with the ordered pairs from the table with a smooth curve and then reflecting it through the origin, we get the graph in Figure 4.38.

x	$f(x) = x^3$
0	0
1	1
2	8
$\frac{1}{2}$	$\frac{1}{8}$

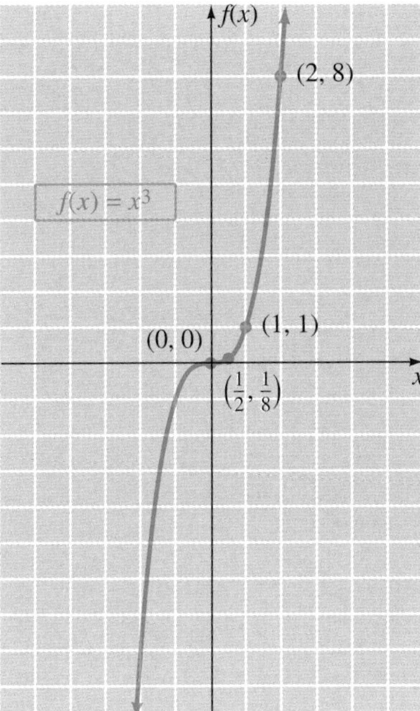

FIGURE 4.38

EXAMPLE 2

Graph $f(x) = x^4$.

Solution

The domain is the set of real numbers. Since $f(0) = 0$, the origin is on the graph. Because $f(-x) = (-x)^4 = x^4 = f(x)$, the graph has y-axis symmetry and we can concentrate our table of values on the positive values of x. If we connect the points associated with the ordered pairs from the table with a smooth curve and then reflect across the vertical axis, we get the graph in Figure 4.39.

x	$f(x) = x^4$
0	0
1	1
2	16
$\frac{1}{2}$	$\frac{1}{16}$

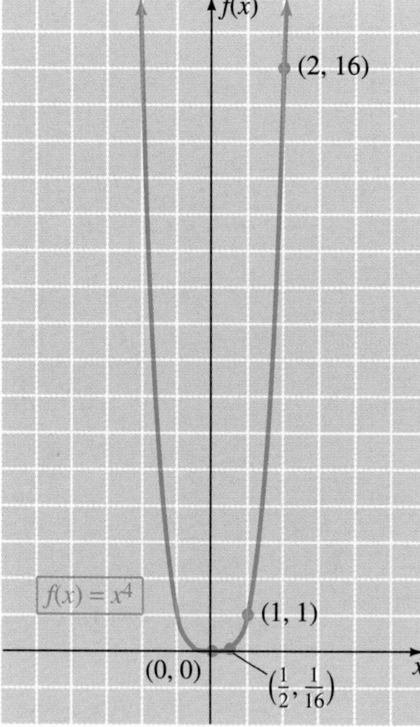

FIGURE 4.39

REMARK The curve in Figure 4.39 is not a parabola, even though it resembles one; this curve is flatter at the bottom and steeper.

EXAMPLE 3

Graph $f(x) = \sqrt{x}$.

Solution

The domain of the function is the set of nonnegative real numbers. Since $f(0) = 0$, the origin is on the graph. Because $f(-x) \neq f(x)$ and $f(-x) \neq -f(x)$, there is no symmetry. So let's set up a table of values using nonnegative values for x.

Plotting the points determined by the table and connecting them with a smooth curve produces Figure 4.40.

x	$f(x) = \sqrt{x}$
0	0
1	1
4	2
9	3

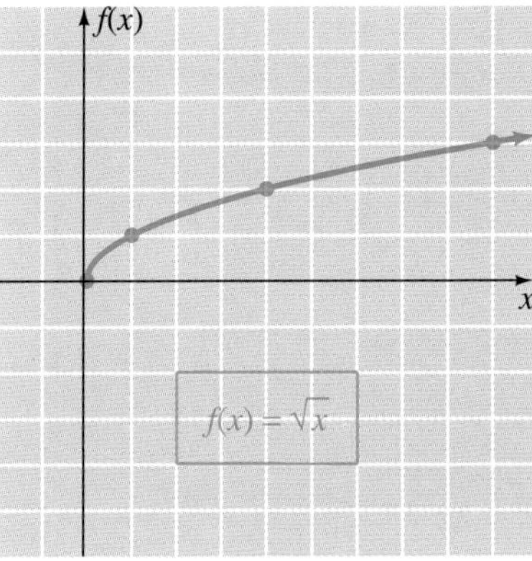

FIGURE 4.40

Sometimes a new function is defined in terms of old functions. In such cases, the definition plays an important role in the study of the new function. Consider the following example.

EXAMPLE 4

Graph $f(x) = |x|$.

Solution

The concept of absolute value is defined for all real numbers by

$$|x| = x \quad \text{if } x \geq 0$$
$$|x| = -x \quad \text{if } x < 0.$$

Therefore, the absolute value function can be expressed

$$f(x) = |x| = \begin{cases} x & \text{if } x \geq 0 \\ -x & \text{if } x < 0. \end{cases}$$

The graph of $f(x) = x$ for $x \geq 0$ is the ray in the first quadrant and the graph of

$f(x) = -x$ for $x < 0$ is the half-line (not including the origin) in the second quadrant, as indicated in Figure 4.41. Note that the graph has y-axis symmetry.

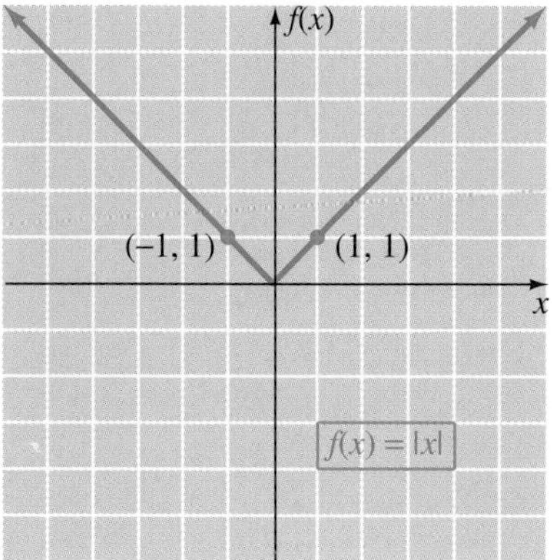

FIGURE 4.41

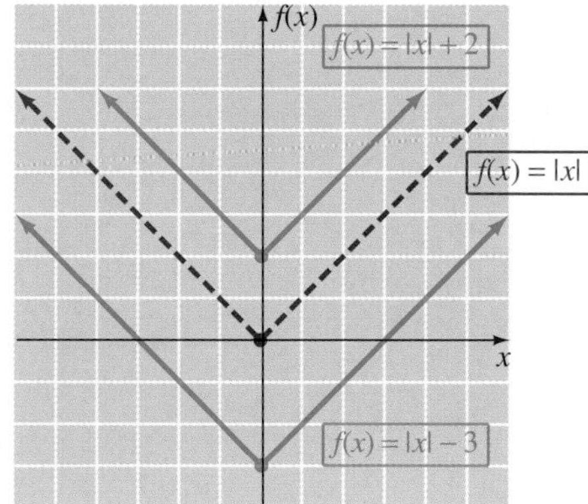

FIGURE 4.42

Translations of the Basic Curves

From our work in Section 4.2, we know that the graph of $f(x) = x^2 + 3$ is the graph of $f(x) = x^2$ moved up three units. Likewise, the graph of $f(x) = x^2 - 2$ is the graph of $f(x) = x^2$ moved down two units. Now let's describe in general the concept of a vertical translation.

Vertical Translation

The graph of $y = f(x) + k$ is the graph of $y = f(x)$ shifted k units upward if $k > 0$ or shifted $|k|$ units downward if $k < 0$.

In Figure 4.42 the graph of $f(x) = |x| + 2$ is obtained by shifting the graph of $f(x) = |x|$ upward two units and the graph of $f(x) = |x| - 3$ is obtained by shifting the graph of $f(x) = |x|$ downward three units. (Remember that $f(x) = |x| - 3$ can be written as $f(x) = |x| + (-3)$.)

Horizontal translations of the basic parabola were also graphed in Section 4.2. For example, the graph of $f(x) = (x - 4)^2$ is the graph of $f(x) = x^2$ shifted four units to the right and the graph of $f(x) = (x + 5)^2$ is the graph of $f(x) = x^2$ shifted five units to the left. The general concept of a horizontal translation can be described as follows.

Horizontal Translation

The graph of $y = f(x - h)$ is the graph of $y = f(x)$ shifted h units to the right if $h > 0$ or shifted $|h|$ units to the left if $h < 0$.

In Figure 4.43 the graph of $f(x) = (x - 3)^3$ is obtained by shifting the graph of $f(x) = x^3$ three units to the right. Likewise, the graph of $f(x) = (x + 2)^3$ is obtained by shifting the graph of $f(x) = x^3$ two units to the left.

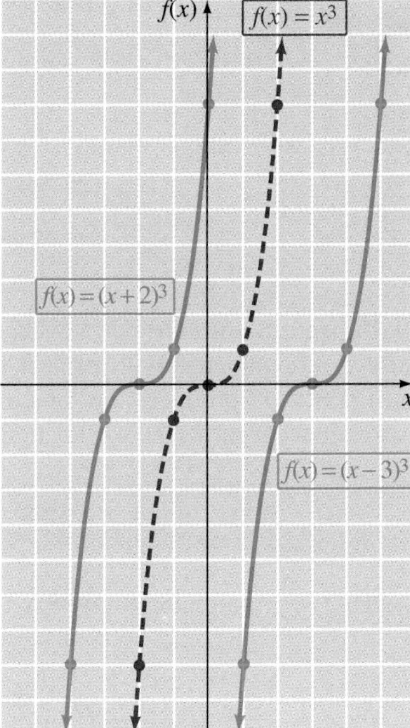

FIGURE 4.43

Reflections of the Basic Curves

From our work in Section 4.2 we know that the graph of $f(x) = -x^2$ is the graph of $f(x) = x^2$ reflected through the x-axis. The general concept of an x-axis reflection can be described as follows.

x-axis Reflection

The graph of $y = -f(x)$ is the graph of $y = f(x)$ reflected through the x-axis.

In Figure 4.44 the graph of $f(x) = -\sqrt{x}$ is obtained by reflecting the graph of $f(x) = \sqrt{x}$ through the x-axis. Reflections are sometimes referred to as **mirror images**. Thus, in Figure 4.44, if we think of the x-axis as a mirror, the graphs of $f(x) = \sqrt{x}$ and $f(x) = -\sqrt{x}$ are mirror images of each other.

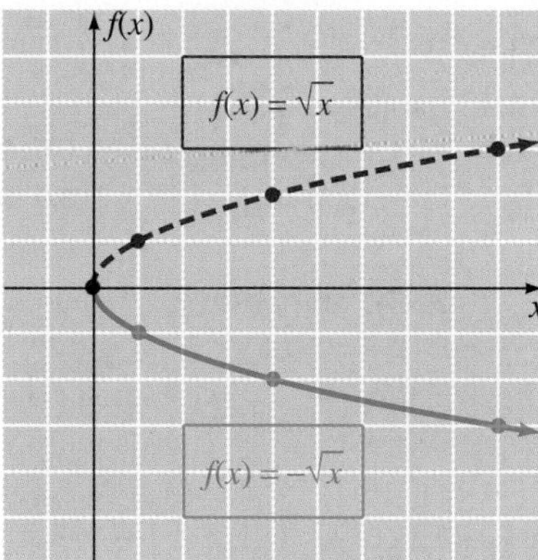

FIGURE 4.44

In Section 4.2 we did not consider a y-axis reflection of the basic parabola $f(x) = x^2$ because it is symmetric with respect to the y-axis. In others words, a y-axis reflection of $f(x) = x^2$ produces the same figure. However, at this time let's describe the general concept of a y-axis reflection.

y-axis Reflection

The graph of $y = f(-x)$ is the graph of $y = f(x)$ reflected through the y-axis.

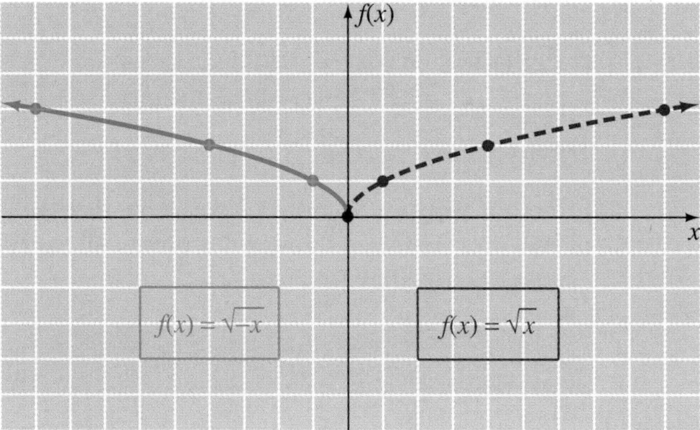

FIGURE 4.45

Now suppose that we want to do a y-axis reflection of $f(x) = \sqrt{x}$. Since $f(x) = \sqrt{x}$ is defined for $x \geq 0$, the y-axis reflection $f(x) = \sqrt{-x}$ is defined for $-x \geq 0$, which is equivalent to $x \leq 0$. Figure 4.45 shows the y-axis reflection of $f(x) = \sqrt{x}$.

Vertical Stretching and Shrinking

Translations and reflections are called **rigid transformations** because the basic shape of the curve being transformed is not changed. In other words, only the positions of the graphs are changed. Now we want to consider some transformations that distort the shape of the original figure somewhat.

In Section 4.2 we graphed the equation $y = 2x^2$ by doubling the y-coordinates of the ordered pairs that satisfy the equation $y = x^2$. We obtained a parabola with its vertex at the origin, symmetric to the y-axis, but *narrower* than the basic parabola. Likewise, we graphed the equation $y = \frac{1}{2}x^2$ by halving the y-coordinates of the ordered pairs that satisfy $y = x^2$. In this case, we obtained a parabola with its vertex at the origin, symmetric to the y-axis, but *wider* than the basic parabola.

The concepts of *narrower* and *wider* can be used to describe parabolas but cannot be used to accurately describe some other curves. Instead, we use the more general concepts of vertical stretching and shrinking.

Vertical Stretching and Shrinking

The graph of $y = cf(x)$ is obtained from the graph of $y = f(x)$ by mulitplying the y-coordinates of $y = f(x)$ by c. If $c > 1$, the graph is said to be *stretched* by a factor of c, and if $0 < c < 1$, the graph is said to be *shrunk* by a factor of c.

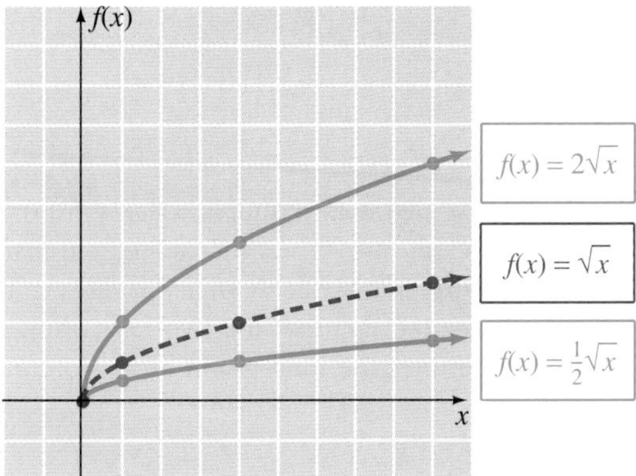

FIGURE 4.46

In Figure 4.46 the graph of $f(x) = 2\sqrt{x}$ is obtained by doubling the y-coordinates of points on the graph of $f(x) = \sqrt{x}$. Likewise, in Figure 4.46, the graph of $f(x) = \frac{1}{2}\sqrt{x}$ is obtained by halving the y-coordinates of points on the graph of $f(x) = \sqrt{x}$.

Successive Transformations

Some curves are the result of performing more than one transformation on a basic curve. Let's consider the graph of a function that involves a stretching, a reflection, a horizontal translation, and a vertical translation of the basic absolute value function.

EXAMPLE 5

Graph $f(x) = -2|x - 3| + 1$.

Solution

This is the basic absolute value curve stretched by a factor of two, reflected through the x-axis, shifted three units to the right, and shifted one unit upward. To sketch the graph we locate the point (3, 1), and then determine a point on each of the rays. The graph is shown in Figure 4.47.

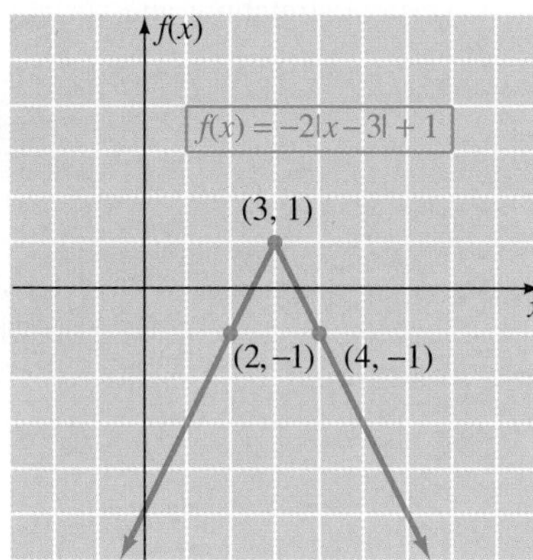

FIGURE 4.47

REMARK Note in Example 5 that we did not sketch the original basic curve $f(x) = |x|$ nor any of the intermediate transformations. However, it is helpful to mentally picture each transformation. This locates the point (3, 1) and establishes the fact that the two rays point downward. Then a point on each ray determines the final graph.

EXAMPLE 6

Graph $f(x) = \sqrt{-3 - x}$.

Solution

This is the basic square root function $f(x) = \sqrt{x}$ shifted three units to the right to produce $f(x) = \sqrt{x - 3}$ and then reflected through the y-axis to produce $f(x) = \sqrt{-x - 3}$. Figure 4.48 shows this graph.

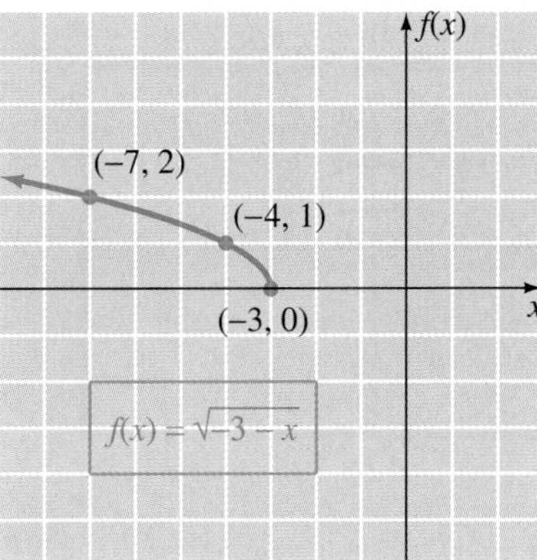

FIGURE 4.48

Now suppose that we want to graph a function such as $f(x) = \dfrac{2x^2}{x^2 + 4}$. Since this is not a basic function that we recognize nor a transformation of a basic function, we need to revert back to our previous graphing experiences. In other words, we need to find the domain, find the intercepts, check for symmetry, check for any restrictions, set up a table of values, plot the points, and sketch the curve. (If you want to do this at this time, you can check your result on page 389.) Furthermore, if the new function is defined in terms of an old function, we may be able to apply the definition of the old function and thereby simplify the new function for graphing purposes. For example, Problem 13 in the next problem set is the function $f(x) = x + |x|$. By applying the definition of absolute value, this function can be simplified. We will leave that for you to do later.

Finally, let's use a graphing utility to give another illustration of the concept of stretching and shrinking a curve.

EXAMPLE 7

If $f(x) = \sqrt{25 - x^2}$, sketch a graph of $y = 2(f(x))$ and $y = \frac{1}{2}(f(x))$.

Solution

If $y = f(x) = \sqrt{25 - x^2}$, then

$$y = 2(f(x)) = 2\sqrt{25 - x^2} \qquad \text{and} \qquad y = \frac{1}{2}(f(x)) = \frac{1}{2}\sqrt{25 - x^2}.$$

Graphing all three of these functions on the same set of axes produces Figure 4.49.

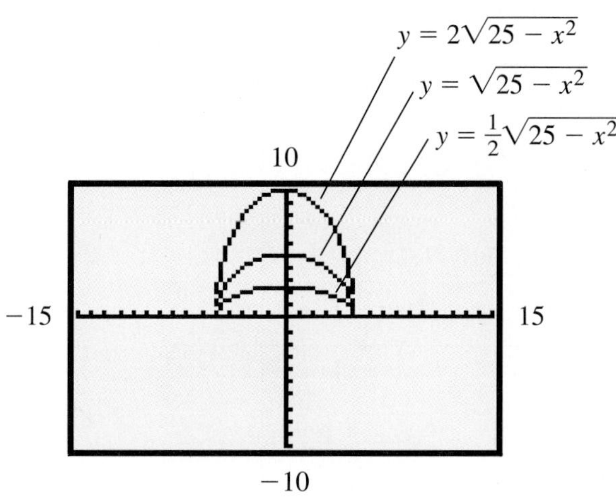

FIGURE 4.49

PROBLEM SET 4.4

For Problems 1–30, graph each of the functions.

1. $f(x) = x^4 + 2$
2. $f(x) = -x^4 - 1$
3. $f(x) = (x - 2)^4$
4. $f(x) = (x + 3)^4 + 1$
5. $f(x) = -x^3$
6. $f(x) = x^3 - 2$
7. $f(x) = (x + 2)^3$
8. $f(x) = (x - 3)^3 - 1$
9. $f(x) = |x - 1| + 2$
10. $f(x) = -|x + 2|$
11. $f(x) = |x + 1| - 3$
12. $f(x) = 2|x|$
13. $f(x) = x + |x|$
14. $f(x) = \dfrac{|x|}{x}$
15. $f(x) = -|x - 2| - 1$
16. $f(x) = 2|x + 1| - 4$
17. $f(x) = x - |x|$
18. $f(x) = |x| - x$
19. $f(x) = -2\sqrt{x}$
20. $f(x) = 2\sqrt{x - 1}$
21. $f(x) = \sqrt{x + 2} - 3$
22. $f(x) = -\sqrt{x + 2} + 2$
23. $f(x) = \sqrt{2 - x}$
24. $f(x) = \sqrt{-1 - x}$
25. $f(x) = -2x^4 + 1$
26. $f(x) = 2(x - 2)^4 - 4$
27. $f(x) = -2x^3$
28. $f(x) = 2x^3 + 3$
29. $f(x) = 3(x - 2)^3 - 1$
30. $f(x) = -2(x + 1)^3 + 2$
31. Suppose that the graph of $y = f(x)$ with a domain of $-2 \le x \le 2$ is shown in Figure 4.50.

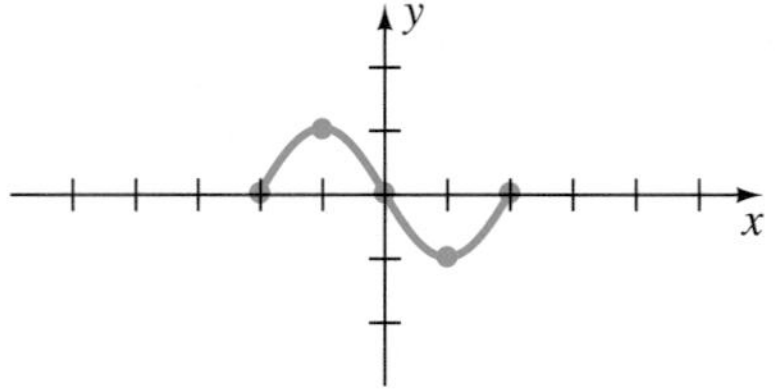

FIGURE 4.50

Sketch the graph of each of the following transformations of $y = f(x)$.

a. $y = f(x) + 3$ **b.** $y = f(x - 2)$

c. $y = -f(x)$ **d.** $y = f(x + 3) - 4$

THOUGHTS into WORDS

32. Are the graphs of the two functions $f(x) = \sqrt{x - 2}$ and $g(x) = \sqrt{2 - x}$ y-axis reflections of each other? Defend your answer.

33. Are the graphs of $f(x) = 2\sqrt{x}$ and $g(x) = \sqrt{2x}$ identical? Defend your answer.

34. Are the graphs of $f(x) = \sqrt{x + 4}$ and $g(x) = \sqrt{-x + 4}$ y-axis reflections of each other? Defend your answer.

GRAPHICS CALCULATOR ACTIVITIES

35. Use your graphics calculator to check your graphs for Problems 13–30.

36. Graph $f(x) = \sqrt{x^2 + 8}$, $f(x) = \sqrt{x^2 + 4}$, and $f(x) = \sqrt{x^2 + 1}$ on the same set of axes. Look at these graphs and predict the graph of $f(x) = \sqrt{x^2 - 4}$. Now graph it with the calculator to test your prediction.

37. For each of the following (a) predict the general shape and location of the graph, and (b) use your calculator to graph the function to check your prediction.

a. $f(x) = \sqrt{x^2}$ **b.** $f(x) = \sqrt{x^3}$

c. $f(x) = |x^2|$ **d.** $f(x) = |x^3|$

38. Graph $f(x) = x^4 + x^3$. Now predict the graph for each of the following and check each prediction with your graphics calculator.

a. $f(x) = x^4 + x^3 - 4$

b. $f(x) = (x - 3)^4 + (x - 3)^3$

c. $f(x) = -x^4 - x^3$

d. $f(x) = x^4 - x^3$

39. Graph $f(x) = \sqrt[3]{x}$. Now predict the graph for each of the following and check each prediction with your graphics calculator.

a. $f(x) = 5 + \sqrt[3]{x}$ **b.** $f(x) = \sqrt[3]{x + 4}$

c. $f(x) = -\sqrt[3]{x}$ **d.** $f(x) = \sqrt[3]{x - 3} - 5$

e. $f(x) = \sqrt[3]{-x}$

4.5 COMBINING FUNCTIONS

In subsequent mathematics courses, it is common to encounter functions that are defined in terms of sums, differences, products, and quotients of simpler functions. For example, if $h(x) = x^2 + \sqrt{x - 1}$, then we may consider the function h as the sum of f and g, where $f(x) = x^2$ and $g(x) = \sqrt{x - 1}$. In general, *if f and g are functions and D is the intersection of their domains*, then the following definitions can be made.

Sum	$(f + g)(x) = f(x) + g(x)$
Difference	$(f - g)(x) = f(x) - g(x)$
Product	$(f \cdot g)(x) = f(x) \cdot g(x)$
Quotient	$\left(\dfrac{f}{g}\right)(x) = \dfrac{f(x)}{g(x)}, \quad g(x) \neq 0$

EXAMPLE 1

If $f(x) = 3x - 1$ and $g(x) = x^2 - x - 2$, find (a) $(f + g)(x)$, (b) $(f - g)(x)$, (c) $(f \cdot g)(x)$, and (d) $(f/g)(x)$. Determine the domain of each.

Solutions

a. $(f + g)(x) = f(x) + g(x) = (3x - 1) + (x^2 - x - 2) = x^2 + 2x - 3$

b.
$$\begin{aligned}(f - g)(x) &= f(x) - g(x)\\ &= (3x - 1) - (x^2 - x - 2)\\ &= 3x - 1 - x^2 + x + 2\\ &= -x^2 + 4x + 1\end{aligned}$$

c.
$$\begin{aligned}(f \cdot g)(x) &= f(x) \cdot g(x)\\ &= (3x - 1)(x^2 - x - 2)\\ &= 3x^3 - 3x^2 - 6x - x^2 + x + 2\\ &= 3x^3 - 4x^2 - 5x + 2\end{aligned}$$

d. $\left(\dfrac{f}{g}\right)(x) = \dfrac{f(x)}{g(x)} = \dfrac{3x - 1}{x^2 - x - 2}$

The domain of both f and g is the set of all real numbers. Therefore, the domain of $f + g$, $f - g$, and $f \cdot g$ is the set of all real numbers. For f/g, the denominator $x^2 - x - 2$ cannot equal zero. Solving $x^2 - x - 2 = 0$ produces

$$\begin{aligned}&(x - 2)(x + 1) = 0\\ &x - 2 = 0 \quad \text{or} \quad x + 1 = 0\\ &\quad\ \ x = 2 \quad \text{or} \quad\quad\ x = -1.\end{aligned}$$

Therefore, the domain for f/g is the set of all real numbers except 2 and -1.

Composition of Functions

Besides adding, subtracting, multiplying, and dividing functions, there is another important operation called *composition*. The composition of two functions can be defined as follows.

DEFINITION 4.2

The **composition** of functions f and g is defined by

$$(f \circ g)(x) = f(g(x))$$

for all x in the domain of g such that $g(x)$ is in the domain of f.

The left side, $(f \circ g)(x)$, of the equation in Definition 4.2 is read *the composition of f and g* and the right side is read *f of g of x*. It may also be helpful for you to have a mental picture of Definition 4.2 as two function machines hooked together to produce another function (called the **composite function**), as illustrated in Figure 4.51. Notice that what comes out of the g function is substituted into the f function. Thus, composition is sometimes called the **substitution of functions**

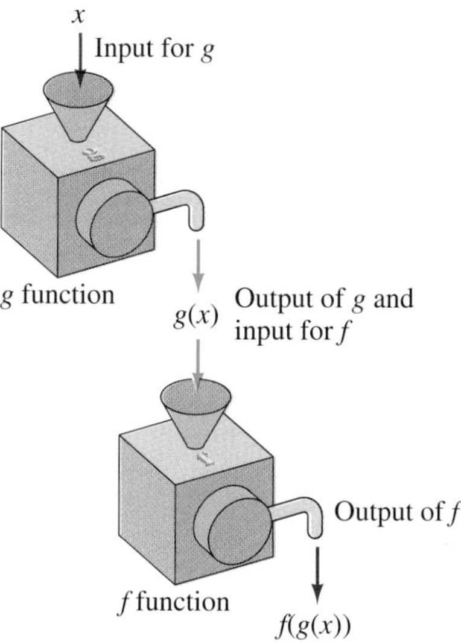

FIGURE 4.51

Figure 4.51 also illustrates the fact that $f \circ g$ is defined *for all x in the domain of g such that $g(x)$ is in the domain of f.* In other words, what comes out of g must be capable of being fed into f. Let's consider some examples.

EXAMPLE 2

If $f(x) = x^2$ and $g(x) = 3x - 4$, find $(f \circ g)(x)$ and determine its domain.

Solution

Apply Definition 4.2 to obtain

$$\begin{aligned}(f \circ g)(x) &= f(g(x)) \\ &= f(3x - 4) \\ &= (3x - 4)^2 \\ &= 9x^2 - 24x + 16.\end{aligned}$$

Because g and f are both defined for all real numbers, so is $f \circ g$.

Definition 4.2, with f and g interchanged, defines the composition of g and f as $(g \circ f)(x) = g(f(x))$.

EXAMPLE 3

If $f(x) = x^2$ and $g(x) = 3x - 4$, find $(g \circ f)(x)$ and determine its domain.

Solution

$$\begin{aligned}(g \circ f)(x) &= g(f(x)) \\ &= g(x^2) \\ &= 3x^2 - 4\end{aligned}$$

Because f and g are defined for all real numbers, so is $g \circ f$.

The results of Examples 2 and 3 demonstrate an important idea, namely, that **the composition of functions is not a commutative operation**. In other words, $f \circ g \neq g \circ f$ for all functions f and g. However, as we will see in the next section, there is a special class of functions for which $f \circ g = g \circ f$.

EXAMPLE 4

If $f(x) = \sqrt{x}$ and $g(x) = 2x - 1$, find $(f \circ g)(x)$ and $(g \circ f)(x)$. Also determine the domain of each composite function.

Solution

$$\begin{aligned}(f \circ g)(x) &= f(g(x)) \\ &= f(2x - 1) \\ &= \sqrt{2x - 1}\end{aligned}$$

The domain and range of g is the set of all real numbers, but the domain of f is all *nonnegative* real numbers. Therefore $g(x)$, which is $2x - 1$, must be nonnegative.

$$\begin{aligned}2x - 1 &\geq 0 \\ 2x &\geq 1 \\ x &\geq \frac{1}{2}\end{aligned}$$

Thus the domain of $f \circ g$ is $D = \left\{x \middle| x \geq \frac{1}{2}\right\}$.

$$\begin{aligned}(g \circ f)(x) &= g(f(x)) \\ &= g\left(\sqrt{x}\right) \\ &= 2\sqrt{x} - 1\end{aligned}$$

The domain and range of f is the set of nonnegative real numbers. The domain of g is the set of all real numbers. Therefore, the domain of $g \circ f$ is $D = \{x | x \geq 0\}$.

EXAMPLE 5

If $f(x) = 2/(x - 1)$ and $g(x) = 1/x$, find $(f \circ g)(x)$ and $(g \circ f)(x)$. Determine the domain for each composite function.

Solution

$$\begin{aligned}(f \circ g)(x) &= f(g(x))\\ &= f\left(\frac{1}{x}\right)\\ &= \frac{2}{\frac{1}{x} - 1} = \frac{2}{\frac{1-x}{x}}\\ &= \frac{2x}{1-x}\end{aligned}$$

The domain of g is all real numbers except zero, and the domain of f is all real numbers except one. Since $g(x)$, which is $1/x$, cannot equal 1,

$$\frac{1}{x} \neq 1$$
$$x \neq 1.$$

Therefore, the domain of $f \circ g$ is $D = \{x | x \neq 0 \text{ and } x \neq 1\}$.

$$\begin{aligned}(g \circ f)(x) &= g(f(x))\\ &= g\left(\frac{2}{x-1}\right)\\ &= \frac{1}{\frac{2}{x-1}}\\ &= \frac{x-1}{2}\end{aligned}$$

The domain of f is all real numbers except 1, and the domain of g is all real numbers except 0. Since $f(x)$, which is $2/(x - 1)$ will never equal 0, the domain of $g \circ f$ is $D = \{x | x \neq 1\}$.

A graphing utility can be used to find the graph of a composite function without actually forming the function algebraically. Let's see how this works.

EXAMPLE 6

If $f(x) = x^3$ and $g(x) = x - 4$, use a graphing utility to obtain the graph of $y = (f \circ g)(x)$ and of $y = (g \circ f)(x)$.

Solution

To find the graph of $y = (f \circ g)(x)$ we can make the following assignments.

$$Y_1 = x - 4$$
$$Y_2 = (Y_1)^3$$

(Note that we have substituted Y_1 for x in $f(x)$ and assigned this expression to Y_2, much the same way as we would do it algebraically.) The graph of $y = (f \circ g)(x)$ is shown in Figure 4.52.

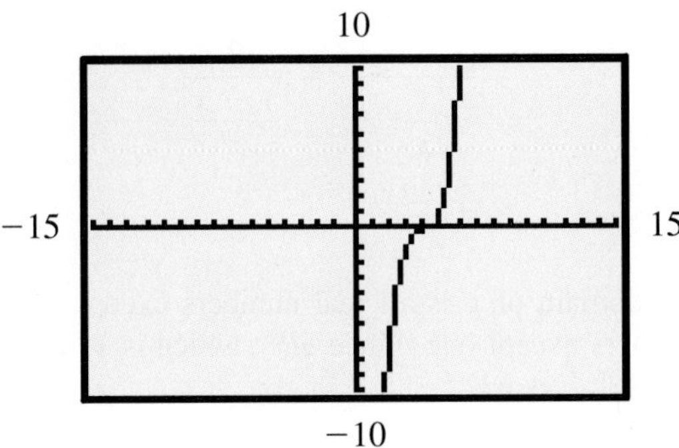

FIGURE 4.52

To find the graph of $y = (g \circ f)(x)$ we can make the following assignments.

$$Y_1 = x^3$$
$$Y_2 = Y_1 - 4$$

The graph of $y = (g \circ f)(x)$ is shown in Figure 4.53.

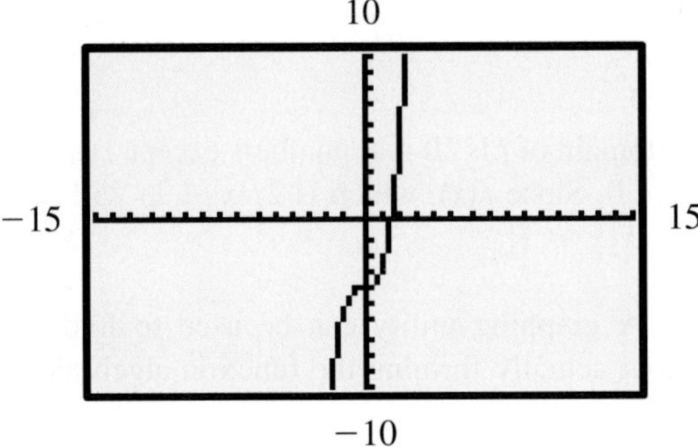

FIGURE 4.53

Take another look at Figures 4.52 and 4.53. Note that in Figure 4.52 the graph of $y = (f \circ g)(x)$ is the basic cubic curve $f(x) = x^3$ shifted 4 units to the right. Likewise, in Figure 4.53 the graph of $y = (g \circ f)(x)$ is the basic cubic curve shifted 4 units downward. These are examples of a more general concept of using composite functions to represent various geometric transformations.

PROBLEM SET 4.5

For Problems 1–8, find $f + g$, $f - g$, $f \cdot g$, and $\frac{f}{g}$.

1. $f(x) = 3x - 4, \quad g(x) = 5x + 2$
2. $f(x) = -6x - 1, \quad g(x) = -8x + 7$
3. $f(x) = x^2 - 6x + 4, \quad g(x) = -x - 1$
4. $f(x) = 2x^2 - 3x + 5, \quad g(x) = x^2 - 4$
5. $f(x) = x^2 - x - 1, \quad g(x) = x^2 + 4x - 5$
6. $f(x) = x^2 - 2x - 24, \quad g(x) = x^2 - x - 30$
7. $f(x) = \sqrt{x - 1}, \quad g(x) = \sqrt{x}$
8. $f(x) = \sqrt{x + 2}, \quad g(x) = \sqrt{3x - 1}$

For Problems 9–26, find $(f \circ g)(x)$ and $(g \circ f)(x)$. Also specify the domain for each.

9. $f(x) = 2x, \quad g(x) = 3x - 1$
10. $f(x) = 4x + 1, \quad g(x) = 3x$
11. $f(x) = 5x - 3, \quad g(x) = 2x + 1$
12. $f(x) = 3 - 2x, \quad g(x) = -4x$
13. $f(x) = 3x + 4, \quad g(x) = x^2 + 1$
14. $f(x) = 3, \quad g(x) = -3x^2 - 1$
15. $f(x) = 3x - 4, \quad g(x) = x^2 + 3x - 4$
16. $f(x) = 2x^2 - x - 1, \quad g(x) = x + 4$
17. $f(x) = \frac{1}{x}, \quad g(x) = 2x + 7$
18. $f(x) = \frac{1}{x^2}, \quad g(x) = x$
19. $f(x) = \sqrt{x - 2}, \quad g(x) = 3x - 1$
20. $f(x) = \frac{1}{x}, \quad g(x) = \frac{1}{x^2}$
21. $f(x) = \frac{1}{x - 1}, \quad g(x) = \frac{2}{x}$
22. $f(x) = \frac{4}{x + 2}, \quad g(x) = \frac{3}{2x}$
23. $f(x) = 2x + 1, \quad g(x) = \sqrt{x - 1}$
24. $f(x) = \sqrt{x + 1}, \quad g(x) = 5x - 2$
25. $f(x) = \frac{1}{x - 1}, \quad g(x) = \frac{x + 1}{x}$
26. $f(x) = \frac{x - 1}{x + 2}, \quad g(x) = \frac{1}{x}$

Solve each of the following problems (Problems 27–32).

27. If $f(x) = 3x - 2$ and $g(x) = x^2 + 1$, find $(f \circ g)(-1)$ and $(g \circ f)(3)$.
28. If $f(x) = x^2 - 2$ and $g(x) = x + 4$, find $(f \circ g)(2)$ and $(g \circ f)(-4)$.
29. If $f(x) = 2x - 3$ and $g(x) = x^2 - 3x - 4$, find $(f \circ g)(-2)$ and $(g \circ f)(1)$.
30. If $f(x) = 1/x$ and $g(x) = 2x + 1$, find $(f \circ g)(1)$ and $(g \circ f)(2)$.
31. If $f(x) = \sqrt{x}$ and $g(x) = 3x - 1$, find $(f \circ g)(4)$ and $(g \circ f)(4)$.
32. If $f(x) = x + 5$ and $g(x) = |x|$, find $(f \circ g)(-4)$ and $(g \circ f)(-4)$.

For Problems 33–38, show that $(f \circ g)(x) = x$ and $(g \circ f) \cdot (x) = x$.

33. $f(x) = 2x, \quad g(x) = \frac{1}{2}x$
34. $f(x) = \frac{3}{4}x, \quad g(x) = \frac{4}{3}x$
35. $f(x) = x - 2, \quad g(x) = x + 2$
36. $f(x) = 2x + 1, \quad g(x) = \frac{x - 1}{2}$
37. $f(x) = 3x + 4, \quad g(x) = \frac{x - 4}{3}$
38. $f(x) = 4x - 3, \quad g(x) = \frac{x + 3}{4}$

THOUGHTS into WORDS

39. Discuss whether or not addition, subtraction, multiplication, and division of functions are commutative operations.

40. Explain why the composition of two functions is not a commutative operation.

MISCELLANEOUS PROBLEMS

41. If $f(x) = 3x - 4$ and $g(x) = ax + b$, find conditions on a and b that will guarantee that $f \circ g = g \circ f$.

42. If $f(x) = x^2$ and $g(x) = \sqrt{x}$, with both having as domain the set of nonnegative real numbers, then show that $(f \circ g)(x) = x$ and $(g \circ f)(x) = x$.

43. If $f(x) = 3x^2 - 2x - 1$ and $g(x) = x$, find $f \circ g$ and $g \circ f$. (Recall that we have previously named $g(x) = x$ the *identity function*.)

44. In Section 4.1, we defined an *even function* to be a function such that $f(-x) = f(x)$ and an *odd function* to be one such that $f(-x) = -f(x)$. Verify that (a) the sum of two even functions is an even function, and (b) the sum of two odd functions is an odd function.

GRAPHICS CALCULATOR ACTIVITIES

45. For each of the following, (a) predict the general shape and location of the graph, and (b) use your calculator to graph the function to check your prediction. (Your knowledge of the graphs of the basic functions that are being added or subtracted should be helpful when making your predictions.)

a. $f(x) = x^4 + x^2$ **b.** $f(x) = x^3 + x^2$

c. $f(x) = x^4 - x^2$ **d.** $f(x) = x^2 - x^4$

e. $f(x) = x^2 - x^3$ **f.** $f(x) = x^3 - x^2$

g. $f(x) = |x| + \sqrt{x}$ **h.** $f(x) = |x| - \sqrt{x}$

46. For each of the following, find the graph of $y = (f \circ g)(x)$ and of $y = (g \circ f)((x)$.

a. $f(x) = x^2$ and $g(x) = x + 5$

b. $f(x) = x^3$ and $g(x) = x + 3$

c. $f(x) = x - 6$ and $g(x) = -x^3$

d. $f(x) = x^2 - 4$ and $g(x) = \sqrt{x}$

e. $f(x) = \sqrt{x}$ and $g(x) = x^2 + 4$

f. $f(x) = \sqrt[3]{x}$ and $g(x) = x^3 - 5$

4.6 INVERSE FUNCTIONS

Recall the *vertical line test*: If each vertical line intersects a graph in no more than one point, then the graph represents a function. There is also a useful distinction between two basic types of functions. Consider the graphs of the two functions in Figure 4.54: (a) $f(x) = 2x - 1$ and (b) $g(x) = x^2$. In part (a), any *horizontal line* will intersect the graph in no more than one point. Therefore, every value of $f(x)$ has only one value of x associated with it. Any function that has this property of

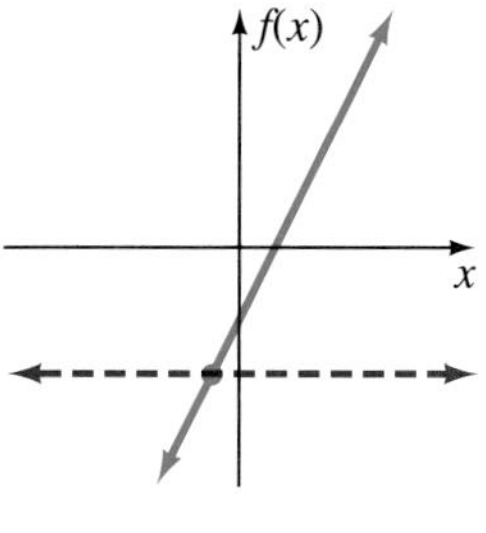

(a) $f(x) = 2x - 1$

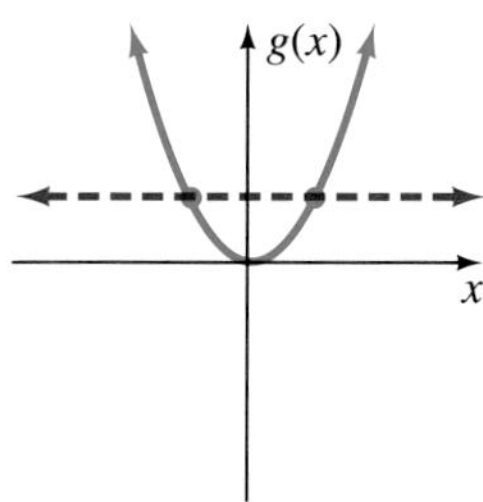

(b) $g(x) = x^2$

FIGURE 4.54

having exactly one value of x associated with each value of $f(x)$ is called a **one-to-one function**. The function $g(x) = x^2$ is not a one-to-one function because the horizontal line in Figure 4.54(b) intersects the parabola in two points.

Stated another way, a function f is said to be one-to-one if $x_1 \neq x_2$ implies that $f(x_1) \neq f(x_2)$. In other words, different values for x always result in different values for $f(x)$. Thus, without a graph, we can show that $f(x) = 2x - 1$ is a one-to-one function as follows: If $x_1 \neq x_2$, then $2x_1 \neq 2x_2$ and therefore $2x_1 - 1 \neq 2x_2 - 1$. Furthermore, we can show that $f(x) = x^2$ is not a one-to-one function because $f(2) = 4$ and $f(-2) = 4$; that is, different values for x produce the same value for $f(x)$.

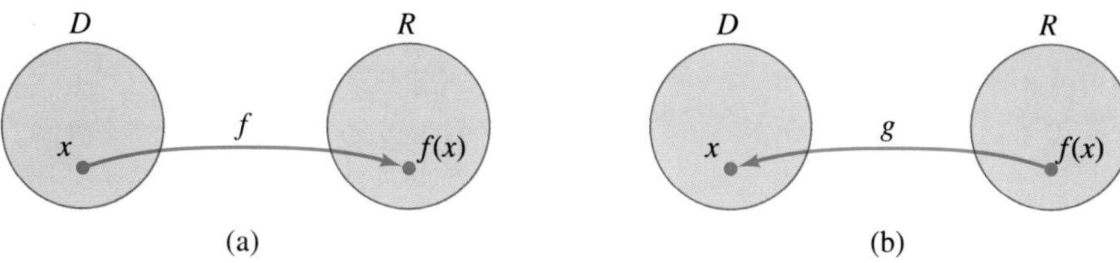

FIGURE 4.55

Now let's consider a one-to-one function f that assigns the value $f(x)$ in its range R to each x in its domain D (Figure 4.55(a)). We can define a new function g that goes from R to D; it assigns $f(x)$ in R back to x in D, as indicated in Figure 4.55(b). The functions f and g are called *inverse functions* of one another. The following definition precisely states this concept.

DEFINITION 4.3

Let f be a one-to-one function with a domain of X and a range of Y. A function g with a domain of Y and a range of X is called the inverse function of f if

$$(f \circ g)(x) = x \quad \text{for every } x \text{ in } Y$$

and

$$(g \circ f)(x) = x \quad \text{for every } x \text{ in } X.$$

In Definition 4.3, note that for f and g to be inverses of each other, the domain of f must equal the range of g and the range of f must equal the domain of g. Furthermore, g must reverse the correspondences given by f, and f must reverse the correspondences given by g. In other words, inverse functions *undo* each other. Let's use Definition 4.3 to verify that two specific functions are inverses of each other.

EXAMPLE 1

Verify that $f(x) = 4x - 5$ and $g(x) = \dfrac{x+5}{4}$ are inverse functions.

Solution

Because the set of real numbers is the domain and range of both functions, we know that the domain of f equals the range of g and the range of f equals the domain of g. Furthermore,

$$\begin{aligned}(f \circ g)(x) &= f(g(x)) \\ &= f\left(\frac{x+5}{4}\right) \\ &= 4\left(\frac{x+5}{4}\right) - 5 = x,\end{aligned}$$

and

$$\begin{aligned}(g \circ f)(x) &= g(f(x)) \\ &= g(4x - 5) \\ &= \frac{4x - 5 + 5}{4} = x.\end{aligned}$$

Therefore, f and g are inverses of each other.

EXAMPLE 2

Verify that $f(x) = x^2 + 1$ for $x \geq 0$ and $g(x) = \sqrt{x-1}$ for $x \geq 1$ are inverse functions.

Solution

First, notice that the domain of f equals the range of g, namely, the set of nonnegative real numbers. Also, the range of f equals the domain of g, namely, the set of real numbers greater than or equal to 1. Furthermore,

$$\begin{aligned}(f \circ g)(x) &= f(g(x)) \\ &= f\left(\sqrt{x-1}\right) \\ &= \left(\sqrt{x-1}\right)^2 + 1 \\ &= x - 1 + 1 = x,\end{aligned}$$

and

$$\begin{aligned}(g \circ f)(x) &= g(f(x)) \\ &= g(x^2 + 1) \\ &= \sqrt{x^2 + 1 - 1} = \sqrt{x^2} = x. \qquad \sqrt{x^2} = x \text{ because } x \geq 1\end{aligned}$$

Therefore, f and g are inverses of each other.

The inverse of a function f is commonly denoted by f^{-1} (read *f inverse* or *the inverse of f*). Do not confuse the -1 in f^{-1} with a negative exponent. The symbol f^{-1} *does not* mean $1/f^1$ but refers to the inverse function of function f.

Remember that a function can also be thought of as a set of ordered pairs no two of which have the same first element. Along those lines, a one-to-one function further requires that no two of the ordered pairs have the same second element. Then, if the components of each ordered pair of a given one-to-one function are interchanged, the resulting function and the given function are inverses of each other. Thus, if

$$f = \{(1, 4), (2, 7), (5, 9)\}$$

then

$$f^{-1} = \{(4, 1), (7, 2), (9, 5)\}$$

Graphically, two functions that are inverses of each other are **mirror images with reference to the line $y = x$**. This is due to the fact that ordered pairs (a, b) and (b, a) are reflections of each other with respect to the line $y = x$, as illustrated in Figure 4.56. (We will have you verify this in the next set of exercises.) Therefore, if the graph of a function f is known, as in Figure 4.57(a), then the graph of f^{-1} can be determined by reflecting f across the line $y = x$ (Figure 4.57(b)).

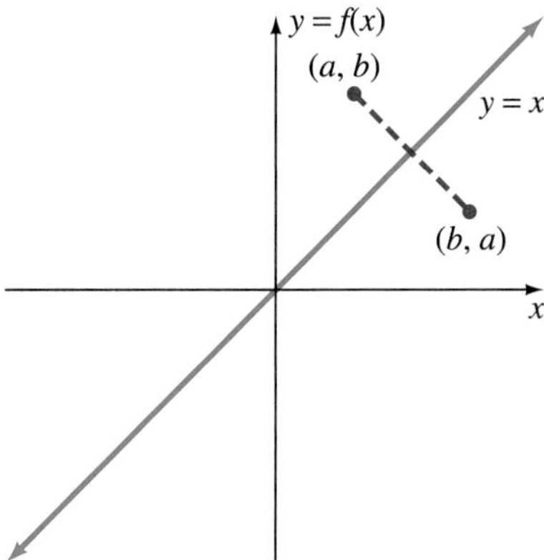

FIGURE 4.56

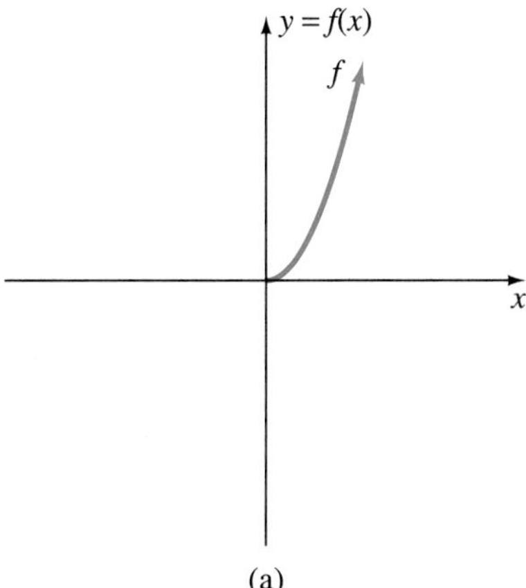

(a)

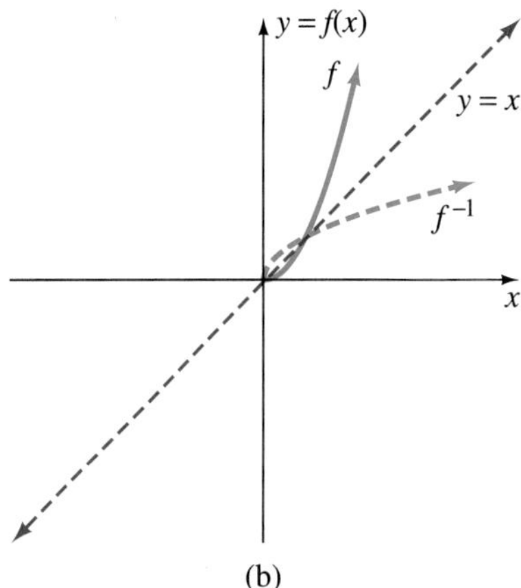

(b)

FIGURE 4.57

Finding Inverse Functions

The idea of inverse functions *undoing each other* provides the basis for an informal approach to finding the inverse of a function. Consider the function

$$f(x) = 2x + 1.$$

To each x this function assigns twice x plus 1. To undo this function, we can subtract 1 and divide by 2. So the inverse is

$$f^{-1}(x) = \frac{x-1}{2}.$$

Now let's verify that f and f^{-1} are indeed inverses of each other.

$$\begin{aligned}(f \circ f^{-1})(x) &= f(f^{-1}(x)) \\ &= f\left(\frac{x-1}{2}\right) \\ &= 2\left(\frac{x-1}{2}\right) + 1 \\ &= x - 1 + 1 = x \\ (f^{-1} \circ f)(x) &= f^{-1}(f(x)) \\ &= f^{-1}(2x+1) \\ &= \frac{2x+1-1}{2} \\ &= \frac{2x}{2} = x\end{aligned}$$

Thus, the inverse of $f(x) = 2x + 1$ is $f^{-1}(x) = \dfrac{x-1}{2}$.

This informal approach may not work very well with more complex functions, but it does emphasize how inverse functions are related to each other. A more formal and systematic technique for finding the inverse of a function can be described as follows.

1. Replace the symbol $f(x)$ by y.
2. Interchange x and y.
3. Solve the equation for y in terms of x.
4. Replace y by the symbol $f^{-1}(x)$.

The following examples illustrate this technique.

EXAMPLE 3

Find the inverse of $f(x) = \dfrac{2}{3}x + \dfrac{3}{5}$.

Solution

Replacing $f(x)$ by y, the equation becomes $y = \frac{2}{3}x + \frac{3}{5}$. Interchanging x and y produces $x = \frac{2}{3}y + \frac{3}{5}$.

Now, solving for y, we obtain

$$\begin{aligned} x &= \frac{2}{3}y + \frac{3}{5} \\ 15(x) &= 15\left(\frac{2}{3}y + \frac{3}{5}\right) \\ 15x &= 10y + 9 \\ 15x - 9 &= 10y \\ \frac{15x - 9}{10} &= y. \end{aligned}$$

Finally, replacing y by $f^{-1}(x)$ we can express the inverse function as

$$f^{-1}(x) = \frac{15x - 9}{10}.$$

The domain of f is equal to the range of f^{-1} (both are the set of real numbers) and the range of f equals the domain of f^{-1} (both are the set of real numbers). Furthermore, we could show that $(f \circ f^{-1})(x) = x$ and $(f^{-1} \circ f)(x) = x$. We leave this for you to complete.

Does $f(x) = x^2 - 2$ have an inverse function? Sometimes a graph of the function helps to answer such a question. In Figure 4.58(a), it should be evident

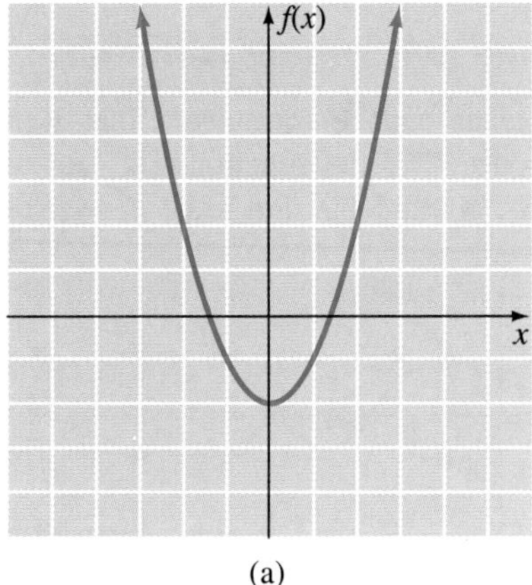

(a)

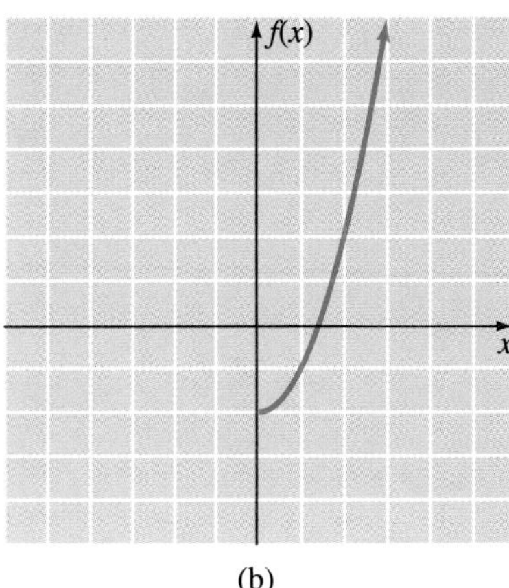

(b)

FIGURE 4.58

that f is not a one-to-one function and therefore cannot have an inverse. However, it should also be apparent from the graph that if we restrict the domain of f to be the nonnegative real numbers, then it is a one-to-one function and should have an inverse (Figure 4.58(b)). The next example illustrates how to find the inverse function.

EXAMPLE 4

Find the inverse of $f(x) = x^2 - 2$, where $x \geq 0$.

Solution

Replacing $f(x)$ by y, the equation becomes

$$y = x^2 - 2, \qquad x \geq 0.$$

Interchanging x and y produces

$$x = y^2 - 2, \qquad y \geq 0.$$

Now let's solve for y; keep in mind that y is to be nonnegative.

$$\begin{aligned} x &= y^2 - 2 \\ x + 2 &= y^2 \\ \sqrt{x + 2} &= y, \quad \text{where } x \geq -2 \end{aligned}$$

Finally, replacing y by $f^{-1}(x)$, the inverse function can be expressed

$$f^{-1}(x) = \sqrt{x + 2}, \qquad x \geq -2.$$

The domain of f equals the range of f^{-1} (both are the nonnegative real numbers), and the range of f equals the domain of f^{-1} (both are the real numbers greater than or equal to -2). It can be shown that $(f \circ f^{-1})(x) = x$ and $(f^{-1} \circ f)(x) = x$. Again, we leave this for you to complete.

Increasing and Decreasing Functions

Some general ideas can be formulated that were specifically illustrated in Example 4. In Figure 4.59 the function f is said to be *increasing* on the intervals $(-\infty, x_1]$ and $[x_2, \infty)$, and f is said to be *decreasing* on the interval $[x_1, x_2]$.

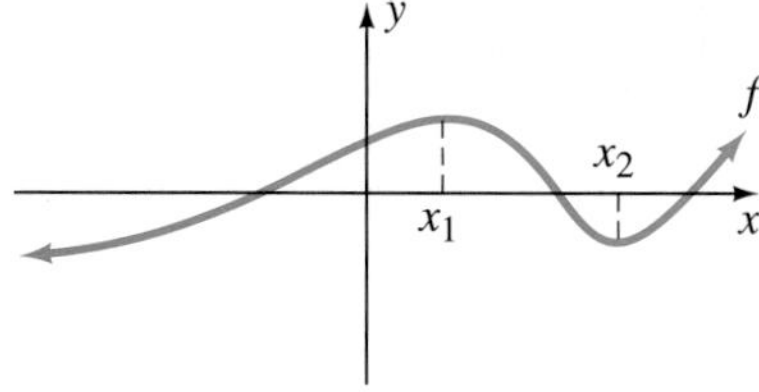

FIGURE 4.59

More specifically, increasing and decreasing functions are defined as follows.

DEFINITION 4.4

Let f be a function, with the interval I a subset of the domain of f. Let x_1 and x_2 be in I. Then

1. f is *increasing on I* if $f(x_1) < f(x_2)$ whenever $x_1 < x_2$.
2. f is *decreasing on I* if $f(x_1) > f(x_2)$ whenever $x_1 < x_2$, and
3. f is *constant on I* if $f(x_1) = f(x_2)$ for every x_1 and x_2.

Apply Definition 4.4 and you will see that the quadratic function $f(x) = x^2$ shown in Figure 4.60 is decreasing on $(-\infty, 0]$ and increasing on $[0, \infty)$. Likewise, the linear function $f(x) = 2x$ in Figure 4.61 is increasing throughout its domain of real numbers, so we say it is increasing on $(-\infty, \infty)$. The function $f(x) = -2x$ in Figure 4.62 is decreasing on $(-\infty, \infty)$. For our purposes in this text, we will rely

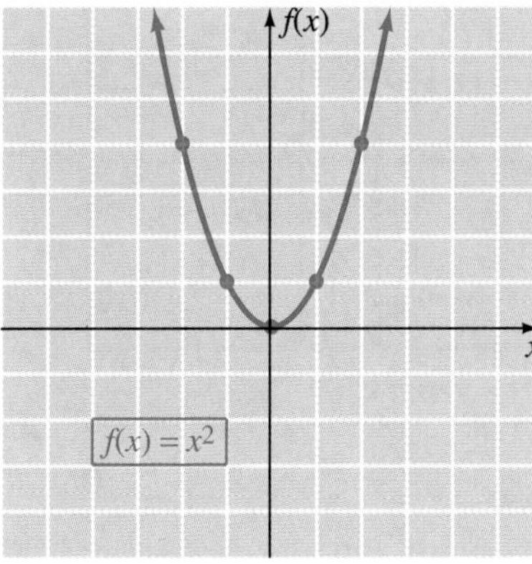

FIGURE 4.60

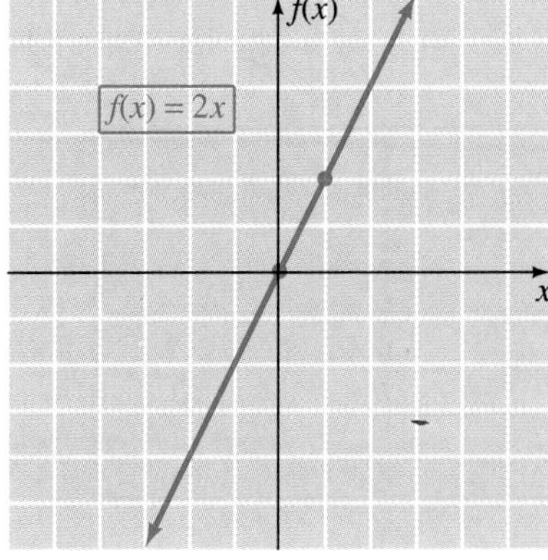

FIGURE 4.61

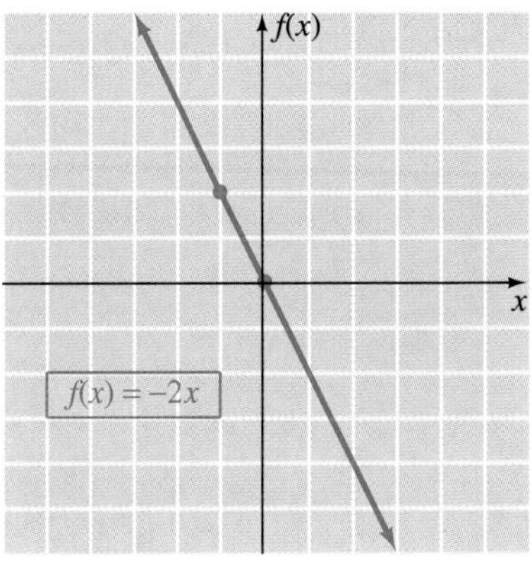

FIGURE 4.62

on our knowledge of the graphs of the functions to determine where functions are increasing and decreasing. More formal techniques for determining where functions increase and decrease will be developed in calculus.

A function that is always increasing (or always decreasing) over its entire domain is one-to-one and so has an inverse. Furthermore, as illustrated by Example 4, even if a function is not one-to-one over its entire domain, it may be so over some subset of the domain. It then has an inverse over this restricted domain.

As functions become more complex, a graphing utility can be used to help with the problems we have discussed in this section. For example, suppose that we want to know if the function $f(x) = \dfrac{3x + 1}{x - 4}$ is a one-to-one function and therefore has an inverse. Using a graphing utility, we can quickly get a sketch of the graph as shown in Figure 4.63. Then by applying the horizontal line test to the graph we feel fairly certain that the function is one-to-one. (Later we will develop some concepts that will allow us to be absolutely certain of this conclusion.)

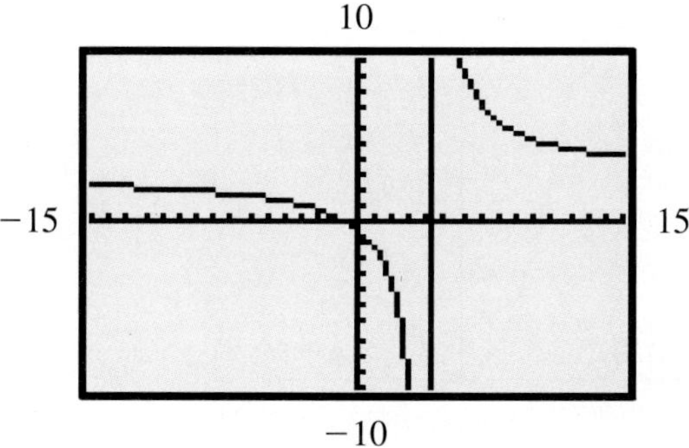

FIGURE 4.63

A graphing utility can also be used to help determine intervals on which a function is increasing or decreasing. For example, to determine such intervals for the function $f(x) = \sqrt{x^2 + 4}$, let's use a graphing utility to get a sketch of the curve as shown in Figure 4.64. From this graph we see that the function is decreasing on the interval $(-\infty, 0]$ and increasing on the interval $[0, \infty)$.

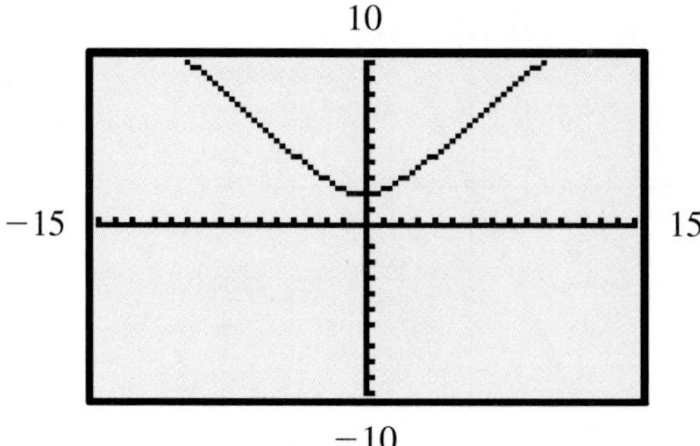

FIGURE 4.64

PROBLEM SET 4.6

For Problems 1–6 (Figures 4.65 through 4.70), determine whether or not the graph represents a one-to-one function.

1. 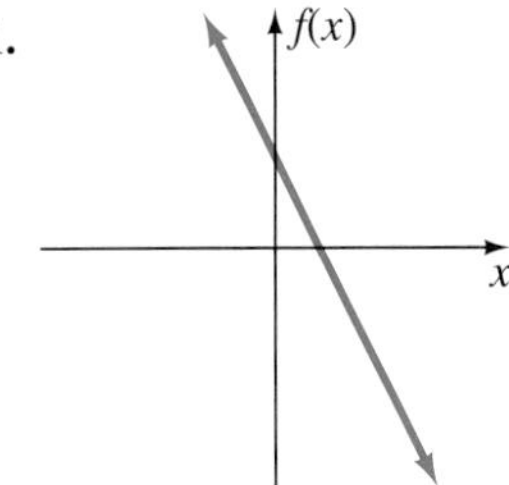

FIGURE 4.65

2. 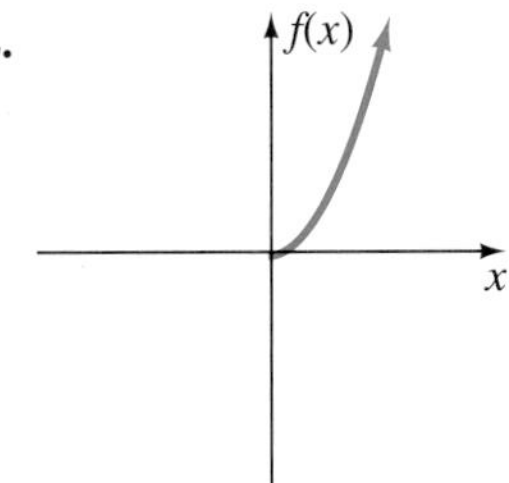

FIGURE 4.66

3. 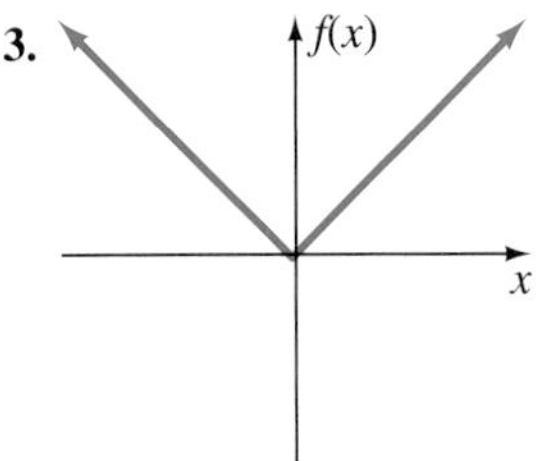

FIGURE 4.67

4. 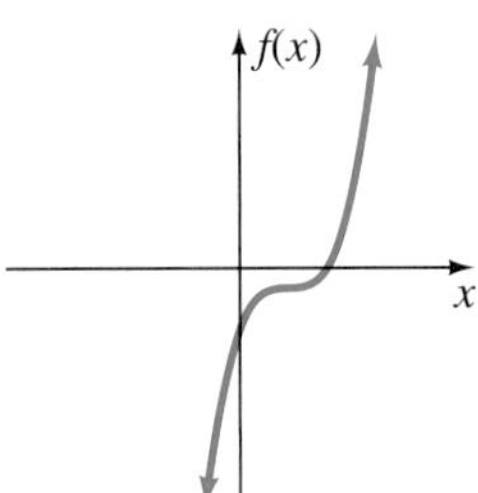

FIGURE 4.68

5. 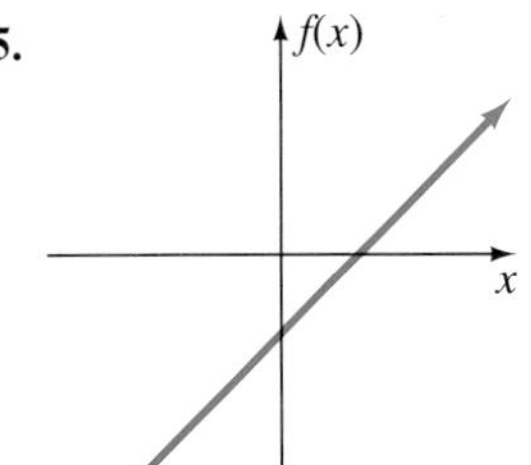

FIGURE 4.69

6. 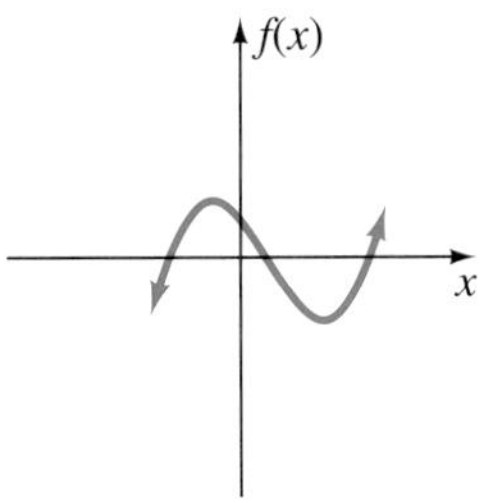

FIGURE 4.70

For Problems 7–14, determine whether the function f is one-to-one.

7. $f(x) = 5x + 4$
8. $f(x) = -3x + 4$
9. $f(x) = x^3$
10. $f(x) = x^5 + 1$
11. $f(x) = |x| + 1$
12. $f(x) = -|x| - 2$
13. $f(x) = -x^4$
14. $f(x) = x^4 + 1$

For each of the functions in Problems 15–18, (a) list the domain and range, (b) form the inverse function f^{-1}, and (c) list the domain and range of f^{-1}.

15. $f = \{(1, 5), (2, 9), (5, 21)\}$
16. $f = \{(1, 1), (4, 2), (9, 3), (16, 4)\}$
17. $f = \{(0, 0), (2, 8), (-1, -1), (-2, -8)\}$
18. $f = \{(-1, 1), (-2, 4), (-3, 9), (-4, 16)\}$

In each of Problems 19–26, verify that the two given functions are inverses of each other.

19. $f(x) = 5x - 9$ and $g(x) = \dfrac{x + 9}{5}$
20. $f(x) = -3x + 4$ and $g(x) = \dfrac{4 - x}{3}$
21. $f(x) = -\dfrac{1}{2}x + \dfrac{5}{6}$ and $g(x) = -2x + \dfrac{5}{3}$
22. $f(x) = x^3 + 1$ and $g(x) = \sqrt[3]{x - 1}$
23. $f(x) = \dfrac{1}{x - 1}$ for $x > 1$ and $g(x) = \dfrac{x + 1}{x}$ for $x > 0$
24. $f(x) = x^2 + 2$ for $x \geq 0$ and $g(x) = \sqrt{x - 2}$ for $x \geq 2$
25. $f(x) = \sqrt{2x - 4}$ for $x \geq 2$ and $g(x) = \dfrac{x^2 + 4}{2}$ for $x \geq 0$
26. $f(x) = x^2 - 4$ for $x \geq 0$ and $g(x) = \sqrt{x + 4}$ for $x \geq -4$

For Problems 27–36, determine if f and g are inverse functions.

27. $f(x) = 3x$ and $g(x) = -\dfrac{1}{3}x$
28. $f(x) = \dfrac{3}{4}x - 2$ and $g(x) = \dfrac{4}{3}x + \dfrac{8}{3}$
29. $f(x) = x^3$ and $g(x) = \sqrt[3]{x}$
30. $f(x) = \dfrac{1}{x + 1}$ and $g(x) = \dfrac{1 - x}{x}$

31. $f(x) = x$ and $g(x) = \frac{1}{x}$

32. $f(x) = \frac{3}{5}x + \frac{1}{3}$ and $g(x) = \frac{5}{3}x - 3$

33. $f(x) = x^2 - 3$ for $x \geq 0$ and $g(x) = \sqrt{x + 3}$ for $x \geq -3$

34. $f(x) = |x - 1|$ for $x \geq 1$ and $g(x) = |x + 1|$ for $x \geq 0$

35. $f(x) = \sqrt{x + 1}$ and $g(x) = x^2 - 1$ for $x \geq 0$

36. $f(x) = \sqrt{2x - 2}$ and $g(x) = \frac{1}{2}x^2 + 1$

For Problems 37–50, (a) find f^{-1} and (b) verify that $(f \circ f^{-1}) \cdot (x) = x$ and $(f^{-1} \circ f)(x) = x$.

37. $f(x) = x - 4$

38. $f(x) = 2x - 1$

39. $f(x) = -3x - 4$

40. $f(x) = -5x + 6$

41. $f(x) = \frac{3}{4}x - \frac{5}{6}$

42. $f(x) = \frac{2}{3}x - \frac{1}{4}$

43. $f(x) = -\frac{2}{3}x$

44. $f(x) = \frac{4}{3}x$

45. $f(x) = \sqrt{x}$ for $x \geq 0$

46. $f(x) = \frac{1}{x}$ for $x \neq 0$

47. $f(x) = x^2 + 4$ for $x \geq 0$

48. $f(x) = x^2 + 1$ for $x \leq 0$

49. $f(x) = 1 + \frac{1}{x}$ for $x > 0$

50. $f(x) = \frac{x}{x + 1}$ for $x > -1$

For Problems 51–58, (a) find f^{-1} and (b) graph f and f^{-1} on the same set of axes.

51. $f(x) = 3x$

52. $f(x) = -x$

53. $f(x) = 2x + 1$

54. $f(x) = -3x - 3$

55. $f(x) = \frac{2}{x - 1}$ for $x > 1$

56. $f(x) = \frac{-1}{x - 2}$ for $x > 2$

57. $f(x) = x^2 - 4$ for $x \geq 0$

58. $f(x) = \sqrt{x - 3}$ for $x \geq 3$

For Problems 59–66, find the intervals on which the given function is increasing and the intervals on which it is decreasing.

59. $f(x) = x^2 + 1$

60. $f(x) = x^3$

61. $f(x) = -3x + 1$

62. $f(x) = (x - 3)^2 + 1$

63. $f(x) = -(x + 2)^2 - 1$

64. $f(x) = x^2 - 2x + 6$

65. $f(x) = -2x^2 - 16x - 35$

66. $f(x) = x^2 + 3x - 1$

MISCELLANEOUS PROBLEMS

67. Explain why every nonconstant linear function has an inverse.

68. The function notation and the operation of composition can be used to find inverses as follows: To find the inverse of $f(x) = 5x + 3$ we know that $f(f^{-1}(x))$ must produce x. Therefore,

$$f(f^{-1}(x)) = 5[f^{-1}(x)] + 3 = x$$
$$5[f^{-1}(x)] = x - 3$$
$$f^{-1}(x) = \frac{x - 3}{5}.$$

Use this approach to find the inverse of each of the following functions.

a. $f(x) = 3x - 9$

b. $f(x) = -2x + 6$

c. $f(x) = -x + 1$

d. $f(x) = 2x$

e. $f(x) = -5x$

f. $f(x) = x^2 + 6$ for $x \geq 0$

69. If $f(x) = 2x + 3$ and $g(x) = 3x - 5$, find

a. $(f \circ g)^{-1}(x)$

b. $(f^{-1} \circ g^{-1})(x)$

c. $(g^{-1} \circ f^{-1})(x)$

GRAPHICS CALCULATOR ACTIVITIES

70. For Problems 37–44, graph the given function, the inverse function that you found, and $f(x) = x$ on the same set of axes. In each case the given function and its inverse should produce graphs that are reflections of each other through the line $f(x) = x$.

71. There is another way that we can use the graphics calculator to help show that two functions are inverses of each other. Suppose that we want to show that $f(x) = x^2 - 2$ for $x \geq 0$ and $g(x) = \sqrt{x + 2}$ for $x \geq -2$ are inverses of each other. Let's make the following assignments for our graphics calculator.

$$\begin{aligned} f&: \quad Y_1 = x^2 - 2 \\ g&: \quad Y_2 = \sqrt{x + 2} \\ f \circ g&: \quad Y_3 = (Y_2)^2 - 2 \\ g \circ f&: \quad Y_4 = \sqrt{Y_1 + 2} \end{aligned}$$

Now we can proceed as follows:

1. Graph $Y_1 = x^2 - 2$ and note that for $x > 0$, the range is greater than or equal to -2.
2. Graph $Y_2 = \sqrt{x + 2}$ and note that for $x \geq -2$, the range is greater than or equal to 0.

   Thus, the domain of f equals the range of g and the range of f equals the domain of g.
3. Graph $Y_3 = (Y_2)^2 - 2$ for $x \geq -2$ and observe the line $y = x$ for $x \geq -2$.
4. Graph $Y_4 = \sqrt{Y_1 + 2}$ for $x \geq 0$ and observe the line $y = x$ for $x \geq 0$.

   Thus, $(f \circ g)(x) = x$ and $(g \circ f)(x) = x$ and the two functions are inverses of each other.

Use this approach to check your answers for Problems 45–50.

72. Use the technique demonstrated in Problem 71 to show that $f(x) = \dfrac{x}{\sqrt{x^2 + 1}}$ and $g(x) = \dfrac{x}{\sqrt{1 - x^2}}$ for $-1 < x < 1$ are inverses of each other.

4.7 DIRECT AND INVERSE VARIATIONS

The amount of simple interest earned by a fixed amount of money invested at a certain rate *varies directly* as the time.

At a constant temperature, the volume of an enclosed gas *varies inversely* as the pressure.

Such statements illustrate two basic types of functional relationships, called **direct** and **inverse variation**, which are widely used, especially in the physical sciences. These relationships can be expressed by equations that determine functions. The purpose of this section is to investigate these special functions.

Direct Variation

The statement, *y varies directly as x* means

$$y = kx$$

where k is a nonzero constant, called the **constant of variation**. The phrase, *y is directly proportional to x* is also used to indicate direct variation; k is then referred to as the **constant of proportionality**.

REMARK Notice that the equation $y = kx$ defines a function and can be written $f(x) = kx$. However, in this section it is more convenient not to use function notation but instead to use variables that are meaningful in terms of the physical entities involved in the particular problem.

Statements that indicate direct variation may also involve powers of a variable. For example, *y varies directly as the square of x* can be written $y = kx^2$. In general, *y varies directly as the nth power of* $x (n > 0)$ means

$$y = kx^n.$$

There are basically three types of problems when dealing with direct variation: (1) translating an English statement into an equation expressing the direct variation, (2) finding the constant of variation from given values of the variables, and (3) finding additional values of the variables once the constant of variation has been determined. Let's consider an example of each of these types of problems.

EXAMPLE 1

Translate the statement, *The tension on a spring varies directly as the distance it is stretched* into an equation using k as the constant of variation.

Solution

Let t represent the tension and d the distance; the equation is

$$t = kd.$$

EXAMPLE 2

If A varies directly as the square of e, and $A = 96$ when $e = 4$, find the constant of variation.

Solution

Since A varies directly as the square of e, we have

$$A = ke^2.$$

Substitute 96 for A and 4 for e to obtain

$$96 = k(4)^2$$
$$96 = 16k$$
$$6 = k.$$

The constant of variation is 6.

EXAMPLE 3

If y is directly proportional to x, and if $y = 6$ when $x = 8$, find the value of y when $x = 24$.

Solution

The statement *y is directly proportional to x* translates into

$$y = kx.$$

Let $y = 6$ and $x = 8$; the constant of variation becomes

$$6 = k(8)$$
$$\frac{6}{8} = k$$
$$\frac{3}{4} = k.$$

So, the specific equation is

$$y = \frac{3}{4}x.$$

Now, let $x = 24$ to obtain

$$y = \frac{3}{4}(24) = 18.$$

Inverse Variation

The second basic type of variation, *inverse variation*, is defined as follows. The statement, *y varies inversely as x*, means

$$y = \frac{k}{x},$$

where k is a nonzero constant, which is again referred to as the constant of variation. The phrase, *y is inversely proportional to x* is also used to express inverse variation. As with direct variation, statements indicating variation may involve powers of x. For example, *y varies inversely as the square of x* can be written $y = k/x^2$. In general, *y varies inversely as the nth power of x* ($n > 0$) means

$$y = \frac{k}{x^n}.$$

The following examples illustrate the three basic kinds of problems that involve inverse variation.

EXAMPLE 4

Translate the statement, *The length of a rectangle of fixed area varies inversely as the width* into an equation using k as the constant of variation.

Solution

Let l represent the length and w the width; the equation is

$$l = \frac{k}{w}.$$

EXAMPLE 5

If y is inversely proportional to x and $y = 14$ when $x = 4$, find the constant of variation.

Solution

Since y is inversely proportional to x, we have

$$y = \frac{k}{x}.$$

Substitute 4 for x and 14 for y to obtain

$$14 = \frac{k}{4}.$$

Solving this equation yields

$$k = 56.$$

The constant of variation is 56.

EXAMPLE 6

The time required for a car to travel a certain distance varies inversely as the rate at which it travels. If it takes 4 hours at 50 miles per hour to travel the distance, how long will it take at 40 miles per hour?

Solution

Let t represent time and r rate. The phrase, *time required . . . varies inversely as the rate*, translates into

$$t = \frac{k}{r}.$$

Substitute 4 for t and 50 for r to produce the constant of variation:

$$4 = \frac{k}{50}$$

$$k = 200.$$

So the specific equation is

$$t = \frac{200}{r}.$$

Now substitute 40 for r to produce

$$t = \frac{200}{40} = 5.$$

It will take 5 hours at 40 miles per hour.

The terms *direct* and *inverse*, as applied to variation, refer to the relative behavior of the variables involved in the equation. That is to say, in *direct variation* ($y = kx$), an assignment of **increasing absolute values for x** produces **increasing absolute values for y**. However, in *inverse variation* ($y = k/x$), an assignment of **increasing absolute values for x** produces **decreasing absolute values for y**.

Joint Variation

Variation may involve more than two variables. The following table illustrates some different types of variation statements and their equivalent algebraic equations that use k as the constant of variation.

VARIATION STATEMENT	ALGEBRAIC EQUATION
1. y *varies jointly* as x and z	$y = kxz$
2. y *varies jointly* as x, z, and w	$y = kxzw$
3. V *varies jointly* as h and the square of r	$V = khr^2$
4. h *varies directly* as V and *inversely* as w	$h = \frac{kV}{w}$
5. y is *directly proportional* to x and *inversely proportional* to the square of z	$y = \frac{kx}{z^2}$
6. y *varies jointly* as w and z, and *inversely* as x	$y = \frac{kwz}{x}$

Statements 1, 2, and 3 illustrate the concept of *joint variation*. Statements 4 and 5 show that both direct and inverse variation may occur in the same problem. Statement 6 combines joint variation with inverse variation.

The final two examples of this section illustrate different kinds of problems involving some of these variation situations.

EXAMPLE 7

The volume of a pyramid varies jointly as its altitude and the area of its base. If a pyramid with an altitude of 9 feet and a base with an area of 17 square feet has a volume of 51 cubic feet, find the volume of a pyramid with an altitude of 14 feet and a base with an area of 45 square feet.

Solution

Let's use some variables as follows.

V = volume

B = area of base

h = altitude

k = constant of variation

The volume varies jointly as the altitude and the area of the base can be represented by the equation

$$V = kBh.$$

Substitute 51 for V, 17 for B, and 9 for h to obtain

$$51 = k(17)(9)$$
$$51 = 153k$$
$$\frac{51}{153} = k$$
$$\frac{1}{3} = k.$$

Therefore, the specific equation is $V = \frac{1}{3}Bh$. Now, substitute 45 for B and 14 for h to obtain

$$V = \frac{1}{3}(45)(14) = (15)(14) = 210.$$

The volume is 210 cubic feet.

EXAMPLE 8

Suppose that y varies jointly as x and z and inversely as w. If $y = 154$ when $x = 6$, $z = 11$, and $w = 3$, find y when $x = 8$, $z = 9$, and $w = 6$.

Solution

The statement, *y varies jointly as x and z and inversely as w*, translates into the equation

$$y = \frac{kxz}{w}.$$

Substitute 154 for y, 6 for x, 11 for z, and 3 for w to produce

$$154 = \frac{(k)(6)(11)}{3}$$
$$154 = 22k$$
$$7 = k.$$

So, the specific equation is

$$y = \frac{7xz}{w}.$$

Now, substitute 8 for x, 9 for z, and 6 for w to obtain

$$y = \frac{7(8)(9)}{6} = 84.$$

PROBLEM SET 4.7

For Problems 1–8, translate each of the statements of variation into an equation; use k as the constant of variation.

1. y varies directly as the cube of x.
2. a varies inversely as the square of b.
3. A varies jointly as l and w.
4. s varies jointly as g and the square of t.
5. At a constant temperature, the volume (V) of a gas varies inversely as the pressure (P).
6. y varies directly as the square of x and inversely as the cube of w.
7. The volume (V) of a cone varies jointly as its height (h) and the square of a radius (r).
8. I is directly proportional to r and t.

For Problems 9–18, find the constant of variation for each of the stated conditions.

9. y varies directly as x, and $y = 72$ when $x = 3$.
10. y varies inversely as the square of x, and $y = 4$ when $x = 2$.
11. A varies directly as the square of r, and $A = 154$ when $r = 7$.
12. V varies jointly as B and h, and $V = 104$ when $B = 24$ and $h = 13$.
13. A varies jointly as b and h, and $A = 81$ when $b = 9$ and $h = 18$.
14. s varies jointly as g and the square of t, and $s = -108$ when $g = 24$ and $t = 3$.
15. y varies jointly as x and z and inversely as w; $y = 154$ when $x = 6$, $z = 11$, and $w = 3$.
16. V varies jointly as h and the square of r, and $V = 1100$ when $h = 14$ and $r = 5$.
17. y is directly proportional to the square of x and inversely proportional to the cube of w, and $y = 18$ when $x = 9$ and $w = 3$.
18. y is directly proportional to x and inversely proportional to the square root of w, and $y = \frac{1}{5}$ when $x = 9$ and $w = 10$.

Solve each of the following problems.

19. If y is directly proportional to x, and $y = 5$ when $x = -15$, find the value of y when $x = -24$.
20. If y is inversely proportional to the square of x, and $y = \frac{1}{8}$ when $x = 4$, find y when $x = 8$.
21. If V varies jointly as B and h, and $V = 96$ when $B = 36$ and $h = 8$, find V when $B = 48$ and $h = 6$.
22. If A varies directly as the square of e, and $A = 150$ when $e = 5$, find A when $e = 10$.
23. The time required for a car to travel a certain distance varies inversely as the rate at which it travels. If it takes 3 hours at 50 miles per hour to travel the distance, how long will it take at 30 miles per hour?
24. The distance that a freely falling body falls varies directly as the square of the time it falls. If a body falls 144 feet in 3 seconds, how far will it fall in 5 seconds?
25. The period (the time required for one complete oscillation) of a simple pendulum varies directly as the square

root of its length. If a pendulum 12 feet long has a period of 4 seconds, find the period of a pendulum of length 3 feet.

26. Suppose the number of days it takes to complete a construction job varies inversely as the number of people assigned to the job. If it takes 7 people 8 days to do the job, how long will it take 10 people to complete the job?

27. The number of days needed to assemble some machines varies directly as the number of machines and inversely as the number of people working. If it takes 4 people 32 days to assemble 16 machines, how many days will it take 8 people to assemble 24 machines?

28. The volume of a gas at a constant temperature varies inversely as the pressure. What is the volume of a gas under a pressure of 25 pounds if the gas occupies 15 cubic centimeters under a pressure of 20 pounds?

29. The volume (V) of a gas varies directly as the temperature (T) and inversely as the pressure (P). If $V = 48$ when $T = 320$ and $P = 20$, find V when $T = 280$ and $P = 30$.

30. The volume of a cylinder varies jointly as its altitutde and the square of the radius of its base. If the volume of a cylinder is 1386 cubic centimeters when the radius of the base is 7 centimeters and its altitude is 9 centimeters, find the volume of a cylinder that has a base of radius 14 centimeters if the altitude of the cylinder is 5 centimeters.

31. The cost of labor varies jointly as the number of workers and the number of days that they work. If it costs $900 to have 15 people work for 5 days, how much will it cost to have 20 people work for 10 days?

32. The cost of publishing pamphlets varies directly as the number of pamphlets produced. If it costs $96 to publish 600 pamphlets, how much does it cost to publish 800 pamphlets?

MISCELLANEOUS PROBLEMS

In the previous problems, we chose numbers to make computations reasonable without the use of a calculator. However, variation-type problems often involve messy computations and the calculator becomes a very useful tool. Use your calculator to help solve the following problems.

33. The simple interest earned by a certain amount of money varies jointly as the rate of interest and the time (in years) that the money is invested.

a. If some money invested at 11% for 2 years earns $385, how much would the same amount earn at 12% for 1 year?

b. If some money invested at 12% for 3 years earns $819, how much would the same amount earn at 14% for 2 years?

c. If some money invested at 14% for 4 years earns $1960, how much would the same amount earn at 15% for 2 years?

34. The period (the time required for one complete oscillation) of a simple pendulum varies directly as the square root of its length. If a pendulum 9 inches long has a period of 2.4 seconds, find the period of a pendulum of length 12 inches. Express the answer to the nearest one-tenth of a second.

35. The volume of a cylinder varies jointly as its altitude and the square of the radius of its base. If the volume of a cylinder is 549.5 cubic meters when the radius of the base is 5 meters and its altitude is 7 meters, find the volume of a cylinder that has a base of radius 9 meters and an altitude of 14 meters.

36. If y is directly proportional to x and inversely proportional to the square of z, and if $y = 0.336$ when $x = 6$ and $z = 5$, find the constant of variation.

37. If y is inversely proportional to the square root of x, and $y = 0.08$ when $x = 225$, find y when $x = 625$.

CHAPTER 4 SUMMARY

The function concept serves as a thread to tie this chapter together.

Function Concept

DEFINITION 4.1

A function f is a correspondence between two sets X and Y that assigns to each element x of set X one and only one element y of set Y. The element y being assigned is called the **image** of x. The set X is called the **domain** of the function and the set of all images is called the **range of the function**.

A function can also be thought of as a set of ordered pairs no two of which have the same first component. If each vertical line intersects a graph in no more than one point, then the graph represents a function.

If no member of the range is assigned to more than one member of the domain, then the function is a **one-to-one function**. If each horizontal line intersects the graph of a function in no more than one point, then the graph represents a one-to-one function.

Graphing Functions

Any function that can be written in the form

$$f(x) = ax + b,$$

where a and b are real numbers, is a **linear function**. The graph of a linear function is a straight line.

Any function that can be written in the form

$$f(x) = ax^2 + bx + c,$$

where a, b, and c are real numbers and $a \neq 0$, is a **quadratic function**. The graph of any quadratic function is a **parabola**, which can be drawn using either one of the following methods.

1. Express the function in the form $f(x) = a(x - h)^2 + k$ and use the values of a, h, and k to determine the parabola.

2. Express the function in the form $f(x) = ax^2 + bx + c$ and use the fact that the vertex is at

$$\left(-\frac{b}{2a}, f\left(-\frac{b}{2a}\right)\right)$$

and the axis of symmetry is

$$x = -\frac{b}{2a}.$$

Another important graphing technique is to be able to recognize equations of the transformations of basic curves. We have worked with the following transformations in this chapter.

Vertical Translation The graph of $y = f(x) + k$ is the graph of $y = f(x)$ shifted k units upward if $k > 0$ or shifted $|k|$ units downward if $k < 0$.

Horizontal Translation The graph of $y = f(x - h)$ is the graph of $y = f(x)$ shifted h units to the right if $h > 0$ or shifted $|h|$ units to the left if $h < 0$.

x-axis Reflection The graph of $y = -f(x)$ is the graph of $y = f(x)$ reflected through the x-axis.

y-axis Reflection The graph of $y = f(-x)$ is the graph of $y = f(x)$ reflected through the y-axis.

Vertical Stretching and Shrinking The graph of $y = cf(x)$ is obtained from the graph of $y = f(x)$ by multiplying the y-coordinates of $y = f(x)$ by c. If $c > 1$, the graph is said to be **stretched** by a factor of c, and if $0 < c < 1$, the graph is said to be **shrunk** by a factor of c.

The following suggestions are made for graphing functions that are unfamiliar.

1. Determine the domain of the function.
2. Find the intercepts.
3. Determine the type of symmetry that the equation exhibits.
4. Set up a table of values that satisfy the equation. The type of symmetry and the domain will affect your choice of values for x in the table.
5. Plot the points associated with the ordered pairs and connect them with a smooth curve. Then, if appropriate, reflect this part of the curve according to the symmetry possessed by the graph.

Operations on Functions

Sum of two functions $(f + g)(x) = f(x) + g(x)$

Difference of two functions $(f - g)(x) = f(x) - g(x)$

Product of two functions $(f \cdot g)(x) = f(x) \cdot g(x)$

Quotient of two functions $\left(\frac{f}{g}\right)(x) = \frac{f(x)}{g(x)}, \quad g(x) \neq 0$

DEFINITION 4.2

The **composition** of functions f and g is defined by

$$(f \circ g)(x) = f(g(x))$$

for all x in the domain of g such that $g(x)$ is in the domain of f.

Remember that the composition of functions is *not a commutative operation.*

Inverse Functions

DEFINITION 4.3

Let f be a one-to-one function with a domain of X and a range of Y. A function g, with a domain of Y and a range of X, is called the **inverse function** of f if

$$(f \circ g)(x) = x \quad \text{for every } x \text{ in } Y$$

and

$$(g \circ f)(x) = x \quad \text{for every } x \text{ in } X.$$

The inverse of a function f is denoted by f^{-1}. Graphically, two functions that are inverses of each other are mirror images with reference to the line $y = x$.

A systematic technique for finding the inverse of a function can be described as follows.

1. Let $y = f(x)$.
2. Interchange x and y.
3. Solve the equation for y in terms of x.
4. The inverse function $f^{-1}(x)$ is determined by the equation in step 3.

Don't forget that the domain of f must equal the range of f^{-1}, and the domain of f^{-1} must equal the range of f.

Increasing and decreasing functions are defined as follows.

DEFINITION 4.4

Let f be a function, with the interval I a subset of the domain of f. Let x_1 and x_2 be in I. Then

1. f is *increasing on I* if $f(x_1) < f(x_2)$ whenever $x_1 < x_2$.
2. f is *decreasing on I* if $f(x_1) > f(x_2)$ whenever $x_1 < x_2$, and
3. f is *constant on I* if $f(x_1) = f(x_2)$ for every x_1 and x_2.

A function that is always increasing or always decreasing over its entire domain is a one-to-one function and therefore has an inverse.

Applications of Functions

Quadratic functions produce parabolas that have either a minimum or a maximum value. Therefore, a real world minimum- or maximum-value problem that can be described by a quadratic function can be solved using the techniques of this chapter.

Relationships that involve direct and inverse variation can be expressed by equations that determine functions. The statement, *y varies directly as x* means

$$y = kx,$$

where k is the constant of variation. The statement, *y varies directly as the nth power of x* ($n > 0$), means

$$y = kx^n.$$

The statement, *y varies inversely as x*, means $y = \dfrac{k}{x}$.

The statement, *y varies inversely as the nth power of x* ($n > 0$), means $y = \dfrac{k}{x^n}$.

The statement, *y varies jointly as x and w*, means $y = kxw$.

CHAPTER 4 REVIEW PROBLEM SET

1. If $f(x)$ is $3x^2 - 2x - 1$, find $f(2)$, $f(-1)$, and $f(-3)$.

2. For each of the following functions, find $\dfrac{f(a + h) - f(a)}{h}$.

a. $f(x) = -5x + 4$ **b.** $f(x) = 2x^2 - x + 4$

c. $f(x) = -3x^2 + 2x - 5$

3. Determine the domain and range of the function $f(x) = x^2 + 5$.

4. Determine the domain of the function $f(x) = \frac{2}{2x^2 + 7x - 4}$.

5. Express the domain of $f(x) = \sqrt{x^2 - 7x + 10}$ using interval notation.

For Problems 6–15, graph each of the functions.

6. $f(x) = -2x + 2$

7. $f(x) = 2x^2 - 1$

8. $f(x) = -\sqrt{x - 2} + 1$

9. $f(x) = x^2 - 8x + 17$

10. $f(x) = -x^3 + 2$

11. $f(x) = 2|x - 1| + 3$

12. $f(x) = -2x^2 - 12x - 19$

13. $f(x) = -\frac{1}{3}x + 1$

14. $f(x) = -\frac{2}{x^2}$

15. $f(x) = 2|x| - x$

16. If $f(x) = 2x + 3$ and $g(x) = x^2 - 4x - 3$, find $f + g$, $f - g$, $f \cdot g$, and f/g.

For Problems 17–20, find $(f \circ g)(x)$ and $(g \circ f)(x)$. Also specify the domain for each.

17. $f(x) = 3x - 9$ and $g(x) = -2x + 7$

18. $f(x) = x^2 - 5$ and $g(x) = 5x - 4$

19. $f(x) = \sqrt{x - 5}$ and $g(x) = x + 2$

20. $f(x) = \frac{1}{x - 3}$ and $g(x) = \frac{1}{x + 2}$

Solve the following two problems.

21. If $f(x) = |x|$ and $g(x) = x^2 - x - 1$, find $(f \circ g)(1)$ and $(g \circ f)(-3)$.

22. Verify that $f(x) = x^2 + 8$ for $x \geq 0$ and $g(x) = \sqrt{x - 8}$ for $x \geq 8$ are inverse functions.

For Problems 23–26, determine if f and g are inverse functions.

23. $f(x) = 7x - 1$ and $g(x) = \frac{x + 1}{7}$

24. $f(x) = -\frac{2}{3}x$ and $g(x) = \frac{3}{2}x$

25. $f(x) = x^2 - 6$ for $x \geq 0$ and $g(x) = \sqrt{x + 6}$ for $x \geq -6$

26. $f(x) = 2 - x^2$ for $x \geq 0$ and $g(x) = \sqrt{2 - x}$ for $x \leq 2$

For Problems 27–30, (a) find f^{-1} and (b) verify that $(f \circ f^{-1})(x) = x$ and $(f^{-1} \circ f)(x) = x$.

27. $f(x) = 4x + 5$

28. $f(x) = -3x - 7$

29. $f(x) = \frac{5}{6}x - \frac{1}{3}$

30. $f(x) = -2 - x^2$ for $x \geq 0$

For Problems 31 and 32, find the intervals on which the function is increasing and the intervals on which it is decreasing.

31. $f(x) = -2x^2 + 16x - 35$

32. $f(x) = 2\sqrt{x - 3}$

33. A group of students is arranging a chartered flight to Europe. The charge per person is \$496 if 100 students go on the flight. If more than 100 students go, the charge per student is reduced by an amount equal to \$4 times the number of students above 100. How many students should the airline try to get in order to maximize their revenue?

34. If y varies directly as x and inversely as w, and if $y = 27$ when $x = 18$ and $w = 6$, find the constant of variation.

35. If y varies jointly as x and the square root of w, and if $y = 140$ when $x = 5$ and $w = 16$, find y when $x = 9$ and $w = 49$.

36. The weight of a body above the surface of the earth varies inversely as the square of its distance from the center of the earth. Assuming the radius of the earth to be 4000 miles, how much would a man weigh 1000 miles above the earth's surface if he weighs 200 pounds on the surface?

Exponential and Logarithmic Functions

In this chapter we will continue our study of exponents in several ways: (1) we will extend the meaning of an exponent; (2) we will work with some exponential functions; (3) we will introduce the concept of a logarithm; (4) we will work some logarithmic functions; and (5) we will use the concepts of exponent and logarithm to expand our problem solving skills. Your calculator will be a valuable tool throughout this chapter.

5

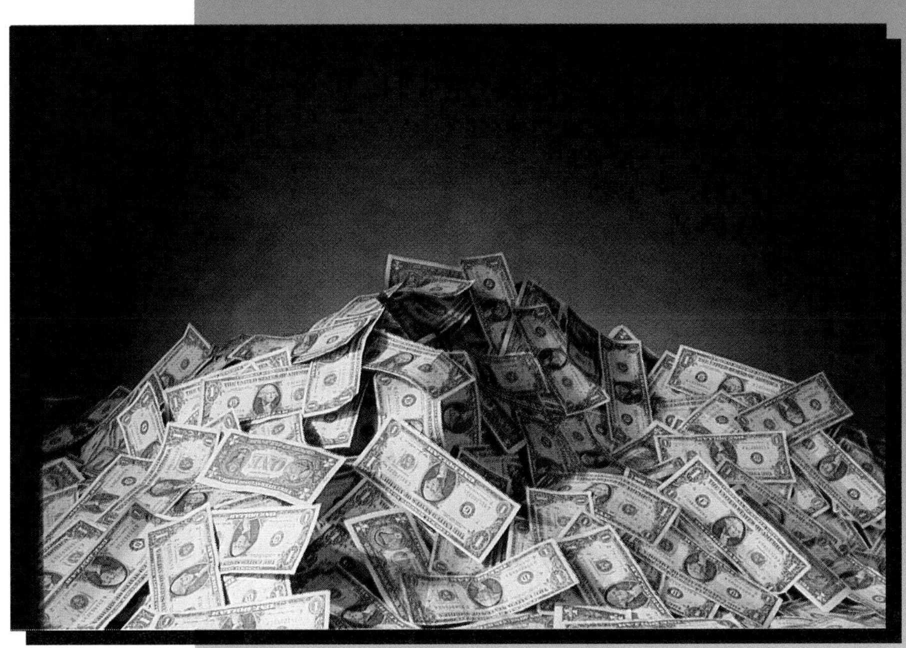

Compound interest is a good illustration of exponential growth.

5.1 EXPONENTS AND EXPONENTIAL FUNCTIONS

In Chapter 1, the expression b^n was defined to mean n factors of b, where n is any positive integer and b is any real number. For example,

$$2^3 = 2 \cdot 2 \cdot 2 = 8, \qquad \left(\frac{1}{3}\right)^4 = \left(\frac{1}{3}\right)\left(\frac{1}{3}\right)\left(\frac{1}{3}\right)\left(\frac{1}{3}\right) = \frac{1}{81},$$

$$(-4)^2 = (-4)(-4) = 16, \qquad -(0.5)^3 = -[(0.5)(0.5)(0.5)] = -0.125.$$

Also in Chapter 1, by defining $b^0 = 1$ and $b^{-n} = 1/b^n$, where n is any positive integer and b is any nonzero real number, we extended the concept of an exponent to include all integers. For example,

$$(0.76)^0 = 1, \qquad 2^{-3} = \frac{1}{2^3} = \frac{1}{8},$$

$$\left(\frac{2}{3}\right)^{-2} = \frac{1}{\left(\frac{2}{3}\right)^2} = \frac{1}{\frac{4}{9}} = \frac{9}{4}, \qquad (0.4)^{-1} = \frac{1}{(0.4)^1} = \frac{1}{0.4} = 2.5.$$

Finally, in Chapter 1 we provided for the use of any rational number as an exponent by defining

$$b^{m/n} = \sqrt[n]{b^m} = \left(\sqrt[n]{b}\right)^m,$$

where n is a positive integer greater than 1 and b is a real number such that $\sqrt[n]{b}$ exists. For example,

$$27^{2/3} = \left(\sqrt[3]{27}\right)^2 = 9, \qquad 16^{1/4} = \sqrt[4]{16^1} = 2,$$

$$\left(\frac{1}{9}\right)^{1/2} = \sqrt{\frac{1}{9}} = \frac{1}{3}, \qquad 32^{-1/5} = \frac{1}{32^{1/5}} = \frac{1}{\sqrt[5]{32}} = \frac{1}{2}.$$

If we were to make a formal extension of the concept of an exponent to include the use of irrational numbers, we would require some ideas from calculus, which is beyond the scope of this text. However, we can give you a brief glimpse at the general idea involved. Consider the number $2^{\sqrt{3}}$. By using the nonterminating and nonrepeating decimal representation 1.73205 . . . for $\sqrt{3}$, we can form the sequence of numbers $2^1, 2^{1.7}, 2^{1.73}, 2^{1.732}, 2^{1.7320}, 2^{1.73205}, \ldots$. It is a reasonable idea that each successive power gets closer to $2^{\sqrt{3}}$. This is precisely what happens if n is irrational and b^n is properly defined by using the concept of a *limit*. Furthermore, this ensures that an expression such as 2^x will yield exactly one value for each value of x.

So from now on we can use any real number as an exponent and the basic properties stated in Chapter 1 can be extended to include all real numbers as exponents. Let's restate those properties with the restriction that the bases a and b are to be positive numbers to avoid expressions such as $(-4)^{1/2}$, which do not represent real numbers.

PROPERTY 5.1

If a and b are positive real numbers and m and n are any real numbers, then the following properties hold.

1. $b^n \cdot b^m = b^{n+m}$ Product of two powers
2. $(b^n)^m = b^{mn}$ Power of a power
3. $(ab)^n = a^n b^n$ Power of a product
4. $\left(\frac{a}{b}\right)^n = \frac{a^n}{b^n}$ Power of a quotient
5. $\frac{b^n}{b^m} = b^{n-m}$ Quotient of two powers

Another property that can be used to solve certain types of equations involving exponents can be stated as follows.

PROPERTY 5.2

If $b > 0$ but $b \neq 1$, and if m and n are real numbers, then

$$b^n = b^m \quad \text{if and only if } n = m.$$

The following examples illustrate the use of Property 5.2.

EXAMPLE 1

Solve $2^x = 32$.

Solution

$$2^x = 32$$

$$2^x = 2^5 \qquad 32 = 2^5$$

$$x = 5 \qquad \text{Apply Property 5.2.}$$

The solution set is $\{5\}$.

EXAMPLE 2

Solve $2^{3x} = \frac{1}{64}$.

Solution

$$2^{3x} = \frac{1}{64}$$

$$2^{3x} = \frac{1}{2^6}$$

$$2^{3x} = 2^{-6}$$
$$3x = -6 \qquad \text{Apply Property 5.2.}$$
$$x = -2$$

The solution set is $\{-2\}$.

EXAMPLE 3

Solve $\left(\frac{1}{5}\right)^{x-4} = \frac{1}{125}$.

Solution

$$\left(\frac{1}{5}\right)^{x-4} = \frac{1}{125}$$
$$\left(\frac{1}{5}\right)^{x-4} = \left(\frac{1}{5}\right)^{3}$$
$$x - 4 = 3 \qquad \text{Apply Property 5.2.}$$
$$x = 7$$

The solution set is $\{7\}$.

EXAMPLE 4

Solve $9^x = 243$.

Solution

$$9^x = 243$$
$$(3^2)^x = 3^5$$
$$3^{2x} = 3^5$$
$$2x = 5 \qquad \text{Apply Property 5.2.}$$
$$x = \frac{5}{2}$$

The solution set is $\left\{\frac{5}{2}\right\}$.

EXAMPLE 5

Solve $(8^{2x})(4^{2x-1}) = 16$.

Solution

$$(8^{2x})(4^{2x-1}) = 16$$
$$(2^3)^{2x}(2^2)^{2x-1} = 2^4$$
$$(2^{6x})(2^{4x-2}) = 2^4$$
$$2^{6x+4x-2} = 2^4$$
$$2^{10x-2} = 2^4$$

$$10x - 2 = 4 \qquad \text{Apply Property 5.2.}$$
$$10x = 6$$
$$x = \frac{6}{10}$$
$$x = \frac{3}{5}$$

The solution set is $\left\{\frac{3}{5}\right\}$.

Exponential Functions

If b is any positive number, then the expression b^x designates exactly one real number for every real value of x. Therefore, the equation $f(x) = b^x$ defines a function whose domain is the set of real numbers. Furthermore, if we place the additional restriction $b \neq 1$, then any equation of the form $f(x) = b^x$ describes a one-to-one function and is called an **exponential function**. This leads to the following definition.

DEFINITION 5.1

If $b > 0$ and $b \neq 1$, then the function f defined by

$$f(x) = b^x,$$

where x is any real number, is called the **exponential function with base** b.

REMARK The function $f(x) = 1^x$ is a constant function and therefore it is not a one-to-one function. Remember from Chapter 4 that one-to-one functions have inverses; this becomes a key issue in a later section.

Now let's consider graphing some exponential functions.

EXAMPLE 6

Graph the function $f(x) = 2^x$.

Solution

Let's set up a table of values. Keep in mind that the domain is the set of real numbers and the equation $f(x) = 2^x$ exhibits no symmetry. Plot these points and connect them with a smooth curve to produce Figure 5.1.

x	2^x
-2	$\frac{1}{4}$
-1	$\frac{1}{2}$
0	1
1	2
2	4
3	8

FIGURE 5.1

In the table for Example 6, we chose integral values for x to keep the computation simple. However, with the use of a calculator we could easily acquire functional values by using nonintegral exponents. Consider the following additional values for $f(x) = 2^x$.

$$f(0.5) \approx 1.41, \qquad f(1.7) \approx 3.25,$$

$$f(-0.5) \approx 0.71, \qquad f(-2.6) \approx 0.16.$$

$\approx$ means *is approximately equal to*

Use your calculator to check these results. Also notice that the points generated by these values do fit the graph in Figure 5.1.

EXAMPLE 7

Graph $f(x) = \left(\frac{1}{2}\right)^x$.

Solution

Again, let's set up a table of values, plot the points, and connect them with a smooth curve. The graph is shown in Figure 5.2.

x	$\left(\frac{1}{2}\right)^x$
-2	4
-1	2
0	1
1	$\frac{1}{2}$
2	$\frac{1}{4}$
3	$\frac{1}{8}$

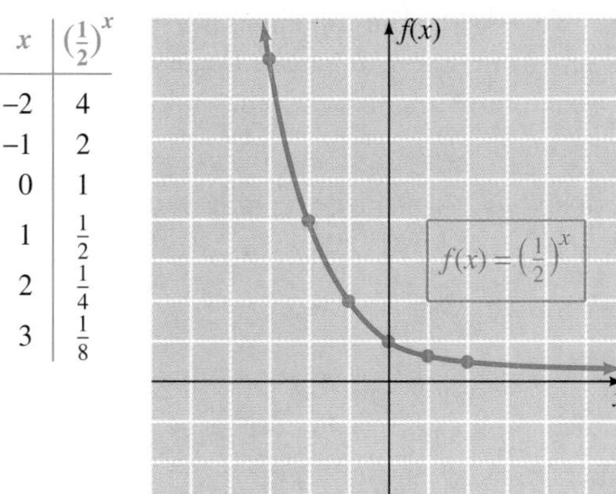

FIGURE 5.2

REMARK Since $\left(\frac{1}{2}\right)^x = 1/2^x = 2^{-x}$, the graphs of $f(x) = 2^x$ and $f(x) = \left(\frac{1}{2}\right)^x$ are reflections of each other across the y-axis. Therefore, Figure 5.2 could been drawn by reflecting Figure 5.1 across the y-axis.

The graphs in Figures 5.1 and 5.2 illustrate a general behavior pattern of exponential functions. That is, if $b > 1$, then the graph of $f(x) = b^x$ goes up to the right, and the function is called an **increasing function**. If $0 < b < 1$, then the graph of $f(x) = b^x$ goes down to the right, and the function is called a **decreasing function**. These facts are illustrated in Figure 5.3. Notice that $b^0 = 1$ for any $b > 0$; thus, **all graphs of $f(x) = b^x$ contain the point (0, 1)**.

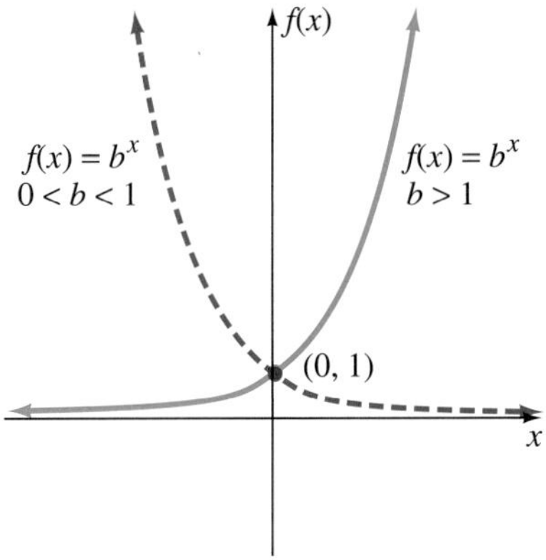

FIGURE 5.3

As you graph exponential functions, don't forget to use your previous graphing experience. For example, consider the following functions.

1. The graph of $f(x) = 2^x + 3$ is the graph of $f(x) = 2^x$ *moved up* 3 *units*.
2. The graph of $f(x) = 2^{x-4}$ is the graph of $f(x) = 2^x$ *moved to the right* 4 *units*.
3. The graph of $f(x) = -2^x$ is the graph of $f(x) = 2^x$ *reflected across the x-axis*.
4. The graph of $f(x) = 2^x + 2^{-x}$ is symmetrical with respect to the y-axis because $f(-x) = 2^{-x} + 2^x = f(x)$.

Furthermore, if you are faced with an exponential function that is not of the form $f(x) = b^x$ nor a variation thereof, don't forget the graphing suggestions offered in Chapter 3. Let's consider one such example.

EXAMPLE 8

Graph $f(x) = 2^{-x^2}$.

Solution

Since $f(-x) = 2^{-(-x)^2} = 2^{-x^2} = f(x)$, we know that this curve is symmetrical with respect to the y-axis. Therefore, let's set up a table of values using nonnegative values for x. Plot these points, connect them with a smooth curve, and reflect this portion of the curve across the y-axis to produce the graph in Figure 5.4.

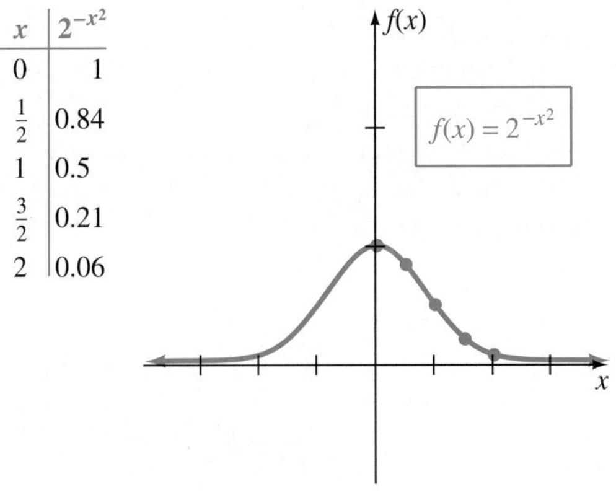

x	2^{-x^2}
0	1
$\frac{1}{2}$	0.84
1	0.5
$\frac{3}{2}$	0.21
2	0.06

FIGURE 5.4

EXAMPLE 9

Use a graphing utility to obtain a graph of $f(x) = 50(2^x)$ and find an approximate value for x when $f(x) = 15{,}000$.

Solution

First, we must find an appropriate viewing rectangle. Since $50(2^{10}) = 51{,}200$, let's set the boundaries so that $0 \leq x \leq 10$ and $0 \leq y \leq 50{,}000$ with a scale of 10,000 on the y-axis. (Certainly other boundaries could be used, but these will give us a graph that we can work with for this problem.) The graph of $f(x) = 50(2^x)$ is shown in Figure 5.5. Now we can use the trace and zoom-in features of the graphing utility to find that $x \approx 8.2$ at $y = 15{,}000$.

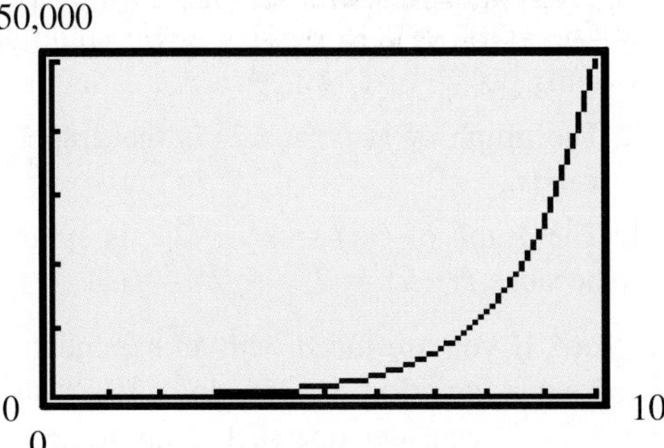

FIGURE 5.5

REMARK In Example 9 we used a graphical approach to solve the equation $50(2^x) = 15{,}000$. In Section 5.5 we will use an algebraic approach for solving that same kind of equation.

PROBLEM SET 5.1

Solve each of the following equations.

1. $3^x = 27$
2. $2^x = 64$
3. $\left(\frac{1}{2}\right)^x = \frac{1}{8}$
4. $\left(\frac{1}{2}\right)^n = 4$
5. $3^{-x} = \frac{1}{81}$
6. $3^{x+1} = 9$
7. $5^{2n-1} = 125$
8. $2^{3-n} = 8$
9. $\left(\frac{2}{3}\right)^t = \frac{9}{4}$
10. $\left(\frac{3}{4}\right)^n = \frac{64}{27}$
11. $4^{3x-1} = 256$
12. $16^x = 64$
13. $4^n = 8$
14. $27^{4x} = 9^{x+1}$
15. $32^x = 16^{1-x}$
16. $\left(\frac{1}{8}\right)^{-2t} = 2^{t+3}$
17. $(2^{2x-1})(2^{x+2}) = 32$
18. $(27)(3^x) = 9^x$
19. $(3^x)(3^{5x}) = 81$
20. $(4^x)(16^{3x-1}) = 8$

Graph each of the following functions.

21. $f(x) = 3^x$
22. $f(x) = \left(\frac{1}{3}\right)^x$
23. $f(x) = 4^x$
24. $f(x) = \left(\frac{1}{4}\right)^x$
25. $f(x) = \left(\frac{2}{3}\right)^x$
26. $f(x) = \left(\frac{3}{2}\right)^x$
27. $f(x) = 2^x + 1$
28. $f(x) = 2^x - 3$
29. $f(x) = 2^{x-1}$
30. $f(x) = 2^{x+2}$
31. $f(x) = -3^x$
32. $f(x) = -2^x$
33. $f(x) = 2^{-x+1}$
34. $f(x) = 2^{-x-2}$
35. $f(x) = 2^x + 2^{-x}$
36. $f(x) = 2^{x^2}$
37. $f(x) = 3^{1-x^2}$
38. $f(x) = 2^{|x|}$
39. $f(x) = 2^{-|x|}$
40. $f(x) = 2^x - 2^{-x}$
41. Graph $f(x) = -\left(\frac{1}{2}\right)^x$. Then on the same set of axes, graph $f(x) = -\left(\frac{1}{2}\right)^x + 2$, $f(x) = -\left(\frac{1}{2}\right)^{x+3}$, and $f(x) = -\left(\frac{1}{2}\right)^{-x}$.

THOUGHTS into WORDS

42. Why is the base of an exponential function restricted to positive numbers not including one?
43. How would you go about graphing the function $f(x) = -\left(\frac{1}{3}\right)^x$?
44. Explain how you would solve the equation

$$(4^{x-1})(8^{2x+3}) = 128.$$

GRAPHICS CALCULATOR ACTIVITIES

45. Use your graphics calculator to check your graphs for Problems 36–41.
46. Graph $f(x) = 4^x$. Where should the graphs of $f(x) = 4^{x-2}$, $f(x) = 4^{x-4}$, and $f(x) = 4^{x+3}$ be located? Graph all three functions on the same set of axes with $f(x) = 4^x$.
47. Graph $f(x) = \left(\frac{1}{4}\right)^x$. Where should the graphs of $f(x) = \left(\frac{1}{4}\right)^x - 2$, $f(x) = \left(\frac{1}{4}\right)^x + 3$, and $f(x) = \left(\frac{1}{4}\right)^x - 4$ be located? Graph all three functions on the same set of axes with $f(x) = \left(\frac{1}{4}\right)^x$.

48. Graph $f(x) = \left(\frac{3}{4}\right)^x$. Now predict the graphs for $f(x) = -\left(\frac{3}{4}\right)^x$, $f(x) = \left(\frac{3}{4}\right)^{-x}$, and $f(x) = -\left(\frac{3}{4}\right)^{-x}$. Graph all three functions on the same set of axes with $f(x) = \left(\frac{3}{4}\right)^x$.

49. Graph $f(x) = (-2)^x$. Explain your result.

50. What is the solution for $3^x = 5$? Do you agree that it is between 1 and 2 since $3^1 = 3$ and $3^2 = 9$? Now graph $f(x) = 3^x - 5$ and use the zoom and trace features of your graphics calculator to find an approximation, to the nearest hundredth, for the x-intercept. You should get an answer of 1.46, to the nearest hundredth. Do you see that this is an approximation for the solution of $3^x = 5$? Try it; raise 3 to the 1.46 power.

Find an approximate solution, to the nearest hundredth, for each of the following equations by graphing the appropriate function and finding the x-intercept.

a. $2^x = 19$ **b.** $3^x = 50$ **c.** $4^x = 47$

d. $5^x = 120$ **e.** $2^x = 1500$ **f.** $3^{x-1} = 34$

5.2 APPLICATIONS OF EXPONENTIAL FUNCTIONS

Many real world situations exhibiting growth or decay can be represented by equations that describe exponential functions. For example, suppose that an economist predicts an annual inflation rate of 5% per year for the next 10 years. This means that an item that presently costs \$8 will cost \$8(105%) = \$8(1.05) = \$8.40 in a year from now. The same item will cost [\$8(105%)] × (105%) = $\$8(1.05)^2$ = \$8.82 in 2 years. In general, the equation

$$P = P_0(1.05)^t$$

yields the predicted price P of an item in t years if the present cost is P_0 and the annual inflation rate is 5%. By using this equation, we can look at some future prices based on the prediction of a 5% inflation rate.

1. A \$3.27 container of hot cocoa mix will cost $\$3.27(1.05)^3 = \$3.79$ in 3 years.
2. A \$4.07 jar of coffee will cost $\$4.07(1.05)^5 = \$5.19$ in 5 years.
3. A \$9500 car will cost $\$9500(1.05)^7 = \$13{,}367$ (to the nearest dollar) in 7 years.

Compound Interest

Compound interest provides another illustration of exponential growth. Suppose that \$500 (called the **principal**) is invested at an interest rate of 8% **compounded annually**. The interest earned the first year is \$500(0.08) = \$40, and this amount is added to the original \$500 to form a new principal of \$540 for the second year. The interest earned during the second year is \$540(0.08) = \$43.20, and this amount is added to \$540 to form a new principal of \$583.20 for the third year. Each year a new principal is formed by reinvesting the interest earned during that year.

In general, suppose that a sum of money P (called the principal) is invested at an interest rate of r percent compounded annually. The interest earned the first year is Pr, and the new principal for the second year is $P + Pr$ or $P(1 + r)$. Note that the new principal for the second year can be found by multiplying the original principal P by $(1 + r)$. In a like fashion, the new principal for the third year can be found by multiplying the previous principal, $P(1 + r)$, by $1 + r$, thus obtaining $P(1 + r)^2$. If this process is continued, then after t years the total amount of money accumulated, A, is given by

$$A = P(1 + r)^t.$$

Consider the following examples of investments made at a certain rate of interest compounded annually.

1. \$750 invested for 5 years at 9% compounded annually produces

$$A = \$750(1.09)^5 = \$1153.97.$$

2. \$1000 invested for 10 years at 11% compounded annually produces

$$A = \$1000(1.11)^{10} = \$2839.42.$$

3. \$5000 invested for 20 years at 12% compounded annually produces

$$A = \$5000(1.12)^{20} = \$48{,}231.47.$$

The compound interest formula can be used to determine what rate of interest is needed to accumulate a certain amount of money based on a given initial investment. The next example illustrates this idea.

EXAMPLE 1

What rate of interest is needed for an investment of \$1000 to yield \$4000 in 10 years if the interest is compounded annually?

Solution

Let's substitute \$1000 for P, \$4000 for A, and 10 years for t in the compound interest formula, and solve for r.

$$\begin{aligned} A &= P(1 + r)^t \\ 4000 &= 1000(1 + r)^{10} \\ 4 &= (1 + r)^{10} \\ 4^{0.1} &= [(1 + r)^{10}]^{0.1} && \text{Raise both sides to the 0.1 power.} \\ 1.148698355 &\approx 1 + r \\ 0.148698355 &\approx r \\ r &= 14.9\% && \text{to the nearest tenth of a percent} \end{aligned}$$

Therefore, a rate of interest of approximately 14.9% is needed. (Perhaps you should check this answer.)

If money invested at a certain rate of interest is to be compounded more than once a year, then the basic formula $A = P(1 + r)^t$ can be adjusted according to the number of compounding periods in a year. For example, for **compounding semiannually**, the formula becomes

$$A = P\left(1 + \frac{r}{2}\right)^{2t}$$

and for **compounding quarterly**, the formula becomes

$$A = P\left(1 + \frac{r}{4}\right)^{4t}.$$

In general, if n represents the number of **compounding periods** in a year, the formula becomes

$$A = P\left(1 + \frac{r}{n}\right)^{nt}.$$

The following examples illustrate the use of the formula.

1. \$750 invested for 5 years at 9% compounded semiannually produces

$$A = \$750\left(1 + \frac{0.09}{2}\right)^{2(5)} = \$750(1.045)^{10} = \$1164.73.$$

2. \$1000 invested for 10 years at 11% compounded quarterly produces

$$A = \$1000\left(1 + \frac{0.11}{4}\right)^{4(10)} = \$1000(1.0275)^{40} = \$2959.87.$$

3. \$5000 invested for 20 years at 12% compounded monthly produces

$$A = \$5000\left(1 + \frac{0.12}{12}\right)^{12(20)} = \$5000(1.01)^{240} = \$54{,}462.77.$$

You may find it interesting to compare these results with those obtained earlier for compounding annually.

Exponential Decay

Suppose that it is estimated that the value of a car depreciates 15% per year for the first 5 years. Therefore, a car that costs \$9500 will be worth \$9500 × (100% − 15%) = \$9500(85%) = \$9500(0.85) = \$8075 in 1 year. In 2 years the value of the car will have depreciated to $9500(0.85)^2$ = \$6864 (to the nearest dollar). The equation

$$V = V_0(0.85)^t$$

yields the value V of a car in t years if the initial cost is V_0 and it depreciates 15% per year. Therefore, we can estimate some car values to the nearest dollar.

1. A \$6900 car will be worth $\$6900(0.85)^3 = \$4237$ in 3 years.
2. A \$10,900 car will be worth $\$10{,}900(0.85)^4 = \$5690$ in 4 years.
3. A \$13,000 car will be worth $\$13{,}000(0.85)^5 = \$5768$ in 5 years.

Another example of exponential decay is associated with radioactive substances. The rate of decay can be described exponentially and is based on the half-life of a substance. The **half-life** of a radioactive substance is the amount of time that it takes for one-half of an initial amount of the substance to disappear as the result of decay. For example, suppose that we have 200 grams of a certain substance that has a half-life of 5 days. After 5 days, $200\left(\frac{1}{2}\right) = 100$ grams remain. After 10 days, $200\left(\frac{1}{2}\right)^2 = 50$ grams remain. After 15 days, $200\left(\frac{1}{2}\right)^3 = 25$ grams remain. In general, after t days, $200\left(\frac{1}{2}\right)^{t/5}$ grams remain.

The previous discussion leads into the following half-life formula. Suppose there is an initial amount, Q_0, of a radioactive substance with a half-life of h. The amount of substance remaining, Q, after a time period of t, is given by the formula

$$Q = Q_0\left(\frac{1}{2}\right)^{t/h}.$$

The units of measure for t and h must be the same.

EXAMPLE 2

Barium-140 has a half-life of 13 days. If there are 500 milligrams of barium initially, how many milligrams remain after 26 days? After 100 days?

Solution

Using $Q_0 = 500$ and $h = 13$, the half-life formula becomes

$$Q = 500\left(\frac{1}{2}\right)^{t/13}.$$

If $t = 26$, then

$$\begin{aligned} Q &= 500\left(\frac{1}{2}\right)^{26/13} \\ &= 500\left(\frac{1}{2}\right)^{2} \\ &= 500\left(\frac{1}{4}\right) \\ &= 125. \end{aligned}$$

So, 125 milligrams remain after 26 days. If $t = 100$, then

$$\begin{aligned} Q &= 500\left(\frac{1}{2}\right)^{100/13} \\ &= 500(0.5)^{100/13} \\ &= 2.4 \quad \text{to the nearest tenth of a milligram.} \end{aligned}$$

So, approximately 2.4 milligrams remain after 100 days.

REMARK The solution to Example 2 clearly demonstrates one facet of the role of the calculator when studying mathematics. We solved the first part of the problem very easily without the calculator but the calculator certainly was helpful for the second part of the problem.

Number e

An interesting situation occurs if we consider the compound interest formula for $P = \$1$, $r = 100\%$, and $t = 1$ year. The formula becomes $A = 1\left(1 + \frac{1}{n}\right)^n$. The following table shows some values, rounded to eight decimal places, of $\left(1 + \frac{1}{n}\right)^n$ for different values of n.

n	$\left(n + \frac{1}{n}\right)^n$
1	2.00000000
10	2.59374246
100	2.70481383
1000	2.71692393
10,000	2.71814593
100,000	2.71826824
1,000,000	2.71828047
10,000,000	2.71828169
100,000,000	2.71828181
1,000,000,000	2.71828183

The table suggests that as n increases, the value of $\left(1 + \frac{1}{n}\right)^n$ gets closer and closer to some fixed number. This does happen and the fixed number is called e. To five decimal places, $e = 2.71828$.

The function defined by the equation $f(x) = e^x$ is the **natural exponential function**. It has a great many real world applications, some of which we will look at in a moment. First, however, let's get a picture of the natural exponential function. Since $2 < e < 3$, the graph of $f(x) = e^x$ must fall between the graphs of $f(x) = 2^x$ and $f(x) = 3^x$. To be more specific, let's use our calculator to determine a table of values. Use the $\boxed{e^x}$ key, and round the results to the nearest tenth to obtain the table. Plot the points determined by this table and connect them with a smooth curve to produce Figure 5.6.

x	$f(x) = e^x$
0	1.0
1	2.7
2	7.4
−1	0.4
−2	0.1

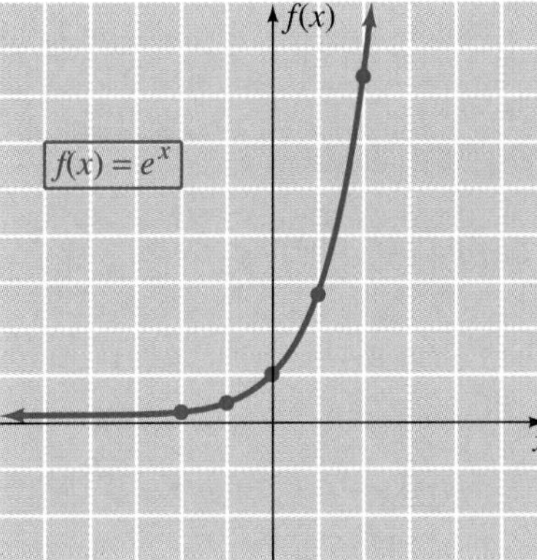

FIGURE 5.6

Back to Compound Interest

Let's return to the concept of compound interest. If the number of compounding periods in a year is increased indefinitely, we arrive at the concept of **compounding continuously**. Mathematically, this can be accomplished by applying the limit concept to the expression

$$P\left(1 + \frac{r}{n}\right)^{nt}.$$

We will not show the details here, but the following result is obtained. The formula

$$A = Pe^{rt}$$

yields the accumulated value (A) of a sum of money (P) that has been invested for t years at a rate of r percent compounded continuously. The following examples illustrate the use of this formula.

1. \$750 invested for 5 years at 9% compounded continuously produces

$$A = 750e^{(0.09)(5)} = 750e^{0.45} = \$1176.23.$$

2. \$1000 invested for 10 years at 11% compounded continuously produces

$$A = 1000e^{(0.11)(10)} = 1000e^{1.1} = \$3004.17.$$

3. \$5000 invested for 20 years at 12% compounded continuously produces

$$A = 5000e^{(0.12)(20)} = 5000e^{2.4} = \$55{,}115.88.$$

Again, you may find it interesting to compare these results with those you obtained earlier using a different number of compounding periods.

Is it better to invest at 6% compounded quarterly or at 5.75% compounded continuously? To answer such a question, we can use the concept of **effective yield** (sometimes called *effective annual rate of interest*). The effective yield of an investment is the simple interest rate that would yield the same amount in one year. Thus, for the *6% compounded quarterly* investment, we can calculate the effective yield as follows.

$$P(1 + r) = P\left(1 + \frac{0.06}{4}\right)^6$$

$$1 + r = \left(1 + \frac{0.06}{4}\right)^4 \qquad \text{Multiply both sides by } \frac{1}{P}.$$

$$1 + r = (1.015)^4$$

$$r = (1.015)^4 - 1$$

$$r \approx 0.0613635506$$

$$r = 6.14\% \quad \text{to the nearest hundredth of a percent}$$

Likewise, for the 5.75% *compounded continuously* investment we can calculate the effective yield as follows.

$$P(1 + r) = Pe^{0.0575}$$

$$1 + r = e^{0.0575}$$

$$r = e^{0.0575} - 1$$

$$r \approx 0.0591852707$$

$$r = 5.92\% \quad \text{to the nearest hundredth of a percent}$$

Therefore, comparing the two effective yields, we see that it is better to invest at 6% compounded quarterly than to invest at 5.75% compounded continuously.

Law of Exponential Growth

The ideas behind compounded continuously carry over to other growth situations. The law of exponential growth

$$Q(t) = Q_0e^{kt}$$

is used as a mathematical model for numerous growth-and-decay applications. In this equation, $Q(t)$ represents the quantity of a given susbstance at any time t; Q_0 is the initial amount of the substance (when $t = 0$); and k is a constant that depends on the particular application. If $k < 0$, then $Q(t)$ decreases as t increases, and we refer to the model as the **law of decay**.

Let's consider some growth-and-decay applications.

EXAMPLE 3

Suppose that in a certain culture the equation $Q(t) = 15000e^{0.3t}$ expresses the number of bacteria present as a function of the time t, where t is expressed in hours. Find (a) the initial number of bacteria, and (b) the number of bacteria after 3 hours.

Solution

a. The initial number of bacteria is produced when $t = 0$.

$$\begin{aligned} Q(0) &= 15000e^{0.3(0)} \\ &= 15000e^{0} \\ &= 15000 \qquad e^0 = 1 \end{aligned}$$

b. $$\begin{aligned} Q(3) &= 15000e^{0.3(3)} \\ &= 15000e^{0.9} = 36894 \quad \text{to the nearest whole number} \end{aligned}$$

There should be approximately 36,894 bacteria present after 3 hours.

EXAMPLE 4

Suppose the number of bacteria present in a certain culture after t minutes is given by the equation $Q(t) = Q_0e^{0.05t}$, where Q_0 represents the initial number of bacteria. If 5000 bacteria are present after 20 minutes, how many bacteria were present initially?

Solution

If 5000 bacteria are present after 20 minutes, then $Q(20) = 5000$.

$$\begin{aligned} 5000 &= Q_0e^{0.05(20)} \\ 5000 &= Q_0e^{1} \\ \frac{5000}{e} &= Q_0 \\ 1839 &= Q_0 \quad \text{to the nearest whole number} \end{aligned}$$

Therefore, there were approximately 1839 bacteria present initially.

EXAMPLE 5

The number of grams of a certain radioactive substance present after t seconds is given by the equation $Q(t) = 200e^{-0.3t}$. How many grams remain after 7 seconds?

Solution

Use $Q(t) = 200e^{-0.3t}$ to obtain

$$\begin{aligned} Q(7) &= 200e^{(-0.3)(7)} \\ &= 200e^{-2.1} = 24.5. \quad \text{to the nearest tenth} \end{aligned}$$

Thus, approximately 24.5 grams remain after 7 seconds.

EXAMPLE 6

Suppose that \$1000 was invested at 6.5% interest compounded continuously. How long would it take for the money to double itself?

Solution

Substitute \$1000 for P and 0.065 for r in the formula $A = Pe^{rt}$ to produce $A = 1000e^{0.065t}$. If we let $y = A$ and $x = t$, we can graph the equation $y = 1000e^{0.065x}$. By letting $x = 20$, we obtain $y = 1000e^{0.065(20)} = 1000e^{1.3} \approx 3670$. Therefore, let's set the boundaries of the viewing rectangle so that $0 \leq x \leq 20$ and $0 \leq y \leq 3700$ with a y-scale of 1000. Then we obtain the graph in Figure 5.7. Now we want to find the value of x so that $y = 2000$. (The money is to double itself.) Using the zoom and trace features of the graphing utility we can determine that an x-value of approximately 10.7 will produce a y-value of 2000. Thus, it will take approximately 10.7 years for the \$1000 investment to double itself.

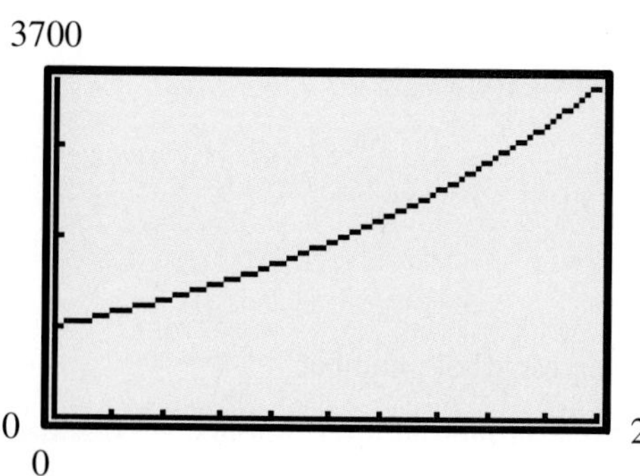

FIGURE 5.7

EXAMPLE 7 Graph the function $y = \frac{1}{\sqrt{2\pi}}e^{-x^2/2}$ and find its maximum value.

Solution

If $x = 0$, then $y = \frac{1}{\sqrt{2\pi}}e^0 = \frac{1}{\sqrt{2\pi}} \approx 0.4$. So let's set the boundaries of the viewing rectangle so that $-5 \leq x \leq 5$ and $0 \leq y \leq 1$ with a y-scale of 0.1; the graph of the function is shown in Figure 5.8. From the graph, we see that the maximum value of the function occurs at $x = 0$, which we have already determined to be approximately 0.4.

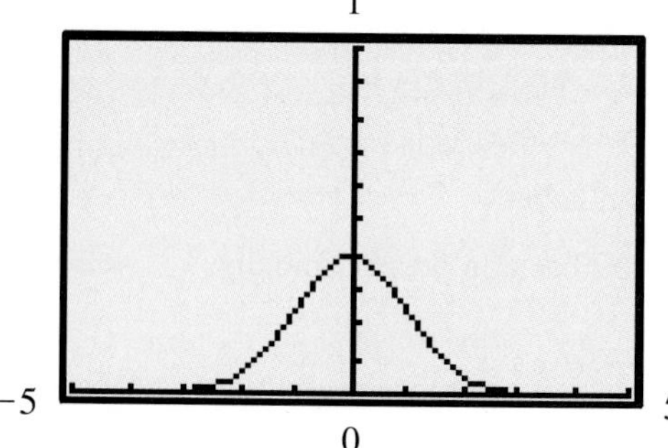

FIGURE 5.8

REMARK The curve in Figure 5.8 is called the **normal distribution curve**. You may want to ask your instructor to explain what it means to assign grades based on the normal distribution curve.

PROBLEM SET 5.2

1. Assuming that the rate of inflation is 7% per year, the equation $P = P_0(1.07)^t$ yields the predicted price P of an item in t years if it presently costs P_0. Find the predicted price of each of the following items for the indicated years ahead.

a. \$.55 can of soup in 3 years

b. \$3.43 container of cocoa mix in 5 years

c. \$1.76 jar of coffee creamer in 4 years

d. \$.44 can of beans and bacon in 10 years

e. \$9000 car in 5 years (to the nearest dollar)

f. \$50,000 house in 8 years (to the nearest dollar)

g. \$500 TV set in 7 years (to the nearest dollar)

2. Suppose that it is estimated that the value of a car depreciates 20% per year for the first 5 years. The equation $A = P_0(0.8)^t$ yields the value (A) of a car after t years if the original price is P_0. Find the value (to the nearest dollar) of each of the following cars after the indicated time.

a. \$9000 car after 4 years

b. \$5295 car after 2 years

c. \$6395 car after 5 years

d. \$15,595 car after 3 years

For Problems 3–14 use the formula

$$A = P\left(1 + \frac{r}{n}\right)^{nt}$$

to find the total amount of money accumulated at the end of the indicated time period for each of the following investments. Estimate to the nearest cent.

3. \$250 for 5 years at 9% compounded annually

4. \$350 for 7 years at 11% compounded annually

5. \$300 for 6 years at 8% compounded semiannually

6. \$450 for 10 years at 10% compounded semiannually

7. \$600 for 12 years at 12% compounded quarterly

8. \$750 for 15 years at 9% compounded quarterly

9. \$1000 for 5 years at 12% compounded monthly

10. \$1250 for 8 years at 9% compounded monthly

11. \$600 for 10 years at $8\frac{1}{2}$% compounded annually

12. \$1500 for 15 years at $9\frac{1}{4}$% compounded semiannually

13. \$8000 for 10 years at 10.5% compounded quarterly

14. \$10,000 for 25 years at 9.25% compounded monthly

For Problems 15–23, use the formula $A = Pe^{rt}$ to find the total amount of money accumulated at the end of the indicated time period by compounding continuously.

15. \$400 for 5 years at 7%

16. \$500 for 7 years at 6%

17. \$750 for 8 years at 8%

18. \$1000 for 10 years at 9%

19. \$2000 for 15 years at 10%

20. \$5000 for 20 years at 11%

21. \$7500 for 10 years at 8.5%

22. \$10,000 for 25 years at 9.25%

23. \$15,000 for 10 years at 7.75%

24. What rate of interest, to the nearest tenth of a percent, compounded annually is needed for an investment of \$200 to grow to \$350 in 5 years?

25. What rate of interest, to the nearest tenth of a percent, compounded quarterly is needed for an investment of \$1500 to grow to \$2700 in 10 years?

26. Find the effective yield, to the nearest tenth of a percent, of an investment at 7.5% compounded monthly.

27. Find the effective yield, to the nearest hundredth of a percent, of an investment at 7.75% compounded continuously.

28. What investment yields the greatest return: 7% compounded monthly or 6.85% compounded continuously?

29. What investment yields the greatest return: 8.25% compounded quarterly or 8.3% compounded semiannually?

30. Suppose that a certain radioactive substance has a half-life of 20 years. If there are presently 2500 milligrams of the substance, how much, to the nearest milligram, will remain after 40 years? After 50 years?

31. Strontium-90 has a half-life of 29 years. If there are 400 grams of strontium initially, how much, to the nearest gram, will remain after 87 years? After 100 years?

32. The half-life of radium is approximately 1600 years. If the present amount of radium in a certain location is 500 grams, how much will remain after 800 years? Express your answer to the nearest gram.

33. Suppose that in a certain culture, the equation $Q(t) = 1000e^{0.4t}$ expresses the number of bacteria present as a function of the time t, where t is expressed in hours. How many bacteria are present at the end of 2 hours? 3 hours? 5 hours?

34. The number of bacteria present at a given time under certain conditions is given by the equation $Q = 5000e^{0.05t}$, where t is expressed in minutes. How many bacteria are present at the end of 10 minutes? 30 minutes? 1 hour?

35. The number of bacteria present in a certain culture after t hours is given by the equation $Q = Q_0e^{0.3t}$, where Q_0 represents the initial number of bacteria. If 6640 bacteria are present after 4 hours, how many bacteria were present initially?

36. The number of grams Q of a certain radioactive substance present after t seconds is given by the equation $Q = 1500e^{-0.4t}$. How many grams remain after 5 seconds? 10 seconds? 20 seconds?

37. The atmospheric pressure, measured in pounds per square inch, is a function of the altitude above sea level. The equation $P(a) = 14.7e^{-0.21a}$, where a is the altitude measured in miles, can be used to approximate atmospheric pressure. Find the atmospheric pressure at each of the following locations.

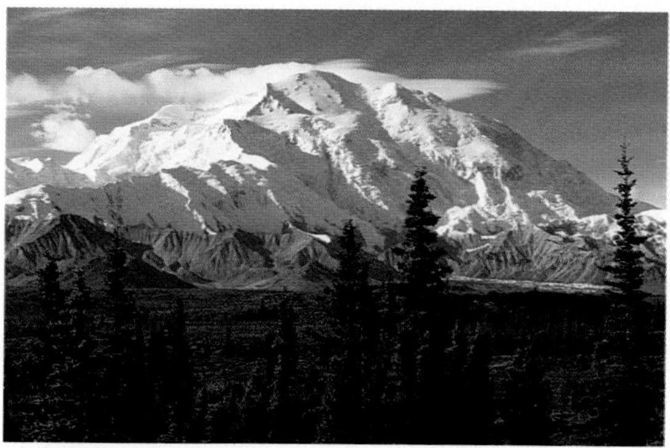

Mount McKinley
Harald Sund/The Image Bank

a. Mount McKinley in Alaska–altitude of 3.85 miles

b. Denver, Colorado–the mile-high city

c. Asheville, North Carolina–altitude of 1985 feet

d. Phoenix, Arizona–altitude of 1090 feet

38. Suppose that the present population of a city is 75,000. Using the equation $P(t) = 75{,}000e^{0.01t}$ to estimate future growth, estimate the population (a) 10 years from now, (b) 15 years from now, and (c) 25 years from now.

For Problems 39–44, graph each of the exponential functions.

39. $f(x) = e^x + 1$

40. $f(x) = e^x - 2$

41. $f(x) = 2e^x$

42. $f(x) = -e^x$

43. $f(x) = e^{2x}$

44. $f(x) = e^{-x}$

MISCELLANEOUS PROBLEMS

45. Complete the following chart that illustrates what happens to \$1000 invested at various rates of interest for different lengths of time but always compounded continuously. Round your answers to the nearest dollar.

\$1000 Compounded continuously

	8%	10%	12%	14%
5 years				
10 years				
15 years				
20 years				
25 years				

46. Complete the following chart that illustrates what happens to \$1000 invested at 12% for different lengths of time and different numbers of compounding periods. Round all of your answers to the nearest dollar.

\$1000 at 12%

	1 year	5 years	10 years	20 years
Compounded annually				
Compounded semiannually				
Compounded quarterly				
Compounded monthly				
Compounded continously				

47. Complete the following chart that illustrates what happens to \$1000 in 10 years based on different rates of interest and different numbers of compounding periods. Round your answers to the nearest dollar.

\$1000 for 10 years

	8%	10%	12%	14%
Compounded annually				
Compounded semiannually				
Compounded quarterly				
Compounded monthly				
Compounded continously				

For Problems 48–52, graph each of the functions.

48. $f(x) = x(2^x)$

49. $f(x) = \dfrac{e^x + e^{-x}}{2}$

50. $f(x) = \dfrac{2}{e^x + e^{-x}}$

51. $f(x) = \dfrac{e^x - e^{-x}}{2}$

52. $f(x) = \dfrac{2}{e^x - e^{-x}}$

GRAPHICS CALCULATOR ACTIVITIES

53. Use your graphics calculator to check your graphs for Problems 48–52.

54. How should the graphs of $f(x) = 2^x$, $f(x) = e^x$, and $f(x) = 3^x$ compare? Graph them on the same set of axes.

55. Graph $f(x) = e^x$. Where should the graphs of $f(x) = e^{x-2}$, $f(x) = e^{x+4}$, and $f(x) = e^{x-6}$ be located? Graph all three functions on the same set of axes.

56. Graph $f(x) = e^x$ again. Now predict the graphs for $f(x) = -e^x$, $f(x) = e^{-x}$, and $f(x) = -e^{-x}$. Graph these three functions on the same set of axes.

57. How do you think the graphs of $f(x) = e^x$, $f(x) = e^{2x}$, and $f(x) = 2e^x$ will compare? Graph them on the same set of axes to see if you were right.

58. Find an approximate solution, to the nearest hundredth, for each of the following equations by graphing the appropriate function and finding the x-intercept.

a. $e^x = 7$ **b.** $e^x = 21$ **c.** $e^x = 53$

d. $2e^x = 60$ **e.** $e^{x+1} = 150$ **f.** $e^{x-2} = 300$

59. Use a graphing approach to argue that it is better to invest money at 6% compounded quarterly than it is at 5.75% compounded continuously.

60. How long will it take \$500 to be worth \$1500 if it is invested at 7.5% interest compounded semiannually?

61. How long will it take \$5000 to triple itself if it is invested at 6.75% interest compounded quarterly?

5.3 LOGARITHMS

In Sections 5.1 and 5.2, we gave meaning to the exponential expressions of the form b^n, where b is any positive real number and n is any real number; we next used exponential expressions of the form b^n to define exponential functions; and then we used exponential functions to help solve problems. In the next three sections we will follow the same basic pattern with respect to a new concept, that of a *logarithm*. Let's begin with the following definition.

DEFINITION 5.2

If r is any positive real number, then the unique exponent t such that $b^t = r$ is called the **logarithm of r with base b** and is denoted by $\log_b r$.

According to Definition 5.2, the logarithm of 16 base 2 is the exponent t such that $2^t = 16$; thus, we can write $\log_2 16 = 4$. Likewise, we can write $\log_{10} 1000 = 3$ because $10^3 = 1000$. In general, Definition 5.2 can be remembered in terms of the statement

$$\log_b r = t \quad \text{is equivalent to } b^t = r.$$

Therefore, we can easily switch back and forth between exponential and logarithmic forms of equations, as the next examples illustrate.

$$\begin{aligned}
\log_2 8 &= 3 && \text{is equivalent to } 2^3 = 8.\\
\log_{10} 100 &= 2 && \text{is equivalent to } 10^2 = 100.\\
\log_3 81 &= 4 && \text{is equivalent to } 3^4 = 81.\\
\log_{10} 0.001 &= -3 && \text{is equivalent to } 10^{-3} = 0.001.\\
2^7 &= 128 && \text{is equivalent to } \log_2 128 = 7.\\
5^3 &= 125 && \text{is equivalent to } \log_5 125 = 3.\\
\left(\frac{1}{2}\right)^4 &= \frac{1}{16} && \text{is equivalent to } \log_{1/2} \frac{1}{16} = 4.\\
10^{-2} &= 0.01 && \text{is equivalent to } \log_{10} 0.01 = -2.
\end{aligned}$$

Some logarithms can be determined by changing to exponential form and using the properties of exponents, as in the next two examples.

EXAMPLE 1

Evaluate $\log_{10} 0.0001$.

Solution

Let $\log_{10} 0.0001 = x$. Then by changing to exponential form we have $10^x = 0.0001$, which can be solved as follows.

$$\begin{aligned}
10^x &= 0.0001\\
10^x &= 10^{-4} && 0.0001 = \frac{1}{10{,}000} = \frac{1}{10^4} = 10^{-4}\\
x &= -4
\end{aligned}$$

Thus, we have $\log_{10} 0.0001 = -4$.

EXAMPLE 2

Evaluate $\log_9\left(\sqrt[5]{27}/3\right)$.

Solution

Let $\log_9\left(\sqrt[5]{27}/3\right) = x$. Then by changing to exponential form we have $9^x = \sqrt[5]{27}/3$, which can be solved as follows.

$$9^x = \frac{(27)^{1/5}}{3}$$
$$(3^2)^x = \frac{(3^3)^{1/5}}{3}$$
$$3^{2x} = \frac{3^{3/5}}{3}$$
$$3^{2x} = 3^{-2/5}$$
$$2x = -\frac{2}{5}$$
$$x = -\frac{1}{5}$$

Therefore, we have $\log_9 \dfrac{\sqrt[5]{27}}{3} = -\dfrac{1}{5}$.

Some equations that involve logarithms can also be solved by changing to exponential form and using our knowledge of exponents.

EXAMPLE 3

Solve $\log_8 x = \frac{2}{3}$.

Solution

Change $\log_8 x = \frac{2}{3}$ to exponential form to obtain

$$8^{2/3} = x.$$

Therefore,

$$x = \left(\sqrt[3]{8}\right)^2 = 2^2 = 4.$$

The solution set is $\{4\}$.

EXAMPLE 4

Solve $\log_b \frac{27}{64} = 3$.

Solution

Change $\log_b \frac{27}{64} = 3$ to exponential form to obtain

$$b^3 = \frac{27}{64}.$$

Therefore,

$$b = \sqrt[3]{\frac{27}{64}} = \frac{3}{4}.$$

The solution set is $\left\{\frac{3}{4}\right\}$.

Properties of Logarithms

There are some properties of logarithms that are a direct consequence of Definition 5.2 and the properties of exponents. For example, by writing the exponential equations $b_1 = b$ and $b_0 = 1$ in logarithmic form, the following property is obtained.

PROPERTY 5.3

For $b > 0$ and $b \neq 1$,

$$\log_b b = 1 \qquad \text{and} \qquad \log_b 1 = 0.$$

Therefore, according to Property 5.3, we can write

$$\log_{10} 10 = 1, \qquad \log_4 4 = 1,$$

$$\log_{10} 1 = 0, \qquad \log_5 1 = 0.$$

Also, from Definition 5.2 we know that $\log_b r$ is the exponent t such that $b^t = r$. Therefore, raising b to the $\log_b r$ power must produce r. This fact is stated in Property 5.4.

PROPERTY 5.4

For $b > 0$, $b \neq 1$, and $r > 0$,

$$b^{\log_b r} = r.$$

Therefore, according to Property 5.4, we can write

$$10^{\log_{10} 72} = 72, \qquad 3^{\log_3 85} = 85, \qquad e^{\log_e 7} = 7.$$

Because a logarithm is by definition an exponent, it is reasonable to predict that logarithms will have some properties that correspond to the basic exponential properties. This is an accurate prediction; these properties provide a basis for computational work with logarithms. Let's state the first of these properties and show how it can be verified by using our knowledge of exponents.

PROPERTY 5.5

For positive numbers b, r, and s, where $b \neq 1$,

$$\log_b rs = \log_b r + \log_b s.$$

To verify Property 5.5, we can proceed as follows. Let $m = \log_b r$ and $n = \log_b s$. Change each of these equations to exponential form:

$m = \log_b r$ becomes $r = b^m$;

$n = \log_b s$ becomes $s = b^n$.

Thus, the product rs becomes

$$rs = b^m \cdot b^n = b^{m+n}.$$

Now, by changing $rs = b^{m+n}$ back to logarithmic form, we obtain

$$\log_b rs = m + n.$$

Replacing m with $\log_b r$ and n with $\log_b s$ yields

$$\log_b rs = \log_b r + \log_b s.$$

The following two examples demonstrate a use of Property 5.5.

EXAMPLE 5

If $\log_2 5 = 2.3219$ and $\log_2 3 = 1.5850$, evaluate $\log_2 15$.

Solution

Because $15 = 5 \cdot 3$, we can apply Property 5.5 as follows.

$$\begin{aligned} \log_2 15 &= \log_2(5 \cdot 3) \\ &= \log_2 5 + \log_2 3 \\ &= 2.3219 + 1.5850 = 3.9069 \end{aligned}$$

EXAMPLE 6

If $\log_{10} 178 = 2.2504$ and $\log_{10} 89 = 1.9494$, evaluate $\log_{10}(178 \cdot 89)$.

Solution

$$\begin{aligned} \log_{10}(178 \cdot 89) &= \log_{10} 178 + \log_{10} 89 \\ &= 2.2504 + 1.9494 = 4.1998 \end{aligned}$$

Since $b^m/b^n = b^{m-n}$, we would expect a corresponding property that pertains to logarithms. Property 5.6 is that property. It can be verified by using an approach similar to the one used for Property 5.5. This verification is left for you to do as an exercise in the next problem set.

PROPERTY 5.6

For positive numbers b, r, and s, where $b \neq 1$,

$$\log_b\left(\frac{r}{s}\right) = \log_b r - \log_b s.$$

Property 5.6 can be used to change a division problem into an equivalent subtraction problem, as the next two examples illustrate.

EXAMPLE 7

If $\log_5 36 = 2.2266$ and $\log_5 4 = 0.8614$, evaluate $\log_5 9$.

Solution

Since $9 = \frac{36}{4}$, we can use Property 5.6 as follows.

$$\begin{aligned} \log_5 9 &= \log_5\left(\frac{36}{4}\right) \\ &= \log_5 36 - \log_5 4 \\ &= 2.2266 - 0.8614 = 1.3652 \end{aligned}$$

EXAMPLE 8

Evaluate $\log_{10}\left(\frac{379}{86}\right)$, given that $\log_{10} 379 = 2.5786$ and $\log_{10} 86 = 1.9345$.

Solution

$$\begin{aligned} \log_{10}\left(\frac{379}{86}\right) &= \log_{10} 379 - \log_{10} 86 \\ &= 2.5786 - 1.9345 = 0.6441 \end{aligned}$$

Another property of exponents states that $(b^n)^m = b^{mn}$. The corresponding property of logarithms is stated in Property 5.7. Again, we leave the verification of this property as an exercise for you to do in the next set of problems.

PROPERTY 5.7

If r is a positive real number, b is a positive real number other than 1, and p is any real number, then

$$\log_b r^p = p(\log_b r).$$

The next two examples demonstrate a use of Property 5.7.

EXAMPLE 9

Evaluate $\log_2 22^{1/3}$, given that $\log_2 22 = 4.4594$.

Solution

$$\log_2 22^{1/3} = \frac{1}{3}\log_2 22 \qquad \text{Property 5.7}$$
$$= \frac{1}{3}(4.4594) = 1.4865$$

EXAMPLE 10

Evaluate $\log_{10}(8540)^{3/5}$, given that $\log_{10} 8540 = 3.9315$.

Solution

$$\log_{10}(8540)^{3/5} = \frac{3}{5}\log_{10} 8540$$
$$= \frac{3}{5}(3.9315) = 2.3589$$

The properties of logarithms can be used to change the forms of various logarithmic expressions, as we will see in the next two examples.

EXAMPLE 11

Express $\log_b \sqrt{xy/z}$ in terms of the logarithms of x, y, and z.

Solution

$$\log_b \sqrt{\frac{xy}{z}} = \log_b\left(\frac{xy}{z}\right)^{1/2}$$
$$= \frac{1}{2}\log_b\left(\frac{xy}{z}\right) \qquad \text{Property 5.7}$$
$$= \frac{1}{2}(\log_b xy - \log_b z) \qquad \text{Property 5.6}$$
$$= \frac{1}{2}(\log_b x + \log_b y - \log_b z) \qquad \text{Property 5.5}$$

EXAMPLE 12

Express $2 \log_b x + 3 \log_b y - 4 \log_b z$ as one logarithm.

Solution

$$2 \log_b x + 3 \log_b y - 4 \log_b z = \log_b x^2 + \log_b y^3 - \log_b z^4$$
$$= \log_b x^2y^3 - \log_b z^4$$
$$= \log_b\left(\frac{x^2y^3}{z^4}\right)$$

Sometimes we need to change from an indicated sum or difference of logarithmic quantities to an indicated product or quotient. This is especially helpful when solving certain kinds of equations that involve logarithms. Note in these next two examples how we can use the properties, along with the process of changing from logarithmic form to exponential form, to solve some equations.

EXAMPLE 13

Solve $\log_{10} x + \log_{10}(x + 9) = 1$.

Solution

$$
\begin{aligned}
\log_{10} x + \log_{10}(x + 9) &= 1 \\
\log_{10}[x(x + 9)] &= 1 && \text{Property 5.5} \\
x(x + 9) &= 10^1 && \text{Change to exponential form.} \\
x^2 + 9x &= 10 \\
x^2 + 9x - 10 &= 0 \\
(x + 10)(x - 1) &= 0
\end{aligned}
$$

$$
\begin{aligned}
x + 10 = 0 \quad &\text{or} \quad x - 1 = 0 \\
x = -10 \quad &\text{or} \quad x = 1
\end{aligned}
$$

Since the left-hand side of the original equation is meaningful only if $x > 0$ and $x + 9 > 0$, the solution -10 must be discarded. Thus, the solution set is $\{1\}$.

EXAMPLE 14

Solve $\log_5(x + 4) - \log_5 x = 2$.

Solution

$$
\begin{aligned}
\log_5(x + 4) - \log_5 x &= 2 \\
\log_5\left(\frac{x + 4}{x}\right) &= 2 && \text{Property 10.6} \\
5^2 &= \frac{x + 4}{x} && \text{Change to exponential form.} \\
25 &= \frac{x + 4}{x} \\
25x &= x + 4 \\
24x &= 4 \\
x &= \frac{4}{24} = \frac{1}{6}
\end{aligned}
$$

The solution set is $\left\{\frac{1}{6}\right\}$.

PROBLEM SET 5.3

Write each of the following in logarithmic form. For example, $2^4 = 16$ becomes $\log_2 16 = 4$.

1. $3^2 = 9$
2. $2^5 = 32$
3. $5^3 = 125$
4. $10^1 = 10$
5. $2^{-4} = \frac{1}{16}$
6. $\left(\frac{2}{3}\right)^{-3} = \frac{27}{8}$
7. $10^{-2} = 0.01$
8. $10^5 = 100{,}000$

Write each of the following in exponential form. For example, $\log_2 8 = 3$ becomes $2^3 = 8$.

9. $\log_2 64 = 6$
10. $\log_3 27 = 3$
11. $\log_{10} 0.1 = -1$
12. $\log_5\left(\frac{1}{25}\right) = -2$
13. $\log_2\left(\frac{1}{16}\right) = -4$
14. $\log_{10} 0.00001 = -5$

Evaluate each of the following.

15. $\log_6 36$
16. $\log_3 243$
17. $\log_5\left(\frac{1}{5}\right)$
18. $\log_4\left(\frac{1}{64}\right)$
19. $\log_{10} 10$
20. $\log_{10} 1$
21. $\log_3\sqrt{3}$
22. $\log_5\sqrt[3]{25}$
23. $\log_3\left(\frac{\sqrt{27}}{3}\right)$
24. $\log_{1/2}\left(\frac{\sqrt[4]{8}}{2}\right)$
25. $\log_{1/4}\left(\frac{\sqrt[4]{32}}{2}\right)$
26. $\log_2\left(\frac{\sqrt[3]{16}}{4}\right)$
27. $10^{\log_{10} 7}$
28. $5^{\log_5 13}$
29. $\log_2(\log_5 5)$
30. $\log_6(\log_2 64)$

Solve each of the following equations.

31. $\log_5 x = 2$
32. $\log_{10} x = 3$
33. $\log_8 t = \frac{5}{3}$
34. $\log_4 m = \frac{3}{2}$
35. $\log_b 3 = \frac{1}{2}$
36. $\log_b 2 = \frac{1}{2}$
37. $\log_{10} x = 0$
38. $\log_{10} x = 1$

Given that $\log_2 5 = 2.3219$ and $\log_2 7 = 2.8074$, evaluate each of the following by using Properties 5.5–5.7.

39. $\log_2 35$
40. $\log_2\left(\frac{7}{5}\right)$
41. $\log_2 125$
42. $\log_2 49$
43. $\log_2 \sqrt{7}$
44. $\log_2 \sqrt[3]{5}$
45. $\log_2 175$
46. $\log_2 56$
47. $\log_2 80$

Given that $\log_8 5 = 0.7740$ and $\log_8 11 = 1.1531$, evaluate each of the following using Properties 5.5–5.7.

48. $\log_8 55$
49. $\log_8\left(\frac{5}{11}\right)$
50. $\log_8 25$
51. $\log_8 \sqrt{11}$
52. $\log_8(5)^{2/3}$
53. $\log_8 88$
54. $\log_8 320$
55. $\log_8\left(\frac{25}{11}\right)$
56. $\log_8\left(\frac{121}{25}\right)$

Express each of the following as the sum or difference of simpler logarithmic quantities. (Assume that all variables represent positive real numbers.) For example,

$$\log_b\left(\frac{x^3}{y^2}\right) = \log_b x^3 - \log_b y^2 = 3\log_b x - 2\log_b y.$$

57. $\log_b xyz$
58. $\log_b\left(\frac{x^2}{y}\right)$
59. $\log_b x^2y^3$
60. $\log_b x^{2/3}y^{3/4}$
61. $\log_b \sqrt{xy}$
62. $\log_b \sqrt[3]{x^2z}$
63. $\log_b \sqrt{\frac{x}{y}}$
64. $\log_b\left[x\left(\sqrt{\frac{x}{y}}\right)\right]$

Express each of the following as a single logarithm. (Assume that all variables represent positive numbers.) For example,

$$3\log_b x + 5\log_b y = \log_b x^3y^5.$$

65. $\log_b x + \log_b y - \log_b z$
66. $2\log_b x - 4\log_b y$

67. $(\log_b x - \log_b y) - \log_b z$

68. $\log_b x - (\log_b y - \log_b z)$

69. $\log_b x + \frac{1}{2}\log_b y$

70. $2 \log_b x + 4 \log_b y - 3 \log_b z$

71. $2 \log_b x + \frac{1}{2}\log_b(x - 1) - 4 \log_b(2x + 5)$

72. $\frac{1}{2}\log_b x - 3 \log_b x + 4 \log_b y$

Solve each of the following equations.

73. $\log_3 x + \log_3 4 = 2$

74. $\log_7 5 + \log_7 x = 1$

75. $\log_{10} x + \log_{10}(x - 21) = 2$

76. $\log_{10} x + \log_{10}(x - 3) = 1$

77. $\log_2 x + \log_2(x - 3) = 2$

78. $\log_3 x + \log_3(x - 2) = 1$

79. $\log_{10}(2x - 1) - \log_{10}(x - 2) = 1$

80. $\log_{10}(9x - 2) = 1 + \log_{10}(x - 4)$

81. $\log_5(3x - 2) = 1 + \log_5(x - 4)$

82. $\log_6 x + \log_6(x + 5) = 2$

83. $\log_8(x + 7) + \log_8 x = 1$

84. $\log_6(x + 1) + \log_6(x - 4) = 2$

85. Verify Property 5.6.

86. Verify Property 5.7.

THOUGHTS into WORDS

87. How would you explain the concept of a logarithm to someone who has never studied algebra?

88. Explain, without using Property 5.4, why $4^{\log_4 9}$ equals 9.

89. In the next section we are going to show that the logarithmic function $f(x) = \log_2 x$ is the inverse of the exponential function $f(x) = 2^x$. From that information how could you sketch a graph of $f(x) = \log_2 x$?

5.4 LOGARITHMIC FUNCTIONS

The concept of a logarithm can now be used to define a logarithmic function.

DEFINITION 5.3

If $b > 0$ and $b \neq 1$, then the function defined by

$$f(x) = \log_b x,$$

where x is any positive real number, is called the **logarithmic function with base b**.

We can obtain the graph of a specific logarithmic function in various ways. For example, we can change the equation $y = \log_2 x$ to the exponential equation $2^y = x$, from which we can determine a table of values. We will instruct you to use this approach to graph some logarithmic functions in the next set of exercises.

We can also obtain the graph of a logarithmic function by setting up a table of values directly from the logarithmic equation. We will demonstrate this approach.

EXAMPLE 1

Graph $f(x) = \log_2 x$.

Solution

Let's choose some values for x where the corresponding values for $\log_2 x$ are easily determined. (Remember that logarithms are defined only for the positive real numbers.) Plot the points determined by the table and connect them with a smooth curve to produce Figure 5.9.

x	$f(x)$	
$\frac{1}{8}$	-3	$\log_2 \frac{1}{8} = -3$ because $2^{-3} = \frac{1}{2^3} = \frac{1}{8}$
$\frac{1}{4}$	-2	
$\frac{1}{2}$	-1	
1	0	$\log_2 1 = 0$ because $2^0 = 1$
2	1	
4	2	
8	3	

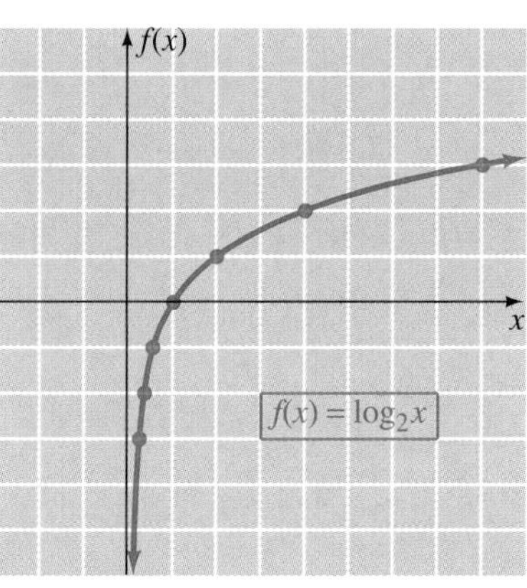

FIGURE 5.9

Now suppose that we consider two functions f and g as follows.

$f(x) = b^x$ Domain: all real numbers
Range: positive real numbers

$g(x) = \log_b x$ Domain: positive real numbers
Range: all real numbers

Furthermore, suppose that we consider the composition of f and g, and the composition of g and f.

$$(f \circ g)(x) = f(g(x)) = f(\log_b x) = b^{\log_b x} = x$$

$$(g \circ f)(x) = g(f(x)) = g(b^x) = \log_b b^x = x \log_b b = x(1) = x$$

Therefore, because the domain of f is the range of g and the range of f is the domain of g, and because $f(g(x)) = x$ and $g(f(x)) = x$, the two functions f and g are *inverses of each other.*

Remember from Chapter 4 that the graphs of a function and its inverse are reflections of each other through the line $y = x$. Thus, the graph of a logarithmic function can be determined by reflecting the graph of its inverse exponential function through the line $y = x$. This idea is illustrated in Figure 5.10, where the graph of $y = 2^x$ has been reflected across the line $y = x$ to produce the graph of $y = \log_2 x$.

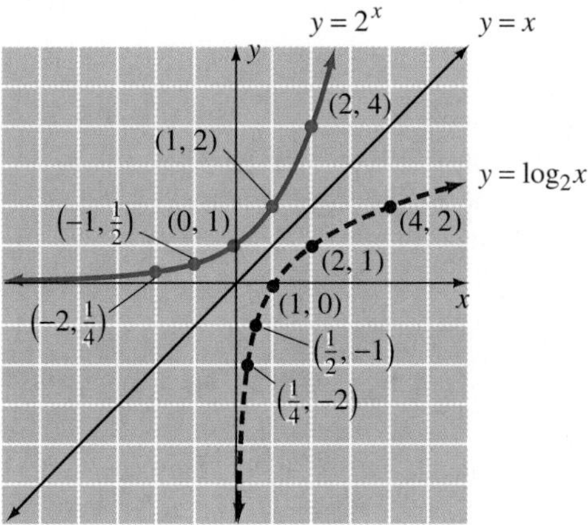

FIGURE 5.10

Figure 5.3 illustrated the general behavior patterns of exponential functions with two graphs. We can now reflect each of these graphs through the line $y = x$ and observe the general behavior patterns of logarithmic functions, as shown in Figure 5.11.

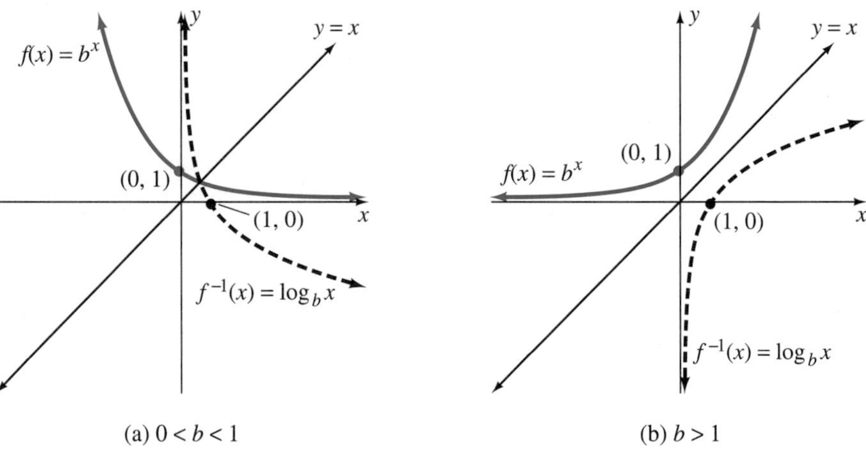

FIGURE 5.11

Finally, when graphing logarithmic functions, don't forget about transformations of the basic curves.

1. The graph of $f(x) = 3 + \log_2 x$ is the graph of $f(x) = \log_2 x$ *moved up three units*. (Since $\log_2 x + 3$ is apt to be confused with $\log_2(x + 3)$, we commonly write $3 + \log_2 x$.)
2. The graph of $f(x) = \log_2(x - 4)$ is the graph of $f(x) = \log_2 x$ *moved four units to the right*.
3. The graph of $f(x) = -\log_2 x$ is the graph of $f(x) = \log_2 x$ *reflected across the x-axis*.

Common Logarithms: Base 10

The properties of logarithms that we discussed in Section 5.3 are true for any valid base. However, since the Hindu-Arabic numeration system that we use is a base 10 system, logarithms to base 10 have historically been used for computational purposes. Base 10 logarithms are called **common logarithms**.

Originally, common logarithms were developed to aid in complicated numerical calculations that involve products, quotients, and powers of real numbers. Today they are seldom used for that purpose because the calculator and computer can much more effectively handle the messy computational problems. However, common logarithms do still occur in applications, so they deserve our attention.

REMARK In Appendix A we have included a short discussion relative to the computational aspects of common logarithms. You may find it interesting to browse through this material. It probably will enhance your appreciation of the calculator.

As we know from earlier work, the definition of a logarithm allows us to evaluate $\log_{10} x$ for values of x that are integral powers of 10. Consider the following examples.

$\log_{10} 1000 = 3$ because $10^3 = 1000$.

$\log_{10} 100 = 2$ because $10^2 = 100$.

$\log_{10} 10 = 1$ because $10^1 = 10$.

$\log_{10} 1 = 0$ because $10^0 = 1$.

$\log_{10} 0.1 = -1$ because $10^{-1} = \frac{1}{10} = 0.1$.

$\log_{10} 0.01 = -2$ because $10^{-2} = \frac{1}{10^2} = 0.01$.

$\log_{10} 0.001 = -3$ because $10^{-3} = \frac{1}{10^3} = 0.001$.

When working exclusively with base-10 logarithms, it is customary to omit writing the numeral 10 to designate the base. Thus, the expression $\log_{10} x$ is written as log x and a statement such as $\log_{10} 1000 = 3$ becomes $\log 1000 = 3$. We will follow this practice from now on in this chapter, but don't forget that the base is understood to be 10.

$$\log_{10} x = \log x$$

To find the common logarithm of a positive number that is not an integral power of 10, we can use an appropriately equipped calculator or a table such as the one that appears in Appendix A. We used a calculator equipped with a common logarithmic function (ordinarily a key labeled [log] is used) to obtain the following results rounded to four decimal places.

$$\log 1.75 = 0.2430$$
$$\log 23.8 = 1.3766$$
$$\log 134 = 2.1271$$
$$\log 0.192 = -0.7167$$
$$\log 0.0246 = -1.6091$$

Be sure that you can use a calculator and obtain these results.

In order to use logarithms to solve problems we sometimes need to be able to determine a number when the logarithm of the number is known. That is to say, we may need to determine x if $\log x$ is known. Let's consider an example.

EXAMPLE 2

Find x if $\log x = 0.2430$.

Solution

If $\log x = 0.2430$, then by changing to exponential form we have $10^{0.2430} = x$; use the [10^x] key to find x.

$$x = 10^{0.2430} \approx 1.749846689$$

Therefore, $x = 1.7498$ rounded to five significant digits.

Be sure that you can use your calculator and obtain the following results. We have rounded the values for x to five significant digits.

If $\log x = 0.7629$, then $x = 10^{0.7629} = 5.7930$.

If $\log x = 1.4825$, then $x = 10^{1.4825} = 30.374$.

If $\log x = 4.0214$, then $x = 10^{4.0214} = 10505$.

If $\log x = -1.5162$, then $x = 10^{-1.5162} = 0.030465$.

If $\log x = -3.8921$, then $x = 10^{-3.8921} = 0.00012820$.

The **common logarithmic function** is defined by the equation $f(x) = \log x$. It should now be a simple matter to set up a table of values and sketch the function. We will have you do this in the next set of exercises. Remember that $f(x) = 10^x$ and $g(x) = \log x$ are inverses of each other. Therefore, we could also get the graph of $g(x) = \log x$ by reflecting the exponential curve $f(x) = 10^x$ across the line $y = x$.

Natural Logarithms–Base e

In many practical applications of logarithms, the number e (remember $e \approx 2.71828$) is used as a base. Logarithms with a base of e are called **natural logarithms** and the symbol $\ln x$ is commonly used instead of $\log_e x$.

$$\log_e x = \ln x$$

Natural logarithms can be found with an appropriately equipped calculator or with a table of natural logarithms. (A table of natural logarithms is provided in Appendix B.) Use a calculator with a natural logarithm function (ordinarily a key labeled [ln x]) to obtain the following results rounded to four decimal places.

$\ln 3.21 = 1.1663$

$\ln 47.28 = 3.8561$

$\ln 842 = 6.7358$

$\ln 0.21 = -1.5606$

$\ln 0.0046 = -5.3817$

$\ln 10 = 2.3026$

Be sure that you can use your calculator to obtain these results. Keep in mind the significance of a statement such as $\ln 3.21 = 1.1663$. By changing to exponential form we are claiming that e raised to the 1.1663 power is approximately 3.21. Using a calculator we obtain $e^{1.1663} = 3.210093293$.

Let's do a few more problems to find x when given $\ln x$. Be sure that you agree with these results.

If $\ln x = 2.4156$, then $x = e^{2.4156} = 11.196$.

If $\ln x = 0.9847$, then $x = e^{0.9847} = 2.6770$.

If $\ln x = 4.1482$, then $e = e^{4.1482} = 63.320$.

If $\ln x = -1.7654$, then $x = e^{-1.7654} = 0.17112$.

The **natural logarithmic function** is defined by the equation $f(x) = \ln x$. It is the inverse of the natural exponential function $f(x) = e^x$. Thus, one way to graph $f(x)$

$= \ln x$ is to reflect the graph of $f(x) = e^x$ across the line $y = x$. We will ask you to do this in the next set of problems.

In Figure 5.12 we have used a graphing utility to sketch the graph of $f(x) = e^x$. Now, based on our previous work with transformations, we should be able to make the following statements.

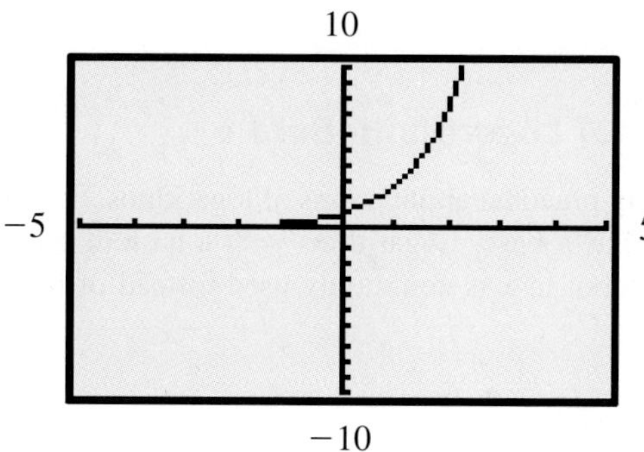

FIGURE 5.12

1. The graph of $f(x) = -e^x$ is the graph of $f(x) = e^x$ reflected through the x-axis.
2. The graph of $f(x) = e^{-x}$ is the graph of $f(x) = e^x$ reflected through the y-axis.
3. The graph of $f(x) = e^x + 4$ is the graph of $f(x) = e^x$ shifted upward 4 units.
4. The graph of $f(x) = e^{x+2}$ is the graph of $f(x) = e^x$ shifted 2 units to the left.

These statements are verified in Figure 5.13, which shows the result of graphing these four functions on the same set of axes using a graphing utility.

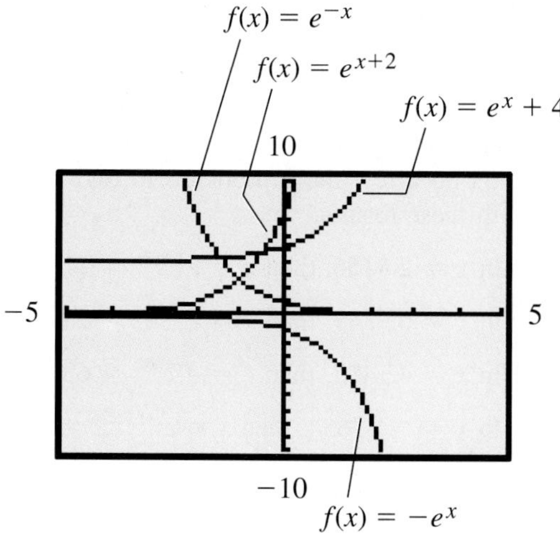

FIGURE 5.13

REMARK At this time we have only used a graphing utility to graph common logarithmic and natural logarithmic functions. In the next section we will see how logarithms with bases other than 10 or e are related to common and natural logarithms. This will provide a way of using a graphing utility to graph a logarithmic function with any valid base.

PROBLEM SET 5.4

For Problems 1–10, use a calculator to find each **common logarithm**. Express answers to four decimal places.

1. log 7.24 **2.** log 2.05 **3.** log 52.23

4. log 825.8 **5.** log 3214.1 **6.** log 14,189

7. log 0.729 **8.** log 0.04376 **9.** log 0.00034

10. log 0.000069

For Problems 11–20, use your calculator to find x when given $\log x$. Express answers to five significant digits.

11. $\log x = 2.6143$ **12.** $\log x = 1.5263$

13. $\log x = 4.9547$ **14.** $\log x = 3.9335$

15. $\log x = 1.9006$ **16.** $\log x = 0.5517$

17. $\log x = -1.3148$ **18.** $\log x = -0.1452$

19. $\log x = -2.1928$ **20.** $\log x = -2.6542$

For Problems 21–30, use your calculator to find each **natural logarithm**. Express answers to four decimal places.

21. ln 5 **22.** ln 18 **23.** ln 32.6

24. ln 79.5 **25.** ln 430 **26.** ln 371.8

27. ln 0.46 **28.** ln 0.524 **29.** ln 0.0314

30. ln 0.008142

For Problems 31–40, use your calculator to find x when given $\ln x$. Express answers to five significant digits.

31. $\ln x = 0.4721$ **32.** $\ln x = 0.9413$

33. $\ln x = 1.1425$ **34.** $\ln x = 2.7619$

35. $\ln x = 4.6873$ **36.** $\ln x = 3.0259$

37. $\ln x = -0.7284$ **38.** $\ln x = -1.6246$

39. $\ln x = -3.3244$ **40.** $\ln x = -2.3745$

41. a. Complete the following table and then graph $f(x) = \log x$. (Express the values for $\log x$ to the nearest tenth.)

x	0.1	0.5	1	2	4	8	10
log x							

b. Complete the following table and express values for 10^x to the nearest tenth.

x	−1	−0.3	0	0.3	0.6	0.9	1
10^x							

Then graph $f(x) = 10^x$ and reflect it across the line $y = x$ to produce the graph for $f(x) = \log x$.

42. a. Complete the following table and then graph $f(x) = \ln x$. (Express the values for $\ln x$ to the nearest tenth.)

x	0.1	0.5	1	2	4	8	10
ln x							

b. Complete the following table and express values for e^x to the nearest tenth.

x	−2.3	−0.7	0	0.7	1.4	2.1	2.3
e^x							

Then graph $f(x) = e^x$ and reflect it across the line $y = x$ to produce the graph for $f(x) = \ln x$.

43. Graph $y = \log_{1/2} x$ by graphing $\left(\frac{1}{2}\right)^y = x$.

44. Graph $y = \log_2 x$ by graphing $2^y = x$.

45. Graph $f(x) = \log_3 x$ by reflecting the graph of $g(x) = 3^x$ across the line $y = x$.

46. Graph $f(x) = \log_4 x$ by reflecting the graph of $g(x) = 4^x$ across the line $y = x$.

For Problems 47–53, graph each of the functions. Remember that the graph of $f(x) = \log_2 x$ is given in Figure 5.9.

47. $f(x) = 3 + \log_2 x$

48. $f(x) = -2 + \log_2 x$

49. $f(x) = \log_2(x + 3)$

50. $f(x) = \log_2(x - 2)$

51. $f(x) = \log_2 2x$

52. $f(x) = -\log_2 x$

53. $f(x) = 2 \log_2 x$

GRAPHICS CALCULATOR ACTIVITIES

54. Graph $f(x) = x$, $f(x) = e^x$, and $f(x) = \ln x$ on the same set of axes.

55. Graph $f(x) = x$, $f(x) = 10^x$, and $f(x) = \log x$ on the same set of axes.

56. Graph $f(x) = \ln x$. How should the graphs of $f(x) = 2 \ln x$, $f(x) = 4 \ln x$, and $f(x) = 6 \ln x$ compare to this basic curve? Graph the three functions on the same set of axes with the graph of $f(x) = \ln x$.

57. Graph $f(x) = \log x$. Now predict the graphs for $f(x) = 3 + \log x$, $f(x) = -2 + \log x$, and $f(x) = -4 + \log x$. Graph them on the same set of axes with the graph of $f(x) = \log x$.

58. For each of the following, (a) predict the general shape and location of the graph, and (b) use your graphics calculator to graph the function to check your prediction.

a. $f(x) = \log x + \ln x$

b. $f(x) = \log x - \ln x$

c. $f(x) = \ln x - \log x$

d. $f(x) = \ln x^2$

5.5 EXPONENTIAL AND LOGARITHMIC EQUATIONS, PROBLEM SOLVING

In Section 5.1 we solved exponential equations such as $3^x = 81$ by expressing both sides of the equation as a power of 3 and then applying the property, *If* $b^n = b^m$, *then* $n = m$. However, if we try this same approach with an equation such as $3^x = 5$, we face the difficulty of expressing 5 as a power of 3. We can solve this type of problem by using the properties of logarithms and the following property of equality.

PROPERTY 5.8

If $x > 0$, $y > 0$, $b > 0$, and $b \neq 1$, then

$$x = y \quad \text{if and only if} \quad \log_b x = \log_b y.$$

Property 5.8 is stated in terms of any valid base b; however, for most applications we use either common logarithms or natural logarithms. Let's consider some examples.

EXAMPLE 1

Solve $3^x = 5$ to the nearest hundredth.

Solution

By using common logarithms, we can proceed as follows.

$$
\begin{aligned}
3^x &= 5 \\
\log 3^x &= \log 5 && \text{Property 5.8} \\
x \log 3 &= \log 5 && \log r^p = p \log r \\
x &= \frac{\log 5}{\log 3} \\
x &= 1.46 && \text{nearest hundredth}
\end{aligned}
$$

Check Since $3^{1.46} \approx 4.972754647$, we say that, to the nearest hundredth, the solution set for $3^x = 5$ is $\{1.46\}$.

A WORD OF CAUTION! The expression $\frac{\log 5}{\log 3}$ means that we must *divide*, not subtract, the logarithms. That is, $\frac{\log 5}{\log 3}$ *does not* mean $\log\left(\frac{5}{3}\right)$. Remember that $\log\left(\frac{5}{3}\right) = \log 5 - \log 3$.

EXAMPLE 2

Solve $e^{x+1} = 5$ to the nearest hundredth.

Solution

Since base e is used in the exponential expression, let's use natural logarithms to help solve this equation.

$$
\begin{aligned}
e^{x+1} &= 5 \\
\ln e^{x+1} &= \ln 5 && \text{Property 5.8} \\
(x+1) \ln e &= \ln 5 && \ln r^p = p \ln r \\
(x+1)(1) &= \ln 5 && \ln e = 1 \\
x &= \ln 5 - 1 \\
x &= 0.61 && \text{nearest hundredth}
\end{aligned}
$$

The solution set is $\{0.61\}$. Check it!

EXAMPLE 3

Solve $2^{3x-2} = 3^{2x+1}$ to the nearest hundredth.

Solution

$$
\begin{aligned}
2^{3x-2} &= 3^{2x+1} \\
\log 2^{3x-2} &= \log 3^{2x+1} \\
(3x-2)\log 2 &= (2x+1)\log 3 \\
3x \log 2 - 2 \log 2 &= 2x \log 3 + \log 3
\end{aligned}
$$

$$
\begin{aligned}
3x \log 2 - 2x \log 3 &= \log 3 + 2 \log 2 \\
x(3 \log 2 - 2 \log 3) &= \log 3 + 2 \log 2 \\
x &= \frac{\log 3 + 2 \log 2}{3 \log 2 - 2 \log 3}. \\
x &= -21.10 \quad \text{nearest hundredth}
\end{aligned}
$$

The solution set is $\{-21.10\}$. Check it!

Logarithmic Equations

In Example 13 of Section 5.3, we solved the logarithmic equation

$$\log_{10} x + \log_{10}(x + 9) = 1$$

by simplifying the left side of the equation to $\log_{10}[x(x + 9)]$ and then changing the equation to exponential form to complete the solution. At this time, we can use Property 5.8 to solve this type of logarithmic equation another way, and we can also expand our equation solving capabilities. Let's consider some examples.

EXAMPLE 4

Solve $\log x + \log(x - 15) = 2$.

Solution

Since $\log 100 = 2$, the given equation becomes

$$\log x + \log(x - 15) = \log 100.$$

Now, simplify the left side and apply Property 5.8 and proceed as follows.

$$
\begin{aligned}
\log[(x)(x - 15)] &= \log 100 \\
x(x - 15) &= 100 \\
x^2 - 15x - 100 &= 0 \\
(x - 20)(x + 5) &= 0 \\
x - 20 = 0 \quad &\text{or} \quad x + 5 = 0 \\
x = 20 \quad &\text{or} \quad x = -5
\end{aligned}
$$

The domain of a logarithmic function must contain only positive numbers, so x and $x - 15$ must be positive in this problem. Therefore, the solution of -5 is discarded and the solution set is $\{20\}$.

EXAMPLE 5

Solve $\ln(x + 2) = \ln(x - 4) + \ln 3$.

Solution

$$
\begin{aligned}
\ln(x + 2) &= \ln(x - 4) + \ln 3 \\
\ln(x + 2) &= \ln[3(x - 4)] \\
x + 2 &= 3(x - 4)
\end{aligned}
$$

$$\begin{aligned} x + 2 &= 3x - 12 \\ 14 &= 2x \\ 7 &= x. \end{aligned}$$

The solution set is $\{7\}$.

EXAMPLE 6

Solve $\log_b(x + 2) + \log_b(2x - 1) = \log_b x$.

Solution

$$\begin{aligned} \log_b(x + 2) + \log_b(2x - 1) &= \log_b x \\ \log_b[(x + 2)(2x - 1)] &= \log_b x \\ (x + 2)(2x - 1) &= x \\ 2x^2 + 3x - 2 &= x \\ 2x^2 + 2x - 2 &= 0 \\ x^2 + x - 1 &= 0 \end{aligned}$$

Use the quadratic formula to obtain

$$x = \frac{-1 \pm \sqrt{1 + 4}}{2} = \frac{-1 \pm \sqrt{5}}{2}.$$

Since $x + 2$, $2x - 1$, and x all have to be positive, the solution of $\left(-1 - \sqrt{5}\right)/2$ has to be discarded and the solution set is

$$\left\{\frac{-1 + \sqrt{5}}{2}\right\}.$$

Problem Solving

In Section 5.2 we used the compound interest formula

$$A = P\left(1 + \frac{r}{n}\right)^{nt}$$

to determine the amount of money (A) accumulated at the end of t years if P dollars is invested at rate r of interest compounded n times per year. Now let's use this formula to solve other types of problems that deal with compound interest.

EXAMPLE 7

How long will it take \$500 to double if it is invested at 12% compounded quarterly?

Solution

To *double* \$500 means that the \$500 will grow into \$1000. We want to find out how long it will take; that is, what is t? Thus,

$$\begin{aligned} 1000 &= 500\left(1 + \frac{0.12}{4}\right)^{4t} \\ &= 500(1 + 0.03)^{4t} \\ &= 500(1.03)^{4t}. \end{aligned}$$

Multiply both sides of $1000 = 500(1.03)^{4t}$ by $\frac{1}{500}$ to yield

$$2 = (1.03)^{4t}.$$

Therefore,

$$\begin{aligned} \log 2 &= \log(1.03)^{4t} && \text{Property 5.8} \\ &= 4t \log 1.03. && \log r^p = p \log r \end{aligned}$$

Now let's solve for t.

$$\begin{aligned} 4t \log 1.03 &= \log 2 \\ t &= \frac{\log 2}{4 \log 1.03} \\ t &= 5.9 \quad \text{nearest tenth} \end{aligned}$$

Therefore, we are claiming that \$500 invested at 12% interest compounded quarterly will double itself in approximately 5.9 years.

Check \$500 invested at 12% compounded quarterly for 5.9 years will produce

$$\begin{aligned} A &= \$500\left(1 + \frac{0.12}{4}\right)^{4(5.9)} \\ &= \$500(1.03)^{23.6} \\ &= \$1004.45. \end{aligned}$$

EXAMPLE 8

Suppose that the number of bacteria present in a certain culture after t minutes is given by the equation $Q(t) = Q_0e^{0.04t}$, where Q_0 represents the initial number of bacteria. How long would it take for the bacteria count to grow from 500 to 2000?

Solution

Substituting into $Q(t) = Q_0e^{0.04t}$ and solving for t, we obtain the following.

$$\begin{aligned} 2000 &= 500e^{0.04t} \\ 4 &= e^{0.04t} \\ \ln 4 &= \ln e^{0.04t} \\ \ln 4 &= 0.04t \ln e \\ \ln 4 &= 0.04t && \ln e = 1 \\ \frac{\ln 4}{0.04} &= t \\ 34.7 &= t && \text{nearest tenth} \end{aligned}$$

It should take approximately 34.7 minutes.

Richter Numbers

Seismologists use the Richter scale to measure and report the magnitude of earthquakes. The equation

$$R = \log \frac{I}{I_0}$$

R is called a Richter number.

Crevice caused by earthquake
Photo by John K. Nakata, #21 U.S.G.S. Photographic Library

compares the intensity I of an earthquake to a minimal or reference intensity I_0. The reference intensity is the smallest earth movement that can be recorded on a seismograph. Suppose that the intensity of an earthquake was determined to be 50,000 times the reference intensity. In this case, $I = 50{,}000\ I_0$ and the Richter number would be calculated as follows.

$$\begin{aligned} R &= \log \frac{50{,}000\ I_0}{I_0} \\ R &= \log 50{,}000 \\ R &\approx 4.698970004 \end{aligned}$$

Thus, a Richter number of 4.7 would be reported. Let's consider two more examples that involve Richter numbers.

EXAMPLE 9

An earthquake in San Francisco in 1989 was reported to have a Richter number of 6.9. How did its intensity compare to the reference intensity?

Solution

$$\begin{aligned} 6.9 &= \log \frac{I}{I_0} \\ 10^{6.9} &= \frac{I}{I_0} \\ I &= (10^{6.9})(I_0) \\ I &\approx 7{,}943{,}282\ I_0 \end{aligned}$$

So its intensity was a little less than 8 million times the reference intensity.

EXAMPLE 10

An earthquake in Iran in 1990 had a Richter number of 7.7. Compare the intensity level of that earthquake to the one in San Francisco (Example 9).

Solution

From Example 9 we have $I = (10^{6.9})(I_0)$ for the earthquake in San Francisco. Then using a Richter number of 7.7 we obtain $I = (10^{7.7})(I_0)$ for the earthquake in Iran.

Therefore, by comparison

$$\frac{(10^{7.7})(I_0)}{(10^{6.9})(I_0)} = 10^{7.7-6.9} = 10^{0.8} \approx 6.3.$$

The earthquake in Iran was about 6 times as intense as the one in San Francisco.

Logarithms with Base Other Than 10 or e

The basic approach of applying Property 5.8 and using either common or natural logarithms can also be used to evaluate a logarithm to some base other than 10 or e. The next example illustrates this idea.

EXAMPLE 11

Evaluate $\log_3 41$.

Solution

Let $x = \log_3 41$. Changing to exponential form, we obtain

$$3^x = 41.$$

Now we can apply Property 5.8.

$$\log 3^x = \log 41$$
$$x \log 3 = \log 41$$
$$x = \frac{\log 41}{\log 3}$$
$$x = 3.3802 \quad \text{rounded to four decimal places}$$

Therefore, we are claiming that 3 raised to the 3.3802 power will produce approximately 41. Check it!

Using the method of Example 11 to evaluate $\log_a r$ produces the following formula, which is often referred to as the **change-of-base formula** for logarithms.

PROPERTY 5.9

If a, b, and r are positive numbers, with $a \neq 1$ and $b \neq 1$, then

$$\log_a r = \frac{\log_b r}{\log_b a}.$$

Property 5.9 provides us with a convenient way to express logarithms with bases other than 10 or e in terms of common or natural logarithms. For example,

$\log_3 41$ is of the form $\log_a r$ with $r = 41$ and $a = 3$. Therefore, in terms of common logarithms (base 10) we have

$$\log_3 41 = \frac{\log_{10} 41}{\log_{10} 3}.$$

Using the abbreviated notation for base 10 logarithms, we have

$$\log_3 41 = \frac{\log 41}{\log 3}.$$

Thus, the following format could be used to evaluate $\log_3 41$.

$$\begin{aligned} \log_3 41 &= \frac{\log 41}{\log 3} \\ &= 3.3802 \quad \text{rounded to four decimal places} \end{aligned}$$

In a similar fashion, we can use natural logarithms to evaluate expressions such as $\log_3 41$.

$$\begin{aligned} \log_3 41 &= \frac{\ln 41}{\ln 3} \\ &= 3.3802 \quad \text{rounded to four decimal places} \end{aligned}$$

Property 5.9 also provides us with another way of solving equations such as $3^x = 5$.

$$\begin{aligned} 3^x &= 5 \\ x &= \log_3 5 && \text{Changed to logarithmic form} \\ x &= \frac{\log 5}{\log 3} && \text{Applied Property 5.9} \\ x &= 1.46 \quad \text{to the nearest hundredth} \end{aligned}$$

Finally, by using Property 5.9, we can obtain a relationship between common and natural logarithms by letting $a = 10$ and $b = e$. Then

$$\log_a r = \frac{\log_b r}{\log_b a}$$

becomes

$$\begin{aligned} \log_{10} r &= \frac{\log_e r}{\log_e 10} \\ \log_e r &= (\log_e 10)(\log_{10} r) \\ \log_e r &= (2.3026)(\log_{10} r). \end{aligned}$$

Thus, the natural logarithm of any positive number is approximately equal to 2.3026 times the common logarithm of the number.

Now we can use a graphing utility to graph logarithmic functions such as $f(x) = \log_2 x$. Using the change-of-base formula, this function can be expressed as $f(x) = \dfrac{\log x}{\log 2}$ or as $f(x) = \dfrac{\ln x}{\ln 2}$. The graph of $f(x) = \log_2 x$ is shown in Figure 5.14.

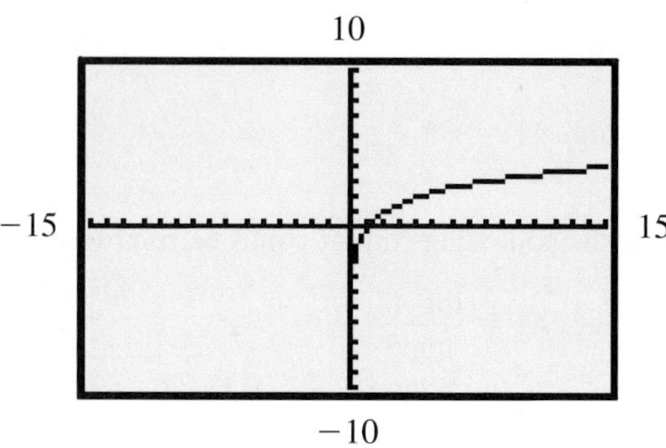

FIGURE 5.14

Finally, let's use a graphical approach to solve an equation that is cumbersome to solve with an algebraic approach.

EXAMPLE 12

Solve the equation $(5^x - 5^{-x})/2 = 3$.

Solution

First, we need to recognize that the solutions for the equation $(5^x - 5^{-x})/2 = 3$ are the x-intercepts of the graph of the equation $y = (5^x - 5^{-x})/2 - 3$. So let's use a graphing utility to obtain the graph of this equation as shown in Figure 5.15. Use the zoom and trace features to determine that the graph crosses the x-axis at approximately 1.13. Thus, the solution set of the original equation is $\{1.13\}$.

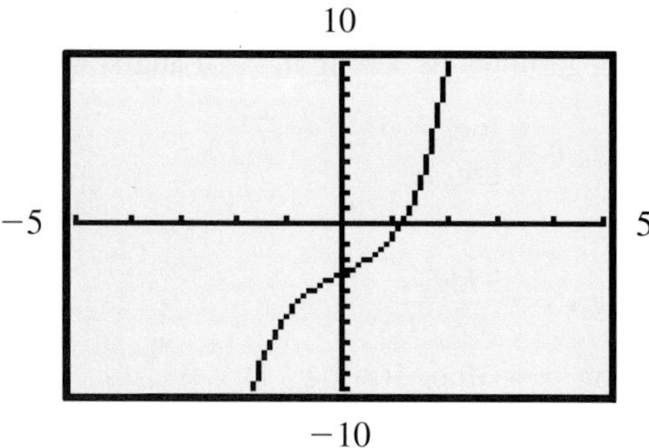

FIGURE 5.15

PROBLEM SET 5.5

Solve each of the following exponential equations and express approximate solutions to the nearest hundredth.

1. $2^x = 9$ **2.** $3^x = 20$ **3.** $5^t = 123$

4. $4^t = 12$ **5.** $2^{x+1} = 7$ **6.** $3^{x-2} = 11$

7. $7^{2t-1} = 35$ **8.** $5^{3t+1} = 9$ **9.** $e^x = 4.1$

10. $e^x = 30$ **11.** $e^{x-1} = 8.2$ **12.** $e^{x-2} = 13.1$

13. $2e^x = 12.4$ **14.** $3e^x - 1 = 17$

15. $3^{x-1} = 2^{x+3}$ **16.** $5^{2x+1} = 7^{x+3}$

17. $5^{x-1} = 2^{2x+1}$ **18.** $3^{2x+1} = 2^{3x+2}$

Solve each of the following logarithmic equations and express irrational solutions in simplest radical form.

19. $\log x + \log(x + 3) = 1$

20. $\log x + \log(x + 21) = 2$

21. $\log(2x - 1) - \log(x - 3) = 1$

22. $\log(3x - 1) = 1 + \log(5x - 2)$

23. $\log(x - 2) = 1 - \log(x + 3)$

24. $\log(x + 1) = \log 3 - \log(2x - 1)$

25. $\log(x + 1) - \log(x + 2) = \log \frac{1}{x}$

26. $\log(x + 2) - \log(2x + 1) = \log x$

27. $\ln(3t - 4) - \ln(t + 1) = \ln 2$

28. $\ln(2t + 5) = \ln 3 + \ln(t - 1)$

29. $\log(x^2) = (\log x)^2$ **30.** $\log\sqrt{x} = \sqrt{\log x}$

Evaluate each of the following logarithms to three decimal places.

31. $\log_3 14$ **32.** $\log_4 94$ **33.** $\log_5 2.1$

34. $\log_6 0.345$ **35.** $\log_7 176$ **36.** $\log_8 296$

37. $\log_9 14.32$ **38.** $\log_7 0.024$

Solve each of the following problems.

39. How long will it take \$1000 to double itself if it is invested at 9% interest compounded semiannually?

40. How long will it take \$750 to be worth \$1000 if it is invested at 12% interest compounded quarterly?

41. How long will it take \$500 to triple itself if it is invested at 9% interest compounded continously?

42. How long will it take \$2000 to double itself if it is invested at 13% interest compounded continuously?

43. At what rate of interest (to the nearest tenth of a percent) compounded annually will an investment of \$200 grow to \$350 in 5 years?

44. At what rate of interest (to the nearest tenth of a percent) compounded continuously will an investment of \$500 grow to \$900 in 10 years?

45. A piece of machinery valued at \$30,000 depreciates at a rate of 10% yearly. How long will it take until the machinery has a value of \$15,000?

46. For a certain strain of bacteria, the number present after t hours is given by the equation $Q = Q_0e^{0.34t}$, where Q_0 represents the initial number of bacteria. How long will it take 400 bacteria to increase to 4000 bacteria?

47. The number of grams of a certain radioactive substance present after t hours is given by the equation $Q = Q_0e^{-0.45t}$, where Q_0 represents the initial number of grams. How long will it take 2500 grams to be reduced to 1250 grams?

48. The atmospheric pressure in pounds per square inch is expressed by the equation $P(a) = 14.7e^{-0.21a}$, where a is the altitude above sea level measured in miles. If the atmospheric pressure in Cheyenne, Wyoming is approximately 11.53 pounds per square inch, find its altitude above sea level. Express your answer to the nearest hundred feet.

49. Suppose you are given the equation $P(t) = P_0e^{0.02t}$ to predict population growth, where P_0 represents an initial population and t is the time in years. How long does this equation predict it will take a city of 50,000 to double its population?

50. In a certain bacterial culture, the equation $Q(t) = Q_0e^{0.4t}$ yields the number of bacteria as a function of the time, where Q_0 is an initial number of bacteria and t is time measured in hours. How long will it take 500 bacteria to increase to 2000?

51. An earthquake in Los Angeles in 1971 had an intensity of approximately five million times the reference intensity. What was the Richter number associated with that earthquake?

52. An earthquake in San Francisco in 1906 was reported to have a Richter number of 8.3. How did its intensity compare to the reference intensity?

53. Calculate how many times more intense an earthquake with a Richter number of 7.3 is than an earthquake with a Richter number of 6.4.

54. Calculate how many times more intense an earthquake with a Richter number of 8.9 is than an earthquake with a Richter number of 6.2.

THOUGHTS into WORDS

55. Explain the concept of a Richter number.

56. Explain how you would solve the equation $7^x = 134$.

57. Explain how you would evaluate $\log_4 79$.

58. How do logarithms with a base of 9 compare to logarithms with a base of 3?

MISCELLANEOUS PROBLEMS

59. Use the approach of Example 11 to develop Property 5.9.

60. Let $r = b$ in Property 5.9 and verify that $\log_a b = \dfrac{1}{\log_b a}$.

61. To solve the equation $(5^x - 5^{-x})/2 = 3$ let's begin as follows.

$$\frac{5^x - 5^{-x}}{2} = 3$$

$$5^x - 5^{-x} = 6$$

$$5^x(5^x - 5^{-x}) = 6(5^x) \qquad \text{Multiply both sides by } 5^x.$$

$$5^{2x} - 1 = 6(5^x)$$

$$5^{2x} - 6(5^x) - 1 = 0$$

This final equation is of quadratic form. Finish the solution and check your answer against the answer in Example 12.

62. Solve the equation $y = (10^x + 10^{-x})/2$ for x in terms of y.

63. Solve the equation $y = (e^x - e^{-x})/2$ for x in terms of y.

GRAPHICS CALCULATOR ACTIVITIES

64. Check your answers for Problems 15–18 by graphing the appropriate function and finding the x-intercept.

65. Graph $f(x) = \log_2 x$. Then predict the graphs for $f(x) = \log_3 x$, $f(x) = \log_4 x$, and $\log_8 x$. Now graph these three functions on the same set of axes with the graph of $f(x) = \log_2 x$.

66. Graph $f(x) = x$, $f(x) = 2^x$, and $f(x) = \log_2 x$ on the same set of axes.

67. Graph $f(x) = x$, $f(x) = \left(\frac{1}{2}\right)^x$, and $f(x) = \log_{1/2} x$ on the same set of axes.

68. Use both a graphical and an algebraic approach to solve the equation $(2^x - 2^{-x})/3 = 4$.

CHAPTER 5 SUMMARY

This chapter can be summarized around three main topics: (1) exponents and exponential functions, (2) logarithms and logarithmic functions, and (3) applications of exponential and logarithmic functions.

Exponents and Exponential Functions

If a and b are positive numbers, and m and n are real numbers, then the following properties hold.

1. $b^n \cdot b^m = b^{n+m}$ — Product of two powers
2. $(b^n)^m = b^{mn}$ — Power of a power
3. $(ab)^n = a^n b^n$ — Power of a product
4. $\left(\frac{a}{b}\right)^n = \frac{a^n}{b^n}$ — Power of a quotient
5. $\frac{b^n}{b^m} = b^{n-m}$ — Quotient of two powers

A function defined by an equation of the form

$$f(x) = b^x, \qquad b > 0 \text{ and } b \neq 1,$$

is called an **exponential function**. Figure 5.16 illustrates the general behavior of the graph of an exponential function of the form $f(x) = b^x$.

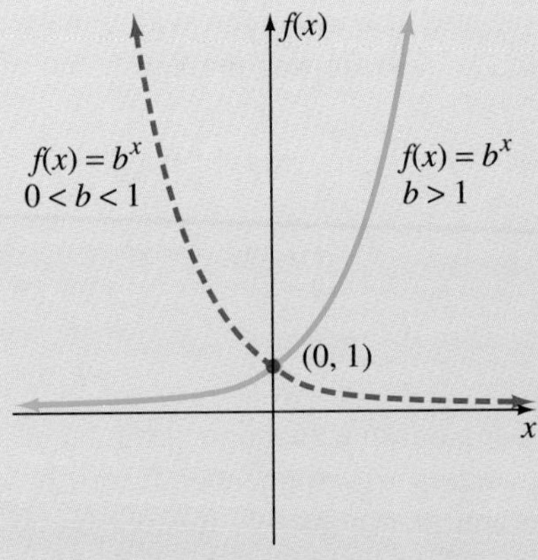

FIGURE 5.16

Logarithms and Logarithmic Functions

If r is any positive real number, then the unique exponent t such that $b^t = r$ is called the **logarithm of r with base b**; it is denoted by $\log_b r$.

The following properties of logarithms are used frequently.

1. $\log_b b = 1$
2. $\log_b 1 = 0$
3. $b^{\log_b r} = r$
4. $\log_b rs = \log_b r + \log_b s$
5. $\log_b\left(\frac{r}{s}\right) = \log_b r - \log_b s$
6. $\log_b(r^p) = p \log_b r$

Logarithms with a base of 10 are called **common logarithms**. The expression $\log_{10} x$ is commonly written $\log x$.

Many calculators are equipped with a common logarithm function. Often a key labeled log is used to find common logarithms.

Natural logarithms are logarithms that have a base of e, where e is an irrational number whose decimal approximation to eight digits is 2.7182818. Natural logarithms are denoted by $\log_e x$ or $\ln x$.

Many calculators are also equipped with a natural logarithmic function. Often a key labeled ln x is used for this purpose.

A function defined by an equation of the form

$$f(x) = \log_b x, \qquad b > 0 \text{ and } b \neq 1,$$

is called a **logarithmic function**.

The graph of a logarithmic function (such as $y = \log_2 x$) can be determined by changing the equation to exponential form ($2^y = x$) and plotting points, or by reflecting the graph of the inverse function ($y = 2^x$) across the line $y = x$. This last approach is based on the fact that exponential and logarithmic functions are inverses of each other.

Figure 5.17 illustrates the general behavior of the graph of a logarithmic function of the form $f(x) = \log_b x$.

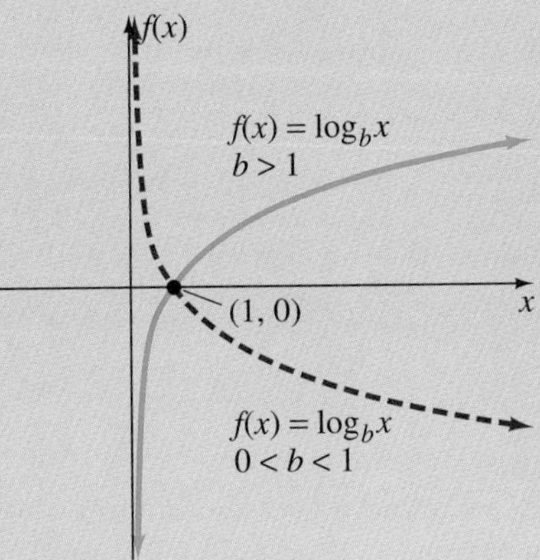

FIGURE 5.17

Applications

The following properties of equality are used frequently when solving exponential and logarithmic equations.

1. If $b > 0$, $b \neq 1$, and m and n are real numbers, then

   $b^n = b^m$ if and only if $n = m$.

2. If $x > 0$, $y > 0$, $b > 0$, and $b \neq 1$, then

   $x = y$ if and only if $\log_b x = \log_b y$.

A general formula for any principal (P) that is compounded n times per year for any number (t) of years at a given rate (r) is

$$A = P\left(1 + \frac{r}{n}\right)^{nt},$$

where A represents the total amount of money accumulated at the end of the t years.

As n gets infinitely large, the value of $[1 + (1/n)]^n$ approaches the number e, where e equals 2.71828 to five decimal places.

The formula

$$A = Pe^{rt}$$

yields the accumulated value (A) of a sum of money (P) that has been invested for t years at a rate of r percent compounded continuously.

The formula

$$Q = Q_0\left(\frac{1}{2}\right)^{t/h}$$

is referred to as the **half-life** formula.

The equation

$$Q(t) = Q_0e^{kt}$$

is used as a mathematical model for exponential growth and decay problems.

The formula

$$R = \log \frac{I}{I_0}$$

yields the Richter number associated with an earthquake.

The formula

$$\log_a r = \frac{\log_b r}{\log_b a}$$

is often called the **change-of-base formula**

CHAPTER 5 REVIEW PROBLEM SET

Evaluate each of the following.

1. $8^{5/3}$ **2.** $-25^{3/2}$ **3.** $(-27)^{4/3}$

4. $\log_6 216$ **5.** $\log_7\left(\frac{1}{49}\right)$ **6.** $\log_2 \sqrt[3]{2}$

7. $\log_2\left(\frac{\sqrt[4]{32}}{2}\right)$ **8.** $\log_{10} 0.00001$ **9.** $\ln e$

10. $7^{\log_7 12}$

Solve each of the following equations. Express approximate solutions to the nearest hundredth.

11. $\log_{10} 2 + \log_{10} x = 1$ **12.** $\log_3 x = -2$

13. $4^x = 128$ **14.** $3^t = 42$

15. $\log_2 x = 3$ **16.** $\left(\frac{1}{27}\right)^{3x} = 3^{2x-1}$

17. $2e^x = 14$ **18.** $2^{2x+1} = 3^{x+1}$

19. $\ln(x + 4) - \ln(x + 2) = \ln x$

20. $\log x + \log(x - 15) = 2$

21. $\log(\log x) = 2$

22. $\log(7x - 4) - \log(x - 1) = 1$

23. $\ln(2t - 1) = \ln 4 + \ln(t - 3)$

24. $64^{2t+1} = 8^{-t+2}$

For Problems 25–28, if $\log 3 = 0.4771$ and $\log 7 = 0.8451$, evaluate each of the following.

25. $\log\left(\frac{7}{3}\right)$

26. $\log 21$

27. $\log 27$

28. $\log 7^{2/3}$

29. Express each of the following as the sum or difference of simpler logarithmic quantities. Assume that all variables represent positive real numbers.

a. $\log_b\left(\dfrac{x}{y^2}\right)$

b. $\log_b \sqrt[4]{xy^2}$

c. $\log_b\left(\dfrac{\sqrt{x}}{y^3}\right)$

30. Express each of the following as a single logarithm. Assume that all variables represent positive real numbers.

a. $3 \log_b x + 2 \log_b y$

b. $\frac{1}{2} \log_b y - 4 \log_b x$

c. $\frac{1}{2}(\log_b x + \log_b y) - 2 \log_b z$

For Problems 31–34, approximate each of the logarithms to three decimal places.

31. $\log_2 3$

32. $\log_3 2$

33. $\log_4 191$

34. $\log_2 0.23$

For Problems 35–42, graph each of the functions.

35. $f(x) = \left(\frac{3}{4}\right)^x$

36. $f(x) = 2^{x+2}$

37. $f(x) = e^{x-1}$

38. $f(x) = -1 + \log x$

39. $f(x) = 3^x - 3^{-x}$

40. $f(x) = e^{-x^2/2}$

41. $f(x) = \log_2(x - 3)$

42. $f(x) = 3 \log_3 x$

For Problems 43–45, use the compound interest formula

$$A = P\left(1 + \frac{r}{n}\right)^{nt}$$

to find the total amount of money accumulated at the end of the indicated time period for each of the investments.

43. \$750 for 10 years at 11% compounded quarterly

44. \$1250 for 15 years at 9% compounded monthly

45. \$2500 for 20 years at 9.5% compounded semiannually

Solve the following problems.

46. How long will it take \$100 to double itself if it is invested at 14% interest compounded annually?

47. How long will it take \$1000 to be worth \$3500 if it is invested at 10.5% interest compounded quarterly?

48. At what rate of interest (to the nearest tenth of a percent) compounded continuously will an investment of \$500 grow to \$1000 in 8 years?

49. Suppose that the present population of a city is 50,000 and suppose that the equation $P(t) = P_0e^{0.02t}$ can be used to estimate future populations, where P_0 represents an initial population. Estimate the population of that city in 10 years, 15 years, and 20 years.

50. The number of bacteria present in a certain culture after t hours is given by the equation $Q = Q_0e^{0.29t}$, where Q_0 represents the initial number of bacteria. How long will it take 500 bacteria to increase to 2000 bacteria?

51. Suppose that a certain radioactive substance has a half-life of 40 days. If there are presently 750 grams of the substance, how much, to the nearest gram, will remain after 100 days?

52. An earthquake occurred in Mexico City in 1985 that had an intensity level about 125,000,000 times the reference intensity. Find the Richter number for that earthquake.

POLYNOMIAL AND RATIONAL FUNCTIONS

In earlier chapters we solved linear and quadratic equations and graphed linear and quadratic functions. In this chapter we will expand our equation solving processes and graphing techniques to include more general polynomial equations and functions. Our knowledge of polynomial functions will then allow us to work with *rational functions*; the function concept will unify the chapter. To facilitate our study in this chapter, we will first review the concept of dividing polynomials; then we will introduce a special division technique called *synthetic division*.

6

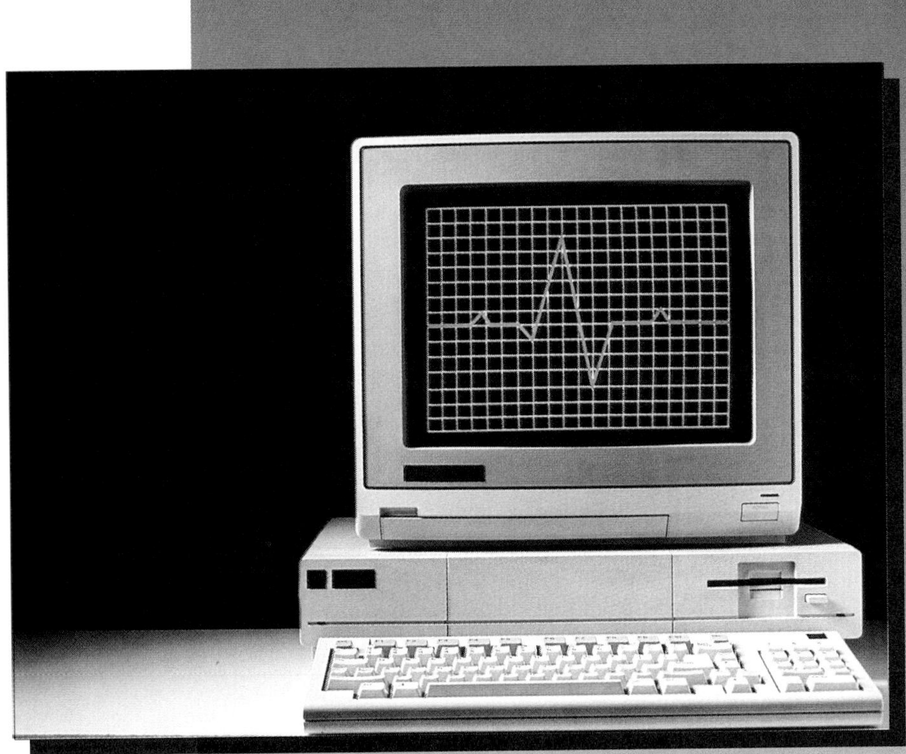

Many problems that involve maximum and minimum values can be solved by using polynomial and rational functions.

6.1 DIVIDING POLYNOMIALS

In Chapter 1 we used the properties

$$\frac{a+b}{c} = \frac{a}{c} + \frac{b}{c} \quad \text{and} \quad \frac{a-b}{c} = \frac{a}{c} - \frac{b}{c}$$

as a basis for dividing a polynomial by a monomial. For example,

$$\frac{18x^3 + 24x^2}{6x} = \frac{18x^3}{6x} + \frac{24x^2}{6x} = 3x^2 + 4x$$

and

$$\frac{35x^2y^3 - 55x^3y^4}{5xy^2} = \frac{35x^2y^3}{5xy^2} - \frac{55x^3y^4}{5xy^2} = 7xy - 11x^2y^2.$$

You may recall from a previous algebra course that the format used to divide a polynomial by a binomial resembles the long division format in arithmetic. Let's work through an example step-by-step.

STEP 1 Use the conventional long division format and arrange both the dividend and the divisor in descending powers of the variable.

$$3x + 1\overline{)3x^3 - 5x^2 + 10x + 1}$$

STEP 2 Find the first term of the quotient by dividing the first term of the dividend by the first term of the divisor.

$$\begin{array}{r} x^2 \\ 3x + 1\overline{)3x^3 - 5x^2 + 10x + 1} \end{array}$$

STEP 3 Multiply the entire divisor by the quotient term in step 2 and place this product in position to be subtracted from the dividend.

$$\begin{array}{r} x^2 \\ 3x + 1\overline{)3x^3 - 5x^2 + 10x + 1} \\ \underline{3x^3 + x^2 } \end{array}$$

STEP 4 Subtract.

$$\begin{array}{r} x^2 \\ 3x + 1\overline{)3x^3 - 5x^2 + 10x + 1} \\ \underline{3x^3 + x^2 } \\ -6x^2 + 10x + 1 \end{array}$$

STEP 5 Repeat steps 2, 3, and 4 and use $-6x^2 + 10x + 1$ as a new dividend.

$$\begin{array}{r} x^2 - 2x \\ 3x + 1\overline{)3x^3 - 5x^2 + 10x + 1} \\ \underline{3x^3 + x^2 } \\ -6x^2 + 10x + 1 \\ \underline{-6x^2 - 2x } \\ 12x + 1 \end{array}$$

STEP 6 Repeat steps 2, 3, and 4 and use $12x + 1$ as a new dividend.

$$\begin{array}{r} x^2 - 2x + 4 \\ 3x + 1 \overline{)3x^3 - 5x^2 + 10x + 1} \\ \underline{3x^3 + x^2 } \\ -6x^2 + 10x + 1 \\ \underline{-6x^2 - 2x } \\ 12x + 1 \\ \underline{12x + 4} \\ -3 \end{array}$$

Therefore $3x^3 - 5x^2 + 10x + 1 = (3x + 1)(x^2 - 2x + 4) + (-3)$, which is of the familiar form

dividend = (divisor)(quotient) + remainder.

This result is commonly called the **division algorithm for polynomials**, which can be stated in general terms as follows.

Division Algorithm for Polynomials

If $f(x)$ and $g(x)$ are polynomials and $g(x) \neq 0$, then unique polynomials $q(x)$ and $r(x)$ exist, such that

$$f(x) = g(x)q(x) + r(x),$$

(Dividend: $f(x)$; Divisor: $g(x)$; Quotient: $q(x)$; Remainder: $r(x)$)

where $r(x) = 0$ or the degree of $r(x)$ is less than the degree of $g(x)$.

Let's consider one more example to illustrate this division process further.

EXAMPLE 1

Divide $t^2 - 3t + 2t^4 - 1$ by $t^2 + 4t$.

Solution

Don't forget to arrange both the dividend and the divisor in descending powers of the variable.

$$\begin{array}{r} 2t^2 - 8t + 33 \\ t^2 + 4t \overline{)2t^4 + 0t^3 + t^2 - 3t - 1} \\ \underline{2t^4 + 8t^3 } \\ -8t^3 + t^2 - 3t - 1 \\ \underline{-8t^3 - 32t^2 } \\ 33t^2 - 3t - 1 \\ \underline{33t^2 + 132t } \\ -135t - 1 \end{array}$$

← Notice the insertion of a t-cubed term with a zero coefficient.

← The division process is completed when the degree of the remainder is less than the degree of the divisor.

Synthetic Division

If the divisor is of the form $x - c$, where c is a constant, then the typical long division algorithm can be simplified to a process called **synthetic division**. First, let's consider another division problem and use the regular division algorithm. Then, in a step-by-step fashion, we will demonstrate some shortcuts that will lead us into the synthetic division procedure. Consider the division problem $(2x^4 + x^3 - 17x^2 + 13x + 2) \div (x - 2)$.

$$
\begin{array}{r|l}
 & 2x^3 + 5x^2 - 7x - 1 \\
\hline
x - 2 & 2x^4 + x^3 - 17x^2 + 13x + 2 \\
 & 2x^4 - 4x^3 \\
\hline
 & \quad 5x^3 - 17x^2 \\
 & \quad 5x^3 - 10x^2 \\
\hline
 & \qquad -7x^2 + 13x \\
 & \qquad -7x^2 + 14x \\
\hline
 & \qquad\qquad -x + 2 \\
 & \qquad\qquad -x + 2 \\
\hline
\end{array}
$$

Because the dividend is written in descending powers of x, the quotient is produced in descending powers of x. In other words, the numerical coefficients are the *key issues*. So let's rewrite the above problem in terms of its coefficients.

$$
\begin{array}{r|rrrrr}
 & 2 & 5 & -7 & -1 & \\
\hline
1 - 2 & 2 & 1 & -17 & 13 & 2 \\
 & \textcircled{2} & -4 & & & \\
\hline
 & & 5 & \textcircled{-17} & & \\
 & & 5 & -10 & & \\
\hline
 & & & -7 & \textcircled{13} & \\
 & & & \textcircled{-7} & 14 & \\
\hline
 & & & & -1 & \textcircled{2} \\
 & & & & \textcircled{-1} & 2 \\
\hline
\end{array}
$$

Now observe that the circled numbers are simply repetitions of the numbers directly above them in the format. Thus, the circled numbers can be omitted and the format will be as follows (disregard the arrows for the moment).

$$
\begin{array}{r|rrrrr}
 & 2 & 5 & -7 & -1 & \\
\hline
1 - 2 & 2 & 1 & -17 & 13 & 2 \\
 & & -4 & & & \\
\hline
 & & 5 & & & \\
 & & & -10 & & \\
\hline
 & & & -7 & & \\
 & & & & 14 & \\
\hline
 & & & & -1 & \\
 & & & & & 2 \\
\end{array}
$$

Next, by moving some numbers up (indicated by the arrows) and by not writing the 1 that is the coefficient of x in the divisor, the following more compact form is obtained.

$$\begin{array}{r|rrrrr} & 2 & 5 & -7 & -1 & \\ \hline -2 & 2 & 1 & -17 & 13 & 2 \\ & & -4 & -10 & 14 & 2 \\ \hline & & 5 & -7 & -1 & \end{array}$$

(1)
(2)
(3)
(4)

Notice that line 4 reveals all of the coefficients of the quotient, (line 1), except for the first coefficient, 2. Thus, we can omit line 1, begin line 4 with the first coefficient, and then use the following form.

$$\begin{array}{r|rrrrr} -2 & 2 & 1 & -17 & 13 & 2 \\ & & -4 & -10 & 14 & 2 \\ \hline & 2 & 5 & -7 & -1 & 0 \end{array}$$

(5)
(6)
(7)

Line 7 contains the coefficients of the quotient, where the zero indicates the remainder. Finally, by changing the constant in the divisor to 2 (instead of -2), which changes the signs of the numbers in line 6, we can *add* the corresponding entries in lines 5 and 6 rather than subtract. Thus, the final synthetic division form for this problem is as follows.

$$\begin{array}{r|rrrrr} 2 & 2 & 1 & -17 & 13 & 2 \\ & & 4 & 10 & -14 & -2 \\ \hline & 2 & 5 & -7 & -1 & 0 \end{array}$$

Now we will consider another problem and indicate a step-by-step procedure for setting up and carrying out the synthetic division process. Suppose that we want to do the following division problem.

$$x + 4\overline{)2x^3 + 5x^2 - 13x - 2}$$

STEP 1 Write the coefficients of the dividend as follows.

$$\overline{)2 \quad 5 \quad -13 \quad -2}$$

STEP 2 In the divisor, use -4 instead of 4 so that later we can add rather than subtract.

$$-4\overline{)2 \quad 5 \quad -13 \quad -2}$$

STEP 3 Bring down the first coefficient of the dividend.

$$\begin{array}{r|rrrr} -4 & 2 & 5 & -13 & -2 \\ & & & & \\ \hline & 2 & & & \end{array}$$

STEP 4 Multiply that first coefficient times the divisor, which yields $2(-4) = -8$; add this result to the second coefficient of the dividend.

$$\begin{array}{r|rrrr} -4 & 2 & 5 & -13 & -2 \\ & 2 & -8 & & \\ \hline & 2 & -3 & & \end{array}$$

STEP 5 Multiply $(-3)(-4)$, which yields 12; add this result to the third coefficient of the dividend.

$$\begin{array}{r|rrrr} -4 & 2 & 5 & -13 & -2 \\ & & -8 & 12 & \\ \hline & 2 & -3 & -1 & \end{array}$$

STEP 6 Multiply $(-1)(-4)$, which yields 4; add this result to the last term of the dividend.

$$\begin{array}{r|rrrr} -4 & 2 & 5 & -13 & -2 \\ & & -8 & 12 & 4 \\ \hline & 2 & -3 & -1 & 2 \end{array}$$

The last row indicates a quotient of $2x^2 - 3x - 1$ and a remainder of 2.

Now let's consider some examples in which we show only the final compact form of synthetic division.

EXAMPLE 2

Find the quotient and remainder for $(x^3 + 8x^2 + 13x - 6) \div (x + 3)$.

Solution

$$\begin{array}{r|rrrr} -3 & 1 & 8 & 13 & -6 \\ & & -3 & -15 & 6 \\ \hline & 1 & 5 & -2 & 0 \end{array}$$

Thus, the quotient is $x^2 + 5x - 2$ and the remainder is zero.

EXAMPLE 3

Find the quotient and the remainder for $(3x^4 + 5x^3 - 29x^2 - 45x + 14) \div (x - 3)$.

Solution

$$\begin{array}{r|rrrrr} 3 & 3 & 5 & -29 & -45 & 14 \\ & & 9 & 42 & 39 & -18 \\ \hline & 3 & 14 & 13 & -6 & -4 \end{array}$$

Thus, the quotient is $3x^3 + 14x^2 + 13x - 6$ and the remainder is -4.

EXAMPLE 4

Find the quotient and remainder for $(4x^4 - 2x^3 + 6x - 1) \div (x - 1)$.

Solution

$$\begin{array}{r|rrrrr} 1 & 4 & -2 & 0 & 6 & -1 \\ & & 4 & 2 & 2 & 8 \\ \hline & 4 & 2 & 2 & 8 & 7 \end{array}$$

Notice that a zero has been inserted as the coefficient of the missing x^2 term.

Thus, the quotient is $4x^3 + 2x^2 + 2x + 8$ and the remainder is 7.

EXAMPLE 5

Find the quotient and remainder for $(x^4 + 16) \div (x + 2)$.

Solution

$$\begin{array}{r|rrrrr} -2 & 1 & 0 & 0 & 0 & 16 \\ & & -2 & 4 & -8 & 16 \\ \hline & 1 & -2 & 4 & -8 & 32 \end{array}$$

Notice that zeros have been inserted as coefficients of the missing terms in the dividend.

Thus, the quotient is $x^3 - 2x^2 + 4x - 8$ and the remainder is 32.

PROBLEM SET 6.1

Find the quotient and remainder for each of the following division problems.

1. $(12x^2 + 7x - 10) \div (3x - 2)$
2. $(20x^2 - 39x + 18) \div (5x - 6)$
3. $(3t^3 + 7t^2 - 10t - 4) \div (3t + 1)$
4. $(4t^3 - 17t^2 + 7t + 10) \div (4t - 5)$
5. $(6x^2 + 19x + 11) \div (3x + 2)$
6. $(20x^2 + 3x - 1) \div (5x + 2)$
7. $(3x^3 + 2x^2 - 5x - 1) \div (x^2 + 2x)$
8. $(4x^3 - 5x^2 + 2x - 6) \div (x^2 - 3x)$
9. $(5y^3 - 6y^2 - 7y - 2) \div (y^2 - y)$
10. $(8y^3 - y^2 - y + 5) \div (y^2 + y)$
11. $(4a^3 - 2a^2 + 7a - 1) \div (a^2 - 2a + 3)$
12. $(5a^3 + 7a^2 - 2a - 9) \div (a^2 + 3a - 4)$
13. $(3x^2 - 2xy - 8y^2) \div (x - 2y)$
14. $(4a^2 - 8ab + 4b^2) \div (a - b)$

Use *synthetic division* to determine the quotient and remainder for each of the following.

15. $(3x^2 + x - 4) \div (x - 1)$
16. $(2x^2 - 5x - 3) \div (x - 3)$
17. $(x^2 + 2x - 10) \div (x - 4)$
18. $(x^2 - 10x + 15) \div (x - 8)$
19. $(4x^2 + 5x - 4) \div (x + 2)$
20. $(5x^2 + 18x - 8) \div (x + 4)$
21. $(x^3 - 2x^2 - x + 2) \div (x - 2)$
22. $(x^3 - 5x^2 + 2x + 8) \div (x + 1)$
23. $(3x^4 - x^3 + 2x^2 - 7x - 1) \div (x + 1)$
24. $(2x^3 - 5x^2 - 4x + 6) \div (x - 2)$

25. $(x^3 - 7x - 6) \div (x + 2)$

26. $(x^3 + 6x^2 - 5x - 1) \div (x - 1)$

27. $(x^4 + 4x^3 - 7x - 1) \div (x - 3)$

28. $(2x^4 + 3x^2 + 3) \div (x + 2)$

29. $(x^3 + 6x^2 + 11x + 6) \div (x + 3)$

30. $(x^3 - 4x^2 - 11x + 30) \div (x - 5)$

31. $(x^5 - 1) \div (x - 1)$

32. $(x^5 - 1) \div (x + 1)$

33. $(x^5 + 1) \div (x - 1)$

34. $(x^5 + 1) \div (x + 1)$

35. $(2x^3 + 3x^2 - 2x + 3) \div \left(x + \frac{1}{2}\right)$

36. $(9x^3 - 6x^2 + 3x - 4) \div \left(x - \frac{1}{3}\right)$

37. $(4x^4 - 5x^2 + 1) \div \left(x - \frac{1}{2}\right)$

38. $(3x^4 - 2x^3 + 5x^2 - x - 1) \div \left(x + \frac{1}{3}\right)$

THOUGHTS into WORDS

39. How would you describe synthetic division to someone who has just completed an elementary algebra course?

40. Why is synthetic division restricted to situations where the divisor is of the form $x - c$?

6.2 REMAINDER AND FACTOR THEOREMS

Let's consider the division algorithm (stated in the previous section) when the dividend, $f(x)$, is divided by a *linear polynomial* of the form $x - c$. Then the division algorithm,

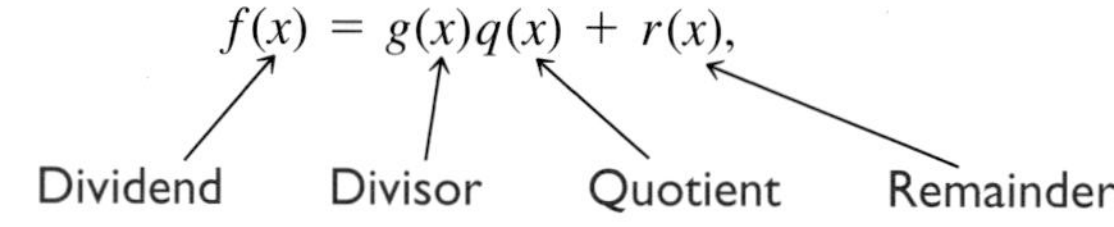

becomes

$$f(x) = (x - c)q(x) + r(x).$$

Because the degree of the remainder, $r(x)$, must be less than the degree of the divisor, $x - c$, the remainder is a constant, Therefore, letting R represent the remainder, we have

$$f(x) = (x - c)q(x) + R.$$

If we evaluate f at c, we obtain

$$\begin{aligned} f(c) &= (c - c)q(c) + R \\ &= 0 \cdot q(c) + R \\ &= R. \end{aligned}$$

In other words, if a polynomial is divided by a linear polynomial of the form $x - c$, then the remainder is the value of the polynomial at c. Let's state this more formally as the **remainder theorem**.

PROPERTY 6.1 Remainder Theorem

If a polynomial $f(x)$ is divided by $x - c$, then the remainder is equal to $f(c)$.

EXAMPLE 1

If $f(x) = x^3 + 2x^2 - 5x - 1$, find $f(2)$ (a) by using synthetic division and the remainder theorem, and (b) by evaluating $f(2)$ directly.

Solutions

a.
$$\begin{array}{r|rrrr} 2 & 1 & 2 & -5 & -1 \\ & & 2 & 8 & 6 \\ \hline & 1 & 4 & 3 & \boxed{5} \end{array} \longleftarrow R = f(2)$$

b. $f(2) = 2^3 + 2(2)^2 - 5(2) - 1 = 8 + 8 - 10 - 1 = 5$

EXAMPLE 2

If $f(x) = x^4 + 7x^3 + 8x^2 + 11x + 5$, find $f(-6)$ (a) by using synthetic divison and the remainder theorem, and (b) by evaluating $f(-6)$ directly.

Solutions

a.
$$\begin{array}{r|rrrrr} -6 & 1 & 7 & 8 & 11 & 5 \\ & & -6 & -6 & -12 & 6 \\ \hline & 1 & 1 & 2 & -1 & \boxed{11} \end{array} \longleftarrow R = f(-6)$$

b. $f(-6) = (-6)^4 + 7(-6)^3 + 8(-6)^2 + 11(-6) + 5$
$= 1296 - 1512 + 288 - 66 + 5 = 11$

In Example 2, notice that finding $f(-6)$ by synthetic division and the remainder theorem involves easier computation than evaluating $f(-6)$ directly. This is often the case.

EXAMPLE 3

Find the remainder when $x^3 + 3x^2 - 13x - 15$ is divided by $x + 1$.

Solution

Let $f(x) = x^3 + 3x^2 - 13x - 15$ and write $x + 1$ as $x - (-1)$ so that we can apply the remainder theorem.

$$f(-1) = (-1)^3 + 3(-1)^2 - 13(-1) - 15 = 0$$

Thus the remainder is zero.

Example 3 illustrates an important special case of the remainder theorem in which the remainder is *zero*. In this case we say that $x + 1$ is a **factor** of $x^3 + 3x^2 - 13x - 15$.

Factor Theorem

A general *factor theorem* can be formulated by considering the equation

$$f(x) = (x - c)q(x) + R.$$

If $x - c$ is a factor of $f(x)$, then the remainder R, which is also $f(c)$, must be zero. Conversely, if $R = f(c) = 0$, then $f(x) = (x - c)q(x)$; in other words, $x - c$ is a factor of $f(x)$. The **factor theorem** can be stated as follows.

PROPERTY 6.2 *Factor Theorem*

A polynomial $f(x)$ has a factor $x - c$ if and only if $f(c) = 0$.

EXAMPLE 4

Is $x - 1$ a factor of $x^3 + 5x^2 + 2x - 8$?

Solution

Let $f(x) = x^3 + 5x^2 + 2x - 8$ and compute $f(1)$ to obtain

$$f(1) = 1^3 + 5(1)^2 + 2(1) - 8 = 0.$$

Therefore, by the factor theorem, $x - 1$ is a factor of $f(x)$.

EXAMPLE 5

Is $x + 3$ a factor of $2x^3 + 5x^2 - 6x - 7$?

Solution

Using synthetic division, we obtain the following.

$$\begin{array}{r|rrrr} -3 & 2 & 5 & -6 & -7 \\ & & -6 & 3 & 9 \\ \hline & 2 & -1 & -3 & \boxed{2} \end{array} \longleftarrow R = f(-3)$$

Since $f(-3) \neq 0$, we know that $x + 3$ is not a factor of the given polynomial.

In Examples 4 and 5 we were concerned only with determining whether a linear polynomial of the form $x - c$ was a factor of another polynomial. For such problems, it is reasonable to compute $f(c)$ either directly or by synthetic division, whichever way seems easier. However, if more information is required, such as the complete factorization of the given polynomial, then the use of synthetic division becomes appropriate, as in the next two examples.

EXAMPLE 6

Show that $x - 1$ is a factor of $x^3 - 2x^2 - 11x + 12$ and find the other linear factors of the polynomial.

Solution

Let's use synthetic division to divide $x^3 - 2x^2 - 11x + 12$ by $x - 1$.

$$\begin{array}{r|rrrr} 1 & 1 & -2 & -11 & 12 \\ & & 1 & -1 & -12 \\ \hline & 1 & -1 & -12 & 0 \end{array}$$

The last line indicates a quotient of $x^2 - x - 12$ and a remainder of zero. The zero remainder means that $x - 1$ is a factor. Furthermore, we can write

$$x^3 - 2x^2 - 11x + 12 = (x - 1)(x^2 - x - 12).$$

The quadratic polynomial $x^2 - x - 12$ can be factored as $(x - 4)(x + 3)$ by using our conventional factoring techniques. Thus we obtain

$$x^3 - 2x^2 - 11x + 12 = (x - 1)(x - 4)(x + 3).$$

Show that $x + 4$ is a factor of $f(x) = x^3 - 5x^2 - 22x + 56$ and complete the factorization of $f(x)$.

Solution

We use synthetic division to divide $x^3 - 5x^2 - 22x + 56$ by $x + 4$.

$$\begin{array}{r|rrrr} -4 & 1 & -5 & -22 & 56 \\ & & -4 & 36 & -56 \\ \hline & 1 & -9 & 14 & 0 \end{array}$$

The last line indicates a quotient of $x^2 - 9x + 14$ and a remainder of zero. The zero remainder means that $x + 4$ is a factor. Furthermore, we can write

$$x^3 - 5x^2 - 22x + 56 = (x + 4)(x^2 - 9x + 14)$$

and then complete the factoring to obtain

$$f(x) = x^3 - 5x^2 - 22x + 56 = (x + 4)(x - 7)(x - 2).$$

The factor theorem also plays a significant role in determining some general factorization ideas, as the last example of this section illustrates.

EXAMPLE 8

Verify that $x + 1$ is a factor of $x^n + 1$ whenever n is an odd positive integer.

Solution

Let $f(x) = x^n + 1$ and compute $f(-1)$ to obtain

$$\begin{aligned} f(-1) &= (-1)^n + 1 \\ &= -1 + 1 && \text{Any odd power of } -1 \text{ is } -1. \\ &= 0. \end{aligned}$$

Since $f(-1) = 0$, we know that $x + 1$ is a factor of $f(x)$.

PROBLEM SET 6.2

For Problems 1–10, find $f(c)$ (a) by using synthetic division and the remainder theorem, and (b) by evaluating $f(c)$ directly.

1. $f(x) = x^2 + x - 8$ and $c = 2$

2. $f(x) = x^3 + x^2 - 2x - 4$ and $c = -1$

3. $f(x) = 3x^3 + 4x^2 - 5x + 3$ and $c = -4$

4. $f(x) = 2x^4 + x^2 + 6$ and $c = 1$

5. $f(x) = x^4 - 2x^3 - 3x^2 + 5x - 1$ and $c = -2$

6. $f(x) = 2x^4 + x^3 - 4x^2 - x + 1$ and $c = 2$

7. $f(t) = 6t^3 - 35t^2 + 8t - 10$ and $c = 6$

8. $f(t) = 2t^5 - 1$ and $c = -2$

9. $f(n) = 3n^4 - 2n^3 + 4n - 1$ and $c = 3$

10. $f(n) = -2n^4 + 4n - 5$ and $c = -3$

For Problems 11–18, find $f(c)$ *either* by using synthetic division and the remainder theorem or by evaluating $f(c)$ directly.

11. $f(x) = 5x^6 - x^3 - 1$ and $c = -1$

12. $f(x) = 2x^3 - 3x^2 - 5x + 4$ and $c = 4$

13. $f(x) = x^4 - 8x^3 + 9x^2 - 15x + 2$ and $c = 7$

14. $f(t) = 5t^3 - 8t^2 + 9t - 4$ and $c = -5$

15. $f(n) = -2n^4 + 2n^2 - n - 5$ and $c = -2$

16. $f(x) = 4x^7 + 3$ and $c = 3$

17. $f(x) = 2x^3 - 5x^2 + 4x - 3$ and $c = \frac{1}{2}$

18. $f(x) = 3x^3 + 4x^2 - 5x - 7$ and $c = -\frac{1}{3}$

For Problems 19–28, use the factor theorem to help answer some questions about factors.

19. Is $x - 2$ a factor of $3x^2 - 4x - 4$?

20. Is $x + 3$ a factor of $6x^2 + 13x - 15$?

21. Is $x + 2$ a factor of $x^3 + x^2 - 7x - 10$?

22. Is $x - 3$ a factor of $2x^3 - 3x^2 - 10x + 3$?

23. Is $x - 1$ a factor of $3x^3 + 5x^2 - x - 2$?

24. Is $x + 4$ a factor of $x^3 - 4x^2 + 2x - 8$?

25. Is $x - 2$ a factor of $x^3 - 8$?

26. Is $x + 2$ a factor of $x^3 + 8$?

27. Is $x - 3$ a factor of $x^4 - 81$?

28. Is $x + 3$ a factor of $x^4 - 81$?

For Problems 29–34, use synthetic division to show that $g(x)$ is a factor of $f(x)$ and complete the factorization of $f(x)$.

29. $g(x) = x + 2$; $f(x) = x^3 + 7x^2 + 4x - 12$

30. $g(x) = x - 1$; $f(x) = 3x^3 + 19x^2 - 38x + 16$

31. $g(x) = x - 3$; $f(x) = 6x^3 - 17x^2 - 5x + 6$

32. $g(x) = x + 2$; $f(x) = 12x^3 + 29x^2 + 8x - 4$

33. $g(x) = x + 1$; $f(x) = x^3 - 2x^2 - 7x - 4$

34. $g(x) = x - 5$; $f(x) = 2x^3 + x^2 - 61x + 30$

For Problems 35–38, find the value(s) of k that make(s) the second polynomial a factor of the first.

35. $x^3 - kx^2 + 5x + k$; $x - 2$

36. $k^2x^4 + 3kx^2 - 4$; $x - 1$

37. $x^3 + 4x^2 - 11x + k$; $x + 2$

38. $kx^3 + 19x^2 + x - 6$; $x + 3$

39. Show that $x + 2$ is a factor of $x^{12} - 4096$.

40. Argue that $f(x) = 2x^4 + x^2 + 3$ has no factor of the form $x - c$, where c is a real number.

41. Verify that $x - 1$ is a factor of $x^n - 1$ for all positive integral values of n.

42. Verify that $x + 1$ is a factor of $x^n - 1$ for all even positive integral values of n.

43. **a.** Verify that $x - y$ is a factor of $x^n - y^n$ whenever n is a positive integer.

b. Verify that $x + y$ is a factor of $x^n - y^n$ whenever n is an even positive integer.

c. Verify that $x + y$ is a factor of $x^n + y^n$ whenever n is an odd positive integer.

MISCELLANEOUS PROBLEMS

The remainder and factor theorems are true for any complex value of c. Therefore, for Problems 44–46, find $f(c)$ (a) by using synthetic division and the remainder theorem, and (b) by evaluating $f(c)$ directly.

44. $f(x) = x^3 - 5x^2 + 2x + 1$ and $c = i$

45. $f(x) = x^2 + 4x - 2$ and $c = 1 + i$

46. $f(x) = x^3 + 2x^2 + x - 2$ and $c = 2 - 3i$

Solve the following problems.

47. Show that $x - 2i$ is a factor of $f(x) = x^4 + 6x^2 + 8$.

48. Show that $x + 3i$ is a factor of $f(x) = x^4 + 14x^2 + 45$.

49. Consider changing the form of the polynomial $f(x) = x^3 + 4x^2 - 3x + 2$ as follows.

$$\begin{aligned} f(x) &= x^3 + 4x^2 - 3x + 2 \\ &= x(x^2 + 4x - 3) + 2 \\ &= x[x(x + 4) - 3] + 2 \end{aligned}$$

The final form, $f(x) = x[x(x + 4) - 3] + 2$, is called the **nested form** of the polynomial. It is particularly well suited to evaluate functional values of f either by hand or with a calculator.

For each of the following, find the indicated functional values, using the nested form of the given polynomial.

a. $f(4)$, $f(-5)$, and $f(7)$ for $f(x) = x^3 + 5x^2 - 2x + 1$

b. $f(3)$, $f(6)$, and $f(-7)$ for $f(x) = 2x^3 - 4x^2 - 3x + 2$

c. $f(4)$, $f(\ 5)$, and $f(-3)$ for $f(x) = -2x^3 + 5x^2 - 6x - 7$

d. $f(5)$, $f(6)$, and $f(-3)$ for $f(x) = x^4 + 3x^3 - 2x^2 + 5x - 1$

6.3 POLYNOMIAL EQUATIONS

In Chapter 2 we solved a large variety of *linear equations* of the form $ax + b = 0$ and *quadratic equations* of the form $ax^2 + bx + c = 0$. Linear and quadratic equations are special cases of a general class of equations we refer to as **polynomial equations**. The equation

$$a_nx^n + a_{n-1}x^{n-1} + \cdots + a_1x + a_0 = 0$$

where the coefficients $a_0, a_1, \ldots, a_n$ are real numbers and n is a positive integer, is called a **polynomial equation of degree n**. The following are examples of polynomial equations.

$\sqrt{2}x - 6 = 0$	Degree 1
$\frac{3}{4}x^2 - \frac{2}{3}x + 5 = 0$	Degree 2
$4x^3 - 3x^2 - 7x - 9 = 0$	Degree 3
$5x^4 - x + 6 = 0$	Degree 4

REMARK The most general polynomial equation allows complex numbers as coefficients. However, for our purposes in this text, we will restrict the coefficients to real numbers. We refer to such equations as **polynomial equations over the reals**.

In general, solving polynomial equations of degree greater than 2 can be very difficult and often requires mathematics beyond the scope of this text. However, there are some general methods for solving polynomial equations that you should know, since there are certain types of polynomial equations that we *can* solve with the techniques available to us at this time.

Let's begin by listing some (previously encountered) polynomial equations and their solution sets.

Equation	Solution set
$3x + 4 = 7$	$\{1\}$
$x^2 + x - 6 = 0$	$\{-3, 2\}$
$2x^3 - 3x^2 - 2x + 3 = 0$	$\left\{-1, 1, \frac{3}{2}\right\}$
$x^4 - 16 = 0$	$\{-2, 2, -2i, 2i\}$

Notice that in each of these examples, the number of solutions corresponds to the degree of the equation. The first-degree equation has one solution, the second-degree equation has two solutions, the third-degree equation has three solutions, and the fourth-degree equation has four solutions. Now consider the equation

$$(x - 4)^2(x + 5)^3 = 0.$$

It can be written

$$(x - 4)(x - 4)(x + 5)(x + 5)(x + 5) = 0,$$

which implies that

$$x - 4 = 0 \quad \text{or} \quad x - 4 = 0 \quad \text{or} \quad x + 5 = 0 \quad \text{or}$$
$$x + 5 = 0 \quad \text{or} \quad x + 5 = 0.$$

Therefore,

$$x = 4 \quad \text{or} \quad x = 4 \quad \text{or} \quad x = -5 \quad \text{or} \quad x = -5 \quad \text{or} \quad x = -5.$$

We say that the solution set of the original equation is $\{-5, 4\}$, but we also say that the equation has a solution of 4 with a **multiplicity of two**, and a solution of -5 with a **multiplicity of three**. Furthermore, notice that the sum of the multiplicities is 5, which agrees with the degree of the equation.

We can state the following general property.

PROPERTY 6.3

A polynomial equation of degree n has n solutions, where any solution of multiplicity p is counted p times.

Finding Rational Solutions

As we stated earlier, solving polynomial equations of degree greater than two can be very difficult. However, *rational solutions* of polynomial equations with integral coefficients can be found by using techniques from this chapter. The following property restricts the possible rational solutions of such an equation.

PROPERTY 6.4 *Rational Root Theorem*

Consider the polynomial equation

$$a_n x^n + a_{n-1}x^{n-1} + \cdots + a_1 x + a_0 = 0,$$

where the coefficients $a_0, a_1, \ldots, a_n$ are integers. If the rational number c/d, reduced to lowest terms, is a solution of the equation, then c is a factor of the constant term a_0, and d is a factor of the leading coefficient a_n.

The *why* behind the rational root theorem is based on some simple factoring ideas, as indicated by the following outline of a proof for the theorem.

Outline of Proof

If c/d is to be a solution, then

$$a_n\left(\frac{c}{d}\right)^n + a_{n-1}\left(\frac{c}{d}\right)^{n-1} + \cdots + a_1\left(\frac{c}{d}\right) + a_0 = 0.$$

Multiply both sides of this equation by d^n and then add $-a_0 d^n$ to both sides to yield

$$a_n c^n + a_{n-1}c^{n-1}d + \cdots + a_1 c d^{n-1} = -a_0 d^n.$$

Because c is a factor of the left side of this equation, c must also be a factor of $-a_0 d^n$. Furthermore, because c/d is in reduced form, c and d have no common factors other than -1 or 1. Thus c must be a factor of a_0. In the same way, from the equation

$$a_{n-1}c^{n-1}d + \cdots + a_1 c d^{n-1} + a_0 d^n = -a_n c^n,$$

we can conclude that d is a factor of the left side and therefore d is also a factor of a_n.

The rational root theorem, synthetic division, the factor theorem, and some previous knowledge about solving linear and quadratic equations all merge to form a basis for finding rational solutions. Let's consider some examples.

EXAMPLE 1

Find all rational solutions of $3x^3 + 8x^2 - 15x + 4 = 0$.

Solution

If c/d is a rational solution, then c must be a factor of 4 and d must be a factor of 3. Therefore, the possible values for c and d are as follows.

For *c*	$\pm 1, \pm 2, \pm 4$
For *d*	$\pm 1, \pm 3$

Thus, the possible values for c/d are

$$\pm 1, \pm\frac{1}{3}, \pm 2, \pm\frac{2}{3}, \pm 4, \pm\frac{4}{3}.$$

By using synthetic division, we can test $x - 1$.

$$\begin{array}{r|rrrr} 1 & 3 & 8 & -15 & 4 \\ & & 3 & 11 & -4 \\ \hline & 3 & 11 & -4 & 0 \end{array}$$

This shows that $x - 1$ is a factor of the given polynomial; therefore, 1 is a rational solution of the equation. Furthermore, the synthetic division result also indicates how to factor the given polynomial.

$$3x^3 + 8x^2 - 15x + 4 = 0$$
$$(x - 1)(3x^2 + 11x - 4) = 0$$

The quadratic factor can be further factored by using our previous techniques.

$$(x - 1)(3x^2 + 11x - 4) = 0$$
$$(x - 1)(3x - 1)(x + 4) = 0$$
$$x - 1 = 0 \quad \text{or} \quad 3x - 1 = 0 \quad \text{or} \quad x + 4 = 0$$
$$x = 1 \quad \text{or} \quad x = \frac{1}{3} \quad \text{or} \quad x = -4$$

Thus, the entire solution set consists of rational numbers and can be listed as $\left\{-4, \frac{1}{3}, 1\right\}$.

In Example 1, we were fortunate that the first time we used synthetic division, we got a rational solution. But this often does not happen; then we would need to conduct a little organized search, as the next example illustrates.

EXAMPLE 2

Find all rational solutions of $3x^3 + 7x^2 - 22x - 8 = 0$.

Solution

If c/d is a rational solution, then c must be a factor of -8 and d must be a factor of 3. Therefore, the possible values for c and d are as follows.

For c $\pm1, \pm2, \pm4, \pm8$

For d $\pm1, \pm3$

Thus, the possible values for c/d are

$$\pm1, \quad \pm\frac{1}{3}, \quad \pm2, \quad \pm\frac{2}{3}, \quad \pm4, \quad \pm\frac{4}{3}, \quad \pm8, \quad \pm\frac{8}{3}.$$

Let's begin our search for rational solutions by trying the integers first.

$$\begin{array}{r|rrrr} 1 & 3 & 7 & -22 & -8 \\ & & 3 & 10 & -12 \\ \hline & 3 & 10 & -12 & \boxed{-20} \end{array}$$

← This indicates that $x - 1$ is not a factor and thus 1 is not a solution.

$$\begin{array}{r|rrrr} -1 & 3 & 7 & -22 & -8 \\ & & -3 & -4 & 26 \\ \hline & 3 & 4 & -26 & \boxed{18} \end{array}$$

← This indicates that -1 is not a solution.

$$\begin{array}{r|rrrr} 2 & 3 & 7 & -22 & -8 \\ & & 6 & 26 & 8 \\ \hline & 3 & 13 & 4 & 0 \end{array}$$

Now we know that $x - 2$ is a factor and we can proceed as follows.

$$3x^3 + 7x^2 - 22x - 8 = 0$$
$$(x - 2)(3x^2 + 13x + 4) = 0$$
$$(x - 2)(3x + 1)(x + 4) = 0$$
$$x - 2 = 0 \quad \text{or} \quad 3x + 1 = 0 \quad \text{or} \quad x + 4 = 0$$
$$x = 2 \quad \text{or} \quad 3x = -1 \quad \text{or} \quad x = -4$$
$$x = 2 \quad \text{or} \quad x = -\frac{1}{3} \quad \text{or} \quad x = -4$$

The solution set is $\left\{-4, -\frac{1}{3}, 2\right\}$.

In Examples 1 and 2, we were solving third-degree equations. Therefore, once we found one linear factor by synthetic division, we were able to factor the remaining quadratic factor in the usual way. However, if the given equation is of degree four or more, we may need to find more than one linear factor by synthetic division, as in the next example.

EXAMPLE 3

Solve $x^4 - 6x^3 + 22x^2 - 30x + 13 = 0$.

Solution

The possible values for c/d are ±1 and ±13. By synthetic division we test 1.

$$\begin{array}{r|rrrrr} 1 & 1 & -6 & 22 & -30 & 13 \\ & & 1 & -5 & 17 & -13 \\ \hline & 1 & -5 & 17 & -13 & 0 \end{array}$$

This indicates that $x - 1$ is a factor of the given polynomial. The bottom line of the synthetic division indicates that the given polynomial can now be factored as follows.

$$\begin{aligned} x^4 - 6x^3 + 22x^2 - 30x + 13 &= 0 \\ (x - 1)(x^3 - 5x^2 + 17x - 13) &= 0 \end{aligned}$$

Therefore,

$$x - 1 = 0 \quad \text{or} \quad x^3 - 5x^2 + 17x - 13 = 0.$$

Now we can use the same approach to look for rational solutions of $x^3 - 5x^2 + 17x - 13 = 0$. The possible values for c/d are, again, ± 1 and ± 13. By synthetic division we test 1 again.

$$\begin{array}{r|rrrr} 1 & 1 & -5 & 17 & -13 \\ & & 1 & -4 & 13 \\ \hline & 1 & -4 & 13 & 0 \end{array}$$

This indicates that $x - 1$ is also a factor of $x^3 - 5x^2 + 17x - 13$, and the other factor is $x^2 - 4x + 13$. Now we can solve the original equation.

$$\begin{aligned} x^4 - 6x^3 + 22x^2 - 30x + 13 &= 0 \\ (x - 1)(x^3 - 5x^2 + 17x - 13) &= 0 \\ (x - 1)(x - 1)(x^2 - 4x + 13) &= 0 \end{aligned}$$

$$x - 1 = 0 \quad \text{or} \quad x - 1 = 0 \quad \text{or} \quad x^2 - 4x + 13 = 0$$

$$x = 1 \quad \text{or} \quad x = 1 \quad \text{or} \quad x^2 - 4x + 13 = 0$$

Use the quadratic formula on $x^2 - 4x + 13 = 0$ to produce

$$x = \frac{4 \pm \sqrt{16 - 52}}{2} = \frac{4 \pm \sqrt{-36}}{2} = \frac{4 \pm 6i}{2} = 2 \pm 3i.$$

Thus, the original equation has a rational solution of 1 with a multiplicity of two and two complex solutions, $2 + 3i$ and $2 - 3i$. We list the solution set as $\{1, 2 \pm 3i\}$.

Example 3 illustrates two general properties. First, notice that the coefficient of x^4 is 1, which forces the possible rational solutions to be integers. In general, **the possible rational solutions of $x^n + a_{n-1}x^{n-1} + \cdots + a_1x + a_0 = 0$ are the integral factors of a_0**. Second, notice that the complex solutions of Example 4 are conjugates of each other. The following general property can be stated.

PROPERTY 6.5

If a polynomial equation with real coefficients has any nonreal complex solutions, they must occur in conjugate pairs.

REMARK The justification for Property 6.5 is based on some properties of conjugates that were presented in Problem 79 of Problem Set 1.8. We will not show the details of such a proof at this time.

Each of Properties 6.3, 6.4, and 6.5 yields some information about the solutions of a polynomial equation. Before we state one more property that will give us some additional information, we need to illustrate two ideas.

In a polynomial that is arranged in descending powers of x, if two successive terms differ in sign, we say that there is a **variation in sign**. Terms with zero coefficients are disregarded when counting sign variations. For example, the polynomial

$$+3x^3 - 2x^2 + 4x + 7$$

has *two* sign variations, whereas the polynomial

$$+x^5 - 4x^3 + x - 5$$

has *three* variations.

Another idea that we need to understand is the fact that the solutions of

$$a_n(-x)^n + a_{n-1}(-x)^{n-1} + \cdots + a_1(-x) + a_0 = 0$$

are the opposites of the solutions of

$$a_nx^n + a_{n-1}x^{n-1} + \cdots + a_1x + a_0 = 0.$$

In other words, if a new equation is formed by replacing x with $-x$ in a given equation, then the solutions of the new equation are the opposites of the solutions of the original equation. For example, the solution set of $x^2 + 7x + 12 = 0$ is $\{-4, -3\}$; the solution set of $(-x)^2 + 7(-x) + 12 = 0$, which simplifies to $x^2 - 7x + 12 = 0$, is $\{3, 4\}$.

Now we can state a property that can help us to determine the nature of the solutions of a polynomial equation without actually solving the equation.

PROPERTY 6.6 *Descartes' Rule of Signs*

Let $a_nx^n + a_{n-1}x^{n-1} + \cdots + a_1x + a_0 = 0$ be a polynomial equation with real coefficients.

1. The number of **positive real solutions** of the given equation is either equal to the number of variations in sign of the

polynomial or less than the number of variations by a positive even integer.

2. The number of **negative real solutions** of the given equation is either equal to the number of variations in sign of the polynomial $a_n(-x)^n + a_{n/1}(-x)^{n-1} + \cdots + a_1(-x) + a_0$ or less than that number of variations by a positive even integer.

Property 6.6, along with Properties 6.3 and 6.5, allows us to acquire some information about the solutions of a polynomial equation without actually solving the equation. Let's consider some equations and indicate how much we know about their solutions without solving them.

$$x^3 + 3x^2 + 5x + 4 = 0$$

1. No variations of sign in $x^3 + 3x^2 + 5x + 4$ means that there are *no positive solutions.*
2. Replace x with $-x$ in the given polynomial to produce $(-x)^3 + 3(-x)^2 + 5(-x) + 4$, which simplifies to $-x^3 + 3x^2 - 5x + 4$. This polynomial contains 3 variations of sign; thus, there are 3 or 1 *negative solutions.*

Conclusion The given equation has either 3 negative real solutions or 1 negative real solution and 2 nonreal complex solutions.

$$2x^4 + 3x^2 - x - 1 = 0$$

1. There is 1 variation of sign in the given polynomial; thus, the equation has 1 *positive solution.*
2. Replace x with $-x$ to produce $2(-x)^4 + 3(-x)^2 - (-x) - 1$, which simplifies to $2x^4 + 3x^2 + x - 1$ and contains 1 variation of sign. Thus, the given equation has 1 *negative solution.*

Conclusion The given equation has 1 positive, 1 negative, and 2 nonreal complex solutions.

$$3x^4 + 2x^2 + 5 = 0$$

1. No variations of sign in the given polynomial means that there are *no positive solutions.*
2. Replace x with $-x$ to produce $3(-x)^4 + 2(-x)^2 + 5$, which simplifies to $3x^4 + 2x^2 + 5$ and still contains no variations of sign. Thus, there are *no negative solutions.*

Conclusion The given equation contains 4 nonreal complex solutions. We also know that these solutions will appear in conjugate pairs.

$$2x^5 - 4x^3 + 2x - 5 = 0$$

1. Three variations of sign in the given polynomial imply that *the number of positive solutions is* 3 *or* 1.
2. Replace x with $-x$ to produce $2(-x)^5 - 4(-x)^3 + 2(-x) - 5 = -2x^5 + 4x^3 - 2x - 5$, which contains 2 variations of sign. Thus *the number of negative solutions is* 2 *or* 0.

Conclusion The given equation has

3 positive and 2 negative solutions, or

3 positive and 2 nonreal complex solutions, or

1 positive, 2 negative, and 2 nonreal complex solutions, or

1 positive and 4 nonreal complex solutions

It should be evident from the previous discussions that sometimes we can truly pinpoint the nature of the solutions of a polynomial equation. However, for some equations (such as the last example), if we use the properties discussed in this section, the best that we can do is to restrict the nature of the solutions to a few possibilities.

Finally, we need to realize that some of the properties presented in these last two sections help us to determine polynomial equations with specified roots. Let's consider some examples.

EXAMPLE 4

Find a polynomial equation with integral coefficients that has the given numbers as solutions and the indicated degree.

a. $1, \frac{1}{2}, -2$; degree 3 **b.** 2 of multiplicity 4; degree 4

c. $1 + i, -3i$; degree 4

Solution

a. If $1, \frac{1}{2}$, and -2 are solutions, then $(x - 1)$, $\left(x - \frac{1}{2}\right)$, and $(x + 2)$ are factors of the polynomial. Thus, the following third-degree polynomial equation can be formed.

$$(x - 1)\left(x - \frac{1}{2}\right)(x + 2) = 0$$

$$(x - 1)(2x - 1)(x + 2) = 0$$

$$2x^3 + x^2 - 5x + 2 = 0$$

b. If 2 is to be a solution with multiplicity 4, then the equation $(x - 2)^4 = 0$ can be formed. Using the binomial expansion pattern we can express the equation as follows.

$$(x - 2)^4 = 0$$

$$x^4 - 8x^3 + 24x^2 - 32x + 16 = 0$$

c. By Property 6.5, if $1 + i$ is a solution, then so is $1 - i$. Likewise, since $-3i$ is a solution, so is $3i$. Therefore, we can form the following equation.

$$[x - (1 + i)][x - (1 - i)](x + 3i)(x - 3i) = 0$$
$$[(x - 1) - i][(x - 1) + i](x^2 + 9) = 0$$
$$[(x - 1)^2 - i^2](x^2 + 9) = 0$$
$$(x^2 - 2x + 1 + 1)(x^2 + 9) = 0$$
$$(x^2 - 2x + 2)(x^2 + 9) = 0$$
$$x^4 - 2x^3 + 11x^2 - 18x + 18 = 0$$

A graphing utility can be very helpful when solving polynomial equations; especially if they are of degree greater than two. Even the search for possible rational solutions can be simplified by looking at a graph. To find the rational solutions of $3x^3 + 8x^2 - 15x + 4 = 0$ (Example 1), we could begin by graphing the equation $y = 3x^3 + 8x^2 - 15x + 4$. This graph is shown in Figure 6.1. From the graph it looks as if 1 and -4 are two of the x-intercepts and therefore solutions of the original equation. Let's check them in the equation.

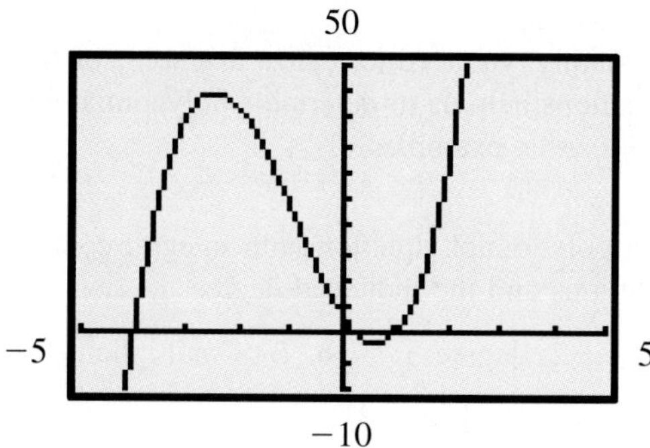

FIGURE 6.1

$$3(1)^3 + 8(1)^2 - 15(1) + 4 = 3 + 8 - 15 + 4 = 0$$
$$3(-4)^3 + 8(-4)^2 - 15(-4) + 4 = -192 + 128 + 60 + 4 = 0$$

Thus, $x - 1$ and $x + 4$ are factors of $3x^3 + 8x^2 - 15x + 4$ and the remaining factor could be found by division. We could then determine the solution set as we did in Example 1. Let's consider an example where we use a graphing utility to approximate the real number solutions of a polynomial equation.

EXAMPLE 5

Find the real number solutions of the equation $x^4 - 2x^3 - 5 = 0$.

Solution

Let's use a graphing utility to get a sketch of the graph of $y = x^4 - 2x^3 - 5$ (Figure 6.2). From this graph we see that one x-intercept is between -1 and -2 and another between 2 and 3. We can use the zoom and trace features to approximate these values at -1.2 and 2.4, to the nearest tenth. Thus, the real solutions for the equation $x^4 - 2x^3 - 5 = 0$ are approximately -1.2 and 2.4 (The other two solutions must be conjugate complex numbers.)

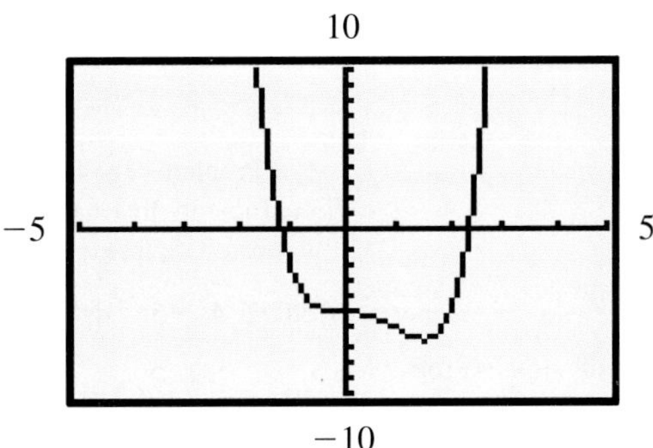

FIGURE 6.2

PROBLEM SET 6.3

Use the rational root theorem and the factor theorem to help solve each of the following equations. Be sure that the number of solutions for each equation agrees with Property 6.3; take into account the multiplicity of solutions.

1. $x^3 + x^2 - 4x - 4 = 0$

2. $x^3 - 2x^2 - 11x + 12 = 0$

3. $6x^3 + x^2 - 10x + 3 = 0$

4. $8x^3 - 2x^2 - 41x - 10 = 0$

5. $3x^3 + 13x^2 - 52x + 28 = 0$

6. $15x^3 + 14x^2 - 3x - 2 = 0$

7. $x^3 - 2x^2 - 7x - 4 = 0$

8. $x^3 - x^2 - 8x + 12 = 0$

9. $x^4 - 4x^3 - 7x^2 + 34x - 24 = 0$

10. $x^4 + 4x^3 - x^2 - 16x - 12 = 0$

11. $x^3 - 10x - 12 = 0$

12. $x^3 - 4x^2 + 8 = 0$

13. $3x^4 - x^3 - 8x^2 + 2x + 4 = 0$

14. $2x^4 + 3x^3 - 11x^2 - 9x + 15 = 0$

15. $6x^4 - 13x^3 - 19x^2 + 12x = 0$

16. $x^3 - x^2 + x - 1 = 0$

17. $x^4 - 3x^3 + 2x^2 + 2x - 4 = 0$

18. $x^4 + x^3 - 3x^2 - 17x - 30 = 0$

19. $2x^5 - 5x^4 + x^3 + x^2 - x + 6 = 0$

20. $4x^4 + 12x^3 + x^2 - 12x + 4 = 0$

Verify that the following equations have no rational solutions.

21. $x^4 - x^3 - 8x^2 - 3x + 1 = 0$

22. $x^4 + 3x - 2 = 0$

23. $2x^4 - 3x^3 + 6x^2 - 24x + 5 = 0$

24. $3x^4 - 4x^3 - 10x^2 + 3x - 4 = 0$

25. $x^5 - 2x^4 + 3x^3 + 4x^2 + 7x - 1 = 0$

26. $x^5 + 2x^4 - 2x^3 + 5x^2 - 2x - 3 = 0$

27. The rational root theorem pertains to polynomial equations with integral coefficients. However, if the coefficients are nonintegral rational numbers, we can first apply the *multiplication property of equality* to produce an equivalent equation with integral coefficients. Use this method to solve each of the following equations.

a. $\frac{1}{10}x^3 + \frac{1}{2}x^2 + \frac{1}{5}x - \frac{4}{5} = 0$

b. $\frac{1}{10}x^3 + \frac{1}{5}x^2 - \frac{1}{2}x - \frac{3}{5} = 0$

c. $x^3 + \frac{9}{2}x^2 - x - 12 = 0$

d. $x^3 - \frac{5}{6}x^2 - \frac{22}{3}x + \frac{5}{2} = 0$

For Problems 28–37, use Descartes' rule of signs (Property 6.6) to determine the possibilities for the nature of the solutions for each of the equations. *Do not solve the equations.*

28. $6x^2 + 7x - 20 = 0$

29. $8x^2 - 14x + 3 = 0$

30. $2x^3 + x - 3 = 0$

31. $4x^3 + 3x + 7 = 0$

32. $3x^3 - 2x^2 + 6x + 5 = 0$

33. $4x^3 + 5x^2 - 6x - 2 = 0$

34. $x^5 - 3x^4 + 5x^3 - x^2 + 2x - 1 = 0$

35. $2x^5 + 3x^3 - x + 1 = 0$

36. $x^5 + 32 = 0$

37. $2x^6 + 3x^4 - 2x^2 - 1 = 0$

For Problems 38–47, find a polynomial equation with integral coefficients that has the given numbers as solutions and the indicated degree.

38. $2, 4, -3$; degree 3

39. $1, -1, 2, -4$; degree 4

40. $-2, \frac{1}{2}, \frac{2}{3}$; degree 3

41. $3, -\frac{2}{3}, \frac{3}{4}$; degree 3

42. 1 of multiplicity 5; degree 5

43. -3 of multiplicity 4; degree 4

44. $3, 2 + 3i$; degree 3

45. $-2, 1 - 4i$; degree 3

46. $1 - i, 2i$; degree 4

47. $-2 + 3i, -i$; degree 4

MISCELLANEOUS PROBLEMS

48. Use the rational root theorem to argue that $\sqrt{2}$ is not a rational number. [*Hint:* The solutions of $x^2 - 2 = 0$ are $\pm\sqrt{2}$.]

49. Use the rational root theorem to argue that $\sqrt{12}$ is not a rational number.

50. Defend the statement, *Every polynomial equation of odd degree with real coefficients has at least one real number solution.*

51. The following synthetic division shows that 2 is a solution of $x^4 + x^3 + x^2 - 9x - 10 = 0$.

$$\begin{array}{r|rrrrr} 2 & 1 & 1 & 1 & -9 & -10 \\ & & 2 & 6 & 14 & 10 \\ \hline & 1 & 3 & 7 & 5 & 0 \leftarrow \end{array}$$

Notice that the new quotient row (indicated by the arrow) consists entirely of nonnegative numbers. This indicates that searching for solutions greater than 2 would be a waste of time, since larger divisors would continue to increase each of the numbers (except the 1 on the far left) in the new quotient row. (Try 3 as a divisor!) Thus, we say that 2 is an **upper bound** for the real number solutions of the given equation.

Now consider the following synthetic division, which shows that -1 is also a solution of $x^4 + x^3 + x^2 - 9x - 10 = 0$.

$$\begin{array}{r|rrrrr} -1 & 1 & 1 & 1 & -9 & -10 \\ & & -1 & 0 & -1 & 10 \\ \hline & 1 & 0 & 1 & -10 & 0 \leftarrow \end{array}$$

The new quotient row (indicated by the arrow) shows that there is no need to look for solutions less than -1 because any divisor less than -1 would increase the size (in absolute value) of each number in the new quotient row (except the 1 on the far left). (Try -2 as a divisor!) Thus, we say that -1 is a **lower bound** for the real number solutions of the given equation.

The following general property can be stated: If $a_nx^n + a_{n/1}x^{n-1} + \cdots + a_1x + a_0 = 0$ is a polynomial equation with real coefficients, and $a_n > 0$, and if the polynomial is divided synthetically by $x - c$, then:

1. If $c > 0$ and all numbers in the new quotient row of the synthetic division are nonnegative, then c is an upper bound for the real number solutions of the given equation.

2. If $c < 0$ and the numbers in the new quotient row alternate in sign (with 0 considered either positive or negative, as needed), then c is a lower bound for the real number solutions of the given equation.

Find the smallest possible integer and the largest negative integer that are upper and lower bounds, respectively, for the real number solutions of each of the following equations. Keep in mind that the integers that serve as bounds do not necessarily have to be solutions of the equation.

a. $x^3 - 3x^2 + 25x - 75 = 0$

b. $x^3 + x^2 - 4x - 4 = 0$

c. $x^4 + 4x^3 - 7x^2 - 22x + 24 = 0$

d. $3x^3 + 7x^2 - 22x - 8 = 0$

e. $x^4 - 2x^3 - 9x^2 + 2x + 8 = 0$

GRAPHICS CALCULATOR ACTIVITIES

52. Suppose that we want to solve the equation $x^3 + 2x^2 - 14x - 40 = 0$. Let's graph the function $f(x) = x^3 + 2x^2 - 14x - 40$. Since the graph has only one x-intercept, the equation must have one real number solution and two nonreal complex solutions. The graph also indicates that the real solution is approximately 4. We can determine that 4 is a solution and then we can proceed to solve the equation using the ideas of this section.

    Solve each of the following equations and use a graphics calculator whenever it seems to be helpful. Express all irrational solutions in lowest radical form.

    a. $x^3 + 2x^2 - 14x - 40 = 0$

    b. $x^3 + x^2 - 7x + 65 = 0$

    c. $x^4 - 6x^3 - 6x^2 + 32x + 24 = 0$

    d. $x^4 + 3x^3 - 39x^2 + 11x + 24 = 0$

    e. $x^3 - 14x^2 + 26x - 24 = 0$

    f. $x^4 + 2x^3 - 3x^2 - 4x + 4 = 0$

53. Use a graphics calculator to help determine the nature of the solutions for each of the following equations. You may also need to use the property stated in Problem 51.

    a. $2x^3 - 3x^2 - 3x + 2 = 0$

    b. $3x^3 + 7x^2 + 8x + 2 = 0$

    c. $2x^4 + 3x^2 + 1 = 0$

    d. $4x^5 - 8x^4 - 5x^3 + 10x^2 + x - 2 = 0$

    e. $x^4 - x^3 + 2x^2 - x - 1 = 0$

    f. $x^5 - x^4 + x^3 - x^2 + x - 3 = 0$

    g. $x^4 - 14x^3 + 23x^2 + 14x - 24 = 0$

    h. $x^3 + 13x^2 - 28x + 30 = 0$

54. Find approximations, to the nearest hundredth, of the real number solutions of each of the following equations.

    a. $x^2 - 4x + 1 = 0$

    b. $3x^3 - 2x^2 + 12x - 8 = 0$

    c. $x^4 - 8x^3 + 14x^2 - 8x + 13 = 0$

    d. $x^4 + 6x^3 - 10x^2 - 22x + 161 = 0$

    e. $7x^5 - 5x^4 + 35x^3 - 25x^2 + 28x - 20 = 0$

6.4 GRAPHING POLYNOMIAL FUNCTIONS

Just as we have a vocabulary to deal with linear, quadratic, and polynomial equations, we also have terms that classify functions. In Chapter 4 we defined a **linear function** by means of the equation

$$f(x) = ax + b$$

and a **quadratic function** by means of the equation

$$f(x) = ax^2 + bx + c.$$

Both of these are special cases of a general class of functions called **polynomial functions**. Any function of the form

$$f(x) = a_n x^n + a_{n-1}x^{n-1} + \cdots + a_1 x + a_0$$

is called a **polynomial function of degree *n***, where a_n is a nonzero real number, $a_{n-1}, \ldots, a_1, a_0$ are real numbers, and n is a nonnegative integer. The following are examples of polynomial functions.

$f(x) = 5x^3 - 2x^2 + x - 4$	Degree 3
$f(x) = -2x^4 - 5x^3 + 3x^2 + 4x - 1$	Degree 4
$f(x) = 3x^5 + 2x^2 - 3$	Degree 5

REMARK Our previous work with polynomial equations is sometimes presented as *finding zeros of polynomial functions*. The *solutions*, or *roots*, of a polynomial equation are also called the **zeros** of the polynomial function. For example, -2 and 2 are solutions of $x^2 - 4 = 0$ and they are zeros of $f(x) = x^2 - 4$. That is, $f(-2) = 0$ and $f(2) = 0$.

For a complete discussion of graphing polynomial functions, we would need some tools from calculus. However, the graphing techniques that we have discussed so far will allow us to graph certain kinds of polynomial functions. For example, polynomial functions of the form

$$f(x) = ax^n$$

are quite easy to graph. We know from our previous work that if $n = 1$, then functions such as $f(x) = 2x$, $f(x) = -3x$, and $f(x) = \frac{1}{2}x$ are lines through the origin that have slopes 2, -3, and $\frac{1}{2}$, respectively.

Furthermore, if $n = 2$ we know that the graphs of functions of the form $f(x) = ax^2$ are parabolas that are symmetrical with respect to the y-axis with vertices at the origin.

We have also previously graphed the special case of $f(x) = ax^n$ where $a = 1$ and $n = 3$, namely, the function $f(x) = x^3$. This graph is shown in Figure 6.3.

From our work with transformations of graphs in Section 4.4, we know that the graphs of functions of the form $f(x) = ax^3$, where $a > 1$ are vertical stretchings of $f(x) = x^3$ and can be easily determined by plotting a few points. Likewise, if $0 < a < 1$, the graph of $f(x) = ax^3$ is a shrinking of $f(x) = x^3$. Furthermore, we know that $f(x) = -x^3$ is an x-axis (also a y-axis) reflection of $f(x) = x^3$. In Figure 6.4 we sketched a graph for $f(x) = \frac{1}{2}x^3$ and $f(x) = -x^3$.

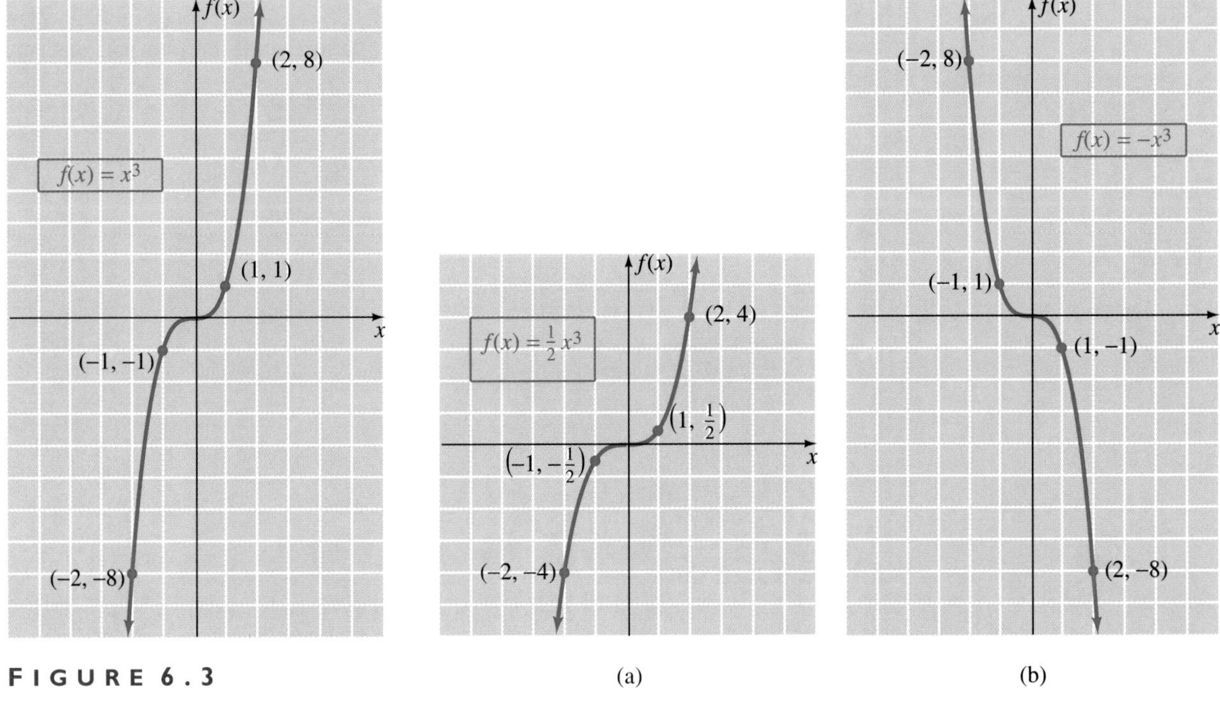

FIGURE 6.3

(a) (b)

FIGURE 6.4

There are two general patterns that emerge from studying functions of the form $f(x) = x^n$. If n is odd and greater than 3, then the graph of $f(x) = x^n$ closely resembles Figure 6.3. For example, the graph of $f(x) = x^5$ is shown in Figure 6.5. Notice that it *flattens out* a little more around the origin than the graph of $f(x) = x^3$, and it increases and decreases more rapidly because of the larger exponent. If

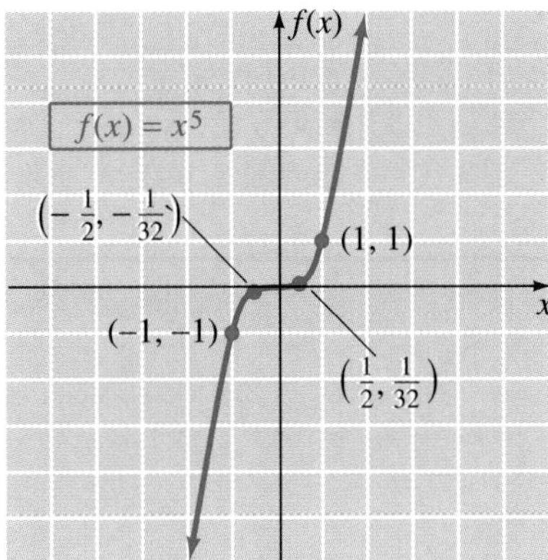

FIGURE 6.5

n is even and greater than 2, then the graphs of $f(x) = x^n$ are not parabolas; they do resemble the basic parabola except they're flatter at the bottom and steeper. Figure 6.6 shows the graph of $f(x) = x^4$.

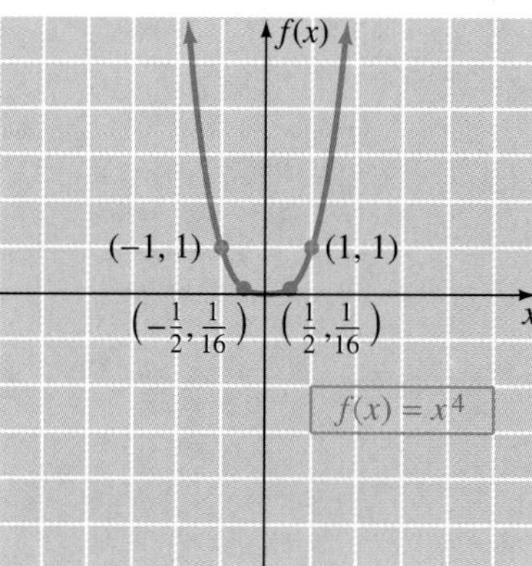

FIGURE 6.6

Graphs of functions of the form $f(x) = ax^n$, where n is an integer greater than 2 and $a \neq 1$, are variations of those shown in Figures 6.3 and 6.6. If n is odd, the curve is symmetrical about the origin; if n is even, the graph is symmetrical about the y-axis.

Transformations of these basic curves are easy to sketch. For example, in Figure 6.7 we translated the graph of $f(x) = x^3$ upward 2 units to produce the graph of $f(x) = x^3 + 2$. In Figure 6.8 we obtained the graph of $f(x) = (x - 1)^5$ by translating the graph of $f(x) = x^5$ one unit to the right. In Figure 6.9 we sketched the graph of $f(x) = -x^4$ as the x-axis reflection of $f(x) = x^4$.

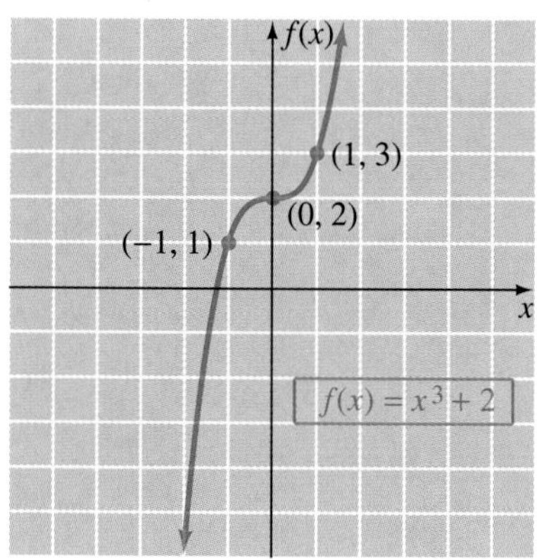

FIGURE 6.7

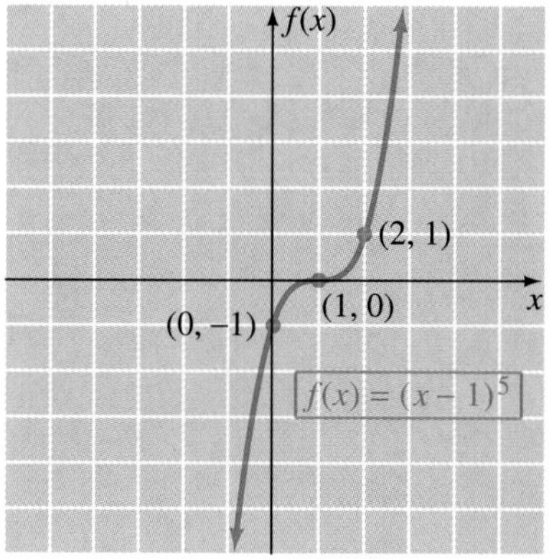

FIGURE 6.8

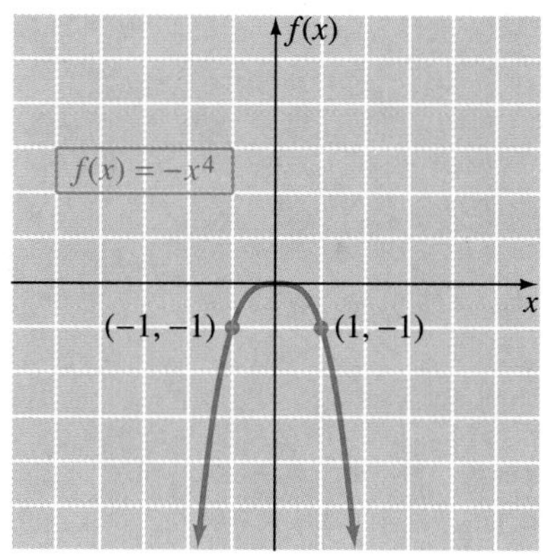

FIGURE 6.9

Graphing Polynomial Functions in Factored Form

As we mentioned earlier, a complete discussion of graphing polynomials of degree greater than 2 requires some tools from calculus. In fact, as the degree increases, the graphs often become more complicated. We do know that polynomial functions produce smooth continuous curves with a number of turning points, as illustrated in Figures 6.10 and 6.11. Some typical graphs of polynomial functions of odd degree are shown in Figure 6.10. As suggested by the graphs, every polynomial function of odd degree has at least *one real zero*, that is, at least one real number c such that $f(c) = 0$. Geometrically, the zeros of the function are the x-intercepts of the graph. In Figure 6.11 we have illustrated some possible graphs of polynomial functions of even degree.

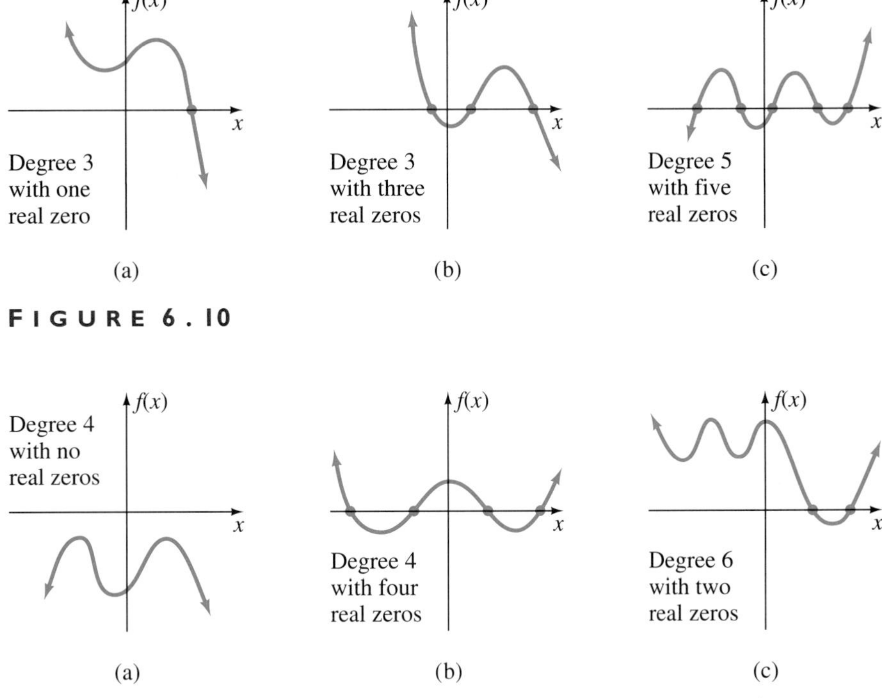

FIGURE 6.10

FIGURE 6.11

As indicated by the graphs in Figures 6.10 and 6.11, polynomial functions usually have **turning points** where the function either changes from increasing to decreasing or from decreasing to increasing. In calculus we are able to verify that *a polynomial function of degree n has at most n − 1 turning points.* Now let's illustrate how this information, along with some other techniques, can be used to graph polynomial functions that are expressed in factored form.

EXAMPLE 1

Graph $f(x) = (x + 2)(x - 1)(x - 3)$.

Solution

First, let's find the x-intercepts (zeros of the function) by setting each factor equal to zero and solving for x.

$$x + 2 = 0 \quad \text{or} \quad x - 1 = 0 \quad \text{or} \quad x - 3 = 0$$
$$x = -2 \quad \text{or} \quad x = 1 \quad \text{or} \quad x = 3$$

Thus, the points $(-2, 0)$, $(1, 0)$, and $(3, 0)$ are on the graph. Second, the points associated with the x-intercepts divide the x-axis into four intervals as we see in Figure 6.12.

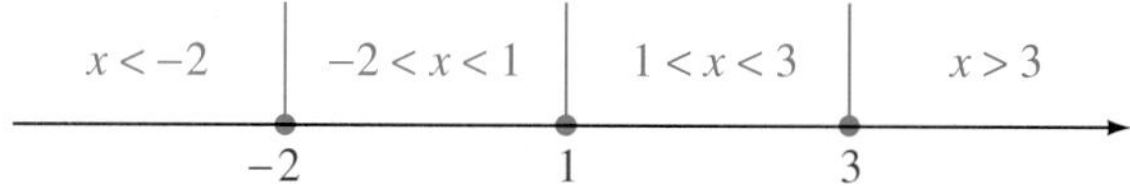

FIGURE 6.12

In each of these intervals, $f(x)$ is either always positive or always negative. That is to say, the graph is either completely above or completely below the x-axis. The sign can be determined by selecting a *test value* for x in each of the intervals. Any additional points that are easily obtained improve the accuracy of the graph. The following table summarizes these results.

INTERVAL	TEST VALUE	SIGN OF $f(x)$	LOCATION OF GRAPH
$x < -2$	$f(-3) = -24$	Negative	Below x-axis
$-2 < x < 1$	$f(0) = 6$	Positive	Above x-axis
$1 < x < 3$	$f(2) = -4$	Negative	Below x-axis
$x > 3$	$f(4) = 18$	Positive	Above x-axis
Additional values: $f(-1) = 8$			

Making use of the x-intercepts and the information in the table, we sketched the graph in Figure 6.13. (The points $(-3, -24)$ and $(4, 18)$ are not shown but they are used to indicate a rapid decrease and increase of the curve in those regions.)

REMARK In Figure 6.13 we indicated turning points of the graph at $(2, -4)$ and $(-1, 8)$. Keep in mind that these are only approximations; again, the tools of calculus are needed to find the exact turning points.

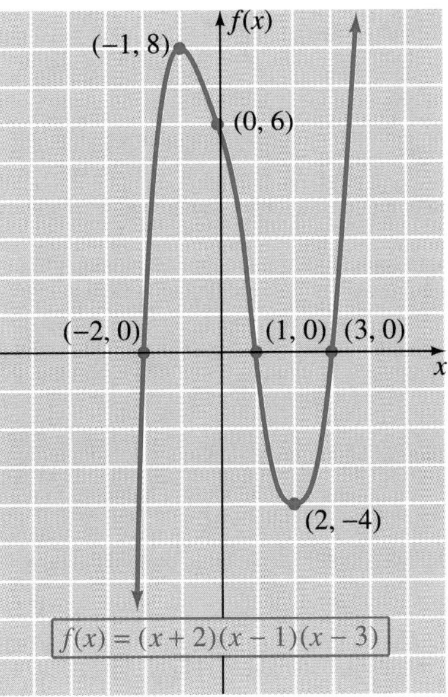

FIGURE 6.13

EXAMPLE 2

Graph $f(x) = -x^4 + 3x^3 - 2x^2$.

Solution

The polynomial can be factored as follows.

$$\begin{aligned} f(x) &= -x^4 + 3x^3 - 2x^2 \\ &= -x^2(x^2 - 3x + 2) \\ &= -x^2(x - 1)(x - 2) \end{aligned}$$

Now we can find the x-intercepts.

$$\begin{array}{ccccc} -x^2 = 0 & \text{or} & x - 1 = 0 & \text{or} & x - 2 = 0 \\ x = 0 & \text{or} & x = 1 & \text{or} & x = 2 \end{array}$$

Thus, the points (0, 0), (1, 0), and (2, 0) are on the graph and divide the x-axis into four intervals (see Figure 6.14).

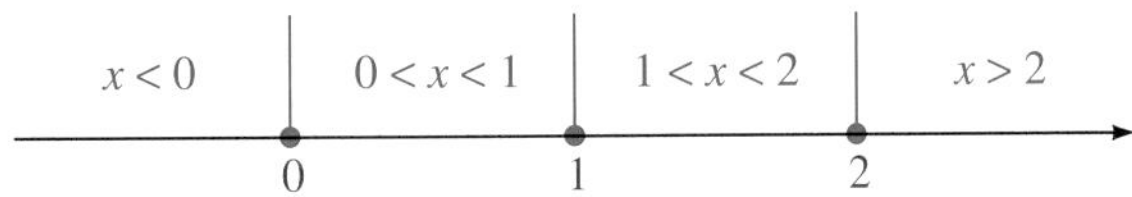

FIGURE 6.14

The following table determines some points and summarizes the sign behavior of $f(x)$.

INTERVAL	TEST VALUE	SIGN OF $f(x)$	LOCATION OF GRAPH
$x < 0$	$f(-1) = -6$	Negative	Below x-axis
$0 < x < 1$	$f\left(\frac{1}{2}\right) = -\frac{3}{16}$	Negative	Below x-axis
$1 < x < 2$	$f\left(\frac{3}{2}\right) = \frac{9}{16}$	Positive	Above x-axis
$x > 2$	$f(3) = -18$	Negative	Below x-axis

Make use of the table and the x-intercepts to graph Figure 6.15.

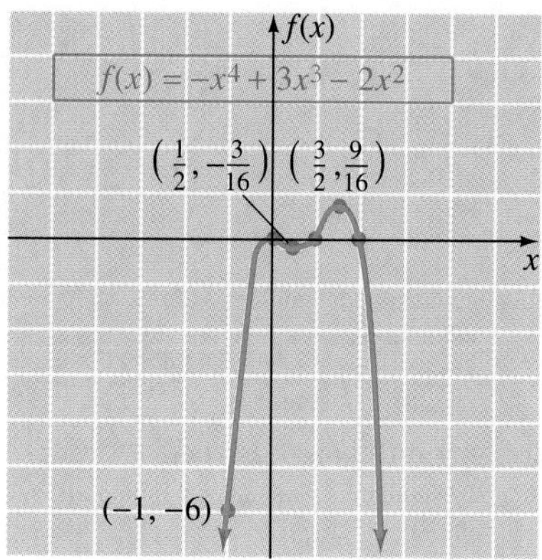

FIGURE 6.15

EXAMPLE 3

Graph $f(x) = x^3 + 3x^2 - 4$.

Solution

By using the rational root theorem, synthetic division, and the factor theorem, we can factor the given polynomial as follows.

$$\begin{aligned} f(x) &= x^3 + 3x^2 - 4 \\ &= (x - 1)(x^2 + 4x + 4) \\ &= (x - 1)(x + 2)^2 \end{aligned}$$

Now we can find the x-intercepts.

$$x - 1 = 0 \quad \text{or} \quad (x + 2)^2 = 0$$

$$x = 1 \qquad \text{or} \qquad x = -2$$

Thus the points $(-2, 0)$ and $(1, 0)$ are on the graph and divide the x-axis into three intervals (see Figure 6.16).

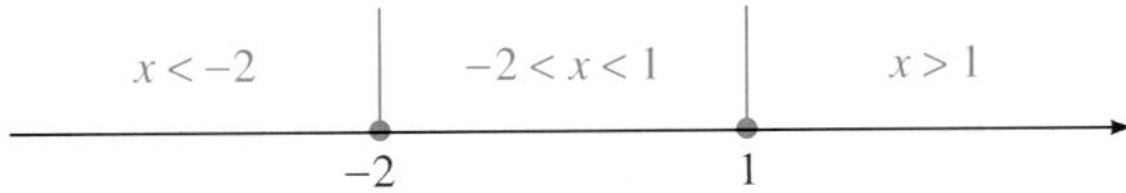

FIGURE 6.16

The following table determines some points and summarizes the sign behavior of $f(x)$.

INTERVAL	TEST VALUE	SIGN OF $f(x)$	LOCATION OF GRAPH
$x < -2$	$f(-3) = -4$	Negative	Below x-axis
$-2 < x < 1$	$f(0) = -4$	Negative	Below x-axis
$x > 1$	$f(2) = 16$	Positive	Above x-axis
Additional values: $f(-1) = -2$; $f(-4) = -20$			

With the results of the table and the x-intercepts, we sketched the graph in Figure 6.17.

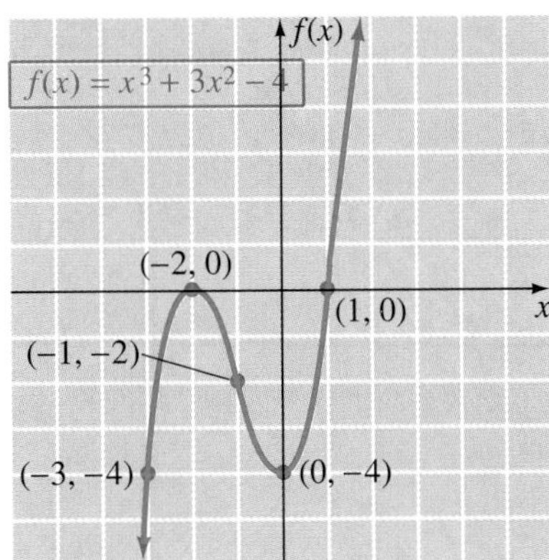

FIGURE 6.17

Finally, let's use a graphical approach to solve a problem involving a polynomial function.

EXAMPLE 4

Suppose that we have a rectangular piece of cardboard that measures 20 inches by 14 inches. From each corner a square piece is cut out and then the flaps are turned up to form an open box (see Figure 6.18). Determine the length of a side of the square pieces to be cut out so that the volume of the box is as large as possible.

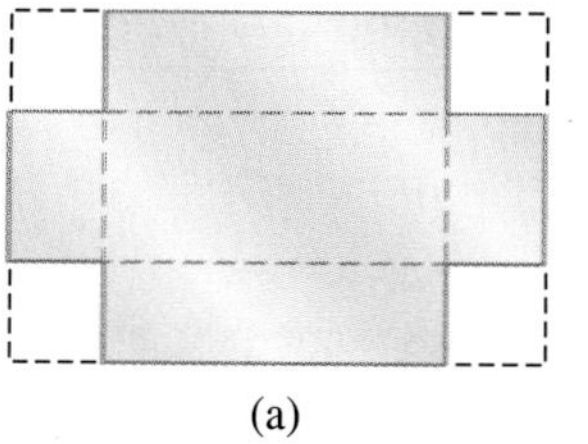

(a)

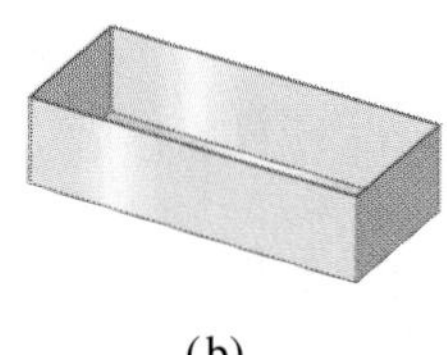

(b)

FIGURE 6.18

Solution

Let x represent the length of a side of the squares to be cut from each corner. Then $20 - 2x$ represents the length of the open box and $14 - 2x$ represents the width. The volume of a rectangular box is given by the formula $V = lwh$. So the volume of this box can be represented by $V = x(20 - 2x)(14 - 2x)$. Now let $y = V$ and graph the function $y = x(20 - 2x)(14 - 2x)$ as shown in Figure 6.19. For this problem we are only interested in the part of the graph between $x = 0$ and $x = 7$ because the length of a side of the squares has to be less than 7 inches for a box to be formed. Figure 6.20 gives us a view of that part of the graph. Now we can use the zoom and trace features to determine that as x equals approximately 2.7 the value of y is a maximum of approximately 339.0. Thus, square pieces of length approximately 2.7 inches on a side should be cut from each corner of the rectangular piece of cardboard. The open box formed will have a volume of approximately 339.0 cubic inches.

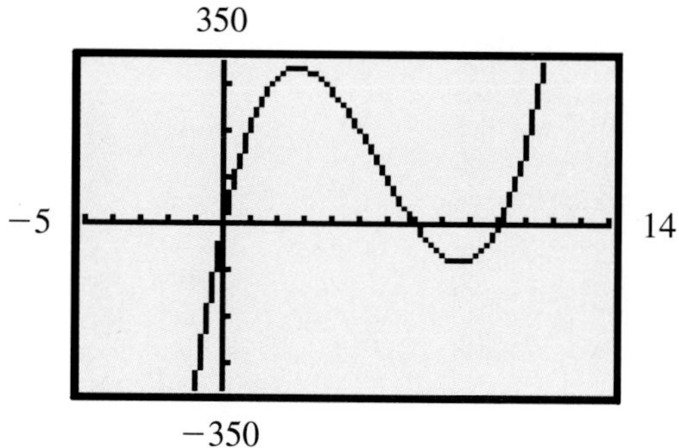

FIGURE 6.19

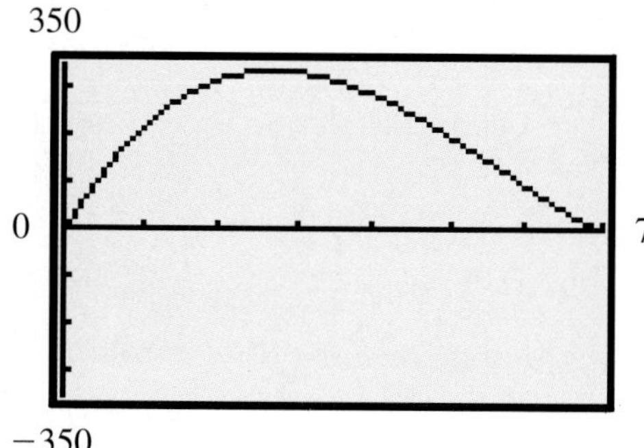

FIGURE 6.20

PROBLEM SET 6.4

Graph each of the following polynomial functions.

1. $f(x) = x^3 - 3$
2. $f(x) = (x + 1)^3$
3. $f(x) = (x - 2)^3 + 1$
4. $f(x) = -(x - 3)^3$
5. $f(x) = x^4 - 2$
6. $f(x) = (x + 3)^4$
7. $f(x) = (x + 1)^4 + 3$
8. $f(x) = -x^5$
9. $f(x) = (x - 1)^5 + 2$
10. $f(x) = -(x - 2)^4$
11. $f(x) = (x - 1)(x + 1)(x - 3)$
12. $f(x) = (x- 2)(x + 1)(x + 3)$
13. $f(x) = (x + 4)(x + 1)(1 - x)$
14. $f(x) = x(x + 2)(2 - x)$
15. $f(x) = -x(x + 3)(x - 2)$
16. $f(x) = -x^2(x - 1)(x + 1)$
17. $f(x) = (x + 3)(x + 1)(x - 1)(x - 2)$
18. $f(x) = (2x - 1)(x - 2)(x - 3)$
19. $f(x) = (x - 1)^2(x + 2)$
20. $f(x) = (x + 2)^3(x - 4)$
21. $f(x) = (x + 1)^2(x - 1)^2$
22. $f(x) = x(x - 2)^2(x + 1)$

Graph each of the following polynomial functions by first factoring the given polynomial. You may need to use some factoring techniques from Chapter 1, as well as the rational root theorem and the factor theorem.

23. $f(x) = x^3 + x^2 - 2x$
24. $f(x) = -x^3 - x^2 + 6x$
25. $f(x) = -x^4 - 3x^3 - 2x^2$
26. $f(x) = x^4 - 6x^3 + 8x^2$
27. $f(x) = x^3 - x^2 - 4x + 4$
28. $f(x) = x^3 + 2x^2 - x - 2$
29. $f(x) = x^3 - 13x + 12$
30. $f(x) = x^3 - x^2 - 9x + 9$
31. $f(x) = x^3 - 2x^2 - 11x + 12$
32. $f(x) = 2x^3 - 3x^2 - 3x + 2$
33. $f(x) = -x^3 + 6x^2 - 11x + 6$
34. $f(x) = x^4 - 5x^2 + 4$

For each of the following, find (a) the y-intercepts, (b) the x-intercepts, and (c) the intervals of x where $f(x) > 0$ and where $f(x) < 0$. *Do not* sketch the graph.

35. $f(x) = (x - 5)(x + 4)(x - 3)$
36. $f(x) = (x + 3)(x - 6)(8 - x)$
37. $f(x) = (x - 4)^2(x + 3)^3$
38. $f(x) = (x + 3)^4(x - 1)^3$
39. $f(x) = (x + 2)^2(x - 1)^3(x - 2)$
40. $f(x) = x(x)^2(x + 4)$
41. $f(x) = (x + 2)^5(x - 4)^2$

THOUGHTS into WORDS

42. Explain the concept of *multiplicity of roots* of an equation.
43. How would you defend the statement that the equation $2x^4 + 3x^3 + x^2 + 5 = 0$ has no positive solutions? Does it have any negative solutions? Defend your answer.
44. How would you go about graphing

$$f(x) = -(x - 1)(x + 2)^3?$$

MISCELLANEOUS PROBLEMS

45. A polynomial function with real coefficients is **continuous everywhere**; that is, its graph has no holes or breaks. This is the basis for the following property: **If $f(x)$ is a polynomial with real coefficents, and if $f(a)$ and $f(b)$ are of opposite sign, then there is at least one real zero between a and b.** This property, along with our previous knowledge of polynomial functions, provides the basis for locating and approximating irrational solutions of a polynomial equation.

Consider the equation $x^3 + 2x - 4 = 0$. Apply Descartes' rule of signs to determine that this equation has 1 positive real solution and 2 nonreal complex solutions. (You may want to confirm this!) The rational root theorem indicates that the only possible positive *rational* solutions are 1, 2, and 4. Use a little more compact format for synthetic division to obtain the following results when testing for 1 and 2 as possible solutions.

	1	0	2	−4
1	1	1	3	−1
2	1	2	6	8

Since $f(1) = -1$ (negative) and $f(2) = 8$ (positive), there must be an *irrational* solution between 1 and 2. Furthermore, since -1 is closer to 0 than 8, our guess is that the solution is closer to 1 than to 2. Let's start looking at 1.0, 1.1, 1.2, etc., until we can clamp the solution between 2 numbers.

	1	0	2	−4
1.0	1	1	3	−1
1.1	1	1.1	3.21	−0.469
1.2	1	1.2	3.44	0.128

A calculator is very helpful at this time.

Since $f(1.1) = -0.469$ and $f(1.2) = 0.128$, the irrational solution must be between 1.1 and 1.2. Furthermore, since 0.128 is closer to 0 than -0.469, our guess is that the solution is closer to 1.2 than to 1.1. Let's start looking at 1.15, 1.16, and so on.

	1	0	2	−4
1.15	1	1.15	3.3225	−0.179
1.16	1	1.16	3.3456	−0.119
1.17	1	1.17	3.3689	−0.058
1.18	1	1.18	3.3924	0.003

Since $f(1.17) = -0.058$ and $f(1.18) = 0.003$, the irrational solution must be between 1.17 and 1.18. Therefore, we can use 1.2 as a rational approximation to the nearest tenth.

For each of the following equations, verify that the equation has exactly one irrational solution, and find an approximation, to the nearest tenth, of that solution.

a. $x^3 + x - 6 = 0$

b. $x^3 - 6x - 4 = 0$

c. $x^3 - 27x + 18 = 0$

d. $x^3 - x^2 - x - 1 = 0$

e. $x^3 - 24x - 32 = 0$

f. $x^3 - 5x^2 + 3 = 0$

GRAPHICS CALCULATOR ACTIVITIES

46. Graph $f(x) = x^3$. Now predict the graphs for $f(x) = x^3 + 2$, $f(x) = -x^3 + 2$, and $f(x) = -x^3 - 2$. Graph these three functions on the same set of axes with the graph of $f(x) = x^3$.

47. Draw a rough sketch of the graphs of the functions $f(x) = x^3 - x^2$, $f(x) = -x^3 + x^2$, and $f(x) = -x^3 - x^2$. Now graph these three functions to check your sketches.

48. Graph $f(x) = x^4 + x^3 + x^2$. What should the graphs of $f(x) = x^4 - x^3 + x^2$ and $f(x) = -x^4 - x^3 - x^2$ look like? Graph them to see if you were right.

49. How should the graphs of $f(x) = x^3$, $f(x) = x^5$, and $f(x) = x^7$ compare? Graph these three functions on the same set of axes.

50. How should the graphs of $f(x) = x^2$, $f(x) = x^4$, and $f(x) = x^6$ compare? Graph these three functions on the same set of axes.

51. For each of the following functions, find the x-intercepts and find the intervals of x where $f(x) > 0$ and those where $f(x) < 0$.

a. $f(x) = x^3 - 3x^2 - 6x + 8$

b. $f(x) = x^3 - 8x^2 - x + 8$

c. $f(x) = x^3 - 7x^2 + 16x - 12$

d. $f(x) = x^3 - 19x^2 + 90x - 72$

e. $f(x) = x^4 + 3x^3 - 3x^2 - 11x - 6$

f. $f(x) = x^4 + 12x^2 - 64$

52. Find the coordinates of the turning points of each of the following graphs. Express x- and y-values to the nearest integer.

a. $f(x) = 2x^3 - 3x^2 - 12x + 40$

b. $f(x) = 2x^3 - 33x^2 + 60x + 1050$

c. $f(x) = -2x^3 - 9x^2 + 24x + 100$

d. $f(x) = x^4 - 4x^3 - 2x^2 + 12x + 3$

e. $f(x) = x^3 - 30x^2 + 288x - 900$

f. $f(x) = x^5 - 2x^4 - 3x^3 - 2x^2 + x - 1$

53. For each of the following functions, find the x-intercepts and find the turning points. Express your answers to the nearest tenth.

a. $f(x) = x^3 + 2x^2 - 3x + 4$

b. $f(x) = 42x^3 - x^2 - 246x - 35$

c. $f(x) = x^4 - 4x^2 - 4$

54. A rectangular piece of cardboard is 13 inches long and 9 inches wide. From each corner a square piece is cut out and then the flaps are turned up to form an open box. Determine the length of a side of the square pieces so that the volume of the box is as large as possible.

55. A company determines that its weekly profit from manufacturing and selling x units of a certain item is given by $P(x) = -x^3 + 3x^2 + 2880x - 500$. What weekly production rate will maximize the profit?

6.5 GRAPHING RATIONAL FUNCTIONS

A function of the form

$$f(x) = \frac{p(x)}{q(x)}, \qquad q(x) \neq 0,$$

where $p(x)$ and $q(x)$ are both polynomial functions, is called a **rational function** The following are examples of rational functions.

$$f(x) = \frac{2}{x - 1} \qquad f(x) = \frac{x}{x - 2}$$

$$f(x) = \frac{x^2}{x^2 - x - 6} \qquad f(x) = \frac{x^3 - 8}{x + 4}$$

In each example, the domain of the rational function is the set of all real numbers except those that make the denominator zero. For example, the domain of $f(x) = \dfrac{2}{x - 1}$ is the set of all real numbers except 1. As you will see, these exclusions from the domain are important numbers from a graphing standpoint. They represent breaks in an otherwise continuous curve.

Let's set the stage for graphing rational functions by considering in detail the function $f(x) = 1/x$. First, note that at $x = 0$, the function is undefined. Second, let's consider an extensive table of values to show some number trends and to build a basis for defining the concept of an *asymptote*.

x	$f(x) = \frac{1}{x}$	
1	1	These values indicate that the value of $f(x)$ is positive and approaches zero from above as x gets larger and larger.
2	0.5	
10	0.1	
100	0.01	
1000	0.001	
0.5	2	These values indicate that $f(x)$ is positive and is getting larger and larger as x approaches zero from the right.
0.1	10	
0.01	100	
0.001	1000	
0.0001	10000	
−0.5	−2	These values indicate that $f(x)$ is negative and is getting smaller and smaller as x approaches zero from the left.
−0.1	−10	
−0.01	−100	
−0.001	−1000	
−0.0001	−10000	
−1	−1	These values indicate that $f(x)$ is negative and approaches zero from below as x gets smaller and smaller.
−2	−0.5	
−10	−0.1	
−100	−0.01	
−1000	−0.001	

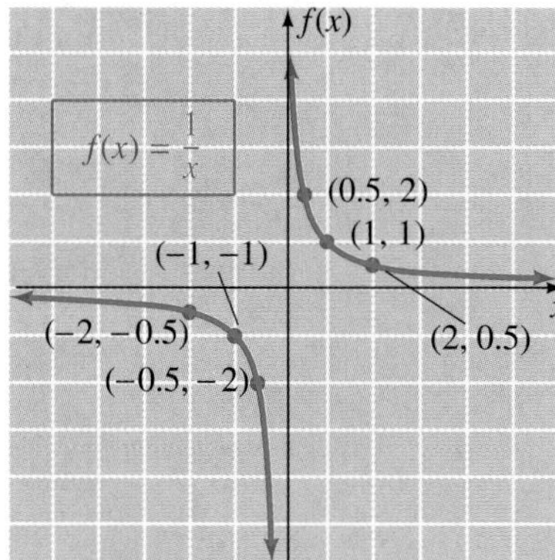

FIGURE 6.21

Using a few points from this table and the patterns we discussed, we have sketched $f(x) = 1/x$ in Figure 6.21. Notice that the graph approaches, but does not touch, either axis. We say that the y-axis (or $f(x)$-axis) is a *vertical asymptote* and the x-axis is a *horizontal asymptote*. In general, the following definitions can be given.

REMARK Observe that the equation $f(x) = 1/x$ exhibits origin symmetry because $f(-x) = -f(x)$. Thus, the graph in Figure 6.21 could have been drawn by first determining the part of the curve in the first quadrant and then reflecting that through the origin.

Vertical Asymptote A line $x = a$ is a **vertical asymptote** for the graph of a function f if it satisfies either of the following two properties.

1. $f(x)$ either increases or decreases without bound as x approaches the number a from the right as in Figure 6.22, or

2. $f(x)$ either increases or decreases without bound as x approaches the number a from the left as in Figure 6.23.

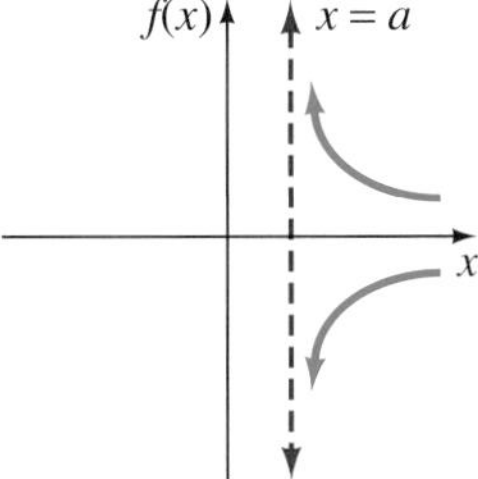

FIGURE 6.22

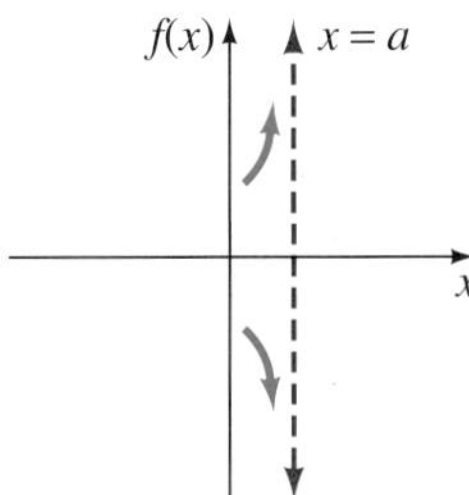

FIGURE 6.23

Horizontal Asymptote A line $y = b$ (or $f(x) = b$) is a **horizontal asymptote** for the graph of a function f if it satisfies either of the following two properties.

1. $f(x)$ approaches the number b from above or below as x gets infinitely small as in Figure 6.24, *or*

2. $f(x)$ approaches the number b from above or below as x gets infinitely large as in Figure 6.25.

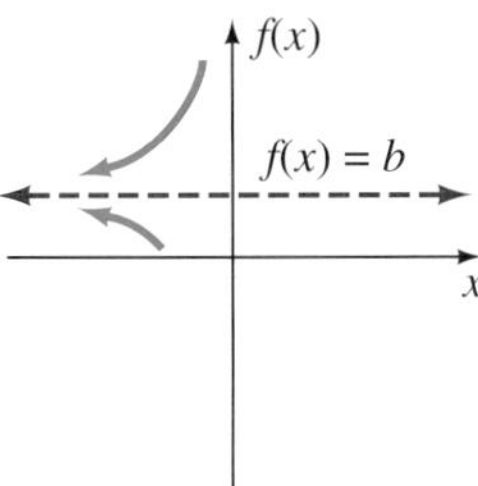

FIGURE 6.24

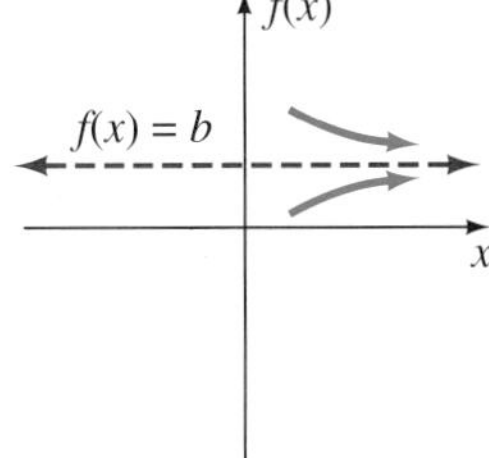

FIGURE 6.25

The following suggestions will help you graph rational functions of the type we are considering in this section.

1. Check for y-axis and origin symmetry.
2. Find any vertical asymptote(s) by setting the denominator equal to zero and solving it for x.
3. Find any horizontal asymptote(s) by studying the behavior of $f(x)$ as x gets infinitely large or as x gets infinitely small.
4. Study the behavior of the graph when it is close to the asymptotes.
5. Plot as many points as necessary to determine the shape of the graph. This may be affected by whether the graph has any symmetry.

Keep these suggestions in mind as you study the following examples.

EXAMPLE 1

Graph $f(x) = \dfrac{-2}{x-1}$.

Solution

Since $x = 1$ makes the denominator zero, the line $x = 1$ is a vertical asymptote; we have indicated this with a dashed line in Figure 6.26. Now let's look for a horizontal asymptote by checking some large and small values of x in the tables that accompany Figure 6.26.

x	$f(x)$	
10	$-\frac{2}{9}$	This portion of the table shows that as x gets very large, the value of $f(x)$ approaches 0 from below.
100	$-\frac{2}{99}$	
1000	$-\frac{2}{999}$	
−10	$\frac{2}{11}$	This portion shows that as x gets very small, the value of $f(x)$ approaches 0 from above.
−100	$\frac{2}{101}$	
−1000	$\frac{2}{1001}$	

Therefore, the x-axis is a horizontal asymptote. Finally, let's check the behavior of the graph near the vertical asymptote.

x	$f(x)$	
2	−2	As x approaches 1 from the right side, the value of $f(x)$ gets smaller and smaller.
1.5	−4	
1.1	−20	
1.01	−200	
1.001	−2000	
0	2	As x approaches 1 from the left side, the value of $f(x)$ gets larger and larger.
0.5	4	
0.9	20	
0.99	200	
0.999	2000	

The graph of $f(x) = \dfrac{-2}{x-1}$ is shown in Figure 6.26.

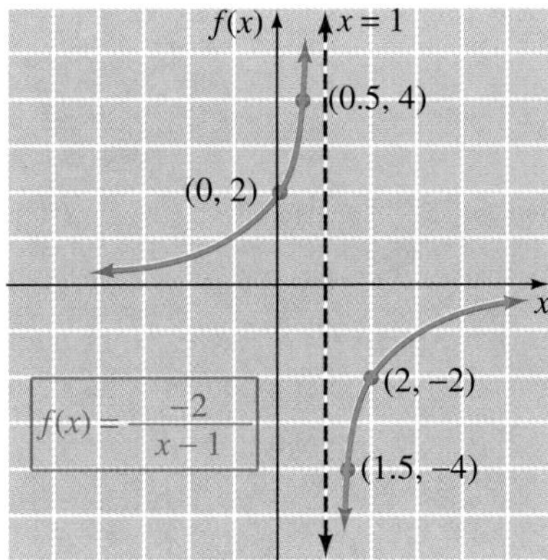

FIGURE 6.26

EXAMPLE 2

Graph $f(x) = \dfrac{x}{x+2}$.

Solution

Since $x = -2$ makes the denominator zero, the line $x = -2$ is a vertical asymptote. To study the behavior of $f(x)$ as x gets very large or very small, let's change the form of the rational expression by dividing both the numerator and the denominator by x.

$$f(x) = \frac{x}{x+2} = \frac{\dfrac{x}{x}}{\dfrac{x+2}{x}} = \frac{1}{\dfrac{x}{x} + \dfrac{2}{x}} = \frac{1}{1 + \dfrac{2}{x}}$$

Now we can see that (1) as x gets larger and larger, the value of $f(x)$ approaches 1 from below, and (2) as x gets smaller and smaller, the value of $f(x)$ approaches 1 from above. (Perhaps you should check these claims by plugging in some values for x.) Thus, the line $f(x) = 1$ is a horizontal asymptote. Drawing the asymptotes (dashed lines) and plotting a few points allows us to complete the graph in Figure 6.27.

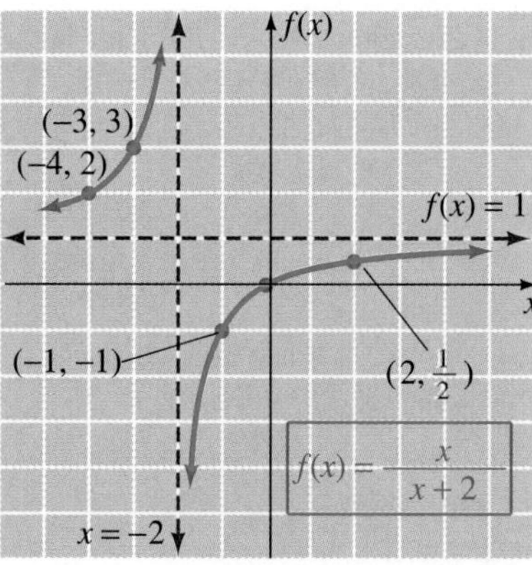

FIGURE 6.27

In the next two examples, pay special attention to the role of symmetry. It will allow us to focus on the portion of a curve in quadrants I and IV and then to reflect that portion of the curve across the vertical axis to complete the graph.

EXAMPLE 3 Graph $f(x) = \frac{2x^2}{x^2 + 4}$.

Solution

First, notice that $f(-x) = f(x)$; therefore, this graph is symmetrical with respect to the y-axis. Second, the denominator $x^2 + 4$ cannot equal zero for any real number x. Thus, there is no vertical asymptote. Third, dividing both numerator and denominator of the rational expression by x^2 produces

$$f(x) = \frac{2x^2}{x^2 + 4} = \frac{\frac{2x^2}{x^2}}{\frac{x^2 + 4}{x^2}} = \frac{2}{\frac{x^2}{x^2} + \frac{4}{x^2}} = \frac{2}{1 + \frac{4}{x^2}}.$$

Now we can see that as x gets larger and larger, the value of $f(x)$ approaches 2 from below. Therefore, the line $f(x) = 2$ is a horizontal asymptote. So we can plot a few points using positive values for x, sketch this part of the curve, and then reflect across the $f(x)$-axis to obtain the complete graph in Figure 6.28.

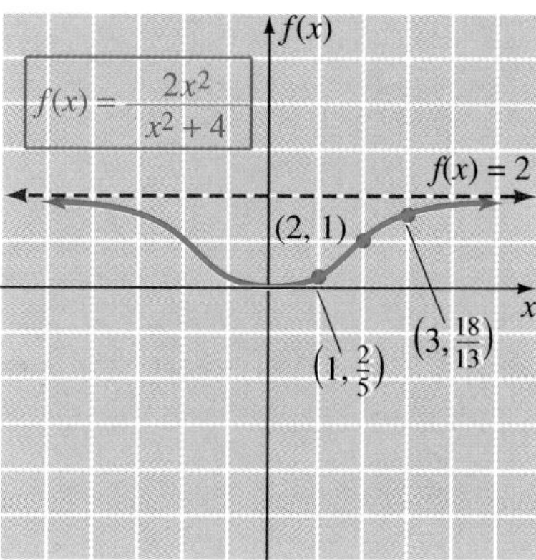

FIGURE 6.28

EXAMPLE 4

Graph $f(x) = \dfrac{3}{x^2 - 4}$.

Solution

First, notice that $f(-x) = f(x)$; therefore, this graph is symmetrical about the $f(x)$-axis. Second, by setting the denominator equal to zero and solving for x, we obtain

$$\begin{aligned} x^2 - 4 &= 0 \\ x^2 &= 4 \\ x &= \pm 2. \end{aligned}$$

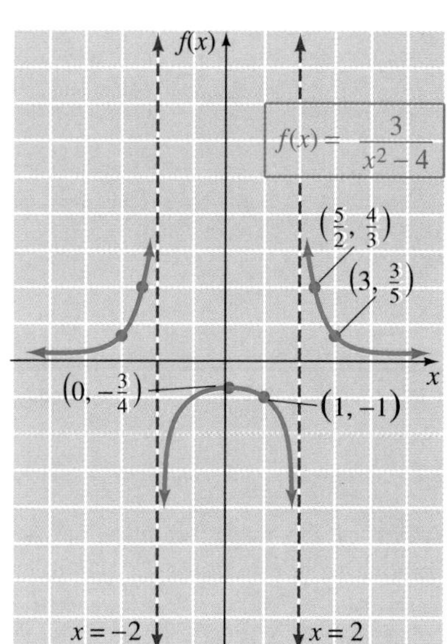

FIGURE 6.29

The lines $x = 2$ and $x = -2$ are vertical asymptotes. Next, we can see that $\dfrac{3}{x^2 - 4}$ approaches zero from above as x gets larger and larger. Finally, we can plot a few points using positive values for x (not 2), sketch this part of the curve, and then reflect it across the $f(x)$-axis to obtain the complete graph in Figure 6.29.

Now suppose that we are going to use a graphing utility to obtain a graph of the function $f(x) = \dfrac{4x^2}{x^4 - 16}$. Before we enter this function into a graphing utility, let's analyze what we know about the graph.

1. Since $f(0) = 0$, the origin is a point on the graph.
2. Since $f(-x) = f(x)$, the graph is symmetric with respect to the y-axis.
3. By setting the denominator equal to zero and solving for x, we can determine the vertical asymptotes.

$$x^4 - 16 = 0$$
$$(x^2 + 4)(x^2 - 4) = 0$$
$$x^2 + 4 = 0 \quad \text{or} \quad x^2 - 4 = 0$$
$$x^2 = -4 \quad \text{or} \quad x^2 = 4$$
$$x = \pm 2i \quad \text{or} \quad x = \pm 2$$

Remember that we are working with ordered pairs of real numbers. Thus, the lines $x = -2$ and $x = 2$ are vertical asymptotes.

4. Divide both the numerator and the denominator of the rational expression by x^4 to produce

$$\frac{4x^2}{x^4 - 16} = \frac{\dfrac{4x^2}{x^4}}{\dfrac{x^4 - 16}{x^4}} = \frac{\dfrac{4}{x^2}}{1 - \dfrac{16}{x^4}}.$$

From the last expression, we see that as $|x|$ gets larger and larger, the value of $f(x)$ approaches zero from above. Therefore, the x-axis is a horizontal asymptote.

Let's enter the function in a graphing utility and obtain a graph as shown in Figure 6.30. Note that the graph is consistent with all of the previous information that we determined before we used the graphing utility. In other words, our knowledge of graphing techniques enhances our use of a graphing utility.

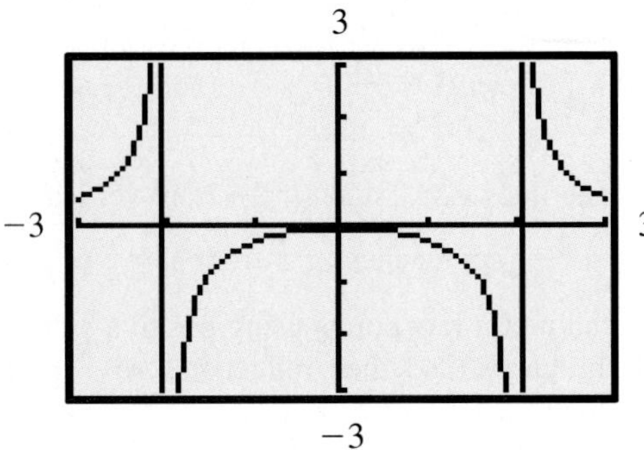

FIGURE 6.30

REMARK In Figure 6.30, the origin is a point of the graph, which is on the horizontal asymptote. More will be said about such situations in the next section.

Back in Problem Set 2.4 you were asked to solve the following problem: How much pure alcohol should be added to 6 liters of a 40% alcohol solution to raise it to a 60% alcohol solution? The answer of 3 liters can be found by solving the following equation, where x represents the amount of pure alcohol to be added.

Pure alcohol to start with	+	Pure alcohol added	=	Pure alcohol in final solution
↓		↓		↓
$0.40(6)$	$+$	x	$=$	$0.60(6 + x)$

Now let's consider this problem in a more general setting. Again, x represents the amount of pure alcohol to be added and the rational expression $\frac{2.4 + x}{6 + x}$ represents the concentration of pure alcohol in the final solution. Let's graph the rational function $y = \frac{2.4 + x}{6 + x}$ as shown in Figure 6.31. For this particular problem x is nonnegative, so we are only interested in the part of the graph that is in the first quadrant. Change the boundaries of the viewing rectangle so that $0 \leq x \leq 15$ and $0 \leq y \leq 2$ to obtain Figure 6.32. Now we are ready to answer questions about this situation.

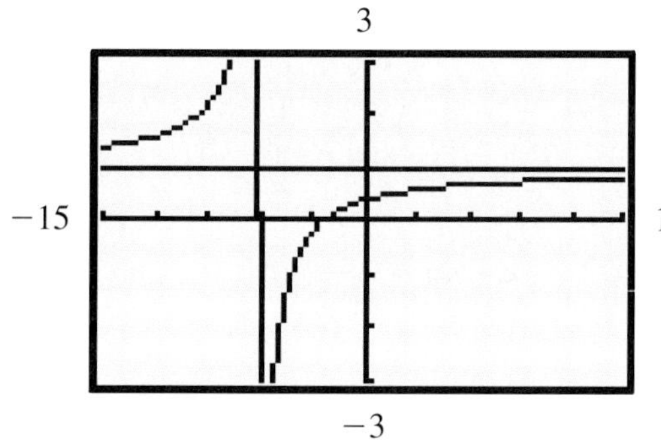

FIGURE 6.31

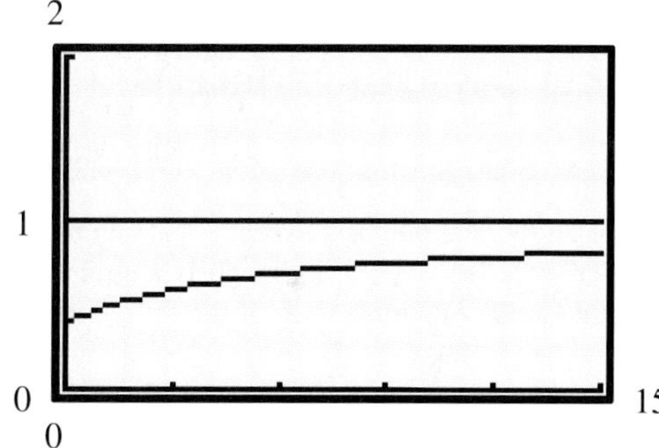

FIGURE 6.32

1. How much pure alcohol needs to be added to raise the 40% solution to a 60% alcohol solution? (*Answer:* Using the trace feature of the graphing utility we find that $y = 0.6$ when $x = 3$. Therefore, 3 liters of pure alcohol need to be added.)
2. How much pure alcohol needs to be added to raise the 40% solution to a 70% alcohol solution? (*Answer:* Using the trace feature we find that $y = 0.7$ when $x = 6$. Therefore, 6 liters of pure alcohol need to be added.)
3. What percent of alcohol do we have if we add 9 liters of pure alcohol to the 6 liters of a 40% solution? (*Answer:* Using the trace feature we find that $y = 0.76$ when $x = 9$. Therefore, adding 9 liters of pure alcohol will give us a 76% alcohol solution.)

PROBLEM SET 6.5

Graph each of the following rational functions.

1. $f(x) = \dfrac{-1}{x}$

2. $f(x) = \dfrac{1}{x^2}$

3. $f(x) = \dfrac{3}{x+1}$

4. $f(x) = \dfrac{-1}{x-3}$

5. $f(x) = \dfrac{2}{(x-1)^2}$

6. $f(x) = \dfrac{-3}{(x+2)^2}$

7. $f(x) = \dfrac{x}{x-3}$

8. $f(x) = \dfrac{2x}{x-1}$

9. $f(x) = \dfrac{-3x}{x+2}$

10. $f(x) = \dfrac{-x}{x+1}$

11. $f(x) = \dfrac{1}{x^2-1}$

12. $f(x) = \dfrac{-2}{x^2-4}$

13. $f(x) = \dfrac{-2}{(x+1)(x-2)}$

14. $f(x) = \dfrac{3}{(x+2)(x-4)}$

15. $f(x) = \dfrac{2}{x^2+x-2}$

16. $f(x) = \dfrac{-1}{x^2+x-6}$

17. $f(x) = \dfrac{x+2}{x}$

18. $f(x) = \dfrac{2x-1}{x}$

19. $f(x) = \dfrac{4}{x^2+2}$

20. $f(x) = \dfrac{4x^2}{x^2+1}$

21. $f(x) = \dfrac{2x^4}{x^4+1}$

22. $f(x) = \dfrac{x^2-4}{x^2}$

MISCELLANEOUS PROBLEMS

23. The rational function $f(x) = \dfrac{(x-2)(x+3)}{x-2}$ has a domain of all the real numbers except 2 and can be simplified to $f(x) = x + 3$. Thus, its graph is a straight line with a hole at (2, 5). Graph each of the following functions.

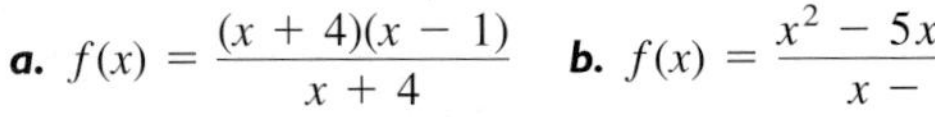

a. $f(x) = \dfrac{(x+4)(x-1)}{x+4}$ **b.** $f(x) = \dfrac{x^2-5x+6}{x-2}$

c. $f(x) = \dfrac{x-1}{x^2-1}$ **d.** $f(x) = \dfrac{x+2}{x^2+6x+8}$

GRAPHICS CALCULATOR ACTIVITIES

24. Use a graphics calculator to check your graphs for Problem 23. What feature of the graph does not show up on the calculator?

25. Each of the following graphs is a transformation of $f(x) = 1/x$. First, predict the general shape and location of the graph and then check your prediction with a graphics calculator.

a. $f(x) = \dfrac{1}{x} - 2$ **b.** $f(x) = \dfrac{1}{x+3}$

c. $f(x) = -\dfrac{1}{x}$ **d.** $f(x) = \dfrac{1}{x-2} + 3$

e. $f(x) = \dfrac{2x+1}{x}$

26. Graph $f(x) = \dfrac{1}{x^2}$. How should the graphs of $f(x) = \dfrac{1}{(x-4)^2}$, $f(x) = \dfrac{1+3x^2}{x^2}$, and $f(x) = -\dfrac{1}{x^2}$ compare to the graph of $f(x) = \dfrac{1}{x^2}$? Graph the three functions on the same set of axes with the graph of $f(x) = \dfrac{1}{x^2}$.

27. Graph $f(x) = \dfrac{1}{x^3}$. How should the graphs of $f(x) = \dfrac{2x^3+1}{x^3}$, $f(x) = \dfrac{1}{(x+2)^3}$, and $f(x) = -\dfrac{1}{x^3}$ compare to the graph of $f(x) = \dfrac{1}{x^3}$? Graph the three functions on the same set of axes with the graph of $f(x) = \dfrac{1}{x^3}$.

28. Use a graphics calculator to check your graphs for Problems 19–22.

29. Graph each of the following functions. Be sure that you get a complete graph for each one. Sketch each graph on a sheet of paper and keep them handy as you study the next section.

a. $f(x) = \dfrac{x^2}{x^2 - x - 2}$ **b.** $f(x) = \dfrac{x}{x^2 - 4}$

c. $f(x) = \dfrac{3x}{x^2 + 1}$ **d.** $f(x) = \dfrac{x^2 - 1}{x - 2}$

30. Suppose that x ounces of pure acid have been added to 14 ounces of a 15% acid solution.

a. Set up the rational expression that represents the concentration of pure acid in the final solution.

b. Graph the rational function that displays the level of concentration.

c. How many ounces of pure acid need to be added to the 14 ounces of a 15% solution to raise it to a 40.5% solution? Check your answer.

d. How many ounces of pure acid need to be added to the 14 ounces of a 15% solution to raise it to a 50% solution? Check your answer.

e. What percent of acid do we obtain if we add 12 ounces of pure acid to the 14 ounces of a 15% solution? Check your answer.

31. Solve the following problem both algebraically and graphically: One solution contains 50% alcohol and another solution contains 80% alcohol. How many liters of each solution should be mixed to produce 10.5 liters of a 70% alcohol solution? Check your answer.

6.6 More on Graphing Rational Functions

The rational functions that we studied in the previous section "behaved rather well." In fact, once we established the vertical and horizontal asymptotes, a little bit of point plotting usually determined the graph rather easily. Such is not always the case with rational functions. In this section we want to investigate some rational functions that behave a little differently.

Since vertical asymptotes occur at values of x where the denominator is zero, there can be no points of a graph on a vertical asymptote. However, recall that horizontal asymptotes are created by the behavior of $f(x)$ as x gets infinitely large or infinitely small. This does not restrict the possibility that for some values of x, there will be points of the graph on the horizontal asymptote. Let's consider some examples.

EXAMPLE 1

Graph $f(x) = \dfrac{x^2}{x^2 - x - 2}$.

Solution

First, let's identify the vertical asymptotes by setting the denominator equal to zero and solving for x.

$$x^2 - x - 2 = 0$$
$$(x - 2)(x + 1) = 0$$
$$x - 2 = 0 \quad \text{or} \quad x + 1 = 0$$
$$x = 2 \quad \text{or} \quad x = -1$$

Thus, the lines $x = 2$ and $x = -1$ are vertical asymptotes. Next, we can divide both the numerator and the denominator of the rational expression by x^2.

$$f(x) = \frac{x^2}{x^2 - x - 2} = \frac{\frac{x^2}{x^2}}{\frac{x^2 - x - 2}{x^2}} = \frac{1}{1 - \frac{1}{x} - \frac{2}{x^2}}$$

Now we can see that as x gets larger and larger, the value of $f(x)$ approaches 1 from above. Thus, the line $f(x) = 1$ is a horizontal asymptote. To determine if any points of the graph are *on* the horizontal asymptote, we can see if the equation

$$\frac{x^2}{x^2 - x - 2} = 1$$

has any solutions.

$$\frac{x^2}{x^2 - x - 2} = 1$$
$$x^2 = x^2 - x - 2$$
$$0 = -x - 2$$
$$x = -2$$

Therefore, the point $(-2, 1)$ is on the graph. Now by drawing the asymptotes, plotting a few points including $(-2, 1)$, and studying the behavior of the function close to the asymptotes, we can sketch the curve in Figure 6.33.

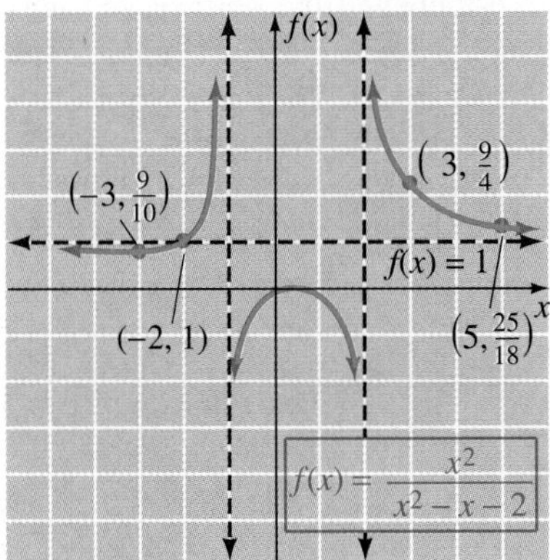

FIGURE 6.33

EXAMPLE 2

Graph $f(x) = \dfrac{x}{x^2 - 4}$.

Solution

First, notice that $f(-x) = -f(x)$; therefore, this graph has origin symmetry. Second, let's identify the vertical asymptotes.

$$x^2 - 4 = 0$$
$$x^2 = 4$$
$$x = \pm 2$$

Thus the lines $x = -2$ and $x = 2$ are vertical asymptotes. Next, by dividing the numerator and the denominator of the rational expression by x^2, we obtain

$$f(x) = \frac{x}{x^2 - 4} = \frac{\frac{x}{x^2}}{\frac{x^2 - 4}{x^2}} = \frac{\frac{1}{x}}{1 - \frac{4}{x^2}}.$$

From this form we can see that as x gets larger and larger, the value of $f(x)$ approaches zero from above. Therefore, the x-axis is a horizontal asymptote. Since $f(0) = 0$, we know that the origin is a point of the graph. Finally, by concentrating our point plotting on positive values of x, we can sketch the portion of the curve to the right of the vertical axis and then use the fact that the graph is symmetric with respect to the origin to complete the graph. Figure 6.34 shows the completed graph.

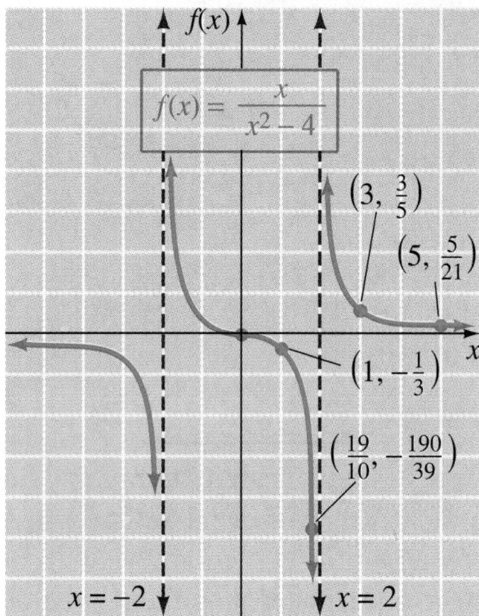

FIGURE 6.34

EXAMPLE 3

Graph $f(x) = \dfrac{3x}{x^2 + 1}$.

Solution

First, observe that $f(-x) = -f(x)$; therefore, this graph is symmetrical with respect to the origin. Second, since $x^2 + 1$ is a positive number for all real number values

of x, there are no vertical asymptotes for this graph. Next, by dividing the numerator and denominator of the rational expression by x^2, we obtain

$$f(x) = \frac{3x}{x^2 + 1} = \frac{\frac{3x}{x^2}}{\frac{x^2 + 1}{x^2}} = \frac{\frac{3}{x}}{1 + \frac{1}{x^2}}.$$

From this form we see that as x gets larger and larger, the value of $f(x)$ approaches zero from above. Thus, the x-axis is a horizontal asymptote. Since $f(0) = 0$, the origin is a point of the graph. Finally, by concentrating our point plotting on positive values of x, we can sketch the portion of the curve to the right of the vertical axis and then use origin symmetry to complete the graph, as shown in Figure 6.35.

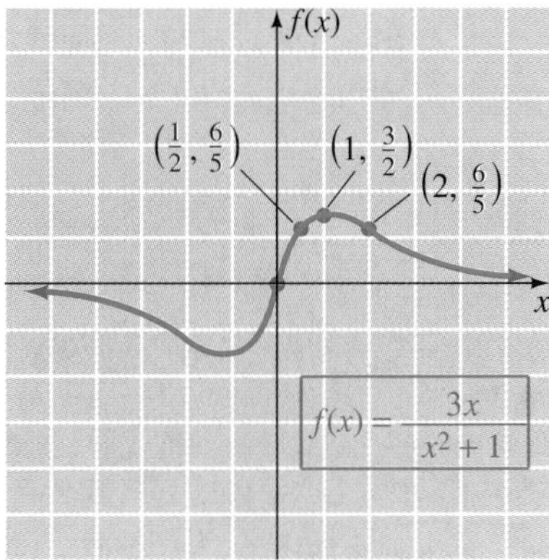

FIGURE 6.35

Oblique Asymptotes

Thus far we have restricted our study of rational functions to those where the degree of the numerator is less than or equal to the degree of the denominator. As our final examples of graphing rational functions, we will consider functions where the degree of the numerator is one greater than the degree of the denominator.

EXAMPLE 4

Graph $f(x) = \frac{x^2 - 1}{x - 2}$.

Solution

First, let's observe that $x = 2$ is a vertical asymptote. Second, since the degree of the numerator is greater than the degree of the denominator, we can change the form of the rational expression by division. We use synthetic division.

$$\begin{array}{r|rrr} 2 & 1 & 0 & -1 \\ & & 2 & 4 \\ \hline & 1 & 2 & 3 \end{array}$$

Therefore, the original function can be rewritten

$$f(x) = \frac{x^2 - 1}{x - 2} = x + 2 + \frac{3}{x - 2}.$$

Now, for very large values of $|x|$, the fraction $\frac{3}{x - 2}$ is close to zero. Therefore, as $|x|$ gets larger and larger, the graph of $f(x) = x + 2 + \frac{3}{x - 2}$ gets closer and closer to the line $f(x) = x + 2$. We call this line an **oblique asymptote** and indicate it with a dashed line in Figure 6.36. Finally, since this is a new situation, it may be necessary to plot a large number of points on both sides of the vertical asymptote, so let's make an extensive table of values. The graph of the function is shown in Figure 6.36.

x	$f(x) = \frac{x^2 - 1}{x - 2}$	
2.1	34.1	These values indicate the behavior of $f(x)$ to the right of the vertical asymptote $x = 2$.
2.5	10.5	
3	8	
4	7.5	
5	8	
6	8.75	
10	12.375	
1.9	−26.1	These values indicate the behavior of $f(x)$ to the left of the vertical asymptote $x = 2$.
1.5	−2.5	
1	0	
0	0.5	
−1	0	
−3	−1.6	
−5	−3.4	
−10	−8.25	

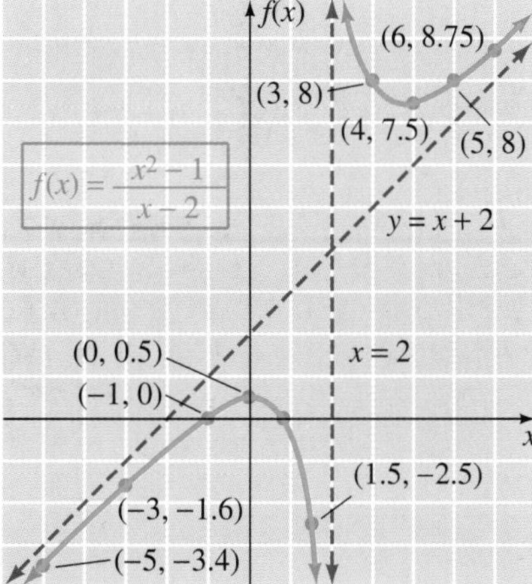

FIGURE 6.36

If the degree of the numerator of a rational function is *exactly one more* than the degree of its denominator, then the graph of the function has an **oblique asymptote**. (If the graph is a line, as is the case with $f(x) = \dfrac{(x-2)(x+1)}{x-2}$, then we consider it to be its own asymptote.) As in Example 4, we find the equation of the oblique asymptote by changing the form of the function using long division. Let's consider another example.

EXAMPLE 5

Graph $f(x) = \dfrac{x^2 - x - 2}{x - 1}$.

Solution

From the given form of the function, we see that $x = 1$ is a vertical asymptote. Then, by factoring the numerator, we can change the form to

$$f(x) = \frac{(x-2)(x+1)}{(x-1)},$$

which indicates x-intercepts of 2 and -1. Then, by long division, we can change the original form of the function to

$$f(x) = x - \frac{2}{x-1},$$

which indicates an oblique asymptote $f(x) = x$. Finally, by plotting a few additional points, we can determine the graph as shown in Figure 6.37.

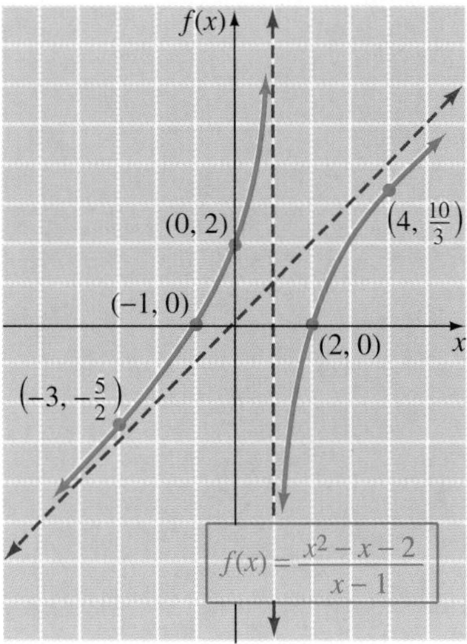

FIGURE 6.37

Finally, let's combine our knowledge of rational functions with the use of a graphing utility to obtain the graph of a fairly complex rational function.

EXAMPLE 6

Graph the rational function $f(x) = \dfrac{x^3 - 2x^2 - x - 1}{x^2 - 36}$.

Solution

Before entering this function into a graphing utility, let's analyze what we know about the graph.

1. Since $f(0) = \frac{1}{36}$, the point $\left(0, \frac{1}{36}\right)$ is on the graph.
2. Since $f(-x) \neq f(x)$ and $f(-x) \neq -f(x)$, there is no symmetry with respect to the origin nor the y-axis.
3. The denominator is zero at $x = \pm 6$. Thus, the lines $x = 6$ and $x = -6$ are vertical asymptotes.
4. Let's change the form of the rational expression by division.

$$\begin{array}{r|l} & x \;-\; 2 \\ \hline x^2 - 36 & x^3 - 2x^2 - x - 1 \\ & x^3 - 36x \\ \hline & -2x^2 + 35x - 1 \\ & -2x^2 + 72 \\ \hline & 35x - 73 \end{array}$$

Thus, the original function can be rewritten as

$$f(x) = x - 2 + \frac{35x - 73}{x^2 - 36}.$$

Therefore, the line $y = x - 2$ is an oblique asymptote. Now let $Y_1 = x - 2$ and $Y_2 = \dfrac{x^3 - 2x^2 - x - 1}{x^2 - 36}$ and use a viewing rectangle where $-15 \leq x \leq 15$ and $-30 \leq y \leq 30$ (Figure 6.38).

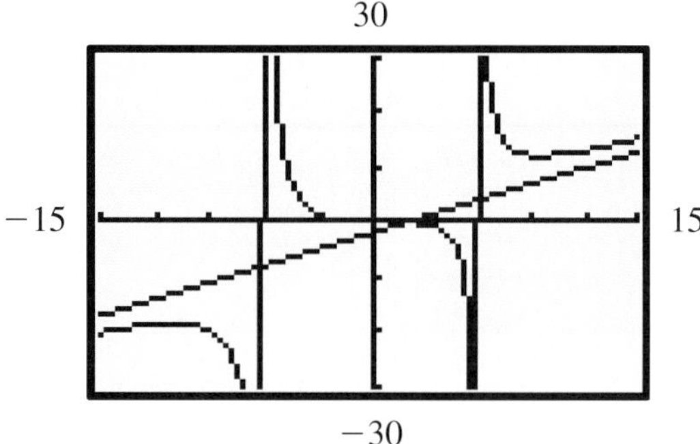

FIGURE 6.38

Note that the graph in Figure 6.38 is consistent with the information that we listed prior to using the graphing utility. Furthermore, notice that the curve does intersect the oblique asymptote. We can use the zoom and trace features of the graphing utility to find that point of intersection or we can do it algebraically as follows: Since $y = \dfrac{x^3 - 2x^2 - x - 1}{x^2 - 36}$ and $y = x - 2$, we can equate the two expressions for y and solve the resulting equation for x.

$$\frac{x^3 - 2x^2 - x - 1}{x^2 - 36} = x - 2$$
$$x^3 - 2x^2 - x - 1 = (x - 2)(x^2 - 36)$$
$$x^3 - 2x^2 - x - 1 = x^3 - 2x^2 - 36x + 72$$
$$35x = 73$$
$$x = \frac{73}{35}$$

If $x = \dfrac{73}{35}$, then $y = x - 2 = \dfrac{73}{35} - 2 = \dfrac{3}{35}$. The point of intersection of the curve and the oblique asymptote is $\left(\dfrac{73}{35}, \dfrac{3}{35}\right)$.

PROBLEM SET 6.6

Graph each of the following rational functions. Check first for symmetry and identify the asymptotes.

1. $f(x) = \dfrac{x^2}{x^2 + x - 2}$
2. $f(x) = \dfrac{x^2}{x^2 + 2x - 3}$
3. $f(x) = \dfrac{2x^2}{x^2 - 2x - 8}$
4. $f(x) = \dfrac{-x^2}{x^2 + 3x - 4}$
5. $f(x) = \dfrac{-x}{x^2 - 1}$
6. $f(x) = \dfrac{2x}{x^2 - 9}$
7. $f(x) = \dfrac{x}{x^2 + x - 6}$
8. $f(x) = \dfrac{-x}{x^2 - 2x - 8}$
9. $f(x) = \dfrac{x^2}{x^2 - 4x + 3}$
10. $f(x) = \dfrac{1}{x^3 + x^2 - 6x}$
11. $f(x) = \dfrac{x}{x^2 + 2}$
12. $f(x) = \dfrac{6x}{x^2 + 1}$
13. $f(x) = \dfrac{-4x}{x^2 + 1}$
14. $f(x) = \dfrac{-5x}{x^2 + 2}$
15. $f(x) = \dfrac{x^2 + 2}{x - 1}$
16. $f(x) = \dfrac{x^2 - 3}{x + 1}$
17. $f(x) = \dfrac{x^2 - x - 6}{x + 1}$
18. $f(x) = \dfrac{x^2 + 4}{x + 2}$
19. $f(x) = \dfrac{x^2 + 1}{1 - x}$
20. $f(x) = \dfrac{x^3 + 8}{x^2}$

THOUGHTS into WORDS

21. Explain the concept of an asymptote.
22. How would you go about graphing $f(x) = \dfrac{x^2 - x - 12}{x - 4}$?
23. How would you go about graphing $f(x) = \dfrac{x^2 - x - 12}{x - 2}$?

GRAPHICS CALCULATOR ACTIVITIES

24. Use your graphics calculator to graph each of the following rational functions. Be sure first to check for symmetry and identify the horizontal and vertical asymptotes.

a. $f(x) = \dfrac{4x^2}{x^2 + x - 2}$ **b.** $f(x) = \dfrac{-2x}{x^2 - 5x - 6}$

c. $f(x) = \dfrac{x^2}{x^2 - 9}$ **d.** $f(x) = \dfrac{x^2 - 4}{x^2 - 9}$

e. $f(x) = \dfrac{x^2 - 9}{x^2 - 4}$ **f.** $f(x) = \dfrac{x^2 + 2x + 1}{x^2 - 5x + 6}$

25. For each of the following, first determine and graph any oblique asymptote; then on the same set of axes graph the function.

a. $f(x) = \dfrac{x^2 - 1}{x - 2}$ **b.** $f(x) = \dfrac{x^2 + 1}{x + 2}$

c. $f(x) = \dfrac{2x^2 + x + 1}{x + 1}$

d. $f(x) = \dfrac{x^2 + 4}{x - 3}$

e. $f(x) = \dfrac{3x^2 - x - 2}{x - 2}$

f. $f(x) = \dfrac{4x^2 + x + 1}{x + 1}$

g. $f(x) = \dfrac{x^3 + x^2 - x - 1}{x^2 + 2x + 3}$

h. $f(x) = \dfrac{x^3 + 2x^2 + x - 3}{x^2 - 4}$

6.7 PARTIAL FRACTIONS

In Chapter 1 we reviewed the process of adding rational expressions. For example,

$$\frac{3}{x-2} + \frac{2}{x+3} = \frac{3(x+3) + 2(x-2)}{(x-2)(x+3)} = \frac{3x + 9 + 2x - 4}{(x-2)(x+3)} = \frac{5x+5}{(x-2)(x+3)}.$$

Now suppose that we want to reverse the process. That is, suppose we are given the rational expression

$$\frac{5x+5}{(x-2)(x+3)}$$

and we want to express it as the sum of two simpler rational expressions, called **partial fractions**. This process, called **partial fraction decomposition**, has several applications in calculus and differential equations. The following property provides the basis for partial fraction decomposition.

PROPERTY 6.7

Let $f(x)$ and $g(x)$ be polynomials with real coefficients, such that the degree of $f(x)$ is less than the degree of $g(x)$. The indicated quotient $f(x)/g(x)$ can be decomposed into partial fractions as follows.

1. If $g(x)$ has a linear factor of the form $ax + b$, then the partial fraction decomposition will contain a term of the form

$$\frac{A}{ax + b}, \quad \text{where } A \text{ is a constant.}$$

2. If $g(x)$ has a linear factor of the form $ax + b$ raised to the kth power, then the partial fraction decomposition will contain terms of the form

$$\frac{A_1}{ax + b} + \frac{A_2}{(ax + b)^2} + \cdots + \frac{A_k}{(ax + b)^k},$$

where $A_1, A_2, \ldots, A_k$ are constants.

3. If $g(x)$ has a quadratic factor of the form $ax^2 + bx + c$, where $b^2 - 4ac < 0$, then the partial fraction decomposition will contain a term of the form

$$\frac{Ax + B}{ax^2 + bx + c}, \quad \text{where } A \text{ and } B \text{ are constants.}$$

4. If $g(x)$ has a quadratic factor of the form $ax^2 + bx + c$ raised to the kth power, where $b^2 - 4ac < 0$, then the partial fraction decomposition will contain terms of the form

$$\frac{A_1x + B_1}{ax^2 + bx + c} + \frac{A_2x + B_2}{(ax^2 + bx + c)^2} + \cdots + \frac{A_kx + B_kx}{(ax^2 + bx + c)^k},$$

where $A_1, A_2, \ldots, A_k$, and $B_1, B_2, \ldots, B_k$ are constants.

Notice that Property 6.7 applies only to **proper fractions**, that is, fractions where the degree of the numerator is less than the degree of the denominator. If the numerator is not of lower degree, we can divide and then apply Property 6.7 to the remainder, which will be a proper fraction. For example,

$$\frac{x^3 - 3x^2 - 3x - 5}{x^2 - 4} = x - 3 + \frac{x - 17}{x^2 - 4}$$

and the proper fraction $\frac{x - 17}{x^2 - 4}$ can be decomposed into partial fractions by applying Property 6.7. Now let's consider some examples to illustrate the four cases in Property 6.7.

EXAMPLE 1

Find the partial fraction decomposition of $\frac{11x + 2}{2x^2 + x - 1}$.

Solution

The denominator can be expressed as $(x + 1)(2x - 1)$. Therefore, according to part 1 of Property 6.7, each of the linear factors produces a partial fraction of the form *constant over linear factor*. In other words, we can write

$$\frac{11x + 2}{(x + 1)(2x - 1)} = \frac{A}{x + 1} + \frac{B}{2x - 1} \qquad \textbf{(1)}$$

for some constants A and B. To find A and B, we multiply both sides of equation (1) by the least common denominator $(x + 1)(2x - 1)$:

$$11x + 2 = A(2x - 1) + B(x + 1). \qquad \textbf{(2)}$$

Equation (2) is an **identity**: *It is true for all values of* x. Therefore, let's choose some convenient values for x that will determine the values for A and B. If we let $x = -1$, then equation (2) becomes an equation only in A.

$$\begin{aligned} 11(-1) + 2 &= A[2(-1) - 1] + B(-1 + 1) \\ -9 &= -3A \\ 3 &= A \end{aligned}$$

If we let $x = \frac{1}{2}$, then equation (2) becomes an equation only in B.

$$\begin{aligned} 11\left(\frac{1}{2}\right) + 2 &= A\left[2\left(\frac{1}{2}\right) - 1\right] + B\left(\frac{1}{2} + 1\right) \\ \frac{15}{2} &= \frac{3}{2}B \\ 5 &= B \end{aligned}$$

Therefore, the given rational expression can now be written

$$\frac{11x + 2}{2x^2 + x - 1} = \frac{3}{x + 1} + \frac{5}{2x - 1}.$$

The key idea in Example 1 is the statement that equation (2) is true for all values of x. If we had chosen *any* two values for x, we still would have been able to determine the values for A and B. For example, letting $x = 1$ and then $x = 2$ produces the equations $13 = A + 2B$ and $24 = 3A + 3B$. Solving this system of two equations in two unknowns produces $A = 3$ and $B = 5$. In Example 1, our choices of letting $x = -1$ and then $x = \frac{1}{2}$ simply eliminated the need for solving a system of equations to find A and B.

EXAMPLE 2

Find the partial fraction decomposition of

$$\frac{-2x^2 + 7x + 2}{x(x - 1)^2}.$$

Solution

Apply part 1 of Property 6.7 to determine that there is a partial fraction of the form A/x corresponding to the factor of x. Next, applying part 2 of Property 6.7 and the squared factor $(x - 1)^2$ gives rise to a sum of partial fractions of the form

$$\frac{B}{x-1} + \frac{C}{(x-1)^2}.$$

Therefore, the complete partial fraction decomposition is of the form

$$\frac{-2x^2 + 7x + 2}{x(x-1)^2} = \frac{A}{x} + \frac{B}{x-1} + \frac{C}{(x-1)^2}. \quad \textbf{(1)}$$

Multiply both sides of equation (1) by $x(x - 1)^2$ to produce

$$-2x^2 + 7x + 2 = A(x-1)^2 + Bx(x-1) + Cx, \quad \textbf{(2)}$$

which is true for all values of x. If we let $x = 1$, then equation (2) becomes an equation only in C.

$$\begin{aligned} -2(1)^2 + 7(1) + 2 &= A(1-1)^2 + B(1)(1-1) + C(1) \\ 7 &= C \end{aligned}$$

If we let $x = 0$, then equation (2) becomes an equation just in A.

$$\begin{aligned} -2(0)^2 + 7(0) + 2 &= A(0-1)^2 + B(0)(0-1) + C(0) \\ 2 &= A \end{aligned}$$

If we let $x = 2$, then equation (2) becomes an equation in A, B, and C.

$$\begin{aligned} -2(2)^2 + 7(2) + 2 &= A(2-1)^2 + B(2)(2-1) + C(2) \\ 8 &= A + 2B + 2C \end{aligned}$$

But since we already know that $A = 2$ and $C = 7$, we can easily determine B.

$$\begin{aligned} 8 &= 2 + 2B + 14 \\ -8 &= 2B \\ -4 &= B \end{aligned}$$

Therefore, the original rational expression can be written

$$\frac{-2x^2 + 7x + 2}{x(x-1)^2} = \frac{2}{x} - \frac{4}{x-1} + \frac{7}{(x-1)^2}.$$

EXAMPLE 3

Find the partial fraction decomposition of

$$\frac{4x^2 + 6x - 10}{(x+3)(x^2 + x + 2)}.$$

Solution

Apply part 1 of Property 6.7 to determine that there is a partial fraction of the form $A/(x + 3)$ that corresponds to the factor $x + 3$. Apply part 3 of Property 6.7 to determine that there is also a partial fraction of the form

$$\frac{Bx + C}{x^2 + x + 2}.$$

Thus, the complete partial fraction decomposition is of the form

$$\frac{4x^2 + 6x - 10}{(x + 3)(x^2 + x + 2)} = \frac{A}{x + 3} + \frac{Bx + C}{x^2 + x + 2}. \tag{1}$$

Multiply both sides of equation (1) by $(x + 3)(x^2 + x + 2)$ to produce

$$4x^2 + 6x - 10 = A(x^2 + x + 2) + (Bx + C)(x + 3), \tag{2}$$

which is true for all values of x. If we let $x = -3$, then equation (2) becomes an equation in A alone.

$$\begin{aligned} 4(-3)^2 + 6(-3) - 10 &= A[(-3)^2 + (-3) + 2] + [B(-3) + C][(-3) + 3] \\ 8 &= 8A \\ 1 &= A \end{aligned}$$

If we let $x = 0$, then equation (2) becomes an equation in A and C.

$$\begin{aligned} 4(0)^2 + 6(0) - 10 &= A(0^2 + 0 + 2) + [B(0) + C](0 + 3) \\ -10 &= 2A + 3C \end{aligned}$$

Since $A = 1$, we obtain the value of C.

$$\begin{aligned} -10 &= 2 + 3C \\ -12 &= 3C \\ -4 &= C \end{aligned}$$

If we let $x = 1$, then equation (2) becomes an equation in A, B, and C.

$$\begin{aligned} 4(1)^2 + 6(1) - 10 &= A(1^2 + 1 + 2) + [B(1) + C](1 + 3) \\ 0 &= 4A + 4B + 4C \end{aligned}$$

But since $A = 1$ and $C = -4$, we obtain the value of B.

$$\begin{aligned} 0 &= 4 + 4B - 16 \\ 12 &= 4B \\ 3 &= B \end{aligned}$$

Therefore, the original rational expression can now be written

$$\frac{4x^2 + 6x - 10}{(x + 3)(x^2 + x + 2)} = \frac{1}{x + 3} + \frac{3x - 4}{x^2 + x + 2}.$$

EXAMPLE 4

Find the partial fraction decomposition of

$$\frac{x^3 + x^2 + x + 3}{(x^2 + 1)^2}.$$

Solution

Apply part 4 of Property 6.7 to determine that the partial fraction decomposition of this fraction is of the form

$$\frac{x^3 + x^2 + x + 3}{(x^2 + 1)^2} = \frac{Ax + B}{x^2 + 1} + \frac{Cx + D}{(x^2 + 1)^2}. \tag{1}$$

Multiply both sides of equation (1) by $(x^2 + 1)^2$ to produce

$$x^3 + x^2 + x + 3 = (Ax + B)(x^2 + 1) + Cx + D, \tag{2}$$

which is true for all values of x. Since equation (2) is an identity, we know that the coefficients of similar terms on both sides of the equation must be equal. Therefore, let's collect similar terms on the right side of equation (2).

$$\begin{aligned} x^3 + x^2 + x + 3 &= Ax^3 + Ax + Bx^2 + B + Cx + D \\ &= Ax^3 + Bx^2 + (A + C)x + B + D \end{aligned}$$

Now we can equate coefficients from both sides:

$$1 = A, \qquad 1 = B, \qquad 1 = A + C, \qquad \text{and} \qquad 3 = B + D.$$

From these equations we can determine that $A = 1$, $B = 1$, $C = 0$, and $D = 2$. Therefore the original rational expression can be written

$$\frac{x^3 + x^2 + x + 3}{(x^2 + 1)^2} = \frac{x + 1}{x^2 + 1} + \frac{2}{(x^2 + 1)^2}.$$

PROBLEM SET 6.7

Find the partial fraction decomposition for each of the following rational expressions.

1. $\frac{11x - 10}{(x - 2)(x + 1)}$
2. $\frac{11x - 2}{(x + 3)(x - 4)}$
3. $\frac{-2x - 8}{x^2 - 1}$
4. $\frac{-2x + 32}{x^2 - 4}$
5. $\frac{20x - 3}{6x^2 + 7x - 3}$
6. $\frac{-2x - 8}{10x^2 - x - 2}$
7. $\frac{x^2 - 18x + 5}{(x - 1)(x + 2)(x - 3)}$
8. $\frac{-9x^2 + 7x - 4}{x^3 - 3x^2 - 4x}$
9. $\frac{-6x^2 + 7x + 1}{x(2x - 1)(4x + 1)}$
10. $\frac{15x^2 + 20x + 30}{(x + 3)(3x + 2)(2x + 3)}$
11. $\frac{2x + 1}{(x - 2)^2}$
12. $\frac{-3x + 1}{(x + 1)^2}$
13. $\frac{-6x^2 + 19x + 21}{x^2(x + 3)}$
14. $\frac{10x^2 - 73x + 144}{x(x - 4)^2}$
15. $\frac{-2x^2 - 3x + 10}{(x^2 + 1)(x - 4)}$
16. $\frac{8x^2 + 15x + 12}{(x^2 + 4)(3x - 4)}$
17. $\frac{3x^2 + 10x + 9}{(x + 2)^3}$
18. $\frac{2x^3 + 8x^2 + 2x + 4}{(x + 1)^2(x^2 + 3)}$
19. $\frac{5x^2 + 3x + 6}{x(x^2 - x + 3)}$
20. $\frac{x^3 + x^2 + 2}{(x^2 + 2)^2}$
21. $\frac{2x^3 + x + 3}{(x^2 + 1)^2}$
22. $\frac{4x^2 + 3x + 14}{x^3 - 8}$

CHAPTER 6 SUMMARY

Two themes unify this chapter: (1) solving polynomial equations and (2) graphing polynomial and rational functions.

Solving Polynomial Equations

The following concepts and properties provide the basis for solving polynomial equations.

1. Synthetic division.
2. The factor theorem: A polynomial $f(x)$ has a factor $x - c$ if and only if $f(c) = 0$.
3. Property 6.3: A polynomial equation of degree n has n solutions, where any solution of multiplicity p is counted p times.
4. The rational root theorem: Consider the polynomial equation

   $$a_n x^n + a_{n-1}x^{n-1} + \cdots + a_1 x + a_0 = 0,$$

   where *the coefficients are integers*. If the rational number c/d, reduced to lowest terms, is a solution of the equation, then c is a factor of the constant term, a_0, and d is a factor of the leading coefficient, a_n.
5. Property 6.5: If a polynomial equation with real coefficients has any nonreal complex solutions, they must occur in conjugate pairs.
6. Descartes' rule of signs: Let $a_n x^n + a_{n-1}x^{n-1} + \cdots + a_1 x + a_0 = 0$ be a polynomial equation with real coefficients.
   - **a.** The number of *positive real solutions* either is equal to the number of sign variations in the given polynomial or is less than that number of sign variations by a positive even integer.
   - **b.** The number of *negative real solutions* either is equal to the number of sign variations in

     $$a_n(-x)^n + a_{n-1}(-x)^{n-1} + \cdots + a_1(-x) + a_0,$$

     or is less than that number of sign variations by a positive even integer.

Graphing Polynomial and Rational Functions

Graphs of polynomial functions of the form $f(x) = ax^n$, where n is an integer greater than 2 and $a \neq 1$, are variations of the graphs shown in Figures 6.3 and 6.6. If n is odd, the curve is symmetrical about the origin, and if n is even, the graph is symmetrical about the vertical axis.

Graphs of polynomial functions of the form $f(x) = ax^n$ can be translated horizontally and vertically and reflected across the x-axis. For example:

1. The graph of $f(x) = 2(x - 4)^3$ is the graph of $f(x) = 2x^3$ moved 4 units to the right.
2. The graph of $f(x) = 3x^4 + 4$ is the graph of $f(x) = 3x^4$ moved up 4 units.
3. The graph of $f(x) = -x^5$ is the graph of $f(x) = x^5$ reflected across the x-axis.

To graph a polynomial function that is expressed in factored form, the following steps are helpful.

1. Find the x-intercepts, which are also called the *zeros* of the polynomial.
2. Use a test value in each of the intervals determined by the x-intercepts to find out whether the function is positive or negative over that interval.
3. Plot any additional points that are needed to determine the graph.

To graph a rational function, the following steps are useful.

1. Check for vertical-axis and origin symmetry.
2. Find any vertical asymptotes by setting the denominator equal to zero and solving it for x.
3. Find any horizontal asymptotes by studying the behavior of $f(x)$ as x gets very large or very small. This may require changing the form of the original rational expression.
4. If the degree of the numerator is one larger than the degree of the denominator, determine the equation of the oblique asymptote.
5. Study the behavior of the graph when it is close to the asymptotic lines.
6. Plot as many points as necessary to determine the graph. This may be affected by whether the graph has any symmetries.

Be sure that you understand the process of partial fraction decomposition outlined in Property 6.7.

CHAPTER 6 REVIEW PROBLEM SET

For Problems 1 and 2, find the quotient and remainder of the division problems.

1. $(6x^3 + 11x^2 - 27x + 32) \div (2x + 7)$
2. $(2a^3 - 3a^2 + 13a - 1) \div (a^2 - a + 6)$

For Problems 3–6, use synthetic division to determine the quotient and remainder.

3. $(3x^3 - 4x^2 + 6x - 2) \div (x - 1)$
4. $(5x^3 + 7x^2 - 9x + 10) \div (x + 2)$

5. $(-2x^4 + x^3 - 2x^2 - x - 1) \div (x + 4)$

6. $(-3x^4 - 5x^2 + 9) \div (x + 3)$

For Problems 7–10, find $f(c)$ either by using synthetic division and the remainder theorem or by evaluating $f(c)$ directly.

7. $f(x) = 4x^5 - 3x^3 + x^2 - 1$ and $c = 1$

8. $f(x) = 4x^3 - 7x^2 + 6x - 8$ and $c = -3$

9. $f(x) = -x^4 + 9x^2 - x - 2$ and $c = -2$

10. $f(x) = x^4 - 9x^3 + 9x^2 - 10x + 16$ and $c = 8$

For Problems 11–14, use the factor theorem to help answer some questions about factors.

11. Is $x + 2$ a factor of $2x^3 + x^2 - 7x - 2$?

12. Is $x - 3$ a factor of $x^4 + 5x^3 - 7x^2 - x + 3$?

13. Is $x - 4$ a factor of $x^5 - 1024$?

14. Is $x + 1$ a factor of $x^5 + 1$?

For Problems 15–18, use the rational root theorem and the factor theorem to help solve each of the equations.

15. $x^3 - 3x^2 - 13x + 15 = 0$

16. $8x^3 + 26x^2 - 17x - 35 = 0$

17. $x^4 - 5x^3 + 34x^2 - 82x + 52 = 0$

18. $x^3 - 4x^2 - 10x + 4 = 0$

For Problems 19 and 20, use Descartes' rule of signs (Property 6.6) to list the possibilities for the nature of the solutions. *Do not* solve the equations.

19. $4x^4 - 3x^3 + 2x^2 + x + 4 = 0$

20. $x^5 + 3x^3 + x + 7 = 0$

For Problems 21–24, graph each of the polynomial functions.

21. $f(x) = -(x - 2)^3 + 3$

22. $f(x) = (x + 3)(x - 1)(3 - x)$

23. $f(x) = x^4 - 4x^2$

24. $f(x) = x^3 - 4x^2 + x + 6$

For Problems 25–28, graph each of the rational functions. Be sure to identify the asymptotes.

25. $f(x) = \dfrac{2x}{x - 3}$

26. $f(x) = \dfrac{-3}{x^2 + 1}$

27. $f(x) = \dfrac{-x^2}{x^2 - x - 6}$

28. $f(x) = \dfrac{x^2 + 3}{x + 1}$

For Problems 29 and 30, find the partial fraction decomposition.

29. $\dfrac{5x^2 - 4}{x^2(x + 2)}$

30. $\dfrac{x^2 - x - 21}{(x^2 + 4)(2x - 1)}$

CUMULATIVE REVIEW PROBLEM SET - CHAPTERS 1–6

For Problems 1–10, evaluate each numerical expression.

1. $\left(\frac{3}{4}\right)^{-3}$

2. $\sqrt[3]{-\frac{8}{27}}$

3. -5^{-2}

4. $8^{4/3}$

5. $9^{-(3/2)}$

6. $\log_4 64$

7. $\log_{10} 0.0001$

8. $\log_2\left(\frac{1}{32}\right)$

9. $(-64)^{2/3}$

10. $\ln e^3$

For Problems 11–33, solve each problem.

11. Express the domain of the function $f(x) = \sqrt{2x^2 + 11x - 6}$ using interval notation.

12. If $f(x) = 3x - 1$ and $g(x) = x^2 - x + 3$, find $(f \circ g)(-2)$ and $(g \circ f)(3)$.

13. If $f(x) = -\dfrac{2}{x}$ and $g(x) = \dfrac{1}{x - 4}$, find $(f \circ g)(x)$ and $(g \circ f)(x)$. Also indicate the domain of each composite function.

14. If $f(x) = -2x + 7$, find the inverse of f.

15. If $f(x) = x^2 + 7x - 2$, find $\dfrac{f(a + h) - f(a)}{h}$.

16. If $f(x) = 2x^4 - 17x^3 - 10x^2 + 11x + 15$, find $f(9)$.

17. Find the quotient for $(3x^5 - 25x^3 - 7x^2 + x + 6) \div (x - 3)$.

18. Is $x + 2$ a factor of $2x^4 + 3x^3 + x^2 + 2x - 16$?

19. Evaluate $\log_2 50$ to the nearest hundredth.

20. Find the center and the length of a radius of the circle $x^2 + y^2 + 6x - 4y + 4 = 0$.

21. Write the equation of the line that contains the points $(-4, 2)$ and $(5, -1)$.

22. Write the equation of the perpendicular bisector of the line segment determined by $(-2, -4)$ and $(6, 2)$.

23. Find the length of the major axis of the ellipse $16x^2 + y^2 = 64$.

24. Find the equations of the asymptotes of the hyperbola $x^2 - 9y^2 = 18$.

25. If y varies directly as x, and if $y = 3$ when $x = 4$, find y when $x = 16$.

26. If y varies inversely as the square root of x, and if $y = \frac{2}{5}$ when $x = 25$, find y when $x = 49$.

27. Find the total amount of money accumulated at the end of 8 years if \$450 is invested at 7% compounded quarterly.

28. How long will it take \$500 to double itself if it is invested at 8% interest compounded continuously?

29. Sandy has a collection of 57 coins worth \$10. They consist of nickels, dimes, and quarters, and the number of quarters is 2 more than three times the number of nickels. How many coins of each kind does she have?

30. A retailer bought a dress for \$75 and wants to sell it at a profit of 40% of the selling price. What price should she ask for the dress?

31. A container has 8 quarts of a 30% alcohol solution. How much pure alcohol should be added to raise it to a 40% solution?

32. Claire rode her bicycle out into the country at a speed of 15 miles per hour and returned along the same route at 10 miles per hour. If the entire trip took $7\frac{1}{2}$ hours, how far out did she ride?

33. Adam can do a job in 2 hours less time than it takes Carl to do the same job. Working together they can do the job in 2 hours and 24 minutes. How long would it take Adam to do the job by himself?

For Problems 34–45, solve each equation.

34. $(2x - 5)(6x + 1) = (3x + 2)(4x - 7)$

35. $(2x + 1)(x - 2) = (3x - 2)(x + 4)$

36. $4x^3 + 20x^2 - 56x = 0$

37. $6x^3 + 17x^2 + x - 10 = 0$

38. $|4x - 3| = 7$

39. $\dfrac{2x - 1}{3} - \dfrac{3x + 2}{4} = -\dfrac{5}{6}$

40. $3^{x-2} = 27^x$

41. $\ln(t + 2) = \ln t + \ln 4$

42. $\log 5 + \log(x - 1) = 1$

43. $x^4 + 3x^2 - 54 = 0$

44. $(2x - 1)(x + 3) = 49$

45. $x^4 - 2x^3 + 2x^2 - 7x + 6 = 0$

For Problems 46–53, solve each inequality and express the solution set using interval notation.

46. $3(x - 1) - 5(x + 2) > 3(x + 4)$

47. $\dfrac{x - 1}{2} + \dfrac{2x + 1}{5} \geq \dfrac{x - 2}{3}$

48. $x^2 - 3x < 18$

49. $(x - 1)(x + 3)(2 - x) \leq 0$

50. $|2x - 1| > 6$

51. $|3x + 2| \leq 8$

52. $\dfrac{4x - 3}{x - 2} \geq 0$

53. $\dfrac{x + 3}{x - 4} < 3$

For Problems 54–64, graph each function.

54. $f(x) = -2x + 4$

55. $f(x) = 2x^2 - 3$

56. $f(x) = 2^x - 3$

57. $f(x) = \log_2(x - 1)$

58. $f(x) = \dfrac{2x}{x + 1}$

59. $f(x) = -|x - 2| + 1$

60. $f(x) = 2\sqrt{x} + 1$

61. $f(x) = 3x^2 + 12x + 9$

62. $f(x) = -(x - 3)^3 + 1$

63. $f(x) = (x + 1)(x - 2)(x - 4)$

64. $f(x) = x^4 - x^2$

TRIGONOMETRIC FUNCTIONS

The word trigonometry derives from two Greek words that mean *measurement of triangles*. Historically, the development of trigonometry began with the study of the various relationships that exist between the angles and sides of triangles. This aspect of trigonometry has many applications in surveying, navigation, carpentry, and the various branches of engineering.

Originally, the trigonometric functions were restricted to domains of angles; a more modern viewpoint allows for domains to be the set of real numbers independent of any angle restriction. This viewpoint has resulted in a greater variety of applications for the trigonometric functions in such areas as light, sound, and electrical wave theories.

Our approach to trigonometry in this text will follow the historical route. That is, we will first introduce the trigonometric functions in terms of angles and then define them with real number domains. Let's begin by reviewing some basic geometric concepts that will be used in our study of trigonometry.

7

Vectors can be used to determine the amount of parallel force needed to keep a tractor from rolling backwards down an incline.

7.1 GEOMETRIC BASIS FOR TRIGONOMETRY

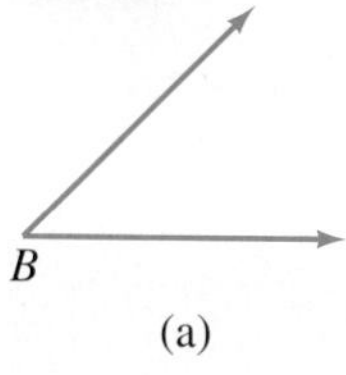

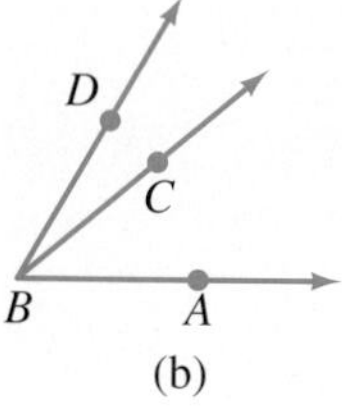

FIGURE 7.1

In geometry a **plane angle** is usually defined as the set of points consisting of two rays that have a common endpoint. The common endpoint is called the **vertex** ofthe angle, and the rays are called the **sides** of the angle. In Figure 7.1(a) the angle can be named by its vertex only; thus, we can refer to $\angle B$. Figure 7.1(b) illustrates three angles that have a common vertex. In such cases, points on the sides of the angles can be used to help name the angles. We can refer to $\angle DBC$, $\angle CBA$, and $\angle DBA$ in Figure 7.1(b).

In trigonometry we define an angle in terms of a ray rotated about its endpoint. Specifically, in Figure 7.2 let's begin with a ray r_1 and rotate it, in a plane, about its endpoint O to a position indicated by the ray r_2. We call r_1 the **initial side**, r_2 the **terminal side**, and O the **vertex** of the angle. If the rotation is in a counterclockwise direction, as indicated by an arrow, then the angle is a **positive angle**. If the rotation is in a clockwise direction, then a **negative angle** is formed. There is no restriction as to the amount of rotation (Figure 7.3). As indicated in parts (a) and (c) of Figure 7.3, different angles can have the same initial and terminal sides (the difference is in the amount of rotation). Any two such angles are called **coterminal**.

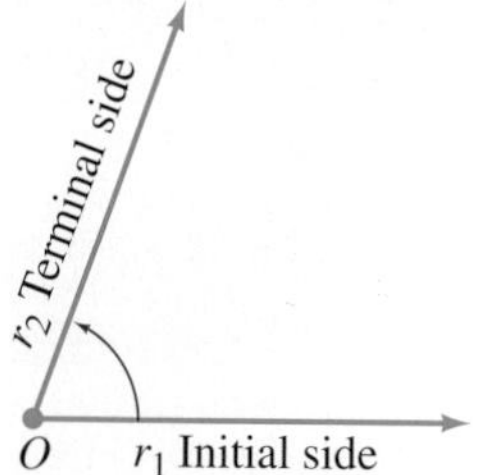

FIGURE 7.2

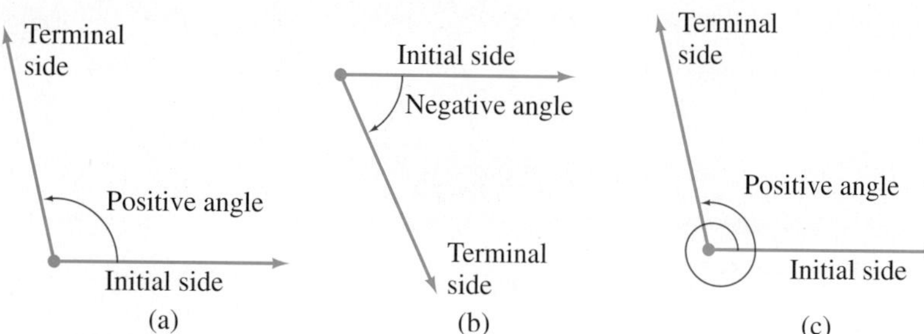

FIGURE 7.3

The size of an angle (amount of rotation from initial side to terminal side)can be described by using *degree measure*. The angle formed by rotating a complete counterclockwise revolution has a measure of 360 degrees, written 360°. Thus, one degree (1°) is $\frac{1}{360}$ of a complete revolution. The diagrams in Figure 7.4 illustrate angles of different degree measure and some commonly used terminology.

Two acute angles are **complementary** if their sum is 90°; one of the angles is referred to as the *complement* of the other. Two positive angles are **supplementary** if their sum is 180°.

The degree system for angle measurement is similar to the hour-minute-second relationship of our time system. Each degree is divided into 60 parts, called **minutes**, and each minute is divided into 60 parts, called **seconds**. Thus, we can speak of an angle as having a measure of 73 degrees, 12 minutes, and 36 seconds, written 73°12′36″.

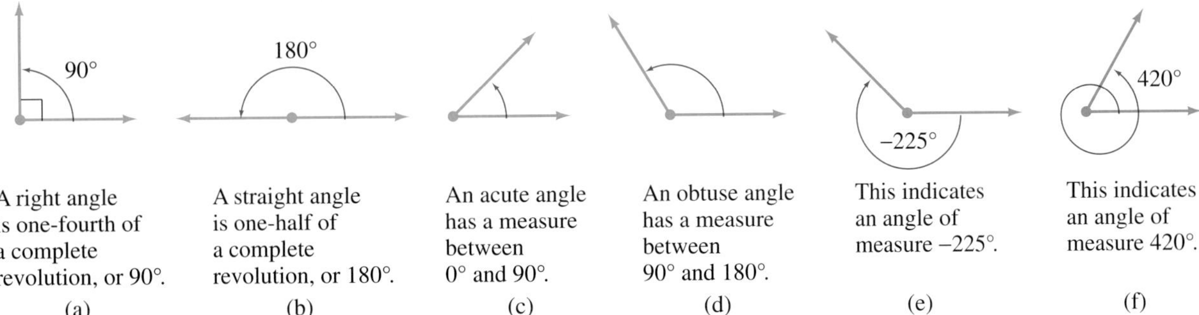

A right angle is one-fourth of a complete revolution, or 90°. (a)

A straight angle is one-half of a complete revolution, or 180°. (b)

An acute angle has a measure between 0° and 90°. (c)

An obtuse angle has a measure between 90° and 180°. (d)

This indicates an angle of measure −225°. (e)

This indicates an angle of measure 420°. (f)

FIGURE 7.4

A calculator displays fractional parts of a degree in decimal form. For example, an angle may have a measure of 73.21°. Some calculators are equipped with a special key sequence that will switch back and forth between the degree-minute-second form and the decimal form. If your calculator is not so equipped, you can proceed as follows.

From degree-minute-second form to decimal form

$$\begin{aligned} 73°12'36'' &= 73° + \left(\frac{12}{60}\right)^\circ + \left(\frac{36}{3600}\right)^\circ \\ &= 73° + \left(\frac{1}{5}\right)^\circ + \left(\frac{1}{100}\right)^\circ \\ &= 73° + (0.2)° + (0.01)° \\ &= 73.21° \end{aligned}$$

If $1° = 60'$ and $1' = 60''$, then $1° = 3600''$.

From decimal form to degree-minute-second form

$$\begin{aligned} 73.21° &= 73° + .21(60') \\ &= 73° + 12.6' \\ &= 73° + 12' + 0.6(60'') \\ &= 73° + 12' + 36'' \\ &= 73°12'36'' \end{aligned}$$

Radian Measure

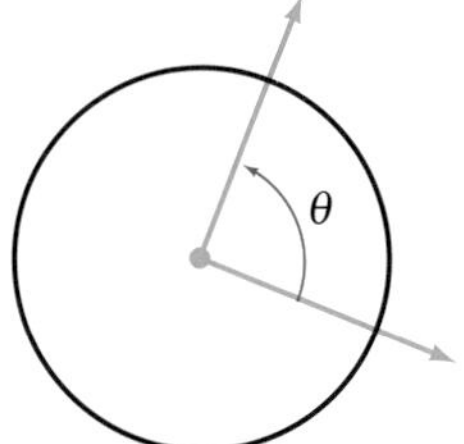

FIGURE 7.5

The *radian* is another basic unit of angle measure that is used extensively in subsequent mathematics courses and in various mathematical applications in the physical sciences. To define a radian, let's first define the concept of a central angle. A **central angle of a circle** is an angle whose vertex is the center of the circle, as illustrated in Figure 7.5. (Lowercase Greek letters are often used to denote angles. The symbol θ, used in Figure 7.5, denotes the letter *theta*.) Using the concept of a central angle, we define a radian as follows.

DEFINITION 7.1

One **radian** is the measure of the central angle of a circle in which the sides of the angle intercept an arc equal in length to the radius of the circle (Figure 7.6).

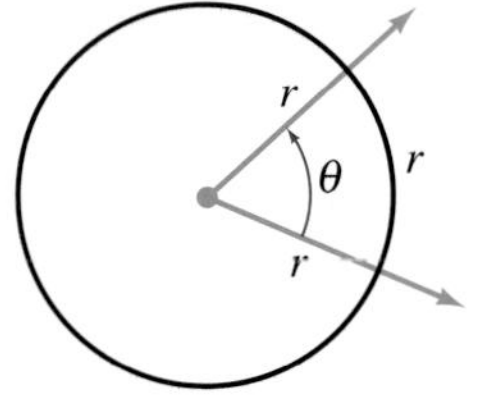

FIGURE 7.6

Therefore, in Figure 7.6, θ is an angle of measure 1 radian. Furthermore, since the circumference of a circle is given by $C = 2\pi r$, and each arc of length r determines an angle of 1 radian, there are $2\pi r/r = 2\pi$ radians in one complete revolution. Thus,

$$2\pi \text{ radians} = 360°$$

or, equivalently,

$$\pi \text{ radians} = 180°.$$

So we have the following two basic relationships between degree and radian measure.

$$1 \text{ radian} = \frac{180}{\pi} \text{ degrees} \qquad \text{and} \qquad 1 \text{ degree} = \frac{\pi}{180} \text{ radians}$$

REMARK Evaluating $\frac{180}{\pi}$, we can determine that 1 radian is approximately 57.3 degrees. This relationship need not be memorized, but it may strengthen your perception of the size of 1 radian.

Sometimes it is necessary to switch back and forth between degree and radian measure. This creates no great difficulty, as illustrated by the following examples.

EXAMPLE 1

Change 150° to radians.

Solution

Since 1 degree $= \frac{\pi}{180}$ radians,

$$\begin{aligned} 150 \text{ degrees} &= 150\left(\frac{\pi}{180}\right) \text{ radians} \\ &= \frac{5\pi}{6} \text{ radians.} \end{aligned}$$

EXAMPLE 2

Change $\frac{3\pi}{4}$ radians to degrees.

Solution

Since 1 radian $= \dfrac{180}{\pi}$ degrees,

$$\frac{3\pi}{4} \text{ radians} = \frac{3\pi}{4}\left(\frac{180}{\pi}\right) \text{ degrees}$$
$$= 135 \text{ degrees.}$$

Some calculators have a [d ↔ r] key that can be used for direct conversion between degrees and radians. If not, you will need to use $\pi/180$ to convert degrees to radians and $180/\pi$ to convert radians to degrees. Therefore,

$$5 \text{ radians} = 5\left(\frac{180}{\pi}\right) = 286.5 \text{ degrees, to the nearest tenth of a degree,}$$

and

$$127.4 \text{ degrees} = 127.4\left(\frac{\pi}{180}\right) = 2.2 \text{ radians, to the nearest tenth of a radian.}$$

DEGREES	RADIANS
30°	$\pi/6$
45°	$\pi/4$
60°	$\pi/3$
90°	$\pi/2$
180°	π
270°	$3\pi/2$
360°	2π

Even though the process of switching between degree and radian measure is quite simple, it would be advantageous for you to know a few basic conversions as listed in the chart at the left. Having these few conversions at your fingertips also provides a basis for expressing the radian measure of many other angles. For example, since $300° = 5(60°) = \pi/3$ radians, then $300° = 5(\pi/3) = 5\pi/3$ radians.

Arc Length

Consider a circle of radius r and a central angle θ, measured in radians that intercept an arc of length s, as in Figure 7.7. Because the arc length s compares to the total circumference of $2\pi r$ as θ radians compare to a complete revolution of 2π radians, we have the proportion

$$\frac{s}{2\pi r} = \frac{\theta}{2\pi}.$$

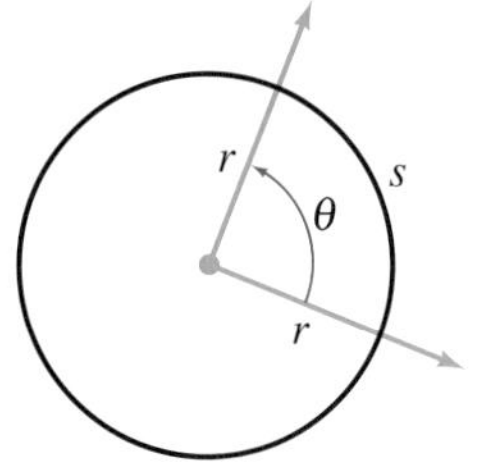

FIGURE 7.7

Solving for s produces

$$(s)(2\pi) = (2\pi r)(\theta)$$

$$s = r\theta.$$

EXAMPLE 3

Find the length of the arc intercepted by a central angle of $\pi/6$ radians if a radius of the circle is 11 inches long.

Solution

Using $s = r\theta$ we obtain

$$\begin{aligned} s &= r\theta \\ &= 11\left(\frac{\pi}{6}\right) \\ &= 5.8 \text{ inches}, \quad \text{to the nearest tenth of an inch.} \end{aligned}$$

EXAMPLE 4

How high will the weight in Figure 7.8 be lifted if the drum is rotated through an angle of 70°?

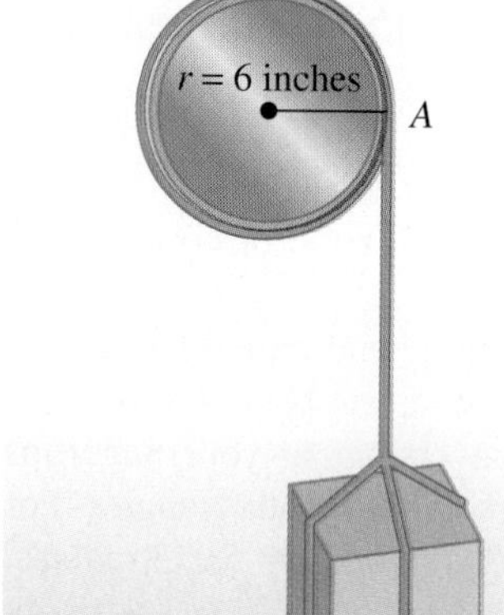

FIGURE 7.8

Solution

First, we need to change 70° to radians.

$$70^\circ = 70\left(\frac{\pi}{180}\right) = \frac{7\pi}{18} \text{ radians}$$

Therefore, point A will move

$$s = r\theta = 6\left(\frac{7\pi}{18}\right) = 7.3 \text{ inches}, \quad \text{to the nearest tenth of an inch.}$$

Thus, the weight will be lifted approximately 7.3 inches.

Triangles

The concepts of linear and angular measurement provide the basis for classifying various geometric figures. Triangles are often classified as in Figure 7.9. It is possible for triangles to fit both classification schemes. For example, an *isosceles right triangle* is a right triangle that has two sides of the same length.

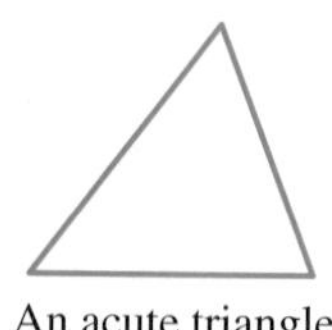

An acute triangle has three acute angles.

An obtuse triangle has one obtuse angle.

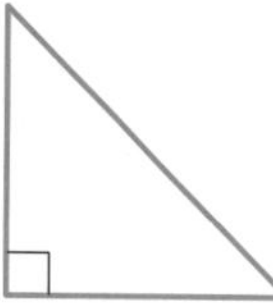

A right triangle has one right angle.

Classification by angles
(a)

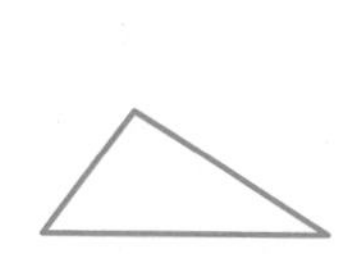

A scalene triangle has no two sides of the same length.

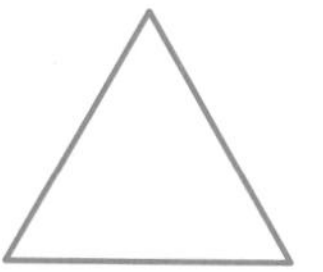

An equilateral triangle has three sides of the same length.

An isosceles triangle has two sides of the same length.

Classification by sides
(b)

FIGURE 7.9

Pythagorean Theorem and Other Right-Triangle Relationships

There are several right-triangle relationships that are used extensively in trigonometry. One of the most important of these relationships is the *Pythagorean theorem.*

Pythagorean Theorem

If, for a right triangle, a and b are measures of the legs and c is the measure of the hypotenuse (Figure 7.10), then

$$a^2 + b^2 = c^2.$$

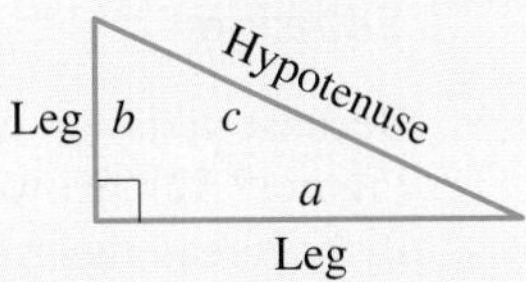

FIGURE 7.10

EXAMPLE 5

A 50-foot rope hangs from the top of a flagpole. When pulled taut to its full length, the rope reaches a point on the ground 18 feet from the base of the pole. Find the height of the pole to the nearest tenth of a foot.

Solution

Let's sketch a figure (Figure 7.11) and record the given information.

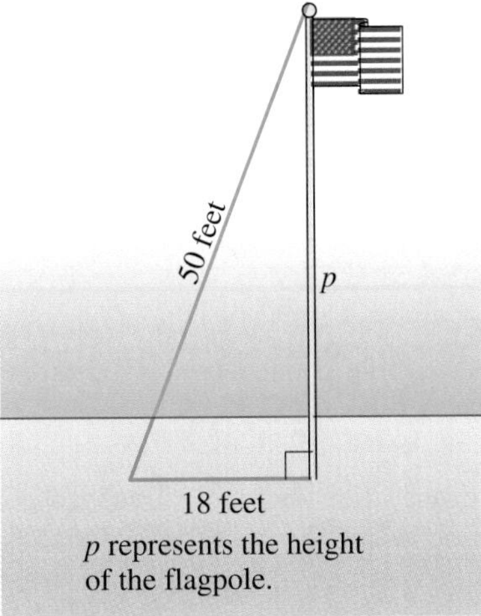

FIGURE 7.11

Using the Pythagorean theorem, we can solve for p as follows.

$$\begin{aligned} p^2 + 18^2 &= 50^2 \\ p^2 + 324 &= 2500 \\ p^2 &= 2176 \\ p &= \pm\sqrt{2176} \approx \pm 46.6 \end{aligned}$$

The negative solution is disregarded, and the height of the flagpole, to the nearest tenth of a foot, is 46.6 feet.

EXAMPLE 6

Find the length of each leg of an isosceles right triangle that has a hypotenuse measuring 6 centimeters.

Solution

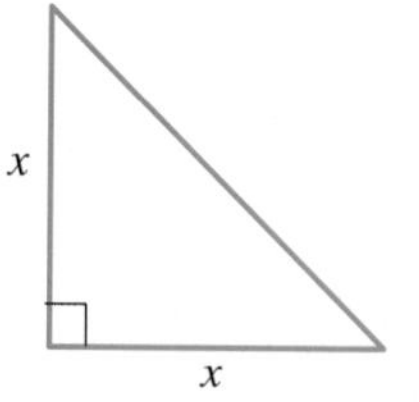

FIGURE 7.12

Because it is an *isosceles right triangle*, both legs are of the same length (see Figure 7.12). So we can use x to represent the length of each leg. Now apply the Pythagorean theorem and proceed as follows.

$$\begin{aligned} x^2 + x^2 &= 6^2 \\ 2x^2 &= 36 \\ x^2 &= 18 \\ x &= \pm\sqrt{18} = \pm 3\sqrt{2} \end{aligned}$$

Therefore, each leg is $3\sqrt{2}$ centimeters long.

Another very useful property comes directly from an equilateral triangle. Consider the equilateral triangle in Figure 7.13(a), which has sides of length s. We know from elementary geometry that an angle bisector of an equilateral triangle is also a perpendicular bisector of the opposite side. Therefore, in Figure 7.13(b) the

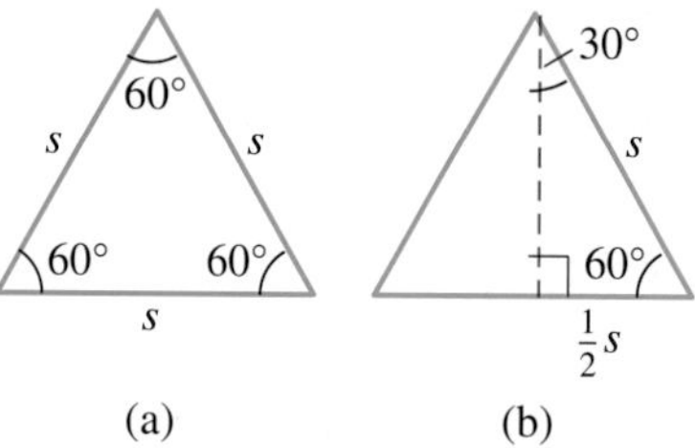

FIGURE 7.13

angle bisector (dashed line segment) divides the equilateral triangle into two right triangles having acute angles of 30° and 60°. The following general property can be stated: **In a 30°–60° right triangle, the length of the side opposite the 30° angle is equal to one-half of the length of the hypotenuse.**

EXAMPLE 7

Suppose that a 20-foot ladder is leaning against a building and makes an angle of 60° with the ground. How far up on the building does the top of the ladder reach? Express your answer to the nearest tenth of a foot.

Solution

Figure 7.14 depicts such a situation.

$$\begin{aligned} h^2 + 10^2 &= 20^2 \\ h^2 + 100 &= 400 \\ h^2 &= 300 \\ h &= \pm\sqrt{300} \approx \pm 17.3 \end{aligned}$$

The top of the ladder touches the building at approximately 17.3 feet from the ground.

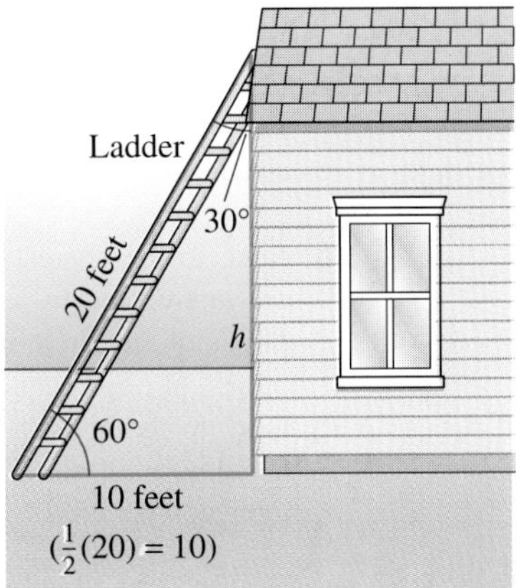

FIGURE 7.14

The two right triangles in Figure 7.15 are used frequently in subsequent trigonometry sections. Notice that in each triangle the hypotenuse is one unit long. We can determine the lengths of the legs for each triangle as we did in Examples 6 and 7. (Remember the dimensions of these right triangles!)

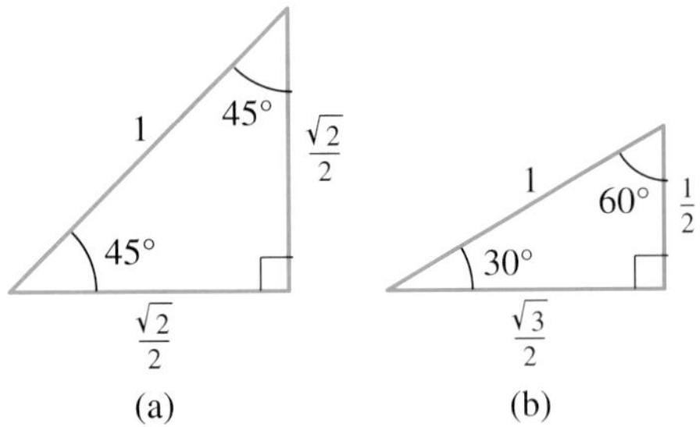

FIGURE 7.15

PROBLEM SET 7.1

1. Find the complement of an angle of measure 38°.

2. Find the supplement of an angle of measure 72°.

3. The difference between two complementary angles is 40°. Find the measure of each angle.

4. Find the measure of an angle that is 10° less than its complement.

5. One of two supplementary angles is eight times as large as the other. Find the measure of each angle.

6. The measures of two supplementary angles are in the ratio of 2 to 7. Find the measure of each angle.

In Problems 7–18, if the measurement is given in degree-minute-second form, change it to decimal form to the nearest one-hundredth of a degree. If the measurement is given in decimal form, change it to degree-minute-second form.

7. 14° 30′ **8.** 62° 15′ **9.** 22.3°

10. 114.6° **11.** 8° 45′ 18″ **12.** 34° 50′ 30″

13. 45.32° **14.** 132.15° **15.** 150° 10′

16. 94° 45′ **17.** 9.13° **18.** 73.47°

In Problems 19–30, change each angle to radians. Do not use a calculator.

19. 10° **20.** 15° **21.** 80°

22. 120° **23.** 150° **24.** 210°

25. 225° **26.** 300° **27.** −30°

28. −330° **29.** −570° **30.** 480°

In Problems 31–42, each angle is expressed in radians. Change each angle to degrees without using a calculator.

31. $\frac{\pi}{9}$ **32.** $\frac{5\pi}{18}$ **33.** $\frac{13\pi}{18}$

34. $\frac{7\pi}{12}$ **35.** $\frac{4\pi}{3}$ **36.** $\frac{7\pi}{4}$

37. $\frac{13\pi}{6}$ **38.** $\frac{17\pi}{6}$ **39.** $-\frac{\pi}{4}$

40. $-\frac{5\pi}{9}$ **41.** $-\frac{7\pi}{6}$ **42.** $-\frac{7\pi}{3}$

In Problems 43–48, each angle is expressed in radians. Use your calculator and change each angle to the nearest tenth of a degree.

43. 2 **44.** 3 **45.** 7

46. 4.1 **47.** −4 **48.** −6.2

In Problems 49–54, use your calculator to help change each angle to the nearest tenth of a radian.

49. 27° **50.** 212° **51.** 14.5°

52. 141.8° **53.** −251.6° **54.** −373.4°

55. Find, to the nearest tenth of an inch, the length of the arc intercepted by a central angle of $2\pi/3$ radians given that a radius of the circle is 22 inches long.

56. Find, to the nearest tenth of a meter, the length of the arc intercepted by a central angle of 130° given that a radius of the circle is 8 meters long.

57. Find, to the nearest tenth of a centimeter, the length of a radius of a circle given that a central angle of 80° intercepts an arc of 25 centimeters.

58. Find, to the nearest tenth of a foot, the length of a radius of a circle given that a central angle of $3\pi/5$ radians intercepts an arc of 12 feet.

59. Find, to the nearest tenth of a degree, the measure of a central angle that intercepts an arc of 7.1 centimeters given that a radius of the circle is 3.2 centimeters long.

60. Refer to Figure 7.16. How much will the weight be lifted if the drum is rotated through an angle of 150°? Express your result to the nearest tenth of an inch.

61. Refer to Figure 7.16. Through what angle (to the nearest tenth of a degree) must the drum be rotated to raise the weight 6 feet?

62. Two pulleys are connected with a belt as indicated in Figure 7.17. If the smaller pulley makes a complete revolution, through what size angle does the larger pulley turn? Express the result to the nearest tenth of a degree.

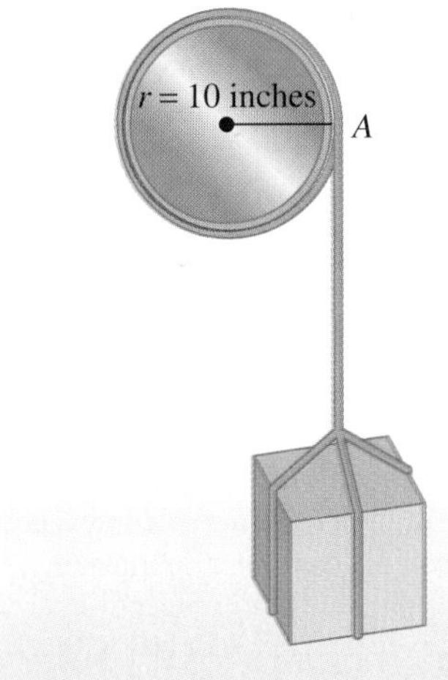

FIGURE 7.16

63. Refer to Figure 7.17. Through what angle does the smaller pulley turn while the larger pulley makes a complete revolution?

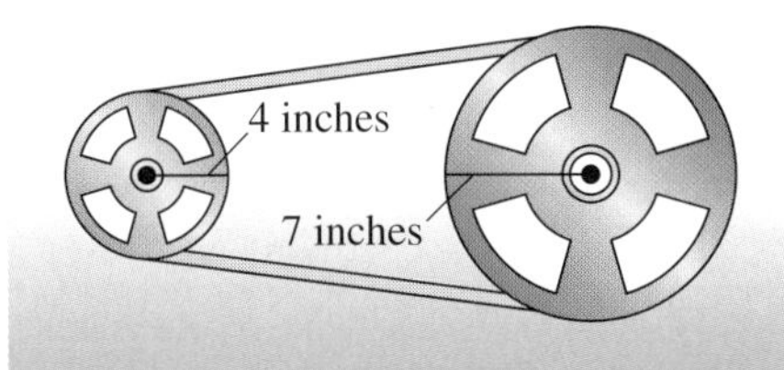

FIGURE 7.17

64. A gas gauge on an automobile is illustrated in Figure 7.18. The scale from E to F is 3.5 inches long. It is an arc of a circle having a radius of 1.5 inches. What angle will the needle make between the empty and full readings? Express the angle to the nearest tenth of a degree.

FIGURE 7.18 FIGURE 7.19

65. A speedometer on an automobile is illustrated in Figure 7.19. An angle of 110° is made as the needle moves from the 55 position to the 0 position. The radius of the circle is 2 inches. Find, to the nearest tenth of an inch, the length of the arc from the 55-reading to the 0-reading.

66. Figure 7.20 depicts the back wheel-drive chain apparatus of a bicycle. How far will the bicycle move forward for each complete revolution of the drive sprocket? Express the result to the nearest inch.

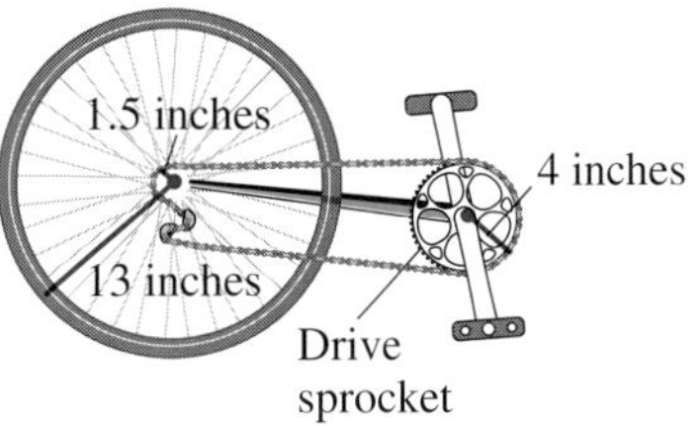

FIGURE 7.20

67. Refer to Figure 7.20. How much rotation of the drive sprocket is needed to move the bicycle forward 50 feet? Express your result to the nearest tenth of a revolution.

For Problems 68–75, we are using Figure 7.21. Express the lengths of the sides in simplest radical form.

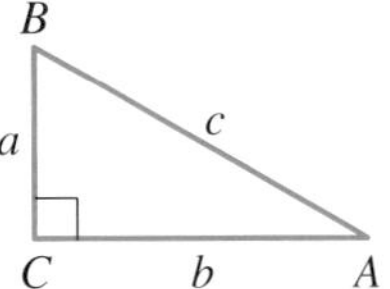

FIGURE 7.21

68. Find c if $a = 5$ inches and $b = 12$ inches.

69. Find b if $c = 25$ feet and $a = 24$ feet.

70. Find b if $c = 12$ meters and $a = 6$ meters.

71. Find a if $c = 8$ centimeters and $b = 4$ centimeters.

72. Find b and c if $A = 30°$ and $a = 3$ yards.

73. Find a and b if $B = 60°$ and $c = 12$ yards.

74. Find c if $A = 45°$ and $a = 4$ meters.

75. Find a and b if $B = 45°$ and $c = 10$ meters.

76. The length of the hypotenuse of an isosceles right triangle is 8 meters. Find, to the nearest tenth of a meter, the length of each leg.

77. The length of the hypotenuse of an isosceles right triangle is 11 centimeters. Find, to the nearest tenth of a centimeter, the length of each leg.

78. An 18-foot ladder resting against a house just reaches a windowsill 16 feet above the ground. How far is the foot of the ladder from the foundation of the house? Express your answer to the nearest tenth of a foot.

79. A 42-foot support wire makes an angle of 60° with the ground and is attached to a telephone pole. Find the distance from the base of the pole to the point on the pole to which the wire is attached. Express your answer to the nearest tenth of a foot.

80. A rectangular plot of ground measures 18 meters by 24 meters. Find the distance, to the nearest meter, from one corner of the plot to the diagonally opposite corner.

81. Consecutive bases of a square-shaped baseball diamond are 90 feet apart. Find the distance, to the nearest foot, from first base diagonally across the diamond to third base.

82. A diagonal of a square parking lot is 50 meters. Find, to the nearest meter, the length of a side of the lot.

83. Find the length of an altitude of an equilateral triangle where each side of the triangle is 6 centimeters long. Express your answer in simplest radical form.

84. Suppose that we are given a cube with edges of length 12 centimeters. Find the length of a diagonal from a lower corner to the diagonally opposite upper corner. Express your answer to the nearest tenth of a centimeter.

85. Suppose that we are given a rectangular box with a length of 8 centimeters, a width of 6 centimeters, and a height of 4 centimeters. Find the length of a diagonal from a lower corner to the diagonally opposite upper corner. Express your answer to the nearest tenth of a centimeter.

THOUGHTS into WORDS

86. How would you explain the difference between degree measure and radian measure?

87. Can an obtuse triangle be an equilateral triangle? Defend your answer.

MISCELLANEOUS PROBLEMS

88. The converse of the Pythagorean theorem is also true:

> If the measures a, b, and c, of the sides of a triangle are such that $a^2 + b^2 = c^2$, then the triangle is a right triangle with a and b the measures of the legs and c the measure of the hypotenuse.

Use the converse of the Pythagorean theorem to determine which of the triangles having sides with the following measures are right triangles.

a. 9, 40, 41 **b.** 20, 48, 52

c. 19, 21, 26 **d.** 32, 37, 49

e. 65, 156, 169 **f.** 21, 72, 75

89. The **angular speed** of a circular wheel that is rotating at a constant rate is the angle generated in one unit of time by a radius. For example, suppose that a wheel is rotating at a rate of 600 revolutions per minute. Since each revolution generates an angle of 2π radians, the angular speed of the wheel is $600(2\pi) = 1200\pi$ radians per minute.

If P is a point on the wheel, the **linear speed** of P is the distance that P travels in one unit of time. For example, suppose that the wheel that is rotating at 600 revolutions per minute has a radius of length 2 feet. The distance that a point P on the wheel travels in a minute is given by $s = r\theta = 2(1200\pi) = 2400\pi$ feet. Thus, the linear speed is 2400π feet per minute. To the nearest foot, this is approximately 7540 feet per minute.

a. A wheel with a radius of 3 inches is rotating at 1500 revolutions per minute. Find the angular speed of the wheel and the linear speed of a point on the wheel.

b. A wheel with a diameter of 5 feet is rotating at 850 revolutions per minute. Find the angular speed of the wheel and the linear speed of a point on the wheel.

c. A car is moving at the rate of 55 miles per hour, and the diameter of each of its wheels is 2.5 feet. Find the number of revolutions per minute that the wheels are rotating and find the angular speed of the wheels.

7.2 TRIGONOMETRIC FUNCTIONS

If a rectangular coordinate system is introduced, then the **standard position** of an angle is obtained by putting the vertex at the origin and having the initial side coincide with the positive side of the x-axis. The angles in Figure 7.22 are each in

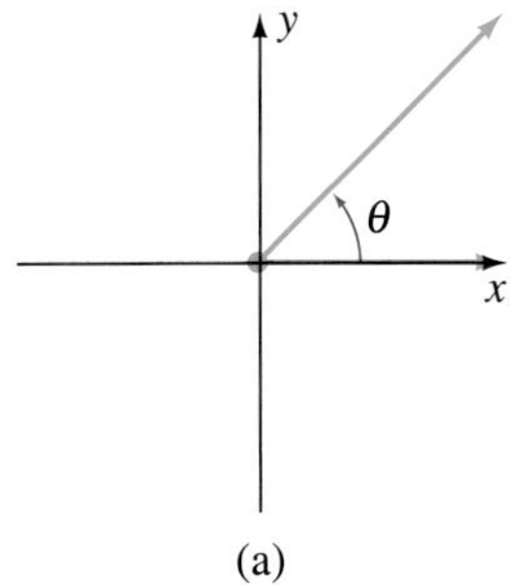

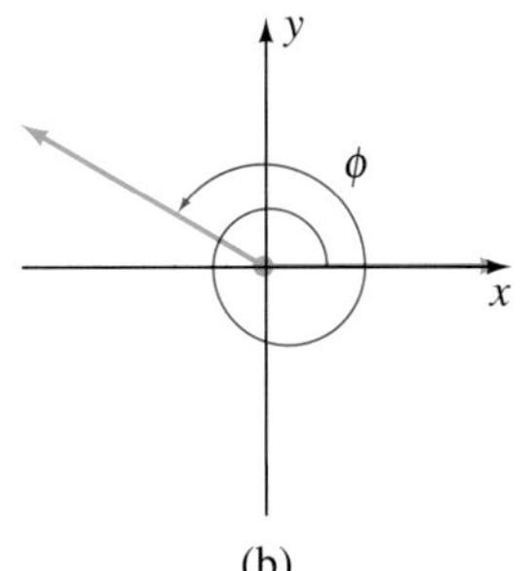

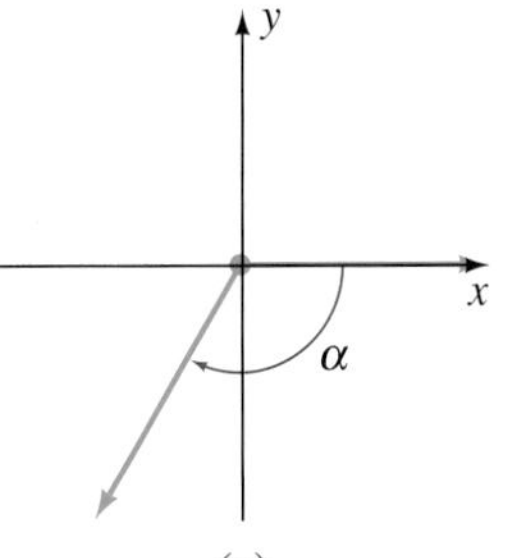

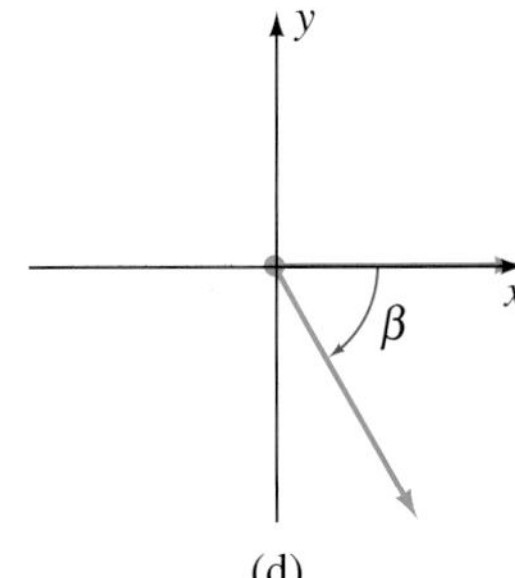

FIGURE 7.22

standard position. Each angle is named with a Greek letter positioned next to the curved arrow. Thus, we can refer to angles θ (theta), ϕ (phi), α (alpha), and β (beta). Angle θ is called a **first-quadrant angle** because its terminal side lies in the first quadrant. Angles ϕ, α, and β are *second-quadrant*, *third-quadrant*, and *fourth-quadrant* angles, respectively. Also note that θ and ϕ are positive angles, whereas α and β are negative angles. If the terminal side of an angle in standard position coincides with a coordinate axis, then the angle is called a **quadrantal angle**. Now the six basic trigonometric functions can be defined as follows.

FIGURE 7.23

DEFINITION 7.2

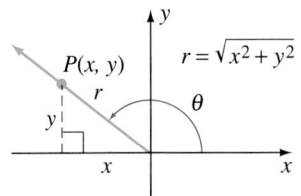

Let θ be an angle in standard position and let $P(x, y)$ be any point (except the origin) on the terminal side of θ (Figure 7.23). The six trigonometric functions are defined as follows.

$$\sin\theta = \frac{y}{r}$$ Read *sine theta*

$$\cos\theta = \frac{x}{r}$$ Read *cosine theta*

$\tan\theta = \frac{y}{x}$ Read *tangent theta*

$\csc\theta = \frac{r}{y}$ Read *cosecant theta*

$\sec\theta = \frac{r}{x}$ Read *secant theta*

$\cot\theta = \frac{x}{y}$ Read *cotangent theta*

These functions are basic to all of trigonometry; **memorize them**.

In Definition 7.2, r is the distance between the origin and point P; *it is always a positive number* and it is determined by $r = \sqrt{x^2 + y^2}$. Recall that a function assigns to each member of a set (the **domain**) a unique member of another set (the **range**). The domain of each of the six trigonometric functions is a set of angles, and Definition 7.2 assigns to each angle (with a few exceptions) a real number determined by the ratios y/r, x/r, y/x, r/y, r/x, and x/y. (The fact that a *unique* **number** is assigned to each angle will be demonstrated a bit later.) Because division by zero is not permitted, $\tan\theta$ and $\sec\theta$ cannot be defined for $x = 0$, and $\csc\theta$ and $\cot\theta$ cannot be defined for $y = 0$. Furthermore, notice that $\csc\theta$, $\sec\theta$, and $\cot\theta$ are the **reciprocals** of $\sin\theta$, $\cos\theta$, and $\tan\theta$, respectively. That is,

$$\csc\theta = \frac{1}{\sin\theta}, \qquad \sin\theta \neq 0$$

$$\sec\theta = \frac{1}{\cos\theta}, \qquad \cos\theta \neq 0$$

$$\cot\theta = \frac{1}{\tan\theta}, \qquad \tan\theta \neq 0.$$

Another very useful relationship that follows directly from Definition 7.2 is

$$\sin^2\theta + \cos^2\theta = 1.$$

The notation $\sin^2\theta$ (usually read *sine squared of theta*) means $(\sin\theta)^2$; therefore, $\sin^2\theta = (\sin\theta)(\sin\theta)$. This relationship can be verified as follows.

$$\begin{aligned}\sin^2\theta + \cos^2\theta &= \left(\frac{y}{r}\right)^2 + \left(\frac{x}{r}\right)^2\\ &= \frac{y^2}{r^2} + \frac{x^2}{r^2}\\ &= \frac{y^2 + x^2}{r^2} = \frac{r^2}{r^2} = 1\end{aligned}$$

The reciprocal relationships and the property $\sin^2\theta + \cos^2\theta = 1$ are called **trigonometric identities**. We will discuss more identities in Chapter 9.

EXAMPLE 1

Find the values of the six trigonometric functions of angle θ if θ is in standard position and the point $(-3, 4)$ is on the terminal side of θ.

Solution

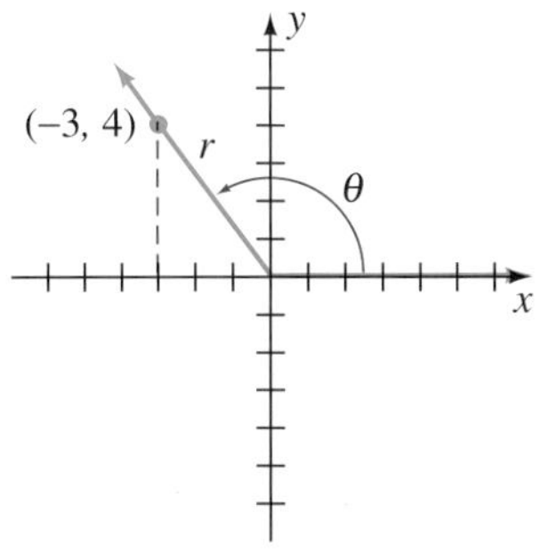

FIGURE 7.24

Figure 7.24 shows θ and the point $(-3, 4)$ on the terminal side of θ. Use $r = \sqrt{x^2 + y^2}$ to obtain

$$\begin{aligned} r &= \sqrt{(-3)^2 + 4^2} \\ &= \sqrt{9 + 16} \\ &= \sqrt{25} \\ &= 5 \end{aligned}$$

Now using $x = -3$, $y = 4$, and $r = 5$, the values of the six trigonometric functions of θ can be determined.

$$\sin\theta = \frac{y}{r} = \frac{4}{5} \qquad \cos\theta = \frac{x}{r} = \frac{-3}{5} = -\frac{3}{5}$$

$$\tan\theta = \frac{y}{x} = \frac{4}{-3} = -\frac{4}{3} \qquad \csc\theta = \frac{r}{y} = \frac{5}{4}$$

$$\sec\theta = \frac{r}{x} = \frac{5}{-3} = -\frac{5}{3} \qquad \cot\theta = \frac{x}{y} = \frac{-3}{4} = -\frac{3}{4}$$

It is important to realize that *any point* (other than the origin) on the terminal side of an angle in standard position can be used to determine the trigonometric functions of the angle. This fact is based on a property of similar triangles illustrated in Figure 7.25. Triangles OQP and $OQ'P'$ are similar triangles, and corresponding sides of similar triangles are proportional. That is, the ratios of corresponding sides are equal. For example, $y/r = y'/r'$ and therefore either of the ratios can be used to determine $\sin\theta$. Similar arguments hold for the remaining five trigonometric functions.

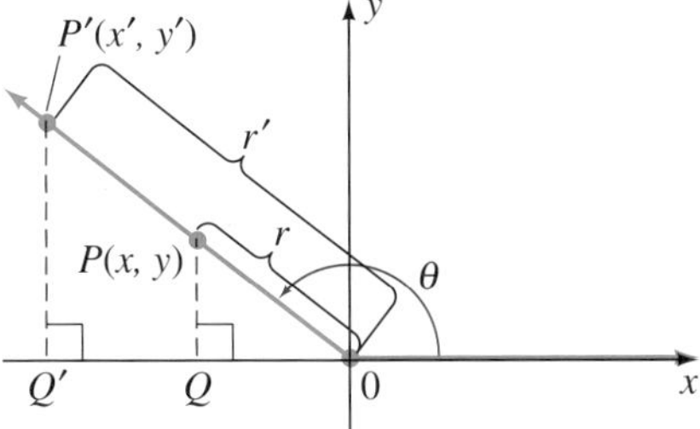

FIGURE 7.25

EXAMPLE 2

Find $\sin\theta$ and $\cos\theta$ if the terminal side of θ lies on the line $y = 2x$ in the third quadrant.

Solution

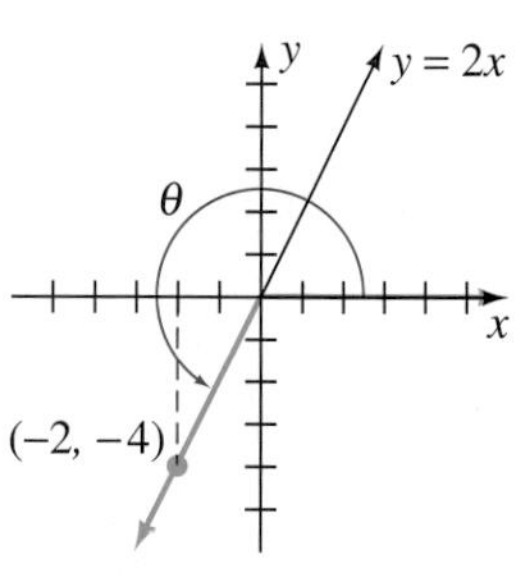

FIGURE 7.26

First let's sketch the line $y = 2x$ (Figure 7.26). The point $(-2, -4)$ on the terminal side of θ can be used to sketch the line. Therefore,

$$\begin{aligned} r &= \sqrt{(-2)^2 + (-4)^2} \\ &= \sqrt{4 + 16} \\ &= \sqrt{20} = 2\sqrt{5}. \end{aligned}$$

Now the values for $\sin\theta$ and $\cos\theta$ can be determined.

$$\sin\theta = \frac{y}{r} = \frac{-4}{2\sqrt{5}} = -\frac{2}{\sqrt{5}} = -\frac{2\sqrt{5}}{5}$$

$$\cos\theta = \frac{x}{r} = \frac{-2}{2\sqrt{5}} = -\frac{1}{\sqrt{5}} = -\frac{\sqrt{5}}{5}$$

EXAMPLE 3

Find $\sin\theta$, $\cos\theta$, and $\tan\theta$ for $\theta = 30°$.

Solution

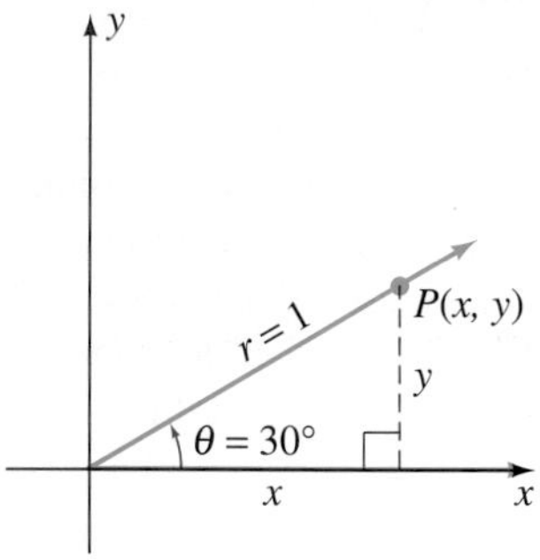

Let's choose a point P on the terminal side of θ so that $r = 1$ (Figure 7.27). Because $\theta = 30°$, the right triangle indicated is a 30°-60° right triangle. Therefore, $y = \frac{1}{2}$ because it is the side opposite the 30° angle. Remember the special 30°-60° right triangle of the previous section, or apply the Pythagorean theorem to determine that $x = \sqrt{3}/2$. Therefore, using $r = 1$, $x = \sqrt{3}/2$, and $y = \frac{1}{2}$, we obtain

FIGURE 7.27

$$\sin 30° = \frac{y}{r} = \frac{\frac{1}{2}}{1} = \frac{1}{2},$$

$$\cos 30° = \frac{x}{r} = \frac{\sqrt{3}/2}{1} = \frac{\sqrt{3}}{2},$$

and

$$\tan 30° = \frac{y}{x} = \frac{\frac{1}{2}}{\sqrt{3}/2} = \frac{1}{\sqrt{3}} = \frac{\sqrt{3}}{3}.$$

Before we consider another example, let's agree on some symbolism. It is customary to omit writing the word *radian* when we use radian measure. For example, an angle θ of radian measure π is usually written as $\theta = \pi$ instead of $\theta = \pi$ radians. However, if the measure is stated in degrees, then be sure to use the degree symbol. In other words, an angle θ of measure 70 degrees is written as $\theta = 70°$ and not as $\theta = 70$.

EXAMPLE 4

Find $\sin\theta$, $\cos\theta$, and $\tan\theta$ for $\theta = \frac{\pi}{4}$.

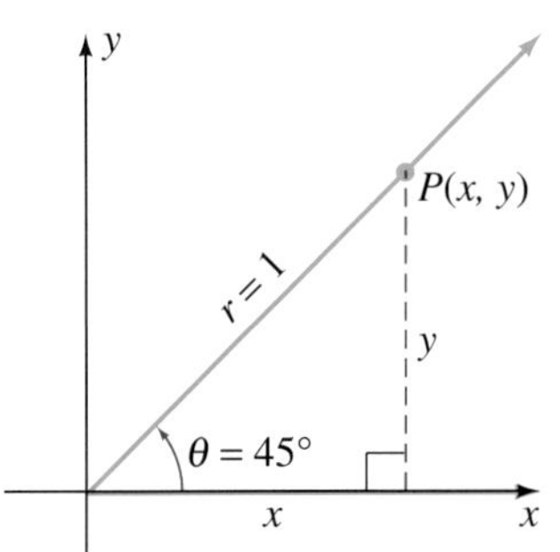

FIGURE 7.28

Solution

Remember that $\pi/4$ radians equals 45°. Therefore, in Figure 7.28 we sketched a 45° angle and chose a point P on the terminal side so that $r = 1$. Since $\theta = 45°$, the indicated right triangle is isosceles with $x = y$. Remember the special isosceles right triangle illustrated in the previous section, or apply the Pythagorean theorem to determine that $x = y = \sqrt{2}/2$. (The values of x and y are both positive since P is in the first quadrant.) Thus, the coordinates of point P are $\left(\sqrt{2}/2, \sqrt{2}/2\right)$ and we can determine the trigonometric functional values of $\pi/4$.

$$\sin \frac{\pi}{4} = \frac{y}{r} = \frac{\sqrt{2}/2}{1} = \frac{\sqrt{2}}{2}$$

$$\cos \frac{\pi}{4} = \frac{x}{r} = \frac{\sqrt{2}/2}{1} = \frac{\sqrt{2}}{2}$$

$$\tan \frac{\pi}{4} = \frac{y}{x} = \frac{\sqrt{2}/2}{\sqrt{2}/2} = 1$$

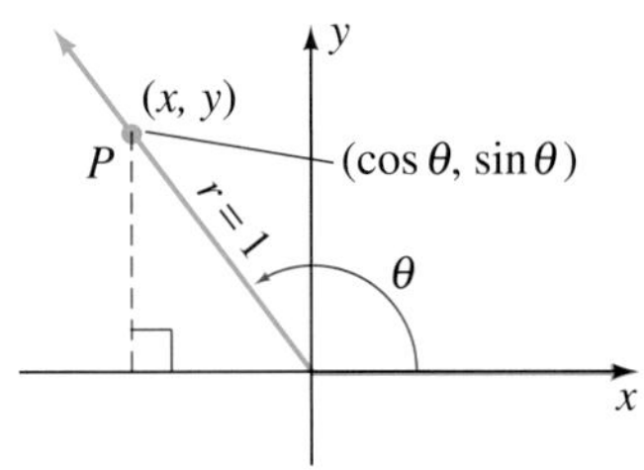

FIGURE 7.29

In Examples 3 and 4 notice the convenience of choosing $r = 1$. In general, by choosing $r = 1$, for any angle θ, we obtain

$$\sin \theta = \frac{y}{r} = \frac{y}{1} = y \qquad \text{and} \qquad \cos \theta = \frac{x}{r} = \frac{x}{1} = x.$$

In other words, if P is a point one unit from the origin on the terminal side of θ, then the coordinates of point P are $(\cos \theta, \sin \theta)$ as indicated in Figure 7.29. Furthermore, the tangent function can be expressed as

$$\tan \theta = \frac{y}{x} = \frac{\sin \theta}{\cos \theta}, \quad \text{for } \cos \theta \neq 0.$$

EXAMPLE 5

Find $\sin \theta$, $\cos \theta$, and $\tan \theta$ for $\theta = 240°$.

Solution

In Figure 7.30 we sketched a 240° angle and indicated a point P on the terminal side so that $r = 1$. The indicated right triangle is a 30°-60° right triangle with the

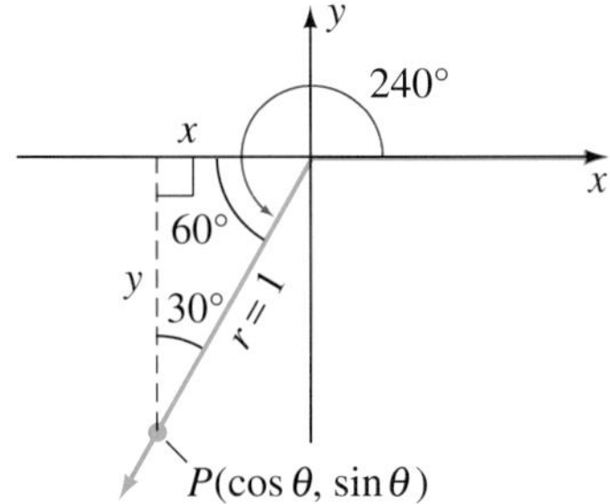

FIGURE 7.30

$30°$ angle at P. Therefore, $x = -\frac{1}{2}$ and $y = -\sqrt{3}/2$ (x and y are both negative because P is in the third quadrant) and the coordinates of P are $\left(-\frac{1}{2}, -\sqrt{3}/2\right)$. Thus, we obtain

$$\sin 240° = y = -\frac{\sqrt{3}}{2},$$

$$\cos 240° = x = -\frac{1}{2},$$

and

$$\tan 240° = \frac{\sin 240°}{\cos 240°} = \frac{-\sqrt{3}/2}{-\frac{1}{2}} = \sqrt{3}.$$

EXAMPLE 6

Find $\sin \theta$, $\cos \theta$, and $\tan \theta$ for $\theta = -240°$.

Solution

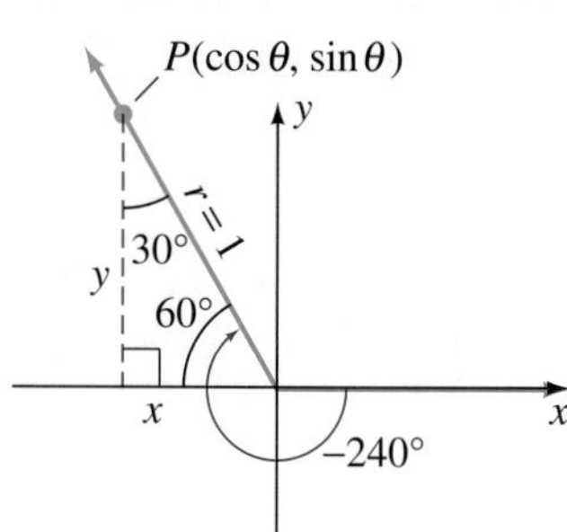

FIGURE 7.31

In Figure 7.31 we sketched an angle of $-240°$ and indicated a point P on the terminal side so that $r = 1$. The indicated right triangle is a 30°-60° right triangle with the 30° angle at P. Therefore, $x = -\frac{1}{2}$ and $y = \sqrt{3}/2$ and we can express the functional values as follows.

$$\sin(-240°) = y = \frac{\sqrt{3}}{2}$$

$$\cos(-240°) = x = -\frac{1}{2}$$

$$\tan(-240°) = \frac{\sin(-240°)}{\cos(-240°)} = \frac{\sqrt{3}/2}{-\frac{1}{2}} = -\sqrt{3}$$

Compare your results for Example 5 and Example 6. Note that $\sin 240° = -\sqrt{3}/2$ and $\sin(-240°) = \sqrt{3}/2$. In other words, $\sin(-240°) = -\sin 240°$. Likewise, observe that $\cos(-240°) = \cos 240°$ and $\tan(-240°) = -\tan 240°$. We will discuss more about some general relationships *of this type* in the next section.

If θ is a quadrantal angle (terminal side lies on an axis), it is quite easy to determine the values of the trigonometric functions, as in the next example.

EXAMPLE 7

Find the values of the six trigonometric functions of θ if $\theta = \pi/2$.

Solution

First let's remember that $\pi/2 = 90°$. Then, as indicated in Figure 7.32, let's choose a point P on the terminal side of a $90°$ angle so that $r = 1$. The coordinates of

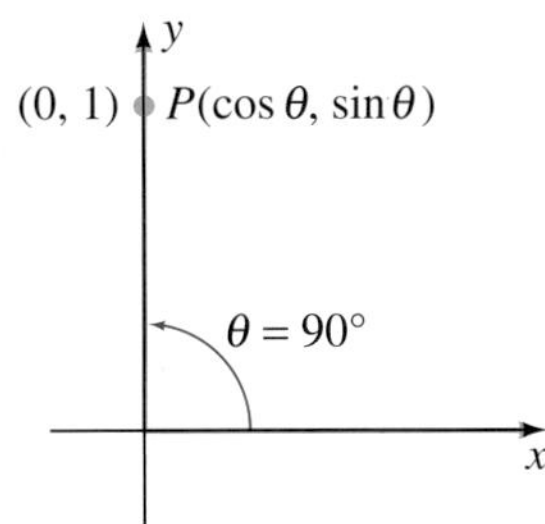

FIGURE 7.32

point P are (0, 1). Therefore,

$$\sin\frac{\pi}{2} = y = 1$$

$$\cos\frac{\pi}{2} = x = 0$$

$$\tan\frac{\pi}{2} = \frac{\sin(\pi/2)}{\cos(\pi/2)} = \frac{1}{0} \qquad \text{Undefined}$$

$$\csc\frac{\pi}{2} = \frac{1}{\sin(\pi/2)} = \frac{1}{1} = 1$$

$$\sec\frac{\pi}{2} = \frac{1}{\cos(\pi/2)} = \frac{1}{0} \qquad \text{Undefined}$$

and

$$\cot\frac{\pi}{2} = \frac{x}{y} = \frac{0}{1} = 0.$$

The reciprocal relationship $\cot\theta = 1/\tan\theta$ cannot be used here because $\tan(\pi/2)$ is undefined.

PROBLEM SET 7.2

For Problems 1–8, point P is on the terminal side of θ, and θ is a positive angle in standard position, less than 360°. Draw θ, and determine the values of the six trigonometric functions of θ.

1. $P(3, -4)$ **2.** $P(-3, -4)$ **3.** $P(-5, 12)$

4. $P(12, 5)$ **5.** $P(1, -1)$ **6.** $P(-1, -1)$

7. $P(-2, -3)$ **8.** $P(3, -2)$

For Problems 9–16, point P is on the terminal side of θ, θ is in standard position, and $0° > \theta > -360°$. Draw θ, and determine the values of the six trigonometric functions of θ.

9. $P(2, 4)$ **10.** $P(1, -3)$ **11.** $P(3, -1)$

12. $P(-2, 2)$ **13.** $P(0, 2)$ **14.** $P(-1, 0)$

15. $P(0, -1)$ **16.** $P(4, 4)$

For Problems 17–34, determine $\sin\theta$, $\cos\theta$, and $\tan\theta$.

17. $\theta = 60°$ **18.** $\theta = 150°$ **19.** $\theta = \frac{3}{4}\pi$

20. $\theta = \frac{7\pi}{6}$ **21.** $\theta = 300°$ **22.** $\theta = 330°$

23. $\theta = -\frac{\pi}{4}$ **24.** $\theta = -\frac{\pi}{3}$ **25.** $\theta = -30°$

26. $\theta = -210°$ **27.** $\theta = 225°$ **28.** $\theta = 315°$

29. $\theta = 390°$ **30.** $\theta = 480°$ **31.** $\theta = 585°$

32. $\theta = 660°$ **33.** $\theta = \frac{23\pi}{6}$ **34.** $\theta = \frac{11\pi}{4}$

35. Complete the following table.

θ	θ in Radians	$\sin\theta$	$\cos\theta$	$\tan\theta$	$\csc\theta$	$\sec\theta$	$\cot\theta$
0°							
30°							
45°							
60°							
90°							
180°							
270°							

36. Find $\sin\theta$ if the terminal side of θ lies on the line $y = x$ in the third quadrant.

37. Find $\cos\theta$ if the terminal side of θ lies on the line $y = -x$ in the second quadrant.

38. Find $\tan\theta$ if the terminal side of θ lies on the line $y = -2x$ in the fourth quadrant.

39. Find $\sin\theta$ if the terminal side of θ lies on the line $y = 3x$ in the first quadrant.

40. If $\sin\theta = -\dfrac{4}{5}$ and the terminal side of θ is in the fourth quadrant, find $\cos\theta$ and $\tan\theta$.

41. If $\cos\theta = -\dfrac{4}{5}$ and the terminal side of θ is in the third quadrant, find $\sin\theta$ and $\cot\theta$.

42. If $\tan\theta = -\dfrac{5}{12}$ and the terminal side of θ is in the second quadrant, find $\sin\theta$ and $\cos\theta$.

43. If $\tan\theta = \dfrac{7}{24}$ and the terminal side of θ is in the first quadrant, find $\sin\theta$ and $\sec\theta$.

44. In which quadrant(s) must the terminal side of θ lie if $\sin\theta$ and $\tan\theta$ are to have the same sign?

45. In which quadrant(s) must the terminal side of θ lie if $\sin\theta$ is negative and $\cos\theta$ is positive?

46. In which quadrant(s) must the terminal side of θ lie if $\sin\theta$, $\cos\theta$, and $\tan\theta$ all have the same sign?

47. In which quadrant(s) must the terminal side of θ lie if $\sin\theta$ and $\cos\theta$ have opposite signs?

For Problems 48–53, determine θ if θ is a positive angle less than 360° and satisfies the stated conditions.

48. $\tan\theta = 1$ and $\sin\theta$ is negative

49. $\cos\theta = \frac{1}{2}$ and $\tan\theta$ is positive

50. $\sin\theta = \sqrt{3}/2$ and $\cos\theta$ is negative

51. $\cos\theta = -\sqrt{3}/2$ and $\sin\theta$ is negative

52. $\cos\theta = -\frac{1}{2}$ and $\tan\theta$ is positive

53. $\sin\theta = -1$ and $\cos\theta = 0$

7.3 TRIGONOMETRIC FUNCTIONS OF ANY ANGLE

Let's begin this section by summarizing some ideas from the previous section and problem set. It is easy to determine the signs (positive or negative) of the trigonometric functions in each of the quadrants. For example, using Definition 7.2 with $r = 1$, we know that $\sin\theta = y$, and therefore $\sin\theta$ is positive in quadrants I and II and negative in quadrants III and IV. Furthermore, because $\csc\theta$ is the reciprocal of $\sin\theta$, its signs will agree with $\sin\theta$. The following chart summarizes the signs of all six trigonometric functions in the four quadrants.

QUADRANT CONTAINING θ	POSITIVE FUNCTIONS	NEGATIVE FUNCTIONS
I	All	None
II	sin, csc	cos, sec, tan, cot
III	tan, cot	sin, csc, cos, sec
IV	cos, sec	sin, csc, tan, cot

Table 7.1 summarizes the trigonometric functional values of some special angles that we have worked with thus far. It will be *very helpful* for you to have these values at your fingertips.

TABLE 7.1

θ	θ in Radians	$\sin\theta$	$\cos\theta$	$\tan\theta$
0°	0	0	1	0
30°	$\frac{\pi}{6}$	$\frac{1}{2}$	$\frac{\sqrt{3}}{2}$	$\frac{\sqrt{3}}{3}$
45°	$\frac{\pi}{4}$	$\frac{\sqrt{2}}{2}$	$\frac{\sqrt{2}}{2}$	1
60°	$\frac{\pi}{3}$	$\frac{\sqrt{3}}{2}$	$\frac{1}{2}$	$\sqrt{3}$
90°	$\frac{\pi}{2}$	1	0	Undefined
180°	π	0	-1	0
270°	$\frac{3\pi}{2}$	-1	0	Undefined

Coterminal Angles

Recall that coterminal angles have the same initial and terminal sides. Therefore, angles in standard position that have the same terminal side are **coterminal**. Coterminal angles differ from each other by a multiple of 360° or 2π radians. For example, angles of 120° and 480° are coterminal as are angles of π and 5π. The angles that are coterminal with any angle θ can be represented by $\theta + 360°n$ or $\theta + 2\pi n$, where n is any integer. Furthermore, it follows from the definitions of the trigonometric functions that **corresponding functions of coterminal angles are equal**. Therefore, using some values from Table 7.1 we can make the following statements.

$$\sin 390° = \sin 30° = \frac{1}{2} \qquad \text{because } 390° \text{ and } 30° \text{ are coterminal}$$

$$\tan 780° = \tan 60° = \sqrt{3} \qquad \text{because } 780° \text{ and } 60° \text{ are coterminal}$$

$$\cos\frac{9\pi}{4} = \cos\frac{\pi}{4} = \frac{\sqrt{2}}{2} \qquad \text{because } \frac{9\pi}{4} \text{ and } \frac{\pi}{4} \text{ are coterminal}$$

$$\sin(-300°) = \sin 60° = \frac{\sqrt{3}}{2} \qquad \text{because } -300° \text{ and } 60° \text{ are coterminal}$$

Reference Angle

The concept of a *reference angle* can be defined as follows.

DEFINITION 7.3

Let θ be any angle in standard position with its terminal side in one of the four quadrants. The **reference angle** associated with θ (we will call it θ') is the acute angle formed by the terminal side of θ and the x-axis.

In Figure 7.33 we have indicated the reference angle θ' for the four different situations in which the terminal side of θ lies in quadrants I, II, III, or IV.

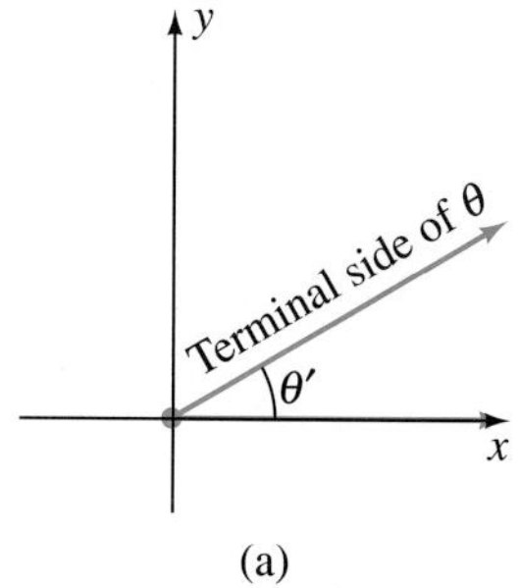

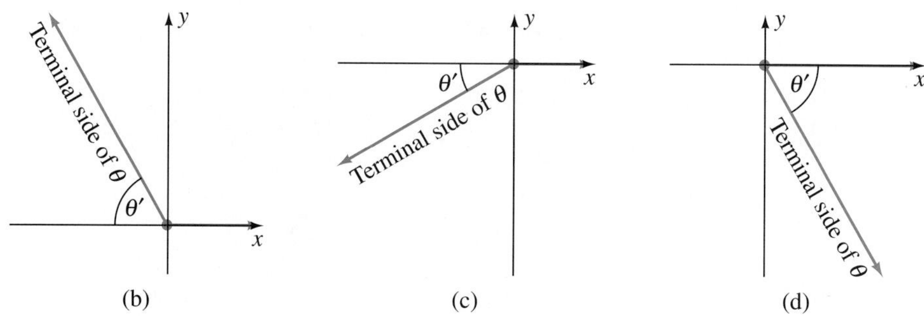

FIGURE 7.33

From our work in the previous section the following fact becomes evident: **The trigonometric functions of any angle θ are equal to those of the reference angle associated with θ, except possibly for the sign. Determine the sign by considering the quadrant in which the terminal side of θ lies.** Let's look at an example.

EXAMPLE 1

Find $\cos\theta$ if $\theta = 225°$.

Solution

In Figure 7.34 we sketched $\theta = 225°$ and indicated its reference angle $\theta' = 45°$. Since the terminal side of θ lies in the third quadrant, $\cos\theta$ is negative. Therefore,

$$\cos 225° = -\cos 45° = -\frac{\sqrt{2}}{2}.$$

FIGURE 7.34

Using the information from Table 7.1 and our knowledge of coterminal angles, reference angles, signs of the trigonometric functions in each of the quadrants, and the reciprocal relationships, we can determine the six trigonometric functions of many special angles.

EXAMPLE 2

Find the six trigonometric functions of θ for $\theta = 510°$.

Solution

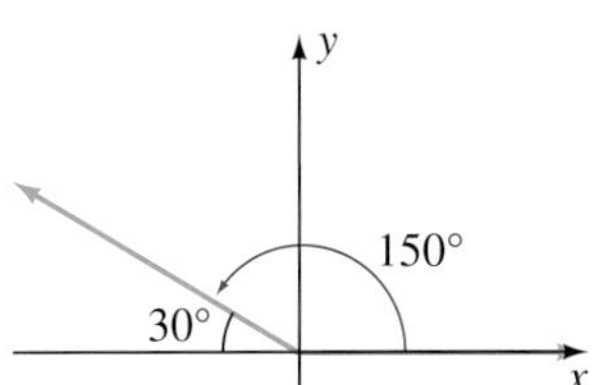

FIGURE 7.35

An angle of 510° is coterminal with an angle of $510° - 360° = 150°$. The reference angle associated with a 150° angle is a 30° angle (Figure 7.35). Because θ is a second-quadrant angle, $\sin\theta$ is positive, $\cos\theta$ is negative, and $\tan\theta$ is negative. Therefore, we obtain

$$\sin 510° = \sin 30° = \frac{1}{2}$$

$$\cos 510° = -\cos 30° = -\frac{\sqrt{3}}{2}$$

$$\tan 510° = -\tan 30° = -\frac{\sqrt{3}}{3}.$$

Using the reciprocal relationships, we obtain

$$\csc 510° = \frac{1}{\sin 510°} = \frac{1}{\frac{1}{2}} = 2$$

$$\sec 510° = \frac{1}{\cos 510°} = \frac{1}{-\sqrt{3}/2} = -\frac{2\sqrt{3}}{3}$$

$$\cot 510° = \frac{1}{\tan 510°} = \frac{1}{-\sqrt{3}/3} = -\sqrt{3}.$$

EXAMPLE 3

Find $\sin\theta$ for $\theta = \dfrac{15\pi}{4}$.

Solution

An angle of $15\pi/4$ radians is coterminal with an angle of $15\pi/4 - 2\pi = 7\pi/4$. The reference angle associated with an angle of $7\pi/4$ radians is an angle of $\pi/4$ radians (Figure 7.36). Because θ is a fourth-quadrant angle, $\sin\theta$ is negative. Therefore,

$$\sin\frac{15\pi}{4} = -\sin\frac{\pi}{4} = -\frac{\sqrt{2}}{2}.$$

FIGURE 7.36

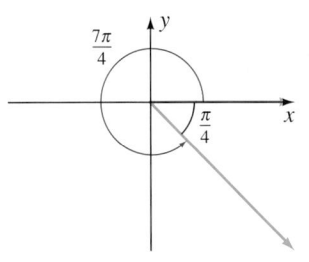

REMARK You may find it easier to solve a problem such as Example 3 by first switching from radian to degree measure. That's acceptable, but it is advantageous in later sections to feel comfortable working with radian measure.

EXAMPLE 4

Find $\cos\theta$ for $\theta = -480°$.

Solution

An angle of $-480°$ is coterminal with an angle of $-480° + 720° = 240°$. The reference angle associated with an angle of $240°$ is a $60°$ angle. (Try to imagine that without drawing a figure.) Because θ is a third-quadrant angle, $\cos\theta$ is negative. Therefore,

$$\cos(-480°) = -\cos 60° = -\frac{1}{2}.$$

Trigonometric Functions of Any Angle

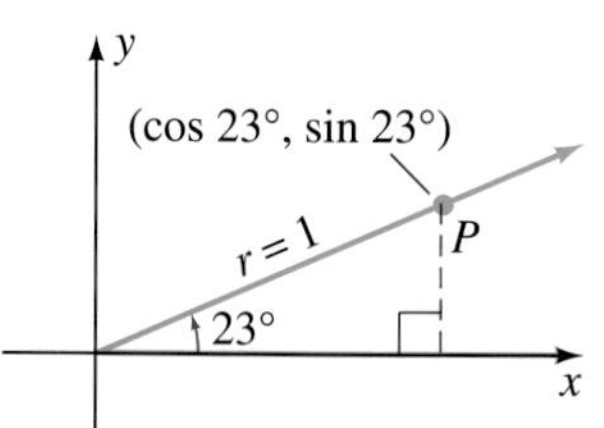

FIGURE 7.37

Until now, we have been looking for *exact values* for the trigonometric functions of some *special angles*. Now suppose we need a value for sin 23°. An approximate value could be found by drawing with a protractor a 23° angle in standard position (Figure 7.37). The ordinate of point P, measured in terms of the unit used for r, is an approximate value for sin 23°. Obviously, this approach would yield a very crude approximation, but it does reemphasize one of the meanings of the trigonometric functions. For our purposes, much better approximations can be found more efficiently by using a table of trigonometric values or a calculator.

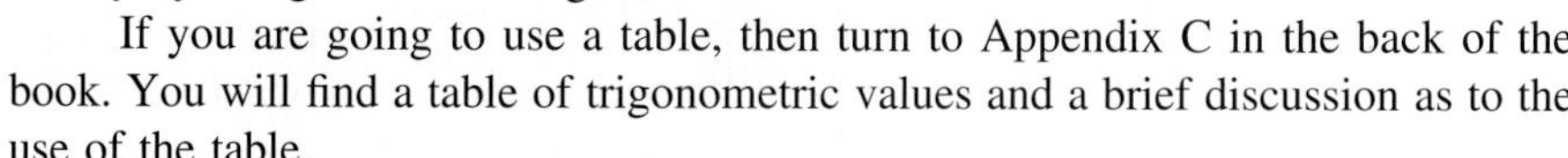

If you are going to use a table, then turn to Appendix C in the back of the book. You will find a table of trigonometric values and a brief discussion as to the use of the table.

The following examples illustrate the use of a calculator to find trigonometric functional values.

EXAMPLE 5

Use a calculator to find the value of

a. sin 23° **b.** cos 212° **c.** tan(−114.2°) **d.** tan 90°

Solutions

First be sure that your calculator is set for degree measure. Your calculator manual will indicate that procedure. Many calculators are automatically set for degree measure when turned on.

a. To find sin 23°, enter the number 23 and press the [SIN] key. The display, to seven decimal places, should read 0.3907311. Therefore, to the nearest ten-thousandth, sin 23° = 0.3907.

b. To find cos 212°, enter the number 212 and press the [COS] key. The display, to seven decimal places, should read −0.8480481. Therefore, to the nearest ten-thousandth, cos 212° = −0.8480.

c. To evaluate tan(−114.2°), enter the number 114.2, press the [+/−] key, and then press the [TAN] key. The display, to seven decimal places, will read 2.2251009. Therefore, to the nearest ten-thousandth, tan(−114.2°) = 2.2251.

d. To attempt to evaluate tan 90°, enter the number 90 and press the TAN key. The display will either blink 9s or give an error message. Either way, it is telling us that tan 90° is undefined.

The trigonometric functions csc θ, sec θ, and cot θ can be evaluated by using the reciprocal relationships, as the next example illustrates.

EXAMPLE 6

Use a calculator to find csc θ, sec θ, and cot θ for $\theta = 57°$.

Solution

Because $\csc 57° = 1/\sin 57°$, we can enter 57, press the SIN key, and then press the 1/x key. This will yield, to the nearest ten-thousandth, $\csc 57° = 1.1924$. In a similar manner, because $\sec 57° = 1/\cos 57°$, we can enter 57, press the COS key, and then press the 1/x key. To the nearest ten-thousandth, we will obtain $\sec 57° = 1.8361$.

Similarly, because $\cot 57° = 1/\tan 57°$, we can enter 57, press the TAN key, and then press the 1/x key. To the nearest ten-thousandth, we will obtain $\cot 57° = 0.6494$.

REMARK We have demonstrated key sequences that work on some calculators. There are other sequences that are used on some calculators. For example, with some calculators, to find sin 23° you first push the SIN key and then enter the number 23. Likewise, some calculators have a x^{-1} key rather than a 1/x key. As always, you need to know how your calculator works.

From Examples 5 and 6 it is evident that using a calculator to evaluate the six trigonometric functions of any angle is very simple. However, we do suggest that you organize your use of the calculator to minimize the chances of making a human error, such as pressing the wrong key. Along this line, we have two specific suggestions to offer at this time. Suppose we want to evaluate $\sin(-23°)$. First, before pressing any keys on your calculator, imagine the terminal side of a $-23°$ angle being in the fourth quadrant. Therefore, $\sin(-23°)$ must be a negative number. Second, after entering -23, check your display to be sure you have entered the correct number. Now, by pressing the SIN key, you should be sure of the result $\sin(-23°) = -0.3907$, to the nearest ten-thousandth. We will use the calculator and involve radian measure in the next chapter.

In Example 6 we used the reciprocal relationships to help determine some trigonometric values. The relationship $\sin^2 \theta + \cos^2 \theta = 1$ can also be used at times for that purpose. For example, suppose that θ is a second-quadrant angle and $\sin \theta = \frac{1}{2}$. We can determine the value for $\cos \theta$ from $\sin^2 \theta + \cos^2 \theta = 1$.

$$\left(\frac{1}{2}\right)^2 + \cos^2 \theta = 1$$

$$\frac{1}{4} + \cos^2\theta = 1$$

$$\cos^2\theta = \frac{3}{4}$$

$$\cos\theta = -\frac{\sqrt{3}}{2}.$$ cos θ is negative since θ is a second-quadrant angle.

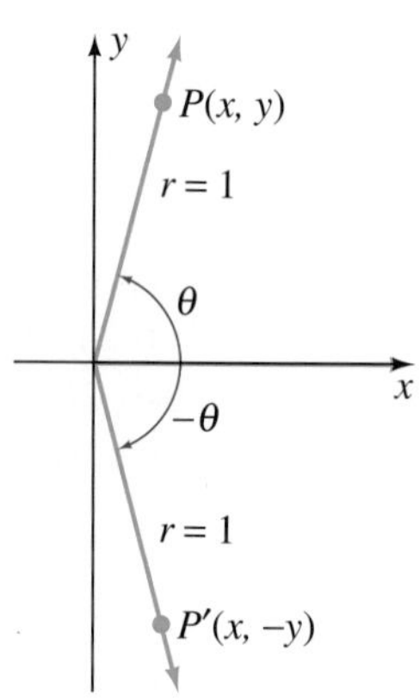

FIGURE 7.38

Some additional relationships are suggested by Figure 7.38. Points P and P' are located one unit from the origin on the terminal sides of θ and $-\theta$, respectively. The points P and P' are x-axis reflections of each other. Therefore, the coordinates of P' are $(x, -y)$. Then the following relationships can be observed.

$$\sin(-\theta) = -y = -\sin\theta$$

$$\cos(-\theta) = x = \cos\theta$$

$$\tan(-\theta) = \frac{-y}{x} = -\frac{y}{x} = -\tan\theta$$

These properties allow us to make statements such as the following.

$$\sin(-30°) = -\sin 30° = -\frac{1}{2}$$

$$\cos(-30°) = \cos 30° = \frac{\sqrt{3}}{2}$$

$$\tan(-30°) = -\tan 30° = -\frac{\sqrt{3}}{3}$$

$$\sin\left(-\frac{\pi}{3}\right) = -\sin\frac{\pi}{3} = -\frac{\sqrt{3}}{2}$$

$$\cos\left(-\frac{\pi}{3}\right) = \cos\frac{\pi}{3} = \frac{1}{2}$$

$$\tan\left(-\frac{\pi}{3}\right) = -\tan\frac{\pi}{3} = -\sqrt{3}$$

PROBLEM SET 7.3

For Problems 1–8, find the quadrant that contains the terminal side of θ if the given conditions are true.

1. $\sin\theta > 0$ and $\cos\theta > 0$

2. $\sin\theta < 0$ and $\cos\theta < 0$

3. $\sin\theta < 0$ and $\cos\theta > 0$

4. $\tan\theta < 0$ and $\cos\theta > 0$

5. $\sin\theta < 0$ and $\cot\theta > 0$

6. $\sec\theta < 0$ and $\tan\theta < 0$

7. $\csc\theta > 0$ and $\cot\theta < 0$

8. $\cos\theta > 0$ and $\cot\theta < 0$

For Problems 9–16, find α such that $0° < \alpha < 360°$ and α is coterminal with θ.

9. $\theta = 510°$ **10.** $\theta = 570°$ **11.** $\theta = 960°$

12. $\theta = 750°$ **13.** $\theta = -60°$ **14.** $\theta = -210°$

15. $\theta = -480°$ **16.** $\theta = -660°$

For Problems 17–22, find α such that $0 < \alpha < 2\pi$ and α is coterminal with θ.

17. $\theta = \frac{7\pi}{2}$ **18.** $\theta = \frac{11\pi}{4}$ **19.** $\theta = \frac{31\pi}{6}$

20. $\theta = \frac{17\pi}{3}$ **21.** $\theta = -\frac{5\pi}{4}$ **22.** $\theta = -\frac{2\pi}{3}$

For Problems 23–34, find the reference angle θ' for each of the given values of θ.

23. $\theta = 265°$ **24.** $\theta = 285.3°$ **25.** $\theta = 431.8°$

26. $\theta = 510°$ **27.** $\theta = -73°$ **28.** $\theta = -190°$

29. $\theta = \frac{5\pi}{4}$ **30.** $\theta = \frac{11\pi}{6}$ **31.** $\theta = \frac{8\pi}{3}$

32. $\theta = \frac{13\pi}{4}$ **33.** $\theta = -\frac{4\pi}{3}$ **34.** $\theta = -\frac{5\pi}{6}$

For Problems 35–66, find exact values. Do not use a calculator or a table.

35. $\sin 120°$ **36.** $\cos 150°$ **37.** $\cos 210°$

38. $\sin 210°$ **39.** $\tan 300°$ **40.** $\tan 315°$

41. $\csc 135°$ **42.** $\sec 240°$ **43.** $\sec 420°$

44. $\csc 480°$ **45.** $\sin(-150°)$ **46.** $\cos(-210°)$

47. $\cos(-300°)$ **48.** $\sin(-390°)$ **49.** $\cot(-930°)$

50. $\cot(-480°)$ **51.** $\sin 630°$ **52.** $\cos 540°$

53. $\cos 315°$ **54.** $\sin(-315°)$ **55.** $\sin \frac{2\pi}{3}$

56. $\cos \frac{3\pi}{4}$ **57.** $\tan \frac{4\pi}{3}$ **58.** $\cot \frac{5\pi}{3}$

59. $\cos \frac{11\pi}{4}$ **60.** $\sin \frac{13\pi}{4}$ **61.** $\cot \frac{13\pi}{3}$

62. $\tan \frac{31\pi}{6}$ **63.** $\sin\left(-\frac{7\pi}{6}\right)$ **64.** $\cos\left(-\frac{5\pi}{3}\right)$

65. $\tan\left(-\frac{3\pi}{2}\right)$ **66.** $\tan(-3\pi)$

For Problems 67–80, use your calculator (or the table in Appendix C) to find approximate values. Express the values to the nearest ten-thousandth.

67. $\sin 75°$ **68.** $\cos 80°$

69. $\tan 256°$ **70.** $\tan 171.4°$

71. $\sin 59.4°$ **72.** $\cos 117.6°$

73. $\cos(-156°)$ **74.** $\sin(-43.7°)$

75. $\sec 15.1°$ **76.** $\csc 114.9°$

77. $\csc(-14.7°)$ **78.** $\cot 214.3°$

79. $\cot 328°$ **80.** $\sec 412.3°$

For Problems 81–88, find the required values using the reciprocal relationships, $\sin^2 \theta + \cos^2 \theta = 1$, $\sin(-\theta) = -\sin \theta$, $\cos(-\theta) = \cos \theta$, and $\tan(-\theta) = -\tan \theta$ as needed.

81. If $\sin \theta = \sqrt{3}/2$ and θ is a second-quadrant angle, find $\cos \theta$.

82. If $\cos \theta = -\frac{1}{2}$ and θ is a third-quadrant angle, find $\sin \theta$.

83. If $\cos \theta = -\sqrt{3}/2$, find $\sec \theta$.

84. If $\sin \theta = -\frac{1}{2}$, find $\csc \theta$.

85. If $\sin \theta = 0.1080$, find $\sin(-\theta)$.

86. If $\cos \theta = 0.2062$, find $\cos(-\theta)$.

87. If $\tan \theta = 1.897$, find $\tan(-\theta)$.

88. If $\sin \theta = \frac{3}{4}$ and θ is a second-quadrant angle, find $\tan \theta$.

THOUGHTS into WORDS

89. Give a step-by-step description of how you would determine sin 855° without using a calculator or a table of trigonometric values.

90. Give a step-by-step description of how you would determine $\cos \frac{17\pi}{3}$ without using a calculator or a table of trigonometric values.

MISCELLANEOUS PROBLEMS

For Problems 91–96, use your calculator to find approximate values to the nearest ten-thousandth.

91. sin 117°6′

92. cos 234°12′

93. tan(−114°48′)

94. sec 221°36′

95. csc 317°54′

96. cot 373°24′

For Problems 97–102, verify each of the statements using a calculator.

97. sin 25° = cos 65°

98. cos 72° = sin 18°

99. sec 10° = csc 80°

100. csc 14.3° = sec 75.7°

101. tan 47° = cot 43°

102. cot 25.7° = tan 64.3°

7.4 RIGHT-TRIANGLE TRIGONOMETRY

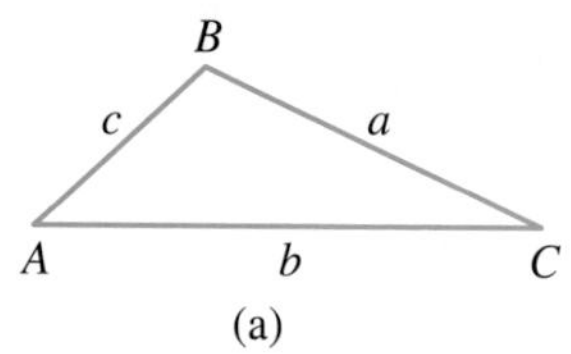

(a)

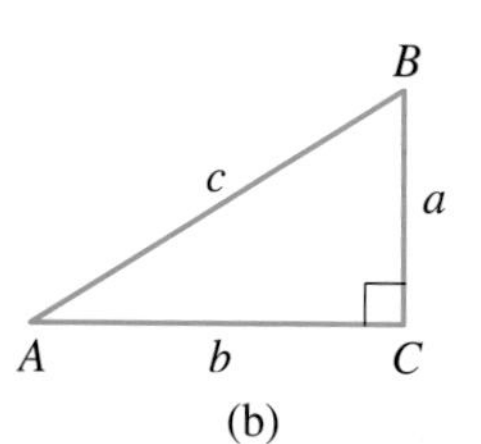

(b)

FIGURE 7.39

In these next three sections we will become involved in the problem solving aspect of trigonometry. More specifically, we will be using triangles to solve problems with a variety of applications. In this context it is customary to designate the vertices and corresponding angles of a triangle by the capital letters A, B, and C. Then the sides opposite the angles A, B, and C are designated by the lowercase letters a, b, and c, respectively (Figure 7.39(a)). If triangle ACB is a right triangle, then we will use C to denote the right angle (Figure 7.39(b)). Furthermore, the sides of a right triangle are often referred to in terms of the acute angles. For example, side a is called the **side opposite** angle A, side b is called the **side adjacent** to angle A, and side c is called the **hypotenuse**. This terminology is summarized in Figure 7.40.

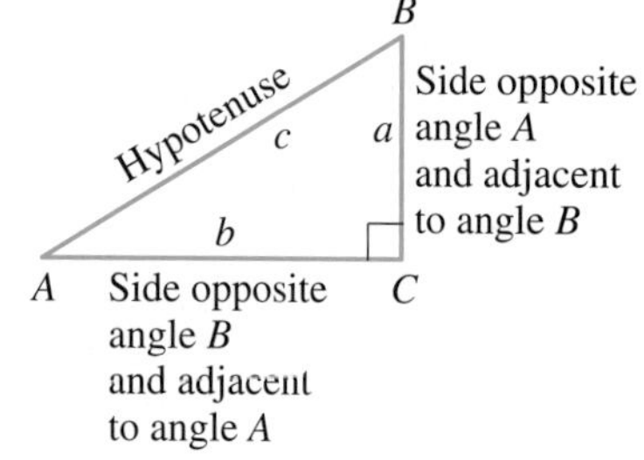

FIGURE 7.40

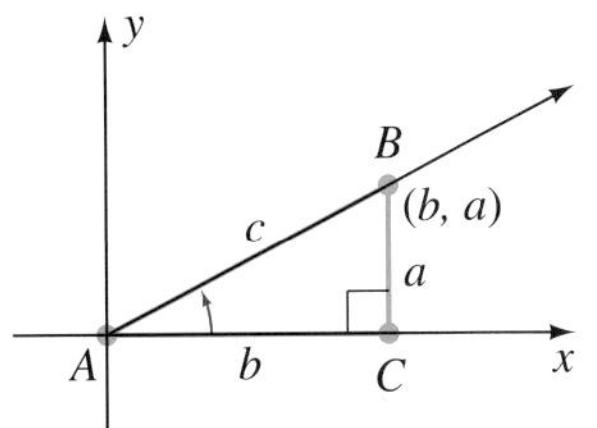

FIGURE 7.41

In Section 7.2 we defined the six basic trigonometric functions in terms of an angle θ in standard position on a Cartesian coordinate system. Now let's consider a special case of this definition, namely, the situation for which $\theta = A$ is an acute angle of a right triangle as indicated in Figure 7.41. The six trigonometric functions can be defined as follows.

$$\sin A = \frac{a}{c} = \frac{\text{side opposite } A}{\text{hypotenuse}} \qquad \csc A = \frac{c}{a} = \frac{\text{hypotenuse}}{\text{side opposite } A}$$

$$\cos A = \frac{b}{c} = \frac{\text{side adjacent } A}{\text{hypotenuse}} \qquad \sec A = \frac{c}{b} = \frac{\text{hypotenuse}}{\text{side adjacent } A}$$

$$\tan A = \frac{a}{b} = \frac{\text{side opposite } A}{\text{side adjacent } A} \qquad \cot A = \frac{b}{a} = \frac{\text{side adjacent } A}{\text{side opposite } A}$$

We could also place right triangle ACB so that angle B is in standard position, and the six trigonometric functions of angle B could be defined in a similar manner. For example,

$$\sin B = \frac{b}{c} = \frac{\text{side opposite } B}{\text{hypotenuse}}.$$

Furthermore, once we have the definitions in terms ofthe ratios of sides of the right triangle, we no longer need to consider the coordinate system. Thus, in right triangle ACB, where C is the right angle, the following statements can be made. Note also that A and B are complementary.

$$\sin A = \frac{a}{c} = \cos B \qquad \tan A = \frac{a}{b} = \cot B \qquad \sec A = \frac{c}{b} = \csc B$$

$$\cos A = \frac{b}{c} = \sin B \qquad \cot A = \frac{b}{a} = \tan B \qquad \csc A = \frac{c}{a} = \sec B$$

Thus, the sine and cosine functions are called **cofunctions**. The tangent and cotangent functions are cofunctions, as are the secant and cosecant functions. In general, **if θ is an acute angle, any trigonometric function of θ is equal to the cofunction of the complement of θ**. We can restate the above relationship in terms of any acute angle θ as follows.

$$\sin \theta = \cos(90° - \theta) \text{ and } \cos \theta = \sin(90° - \theta)$$

$$\tan \theta = \cot(90° - \theta) \text{ and } \cot \theta = \tan(90° - \theta)$$

$$\sec \theta = \csc(90° - \theta) \text{ and } \csc \theta = \sec(90° - \theta)$$

We refer to a triangle as having six parts–three sides and three angles. The phrase *solving a triangle* refers to finding the values of all six parts. Solving a right triangle can be analyzed as follows.

1. If an acute angle and the length of one of the sides is known, then the other acute angle is the complement of the one given. The lengths of the other sides can be determined from equations involving the appropriate trigonometric functions.
2. If the lengths of the two sides are given, the third side can be found by using the Pythagorean theorem. The two acute angles can be determined from equations involving the appropriate trigonometric functions.

Let's consider two examples of solving right triangles.

EXAMPLE 1

Solve the right tirangle in Figure 7.42.

Solution

Because A and B are complementary angles, $B = 90° - 34° = 56°$. Using the tangent function, a can be determined as follows.

$$\tan 34° = \frac{a}{8.1}$$

$$a = 8.1 \tan 34° = 5.5 \text{ centimeters}, \quad \text{to the nearest tenth}$$

FIGURE 7.42

Using the cosine function, c can be determined as follows.

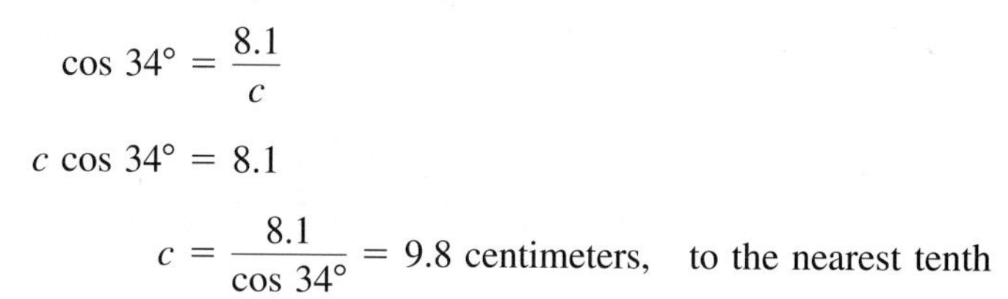

$$\cos 34° = \frac{8.1}{c}$$

$$c \cos 34° = 8.1$$

$$c = \frac{8.1}{\cos 34°} = 9.8 \text{ centimeters}, \quad \text{to the nearest tenth}$$

Before we consider the next right-triangle problem, we should expand our use of the calculator. We need to be able to find the measure of an acute angle when given a trigonometric functional value of the angle. Consider the following examples.

EXAMPLE 2

Find the measure of an acute angle θ when given the following.

a. $\sin \theta = 0.2706$ **b.** $\cos \theta = 0.9449$

c. $\tan \theta = 3.8947$ **d.** $\csc \theta = 1.249$

Solutions

First be sure that your calculator is in the degree mode.

a. Enter the number 0.2706 and press the SIN⁻¹ key. Your result should be $\theta = 15.7°$, to the nearest tenth of a degree. (Some calculators require that you press the SIN⁻¹ key first and then enter the number.)

b. Enter the number 0.9449 and press the COS^{-1} key. Your result should be $\theta = 19.1°$, to the nearest tenth of a degree.

c. Enter the number 3.8947 and press the TAN^{-1} key. Your result should be $\theta = 75.6°$, to the nearest tenth of a degree.

d. Enter the number 1.249 and press the 1/x and SIN^{-1} keys in that order. Your result should be $\theta = 53.2°$, to the nearest tenth of a degree.

REMARK Some calculators have one inverse key labeled INV. Thus, for part (a) you would enter the number and press the INV and SIN keys in that order. We define and discuss the inverse trigonometric functions in the next chapter.

Now let's return to solving some problems with right triangles.

EXAMPLE 3

Solve right triangle ACB where $C = 90°$, $a = 9.4$ meters, and $b = 12.6$ meters.

Solution

Let's sketch the figure and record the known facts (Figure 7.43). Using the Pythagorean theorem, c can be obtained as follows.

$$\begin{aligned} c &= \sqrt{a^2 + b^2} \\ &= \sqrt{(9.4)^2 + (12.6)^2} \\ &= \sqrt{247.12} = 15.7 \text{ meters}, \quad \text{to the nearest tenth of a meter} \end{aligned}$$

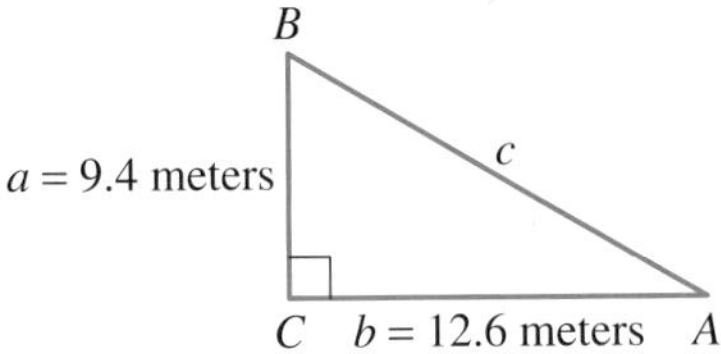

FIGURE 7.43

Angles A and B can be found as follows.

$$\tan A = \frac{9.4}{12.6}$$

$$A = 36.7°, \quad \text{to the nearest tenth of a degree}$$

$$\tan B = \frac{12.6}{9.4}$$

$$B = 53.3°, \quad \text{to the nearest tenth of a degree}$$

Notice in Example 3 that after finding angle A to be 36.7°, we did not subtract this value from 90° to find angle B. In so doing, an error in calculating angle A would produce a corresponding error in angle B. Instead we would suggest finding A and B from the given information and then using the complementary relationship for checking purposes.

Right-triangle trigonometry has a variety of applications. Let's consider a few examples.

EXAMPLE 4

Suppose that a 20-foot ladder is placed against a building so that its lower end is 5 feet from the base of the building (Figure 7.44). What angle does the ladder make with the ground?

FIGURE 7.44

Solution

Using the cosine function, angle A can be determined as follows.

$$\cos A = \frac{5}{20} = 0.2500$$

$$A = 75.5°, \quad \text{to the nearest tenth of a degree}$$

In Figure 7.45 we indicate some terminology that is commonly used in *line-of-sight* type problems. Angles of elevation and depression are measured with reference to a horizontal line. If the object being sighted is above the observer, then the angle formed by the line of sight and the horizontal line is called an **angle of elevation** (Figure 7.45(a)). If the object being sighted is below the observer, then the angle formed by the line of sight and the horizontal line is called an **angle of depression** (Figure 7.45(b)).

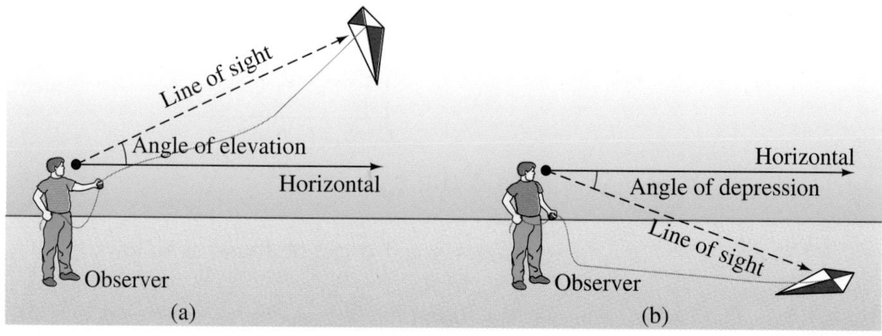

FIGURE 7.45

EXAMPLE 5

At a point 42 meters from the base of a smokestack, the angle of elevation to the top of the stack is 47°. Find the height of the smokestack to the nearest meter.

Solution

Let's sketch a figure and record the given information (Figure 7.46). Let h represent the height of the stack and use the tangent function to find h as follows.

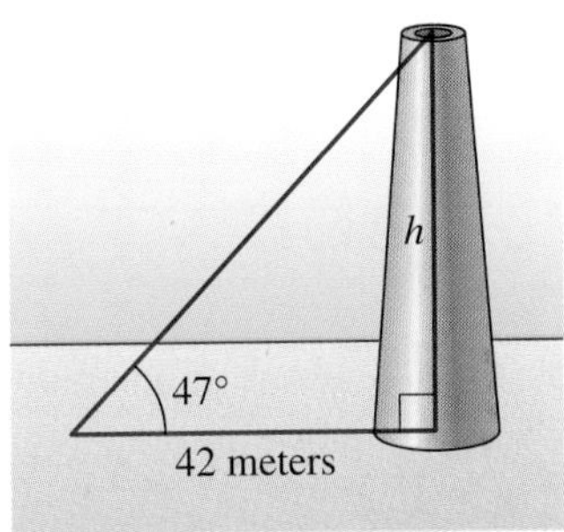

FIGURE 7.46

$$\tan 47° = \frac{h}{42}$$

$$h = 42 \tan 47° = 45, \quad \text{to the nearest meter}$$

The height of the smokestack is approximately 45 meters.

EXAMPLE 6

From the top of a building 350 feet tall, the angle of depression to a special landmark is 71.5°. How far from the base of the building is the landmark?

Solution

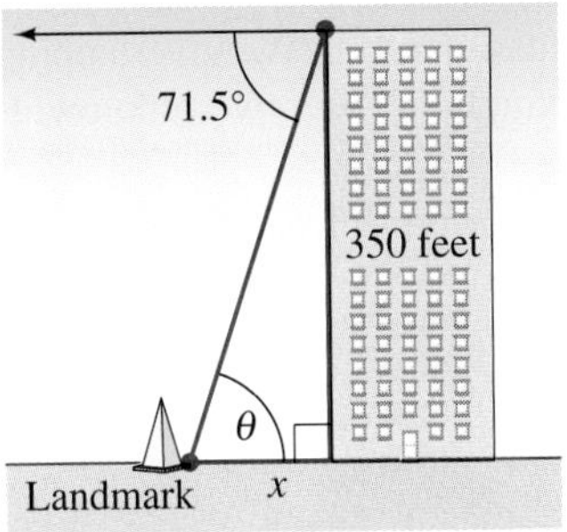

Figure 7.47 depicts the situation described in the problem. Angle θ and the given angle of depression are alternate interior angles; thus, $\theta = 71.5°$. Use the tangent function of θ to find x as follows.

$$\tan 71.5° = \frac{350}{x}$$

$$x \tan 71.5° = 350$$

$$x = \frac{350}{\tan 71.5°} = 117 \text{ feet}, \quad \text{to the nearest foot}$$

FIGURE 7.47

The landmark is approximately 117 feet from the base of the building.

EXAMPLE 7

A TV tower stands on the top of the One Shell Plaza building in Houston, Texas. From a point 3000 feet from the base of the building, the angle of elevation to the top of the TV tower is 18.4°. If the building itself is 714 feet tall, how tall is the TV tower?

Solution

Figure 7.48 depicts the situation described in the problem. Use the tangent function to proceed as follows.

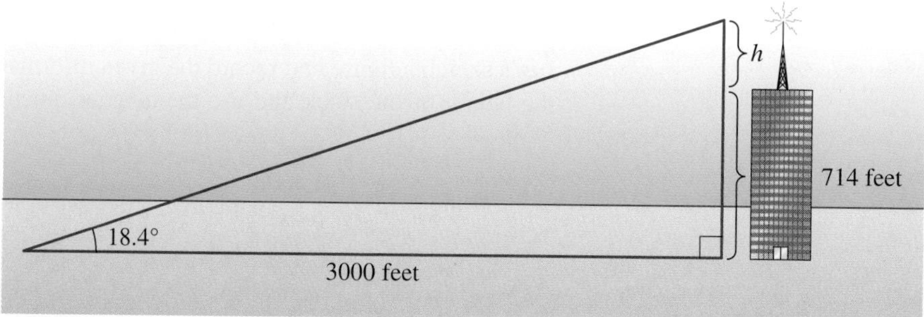

FIGURE 7.48

$$\tan 18.4° = \frac{h + 714}{3000}$$

$$h + 714 = 3000 \tan 18.4°$$

$$h = 3000 \tan 18.4° - 714 = 284 \text{ feet}, \quad \text{to the nearest foot}$$

The TV tower is approximately 284 feet tall.

REMARK Often an important part of the problem solving process is the sketching of a meaningful figure that can be used to record the given information and help in the analysis of the problem. Our sketches were done by professional artists for aesthetic purposes. Your sketches can be very roughly drawn as long as they depict the situation in a way that helps you analyze the problem.

Our final example of this section illustrates a situation where two overlapping right triangles are used to generate a system of two equations with two unknowns.

EXAMPLE 8

In Figure 7.49 a tree is located on the opposite side of a pond from points A and B. From point B the angle of elevation to the top of the tree is 35°. From point A

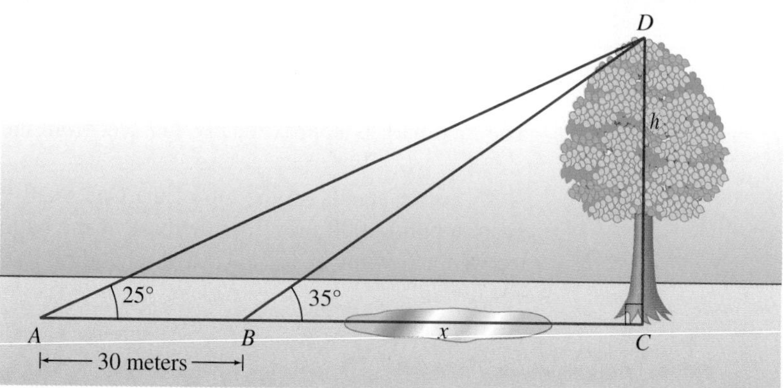

FIGURE 7.49

the angle of elevation to the top of the tree is 25°. If points A and B are 30 meters apart, what is the height of the tree to the nearest tenth of a meter?

Solution

Let x represent the distance between B and C, and h the height of the tree. Using right triangle BCD, we have $\tan 35° = \frac{h}{x}$ where

$$h = x \tan 35°.$$

Using right triangle ACD, we have $\tan 25° = \frac{h}{x + 30}$ where

$$h = (x + 30)\tan 25°.$$

Equating the two values for h and solving for x produces

$$x \tan 35° = (x + 30)\tan 25°$$

$$x \tan 35° = x \tan 25° + 30 \tan 25°$$

$$x \tan 35° - x \tan 25° = 30 \tan 25°$$

$$x(\tan 35° - \tan 25°) = 30 \tan 25°$$

$$x = \frac{30 \tan 25°}{\tan 35° - \tan 25°} = 59.81, \quad \text{to the nearest hundredth.}$$

Substituting 59.81 for x in $h = x \tan 35°$ produces

$$h = 59.81 \tan 35° = 41.9, \quad \text{to the nearest tenth.}$$

The tree is approximately 41.9 meters tall.

PROBLEM SET 7.4

For Problems 1–14, find an acute angle θ, to the nearest tenth of a degree, that satisfies the given conditions.

1. $\sin \theta = 0.2233$
2. $\cos \theta = 0.7902$
3. $\cos \theta = 0.4051$
4. $\sin \theta = 0.8281$
5. $\tan \theta = 0.3365$
6. $\tan \theta = 1.235$
7. $\cot \theta = 0.5704$
8. $\cot \theta = 6.940$
9. $\sin \theta = 0.5182$
10. $\cos \theta = 0.7768$
11. $\tan \theta = 6.400$
12. $\tan \theta = 0.4937$
13. $\sec \theta = 3.103$
14. $\csc \theta = 1.354$

In Problems 15–24, we refer to Figure 7.50. Solve each of the right triangles and express lengths of sides to the nearest unit and angles to the nearest degree.

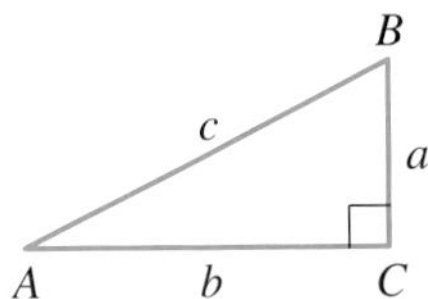

FIGURE 7.50

15. $A = 37°$ and $b = 14$
16. $A = 58°$ and $a = 19$
17. $B = 23°$ and $b = 12$
18. $B = 42°$ and $a = 9$

19. $A = 67°$ and $c = 26$

20. $B = 19°$ and $c = 34$

21. $a = 5$ and $b = 12$

22. $a = 24$ and $c = 25$

23. $b = 12$ and $c = 29$

24. $a = 18$ and $b = 14$

25. A 30-foot ladder, leaning against the side of a building, makes a 50° angle with the ground. How far up on the building does the top of the ladder reach? Express your answer to the nearest tenth of a foot.

26. For safety purposes, a manufacturer recommends that the maximum angle made by a ladder with the ground as it leans against a building be 70°. What is the maximum height, to the nearest tenth of a foot, that the top of a 24-foot ladder can reach? To the nearest tenth of a foot, determine the distance from the bottom of the ladder to the base of the building.

27. From a point 50 meters from the base of a fir tree, the angle of elevation to the top of the tree is 61.5°. What is the height of the tree to the nearest tenth of a meter?

28. Bill is standing on top of a 175-foot cliff overlooking a lake. The measurement of the angle of depression to a boat on the lake is 29°. How far is the boat from the base of the cliff? Express your answer to the nearest foot.

29. From a point 2156 feet from the base of the Sears Tower in Chicago, Illinois the angle of elevation to the top of the tower is 34°. What is the height of the Sears Tower to the nearest foot?

30. A radar station is tracking a missile. The angle of elevation is 22.6° and the distance of the line of sight is 36.8 kilometers (see Figure 7.51). What is the altitude and the horizontal range of the missile to the nearest tenth of a kilometer?

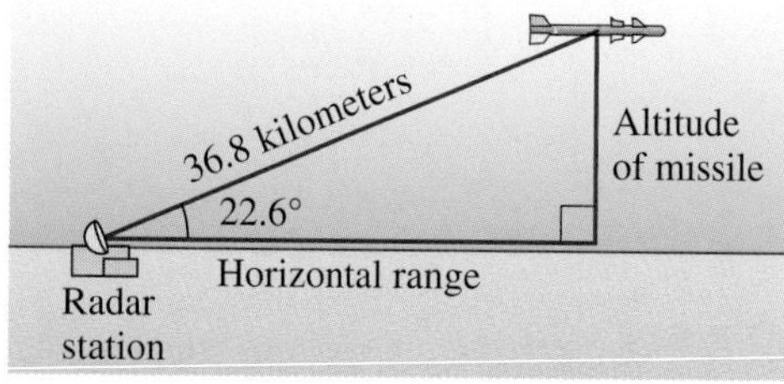

FIGURE 7.51

31. A person wishing to know the width of a river walks 50 yards downstream from a point directly across from a tree on the opposite bank. At this point the angle between the river bank and the line of sight to the tree is 40°. What is the width of the river to the nearest yard?

32. A diagonal of a rectangle is 17 centimeters long and makes an angle of 27° with a side of the rectangle. What is the length and width of the rectangle to the nearest centimeter?

33. Use the information in Figure 7.52 to compute the height of the Gateway Arch in St. Louis, Missouri to the nearest foot.

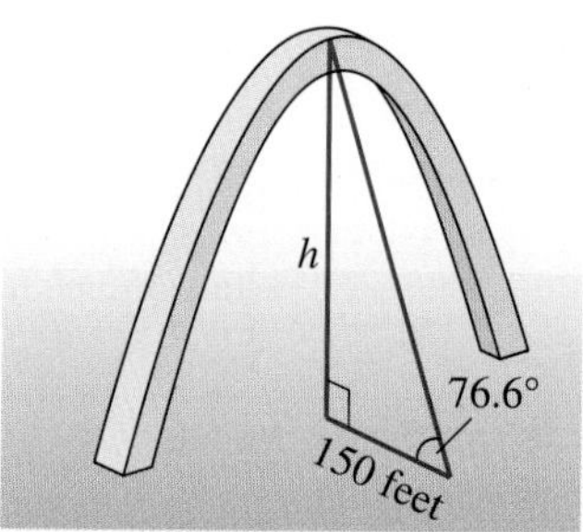

FIGURE 7.52

34. From the top of the Ala Moana Hotel in Honolulu, Hawaii the angle of depression to a landmark on the ground is 57.3°. If the height of the hotel is 390 feet, determine the distance that the landmark is from the hotel, to the nearest foot.

35. A 35-foot-long, upward moving escalator has an angle of elevation of 34°. What is the vertical distance between the floors? Express your answer to the nearest tenth of a foot.

36. The lengths of the three sides of an isosceles triangle are 18 centimeters, 18 centimeters, and 12 centimeters. What is the measure of each of the three angles to the nearest tenth of a degree?

37. A TV tower stands on top of the Empire State Building. From a point 1150 feet from the base of the building, the angle of elevation to the top of the TV tower is 52°. If the building is 1250 feet tall, what is the height of the TV tower to the nearest foot?

38. In Figure 7.53, determine the length of x to the nearest tenth of a meter.

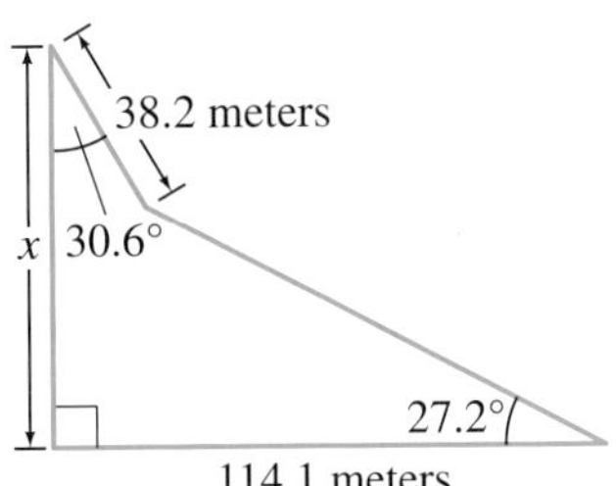

FIGURE 7.53

39. The length of each blade of a pair of shears from the pivot to the point is 8 inches. To the nearest tenth of a degree, what angle do the blades make with each other when the points of the open shears are 6 inches apart?

40. Two people standing 200 feet apart are in line with the base of a tower. The angle of elevation to the top of the tower from one person is 30° and from the other person is 60°. How far is the tower from each person? (Since the angles are 30° and 60°, try doing this problem without your calculator.)

41. Two buildings are separated by an alley that is 15 feet wide. From a second-floor window in one of the buildings, it can be determined that the angle of elevation to the top of the other building is 75° and the angle of depression to the bottom of the building is 50°. Find the height, to the nearest foot, of the building for which these measurements were taken.

42. Two buildings are separated by an alley. From a window 66 feet above the ground in one of the buildings it can be observed that the angle of elevation to the top of the other building is 52°, and the angle of depression to the bottom of the building is 68°. Find the height, to the nearest foot, of the building for which these angle measurements were taken.

43. A TV antenna was placed on top of a building. From a point on the ground 75 feet from the base of the building, a person determines the angle of elevation to the bottom of the antenna to be 48°, and the angle of elevation to the top of the antenna to be 54°. What is the height of the antenna to the nearest foot?

44. Find h, to the nearest tenth of a meter, in Figure 7.54.

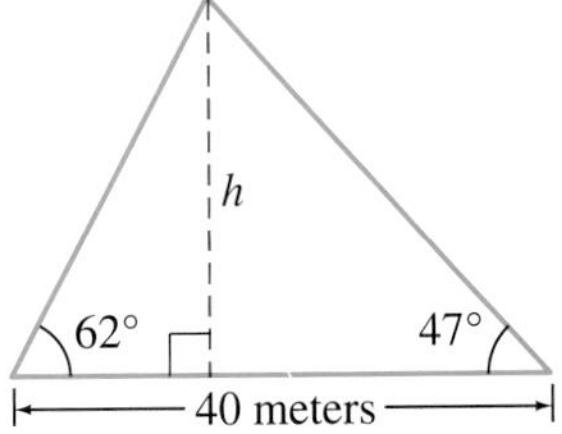

FIGURE 7.54

45. The right circular cone in Figure 7.55 has a radius of 3 feet and a volume of 50 cubic feet. Find the measure of angle θ to the nearest tenth of a degree. The volume of a right circular cone is given by the formula $V = \frac{1}{3}\pi r^2 h$.

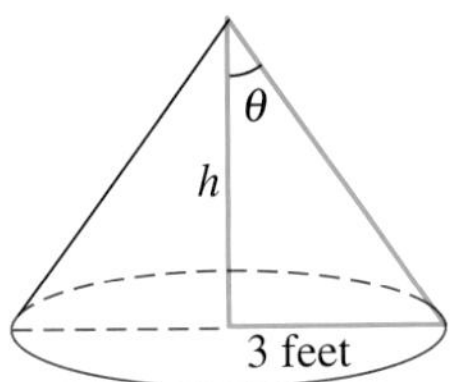

FIGURE 7.55

46. Suppose that a spacelab is circling the earth at an altitude of 400 miles as shown in Figure 7.56. The angle θ in the figure is 65.5°. Use this information to estimate the radius of the earth to the nearest mile.

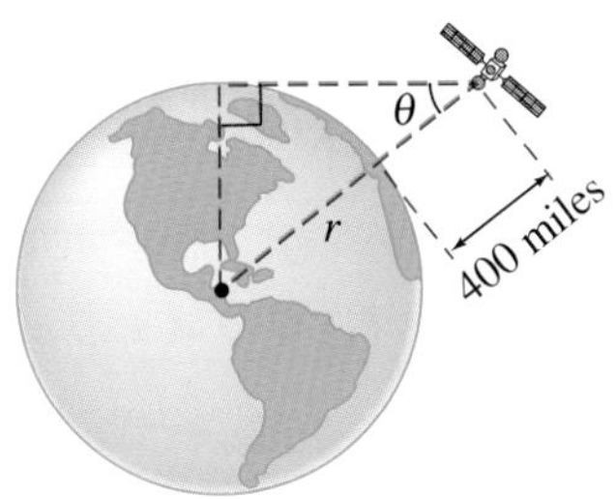

FIGURE 7.56

47. A draw bridge is 160 feet long. As shown in Figure 7.57, the two sections of the bridge can be lifted upward to an angle of 40°. If the water level is 20 feet below the bridge, find the distance d between the end of a section and the water level when the bridge is fully open. Express the answer to the nearest foot.

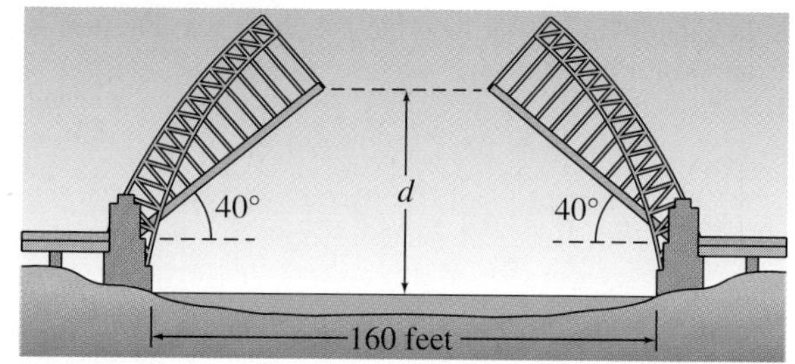

FIGURE 7.57

7.5 SOLVING OBLIQUE TRIANGLES: LAW OF COSINES

In the previous section we solved problems by using the trigonometric functions of the acute angles of right triangles. Now we want to expand our problem solving capabilities to include any kind of triangle. Any triangle that is not a right triangle is called an **oblique triangle**.

In elementary geometry you studied the concept of congruence as it pertains to geometric figures. Two geometric figures are said to be *congruent* if they have exactly the same shape and size; that is to say, they can be made to coincide. You also discovered that there are certain conditions that determine the *congruence of triangles*. For example, the SAS condition states that if two sides and the included angle of one triangle are equal in measure, respectively, to two sides and the included angle of another triangle, then the triangles are *congruent*. In other words, the lengths of two sides and the measure of the angle included by those two sides determines the exact shape and size of the triangle. Example 1 will illustrate this idea.

EXAMPLE 1

Use the information given in Figure 7.58 to find the length of $\overline{BC}$.

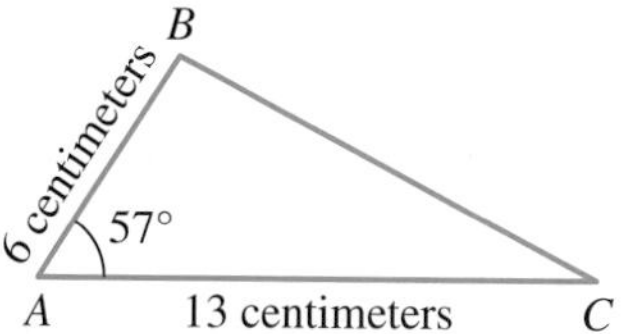

FIGURE 7.58

Solution

Draw a line segment $\overline{BD}$ perpendicular to $\overline{AC}$ to form two right triangles as indicated in Figure 7.59. From triangle ADB the values of x and h can be found as follows.

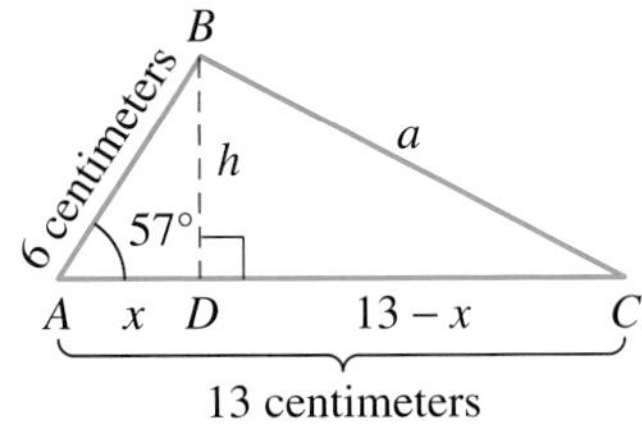

FIGURE 7.59

$$\cos 57° = \frac{x}{6} \quad \text{and} \quad \sin 57° = \frac{h}{6}$$

$$x = 6 \cos 57° = 3.3 \quad \text{and} \quad h = 6 \sin 57° = 5.0, \quad \text{to the nearest tenth}$$

Now, apply the Pythagorean theorem to triangle CDB to obtain

$$\begin{aligned} a^2 &= h^2 + (13 - x)^2 \\ &= (5.0)^2 + (9.7)^2 \qquad 13 - x = 13 - 3.3 = 9.7 \\ &= 25.0 + 94.09 = 119.09. \end{aligned}$$

Thus,

$$a = \sqrt{119.09} = 11 \text{ centimeters}, \quad \text{to the nearest centimeter.}$$

Without carrying out the details, it should be evident that the measures of angles B and C in Figure 7.59 are also determined by the given information. In other words, knowing the lengths of two sides and the measure of the angle included by those two sides does determine the remaining parts of the triangle.

The steps we followed to solve Example 1 point to a general formula for solving like problems. Let's consider a triangle as in Figure 7.60. By drawing $\overline{BD}$ perpendicular to $\overline{AC}$, we can form two right triangles. From triangle ADB we obtain the values of x and h as follows.

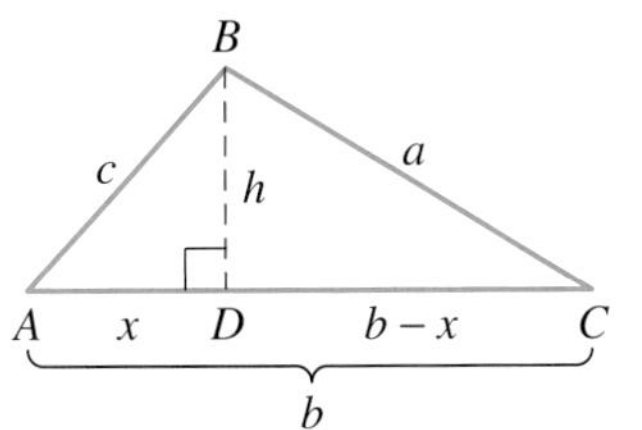

FIGURE 7.60

$$\cos A = \frac{x}{c} \qquad \sin A = \frac{h}{c}$$

$$x = c \cos A \qquad h = c \sin A$$

Use the Pythagorean theorem and triangle CDB to obtain

$$\begin{aligned} a^2 &= h^2 + (b - x)^2 \\ &= h^2 + b^2 - 2bx + x^2. \end{aligned}$$

Now we can substitute $c \cos A$ for x and $c \sin A$ for h and simplify.

$$\begin{aligned} a^2 &= h^2 + b^2 - 2bx + x^2 \\ &= (c \sin A)^2 + b^2 - 2b(c \cos A) + (c \cos A)^2 \\ &= c^2 \sin^2 A + b^2 - 2bc \cos A + c^2 \cos^2 A \\ &= c^2 \sin^2 A + c^2 \cos^2 A + b^2 - 2bc \cos A \\ &= c^2(\sin^2 A + \cos^2 A) + b^2 - 2bc \cos A \\ &= c^2 + b^2 - 2bc \cos A \end{aligned}$$

Remember that $\sin^2 A + \cos^2 A = 1$.

A similar type of development could be used to show that $b^2 = a^2 + c^2 - 2ac \cos B$ and $c^2 = a^2 + b^2 - 2ab \cos C$. We refer to these three relationships as the *law of cosines*, which we formally state as follows.

Law of Cosines

In any triangle ABC having sides of length a, b, and c, the following relationships are true.

$$a^2 = b^2 + c^2 - 2bc \cos A$$

$$b^2 = a^2 + c^2 - 2ac \cos B$$

$$c^2 = a^2 + b^2 - 2ab \cos C$$

REMARK You should realize that the development based on Figure 7.60 assumes angle A to be an acute angle. A similar type of development follows if A is an obtuse angle. This development will be an exercise in the next set of problems.

Using the appropriate part of the law of cosines, problems like Example 1 become easy to solve. Let's consider another example of that type.

EXAMPLE 2

Using the information given in Figure 7.61, find the value of c to the nearest tenth of a meter.

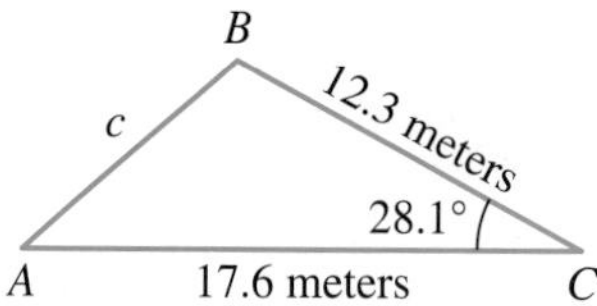

FIGURE 7.61

Solution

Using $c^2 = a^2 + b^2 - 2ab \cos C$, we obtain

$$c^2 = (12.3)^2 + (17.6)^2 - 2(12.3)(17.6)\cos 28.1° = 79.12, \quad \text{to the nearest hundredth.}$$

Therefore,

$$c = \sqrt{79.12} = 8.9 \text{ meters}, \quad \text{to the nearest tenth of a meter.}$$

SSS Condition from Elementary Geometry

Refer to elementary geometry and you may recall the SSS property of congruence. It stated that if the lengths of three sides of one triangle are equal to the lengths of three sides of another triangle, then the triangles are congruent. In other words, if we know the lengths of three sides of a triangle, we can determine its exact shape and size. Furthermore, from the law of cosines we see that cos A, cos B, and cos C can each be expressed in terms of the lengths of the sides as follows.

$$\cos A = \frac{b^2 + c^2 - a^2}{2bc} \qquad \cos B = \frac{a^2 + c^2 - b^2}{2ac} \qquad \cos C = \frac{a^2 + b^2 - c^2}{2ab}$$

Therefore, it becomes evident that the law of cosines can also be used to find the size of the angles of a triangle when the lengths of the three sides are given, as in the next example.

EXAMPLE 3

Find the measure of each angle of a triangle that has sides of length 9 feet, 15 feet, and 19 feet.

Solution

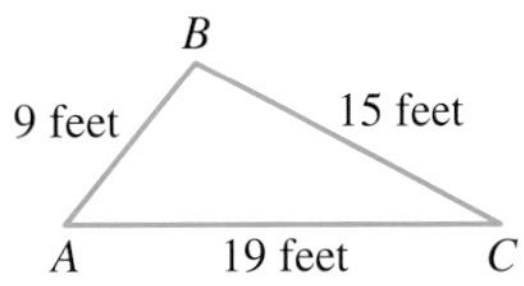

Let's sketch and label a triangle to organize our use of the law of cosines (Figure 7.62).

FIGURE 7.62

$$\cos A = \frac{b^2 + c^2 - a^2}{2bc}$$
$$= \frac{19^2 + 9^2 - 15^2}{2(19)(9)} = \frac{217}{342}$$

Therefore, $A = 50.6°$, to the nearest tenth of a degree.

$$\cos B = \frac{a^2 + c^2 - b^2}{2ac}$$
$$= \frac{15^2 + 9^2 - 19^2}{2(15)(9)} = -\frac{55}{270}$$

Therefore, $B = 101.8°$, to the nearest tenth of a degree.

$$\cos C = \frac{a^2 + b^2 - c^2}{2ab}$$
$$= \frac{15^2 + 19^2 - 9^2}{2(15)(19)} = \frac{505}{570}$$

Therefore, $C = 27.6°$, to the nearest tenth of a degree. As a partial check, we see that $A + B + C = 50.6° + 101.8° + 27.6° = 180°$.

EXAMPLE 4

A vertical pole 50 feet tall stands on a hillside that makes an angle of 25° with the horizontal. A support wire is attached to the top of the pole and to a point on the hillside 35 feet down from the base of the pole. Find the length of the support wire.

Solution

First, let's make a sketch and record the data from the problem (Figure 7.63). From the figure we see that $\angle ABD = 90° - 25° = 65°$. Therefore, $\angle CBE = 180 - 65° = 115°$. Now use triangle EBC and the law of cosines to find the length of the support wire $\overline{EC}$, denoted EC.

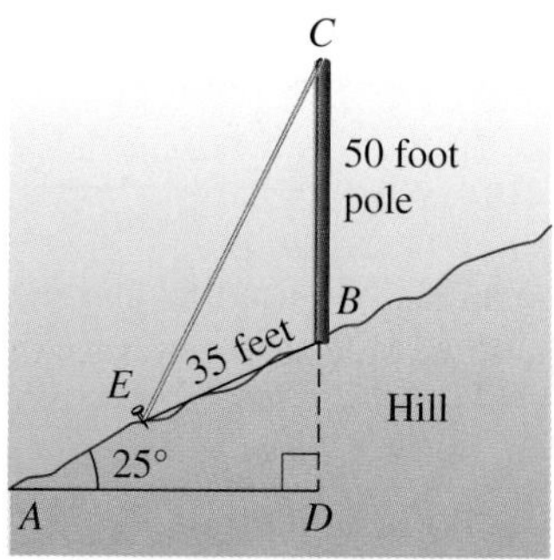

FIGURE 7.63

$$(EC)^2 = (35)^2 + (50)^2 - 2(35)(50)\cos 115°$$
$$= 1225 + 2500 - (-1479.16)$$
$$= 5204.16, \quad \text{to the nearest hundredth}$$

Therefore,

$$EC = \sqrt{5204.16} = 72.1, \quad \text{to the nearest tenth of a foot.}$$

One final comment about the material in this section: Keep in mind that we developed the law of cosines by partitioning an oblique triangle into two right triangles and then using our knowledge of right triangles. Therefore, if we should forget the law of cosines, all is not lost. We can solve a specific problem by forming right triangles and using our knowledge of right-triangle trigonometry as we did in Example 1 of this section.

PROBLEM SET 7.5

Each of the Problems 1–15 refers to triangle ABC. Express measures of angles to the nearest tenth of a degree and lengths of sides to the nearest tenth of a unit.

1. Find a when $b = 8$ centimeters, $c = 12$ centimeters, and $A = 53°$.

2. Find b when $c = 13$ meters, $a = 10$ meters, and $B = 22°$.

3. Find c when $a = 11.6$ feet, $b = 5.1$ feet, and $C = 85°$.

4. Find a when $b = 21.4$ yards, $c = 15.1$ yards, and $A = 74°$.

5. Find b when $a = 27$ centimeters, $c = 21$ centimeters, and $B = 112°$.

6. Find A when $a = 14$ centimeters, $b = 18$ centimeters, and $c = 12$ centimeters.

7. Find B when $a = 17$ feet, $b = 25$ feet, and $c = 17$ feet.

8. Find C when $a = 14.6$ meters, $b = 11.2$ meters, and $c = 4.1$ meters.

9. Find A when $a = 8.3$ centimeters, $b = 16.4$ centimeters, and $c = 11.8$ centimeters.

10. Find B when $a = 7.2$ feet, $b = 11.4$ feet, and $c = 5.1$ feet.

11. Find A, B, and C when $a = 7$ yards, $b = 10$ yards, and $c = 4$ yards.

12. Find A, B, and C when $a = 7.1$ centimeters, $b = 17.8$ centimeters, and $c = 12.5$ centimeters.

13. Find b to the nearest tenth of a foot, and A and C to the nearest tenth of a degree when $a = 41$ feet, $c = 32$ feet, and $B = 100°$.

14. Find c to the nearest tenth of an inch, and A and B to the nearest tenth of a degree when $a = 8.1$ inches, $b = 14.3$ inches, and $C = 12°$.

15. Find C when $a = 24$ centimeters, $b = 7$ centimeters, and $c = 25$ centimeters.

16. The diagonals of a parallelogram are 30 centimeters and 22 centimeters long and intersect at an angle of 67°. Find the lengths of the sides of the parallelogram to the nearest tenth of a centimeter. (Remember that the diagonals of a parallelogram bisect each other.)

17. A parallelogram has sides of lengths 25 centimeters and 40 centimeters and one of the angles has measure 62°. Find the length of each diagonal to the nearest tenth of a centimeter.

18. A triangular plot of ground measures 50 meters by 65 meters by 80 meters. Find, to the nearest tenth of a degree, the measure of the angle opposite the longest side.

19. City A is located 14 miles due north of city B. City C is located 17 miles from city B on a line that is 13° north

of west from B. Find the distance between city A and city C to the nearest mile.

20. A plane travels 250 miles due east after take-off, then adjusts its course 15° southward and flies another 175 miles. How far is it from the point of departure? Express your answer to the nearest tenth of a mile.

21. To measure the length of a pond, a surveyor determines the measurements indicated in Figure 7.64. Find the length of the pond to the nearest meter.

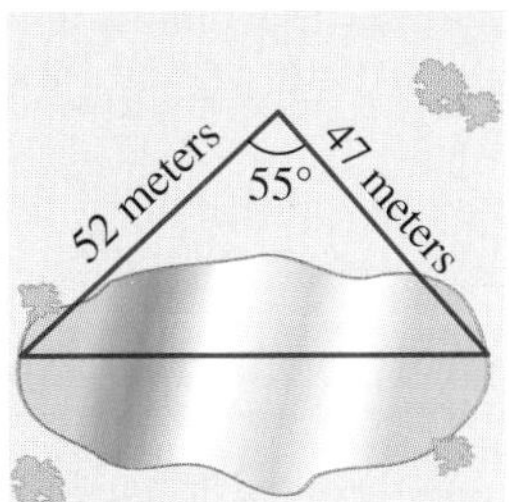

FIGURE 7.64

Problems 22–24 refer to baseball fields that consist of the 90-foot square diamond and the surrounding outfield as indicated in Figure 7.65.

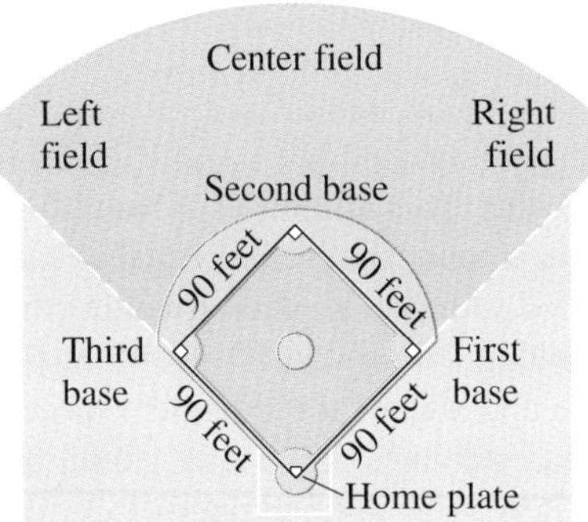

FIGURE 7.65

22. The pitcher's mound is located 60.5 feet from home plate on the diagonal connecting home plate to second base. Find the distance, to the nearest tenth of a foot, between the pitcher's mound and first base.

23. Suppose that the center fielder is standing in straightaway center field 375 feet from home plate. How far is it from where he is standing to third base? Express your answer to the nearest foot.

24. Suppose the left fielder is standing 320 feet from home plate and the line segment that connects her to home plate bisects the line segment connecting second base and third base. How far is she from second base? Express your answer to the nearest foot.

25. Figure 7.66 shows a rooftop with measurements as indicated. Find, to the nearest tenth of a degree, the measure of angle θ.

FIGURE 7.66

26. A solar collector is attached to a roof as indicated in Figure 7.67. Find the length of the collector to the nearest tenth of a foot.

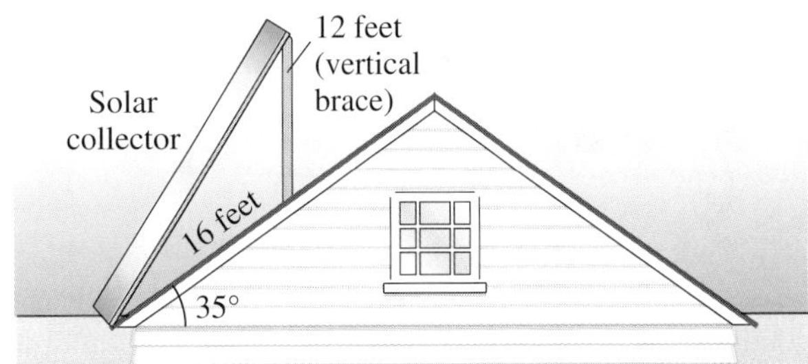

FIGURE 7.67

27. Two airplanes leave an airport at the same time, one going west at 375 miles per hour and the other going northeast at 425 miles per hour. How far apart are they two hours after departure? Express your answer to the nearest mile.

28. Two points A and B are on opposite sides of a building. In order to find the distance between the points, Bob chooses a point C that is 200 feet from A and 175 feet from B, and then determines that angle ACB has a measure of 110°. Find, to the nearest foot, the distance between A and B.

29. The rectangular box in Figure 7.68 is 20 inches long, 12 inches wide, and 10 inches deep. The angle θ is formed by a diagonal of the base and a diagonal of the 12-inch by 10-inch side. Find the measure of angle θ to the nearest tenth of a degree.

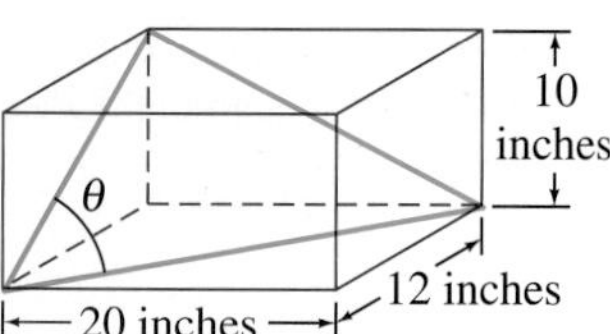

FIGURE 7.68

30. Verify that given the triangle in Figure 7.69 where A is an obtuse angle, $a^2 = b^2 + c^2 - 2bc \cos A$.

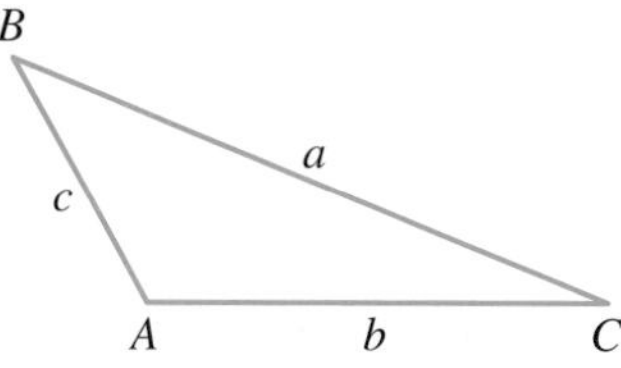

FIGURE 7.69

THOUGHTS into WORDS

31. Suppose that the lengths of the sides of a triangle are given as $a = 7.3$ inches, $b = 14.1$ inches, and $c = 5.2$ inches. Using a calculator to find the measure of angle A, we keep getting an error message. Explain the reason for this.

32. One form of the law of cosines is symbolically stated as

$$\cos A = \frac{b^2 + c^2 - a^2}{2bc}.$$

Express this form in your own words.

7.6 LAW OF SINES

From elementary geometry you may recall the ASA property of congruence. It states that if two angles and the included side of one triangle are equal in measure, respectively, to two angles and the included side of another triangle, then the triangles are congruent. That is to say, knowing the measures of two angles and the included side determines the exact shape and size of the triangle. Furthermore, knowing the size of two angles of a triangle determines the size of the third angle. Therefore, the ASA property can also be stated as an AAS property.

In the previous section we found that the law of cosines can be used to solve triangles when the given information fits the SAS or SSS properties. However, the law of cosines is not useful in the ASA or AAS situations. Instead we need another set of properties, which we refer to as the *law of sines.*

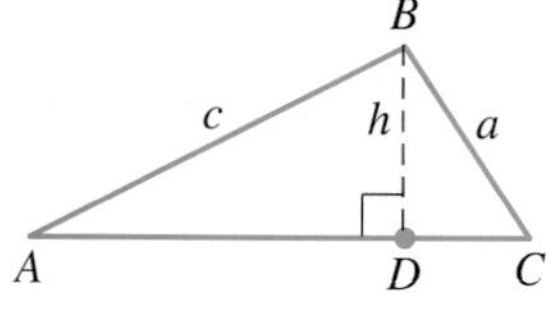

FIGURE 7.70

Let's consider triangle ACB in Figure 7.70 with $\overline{BD}$ drawn perpendicular to $\overline{AC}$. Using right triangle ADB we obtain

$$\sin A = \frac{h}{c}$$

$$h = c \sin A.$$

Using right triangle CDB, we obtain

$$\sin C = \frac{h}{a}$$

$$h = a \sin C.$$

Equating the two expressions for h produces

$$a \sin C = c \sin A,$$

which can be written

$$\frac{a}{\sin A} = \frac{c}{\sin C}.$$

Returning to triangle ACB in Figure 7.70 we could also draw a line segment from vertex C perpendicular to AB. Similar reasoning would result in the relationship

$$\frac{a}{\sin A} = \frac{b}{\sin B}.$$

Therefore, the law of sines can be stated as follows.

Law of Sines

In any triangle ABC, having sides of lengths a, b, and c, the following relationships are true.

$$\frac{a}{\sin A} = \frac{b}{\sin B} = \frac{c}{\sin C}$$

REMARK The previous development of the law of sines was based on Figure 7.70, which has angle A as an acute angle. A similar type of development follows if A is an obtuse angle. The details of this will be covered in the next set of problems.

EXAMPLE 1

Suppose that in triangle ACB, $A = 71°$, $C = 40°$, and $a = 19$ centimeters. Find c to the nearest tenth of a centimeter.

Solution

Let's sketch a triangle and record the given information (Figure 7.71). From the law of sines $\frac{a}{\sin A} = \frac{c}{\sin C}$. We can substitute 19 for a, 71° for A, 40° for C, and solve for c.

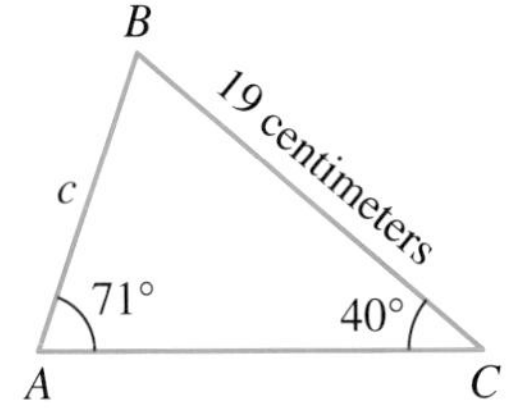

FIGURE 7.71

$$\frac{a}{\sin A} = \frac{c}{\sin C}$$

$$\frac{19}{\sin 71°} = \frac{c}{\sin 40°}$$

$$c \sin 71° = 19 \sin 40°$$

$$c = \frac{19 \sin 40°}{\sin 71°} = 12.9 \text{ centimeters}, \quad \text{to the nearest tenth of a centimeter}$$

EXAMPLE 2

Two points A and B are on opposite sides of a river. Point C is located on the same side of the river and 350 feet from A. In triangle ACB, $C = 52°$ and $A = 67°$. Find the distance between A and B to the nearest foot. (See Figure 7.72.)

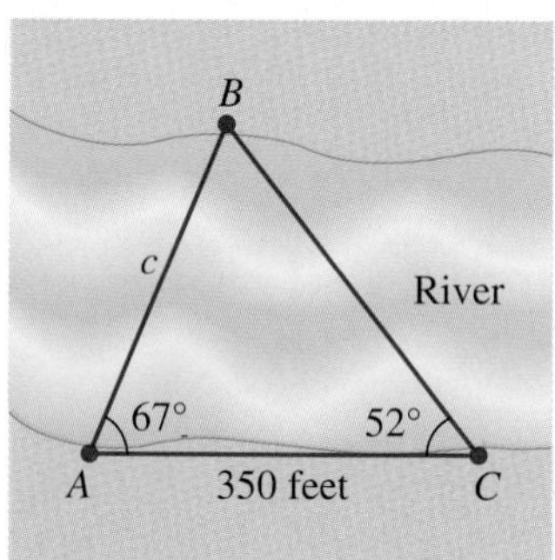

FIGURE 7.72

Solution

Because $A + B + C = 180°$, $A = 67°$, and $C = 52°$, we have

$$67° + B + 52° = 180°$$

$$B + 119° = 180°$$

$$B = 61°.$$

Now using $\frac{b}{\sin B} = \frac{c}{\sin C}$, we obtain

$$\frac{350}{\sin 61°} = \frac{c}{\sin 52°}$$

$$c \sin 61° = 350 \sin 52°$$

$$c = \frac{350 \sin 52°}{\sin 61°} = 315 \text{ feet}, \quad \text{to the nearest foot.}$$

The next example illustrates a situation where both the law of cosines and the law of sines can be used to find the measure of an angle, *but* one approach may be preferable to the other. Let's solve the problem both ways to illustrate our point.

EXAMPLE 3

In Figure 7.73, $A = 23.1°$, $a = 14$ yards, $b = 21$ yards, and $c = 8$ yards. Find B to the nearest tenth of a degree.

8 yards B 14 yards
23.1°
A 21 yards C

FIGURE 7.73

Solution A

Using the law of cosines we obtain

$$\cos B = \frac{a^2 + c^2 - b^2}{2ac}$$
$$= \frac{14^2 + 8^2 - 21^2}{2(14)(8)} = -\frac{181}{224}.$$

Therefore,

$$B = 143.9°, \quad \text{to the nearest tenth of a degree.}$$

Solution B

Using the law of sines we obtain

$$\frac{a}{\sin A} = \frac{b}{\sin B}$$
$$\frac{14}{\sin 23.1°} = \frac{21}{\sin B}$$
$$14 \sin B = 21 \sin 23.1°$$
$$\sin B = \frac{21 \sin 23.1°}{14}.$$

With a calculator we obtain $B = 36.1°$, to the nearest tenth of a degree. But remember that the sine function is also positive in the second quadrant. Therefore, B might be an obtuse angle with a measure of $180° - 36.1° = 143.9°$.

Note that the law of sines does not indicate which value of B should be used for a particular problem. However, in this problem our sketch of the triangle in Figure 7.73 clearly indicates that B is an obtuse angle and, thus, we need to use $B = 143.9°$.

Relative to Solution B, situations do arise where a rough sketch of the triangle may not be sufficient to figure out which value of an angle determined by the law of sines is to be used. Therefore, in general, we suggest that you use the law of cosines whenever possible.

SSA Situation (The Ambiguous Case)

Thus far, the law of cosines and the law of sines have provided us with methods for solving triangles when the given information fits the SAS, SSS, ASA, or AAS patterns. But what happens if the given information fits the SSA pattern? In other words, how do we solve a triangle if we know the lengths of two sides and the measure of an angle opposite one of those two sides? This question is a little more difficult to answer because the given information of *the lengths of two sides and the measure of an angle opposite one of those two sides* may determine two triangles, one triangle, or no triangle at all. Let's see why this is true. Consider triangle ACB and assume that we are given A, a, and c. The situations shown in Figure 7.74 may exist.

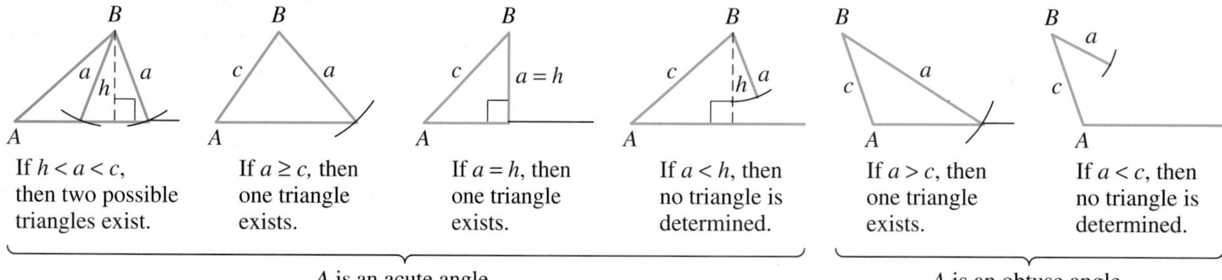

If $h < a < c$, then two possible triangles exist.
If $a \geq c$, then one triangle exists.
If $a = h$, then one triangle exists.
If $a < h$, then no triangle is determined.
If $a > c$, then one triangle exists.
If $a < c$, then no triangle is determined.
A is an acute angle.
A is an obtuse angle.

FIGURE 7.74

Fortunately, it is not necessary to memorize all of the previous possibilities; each problem can be analyzed on an individual basis. Frequently, by making a careful sketch and using some common sense, we can determine which situation exists. The important issue is that we are alert to the fact that when the given information is of the SSA pattern, then various possibilities might exist. Let's analyze a few problems.

EXAMPLE 4

Suppose we are given $A = 57°$, $a = 17$ feet, and $c = 14$ feet. How many triangles exist that satisfy these conditions? Find C for each such triangle.

Solution

Let's make a careful sketch (Figure 7.75). Because $17 > 14$ there is only one possible triangle determined. We can use the law of sines to determine C.

FIGURE 7.75

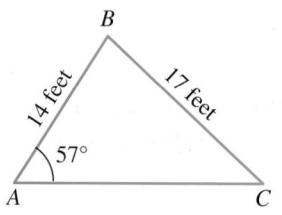

$$\frac{a}{\sin A} = \frac{c}{\sin C}$$

$$\frac{17}{\sin 57°} = \frac{14}{\sin C}$$

$$17 \sin C = 14 \sin 57°$$

$$\sin C = \frac{14 \sin 57°}{17}$$

Therefore,

$$C = 43.7° \quad \text{or} \quad C = 136.3°, \quad \text{to the nearest tenth of a degree.}$$

Either from the figure or from the fact that $57° + 136.3° > 180°$, we know that the solution $136.3°$ must be discarded and $C = 43.7°$.

EXAMPLE 5

Suppose we are given $A = 43.2°$, $a = 7.7$ meters, and $c = 9.1$ meters. How many triangles exist that satisfy these conditions? Find C for each such triangle.

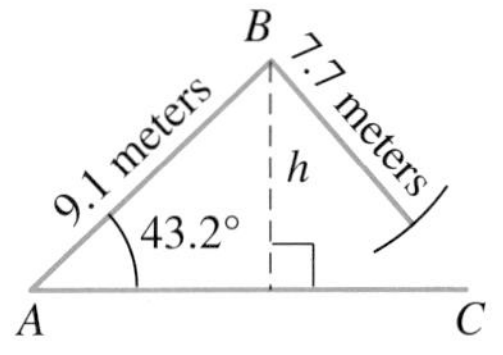

FIGURE 7.76

Solution

As we attempt to sketch a triangle for this situation (Figure 7.76), we might not be able to tell whether two, one, or no triangles exist. Therefore, let's find an approximate value for h.

$$\sin 43.2° = \frac{h}{9.1}$$

$$h = 9.1 \sin 43.2° = 6.2 \text{ meters}, \quad \text{to the nearest tenth of a meter}$$

Now, because $h < a < c$, we know that two triangles exist as indicated in Figure 7.77. Using the law of sines we can find the two possible values for C as follows.

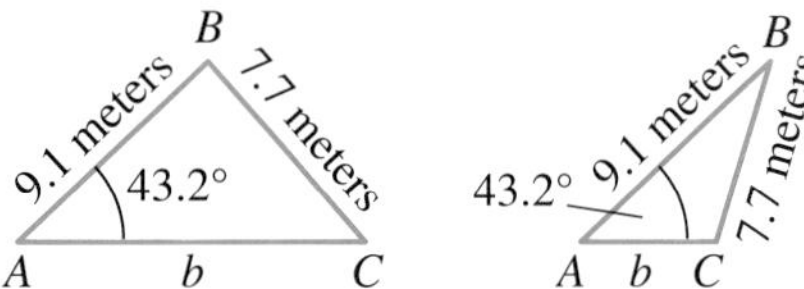

FIGURE 7.77

$$\frac{a}{\sin A} = \frac{c}{\sin C}$$

$$\frac{7.7}{\sin 43.2°} = \frac{9.1}{\sin C}$$

$$7.7 \sin C = 9.1 \sin 43.2°$$

$$\sin C = \frac{9.1 \sin 43.2°}{7.7}$$

Therefore,

$C = 54.0°$ or $C = 180° - 54.0° = 126.0°$, to the nearest tenth of a degree.

Each of the two values for angle C in Example 5 produces a different value for the length of side b as indicated by the two triangles in Figure 7.77. If $C = 54.0°$, then $B = 180° - (43.2° + 54.0°) = 82.8°$. Then b can be determined by using the law of sines.

$$\frac{b}{\sin 82.8°} = \frac{7.7}{\sin 43.2°}$$

$$b \sin 43.2° = 7.7 \sin 82.8°$$

$$b = \frac{7.7 \sin 82.8^\circ}{\sin 43.2^\circ}$$

$$b = 11.2 \text{ meters}, \quad \text{to the nearest tenth of a meter}$$

If $C = 126.0^\circ$, then $B = 180^\circ - (43.2^\circ + 126.0^\circ) = 10.8^\circ$. Then b can be determined as follows.

$$\frac{b}{\sin 10.8^\circ} = \frac{7.7}{\sin 43.2^\circ}$$

$$b \sin 43.2^\circ = 7.7 \sin 10.8^\circ$$

$$b = \frac{7.7 \sin 10.8^\circ}{\sin 43.2^\circ}$$

$$b = 2.1 \text{ meters}, \quad \text{to the nearest tenth of a meter}$$

EXAMPLE 6

Suppose we are given $A = 68^\circ$, $a = 22$ inches, and $c = 25$ inches. How many triangles exist that satisfy these conditions? Find C for each such triangle.

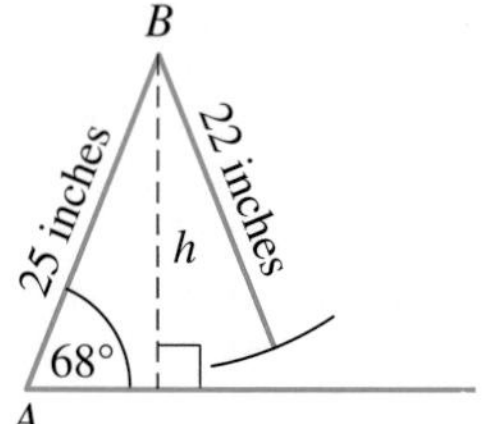

FIGURE 7.78

Solution

As we attempt to sketch a triangle with these conditions (Figure 7.78), we realize that we may not be able to tell whether two, one, or no triangles exist. Therefore, let's find an approximate value for h.

$$\sin 68^\circ = \frac{h}{25}$$

$$h = 25 \sin 68^\circ \approx 23.2 \text{ inches.}$$

Because $a < h$, no triangle exists that satisfies these conditions.

In Example 6, had we attempted to find c using the law of sines, the following situation would have arisen.

$$\frac{a}{\sin A} = \frac{c}{\sin C}$$

$$\frac{22}{\sin 68^\circ} = \frac{25}{\sin C}$$

$$22 \sin C = 25 \sin 68^\circ$$

$$\sin C = \frac{25 \sin 68^\circ}{22}$$

Using a calculator at this stage produces an error message. The reason for this message can be found by evaluating $(25 \sin 68^\circ)/22$. Its value is approximately 1.054, and we know that the sine of an angle cannot exceed 1. Thus, we conclude that the original set of conditions in this problem does not determine a triangle.

PROBLEM SET 7.6

Each of Problems 1–12 refers to triangle ABC. Express measures of angles to the nearest tenth of a degree and lengths of sides to the nearest tenth of a unit.

1. Find c when $A = 64°$, $C = 47°$, and $a = 17$ centimeters.
2. Find c when $A = 28°$, $C = 61°$, and $a = 6$ feet.
3. Find a when $A = 20.4°$, $B = 31.2°$, and $b = 25$ meters.
4. Find b when $B = 115°$, $C = 32°$, and $c = 6.1$ yards.
5. Find B and then b when $A = 41°$, $C = 37°$, and $a = 14$ centimeters.
6. Find C and then c when $A = 26.3°$, $B = 94.5°$, and $a = 9.2$ feet.
7. Find B and then b when $A = 132°$, $C = 17°$, and $a =$ 75 miles.
8. Find B and C when $A = 34.1°$, $a = 23$ feet, $b = 35$ feet, and $c = 17$ feet.
9. Find A and C when $B = 71.7°$, $a = 15$ miles, $b = 17$ miles, and $c = 14$ miles.
10. Find A and B when $C = 136°$, $a = 6$ kilometers, $b = 8$ kilometers, and $c = 13$ kilometers.
11. Find b and c to the nearest tenth of a foot when $A = 54°$, $C = 33°$, and $a = 28$ feet.
12. Find a and c to the nearest tenth of a yard when $A =$ 144°, $B = 19°$, and $b = 12$ yards.

In Problems 13–20, first decide whether two, one, or no triangles are determined by the given information. If one or two triangles are determined, calculate all possible values for the indicated part. Express measures of angles to the nearest tenth of a degree and lengths of sides to the nearest tenth of a unit.

13. Find C when $A = 59°$, $a = 14$ centimeters, and $c = 9$ centimeters.
14. Find C when $A = 17°$, $a = 7$ feet, and $c = 22$ feet.
15. Find C when $A = 53.1°$, $a = 10$ meters, and $c = 14$ meters.
16. Find C when $A = 119°$, $a = 25$ yards, and $c = 12$ yards.
17. Find B, C, and b when $A = 28°$, $a = 19$ miles, and $c = 32$ miles.
18. Find B and C when $A = 55°$, $a = 31$ feet, and $c = 34$ feet.
19. Find C when $A = 30°$, $a = 21$ centimeters, and $c = 42$ centimeters.
20. Find C and b when $A = 124°$, $a = 21$ yards, and $c =$ 27 yards.
21. Verify that $\frac{a}{\sin A} = \frac{c}{\sin C}$ for a triangle in which A is an obtuse angle.
22. Two points A and B are located on opposite sides of a river. Point C is located 275 meters from B on the same side of the river. In triangle ABC, $B = 71°$ and $C = 46°$. Find the distance between A and B to the nearest meter.
23. In Figure 7.79, a surveyor is standing at point A wanting to locate point C such that A, B, and C are on a straight line. However, his line of sight toward C is blocked by a building. Therefore, he determines angle ABP to be 140°, the distance from B to P to be 75 feet, and angle BPC to be 58°. What is the distance from P to C to the nearest tenth of a foot?

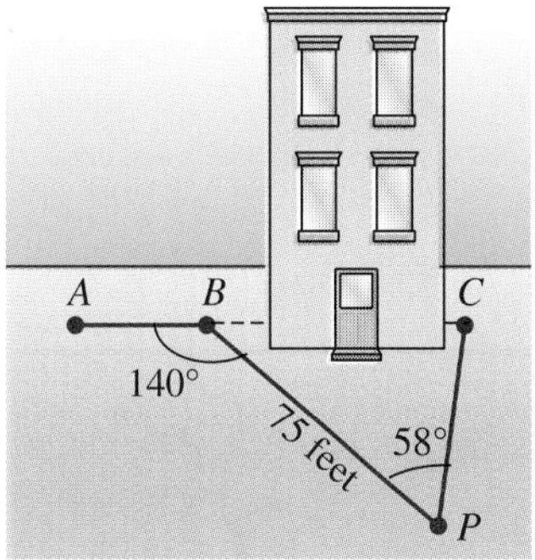

FIGURE 7.79

24. In Figure 7.80, one end of a 20-foot plank is placed on the ground at a point 8 feet from the start of a 41° incline. The other end of the plank rests on the incline. How far

up the incline does the plank extend? Express your answer to the nearest tenth of a foot.

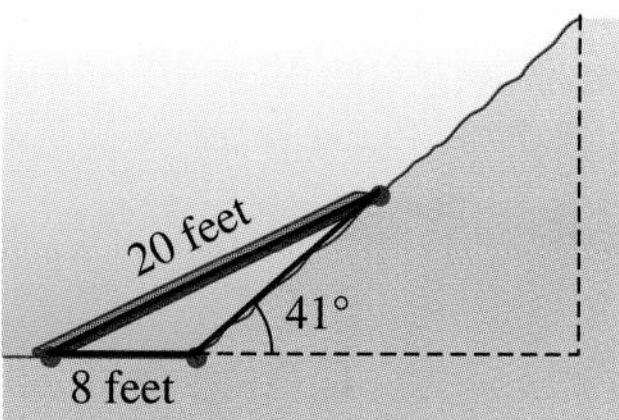

FIGURE 7.80

25. A 200-foot TV antenna is positioned at the top of a hill that has an incline of 23° with the horizontal. How far down the hill will a 150-foot support cable extend if it is attached halfway up the antenna? Express your answer to the nearest foot.

26. A building 55 feet tall is on top of a hill. A surveyor standing on the hillside determines that the angle of elevation to the top of the building is 42° and to the bottom of the building is 19°. How far is the surveyor from the bottom of the building? Express your answer to the nearest tenth of a foot. (Remember that an angle of elevation is measured with reference to a horizontal line.)

27. A triangular lot faces two streets that meet at an angle of 82° (see Figure 7.81). The two sides of the lot facing the streets are each 175 feet long. Find the length of the third side to the nearest tenth of a foot.

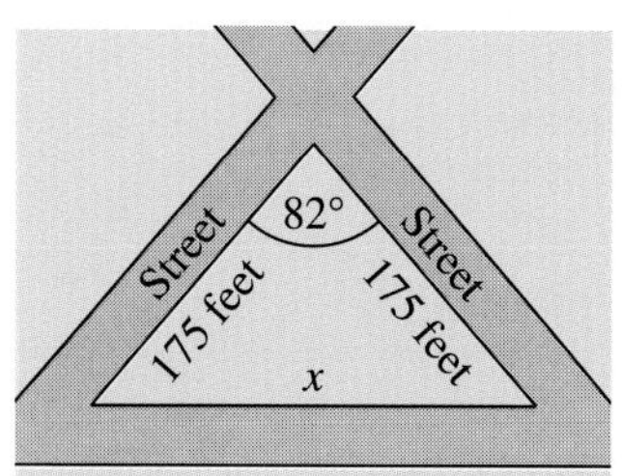

FIGURE 7.81

28. Two people standing 75 feet apart are in line with the base of a tower. The angle of elevation to the top of the tower from one person is 41.2° and from the other person is 32.6°. Find the height of the tower to the nearest tenth of a foot.

29. A triangular lot is bounded by three straight streets: Washington, Jefferson, and Monroe. Washington and Monroe Streets intersect at an angle of 65°. The side of the lot along Washington is 250 meters long and the side along Jefferson is 235 meters long. Find the length of the side of the lot along Monroe Street. Express the length to the nearest meter.

30. Two landmarks, A and B, are on the same side of a river and 750 yards apart. A surveyor stands at point C on the opposite side of the river, thus forming triangle ABC. If $C = 57.2°$ and the distance between A and C is 600 yards, find the distance between B and C to the nearest yard.

31. A solar collector that is 10 feet wide is attached to a roof as indicated in Figure 7.82. The roof makes an angle of 35° with the horizontal and the collector makes an angle of 50° with the horizontal. Find the length of the vertical brace to the nearest tenth of a foot.

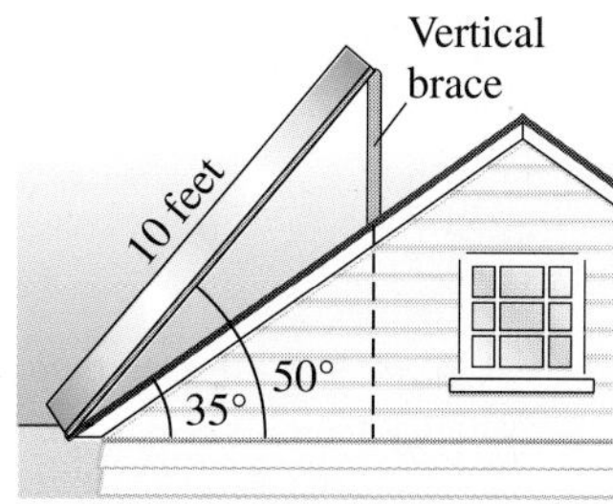

FIGURE 7.82

32. An office building is located at the top of a hill. From the base of the hill to the top of the office building, the angle of elevation is 40°. From a point 150 feet from the base of the hill, the angle of elevation to the top of the building is 32°. The hill rises at a 25° angle with the horizontal. Find the height of the office building to the nearest tenth of a foot. (See Figure 7.83.)

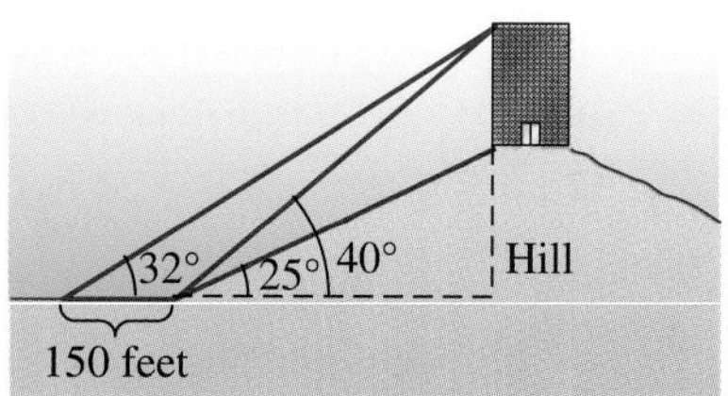

FIGURE 7.83

33. Two people standing 1375 yards apart observe a helicopter hovering in the sky as in Figure 7.84. From the position of one person the angle of elevation to the helicopter is 41° and from the other person the angle of elevation to the helicopter is 32°. Find, to the nearest yard, the height of the helicopter.

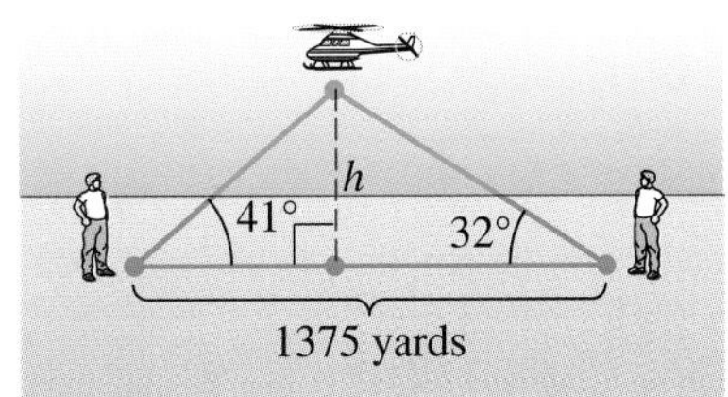

FIGURE 7.84

MISCELLANEOUS PROBLEMS

34. Consider the triangle in Figure 7.85. The area (K) of the triangle is given by $K = \frac{1}{2}bh$. Furthermore, $\sin C = h/a$ or $h = a \sin C$. Therefore, substituting $a \sin C$ for h in the formula $K = \frac{1}{2}bh$ produces

$$K = \frac{1}{2}ab \sin C.$$

This formula can be used to find the area of a triangle if the measurements of two sides and the included angle are known. Use the formula to help solve the following problems.

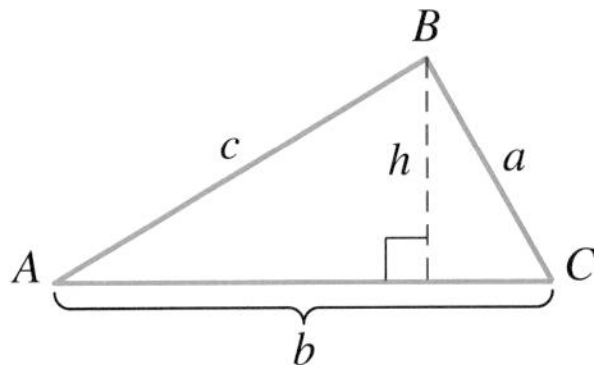

FIGURE 7.85

a. Find the area of a triangle if two sides are 18 meters and 24 meters long, and the included angle has a measure of 47°. Express the area to the nearest square meter.

b. Two sides of a triangular plot of ground are 75 yards and 90 yards long, and the angle included by those two sides is 65°. Find the area of the plot to the nearest square yard.

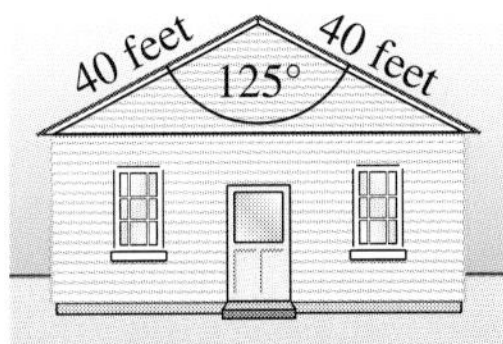

FIGURE 7.86

c. The gable end of a house is shown in Figure 7.86. Find the area of the shaded triangle, to the nearest square foot.

35. From the law of sines we know that $a/(\sin A) = b/\sin B$ or $b = (a \sin B)/\sin A$. Substituting this expression for b in $K = \frac{1}{2}ab \sin C$ produces

$$K = \frac{a^2 \sin B \sin C}{2 \sin A}.$$

This formula can be used to find the area of a triangle if the measurements of two angles and the included side are known. Use it to help solve the following problems.

a. Find the area of a triangle if two angles have measures of 43° and 61°, and the included side is 14 centimeters long. Express the area to the nearest square centimeter.

b. A triangular plot of ground has two angles that measure 50° each. The side included between the two angles is 80 feet long. Find the area of the plot to the nearest square foot.

c. The dimensions of a triangle are indicated in Figure 7.87. Find the area to the nearest square inch.

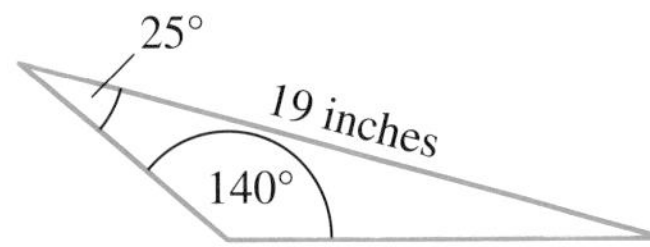

FIGURE 7.87

36. There is yet another formula, commonly referred to as Heron's formula, that can be used to find the area of a triangle if the lengths of the three sides are known. We will not show the development of the formula but simply state it so that you can use it to solve some problems. The area of a triangle is given by

$$K = \sqrt{s(s - a)(s - b)(s - c)},$$

where a, b, and c are the lengths of the three sides and $s = \frac{1}{2}(a + b + c)$.

a. Find the area, to the nearest square centimeter, of a triangle that measures 14 centimeters by 16 centimeters by 18 centimeters.

b. Find the area, to the nearest square yard, of a triangular plot of ground that measures 45 yards by 60 yards by 75 yards.

c. Find the area of an equilateral triangle, of which each side is 18 inches long. Express the area to the nearest square inch.

7.7 VECTORS

In the previous three sections we solved numerous problems involving the lengths of sides of triangles. It was not necessary to specify any sort of direction in which the measurements were made. In other words, the sides of the triangles had magnitude (length), but no direction.

Many quantities, however, can be described completely by specifying both magnitude and direction. For example, we speak of a car traveling north at 55 miles per hour or an airplane flying east at 350 miles per hour. Another example is force that is exerted with a given magnitude and in a given direction–30 pounds exerted at a 30° angle to the horizontal. Quantities that have both magnitude and direction are called **vector quantities**.

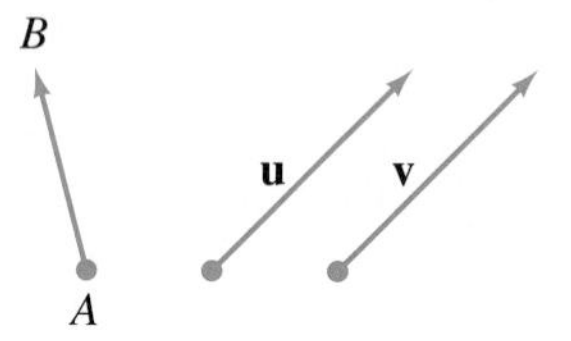

FIGURE 7.88

Geometrically, vector quantities can be represented by directed line segments called **vectors**. In Figure 7.88, a vector was drawn from point A (called the **initial point**) to point B (called the **terminal point**) and is denoted by $\overrightarrow{AB}$. Boldface letters are also commonly used to name vectors. For example, vectors $\mathbf{u}$ and $\mathbf{v}$ are also represented in Figure 7.88. (Since handwritten boldface letters are difficult to distinguish you might use $\vec{u}$ and $\vec{v}$.) The length of a directed line segment represents the *magnitude* of the vector and is denoted by $|\overrightarrow{AB}|$ or $|\mathbf{v}|$. Vectors that have the same magnitude and same direction are considered **equal vectors**. In Figure 7.88, $\mathbf{u}$ and $\mathbf{v}$ are equal vectors and we write $\mathbf{u} = \mathbf{v}$. The vector quantities (car traveling north at 55 miles per hour, airplane flying east at 350 miles per hour, and a force of 30 pounds acting at a 30° angle to the horizontal) are represented in Figure 7.89.

$|\mathbf{u}| = 55$ mph

$|\mathbf{v}| = 350$ mph

$|\mathbf{w}| = 30$ lb

30°

FIGURE 7.89

Vector Addition

The *sum* (also called the *resultant*) of two vectors **u** and **v** is pictured in Figure 7.90. The initial point of **v** is placed at the terminal point of **u** and the vector from the initial point of **u** to the terminal point of **v** is the sum **u** + **v**.

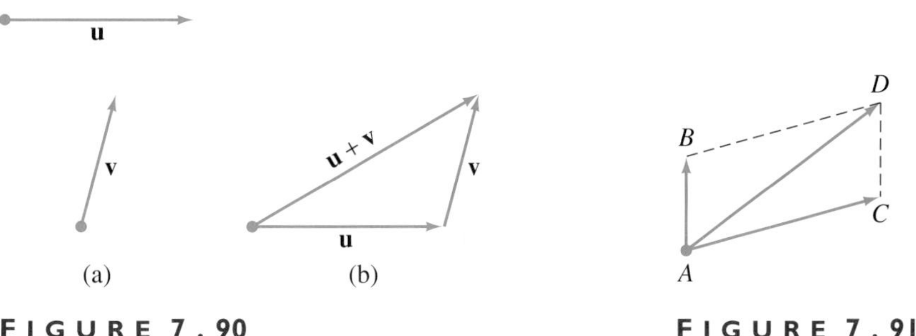

FIGURE 7.90

FIGURE 7.91

Another way to describe the addition of vectors is known as the parallelogram law. Consider two vectors $\overrightarrow{AB}$ and $\overrightarrow{AC}$ having a common initial point as in Figure 7.91. Complete the parallelogram $ACDB$. Since $\overrightarrow{AB} = \overrightarrow{CD}$, the diagonal vector $\overrightarrow{AD}$ represents the sum of $\overrightarrow{AC}$ and $\overrightarrow{AB}$. So we can write $\overrightarrow{AC} + \overrightarrow{AB} = \overrightarrow{AD}$. (If $\overrightarrow{AC}$ and $\overrightarrow{AB}$ are perpendicular, then $\overrightarrow{AD}$ will be the diagonal of a rectangle.) If $\overrightarrow{AC}$ and $\overrightarrow{AB}$ represent two physical forces acting on some object at A, then it can be shown by physical experiments that $\overrightarrow{AD}$ represents the resultant force, that is, the single force that produces the same effect as the two combined forces. In other words, vector addition was defined to be consistent with the action of vector quantities being represented.

Scalar Multiplication and Vector Subtraction

If **u** is a vector, then 2**u** is the vector in the same direction as **u** but twice as long; −3**u** is three times as long as **u** but in the opposite direction (Figure 7.92). In general, if k is a real number (called a scalar) and **u** is any vector, then k**u** (called the *scalar multiple* of **u**) is a vector whose magnitude is $|k|$ times the magnitude

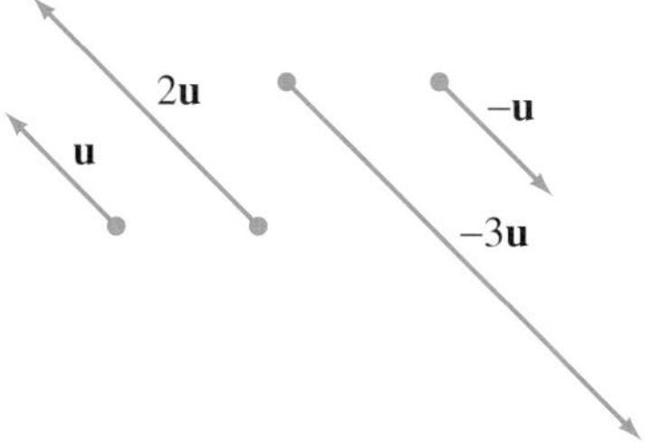

FIGURE 7.92

of **u** and whose direction is the same as **u** if $k > 0$, and opposite that of **u** if $k < 0$. In particular, $(-1)\mathbf{u}$ (usually written as $-\mathbf{u}$) has the same magnitude as **u** but is directed in the opposite direction of **u**, as indicated in Figure 7.92. The vector $-\mathbf{u}$ is called the **opposite** or the **negative of u** and when added to **u** produces the zero vector denoted by **0**. So we can write $\mathbf{u} + (-\mathbf{u}) = \mathbf{0}$. (The **zero vector** is interpreted as a point having a magnitude of 0 and no direction.) The zero vector is the *identity element* for vector addition, that is, $\mathbf{u} + \mathbf{0} = \mathbf{0} + \mathbf{u} = \mathbf{u}$. Finally, vector subtraction is defined by $\mathbf{u} - \mathbf{v} = \mathbf{u} + (-\mathbf{v})$.

EXAMPLE 1

Using the two vectors in Figure 7.93(a), draw $3\mathbf{u} + 2\mathbf{v}$ and $\mathbf{u} - 3\mathbf{v}$.

Solution

Position $3\mathbf{u}$ (a vector in the same direction as **u** but three times as long) and $2\mathbf{v}$ (a vector in the same direction as **v** but twice as long) so that they share a common initial point. Then complete the parallelogram and draw $3\mathbf{u} + 2\mathbf{v}$, as in Figure 7.93(b).

Now to draw $\mathbf{u} - 3\mathbf{v}$: Position **u** and $-3\mathbf{v}$ (a vector in the opposite direction of **v** and three times as long) with a common initial point. Then complete the parallelogram and draw $\mathbf{u} - 3\mathbf{v}$, as in Figure 7.93(c).

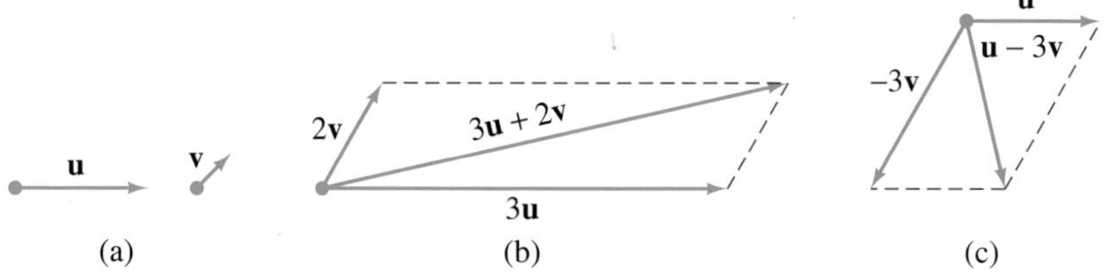

FIGURE 7.93

Vectors and Problem Solving

Vectors can be used to solve a variety of problems from diverse areas. In this brief introduction to vectors, we simply want to show you a few such applications. Some of these problems can be solved using only right-triangle trigonometry, but others require the use of the law of cosines and the law of sines as applied to oblique triangles.

PROBLEM 1

Suppose that an airplane flying at 300 miles per hour is headed due north but is blown off course by the wind that is blowing at 30 miles per hour from the west. By what angle is the plane diverted from its intended path and what is its actual ground speed?

Solution

The velocity of the plane and the wind velocity can be represented by two vectors as shown in Figure 7.94. The actual path of the plane is represented by the resultant

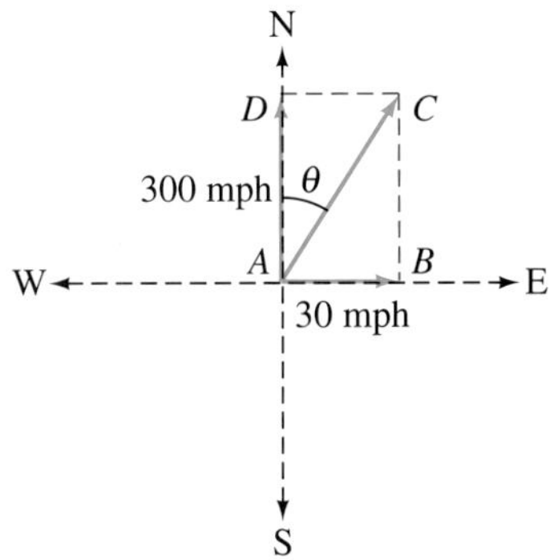

FIGURE 7.94

$\overrightarrow{AC}$, the diagonal of the rectangle. Angle θ is the angle at which the plane is blown off the intended course. Because DC is also 30, we can find θ by using the tangent function.

$$\tan\theta = \frac{30}{300}$$

Therefore,

$$\theta = 5.7°, \quad \text{to the nearest tenth of a degree.}$$

The magnitude of $\overrightarrow{AC}$ represents the ground speed of the plane.

$$|\overrightarrow{AC}|^2 = 300^2 + 30^2 = 90{,}900$$

Therefore,

$$|\overrightarrow{AC}| = \sqrt{90{,}900} \approx 301.5 \text{ miles per hour.}$$

Instead of adding two vectors to form a resultant vector, it is sometimes necessary to reverse the process and to find two vectors whose sum is a given vector. The two vectors that we find are called **components** of the given vector. The components are especially easy to find if they are to be perpendicular.

PROBLEM 2

Using a rope to pull a sled with a child on it, a man applies a force of 90 pounds. If the rope makes a 40° angle with the horizontal (Figure 7.95), find the horizontal

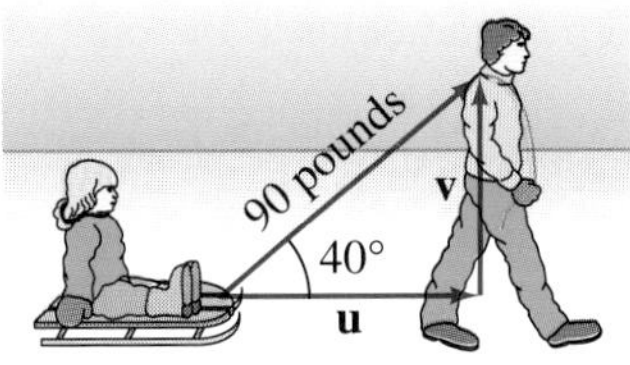

FIGURE 7.95

force that tends to move the sled along the ground, and the vertical force that tends to lift the sled vertically.

Solution

The vectors **u** and **v** in Figure 7.95 represent the horizontal and vertical components, respectively, of the given force of 90 pounds.

$$\cos 40^\circ = \frac{|\mathbf{u}|}{90}$$

$$|\mathbf{u}| = 90 \cos 40^\circ = 68.9, \quad \text{to the nearest tenth of a pound}$$

$$\sin 40^\circ = \frac{|\mathbf{v}|}{90}$$

$$|\mathbf{v}| = 90 \sin 40^\circ = 57.9, \quad \text{to the nearest tenth of a pound}$$

Suppose that a spherical object weighing 250 pounds is placed on an inclined ramp, as indicated in Figure 7.96. Gravity pulls directly downward with a force of

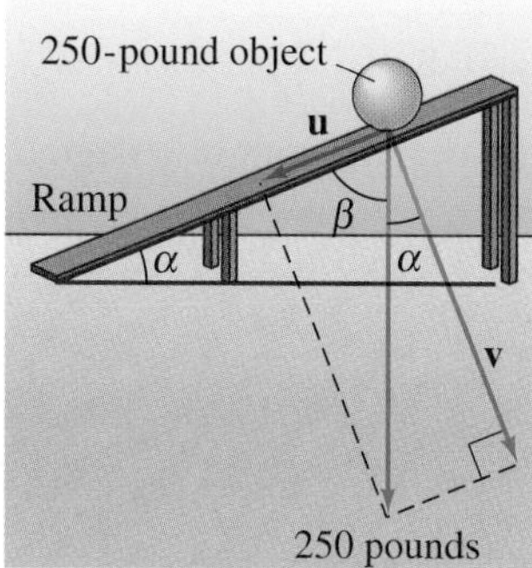

FIGURE 7.96

250 pounds. Part of this force (represented by **u**) tends to pull the object down the ramp, and another part (represented by **v**) presses the object against the ramp on a line perpendicular to the ramp. The angle of the ramp, α, is also the angle between two of the vectors, since they are both complementary to angle β. Therefore, if the angle of the ramp is known, then the 250-pound resultant vector, which is a diagonal of the rectangle, can be *resolved into two components*–one directed down the ramp and the other directed perpendicular to the ramp. Let's consider a specific problem.

PROBLEM 3

A 250-pound spherical lead ball is placed on a ramp that has an incline of 22° with the horizontal. How much force is pulling down the ramp and how much force is pressing on a line perpendicular against the ramp?

Solution

We can use Figure 7.96 with $\alpha = 22^\circ$.

$$\cos 22^\circ = \frac{|\mathbf{v}|}{250}$$

$$|\mathbf{v}| = 250 \cos 22^\circ = 231.8, \quad \text{to the nearest tenth}$$

$$\sin 22^\circ = \frac{|\mathbf{u}|}{250}$$

$$|\mathbf{u}| = 250 \sin 22^\circ = 93.7, \quad \text{to the nearest tenth}$$

Therefore, there is a force of approximately 93.7 pounds acting down the ramp and 231.8 pounds acting against the ramp.

In certain navigation problems, the *direction* or *bearing* of a ship may be given by stating the size of an acute angle that is measured with respect to a north-south line. For example, a bearing of N 40°E denotes an angle whose one side points north and whose other side points 40° east of north, as indicated in Figure 7.97. In the same figure, we indicated a bearing of S 70°W, that is, 70° west of south.

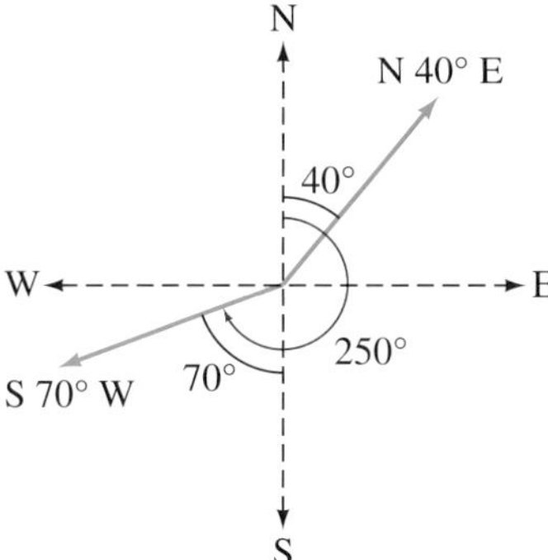

FIGURE 7.97

Vectors and navigation
Mark Kelley/Stock Boston

In air navigation, directions and bearings are often specified by measuring from a north line in a *clockwise* direction. (The angles are stated as positive angles even though they are measured in a clockwise direction.) Thus, in Figure 7.97 the bearing of S 70°W could also be expressed as a bearing or direction of 250°.

PROBLEM 4

If a ship sails 25 miles in the direction N 40°E and then 60 miles east, what is its distance and bearing with respect to its starting point?

Solution

The route of the ship is indicated in Figure 7.98. The measure of $\angle ABD$ is $90° - 40° = 50°$. Therefore, the measure of $\angle ABC$ is $180° - 50° = 130°$. Then the magnitude of $\overrightarrow{AC}$ can be found by using the law of cosines.

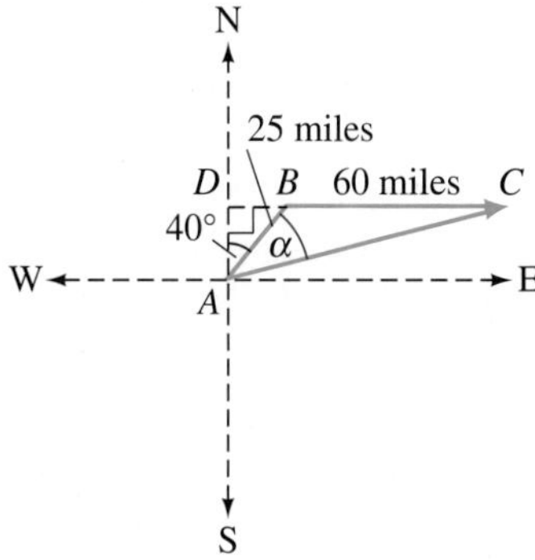

FIGURE 7.98

$$\begin{aligned}|\overrightarrow{AC}|^2 &= 25^2 + 60^2 - 2(25)(60)\cos 130° \\ &= 6153.36, \quad \text{to the nearest hundredth}\end{aligned}$$

Therefore,

$$|\overrightarrow{AC}| = \sqrt{6153.36} \approx 78.44 \text{ miles.}$$

Then the measure of angle α can be found by the law of sines.

$$\frac{\sin \alpha}{60} = \frac{\sin 130°}{78.44}$$

$$78.44 \sin \alpha = 60 \sin 130°$$

$$\sin \alpha = \frac{60 \sin 130°}{78.44}$$

$$\alpha = 35.9°, \quad \text{to the nearest tenth of a degree}$$

From Figure 7.98 we see that the bearing of $\overrightarrow{AC}$ is stated in terms of $\alpha + 40°$ and, therefore, is N 75.9°E.

PROBLEM SET 7.7

For Problems 1–4, use the vectors **u** and **v** in Figure 7.99 to draw the indicated vectors.

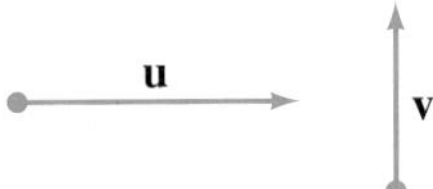

FIGURE 7.99

1. **u** + **v**
2. **u** − **v**
3. 2**u** + 3**v**
4. **u** − 2**v**

For Problems 5–10, use the vectors **u** and **v** in Figure 7.100 to draw the indicated vectors.

FIGURE 7.100

5. **u** − **v**
6. **u** + **v**
7. **v** + 2**u**
8. 2**u** + **v**
9. 2**u** − **v**
10. 2**v** − **u**

11. Suppose that an airplane flying at 250 miles per hour is headed due south but is blown off course by the wind, which is blowing at 40 miles per hour from the east. By what angle is the plane blown off its intended path and what is its actual ground speed?

12. A boat that can travel 15 miles per hour in still water attempts to go directly across a river that is flowing at 4 miles per hour. By what angle is the boat pushed off its intended path?

13. A river flows due south at 125 feet per minute. In what direction must a motor boat, which can travel 500 feet per minute, be headed so that it actually travels due east?

14. A balloon is rising at 5 feet per second while a wind is blowing at 15 feet per second. Find the speed of the balloon and the angle it makes with the horizontal.

15. Two forces of 50 pounds and 75 pounds act on an object at right angles. Find the magnitude of the resultant force and the angle it makes with the greater force.

16. Two forces of 4 kilograms and 9 kilograms act on an object at right angles. Find the magnitude of the resultant force and the angle it makes with the smaller force.

17. An airplane heads in a direction of 80° at 200 kilometers per hour. A wind blowing in the direction of 170° forces the plane onto a course that is due east. Find the speed of the wind.

18. An airplane heads in a direction of 265° at a speed of 330 miles per hour. A wind blowing in the direction of 355° forces the plane onto a course that is due west. Find the speed of the wind.

19. A force of 45 pounds is acting on an object at a 20° angle to the horizontal. Find the horizontal and vertical components of the force.

20. An airplane is flying at 275 miles per hour in a direction of 235°. Find the westerly and southerly components of its velocity.

21. A boy exerts a force of 20 pounds to pull a wagon filled with toys. The handle of the wagon makes a 35° angle with the horizontal. Find the horizontal and vertical components of the 20-pound force (see Figure 7.101).

FIGURE 7.101

22. An automobile weighing 3200 pounds is parked on a ramp that makes a 15° angle with the horizontal. How much parallel force is exerted to the ramp and how much perpendicular force is exerted to the ramp?

23. A small car weighing 1750 pounds is parked on a ramp that makes a 27° angle with the horizontal. How much parallel force must be applied to the ramp to keep the car from rolling down the ramp?

24. A force of 30 pounds is needed to hold a barrel in place on a ramp that makes an angle of 32° with the horizontal. Find the weight of the barrel.

25. A force of 52.1 pounds is needed to keep a 75-pound lead ball from rolling down a ramp. Find the angle that the ramp makes with the horizontal.

26. A weight of 60 pounds is held by two wires as indicated in Figure 7.102. Find the horizontal component of force F_1.

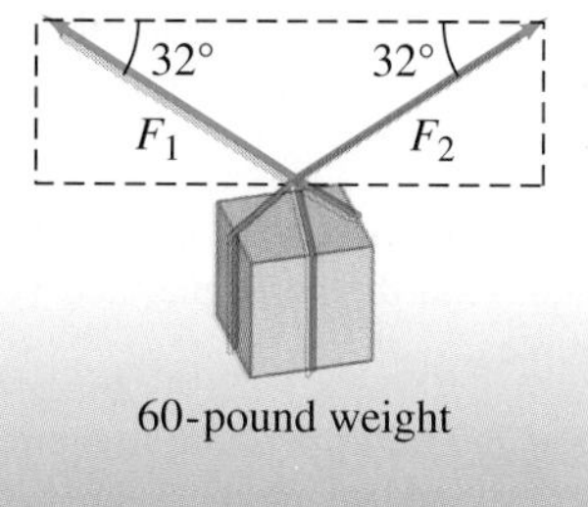

FIGURE 7.102

27. Forces of 220 kilograms and 175 kilograms act on an object. The angle between the forces is 65°. Find the magnitude of the resultant of the two forces and also find the angle that the resultant makes with the greater force.

28. Forces of 105 pounds and 85 pounds act on an object. The angle between the forces is 42°. Find the magnitude of the resultant of the two forces and also find the angle that the resultant makes with the lesser force.

29. If a ship sails 35 miles in the direction N 32°W and then 70 miles west, what are its distance and bearing with respect to its starting point?

30. If a ship sails 50 miles in the direction S 47°E and then 80 miles east, what are its distance and bearing with respect to its starting point?

31. Suppose that a boat travels 70 miles in the direction N 50° E and then travels 85 miles in the direction S 65°E. What are its distance and bearing with respect to its starting point?

32. Suppose that a boat travels 40 miles in the direction S 27°W and then 90 miles east. What are its distance and bearing with respect to its starting point?

MISCELLANEOUS PROBLEMS

33. Give a geometric argument to show that vector addition is a commutative operation.

34. Give a geometric argument to show that vector addition is an associative operation.

35. Give a geometric argument to show that $k(\mathbf{u} + \mathbf{v}) = k\mathbf{u} + k\mathbf{v}$ for any vectors $\mathbf{u}$ and $\mathbf{v}$, and any real number k.

CHAPTER 7 SUMMARY

Section 7.1 contains a brief review of some geometric concepts that are used in the study of trigonometry. Be sure that you know the meanings of the following geometric concepts: plane angle, vertex, initial side, terminal side, positive angle, negative angle, coterminal angles, degree measure, complementary angles, supplementary angles, radian measure, central angle, arc length, right triangle, scalene triangle, isosceles triangle, isosceles right triangle, and equilateral triangle.

The following important relationships are contained in Section 7.1.

1 radian $= 180/\pi$ degrees; therefore, to change from radians to degrees, multiply by $180/\pi$.

1 degree $= \pi/180$ radians; therefore, to change from degrees to radians, multiply by $\pi/180$.

The length of an arc s intercepted by a central angle θ in a circle of radius r is given by $s = r\theta$, where θ is expressed in radians.

The Pythagorean theorem states that if a and b are the measures of the legs of a right triangle, and c is the measure of the hypotenuse, then $a^2 + b^2 = c^2$.

In a 30°-60° right triangle, the length of the side opposite the 30° angle is equal to one-half of the length of the hypotenuse.

The defining and evaluating of trigonometric functions is the main theme of Sections 7.2 and 7.3.

If θ is an angle in standard position and $P(x, y)$ is *any point* on the terminal side of θ (see Figure 7.103), then

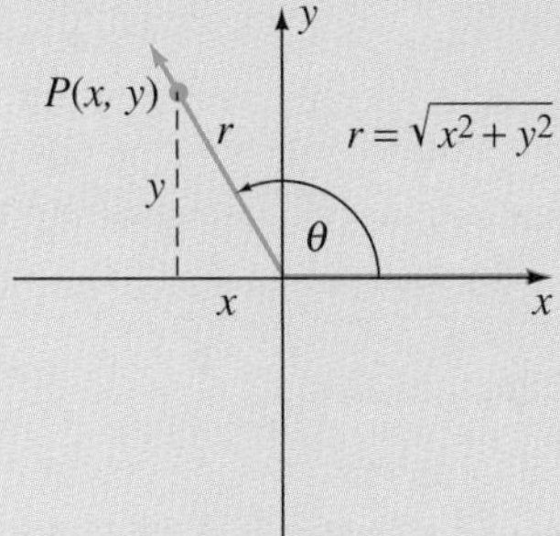

FIGURE 7.103

$$\sin \theta = \frac{y}{r} \qquad \csc \theta = \frac{r}{y}$$

$$\cos \theta = \frac{x}{r} \qquad \sec \theta = \frac{r}{x}$$

$$\tan \theta = \frac{y}{x} \qquad \cot \theta = \frac{x}{y}.$$

If $r = 1$, then $\sin \theta = y$ and $\cos \theta = x$.

If θ is any angle in standard position with its terminal side in one of the four quadrants, then the **reference angle** of θ (denoted θ') is the acute angle formed by the terminal side of θ and the x-axis.

The trigonometric functions of any angle θ are equal to those of the reference angle associated with θ, except possibly for the sign. The sign can be determined by considering the quadrant in which the terminal side of θ lies.

You should be able to find, without a calculator or a table, the exact trigonometric functional values of any angle that has a reference angle of 30°, 45°, or 60°.

Be sure that you have the following relationships at your fingertips.

$$\csc \theta = \frac{1}{\sin \theta} \qquad \sec \theta = \frac{1}{\cos \theta} \qquad \cot \theta = \frac{1}{\tan \theta}$$

$$\tan \theta = \frac{\sin \theta}{\cos \theta} \qquad \cot \theta = \frac{\cos \theta}{\sin \theta} \qquad \sin^2 \theta + \cos^2 \theta = 1$$

$$\sin(-\theta) = -\sin \theta \qquad \cos(-\theta) = \cos \theta \qquad \tan(-\theta) = -\tan \theta$$

Problem solving is the central theme of Sections 7.4, 7.5, and 7.6.

Solving a right triangle can be analyzed as follows.

1. If an acute angle and the length of one side are given, then the other acute angle is the complement of the given one, and the lengths of the other two sides can be found by solving equations involving the appropriate trigonometric functions.
2. If the lengths of two sides are given, the third side can be found by using the Pythagorean theorem. The two acute angles can be determined from equations involving the appropriate trigonometric functions.

The following relationships are referred to as the **law of cosines**.

$$a^2 = b^2 + c^2 - 2bc \cos A$$

$$b^2 = a^2 + c^2 - 2ac \cos B$$

$$c^2 = a^2 + b^2 - 2ab \cos C$$

If two sides and the included angle (SAS) of a triangle are known or if three sides (SSS) are known, then the law of cosines can be used to solve the triangle.

The following relationships are referred to as the **law of sines**.

$$\frac{a}{\sin A} = \frac{b}{\sin B} = \frac{c}{\sin C}$$

If two angles and a side of a triangle are known, then the law of sines can be used to solve the triangle.

If, in a triangle, two sides and an angle opposite one of them are known, then two, one, or no triangles may be determined.

Vectors, as directed line segments, play an important role in many applications because they can be used to represent quantities that have both magnitude and direction. We call two vectors equal if they have the same magnitude and same direction. Vectors can be added and multiplied by real numbers called scalars. Problems 1–4 of Section 7.7 provide a good review of some of the applications of vectors.

CHAPTER 7 REVIEW PROBLEM SET

1. The measures of two complementary angles are in the ratio of 7 to 11. What is the measure of each angle?

2. Change 35°17′ and 82°15′36″ to decimal form and express each to the nearest one-hundredth of a degree.

3. Change 93.35° and 163.27° to degree-minute-second form.

4. Without using a calculator, change each of the following to radians.

a. 420° **b.** 570° **c.** −45°

5. Without using a calculator, change each of the following to degrees.

a. $\frac{7\pi}{6}$ **b.** $-\frac{4\pi}{3}$ **c.** $\frac{17\pi}{4}$

6. Find, to the nearest tenth of a centimeter, the length of the arc intercepted by a central angle of $4\pi/3$ radians if a radius of the circle is 17 centimeters long.

7. Find, to the nearest tenth of an inch, the length of the arc intercepted by a central angle of 130° if a radius of the circle is 14 inches long.

8. A rectangular floor measures 22 feet by 13 feet. Find the distance, to the nearest tenth of a foot, from one corner of the floor to the diagonally opposite corner.

9. A diagonal of a square plot of ground is 46 meters long. Find, to the nearest tenth of a meter, the length of a side of the square plot.

10. A 50-foot support wire makes an angle of 60° with the ground and is attached to a telephone pole. Find the distance from the base of the pole to the point on the pole where the support wire is attached. Express your answer to the nearest tenth of a foot.

For Problems 11–14, point P is a point on the terminal side of θ, and θ is in standard position. Find $\sin\theta$, $\cos\theta$, and $\tan\theta$.

11. $P(-6, 8)$
12. $P(1, -3)$
13. $P(-2, -4)$
14. $P(-5, -12)$

For Problems 15–26, find exact values without using a calculator or a table.

15. $\cos(-150°)$
16. $\sin(-330°)$
17. $\csc 45°$
18. $\sec 120°$
19. $\tan 675°$
20. $\cot 480°$
21. $\sin\frac{7\pi}{3}$
22. $\cos\frac{13\pi}{3}$
23. $\sec\left(-\frac{3\pi}{2}\right)$
24. $\csc(-\pi)$
25. $\cot\frac{7\pi}{4}$
26. $\tan\frac{5\pi}{4}$

27. If $\sin\theta > 0$ and $\tan\theta < 0$, what quadrant contains the terminal side of θ?

28. If $\sec\theta > 0$ and $\csc\theta < 0$, what quadrant contains the terminal side of θ?

29. Find $\tan\theta$ if the terminal side of θ lies in the fourth quadrant on the line $y = -5x$.

30. Find $\cos\theta$ if $\sin\theta = -5/13$ and θ is a fourth-quadrant angle.

31. Find $\sin(-\theta)$ if $\sin\theta = 0.7986$.

32. Find $\cos(-\theta)$ if $\cos\theta = -0.4067$.

33. Find θ, such that $0° < \theta < 90°$ if $\sin\theta = 0.7615$. Express your answer to the nearest tenth of a degree.

34. Find θ, such that $0° < \theta < 90°$ if $\cos\theta = 0.3697$. Express your answer to the nearest tenth of a degree.

35. Find θ, such that $0° < \theta < 90°$ if $\tan\theta = 13.6174$. Express your answer to the nearest tenth of a degree.

36. The sides of a right triangle are of length 8 feet, 15 feet, and 17 feet. Find the measure of the angle opposite the 15-foot side. Express your answer to the nearest tenth of a degree.

37. From a point 80 feet from the base of a tree, the angle of elevation to the top of the tree is 28°. Find the height of the tree to the nearest tenth of a foot.

38. A 50-foot antenna is positioned on top of a building. From a point on the ground the angle of elevation to the top of the antenna is 51° and the angle of elevation to the bottom of the antenna is 42°. Find the height of the building to the nearest foot.

39. In triangle ABC, if $A = 74°$, $c = 19$ miles, and $b = 27$ miles, what is the length of a to the nearest tenth of a mile?

40. In triangle ABC, $A = 118.2°$, $C = 17.3°$, and $c = 56$ yards. Find b to the nearest tenth of a yard.

41. If $A = 29°$, $a = 19$ meters, and $c = 37$ meters, how many triangles are determined?

42. In Figure 7.104, find x to the nearest tenth of a foot.

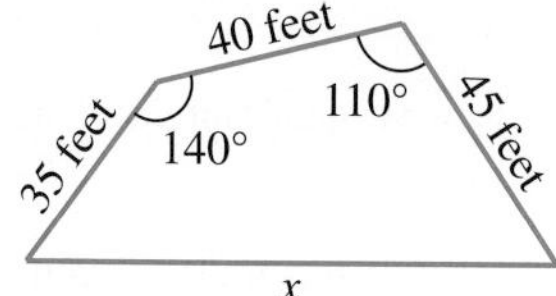

FIGURE 7.104

43. A triangular lot measures 52 yards by 47 yards by 85 yards. Find, to the nearest tenth of a degree, the measure of the angle opposite the longest side.

44. A plane travels 350 miles due east after take-off, then adjusts its course 20° northward and flies another 175 miles. How far is it from the point of departure? Express your answer to the nearest mile.

45. Two points A and B are located on opposite sides of a river. Point C is located 350 yards from B and on the same side of the river. In $\triangle ABC$, $B = 69°$ and $C = 43°$. Find the distance between A and B to the nearest yard.

46. Two people standing 60 feet apart are in line with the base of a tower. The angle of elevation to the top of the tower from one person is 39.6° and from the other person is 27.4°. Find the height of the tower to the nearest tenth of a foot.

Use vectors to help solve Problems 47–50.

47. A lawnmower is being pushed by applying a force of 45 pounds. The handle of the mower makes an angle of 40° with the horizontal. How much force is being applied horizontally and how much force is being applied downward?

48. A car weighing 2750 pounds is parked on a ramp that makes a 32° angle with the horizontal. How much force must be applied to keep the car from rolling down the ramp?

49. Forces of 130 pounds and 165 pounds act on an object. The angle between the forces is 28°. Find the magnitude of the resultant of the two forces and also find the angle that the resultant makes with the greater force.

50. A ship sails 70 miles west and then 110 miles in the direction N 20°W. What are its distance and bearing with respect to its starting point?

GRAPHING TRIGONOMETRIC FUNCTIONS

What are the graphs of the basic trigonometric functions? Can we apply our knowledge of graphing variations of basic curves to the graphing of trigonometric functions? Do the basic trigonometric functions have inverses? How does the polar coordinate system compare to the rectangular coordinate system? The focus of this chapter is to provide answers to these questions.

8

The sine function is sometimes used to describe the predator-prey relationship in a balanced ecological system.

8.1 UNIT CIRCLE: SINE AND COSINE CURVES

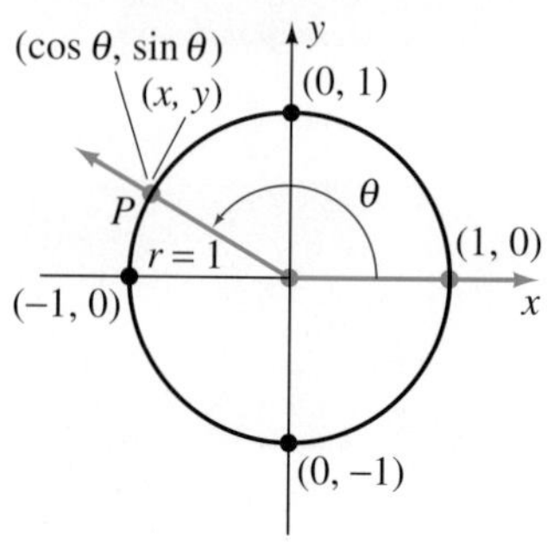

FIGURE 8.1

Consider the circle $x^2 + y^2 = 1$, commonly referred to as the **unit circle**, with its center at the origin and a radius of length one unit (Figure 8.1). The terminal side of any angle in standard position intersects the unit circle at a point $P(x, y)$ such that $r = 1$. (Note that this is equivalent to choosing a point on the terminal side such that $r = 1$ as we did in Chapter 7.) Again, observe that $\sin\theta = y$ and $\cos\theta = x$; so the coordinates of point P are $(\cos\theta, \sin\theta)$. In other words, the trigonometric functions $\sin\theta$ and $\cos\theta$ can be thought of as the **ordinate** and **abscissa** values, respectively, of the point of intersection of the terminal side of θ and the unit circle.

The formula for arc length, $s = r\theta$, which we developed in Section 7.1, takes on special significance when applied to the unit circle. In Figure 8.2 consider a central angle θ in standard position that intercepts an arc s on the unit circle. Since $r = 1$, the formula $s = r\theta$ becomes $s = \theta$. That is to say, numerically the length of the arc from $(1, 0)$ to P equals the measure of the angle θ in radians. Consider the following examples.

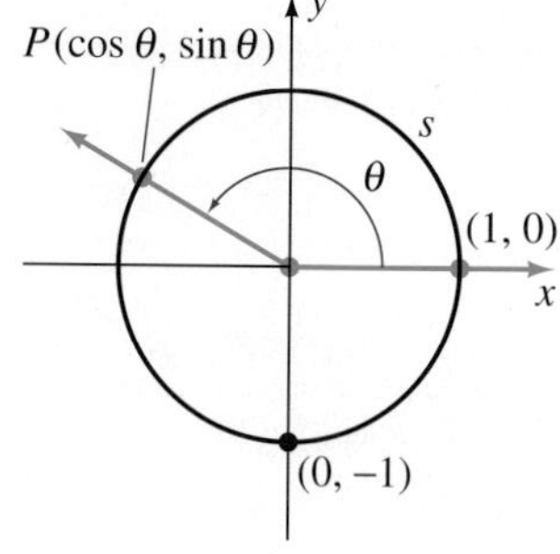

FIGURE 8.2

$\theta = \dfrac{7\pi}{6}$,	then $s = \dfrac{7\pi}{6}$ units	Figure 8.3(a)
$\theta = \dfrac{7\pi}{3}$,	then $s = \dfrac{7\pi}{3}$ units	Figure 8.3(b)
$\theta = -\dfrac{3\pi}{4}$,	then $s = -\dfrac{3\pi}{4}$ units	Figure 8.3(c)

In general, to each real number that represents the radian measure of a central angle θ in standard position, we can associate a real number s that represents the length of the arc intercepted by the angle on the unit circle. If θ is positive, then the arc is measured from $(1, 0)$ in a counterclockwise direction to P, where P is the point of intersection of the terminal side of θ and the unit circle. If θ is negative,

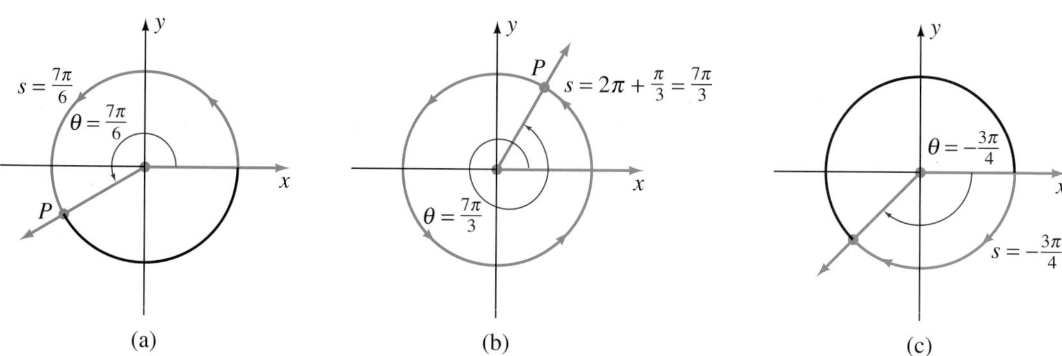

FIGURE 8.3

the arc is measured from (1, 0) in a clockwise direction to P. If $\theta = 0$, then $s = 0$. The point $P(x, y)$ is referred to as *the point on the unit circle that corresponds to s.*

An important consequence of the previous discussion is that we can now consider the trigonometric functions by using a domain of real numbers independent of any reference to angles. In such a setting, we often refer to the trigonometric functions as the **circular functions. If s is a real number and $P(x, y)$ is the point on the unit circle that corresponds to s, then**

$$\sin s = y \qquad \csc s = \frac{1}{y} \quad \text{if } y \neq 0$$

$$\cos s = x \qquad \sec s = \frac{1}{x} \quad \text{if } x \neq 0$$

$$\tan s = \frac{y}{x} \quad \text{if } x \neq 0 \qquad \cot s = \frac{x}{y} \quad \text{if } y \neq 0.$$

EXAMPLE 1

Determine the circular functions $\sin s$, $\cos s$, and $\tan s$ for

a. $s = \pi$ **b.** $s = -\dfrac{3\pi}{2}$ **c.** $s = \dfrac{\pi}{4}$ **d.** $s = \dfrac{5\pi}{4}$

Solutions

a. Since the circumference of the unit circle is 2π, the point that corresponds to $s = \pi$ is $(-1, 0)$ as indicated in Figure 8.4. Therefore,

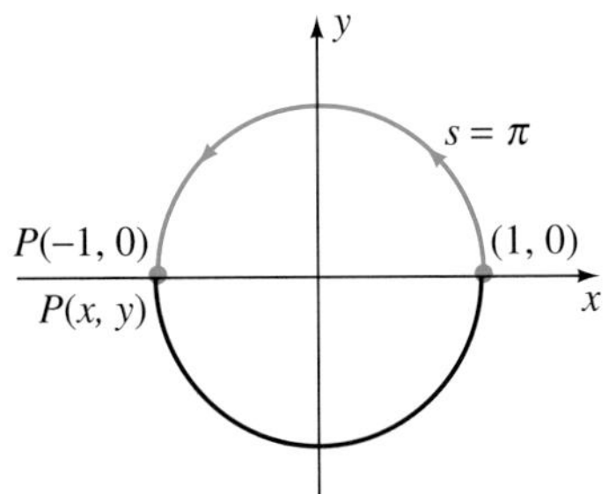

FIGURE 8.4

$$\sin \pi = y = 0$$

$$\cos \pi = x = -1$$

and

$$\tan \pi = \frac{y}{x} = \frac{0}{-1} = 0.$$

b. Since $-(3\pi/2) = \frac{3}{4}(-2\pi)$, the point $P(x, y)$ that corresponds to $s = -(3\pi/2)$ is (0, 1) as indicated in Figure 8.5. Therefore,

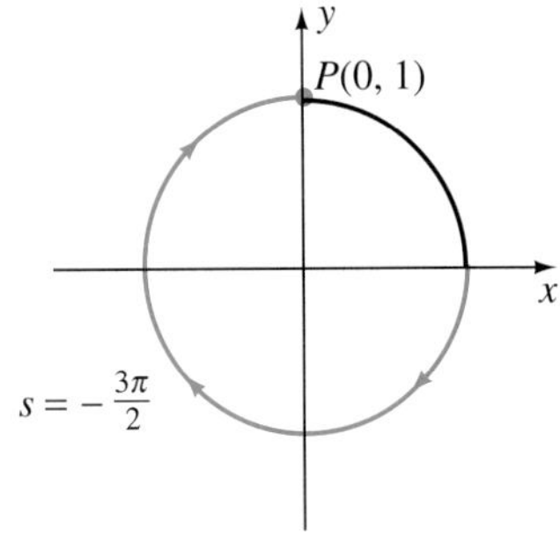

FIGURE 8.5

$$\sin\left(-\frac{3\pi}{2}\right) = y = 1$$

$$\cos\left(-\frac{3\pi}{2}\right) = x = 0$$

$$\tan\left(-\frac{3\pi}{2}\right) = \frac{y}{x} = \frac{1}{0}. \qquad \text{Undefined}$$

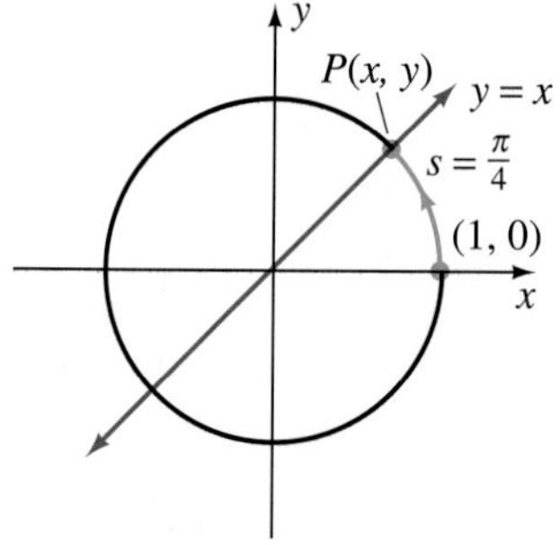

FIGURE 8.6

c. Since $\pi/4 = \frac{1}{8}(2\pi)$, the point $P(x, y)$ must lie on the line $y = x$ that bisects the first quadrant (Figure 8.6). Therefore, substituting x for y in the equation $x^2 + y^2 = 1$, we obtain

$$x^2 + x^2 = 1$$
$$2x^2 = 1$$
$$x^2 = \frac{1}{2}$$
$$x = \sqrt{\frac{1}{2}} = \frac{\sqrt{2}}{2}.$$

x and y are positive because P is in the first quadrant.

The coordinates of P are $\left(\sqrt{2}/2, \sqrt{2}/2\right)$ and the circular functions are determined.

$$\sin\frac{\pi}{4} = y = \frac{\sqrt{2}}{2}$$

$$\cos\frac{\pi}{4} = x = \frac{\sqrt{2}}{2}$$

$$\tan\frac{\pi}{4} = \frac{y}{x} = \frac{\sqrt{2}/2}{\sqrt{2}/2} = 1$$

d. Let's refer back to Figure 8.6. The point of intersection in the third quadrant of the line $y = x$ and the unit cicle is the point on the unit circle that corresponds to $s = 5\pi/4$. This point is the reflection of $P\left(\sqrt{2}/2, \sqrt{2}/2\right)$ through the origin. Thus, its coordinates are $\left(-\sqrt{2}/2, -\sqrt{2}/2\right)$. Therefore, the trigonometric functions can be determined.

$$\sin\frac{5\pi}{4} = y = -\frac{\sqrt{2}}{2}$$

$$\cos\frac{5\pi}{4} = x = -\frac{\sqrt{2}}{2}$$

$$\tan\frac{5\pi}{4} = \frac{y}{x} = \frac{-\frac{\sqrt{2}}{2}}{-\frac{\sqrt{2}}{2}} = 1$$

In Example 1 we determined the values of some trigonometric (circular) functions without any reference to angles, but don't get the wrong idea. We are not trying to eliminate completely the relationships between angles and trigonometric functions. On the contrary, there are many applications of trigonometry that are based on an angle definition of the trigonometric functions. However, there are also numerous applications, especially in the calculus, that use the trigonometric (circular) functions without any reference to angles. So we need to be able to work with these functions in different settings.

Sine Curve

As we graph the trigonometric functions with an xy-coordinate system, we will use equations of the type $f(x) = \sin x$, $f(x) = \cos x$, and $f(x) = \tan x$, and so on. Be careful not to confuse the use of the variable x in these equations with the x we employed earlier in the unit circle approach; in the unit circle approach, x denoted the x-coordinate of the point P on the unit circle. What is the graph of the function $f(x) = \sin x$? We could begin by making a fairly extensive table of values and then plotting the corresponding points. However, in this case our previous work with the sine function gives us some initial guidance. For example, we know that the sine function behaves as follows.

AS x INCREASES	THE SINE FUNCTION
From 0 to $\frac{\pi}{2}$	Increases from 0 to 1
From $\frac{\pi}{2}$ to π	Decreases from 1 to 0
From π to $\frac{3\pi}{2}$	Decreases from 0 to -1
From $\frac{3\pi}{2}$ to 2π	Increases from -1 to 0

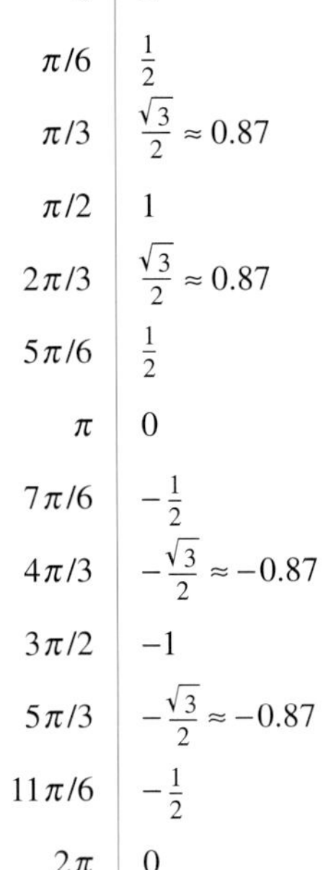

x	$f(x) = \sin x$
0	0
$\pi/6$	$\frac{1}{2}$
$\pi/3$	$\frac{\sqrt{3}}{2} \approx 0.87$
$\pi/2$	1
$2\pi/3$	$\frac{\sqrt{3}}{2} \approx 0.87$
$5\pi/6$	$\frac{1}{2}$
π	0
$7\pi/6$	$-\frac{1}{2}$
$4\pi/3$	$-\frac{\sqrt{3}}{2} \approx -0.87$
$3\pi/2$	-1
$5\pi/3$	$-\frac{\sqrt{3}}{2} \approx -0.87$
$11\pi/6$	$-\frac{1}{2}$
2π	0

Furthermore, we know from our work with coterminal angles (or the unit circle) that the sine function will repeat itself every 2π radians. More formally, we say that the sine function has a *period* of 2π. (A precise definition of the term *period* is given in the next section.) Now with these ideas in mind and our knowledge of some specific angles, let's set up a table of values that allows x to vary from 0 to 2π at intervals of $\pi/6$. (For graphing purposes, we will use decimal approximations of the radical expressions such as $\sqrt{3}/2$.) Plotting the points $(x, f(x))$, determined by the table, and connecting them with a smooth curve produces Figure 8.7.

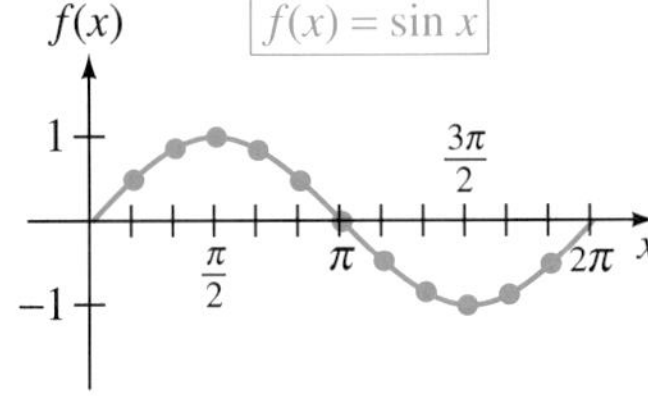

FIGURE 8.7

Now we can use the fact that the sine function has a period of 2π and sketch the curve from -4π to 3π as in Figure 8.8. The curve continues to repeat itself every 2π units in both directions indefinitely.

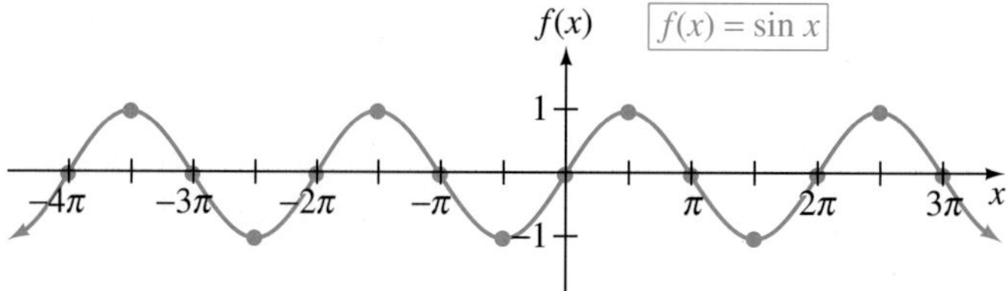

FIGURE 8.8

Cosine Curve

As you might expect from the definitions of the sine and cosine functions, their graphs are very similar. The cosine function also has a period of 2π; that is to say, the graph repeats itself every 2π units. In Figure 8.9 we graphed $f(x) = \cos x$, for x between 0 and 2π, by plotting the points determined by the table.

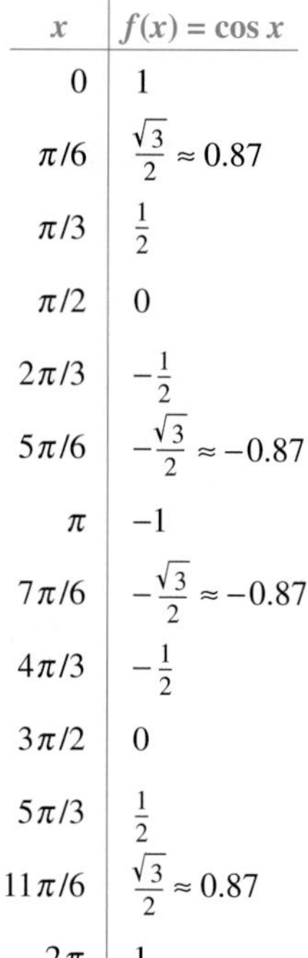

x	$f(x) = \cos x$
0	1
$\pi/6$	$\frac{\sqrt{3}}{2} \approx 0.87$
$\pi/3$	$\frac{1}{2}$
$\pi/2$	0
$2\pi/3$	$-\frac{1}{2}$
$5\pi/6$	$-\frac{\sqrt{3}}{2} \approx -0.87$
π	-1
$7\pi/6$	$-\frac{\sqrt{3}}{2} \approx -0.87$
$4\pi/3$	$-\frac{1}{2}$
$3\pi/2$	0
$5\pi/3$	$\frac{1}{2}$
$11\pi/6$	$\frac{\sqrt{3}}{2} \approx 0.87$
2π	1

FIGURE 8.9

Figure 8.10 depicts the cosine curve for $-2\pi \le x \le 4\pi$. Note that the cosine curve is identical to the sine curve, but moved $\pi/2$ units to the left. In other words,

$$\sin\left(x + \frac{\pi}{2}\right) = \cos x.$$

We will verify this relationship in the next chapter.

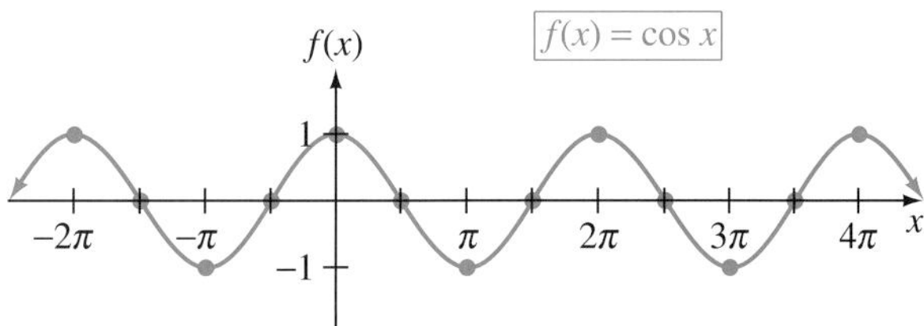

FIGURE 8.10

Variations of Sine and Cosine Curves

Let's briefly review a few ideas from our previous graphing experiences. Recall the following ideas from graphing parabolas.

1. The graph of $f(x) = x^2 + 2$ is the graph of $f(x) = x^2$ *moved up* 2 *units.* ($f(x) = x^2 + 2$ could be written as $f(x) = 2 + x^2$.)
2. The graph of $f(x) = (x - 3)^2$ is the graph of $f(x) = x^2$ *moved* 3 *units to the right.*
3. The graph of $f(x) = -x^2$ is the graph of $f(x) = x^2$ *reflected across the x-axis.*

Now let's apply these same ideas to the sine and cosine graphs. Furthermore, since we now know the general shapes of the sine and cosine curves, it is no longer necessary to plot so many points.

EXAMPLE 2

Graph $f(x) = 1 + \sin x$ for $0 \le x \le 2\pi$.

Solution

The graph of $f(x) = 1 + \sin x$ is the graph of $f(x) = \sin x$ *moved up* 1 *unit* (Figure 8.11).

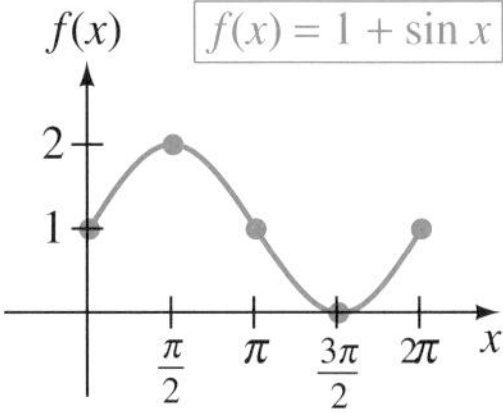

FIGURE 8.11

EXAMPLE 3

Graph $f(x) = \sin\left(x - \dfrac{\pi}{2}\right)$ for $\dfrac{\pi}{2} \le x \le \dfrac{5\pi}{2}$.

Solution

The graph of

$$f(x) = \sin\left(x - \frac{\pi}{2}\right)$$

is the graph of $f(x) = \sin x$ *moved* $\pi/2$ *units to the right*. Thus, in the interval $\pi/2 \le x \le 5\pi/2$ we get Figure 8.12.

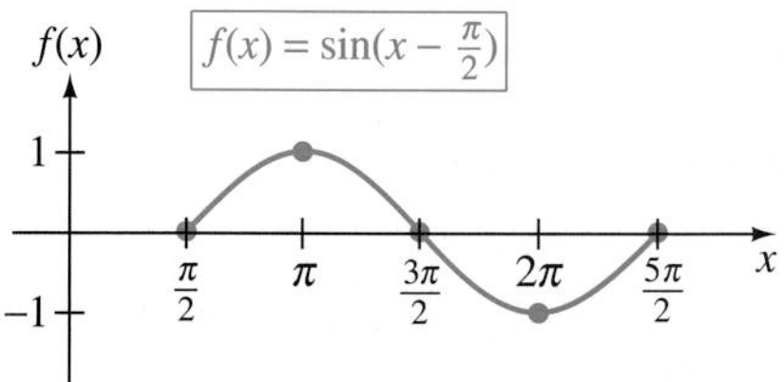

FIGURE 8.12

EXAMPLE 4

Graph $f(x) = -\cos x$ for $0 \le x \le 2\pi$.

Solution

The graph of $f(x) = -\cos x$ is the graph of $f(x) = \cos x$ *reflected across the x-axis* (Figure 8.13).

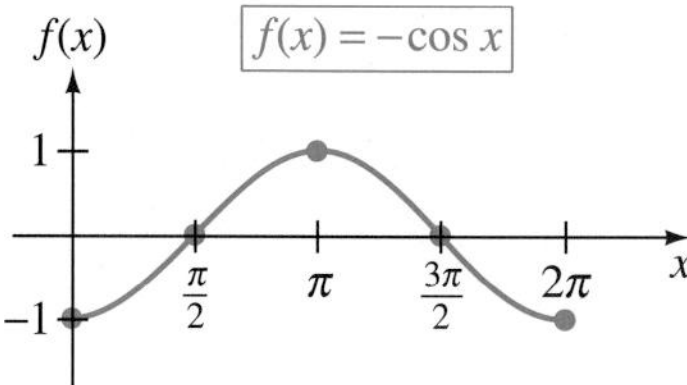

FIGURE 8.13

In Chapter 7 we established that $\sin(-x) = -\sin x$. Therefore, the graph of $f(x) = \sin(-x)$ is the same as the graph of $f(x) = -\sin x$ as shown in Figure 8.14 using a graphing utility. Also, since $\cos(-x) = \cos x$, the graph of $f(x) = \cos(-x)$ is the same as the graph of $f(x) = \cos x$ as shown in Figure 8.15.

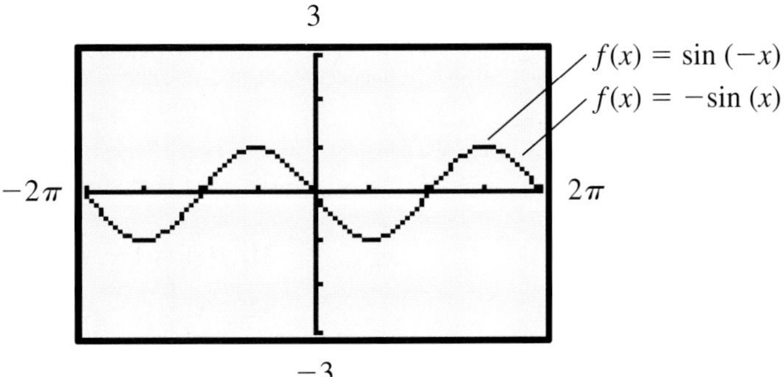

FIGURE 8.14

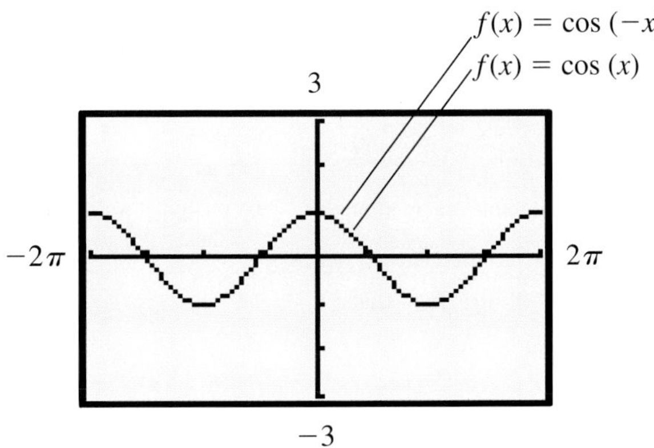

FIGURE 8.15

PROBLEM SET 8.1

For Problems 1–12, sketch the unit circle and indicate the given arc s along with its initial and terminal points. Then determine the exact values of the six circular functions of s without using a calculator or a table.

1. $s = \frac{3\pi}{4}$ **2.** $s = \frac{7\pi}{4}$ **3.** $s = \frac{3\pi}{2}$

4. $s = 2\pi$ **5.** $s = -\frac{\pi}{4}$ **6.** $s = -\pi$

7. $s = -3\pi$ **8.** $s = \frac{9\pi}{4}$ **9.** $s = \frac{5\pi}{2}$

10. $s = -\frac{5\pi}{4}$ **11.** $s = -\frac{11\pi}{4}$ **12.** $s = -\frac{15\pi}{4}$

For Problems 13–34, graph each of the functions in the indicated interval.

13. $f(x) = 2 + \sin x, \quad -2\pi \leq x \leq 2\pi$

14. $f(x) = -1 + \sin x, \quad -2\pi \leq x \leq 2\pi$

15. $f(x) = 3 + \cos x, \quad -2\pi \leq x \leq 2\pi$

16. $f(x) = -2 + \cos x, \quad -2\pi \leq x \leq 2\pi$

17. $f(x) = -\sin x, \quad 0 \leq x \leq 2\pi$

18. $f(x) = 1 - \sin x, \quad 0 \leq x \leq 2\pi$

19. $f(x) = 1 - \cos x, \quad 0 \leq x \leq 2\pi$

20. $f(x) = -2 - \cos x, \quad 0 \leq x \leq 2\pi$

21. $f(x) = \sin\left(x + \frac{\pi}{2}\right), \quad -\frac{\pi}{2} \le x \le \frac{3\pi}{2}$

22. $f(x) = \sin(x - \pi), \quad \pi \le x \le 3\pi$

23. $f(x) = \cos\left(x - \frac{\pi}{2}\right), \quad \frac{\pi}{2} \le x \le \frac{5\pi}{2}$

24. $f(x) = \cos(x + \pi), \quad -\pi \le x \le \pi$

25. $f(x) = 1 + \sin(x - \pi), \quad \pi \le x \le 3\pi$

26. $f(x) = -2 + \sin\left(x - \frac{\pi}{2}\right), \quad \frac{\pi}{2} \le x \le \frac{5\pi}{2}$

27. $f(x) = -1 + \cos(x + \pi), \quad -\pi \le x \le \pi$

28. $f(x) = 1 + \cos\left(x - \frac{\pi}{2}\right), \quad \frac{\pi}{2} \le x \le \frac{5\pi}{2}$

29. $f(x) = -\sin(x + \pi), \quad -\pi \le x \le \pi$

30. $f(x) = -\cos\left(x + \frac{\pi}{2}\right), \quad -\frac{\pi}{2} \le x \le \frac{3\pi}{2}$

31. $f(x) = 1 - \sin(x - \pi), \quad \pi \le x \le 3\pi$

32. $f(x) = 2 - \cos(x + \pi), \quad -\pi \le x \le \pi$

33. $f(x) = \sin(-x), \quad -2\pi \le x \le 2\pi$ [*Hint:* Don't forget that $\sin(-x) = -\sin x$.]

34. $f(x) = \cos(-x), \quad -2\pi \le x \le 2\pi$ [*Hint:* Don't forget that $\cos(-x) = \cos x$.]

THOUGHTS into WORDS

35. **a.** Give an *angle interpretation* of the fact that $\sin 45° = \sqrt{2}/2$.

b. Give an *arc length interpretation* of the fact that $\sin(\pi/4) = \sqrt{2}/2$.

36. Explain how you would graph the function $f(x) = -1 - \sin x$.

MISCELLANEOUS PROBLEMS

For Problems 37–40, graph each of the functions in the interval $0 \le x \le 2\pi$.

37. $f(x) = 2 \sin x$

38. $f(x) = 3 \sin x$

39. $f(x) = \frac{1}{2} \sin x$

40. $f(x) = -2 \sin x$

41. Graph $f(x) = \sin 2x$ in the interval $0 \le x \le 2\pi$.

42. Graph $f(x) = \sin 4x$ in the interval $0 \le x \le 2\pi$.

43. Graph $f(x) = \sin \frac{1}{2}x$ in the interval $0 \le x \le 4\pi$.

GRAPHICS CALCULATOR ACTIVITIES

Set your graphics calculator to graph trigonometric functions using radian measure. If necessary, consult your user's manual for specific instructions. Now graph the functions $f(x) = \sin x$ and $f(x) = \cos x$ for $0 \le x \le 2\pi$. Be sure that your graphs agree with Figures 8.7 and 8.9. Then graph the functions in Examples 2, 3, and 4 of this section. Be sure to set the domain and range so that your graphs look like Figures 8.11, 8.12, and 8.13.

The following problems are designed to reinforce some concepts of this section and to lay groundwork for material in the next section.

44. Graph $f(x) = \sin x$, $f(x) = \sin 2x$, and $f(x) = \sin 4x$ on the same set of axes. What effect do the 2 and 4 seem to have on the graphs? Predict the graph of $f(x) = \sin 8x$. Then graph $f(x) = \sin x$ and $f(x) = \sin 8x$ on the same set of axes.

45. Graph $f(x) = \sin x$, $f(x) = 2 \sin x$, $f(x) = 3 \sin x$, and $f(x) = 4 \sin x$ on the same set of axes. What effect do

the 2, 3, and 4 seem to have on the graphs? Predict the graph of $f(x) = 5 \sin x$. Now graph $f(x) = \sin x$ and $f(x) = 5 \sin x$ on the same set of axes.

46. Graph $f(x) = \sin x$ and $f(x) = \sin[x - (\pi/2)]$ on the same set of axes. First, predict the graph of $f(x) = \sin[x + (\pi/2)]$, and then graph $f(x) = \sin x$ and $f(x) = \sin[x + (\pi/2)]$ on the same set of axes.

47. Use your graphics calculator to demonstrate the truth of each of the following.

a. $\sin[x + (\pi/2)] = \cos x$

b. $\sin(-x) = -\sin x$

c. $\cos(-x) = \cos x$

48. Use your graphics calculator to determine what relationship holds between $\cos[x + (\pi/2)]$ and $\sin x$.

8.2 More on Graphing: Period, Amplitude, and Phase Shift

Let's begin this section by summarizing some of our ideas from Section 8.1 that pertain to the graphing of variations of the sine and cosine curves. We found that both the sine and cosine functions have a period of 2π. This means that their graphs repeat themselves every 2π units. The following general definition is helpful.

DEFINITION 8.1

A function f is called **periodic** if there exists a positive real number p such that

$$f(x + p) = f(x)$$

for all x in the domain of f. The smallest value for p is called the **period** of the function.

From our work with the sine and cosine functions, it is evident that 2π is the smallest positive number such that $\sin(x + 2\pi) = \sin x$ and $\cos(x + 2\pi) = \cos x$. Thus, we conclude that both the sine and cosine functions are periodic with a period of 2π.

Now suppose we consider the graph of a function that has an equation of the form $f(x) = \sin bx$, where $b > 0$. One cycle of the graph is completed as bx increases from 0 to 2π. When $bx = 0$, $x = 0$; and when $bx = 2\pi$, $x = 2\pi/b$. A similar line of reasoning holds for $f(x) = \cos bx$ and therefore we can state that **the period of $f(x) = \sin bx$ and also of $f(x) = \cos bx$, where $b > 0$, is $2\pi/b$.** If $b < 0$, then we can first apply the appropriate property, $\sin(-x) = -\sin x$ or $\cos(-x) = \cos x$. For example, to graph $f(x) = \sin(-3x)$ we can first change to $f(x) = -\sin 3x$ and then proceed.

EXAMPLE 1

Find the period of $f(x) = \cos 3x$ and sketch the graph for one period beginning at $x = 0$.

Solution

The period of $f(x) = \cos 3x$ is $2\pi/3$. Therefore, let's divide the interval from 0 to $2\pi/3$ on the x-axis into four equal subintervals. Each subinterval will be of length

$$\frac{2\pi/3 - 0}{4} = \frac{2\pi/3}{4} = \frac{2\pi}{12} = \frac{\pi}{6}.$$

Then we can plot points determined by the endpoint values of the subintervals and sketch the curve as in Figure 8.16.

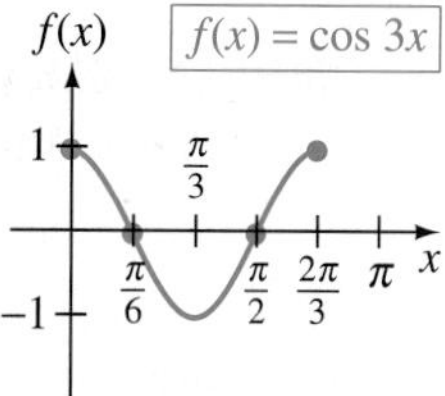

FIGURE 8.16

EXAMPLE 2

Find the period of $f(x) = \sin \pi x$ and sketch the graph for one period beginning at $x = 0$.

Solution

The period of $f(x) = \sin \pi x$ is $2\pi/\pi = 2$. Again, by dividing the interval from 0 to 2 into four equal subintervals, we can plot points determined by the endpoint values of the subintervals and sketch the curve as in Figure 8.17.

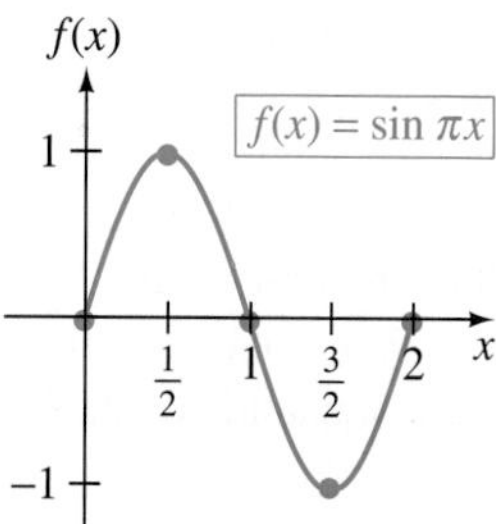

FIGURE 8.17

Amplitude

Consider the graph of $f(x) = 2 \sin x$. It should be evident that for each x-coordinate, the y-coordinate is twice that of the corresponding y-coordinate on the basic sine graph. Likewise, the corresponding y-coordinates of $f(x) = 3 \sin x$ are three times those of the basic sine graph. Thus, in Figure 8.18 we have sketched the graphs of $f(x) = 2 \sin x$ and $f(x) = 3 \sin x$ in the interval from 0 to 2π.

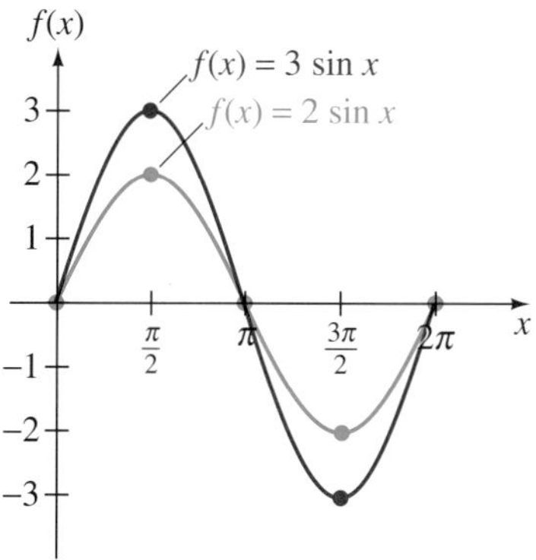

FIGURE 8.18

The maximum functional value attained by $f(x) = 2 \sin x$ is 2 and by $f(x) = 3 \sin x$ is 3. Each of these maximum functional values is called the **amplitude** of the graph. In general, we can state that **the amplitude of the graph of $f(x) = a \sin x$ or $f(x) = a \cos x$ is $|a|$**. For example, the amplitude of $f(x) = \frac{1}{2} \sin x$ is $\left|\frac{1}{2}\right| = \frac{1}{2}$ and the amplitude of $f(x) = -2 \cos x$ is $|-2| = 2$. Together the concepts of period and amplitude help us graph functions of the form $f(x) = a \sin bx$ or $f(x) = a \cos bx$.

EXAMPLE 3

Find the period and amplitude of $f(x) = -3 \sin \frac{1}{2}x$, and sketch the curve for one period beginning at $x = 0$.

Solution

The period is $2\pi/(1/2) = 4\pi$ and the amplitude is $|-3| = 3$. The curve $f(x) = -3 \sin \frac{1}{2}x$ is an x-axis reflection of $f(x) = 3 \sin \frac{1}{2}x$. Thus, we obtain Figure 8.19.

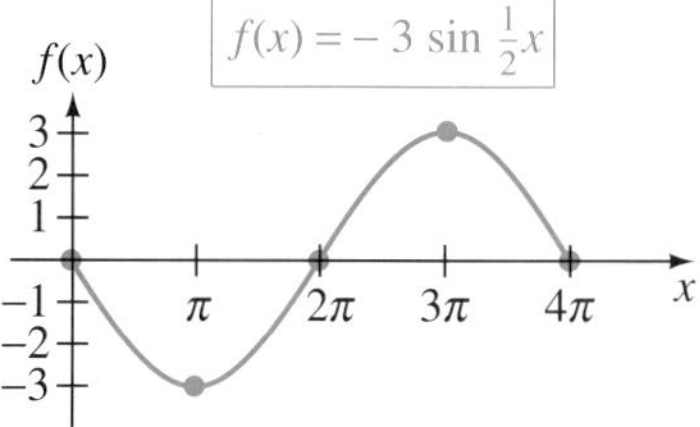

FIGURE 8.19

Phase Shift

In the previous section we found that the graph of $f(x) = \sin(x - \pi/2)$ is the basic sine curve shifted $\pi/2$ units to the right. Likewise, the graph of $f(x) = \cos(x + \pi)$ is the basic cosine curve shifted π units to the left. Each of the numbers $\pi/2$ and π, which represent the amount of shift, is called the **phase shift** of the graph. In general, **the phase shift of $f(x) = \sin(x - c)$ or $f(x) = \cos(x - c)$ is $|c|$. If c is positive, the shift is to the right and if c is negative, the shift is to the left.**

Now let's pull together the concepts of period, amplitude, and phase shift in one very useful property.

PROPERTY 8.1

Consider the functions $f(x) = a \sin b(x - c)$ and $f(x) = a \cos b(x - c)$, where $b > 0$.

1. The period of both curves is $2\pi/b$.
2. The amplitude of both curves is $|a|$.
3. The phase shift of both curves is $|c|$. The shift is to the right if c is positive and to the left if c is negative.

EXAMPLE 4

Find the period, amplitude, and phase shift of $f(x) = 2 \sin(x - \pi)$, and sketch the curve.

Solution

By applying Property 8.1 we can determine the following information directly from the equation.

$$f(x) = 2 \sin 1(x - \pi)$$

Amplitude is $|2| = 2$

Period is $2\pi/1 = 2\pi$

Phase shift is $|\pi| = \pi$ units to the right

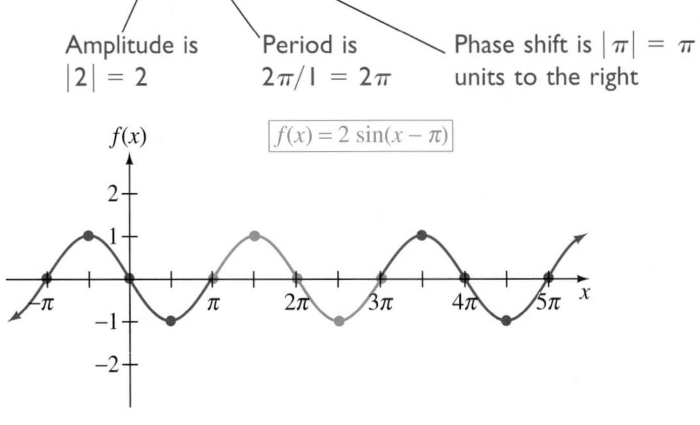

FIGURE 8.20

Thus, one complete cycle of a sine curve having an amplitude of 2 is contained in the interval from $x = \pi$ to $x = \pi + 2\pi = 3\pi$. The curve repeats itself every 2π units in both directions (Figure 8.20).

EXAMPLE 5

Find the period, amplitude, and phase shift of $f(x) = \frac{1}{2}\sin(2x + \pi)$, and sketch the curve.

Solution

First, let's change the form of the equation so that we can apply Property 8.1.

$$\begin{aligned} f(x) &= \frac{1}{2}\sin(2x + \pi) \\ &= \frac{1}{2}\sin 2\left(x + \frac{\pi}{2}\right) \\ &= \frac{1}{2}\sin 2\left(x - \left(-\frac{\pi}{2}\right)\right) \end{aligned}$$

Now we can obtain some information about the graph as follows.

$$f(x) = \frac{1}{2}\sin 2\left(x - \left(-\frac{\pi}{2}\right)\right)$$

Amplitude is $\left|\frac{1}{2}\right| = \frac{1}{2}$

Period is $2\pi/2 = \pi$

Phase shift is $|-\pi/2| = \pi/2$ units to the left

Therefore, one complete cycle of the sine curve having an amplitude of $\frac{1}{2}$ is contained in the interval from $x = -\pi/2$ to $x = -\pi/2 + \pi = \pi/2$. The curve repeats itself every π units in both directions (Figure 8.21).

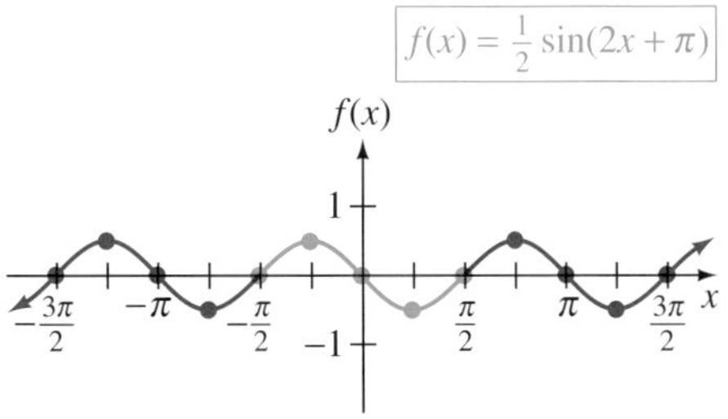

FIGURE 8.21

EXAMPLE 6

Find the period, amplitude, and phase shift of

$$f(x) = 3\cos\left(2x - \frac{\pi}{2}\right)$$

and sketch the curve.

Solution

The given equation can be written as

$$\begin{aligned} f(x) &= 3\cos\left(2x - \frac{\pi}{2}\right) \\ &= 3\cos 2\left(x - \frac{\pi}{4}\right). \end{aligned}$$

From this form we can obtain the following information.

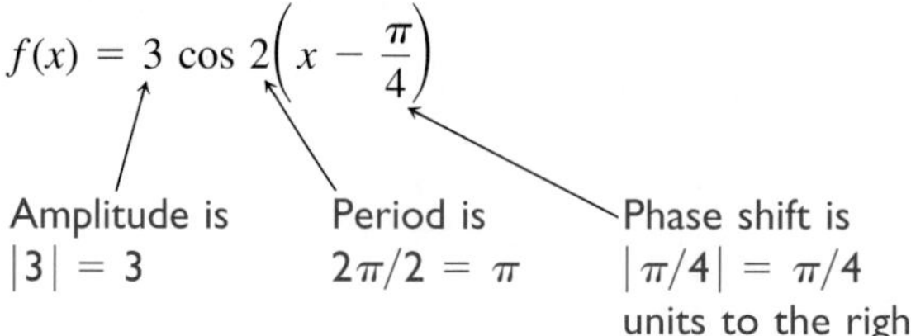

Therefore, one complete cycle of a cosine curve having an amplitude of 3 is contained in the interval from $x = \pi/4$ to $x = \pi/4 + \pi = 5\pi/4$. The curve repeats itself every π units in both directions (Figure 8.22).

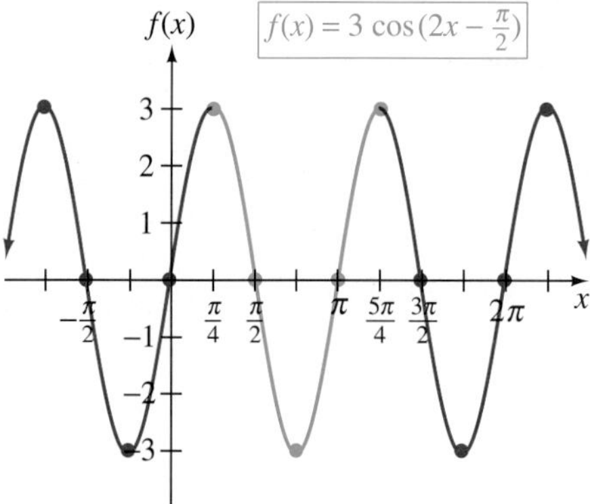

FIGURE 8.22

Sometimes a mathematical application involves a function that is the sum (or difference) of two or more functions. In such situations a graph of the function is often helpful when solving a problem. For example, suppose that we need the graph of the function $f(x) = \sin x + \cos x$ for $0 \le x \le 2\pi$. Certainly we could set up a table of values and plot some points, but it would take a large number of points to determine an accurate graph. Another approach would be to graph the sine curve and cosine curve on the same set of axes and then, with a ruler and/or a compass, graphically add y-coordinates for a number of x-values. Again, this approach would require quite a number of points in order to obtain an accurate graph. By far the most efficient way of obtaining the graph is by using a graphing utility as we will see in the following example.

EXAMPLE 7

Use a graphing utility to graph the function $f(x) = \sin x + \cos x$ for $0 \le x \le 2\pi$.

Solution

Let's set a viewing rectangle so that $0 \le x \le 2\pi$ and $-2 \le y \le 2$. Then we can make the following assignments.

$$Y_1 = \sin x \qquad Y_2 = \cos x \qquad Y_3 = \sin x + \cos x$$

(Remember that we could also let $Y_3 = Y_1 + Y_2$.) Now the graphing utility can be set to produce all three graphs on the same set of axes as in Figure 8.23(a) or only the final graph of $f(x) = \sin x + \cos x$ as in Figure 8.23(b). Watch your graphing utility produce Figure 8.23(a) to give you a dynamic view of the addition of y-coordinates.

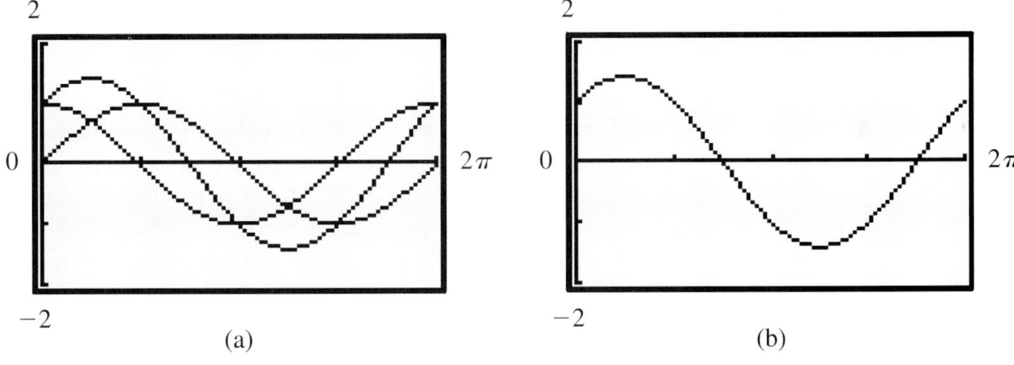

FIGURE 8.23

Many phenomena in our world behave in a cyclic or rhythmic manner. Such behavior can often be described mathematically by using variations of the sine and cosine curves. Such phenomena occur in a variety of applications ranging from alternating current in electricity to the sound waves generated by a vibrating tuning fork. There is a large class of problems known as simple harmonic motion problems. An example of simple harmonic motion is illustrated in Figure 8.24. Suppose that a weight W is attached to a spring. If the weight is pulled down and then released, it will oscillate up and down about the rest or equilibrium point marked 0 in Figure 8.24. Vibratory motion of this type is called **simple harmonic motion**.

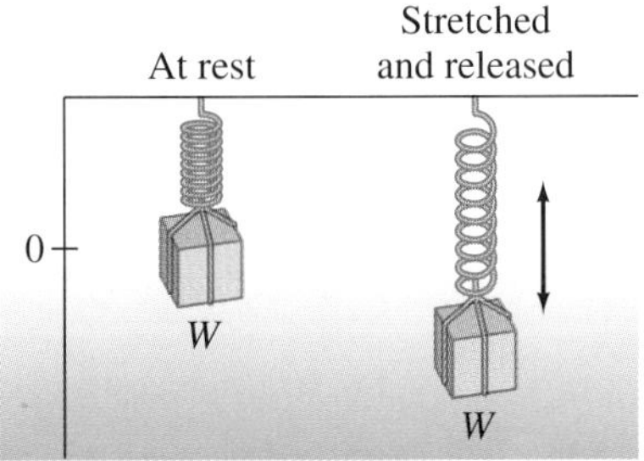

FIGURE 8.24

An example of harmonic motion
Dean Abramson/Stock Boston

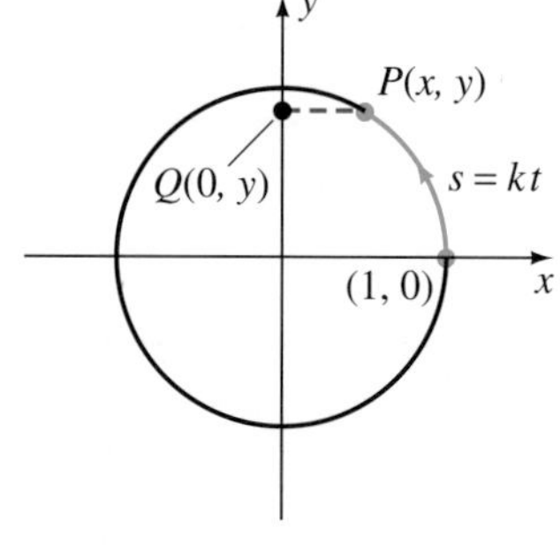

FIGURE 8.25

To obtain a mathematical model of simple harmonic motion, let's use the unit circle in Figure 8.25. Consider a point $P(x, y)$ moving at a constant rate of k radians per unit of time in a counterclockwise direction around the circle. Using $(1, 0)$ as the initial position of P (when $t = 0$), then the arc distance s is given by $s = kt$ after t units of time. So, $y = \sin s = \sin kt$. Now consider the point Q, which is the projection of P on the vertical axis. As P moves around the circle, point Q oscillates up and down (simple harmonic motion). At any time t, the distance of point Q from the center of the circle is given by $y = \sin kt$.

PROBLEM SET 8.2

For Problems 1–18, graph the given function in the indicated interval.

1. $f(x) = \cos 2x, \quad 0 \le x \le 2\pi$

2. $f(x) = \cos \pi x, \quad 0 \le x \le 2$

3. $f(x) = \sin 3x, \quad -\frac{2\pi}{3} \le x \le \frac{2\pi}{3}$

4. $f(x) = \sin 2x, \quad -2\pi \le x \le 2\pi$

5. $f(x) = 2 \cos x, \quad 0 \le x \le 2\pi$

6. $f(x) = -3 \cos x, \quad 0 \le x \le 2\pi$

7. $f(x) = \frac{1}{2} \sin x, \quad -2\pi \le x \le 2\pi$

8. $f(x) = \frac{1}{2} \cos x, \quad -2\pi \le x \le 2\pi$

9. $f(x) = 2 \sin \frac{1}{2}x, \quad 0 \le x \le 4\pi$

10. $f(x) = 3 \cos \frac{1}{2}x, \quad 0 \le x \le 4\pi$

11. $f(x) = \cos\left(x + \frac{\pi}{2}\right), \quad -\frac{\pi}{2} \le x \le \frac{3\pi}{2}$

12. $f(x) = \sin(x - \pi), \quad \pi \le x \le 3\pi$

13. $f(x) = -\sin\left(x - \frac{\pi}{2}\right), \quad \frac{\pi}{2} \le x \le \frac{5\pi}{2}$

14. $f(x) = -\cos(x - \pi), \quad \pi \le x \le 3\pi$

15. $f(x) = \sin(-2x), \quad 0 \le x \le 2\pi$

16. $f(x) = \sin(-\pi x), \quad 0 \le x \le 4$

17. $f(x) = \cos(-\pi x), \quad 0 \le x \le 4$

18. $f(x) = \cos(-2x), \quad 0 \le x \le 2\pi$

For Problems 19–40, find the period, amplitude, and phase shift of the given function and draw the graph of the function.

19. $f(x) = 3 \sin\left(x + \frac{\pi}{2}\right)$

20. $f(x) = \frac{1}{2} \sin\left(x - \frac{\pi}{2}\right)$

21. $f(x) = 2 \cos(x - \pi)$

22. $f(x) = 3 \cos\left(x + \frac{\pi}{2}\right)$

23. $f(x) = \frac{1}{2} \cos\left(x + \frac{\pi}{4}\right)$

24. $f(x) = 2 \sin\left(x - \frac{\pi}{3}\right)$

25. $f(x) = -2 \sin(x + \pi)$

26. $f(x) = -3 \cos(x + \pi)$

27. $f(x) = 2 \sin 2(x - \pi)$

28. $f(x) = 3 \cos 2\left(x - \frac{\pi}{2}\right)$

29. $f(x) = \frac{1}{2} \cos 3(x + \pi)$

30. $f(x) = 2 \sin 3(x - \pi)$

31. $f(x) = \sin(2x - \pi)$

32. $f(x) = \cos\left(2x + \frac{\pi}{2}\right)$

33. $f(x) = 2 \cos(3x - \pi)$

34. $f(x) = 4 \sin(2x + \pi)$

35. $f(x) = \frac{1}{2} \sin(3x - \pi)$

36. $f(x) = 2 \sin\left(\frac{1}{2}x - \frac{\pi}{2}\right)$

37. $f(x) = 2 \cos\left(\frac{1}{2}x + \frac{\pi}{2}\right)$

38. $f(x) = -2 \sin(2x - \pi)$

39. $f(x) = -3 \sin(2x + \pi)$

40. $f(x) = -\frac{1}{2} \cos\left(2x - \frac{\pi}{2}\right)$

MISCELLANEOUS PROBLEMS

Graph each of the following functions.

41. $f(x) = 1 + 2 \sin\left(x - \frac{\pi}{2}\right)$

42. $f(x) = 2 - 3 \cos\left(x + \frac{\pi}{2}\right)$

43. $f(x) = -1 + 2 \cos(2x + \pi)$

44. $f(x) = -2 + 2 \sin\left(2x - \frac{\pi}{2}\right)$

GRAPHICS CALCULATOR ACTIVITIES

45. Use your graphics calculator to check your graphs for Problems 13–24 and Problems 41–44.

46. For each of the following, (a) predict the general shape and location of the graph, and (b) graph the function with your graphics calculator to check your prediction.

a. $f(x) = 2 \sin x + \cos x$ **b.** $f(x) = \sin 2x + \cos x$

c. $f(x) = \sin x + \cos 2x$ **d.** $f(x) = \sin x - \cos x$

e. $f(x) = \cos x - \sin x$ **f.** $f(x) = 2 \cos x + \sin x$

47. Graph $f(x) = |x|$, $f(x) = -|x|$, and $f(x) = x \sin x$ on the same set of axes. The graph of $f(x) = x \sin x$ is called a **damped sine wave** and the x factor is called the **damping factor**. Look at your display and note that the curve for $f(x) = x \sin x$ appears to be clamped between $f(x) = |x|$ and $f(x) = -|x|$. Construct an algebraic argument to verify that $-|x| \le x \sin x \le |x|$.

48 How do you think the graph of $f(x) = x \cos x$ compares to the graph of $f(x) = x \sin x$? Graph $f(x) = |x|$, $f(x) = -|x|$, $f(x) = x \sin x$, and $f(x) = x \cos x$ on the same set of axes to check your prediction.

8.3 GRAPHING THE OTHER BASIC TRIGONOMETRIC FUNCTIONS

Recall that $\tan x = \sin x/\cos x$ and therefore, it is not defined for those values of x where $\cos x = 0$ (namely, at $x = \pm\pi/2, \pm 3\pi/2, \pm 5\pi/2$, etc.). Thus, to graph $f(x) = \tan x$, we will first use vertical dashed lines to indicate those values of x where the tangent function is undefined (Figure 8.26).

x	$f(x) = \tan x$
$-\pi/3$	$-\sqrt{3} \approx -1.73$
$-\pi/4$	-1
$-\pi/6$	$-\frac{\sqrt{3}}{3} \approx -0.58$
0	0
$\pi/6$	$\frac{\sqrt{3}}{3} \approx 0.58$
$\pi/4$	1
$\pi/3$	$\sqrt{3} \approx 1.73$

FIGURE 8.26

Next, let's consider some functional values in the interval between $x = -\pi/2$ and $x = \pi/2$ as indicated in the table. The points determined by these values are plotted in Figure 8.26. Notice that $\tan x$ is getting larger as x increases from $\pi/6$ to $\pi/4$ to $\pi/3$. By choosing more values of x between $\pi/3$ and $\pi/2$, and using a calculator, we obtain the following results.

$\tan 1.1 = 2.0 \qquad \tan 1.2 = 2.6 \qquad \tan 1.3 = 3.6$

$\tan 1.4 = 5.8 \qquad \tan 1.5 = 14.1 \qquad \tan 1.55 = 48.1$

(Perhaps you should check these values with your calculator. Don't forget to put the calculator in the radian mode.) As x approaches $\pi/2$ from the left side, $\tan x$ is getting larger and larger. The line $x = \pi/2$ is an asymptote. Likewise, the line $x = -\pi/2$ is an asymptote and the graph of $f(x) = \tan x$ for $-\pi/2 < x < \pi/2$ is shown in Figure 8.27.

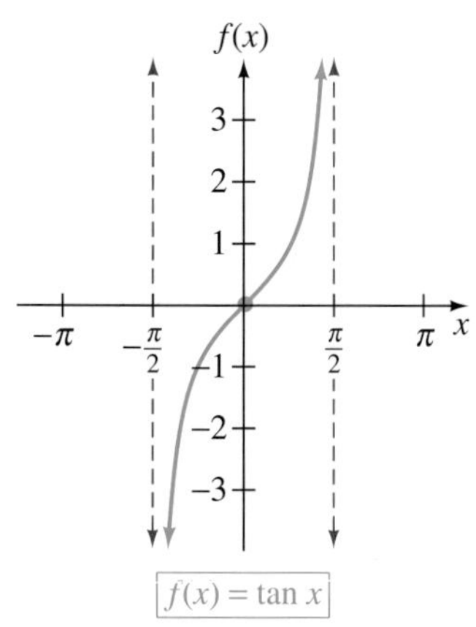

FIGURE 8.27

Using a similar approach we can graph $f(x) = \tan x$ in the intervals $\pi/2 < x < 3\pi/2$ and $-3\pi/2 < x < -\pi/2$ to obtain Figure 8.28. Note that the tangent function has a period of π.

We can graph the cotangent function in much the same manner that we graphed the tangent function. First, because $\cot x = \cos x/\sin x$, where $\sin x \neq 0$, we know that the cotangent function is not defined at $x = 0, \pm\pi, \pm 2\pi$, etc. These values locate the vertical asymptotes as indicated in Figure 8.29. As the reciprocal of the tangent function, the cotangent function also has a period of π. Finally, by

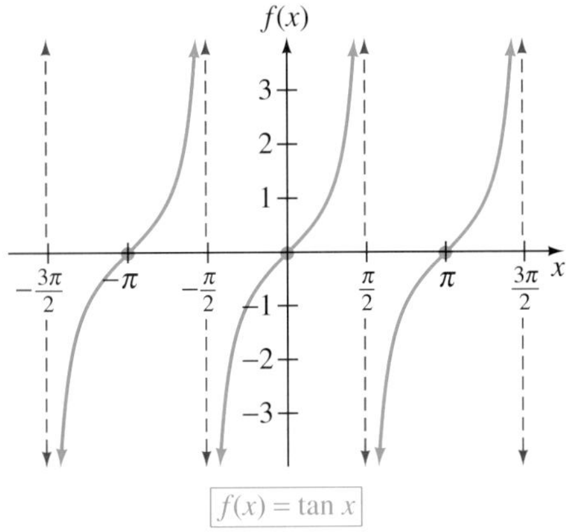

FIGURE 8.28

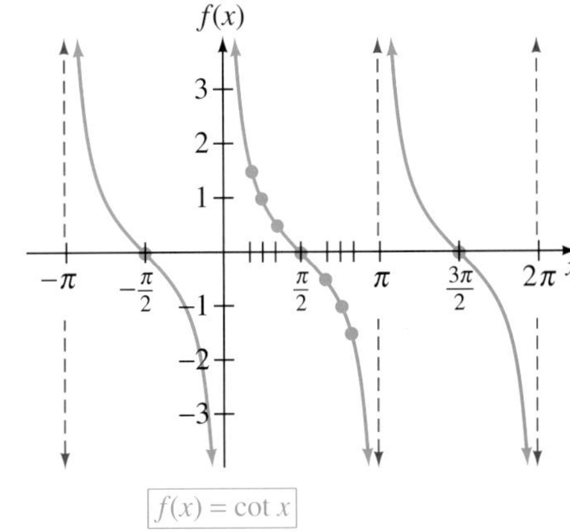

FIGURE 8.29

plotting a few points, $(\pi/6, \sqrt{3})$, $(\pi/4, 1)$, $(\pi/3, \sqrt{3}/3)$, $(2\pi/3, -\sqrt{3}/3)$, $(3\pi/4, -1)$, and $(5\pi/6, -\sqrt{3})$, in the interval $0 < x < \pi$, we can sketch the cotangent curve as in Figure 8.29.

Variations of Tangent and Cotangent Curves

We have seen that both the tangent and cotangent functions have period π; that is, they repeat themselves every π units. This also means that their asymptotes are π units apart. Furthermore, because the tangent and cotangent functions increase and decrease without bound, the concept of amplitude has no meaning. However, to see the effect of a number a on the graph of a function of the form $f(x) = a \tan x$ or $f(x) = a \cot x$, the graphs of one period of $f(x) = 3 \tan x$ and $f(x) = \frac{1}{2} \cot x$ are shown in Figures 8.30(a) and (b). Notice how the coefficient 3 in $f(x) =$

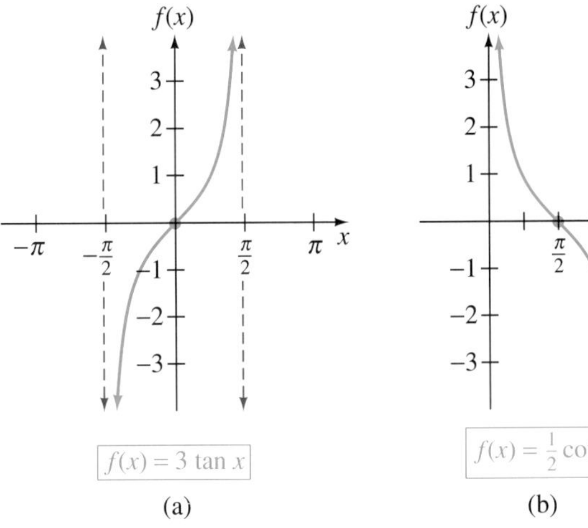

FIGURE 8.30

3 tan x appears to *stretch* the tangent curve and the coefficient $\frac{1}{2}$ in $f(x) = \frac{1}{2}\cot x$ appears to *shrink* the cotangent curve. These facts need to be kept in mind as we use the general equations $f(x) = a \tan b(x - c)$ and $f(x) = a \cot b(x - c)$, $b > 0$, to study variations of the basic tangent and cotangent curves. The number a effects ordinate values but has no significance in terms of amplitude. In each case, *the period is determined by* π/b *and the phase shift is again* $|c|$. Let's consider some examples.

EXAMPLE 1

Find the period and phase shift of $f(x) = 2\tan\left(x - \frac{\pi}{4}\right)$, and sketch the curve.

Solution

From the equation we can determine the period and the phase shift.

$$f(x) = 2 \tan 1\left(x - \frac{\pi}{4}\right)$$

Period is $\pi/1 = \pi$ (from the coefficient 1)

Phase shift is $\pi/4$ units to the right (from $\frac{\pi}{4}$)

Let's shift the asymptotes $x = -\pi/2$ and $x = \pi/2$ of $g(x) = \tan x$ to the right $\pi/4$ units; so we have asymptotes at $x = -\pi/4$ and $x = 3\pi/4$ as shown in Figure 8.31. The curve crosses the x-axis at $x = \pi/4$ since $f(\pi/4) = 0$. The points $(0, -2)$ and $(\pi/2, 2)$ help determine the shape of the curve. Figure 8.31 shows the graph through three periods.

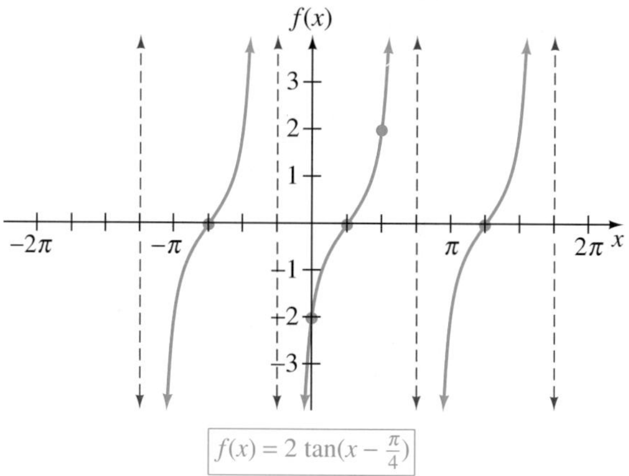

FIGURE 8.31

Before we consider the next example, let's make an important observation. Remember that two asymptotes of the curve $f(x) = \cot x$ are located at $x = 0$ and at $x = \pi$. Now consider the function $g(x) = \cot 2x$. If $2x = 0$, then $x = 0$; and if

$2x = \pi$, then $x = \pi/2$. Therefore, two asymptotes of the curve $g(x) = \cot 2x$ are located at $x = 0$ and $x = \pi/2$. Keep this in mind as we consider the next example.

EXAMPLE 2

Find the period and phase shift of $f(x) = \frac{1}{2}\cot(2x + \pi)$, and sketch the curve.

Solution

First, let's change the form of the equation so that it yields some useful information.

$$f(x) = \frac{1}{2}\cot(2x + \pi)$$

$$= \frac{1}{2}\cot 2\left(x + \frac{\pi}{2}\right)$$

Period is $\pi/2$ — Phase shift is $\pi/2$ units to the left

Let's shift the asymptotes of $g(x) = \cot 2x$, $x = 0$ and $x = \pi/2$, to the left $\pi/2$ units; so we have asymptotes at $x = -\pi/2$ and at $x = 0$, as shown in Figure 8.32. The curve crosses the x-axis at $x = -\pi/4$ since $f(-\pi/4) = 0$. The points $\left(-\pi/8, -\frac{1}{2}\right)$ and $\left(-3\pi/8, \frac{1}{2}\right)$ help determine the shape of the curve. Figure 8.32 shows the graph through three periods.

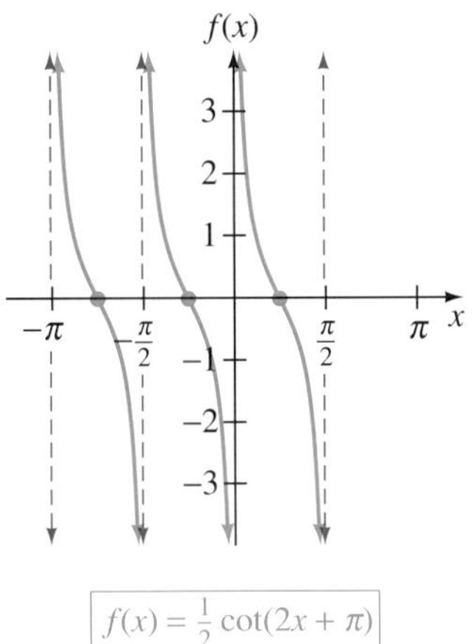

FIGURE 8.32

Cosecant and Secant Curves

The graphs of $f(x) = \csc x$ and $f(x) = \sec x$ can be sketched easily by using the reciprocal relationships $\csc x = 1/\sin x$ and $\sec x = 1/\cos x$. In Figures 8.33 and 8.34 we first drew the sine and cosine curves (dashed curves) and then used those

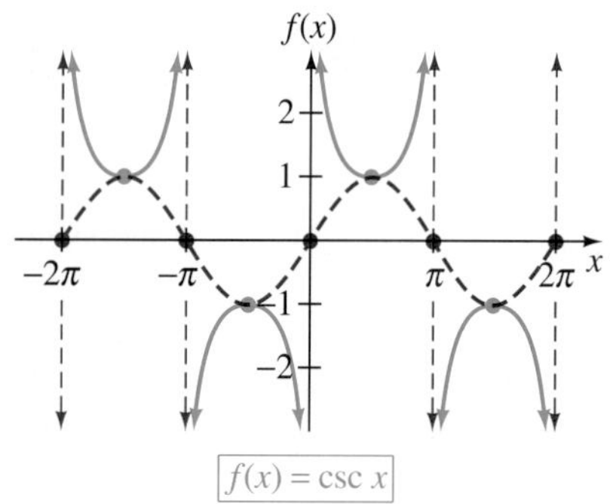

FIGURE 8.33

FIGURE 8.34

curves to help sketch the cosecant and secant curves, respectively. The following general properties should be noted.

1. Because all functional values of sin x and cos x are between ± 1 inclusive, we know that allfunctional values for csc x and sec x are greater than or equal to 1, or less than or equal to -1.
2. For graphing $f(x) = \csc x$, vertical asymptotes exist at $x = 0, \pm\pi, \pm 2\pi$, etc.
3. For graphing $f(x) = \sec x$, vertical asymptotes exist at $x = \pm\pi/2, \pm 3\pi/2, \pm 5\pi/2$, etc.
4. Both the cosecant and secant functions have a period of 2π.

The two equations $f(x) = a \csc b(x - c)$ and $f(x) = a \sec b(x - c)$ are used to express variations of the cosecant and secant curves, respectively. As with tangent and cotangent curves, the concept of amplitude has no meaning with cosecant and secant curves. Thus, the number a simply effects ordinate values but has no significance relative to amplitude. **For both $f(x) = a \csc b(x - c)$ and $f(x) = a \sec b(x - c)$, where $b > 0$, the period is determined by $2\pi/b$ and the phase shift is again $|c|$.**

In Figure 8.33, notice that three asymptotes of the curve $f(x) = \csc x$ are located at $x = 0$, $x = \pi$, and $x = 2\pi$. Now consider the function $g(x) = \csc 2x$.

If $2x = 0$, then $x = 0$.

If $2x = \pi$, then $x = \pi/2$.

If $2x = 2\pi$, then $x = \pi$.

So, the curve $g(x) = \csc 2x$ has three asymptotes at $x = 0$, $x = \pi/2$, and $x = \pi$.

EXAMPLE 3

Find the period and phase shift of $f(x) = \csc 2[x - (\pi/4)]$, and sketch the curve.

Solution

From the equation we can determine the period and phase shift.

$$f(x) = \csc 2\left(x - \frac{\pi}{4}\right)$$

Period is $2\pi/2 = \pi$

Phase shift is $\pi/4$ units to the right

Let's shift the asymptotes, $x = 0$, $x = \pi/2$, and $x = \pi$ of $g(x) = \csc 2x$, to the right $\pi/4$ units. Therefore, we have asymptotes at $x = \pi/4$, $x = 3\pi/4$, and $x = 5\pi/4$ as indicated in Figure 8.35. The points $(\pi/2, 1)$ and $(\pi, -1)$ determine the turning points. One period of the graph is shown in Figure 8.35.

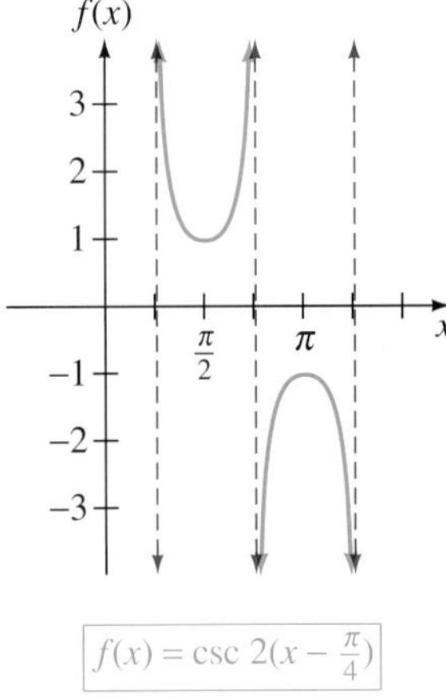

FIGURE 8.35

EXAMPLE 4

Use a graphing utility to graph the function $f(x) = \csc x + \sec x$ for $0 \leq x \leq 2\pi$.

Solution

Before we use a graphing utility, let's go back and take another look at the cosecant and secant curves between $x = 0$ and $x = 2\pi$ in Figures 8.33 and 8.34, respectively. From these graphs we can tell that the graph of the sum of these two functions will have vertical asymptotes at $x = 0$, $x = \pi/2$, $x = \pi$, $x = 3\pi/2$, and $x = 2\pi$. Furthermore, between $x = 0$ and $x = \pi/2$, the graph should be entirely above the x-axis, but between $x = \pi/2$ and $x = \pi$, the graph will appear both above and below the x-axis. Then between $x = \pi$ and $x = 3\pi/2$, the graph will be below the x-axis, but between $x = 3\pi/2$ and $x = 2\pi$, the graph will be both above and below the x-axis.

Now let's use a graphing utility to obtain the graph of $f(x) = \csc x + \sec x$. Figure 8.36 shows this display where we used a viewing rectangle such that $0 \leq x \leq 2\pi$ and $-5 \leq y \leq 5$.

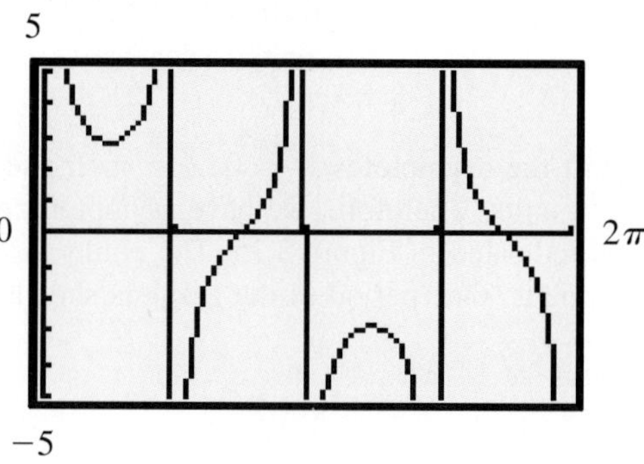

FIGURE 8.36

PROBLEM SET 8.3

For Problems 1–22, graph each of the functions in the indicated interval.

1. $f(x) = \tan x, \quad 0 \leq x \leq \pi$
2. $f(x) = -\tan x, \quad -\frac{\pi}{2} \leq x \leq \frac{\pi}{2}$
3. $f(x) = -2 \tan x, \quad -\frac{\pi}{2} \leq x \leq \frac{\pi}{2}$
4. $f(x) = \frac{1}{2} \tan x, \quad -\frac{\pi}{2} \leq x \leq \frac{\pi}{2}$
5. $f(x) = 2 + \tan x, \quad -\frac{\pi}{2} \leq x \leq \frac{\pi}{2}$
6. $f(x) = -1 + \tan x, \quad -\frac{\pi}{2} \leq x \leq \frac{\pi}{2}$
7. $f(x) = \tan(-x), \quad -\frac{\pi}{2} \leq x \leq \frac{\pi}{2}$
8. $f(x) = \tan 2x, \quad -\frac{\pi}{4} \leq x \leq \frac{\pi}{4}$
9. $f(x) = 1 + \cot x, \quad 0 \leq x \leq 2\pi$
10. $f(x) = -2 + \cot x, \quad 0 \leq x \leq 2\pi$
11. $f(x) = \cot(-x), \quad 0 \leq x \leq \pi$
12. $f(x) = -\cot x, \quad 0 \leq x \leq \pi$
13. $f(x) = \cot 3x, \quad 0 \leq x \leq \pi$
14. $f(x) = \cot \pi x, \quad -1 \leq x \leq 1$
15. $f(x) = \csc 2x, \quad 0 \leq x \leq \pi$
16. $f(x) = -\csc x, \quad 0 \leq x \leq 2\pi$
17. $f(x) = 3 \csc \pi x, \quad 0 \leq x \leq 2$
18. $f(x) = 1 + \csc x, \quad 0 \leq x \leq 2\pi$
19. $f(x) = -\sec x, \quad -\frac{\pi}{2} \leq x \leq \frac{3\pi}{2}$
20. $f(x) = \sec 3x, \quad -\frac{\pi}{6} \leq x \leq \frac{\pi}{2}$
21. $f(x) = \sec(-x), \quad -\frac{\pi}{2} \leq x \leq \frac{3\pi}{2}$
22. $f(x) = 1 + \sec x, \quad -\frac{\pi}{2} \leq x \leq \frac{3\pi}{2}$

For Problems 23–36, find the period and phase shift of the given function, and sketch the graph through two periods.

23. $f(x) = \tan\left(x + \frac{\pi}{2}\right)$

24. $f(x) = \tan(x - \pi)$

25. $f(x) = 2\tan\left(x - \frac{\pi}{4}\right)$

26. $f(x) = 3\tan\left(x + \frac{\pi}{4}\right)$

27. $f(x) = \tan(2x + \pi)$

28. $f(x) = \tan(3x - \pi)$

29. $f(x) = -2\tan\left(x - \frac{\pi}{2}\right)$

30. $f(x) = -\tan(x + \pi)$

31. $f(x) = \tan(3x - 2\pi)$

32. $f(x) = \tan(2x + 3\pi)$

33. $f(x) = \frac{1}{2}\cot\left(x - \frac{\pi}{4}\right)$

34. $f(x) = -\frac{1}{2}\cot\left(x + \frac{\pi}{2}\right)$

35. $f(x) = \cot\frac{1}{2}\left(x + \frac{\pi}{2}\right)$

36. $f(x) = \cot(2x - 2\pi)$

For Problems 37–46, find the period and phase shift of the given function, and sketch the graph through one period.

37. $f(x) = \csc\left(x - \frac{\pi}{2}\right)$

38. $f(x) = \csc(x + \pi)$

39. $f(x) = -\csc\left(x + \frac{\pi}{4}\right)$

40. $f(x) = -\csc\left(x - \frac{\pi}{2}\right)$

41. $f(x) = \csc(2x + \pi)$

42. $f(x) = \csc\frac{1}{2}\left(x + \frac{\pi}{2}\right)$

43. $f(x) = \sec(x + \pi)$

44. $f(x) = 2\sec\left(x - \frac{\pi}{2}\right)$

45. $f(x) = -2\sec\left(x - \frac{\pi}{4}\right)$

46. $f(x) = \sec(3x - 3\pi)$

THOUGHTS into WORDS

47. Give a step-by-step description of how you would graph the function $f(x) = 2\tan(2x - \pi)$.

48. Explain in your own words the concepts of period, amplitude, and phase shift.

GRAPHICS CALCULATOR ACTIVITIES

49. Be sure that you agree with each of the graphs in Figures 8.30–8.35.

50. How should the graphs of $f(x) = 2\tan x$ and $f(x) = 3\tan x$ compare to the graph of $f(x) = \tan x$? Graph all three functions on the same set of axes to check your prediction.

51. How should the graph of $f(x) = -\tan x$ compare to the graph of $f(x) = \tan x$? Graph both functions on the same set of axes to check your prediction.

52. For each of the following, (a) predict how the graph of the given function will compare to the graph of $f(x) = \csc x$, and (b) graph the given function on a set of axes with the graph of $f(x) = \csc x$ to check your prediction.

a. $f(x) = -\csc x$

b. $f(x) = \csc(2x)$

c. $f(x) = 1 - \csc x$

d. $f(x) = 1 + \csc x$

e. $f(x) = 2\csc x$

f. $f(x) = 2 + \csc(x - 2)$

53. For each of the following, (a) predict the general shape and location of the graph, and (b) use your graphics calculator to graph the function to check your prediction.

a. $f(x) = \sin x + \tan x$

b. $f(x) = \sin x - \tan x$

c. $f(x) = \tan x - \sin x$

d. $f(x) = \tan x + \cot x$

e. $f(x) = \sin x + \csc x$

f. $f(x) = \cos x + \sec x$

8.4 INVERSE TRIGONOMETRIC FUNCTIONS

Before we introduce the inverse trigonometric functions, let's review some general ideas that pertain to inverse functions.

1. If the components of each ordered pair of a given one-to-one function are interchanged, the resulting function and the given function are called **inverses** of each other. (Remember that only one-to-one functions have inverses.)
2. Symbolically, the inverse of function f is denoted by f^{-1} (read *f inverse* or *the inverse of f*).
3. If f and g are inverses, then $f(g(x)) = x$ for all x in the domain of g and $g(f(x)) = x$ for all x in the domain of f.
4. Graphically, two functions that are inverses of each other are mirror images with reference to the line $y = x$.

It is evident that the sine function over the domain of all real numbers is not a one-to-one function. For example, suppose that we consider the solutions for $\sin x = \frac{1}{2}$. Certainly, $\pi/6$ is a solution, but there are infinitely many more solutions, such as $5\pi/6$, $13\pi/6$, $17\pi/6$, $-7\pi/6$, $-11\pi/6$, etc. as indicated in Figure 8.37. However, we can form a one-to-one function from the sine function, and not eliminate any values from its range, by restricting the domain to the interval $-\pi/2 \leq x \leq \pi/2$. Therefore, we have a *new* function defined by the equation $y = \sin x$ having domain $-\pi/2 \leq x \leq \pi/2$ and range $-1 \leq y \leq 1$ (Figure 8.38).

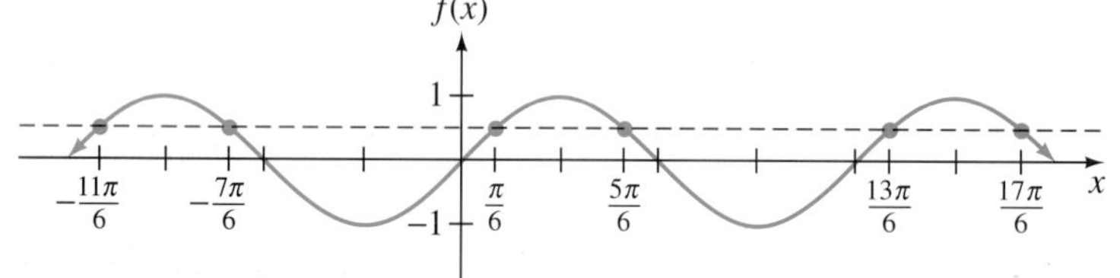

FIGURE 8.37

FIGURE 8.38

Now the inverse sine function can be defined as follows.

DEFINITION 8.2

The **inverse sine function** is defined by

$$y = \sin^{-1} x \quad \text{if and only if } x = \sin y$$

where $-1 \leq x \leq 1$ and $-\pi/2 \leq y \leq \pi/2$.

In Definition 8.2 the equation $y = \sin^{-1} x$ can be read *y is the angle whose sine is x*. Therefore, $y = \sin^{-1} \frac{1}{2}$ means *y is the angle between $-\pi/2$ and $\pi/2$, inclusive, whose sine is $\frac{1}{2}$*; thus, $y = \pi/6$. (The angle could also be expressed as 30°.)

EXAMPLE 1

Solve $y = \sin^{-1}\left(-\frac{\sqrt{2}}{2}\right)$ for y, where $-\frac{\pi}{2} \le y \le \frac{\pi}{2}$.

Solution

From our previous work with special angles, we should recall that the angle between $-\pi/2$ and $\pi/2$, inclusive, whose sine equals $-\sqrt{2}/2$, is $-\pi/4$. Therefore, $y = -\pi/4$.

EXAMPLE 2

Use a calculator and solve each of the following for y.

a. $y = \sin^{-1}(0.7256), \qquad -\frac{\pi}{2} \le y \le \frac{\pi}{2}$

b. $y = \sin^{-1}(-0.3402), \qquad -\frac{\pi}{2} \le y \le \frac{\pi}{2}$

c. $y = \sin^{-1}(0.6378), \qquad -90° \le y \le 90$

Solution

For parts (a) and (b), be sure that your calculator is set for radian measure.

a. Enter 0.7256 and press the [SIN⁻¹] key to obtain $y = 0.812$, to three decimal places. (Remember that some calculators require that you press the [SIN⁻¹] key first and then enter the number.)

b. Enter -0.3402 and press the [SIN⁻¹] key to obtain $y = -0.347$, to three decimal places.

c. Set your calculator for degree measure. Then enter 0.6378 and press the [SIN⁻¹] key to obtain $y = 39.6°$, to the nearest tenth of a degree.

REMARK Note that your calculator has been designed to yield values that agree with the range of the inverse sine function. In other words, your calculator will produce inverse sine values between $-\pi/2$ and $\pi/2$, inclusive, or between $-90°$ and $90°$, inclusive.

EXAMPLE 3

Evaluate $\cos\left(\sin^{-1}\left(-\frac{1}{2}\right)\right)$.

Solution

The expression $\cos\left(\sin^{-1}\left(-\frac{1}{2}\right)\right)$ means *the cosine of the angle between $-\pi/2$ and $\pi/2$, inclusive, whose sine is $-\frac{1}{2}$*. From our previous work with special angles we know that the angle between $-\pi/2$ and $\pi/2$, inclusive, whose sine is $-\frac{1}{2}$, is $-\pi/6$. In addition we know that $\cos(-\pi/6) = \sqrt{3}/2$. Therefore,

$$\cos\left(\sin^{-1}\left(-\frac{1}{2}\right)\right) = \frac{\sqrt{3}}{2}.$$

EXAMPLE 4

Graph $y = \sin^{-1} x$.

Solution

A table of values can be easily formed from our previous work with special angles. Plotting the points determined by the table and connecting them with a smooth curve produces Figure 8.39.

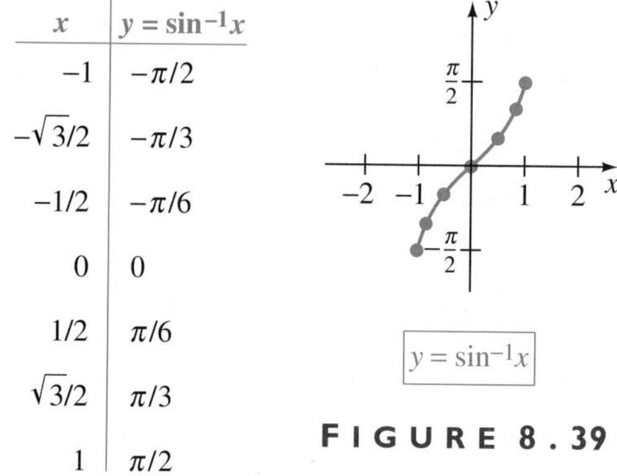

x	$y = \sin^{-1}x$
-1	$-\pi/2$
$-\sqrt{3}/2$	$-\pi/3$
$-1/2$	$-\pi/6$
0	0
$1/2$	$\pi/6$
$\sqrt{3}/2$	$\pi/3$
1	$\pi/2$

FIGURE 8.39

Remember that for two functions f and g to be inverses of each other, the following two conditions must be satisfied: (1) $f(g(x)) = x$ for all x in the domain of g, and (2) $g(f(x)) = x$ for all x in the domain of f. Let's show that the two functions

$$f(x) = \sin x \quad \text{for } -\frac{\pi}{2} \le x \le \frac{\pi}{2}$$

and

$$g(x) = \sin^{-1} x \quad \text{for } -1 \le x \le 1$$

satisfy these conditions.

$$f(g(x)) = f(\sin^{-1} x) = \sin(\sin^{-1} x) = x \quad \text{for } -1 \le x \le 1$$

and

$$g(f(x)) = g(\sin x) = \sin^{-1}(\sin x) = x \quad \text{for } -\frac{\pi}{2} \le x \le \frac{\pi}{2}$$

The fact that $f(x) = \sin x$ and $g(x) = \sin^{-1} x$ are inverses could also be used for graphing purposes. That is, their graphs are reflections of each other through the line $y = x$. You will use this idea to graph $f(x) = \sin^{-1} x$ in the next set of problems.

The inverse sine function is also called the *arcsine function* and the notation *arcsin x* can be used in place of $\sin^{-1} x$. The *arc* vocabulary refers to the fact that $y = \arcsin x$ means $x = \sin y$; that is, y is a real number whose sine is x, which can be interpreted geometrically as arc-length on a unit circle. Therefore, $\arcsin \frac{1}{2}$ can also refer to the arc indicated in Figure 8.40(a), which is $\pi/6$ units long. Likewise, arcsin(−0.9440) can refer to the arc indicated in Figure 8.40(b), which is −1.235 units long.

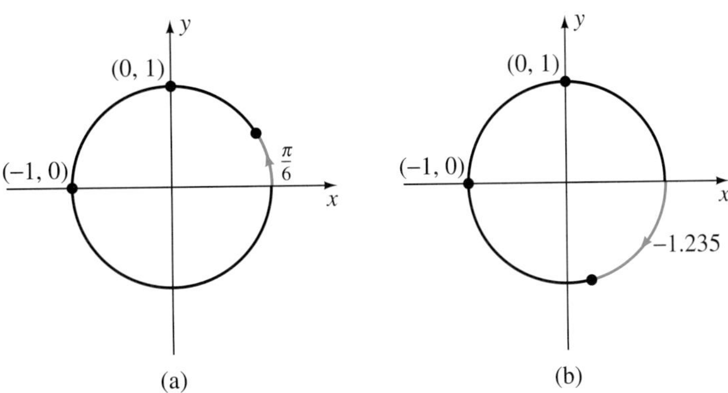

FIGURE 8.40

Inverse Cosine Function

The other trigonometric functions can also be used to introduce inverse functions. In each case a restriction needs to be placed on the original domain to create a one-to-one function that contains the entire range of the original function. Then a corresponding inverse function can be defined.

By restricting the domain of the cosine function to real numbers between 0 and π, inclusive, a one-to-one function with a range between −1 and 1, inclusive, is obtained. Then the following definition creates the inverse cosine function.

DEFINITION 8.3

The *inverse cosine function* or **arccosine function** is defined by

$$y = \cos^{-1} x = \arccos x \quad \text{if and only if } x = \cos y$$

where $-1 \leq x \leq 1$ and $0 \leq y \leq \pi$.

EXAMPLE 5

Solve $y = \cos^{-1}\left(-\frac{\sqrt{3}}{2}\right)$ for y, where $0 \leq y \leq \pi$.

Solution

The expression $\cos^{-1}(-\sqrt{3}/2)$ can be interpreted as *the angle whose cosine is* $-\sqrt{3}/2$. From our previous work with special angles we know that $y = 5\pi/6$.

EXAMPLE 6

Use a calculator and solve each of the following for y.

a. $y = \cos^{-1}(0.3214), \quad 0 \le y \le \pi$

b. $y = \arccos(0.7914), \quad 0 \le y \le \pi$

c. $y = \cos^{-1}(-0.7120), \quad 0° \le y \le 180°$

Solution

For parts (a) and (b), be sure that your calculator is set for radian measure.

a. Enter 0.3214 and press [COS⁻¹] to obtain $y = 1.244$, to three decimal places.

b. Enter 0.7914 and press [COS⁻¹] to obtain $y = 0.658$, to three decimal places.

c. Set your calculator for degree measure. Then enter -0.7120 and press [COS⁻¹] to obtain $y = 135.4°$, to the nearest tenth of a degree.

EXAMPLE 7

Evaluate $\sin\left(\cos^{-1} \frac{1}{2}\right)$.

Solution

The expression $\sin\left(\cos^{-1} \frac{1}{2}\right)$ means *the sine of the angle between* 0 *and* π, *inclusive, whose cosine is* $\frac{1}{2}$. We know that $\pi/3$ is the angle whose cosine is $\frac{1}{2}$ and we know that $\sin(\pi/3) = \sqrt{3}/2$. Therefore, $\sin\left(\cos^{-1} \frac{1}{2}\right) = \sqrt{3}/2$.

Inverse Tangent Function

By restricting the domain of the tangent function to real numbers between $-\pi/2$ and $\pi/2$ ($-\pi/2$ and $\pi/2$ are not included since the tangent is undefined at those values), a one-to-one function is obtained. Therefore, the inverse tangent function can be defined.

DEFINITION 8.4

The *inverse tangent function* or **arctangent function** is defined by

$$y = \tan^{-1} x = \arctan x \quad \text{if and only if } x = \tan y$$

where $-\infty < x < \infty$ and $-\pi/2 < y < \pi/2$.

EXAMPLE 8

Solve $y = \tan^{-1}\left(-\dfrac{\sqrt{3}}{3}\right)$ for y, where $-90° < y < 90°$.

Solution

The expression $\tan^{-1}(-\sqrt{3}/3)$ can be interpreted as *the angle between* $-90°$ *and* $90°$ *whose tangent is* $-\sqrt{3}/3$. Therefore, from our previous work with special angles, we know that $y = -30°$.

EXAMPLE 9

Find an exact value for $\sin\left(\tan^{-1} \frac{4}{3}\right)$ without using a calculator or a table.

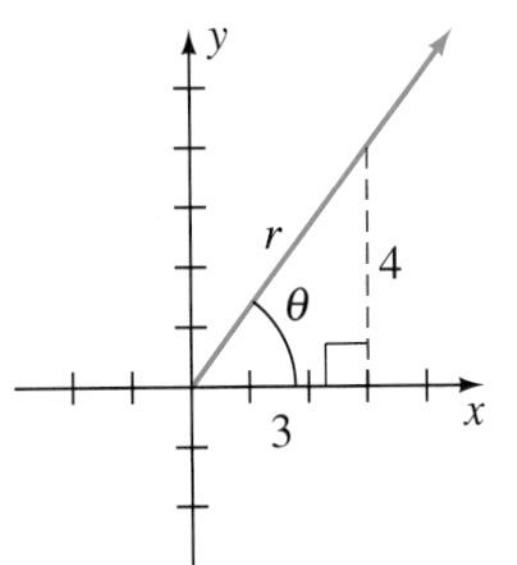

FIGURE 8.41

Solution

Let $\theta = \tan^{-1} \frac{4}{3}$, that is, θ is the angle between $-\pi/2$ and $\pi/2$ whose tangent is $\frac{4}{3}$. Therefore, θ is a first-quadrant angle as indicated in Figure 8.41. From the right triangle formed, we obtain

$$r^2 = 3^2 + 4^2 = 9 + 16 = 25$$

$$r = 5.$$

Therefore, $\sin \theta = \frac{4}{5}$, and we have $\sin\left(\tan^{-1} \frac{4}{3}\right) = \frac{4}{5}$.

As a final example of this section, let's observe a graphing utility perform some transformations with the basic inverse sine curve.

EXAMPLE 10

Use a graphing utility to graph $f(x) = \sin^{-1} x$, $f(x) = -\sin^{-1} x$, $f(x) = -\sin^{-1}(x - 3)$, and $f(x) = -\sin^{-1}(x - 3) + 2$ on the same set of axes.

Solution

Let's make the following assignments.

$$Y_1 = \sin^{-1} x, \qquad Y_2 = -\sin^{-1} x,$$

$$Y_3 = -\sin^{-1}(x - 3) \qquad Y_4 = -\sin^{-1}(x - 3) + 2$$

Now by activating the graphing feature of the utility, we can observe the following sequence of graphs being produced in this order.

1. The graph of $f(x) = \sin^{-1} x$ will appear.
2. The graph of $f(x) = -\sin^{-1} x$, which is the x-axis reflection of $f(x) = \sin^{-1} x$, will appear.
3. The graph of $f(x) = -\sin^{-1}(x - 3)$, which is a 3-unit horizontal shift to the right of the graph of $f(x) = -\sin^{-1} x$, will appear.
4. The graph of $f(x) = -\sin^{-1}(x - 3) + 2$, which is a 2-unit vertical shift upward of the graph of $f(x) = -\sin^{-1}(x - 3)$, will appear.

Figure 8.42 shows the graph of all four of the functions on the same set of axes.

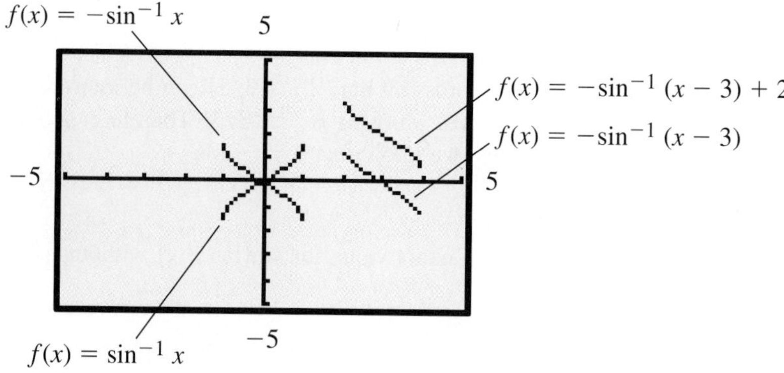

FIGURE 8.42

PROBLEM SET 8.4

For Problems 1–14, solve for y and express y in radian measure. Do not use a calculator or a table.

1. $y = \sin^{-1} \frac{\sqrt{2}}{2}$
2. $y = \sin^{-1} \frac{\sqrt{3}}{2}$
3. $y = \sin^{-1}\left(-\frac{\sqrt{3}}{2}\right)$
4. $y = \sin^{-1} 1$
5. $y = \cos^{-1} \frac{1}{2}$
6. $y = \cos^{-1}\left(-\frac{1}{2}\right)$
7. $y = \cos^{-1} \frac{\sqrt{3}}{2}$
8. $y = \cos^{-1} 0$
9. $y = \arctan 1$
10. $y = \arctan(-1)$
11. $y = \arctan \sqrt{3}$
12. $y = \arctan 0$
13. $y = \arcsin(-1)$
14. $y = \arccos(-1)$

For Problems 15–22, solve for y and express y in degree measure. Do not use a calculator or a table.

15. $y = \tan^{-1} \frac{\sqrt{3}}{3}$
16. $y = \tan^{-1}\left(-\frac{\sqrt{3}}{3}\right)$
17. $y = \cos^{-1}\left(-\frac{\sqrt{2}}{2}\right)$
18. $y = \cos^{-1}\left(-\frac{\sqrt{3}}{2}\right)$
19. $y = \sin^{-1} 0$
20. $y = \sin^{-1}\left(-\frac{1}{2}\right)$
21. $y = \sin^{-1}\left(-\frac{\sqrt{2}}{2}\right)$
22. $y = \tan^{-1}(-\sqrt{3})$

For Problems 23–34, use a calculator or the table in the back of the book to solve for y. Express answers in radians to three decimal places.

23. $y = \sin^{-1}(0.3578)$
24. $y = \sin^{-1}(0.8629)$
25. $y = \arcsin(-0.9142)$
26. $y = \arcsin(-0.1654)$
27. $y = \arccos(0.5894)$
28. $y = \arccos(0.0428)$
29. $y = \cos^{-1}(-0.4162)$
30. $y = \cos^{-1}(-0.8894)$
31. $y = \tan^{-1}(8.6214)$
32. $y = \tan^{-1}(0.9145)$
33. $y = \arctan(-0.1986)$
34. $y = \arctan(-56.2413)$

For Problems 35–46, use a calculator or the table in the back of the book to solve for y. Express answers to the nearest tenth of a degree.

35. $y = \sin^{-1}(0.4310)$
36. $y = \sin^{-1}(0.7214)$
37. $y = \sin^{-1}(-0.8214)$
38. $y = \sin^{-1}(-0.2318)$
39. $y = \cos^{-1}(0.2644)$
40. $y = \cos^{-1}(0.8419)$
41. $y = \cos^{-1}(-0.1620)$
42. $y = \cos^{-1}(-0.6217)$
43. $y = \tan^{-1}(14.2187)$
44. $y = \tan^{-1}(0.9854)$
45. $y = \tan^{-1}(-8.2176)$
46. $y = \tan^{-1}(-21.1765)$

For Problems 47–70, evaluate each expression without using a calculator or a table.

47. $\sin\left(\cos^{-1}\left(-\frac{1}{2}\right)\right)$

48. $\cos\left(\sin^{-1}\frac{1}{2}\right)$

49. $\cos(\sin^{-1} 1)$

50. $\sin(\cos^{-1}(-1))$

51. $\tan\left(\sin^{-1}\frac{\sqrt{2}}{2}\right)$

52. $\tan\left(\cos^{-1}\left(-\frac{\sqrt{3}}{2}\right)\right)$

53. $\sin\left(\tan^{-1}\sqrt{3}\right)$

54. $\cos\left(\tan^{-1}\left(-\frac{\sqrt{3}}{3}\right)\right)$

55. $\sec\left(\sin^{-1}\frac{\sqrt{2}}{2}\right)$

56. $\cos(\cos^{-1} 0)$

57. $\cos\left(\arcsin\frac{4}{5}\right)$

58. $\cos\left(\arcsin\frac{5}{13}\right)$

59. $\sin\left(\arctan\frac{3}{4}\right)$

60. $\cos\left(\arctan\left(-\frac{4}{3}\right)\right)$

61. $\tan\left(\sin^{-1}\left(-\frac{4}{5}\right)\right)$

62. $\tan\left(\cos^{-1}\left(-\frac{5}{13}\right)\right)$

63. $\cos\left(\sin^{-1}\frac{2}{3}\right)$

64. $\sin\left(\tan^{-1}\left(-\frac{2}{3}\right)\right)$

65. $\tan\left(\cos^{-1}\left(-\frac{1}{3}\right)\right)$

66. $\tan\left(\sin^{-1}\frac{2}{5}\right)$

67. $\sec\left(\sin^{-1}\left(-\frac{3}{4}\right)\right)$

68. $\csc\left(\cos^{-1}\left(-\frac{2}{3}\right)\right)$

69. $\cot\left(\cos^{-1}\left(-\frac{3}{7}\right)\right)$

70. $\sec\left(\sin\left(-\frac{1}{5}\right)\right)$

71. Graph $y = \sin^{-1} x$ by reflecting $y = \sin x$, where $-\pi/2 \le x \le \pi/2$, across the line $y = x$.

For Problems 72–81, graph each function.

72. $y = \cos^{-1} x$

73. $y = \tan^{-1} x$

74. $y = 2\sin^{-1} x$

75. $y = 2 + \cos^{-1} x$

76. $y = \cos^{-1}(x - 1)$

77. $y = \sin^{-1} 2x$

78. $y = \sin(\sin^{-1} x)$

79. $y = \sin(\cos^{-1} x)$

80. $y = -\sin^{-1} x$

81. $y = -2\cos^{-1} x$

GRAPHICS CALCULATOR ACTIVITIES

82. Use your graphics calculator to check your graphs for Problems 72–81.

83. How should the graphs of $f(x) = 2\sin^{-1} x$ and $f(x) = 3\sin^{-1} x$ compare to the graph of $f(x) = \sin^{-1} x$? Graph all three functions on the same set of axes to check your prediction.

84. How should the graphs of $f(x) = 1 + \sin^{-1} x$ and $f(x) = -1 + \sin^{-1} x$ compare to the graph of $f(x) = \sin^{-1} x$? Graph all three functions on the same set of axes to check your prediction.

85. How should the graph of $f(x) = 1 - \cos^{-1} x$ compare to the graph of $f(x) = \cos^{-1} x$? Graph both functions on the same set of axes to check your prediction.

86. How should the graphs of $f(x) = \cos^{-1}(x + 2)$ and $f(x) = \cos^{-1}(x - 2)$ compare to the graph of $f(x) = \cos^{-1} x$? Graph all three functions on the same set of axes to check your prediction.

87. a. Evaluate $\left(\sin^{-1}\frac{\sqrt{2}}{2} + \cos^{-1}\frac{\sqrt{2}}{2}\right)$, $\left(\sin^{-1}\frac{1}{2} + \cos^{-1}\frac{1}{2}\right)$, and $\left(\sin^{-1}\frac{\sqrt{3}}{2} + \cos^{-1}\frac{\sqrt{3}}{2}\right)$.

b. Graph $f(x) = \sin^{-1} x$ and $f(x) = \cos^{-1} x$ on the same set of axes.

c. What do you think the graph of $f(x) = \sin^{-1} x + \cos^{-1} x$ should be?

d. Graph $f(x) = \sin^{-1} x + \cos^{-1} x$.

88. For each of the following, (a) predict the general shape and location of the graph, and (b) graph the function to check your prediction.

a. $f(x) = \cos(\sin^{-1} x)$

b. $f(x) = \sin^{-1} x - \cos^{-1} x$

c. $f(x) = \cos^{-1} x - \sin^{-1} x$

d. $f(x) = \sin^{-1} x + \tan^{-1} x$

e. $f(x) = \cos^{-1} x + \tan^{-1} x$

8.5 POLAR COORDINATE SYSTEM

Some problems in analytic geometry, especially those involving motion about a point, are difficult to solve using the rectangular coordinate system. In fact, unwieldy equations such as $x^2 + y^2 - 2x = 2\sqrt{x^2 + y^2}$ are generated from relatively simple motion problems. However, this same equation can be transformed into a much more workable form using the variables r and θ of a plotting system called the **polar coordinate system**.

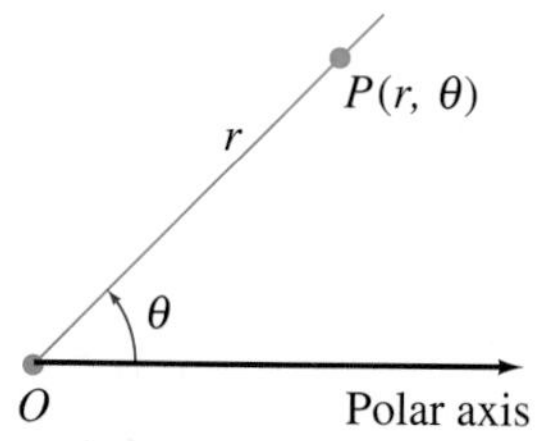

FIGURE 8.43

To set up the polar coordinate system in a plane, we begin with a fixed point O (called the *origin* or *pole*) and a directed half-line called the **polar axis** with its endpoint at O (Figure 8.43). The polar axis is usually drawn horizontally and is directed to the right. To each point P in the plane we assign **polar coordinates** (r, θ) as indicated in Figure 8.43. The angle θ has the polar axis as its initial side and the half-line OP as its terminal side. As usual, θ is considered positive if the angle is generated by a counterclockwise rotation of the polar axis and negative if the rotation is clockwise. Either radians or degrees can be used to express the measure of θ. The number r indicates the distance from the pole to the point P. If r is positive, then the distance is measured along the terminal side of θ. If r is negative, then the distance is measured along the half-line from O in the opposite direction of the terminal side of θ. The points associated with the ordered pairs $(5, \pi/3)$, $(-5, \pi/3)$, $(4, -30°)$, and $(-4, -135°)$ are plotted in Figure 8.44.

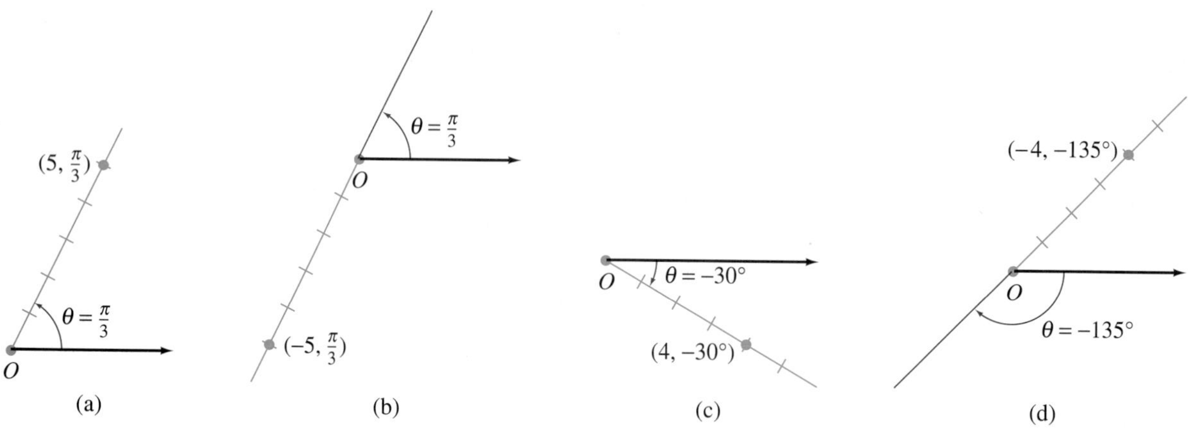

FIGURE 8.44

In Figure 8.45 we drew a model of polar coordinate graph paper. The concentric circles with point O as a common center and the rays emanating from O at intervals that correspond to some special angles facilitate the plotting of points. It should be evident that the polar coordinates of a point are not unique. Note in Figure 8.45 that we have assigned five different ordered pairs to the point P. Actually, every point has infinitely many ordered pairs assigned to it. The pole has polar coordinates $(0, \theta)$ for any angle θ.

Circular motion about a point
Chromasohm/Sohm/Stock Boston

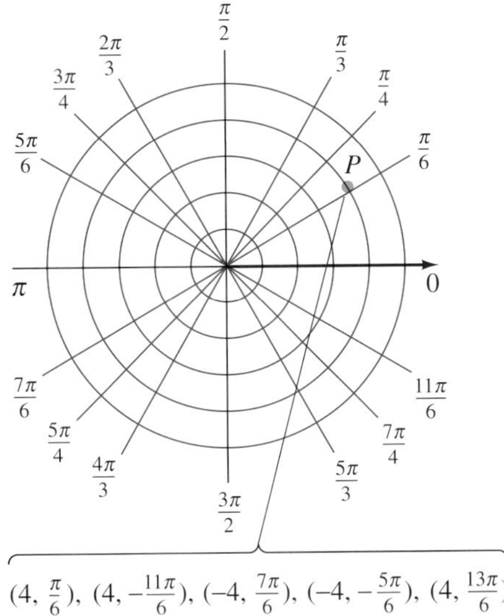

FIGURE 8.45

Relationships Between Rectangular and Polar Coordinates

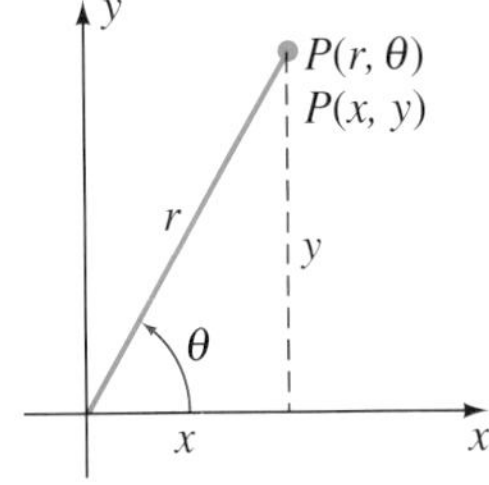

FIGURE 8.46

In Figure 8.46 we superimposed an xy-plane on an $r\theta$-plane where the positive x-axis coincides with the polar axis. The following polar-rectangular relationships can be easily deduced.

$$x = r\cos\theta \qquad y = r\sin\theta \qquad \tan\theta = \frac{y}{x} \qquad r^2 = x^2 + y^2$$

EXAMPLE 1

Find the rectangular coordinates of the point P whose polar coordinates are $(4, 210°)$.

Solution

Let's substitute 4 for r and 210° for θ in the equations $x = r \cos \theta$ and $y = r \sin \theta$.

$$x = 4 \cos 210° = 4\left(-\frac{\sqrt{3}}{2}\right) = -2\sqrt{3}$$

$$y = 4 \sin 210° = 4\left(-\frac{1}{2}\right) = -2$$

Thus, the rectangular coordinates of point P are $\left(-2\sqrt{3}, -2\right)$.

EXAMPLE 2

Suppose that point P has rectangular coordinates $(3, -3)$. Find polar coordinates (r, θ), such that $r > 0$ and $0 \leq \theta < 2\pi$, for point P.

Solution

Since r is to be positive, we can use $r = \sqrt{x^2 + y^2}$.

$$r = \sqrt{3^2 + (-3)^2} = \sqrt{9 + 9} = \sqrt{18} = 3\sqrt{2}$$

Also, we can use the equation $\tan \theta = y/x$.

$$\tan \theta = \frac{-3}{3} = -1$$

Since $\tan \theta = -1$ and the point $(3, -3)$ lies in the fourth quadrant, it follows that

$$\theta = \frac{7\pi}{4}.$$

Thus, $\left(3\sqrt{2}, 7\pi/4\right)$ are polar coordinates for point P.

The polar-rectangular relationships also provide the basis for changing polar equations to equations in rectangular form and vice versa. The next two examples illustrate this process.

EXAMPLE 3

Change $x^2 + y^2 - 2x = 2\sqrt{x^2 + y^2}$ to polar form.

Solution

Substituting r^2 for $x^2 + y^2$, $r \cos \theta$ for x, and r for $\sqrt{x^2 + y^2}$, the equation $x^2 + y^2 - 2x = 2\sqrt{x^2 + y^2}$ becomes

$$r^2 - 2r \cos \theta = 2r.$$

This equation simplifies to

$$r^2 - 2r \cos \theta - 2r = 0$$

$$r(r - 2\cos\theta - 2) = 0$$

$$r = 0 \qquad \text{or} \qquad r - 2\cos\theta - 2 = 0.$$

The graph of $r = 0$ is the pole, and since the pole is also included in the graph of $r - 2\cos\theta - 2 = 0$ (let $\theta = \pi$), we can discard $r = 0$ and keep only

$$r - 2\cos\theta - 2 = 0.$$

This could also be written as

$$r = 2 + 2\cos\theta.$$

Notice in Example 3 that the original complicated equation in rectangular form produced a fairly simple polar equation. Furthermore, in the next section we will see that the polar equation $r = 2 + 2\cos\theta$ is easy to graph. So in a case like this, changing from rectangular form to polar form can be very beneficial.

EXAMPLE 4

Change $r = \cos\theta + \sin\theta$ to rectangular form.

Solution

Substituting x/r for $\cos\theta$ and y/r for $\sin\theta$, the given equation $r = \cos\theta + \sin\theta$ becomes

$$r = \frac{x}{r} + \frac{y}{r}.$$

Now we can multiply both sides by r. This, in effect, adds $r = 0$ (the pole) to the graph. But the pole is already a part of the graph of $r = \cos\theta + \sin\theta$ (let $\theta = 3\pi/4$), so an equivalent equation

$$r^2 = x + y$$

is produced. Finally, by substituting $x^2 + y^2$ for r^2, we obtain

$$x^2 + y^2 = x + y,$$

which can be written

$$x^2 + y^2 - x - y = 0.$$

In Example 4 the switch from polar form to rectangular form produced an equation ($x^2 + y^2 - x - y = 0$) that should look familiar to you. Its graph is a circle and, by completing the square, we could find its center and the length of a radius. In other words, in this case the switch from polar to rectangular form was beneficial.

Together Examples 3 and 4 illustrate that for some problems the rectangular system is most appropriate; however, for other problems it may be easier to use the polar coordinate system. Having both systems provides us with more flexibility to solve problems.

Graphing Polar Equations

In Chapter 3, when introducing the rectangular coordinate system, we mentioned that there are basically two kinds of problems to solve in analytic geometry, namely, (1) given an algebraic equation, find its geometric graph, and (2) given a set of conditions pertaining to a geometric figure, find its algebraic equation. The polar coordinate system provides another basis for solving those same two kinds of problems. However, in this brief introduction to the polar coordinate system, we will limit our study to problems of type (1), that is, to sketching the graph of a given polar equation.

A **polar equation** is an equation involving the variables r and θ. An ordered pair (a, b) is said to be a *solution* of a polar equation if a substituted for r and b substituted for θ produce a true numerical statement. For example, $(1, \pi/6)$ is a solution of $r = 2 \sin \theta$, because 1 substituted for r and $\pi/6$ substituted for θ produces the true numerical statement $1 = 2\left(\frac{1}{2}\right)$. The *graph* of a polar equation is the set of all points (in the $r\theta$-plane) that correspond to the set of all solutions of the equation.

EXAMPLE 5

Graph the polar equation $r \cos \theta = 2$.

Solution

Let's change the form of the given equation by solving for r.

$$r \cos \theta = 2 \quad \text{therefore, } r = \frac{2}{\cos \theta}, \qquad \cos \theta \neq 0$$

For each value assigned to θ, starting with 0 and using special positive angles, r takes on a corresponding value. (Since $\cos \theta$ cannot equal zero, θ cannot equal

r	θ
2	0
$\frac{4\sqrt{3}}{3} \approx 2.3$	$\pi/6$
$2\sqrt{2} \approx 2.8$	$\pi/4$
4	$\pi/3$
-4	$2\pi/3$
$-2\sqrt{2} \approx -2.8$	$3\pi/4$
$\frac{-4\sqrt{3}}{3} \approx -2.3$	$5\pi/6$
-2	π

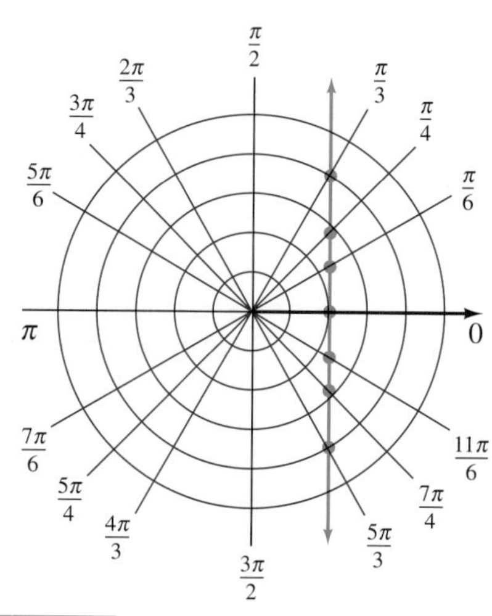

$r \cos \theta = 2$

FIGURE 8.47

$\pi/2$.) The table accompanying Figure 8.47 contains eight solutions of the equation. Plotting the points associated with the ordered pairs (r, θ) and connecting them produces the line in Figure 8.47.

Notice that the last entry in the table for Example 5, $(-2, \pi)$, determines the same point as the first entry, $(2, 0)$. This fact alerted us to the realization that in this case there was no need to allow θ to vary from π to 2π because the same points would be determined again. For example, if $\theta = 7\pi/6$ we get the ordered pair $\left(-4\sqrt{3}/3,\ 7\pi/6\right)$, which determines the same point as $\left(4\sqrt{3}/3,\ \pi/6\right)$. In other words, by paying special attention to the nature of the trigonometric functions, we can often circumvent the need for a large table of values.

Many graphing utilities are designed to graph polar equations, but there may be a difference in the way this is accomplished. For example, some graphing utilities have a special mode so that polar equations can be handled in much the same way as polynomial equations. However, some graphing utilities use a *parametric mode* for graphing polar equations. In such cases, the user must first change a polar equation into equivalent parametric equations. (Problems 95 and 96 in the next problem set are designed to help you with a graphing utility that uses the parametric approach.) You will need to consult your user's manual for specific instructions relative to your graphing utility. Figure 8.48 shows the graph of the polar equation $r = 4/(1 + \cos\theta)$ using a graphing utility.

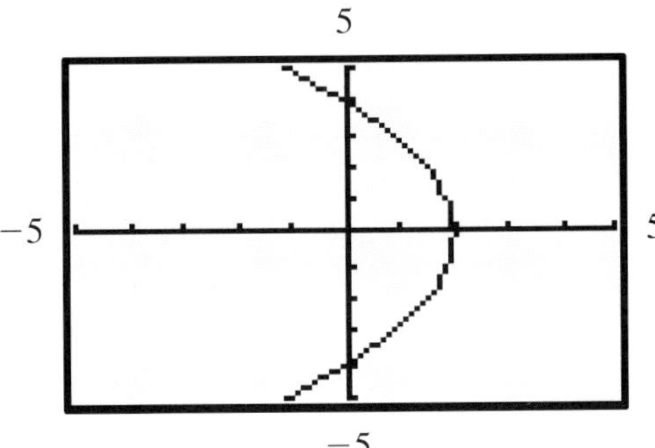

FIGURE 8.48

EXAMPLE 6

Graph the polar equation $r = 4 \sin \theta$.

Solution

Let's set up a table of values, plot the points associated with the ordered pairs, and draw the graph (Figure 8.49).

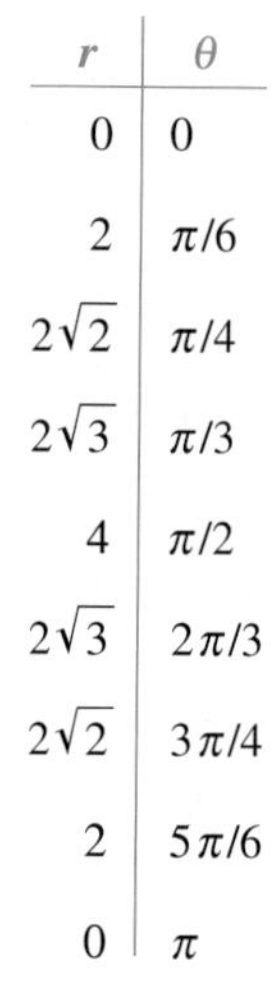

r	θ
0	0
2	$\pi/6$
$2\sqrt{2}$	$\pi/4$
$2\sqrt{3}$	$\pi/3$
4	$\pi/2$
$2\sqrt{3}$	$2\pi/3$
$2\sqrt{2}$	$3\pi/4$
2	$5\pi/6$
0	π

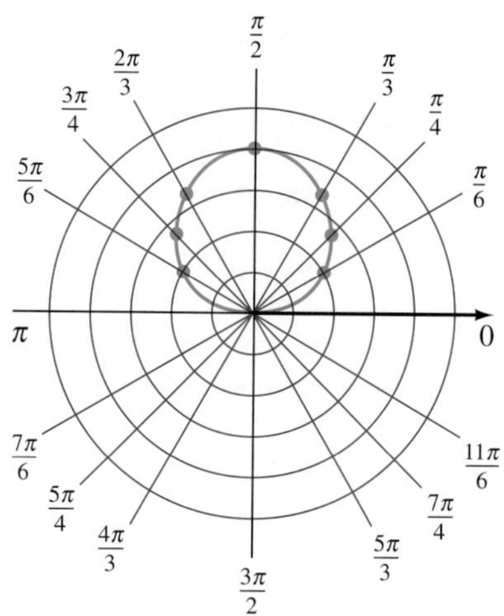

$r = 4 \sin \theta$

FIGURE 8.49

Examples 5 and 6 illustrate the fact that the graphs of some polar equations are familiar geometric figures. In fact, the polar equation $r \cos \theta = 2$ of Example 5 can be easily changed to the rectangular form $x = 2$, where the graph in Figure 8.47 is obvious. Likewise, the polar equation $r = 4 \sin \theta$ of Example 6 can be changed to the rectangular form $x^2 + y^2 = 4y$, which is equivalent to $x^2 + (y - 2)^2 = 4$. So in the xy-plane it is a circle with its center at (0, 2) that has a radius of 2 units. This agrees with our graph in Figure 8.49. However, at this time our primary objective is to give you some practice graphing polar equations by plotting a sufficient number of points to determine the figure. This will help you in the next section when we encounter some not so familiar geometric figures.

PROBLEM SET 8.5

For Problems 1–12, plot the indicated ordered pairs as points in a polar coordinate system.

1. $A\left(3, \frac{\pi}{4}\right)$

2. $B\left(4, \frac{\pi}{3}\right)$

3. $C\left(-2, \frac{2\pi}{3}\right)$

4. $D\left(-3, \frac{5\pi}{6}\right)$

5. $E\left(5, -\frac{3\pi}{4}\right)$

6. $F\left(5, -\frac{5\pi}{4}\right)$

7. $G\left(-4, -\frac{\pi}{6}\right)$

8. $H\left(-4, -\frac{23\pi}{6}\right)$

9. $I(5, 270°)$

10. $J(4, -180°)$

11. $K(-2, -510°)$

12. $L(-2, 150°)$

For Problems 13–30, find the rectangular coordinates of the points whose polar coordinates are given.

13. $(3, 30°)$ **14.** $(6, 150°)$ **15.** $(-4, 225°)$

16. $(-2, 315°)$ **17.** $(2, 420°)$ **18.** $(5, 570°)$

19. $\left(-3, \frac{\pi}{3}\right)$ **20.** $\left(-7, \frac{5\pi}{6}\right)$ **21.** $\left(4, \frac{4\pi}{3}\right)$

22. $\left(6, \frac{5\pi}{3}\right)$ **23.** $\left(1, -\frac{2\pi}{3}\right)$ **24.** $\left(3, -\frac{11\pi}{6}\right)$

25. $\left(-7, \frac{9\pi}{4}\right)$ **26.** $\left(-4, \frac{11\pi}{4}\right)$

27. $\left(-2, -\frac{17\pi}{6}\right)$ **28.** $\left(-1, -\frac{11\pi}{3}\right)$

29. $\left(-3, -\frac{3\pi}{2}\right)$ **30.** $\left(8, -\frac{5\pi}{2}\right)$

For Problems 31–40, the rectangular coordinates of a point P are given. Find a pair of polar coordinates (r, θ) for P such that $r > 0$ and $0 \le \theta < 2\pi$.

31. $(-\sqrt{2}, \sqrt{2})$ **32.** $(2\sqrt{2}, -2\sqrt{2})$

33. $\left(-\frac{5\sqrt{3}}{2}, -\frac{5}{2}\right)$ **34.** $(-3\sqrt{3}, 3)$

35. $(3, -3\sqrt{3})$ **36.** $\left(-\frac{1}{2}, -\frac{\sqrt{3}}{2}\right)$

37. $(-4, 0)$ **38.** $(0, -3)$

39. $\left(\frac{3\sqrt{3}}{2}, \frac{3}{2}\right)$ **40.** $(\sqrt{3}, 1)$

For Problems 41–46, the rectangular coordinates of a point P are given. Find a pair of polar coordinates (r, θ) for P such that $r < 0$ and $0 \le \theta < 2\pi$.

41. $(\sqrt{2}, \sqrt{2})$ **42.** $\left(-\frac{3\sqrt{2}}{2}, \frac{3\sqrt{2}}{2}\right)$

43. $(2, -2\sqrt{3})$ **44.** $\left(\frac{1}{2}, \frac{\sqrt{3}}{2}\right)$

45. $\left(-\frac{5\sqrt{3}}{2}, \frac{5}{2}\right)$ **46.** $(-\sqrt{3}, -1)$

For Problems 47–52, the rectangular coordinates of a point P are given. Find a pair of polar coordinates (r, θ) for P such that $r > 0$ and $0° \le \theta < 360°$. Express θ to the nearest tenth of a degree.

47. $(3, 2)$ **48.** $(2, 5)$ **49.** $(-4, 3)$

50. $(6, -2)$ **51.** $(-4, -1)$ **52.** $(-3, -4)$

For Problems 53–64, change each equation to polar form.

53. $y = 2$ **54.** $x = 7$

55. $3x - 2y = 4$ **56.** $5x + 4y = 10$

57. $y = x$ **58.** $y = -2x$

59. $x^2 + y^2 - 8x = 0$ **60.** $x^2 + y^2 + 6y = 0$

61. $x^2 + y^2 + x = \sqrt{x^2 + y^2}$

62. $x^2 + y^2 - 2y = 2\sqrt{x^2 + y^2}$

63. $x^2 = 4y$ **64.** $y^2 = x$

For Problems 65–76, change each polar equation to rectangular form.

65. $r \sin \theta = -4$ **66.** $r \cos \theta = 6$

67. $r - 3 \cos \theta = 0$ **68.** $r = 2 \sin \theta$

69. $r = 2 \cos \theta + 3 \sin \theta$ **70.** $r = 3 \cos \theta - 4 \sin \theta$

71. $r(\sin \theta + 4 \cos \theta) = 5$

72. $r(2 \sin \theta - 3 \cos \theta) = -4$

73. $r = \frac{4}{2 + \cos \theta}$ **74.** $r = \frac{5}{2 - 3 \sin \theta}$

75. $r = 2 + 3 \sin \theta$ **76.** $r = -3 - 2 \cos \theta$

For Problems 77–92, sketch the graph of each of the polar equations. These graphs should be the familiar figures: straight line, circle, parabola, ellipse, or hyperbola.

77. $r = 4$ **78.** $r = -3$

79. $\theta = \frac{\pi}{6}$ **80.** $\theta = -\frac{\pi}{4}$

81. $r = 4 \sin \theta$ **82.** $r = -3 \cos \theta$

83. $r \sin \theta = 3$ **84.** $r \cos \theta = -2$

85. $r = 3 \cos \theta + 4 \sin \theta$ **86.** $r = 2 \cos \theta - 3 \sin \theta$

87. $r = \frac{4}{1 + \sin \theta}$ **88.** $r = \frac{3}{1 - \sin \theta}$

89. $r = \dfrac{5}{3 + 2\cos\theta}$ **90.** $r = \dfrac{5}{3 - 2\cos\theta}$ **91.** $r = \dfrac{5}{2 + 3\sin\theta}$ **92.** $r = \dfrac{5}{2 - 3\cos\theta}$

MISCELLANEOUS PROBLEMS

93. The formula $d = \sqrt{(r_1)^2 + (r_2)^2 - 2r_1r_2\cos(\theta_2 - \theta_1)}$ can be used to find the distance between two points (r_1, θ_1) and (r_2, θ_2) in the polar coordinate system. Use Figure 8.50 and the law of cosines to develop the formula.

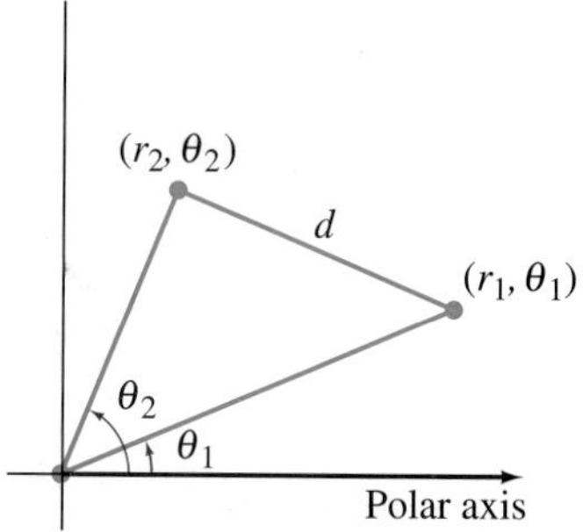

FIGURE 8.50

94. Use the distance formula from Problem 93 to find the distance between each of the following pairs of points.

a. $\left(4, \dfrac{\pi}{2}\right)$ and $\left(3, \dfrac{\pi}{6}\right)$

b. $\left(6, \dfrac{3\pi}{4}\right)$ and $\left(8, \dfrac{\pi}{4}\right)$

c. $\left(10, \dfrac{7\pi}{6}\right)$ and $\left(2, \dfrac{2\pi}{3}\right)$

d. $\left(3, \dfrac{5\pi}{6}\right)$ and $\left(6, \dfrac{\pi}{6}\right)$

GRAPHICS CALCULATOR ACTIVITIES

95. Sometimes when using the rectangular coordinate system, the xy-coordinates of points on a curve are difficult to specify using only one equation involving x and y. It may be easier to express both x and y in terms of a third variable. For example, the x- and y-coordinates may be related to a time element t; so both x and y can be expressed as functions of t. In general, the equations $x = f(t)$ and $y = g(t)$ are called **parametric equations** with a parameter of t.

To graph a curve represented by parametric equations, we can allow the parameter to vary within the boundaries set by the problem and determine a set of ordered pairs (x, y). For example, suppose that we want to graph the curve determined by the parametric equations $x = t - 1$ and $y = t^2 - 6t + 8$, where t is any real number between 0 and 5, inclusive. Let's set up a table of values that allows t to vary from 0 to 5, inclusive. (We will use only whole numbers for t to keep the computation simple.) Plotting the points (x, y) and connecting them with a smooth curve produces Figure 8.51. The curve in this figure appears to be a portion of a parabola. We can verify this by **eliminating the parameter** t as follows. Solve the equation $x = t - 1$ for t.

$$x = t - 1$$
$$x + 1 = t$$

Now substitute $x + 1$ for t in the other equation.

$$y = t^2 - 6t + 8$$
$$y = (x + 1)^2 - 6(x + 1) + 8$$
$$y = x^2 + 2x + 1 - 6x - 6 + 8$$
$$y = x^2 - 4x + 3$$

From our previous graphing experiences, we know that $y = x^2 - 4x + 3$ does represent a parabola. So the curve in the figure is a portion of the parabola between and including the points $(-1, 8)$ and $(4, 3)$.

t	$x = t - 1$	$y = t^2 - 6t + 8$
0	−1	8
1	0	3
2	1	0
3	2	−1
4	3	0
5	4	3

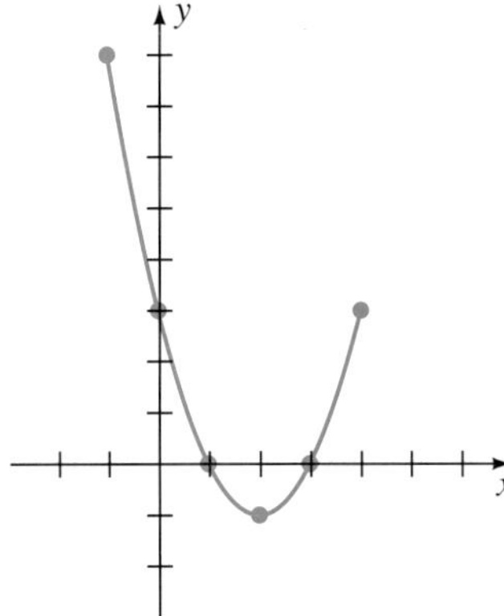

FIGURE 8.51

For each of the following, (a) use your graphics calculator to graph the curve represented by the parametric equations, and (b) eliminate the parameter to form an equation in terms of x and y. (You may need to refer to the user's manual for specific instructions for graphing parametric equations.)

a. $x = t - 2,\ y = 3t + 1;\quad 0 \le t \le 4$

b. $x = t + 1,\ y = -2t + 2;\quad -2 \le t \le 3$

c. $x = t + 2,\ y = t^2 + 4;\quad -3 \le t \le 2$

d. $x = t - 1,\ y = -t^2 + 2;\quad -2 \le t \le 3$

e. $x = 2 \cos t,\ y = 2 \sin t;\quad 0 \le t \le 2\pi$

f. $x = 4 \cos t,\ y = 2 \sin t;\quad 0 \le t \le 2\pi$

g. $x = 2 \sec t,\ y = \tan t;\quad 0 \le t \le 2\pi$

h. $x = e^t,\ y = e^{-2t};\quad -10 \le t \le 10$

96. Some graphics calculators use parametric equations to graph polar equations. This can be done because the polar equation $r = f(\theta)$ can be defined parametrically by the equations $x = f(\theta) \cos \theta$ and $y = f(\theta) \sin \theta$. Thus, to graph a polar equation such as $r = 4 \sin \theta$ (Problem 81), we can graph the parametric equations $x = f(\theta) \cos \theta = 4 \sin \theta \cos \theta$ and $y = f(\theta) \sin \theta = 4 \sin \theta \sin \theta = 4 \sin^2 \theta$. (Your calculator may use the variable t instead of θ.)

For Problems 81–92, express each polar equation as parametric equations and then use your graphics calculator to graph the parametric equations. These graphs should agree with those you obtained earlier without using the graphics calculator.

8.6 More on Graphing Polar Equations

In the previous section we graphed some polar equations by plotting a sufficient number of points to determine the curve. Now let's discuss how the concept of **symmetry** can decrease the number of points that we need to plot and increase our efficiency in graphing polar equations.

In Figure 8.52(a) we indicated that the polar axis reflection of point (r, θ) can be named $(r, -\theta)$ or $(-r, \pi - \theta)$. Likewise, in parts (b) and (c) of Figure 8.52 we indicated the $(\pi/2)$-axis reflection and the pole reflection of (r, θ). From this information the following tests for symmetry can be stated.

Polar axis A polar equation exhibits polar-axis symmetry if replacing θ by $-\theta$ or replacing r by $-r$ and θ by $\pi - \theta$ produces an equivalent equation.

$\pi/2$-axis A polar equation exhibits $(\pi/2)$-axis symmetry if replacing r by $-r$ and θ by $-\theta$ or replacing θ by $\pi - \theta$ produces an equivalent equation.

Pole A polar equation exhibits pole symmetry if replacing r by $-r$ or replacing θ by $\pi + \theta$ produces an equivalent equation.

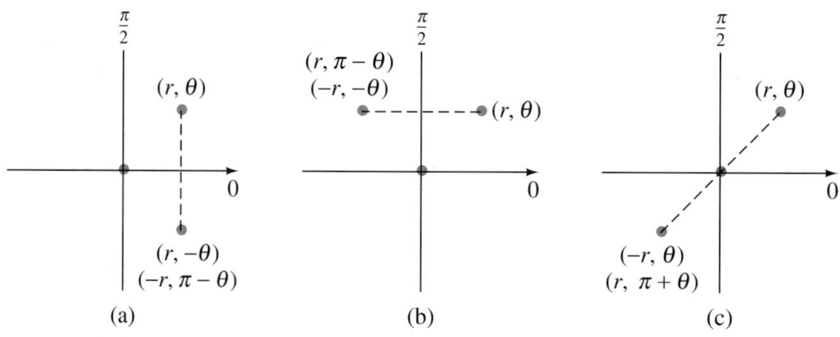

FIGURE 8.52

A few comments about the tests for symmetry should be made. We refer to polar-axis symmetry, but technically we mean symmetry with respect to the line determined by the polar axis. Likewise, $(\pi/2)$-axis symmetry means symmetry with respect to the line determined by the $(\pi/2)$-axis. Also, note that we have stated more than one test for each kind of symmetry. This is due to the fact that different polar equations produce the same set of points. For example, $r = 2$ and $r = -2$ would both produce a circle of radius 2 with the center at the pole. Finally, we suggest that as you begin to use the tests for symmetry you retain a mental picture of Figure 8.52. It may help you recall the specific tests.

EXAMPLE 1

Graph $r = 2 + 2 \cos \theta$.

Solution

First, since $\cos(-\theta) = \cos \theta$, we know that replacing θ by $-\theta$ will produce an equivalent equation. Thus, this curve is symmetric with respect to the polar axis. So in our table of values we can let θ vary from 0 to π. Then $\cos \theta$ decreases from 1 to -1 and $2 + 2 \cos \theta$ decreases from 4 to 0. Plotting the points represented in the table and connecting them with a smooth curve produces the upper half of Figure 8.53. Then reflecting this across the polar axis completes the figure.

r	θ
4	0
$2+\sqrt{3}\approx 3.7$	$\pi/6$
$2+\sqrt{2}\approx 3.4$	$\pi/4$
3	$\pi/3$
2	$\pi/2$
1	$2\pi/3$
$2-\sqrt{2}\approx 0.6$	$3\pi/4$
$2-\sqrt{3}\approx 0.3$	$5\pi/6$
0	π

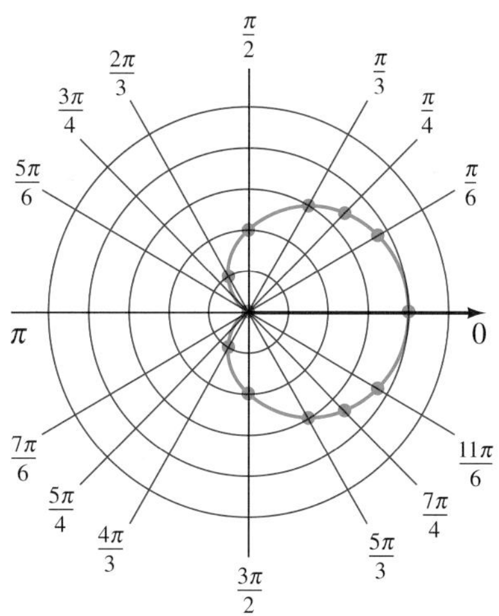

$r = 2 + 2\cos\theta$

FIGURE 8.53

The heart-shaped graph in Figure 8.53 is called a cardioid. In general, the graph of a polar equation of the form

$$r = a(1 \pm \cos\theta) \qquad \text{or} \qquad r = a(1 \pm \sin\theta),$$

where a is a nonzero real number, is called a **cardioid**. Therefore, by recognizing the general form of the equation of a cardioid, we can sketch a rough graph of one by simply plotting points for $\theta = 0$, $\pi/2$, π, and $3\pi/2$. Let's consider an example.

EXAMPLE 2

Sketch the graph of $r = 2 - 2\sin\theta$.

Solution

Letting $\theta = 0$, $\pi/2$, π, and $3\pi/2$ produces the ordered pairs $(2, 0)$, $(0, \pi/2)$, $(2, \pi)$, and $(4, 3\pi/2)$. These points along with the knowledge that the graph is a cardioid allow us to sketch a rough graph, as in Figure 8.54.

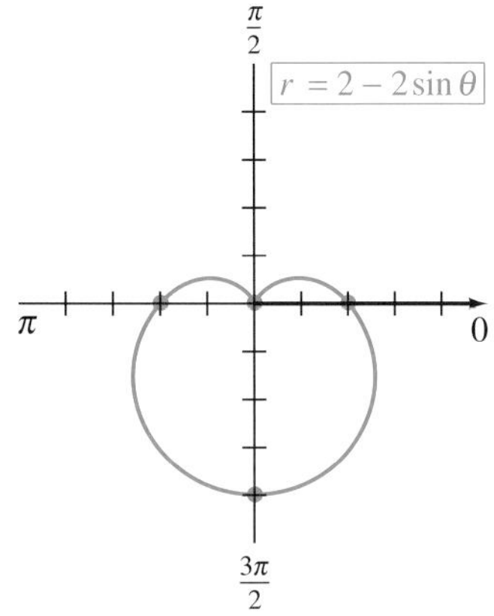

FIGURE 8.54

The graph of a polar equation of the form

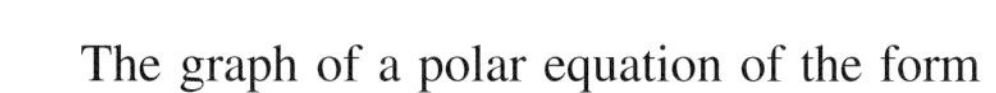

$$r = a \pm b\cos\theta \qquad \text{or} \qquad r = a \pm b\sin\theta,$$

where $a \neq b$, is called a **limacon**. The graph of a limacon is similar in shape to a cardioid, but depending upon the relative sizes of a and b, it may contain an additional loop, as illustrated by the next example.

EXAMPLE 3

Graph $r = 2 + 4 \cos \theta$.

Solution

Again, since $\cos(-\theta) = \cos \theta$, this equation exhibits polar-axis symmetry. So in the table we have allowed θ to vary from 0 to π. Notice in the table that $r = 0$ when $\theta = 2\pi/3$. Then r becomes negative for $2\pi/3 < \theta \leq \pi$. So the points from the table determine the upper half of the large loop and the lower half of the small loop in Figure 8.55. Then because of symmetry, the complete figure is determined.

r	θ
6	0
$2 + 2\sqrt{3} \approx 5.4$	$\pi/6$
$2 + 2\sqrt{2} \approx 4.8$	$\pi/4$
4	$\pi/3$
2	$\pi/2$
0	$2\pi/3$
$2 - 2\sqrt{2} \approx -0.8$	$3\pi/4$
$2 - 2\sqrt{3} \approx -1.4$	$5\pi/6$
-2	π

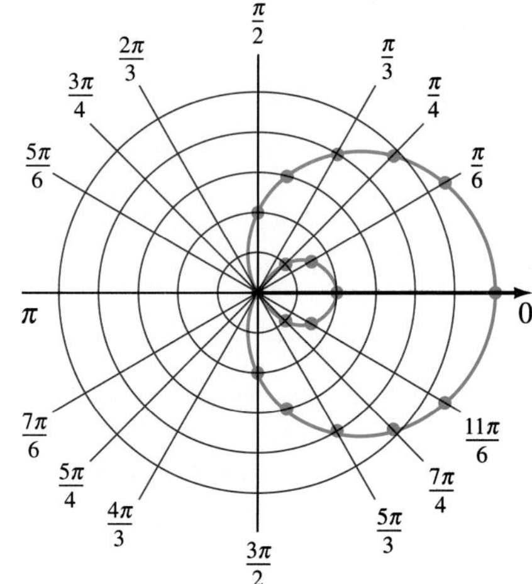

FIGURE 8.55

EXAMPLE 4

Graph $r = 5 \sin 2\theta$.

Solution

To test for symmetry, it might be easier to replace $\sin 2\theta$ by $2 \sin \theta \cos \theta$. Then the given equation becomes $r = 10 \sin \theta \cos \theta$. Using the identities $\sin(\pi - \theta) = \sin \theta$, $\cos(\pi - \theta) = -\cos \theta$, $\sin(\pi + \theta) = -\sin \theta$, and $\cos(\pi + \theta) = -\cos \theta$, we can verify that this curve is symmetric with respect to the polar axis, the $(\pi/2)$-axis, and the pole. (The details of applying the tests for symmetry we shall leave for you to complete.) Thus, we can concentrate on values of θ from 0 to $\pi/2$. As θ increases from 0 to $\pi/4$, the value of r increases from 0 to 5. Then as θ continues to increase from $\pi/4$ to $\pi/2$, the value of r decreases from 5 to 0. By keeping these facts in mind and plotting the points $(5\sqrt{3}/2, \pi/6)$, $(5, \pi/4)$, and $(5\sqrt{3}/2, \pi/3)$, we can sketch the upper right-hand part of Figure 8.56. Then the concept of symmetry allows us to complete the figure, which is called a **four-leafed rose**.

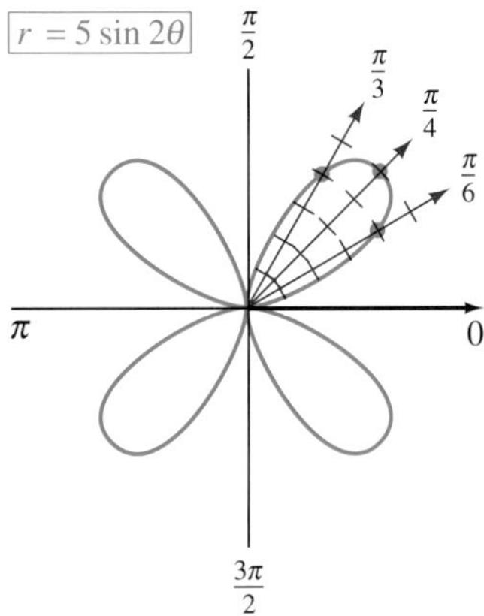

FIGURE 8.56

Another interesting type of curve is produced by equations of the form $r = a\theta$. These curves *wind around the pole* infinitely many times in such a way that r increases (or decreases) steadily as θ increases (or decreases). They are called **Archimedean spirals**. Let's consider one specific example.

EXAMPLE 5

Sketch the curve $r = \theta$ for $\theta \geq 0$.

Solution

A reasonably accurate sketch can be obtained by plotting some points on the axes and using the fact that r increases steadily as θ increases. In Figure 8.57 we plotted the points $(0, 0)$, $(\pi/2, \pi/2)$, (π, π), $(3\pi/2, 3\pi/2)$, $(2\pi, 2\pi)$, $(5\pi/2, 5\pi/2)$, $(3\pi, 3\pi)$, $(7\pi/2, 7\pi/2)$, and $(4\pi, 4\pi)$, and sketched the curve.

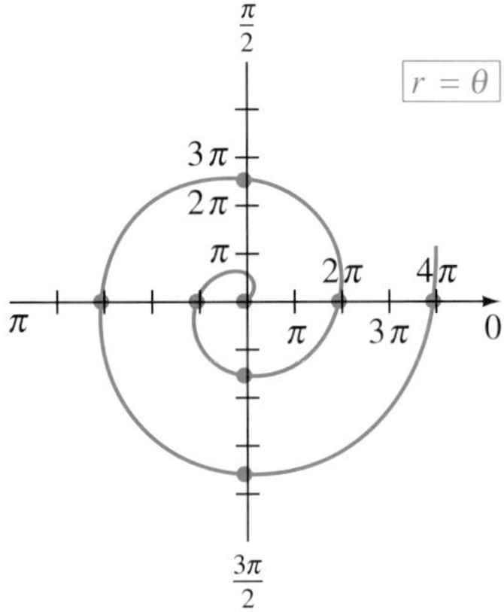

FIGURE 8.57

A graphing utility provides us with an efficient way of studying many variations of cardioids, limacons, and other polar curves. The graphics calculator activities in the next problem set give you the opportunity to investigate some of these polar curves.

As we demonstrated in the previous section, watching your graphing utility produce a particular curve can give you a very dynamic view of graphing. Let's illustrate this again by using a graphing utility to produce the graph of $r = \theta$ for $0 \leq \theta \leq 8\pi$. Enter this equation and activate the graphing feature so that you can observe the spiral in Figure 8.58 being produced.

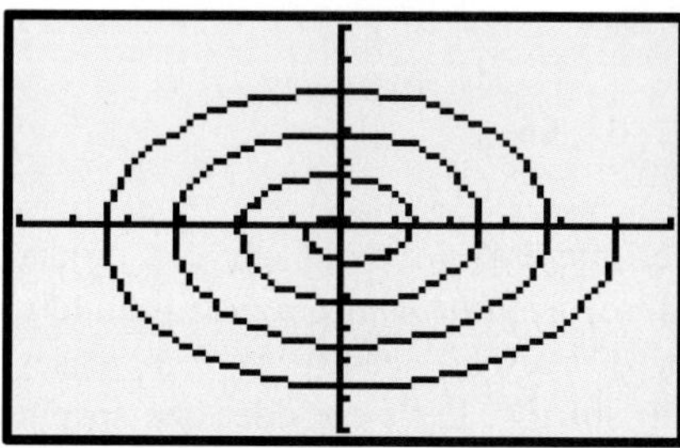

FIGURE 8.58

The spiral in Figure 8.58 was produced from the pole outward in a counterclockwise direction. Now suppose that we graph $r = \theta$ for $-8\pi \leq \theta \leq 0$. This curve, as shown in Figure 8.59, can be considered a clockwise spiral as θ varies from 0 through -8π. (Using a graphing utility, the curve in Figure 8.59 was actually produced in a counterclockwise direction starting at -8π and spiraling inward toward the pole.)

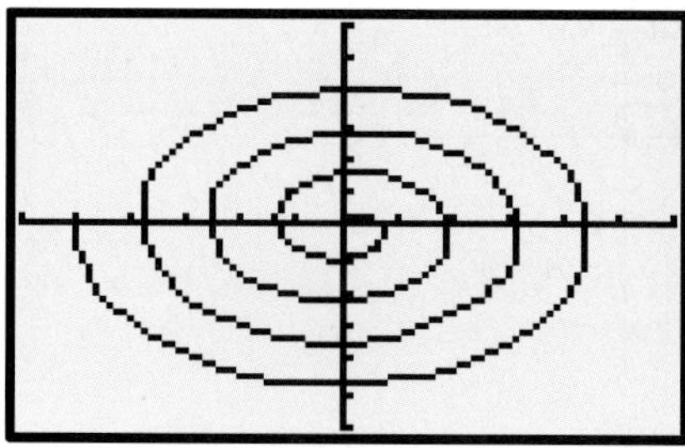

FIGURE 8.59

PROBLEM SET 8.6

For Problems 1–16, determine the symmetry (polar axis, $\pi/2$-axis, pole, or none) that each of the following equations exhibits. Do not sketch the graphs.

1. $r \cos \theta = -6$
2. $r \sin \theta = 8$
3. $r = \dfrac{3}{1 - \sin \theta}$
4. $r = \dfrac{2}{1 + \cos \theta}$
5. $r = \dfrac{4}{2 + 3 \cos \theta}$
6. $r = \dfrac{3}{3 - 2 \sin \theta}$
7. $r = 4 \sin \theta$
8. $r = 6 \cos \theta$
9. $r = 3 \cos \theta + 2 \sin \theta$
10. $r = 5 \sin \theta + 3 \cos \theta$
11. $r = \sec \theta + 2$
12. $r = \csc \theta - 3$
13. $r = 10 \tan \theta \sin \theta$
14. $r = 4 \cot \theta \cos \theta$
15. $r^2 = \sin 2\theta$
16. $r^2 = \cos 2\theta$

For Problems 17–40, graph each of the following polar equations.

17. $r = 3 + 3 \sin \theta$
18. $r = 2 - 2 \cos \theta$
19. $r = 1 - \cos \theta$
20. $r = 3 - 3 \sin \theta$
21. $r = 2 + 4 \sin \theta$
22. $r = 3 - 4 \sin \theta$
23. $r = 4 - 2 \cos \theta$
24. $r = 4 + 2 \cos \theta$
25. $r = 2 - 4 \cos \theta$
26. $r = 1 - 3 \cos \theta$
27. $r = 3 + \sin \theta$
28. $r = 3 - \sin \theta$
29. $r = 4 \cos 2\theta$
30. $r = 3 \sin 2\theta$
31. $r = 3 \sin 3\theta$
32. $r = 3 \cos 2\theta$
33. $r^2 = 9 \cos 2\theta$
34. $r^2 = -16 \cos 2\theta$
35. $r^2 = -16 \sin 2\theta$
36. $r^2 = 9 \sin 2\theta$
37. $r = 3 \sin \theta \tan \theta$
38. $r = 2 \cos \theta \cot \theta$
39. $r = \theta, \quad \theta \leq 0$
40. $r = 2\theta, \quad \theta \geq 0$

GRAPHICS CALCULATOR ACTIVITIES

41. Use your graphics calculator and be sure that you agree with the graphs in Examples 1–5 of this section.

42. Use your graphics calculator to check your graphs for Problems 17–40.

43. Graph $r = 1 + 2 \cos t$, $r = 2 + 4 \cos t$, and $r = 3 + 7 \cos t$ on the same set of axes. Then predict the graph of $r = 1 + 5 \cos t$ and check your prediction.

44. Graph $r = 1 + \sin t$, $r = 1 - \sin t$, and $r = 2 + 2 \sin t$ on the same set of axes. Then predict the graph of $r = 4 - 4 \sin t$ and check your prediction.

45. Set your calculator so that $-2 \leq x \leq 2$ and $-2 \leq y \leq 2$. Graph $r = \sin 2t$ and $r = \sin 3t$ on the same set of axes. Then predict the graphs of $r = \sin 4t$ and $r = \sin 5t$ and check your predictions.

46. Keep your calculator set so that $-2 \leq x \leq 2$ and $-2 \leq y \leq 2$. Graph $r = \cos 2t$ and $r = \cos 3t$ on the same set of axes. Then predict the graphs of $r = \cos 4t$ and $r = \cos 5t$ and check your predictions.

47. Set your calculator so that $-5 \leq x \leq 5$ and $-10 \leq y \leq 10$. Graph $r = \sin t \tan t$ and $r = 2 \sin t \tan t$ on the same set of axes. Then predict the graphs of $r = 3 \sin t \tan t$ and $r = -2 \sin t \tan t$ and check your predictions.

CHAPTER 8 SUMMARY

The basic trigonometric functions can also be interpreted as *circular functions*. Perhaps another reading of the first part of Section 8.1 would help your understanding of the circular functions.

The following ideas pertaining to the sine and cosine functions are helpful for graphing purposes.

1. Both $\sin x$ and $\cos x$ are bounded above by 1 and below by -1.
2. Both $\sin x$ and $\cos x$ have period 2π.
3. Through one period of 2π the sine and cosine functions vary as follows.

AS x INCREASES	THE SINE FUNCTION	THE COSINE FUNCTION
From 0 to $\frac{\pi}{2}$	Increases from 0 to 1	Decreases from 1 to 0
From $\frac{\pi}{2}$ to π	Decreases from 1 to 0	Decreases from 0 to -1
From π to $\frac{3\pi}{2}$	Decreases from 0 to -1	Increases from -1 to 0
From $\frac{3\pi}{2}$ to 2π	Increases from -1 to 0	Increases from 0 to 1

The following are examples of variations of the basic curves $f(x) = \sin x$ and $f(x) = \cos x$.

1. The graph of $f(x) = 1 + \sin x$ is the graph of $f(x) = \sin x$ *moved up 1 unit.*
2. The graph of $f(x) = \sin(x - \pi/2)$ is the graph of $f(x) = \sin x$ *moved to the right $\pi/2$ units.*
3. The graph of $f(x) = -\cos x$ is the graph of $f(x) = \cos x$ *reflected across the x-axis.*

A function f is called **periodic** if there exists a positive real number p such that $f(x + p) = f(x)$, for all x in the domain of f. The smallest value for p is called the **period** of the function.

The following information about equations of the form $f(x) = a \sin b(x - c)$ or $f(x) = a \cos b(x - c)$, where $b > 0$, is very useful for sketching their graphs.

1. The period of both curves is $2\pi/b$.
2. The amplitude of both curves is $|a|$.
3. The phase shift of both curves is $|c|$. The shift is to the right if c is positive and to the left if c is negative.

The tangent curve has vertical asymptotes at $x = \pm\pi/2, \pm 3\pi/2, \pm 5\pi/2$, etc. Refer to Figure 8.28 to review the shape of the tangent curve. It has period π.

The cotangent curve has vertical asymptotes at $x = 0, \pm\pi, \pm 2\pi$, etc. Refer to Figure 8.29 to review the shape of the cotangent curve. It has period π.

The following information about equations of the form $f(x) = a \tan b(x - c)$ or $f(x) = a \cot b(x - c)$, where $b > 0$, is very useful for sketching their graphs.

1. The number a effects ordinate values but has no significance in terms of amplitude.
2. The period of both curves is π/b.
3. The phase shift of both curves is $|c|$.

The graphs of $f(x) = \csc x$ and $f(x) = \sec x$ are shown in Figures 8.33 and 8.34, respectively. Both curves have period 2π.

The following information about equations of the form $f(x) = a \csc b(x - c)$ or $f(x) = a \sec b(x - c)$, where $b > 0$, is very useful for sketching their graphs.

1. The number a effects ordinate values but has no significance in terms of amplitude.
2. The period of both curves is $2\pi/b$.
3. The phase shift of both curves is $|c|$.

Definitions 8.2–8.4 form the basis for working with the inverse sine, inverse cosine, and inverse tangent functions. Know those definitions.

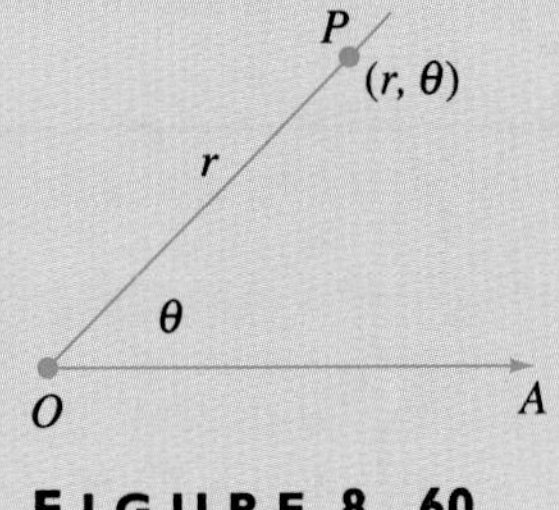

FIGURE 8.60

The polar coordinate system provides another way of naming points in a coordinatized plane, which is illustrated in Figure 8.60. The polar coordinates (r, θ) of a point P measure its distance r from a fixed point O and the angle θ that ray OP makes with a horizontal ray OA directed to the right.

The point O is called the **pole** and the ray OA is called the **polar axis**.

The following equations express relationships between the polar coordinates (r, θ) and the rectangular coordinates (x, y).

$$x = r\cos\theta \qquad y = r\sin\theta \qquad \tan\theta = \frac{y}{x} \qquad r^2 = x^2 + y^2$$

The graph of a polar equation of the form

$$r = a(1 + \cos\theta), \qquad r = a(1 - \cos\theta),$$
$$r = a(1 + \sin\theta), \qquad \text{or} \qquad r = a(1 - \sin\theta)$$

is a **cardioid**.

The graph of a polar equation of the form

$$r = a \pm b\cos\theta \qquad \text{or} \qquad r = a \pm b\sin\theta,$$

where $a \neq b$, is a **limacon**.

CHAPTER 8 REVIEW PROBLEM SET

For Problems 1–6, find exact values without using a calculator or a table.

1. $\sin\left(-\frac{5\pi}{6}\right)$

2. $\tan\left(\frac{9\pi}{4}\right)$

3. $\sin\left(\cos^{-1}\frac{\sqrt{2}}{2}\right)$

4. $\cos\left(\sin^{-1}\left(-\frac{12}{13}\right)\right)$

5. $\tan\left(\arcsin\frac{1}{2}\right)$

6. $\sin\left(\arctan\left(-\frac{2}{3}\right)\right)$

For Problems 7–10, solve for y and express y in radians. Do not use a calculator or a table.

7. $y = \tan^{-1}\left(-\frac{\sqrt{3}}{3}\right)$

8. $y = \cos^{-1}\left(-\frac{\sqrt{3}}{2}\right)$

9. $y = \arcsin\left(-\frac{1}{2}\right)$

10. $y = \arctan\left(-\sqrt{3}\right)$

For Problems 11–14, solve for y and express y in degrees. Do not use a calculator or a table.

11. $y = \sin^{-1}\frac{\sqrt{3}}{2}$

12. $y = \cos^{-1}\left(-\frac{1}{2}\right)$

13. $y = \tan^{-1}(-1)$

14. $y = \sin^{-1}\left(-\frac{1}{2}\right)$

For Problems 15–18, use a calculator or the table in the back of the book to solve for y. Express y in radians to three decimal places.

15. $y = \cos^{-1}(-0.5724)$

16. $y = \sin^{-1}(-0.7219)$

17. $y = \arctan(-71.2134)$

18. $y = \arcsin(0.9417)$

For Problems 19–22, use a calculator or the table in the back of the book to solve for y. Express y to the nearest tenth of a degree.

19. $y = \cos^{-1}(0.2479)$

20. $y = \sin^{-1}(-0.4100)$

21. $y = \tan^{-1}(-9.2147)$

22. $y = \cos^{-1}(-0.5628)$

For Problems 23–36, find the period, amplitude (if it exists), and phase shift for each graph. *Do not* graph the functions.

23. $f(x) = 4\sin\left(x + \frac{\pi}{4}\right)$

24. $f(x) = -3\cos 2\left(x - \frac{\pi}{3}\right)$

25. $f(x) = 2\tan 2(x + \pi)$

26. $f(x) = \sin(3x - \pi)$

27. $f(x) = -2\sin(\pi x + \pi)$

28. $f(x) = 2\cos\left(\pi x - \frac{\pi}{2}\right)$

29. $f(x) = -4\cos(-3x)$

30. $f(x) = 5\sin(-2x)$

31. $f(x) = 5\cot(3x - \pi)$

32. $f(x) = -4\cot(4x + \pi)$

33. $f(x) = \csc\left(4x + \frac{\pi}{2}\right)$

34. $f(x) = 2\csc\left(3x - \frac{\pi}{4}\right)$

35. $f(x) = 2\sec 3(x - 2)$

36. $f(x) = 3\sec(2x - \pi)$

For Problems 37–50, graph the given function in the indicated interval.

37. $f(x) = -\sin 2x, \quad -2\pi \le x \le 2\pi$

38. $f(x) = -1 + \cos x, \quad -2\pi \le x \le 2\pi$

39. $f(x) = \tan \pi x, \quad -\frac{3}{2} \le x \le \frac{3}{2}$

40. $f(x) = 1 + \cos(x - \pi), \quad \pi \le x \le 3\pi$

41. $f(x) = \csc \frac{1}{2}x, \quad 0 \le x \le 4\pi$

42. $f(x) = 1 - \sec x, \quad -\frac{\pi}{2} \le x \le \frac{3\pi}{2}$

43. $f(x) = 2\sin\left(x - \frac{\pi}{2}\right), \quad -\frac{\pi}{2} \le x \le \frac{5\pi}{2}$

44. $f(x) = \cos\left(2x + \frac{\pi}{2}\right), \quad -\pi \le x \le \pi$

45. $f(x) = \tan\left(x - \frac{\pi}{4}\right), \quad -\frac{\pi}{4} \le x \le \frac{7\pi}{4}$

46. $f(x) = 1 + \cot 2x, \quad 0 \le x \le \pi$

47. $f(x) = 2 - \csc 2x, \quad 0 \le x \le 2\pi$

48. $f(x) = \sec \pi\left(x - \frac{1}{2}\right), \quad 0 \le x \le 2$

49. $f(x) = -\cot \pi(x + 1), \quad -2 \le x \le 0$

50. $f(x) = \cot \frac{1}{2}\left(x - \frac{\pi}{4}\right), \quad \frac{\pi}{4} \le x \le \frac{9\pi}{4}$

For Problems 51–56, sketch the graph of each of the polar equations.

51. $r = \dfrac{2}{1 + \cos\theta}$

52. $r = 2\cos\theta$

53. $r = 1 + \cos\theta$

54. $r = 1 - \sin\theta$

55. $r = 2 - 4\sin\theta$

56. $r = 3 - 2\cos\theta$

For Problems 57–60, if the equation is given in rectangular form, change it to polar form. If the equation is given in polar form, change it to rectangular form. Also identify each curve using either of the two forms.

57. $r = 1 - \cos\theta$

58. $y = -\frac{1}{3}x^2$

59. $r = 3\sin\theta$

60. $x^2 + y^2 - 3y = 2\sqrt{x^2 + y^2}$

Trigonometric Identities and Equations

In algebra, a statement such as $3x + 4 = 7$ is called an *equation*, or more specifically, a *conditional equation*. Solving an algebraic equation refers to the process of finding, from a set of potential replacements, those values for the variable that will make a true statement. Likewise, in trigonometry we are confronted with *trigonometric equations* such as $2 \sin \theta = 1$. Solving a trigonometric equation also refers to the process of finding replacements for the variables that will make a true statement.

The algebraic equation

$$\frac{1}{x} + \frac{2}{x} = \frac{3}{x}$$

is called an *identity* because it is true for all replacements for x in which both sides of the equation are defined. Similarly, in trigonometry a statement such as $\csc \theta = 1/\sin \theta$ is called a **trigonometric identity** because it is true for all values of θ where both sides of the equation are defined. The major emphasis of this chapter is on solving trigonometric equations and verifying trigonometric identities.

9

We can describe the intensity of the sun on clear and partly cloudy days with different trigonometric equations.

9.1 TRIGONOMETRIC IDENTITIES

In Chapter 7 we frequently referred to the reciprocal relationships. These are actually trigonometric identities and can be verified using the definitions of the trigonometric functions (Definition 7.2). For example,

$$\sin\theta = \frac{y}{r} \quad \text{and} \quad \csc\theta = \frac{r}{y}.$$

Therefore, because

$$\frac{r}{y} = \frac{1}{y/r},$$

the identity

$$\csc\theta = \frac{1}{\sin\theta}$$

is established. In a like manner, the identities

$$\sec\theta = \frac{1}{\cos\theta} \quad \text{and} \quad \cot\theta = \frac{1}{\tan\theta}$$

can be verified. Remember that an identity such as $\csc\theta = 1/\sin\theta$ is true for all values of θ where both sides of the equation are defined.

Again using the definitions of the trigonometric functions we can show that

$$\frac{\sin\theta}{\cos\theta} = \frac{y/r}{x/r} = \left(\frac{y}{r}\right)\left(\frac{r}{x}\right) = \frac{y}{x} = \tan\theta.$$

Therefore, the identity

$$\tan\theta = \frac{\sin\theta}{\cos\theta}$$

is established. Similarly, the identity

$$\cot\theta = \frac{\cos\theta}{\sin\theta}$$

can be verified.

In Chapter 7 we verified the identity

$$\sin^2\theta + \cos^2\theta = 1.$$

From this identity we can develop two additional identities as follows: Divide both sides of $\sin^2\theta + \cos^2\theta = 1$ by $\cos^2\theta$ and simplify to obtain

$$\frac{\sin^2\theta}{\cos^2\theta} + \frac{\cos^2\theta}{\cos^2\theta} = \frac{1}{\cos^2\theta}$$

$$\tan^2\theta + 1 = \sec^2\theta.$$

Likewise, by dividing both sides of $\sin^2\theta + \cos^2\theta = 1$ by $\sin^2\theta$ and simplifying, we obtain

$$\frac{\sin^2\theta}{\sin^2\theta} + \frac{\cos^2\theta}{\sin^2\theta} = \frac{1}{\sin^2\theta}$$

$$1 + \cot^2\theta = \csc^2\theta.$$

Let's pause for a moment and list the identities discussed thus far.

$\csc\theta = \dfrac{1}{\sin\theta}$	$\sec\theta = \dfrac{1}{\cos\theta}$	$\cot\theta = \dfrac{1}{\tan\theta}$
$\tan\theta = \dfrac{\sin\theta}{\cos\theta}$	$\cot\theta = \dfrac{\cos\theta}{\sin\theta}$	
$\sin^2\theta + \cos^2\theta = 1$	$1 + \tan^2\theta = \sec^2\theta$	$1 + \cot^2\theta = \csc^2\theta$

You should know this list, which contains the **basic** or **fundamental identities** of trigonometry.

The basic identities are used throughout the remainder of this chapter to (1) simplify trigonometric expressions, (2) verify additional identities, (3) determine the other functional values from a given value, (4) derive other important formulas, and (5) aid in the solving of trigonometric equations. Let's first consider simplifying some trigonometric expressions.

EXAMPLE 1

Simplify $\sin\theta\cot\theta$.

Solution

Replace $\cot\theta$ with $\cos\theta/\sin\theta$ and proceed as follows.

$$\sin\theta\cot\theta = \sin\theta\left(\frac{\cos\theta}{\sin\theta}\right) = \cos\theta$$

Therefore, $\sin\theta\cot\theta$ simplifies to $\cos\theta$.

EXAMPLE 2

Simplify $\sec\theta\cot\theta$.

Solution

Replace $\sec\theta$ with $1/\cos\theta$ and $\cot\theta$ with $\cos\theta/\sin\theta$ and proceed as follows.

$$\sec\theta\cot\theta = \left(\frac{1}{\cos\theta}\right)\left(\frac{\cos\theta}{\sin\theta}\right)$$

$$= \frac{1}{\sin\theta} = \csc\theta$$

Therefore, $\sec\theta\cot\theta$ simplifies to $\csc\theta$.

EXAMPLE 3

Simplify $\dfrac{\sin\theta}{\csc\theta} + \dfrac{\cos\theta}{\sec\theta}$.

Solution

$$\frac{\sin\theta}{\csc\theta} + \frac{\cos\theta}{\sec\theta} = \frac{\sin\theta}{1/\sin\theta} + \frac{\cos\theta}{1/\cos\theta}$$
$$= \sin^2\theta + \cos^2\theta = 1$$

Therefore, $\dfrac{\sin\theta}{\csc\theta} + \dfrac{\cos\theta}{\sec\theta} = 1$.

From Examples 1, 2, and 3 we see that the end result of simplifying a trigonometric expression may be a constant as in Example 3 or a simpler trigonometric expression as in Examples 1 and 2. In Example 2, whether we use the final result of $1/\sin\theta$ or $\csc\theta$ depends on the context of the problem in which the simplifying occurs.

Verifying Identities

The process of verifying trigonometric identities is the same as simplifying trigonometric expressions except that we know the desired result in advance. Consider the following examples and be sure that you can supply reasons for all of the steps.

EXAMPLE 4

Verify the identity $\cos x + \cos x \tan^2 x = \sec x$.

Solution

Let's simplify the left side.

$$\cos x + \cos x \tan^2 x = \cos x(1 + \tan^2 x)$$
$$= \cos x \sec^2 x$$
$$= (\cos x)\left(\frac{1}{\cos^2 x}\right)$$
$$= \frac{1}{\cos x} = \sec x$$

Therefore, we have verified that $\cos x + \cos x \tan^2 x = \sec x$.

EXAMPLE 5

Verify the identity $1/\sec^2 x = (1 + \sin x)(1 - \sin x)$.

Solution

By finding the indicated product on the right side, we can proceed as follows.

$$(1 + \sin x)(1 - \sin x) = 1 - \sin^2 x$$
$$= \cos^2 x$$
$$= \frac{1}{\sec^2 x}$$

Therefore, we have verified that $1/\sec^2 x = (1 + \sin x)(1 - \sin x)$.

Notice that in Example 4 we transformed the left side into the right side, but in Example 5 we transformed the right side into the left side. In general, we suggest that you attempt to transform the more complicated side into the other side. In some examples, like Example 6, the choice of either side will result in the same amount of work.

EXAMPLE 6

Verify the identity $\sec x - \cos x = \sin x \tan x$.

Solution A

We can transform the left side into the right side as follows.

$$\begin{aligned} \sec x - \cos x &= \frac{1}{\cos x} - \cos x \\ &= \frac{1 - \cos^2 x}{\cos x} \\ &= \frac{\sin^2 x}{\cos x} \\ &= \sin x\left(\frac{\sin x}{\cos x}\right) \\ &= \sin x \tan x \end{aligned}$$

Solution B

We can transform the right side into the left side as follows.

$$\begin{aligned} \sin x \tan x &= \sin x\left(\frac{\sin x}{\cos x}\right) \\ &= \frac{\sin^2 x}{\cos x} \\ &= \frac{1 - \cos^2 x}{\cos x} \\ &= \frac{1}{\cos x} - \cos x \\ &= \sec x - \cos x \end{aligned}$$

A word of caution is in order before we consider additional identities. The transformation of the left side into the right side or the right side into the left side is an acceptable procedure for verifying an identity. However, *do not* assume the truth of the identity at the beginning and apply properties of equality to both sides.

EXAMPLE 7

Verify the identity

$$\frac{\cos x}{1 + \sin x} + \frac{\cos x}{1 - \sin x} = 2 \sec x.$$

Solution

Adding the two fractions on the left, we can transform the left side into the right side as follows.

$$\frac{\cos x}{1 + \sin x} + \frac{\cos x}{1 - \sin x} = \frac{\cos x(1 - \sin x) + \cos x(1 + \sin x)}{(1 + \sin x)(1 - \sin x)}$$
$$= \frac{\cos x - \cos x \sin x + \cos x + \cos x \sin x}{(1 + \sin x)(1 - \sin x)}$$
$$= \frac{2 \cos x}{(1 + \sin x)(1 - \sin x)}$$
$$= \frac{2 \cos x}{1 - \sin^2 x} = \frac{2 \cos x}{\cos^2 x}$$
$$= \frac{2}{\cos x} = 2 \sec x$$

EXAMPLE 8

Verify the identity

$$\frac{\cos x + \tan x}{\sin x \cos x} = \csc x + \sec^2 x.$$

Solution

Let's apply the property of fractions $(a + c)/b = (a/b) + (c/b)$ to the left side and then proceed as follows.

$$\frac{\cos x + \tan x}{\sin x \cos x} = \frac{\cos x}{\sin x \cos x} + \frac{\tan x}{\sin x \cos x}$$
$$= \frac{1}{\sin x} + \frac{\frac{\sin x}{\cos x}}{\sin x \cos x}$$
$$= \frac{1}{\sin x} + \frac{\sin x}{\sin x \cos^2 x}$$
$$= \frac{1}{\sin x} + \frac{1}{\cos^2 x}$$
$$= \csc x + \sec^2 \mathrm{x}$$

EXAMPLE 9

Verify the identity

$$\frac{\cos x}{1 - \sin x} = \frac{1 + \sin x}{\cos x}.$$

Solution

We can apply the property of fractions $a/b = ac/bc$ and multiply both the numerator and the denominator of the left side by $1 + \sin x$.

$$\frac{\cos x}{1 - \sin x} = \frac{\cos x}{1 - \sin x} \cdot \frac{1 + \sin x}{1 + \sin x}$$
$$= \frac{\cos x(1 + \sin x)}{1 - \sin^2 x}$$
$$= \frac{\cos x(1 + \sin x)}{\cos^2 x}$$
$$= \frac{1 + \sin x}{\cos x}$$

We cannot outline for you a specific procedure that will guarantee success at verifying identities. However, we do offer the following suggestions.

1. Know the basic identities listed at the beginning of this section. You must have these at your fingertips.
2. Attempt to transform the more complicated side into the other side.
3. Keep in mind that many properties from algebra apply to trigonometric expressions. Similar terms can be combined. Trigonometric expressions can be multiplied and factored. Trigonometric fractions can be simplified, added, subtracted, multiplied, and divided using the same properties that we use with algebraic fractions.
4. Since 1 can be substituted for $\sin^2 \theta + \cos^2 \theta$, it is often helpful to simplify in terms of the sine and cosine functions.

Finally, let's consider an example that illustrates how the basic identities can be used to find the other functional values when given one specific functional value.

EXAMPLE 10

If $\sin \theta = \frac{3}{5}$ and $\cos \theta < 0$, find the values of the other trigonometric functions.

Solution

Using $\csc \theta = 1/\sin \theta$, we obtain

$$\csc \theta = \frac{1}{\frac{3}{5}} = \frac{5}{3}.$$

Substituting $\frac{3}{5}$ for $\sin \theta$ in the identity $\sin^2 \theta + \cos^2 \theta = 1$, and solving for $\cos \theta$, we obtain

$$\sin^2 \theta + \cos^2 \theta = 1$$
$$\left(\frac{3}{5}\right)^2 + \cos^2 \theta = 1$$
$$\cos^2 \theta = 1 - \frac{9}{25}$$
$$\cos^2 \theta = \frac{16}{25}$$
$$\cos \theta = -\frac{4}{5}.$$

Remember we were given that $\cos \theta < 0$.

Using $\sec\theta = 1/\cos\theta$, we obtain

$$\sec\theta = \frac{1}{-\frac{4}{5}} = -\frac{5}{4}.$$

Finally, using $\tan\theta = \sin\theta/\cos\theta$ and $\cot\theta = 1/\tan\theta$, we obtain

$$\tan\theta = \frac{\frac{3}{5}}{-\frac{4}{5}} = \left(\frac{3}{5}\right)\left(-\frac{5}{4}\right) = -\frac{3}{4}$$

and

$$\cot\theta = \frac{1}{\tan\theta} = \frac{1}{-\frac{3}{4}} = -\frac{4}{3}.$$

You should recognize that a problem such as Example 10 can be worked in many different ways. For example, after we found that $\csc\theta = \frac{5}{3}$, the identity $1 + \cot^2\theta = \csc^2\theta$ could be used to determine the value of $\cot\theta$. Likewise, after finding that $\sec\theta = -\frac{5}{4}$, we could use $1 + \tan^2\theta = \sec^2\theta$ to determine the value of $\tan\theta$. Don't forget that the entire problem could be done by applying the definitions of the trigonometric functions as we did in Chapter 7. However, at this time we would prefer to use the basic identities as much as possible.

A graphing utility can be used to provide visual support for a claim that a particular equation is or is not an identity. In Example 4 we verified that the equation $\cos x + \cos x \tan^2 x = \sec x$ is an identity by simplifying $\cos x + \cos x \tan^2 x$ to $\sec x$. Now let's use a graphing utility to give visual support to this claim. By letting $Y_1 = \cos x + \cos x \tan^2 x$, $Y_2 = \sec x$, and graphing both of these equations on the same set of axes we obtain Figure 9.1. The graphs appear to be identical, which supports the claim that $\cos x + \cos x \tan^2 x = \sec x$ is an identity.

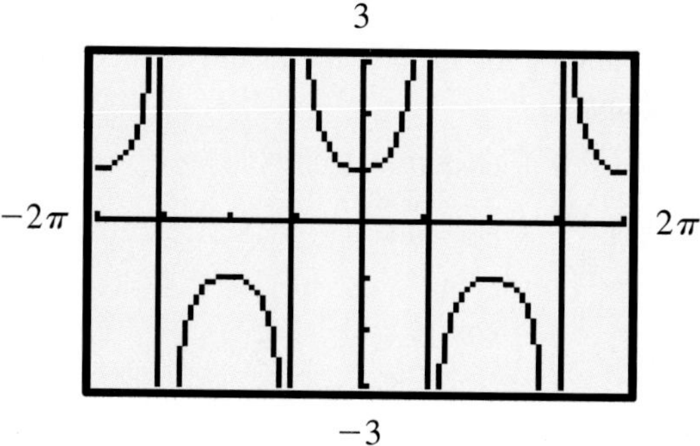

FIGURE 9.1

Is $(\sin x - \cos x)^2 = 1 + 2 \sin x \cos x$ an identity? First, let's use a graphing utility to obtain the graphs of $Y_1 = (\sin x - \cos x)^2$ and $Y_2 = 1 + 2 \sin x \cos x$ on the same set of axes as shown in Figure 9.2. Since two different graphs are produced, we know that the equation $(\sin x - \cos x)^2 = 1 + 2 \sin x \cos x$ is not an identity. Furthermore, the points of intersection of the two graphs seem to indicate that the equation is satisfied by the values $n(\pi/2)$, where n is any integer. We will discuss using an algebraic approach to solving such equations in the next section.

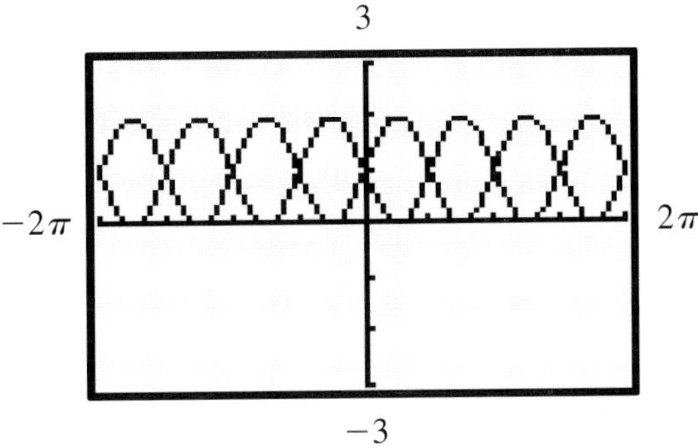

FIGURE 9.2

PROBLEM SET 9.1

1. Use Definition 7.2 and prove that $\sec \theta = \dfrac{1}{\cos \theta}$ and $\cot \theta = \dfrac{1}{\tan \theta}$.

2. Use Definition 7.2 and prove that $\dfrac{\cos \theta}{\sin \theta} = \cot \theta$.

For Problems 3–12, use the basic trigonometric identities listed at the beginning of this section to help find the remaining five trigonometric functional values.

3. $\sin \theta = \frac{4}{5}$ and the terminal side of θ lies in the first quadrant.

4. $\cos \theta = -\frac{5}{13}$ and the terminal side of θ lies in the second quadrant.

5. $\tan \theta = \frac{12}{5}$ and the terminal side of θ lies in the third quadrant.

6. $\sin \theta = -\frac{8}{17}$ and the terminal side of θ lies in the fourth quadrant.

7. $\sin \theta = \frac{4}{5}$ and $\cos \theta < 0$

8. $\csc \theta = -\frac{5}{4}$ and $\sec \theta < 0$

9. $\tan \theta = -\frac{1}{2}$ and $\cos \theta > 0$

10. $\sec \theta = 3$ and $\sin \theta < 0$

11. $\csc \theta = \frac{3}{2}$ and $\sec \theta < 0$

12. $\cot \theta = \frac{1}{3}$ and $\csc \theta < 0$

For Problems 13–24, simplify the given trigonometric expression to a single trigonometric function or a constant.

13. $\cos \theta \tan \theta$

14. $\dfrac{\tan \theta}{\sin \theta}$

15. $\cos x \csc x$

16. $\sec x - \sin x \tan x$

17. $(\cos^2 x - 1)(\tan^2 x + 1)$

18. $\sin x + \sin x \cot^2 x$

19. $(\cos^2 x)(1 + \tan^2 x)$

20. $(1 - \sin^2 x)\sec^2 x$

21. $\cos \theta + \tan \theta \sin \theta$

22. $\dfrac{\sec\theta - \cos\theta}{\tan\theta}$

23. $\dfrac{\tan x \sin x}{\sec^2 x - 1}$

24. $\tan x(\sin x + \cot x \cos x)$

For Problems 25–60, verify each of the following identities.

25. $\sin\theta \sec\theta = \tan\theta$

26. $\cos\theta \tan\theta \csc\theta = 1$

27. $\sin\theta + \sin\theta \tan^2\theta = \tan\theta \sec\theta$

28. $\cos\theta + \cos\theta \cot^2\theta = \cot\theta \csc\theta$

29. $\dfrac{\sin x + \cos x}{\cos x} = 1 + \tan x$

30. $\dfrac{\sin x + \tan x}{\sin x} = 1 + \sec x$

31. $\dfrac{\tan x + \cos x}{\sin x} = \sec x + \cot x$

32. $\csc x - \sin x = \cos x \cot x$

33. $\csc x \sec x = \tan x + \cot x$

34. $\dfrac{(\tan x)(1 + \cot^2 x)}{1 + \tan^2 x} = \cot x$

35. $\tan x = \dfrac{\cot x(1 + \tan^2 x)}{1 + \cot^2 x}$

36. $\dfrac{\cot x \cos x}{\csc^2 x - 1} = \sin x$

37. $\sin x(\csc x - \sin x) = \cos^2 x$

38. $1 - 2\sin^2\theta = 2\cos^2\theta - 1$

39. $2\sec^2\theta - 1 = 1 + 2\tan^2\theta$

40. $\cos^2\theta - \sin^2\theta = 1 - 2\sin^2\theta$

41. $\cos^2\theta - \sin^2\theta = 2\cos^2\theta - 1$

42. $\dfrac{1 + \cos x}{\sin x} + \dfrac{\sin x}{1 + \cos x} = 2\csc x$

43. $\dfrac{1}{1 - \cos x} + \dfrac{1}{1 + \cos x} = 2\csc^2 x$

44. $(\sec x - \tan x)(\csc x + 1) = \cot x$

45. $(\cos x - \sin x)(\cos x + \sin x) = 1 - 2\sin^2 x$

46. $\dfrac{1 + \sec x}{\sin x + \tan x} = \csc x$

47. $\cos^2 x = \dfrac{\csc^2 x - \cot^2 x}{\sec^2 x}$

48. $\dfrac{\cos x + \tan x}{\sin x \cos x} = \csc x + \sec^2 x$

49. $\sin^4 x - \cos^4 x = 1 - 2\cos^2 x$

50. $\tan^4 x - \sec^4 x = 1 - 2\sec^2 x$

51. $(\sin x - \cos x)^2 = 1 - 2\sin x \cos x$

52. $1 - \sin x = \dfrac{\cot x - \cos x}{\cot x}$

53. $1 + \tan x = \dfrac{\sec^2 x + 2\tan x}{1 + \tan x}$

54. $\dfrac{\sin x}{\sin x + \cos x} = \dfrac{\tan x}{1 + \tan x}$

55. $\dfrac{\sin x}{1 + \cos x} = \dfrac{1 - \cos x}{\sin x}$

56. $\dfrac{\tan x}{\sec x - 1} = \dfrac{\sec x + 1}{\tan x}$

57. $\dfrac{\csc x - 1}{\cot x} = \dfrac{\cot x}{\csc x + 1}$

58. $(\tan x - \sec x)^2 = \dfrac{1 - \sin x}{1 + \sin x}$

59. $\dfrac{1}{\tan x + \cot x} = \sin x \cos x$

60. $\dfrac{\sin x + \cos x}{\sin x - \cos x} = \dfrac{\sec x + \csc x}{\sec x - \csc x}$

GRAPHICS CALCULATOR ACTIVITIES

61. To demonstrate graphically that $\sin^2 x + \cos^2 x = 1$ is an identity, we need to show that $y = \sin^2 x + \cos^2 x$ and $y = 1$ produce identical graphs. In other words, the graph of $y = \sin^2 x + \cos^2 x$ should be the horizontal line $y = 1$.

Use your graphics calculator to demonstrate that each of the following is an identity.

a. $\sin^2 x + \cos^2 x = 1$ **b.** $1 + \tan^2 x = \sec^2 x$

c. $1 + \cot^2 x = \csc^2 x$

62. Use your graphics calculator to demonstrate the validity of each of the identities in Examples 4–9 of this section.

63. Another technique for verifying that an equation $p = q$ is an identity is to transform the left side p into another expression s, making sure that all steps taken are *reversible*. Therefore, $p = s$ is an identity. Then if the right side q of the original equation can also be transformed to s using reversible steps, we have $q = s$ is an identity. Thus, $p = s$ and $s = q$ implies that $p = q$ is an identity.

Use this approach to verify each of the following identities. Then use your graphics calculator to demonstrate the validity of each identity.

a. $\sec\theta + \csc\theta - \cos\theta - \sin\theta = \sin\theta\tan\theta + \cos\theta\cot\theta$

b. $(1 - \tan^2 x)^2 = \sec^4 x - 4\tan^2 x$

c. $(\tan x - \sec x)^2 = \dfrac{1 - \sin x}{1 + \sin x}$

d. $\dfrac{\sin\theta}{1 - \cos\theta} = \csc\theta + \cot\theta$

e. $\dfrac{1}{\csc x - \cot x} = \csc x + \cot x$

9.2 TRIGONOMETRIC EQUATIONS

As we stated in the introductory paragraph of this chapter, solving a conditional trigonometric equation such as $2\sin x = 1$ refers to the process of finding the values for the variable x that will make a true numerical statement. Because the trigonometric functions are periodic, most trigonometric equations have infinitely many solutions. However, once the solutions within one period have been found, the remainder of the solutions are easily determined. For example, $\pi/6$ and $5\pi/6$ are the solutions between 0 and 2π (remember that 2π is the period of the sine function) that satisfy the equation $\sin x = \frac{1}{2}$. Then by adding multiples of 2π to each of these, all of the solutions can be represented by $\pi/6 + 2\pi n$ and $5\pi/6 + 2\pi n$, where n is an integer. The expressions $\pi/6 + 2\pi n$ and $5\pi/6 + 2\pi n$ are referred to as the *general solutions* of the equation. Using degree measure, the general solutions could be represented by $30° + n \cdot 360°$ and $150° + n \cdot 360°$, where n is an integer.

Solving trigonometric equations requires the use of many of the techniques used to solve algebraic equations. The following examples will illustrate some of those techniques.

EXAMPLE 1

Solve $2\cos\theta + 1 = 0$ if $0° \le \theta < 360°$.

Solution

$$2\cos\theta + 1 = 0$$

$$2\cos\theta = -1 \qquad \text{Add } -1 \text{ to both sides.}$$

$$\cos\theta = -\frac{1}{2} \qquad \text{Multiply both sides by } \frac{1}{2}.$$

From our work with special angles, we know that $\cos 120° = -\frac{1}{2}$ and $\cos 240° = -\frac{1}{2}$. The solutions are 120° and 240°.

EXAMPLE 2

Solve $\sin x \cos x = 0$ if $0 \le x < 2\pi$.

Solution

By applying the property, if $ab = 0$, then $a = 0$ or $b = 0$, we can proceed as follows.

$$\sin x \cos x = 0$$

$$\sin x = 0 \quad \text{or} \quad \cos x = 0$$

We know that $\sin 0 = 0$, $\sin \pi = 0$, $\cos(\pi/2) = 0$, and $\cos(3\pi/2) = 0$. Therefore, the solutions are 0, $\pi/2$, π, and $3\pi/2$.

In Example 1, because the statement of the problem contained degree measure ($0° \le \theta < 360°$), we expressed the solutions in degrees. Likewise, in Example 2 the statement $0 \le x < 2\pi$ implies the use of real numbers or radian measure.

EXAMPLE 3

Find the general solutions for $\sin x \tan x = \sin x$.

Solution

$$\sin x \tan x = \sin x$$

$$\sin x \tan x - \sin x = 0 \qquad \text{Add } -\sin x \text{ to both sides.}$$

$$\sin x(\tan x - 1) = 0 \qquad \text{Factor the left side.}$$

$$\sin x = 0 \quad \text{or} \quad \tan x - 1 = 0 \qquad \text{Apply: if } ab = 0, \text{ then } a = 0 \text{ or } b = 0.$$

$$\sin x = 0 \quad \text{or} \quad \tan x = 1$$

Since the sine function has a period of 2π, it is sufficient to find the solutions of $\sin x = 0$ for $0 \le x < 2\pi$. Those solutions are 0 and π. The general expression $n\pi$, where n is an integer, will generate all of the solutions for $\sin x = 0$.

The tangent function has period π, so it is sufficient to find the solutions of $\tan x = 1$ for $0 \le x < \pi$. The only solution is $\pi/4$. The general expression $\pi/4 + n\pi$, where n is an integer, will generate all of the solutions for $\tan x = 1$.

Therefore, the solutions for $\sin x \tan x = \sin x$ can be represented by

$$n\pi \quad \text{and} \quad \frac{\pi}{4} + n\pi, \quad \text{where } n \text{ is an integer.}$$

Note that in Example 3 we *did not* begin by dividing both sides of the original equation by $\sin x$. Doing so would cause us to lose the solutions for $\sin x = 0$. As in algebra, we want to avoid dividing both sides of an equation by a variable, or an expression that contains a variable.

EXAMPLE 4

Solve $2 \sin^2 x + \sin x - 1 = 0$, if $0 \le x < 2\pi$.

Solution

Factoring the left side, we can proceed as follows.

$$2\sin^2 x + \sin x - 1 = 0$$
$$(2\sin x - 1)(\sin x + 1) = 0$$
$$2\sin x - 1 = 0 \quad \text{or} \quad \sin x + 1 = 0$$
$$2\sin x = 1 \quad \text{or} \quad \sin x = -1$$
$$\sin x = \frac{1}{2} \quad \text{or} \quad \sin x = -1$$

If $\sin x = \frac{1}{2}$, then $x = \pi/6$ or $5\pi/6$. If $\sin x = -1$, then $x = 3\pi/2$. The solutions are $\pi/6$, $5\pi/6$, and $3\pi/2$. (Perhaps you should check these solutions by substituting them back into the original equation.)

Sometimes it is necessary to use one of the basic identities to make a substitution, as in the next example.

EXAMPLE 5

Solve $\sec^2 x + \tan^2 x = 3$ if $0 \le x < 2\pi$.

Solution

Using the identity $1 + \tan^2 x = \sec^2 x$, we can substitute $1 + \tan^2 x$ for $\sec^2 x$ in the given equation and proceed as follows.

$$\sec^2 x + \tan^2 x = 3$$
$$(1 + \tan^2 x) + \tan^2 x = 3$$
$$2\tan^2 x = 2$$
$$\tan^2 x = 1$$
$$\tan x = \pm 1$$

If $\tan x = 1$, then $x = \pi/4$ or $5\pi/4$. If $\tan x = -1$, then $x = 3\pi/4$ or $7\pi/4$. The solutions are $\pi/4$, $3\pi/4$, $5\pi/4$, and $7\pi/4$.

Recall from algebra that squaring both sides of an equation may produce some extraneous solutions. Therefore, we have learned that potential solutions *must be checked* if the squaring property is applied.

EXAMPLE 6

Solve $\sin x + \cos x = \sqrt{2}$ if $0 \le x < 2\pi$.

Solution

$$\sin x + \cos x = \sqrt{2}$$
$$\sin x = \sqrt{2} - \cos x$$
$$\sin^2 x = 2 - 2\sqrt{2}\cos x + \cos^2 x \qquad \text{Square both sides.}$$
$$1 - \cos^2 x = 2 - 2\sqrt{2}\cos x + \cos^2 x \qquad \text{Substitute } 1 - \cos^2 x \text{ for } \sin^2 x.$$
$$0 = 2\cos^2 x - 2\sqrt{2}\cos x + 1$$

Now we can use the quadratic formula to solve for $\cos x$.

$$\cos x = \frac{2\sqrt{2} \pm \sqrt{8-8}}{4} = \frac{2\sqrt{2}}{4} = \frac{\sqrt{2}}{2}$$

If $\cos x = \sqrt{2}/2$, then $x = \pi/4$ or $7\pi/4$.

Check

$$\sin x + \cos x = \sqrt{2} \qquad \sin x + \cos x = \sqrt{2}$$

$$\sin\frac{\pi}{4} + \cos\frac{\pi}{4} \stackrel{?}{=} \sqrt{2} \qquad \sin\frac{7\pi}{4} + \cos\frac{7\pi}{4} \stackrel{?}{=} \sqrt{2}$$

$$\frac{\sqrt{2}}{2} + \frac{\sqrt{2}}{2} \stackrel{?}{=} \sqrt{2} \qquad -\frac{\sqrt{2}}{2} + \frac{\sqrt{2}}{2} \stackrel{?}{=} \sqrt{2}$$

$$\frac{2\sqrt{2}}{2} = \sqrt{2} \qquad 0 \neq \sqrt{2}$$

The only solution is $\pi/4$.

Thus far in this section we have been able to determine solutions without the use of a calculator or a table. Now let's consider two examples where we can obtain approximate solutions using a calculator or a table.

EXAMPLE 7

Approximate, to the nearest hundredth of a radian, the solutions for

$$5\sin^2 x + 7\sin x - 6 = 0, \qquad 0 \le x < 2\pi.$$

Solution

Factoring the left side of the equation, we can proceed as follows.

$$5\sin^2 x + 7\sin x - 6 = 0$$

$$(\sin x + 2)(5\sin x - 3) = 0$$

$$\sin x + 2 = 0 \quad \text{or} \quad 5\sin x - 3 = 0$$

$$\sin x = -2 \quad \text{or} \quad 5\sin x = 3$$

$$\sin x = -2 \quad \text{or} \quad \sin x = \frac{3}{5}$$

The equation $\sin x = -2$ produces no solutions because sine values must be between -1 and 1, inclusive.

If $\sin x = \frac{3}{5} = 0.6$, then either from the table in Appendix C or by using a calculator, we can determine that $x = 0.64$, to the nearest hundredth of a radian. Because the sine function is also positive in the second quadrant, x can be $\pi - 0.64 = 3.14 - 0.64 = 2.50$, to the nearest hundredth of a radian. Therefore, the approximate solutions of the original equation are 0.64 and 2.50.

EXAMPLE 8

Approximate, to the nearest tenth of a degree, the solutions for

$$\tan^2 \theta + 2 \tan \theta - 1 = 0, \qquad 0° \le \theta < 360°.$$

Solution

Using the quadratic formula we can determine approximate values for $\tan \theta$.

$$\begin{aligned}\tan \theta &= \frac{-2 \pm \sqrt{4 + 4}}{2} \\ &= \frac{-2 \pm \sqrt{8}}{2} \\ &\approx \frac{-2 \pm 2.8284}{2}\end{aligned}$$

$$\begin{aligned}\tan \theta &\approx \frac{-2 + 2.8284}{2} \\ &\approx 0.4142\end{aligned} \qquad \text{or} \qquad \begin{aligned}\tan \theta &\approx \frac{-2 - 2.8284}{2} \\ &\approx -2.4142\end{aligned}$$

Now, use a calculator to find the approximate solutions of the original equation. Tan $\theta = 0.4142$ implies that $\theta = 22.5°$. Since the tangent is also positive in the third quadrant, another approximate solution is $180° + 22.5° = 202.5°$. To solve $\tan \theta = -2.4142$, consider the reference angle θ' such that $\tan \theta' = 2.4142$. Solving this equation for θ' we find that $\theta' = 67.5°$. Since the tangent function is negative in the second and fourth quadrants, we obtain

$$\theta = 180° - 67.5° = 112.5° \qquad \text{or} \qquad \theta = 360° - 67.5° = 292.5°.$$

Therefore, the approximate solutions of the original equation are 22.5°, 112.5°, 202.5°, and 292.5°.

A graphing utility can be a very useful tool for solving trigonometric equations. Let's demonstrate this by checking our answers for Example 7 using a graphing utility. First, we can graph the equation $y = 5 \sin^2 x + 7 \sin x - 6$ in the interval $0 \le x < 2\pi$ as shown in Figure 9.3. Then we can use the trace and zoom features

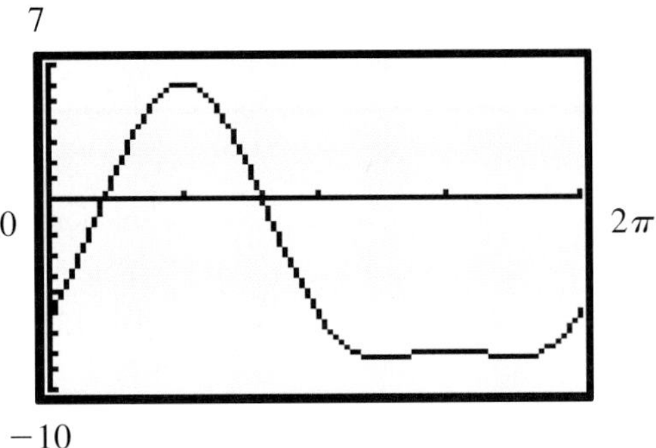

FIGURE 9.3

to approximate the x-intercepts of the graph. We will find that, to the nearest hundredth, the x-intercepts are 0.64 and 2.50. These answers agree with our solutions in Example 7.

If you have access to a graphing utility, we would suggest that you check our answers for Example 8. You can find the solutions in radians and then convert to degrees to see if you agree with the four solutions given.

PROBLEM SET 9.2

Solve each of the following equations for θ, if $0° \leq \theta < 360°$. Do not use a calculator or a table.

1. $2 \sin \theta = \sqrt{3}$
2. $2 \cos \theta + 1 = 0$
3. $2 \cos \theta + 2 = 0$
4. $2 \sin \theta + \sqrt{2} = 0$
5. $3 \tan \theta + 3\sqrt{3} = 0$
6. $\tan^2 \theta = 3$
7. $2 \sin \theta = \sin \theta - 1$
8. $3 \cos \theta + 1 = \cos \theta + 3$

Solve each of the following equations for x, if $0 \leq x < 2\pi$. Do not use a calculator or a table.

9. $2 \sin x + \sqrt{3} = 0$
10. $-2 \cos x = \sqrt{2}$
11. $3 \cos x - 2 - \cos x = 0$
12. $\sin^2 x - 1 = 0$
13. $(2 \sin x + 1)(\tan x - 1) = 0$
14. $(\cos x + 1)(\sec x - 1) = 0$
15. $3 \sin x + 5 = 0$
16. $\tan x \sin x = 0$

Solve each of the following equations. If the variable is θ, find all solutions such that $0° \leq \theta < 360°$. If the variable is x, find all solutions such that $0 \leq x < 2\pi$. Do not use a calculator or a table.

17. $2 \sin^2 x = \sin x$
18. $\sin x \tan^2 x = \sin x$
19. $\tan^2 \theta - \tan \theta = 0$
20. $2 \tan \theta \sec \theta - \tan \theta = 0$
21. $2 \cos^3 \theta = \cos \theta$
22. $2 \cos x = \cos x \csc x$
23. $2 \cos^2 x + 3 \cos x + 1 = 0$
24. $2 \sin^2 x - \sin x - 1 = 0$
25. $\sec^2 \theta - \sec \theta - 2 = 0$
26. $\sin \theta \cos \theta - \cos \theta + \sin \theta - 1 = 0$
27. $2 \sin^2 x - \cos x - 1 = 0$
28. $2 \cos^2 x - \sin x - 1 = 0$
29. $\sin^2 x + \cos x = 1$
30. $2 \cos^2 x + \sin x - 2 = 0$
31. $\sin x = 1 - \cos x$
32. $\tan x + 1 = \sec x$
33. $\tan x = \cot x$
34. $\sin x - \cos x = 1$

Find *all* solutions of each of the following equations. If the variable is θ, express the solutions in degrees and if the variable is x, express the solutions in radians. Do not use a calculator or a table.

35. $2 \cos \theta = \sqrt{3}$
36. $\sin \theta + 1 = 0$
37. $2 \sin x + \sqrt{3} = 0$
38. $2 \cos x - 1 = 0$
39. $\tan x + 1 = 0$
40. $\cot x - 1 = 0$
41. $\sec^2 \theta = 4$
42. $\csc^2 \theta = 4$
43. $\cot^2 x - \cot x = 0$
44. $\sec^2 x = \sec x$
45. $\csc^2 x - \csc x - 2 = 0$
46. $(\tan x - 1)(\tan x - \sqrt{3}) = 0$

Use your calculator or the table in Appendix C to help find approximate solutions for θ, where $0° \leq \theta < 360°$. Express the solutions to the nearest tenth of a degree.

47. $\sin \theta = -0.2157$
48. $\sin \theta = 0.8217$
49. $\cos \theta = -0.6427$
50. $\cos \theta = -0.2179$
51. $\tan \theta = -3.1426$
52. $\tan \theta = 14.2789$
53. $(3 \sin \theta - 1)(2 \sin \theta + 3) = 0$

54. $(4 \sin \theta - 1)(\sin \theta + 2) = 0$

55. $6 \cos^2 \theta - 13 \cos \theta + 6 = 0$

56. $12 \cos^2 \theta - 13 \cos \theta + 3 = 0$

57. $\sin^2 \theta - 4 \sin \theta + 1 = 0$

58. $\cos^2 \theta - 3 \cos \theta + 1 = 0$

Use your calculator or the table in Appendix C to help find approximate solutions for x, where $0 \le x < 2\pi$. Express the solutions to the nearest hundredth of a radian.

59. $\sin x = -0.7126$ 60. $\sin x = 0.2314$

61. $\cos x = -0.8214$ 62. $\cos x = -0.1429$

63. $\tan x = -1.2784$ 64. $\tan x = 9.1275$

65. $4 \sin^2 x + 11 \sin x - 3 = 0$

66. $12 \cos^2 x + 5 \cos x - 3 = 0$

67. $\sin^2 x - 3 \sin x - 2 = 0$

68. $\sin^2 x - \sin x - 1 = 0$

69. $\cos^2 x + \cos x - 1 = 0$

70. $2 \cos^2 x - 3 \cos x - 1 = 0$

THOUGHTS into WORDS

71. Is $\sin x + \cos x = 1$ a conditional equation or an identity? Defend your answer.

72. Give a step-by-step description of how you would find all solutions for the equation $3 \tan^4 x = 1 + \sec^2 x$.

GRAPHICS CALCULATOR ACTIVITIES

73. Use a graphics calculator and check your solutions for Problems 59–70.

74. Use your graphics calculator to approximate the solutions for each of the following equations. Express your answers to the nearest tenth.

a. $\cos x = x$ **b.** $\sin x = \frac{1}{2}x$ **c.** $\cos x = -2x$

75. Use your graphics calculator to determine the number of solutions for the equation $\cos x = -0.1x$.

9.3 SUM AND DIFFERENCE FORMULAS

We offer the following suggestions before you study these next two sections. First, read each section and look for the big ideas; don't be concerned about the details of proofs and worked-out examples. Then read the section again and pay attention to the details. This material is not difficult, but it is messy because of an abundance of formulas. However, you will soon see that there are a few very key formulas from which the others are easily derived.

The following identities were established in Chapter 7 and will be used at times in this section.

$$\sin(-\theta) = -\sin \theta \qquad \cos(-\theta) = \cos \theta \qquad \tan(-\theta) = -\tan \theta$$

Does $\cos(\alpha - \beta) = \cos \alpha - \cos \beta$? The following example gives an immediate response to this equation.

EXAMPLE 1

Evaluate $\cos(\alpha - \beta)$ and $\cos \alpha - \cos \beta$ for $\alpha = 90°$ and $\beta = 60°$.

Solution

$$\cos(\alpha - \beta) = \cos(90° - 60°) = \cos 30° = \frac{\sqrt{3}}{2}$$

$$\cos \alpha - \cos \beta = \cos 90° - \cos 60° = 0 - \frac{1}{2} = -\frac{1}{2}$$

In general, $\cos(\alpha - \beta) \neq \cos \alpha - \cos \beta$. Additional examples would demonstrate that $\sin(\alpha - \beta) \neq \sin \alpha - \sin \beta$, $\cos 2\alpha \neq 2 \cos \alpha$, $\sin \frac{1}{2}\alpha \neq \frac{1}{2} \sin \alpha$, and so on. We need special formulas (identities) for these situations.

To develop a formula for $\cos(\alpha - \beta)$, let's consider two angles α and β, the angle representing $\alpha - \beta$, and the unit circle, as in Figure 9.4. The terminal side of α intersects the unit circle at the point $(\cos \alpha, \sin \alpha)$ and the terminal side of β intersects the unit circle at $(\cos \beta, \sin \beta)$. The distance d between the two points is given by

$$d = \sqrt{(\cos \alpha - \cos \beta)^2 + (\sin \alpha - \sin \beta)^2}.$$

Therefore,

$$\begin{aligned} d^2 &= (\cos \alpha - \cos \beta)^2 + (\sin \alpha - \sin \beta)^2 \\ &= \cos^2 \alpha - 2 \cos \alpha \cos \beta + \cos^2 \beta + \sin^2 \alpha - 2 \sin \alpha \sin \beta + \sin^2 \beta. \end{aligned}$$

Since $\cos^2 \alpha + \sin^2 \alpha = 1$ and $\cos^2 \beta + \sin^2 \beta = 1$, we obtain

$$d^2 = 2 - 2 \cos \alpha \cos \beta - 2 \sin \alpha \sin \beta.$$

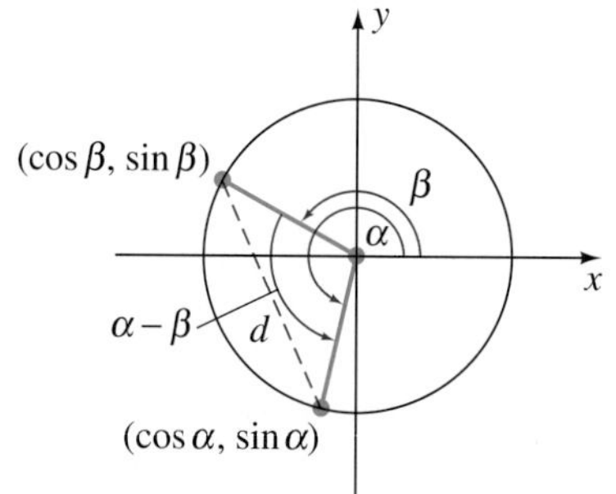

FIGURE 9.4

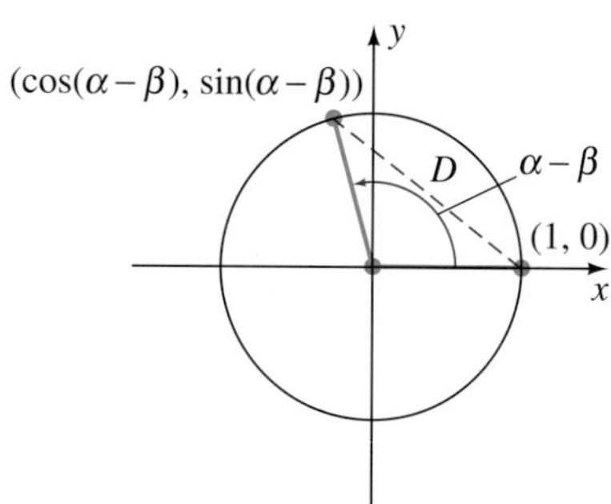

FIGURE 9.5

Now, looking back at Figure 9.4, let's construct angle $(\alpha - \beta)$ in standard position as indicated in Figure 9.5. The terminal side of $(\alpha - \beta)$ intersects the unit circle at the point $(\cos(\alpha - \beta), \sin(\alpha - \beta))$. The distance D between this point and $(1, 0)$ is given by

$$D = \sqrt{[\cos(\alpha - \beta) - 1]^2 + [\sin(\alpha - \beta) - 0]^2}.$$

Therefore,

$$\begin{aligned} D^2 &= [\cos(\alpha - \beta) - 1]^2 + [\sin(\alpha - \beta) - 0]^2 \\ &= \cos^2(\alpha - \beta) - 2\cos(\alpha - \beta) + 1 + \sin^2(\alpha - \beta). \end{aligned}$$

Since $\cos^2(\alpha - \beta) + \sin^2(\alpha - \beta) = 1$, we obtain

$$D^2 = 2 - 2\cos(\alpha - \beta).$$

The isosceles triangles formed in Figures 9.4 and 9.5 are congruent; thus, $d = D$ and we can equate the expressions for D^2 and d^2.

$$2 - 2\cos(\alpha - \beta) = 2 - 2\cos\alpha\cos\beta - 2\sin\alpha\sin\beta$$

Subtract 2 from both sides and then divide both sides by -2 to produce the following important identity.

$$\cos(\alpha - \beta) = \cos\alpha\cos\beta + \sin\alpha\sin\beta$$

The previous development assumes that α and β are positive angles with $\alpha > \beta$. However, the identity holds for all angles measured in radians or degrees, and, in fact, is true for all real numbers.

EXAMPLE 2

Find an exact value for $\cos 15°$.

Solution

Let $\alpha = 45°$ and $\beta = 30°$. Therefore,

$$\begin{aligned} \cos 15° = \cos(45° - 30°) &= \cos 45° \cos 30° + \sin 45° \sin 30° \\ &= \left(\frac{\sqrt{2}}{2}\right)\left(\frac{\sqrt{3}}{2}\right) + \left(\frac{\sqrt{2}}{2}\right)\left(\frac{1}{2}\right) \\ &= \frac{\sqrt{6}}{4} + \frac{\sqrt{2}}{4} \\ &= \frac{\sqrt{6} + \sqrt{2}}{4}. \end{aligned}$$

If in the formula for $\cos(\alpha - \beta)$ we replace β with $-\beta$, we obtain

$$\cos(\alpha - (-\beta)) = \cos\alpha\cos(-\beta) + \sin\alpha\sin(-\beta).$$

Using the identities $\cos(-\beta) = \cos\beta$ and $\sin(-\beta) = -\sin\beta$, the following sum formula is produced.

$$\cos(\alpha + \beta) = \cos\alpha\cos\beta - \sin\alpha\sin\beta$$

EXAMPLE 3

Find an exact value for $\cos(5\pi/12)$.

Solution

Let $\alpha = \pi/4$ and $\beta = \pi/6$. Therefore,

$$\begin{aligned}\cos\frac{5\pi}{12} = \cos\left(\frac{\pi}{4}+\frac{\pi}{6}\right) &= \cos\frac{\pi}{4}\cos\frac{\pi}{6} - \sin\frac{\pi}{4}\sin\frac{\pi}{6}\\ &= \left(\frac{\sqrt{2}}{2}\right)\left(\frac{\sqrt{3}}{2}\right) - \left(\frac{\sqrt{2}}{2}\right)\left(\frac{1}{2}\right)\\ &= \frac{\sqrt{6}}{4} - \frac{\sqrt{2}}{4} = \frac{\sqrt{6}-\sqrt{2}}{4}.\end{aligned}$$

Before we develop formulas for $\sin(\alpha + \beta)$ and $\sin(\alpha - \beta)$, let's consider some identities involving a 90° angle.

$$\begin{aligned}\cos(90^\circ - \alpha) &= \cos 90^\circ \cos\alpha + \sin 90^\circ \sin\alpha\\ &= 0\cdot\cos\alpha + 1\cdot\sin\alpha\\ &= \sin\alpha\end{aligned}$$

Therefore,

$$\cos(90^\circ - \alpha) = \sin\alpha.$$

Now substitute $90^\circ - \alpha$ for α in the previous identity to obtain

$$\begin{aligned}\cos[90^\circ - (90^\circ - \alpha)] &= \sin(90^\circ - \alpha)\\ \cos(90^\circ - 90^\circ + \alpha) &= \sin(90^\circ - \alpha)\\ \cos\alpha &= \sin(90^\circ - \alpha).\end{aligned}$$

$$\sin(90^\circ - \alpha) = \cos\alpha$$

An identity involving $\tan(90^\circ - \alpha)$ follows directly from the sine and cosine relationships.

$$\tan(90^\circ - \alpha) = \frac{\sin(90^\circ - \alpha)}{\cos(90^\circ - \alpha)} = \frac{\cos\alpha}{\sin\alpha} = \cot\alpha$$

$$\tan(90^\circ - \alpha) = \cot\alpha$$

In Chapter 7 we verified that in a right triangle ACB, where A and B are the acute angles, any functional value of A equals the value of the cofunction of B. For example, $\sin A = \cos B$. The three previously stated identities are generalizations of this idea. For example, $\sin \alpha = \cos(90° - \alpha)$ for *any angle* α.

Now let's use the identity $\sin \alpha = \cos(90° - \alpha)$ to develop a formula for $\sin(\alpha + \beta)$.

$$\begin{aligned}\sin(\alpha + \beta) &= \cos[90° - (\alpha + \beta)] \\ &= \cos[(90° - \alpha) - \beta] \\ &= \cos(90° - \alpha) \cos \beta + \sin(90° - \alpha) \sin \beta \\ &= \sin \alpha \cos \beta + \cos \alpha \sin \beta\end{aligned}$$

$$\sin(\alpha + \beta) = \sin \alpha \cos \beta + \cos \alpha \sin \beta$$

Substitute $-\beta$ for β into the formula for $\sin(\alpha + \beta)$ to produce

$$\begin{aligned}\sin(\alpha + (-\beta)) &= \sin \alpha \cos(-\beta) + \cos \alpha \sin(-\beta) \\ &= \sin \alpha \cos \beta - \cos \alpha \sin \beta.\end{aligned}$$

$$\sin(\alpha - \beta) = \sin \alpha \cos \beta - \cos \alpha \sin \beta$$

EXAMPLE 4

Find an exact value for $\sin(\pi/12)$.

Solution

Let $\alpha = \pi/3$ and $\beta = \pi/4$. Therefore,

$$\begin{aligned}\sin \frac{\pi}{12} = \sin\left(\frac{\pi}{3} - \frac{\pi}{4}\right) &= \sin \frac{\pi}{3} \cos \frac{\pi}{4} - \cos \frac{\pi}{3} \sin \frac{\pi}{4} \\ &= \left(\frac{\sqrt{3}}{2}\right)\left(\frac{\sqrt{2}}{2}\right) - \left(\frac{1}{2}\right)\left(\frac{\sqrt{2}}{2}\right) \\ &= \frac{\sqrt{6}}{4} - \frac{\sqrt{2}}{4} = \frac{\sqrt{6} - \sqrt{2}}{4}.\end{aligned}$$

As you might expect, a formula for $\tan(\alpha + \beta)$ follows directly from the sine and cosine relationships.

$$\tan(\alpha + \beta) = \frac{\sin(\alpha + \beta)}{\cos(\alpha + \beta)} = \frac{\sin \alpha \cos \beta + \cos \alpha \sin \beta}{\cos \alpha \cos \beta - \sin \alpha \sin \beta}$$

Divide both numerator and denominator by $\cos \alpha \cos \beta$ to produce

$$\begin{aligned}\tan(\alpha + \beta) &= \frac{\dfrac{\sin \alpha \cos \beta}{\cos \alpha \cos \beta} + \dfrac{\cos \alpha \sin \beta}{\cos \alpha \cos \beta}}{\dfrac{\cos \alpha \cos \beta}{\cos \alpha \cos \beta} - \dfrac{\sin \alpha \sin \beta}{\cos \alpha \cos \beta}} \\ &= \frac{\tan \alpha + \tan \beta}{1 - \tan \alpha \tan \beta}.\end{aligned}$$

$$\tan(\alpha + \beta) = \frac{\tan \alpha + \tan \beta}{1 - \tan \alpha \tan \beta}$$

If in the formula for $\tan(\alpha + \beta)$ we replace β with $-\beta$, the following result will be obtained. (We will leave the details for you as an exercise.)

$$\tan(\alpha - \beta) = \frac{\tan \alpha - \tan \beta}{1 + \tan \alpha \tan \beta}$$

EXAMPLE 5

Find an exact value for $\tan 195°$.

Solution

Let $\alpha = 135°$ and $\beta = 60°$. Therefore,

$$\begin{aligned} \tan 195° = \tan(135° + 60°) &= \frac{\tan 135° + \tan 60°}{1 - \tan 135° \tan 60°} \\ &= \frac{-1 + \sqrt{3}}{1 - (-1)(\sqrt{3})} \\ &= \frac{-1 + \sqrt{3}}{1 + \sqrt{3}} \\ &= \frac{-1 + \sqrt{3}}{1 + \sqrt{3}} \cdot \frac{1 - \sqrt{3}}{1 - \sqrt{3}} = \frac{-1 + 2\sqrt{3} - 3}{1 - 3} \\ &= \frac{-4 + 2\sqrt{3}}{-2} = 2 - \sqrt{3}. \end{aligned}$$

EXAMPLE 6

Given that $\sin \alpha = \frac{3}{5}$ with α in the first quadrant, and $\cos \beta = -\frac{7}{25}$ with β in the second quadrant, find $\cos(\alpha - \beta)$, $\sin(\alpha + \beta)$, and $\tan(\alpha + \beta)$.

Solution

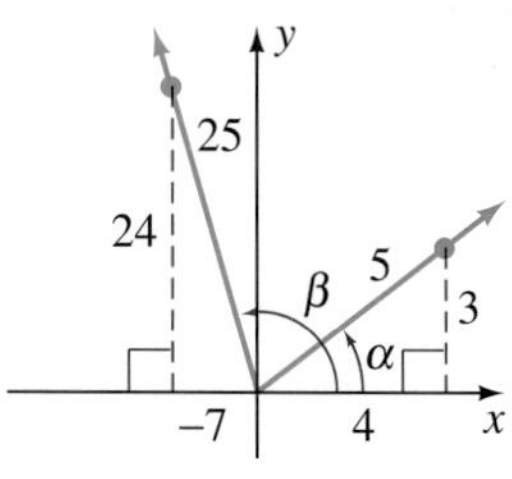

FIGURE 9.6

If $\sin \alpha = \frac{3}{5}$ and α is in the first quadrant, then $\cos \alpha = \frac{4}{5}$ and $\tan \alpha = \frac{3}{4}$ (Figure 9.6). Likewise, if $\cos \beta = -\frac{7}{25}$ and β is in the second quadrant, then $\sin \beta = \frac{24}{25}$ and $\tan \beta = -\frac{24}{7}$.

$$\begin{aligned} \cos(\alpha - \beta) &= \cos \alpha \cos \beta + \sin \alpha \sin \beta \\ &= \left(\frac{4}{5}\right)\left(-\frac{7}{25}\right) + \left(\frac{3}{5}\right)\left(\frac{24}{25}\right) \\ &= -\frac{28}{125} + \frac{72}{125} = \frac{44}{125} \end{aligned}$$

$$\begin{aligned}\sin(\alpha + \beta) &= \sin \alpha \cos \beta + \cos \alpha \sin \beta \\ &= \left(\frac{3}{5}\right)\left(-\frac{7}{25}\right) + \left(\frac{4}{5}\right)\left(\frac{24}{25}\right) \\ &= -\frac{21}{125} + \frac{96}{125} = \frac{75}{125} = \frac{3}{5}\end{aligned}$$

$$\begin{aligned}\tan(\alpha + \beta) &= \frac{\tan \alpha + \tan \beta}{1 - \tan \alpha \tan \beta} \\ &= \frac{\frac{3}{4} + \left(-\frac{24}{7}\right)}{1 - \left(\frac{3}{4}\right)\left(-\frac{24}{7}\right)} \\ &= \frac{-\frac{75}{28}}{1 + \frac{72}{28}} = \frac{-\frac{75}{28}}{\frac{100}{28}} = -\frac{75}{100} = -\frac{3}{4}\end{aligned}$$

EXAMPLE 7

Evaluate $\sin\left(\tan^{-1} \frac{1}{2} + \cos^{-1} \frac{3}{5}\right)$.

Solution

Let $\alpha = \tan^{-1} \frac{1}{2}$ and $\beta = \cos^{-1} \frac{3}{5}$. From the definitions of the inverse functions in Chapter 8, and since $\tan \alpha$ and $\cos \beta$ are positive, α and β can be considered acute angles of the right triangles in Figure 9.7. Therefore,

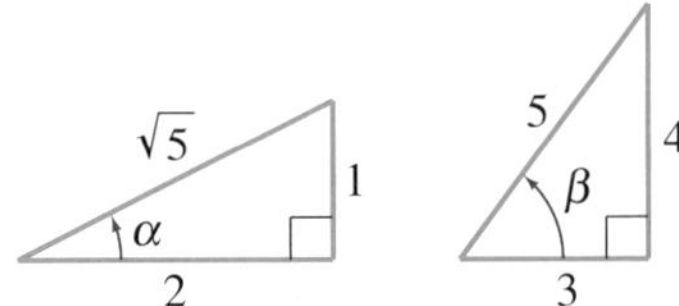

FIGURE 9.7

$$\begin{aligned}\sin(\alpha + \beta) &= \sin \alpha \cos \beta + \cos \alpha \sin \beta \\ &= \left(\frac{\sqrt{5}}{5}\right)\left(\frac{3}{5}\right) + \left(\frac{2\sqrt{5}}{5}\right)\left(\frac{4}{5}\right) \\ &= \frac{3\sqrt{5}}{25} + \frac{8\sqrt{5}}{25} = \frac{11\sqrt{5}}{25}.\end{aligned}$$

Thus, $\sin\left(\tan^{-1} \frac{1}{2} + \cos^{-1} \frac{3}{5}\right) = 11\sqrt{5}/25$.

EXAMPLE 8

Use a graphing utility to demonstrate that $\cos[(\pi/2) - x] = \sin x$ is an identity.

Solution

Let $Y_1 = \cos[(\pi/2) - x]$ and $Y_2 = \sin x$ and graph both of these curves on the same set of axes (Figure 9.8). Since their graphs appear to be identical, this visually demonstrates that $\cos[(\pi/2) - x] = \sin x$.

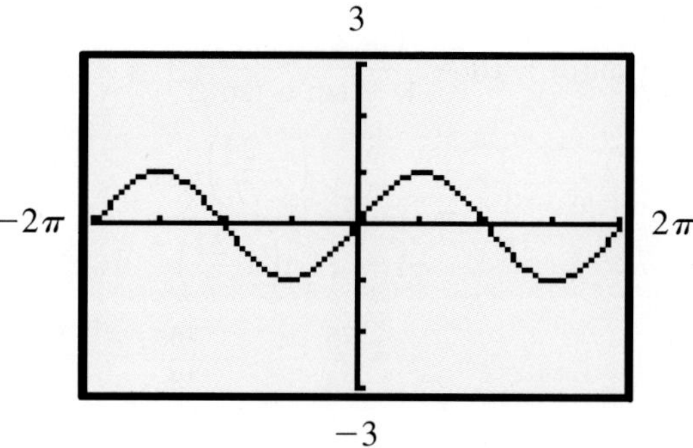

FIGURE 9.8

A graphing utility can also be used to *predict* possible identities. For example, suppose that we graph the function $f(x) = \sin[x + (3\pi/2)]$ as shown in Figure 9.9. This curve appears to be the x-axis reflection of the cosine curve; that is, the graph of $f(x) = -\cos x$. So let's graph $f(x) = -\cos x$ on the same set of axes and we do obtain the same graph as in Figure 9.9. Thus, from a graphing viewpoint we have demonstrated the identity $\sin[x + (3\pi/2)] = -\cos x$. Using the sum identity, we can verify this identity as follows.

$$\begin{aligned}\sin\left(x + \frac{3\pi}{2}\right) &= \sin x \cos \frac{3\pi}{2} + \cos x \sin \frac{3\pi}{2} \\ &= (\sin x)(0) + (\cos x)(-1) \\ &= -\cos x\end{aligned}$$

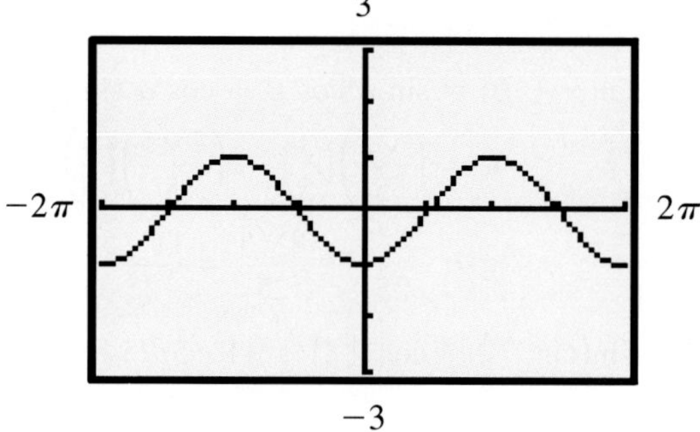

FIGURE 9.9

If you are following the suggestion offered at the beginning of this section, it is now time to go back and fill in some details. Keep the following continuity pattern in mind.

The first formula derived was

$$\cos(\alpha - \beta) = \cos \alpha \cos \beta + \sin \alpha \sin \beta. \quad \textbf{(1)}$$

Then substituting $-\beta$ for β produced

$$\cos(\alpha + \beta) = \cos \alpha \cos \beta - \sin \alpha \sin \beta. \quad \textbf{(2)}$$

Then applying $\cos(\alpha - \beta)$ to $\cos(90° - \alpha)$ produced

$$\cos(90° - \alpha) = \sin \alpha. \quad \textbf{(3)}$$

Then substituting $(\alpha + \beta)$ for α in equation (3) produced

$$\sin(\alpha + \beta) = \sin \alpha \cos \beta + \cos \alpha \sin \beta. \quad \textbf{(4)}$$

Then substituting $-\beta$ for β produced

$$\sin(\alpha - \beta) = \sin \alpha \cos \beta - \cos \alpha \sin \beta. \quad \textbf{(5)}$$

Then using the formulas for $\sin(\alpha + \beta)$ and $\cos(\alpha + \beta)$, we obtained

$$\tan(\alpha + \beta) = \frac{\tan \alpha + \tan \beta}{1 - \tan \alpha \tan \beta}. \quad \textbf{(6)}$$

Finally, substituting $-\beta$ for β produced

$$\tan(\alpha - \beta) = \frac{\tan \alpha - \tan \beta}{1 + \tan \alpha \tan \beta}. \quad \textbf{(7)}$$

PROBLEM SET 9.3

For Problems 1–14, find the exact values without using a table or a calculator.

1. $\sin 15°$ **2.** $\tan 15°$ **3.** $\tan 75°$

4. $\sin 75°$ **5.** $\sin 105°$ **6.** $\cos 105°$

7. $\cos 195°$ **8.** $\sin 195°$ **9.** $\tan 255°$

10. $\cos 345°$ **11.** $\cos \frac{\pi}{12}$ **12.** $\cos \frac{5\pi}{12}$

13. $\sin \frac{7\pi}{12}$ **14.** $\sin \frac{11\pi}{12}$

15. Given that $\cos \alpha = \frac{3}{5}$ with α in the first quadrant, and $\sin \beta = \frac{15}{17}$ with β in the second quadrant, find $\sin(\alpha - \beta)$ and $\tan(\alpha + \beta)$.

16. If α and β are acute angles such that $\cos \alpha = \frac{4}{5}$ and $\tan \beta = \frac{8}{15}$, find $\cos(\alpha + \beta)$ and $\sin(\alpha + \beta)$.

17. If $\cos \alpha = -\frac{3}{5}$ and $\tan \beta = \frac{8}{15}$, where α is a second-quadrant angle and β is a third-quadrant angle, find $\sin(\alpha + \beta)$ and $\cos(\alpha - \beta)$.

18. If $\tan \alpha = -\frac{4}{3}$ and $\sin \beta = -\frac{3}{5}$, where α is a second-quadrant angle and β is a fourth-quadrant angle, find $\sin(\alpha - \beta)$ and $\tan(\alpha + \beta)$.

19. Given that $\tan\alpha = \frac{8}{15}$ with α in the first quadrant, and $\cos\beta = \frac{7}{25}$ with β in the fourth quadrant, find $\sin(\alpha - \beta)$ and $\cos(\alpha - \beta)$.

20. Given that $\tan\alpha = -\frac{2}{3}$ with α in the second quadrant, and $\tan\beta = \frac{3}{5}$ with β in the third quadrant, find $\tan(\alpha + \beta)$ and $\tan(\alpha - \beta)$.

For Problems 21–26, find exact values without using a calculator or a table.

21. $\sin\left(\tan^{-1}\frac{3}{4} + \cos^{-1}\frac{24}{25}\right)$

22. $\cos\left(\tan^{-1}\frac{7}{24} - \sin^{-1}\frac{4}{5}\right)$

23. $\tan\left(\arcsin\frac{15}{17} + \arccos\frac{4}{5}\right)$

24. $\tan\left[\arcsin\frac{3}{5} + \arccos\left(-\frac{3}{5}\right)\right]$

25. $\cos\left(\tan^{-1}\frac{1}{3} + \cos^{-1}\frac{8}{17}\right)$

26. $\sin\left(\tan^{-1}\frac{1}{2} - \cos^{-1}\frac{4}{5}\right)$

Verify each of the identities in Problems 27–38.

27. $\sin(\alpha + 90°) = \cos\alpha$ **28.** $\cos(\alpha + 90°) = -\sin\alpha$

29. $\sin(\alpha + \pi) = -\sin\alpha$ **30.** $\cos(\alpha - \pi) = -\cos\alpha$

31. $\tan(\alpha + \pi) = \tan\alpha$ **32.** $\tan(\alpha - \pi) = \tan\alpha$

33. $\sin(\alpha + 45°) = \frac{\sqrt{2}}{2}(\sin\alpha + \cos\alpha)$

34. $\cos(\alpha - 45°) = \frac{\sqrt{2}}{2}(\cos\alpha + \sin\alpha)$

35. $\tan\left(\alpha + \frac{\pi}{4}\right) = \frac{1 + \tan\alpha}{1 - \tan\alpha}$

36. $\tan\left(\alpha - \frac{\pi}{4}\right) = \frac{\tan\alpha - 1}{\tan\alpha + 1}$

37. $\sin(\alpha + 270°) = -\cos\alpha$

38. $\cos(\alpha - 270°) = -\sin\alpha$

39. Develop the formula for $\tan(\alpha - \beta)$ by replacing β with $-\beta$ in the formula for $\tan(\alpha + \beta)$.

40. Derive the formula $\cot(\alpha + \beta) = \dfrac{\cot\alpha\cot\beta - 1}{\cot\alpha + \cot\beta}$.

41. Derive the formula $\cot(\alpha - \beta) = \dfrac{\cot\alpha\cot\beta + 1}{\cot\beta - \cot\alpha}$.

42. If α and β are complementary angles, verify that $\sin^2\alpha + \sin^2\beta = 1$.

MISCELLANEOUS PROBLEMS

43. Use the identities

$$\cos(\alpha + \beta) = \cos\alpha\cos\beta - \sin\alpha\sin\beta$$
$$\cos(\alpha - \beta) = \cos\alpha\cos\beta + \sin\alpha\sin\beta$$

to produce the following *product identities*.

a. $\cos\alpha\cos\beta = \frac{1}{2}[\cos(\alpha + \beta) + \cos(\alpha - \beta)]$

b. $\sin\alpha\sin\beta = \frac{1}{2}[\cos(\alpha - \beta) - \cos(\alpha + \beta)]$

44. Use the identities

$$\sin(\alpha + \beta) = \sin\alpha\cos\beta + \cos\alpha\sin\beta$$
$$\sin(\alpha - \beta) = \sin\alpha\cos\beta - \cos\alpha\sin\beta$$

to produce the following *product identities*.

a. $\sin\alpha\cos\beta = \frac{1}{2}[\sin(\alpha + \beta) + \sin(\alpha - \beta)]$

b. $\cos\alpha\sin\beta = \frac{1}{2}[\sin(\alpha + \beta) - \sin(\alpha - \beta)]$

45. Use the identities established in Problems 43 and 44 to express each of the following products as a sum or difference.

a. $\cos 3\theta\cos 2\theta$ **b.** $\cos 2\theta\cos 4\theta$

c. $\sin 3\theta\sin\theta$ **d.** $\sin 4\theta\cos 3\theta$

e. $2\sin 9\theta\cos 3\theta$ **f.** $3\cos\theta\sin 2\theta$

46. Use the identities from parts (a) and (b) of Problem 43 to help verify the following *sum* and *difference identities*. [*Hint:* Let $A = \alpha + \beta$ and $B = \alpha - \beta$.]

a. $\cos A + \cos B = 2\cos\frac{A + B}{2}\cos\frac{A - B}{2}$

b. $\cos B - \cos A = 2\sin\frac{A + B}{2}\sin\frac{A - B}{2}$

47. Use the identities from parts (a) and (b) of Problem 44 to help verify the following *sum* and *difference identities.*

a. $\sin A + \sin B = 2 \sin \frac{A + B}{2} \cos \frac{A - B}{2}$

b. $\sin A - \sin B = 2 \cos \frac{A + B}{2} \sin \frac{A - B}{2}$

48. Use the identities established in Problems 46 and 47 to help write each of the following as a product.

a. $\cos 4\theta + \cos 2\theta$ **b.** $\cos 6\theta - \cos 2\theta$

c. $\cos \theta - \cos 5\theta$ **d.** $\sin 6\theta + \sin 2\theta$

e. $\sin 3\theta - \sin \theta$ **f.** $\sin 2\theta - \sin 4\theta$

GRAPHICS CALCULATOR ACTIVITIES

49. We want to decide if each of the following represents a conditional equation or an identity. Therefore, let's (a) try some specific examples and from these results make a prediction, and (b) use a graphics calculator to test the predictions.

a. $\sin 2x = 2 \sin x$ **b.** $\sin 2x = 2 \sin x \cos x$

c. $\cos 2x = 2 \cos^2 x - 1$

d. $\cos 2x = \cos^2 x - \sin^2 x$

e. $\cos 2x = 2 \cos x$ **f.** $\cos 2x = 1 - 2 \sin^2 x$

g. $\sin 3x = 3 \sin x \cos x$

h. $\sin 3x = 3 \sin x - 4 \sin^3 x$

9.4 MULTIPLE AND HALF-ANGLE FORMULAS

Again let's emphasize the point that $\sin 2\alpha \neq 2 \sin \alpha$. For example, $\sin 2(30°) = \sin 60° = \sqrt{3}/2$, but $2 \sin 30° = 2\left(\frac{1}{2}\right) = 1$. Thus, *multiple angle formulas* are needed. As you might expect, the multiple angle formulas are nothing more than special cases of the sum formulas we developed in the previous section. For example, in the formula for $\sin(\alpha + \beta)$, if we let $\alpha = \beta$ we obtain the following.

$$\sin(\alpha + \beta) = \sin \alpha \cos \beta + \cos \alpha \sin \beta$$

becomes

$$\sin(\alpha + \alpha) = \sin \alpha \cos \alpha + \cos \alpha \sin \alpha,$$

or, equivalently,

$$\sin 2\alpha = 2 \sin \alpha \cos \alpha.$$

EXAMPLE 1

Find $\sin 2\alpha$ if $\sin \alpha = \frac{4}{5}$ and α is in the first quadrant.

Solution

If $\sin \alpha = \frac{4}{5}$ and α is in the first quadrant, then $\cos \alpha = \frac{3}{5}$. Therefore,

$$\begin{aligned} \sin 2\alpha &= 2 \sin \alpha \cos \alpha \\ &= 2\left(\frac{4}{5}\right)\left(\frac{3}{5}\right) = \frac{24}{25}. \end{aligned}$$

Substituting α for β in the formula for $\cos(\alpha + \beta)$ produces

$$\begin{aligned} \cos(\alpha + \beta) &= \cos \alpha \cos \beta - \sin \alpha \sin \beta \\ \cos(\alpha + \alpha) &= \cos \alpha \cos \alpha - \sin \alpha \sin \alpha \\ \cos 2\alpha &= \cos^2 \alpha - \sin^2 \alpha. \end{aligned}$$

Two other formulas for $\cos 2\alpha$ can be obtained by using the basic identity $\sin^2 \alpha + \cos^2 \alpha = 1$. Substituting $1 - \sin^2 \alpha$ for $\cos^2 \alpha$ produces

$$\begin{aligned} \cos^2 \alpha - \sin^2 \alpha &= (1 - \sin^2 \alpha) - \sin^2 \alpha \\ &= 1 - 2 \sin^2 \alpha. \end{aligned}$$

Similarly, substituting $1 - \cos^2 \alpha$ for $\sin^2 \alpha$ produces

$$\begin{aligned} \cos^2 \alpha - \sin^2 \alpha &= \cos^2 \alpha - (1 - \cos^2 \alpha) \\ &= 2 \cos^2 \alpha - 1. \end{aligned}$$

$$\cos 2\alpha = \cos^2 \alpha - \sin^2 \alpha \quad \text{or} \quad \cos 2\alpha = 1 - 2 \sin^2 \alpha$$
$$\text{or} \quad \cos 2\alpha = 2 \cos^2 \alpha - 1$$

Substituting α for β in the formula for $\tan(\alpha + \beta)$ produces

$$\begin{aligned} \tan(\alpha + \beta) &= \frac{\tan \alpha + \tan \beta}{1 - \tan \alpha \tan \beta} \\ \tan(\alpha + \alpha) &= \frac{\tan \alpha + \tan \alpha}{1 - \tan \alpha \tan \alpha}. \end{aligned}$$

$$\tan 2\alpha = \frac{2 \tan \alpha}{1 - \tan^2 \alpha}$$

EXAMPLE 2

Find $\sin 2\theta$, $\cos 2\theta$, and $\tan 2\theta$ if $\sin \theta = -\frac{5}{13}$ and θ is a fourth-quadrant angle.

Solution

Figure 9.10 depicts the situation.

Therefore, $\cos \theta = \frac{12}{13}$ and $\tan \theta = -\frac{5}{12}$. Now we can use the double-angle identities.

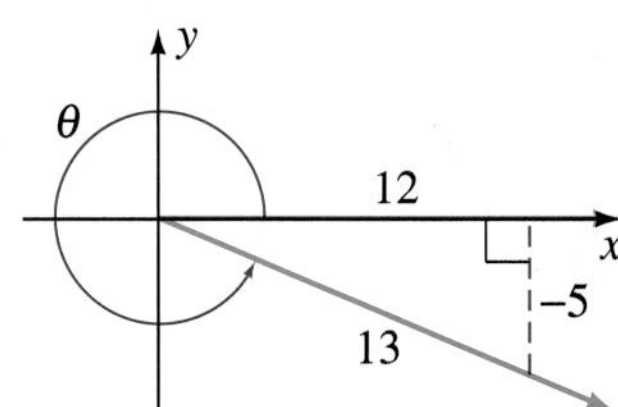

FIGURE 9.10

$$\sin 2\theta = 2 \sin \theta \cos \theta$$
$$= 2\left(-\frac{5}{13}\right)\left(\frac{12}{13}\right) = -\frac{120}{169}$$
$$\cos 2\theta = \cos^2 \theta - \sin^2 \theta$$
$$= \left(\frac{12}{13}\right)^2 - \left(-\frac{5}{13}\right)^2$$
$$= \frac{144}{169} - \frac{25}{169} = \frac{119}{169}$$
$$\tan 2\theta = \frac{2 \tan \theta}{1 - \tan^2 \theta}$$
$$= \frac{2\left(-\frac{5}{12}\right)}{1 - \left(-\frac{5}{12}\right)^2}$$
$$= \frac{-\frac{10}{12}}{1 - \frac{25}{144}}$$
$$= \frac{-\frac{5}{6}}{\frac{119}{144}} = \left(-\frac{5}{6}\right)\left(\frac{144}{119}\right) = -\frac{120}{119}$$

REMARK In Example 2, after finding the values for $\sin 2\theta$ and $\cos 2\theta$, the identity

$$\tan 2\theta = \frac{\sin 2\theta}{\cos 2\theta}$$

could also be used to find the value of $\tan 2\theta$.

The double-angle identities provide a broader base for proving identities and solving trigonometric equations, as the next examples illustrate.

EXAMPLE 3

Verify the identity $\cos 2\alpha = \cos^4 \alpha - \sin^4 \alpha$.

Solution

The right side can be factored as the difference of two squares.

$$\cos^4 \alpha - \sin^4 \alpha = (\cos^2 \alpha + \sin^2 \alpha)(\cos^2 \alpha - \sin^2 \alpha)$$

We know that $\cos^2 \alpha + \sin^2 \alpha = 1$ and $\cos^2 \alpha - \sin^2 \alpha = \cos 2\alpha$. Therefore,

$$\cos^4 \alpha - \sin^4 \alpha = 1 \cdot \cos 2\alpha = \cos 2\alpha.$$

EXAMPLE 4

Solve the equation $\cos 2\theta - \cos\theta = 0$, where $0° \le \theta < 360°$.

Solution

We can substitute $2\cos^2\theta - 1$ for $\cos 2\theta$ and proceed as follows.

$$\begin{aligned}\cos 2\theta - \cos\theta &= 0\\ (2\cos^2\theta - 1) - \cos\theta &= 0\\ 2\cos^2\theta - \cos\theta - 1 &= 0\\ (2\cos\theta + 1)(\cos\theta - 1) &= 0\end{aligned}$$

$$\begin{aligned}2\cos\theta + 1 &= 0 &\quad\text{or}\quad \cos\theta - 1 &= 0\\ 2\cos\theta &= -1 &\quad\text{or}\quad \cos\theta &= 1\\ \cos\theta &= -\frac{1}{2} &\quad\text{or}\quad \cos\theta &= 1\end{aligned}$$

If $\cos\theta = -\frac{1}{2}$, then $\theta = 120°$ or $240°$. If $\cos\theta = 1$, then $\theta = 0°$. The solutions are $0°$, $120°$, and $240°$.

EXAMPLE 5

Express $\sin 3\theta$ in terms of $\sin\theta$.

Solution

$$\begin{aligned}\sin 3\theta &= \sin(2\theta + \theta)\\ &= \sin 2\theta\cos\theta + \cos 2\theta\sin\theta\\ &= (2\sin\theta\cos\theta)\cos\theta + (1 - 2\sin^2\theta)\sin\theta\\ &= 2\sin\theta\cos^2\theta + \sin\theta - 2\sin^3\theta\\ &= 2\sin\theta(1 - \sin^2\theta) + \sin\theta - 2\sin^3\theta\\ &= 2\sin\theta - 2\sin^3\theta + \sin\theta - 2\sin^3\theta\\ &= -4\sin^3\theta + 3\sin\theta\end{aligned}$$

Now let's change the form of the identity $\cos 2\alpha = 1 - 2\sin^2\alpha$ by solving for $\sin^2\alpha$.

$$\begin{aligned}\cos 2\alpha &= 1 - 2\sin^2\alpha\\ 2\sin^2\alpha &= 1 - \cos 2\alpha\\ \sin^2\alpha &= \frac{1 - \cos 2\alpha}{2}\end{aligned}$$

In a similar fashion, the identity $\cos 2\alpha = 2\cos^2\alpha - 1$ can be written as $\cos^2\alpha = (1 + \cos 2\alpha)/2$. Then using the fact that $\tan^2\alpha = (\sin^2\alpha)/(\cos^2\alpha)$ we can express $\tan^2\alpha$ as $\tan^2\alpha = \dfrac{1 - \cos 2\alpha}{1 + \cos 2\alpha}$. Thus, the following three identities are formed.

$$\sin^2 \alpha = \frac{1 - \cos 2\alpha}{2} \quad \cos^2 \alpha = \frac{1 + \cos 2\alpha}{2} \quad \tan^2 \alpha = \frac{1 - \cos 2\alpha}{1 + \cos 2\alpha}$$

These identities are used in calculus to change the form of some expressions involving powers of trigonometric functions as in the next example.

EXAMPLE 6 Express $\sin^4 \alpha$ as an expression involving no powers of trigonometric functions.

Solution

$$\begin{aligned}
\sin^4 \alpha &= (\sin^2 \alpha)^2 \\
&= \left(\frac{1 - \cos 2\alpha}{2}\right)^2 && \text{Substitute } \frac{1 - \cos 2\alpha}{2} \text{ for } \sin^2 \alpha. \\
&= \frac{1 - 2\cos 2\alpha + \cos^2 2\alpha}{4} \\
&= \frac{1}{4} - \frac{1}{2}\cos 2\alpha + \frac{1}{4}\cos^2 2\alpha \\
&= \frac{1}{4} - \frac{1}{2}\cos 2\alpha + \frac{1}{4}\left(\frac{1 + \cos 4\alpha}{2}\right) && \text{Substitute } \frac{1 + \cos 4\alpha}{2} \text{ for } \cos^2 2\alpha. \\
&= \frac{1}{4} - \frac{1}{2}\cos 2\alpha + \frac{1}{8} + \frac{1}{8}\cos 4\alpha \\
&= \frac{3}{8} - \frac{1}{2}\cos 2\alpha + \frac{1}{8}\cos 4\alpha
\end{aligned}$$

REMARK In Example 6 we changed a simple-looking expression, $\sin^4 \alpha$, to a more complicated expression, $\frac{3}{8} - \frac{1}{2}\cos 2\alpha + \frac{1}{8}\cos 4\alpha$. For some purposes in calculus, the multiple angle expressions can be handled more easily than powers of a trigonometric function.

Half-Angle Formulas

By substituting $\alpha/2$ for α in the three identities in the previous box, and then solving for $\sin(\alpha/2)$, $\cos(\alpha/2)$, and $\tan(\alpha/2)$ we obtain the following three half-angle formulas.

$$\sin \frac{\alpha}{2} = \pm\sqrt{\frac{1 - \cos \alpha}{2}} \qquad \cos \frac{\alpha}{2} = \pm\sqrt{\frac{1 + \cos \alpha}{2}}$$

$$\tan \frac{\alpha}{2} = \pm\sqrt{\frac{1 - \cos \alpha}{1 + \cos \alpha}}$$

In the formula for $\sin(\alpha/2)$ and $\cos(\alpha/2)$, the choice of the plus or minus sign is determined by the quadrant in which $\alpha/2$ lies. For example, if $\alpha/2$ is in the first or second quadrant, then we would use

$$\sin \frac{\alpha}{2} = \sqrt{\frac{1 - \cos \alpha}{2}}.$$

However, if $\alpha/2$ is a third- or fourth-quadrant angle, then we would use

$$\sin \frac{\alpha}{2} = -\sqrt{\frac{1 - \cos \alpha}{2}}.$$

An alternative form for the $\tan(\alpha/2)$ formula can be obtained by multiplying the radicand by a form of one, namely, $(1 - \cos \alpha)/(1 - \cos \alpha)$.

$$\begin{aligned} \tan \frac{\alpha}{2} &= \pm\sqrt{\frac{1 - \cos \alpha}{1 + \cos \alpha} \cdot \frac{1 - \cos \alpha}{1 - \cos \alpha}} \\ &= \pm\sqrt{\frac{(1 - \cos \alpha)^2}{1 - \cos^2 \alpha}} \\ &= \pm\sqrt{\frac{(1 - \cos \alpha)^2}{\sin^2 \alpha}} = \frac{1 - \cos \alpha}{\sin \alpha} \end{aligned}$$

We no longer need the $\pm$ sign because $1 - \cos \alpha$ is never negative, and $\sin \alpha$ and $\tan(\alpha/2)$ will always agree in sign. For example, if $0 < \alpha < \pi$, then $0 < \alpha/2 < \pi/2$ and therefore both $\sin \alpha$ and $\tan(\alpha/2)$ are positive. If $\pi < \alpha < 2\pi$, then $\pi/2 < \alpha/2 < \pi$ and both $\sin \alpha$ and $\tan(\alpha/2)$ are negative. Furthermore, the form of $(1 - \cos \alpha)/\sin \alpha$ can be changed as follows.

$$\begin{aligned} \frac{1 - \cos \alpha}{\sin \alpha} \cdot \frac{1 + \cos \alpha}{1 + \cos \alpha} &= \frac{1 - \cos^2 \alpha}{\sin \alpha(1 + \cos \alpha)} \\ &= \frac{\sin^2 \alpha}{\sin \alpha(1 + \cos \alpha)} \\ &= \frac{\sin \alpha}{1 + \cos \alpha} \end{aligned}$$

Therefore, either of the following two identities can be used.

$$\tan \frac{\alpha}{2} = \frac{1 - \cos \alpha}{\sin \alpha} \quad \text{or} \quad \tan \frac{\alpha}{2} = \frac{\sin \alpha}{1 + \cos \alpha}$$

EXAMPLE 7

If $\cos \alpha = -\frac{4}{5}$ and α is in the third quadrant, find $\sin(\alpha/2)$, $\cos(\alpha/2)$, and $\tan(\alpha/2)$.

Solution

Because α is in the third quadrant, $\alpha/2$ is in the second quadrant; thus, $\sin(\alpha/2)$

is positive and $\cos(\alpha/2)$ is negative.

$$\sin\frac{\alpha}{2} = \sqrt{\frac{1-\cos\alpha}{2}} = \sqrt{\frac{1-\left(-\frac{4}{5}\right)}{2}} = \sqrt{\frac{1+\frac{4}{5}}{2}}$$

$$= \sqrt{\frac{\frac{9}{5}}{2}} = \sqrt{\frac{9}{10}} = \frac{3\sqrt{10}}{10}$$

$$\cos\frac{\alpha}{2} = -\sqrt{\frac{1+\cos\alpha}{2}} = -\sqrt{\frac{1+\left(-\frac{4}{5}\right)}{2}} = -\sqrt{\frac{\frac{1}{5}}{2}}$$

$$= -\sqrt{\frac{1}{10}} = -\frac{\sqrt{10}}{10}$$

If $\cos\alpha = -\frac{4}{5}$ and α is in the third quadrant, then $\sin\alpha = -\frac{3}{5}$. Therefore,

$$\tan\frac{\alpha}{2} = \frac{1-\cos\alpha}{\sin\alpha} = \frac{1-\left(-\frac{4}{5}\right)}{-\frac{3}{5}} = \frac{\frac{9}{5}}{-\frac{3}{5}} = -3.$$

EXAMPLE 8

Find an exact value for $\tan 67.5°$.

Solution

$$\tan 67.5° = \tan\frac{1}{2}(135°)$$

$$= \frac{1-\cos 135°}{\sin 135°}$$

$$= \frac{1-\left(-\frac{\sqrt{2}}{2}\right)}{\frac{\sqrt{2}}{2}} = \frac{1+\frac{\sqrt{2}}{2}}{\frac{\sqrt{2}}{2}}$$

$$= \frac{\frac{2+\sqrt{2}}{2}}{\frac{\sqrt{2}}{2}} = \frac{2+\sqrt{2}}{\sqrt{2}} = 1+\sqrt{2}$$

EXAMPLE 9

Verify the identity $2\sin^2(x/2)\tan x = \tan x - \sin x$.

Solution

Let's simplify the left side of the equation.

$$\begin{aligned} 2\sin^2\frac{x}{2}\tan x &= 2\left(\pm\sqrt{\frac{1-\cos x}{2}}\right)^2\tan x \\ &= 2\left(\frac{1-\cos x}{2}\right)\tan x \\ &= (1-\cos x)\tan x \\ &= \tan x - \tan x\cos x \\ &= \tan x - \left(\frac{\sin x}{\cos x}\right)\cos x \\ &= \tan x - \sin x \end{aligned}$$

Finally, let's consider three different approaches for solving an equation that involves a half-angle formula.

EXAMPLE 10

Solve $4\cos^2(x/2) = 1$, where $0 \le x < 2\pi$.

Solution A

First, let's recognize that if $0 \le x < 2\pi$, then $0 \le x/2 < \pi$. In other words, we want to find the solutions for $x/2$ that are greater than or equal to 0 and less than π.

$$\begin{aligned} 4\cos^2\frac{x}{2} &= 1 \\ \cos^2\frac{x}{2} &= \frac{1}{4} \\ \cos\frac{x}{2} = \frac{1}{2} \quad &\text{or} \quad \cos\frac{x}{2} = -\frac{1}{2} \end{aligned}$$

If $\cos(x/2) = \frac{1}{2}$, then $x/2 = \pi/3$, and therefore $x = 2\pi/3$. If $\cos(x/2) = -\frac{1}{2}$, then $x/2 = 2\pi/3$, and therefore $x = 4\pi/3$. The solutions are $2\pi/3$ and $4\pi/3$.

Solution B

Using the half-angle formula, we can substitute $(1 + \cos x)/2$ for $\cos^2(x/2)$.

$$\begin{aligned} 4\cos^2\frac{x}{2} &= 1 \\ 4\left(\frac{1+\cos x}{2}\right) &= 1 \\ 2(1+\cos x) &= 1 \\ 1+\cos x &= \frac{1}{2} \\ \cos x &= -\frac{1}{2} \end{aligned}$$

If $\cos x = -\frac{1}{2}$, then $x = 2\pi/3$ or $x = 4\pi/3$. Thus the solutions are $2\pi/3$ and $4\pi/3$.

Solution C

Using a graphing utility with a viewing rectangle such that $0 \le x \le 2\pi$ and $-3 \le y \le 3$, we can graph the function $f(x) = 4\cos^2(x/2) - 1$ and obtain Figure 9.11.

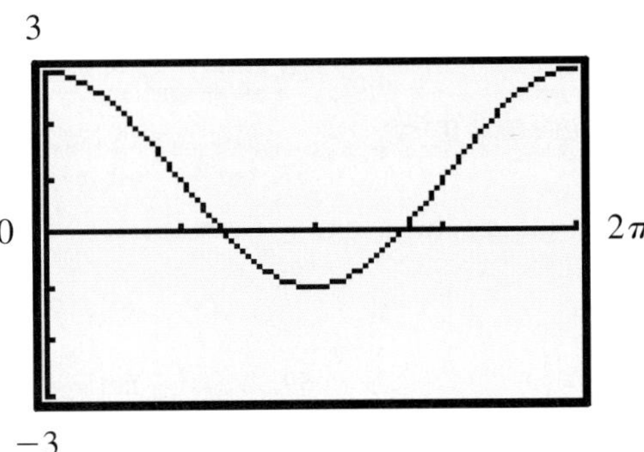

FIGURE 9.11

Now we can use the trace and zoom features to approximate the x-intercepts that are the solutions to the given equation. To the nearest hundredth, we find these to be 2.09 and 4.19. These answers agree with the previous answers of $2\pi/3$ and $4\pi/3$.

PROBLEM SET 9.4

For Problems 1–8, find the exact values for $\sin 2\theta$, $\cos 2\theta$, and $\tan 2\theta$. Do not use a calculator or a table.

1. $\cos\theta = \frac{4}{5}$ and θ is a first-quadrant angle.
2. $\sin\theta = -\frac{4}{5}$ and θ is a third-quadrant angle.
3. $\tan\theta = -\frac{12}{5}$ and θ is a second-quadrant angle.
4. $\cot\theta = \frac{12}{5}$ and θ is a first-quadrant angle.
5. $\sin\theta = -\frac{7}{25}$ and θ is a fourth-quadrant angle.
6. $\cos\theta = \frac{15}{17}$ and θ is a fourth-quadrant angle.
7. $\tan\theta = \frac{1}{2}$ and θ is a first-quadrant angle.
8. $\tan\theta = -\frac{3}{2}$ and θ is a second-quadrant angle.

For Problems 9–20, use the half-angle formulas to find exact values. Do not use a calculator or a table.

9. $\sin 15^\circ$
10. $\cos 15^\circ$
11. $\tan 15^\circ$
12. $\sin 67.5^\circ$
13. $\tan 157.5^\circ$
14. $\tan 22.5^\circ$
15. $\cos\frac{3\pi}{8}$
16. $\sin\frac{5\pi}{12}$
17. $\tan\frac{5\pi}{12}$
18. $\tan\frac{7\pi}{12}$
19. $\cos\frac{7\pi}{12}$
20. $\cos\frac{5\pi}{8}$

For Problems 21–28, find the exact values for $\sin(\theta/2)$, $\cos(\theta/2)$, and $\tan(\theta/2)$. Do not use a calculator or a table.

21. $\sin\theta = \frac{3}{5}$ and $0^\circ < \theta < 90^\circ$
22. $\sin\theta = \frac{4}{5}$ and $90^\circ < \theta < 180^\circ$

23. $\cos\theta = -\frac{3}{5}$ and $180° < \theta < 270°$

24. $\tan\theta = -\frac{5}{12}$ and $270° < \theta < 360°$

25. $\tan\theta = -\frac{12}{5}$ and $90° < \theta < 180°$

26. $\cos\theta = \frac{1}{3}$ and $0° < \theta < 90°$

27. $\sec\theta = \frac{3}{2}$ and $270° < \theta < 360°$

28. $\sec\theta = -\frac{4}{3}$ and $180° < \theta < 270°$

For Problems 29–40, solve each equation for θ, where $0° \leq \theta < 360°$. Do not use a calculator or a table.

29. $\sin 2\theta = \sin\theta$

30. $\cos 2\theta + \cos\theta = 0$

31. $\cos 2\theta + 3\sin\theta - 2 = 0$

32. $\tan 2\theta = \tan\theta$

33. $2 - \cos^2\theta = 4\sin^2\frac{\theta}{2}$

34. $\sin 2\theta \sin\theta + \cos\theta = 0$

35. $\cos\theta = \cos\frac{\theta}{2}$

36. $\sin\frac{\theta}{2} + \cos\theta = 1$

37. $\sin 4\theta = \sin 2\theta$

38. $\cos 4\theta = \cos 2\theta$

39. $\tan\frac{1}{2}\theta = \sin\theta$

40. $\sin\frac{1}{2}\theta = -\cos\frac{1}{2}\theta$

For Problems 41–50, solve each equation for x, where $0 \leq x < 2\pi$. Do not use a calculator or a table.

41. $\cos x = \sin 2x$

42. $\sin 2x + \sqrt{2}\cos x = 0$

43. $\cos 2x - 3\sin x - 2 = 0$

44. $\sin 2x - 2\cos x + \sin x - 1 = 0$

45. $\sin\frac{x}{2} + \cos x = 1$

46. $\cos 2x = 1 - \sin x$

47. $\tan 2x - 2\cos x = 0$

48. $\tan 2x + \sec 2x = 1$

49. $2 - \sin^2 x = 2\cos^2\frac{x}{2}$

50. $\cos\frac{x}{2} + \cos x = 0$

For Problems 51–70, verify each identity.

51. $\dfrac{\sin 2\theta}{1 - \cos 2\theta} = \cot\theta$

52. $(\sin\theta + \cos\theta)^2 - \sin 2\theta = 1$

53. $\dfrac{\sin 2\theta \sin\theta}{2\cos\theta} + \cos^2\theta = 1$

54. $\csc 2\theta = \dfrac{1}{2}\sec\theta\csc\theta$

55. $2\sin^2 x = \tan x \sin 2x$

56. $\dfrac{1 - \tan^2 x}{1 + \tan^2 x} = \cos 2x$

57. $\tan\dfrac{x}{2} = \csc x - \cot x$

58. $\sin^2\dfrac{x}{2} = \dfrac{\sec x - 1}{2\sec x}$

59. $2\cos^2\dfrac{x}{2}\tan x = \tan x + \sin x$

60. $\dfrac{\tan x}{1 + \tan^2 x} = \dfrac{1}{2}\sin 2x$

61. $\cot\theta \sin 2\theta = 1 + \cos 2\theta$

62. $\dfrac{\sin 2\theta}{\sin\theta} - \dfrac{\cos 2\theta}{\cos\theta} = \sec\theta$

63. $\sin 4\theta = 4\cos\theta\sin\theta(1 - 2\sin^2\theta)$

64. $\cos^4\theta - \sin^4\theta = \cos 2\theta$

65. $\sec 2\theta = \dfrac{\sec^2\theta}{2 - \sec^2\theta}$

66. $\cot 2\theta = \dfrac{\cot^2\theta - 1}{2\cot\theta}$

67. $2\sin^2 2\theta + \cos 4\theta = 1$

68. $\cos^2\theta\tan^2\theta = \dfrac{1}{2} - \dfrac{1}{2}\cos 2\theta$

69. $\sin^2\theta\cos^2\theta = \dfrac{1}{8} - \dfrac{1}{8}\cos 4\theta$

70. $\cos^4\theta = \dfrac{3}{8} + \dfrac{1}{2}\cos 2\theta + \dfrac{1}{8}\cos 4\theta$

71. Express $\cos 3\theta$ in terms of $\cos\theta$.

72. Express $\cos 4\theta$ in terms of $\cos\theta$.

73. Express $\cos 6\theta$ in terms of $\cos\theta$.

THOUGHTS into WORDS

74. How would you convince someone that $\sin \frac{1}{2}x = \frac{1}{2} \sin x$ is not an identity? Does $\sin \frac{1}{2}x = \frac{1}{2} \sin x$ for any values of x? Defend your answer.

75. Describe how you would solve the equation $\sin \frac{1}{2}\theta = \frac{1}{2}$, where $0° \le \theta < 360°$.

GRAPHICS CALCULATOR ACTIVITIES

Use a graphics calculator to help solve each of the following equations. Express your answers to the nearest tenth.

76. $\sin \frac{1}{2}x = x + 1$

77. $\cos \frac{1}{2}x = x$

78. $\sin 2x = x - 1$

79. $\cos \frac{1}{2}x = -2x$

9.5 TRIGONOMETRIC FORM OF COMPLEX NUMBERS

Recall that any number of the form $a + bi$, where a and b are real numbers and i is the imaginary unit with the property $i^2 = -1$, is called a **complex number**. As we have seen, real numbers can be represented geometrically by points on a coordinate line. Complex numbers can also be represented geometrically, but by points in a plane. Every complex number $a + bi$ determines an ordered pair of real numbers (a, b), where a is called the **real part** and b the **imaginary part** of the complex number. Therefore, using a rectangular coordinate system with the horizontal axis representing the **real axis** and the vertical axis representing the **imaginary axis**, we can associate points of the plane with complex numbers as illustrated in Figure 9.12. Note that the complex number $5 + 2i$ is represented by the point $(5, 2)$, the number $-6 + i$ is represented by the point $(-6, 1)$, and so on. A coordinate plane with complex numbers assigned to the points in this manner is called a **complex plane**. In this context, the form $a + bi$ of a complex number is referred to as the **rectangular form** of the number.

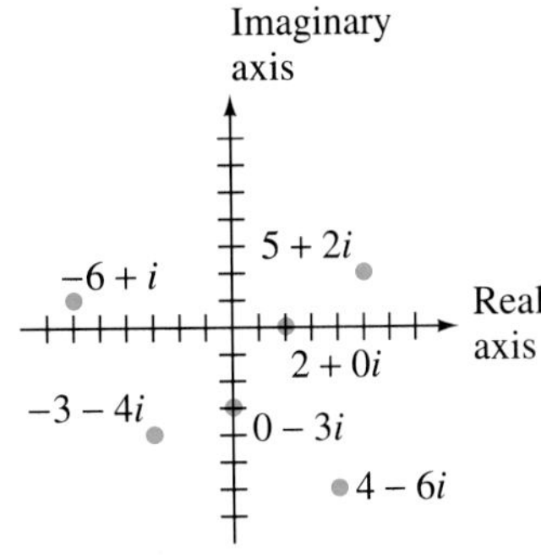

FIGURE 9.12

The absolute value of a real number can be geometrically interpreted as the distance between 0 and the number on the real number line. It is natural, therefore, to interpret the absolute value of a complex number as the distance between the origin and the point representing the complex number in the complex plane. Specifically, the absolute value of a complex number is defined as follows.

DEFINITION 9.1

The **absolute value** of a complex number $a + bi$ is denoted by $|a + bi|$ and it is defined by

$$|a + bi| = \sqrt{a^2 + b^2}.$$

EXAMPLE 1

Compute

a. $|-5 + 12i|$, **b.** $|2 - 4i|$, **c.** $|3i|$.

Solutions

a. $|-5 + 12i| = \sqrt{(-5)^2 + 12^2} = \sqrt{25 + 144} = \sqrt{169} = 13$

b. $|2 - 4i| = \sqrt{2^2 + (-4)^2} = \sqrt{4 + 16} = \sqrt{20} = 2\sqrt{5}$

c. $|3i| = |0 + 3i| = \sqrt{0^2 + 3^2} = \sqrt{0 + 9} = 3$

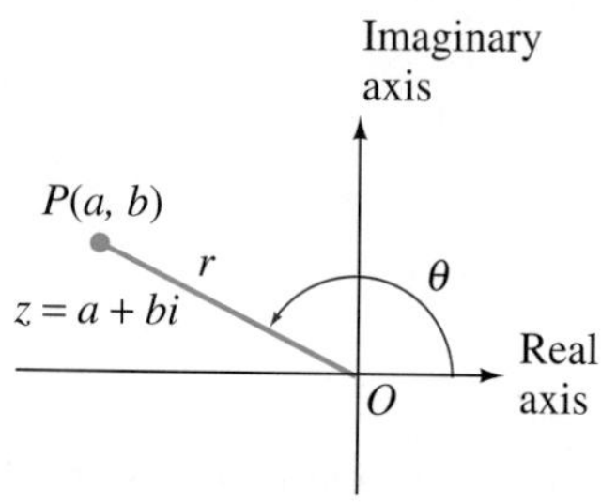

FIGURE 9.13

The geometric representation of a complex number as a point in a coordinate plane provides the basis for representing a complex number using trigonometric functions. Consider the complex number $z = a + bi$ as illustrated in Figure 9.13. Let θ be an angle in standard position whose terminal side is OP and let $r = |z|$; that is, $r = \sqrt{a^2 + b^2}$. By the definition of the sine and cosine functions we have

$$\sin\theta = \frac{b}{r} \qquad \text{or} \qquad b = r\sin\theta$$

and

$$\cos\theta = \frac{a}{r} \qquad \text{or} \qquad a = r\cos\theta.$$

Therefore, the complex number $a + bi$ can be expressed as

$$a + bi = (r\cos\theta) + (r\sin\theta)i,$$

which can be written in *trigonometric form* as

$$z = a + bi = r(\cos\theta + i\sin\theta).$$

The form $r(\cos\theta + i\sin\theta)$ is also called the **polar form** of a complex number. Remember that r is the absolute value of the complex number and therefore it is always nonnegative. Angle θ can be expressed in either radian or degree measure. Furthermore, θ is not uniquely determined since $\theta \pm 2k\pi$ will also do for k any integer. We will usually take the smallest positive angle for θ when writing complex numbers in trigonometric form.

REMARK Traditionally the number r in the trigonometric form of a complex number is also called the *modulus* of the number, and angle θ is called the *argument*. Also, the expression $r(\cos\theta + i\sin\theta)$ can be abbreviated as r cis θ. However, we will not use this terminology or symbolism in this brief introduction of the complex numbers in trigonometric form.

Since a complex number can be written in either rectangular form $(a + bi)$ or trigonometric form $(r(\cos\theta + i\sin\theta))$, we need to be able to switch back and forth between the two forms. Let's consider some examples.

EXAMPLE 2

Express the following complex numbers in trigonometric form, where $0 \le \theta < 2\pi$.

a. $3 + 3i$ **b.** $-2\sqrt{3} - 2i$

Solutions

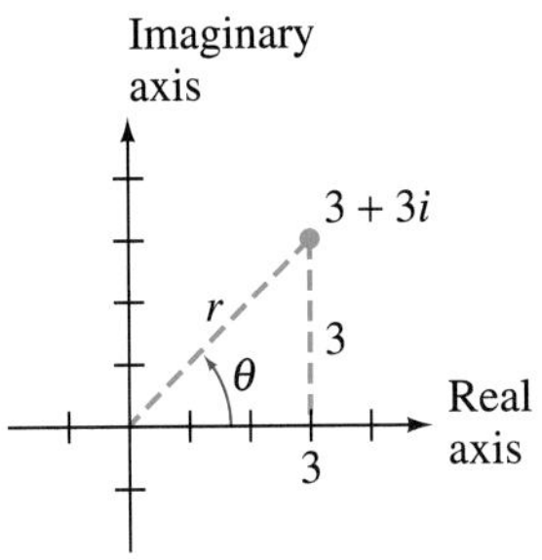

FIGURE 9.14

a. The complex number $3 + 3i$ is represented geometrically in Figure 9.14. Since the right triangle indicated in the figures is isosceles, $\theta = \pi/4$. Then using either the Pythagorean theorem or the distance formula, we can find the value of r.

$$r = \sqrt{3^2 + 3^2} = \sqrt{18} = 3\sqrt{2}.$$

Therefore,

$$3 + 3i = 3\sqrt{2}\left(\cos\frac{\pi}{4} + i\sin\frac{\pi}{4}\right).$$

b. The complex number $-2\sqrt{3} - 2i$ is graphed in Figure 9.15.

$$r = \sqrt{\left(-2\sqrt{3}\right)^2 + (-2)^2} = \sqrt{12 + 4} = \sqrt{16} = 4.$$

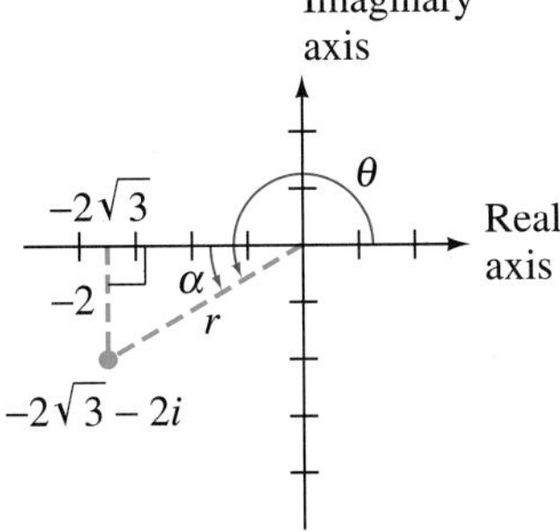

FIGURE 9.15

Now we should recognize that $\alpha = \pi/6$ and thus $\theta = \pi + \pi/6 = 7\pi/6$. Therefore,

$$-2\sqrt{3} - 2i = 4\left(\cos\frac{7\pi}{6} + i\sin\frac{7\pi}{6}\right).$$

EXAMPLE 3

Express the following complex numbers in trigonometric form, where $0° \le \theta < 360°$.

a. $0 - 4i$ **b.** $5 + 2i$

Solutions

a. The complex number $0 - 4i$ is graphed in Figure 9.16. It is evident from the figure that $r = 4$ and $\theta = 270°$. Therefore,

$$0 - 4i = 4(\cos 270° + i \sin 270°).$$

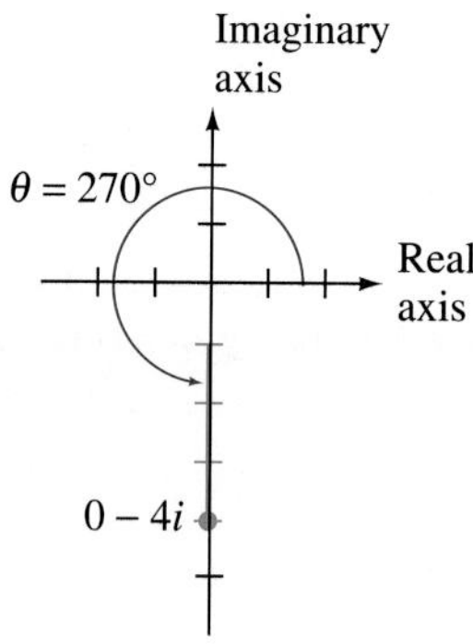

FIGURE 9.16

b. The complex number $5 + 2i$ is represented in Figure 9.17.

$$r = \sqrt{5^2 + 2^2} = \sqrt{29}$$

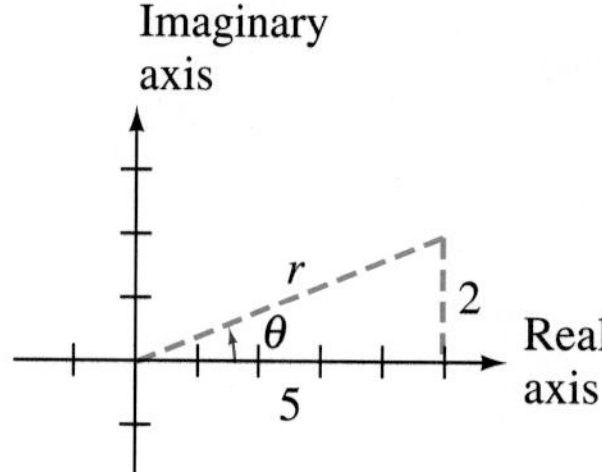

FIGURE 9.17

From the figure we see that $\tan \theta = \frac{2}{5} = 0.4$. Using a calculator we can find that $\theta = 21.8°$, to the nearest tenth of a degree. Therefore,

$$5 + 2i = \sqrt{29}(\cos 21.8° + i \sin 21.8°).$$

EXAMPLE 4

Change the complex number $8(\cos 60° + i \sin 60°)$ to $a + bi$ form.

Solution

$$\begin{aligned} 8(\cos 60° + i \sin 60°) &= 8\left(\frac{1}{2} + i\frac{\sqrt{3}}{2}\right) \\ &= 4 + 4\sqrt{3}i \end{aligned}$$

Multiplying Complex Numbers in Trigonometric Form

In Section 1.8 we discussed the multiplication of complex numbers in the form $a + bi$. For example, the product of $2 + 2i$ and $-3 + 3i$ was handled as follows.

$$\begin{aligned}(2 + 2i)(-3 + 3i) &= 2(-3 + 3i) + 2i(-3 + 3i) \\ &= -6 + 6i - 6i + 6i^2 \\ &= -6 + 6(-1)i \\ &= -12 + 0i\end{aligned}$$

Let's consider the product of two complex numbers using their trigonometric forms. Let

$$z_1 = r_1(\cos\theta_1 + i\sin\theta_1) \qquad \text{and} \qquad z_2 = r_2(\cos\theta_2 + i\sin\theta_2)$$

be the two numbers. Therefore,

$$\begin{aligned}z_1 z_2 &= [r_1(\cos\theta_1 + i\sin\theta_1)][r_2(\cos\theta_2 + i\sin\theta_2)] \\ &= r_1 r_2[\cos\theta_1\cos\theta_2 + i\cos\theta_1\sin\theta_2 + i\sin\theta_1\cos\theta_2 + i^2\sin\theta_1\sin\theta_2] \\ &= r_1 r_2[\cos\theta_1\cos\theta_2 - \sin\theta_1\sin\theta_2 + i(\cos\theta_1\sin\theta_2 + \sin\theta_1\cos\theta_2)].\end{aligned}$$

Now applying the sum formulas for $\cos(\theta_1 + \theta_2)$ and $\sin(\theta_1 + \theta_2)$, we obtain the following description for multiplying complex numbers in trigonometric form.

$$z_1 z_2 = r_1 r_2[\cos(\theta_1 + \theta_2) + i\sin(\theta_1 + \theta_2)]$$

EXAMPLE 5

Find the product of $(2 + 2i)$ and $(-3 + 3i)$ by using their trigonometric forms.

Solution

Let $z_1 = 2 + 2i$ and $z_2 = -3 + 3i$. Their trigonometric forms are as follows.

$$z_1 = 2 + 2i = 2\sqrt{2}\left(\cos\frac{\pi}{4} + i\sin\frac{\pi}{4}\right)$$

$$z_2 = -3 + 3i = 3\sqrt{2}\left(\cos\frac{3\pi}{4} + i\sin\frac{3\pi}{4}\right)$$

Therefore,

$$\begin{aligned}z_1 z_2 &= \left(2\sqrt{2}\right)\left(3\sqrt{2}\right)\left[\cos\left(\frac{\pi}{4} + \frac{3\pi}{4}\right) + i\sin\left(\frac{\pi}{4} + \frac{3\pi}{4}\right)\right] \\ &= 12[\cos\pi + i\sin\pi] \\ &= 12[-1 + i(0)] \\ &= -12 + 0i. \qquad \text{This agrees with our earlier result!}\end{aligned}$$

Example 5 demonstrates that multiplying complex numbers in trigonometric form is quite easy. Basically, we multiply absolute values and add angles. Most of us would probably agree that multiplying complex numbers in the $a + bi$ form is not very difficult; therefore, it may seem as if the trigonometric form has not provided us with much extra firepower. This is indeed true at this time; however, in the next section we will find that the trigonometric form is very convenient for finding powers and roots of complex numbers.

PROBLEM SET 9.5

For Problems 1–12, plot each complex number and find its absolute value.

1. $3 + 4i$ **2.** $-4 + 3i$ **3.** $-5 - 12i$

4. $12 - 5i$ **5.** $0 - 5i$ **6.** $-4 + 0i$

7. $1 - 2i$ **8.** $-1 - i$ **9.** $-2 + 3i$

10. $3 - 2i$ **11.** $\frac{3}{5} - \frac{4}{5}i$ **12.** $-\frac{5}{13} - \frac{12}{13}i$

For Problems 13–22, express each complex number in trigonometric form, where $0 \leq \theta < 2\pi$.

13. $-2 + 2i$ **14.** $-4 - 4i$ **15.** $0 - 3i$

16. $-4 + 0i$ **17.** $2\sqrt{3} - 2i$ **18.** $-3\sqrt{3} + 3i$

19. $-1 - i$ **20.** $1 - i$ **21.** $-1 + \sqrt{3}i$

22. $2 - 2\sqrt{3}i$

For Problems 23–32, express each complex number in trigonometric form, where $0 \leq \theta < 360°$.

23. $5 - 5i$ **24.** $6 + 6i$

25. $-2 + 0i$ **26.** $0 + 7i$

27. $\sqrt{3} + i$ **28.** $-\sqrt{3} - i$

29. $-4 - 4\sqrt{3}i$ **30.** $5 - 5\sqrt{3}i$

31. $6\sqrt{3} - 6i$ **32.** $-7\sqrt{3} + 7i$

For Problems 33–40, express each complex number in trigonometric form, where $0° \leq \theta < 360°$. Express θ to the nearest tenth of a degree.

33. $2 + 3i$ **34.** $4 - 3i$ **35.** $-2 - i$

36. $-5 + 3i$ **37.** $4 - i$ **38.** $2 + i$

39. $-2 + 4i$ **40.** $-6 - 3i$

For Problems 41–50, change the given complex number from trigonometric form to $a + bi$ form.

41. $4(\cos 30° + i \sin 30°)$ **42.** $5(\cos 120° + i \sin 120°)$

43. $3\left(\cos \frac{5\pi}{4} + i \sin \frac{5\pi}{4}\right)$ **44.** $1\left(\cos \frac{11\pi}{6} + i \sin \frac{11\pi}{6}\right)$

45. $2\left(\cos \frac{4\pi}{3} + i \sin \frac{4\pi}{3}\right)$ **46.** $\frac{1}{2}\left(\cos \frac{5\pi}{6} + i \sin \frac{5\pi}{6}\right)$

47. $\frac{2}{3}\left(\cos \frac{5\pi}{3} + i \sin \frac{5\pi}{3}\right)$ **48.** $7(\cos \pi + i \sin \pi)$

49. $6(\cos 0° + i \sin 0°)$ **50.** $8(\cos 60° + i \sin 60°)$

For Problems 51–60, find the product z_1z_2 by using the trigonometric form of the numbers. Express the final result in $a + bi$ form. Check by using the methods of Section 1.8.

51. $z_1 = \sqrt{3} + i, \quad z_2 = -2\sqrt{3} - 2i$

52. $z_1 = -3\sqrt{3} - 3i, \quad z_2 = 2\sqrt{3} + 2i$

53. $z_1 = 5\sqrt{3} + 5i, \quad z_2 = 6\sqrt{3} + 6i$

54. $z_1 = 1 + \sqrt{3}i, \quad z_2 = -\frac{1}{2} - \frac{\sqrt{3}}{2}i$

55. $z_1 = \frac{1}{2} + \frac{\sqrt{3}}{2}i, \quad z_2 = -\frac{3}{2} - \frac{3\sqrt{3}}{2}i$

56. $z_1 = -1 + i, \quad z_2 = 1 + i$

57. $z_1 = -2 - 2\sqrt{3}i, \quad z_2 = 0 + 5i$

58. $z_1 = 0 + 4i, \quad z_2 = 0 - 7i$

59. $z_1 = 8 + 0i, \quad z_2 = 0 - 3i$

60. $z_1 = -\frac{5\sqrt{2}}{2} + \frac{5\sqrt{2}}{2}i, \quad z_2 = -3 + 0i$

For Problems 61–66, find the product z_1z_2 by using the trigonometric form of the numbers. Express the final results in trigonometric form.

61. $z_1 = 5(\cos 20° + i \sin 20°)$,
$z_2 = 4(\cos 55° + i \sin 55°)$

62. $z_1 = 3(\cos 110° + i \sin 110°)$,
$z_2 = (\cos 28° + i \sin 28°)$

63. $z_1 = \sqrt{2}(\cos 120° + i \sin 120°)$,
$z_2 = 3\sqrt{2}(\cos 310° + i \sin 310°)$

64. $z_1 = 2\sqrt{3}(\cos 260° + i \sin 260°)$,
$z_2 = 4\sqrt{3}(\cos 320° + i \sin 320°)$

65. $z_1 = 5\left(\cos \frac{3\pi}{5} + i \sin \frac{3\pi}{5}\right)$,
$z_2 = 7\left(\cos \frac{\pi}{2} + i \sin \frac{\pi}{2}\right)$

66. $z_1 = 6\left(\cos \frac{3\pi}{4} + i \sin \frac{3\pi}{4}\right)$,
$z_2 = 4\left(\cos \frac{2\pi}{3} + i \sin \frac{2\pi}{3}\right)$

MISCELLANEOUS PROBLEMS

67. If $z_1 = r_1(\cos \theta_1 + i \sin \theta_1)$ and $z_2 = r_2(\cos \theta_2 + i \sin \theta_2)$, then verify that $z_1/z_2 = (r_1/r_2)[\cos(\theta_1 - \theta_2) + i \sin(\theta_1 - \theta_2)]$.

For Problems 68–73, find the quotient z_1/z_2 using the trigonometric form of the numbers (see Problem 67). Express the final quotient in $a + bi$ form.

68. $z_1 = 1 + i, \quad z_2 = 0 + i$

69. $z_1 = 2 - 2i, \quad z_2 = 0 + 3i$

70. $z_1 = -1 + i, \quad z_2 = 1 + i$

71. $z_1 = 1 - i, \quad z_2 = -1 - i$

72. $z_1 = -1 + \sqrt{3}i, \quad z_2 = -1 - \sqrt{3}i$

73. $z_1 = 3 + 3\sqrt{3}i, \quad z_2 = -\frac{3\sqrt{3}}{2} - \frac{3}{2}i$

9.6 POWERS AND ROOTS OF COMPLEX NUMBERS

By repeated application of the principle for multiplying complex numbers in trigonometric form, the powers of a complex number $z = r(\cos \theta + i \sin \theta)$ can be easily obtained.

$$z^2 = z \cdot z = [r(\cos \theta + i \sin \theta)][r(\cos \theta + i \sin \theta)] = r^2(\cos 2\theta + i \sin 2\theta)$$

$$z^3 = z^2 \cdot z = [r^2(\cos 2\theta + i \sin 2\theta)][r(\cos \theta + i \sin \theta)] = r^3(\cos 3\theta + i \sin 3\theta)$$

$$z^4 = z^3 \cdot z = [r^3(\cos 3\theta + i \sin 3\theta)][r(\cos \theta + i \sin \theta)] = r^4(\cos 4\theta + i \sin 4\theta)$$

In general, the following result, named after the French mathematician Abraham De Moivre (1667–1754), can be stated.

De Moivre's Theorem

For every positive integer n,

$$[r(\cos \theta + i \sin \theta)]^n = r^n(\cos n\theta + i \sin n\theta).$$

A complete proof of De Moivre's theorem can be given by using the principle of mathematical induction we discuss in Chapter 13.

EXAMPLE 1

Find $(1 + i)^{16}$.

Solution

First, let's change $1 + i$ to trigonometric form.

$$1 + i = \sqrt{2}\left(\cos\frac{\pi}{4} + i\sin\frac{\pi}{4}\right)$$

Next, we can apply De Moivre's theorem.

$$\begin{aligned}(1 + i)^{16} &= (2^{1/2})^{16}\left[\cos 16\left(\frac{\pi}{4}\right) + i\sin 16\left(\frac{\pi}{4}\right)\right]\\ &= 2^8[\cos 4\pi + i\sin 4\pi]\\ &= 256[\cos 4\pi + i\sin 4\pi]\end{aligned}$$

Finally, we can change back to $a + bi$ form.

$$\begin{aligned}(1 + i)^{16} &= 256[\cos 4\pi + i\sin 4\pi]\\ &= 256(1) + 256(i)(0)\\ &= 256 + 0i\end{aligned}$$

EXAMPLE 2

Find $\left(-\frac{\sqrt{3}}{2} - \frac{1}{2}i\right)^{20}$.

Solution

$$-\frac{\sqrt{3}}{2} - \frac{1}{2}i = 1(\cos 210° + i\sin 210°)$$

Therefore,

$$\begin{aligned}\left(-\frac{\sqrt{3}}{2} - \frac{1}{2}i\right)^{20} &= 1^{20}[\cos 20(210°) + i\sin 20(210°)]\\ &= 1[\cos 4200° + i\sin 4200°]\\ &= 1[\cos 240° + i\sin 240°]\\ &= 1\left(-\frac{1}{2}\right) + (1)(i)\left(-\frac{\sqrt{3}}{2}\right)\\ &= -\frac{1}{2} - \frac{\sqrt{3}}{2}i.\end{aligned}$$

Finding Roots of Complex Numbers

Now suppose that we wanted to find the two square roots of the complex number $2 + 2\sqrt{3}i$. First, let's express $2 + 2\sqrt{3}i$ in trigonometric form.

$$2 + 2\sqrt{3}i = 4\left(\cos\frac{\pi}{3} + i\sin\frac{\pi}{3}\right)$$

If $s(\cos\alpha + i\sin\alpha)$ is one of the square roots, then

$$[s(\cos\alpha + i\sin\alpha)]^2 = 4\left(\cos\frac{\pi}{3} + i\sin\frac{\pi}{3}\right).$$

Now we can apply De Moivre's theorem to the left side.

$$s^2(\cos 2\alpha + i\sin 2\alpha) = 4\left(\cos\frac{\pi}{3} + i\sin\frac{\pi}{3}\right)$$

Since equal complex numbers have equal absolute values, we obtain

$$s^2 = 4$$
$$s = 2. \qquad \text{s must be nonnegative.}$$

It also follows that $\cos 2\alpha = \cos(\pi/3)$ and $\sin 2\alpha = \sin(\pi/3)$. For this to be true where $0 \le \alpha < 2\pi$, we have

$$2\alpha = \frac{\pi}{3} \qquad \text{or} \qquad 2\alpha = \frac{\pi}{3} + 2\pi = \frac{7\pi}{3}.$$

Therefore, $\alpha = \pi/6$ or $\alpha = 7\pi/6$. Thus, the two square roots are as follows.

$$\begin{aligned} 2\left(\cos\frac{\pi}{6} + i\sin\frac{\pi}{6}\right) &= 2\left(\frac{\sqrt{3}}{2}\right) + 2i\left(\frac{1}{2}\right) \\ &= \sqrt{3} + i \end{aligned}$$

$$\begin{aligned} 2\left(\cos\frac{7\pi}{6} + i\sin\frac{7\pi}{6}\right) &= 2\left(-\frac{\sqrt{3}}{2}\right) + 2i\left(-\frac{1}{2}\right) \\ &= -\sqrt{3} - i \end{aligned}$$

Let's check both of these roots.

$$\begin{aligned} \left(\sqrt{3} + i\right)^2 &= 3 + 2\sqrt{3}i + i^2 \\ &= 3 + 2\sqrt{3}i - 1 \\ &= 2 + 2\sqrt{3}i \end{aligned}$$

$$\begin{aligned} \left(-\sqrt{3} - i\right)^2 &= 3 + 2\sqrt{3}i + i^2 \\ &= 3 + 2\sqrt{3}i - 1 \\ &= 2 + 2\sqrt{3}i \end{aligned}$$

The previous example can be generalized as follows. If the complex number w is an nth root of the complex number z, then $w^n = z$. Substituting the trigonometric form $s(\cos\alpha + i\sin\alpha)$ for w and $r(\cos\theta + i\sin\theta)$ for z, we obtain

$$[s(\cos\alpha + i\sin\alpha)]^n = r(\cos\theta + i\sin\theta).$$

Now applying De Moivre's theorem to the left side, we obtain

$$s^n(\cos n\alpha + i\sin n\alpha) = r(\cos\theta + i\sin\theta). \tag{1}$$

Equal complex numbers have equal absolute values. Therefore, $s^n = r$ and because s and r are nonnegative,

$$s = \sqrt[n]{r}.$$

Furthermore, for equation (1) to hold,

$$\cos n\alpha + i \sin n\alpha = \cos \theta + i \sin \theta.$$

It follows that

$$\cos n\alpha = \cos \theta \qquad \text{and} \qquad \sin n\alpha = \sin \theta.$$

Since both the sine and cosine functions have a period of 2π, the last two equations are true if and only if $n\alpha$ and θ differ by a multiple of 2π. Therefore,

$$n\alpha = \theta + 2\pi k, \quad \text{where } k \text{ is an integer.}$$

Solving for α, we obtain

$$\alpha = \frac{\theta + 2\pi k}{n}.$$

Substituting $\sqrt[n]{r}$ for s and $(\theta + 2\pi k)/n$ for α in the trigonometric form for w, we obtain

$$w = \sqrt[n]{r}\left[\cos\left(\frac{\theta + 2\pi k}{n}\right) + i \sin\left(\frac{\theta + 2\pi k}{n}\right)\right].$$

If we let $k = 0, 1, 2, \ldots, n - 1$ successively, we obtain n distinct values for w, that is, n distinct nth roots of z. Furthermore, no other value of k will produce a new value for w. For example, if $k = n$, then

$$\alpha = \frac{\theta + 2\pi n}{n} = \frac{\theta}{n} + 2\pi,$$

which produces the same value for w as when $k = 0$. Similarly, $k = n + 1$ yields the same value for w as $k = 1$, and so on. Likewise, negative values of k will merely produce repeat values for w.

The previous discussion has verified the following property.

PROPERTY 9.1

If $z = r(\cos \theta + i \sin \theta)$ is any nonzero complex number and n is any positive integer, then z has precisely n distinct nth roots that are given by

$$\sqrt[n]{r}\left[\cos\left(\frac{\theta + 2\pi k}{n}\right) + i \sin\left(\frac{\theta + 2\pi k}{n}\right)\right],$$

where $k = 0, 1, 2, \ldots, n - 1$.

Property 9.1 can also be stated in terms of degree measure. It then becomes

$$\sqrt[n]{r}\left[\cos\left(\frac{\theta + k \cdot 360°}{n}\right) + i \sin\left(\frac{\theta + k \cdot 360°}{n}\right)\right],$$

where $k = 0, 1, 2, \ldots, n - 1$.

Now let's use the degree interpretation of Property 9.1 to find the two square roots of $2 + 2\sqrt{3}i$.

EXAMPLE 3

Find the two square roots of $2 + 2\sqrt{3}i$.

Solution

First, let's express the given number in trigonometric form.

$$2 + 2\sqrt{3}i = 4(\cos 60° + i \sin 60°)$$

With $n = 2$, the square roots are given by

$$\sqrt{4}\left[\cos\left(\frac{60° + k \cdot 360°}{2}\right) + i \sin\left(\frac{60° + k \cdot 360°}{2}\right)\right],$$

which simplifies to

$$2[\cos(30° + k \cdot 180°) + i \sin(30° + k \cdot 180°)].$$

Substituting 0 and 1 for k yields the following square roots.

$$2(\cos 30° + i \sin 30°) = 2\left(\frac{\sqrt{3}}{2}\right) + 2i\left(\frac{1}{2}\right) = \sqrt{3} + i$$

$$2(\cos 210° + i \sin 210°) = 2\left(-\frac{\sqrt{3}}{2}\right) + 2i\left(-\frac{1}{2}\right) = -\sqrt{3} - i$$

EXAMPLE 4

Find the four fourth roots of $-8 - 8\sqrt{3}i$.

Solution

First, let's express the given number in trigonometric form.

$$-8 - 8\sqrt{3}i = 16(\cos 240° + i \sin 240°)$$

With $n = 4$, the fourth roots are given by

$$\sqrt[4]{16}\left[\cos\left(\frac{240° + k \cdot 360°}{4}\right) + i \sin\left(\frac{240° + k \cdot 360°}{4}\right)\right],$$

which simplifies to

$$2[\cos(60° + k \cdot 90°) + i \sin(60° + k \cdot 90°)].$$

Substituting 0, 1, 2, and 3 for k yields the following fourth roots.

$$2(\cos 60° + i \sin 60°) = 2\left(\frac{1}{2}\right) + 2i\left(\frac{\sqrt{3}}{2}\right) = 1 + \sqrt{3}i$$

$$2(\cos 150° + i \sin 150°) = 2\left(-\frac{\sqrt{3}}{2}\right) + 2i\left(\frac{1}{2}\right) = -\sqrt{3} + i$$

$$2(\cos 240° + i \sin 240°) = 2\left(-\frac{1}{2}\right) + 2i\left(-\frac{\sqrt{3}}{2}\right) = -1 - \sqrt{3}i$$

$$2(\cos 330° + i \sin 330°) = 2\left(\frac{\sqrt{3}}{2}\right) + 2i\left(-\frac{1}{2}\right) = \sqrt{3} - i$$

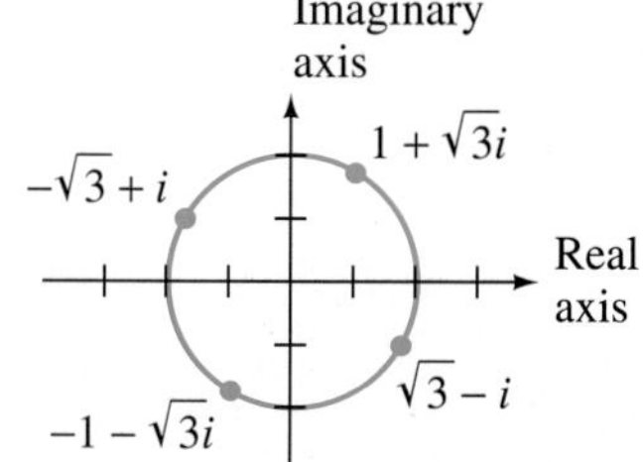

FIGURE 9.18

Note in Example 4 that each root has an absolute value of 2. Therefore, geometrically they are all located on a circle with its center at the origin and having a radius of length 2 units as illustrated in Figure 9.18. Furthermore, the points associated with the roots are equally spaced around the circle. In general, the nth roots of a complex number $z = r(\cos \theta + i \sin \theta)$ are equally spaced around a circle with a radius of length $\sqrt[n]{r}$ and its center at the origin.

EXAMPLE 5

Find the three cube roots of $8i$.

Solution

In trigonometric form, $0 + 8i = 8[\cos(\pi/2) + i \sin(\pi/2)]$. With $n = 3$, the cube roots are given by

$$\sqrt[3]{8}\left[\cos\left(\frac{\pi/2 + 2\pi k}{3}\right) + i \sin\left(\frac{\pi/2 + 2\pi k}{3}\right)\right],$$

which simplifies to

$$2\left[\cos\left(\frac{\pi}{6} + \frac{2\pi k}{3}\right) + i \sin\left(\frac{\pi}{6} + \frac{2\pi k}{3}\right)\right].$$

Substituting 0, 1, and 2 for k yields the following cube roots.

$$2\left(\cos \frac{\pi}{6} + i \sin \frac{\pi}{6}\right) = 2\left(\frac{\sqrt{3}}{2}\right) + 2i\left(\frac{1}{2}\right) = \sqrt{3} + i$$

$$2\left(\cos \frac{5\pi}{6} + i \sin \frac{5\pi}{6}\right) = 2\left(-\frac{\sqrt{3}}{2}\right) + 2i\left(\frac{1}{2}\right) = -\sqrt{3} + i$$

$$2\left(\cos \frac{3\pi}{2} + i \sin \frac{3\pi}{2}\right) = 2(0) + 2i(-1) = 0 - 2i$$

EXAMPLE 6

Find the five fifth roots of -243.

Solution

In trigonometric form, $-243 + 0i = 243(\cos 180° + i \sin 180°)$. With $n = 5$, the fifth roots are given by

$$\sqrt[5]{243}\left[\cos\left(\frac{180° + k \cdot 360°}{5}\right) + i \sin\left(\frac{180° + k \cdot 360°}{5}\right)\right],$$

which simplifies to

$$3[\cos(36° + k \cdot 72°) + i \sin(36° + k \cdot 72°)].$$

Substituting 0, 1, 2, 3, and 4 for k produces the following fifth roots.

$$3(\cos 36° + i \sin 36°)$$

$$3(\cos 108° + i \sin 108°)$$

$$3(\cos 180° + i \sin 180°) = -3 + 0i$$

$$3(\cos 252° + i \sin 252°)$$

$$3(\cos 324° + i \sin 324°)$$

There is one exact root, namely, $-3 + 0i$. Approximations for the other four roots could be obtained by using a calculator or a table. We will leave them in trigonometric form.

PROBLEM SET 9.6

For Problems 1–16, use De Moivre's theorem to find the indicated powers. Express results in $a + bi$ form.

1. $(1 + i)^{20}$

2. $(1 - i)^{12}$

3. $(-1 + i)^{10}$

4. $(-2 - 2i)^4$

5. $(3 + 3i)^5$

6. $(\sqrt{3} + i)^7$

7. $(-1 + \sqrt{3}i)^4$

8. $(-2\sqrt{3} + 2i)^5$

9. $\left(\frac{\sqrt{3}}{2} - \frac{1}{2}i\right)^{14}$

10. $\left(\frac{1}{2} - \frac{\sqrt{3}}{2}i\right)^{11}$

11. $\left(-\frac{\sqrt{2}}{2} + \frac{\sqrt{2}}{2}i\right)^{15}$

12. $\left(-\frac{\sqrt{2}}{2} - \frac{\sqrt{2}}{2}i\right)^{13}$

13. $[2(\cos 15° + i \sin 15°)]^4$

14. $[2(\cos 50° + i \sin 50°)]^6$

15. $\left(\cos \frac{\pi}{8} + i \sin \frac{\pi}{8}\right)^{10}$

16. $\left(\cos \frac{\pi}{12} + i \sin \frac{\pi}{12}\right)^8$

For Problems 17–32, find the indicated roots. Express the roots in $a + bi$ form if they are exact. Otherwise, leave them in trigonometric form.

17. The three cube roots of 8

18. The three cube roots of -27

19. The two square roots of $-16i$

20. The two square roots of $9i$

21. The four fourth roots of $-8 + 8\sqrt{3}i$

22. The four fourth roots of $-8 - 8\sqrt{3}i$

23. The five fifth roots of $1 + i$

24. The five fifth roots of $1 - i$

25. The six sixth roots of 1

26. The eight eighth roots of 1

27. The three cube roots of $-1 + \sqrt{3}i$

28. The three cube roots of $1 - \sqrt{3}i$

29. The five fifth roots of $-\sqrt{2} + \sqrt{2}i$

30. The five fifth roots of $\sqrt{2} - \sqrt{2}i$

31. The two square roots of $\frac{9}{2} + \frac{9\sqrt{3}}{2}i$

32. The two square roots of $-2 - 2\sqrt{3}i$

MISCELLANEOUS PROBLEMS

For Problems 33–40, solve each of the equations using any approach that you think is appropriate.

33. $x^4 - 16 = 0$

34. $x^3 - 27 = 0$

35. $x^4 + 16 = 0$

36. $x^3 + 27 = 0$

37. $x^5 - 1 = 0$

38. $x^5 + 1 = 0$

39. $x^6 + 1 = 0$

40. $x^6 - 1 = 0$

CHAPTER 9 SUMMARY

You should have the following basic trigonometric identities at your fingertips.

$$\csc\theta = \frac{1}{\sin\theta} \qquad \sec\theta = \frac{1}{\cos\theta} \qquad \cot\theta = \frac{1}{\tan\theta}$$

$$\tan\theta = \frac{\sin\theta}{\cos\theta} \qquad \cot\theta = \frac{\cos\theta}{\sin\theta} \qquad \sin^2\theta + \cos^2\theta = 1$$

$$1 + \tan^2\theta = \sec^2\theta \qquad 1 + \cot^2\theta = \csc^2\theta$$

These identities can be used to (1) determine other functional values from a given value, (2) simplify trigonometric expressions, (3) verify additional identities (formulas), and (4) help solve trigonometric equations.

Many of the techniques used to solve algebraic equations (such as factoring and applying the property, if $ab = 0$, then $a = 0$ or $b = 0$) also apply to the solving of trigonometric equations.

The following important identities were verified in this chapter.

$$\sin(-\theta) = -\sin\theta$$

$$\cos(-\theta) = \cos\theta$$

$$\tan(-\theta) = -\tan\theta$$

$$\cos(\alpha - \beta) = \cos\alpha\cos\beta + \sin\alpha\sin\beta$$

$\cos(\alpha + \beta) = \cos\alpha \cos\beta - \sin\alpha \sin\beta$

$\cos(90° - \alpha) = \sin\alpha$

$\sin(90° - \alpha) = \cos\alpha$

$\tan(90° - \alpha) = \cot\alpha$

$\sin(\alpha + \beta) = \sin\alpha \cos\beta + \cos\alpha \sin\beta$

$\sin(\alpha - \beta) = \sin\alpha \cos\beta - \cos\alpha \sin\beta$

$$\tan(\alpha + \beta) = \frac{\tan\alpha + \tan\beta}{1 - \tan\alpha \tan\beta}$$

$$\tan(\alpha - \beta) = \frac{\tan\alpha - \tan\beta}{1 + \tan\alpha \tan\beta}$$

$\sin 2\alpha = 2\sin\alpha \cos\alpha$

$\cos 2\alpha = \cos^2\alpha - \sin^2\alpha = 1 - 2\sin^2\alpha = 2\cos^2\alpha - 1$

$$\tan 2\alpha = \frac{2\tan\alpha}{1 - \tan^2\alpha}$$

$$\sin^2\alpha = \frac{1 - \cos 2\alpha}{2}$$

$$\cos^2\alpha = \frac{1 + \cos 2\alpha}{2}$$

$$\tan^2\alpha = \frac{1 - \cos 2\alpha}{1 + \cos 2\alpha}$$

$$\sin\frac{\alpha}{2} = \pm\sqrt{\frac{1 - \cos\alpha}{2}}$$

$$\cos\frac{\alpha}{2} = \pm\sqrt{\frac{1 + \cos\alpha}{2}}$$

$$\tan\frac{\alpha}{2} = \frac{1 - \cos\alpha}{\sin\alpha} = \frac{\sin\alpha}{1 + \cos\alpha}$$

The following relationships are apparent from Figure 9.19.

$b = r\sin\theta$

$a = r\cos\theta$

$r = \sqrt{a^2 + b^2}$

$a + bi = r(\cos\theta + i\sin\theta)$ Trigonometric form of a complex number

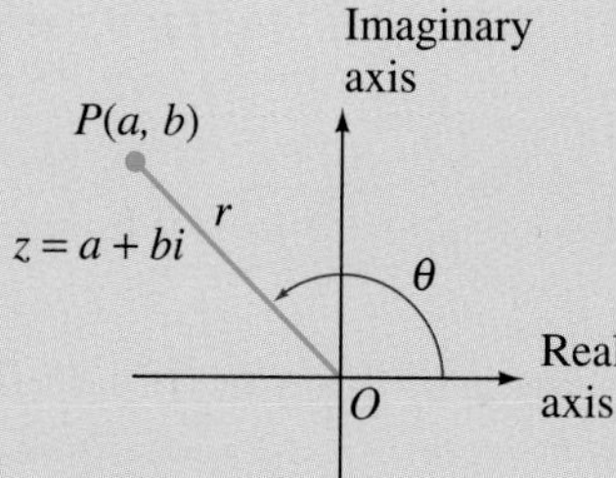

FIGURE 9.19

The product of two complex numbers z_1 and z_2, expressed in trigonometric form, is given by

$$z_1z_2 = r_1r_2[\cos(\theta_1 + \theta_2) + i\sin(\theta_1 + \theta_2)].$$

De Moivre's theorem is the basis for raising a complex number to a power.

$$[r(\cos\theta + i\sin\theta)]^n = r^n(\cos n\theta + i\sin n\theta)$$

The nth roots of a complex number $z = r(\cos\theta + i\sin\theta)$ are given by

$$\sqrt[n]{r}\left[\cos\left(\frac{\theta + 2\pi k}{n}\right) + i\sin\left(\frac{\theta + 2\pi k}{n}\right)\right],$$

where $k = 0, 1, 2, \ldots, n - 1$.

CHAPTER 9 REVIEW PROBLEM SET

1. If $\tan\theta = -\frac{5}{12}$ and $\cos\theta > 0$, find $\sin\theta$, $\sec\theta$, and $\cot\theta$.

2. If $\cos\alpha = -\frac{3}{5}$ and $\tan\beta = \frac{5}{12}$, where α is a third-quadrant angle and where β is a first-quadrant angle, find $\sin(\alpha + \beta)$, $\cos(\alpha - \beta)$, and $\tan(\alpha + \beta)$.

3. If $\sin\theta = \frac{5}{13}$ and $90° < \theta < 180°$, find $\sin 2\theta$, $\cos 2\theta$, and $\tan 2\theta$.

4. If $\cos x = \frac{4}{5}$ and $\frac{3\pi}{2} < x < 2\pi$, find $\sin\frac{x}{2}$, $\cos\frac{x}{2}$, and $\tan\frac{x}{2}$.

For Problems 5–10, find exact values. Do not use a calculator or a table.

5. $\sin 165°$ **6.** $\cos 75°$ **7.** $\sin\frac{7\pi}{12}$

8. $\tan\frac{\pi}{8}$ **9.** $\cos\left[\sin^{-1}\frac{3}{5} + \tan^{-1}\left(-\frac{4}{3}\right)\right]$

10. $\tan\left(\arcsin\frac{4}{5} - \arccos\frac{12}{13}\right)$

For Problems 11–24, solve each of the equations. If the variable is θ, find all solutions such that $0° \le \theta < 360°$. If the variable is x, find all solutions such that $0 \le x < 2\pi$. Do not use a calculator or a table.

11. $\sin^2\theta - \sin\theta = 0$

12. $2\cos^2\theta + 5\sin\theta - 4 = 0$

13. $4\sin^2\theta - 4\sin\theta + 1 = 0$

14. $\tan\theta = 2\cos\theta\tan\theta$

15. $\cos 2\theta + 3\cos\theta + 2 = 0$

16. $2\cos^2\frac{\theta}{2} - 3\cos\theta = 0$

17. $\cos\theta - \sqrt{3}\sin\theta = 1$

18. $\tan 2\theta + 2\sin\theta = 0$

19. $2\cos x + \tan x = \sec x$

20. $\cos 2x \sin x - \cos 2x = 0$

21. $2\sec x \sin x + 2 = 4\sin x + \sec x$

22. $\sin 2x = \cos 2x$

23. $\cos\frac{x}{2} = \frac{\sqrt{3}}{2}$

24. $\sin 3x + \sin x = 0$

25. Find *all* solutions of $\sec x - 1 = \tan x$ using radian measure.

26. Find *all* solutions of $2\sin\theta\tan\theta + \tan\theta - 2\sin\theta - 1 = 0$ using degree measure.

27. Solve $8\sin^2\theta + 13\sin\theta - 6 = 0$ for θ, where $0° \le \theta < 360°$. Express the solutions to the nearest tenth of a degree.

28. Solve $2\sin^2\theta - 3\sin\theta - 1 = 0$ for θ, where $0° \le \theta < 360°$. Express the solutions to the nearest tenth of a degree.

29. Solve $2\tan^2 x + 5\tan x - 12 = 0$ for x, where $0 \le x < 2\pi$. Express the solutions to the nearest hundredth of a radian.

30. Solve $2\cos^2(x/2) = \cos^2 x$ for x, where $0 \le x < 2\pi$. Express the solutions to the nearest hundredth of a radian.

For Problems 31–44, verify each of the identities.

31. $(\cot^2 x + 1)(1 - \cos^2 x) = 1$

32. $\tan x = \frac{\tan x - 1}{1 - \cot x}$

33. $\frac{1}{1 - \sin x} + \frac{1}{1 + \sin x} = 2\sec^2 x$

34. $\sin(\theta + 270°) = -\cos\theta$

35. $\tan(x + \pi) = \tan x$

36. $\cos\left(x + \frac{\pi}{4}\right) = \frac{\sqrt{2}}{2}(\cos x - \sin x)$

37. $\frac{\sin(\alpha - \beta)}{\cos(\alpha + \beta)} = \frac{\tan\alpha - \tan\beta}{1 - \tan\alpha\tan\beta}$

38. $4\sin^2\frac{\theta}{2}\cos^2\frac{\theta}{2} = \sin^2\theta$

39. $\frac{\cos x}{1 + \sin x} = \sec x - \tan x$

40. $\sin x + \cos x = \frac{1 + \cot x}{\csc x}$

41. $2\cot 2\theta = \cot\theta - \tan\theta$

42. $\cos\theta\sin 2\theta = 2\sin\theta - 2\sin^3\theta$

43. $1 - \frac{1}{2}\sin 2x = \frac{\sin^3 x + \cos^3 x}{\sin x + \cos x}$

44. $\tan\frac{\theta}{2} = \csc\theta - \cot\theta$

45. Plot the complex number $2 - 4i$ and find its absolute value.

46. Express the complex number $\sqrt{3} - i$ in trigonometric form, where $0 \le \theta < 2\pi$.

47. Express the complex number $-3\sqrt{2} - 3\sqrt{2}i$ in trigonometric form, where $0° \le \theta < 360°$.

48. Express the complex number $5[\cos(3\pi/2) + i\sin(3\pi/2)]$ in $a + bi$ form.

49. Express the complex number $8(\cos 300° + i\sin 300°)$ in $a + bi$ form

For Problems 50–52, use De Moivre's theorem to find the indicated powers. Express results in $a + bi$ form.

50. $(-1 - i)^8$

51. $\left(1 - \sqrt{3}i\right)^{10}$

52. $\left(\frac{\sqrt{2}}{2} + \frac{\sqrt{2}}{2}i\right)^{17}$

For Problems 53–55, find the indicated roots and express them in $a + bi$ form.

53. The three cube roots of $-27i$.

54. The four fourth roots of $-2 + 2\sqrt{3}i$

55. The two square roots of $8 - 8\sqrt{3}i$

CUMULATIVE REVIEW PROBLEM SET - CHAPTERS 7–9

For Problems 1–16, find exact values without using a calculator or a table.

1. $\sin 120°$ **2.** $\cos 330°$ **3.** $\tan \frac{5\pi}{4}$

4. $\sec 570°$ **5.** $\csc \frac{11\pi}{3}$ **6.** $\cot(-300°)$

7. $\cos\left(-\frac{5\pi}{6}\right)$ **8.** $\sin\left[\cos^{-1}\left(-\frac{\sqrt{3}}{2}\right)\right]$

9. $\cos\left[\sin^{-1}\left(-\frac{1}{2}\right)\right]$ **10.** $\cos\left[\tan^{-1}\left(-\sqrt{3}\right)\right]$

11. $\sin\left[\arccos\left(-\frac{3}{5}\right)\right]$ **12.** $\tan\left[\arcsin\left(-\frac{4}{5}\right)\right]$

13. $\sin 75°$ **14.** $\cos 22.5°$

15. $\tan\left[\cos^{-1}\left(-\frac{3}{5}\right) + \sin^{-1}\left(-\frac{12}{13}\right)\right]$

16. $\cos\left[\sin^{-1}\left(-\frac{5}{13}\right) - \tan^{-1}\left(\frac{3}{4}\right)\right]$

17. Find $\sec \theta$ if $\sin \theta = -\frac{7}{25}$ and θ is a third-quadrant angle.

18. Find $\csc \theta$ if the terminal side of θ lies in the second quadrant on the line $y = -\frac{1}{2}x$.

For Problems 19–26, solve each of the equations. If the variable is θ, find all solutions such that $0° \leq \theta < 360°$. If the variable is x, find all solutions such that $0 \leq x < 2\pi$. Do not use a calculator or a table.

19. $2 \sin^2 \theta + 7 \sin \theta - 4 = 0$

20. $\cos x \tan x - \cos x + 2 \tan x - 2 = 0$

21. $2 \cos^2 x + \sin x - 2 = 0$

22. $\tan \theta + 1 = \sec \theta$

23. $\cos 2x + \cos x = 0$ **24.** $\cos \frac{x}{2} + \cos x = 0$

25. $\tan 2\theta - \sqrt{3} = 0$ **26.** $\sin \frac{\theta}{2} + \cos \theta = 1$

27. Solve $12 \sin^2 \theta + \sin \theta - 6 = 0$, where $0° \leq \theta < 360°$. Express solutions to the nearest tenth of a degree.

28. Solve $\cos^2 x + 3 \cos x - 3 = 0$, where $0 \leq x < 2\pi$. Express solutions to the nearest hundredth of a radian.

29. From a point 65 feet from the base of a building, the angle of elevation to the top of the building is 51°. Find the height of the building to the nearest foot.

30. One leg of a right triangle is 5 centimeters long and the hypotenuse is 9 centimeters long. Find the measure, to the nearest tenth of a degree, of the angle opposite the 5-foot leg.

31. The lengths of the three sides of a triangle are 7 yards, 13 yards, and 18 yards. Find the measure, to the nearest tenth of a degree, of the angle opposite the longest side.

32. In $\triangle ABC$, $A = 37°$, $c = 10$ inches, and $b = 6$ inches. Find a to the nearest tenth of an inch.

33. In $\triangle ABC$, $A = 75°$, $B = 68°$, and $b = 14$ meters. Find c to the nearest tenth of a meter.

34. Two people standing 35 feet apart are in line with the base of a tower. The angle of elevation to the top of the tower from one person is 47° and from the other person is 27°. Find the height of the tower to the nearest foot.

35. A plane flies 265 miles due east after take-off, then adjusts its course 25° southward and flies another 450 miles. How far is it from point of departure? Express the answer to the nearest mile.

36. A car weighing 3200 pounds is parked on a ramp that makes a 25° angle with the horizontal. How much force must be applied to keep the car from rolling down the ramp? Express the answer to the nearest pound.

37. Change $r = 2 \cos \theta - 3 \sin \theta$ to an equation in rectangular form.

38. Change $x^2 + y^2 - 4x = 0$ to an equation in polar form.

39. Find the absolute value of the complex number $6 - 2i$.

40. Express the complex number $3 - 3i\sqrt{3}$ in trigonometric form, where $0 \leq \theta < 2\pi$.

41. Change the complex number $12(\cos 225° + i \sin 225°)$ to $a + bi$ form.

42. Find the indicated power $(1 - i)^6$ and express the result in $a + bi$ form.

43. Find the four fourth roots of $-8 - 8\sqrt{3}i$.

44. Find the period, amplitude, and phase shift for $f(x) = -2\sin(3x - 2\pi)$.

45. Find the period, amplitude, and phase shift for $f(x) = -\cos(\pi x + \pi)$.

46. Find the period and phase shift for

$$f(x) = 3\tan\left(2x + \frac{\pi}{2}\right).$$

For Problems 47–52, graph each function in the indicated interval.

47. $f(x) = -2 - \sin x, \quad 0 \le x \le 2\pi$

48. $f(x) = -\cos 2x, \quad 0 \le x \le 2\pi$

49. $f(x) = 2\sin\left(x + \frac{\pi}{2}\right), \quad -\frac{\pi}{2} \le x \le \frac{3\pi}{2}$

50. $f(x) = -\frac{1}{2}\cos\left(2x - \frac{\pi}{2}\right), \quad \frac{\pi}{4} \le x \le \frac{9\pi}{4}$

51. $f(x) = \tan\left(x + \frac{\pi}{4}\right), \quad -\frac{3\pi}{4} \le x \le \frac{\pi}{4}$

52. $f(x) = 1 + \csc 2x, \quad 0 \le x \le \pi$

53. Graph the polar equation $r = 1 + 3\cos\theta$.

54. Graph the polar equation $r = 2 - 2\sin\theta$.

For Problems 55–60, verify each of the identities.

55. $\sin x \tan x = \sec x - \cos x$

56. $\dfrac{\tan x}{\sec x - 1} = \dfrac{\sec x + 1}{\tan x}$

57. $\tan^4 x - \sec^4 x = 1 - 2\sec^2 x$

58. $\tan(x + \pi) = \tan x$

59. $\sin 3x = -4\sin^3 x + 3\sin x$

60. $\dfrac{\sin 2x}{\sin x} - \dfrac{\cos 2x}{\cos x} = \sec x$

SYSTEMS OF EQUATIONS

In this chapter we will begin by reviewing some techniques for solving systems of linear equations involving two or three variables. Then, since many applications of mathematics require the use of large numbers of variables and equations, we will introduce some additional techniques for solving such extensive systems. These new techniques also form a basis for solving systems by using a computer.

10

A system of two linear equations in two variables can be used to approximate the effect of the jet stream on airline schedules.

10.1 SYSTEMS OF TWO LINEAR EQUATIONS IN TWO VARIABLES

In Chapter 3 we stated that any equation of the form $Ax + By = C$, where A, B, and C are real numbers (A and B not both zero), is a **linear equation** in the two variables x and y, and its graph is a straight line. Two linear equations in two variables considered together form a **system of two linear equations in two variables**, as illustrated by the following examples.

$$\begin{pmatrix} x + y = 6 \\ x - y = 2 \end{pmatrix}, \quad \begin{pmatrix} 3x + 2y = 1 \\ 5x - 2y = 23 \end{pmatrix}, \quad \begin{pmatrix} 4x - 5y = 21 \\ -3x + y = -7 \end{pmatrix}$$

To solve a system (such as any of these three examples) means to find all of the ordered pairs that simultaneously satisfy both equations in the system. For example, if we graph the two equations $x + y = 6$ and $x - y = 2$ on the same set of axes, as in Figure 10.1, then the ordered pair associated with the point of intersection of

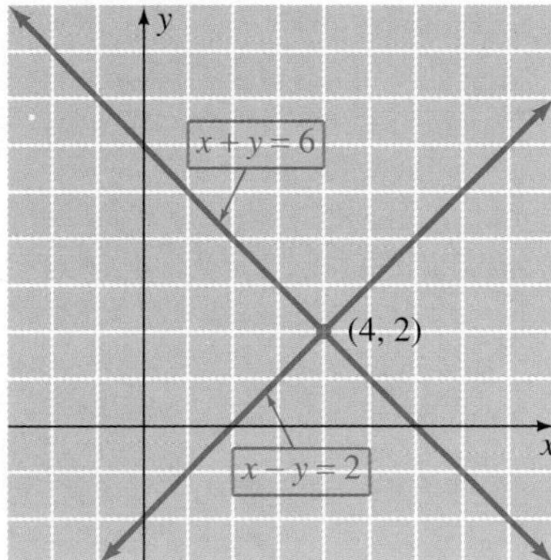

FIGURE 10.1

the two lines is the **solution of the system**. Thus, we say that $\{(4, 2)\}$ is the solution set of the system

$$\begin{pmatrix} x + y = 6 \\ x - y = 2 \end{pmatrix}.$$

To check the solution, we substitute 4 for x and 2 for y in the two equations:

$x + y = 6$ becomes $4 + 2 = 6$, a true statement;

$x - y = 2$ becomes $4 - 2 = 2$, a true statement.

Because the graph of a linear equation in two variables is astraight line, there are three possible situations that can occur when solving a system of two linear equations in two variables. Each situation is shown in Figure 10.2.

CASE 1 The graphs of the two equations are two lines intersecting in *one* point. There is exactly one solution and the system is called a **consistent system**.

CASE 2 The graphs of the two equations are parallel lines. There is *no solution* and the system is called an **inconsistent system**.

CASE 3 The graphs of the two equations are the same line and there are *infinitely many solutions* of the system. Any pair of real numbers that satisfies one of the equations will also satisfy the other equation, and we say that the equations are **dependent**.

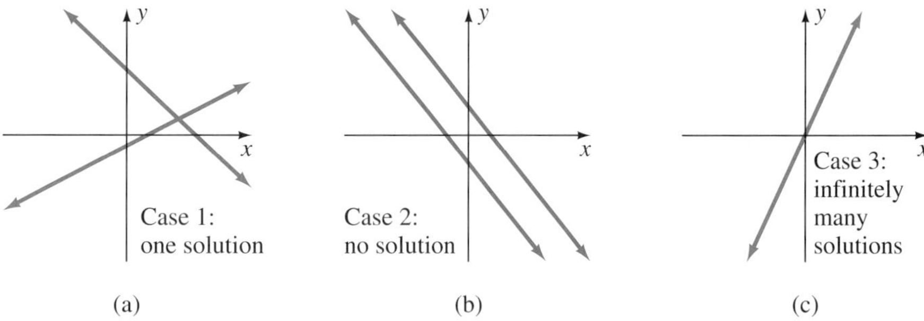

FIGURE 10.2

Thus, as we solve a system of two linear equations in two variables, we can expect one of three things: The system will have either *no* solutions, *one* ordered pair as a solution, or *infinitely many* ordered pairs as solutions.

The Substitution Method

Solving specific systems of equations by graphing requires accurate graphs. However, unless the solutions are integers, it is difficult to obtain exact solutions from a graph. Therefore we will consider some other techniques for solving systems of equations.

The **substitution method**, which works especially well with systems of two equations in two unknowns, can be described as follows.

STEP 1 Solve one of the equations for one variable in terms of the other. (If possible, make a choice that will avoid fractions.)

STEP 2 Substitute the expression obtained in step 1 into the other equation, producing an equation in one variable.

STEP 3 Solve the equation obtained in step 2.

STEP 4 Use the solution obtained in step 3, along with the expression obtained in step 1, to determine the solution of the system.

EXAMPLE 1

Solve the system

$$\begin{pmatrix} x - 3y = -25 \\ 4x + 5y = 19 \end{pmatrix}.$$

Solution

Solve the first equation for x in terms of y to produce

$$x = 3y - 25.$$

Substitute $3y - 25$ for x in the second equation and we can now solve for y.

$$\begin{aligned} 4x + 5y &= 19 \\ 4(3y - 25) + 5y &= 19 \\ 12y - 100 + 5y &= 19 \\ 17y &= 119 \\ y &= 7 \end{aligned}$$

Next, substitute 7 for y in the equation $x = 3y - 25$ to obtain

$$x = 3(7) - 25 = -4.$$

The solution set of the given system is $\{(-4, 7)\}$. (Perhaps you should check this solution in both of the original equations.)

EXAMPLE 2

Solve the system $\begin{pmatrix} 5x + 9y = -2 \\ 2x + 4y = -1 \end{pmatrix}$.

Solution

A glance at the system should tell us that solving either equation for either variable will produce a fractional form. So let's just use the first equation and solve for x in terms of y.

$$\begin{aligned} 5x + 9y &= -2 \\ 5x &= -9y - 2 \\ x &= \frac{-9y - 2}{5} \end{aligned}$$

Now we can substitute this value for x into the second equation and solve for y.

$$\begin{aligned} 2x + 4y &= -1 \\ 2\left(\frac{-9y - 2}{5}\right) + 4y &= -1 \\ 2(-9y - 2) + 20y &= -5 \qquad \text{Multiplied both sides by 5.} \\ -18y - 4 + 20y &= -5 \\ 2y - 4 &= -5 \\ 2y &= -1 \\ y &= -\frac{1}{2} \end{aligned}$$

Now we can substitute $-\frac{1}{2}$ for y in $x = \dfrac{-9y - 2}{5}$.

$$x = \frac{-9\left(-\frac{1}{2}\right) - 2}{5} = \frac{\frac{9}{2} - 2}{5} = \frac{1}{2}.$$

The solution set is $\left\{\left(\frac{1}{2}, -\frac{1}{2}\right)\right\}$.

EXAMPLE 3

Solve the system

$$\begin{pmatrix} 6x - 4y = 18 \\ y = \frac{3}{2}x - \frac{9}{2} \end{pmatrix}.$$

Solution

The second equation is given in appropriate form to begin the substitution process. Substitute $\frac{3}{2}x - \frac{9}{2}$ for y in the first equation to yield

$$\begin{aligned} 6x - 4y &= 18 \\ 6x - 4\left(\frac{3}{2}x - \frac{9}{2}\right) &= 18 \\ 6x - 6x + 18 &= 18 \\ 18 &= 18. \end{aligned}$$

Obtaining the true numerical statement $18 = 18$ indicates that the system has infinitely many solutions. Any ordered pair that satisfies one of the equations will also satisfy the other equation. Thus, in the second equation of the original system, if we let $x = k$, then $y = \frac{3}{2}k - \frac{9}{2}$. So the solution set can be expressed $\left\{\left(k, \frac{3}{2}k - \frac{9}{2}\right) \middle| k \text{ is a real number}\right\}$. If some specific solutions are needed, they can be generated by the ordered pair $\left(k, \frac{3}{2}k - \frac{9}{2}\right)$. For example, if we let $k = 1$, then we get $\frac{3}{2}(1) - \frac{9}{2} = -\frac{6}{2} = -3$. Thus, the ordered pair $(1, -3)$ is a member of the solution set of the given system.

Elimination-by-Addition Method

Now let's consider the **elimination-by-addition method** for solving a system of equations. This is a very important method since it is the basis for developing other techniques for solving systems that contain many equations and variables. The method involves the replacement of systems of equations with *simpler equivalent systems* until we obtain a system where the solutions are obvious. **Equivalent systems of equations are systems that have exactly the same solution set.** The following operations or transformations can be applied to a system of equations to produce an equivalent system.

1. Any two equations of the system can be interchanged.
2. Both sides of any equation of the system can be multiplied by any nonzero real number.

3. Any equation of the system can be replaced by the sum of that equation and a nonzero multiple of another equation.

EXAMPLE 4

Solve the system

$$\begin{pmatrix} 3x + 5y = -9 \\ 2x - 3y = 13 \end{pmatrix}. \quad \begin{matrix} \textbf{(1)} \\ \textbf{(2)} \end{matrix}$$

Solution

We can replace the given system with an equivalent system by multiplying equation (2) by -3.

$$\begin{pmatrix} 3x + 5y = -9 \\ -6x + 9y = -39 \end{pmatrix} \quad \begin{matrix} \textbf{(3)} \\ \textbf{(4)} \end{matrix}$$

Now let's replace equation (4) with an equation formed by multiplying equation (3) by 2 and adding this result to equation (4).

$$\begin{pmatrix} 3x + 5y = -9 \\ 19y = -57 \end{pmatrix}. \quad \begin{matrix} \textbf{(5)} \\ \textbf{(6)} \end{matrix}$$

From equation (6) we can easily determine that $y = -3$. Then substituting -3 for y in equation (5) produces

$$\begin{aligned} 3x + 5(-3) &= -9 \\ 3x - 15 &= -9 \\ 3x &= 6 \\ x &= 2. \end{aligned}$$

The solution set for the given system is $\{(2, -3)\}$.

REMARK We are using a format for the elimination-by-addition method that highlights the use of equivalent systems. In Section 10.3, this format will lead naturally to an approach using matrices. Thus, it is beneficial to stress the use of equivalent systems at this time.

EXAMPLE 5

Solve the system

$$\begin{pmatrix} \frac{1}{2}x + \frac{2}{3}y = -4 \\ \frac{1}{4}x - \frac{3}{2}y = 20 \end{pmatrix}. \quad \begin{matrix} \textbf{(7)} \\ \textbf{(8)} \end{matrix}$$

Solution

The given system can be replaced with an equivalent system by multiplying equation (7) by 6 and equation (8) by 4.

$$\begin{pmatrix} 3x + 4y = -24 \\ x - 6y = 80 \end{pmatrix} \quad \begin{matrix} \textbf{(9)} \\ \textbf{(10)} \end{matrix}$$

Now let's exchange equations (9) and (10).

$$\begin{pmatrix} x - 6y = 80 \\ 3x + 4y = -24 \end{pmatrix} \quad \begin{matrix} \textbf{(11)} \\ \textbf{(12)} \end{matrix}$$

We can replace equation (12) with an equation formed by multiplying equation (11) by -3 and adding this result to equation (12).

$$\begin{pmatrix} x - 6y = 80 \\ 22y = -264 \end{pmatrix} \quad \begin{matrix} \textbf{(13)} \\ \textbf{(14)} \end{matrix}$$

From equation (14) we can determine that $y = -12$. Then substituting -12 for y in equation (13) produces

$$x - 6(-12) = 80$$
$$x + 72 = 80$$
$$x = 8.$$

The solution set of the given system is $\{(8, -12)\}$. (Check this!)

EXAMPLE 6

Solve the system

$$\begin{pmatrix} x - 4y = 9 \\ x - 4y = 3 \end{pmatrix}. \quad \begin{matrix} \textbf{(15)} \\ \textbf{(16)} \end{matrix}$$

Solution

We can replace equation (16) with an equation formed by multiplying equation (15) by -1 and adding this result to equation (16).

$$\begin{pmatrix} x - 4y = 9 \\ 0 = -6 \end{pmatrix} \quad \begin{matrix} \textbf{(17)} \\ \textbf{(18)} \end{matrix}$$

The statement $0 = -6$ is a contradiction and therefore the original system is *inconsistent*; it has no solution. The solution set is $\varnothing$.

Both the elimination-by-addition and substitution methods can be used to obtain exact solutions for any system of two linear equations in two unknowns. Sometimes the issue is one of deciding which method to use on a particular system. Some systems lend themselves to one or the other of the methods by the original format of the equations. We will illustrate this idea in a moment when we solve some word problems.

Using Systems to Solve Problems

Many word problems that we solved earlier in this text with one variable and one equation can also be solved by using a system of two linear equations in two variables. In fact, in many of these problems you may find it more natural to use two variables and two equations.

The two-variable expression $10t + u$ can be used to represent any two-digit whole number. The t represents the tens digit and the u represents the units digit.

For example, if $t = 4$ and $u = 8$, then $10t + u$ becomes $10(4) + 8 = 48$. Now let's use this general representation for a two-digit number to help solve a problem.

PROBLEM 1

The units digit of a two-digit number is one more than twice the tens digit. The number with the digits reversed is 45 larger than the original number. Find the original number.

Solution

Let u represent the units digit of the original number and let t represent the tens digit. Then $10t + u$ represents the original number and $10u + t$ represents the new number with the digits reversed. The problem translates into the following system.

$$\begin{pmatrix} u = 2t + 1 \\ 10u + t = 10t + u + 45 \end{pmatrix}$$

← The units digit is one more than twice the tens digit.
← The number with the digits reversed is 45 larger than the original number.

Simplify the second equation and the system becomes

$$\begin{pmatrix} u = 2t + 1 \\ 9u - 9t = 45 \end{pmatrix}.$$

Because of the form of the first equation, this system lends itself to solution by the substitution method. Substitute $2t + 1$ for u in the second equation to produce

$$\begin{aligned} 9(2t + 1) - 9t &= 45 \\ 18t + 9 - 9t &= 45 \\ 9t &= 36 \\ t &= 4. \end{aligned}$$

Now substitute 4 for t in the equation $u = 2t + 1$ to get

$$u = 2(4) + 1 = 9.$$

The tens digit is 4 and the units digit is 9, so the number is 49.

PROBLEM 2

Lucinda invested \$950, part of it at 11% interest and the remainder at 12%. Her total yearly income from the two investments was \$111.50. How much did she invest at each rate?

Solution

Let x represent the amount invested at 11% and y the amount invested at 12%. The problem translates into the following system.

$$\begin{pmatrix} x + y = 950 \\ 0.11x + 0.12y = 111.50 \end{pmatrix}$$

← The two investments total \$950.
← The yearly interest from the two investments totals \$111.50.

Multiply the second equation by 100 to produce an equivalent system.

$$\begin{pmatrix} x + y = 950 \\ 11x + 12y = 11150 \end{pmatrix}$$

Since neither equation is solved for one variable in terms of the other, let's use the elimination-by-addition method to solve the system. The second equation can be replaced by an equation formed by multiplying the first equation by -11 and adding this result to the second equation.

$$\begin{pmatrix} x + y = 950 \\ y = 700 \end{pmatrix}$$

Now we substitute 700 for y in the equation $x + y = 950$.

$$x + 700 = 950$$
$$x = 250$$

Therefore, Lucinda must have invested \$250 at 11% and \$700 at 12%.

In our final example of this section, we will use a graphing utility to help solve a system of equations.

EXAMPLE 7

Solve the system $\begin{pmatrix} 1.14x + 2.35y = -7.12 \\ 3.26x - 5.05y = 26.72 \end{pmatrix}$.

Solution

First, we need to solve each equation for y in terms of x. Thus, the system becomes

$$\begin{pmatrix} y = \dfrac{-7.12 - 1.14x}{2.35} \\ y = \dfrac{3.26x - 26.72}{5.05} \end{pmatrix}.$$

Now we can enter both of these equations into a graphing utility and obtain Figure 10.3. In this figure it appears that the point of intersection is at approximately

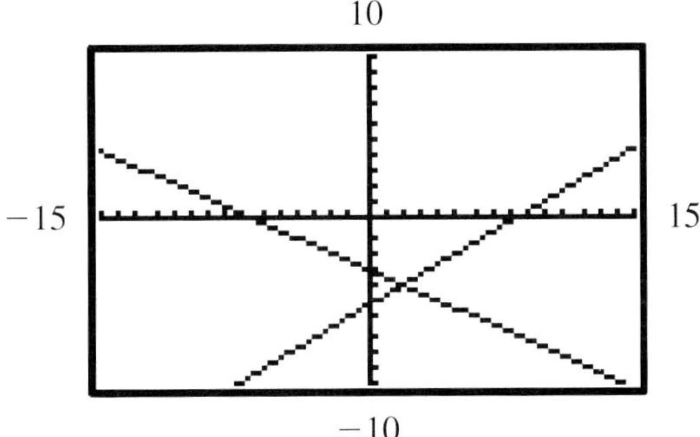

FIGURE 10.3

$x = 2$ and $y = -4$. By direct substitution into the given equations we can verify that the point of intersection is exactly $(2, -4)$.

PROBLEM SET 10.1

Solve each of the following systems by using the substitution method.

1. $\begin{pmatrix} x + y = 16 \\ y = x + 2 \end{pmatrix}$

2. $\begin{pmatrix} 2x + 3y = -5 \\ y = 2x + 9 \end{pmatrix}$

3. $\begin{pmatrix} x = 3y - 25 \\ 4x + 5y = 19 \end{pmatrix}$

4. $\begin{pmatrix} 3x - 5y = 25 \\ x = y + 7 \end{pmatrix}$

5. $\begin{pmatrix} y = \frac{2}{3}x - 1 \\ 5x - 7y = 9 \end{pmatrix}$

6. $\begin{pmatrix} y = \frac{3}{4}x + 5 \\ 4x - 3y = -1 \end{pmatrix}$

7. $\begin{pmatrix} a = 4b + 13 \\ 3a + 6b = -33 \end{pmatrix}$

8. $\begin{pmatrix} 9a - 2b = 28 \\ b = -3a + 1 \end{pmatrix}$

9. $\begin{pmatrix} 2x - 3y = 4 \\ y = \frac{2}{3}x - \frac{4}{3} \end{pmatrix}$

10. $\begin{pmatrix} t + u = 11 \\ t = u + 7 \end{pmatrix}$

11. $\begin{pmatrix} u = t - 2 \\ t + u = 12 \end{pmatrix}$

12. $\begin{pmatrix} y = 5x - 9 \\ 5x - y = 9 \end{pmatrix}$

13. $\begin{pmatrix} 4x + 3y = -7 \\ 3x - 2y = 16 \end{pmatrix}$

14. $\begin{pmatrix} 5x - 3y = -34 \\ 2x + 7y = -30 \end{pmatrix}$

15. $\begin{pmatrix} 5x - y = 4 \\ y = 5x + 9 \end{pmatrix}$

16. $\begin{pmatrix} 2x + 3y = 3 \\ 4y - 9y = -4 \end{pmatrix}$

17. $\begin{pmatrix} 4x - 5y = 3 \\ 8x + 15y = -24 \end{pmatrix}$

18. $\begin{pmatrix} 4x + y = 9 \\ y = 15 - 4x \end{pmatrix}$

Solve each of the following systems by using the elimination-by-addition method.

19. $\begin{pmatrix} 3x + 2y = 1 \\ 5x - 2y = 23 \end{pmatrix}$

20. $\begin{pmatrix} 4x + 3y = -22 \\ 4x - 5y = 26 \end{pmatrix}$

21. $\begin{pmatrix} x - 3y = -22 \\ 2x + 7y = 60 \end{pmatrix}$

22. $\begin{pmatrix} 6x - y = 3 \\ 5x + 3y = -9 \end{pmatrix}$

23. $\begin{pmatrix} 4x - 5y = 21 \\ 3x + 7y = -38 \end{pmatrix}$

24. $\begin{pmatrix} 5x - 3y = -34 \\ 2x + 7y = -30 \end{pmatrix}$

25. $\begin{pmatrix} 5x - 2y = 19 \\ 5x - 2y = 7 \end{pmatrix}$

26. $\begin{pmatrix} 3a - 2b = 5 \\ 2a + 7b = 9 \end{pmatrix}$

27. $\begin{pmatrix} 6a - 3b = 4 \\ 5a + 2b = -1 \end{pmatrix}$

28. $\begin{pmatrix} 7x + 2y = 11 \\ 7x + 2y = -4 \end{pmatrix}$

29. $\begin{pmatrix} \frac{2}{3}s + \frac{1}{4}t = -1 \\ \frac{1}{2}s - \frac{1}{3}t = -7 \end{pmatrix}$

30. $\begin{pmatrix} \frac{1}{4}s - \frac{2}{3}t = -3 \\ \frac{1}{3}s + \frac{1}{3}t = 7 \end{pmatrix}$

31. $\begin{pmatrix} \frac{x}{2} - \frac{2y}{5} = \frac{-23}{60} \\ \frac{2x}{3} + \frac{y}{4} = \frac{-1}{4} \end{pmatrix}$

32. $\begin{pmatrix} \frac{2x}{3} - \frac{y}{2} = \frac{3}{5} \\ \frac{x}{4} + \frac{y}{2} = \frac{7}{80} \end{pmatrix}$

33. $\begin{pmatrix} \frac{4x}{5} - \frac{3y}{2} = \frac{1}{5} \\ -2x + y = -1 \end{pmatrix}$

34. $\begin{pmatrix} \frac{3x}{2} - \frac{2y}{7} = -1 \\ 4x + y = 2 \end{pmatrix}$

Solve each of the following systems by either the substitution method or the elimination-by-addition method, whichever seems more appropriate.

35. $\begin{pmatrix} 5x - y = -22 \\ 2x + 3y = -2 \end{pmatrix}$

36. $\begin{pmatrix} 4x + 5y = -41 \\ 3x - 2y = 21 \end{pmatrix}$

37. $\begin{pmatrix} x = 3y - 10 \\ x = -2y + 15 \end{pmatrix}$

38. $\begin{pmatrix} y = 4x - 24 \\ 7x + y = 42 \end{pmatrix}$

39. $\begin{pmatrix} 3x - 5y = 9 \\ 6x - 10y = -1 \end{pmatrix}$

40. $\begin{pmatrix} y = \frac{2}{5}x - 3 \\ 4x - 7y = 33 \end{pmatrix}$

41. $\begin{pmatrix} \frac{1}{2}x - \frac{2}{3}y = 22 \\ \frac{1}{2}x + \frac{1}{4}y = 0 \end{pmatrix}$

42. $\begin{pmatrix} \frac{2}{5}x - \frac{1}{3}y = -9 \\ \frac{3}{4}x + \frac{1}{3}y = -14 \end{pmatrix}$

43. $\begin{pmatrix} t = 2u + 2 \\ 9u - 9t = -45 \end{pmatrix}$

44. $\begin{pmatrix} 9u - 9t = 36 \\ u = 2t + 1 \end{pmatrix}$

45. $\begin{pmatrix} x + y = 1000 \\ 0.12x + 0.14y = 136 \end{pmatrix}$

46. $\begin{pmatrix} x + y = 10 \\ 0.3x + 0.7y = 4 \end{pmatrix}$

47. $\begin{pmatrix} y = 2x \\ 0.09x + 0.12y = 132 \end{pmatrix}$

48. $\begin{pmatrix} y = 3x \\ 0.1x + 0.11y = 64.5 \end{pmatrix}$

49. $\begin{pmatrix} x + y = 10.5 \\ 0.5x + 0.8y = 7.35 \end{pmatrix}$

50. $\begin{pmatrix} 2x + y = 7.75 \\ 3x + 2y = 12.5 \end{pmatrix}$

Solve each of the following problems by using a system of equations.

51. The sum of two numbers is 53 and their difference is 19. Find the numbers.

52. The sum of two numbers is -3 and their difference is 25. Find the numbers.

53. Find two numbers such that one of them is 3 times the other and their difference is 10.

54. One number is 2 less than 3 times the other number. Find the two numbers if their sum is 26.

55. The tens digit of a two-digit number is 1 more than 3 times the units digit. If the sum of the digits is 9, find the number.

56. The units digit of a two-digit number is 1 less than twice the tens digit. The sum of the digits is 8. Find the number.

57. The sum of the digits of a two-digit number is 7. If the digits are reversed, the newly formed number is 9 larger than the original number. Find the original number.

58. The units digit of a two-digit number is 1 less than twice the tens digit. If the digits are reversed, the newly formed number is 27 larger than the original number. Find the original number.

59. A motel rents double rooms at $32 per day and single rooms at $26 per day. If 23 rooms were rented one day for a total of $688, how many rooms of each kind were rented?

60. An apartment complex rents one-bedroom apartments for $325 per month and two-bedroom apartments for $375 per month. One month the number of one-bedroom apartments was twice the number of two-bedroom apartments. If the total income for that month was $12,300, how many apartments of each kind were rented?

61. The income from a student production was $10,000. The price of a student ticket was $3 and nonstudent tickets were sold at $5 each. Three thousand tickets were sold. How many tickets of each kind were sold?

62. Eric bought 50 stamps for $12.40. Some of them were 29-cent stamps and the rest were 22-cent stamps. How many of each kind did he buy?

63. Melinda invested three times as much money at 11% yearly interest as she did at 9%. Her total yearly interest from the two investments was $210. How much did she invest at each rate?

64. Sam invested $1950, part of it at 10% and the rest at 12% yearly interest. The yearly income on the 12% investment was $6 less than twice the income from the 10% investment. How much did he invest at each rate?

65. One solution contains 40% alcohol and another solution contains 60% alcohol. How many liters of each solution should be mixed to produce 20 liters of a 52% alcohol solution?

66. One solution contains 30% alcohol and a second solution contains 70% alcohol. How many liters of each solution should be mixed to make 10 liters containing 40% alcohol?

67. Bill bought 4 tennis balls and 3 golf balls for a total of $10.25. Bret went into the same store and bought 2 tennis balls and 5 golf balls for $11.25. What was the price for a tennis ball and the price for a golf ball?

68. Six cans of pop and 2 bags of potato chips cost $5.12. At the same prices, 8 cans of pop and 5 bags of potato chips cost $9.86. Find the price per can of pop and the price per bag of potato chips.

69. A cash drawer contains only five- and ten-dollar bills. There are 12 more five-dollar bills than there are ten-dollar bills. If the drawer contains $330, find the number of each kind of bill.

70. Brad has a collection of dimes and quarters totaling $47.50. The number of quarters is ten more than twice the number of dimes. How many coins of each kind does he have?

MISCELLANEOUS PROBLEMS

A system such as

$$\begin{pmatrix} \frac{2}{x} + \frac{3}{y} = \frac{19}{15} \\ -\frac{2}{x} + \frac{1}{y} = -\frac{7}{15} \end{pmatrix}$$

is not a linear system, but it can be solved using the elimination-by-addition method as follows. Add the first equation to the second to produce the equivalent system.

$$\begin{pmatrix} \frac{2}{x} + \frac{3}{y} = \frac{19}{15} \\ \frac{4}{y} = \frac{12}{15} \end{pmatrix}$$

Now solve $\frac{4}{y} = \frac{12}{15}$ to produce $y = 5$. Substitute 5 for y in the first equation and solve for x to produce

$$\frac{2}{x} + \frac{3}{5} = \frac{19}{15}$$
$$\frac{2}{x} = \frac{10}{15}$$
$$10x = 30$$
$$x = 3.$$

The solution set of the original system is $\{(3, 5)\}$.

Solve each of the following systems in Problems 71–76.

71. $\begin{pmatrix} \frac{1}{x} + \frac{2}{y} = \frac{7}{12} \\ \frac{3}{x} - \frac{2}{y} = \frac{5}{12} \end{pmatrix}$

72. $\begin{pmatrix} \frac{3}{x} + \frac{2}{y} = 2 \\ \frac{2}{x} - \frac{3}{y} = \frac{1}{4} \end{pmatrix}$

73. $\begin{pmatrix} \frac{3}{x} - \frac{2}{y} = \frac{13}{6} \\ \frac{2}{x} + \frac{3}{y} = 0 \end{pmatrix}$

74. $\begin{pmatrix} \frac{4}{x} + \frac{1}{y} = 11 \\ \frac{3}{x} - \frac{5}{y} = -9 \end{pmatrix}$

75. $\begin{pmatrix} \frac{5}{x} - \frac{2}{y} = 23 \\ \frac{4}{x} + \frac{3}{y} = \frac{23}{2} \end{pmatrix}$

76. $\begin{pmatrix} \frac{2}{x} - \frac{7}{y} = \frac{9}{10} \\ \frac{5}{x} + \frac{4}{y} = -\frac{41}{20} \end{pmatrix}$

77. Consider the linear system $\begin{pmatrix} a_1x + b_1y = c_1 \\ a_2x + b_2y = c_2 \end{pmatrix}$.

a. Prove that this system has exactly one solution if and only if $\frac{a_1}{a_2} \neq \frac{b_1}{b_2}$.

b. Prove that this system has no solutions if and only if $\frac{a_1}{a_2} = \frac{b_1}{b_2} \neq \frac{c_1}{c_2}$.

c. Prove that this system has infinitely many solutions if and only if $\frac{a_1}{a_2} = \frac{b_1}{b_2} = \frac{c_1}{c_2}$.

78. Use the results from Problem 77 to determine whether each of the following systems is consistent, inconsistent, or dependent.

a. $\begin{pmatrix} 5x + y = 9 \\ x - 5y = 4 \end{pmatrix}$ **b.** $\begin{pmatrix} 3x - 2y = 14 \\ 2x + 3y = 9 \end{pmatrix}$

c. $\begin{pmatrix} x - 7y = 4 \\ x - 7y = 9 \end{pmatrix}$ **d.** $\begin{pmatrix} 3x - 5y = 10 \\ 6x - 10y = 1 \end{pmatrix}$

e. $\begin{pmatrix} 3x + 6y = 2 \\ \frac{3}{5}x + \frac{6}{5}y = \frac{2}{5} \end{pmatrix}$ **f.** $\begin{pmatrix} \frac{2}{3}x - \frac{3}{4}y = 2 \\ \frac{1}{2}x + \frac{2}{5}y = 9 \end{pmatrix}$

g. $\begin{pmatrix} 7x + 9y = 14 \\ 8x - 3y = 12 \end{pmatrix}$ **h.** $\begin{pmatrix} 4x - 5y = 3 \\ 12x - 15y = 9 \end{pmatrix}$

GRAPHICS CALCULATOR ACTIVITIES

79. Use your graphics calculator to help determine whether each of the systems of equations in Problem 78 is consistent, inconsistent, or dependent.

80. Use your graphics calculator to help determine the solution set for each of the following systems. Be sure to check your answers.

a. $\begin{pmatrix} y = 3x - 1 \\ y = 9 - 2x \end{pmatrix}$ **b.** $\begin{pmatrix} 5x + y = -9 \\ 3x - 2y = 5 \end{pmatrix}$

c. $\begin{pmatrix} 4x - 3y = 18 \\ 5x + 6y = 3 \end{pmatrix}$ **d.** $\begin{pmatrix} 2x - y = 20 \\ 7x + y = 79 \end{pmatrix}$

e. $\begin{pmatrix} 13x - 12y = 37 \\ 15x + 13y = -11 \end{pmatrix}$ **f.** $\begin{pmatrix} 1.98x + 2.49y = 13.92 \\ 1.19x + 3.45y = 16.18 \end{pmatrix}$

10.2 SYSTEMS OF THREE LINEAR EQUATIONS IN THREE VARIABLES

Consider a linear equation in three variables x, y, and z, such as $3x - 2y + z = 7$. Any ordered triple (x, y, z) that makes the equation a true numerical statement is said to be a solution of the equation. For example, the ordered triple $(2, 1, 3)$ is a solution because $3(2) - 2(1) + 3 = 7$. However, the ordered triple $(5, 2, 4)$ is not a solution because $3(5) - 2(2) + 4 \neq 7$. There are infinitely many solutions in the solution set.

> **REMARK** The idea of a linear equation is generalized to include equations of more than two variables. Thus, an equation such as $5x - 2y + 9z = 8$ is called a *linear equation in three variables*; the equation $5x - 7y + 2z - 11w = 1$ is called a *linear equation in four variables*, and so on.

To *solve* a system of three linear equations in three variables, such as

$$\begin{pmatrix} 3x - y + 2z = 13 \\ 4x + 2y + 5z = 30 \\ 5x - 3y - z = 3 \end{pmatrix},$$

means to find all of the ordered triples that satisfy all three equations. In other words, the solution set of the system is the intersection of the solution sets of all three equations in the system.

The graph of a linear equation in three variables is a *plane*, not a line. In fact, graphing equations in three variables requires the use of a three-dimensional coordinate system. Thus, using a graphing approach to solve systems of three linear equations in three variables is not at all practical. However, a simple graphic analysis does provide us with some direction as to what we can expect as we begin solving such systems.

In general, because each linear equation in three variables produces a plane, a system of three such equations produces three planes. There are various ways that three planes can be related. For example, they may be mutually parallel, or two of the planes may be parallel with the third one intersecting the other two. (You may want to analyze all of the other possibilities for the three planes!) However, for our purposes at this time we need to realize that from a solution set viewpoint, a system of three linear equations in three variables produces one of the following possibilities.

1. There is *one ordered triple* that satisfies all three equations. The three planes have a common point of intersection as indicated in Figure 10.4.

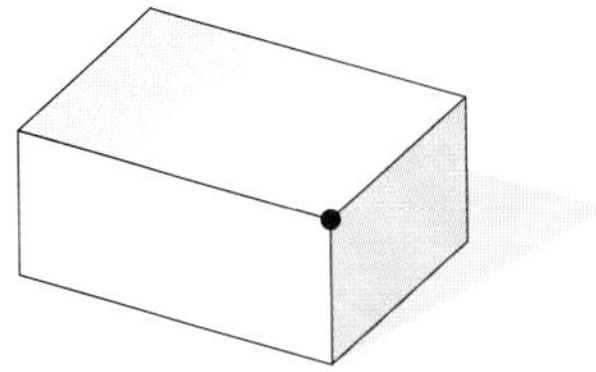

FIGURE 10.4

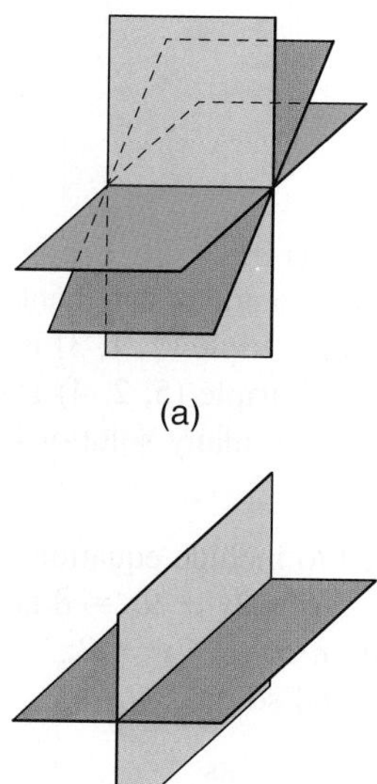

FIGURE 10.5

2. There are *infinitely many ordered triples* in the solution set, all of which are coordinates of *points on a line* common to the three planes. This can happen if the three planes have a common line of intersection (Figure 10.5(a)), or if two of the planes coincide and the third plane intersects them (Figure 10.5(b)).

3. There are *infinitely many ordered triples* in the solution set, all of which are coordinates of *points on a plane*. This can happen if the three planes coincide, as illustrated in Figure 10.6.

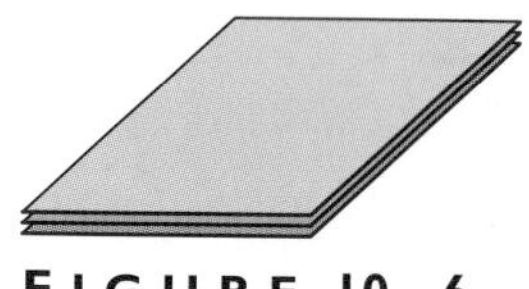

FIGURE 10.6

4. The solution set is *empty*; thus we write $\varnothing$. This can happen in various ways, as illustrated in Figure 10.7. Notice that in each situation there are no points common to all three planes.

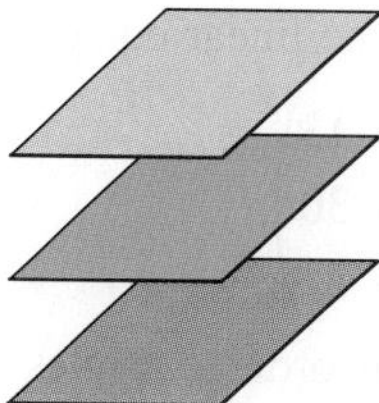

(a) Three parallel planes

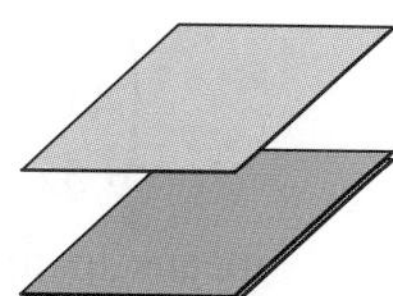

(b) Two planes coincide and the third one is parallel to the coinciding planes.

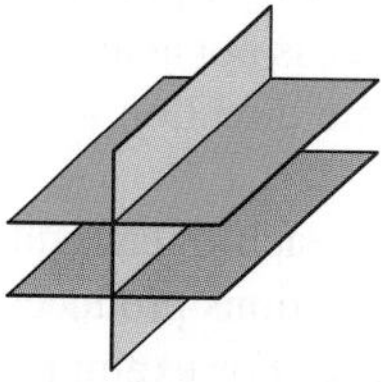

(c) Two planes are parallel and the third intersects them in parallel lines.

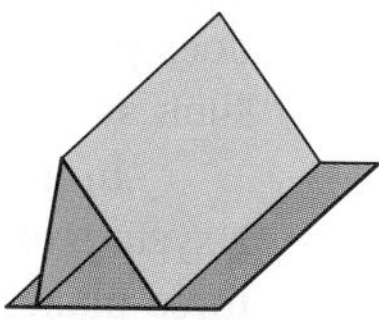

(d) No two planes are parallel, but two of them intersect in a line that is parallel to the third plane.

FIGURE 10.7

Now that we know what possibilities exist, let's consider finding the solution sets for some systems. Our approach will be the elimination-by-addition method, whereby systems are replaced with equivalent systems until a system is obtained where we can easily determine the solution set. The details of this approach will become apparent as we work a few examples.

EXAMPLE 1

Solve the system

$$\begin{pmatrix} 4x - 3y - 2z = 5 \\ 5y + z = -11 \\ 3z = 12 \end{pmatrix}. \quad \begin{matrix} (1) \\ (2) \\ (3) \end{matrix}$$

Solution

The given form of this system makes it easy to solve. From equation (3) we obtain $z = 4$. Then substituting 4 for z in equation (2), we get

$$\begin{aligned} 5y + 4 &= -11 \\ 5y &= -15 \\ y &= -3. \end{aligned}$$

Finally, substituting 4 for z and -3 for y in equation (1) yields

$$\begin{aligned} 4x - 3(-3) - 2(4) &= 5 \\ 4x + 1 &= 5 \\ 4x &= 4 \\ x &= 1. \end{aligned}$$

Thus, the solution set of the given system is $\{(1, -3, 4)\}$.

EXAMPLE 2

Solve the system

$$\begin{pmatrix} x - 2y + 3z = 22 \\ 2x - 3y - z = 5 \\ 3x + y - 5z = -32 \end{pmatrix}. \quad \begin{matrix} (4) \\ (5) \\ (6) \end{matrix}$$

Solution

Equation (5) can be replaced with the equation formed by multiplying equation (4) by -2 and adding this result to equation (5). Equation (6) can be replaced with the equation formed by multiplying equation (4) by -3 and adding this result to equation (6). The following equivalent system is produced, in which equations (8) and (9) contain only the two variables y and z.

$$\begin{pmatrix} x - 2y + 3z = 22 \\ y - 7z = -39 \\ 7y - 14z = -98 \end{pmatrix} \quad \begin{matrix} (7) \\ (8) \\ (9) \end{matrix}$$

Equation (9) can be replaced with the equation formed by multiplying equation (8) by -7 and adding this result to equation (9). This produces the following equivalent system.

$$\begin{pmatrix} x - 2y + 3z = 22 \\ y - 7z = -39 \\ 35z = 175 \end{pmatrix} \quad \begin{matrix} (10) \\ (11) \\ (12) \end{matrix}$$

From equation (12) we obtain $z = 5$. Then, substituting 5 for z in equation (11) we obtain

$$\begin{aligned} y - 7(5) &= -39 \\ y - 35 &= -39 \\ y &= -4. \end{aligned}$$

Finally, substituting -4 for y and 5 for z in equation (10) produces

$$\begin{aligned} x - 2(-4) + 3(5) &= 22 \\ x + 8 + 15 &= 22 \\ x + 23 &= 22 \\ x &= -1. \end{aligned}$$

The solution set of the original system is $\{(-1, -4, 5)\}$. (Perhaps you should check this ordered triple in all three of the original equations.)

EXAMPLE 3

Solve the system

$$\begin{pmatrix} 3x - y + 2z = 13 \\ 5x - 3y - z = 3 \\ 4x + 2y + 5z = 30 \end{pmatrix}. \quad \begin{matrix} \textbf{(13)} \\ \textbf{(14)} \\ \textbf{(15)} \end{matrix}$$

Solution

Equation (14) can be replaced with the equation formed by multiplying equation (13) by -3 and adding this result to equation (14). Equation (15) can be replaced with the equation formed by multiplying equation (13) by 2 and adding this result to equation (15). Thus, we produce the following equivalent system in which equations (17) and (18) contain only the two variables x and z.

$$\begin{pmatrix} 3x - y + 2z = 13 \\ -4x - 7z = -36 \\ -10x + 9z = 56 \end{pmatrix} \quad \begin{matrix} \textbf{(16)} \\ \textbf{(17)} \\ \textbf{(18)} \end{matrix}$$

Now, if we multiply equation (17) by 5 and equation (18) by 2, we get the following equivalent system.

$$\begin{pmatrix} 3x - y + 2z = 13 \\ -20x - 35z = -180 \\ 20x + 18z = 112 \end{pmatrix} \quad \begin{matrix} \textbf{(19)} \\ \textbf{(20)} \\ \textbf{(21)} \end{matrix}$$

Equation (21) can be replaced with the equation formed by adding equation (20) to equation (21).

$$\begin{pmatrix} 3x - y + 2z = 13 \\ -20x - 35z = -180 \\ -17z = -68 \end{pmatrix} \quad \begin{matrix} \textbf{(22)} \\ \textbf{(23)} \\ \textbf{(24)} \end{matrix}$$

From equation (24), we obtain $z = 4$. Then we can substitute 4 for z in equation (23).

$$\begin{aligned} -20x - 35(4) &= -180 \\ -20x - 140 &= -180 \\ -20x &= -40 \\ x &= 2 \end{aligned}$$

Now we can substitute 2 for x and 4 for z in equation (22).

$$\begin{aligned} 3(2) - y + 2(4) &= 13 \\ 6 - y + 8 &= 13 \\ -y + 14 &= 13 \\ -y &= -1 \\ y &= 1 \end{aligned}$$

The solution set of the original system is $\{(2, 1, 4)\}$.

EXAMPLE 4

Solve the system

$$\begin{pmatrix} 2x + 3y + z = 14 \\ 3x - 4y - 2z = -30 \\ 5x + 7y + 3z = 32 \end{pmatrix}. \qquad \begin{matrix} \textbf{(25)} \\ \textbf{(26)} \\ \textbf{(27)} \end{matrix}$$

Solution

Equation (26) can be replaced with the equation formed by multiplying equation (25) by 2 and adding this result to equation (26). Equation (27) can be replaced with the equation formed by multiplying equation (25) by -3 and adding this result to equation (27). The following equivalent system is produced, in which equations (29) and (30) contain only the two variables x and y.

$$\begin{pmatrix} 2x + 3y + z = 14 \\ 7x + 2y = -2 \\ -x - 2y = -10 \end{pmatrix} \qquad \begin{matrix} \textbf{(28)} \\ \textbf{(29)} \\ \textbf{(30)} \end{matrix}$$

Now equation (30) can be replaced with the equation formed by adding equation (29) to equation (30).

$$\begin{pmatrix} 2x + 3y + z = 14 \\ 7x + 2y = -2 \\ 6x = -12 \end{pmatrix} \qquad \begin{matrix} \textbf{(31)} \\ \textbf{(32)} \\ \textbf{(33)} \end{matrix}$$

From equation (33) we obtain $x = -2$. Then, substituting -2 for x in equation (32) we obtain

$$\begin{aligned} 7(-2) + 2y &= -2 \\ 2y &= 12 \\ y &= 6. \end{aligned}$$

Finally, substituting 6 for y and -2 for x in equation (31) produces

$$\begin{aligned} 2(-2) + 3(6) + z &= 14 \\ 14 + z &= 14 \\ z &= 0. \end{aligned}$$

The solution set of the original system is $\{(-2, 6, 0)\}$.

The ability to solve systems of three linear equations in three unknowns enhances our problem solving capabilities. Let's conclude this section with a problem that we can solve using such a system.

PROBLEM 1

A small company that manufactures sporting equipment produces three different styles of golf shirts. Each style of shirt requires the services of three departments, as indicated by the following table.

	STYLE *A*	STYLE *B*	STYLE *C*
Cutting department	0.1 hr	0.1 hr	0.3 hr
Sewing department	0.3 hr	0.2 hr	0.4 hr
Packaging department	0.1 hr	0.2 hr	0.1 hr

The cutting, sewing, and packaging departments have available a maximum of 340, 580, and 255 work-hours per week, respectively. How many of each style of golf shirt should be produced each week so that the company is operating at full capacity?

Solution

Let a represent the number of shirts of style A produced per week, b the number of style B per week, and c the number of style C per week. Then the problem translates into the following system of equations.

$$\begin{pmatrix} 0.1a + 0.1b + 0.3c = 340 \\ 0.3a + 0.2b + 0.4c = 580 \\ 0.1a + 0.2b + 0.1c = 255 \end{pmatrix} \begin{matrix} \longleftarrow \text{Cutting department} \\ \longleftarrow \text{Sewing department} \\ \longleftarrow \text{Packaging department} \end{matrix}$$

Solving this system (we will leave the details for you to carry out) produces $a = 500$, $b = 650$, and $c = 750$. Thus, the company should produce 500 golf shirts of style A, 650 of style B, and 750 of style C per week.

PROBLEM SET 10.2

Solve each of the following systems.

1. $\begin{pmatrix} 2x - 3y + 4z = 10 \\ 5y - 2z = -16 \\ 3z = 9 \end{pmatrix}$

2. $\begin{pmatrix} -3x + 2y + z = -9 \\ 4x \quad - 3z = 18 \\ 4z = -8 \end{pmatrix}$

3. $\begin{pmatrix} x + 2y - 3z = 2 \\ 3y - z = 13 \\ 3y + 5z = 25 \end{pmatrix}$

4. $\begin{pmatrix} 2x + 3y - 4z = -10 \\ 2y + 3z = 16 \\ 2y - 5z = -16 \end{pmatrix}$

5. $\begin{pmatrix} 3x + 2y - 2z = 14 \\ x \quad - 6z = 16 \\ 2x \quad + 5z = -2 \end{pmatrix}$

6. $\begin{pmatrix} 3x + 2y - z = -11 \\ 2x - 3y \quad = -1 \\ 4x + 5y \quad = -13 \end{pmatrix}$

7. $\begin{pmatrix} x - 2y + 3z = 7 \\ 2x + y + 5z = 17 \\ 3x - 4y - 2z = 1 \end{pmatrix}$

8. $\begin{pmatrix} x - 2y + z = -4 \\ 2x + 4y - 3z = -1 \\ -3x - 6y + 7z = 4 \end{pmatrix}$

9. $\begin{pmatrix} 2x - y + z = 0 \\ 3x - 2y + 4z = 11 \\ 5x + y - 6z = -32 \end{pmatrix}$

10. $\begin{pmatrix} 2x - y + 3z = -14 \\ 4x + 2y - z = 12 \\ 6x - 3y + 4z = -22 \end{pmatrix}$

11. $\begin{pmatrix} 3x + 2y - z = -11 \\ 2x - 3y + 4z = 11 \\ 5x + y - 2z = -17 \end{pmatrix}$

12. $\begin{pmatrix} 9x + 4y - z = 0 \\ 3x - 2y + 4z = 6 \\ 6x - 8y - 3z = 3 \end{pmatrix}$

13. $\begin{pmatrix} 2x + 3y - 4z = -10 \\ 4x - 5y + 3z = 2 \\ 2y + z = 8 \end{pmatrix}$

14. $\begin{pmatrix} x + 2y - 3z = 2 \\ 3x - z = -8 \\ 2x - 3y + 5z = -9 \end{pmatrix}$

15. $\begin{pmatrix} 3x + 2y - 2z = 14 \\ 2x - 5y + 3z = 7 \\ 4x - 3y + 7z = 5 \end{pmatrix}$

16. $\begin{pmatrix} 4x + 3y - 2z = -11 \\ 3x - 7y + 3z = 10 \\ 9x - 8y + 5z = 9 \end{pmatrix}$

17. $\begin{pmatrix} 2x - 3y + 4z = -12 \\ 4x + 2y - 3z = -13 \\ 6x - 5y + 7z = -31 \end{pmatrix}$

18. $\begin{pmatrix} 3x + 5y - 2z = -27 \\ 5x - 2y + 4z = 27 \\ 7x + 3y - 6z = -55 \end{pmatrix}$

19. $\begin{pmatrix} 5x - 3y - 6z = 22 \\ x - y + z = -3 \\ -3x + 7y - 5z = 23 \end{pmatrix}$

20. $\begin{pmatrix} 4x + 3y - 5z = -29 \\ 3x - 7y - z = -19 \\ 2x + 5y + 2z = -10 \end{pmatrix}$

Solve each of the following problems by setting up and solving a system of three linear equations in three variables.

21. The sum of three numbers is 43. The sum of the two smaller numbers is three larger than the largest number. Twice the smallest number plus three times the second number plus four times the largest number equals 141. Find the numbers.

22. The sum of three numbers is 20. The sum of the first and third numbers is two more than twice the second number. The third number minus the first yields three times the second number. Find the numbers.

23. A box contains \$7.15 in nickels, dimes, and quarters. There are 42 coins in all and the sum of the numbers of nickels and dimes is two less than the number of quarters. How many coins of each kind are there?

24. A handful of 65 coins consists of pennies, nickels, and dimes. The number of nickels is four less than twice the number of pennies, and there are 13 more dimes than nickels. How many coins of each kind are there?

25. The measure of the largest angle of a triangle is twice the smallest angle. The sum of the smallest and the largest angle is twice the other angle. Find the measure of each angle.

26. The perimeter of a triangle is 45 centimeters. The longest side is 4 centimeters less than twice the shortest side. The sum of the lengths of the shortest and longest sides is 7 centimeters less than three times the length of the remaining side. Find the lengths of all three sides of the triangle.

27. Part of \$3000 is invested at 12%, another part at 13%, and the remainder at 14% yearly interest. The total yearly income from the three investments is \$400. The sum of the amounts invested at 12% and 13% equals the amount invested at 14%. How much is invested at each rate?

28. Different amounts of money are invested at 10%, 11%, and 12% yearly interest. The amount invested at 11% is \$300 more than what is invested at 10%, and the total yearly income from all three investments is \$324. If a total of \$2900 is invested, find the amount invested at each rate.

29. A small company makes three different types of bird houses. Each type requires the services of three different departments according to the following table.

	TYPE *A*	TYPE *B*	TYPE *C*
Cutting department	0.1 hr	0.2 hr	0.1 hr
Finishing department	0.4 hr	0.4 hr	0.3 hr
Assembly department	0.2 hr	0.1 hr	0.3 hr

The cutting, finishing, and assembly departments have available a maximum of 35, 95, and 62.5 work-hours per week, respectively. How many bird houses of each type should be made per week so that the company is operating at full capacity?

30. A certain diet consists of dishes *A*, *B*, and *C*. Each serving of *A* has 1 gram of fat, 2 grams of carbohydrate, and 4 grams of protein. Each serving of *B* has 2 grams of fat, 1 gram of carbohydrate, and 3 grams of protein. Each serving of *C* has 2 grams of fat, 4 grams of carbohydrate, and 3 grams of protein. The diet allows 15 grams of fat, 24 grams of carbohydrate, and 30 grams of protein. How many servings of each dish can be eaten?

31. Recall that one form of the equation of a circle is $x^2 + y^2 + Dx + Ey + F = 0$. Find the equation of the circle that passes through the points $(-3, 1)$, $(7, 1)$, and $(-7, 5)$.

10.3 MATRIX APPROACH TO SOLVING SYSTEMS

In the first two sections of this chapter, we found that the techniques of substitution and elimination by addition worked effectively with two equations and two unknowns, but they started to get a bit cumbersome with three equations and three unknowns. Therefore, we shall now begin to analyze some techniques that lend themselves to use with larger systems of equations. Furthermore, some of these techniques form the basis for using a computer to solve systems. Even though these techniques are primarily designed for large systems of equations, we shall study them in the context of small systems so we won't become too bogged down with the computational aspects of the techniques.

Matrices

A **matrix** is an array of numbers arranged in horizontal rows and vertical columns and enclosed in brackets. For example, the matrix

$$\text{2 rows} \begin{matrix}\longrightarrow \\ \longrightarrow\end{matrix} \begin{bmatrix} 2 & 3 & -1 \\ -4 & 7 & 12 \end{bmatrix}$$

$$\uparrow \quad \uparrow \quad \uparrow$$

3 columns

has 2 rows and 3 columns and is called a 2×3 (read *two by three*) matrix. Each number in a matrix is called an **element** of the matrix. Some additional examples of matrices (*matrices* is the plural of matrix) are as follows.

$$\underset{3 \times 2}{\begin{bmatrix} 2 & 1 \\ 1 & -4 \\ \frac{1}{2} & \frac{2}{3} \end{bmatrix}} \qquad \underset{2 \times 2}{\begin{bmatrix} 17 & 18 \\ -14 & 16 \end{bmatrix}} \qquad \underset{1 \times 2}{[7 \quad 14]} \qquad \underset{4 \times 1}{\begin{bmatrix} 3 \\ -2 \\ 1 \\ 19 \end{bmatrix}}$$

In general, a matrix of m rows and n columns is called a matrix of **dimension $m \times n$** or **order $m \times n$**.

With every system of linear equations we can associate a matrix that consists of the coefficients and constant terms. For example, with the system

$$\begin{pmatrix} a_1x + b_1y + c_1z = d_1 \\ a_2x + b_2y + c_2z = d_2 \\ a_3x + b_3y + c_3z = d_3 \end{pmatrix}.$$

we can associate the matrix

$$\left[\begin{array}{ccc:c} a_1 & b_1 & c_1 & d_1 \\ a_2 & b_2 & c_2 & d_2 \\ a_3 & b_3 & c_3 & d_3 \end{array}\right],$$

which is commonly called the **augmented matrix** of the system of equations. The dashed line simply separates the coefficients from the constant terms and reminds us that we are working with an augmented matrix.

On pages 417 and 418 we listed the operations or transformations that can be applied to a system of equations to produce an equivalent system. Since augmented matrices are essentially abbreviated forms of systems of linear equations, there are analogous transformations that can be applied to augmented matrices. These transformations are usually referred to as **elementary row operations** and can be stated as follows.

For any augmented matrix of a system of linear equations, the following elementary row operations will produce a matrix of an equivalent system:

1. Any two rows of the matrix can be interchanged.
2. Any row of the matrix can be multiplied by a nonzero real number.
3. Any row of the matrix can be replaced by the sum of a nonzero multiple of another row plus that row.

Let's illustrate the use of augmented matrices and elementary row operations to solve a system of two linear equations in two variables.

EXAMPLE 1

Solve the system

$$\begin{pmatrix} x - 3y = -17 \\ 2x + 7y = 31 \end{pmatrix}.$$

Solution

The augmented matrix of the system is

$$\left[\begin{array}{cc:c} 1 & -3 & -17 \\ 2 & 7 & 31 \end{array}\right].$$

We would like to change this matrix to one of the form

$$\left[\begin{array}{cc:c} 1 & 0 & a \\ 0 & 1 & b \end{array}\right],$$

where we can easily determine that the solution is $x = a$ and $y = b$. Let's begin by adding -2 times row 1 to row 2 to produce a new row 2.

$$\left[\begin{array}{cc:c} 1 & -3 & -17 \\ 0 & 13 & 65 \end{array}\right]$$

Now we can multiply row 2 by $\frac{1}{13}$.

$$\left[\begin{array}{cc:c} 1 & -3 & -17 \\ 0 & 1 & 5 \end{array}\right]$$

Finally, we can add 3 times row 2 to row 1 to produce a new row 1.

$$\left[\begin{array}{cc:c} 1 & 0 & -2 \\ 0 & 1 & 5 \end{array}\right]$$

From this last matrix we see that $x = -2$ and $y = 5$. In other words, the solution set of the original system is $\{(-2, 5)\}$.

It may seem as though the matrix approach does not provide us with much extra power for solving systems of two linear equations in two unknowns. However, as the systems become larger, the compactness of the matrix approach becomes more convenient. Let's consider a system of three equations in three variables.

EXAMPLE 2

Solve the system

$$\begin{pmatrix} x + 2y - 3z = 15 \\ -2x - 3y + z = -15 \\ 4x + 9y - 4z = 49 \end{pmatrix}.$$

Solution

The augmented matrix of this system is

$$\begin{bmatrix} 1 & 2 & -3 & \vdots & 15 \\ -2 & -3 & 1 & \vdots & -15 \\ 4 & 9 & -4 & \vdots & 49 \end{bmatrix}.$$

If this system has a unique solution, then we will be able to change the augmented matrix to the form

$$\begin{bmatrix} 1 & 0 & 0 & \vdots & a \\ 0 & 1 & 0 & \vdots & b \\ 0 & 0 & 1 & \vdots & c \end{bmatrix}$$

where we will be able to read the solution $x = a$, $y = b$, and $z = c$.

Add 2 times row 1 to row 2 to produce a new row 2. Likewise, add -4 times row 1 to row 3 to produce a new row 3.

$$\begin{bmatrix} 1 & 2 & -3 & \vdots & 15 \\ 0 & 1 & -5 & \vdots & 15 \\ 0 & 1 & 8 & \vdots & -11 \end{bmatrix}$$

Now add -2 times row 2 to row 1 to produce a new row 1. Also, add -1 times row 2 to row 3 to produce a new row 3.

$$\begin{bmatrix} 1 & 0 & 7 & \vdots & -15 \\ 0 & 1 & -5 & \vdots & 15 \\ 0 & 0 & 13 & \vdots & -26 \end{bmatrix}$$

Now let's multiply row 3 by $\frac{1}{13}$.

$$\begin{bmatrix} 1 & 0 & 7 & \vdots & -15 \\ 0 & 1 & -5 & \vdots & 15 \\ 0 & 0 & 1 & \vdots & -2 \end{bmatrix}$$

Finally, we can add -7 times row 3 to row 1 to produce a new row 1, and add 5 times row 3 to row 2 for a new row 2.

$$\begin{bmatrix} 1 & 0 & 0 & \vdots & -1 \\ 0 & 1 & 0 & \vdots & 5 \\ 0 & 0 & 1 & \vdots & -2 \end{bmatrix}$$

From this last matrix we can see that the solution set of the original system is $\{(-1, 5, -2)\}$.

The final matrices of Examples 1 and 2,

$$\begin{bmatrix} 1 & 0 & \vdots & -2 \\ 0 & 1 & \vdots & 5 \end{bmatrix} \quad \text{and} \quad \begin{bmatrix} 1 & 0 & 0 & \vdots & -1 \\ 0 & 1 & 0 & \vdots & 5 \\ 0 & 0 & 1 & \vdots & -2 \end{bmatrix},$$

are said to be in **reduced echelon form**. In general, a matrix is in reduced echelon form if the following conditions are satisfied.

1. Reading from left to right, the first nonzero entry of each row is 1.
2. In the *column* containing the leftmost 1 of a row, all the remaining entries are zeros.
3. The leftmost 1 of any row is to the right of the leftmost 1 of the preceding row.
4. Rows containing only zeros are below all the rows containing nonzero entries.

In addition to the final matrices of Examples 1 and 2, the following are also in reduced echelon form.

$$\left[\begin{array}{cc|c} 1 & 2 & -3 \\ 0 & 0 & 0 \end{array}\right], \quad \left[\begin{array}{ccc|c} 1 & 0 & -2 & 5 \\ 0 & 1 & 4 & 7 \\ 0 & 0 & 0 & 0 \end{array}\right], \quad \left[\begin{array}{cccc|c} 1 & 0 & 0 & 0 & 8 \\ 0 & 1 & 0 & 0 & -9 \\ 0 & 0 & 1 & 0 & -2 \\ 0 & 0 & 0 & 1 & 12 \end{array}\right]$$

In contrast the following matrices are *not* in reduced echelon form for the reason indicated below each matrix.

$$\left[\begin{array}{ccc|c} 1 & 0 & 0 & 11 \\ 0 & 3 & 0 & -1 \\ 0 & 0 & 1 & -2 \end{array}\right]$$

Violates condition 1

$$\left[\begin{array}{ccc|c} 1 & 2 & -3 & 5 \\ 0 & 1 & 7 & 9 \\ 0 & 0 & 1 & -6 \end{array}\right]$$

Violates condition 2

$$\left[\begin{array}{ccc|c} 1 & 0 & 0 & 7 \\ 0 & 0 & 1 & -8 \\ 0 & 1 & 0 & 14 \end{array}\right]$$

Violates condition 3

$$\left[\begin{array}{cccc|c} 1 & 0 & 0 & 0 & -1 \\ 0 & 0 & 0 & 0 & 0 \\ 0 & 0 & 1 & 0 & 7 \\ 0 & 0 & 0 & 0 & 0 \end{array}\right]$$

Violates condition 4

Once we have an augmented matrix in reduced echelon form, it is easy to determine the solution set of the system. Furthermore, the procedure for changing a given augmented matrix to reduced echelon form can be described in a very systematic way. For example, if an augmented matrix of a system of three linear equations in three unknowns has a unique solution, it can be changed to reduced echelon form as follows.

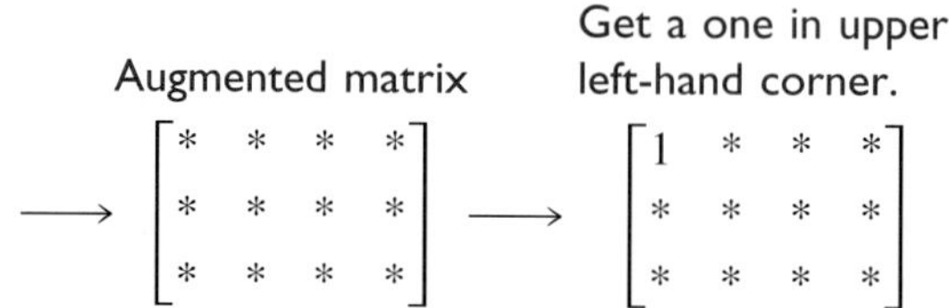
Augmented matrix
Get a one in upper left-hand corner.

Get zeros in first column beneath the one.

$$\longrightarrow \begin{bmatrix} 1 & * & * & * \\ 0 & * & * & * \\ 0 & * & * & * \end{bmatrix}$$

Get a one in the second row/second column position.

$$\longrightarrow \begin{bmatrix} 1 & * & * & * \\ 0 & 1 & * & * \\ 0 & * & * & * \end{bmatrix}$$

Get zeros above and below the one in the second column.

$$\longrightarrow \begin{bmatrix} 1 & 0 & * & * \\ 0 & 1 & * & * \\ 0 & 0 & * & * \end{bmatrix}$$

Get a one in the third row/third column position.

$$\longrightarrow \begin{bmatrix} 1 & 0 & * & * \\ 0 & 1 & * & * \\ 0 & 0 & 1 & * \end{bmatrix}$$

Get zeros above the one in the third column.

$$\begin{bmatrix} 1 & 0 & 0 & * \\ 0 & 1 & 0 & * \\ 0 & 0 & 1 & * \end{bmatrix}$$

We can identify inconsistent and dependent systems while we are changing a matrix to reduced echelon form. We will show some examples of such cases in a moment, but first let's consider another example of a system of three linear equations in three unknowns where there's a unique solution.

EXAMPLE 3

Solve the system

$$\begin{pmatrix} 2x + 4y - 5z = 37 \\ x + 3y - 4z = 29 \\ 5x - y + 3z = -20 \end{pmatrix}.$$

Solution

The augmented matrix

$$\left[\begin{array}{ccc|c} 2 & 4 & -5 & 37 \\ 1 & 3 & -4 & 29 \\ 5 & -1 & 3 & -20 \end{array}\right]$$

does not have a one in the upper left-hand corner, but this can be remedied by exchanging rows 1 and 2.

$$\left[\begin{array}{ccc|c} 1 & 3 & -4 & 29 \\ 2 & 4 & -5 & 37 \\ 5 & -1 & 3 & -20 \end{array}\right]$$

Now we can get zeros in the first column beneath the one by adding -2 times row

1 to row 2 and by adding -5 times row 1 to row 3.

$$\left[\begin{array}{rrr|r} 1 & 3 & -4 & 29 \\ 0 & -2 & 3 & -21 \\ 0 & -16 & 23 & -165 \end{array}\right]$$

Next, we can get a one for the first nonzero entry of the second row by multiplying the second row by $-\frac{1}{2}$.

$$\left[\begin{array}{rrr|r} 1 & 3 & -4 & 29 \\ 0 & 1 & -\frac{3}{2} & \frac{21}{2} \\ 0 & -16 & 23 & -165 \end{array}\right]$$

Now we can get zeros above and below the one in the second column by adding -3 times row 2 to row 1 and by adding 16 times row 2 to row 3.

$$\left[\begin{array}{rrr|r} 1 & 0 & \frac{1}{2} & -\frac{5}{2} \\ 0 & 1 & -\frac{3}{2} & \frac{21}{2} \\ 0 & 0 & -1 & 3 \end{array}\right]$$

Next, we can get a one as the first nonzero entry of the third row by multiplying the third row by -1.

$$\left[\begin{array}{rrr|r} 1 & 0 & \frac{1}{2} & -\frac{5}{2} \\ 0 & 1 & -\frac{3}{2} & \frac{21}{2} \\ 0 & 0 & 1 & -3 \end{array}\right]$$

Finally, we can get zeros above the one in the third column by adding $-\frac{1}{2}$ times row 3 to row 1 and by adding $\frac{3}{2}$ times row 3 to row 2.

$$\left[\begin{array}{rrr|r} 1 & 0 & 0 & -1 \\ 0 & 1 & 0 & 6 \\ 0 & 0 & 1 & -3 \end{array}\right]$$

From this last matrix, we see that the solution set of the original system is $\{(-1, 6, -3)\}$.

Example 3 illustrates that even though the process of changing to reduced echelon form can be systematically described, it can involve some rather messy calculations. However, with the aid of a computer, such calculations are not troublesome. For our purposes in this text, the examples and problems involve systems that minimize messy calculations. This will allow us to concentrate on the procedures.

We want to call your attention to another issue in the solution of Example 3. Consider the matrix

$$\left[\begin{array}{rrr|r} 1 & 3 & -4 & 29 \\ 0 & 1 & -\frac{3}{2} & \frac{21}{2} \\ 0 & -16 & 23 & -165 \end{array}\right]$$

which is obtained about halfway through the solution. At this step it seems evident that the calculations are getting a little messy. Therefore, instead of continuing toward the reduced echelon form, let's add 16 times row 2 to row 3 to produce a new row 3.

$$\left[\begin{array}{ccc|c} 1 & 3 & -4 & 29 \\ 0 & 1 & -\frac{3}{2} & \frac{21}{2} \\ 0 & 0 & -1 & 3 \end{array}\right].$$

The system represented by this matrix is

$$\begin{pmatrix} x + 3y - 4z = 29 \\ y - \frac{3}{2}z = \frac{21}{2} \\ -z = 3 \end{pmatrix};$$

it is said to be in **triangular form**. The last equation determines the value for z; then we can use the process of back-substitution to determine the values for y and x.

Finally, let's consider two examples to illustrate what happens when we use the matrix approach on inconsistent and dependent systems.

EXAMPLE 4

Solve the system

$$\begin{pmatrix} x - 2y + 3z = 3 \\ 5x - 9y + 4z = 2 \\ 2x - 4y + 6z = -1 \end{pmatrix}.$$

Solution

The augmented matrix of the system is

$$\left[\begin{array}{ccc|c} 1 & -2 & 3 & 3 \\ 5 & -9 & 4 & 2 \\ 2 & -4 & 6 & -1 \end{array}\right].$$

We can get zeros below the one in the first column by adding -5 times row 1 to row 2 and by adding -2 times row 1 to row 3.

$$\left[\begin{array}{ccc|c} 1 & -2 & 3 & 3 \\ 0 & 1 & -11 & -13 \\ 0 & 0 & 0 & -7 \end{array}\right]$$

At this step we can stop, because the bottom row of the matrix represents the statement $0(x) + 0(y) + 0(z) = -7$, which is obviously false for all values of x, y, and z. Thus, the original system is inconsistent; its solution set is $\varnothing$.

EXAMPLE 5

Solve the system

$$\begin{pmatrix} x + 2y + 2z = 9 \\ x + 3y - 4z = 5 \\ 2x + 5y - 2z = 14 \end{pmatrix}.$$

Solution

The augmented matrix of the system is

$$\left[\begin{array}{ccc|c} 1 & 2 & 2 & 9 \\ 1 & 3 & -4 & 5 \\ 2 & 5 & -2 & 14 \end{array}\right].$$

We can get zeros in the first column below the one in the upper left-hand corner by adding -1 times row 1 to row 2 and by adding -2 times row 1 to row 3.

$$\left[\begin{array}{ccc|c} 1 & 2 & 2 & 9 \\ 0 & 1 & -6 & -4 \\ 0 & 1 & -6 & -4 \end{array}\right]$$

Now we can get zeros in the second column above and below the one in the second row by adding -2 times row 2 to row 1 and by adding -1 times row 2 to row 3.

$$\left[\begin{array}{ccc|c} 1 & 0 & 14 & 17 \\ 0 & 1 & -6 & -4 \\ 0 & 0 & 0 & 0 \end{array}\right]$$

The bottom row of zeros represents the statement $0(x) + 0(y) + 0(z) = 0$, which is true for all values of x, y, and z. The second row represents the statement $y - 6z = -4$, which can be rewritten $y = 6z - 4$. The top row represents the statement $x + 14z = 17$, which can be rewritten $x = -14z + 17$. Therefore, if we let $z = k$, where k is any real number, the solution set of infinitely many ordered triples can be represented by $\{(-14k + 17, 6k - 4, k) | k \text{ is a real number}\}$. Specific solutions can be generated by letting k take on a value. For example if $k = 2$, then $6k - 4$ becomes $6(2) - 4 = 8$ and $-14k + 17$ becomes $-14(2) + 17 = -11$. Thus the ordered triple $(-11, 8, 2)$ is a member of the solution set.

PROBLEM SET 10.3

For Problems 1–10, indicate whether each matrix is in reduced echelon form.

1. $\left[\begin{array}{cc|c} 1 & 0 & -4 \\ 0 & 1 & 14 \end{array}\right]$

2. $\left[\begin{array}{cc|c} 1 & 2 & 8 \\ 0 & 0 & 0 \end{array}\right]$

3. $\left[\begin{array}{ccc|c} 1 & 0 & 2 & 5 \\ 0 & 1 & 3 & 7 \\ 0 & 0 & 0 & 0 \end{array}\right]$

4. $\left[\begin{array}{ccc|c} 1 & 0 & 0 & 5 \\ 0 & 3 & 0 & 8 \\ 0 & 0 & 1 & -11 \end{array}\right]$

5. $\left[\begin{array}{cccc|c} 1 & 0 & 0 & 17 \\ 0 & 0 & 0 & 0 \\ 0 & 1 & 0 & -14 \end{array}\right]$

6. $\left[\begin{array}{ccc|c} 1 & 0 & 0 & -7 \\ 0 & 1 & 0 & 0 \\ 0 & 0 & 1 & 9 \end{array}\right]$

7. $\left[\begin{array}{ccc|c} 1 & 1 & 0 & -3 \\ 0 & 1 & 2 & 5 \\ 0 & 0 & 1 & 7 \end{array}\right]$

8. $\left[\begin{array}{ccc|c} 1 & 0 & 3 & 8 \\ 0 & 1 & 2 & -6 \\ 0 & 0 & 0 & 0 \end{array}\right]$

9. $\left[\begin{array}{cccc|c} 1 & 0 & 0 & 3 & 4 \\ 0 & 1 & 0 & 5 & -3 \\ 0 & 0 & 1 & -1 & 7 \\ 0 & 0 & 0 & 0 & 0 \end{array}\right]$

10. $\left[\begin{array}{cccc|c} 1 & 0 & 0 & 0 & 2 \\ 0 & 0 & 1 & 0 & 4 \\ 0 & 1 & 0 & 0 & -3 \\ 0 & 0 & 0 & 1 & 9 \end{array}\right]$

Use a matrix approach to solve each of the following systems.

11. $\begin{pmatrix} x - 3y = 14 \\ 3x + 2y = -13 \end{pmatrix}$

12. $\begin{pmatrix} x + 5y = -18 \\ -2x + 3y = -16 \end{pmatrix}$

13. $\begin{pmatrix} 3x - 4y = 33 \\ x + 7y = -39 \end{pmatrix}$

14. $\begin{pmatrix} 2x + 7y = -55 \\ x - 4y = 25 \end{pmatrix}$

15. $\begin{pmatrix} x - 6y = -2 \\ 2x - 12y = 5 \end{pmatrix}$

16. $\begin{pmatrix} 2x - 3y = -12 \\ 3x + 2y = 8 \end{pmatrix}$

17. $\begin{pmatrix} 3x - 5y = 39 \\ 2x + 7y = -67 \end{pmatrix}$

18. $\begin{pmatrix} 3x + 9y = -1 \\ x + 3y = 10 \end{pmatrix}$

19. $\begin{pmatrix} x - 2y - 3z = -6 \\ 3x - 5y - z = 4 \\ 2x + y + 2z = 2 \end{pmatrix}$

20. $\begin{pmatrix} x + 3y - 4z = 13 \\ 2x + 7y - 3z = 11 \\ -2x - y + 2z = -8 \end{pmatrix}$

21. $\begin{pmatrix} -2x - 5y + 3z = 11 \\ x + 3y - 3z = -12 \\ 3x - 2y + 5z = 31 \end{pmatrix}$

22. $\begin{pmatrix} -3x + 2y + z = 17 \\ x - y + 5z = -2 \\ 4x - 5y - 3z = -36 \end{pmatrix}$

23. $\begin{pmatrix} x - 3y - z = 2 \\ 3x + y - 4z = -18 \\ -2x + 5y + 3z = 2 \end{pmatrix}$

24. $\begin{pmatrix} x - 4y + 3z = 16 \\ 2x + 3y - 4z = -22 \\ -3x + 11y - z = -36 \end{pmatrix}$

25. $\begin{pmatrix} x - y + 2z = 1 \\ -3x + 4y - z = 4 \\ -x + 2y + 3z = 6 \end{pmatrix}$

26. $\begin{pmatrix} x + 2y - 5z = -1 \\ 2x + 3y - 2z = 2 \\ 3x + 5y - 7z = 4 \end{pmatrix}$

27. $\begin{pmatrix} -2x + y + 5z = -5 \\ 3x + 8y - z = -34 \\ x + 2y + z = -12 \end{pmatrix}$

28. $\begin{pmatrix} 4x - 10y + 3z = -19 \\ 2x + 5y - z = -7 \\ x - 3y - 2z = -2 \end{pmatrix}$

29. $\begin{pmatrix} 2x + 3y - z = 7 \\ 3x + 4y + 5z = -2 \\ 5x + y + 3z = 13 \end{pmatrix}$

30. $\begin{pmatrix} 4x + 3y - z = 0 \\ 3x + 2y + 5z = 6 \\ 5x - y - 3z = 3 \end{pmatrix}$

Subscript notation is frequently used for working with larger systems of equations. Use a matrix approach to solve each of the following systems. Express the solutions as 4-tuples of the form (x_1, x_2, x_3, x_4).

31. $\begin{pmatrix} x_1 - 3x_2 - 2x_3 + x_4 = -3 \\ -2x_1 + 7x_2 + x_3 - 2x_4 = -1 \\ 3x_1 - 7x_2 - 3x_3 + 3x_4 = -5 \\ 5x_1 + x_2 + 4x_3 - 2x_4 = 18 \end{pmatrix}$

32. $\begin{pmatrix} x_1 - 2x_2 + 2x_3 - x_4 = -2 \\ -3x_1 + 5x_2 - x_3 - 3x_4 = 2 \\ 2x_1 + 3x_2 + 3x_3 + 5x_4 = -9 \\ 4x_1 - x_2 - x_3 - 2x_4 = 8 \end{pmatrix}$

33. $\begin{pmatrix} x_1 + 3x_2 - x_3 + 2x_4 = -2 \\ 2x_1 + 7x_2 + 2x_3 - x_4 = 19 \\ -3x_1 - 8x_2 + 3x_3 + x_4 = -7 \\ 4x_1 + 11x_2 - 2x_3 - 3x_4 = 19 \end{pmatrix}$

34. $\begin{pmatrix} x_1 + 2x_2 - 3x_3 + x_4 = -2 \\ -2x_1 - 3x_2 + x_3 - x_4 = 5 \\ 4x_1 + 9x_2 - 2x_3 - 2x_4 = -28 \\ -5x_1 - 9x_2 + 2x_3 - 3x_4 = 14 \end{pmatrix}$

Each matrix in Problems 35–42 is the reduced echelon matrix for a system with variables x_1, x_2, x_3, and x_4. Find the solution set of each system.

35. $\left[\begin{array}{cccc|c} 1 & 0 & 0 & 0 & -2 \\ 0 & 1 & 0 & 0 & 4 \\ 0 & 0 & 1 & 0 & -3 \\ 0 & 0 & 0 & 1 & 0 \end{array}\right]$

36. $\left[\begin{array}{cccc|c} 1 & 0 & 0 & 0 & 0 \\ 0 & 1 & 0 & 0 & -5 \\ 0 & 0 & 1 & 0 & 0 \\ 0 & 0 & 0 & 1 & 4 \end{array}\right]$

37. $\left[\begin{array}{cccc|c} 1 & 0 & 0 & 0 & -8 \\ 0 & 1 & 0 & 0 & 5 \\ 0 & 0 & 1 & 0 & -2 \\ 0 & 0 & 0 & 0 & 1 \end{array}\right]$

38. $\left[\begin{array}{cccc|c} 1 & 0 & 0 & 0 & 2 \\ 0 & 1 & 0 & 2 & -3 \\ 0 & 0 & 1 & 3 & 4 \\ 0 & 0 & 0 & 0 & 0 \end{array}\right]$

39. $\left[\begin{array}{cccc|c} 1 & 0 & 0 & 3 & 5 \\ 0 & 1 & 0 & 0 & -1 \\ 0 & 0 & 1 & 4 & 2 \\ 0 & 0 & 0 & 0 & 0 \end{array}\right]$

40. $\left[\begin{array}{cccc|c} 1 & 3 & 0 & 0 & 0 \\ 0 & 0 & 1 & 0 & 0 \\ 0 & 0 & 0 & 0 & 1 \\ 0 & 0 & 0 & 0 & 0 \end{array}\right]$

41. $\left[\begin{array}{cccc|c} 1 & 3 & 0 & 0 & 9 \\ 0 & 0 & 1 & 0 & 2 \\ 0 & 0 & 0 & 1 & -3 \\ 0 & 0 & 0 & 0 & 0 \end{array}\right]$

42. $\left[\begin{array}{cccc|c} 1 & 0 & 0 & 0 & 7 \\ 0 & 1 & 0 & 0 & -3 \\ 0 & 0 & 1 & -2 & 5 \\ 0 & 0 & 0 & 0 & 0 \end{array}\right]$

MISCELLANEOUS PROBLEMS

For Problems 43–48, change each augmented matrix of the system to reduced echelon form and then indicate the solutions of the system.

43. $\begin{pmatrix} x - 2y + 3z = 4 \\ 3x - 5y - z = 7 \end{pmatrix}$

44. $\begin{pmatrix} x + 3y - 2z = -1 \\ -2x - 5y + 7z = 4 \end{pmatrix}$

45. $\begin{pmatrix} 2x - 4y + 3z = 8 \\ 3x + 5y - z = 7 \end{pmatrix}$

46. $\begin{pmatrix} 3x + 6y - z = 9 \\ 2x - 3y + 4z = 1 \end{pmatrix}$

47. $\begin{pmatrix} x - 2y + 4z = 9 \\ 2x - 4y + 8z = 3 \end{pmatrix}$

48. $\begin{pmatrix} x + y - 2z = -1 \\ 3x + 3y - 6z = -3 \end{pmatrix}$

GRAPHICS CALCULATOR ACTIVITIES

49. If your graphics calculator has the capability of manipulating matrices, this is a good time to become familiar with those operations. You may need to refer to your calculator manual for the technical instructions. To begin the familiarization process, load your calculator with the three augmented matrices in Examples 1, 2, and 3. Then for each one, carry out the row operations as described in the text.

10.4 DETERMINANTS

Before we introduce the concept of a determinant, let's agree upon some convenient new notation. A **general $m \times n$ (m-by-n) matrix** can be represented by

$$A = \begin{bmatrix} a_{11} & a_{12} & a_{13} & \dots & a_{1n} \\ a_{21} & a_{22} & a_{23} & \dots & a_{2n} \\ \vdots & \vdots & \vdots & & \vdots \\ a_{m1} & a_{m2} & a_{m3} & \dots & a_{mn} \end{bmatrix},$$

where the double subscripts are used to identify the number of the row and the number of the column, in that order. For example, a_{23} is the entry at the intersection of the second row and the third column. In general, the entry at the intersection of row i and column j is denoted by a_{ij}.

A square matrix is one that has the same number of rows as columns. Each square matrix A with real number entries can be associated with a real number called the determinant of the matrix, denoted by $|A|$. We will first define $|A|$ for a 2×2 matrix.

DEFINITION 10.1

If $A = \begin{bmatrix} a_{11} & a_{12} \\ a_{21} & a_{22} \end{bmatrix}$, then

$$|A| = \begin{vmatrix} a_{11} & a_{12} \\ a_{21} & a_{22} \end{vmatrix} = a_{11}a_{22} - a_{12}a_{21}.$$

EXAMPLE 1

If $A = \begin{bmatrix} 3 & -2 \\ 5 & 8 \end{bmatrix}$, find $|A|$.

Solution

Use Definition 10.1 to obtain

$$\begin{aligned} |A| = \begin{vmatrix} 3 & -2 \\ 5 & 8 \end{vmatrix} &= 3(8) - (-2)(5) \\ &= 24 + 10 \\ &= 34. \end{aligned}$$

Finding the determinant of a square matrix is commonly called evaluating the determinant, and the matrix notation is often omitted.

EXAMPLE 2

Evaluate $\begin{vmatrix} -3 & 6 \\ 2 & 8 \end{vmatrix}$.

Solution

$$\begin{aligned} \begin{vmatrix} -3 & 6 \\ 2 & 8 \end{vmatrix} &= (-3)(8) - (6)(2) \\ &= -24 - 12 \\ &= -36 \end{aligned}$$

To define determinants of 3×3 and larger square matrices, it is convenient to introduce some additional terminology.

DEFINITION 10.2

If A is a 3×3 matrix, then the **minor** (denoted by M_{ij}) of the a_{ij} element is the determinant of the 2×2 matrix obtained by deleting row i and column j of A.

EXAMPLE 3

If $A = \begin{bmatrix} 2 & 1 & 4 \\ -6 & 3 & -2 \\ 4 & 2 & 5 \end{bmatrix}$, find (a) M_{11} and (b) M_{23}.

Solutions

a. To find M_{11} we first delete row 1 and column 1 of A.

$$\begin{bmatrix} 2 & 1 & 4 \\ -6 & 3 & -2 \\ 4 & 2 & 5 \end{bmatrix}$$

Thus,

$$M_{11} = \begin{vmatrix} 3 & -2 \\ 2 & 5 \end{vmatrix} = 3(5) - (-2)(2) = 19.$$

b. To find M_{23} we first delete row 2 and column 3 of A.

$$\begin{bmatrix} 2 & 1 & 4 \\ -6 & 3 & -2 \\ 4 & 2 & 5 \end{bmatrix}$$

Thus,

$$M_{23} = \begin{vmatrix} 2 & 1 \\ 4 & 2 \end{vmatrix} = 2(2) - (1)(4) = 0.$$

The following definition will also be used.

DEFINITION 10.3

If A is a 3×3 matrix, then the **cofactor** (denoted by C_{ij}) of the element a_{ij} is defined by

$$C_{ij} = (-1)^{i+j} M_{ij}.$$

According to Definition 10.3, to find the cofactor of any element a_{ij} of a square matrix A, we find the minor of a_{ij} and multiply it by 1 if $i + j$ is even, or multiply it by -1 if $i + j$ is odd.

EXAMPLE 4

If $A = \begin{bmatrix} 3 & 2 & -4 \\ 1 & 5 & 4 \\ 2 & -3 & 1 \end{bmatrix}$, find C_{32}.

Solution

First, let's find M_{32} by deleting row 3 and column 2 of A.

$$\begin{bmatrix} 3 & 2 & -4 \\ 1 & 5 & 4 \\ 2 & -3 & 1 \end{bmatrix}$$

Thus,

$$M_{32} = \begin{vmatrix} 3 & -4 \\ 1 & 4 \end{vmatrix} = 3(4) - (-4)(1) = 16.$$

Therefore,

$$C_{32} = (-1)^{3+2} M_{32} = (-1)^5(16) = -16.$$

The concept of a cofactor can be used to define the determinant of a 3×3 matrix as follows.

DEFINITION 10.4

If $A = \begin{bmatrix} a_{11} & a_{12} & a_{13} \\ a_{21} & a_{22} & a_{23} \\ a_{31} & a_{32} & a_{33} \end{bmatrix}$, then

$$|A| = a_{11}C_{11} + a_{21}C_{21} + a_{31}C_{31}.$$

Definition 10.4 simply states that the determinant of a 3×3 matrix can be found by multiplying each element of the first column by its corresponding cofactor and then adding the three results. Let's illustrate this procedure.

EXAMPLE 5

Find $|A|$ if $A = \begin{bmatrix} -2 & 1 & 4 \\ 3 & 0 & 5 \\ 1 & -4 & -6 \end{bmatrix}$.

Solution

$$\begin{aligned} |A| &= a_{11}C_{11} + a_{21}C_{21} + a_{31}C_{31} \\ &= (-2)(-1)^{1+1} \begin{vmatrix} 0 & 5 \\ -4 & -6 \end{vmatrix} + (3)(-1)^{2+1} \begin{vmatrix} 1 & 4 \\ -4 & -6 \end{vmatrix} + (1)(-1)^{3+1} \begin{vmatrix} 1 & 4 \\ 0 & 5 \end{vmatrix} \end{aligned}$$

$$= (-2)(1)(20) + (3)(-1)(10) + (1)(1)(5)$$
$$= -40 - 30 + 5$$
$$= -65$$

When we use Definition 10.4, we often say that *the determinant is being expanded about the first column*. It can be shown that **any row or column can be used to expand a determinant**. For example, for matrix A in Example 5, the expansion of the determinant about the *second row* is as follows.

$$\begin{vmatrix} -2 & 1 & 4 \\ 3 & 0 & 5 \\ 1 & -4 & -6 \end{vmatrix} = (3)(-1)^{2+1}\begin{vmatrix} 1 & 4 \\ -4 & -6 \end{vmatrix} + (0)(-1)^{2+2}\begin{vmatrix} -2 & 4 \\ 1 & -6 \end{vmatrix} + (5)(-1)^{2+3}\begin{vmatrix} -2 & 1 \\ 1 & -4 \end{vmatrix}$$
$$= (3)(-1)(10) + (0)(1)(8) + (5)(-1)(7)$$
$$= -30 + 0 - 35$$
$$= -65$$

Notice that when we expanded about the second row, the computation was simplified by the presence of a zero. In general, it is helpful to expand about the row or column that contains the most zeros.

The concepts of minor and cofactor have been defined in terms of 3×3 matrices. Analogous definitions can be given for any square matrix (that is, any $n \times n$ matrix with $n \geq 2$), and the determinant can then be expanded about any row or column. Certainly as the matrices become larger than 3×3, the computations get more and more tedious. We will concentrate most of our efforts in this text on 2×2 and 3×3 matrices.

Properties of Determinants

Determinants have several interesting properties, some of which are primarily important from a theoretical standpoint. But some of the properties are also very useful when evaluating determinants. We will state these properties for square matrices in general, but we will use 2×2 or 3×3 matrices as examples. We can demonstrate some of the proofs of these properties by evaluating the determinants involved, and some of the proofs for 3×3 matrices will be left for you to verify in the next problem set.

PROPERTY 10.1

If any row (or column) of a square matrix A contains only zeros, then $|A| = 0$.

If every element of a row (or column) of a square matrix A is 0, then it should be evident that expanding the determinant about that row (or column) of zeros will produce 0.

PROPERTY 10.2

If square matrix B is obtained from square matrix A by interchanging two rows (or two columns), then $|B| = -|A|$.

Property 10.2 states that interchanging two rows (or columns) changes the sign of the determinant. As an example of this property suppose that

$$A = \begin{bmatrix} 2 & 5 \\ -1 & 6 \end{bmatrix}$$

and that rows 1 and 2 are interchanged to form

$$B = \begin{bmatrix} -1 & 6 \\ 2 & 5 \end{bmatrix}.$$

Calculating $|A|$ and $|B|$ we obtain

$$|A| = \begin{vmatrix} 2 & 5 \\ -1 & 6 \end{vmatrix} = 2(6) - (5)(-1) = 17$$

and

$$|B| = \begin{vmatrix} -1 & 6 \\ 2 & 5 \end{vmatrix} = (-1)(5) - (6)(2) = -17.$$

PROPERTY 10.3

If square matrix B is obtained from square matrix A by multiplying each element of any row (or column) of A by some real number k, then $|B| = k|A|$.

Property 10.3 states that multiplying any row (or column) by a factor of k affects the value of the determinant by a factor of k. As an example of this property, suppose that

$$A = \begin{bmatrix} 1 & -2 & 8 \\ 2 & 1 & 12 \\ 3 & 2 & -16 \end{bmatrix}$$

and that B is formed by multiplying each element of the third column by $\frac{1}{4}$.

$$B = \begin{bmatrix} 1 & -2 & 2 \\ 2 & 1 & 3 \\ 3 & 2 & -4 \end{bmatrix}$$

Now let's calculate $|A|$ and $|B|$ by expanding about the third column in each case.

$$|A| = \begin{vmatrix} 1 & -2 & 8 \\ 2 & 1 & 12 \\ 3 & 2 & -16 \end{vmatrix} = (8)(-1)^{1+3}\begin{vmatrix} 2 & 1 \\ 3 & 2 \end{vmatrix} + (12)(-1)^{2+3}\begin{vmatrix} 1 & -2 \\ 3 & 2 \end{vmatrix} + (-16)(-1)^{3+3}\begin{vmatrix} 1 & -2 \\ 2 & 1 \end{vmatrix}$$
$$= (8)(1)(1) + (12)(-1)(8) + (-16)(1)(5)$$
$$= -168$$

$$|B| = \begin{vmatrix} 1 & -2 & 2 \\ 2 & 1 & 3 \\ 3 & 2 & -4 \end{vmatrix} = (2)(-1)^{1+3}\begin{vmatrix} 2 & 1 \\ 3 & 2 \end{vmatrix} + (3)(-1)^{2+3}\begin{vmatrix} 1 & -2 \\ 3 & 2 \end{vmatrix} + (-4)(-1)^{3+3}\begin{vmatrix} 1 & -2 \\ 2 & 1 \end{vmatrix}$$
$$= (2)(1)(1) + (3)(-1)(8) + (-4)(1)(5)$$
$$= -42$$

We see that $|B| = \frac{1}{4}|A|$. This example also illustrates the usual computational use of Property 10.3: We can factor out a common factor from a row or column and then adjust the value of the determinant by that factor. For example,

$$\begin{vmatrix} 2 & 6 & 8 \\ -1 & 2 & 7 \\ 5 & 2 & 1 \end{vmatrix} = 2\begin{vmatrix} 1 & 3 & 4 \\ -1 & 2 & 7 \\ 5 & 2 & 1 \end{vmatrix}.$$

Factor a 2 from the top row.

PROPERTY 10.4

If square matrix B is obtained from square matrix A by adding k times a row (or column) of A to another row (or column) of A, then $|B| = |A|$.

Property 10.4 states that **adding the product of k times a row (or column) to another row (or column) does not affect the value of the determinant**. As an example of this property, suppose that

$$A = \begin{bmatrix} 1 & 2 & 4 \\ 2 & 4 & 7 \\ -1 & 3 & 5 \end{bmatrix}.$$

Now let's form B by replacing row 2 with the result of adding -2 times row 1 to row 2.

$$B = \begin{bmatrix} 1 & 2 & 4 \\ 0 & 0 & -1 \\ -1 & 3 & 5 \end{bmatrix}.$$

Next, let's evaluate $|A|$ and $|B|$ by expanding about the second row in each case.

$$|A| = \begin{vmatrix} 1 & 2 & 4 \\ 2 & 4 & 7 \\ -1 & 3 & 5 \end{vmatrix} = (2)(-1)^{2+1}\begin{vmatrix} 2 & 4 \\ 3 & 5 \end{vmatrix} + (4)(-1)^{2+2}\begin{vmatrix} 1 & 4 \\ -1 & 5 \end{vmatrix} + (7)(-1)^{2+3}\begin{vmatrix} 1 & 2 \\ -1 & 3 \end{vmatrix}$$

$$= 2(-1)(-2) + (4)(1)(9) + (7)(-1)(5)$$

$$= 5$$

$$|B| = \begin{vmatrix} 1 & 2 & 4 \\ 0 & 0 & -1 \\ -1 & 3 & 5 \end{vmatrix} = (0)(-1)^{2+1}\begin{vmatrix} 2 & 4 \\ 3 & 5 \end{vmatrix} + (0)(-1)^{2+2}\begin{vmatrix} 1 & 4 \\ -1 & 5 \end{vmatrix} + (-1)(-1)^{2+3}\begin{vmatrix} 1 & 2 \\ -1 & 3 \end{vmatrix}$$

$$= 0 + 0 + (-1)(-1)(5)$$

$$= 5$$

Notice that $|B| = |A|$. Furthermore, notice that because of the zeros in the second row, evaluating $|B|$ is much easier than evaluating $|A|$. Property 10.4 can often be used to obtain some zeros before evaluating a determinant.

A word of caution is in order at this time. Be careful not to confuse Properties 10.2, 10.3, and 10.4 with the three elementary row transformations of augmented matrices that were used in Section 10.3. The statements of the two sets of properties do resemble each other, but the properties pertain to *two different concepts*, so be sure to keep them separate.

One final property of determinants should be mentioned.

PROPERTY 10.5

If two rows (or columns) of a square matrix A are identical, then $|A| = 0$.

Property 10.5 is a direct consequence of Property 10.2. Suppose that A is a square matrix (any size) with two identical rows. Square matrix B can be formed from A by interchanging the two identical rows. Since identical rows were interchanged, $|B| = |A|$. *But* by Property 10.2, $|B| = -|A|$. For both of these statements to hold, $|A| = 0$.

Let's conclude this section by evaluating a 4×4 determinant using Properties 10.3 and 10.4 to facilitate the computation.

EXAMPLE 6

Evaluate $\begin{vmatrix} 6 & 2 & 1 & -2 \\ 9 & -1 & 4 & 1 \\ 12 & -2 & 3 & -1 \\ 0 & 0 & 9 & 3 \end{vmatrix}$.

Solution

First, let's add -3 times the fourth column to the third column.

$$\begin{vmatrix} 6 & 2 & 7 & -2 \\ 9 & -1 & 1 & 1 \\ 12 & -2 & 6 & -1 \\ 0 & 0 & 0 & 3 \end{vmatrix}$$

Now if we expand about the fourth row, we get only one nonzero product.

$$(3)(-1)^{4+4}\begin{vmatrix} 6 & 2 & 7 \\ 9 & -1 & 1 \\ 12 & -2 & 6 \end{vmatrix}$$

Factoring a 3 out of the first column of the 3×3 determinant, we obtain

$$(3)(-1)^{8}(3)\begin{vmatrix} 2 & 2 & 7 \\ 3 & -1 & 1 \\ 4 & -2 & 6 \end{vmatrix}.$$

Now working with the 3×3 determinant, we can first add column 3 to column 2 and then add -3 times column 3 to column 1.

$$(3)(-1)^{8}(3)\begin{vmatrix} -19 & 9 & 7 \\ 0 & 0 & 1 \\ -14 & 4 & 6 \end{vmatrix}.$$

Finally, by expanding this 3×3 determinant about the second row, we obtain

$$(3)(-1)^{8}(3)(1)(-1)^{2+3}\begin{vmatrix} -19 & 9 \\ -14 & 4 \end{vmatrix}.$$

Our final result is

$$(3)(-1)^{8}(3)(1)(-1)^{5}(50) = -450.$$

PROBLEM SET 10.4

Evaluate each of the following 2×2 determinants by using Definition 10.1.

1. $\begin{vmatrix} 4 & 3 \\ 2 & 7 \end{vmatrix}$

2. $\begin{vmatrix} 3 & 5 \\ 6 & 4 \end{vmatrix}$

3. $\begin{vmatrix} -3 & 2 \\ 7 & 5 \end{vmatrix}$

4. $\begin{vmatrix} 5 & 3 \\ 6 & -1 \end{vmatrix}$

5. $\begin{vmatrix} 2 & -3 \\ 8 & -2 \end{vmatrix}$

6. $\begin{vmatrix} -5 & 5 \\ -6 & 2 \end{vmatrix}$

7. $\begin{vmatrix} -2 & -3 \\ -1 & -4 \end{vmatrix}$

8. $\begin{vmatrix} -4 & -3 \\ -5 & -7 \end{vmatrix}$

9. $\begin{vmatrix} \frac{1}{2} & \frac{1}{3} \\ -3 & -6 \end{vmatrix}$

10. $\begin{vmatrix} \frac{2}{3} & \frac{3}{4} \\ 8 & 6 \end{vmatrix}$

11. $\begin{vmatrix} \frac{1}{2} & \frac{2}{3} \\ \frac{3}{4} & -\frac{1}{3} \end{vmatrix}$

12. $\begin{vmatrix} \frac{2}{3} & \frac{1}{5} \\ -\frac{1}{4} & \frac{3}{2} \end{vmatrix}$

Evaluate each of the following 3×3 determinants. Use the properties of determinants to your advantage.

13. $\begin{vmatrix} 1 & 2 & -1 \\ 3 & 1 & 2 \\ 2 & 4 & 3 \end{vmatrix}$

14. $\begin{vmatrix} 1 & -2 & 1 \\ 2 & 1 & -1 \\ 3 & 2 & 4 \end{vmatrix}$

15. $\begin{vmatrix} 1 & -4 & 1 \\ 2 & 5 & -1 \\ 3 & 3 & 4 \end{vmatrix}$

16. $\begin{vmatrix} 3 & -2 & 1 \\ 2 & 1 & 4 \\ -1 & 3 & 5 \end{vmatrix}$

17. $\begin{vmatrix} 6 & 12 & 3 \\ -1 & 5 & 1 \\ -3 & 6 & 2 \end{vmatrix}$

18. $\begin{vmatrix} 2 & 35 & 5 \\ 1 & -5 & 1 \\ -4 & 15 & 2 \end{vmatrix}$

19. $\begin{vmatrix} 2 & -1 & 3 \\ 0 & 3 & 1 \\ 1 & -2 & -1 \end{vmatrix}$

20. $\begin{vmatrix} 2 & -17 & 3 \\ 0 & 5 & 1 \\ 1 & -3 & -1 \end{vmatrix}$

21. $\begin{vmatrix} -3 & -2 & 1 \\ 5 & 0 & 6 \\ 2 & 1 & -4 \end{vmatrix}$

22. $\begin{vmatrix} -5 & 1 & -1 \\ 3 & 4 & 2 \\ 0 & 2 & -3 \end{vmatrix}$

23. $\begin{vmatrix} 3 & -4 & -2 \\ 5 & -2 & 1 \\ 1 & 0 & 0 \end{vmatrix}$

24. $\begin{vmatrix} -6 & 5 & 3 \\ 2 & 0 & -1 \\ 4 & 0 & 7 \end{vmatrix}$

25. $\begin{vmatrix} 24 & -1 & 4 \\ 40 & 2 & 0 \\ -16 & 6 & 0 \end{vmatrix}$

26. $\begin{vmatrix} 2 & -1 & 3 \\ 0 & 3 & 1 \\ 4 & -8 & -4 \end{vmatrix}$

27. $\begin{vmatrix} 2 & 3 & -4 \\ 4 & 6 & -1 \\ -6 & 1 & -2 \end{vmatrix}$

28. $\begin{vmatrix} 1 & 2 & -3 \\ -3 & -1 & 1 \\ 4 & 5 & 4 \end{vmatrix}$

Evaluate each of the following 4×4 determinants. Use the properties of determinants to your advantage.

29. $\begin{vmatrix} 1 & -2 & 3 & 2 \\ 2 & -1 & 0 & 4 \\ -3 & 4 & 0 & -2 \\ -1 & 1 & 1 & 5 \end{vmatrix}$

30. $\begin{vmatrix} 1 & 2 & 5 & 7 \\ -6 & 3 & 0 & 9 \\ -3 & 5 & 2 & 7 \\ 2 & 1 & 4 & 3 \end{vmatrix}$

31. $\begin{vmatrix} 3 & -1 & 2 & 3 \\ 1 & 0 & 2 & 1 \\ 2 & 3 & 0 & 1 \\ 5 & 2 & 4 & -5 \end{vmatrix}$

32. $\begin{vmatrix} 1 & 2 & 0 & 0 \\ 3 & -1 & 4 & 5 \\ -2 & 4 & 1 & 6 \\ 2 & -1 & -2 & -3 \end{vmatrix}$

Use the appropriate property of determinants from this section to justify each of the following true statements. *Do not* evaluate the determinants.

33. $(-4)\begin{vmatrix} 2 & 1 & -1 \\ 3 & 2 & 1 \\ 2 & 1 & 3 \end{vmatrix} = \begin{vmatrix} 2 & -4 & -1 \\ 3 & -8 & 1 \\ 2 & -4 & 3 \end{vmatrix}$

34. $\begin{vmatrix} 1 & -2 & 3 \\ 4 & -6 & -8 \\ 0 & 2 & 7 \end{vmatrix} = (-2)\begin{vmatrix} 1 & -2 & 3 \\ -2 & 3 & 4 \\ 0 & 2 & 7 \end{vmatrix}$

35. $\begin{vmatrix} 4 & 7 & 9 \\ 6 & -8 & 2 \\ 4 & 3 & -1 \end{vmatrix} = -\begin{vmatrix} 4 & 9 & 7 \\ 6 & 2 & -8 \\ 4 & -1 & 3 \end{vmatrix}$

36. $\begin{vmatrix} 3 & -1 & 4 \\ 5 & 2 & 7 \\ 3 & -1 & 4 \end{vmatrix} = 0$

37. $\begin{vmatrix} 1 & 3 & 4 \\ -2 & 5 & 7 \\ -3 & -1 & 2 \end{vmatrix} = \begin{vmatrix} 1 & 3 & 4 \\ -2 & 5 & 7 \\ 0 & 8 & 14 \end{vmatrix}$

38. $\begin{vmatrix} 3 & 2 & 0 \\ 1 & 4 & 1 \\ -4 & 9 & 2 \end{vmatrix} = \begin{vmatrix} 3 & 2 & -3 \\ 1 & 4 & 0 \\ -4 & 9 & 6 \end{vmatrix}$

39. $\begin{vmatrix} 6 & 2 & 2 \\ 3 & -1 & 4 \\ 9 & -3 & 6 \end{vmatrix} = 6\begin{vmatrix} 2 & 2 & 1 \\ 1 & -1 & 2 \\ 3 & -3 & 3 \end{vmatrix} = 18\begin{vmatrix} 2 & 2 & 1 \\ 1 & -1 & 2 \\ 1 & -1 & 1 \end{vmatrix}$

40. $\begin{vmatrix} 2 & 1 & -3 \\ 0 & 2 & -4 \\ -5 & 1 & 3 \end{vmatrix} = -\begin{vmatrix} 2 & 1 & -3 \\ -5 & 1 & 3 \\ 0 & 2 & -4 \end{vmatrix}$

41. $\begin{vmatrix} 2 & -3 & 2 \\ 1 & -4 & 1 \\ 7 & 8 & 7 \end{vmatrix} = 0$

42. $\begin{vmatrix} 3 & 1 & 2 \\ -4 & 5 & -1 \\ 2 & -2 & -4 \end{vmatrix} = \begin{vmatrix} 3 & 1 & 0 \\ -4 & 5 & -11 \\ 2 & -2 & 0 \end{vmatrix}$

MISCELLANEOUS PROBLEMS

For Problems 43–45, use

$$A = \begin{bmatrix} a_{11} & a_{12} & a_{13} \\ a_{21} & a_{22} & a_{23} \\ a_{31} & a_{32} & a_{33} \end{bmatrix}$$

as a general representation for any 3×3 matrix.

43. Verify Property 10.2 for 3×3 matrices.

44. Verify Property 10.3 for 3×3 matrices.

45. Verify Property 10.4 for 3×3 matrices.

46. If

$$A = \begin{bmatrix} a_{11} & a_{12} & a_{13} & a_{14} \\ 0 & a_{22} & a_{23} & a_{24} \\ 0 & 0 & a_{33} & a_{34} \\ 0 & 0 & 0 & a_{44} \end{bmatrix}$$

then show that $|A| = a_{11}a_{22}a_{33}a_{44}$.

GRAPHICS CALCULATOR ACTIVITIES

47. Use a graphics calculator to check your answers for problems 29–32.

48. Let matrix

$$A = \begin{bmatrix} 2 & 5 & 7 & 9 \\ -4 & 6 & 2 & 4 \\ 6 & 9 & 12 & 3 \\ 5 & 4 & -2 & 8 \end{bmatrix}.$$

Form matrix B by interchanging rows 1 and 3 of matrix A. Now use your calculator to show that $|B| = -|A|$.

49. Let matrix

$$A = \begin{bmatrix} 2 & 1 & 7 & 6 & 8 \\ 3 & -2 & 4 & 5 & -1 \\ 6 & 7 & 9 & 12 & 13 \\ -4 & -7 & 6 & 2 & 1 \\ 9 & 8 & 12 & 14 & 17 \end{bmatrix}.$$

Form matrix B by multiplying each element of the second row of A by 3. Now use your calculator to show that $|B| = 3|A|$.

50. Let matrix

$$A = \begin{bmatrix} 4 & 3 & 2 & 1 & 5 & -3 \\ 5 & 2 & 7 & 8 & 6 & 3 \\ 0 & 9 & 1 & 4 & 7 & 2 \\ 4 & 3 & 2 & 1 & 5 & -3 \\ -4 & -6 & 7 & 12 & 11 & 9 \\ 5 & 8 & 6 & -3 & 2 & -1 \end{bmatrix}.$$

Use your calculator to show that $|A| = 0$.

10.5 CRAMER'S RULE

Determinants provide the basis for another method of solving linear systems. Consider the following linear system of two equations and two unknowns:

$$\begin{pmatrix} a_1x + b_1y = c_1 \\ a_2x + b_2y = c_2 \end{pmatrix}.$$

The augmented matrix of this system is

$$\left[\begin{array}{cc|c} a_1 & b_1 & c_1 \\ a_2 & b_2 & c_2 \end{array}\right].$$

Using the elementary row transformations of augmented matrices, we can change this matrix to the following reduced echelon form. (The details of this are left for you to do as an exercise.)

$$\left[\begin{array}{cc|c} 1 & 0 & \dfrac{c_1b_2 - c_2b_1}{a_1b_2 - a_2b_1} \\ 0 & 1 & \dfrac{a_1c_2 - a_2c_1}{a_1b_2 - a_2b_1} \end{array}\right], \qquad a_1b_2 - a_2b_1 \neq 0$$

The solution for x and y can be expressed in determinant form as follows.

$$x = \frac{c_1b_2 - c_2b_1}{a_1b_2 - a_2b_1} = \frac{\begin{vmatrix} c_1 & b_1 \\ c_2 & b_2 \end{vmatrix}}{\begin{vmatrix} a_1 & b_1 \\ a_2 & b_2 \end{vmatrix}}$$

$$y = \frac{a_1c_2 - a_2c_1}{a_1b_2 - a_2b_1} = \frac{\begin{vmatrix} a_1 & c_1 \\ a_2 & c_2 \end{vmatrix}}{\begin{vmatrix} a_1 & b_1 \\ a_2 & b_2 \end{vmatrix}}$$

This method of using determinants to solve a system of two linear equations in two variables is called **Cramer's rule** and can be stated as follows.

Cramer's Rule (2 × 2 case)

Given the system

$$\begin{pmatrix} a_1x + b_1y = c_1 \\ a_2x + b_2y = c_2 \end{pmatrix},$$

with

$$D = \begin{vmatrix} a_1 & b_1 \\ a_2 & b_2 \end{vmatrix} \neq 0,$$

$$D_x = \begin{vmatrix} c_1 & b_1 \\ c_2 & b_2 \end{vmatrix}, \quad \text{and} \quad D_y = \begin{vmatrix} a_1 & c_1 \\ a_2 & c_2 \end{vmatrix},$$

then the solution for this system is given by

$$x = \frac{D_x}{D} \quad \text{and} \quad y = \frac{D_y}{D}.$$

Notice that the elements of D are the coefficients of the variables in the given system. In D_x the coefficients of x are replaced by the corresponding constants, and in D_y the coefficients of y are replaced by the corresponding constants. Let's illustrate the use of Cramer's rule to solve some systems.

EXAMPLE 1

Solve the system $\begin{pmatrix} 6x + 3y = 2 \\ 3x + 2y = -4 \end{pmatrix}$.

Solution

The system is in the proper form to apply Cramer's rule. So let's determine D, D_x, and D_y.

$$D = \begin{vmatrix} 6 & 3 \\ 3 & 2 \end{vmatrix} = 12 - 9 = 3$$

$$D_x = \begin{vmatrix} 2 & 3 \\ -4 & 2 \end{vmatrix} = 4 + 12 = 16$$

$$D_y = \begin{vmatrix} 6 & 2 \\ 3 & -4 \end{vmatrix} = -24 - 6 = -30$$

Therefore,

$$x = \frac{D_x}{D} = \frac{16}{3}$$

and

$$y = \frac{D_y}{D} = \frac{-30}{3} = -10.$$

The solution set is $\left\{\left(\frac{16}{3}, -10\right)\right\}$.

EXAMPLE 2

Solve the system

$$\begin{pmatrix} y = -2x - 2 \\ 4x - 5y = 17 \end{pmatrix}.$$

Solution

To begin, we must change the form of the first equation so that the system fits the form given in Cramer's rule. The equation $y = -2x - 2$ can be rewritten $2x + y = -2$. The system now becomes

$$\begin{pmatrix} 2x + y = -2 \\ 4x - 5y = 17 \end{pmatrix},$$

and we can proceed to determine D, D_x, and D_y.

$$D = \begin{vmatrix} 2 & 1 \\ 4 & -5 \end{vmatrix} = -10 - 4 = -14$$

$$D_x = \begin{vmatrix} -2 & 1 \\ 17 & -5 \end{vmatrix} = 10 - 17 = -7$$

$$D_y = \begin{vmatrix} 2 & -2 \\ 4 & 17 \end{vmatrix} = 34 - (-8) = 42$$

Thus,

$$x = \frac{D_x}{D} = \frac{-7}{-14} = \frac{1}{2} \qquad \text{and} \qquad y = \frac{D_y}{D} = \frac{42}{-14} = -3.$$

The solution set is $\left\{\left(\frac{1}{2}, -3\right)\right\}$, which can be verified, as always, by substituting back into the original equations.

EXAMPLE 3

Solve the system

$$\begin{pmatrix} \frac{1}{2}x + \frac{2}{3}y = -4 \\ \frac{1}{4}x - \frac{3}{2}y = 20 \end{pmatrix}.$$

Solution

With such a system either we can first produce an equivalent system with integral coefficients and then apply Cramer's rule, or we can apply the rule immediately. Let's avoid some work with fractions by multiplying the first equation by 6 and the second equation by 4, to produce the following equivalent system.

$$\begin{pmatrix} 3x + 4y = -24 \\ x - 6y = 80 \end{pmatrix}.$$

Now we can proceed as before.

$$D = \begin{vmatrix} 3 & 4 \\ 1 & -6 \end{vmatrix} = -18 - 4 = -22;$$

$$D_x = \begin{vmatrix} -24 & 4 \\ 80 & -6 \end{vmatrix} = 144 - 320 = -176$$

$$D_y = \begin{vmatrix} 3 & -24 \\ 1 & 80 \end{vmatrix} = 240 - (-24) = 264.$$

Therefore,

$$x = \frac{D_x}{D} = \frac{-176}{-22} = 8 \qquad \text{and} \qquad y = \frac{D_y}{D} = \frac{264}{-22} = -12.$$

The solution set is $\{(8, -12)\}$.

In the statement of Cramer's rule, the condition was imposed that $D \neq 0$. If $D = 0$ and either D_x or D_y (or both) is nonzero, then the system is inconsistent and has no solution. If $D = 0$, $D_x = 0$, and $D_y = 0$, then the equations are dependent and there are infinitely many solutions.

Cramer's Rule Extended

Without showing the details, we will simply state that Cramer's rule also applies to solving systems of three linear equations in three variables. It can be stated as follows.

Cramer's Rule (3 × 3 cases)

Given the system

$$\begin{pmatrix} a_1x + b_1y + c_1z = d_1 \\ a_2x + b_2y + c_2z = d_2 \\ a_3x + b_3y + c_3z = d_3 \end{pmatrix},$$

with

$$D = \begin{vmatrix} a_1 & b_1 & c_1 \\ a_2 & b_2 & c_2 \\ a_3 & b_3 & c_3 \end{vmatrix} \neq 0, \qquad D_x = \begin{vmatrix} d_1 & b_1 & c_1 \\ d_2 & b_2 & c_2 \\ d_3 & b_3 & c_3 \end{vmatrix},$$

$$D_y = \begin{vmatrix} a_1 & d_1 & c_1 \\ a_2 & d_2 & c_2 \\ a_3 & d_3 & c_3 \end{vmatrix}, \qquad D_z = \begin{vmatrix} a_1 & b_1 & d_1 \\ a_2 & b_2 & d_2 \\ a_3 & b_3 & d_3 \end{vmatrix},$$

then

$$x = \frac{D_x}{D}, \quad y = \frac{D_y}{D}, \quad \text{and} \quad z = \frac{D_z}{D}.$$

Again, notice the restriction that $D \neq 0$. If $D = 0$ and at least one of D_x, D_y, and D_z is not zero, then the system is inconsistent. If D, D_x, D_y, and D_z are all zero, then the equations are dependent and there are infinitely many solutions.

EXAMPLE 4

Solve the system

$$\begin{pmatrix} x - 2y + z = -4 \\ 2x + y - z = 5 \\ 3x + 2y + 4z = 3 \end{pmatrix}.$$

Solution

We will simply indicate the values of D, D_x, D_y, and D_z, and leave the computations for you to check.

$$D = \begin{vmatrix} 1 & -2 & 1 \\ 2 & 1 & -1 \\ 3 & 2 & 4 \end{vmatrix} = 29 \qquad D_x = \begin{vmatrix} -4 & -2 & 1 \\ 5 & 1 & -1 \\ 3 & 2 & 4 \end{vmatrix} = 29$$

$$D_y = \begin{vmatrix} 1 & -4 & 1 \\ 2 & 5 & -1 \\ 3 & 3 & 4 \end{vmatrix} = 58 \qquad D_z = \begin{vmatrix} 1 & -2 & -4 \\ 2 & 1 & 5 \\ 3 & 2 & 3 \end{vmatrix} = -29$$

Therefore,

$$x = \frac{D_x}{D} = \frac{29}{29} = 1,$$

$$y = \frac{D_y}{D} = \frac{58}{29} = 2,$$

and

$$z = \frac{D_z}{D} = \frac{-29}{29} = -1.$$

The solution set is $\{(1, 2, -1)\}$. (Be sure to check it!)

EXAMPLE 5

Solve the system

$$\begin{pmatrix} x + 3y - z = 4 \\ 3x - 2y + z = 7 \\ 2x + 6y - 2z = 1 \end{pmatrix}.$$

Solution

$$D = \begin{vmatrix} 1 & 3 & -1 \\ 3 & -2 & 1 \\ 2 & 6 & -2 \end{vmatrix} = 2\begin{vmatrix} 1 & 3 & -1 \\ 3 & -2 & 1 \\ 1 & 3 & -1 \end{vmatrix} = 2(0) = 0$$

$$D_x = \begin{vmatrix} 4 & 3 & -1 \\ 7 & -2 & 1 \\ 1 & 6 & -2 \end{vmatrix} = -7$$

Therefore, since $D = 0$ and at least one of D_x, D_y, and D_z is not zero, the system is inconsistent. The solution set is $\varnothing$.

Example 5 illustrates why D should be determined first. Once we found that $D = 0$ and $D_x \neq 0$, we knew that the system was inconsistent and there was no need to find D_y, and D_z.

Finally, it should be noted that Cramer's rule can be extended to systems of n linear equations in n variables; however, that method is not considered to be a very efficient way of solving a large system of linear equations.

PROBLEM SET 10.5

Use Cramer's rule to find the solution set for each of the following systems. If the equations are dependent, simply indicate that there are infinitely many solutions.

1. $\begin{pmatrix} 2x - y = -2 \\ 3x + 2y = 11 \end{pmatrix}$

2. $\begin{pmatrix} 3x + y = -9 \\ 4x - 3y = 1 \end{pmatrix}$

3. $\begin{pmatrix} 5x + 2y = 5 \\ 3x - 4y = 29 \end{pmatrix}$

4. $\begin{pmatrix} 4x - 7y = -23 \\ 2x + 5y = -3 \end{pmatrix}$

5. $\begin{pmatrix} 5x - 4y = 14 \\ -x + 2y = -4 \end{pmatrix}$

6. $\begin{pmatrix} -x + 2y = 10 \\ 3x - y = -10 \end{pmatrix}$

7. $\begin{pmatrix} y = 2x - 4 \\ 6x - 3y = 1 \end{pmatrix}$

8. $\begin{pmatrix} -3x - 4y = 14 \\ -2x + 3y = -19 \end{pmatrix}$

9. $\begin{pmatrix} -4x + 3y = 3 \\ 4x - 6y = -5 \end{pmatrix}$

10. $\begin{pmatrix} x = 4y - 1 \\ 2x - 8y = -2 \end{pmatrix}$

11. $\begin{pmatrix} 9x - y = -2 \\ 8x + y = 4 \end{pmatrix}$

12. $\begin{pmatrix} 6x - 5y = 1 \\ 4x - 7y = 2 \end{pmatrix}$

13. $\begin{pmatrix} -\frac{2}{3}x + \frac{1}{2}y = -7 \\ \frac{1}{3}x - \frac{3}{2}y = 6 \end{pmatrix}$

14. $\begin{pmatrix} \frac{1}{2}x + \frac{2}{3}y = -6 \\ \frac{1}{4}x - \frac{1}{3}y = -1 \end{pmatrix}$

15. $\begin{pmatrix} 2x + 7y = -1 \\ x = 2 \end{pmatrix}$

16. $\begin{pmatrix} 5x - 3y = 2 \\ y = 4 \end{pmatrix}$

17. $\begin{pmatrix} x - y + 2z = -8 \\ 2x + 3y - 4z = 18 \\ -x + 2y - z = 7 \end{pmatrix}$

18. $\begin{pmatrix} x - 2y + z = 3 \\ 3x + 2y + z = -3 \\ 2x - 3y - 3z = -5 \end{pmatrix}$

19. $\begin{pmatrix} 2x - 3y + z = -7 \\ -3x + y - z = -7 \\ x - 2y - 5z = -45 \end{pmatrix}$

20. $\begin{pmatrix} 3x - y - z = 18 \\ 4x + 3y - 2z = 10 \\ -5x - 2y + 3z = -22 \end{pmatrix}$

21. $\begin{pmatrix} 4x + 5y - 2z = -14 \\ 7x - y + 2z = 42 \\ 3x + y + 4z = 28 \end{pmatrix}$

22. $\begin{pmatrix} -5x + 6y + 4z = -4 \\ -7x - 8y + 2z = -2 \\ 2x + 9y - z = 1 \end{pmatrix}$

23. $\begin{pmatrix} 2x - y + 3z = -17 \\ 3y + z = 5 \\ x - 2y - z = -3 \end{pmatrix}$

24. $\begin{pmatrix} 2x - y + 3z = -5 \\ 3x + 4y - 2z = -25 \\ -x + z = 6 \end{pmatrix}$

25. $\begin{pmatrix} x + 3y - 4z = -1 \\ 2x - y + z = 2 \\ 4x + 5y - 7z = 0 \end{pmatrix}$

26. $\begin{pmatrix} x - 2y + z = 1 \\ 3x + y - z = 2 \\ 2x - 4y + 2z = -1 \end{pmatrix}$

27. $\begin{pmatrix} 3x - 2y - 3z = -5 \\ x + 2y + 3z = -3 \\ -x + 4y - 6z = 8 \end{pmatrix}$

28. $\begin{pmatrix} 3x - 2y + z = 11 \\ 5x + 3y = 17 \\ x + y - 2z = 6 \end{pmatrix}$

29. $\begin{pmatrix} x - 2y + 3z = 1 \\ -2x + 4y - 3z = -3 \\ 5x - 6y + 6z = 10 \end{pmatrix}$

30. $\begin{pmatrix} 2x - y + 2z = -1 \\ 4x + 3y - 4z = 2 \\ x + 5y - z = 9 \end{pmatrix}$

31. $\begin{pmatrix} -x - y + 3z = -2 \\ -2x + y + 7z = 14 \\ 3x + 4y - 5z = 12 \end{pmatrix}$

32. $\begin{pmatrix} -2x + y - 3z = -4 \\ x + 5y - 4z = 13 \\ 7x - 2y - z = 37 \end{pmatrix}$

THOUGHTS into WORDS

33. Explain the difference between a matrix and a determinant.

34. Give a step-by-step description of how you would solve the system

$$\begin{pmatrix} 2x - y + 3z = 31 \\ x - 2y - z = 8 \\ 3x + 5y + 8z = 35 \end{pmatrix}.$$

35. Give a step-by-step description of how you would find the value of x in the solution for the system

$$\begin{pmatrix} x + 5y - z = -9 \\ 2x - y + z = 11 \\ -3x - 2y + 4z = 20 \end{pmatrix}.$$

MISCELLANEOUS PROBLEMS

36. A linear system in which the constant terms are all zero is called a **homogeneous system**.

a. Verify that for a 3 × 3 homogeneous system, if $D \neq 0$, then (0, 0, 0) is the only solution for the system.

b. Verify that for a 3 × 3 homogeneous system, if $D = 0$, then the equations are dependent.

For Problems 37–40, solve each of the homogeneous systems (see problem 36). If the equations are dependent, indicate that the system has infinitely many solutions.

37. $\begin{pmatrix} x - 2y + 5z = 0 \\ 3x + y - 2z = 0 \\ 4x - y + 3z = 0 \end{pmatrix}$

38. $\begin{pmatrix} 2x - y + z = 0 \\ 3x + 2y + 5z = 0 \\ 4x - 7y + z = 0 \end{pmatrix}$

39. $\begin{pmatrix} 3x + y - z = 0 \\ x - y + 2z = 0 \\ 4x - 5y - 2z = 0 \end{pmatrix}$

40. $\begin{pmatrix} 2x - y + 2z = 0 \\ x + 2y + z = 0 \\ x - 3y + z = 0 \end{pmatrix}$

GRAPHICS CALCULATOR ACTIVITIES

41. Use determinants and your calculator to help solve each of the following systems.

a. $\begin{pmatrix} 4x - 3y + z = 10 \\ 8x + 5y - 2z = -6 \\ -12x - 2y + 3z = -2 \end{pmatrix}$

b. $\begin{pmatrix} 2x + y - z + w = -4 \\ x + 2y + 2z - 3w = 6 \\ 3x - y - z + 2w = 0 \\ 2x + 3y + z + 4w = -5 \end{pmatrix}$

c. $\begin{pmatrix} x - 2y + z - 3w = 4 \\ 2x + 3y - z - 2w = -4 \\ 3x - 4y + 2z - 4w = 12 \\ 2x - y - 3z + 2w = -2 \end{pmatrix}$

d. $\begin{pmatrix} 1.98x + 2.49y + 3.45z = 80.10 \\ 2.15x + 3.20y + 4.19z = 97.16 \\ 1.49x + 4.49y + 2.79z = 83.92 \end{pmatrix}$

CHAPTER 10 SUMMARY

The primary focus of this entire chapter is the development of different techniques for solving systems of linear equations.

Substitution Method

With the aid of an example, we can describe the substitution method as follows. Suppose we want to solve the system

$$\begin{pmatrix} x - 2y = 22 \\ 3x + 4y = -24 \end{pmatrix}.$$

STEP 1 Solve the first equation for x in terms of y.

$$x - 2y = 22$$
$$x = 2y + 22$$

STEP 2 Substitute $2y + 22$ for x in the second equation.

$$3(2y + 22) + 4y = -24$$

STEP 3 Solve the equation obtained in step 2.

$$6y + 66 + 4y = -24$$
$$10y + 66 = -24$$
$$10y = -90$$
$$y = -9$$

STEP 4. Substitute -9 for y in the equation of step 1.

$$x = 2(-9) + 22 = 4$$

The solution set is $\{(4, -9)\}$.

Elimination-by-Addition Method

This method allows us to replace systems of equations with *simpler equivalent systems* until we obtain a system where we can easily determine the solution. The following operations produce equivalent systems.

1. Any two equations of a system can be interchanged.
2. Both sides of any equation of the system can be multiplied by any nonzero real number.
3. Any equation of the system can be replaced by the sum of a nonzero multiple of another equation plus that equation.

For example, through a sequence of operations, we can transform the system

$$\begin{pmatrix} 5x + 3y = -28 \\ \frac{1}{2}x - y = -8 \end{pmatrix}$$

to the equivalent system

$$\begin{pmatrix} x - 2y = -16 \\ 13y = 52 \end{pmatrix},$$

where we can easily determine the solution set $\{(-8, 4)\}$.

Matrix Approach

We can change the augmented matrix of a system to reduced echelon form by applying the following elementary row operations.

1. Any two rows of the matrix can be interchanged.
2. Any row of the matrix can be multiplied by a nonzero real number.
3. Any row of the matrix can be replaced by the sum of a nonzero multiple of another row plus that row.

For example, the augmented matrix of the system

$$\begin{pmatrix} x - 2y + 3z = 4 \\ 2x + y - 4z = 3 \\ -3x + 4y - z = -2 \end{pmatrix}$$

is

$$\left[\begin{array}{rrr|r} 1 & -2 & 3 & 4 \\ 2 & 1 & -4 & 3 \\ -3 & 4 & -1 & -2 \end{array}\right].$$

We can change this matrix to the reduced echelon form

$$\left[\begin{array}{ccc|c} 1 & 0 & 0 & 4 \\ 0 & 1 & 0 & 3 \\ 0 & 0 & 1 & 2 \end{array}\right],$$

where the solution set of $\{(4, 3, 2)\}$ is obvious.

Cramer's Rule

Cramer's rule for solving systems of linear equations involves the use of determinants. It is stated for the 2×2 case on page 633 and for the 3×3 case on page 636. For example, the solution set of the system

$$\begin{pmatrix} 3x - y - z = 2 \\ 2x + y + 3z = 9 \\ -x + 5y - 6z = -29 \end{pmatrix}$$

is determined by

$$x = \frac{\begin{vmatrix} 2 & -1 & -1 \\ 9 & 1 & 3 \\ -29 & 5 & -6 \end{vmatrix}}{\begin{vmatrix} 3 & -1 & -1 \\ 2 & 1 & 3 \\ -1 & 5 & -6 \end{vmatrix}} = \frac{-83}{-83} = 1,$$

$$y = \frac{\begin{vmatrix} 3 & 2 & -1 \\ 2 & 9 & 3 \\ -1 & -29 & -6 \end{vmatrix}}{\begin{vmatrix} 3 & -1 & -1 \\ 2 & 1 & 3 \\ -1 & 5 & -6 \end{vmatrix}} = \frac{166}{-83} = -2,$$

and

$$z = \frac{\begin{vmatrix} 3 & -1 & 2 \\ 2 & 1 & 9 \\ -1 & 5 & -29 \end{vmatrix}}{\begin{vmatrix} 3 & -1 & -1 \\ 2 & 1 & 3 \\ -1 & 5 & -6 \end{vmatrix}} = \frac{-249}{-83} = 3.$$

CHAPTER 10 REVIEW PROBLEM SET

Solve each of the following systems by using the *substitution* method.

1. $\begin{pmatrix} 3x - y = 16 \\ 5x + 7y = -34 \end{pmatrix}$

2. $\begin{pmatrix} 6x + 5y = -21 \\ x - 4y = 11 \end{pmatrix}$

3. $\begin{pmatrix} 2x - 3y = 12 \\ 3x + 5y = -20 \end{pmatrix}$

4. $\begin{pmatrix} 5x + 8y = 1 \\ 4x + 7y = -2 \end{pmatrix}$

Solve each of the following systems by using the *elimination-by-addition* method.

5. $\begin{pmatrix} 4x - 3y = 34 \\ 3x + 2y = 0 \end{pmatrix}$

6. $\begin{pmatrix} \frac{1}{2}x - \frac{2}{3}y = 1 \\ \frac{3}{4}x + \frac{1}{6}y = -1 \end{pmatrix}$

7. $\begin{pmatrix} 2x - y + 3z = -19 \\ 3x + 2y - 4z = 21 \\ 5x - 4y - z = -8 \end{pmatrix}$

8. $\begin{pmatrix} 3x + 2y - 4z = 4 \\ 5x + 3y - z = 2 \\ 4x - 2y + 3z = 11 \end{pmatrix}$

Solve each of the following systems by *changing the augmented matrix to reduced echelon form.*

9. $\begin{pmatrix} x - 3y = 17 \\ -3x + 2y = -23 \end{pmatrix}$

10. $\begin{pmatrix} 2x + 3y = 25 \\ 3x - 5y = -29 \end{pmatrix}$

11. $\begin{pmatrix} x - 2y + z = -7 \\ 2x - 3y + 4z = -14 \\ -3x + y - 2z = 10 \end{pmatrix}$

12. $\begin{pmatrix} -2x - 7y + z = 9 \\ x + 3y - 4z = -11 \\ 4x + 5y - 3z = -11 \end{pmatrix}$

Solve each of the following systems by using *Cramer's rule.*

13. $\begin{pmatrix} 5x + 3y = -18 \\ 4x - 9y = -3 \end{pmatrix}$

14. $\begin{pmatrix} 0.2x + 0.3y = 2.6 \\ 0.5x - 0.1y = 1.4 \end{pmatrix}$

15. $\begin{pmatrix} 2x - 3y - 3z = 25 \\ 3x + y + 2z = -5 \\ 5x - 2y - 4z = 32 \end{pmatrix}$

16. $\begin{pmatrix} 3x - y + z = -10 \\ 6x - 2y + 5z = -35 \\ 7x + 3y - 4z = 19 \end{pmatrix}$

Solve each of the following systems by using the method you think is most appropriate.

17. $\begin{pmatrix} 4x + 7y = -15 \\ 3x - 2y = 25 \end{pmatrix}$

18. $\begin{pmatrix} \frac{3}{4}x - \frac{1}{2}y = -15 \\ \frac{2}{3}x + \frac{1}{4}y = -5 \end{pmatrix}$

19. $\begin{pmatrix} x + 4y = 3 \\ 3x - 2y = 1 \end{pmatrix}$

20. $\begin{pmatrix} 7x - 3y = -49 \\ y = \frac{3}{5}x - 1 \end{pmatrix}$

21. $\begin{pmatrix} x - y - z = 4 \\ -3x + 2y + 5z = -21 \\ 5x - 3y - 7z = 30 \end{pmatrix}$

22. $\begin{pmatrix} 2x - y + z = -7 \\ -5x + 2y - 3z = 17 \\ 3x + y + 7z = -5 \end{pmatrix}$

23. $\begin{pmatrix} 3x - 2y - 5z = 2 \\ -4x + 3y + 11z = 3 \\ 2x - y + z = -1 \end{pmatrix}$

24. $\begin{pmatrix} 7x - y + z = -4 \\ -2x + 9y - 3z = -50 \\ x - 5y + 4z = 42 \end{pmatrix}$

Evaluate each of the following determinants.

25. $\begin{vmatrix} -2 & 6 \\ 3 & 8 \end{vmatrix}$

26. $\begin{vmatrix} 5 & -4 \\ 7 & -3 \end{vmatrix}$

27. $\begin{vmatrix} 2 & 3 & -1 \\ 3 & 4 & -5 \\ 6 & 4 & 2 \end{vmatrix}$

28. $\begin{vmatrix} 3 & -2 & 4 \\ 1 & 0 & 6 \\ 3 & -3 & 5 \end{vmatrix}$

29. $\begin{vmatrix} 5 & 4 & 3 \\ 2 & -7 & 0 \\ 3 & -2 & 0 \end{vmatrix}$

30. $\begin{vmatrix} 5 & -4 & 2 & 1 \\ 3 & 7 & 6 & -2 \\ 2 & 1 & -5 & 0 \\ 3 & -2 & 4 & 0 \end{vmatrix}$

Solve each of the following problems by setting up and solving a system of linear equations.

31. The sum of the digits of a two-digit number is 9. If the digits are reversed, the newly formed number is 45 less than the original number. Find the original number.

32. Sara invested \$2500, part of it at 10% and the rest at 12% yearly interest. The yearly income on the 12% investment was \$102 more than the income on the 10% investment. How much money did she invest at each rate?

33. A box contains \$17.70 in nickels, dimes, and quarters. The number of dimes is eight less than twice the number of nickels. The number of quarters is two more than the sum of the number of nickels and dimes. How many coins of each kind are there in the box?

34. The measure of the largest angle of a triangle is 10° more than four times the smallest angle. The sum of the smallest and largest angles is three times the measure of the other angle. Find the measure of each angle of the triangle.

ALGEBRA OF MATRICES

In Section 10.3, matrices were used strictly as a device to help solve systems of linear equations. Our primary objective was the development of techniques for solving systems of equations, not the study of matrices. However, matrices can be studied from an algebraic viewpoint, much as we study the set of real numbers. That is, we can define certain operations on matrices and verify properties of those operations. This algebraic approach to matrices is the focal point of this chapter. In order to get a simplified view of the algebra of matrices, we will begin by studying 2×2 matrices, and then later we will expand to $m \times n$ matrices. As a bonus, another technique for solving systems of equations will emerge from our study.

11

Matrices can be used to organize and manipulate data that determine population trends over a certain time interval.

11.1 ALGEBRA OF 2 × 2 MATRICES

Throughout these next two sections, we will be working primarily with 2 × 2 matrices; therefore, any reference to matrices means 2 × 2 matrices unless stated otherwise. The following general 2 × 2 matrix notation will be used frequently.

$$A = \begin{bmatrix} a_{11} & a_{12} \\ a_{21} & a_{22} \end{bmatrix} \qquad B = \begin{bmatrix} b_{11} & b_{12} \\ b_{21} & b_{22} \end{bmatrix} \qquad C = \begin{bmatrix} c_{11} & c_{12} \\ c_{21} & c_{22} \end{bmatrix}$$

Two matrices are **equal** if and only if all elements in corresponding positions are equal. Thus, $A = B$ if and only if $a_{11} = b_{11}$, $a_{12} = b_{12}$, $a_{21} = b_{21}$, and $a_{22} = b_{22}$.

Addition of Matrices

To add two matrices, we add the elements that appear in corresponding positions. Therefore, the sum of A and B is defined as follows.

DEFINITION 11.1

$$A + B = \begin{bmatrix} a_{11} & a_{12} \\ a_{21} & a_{22} \end{bmatrix} + \begin{bmatrix} b_{11} & b_{12} \\ b_{21} & b_{22} \end{bmatrix} = \begin{bmatrix} a_{11} + b_{11} & a_{12} + b_{12} \\ a_{21} + b_{21} & a_{22} + b_{22} \end{bmatrix}$$

For example,

$$\begin{bmatrix} 2 & -1 \\ -3 & 4 \end{bmatrix} + \begin{bmatrix} -5 & 4 \\ -1 & 7 \end{bmatrix} = \begin{bmatrix} -3 & 3 \\ -4 & 11 \end{bmatrix}.$$

It is not difficult to show that **the commutative and associative properties are valid for the addition of matrices**. Thus, we can state that

$$A + B = B + A \qquad \text{and} \qquad (A + B) + C = A + (B + C).$$

Since

$$\begin{bmatrix} a_{11} & a_{12} \\ a_{21} & a_{22} \end{bmatrix} + \begin{bmatrix} 0 & 0 \\ 0 & 0 \end{bmatrix} = \begin{bmatrix} a_{11} & a_{12} \\ a_{21} & a_{22} \end{bmatrix},$$

we see that $\begin{bmatrix} 0 & 0 \\ 0 & 0 \end{bmatrix}$, which is called the **zero matrix** and represented by O, is the additive identity element. Thus, we can state that

$$A + O = O + A = A.$$

Since every real number has an additive inverse, it follows that any matrix A has an additive inverse, $-A$, that is formed by taking the additive inverse of each element of A. For example, if

$$A = \begin{bmatrix} 4 & -2 \\ -1 & 0 \end{bmatrix} \quad \text{then } -A = \begin{bmatrix} -4 & 2 \\ 1 & 0 \end{bmatrix}$$

and

$$A + (-A) = \begin{bmatrix} 4 & -2 \\ -1 & 0 \end{bmatrix} + \begin{bmatrix} -4 & 2 \\ 1 & 0 \end{bmatrix} = \begin{bmatrix} 0 & 0 \\ 0 & 0 \end{bmatrix}.$$

In general, we can state that every matrix A has an additive inverse $-A$ such that

$$A + (-A) = (-A) + A = O.$$

Subtraction of Matrices

Again, paralleling the algebra of real numbers, *subtraction* of matrices can be defined in terms of *adding the additive inverse*. Therefore, we can define subtraction as follows.

DEFINITION 11.2

$$A - B = A + (-B)$$

For example,

$$\begin{bmatrix} 2 & -7 \\ -6 & 5 \end{bmatrix} - \begin{bmatrix} 3 & 4 \\ -2 & -1 \end{bmatrix} = \begin{bmatrix} 2 & -7 \\ -6 & 5 \end{bmatrix} + \begin{bmatrix} -3 & -4 \\ 2 & 1 \end{bmatrix}$$
$$= \begin{bmatrix} -1 & -11 \\ -4 & 6 \end{bmatrix}.$$

Scalar Multiplication

When we work with matrices, we commonly refer to a single real number as a scalar to distinguish it from a matrix. Then the product of a scalar and a matrix (often referred to as **scalar multiplication**) can be formed by multiplying each element of the matrix by the scalar. For example,

$$3\begin{bmatrix} -4 & -6 \\ 1 & -2 \end{bmatrix} = \begin{bmatrix} 3(-4) & 3(-6) \\ 3(1) & 3(-2) \end{bmatrix} = \begin{bmatrix} -12 & -18 \\ 3 & -6 \end{bmatrix}.$$

In general, scalar multiplication can be defined as follows.

DEFINITION 11.3

$$kA = k\begin{bmatrix} a_{11} & a_{12} \\ a_{21} & a_{22} \end{bmatrix} = \begin{bmatrix} ka_{11} & ka_{12} \\ ka_{21} & ka_{22} \end{bmatrix},$$

where k is any real number.

EXAMPLE 1

If $A = \begin{bmatrix} -4 & 3 \\ 2 & -5 \end{bmatrix}$ and $B = \begin{bmatrix} 2 & -3 \\ 7 & -6 \end{bmatrix}$, find

a. $-2A$, **b.** $3A + 2B$, and **c.** $A - 4B$.

Solutions

a. $-2A = -2\begin{bmatrix} -4 & 3 \\ 2 & -5 \end{bmatrix} = \begin{bmatrix} 8 & -6 \\ -4 & 10 \end{bmatrix}$

b.
$$\begin{aligned} 3A + 2B &= 3\begin{bmatrix} -4 & 3 \\ 2 & -5 \end{bmatrix} + 2\begin{bmatrix} 2 & -3 \\ 7 & -6 \end{bmatrix} \\ &= \begin{bmatrix} -12 & 9 \\ 6 & -15 \end{bmatrix} + \begin{bmatrix} 4 & -6 \\ 14 & -12 \end{bmatrix} \\ &= \begin{bmatrix} -8 & 3 \\ 20 & -27 \end{bmatrix} \end{aligned}$$

c.
$$\begin{aligned} A - 4B &= \begin{bmatrix} -4 & 3 \\ 2 & -5 \end{bmatrix} - 4\begin{bmatrix} 2 & -3 \\ 7 & -6 \end{bmatrix} \\ &= \begin{bmatrix} -4 & 3 \\ 2 & -5 \end{bmatrix} - \begin{bmatrix} 8 & -12 \\ 28 & -24 \end{bmatrix} \\ &= \begin{bmatrix} -4 & 3 \\ 2 & -5 \end{bmatrix} + \begin{bmatrix} -8 & 12 \\ -28 & 24 \end{bmatrix} \\ &= \begin{bmatrix} -12 & 15 \\ -26 & 19 \end{bmatrix} \end{aligned}$$

The following properties, which are easy to check, pertain to scalar multiplication and matrix addition (where k and l represent any real numbers).

$$k(A + B) = kA + kB$$
$$(k + l)A = kA + lA$$
$$(kl)A = k(lA)$$

Multiplication of Matrices

At this time, it probably would seem quite natural to define matrix multiplication by multiplying corresponding elements of two matrices. However, it turns out that such a definition does not have many worthwhile applications. Therefore we use a special type of **matrix multiplication**, sometimes referred to as *row-by-column* multiplication. We will state the definition, paraphrase what it says, and then use some examples to be sure of its meaning.

DEFINITION 11.4

$$AB = \begin{bmatrix} a_{11} & a_{12} \\ a_{21} & a_{22} \end{bmatrix} \begin{bmatrix} b_{11} & b_{12} \\ b_{21} & b_{22} \end{bmatrix}$$
$$= \begin{bmatrix} a_{11}b_{11} + a_{12}b_{21} & a_{11}b_{12} + a_{12}b_{22} \\ a_{21}b_{11} + a_{22}b_{21} & a_{21}b_{12} + a_{22}b_{22} \end{bmatrix}$$

Notice the row-by-column pattern of Definition 11.4. We multiply the rows of A times the columns of B in a pairwise entry fashion, adding the results. For example, the element in the first row and second column of the product is obtained by multiplying the elements of the first row of A times the elements of the second column of B and adding the results.

$$\begin{bmatrix} \boxed{a_{11} \quad a_{12}} \\ a_{21} \quad a_{22} \end{bmatrix} \begin{bmatrix} b_{11} & \boxed{b_{12}} \\ b_{21} & \boxed{b_{22}} \end{bmatrix} = \begin{bmatrix} & a_{11}b_{12} + a_{12}b_{22} \\ & \end{bmatrix}$$

Now let's look at some specific examples.

EXAMPLE 2

If $A = \begin{bmatrix} -2 & 1 \\ 4 & 5 \end{bmatrix}$ and $B = \begin{bmatrix} 3 & -2 \\ -1 & 7 \end{bmatrix}$, find (a) AB and (b) BA.

Solutions

a. $AB = \begin{bmatrix} -2 & 1 \\ 4 & 5 \end{bmatrix} \begin{bmatrix} 3 & -2 \\ -1 & 7 \end{bmatrix}$

$$= \begin{bmatrix} (-2)(3) + (1)(-1) & (-2)(-2) + (1)(7) \\ (4)(3) + (5)(-1) & (4)(-2) + (5)(7) \end{bmatrix}$$
$$= \begin{bmatrix} -7 & 11 \\ 7 & 27 \end{bmatrix}$$

b. $$BA = \begin{bmatrix} 3 & -2 \\ -1 & 7 \end{bmatrix}\begin{bmatrix} -2 & 1 \\ 4 & 5 \end{bmatrix}$$
$$= \begin{bmatrix} (3)(-2) + (-2)(4) & (3)(1) + (-2)(5) \\ (-1)(-2) + (7)(4) & (-1)(1) + (7)(5) \end{bmatrix} = \begin{bmatrix} -14 & -7 \\ 30 & 34 \end{bmatrix}$$

Already from Example 2 we see that **matrix multiplication is not a commutative operation**.

EXAMPLE 3

If $A = \begin{bmatrix} 2 & -6 \\ -3 & 9 \end{bmatrix}$ and $B = \begin{bmatrix} -3 & 6 \\ -1 & 2 \end{bmatrix}$, find AB.

Solution

Once you feel comfortable with Definition 11.4, the additions can be done mentally.

$$AB = \begin{bmatrix} 2 & -6 \\ -3 & 9 \end{bmatrix}\begin{bmatrix} -3 & 6 \\ -1 & 2 \end{bmatrix} = \begin{bmatrix} 0 & 0 \\ 0 & 0 \end{bmatrix}.$$

Example 3 illustrates that the product of two matrices can be the zero matrix even though neither of the two matrices is the zero matrix. This is different from the property of real numbers that states $ab = 0$ *if and only if* $a = 0$ *or* $b = 0$.

As we illustrated and stated earlier, matrix multiplication is *not* a commutative operation. However, it is an **associative operation** and it does abide by two **distributive properties**. These properties can be stated as follows.

$$(AB)C = A(BC)$$
$$A(B + C) = AB + AC$$
$$(B + C)A = BA + CA$$

We will have you verify these properties in the next set of problems.

PROBLEM SET 11.1

For Problems 1–12, compute the indicated matrix using the following matrices:

$A = \begin{bmatrix} 1 & -2 \\ 3 & 4 \end{bmatrix}$, $B = \begin{bmatrix} 2 & -3 \\ 5 & -1 \end{bmatrix}$, $C = \begin{bmatrix} 0 & 6 \\ -4 & 2 \end{bmatrix}$, $D = \begin{bmatrix} -2 & 3 \\ 5 & -4 \end{bmatrix}$, and $E = \begin{bmatrix} 2 & 5 \\ 7 & 3 \end{bmatrix}$.

1. $A + B$

2. $B - C$

3. $3C + D$

4. $2D - E$

5. $4A - 3B$

6. $2B + 3D$

7. $(A - B) - C$

8. $B - (D - E)$

9. $2D - 4E$

10. $3A - 4E$

11. $B - (D + E)$

12. $A - (B + C)$

For Problems 13–26, compute AB and BA.

13. $A = \begin{bmatrix} 1 & -1 \\ 2 & -2 \end{bmatrix}$, $B = \begin{bmatrix} 3 & -4 \\ -1 & 2 \end{bmatrix}$

14. $A = \begin{bmatrix} -3 & 4 \\ 2 & 1 \end{bmatrix}$, $B = \begin{bmatrix} -2 & 5 \\ 6 & -1 \end{bmatrix}$

15. $A = \begin{bmatrix} 1 & -3 \\ -4 & 6 \end{bmatrix}$, $B = \begin{bmatrix} 7 & -3 \\ 4 & 5 \end{bmatrix}$

16. $A = \begin{bmatrix} 5 & 0 \\ -2 & 3 \end{bmatrix}$, $B = \begin{bmatrix} -3 & 6 \\ 4 & 1 \end{bmatrix}$

17. $A = \begin{bmatrix} 2 & -4 \\ 1 & -2 \end{bmatrix}$, $B = \begin{bmatrix} 1 & -2 \\ -3 & 6 \end{bmatrix}$

18. $A = \begin{bmatrix} 1 & 2 \\ 1 & 2 \end{bmatrix}$, $B = \begin{bmatrix} 2 & 2 \\ -1 & -1 \end{bmatrix}$

19. $A = \begin{bmatrix} -3 & -2 \\ -4 & -1 \end{bmatrix}$, $B = \begin{bmatrix} 2 & -1 \\ 4 & 5 \end{bmatrix}$

20. $A = \begin{bmatrix} -2 & 3 \\ -1 & 7 \end{bmatrix}$, $B = \begin{bmatrix} -1 & -3 \\ -5 & -7 \end{bmatrix}$

21. $A = \begin{bmatrix} 2 & -1 \\ -5 & 3 \end{bmatrix}$, $B = \begin{bmatrix} 3 & 1 \\ 5 & 2 \end{bmatrix}$

22. $A = \begin{bmatrix} -8 & -5 \\ 3 & 2 \end{bmatrix}$, $B = \begin{bmatrix} -2 & -5 \\ 3 & 8 \end{bmatrix}$

23. $A = \begin{bmatrix} \frac{1}{2} & -\frac{1}{3} \\ \frac{1}{3} & \frac{1}{4} \end{bmatrix}$, $B = \begin{bmatrix} 4 & -6 \\ 6 & -4 \end{bmatrix}$

24. $A = \begin{bmatrix} \frac{1}{3} & -\frac{1}{2} \\ \frac{3}{2} & -\frac{2}{3} \end{bmatrix}$, $B = \begin{bmatrix} -6 & -18 \\ 12 & -12 \end{bmatrix}$

25. $A = \begin{bmatrix} 5 & 6 \\ 2 & 3 \end{bmatrix}$, $B = \begin{bmatrix} 1 & -2 \\ -\frac{2}{3} & \frac{5}{3} \end{bmatrix}$

26. $A = \begin{bmatrix} -3 & -5 \\ 2 & 4 \end{bmatrix}$, $B = \begin{bmatrix} -2 & -\frac{5}{2} \\ 1 & \frac{3}{2} \end{bmatrix}$

For Problems 27–30, use the following matrices.

$$A = \begin{bmatrix} -2 & 3 \\ 5 & 4 \end{bmatrix}, \quad B = \begin{bmatrix} 0 & 1 \\ 1 & 0 \end{bmatrix},$$

$$C = \begin{bmatrix} 1 & 0 \\ 1 & 0 \end{bmatrix}, \quad D = \begin{bmatrix} 1 & 1 \\ 1 & 1 \end{bmatrix},$$

$$I = \begin{bmatrix} 1 & 0 \\ 0 & 1 \end{bmatrix}$$

27. Compute AB and BA.

28. Compute AC and CA.

29. Compute AD and DA.

30. Compute AI and IA.

For Problems 31–34, use the following matrices.

$$A = \begin{bmatrix} 2 & 4 \\ 5 & -3 \end{bmatrix}, \quad B = \begin{bmatrix} -2 & 3 \\ -1 & 2 \end{bmatrix},$$

$$C = \begin{bmatrix} 2 & 1 \\ 3 & 7 \end{bmatrix}$$

31. Show that $(AB)C = A(BC)$.

32. Show that $A(B + C) = AB + AC$.

33. Show that $(A + B)C = AC + BC$.

34. Show that $(3 + 2)A = 3A + 2A$.

For Problems 35–43, use the following matrices.

$$A = \begin{bmatrix} a_{11} & a_{12} \\ a_{21} & a_{22} \end{bmatrix}, \quad B = \begin{bmatrix} b_{11} & b_{12} \\ b_{21} & b_{22} \end{bmatrix},$$

$$C = \begin{bmatrix} c_{11} & c_{12} \\ c_{21} & c_{22} \end{bmatrix}, \quad O = \begin{bmatrix} 0 & 0 \\ 0 & 0 \end{bmatrix}$$

35. Show that $A + B = B + A$.

36. Show that $(A + B) + C = A + (B + C)$.

37. Show that $A + (-A) = O$.

38. Show that $k(A + B) = kA + kB$ for any real number k.

39. Show that $(k + l)A = kA + lA$ for any real numbers k and l.

40. Show that $(kl)A = k(lA)$ for any real numbers k and l.

41. Show that $(AB)C = A(BC)$.

42. Show that $A(B + C) = AB + AC$.

43. Show that $(A + B)C = AC + BC$.

MISCELLANEOUS PROBLEMS

44. If $A = \begin{bmatrix} 2 & 0 \\ 0 & 3 \end{bmatrix}$, calculate A^2 and A^3, where A^2 means AA and A^3 means AAA.

45. If $A = \begin{bmatrix} 1 & -1 \\ 2 & 3 \end{bmatrix}$, calculate A^2 and A^3.

46. Does $(A + B)(A - B) = A^2 - B^2$ for all 2×2 matrices? Defend your answer.

GRAPHICS CALCULATOR ACTIVITIES

47. Use your calculator to check the answers to all three parts of Example 1.

48. Use a calculator to check your answers for Problems 21–26.

49. Use the following matrices.

$$A = \begin{bmatrix} 7 & -4 \\ 6 & 9 \end{bmatrix}, \quad B = \begin{bmatrix} -3 & 8 \\ -5 & 7 \end{bmatrix}, \quad C = \begin{bmatrix} 8 & -2 \\ 4 & -7 \end{bmatrix}$$

a. Show that $(AB)C = A(BC)$.

b. Show that $A(B + C) = AB + AC$.

c. Show that $(B + C)A = BA + CA$.

11.2 MULTIPLICATIVE INVERSES

We know that 1 is a multiplicative identity element for the set of real numbers. That is, $a(1) = 1(a) = a$ for any real number a. Is there a multiplicative identity element for 2×2 matrices? Yes, the matrix

$$I = \begin{bmatrix} 1 & 0 \\ 0 & 1 \end{bmatrix}$$

is the **multiplicative identity element**, because

$$\begin{bmatrix} 1 & 0 \\ 0 & 1 \end{bmatrix} \begin{bmatrix} a_{11} & a_{12} \\ a_{21} & a_{22} \end{bmatrix} = \begin{bmatrix} a_{11} & a_{12} \\ a_{21} & a_{22} \end{bmatrix}$$

and

$$\begin{bmatrix} a_{11} & a_{12} \\ a_{21} & a_{22} \end{bmatrix} \begin{bmatrix} 1 & 0 \\ 0 & 1 \end{bmatrix} = \begin{bmatrix} a_{11} & a_{12} \\ a_{21} & a_{22} \end{bmatrix}.$$

Therefore, we can state that

$$AI = IA = A$$

for all 2×2 matrices.

Again, refer to the real numbers, where every nonzero real number a has a multiplicative inverse $1/a$ such that $a(1/a) = (1/a)a = 1$. Does every 2×2 matrix have a multiplicative inverse? To help answer this question, let's think about finding the multiplicative inverse (if one exists) for a specific matrix. This should give us some clues about a general approach.

EXAMPLE 1

Find the multiplicative inverse of $A = \begin{bmatrix} 3 & 5 \\ 2 & 4 \end{bmatrix}$.

Solution

We are looking for a matrix A^{-1} such that $AA^{-1} = A^{-1}A = I$. In other words, we want to solve the following matrix equation

$$\begin{bmatrix} 3 & 5 \\ 2 & 4 \end{bmatrix} \begin{bmatrix} x & y \\ z & w \end{bmatrix} = \begin{bmatrix} 1 & 0 \\ 0 & 1 \end{bmatrix}.$$

We need to multiply the two matrices on the left side of this equation and then set the elements of the product matrix equal to the corresponding elements of the identity matrix. We obtain the following system of equations.

$$\begin{pmatrix} 3x + 5z = 1 \\ 3y + 5w = 0 \\ 2x + 4z = 0 \\ 2y + 4w = 1 \end{pmatrix} \quad \begin{matrix} \textbf{(1)} \\ \textbf{(2)} \\ \textbf{(3)} \\ \textbf{(4)} \end{matrix}$$

Solving equations (1) and (3) simultaneously produces values for x and z as follows.

$$x = \frac{\begin{vmatrix} 1 & 5 \\ 0 & 4 \end{vmatrix}}{\begin{vmatrix} 3 & 5 \\ 2 & 4 \end{vmatrix}} = \frac{1(4) - 5(0)}{3(4) - 5(2)} = \frac{4}{2} = 2$$

$$z = \frac{\begin{vmatrix} 3 & 1 \\ 2 & 0 \end{vmatrix}}{\begin{vmatrix} 3 & 5 \\ 2 & 4 \end{vmatrix}} = \frac{3(0) - 1(2)}{3(4) - 5(2)} = \frac{-2}{2} = -1$$

Likewise, solving equations (2) and (4) simultaneously produces values for y and w.

$$y = \frac{\begin{vmatrix} 0 & 5 \\ 1 & 4 \end{vmatrix}}{\begin{vmatrix} 3 & 5 \\ 2 & 4 \end{vmatrix}} = \frac{0(4) - 5(1)}{3(4) - 5(2)} = \frac{-5}{2} = -\frac{5}{2}$$

$$w = \frac{\begin{vmatrix} 3 & 0 \\ 2 & 1 \end{vmatrix}}{\begin{vmatrix} 3 & 5 \\ 2 & 4 \end{vmatrix}} = \frac{3(1) - 0(2)}{3(4) - 5(2)} = \frac{3}{2}$$

Therefore,

$$A^{-1} = \begin{bmatrix} x & y \\ z & w \end{bmatrix} = \begin{bmatrix} 2 & -\frac{5}{2} \\ -1 & \frac{3}{2} \end{bmatrix}.$$

To check this, we perform the following multiplication.

$$\begin{bmatrix} 3 & 5 \\ 2 & 4 \end{bmatrix} \begin{bmatrix} 2 & -\frac{5}{2} \\ -1 & \frac{3}{2} \end{bmatrix} = \begin{bmatrix} 2 & -\frac{5}{2} \\ -1 & \frac{3}{2} \end{bmatrix} \begin{bmatrix} 3 & 5 \\ 2 & 4 \end{bmatrix} = \begin{bmatrix} 1 & 0 \\ 0 & 1 \end{bmatrix}.$$

Now let's use the approach in Example 1 on the general matrix

$$A = \begin{bmatrix} a_{11} & a_{12} \\ a_{21} & a_{22} \end{bmatrix}.$$

We want to find

$$A^{-1} = \begin{bmatrix} x & y \\ z & w \end{bmatrix}$$

such that $AA^{-1} = I$. Therefore, we need to solve the matrix equation

$$\begin{bmatrix} a_{11} & a_{12} \\ a_{21} & a_{22} \end{bmatrix} \begin{bmatrix} x & y \\ z & w \end{bmatrix} = \begin{bmatrix} 1 & 0 \\ 0 & 1 \end{bmatrix}$$

for x, y, z, and w. Once again, we multiply the two matrices on the left side of the equation and set the elements of this product matrix equal to the corresponding elements of the identity matrix. We then obtain the following system of equations.

$$\begin{pmatrix} a_{11}x + a_{12}z = 1 \\ a_{11}y + a_{12}w = 0 \\ a_{21}x + a_{22}z = 0 \\ a_{21}y + a_{22}w = 1 \end{pmatrix}$$

Solving this system produces

$$x = \frac{a_{22}}{a_{11}a_{22} - a_{12}a_{21}} \qquad y = \frac{-a_{12}}{a_{11}a_{22} - a_{12}a_{21}}$$

$$z = \frac{-a_{21}}{a_{11}a_{22} - a_{12}a_{21}} \qquad w = \frac{a_{11}}{a_{11}a_{22} - a_{12}a_{21}}.$$

Notice that the number in each denominator, $a_{11}a_{22} - a_{12}a_{21}$, is the determinant of the matrix A. Thus, if $|A| \neq 0$,

$$A^{-1} = \frac{1}{|A|} \begin{bmatrix} a_{22} & -a_{12} \\ -a_{21} & a_{11} \end{bmatrix}.$$

Matrix multiplication will show that $AA^{-1} = A^{-1}A = I$. **If $|A| = 0$, then the matrix A has no multiplicative inverse.**

EXAMPLE 2

Find A^{-1} if $A = \begin{bmatrix} 3 & 5 \\ -2 & -4 \end{bmatrix}$.

Solution

First, let's find $|A|$.

$$|A| = (3)(-4) - (5)(-2) = -2$$

Therefore,

$$A^{-1} = \frac{1}{-2}\begin{bmatrix} -4 & -5 \\ 2 & 3 \end{bmatrix} = -\frac{1}{2}\begin{bmatrix} -4 & -5 \\ 2 & 3 \end{bmatrix} = \begin{bmatrix} 2 & \frac{5}{2} \\ -1 & -\frac{3}{2} \end{bmatrix}.$$

It is easily checked that $AA^{-1} = A^{-1}A = I$.

EXAMPLE 3

Find A^{-1} if $A = \begin{bmatrix} 8 & -2 \\ -12 & 3 \end{bmatrix}$.

Solution

$$|A| = (8)(3) - (-2)(-12) = 0$$

Therefore, A has no multiplicative inverse.

More About Multiplication of Matrices

Thus far we have found the product only of 2×2 matrices. The *row-by-column* multiplication pattern can be applied to many different kinds of matrices, which we shall see in the next section. For now let's find the product of a 2×2 matrix and a 2×1 matrix, with the 2×2 matrix on the left, as follows.

$$\begin{bmatrix} a_{11} & a_{12} \\ a_{21} & a_{22} \end{bmatrix}\begin{bmatrix} b_{11} \\ b_{21} \end{bmatrix} = \begin{bmatrix} a_{11}b_{11} + a_{12}b_{21} \\ a_{21}b_{11} + a_{22}b_{21} \end{bmatrix}$$

Notice that the product matrix is a 2×1 matrix. The following example illustrates this pattern.

$$\begin{bmatrix} -2 & 3 \\ 1 & -4 \end{bmatrix}\begin{bmatrix} 5 \\ 7 \end{bmatrix} = \begin{bmatrix} (-2)(5) + (3)(7) \\ (1)(5) + (-4)(7) \end{bmatrix} = \begin{bmatrix} 11 \\ -23 \end{bmatrix}$$

Back to Solving Systems of Equations

The linear system of equations

$$\begin{pmatrix} a_{11}x + a_{12}y = d_1 \\ a_{21}x + a_{22}y = d_2 \end{pmatrix}$$

can be represented by the matrix equation

$$\begin{bmatrix} a_{11} & a_{12} \\ a_{21} & a_{22} \end{bmatrix} \begin{bmatrix} x \\ y \end{bmatrix} = \begin{bmatrix} d_1 \\ d_2 \end{bmatrix}.$$

If we let

$$A = \begin{bmatrix} a_{11} & a_{12} \\ a_{21} & a_{22} \end{bmatrix}, \quad X = \begin{bmatrix} x \\ y \end{bmatrix}, \quad \text{and} \quad B = \begin{bmatrix} d_1 \\ d_2 \end{bmatrix},$$

then the previous matrix equation can be written $AX = B$.

If A^{-1} exists, then we can multiply both sides of $AX = B$ by A^{-1} (on the left) and simplify as follows.

$$\begin{aligned} AX &= B \\ A^{-1}(AX) &= A^{-1}(B) \\ (A^{-1}A)X &= A^{-1}B \\ IX &= A^{-1}B \\ X &= A^{-1}B \end{aligned}$$

Therefore, the product $A^{-1}B$ is the solution of the system.

EXAMPLE 4

Solve the system

$$\begin{pmatrix} 5x + 4y = 10 \\ 6x + 5y = 13 \end{pmatrix}.$$

Solution

By letting

$$A = \begin{bmatrix} 5 & 4 \\ 6 & 5 \end{bmatrix}, \quad X = \begin{bmatrix} x \\ y \end{bmatrix}, \quad \text{and} \quad B = \begin{bmatrix} 10 \\ 13 \end{bmatrix},$$

the given system can be represented by the matrix equation $AX = B$. From our previous discussion, we know that the solution of this equation is $X = A^{-1}B$, so we need to find A^{-1} and the product $A^{-1}B$.

$$A^{-1} = \frac{1}{|A|}\begin{bmatrix} 5 & -4 \\ -6 & 5 \end{bmatrix} = \frac{1}{1}\begin{bmatrix} 5 & -4 \\ -6 & 5 \end{bmatrix} = \begin{bmatrix} 5 & -4 \\ -6 & 5 \end{bmatrix}$$

Therefore,

$$A^{-1}B = \begin{bmatrix} 5 & -4 \\ -6 & 5 \end{bmatrix} \begin{bmatrix} 10 \\ 13 \end{bmatrix} = \begin{bmatrix} -2 \\ 5 \end{bmatrix}.$$

The solution set of the given system is $\{(-2, 5)\}$.

EXAMPLE 5

Solve the system

$$\begin{pmatrix} 3x - 2y = 9 \\ 4x + 7y = -17 \end{pmatrix}.$$

Solution

If we let

$$A = \begin{bmatrix} 3 & -2 \\ 4 & 7 \end{bmatrix}, \quad X = \begin{bmatrix} x \\ y \end{bmatrix}, \quad \text{and} \quad B = \begin{bmatrix} 9 \\ -17 \end{bmatrix},$$

then the system is represented by $AX = B$, where $X = A^{-1}B$ and

$$A^{-1} = \frac{1}{|A|}\begin{bmatrix} 7 & 2 \\ -4 & 3 \end{bmatrix} = \frac{1}{29}\begin{bmatrix} 7 & 2 \\ -4 & 3 \end{bmatrix} = \begin{bmatrix} \frac{7}{29} & \frac{2}{29} \\ -\frac{4}{29} & \frac{3}{29} \end{bmatrix}.$$

Therefore,

$$A^{-1}B = \begin{bmatrix} \frac{7}{29} & \frac{2}{29} \\ -\frac{4}{29} & \frac{3}{29} \end{bmatrix}\begin{bmatrix} 9 \\ -17 \end{bmatrix} = \begin{bmatrix} 1 \\ -3 \end{bmatrix}.$$

The solution set of the given system is $\{(1, -3)\}$.

This technique of using matrix inverses to solve systems of linear equations is especially useful when there are many systems to be solved that have the same coefficients but different constant terms.

PROBLEM SET 11.2

Find the multiplicative inverse (if it exists) of each of the following matrices.

1. $\begin{bmatrix} 5 & 7 \\ 2 & 3 \end{bmatrix}$

2. $\begin{bmatrix} 3 & 4 \\ 2 & 3 \end{bmatrix}$

3. $\begin{bmatrix} 3 & 8 \\ 2 & 5 \end{bmatrix}$

4. $\begin{bmatrix} 2 & 9 \\ 3 & 13 \end{bmatrix}$

5. $\begin{bmatrix} -1 & 2 \\ 3 & 4 \end{bmatrix}$

6. $\begin{bmatrix} 1 & -2 \\ 4 & -3 \end{bmatrix}$

7. $\begin{bmatrix} -2 & -3 \\ 4 & 6 \end{bmatrix}$

8. $\begin{bmatrix} 5 & -1 \\ 3 & 4 \end{bmatrix}$

9. $\begin{bmatrix} -3 & 2 \\ -4 & 5 \end{bmatrix}$

10. $\begin{bmatrix} 3 & -4 \\ 6 & -8 \end{bmatrix}$

11. $\begin{bmatrix} 0 & 1 \\ 5 & 3 \end{bmatrix}$

12. $\begin{bmatrix} -2 & 0 \\ -3 & 5 \end{bmatrix}$

13. $\begin{bmatrix} -2 & -3 \\ -1 & -4 \end{bmatrix}$

14. $\begin{bmatrix} -2 & -5 \\ -3 & -6 \end{bmatrix}$

15. $\begin{bmatrix} -2 & 5 \\ -3 & 6 \end{bmatrix}$

16. $\begin{bmatrix} -3 & 4 \\ 1 & -2 \end{bmatrix}$

17. $\begin{bmatrix} 1 & 1 \\ 1 & -1 \end{bmatrix}$

18. $\begin{bmatrix} 1 & -1 \\ 1 & 1 \end{bmatrix}$

For Problems 19–26, compute AB.

19. $A = \begin{bmatrix} 4 & 3 \\ 2 & 5 \end{bmatrix}, \quad B = \begin{bmatrix} 3 \\ 6 \end{bmatrix}$

20. $A = \begin{bmatrix} 5 & -2 \\ 3 & 1 \end{bmatrix}, \quad B = \begin{bmatrix} 5 \\ 8 \end{bmatrix}$

21. $A = \begin{bmatrix} -3 & -4 \\ 2 & 1 \end{bmatrix}, \quad B = \begin{bmatrix} 4 \\ -3 \end{bmatrix}$

22. $A = \begin{bmatrix} 5 & 2 \\ -1 & -3 \end{bmatrix}, \quad B = \begin{bmatrix} 3 \\ -5 \end{bmatrix}$

23. $A = \begin{bmatrix} -4 & 2 \\ 7 & -5 \end{bmatrix}$, $B = \begin{bmatrix} -1 \\ -4 \end{bmatrix}$

24. $A = \begin{bmatrix} 0 & -3 \\ 2 & 9 \end{bmatrix}$, $B = \begin{bmatrix} -3 \\ -6 \end{bmatrix}$

25. $A = \begin{bmatrix} -2 & -3 \\ -5 & -6 \end{bmatrix}$, $B = \begin{bmatrix} 5 \\ -2 \end{bmatrix}$

26. $A = \begin{bmatrix} -3 & -5 \\ 4 & -7 \end{bmatrix}$, $B = \begin{bmatrix} -3 \\ -10 \end{bmatrix}$

Use the method of matrix inverses to solve each of the following systems.

27. $\begin{pmatrix} 2x + 3y = 13 \\ x + 2y = 8 \end{pmatrix}$

28. $\begin{pmatrix} 3x + 2y = 10 \\ 7x + 5y = 23 \end{pmatrix}$

29. $\begin{pmatrix} 4x - 3y = -23 \\ -3x + 2y = 16 \end{pmatrix}$

30. $\begin{pmatrix} 6x - y = -14 \\ 3x + 2y = -17 \end{pmatrix}$

31. $\begin{pmatrix} x - 7y = 7 \\ 6x + 5y = -5 \end{pmatrix}$

32. $\begin{pmatrix} x + 9y = -5 \\ 4x - 7y = -20 \end{pmatrix}$

33. $\begin{pmatrix} 3x - 5y = 2 \\ 4x - 3y = -1 \end{pmatrix}$

34. $\begin{pmatrix} 5x - 2y = 6 \\ 7x - 3y = 8 \end{pmatrix}$

35. $\begin{pmatrix} y = 19 - 3x \\ 9x - 5y = 1 \end{pmatrix}$

36. $\begin{pmatrix} 4x + 3y = 31 \\ x = 5y + 2 \end{pmatrix}$

37. $\begin{pmatrix} 3x + 2y = 0 \\ 30x - 18y = -19 \end{pmatrix}$

38. $\begin{pmatrix} 12x + 30y = 23 \\ 12x - 24y = -13 \end{pmatrix}$

39. $\begin{pmatrix} \frac{1}{3}x + \frac{3}{4}y = -18 \\ \frac{2}{3}x + \frac{1}{5}y = -2 \end{pmatrix}$

40. $\begin{pmatrix} \frac{3}{2}x + \frac{1}{6}y = 11 \\ \frac{2}{3}x - \frac{1}{4}y = 1 \end{pmatrix}$

THOUGHTS into WORDS

41. Describe how to solve the system $\begin{pmatrix} x - 2y = -10 \\ 3x + 5y = 14 \end{pmatrix}$ using each of the following techniques.

a. substitution method

b. elimination-by-addition method

c. reduced echelon form of the augmented matrix

d. determinants

e. matrix inverse method

GRAPHICS CALCULATOR ACTIVITIES

42. Use your calculator to find the multiplicative inverse (if it exists) of each of the following matrices. Be sure to check your answers by showing that $A^{-1}A = I$.

a. $\begin{bmatrix} 7 & 6 \\ 8 & 7 \end{bmatrix}$ **b.** $\begin{bmatrix} -12 & 5 \\ -19 & 8 \end{bmatrix}$

c. $\begin{bmatrix} -7 & 9 \\ 6 & -8 \end{bmatrix}$ **d.** $\begin{bmatrix} -6 & -11 \\ -4 & -8 \end{bmatrix}$

e. $\begin{bmatrix} 13 & 12 \\ 4 & 4 \end{bmatrix}$ **f.** $\begin{bmatrix} 15 & -8 \\ -9 & 5 \end{bmatrix}$

g. $\begin{bmatrix} 9 & 36 \\ 3 & 12 \end{bmatrix}$ **h.** $\begin{bmatrix} 1.2 & 1.5 \\ 7.6 & 4.5 \end{bmatrix}$

43. Use your calculator to find the multiplicative inverse of $\begin{bmatrix} \frac{1}{2} & \frac{2}{5} \\ \frac{3}{4} & \frac{1}{4} \end{bmatrix}$. What difficulty did you encounter?

44. Use your calculator and the method of matrix inverses to solve each of the following systems. Be sure to check your solutions.

a. $\begin{pmatrix} 5x + 7y = 82 \\ 7x + 10y = 116 \end{pmatrix}$ **b.** $\begin{pmatrix} 9x - 8y = -150 \\ -10x + 9y = 168 \end{pmatrix}$

c. $\begin{pmatrix} 15x - 8y = -15 \\ -9x + 5y = 12 \end{pmatrix}$ **d.** $\begin{pmatrix} 1.2x + 1.5y = 5.85 \\ 7.6x + 4.5y = 19.55 \end{pmatrix}$

e. $\begin{pmatrix} 12x - 7y = -34.5 \\ 8x + 9y = 79.5 \end{pmatrix}$ **f.** $\begin{pmatrix} \frac{3x}{2} + \frac{y}{6} = 11 \\ \frac{2x}{3} - \frac{y}{4} = 1 \end{pmatrix}$ **g.** $\begin{pmatrix} 114x + 129y = 2832 \\ 127x + 214y = 4139 \end{pmatrix}$ **h.** $\begin{pmatrix} \frac{x}{2} + \frac{2y}{5} = 14 \\ \frac{3x}{4} + \frac{y}{4} = 14 \end{pmatrix}$

11.3 $m \times n$ MATRICES

Now let's see how much of the algebra of 2×2 matrices extends to $m \times n$ matrices, that is, to matrices of any dimension. In Section 10.4 we represented a general $m \times n$ matrix by

$$A = \begin{bmatrix} a_{11} & a_{12} & a_{13} & \cdots & a_{1n} \\ a_{21} & a_{22} & a_{23} & \cdots & a_{2n} \\ \vdots & \vdots & \vdots & & \vdots \\ a_{m1} & a_{m2} & a_{m3} & \cdots & a_{mn} \end{bmatrix}.$$

We denote the element at the intersection of row i and column j by a_{ij}. It is also customary to denote a matrix A with the abbreviated notation (a_{ij}).

Addition of matrices can be extended to matrices of any dimensions by the following definition.

DEFINITION 11.5

Let $A = (a_{ij})$ and $B = (b_{ij})$ be two matrices *of the same dimension*. Then

$$A + B = (a_{ij}) + (b_{ij}) = (a_{ij} + b_{ij}).$$

Definition 11.5 states that to add two matrices, we add the elements that appear in corresponding positions in the matrices. The matrices must be of the same dimension for this to work. An example of the sum of two 3×2 matrices is

$$\begin{bmatrix} 3 & 2 \\ 4 & -1 \\ -3 & 8 \end{bmatrix} + \begin{bmatrix} -2 & 1 \\ -3 & -7 \\ 5 & 9 \end{bmatrix} = \begin{bmatrix} 1 & 3 \\ 1 & -8 \\ 2 & 17 \end{bmatrix}.$$

The **commutative** and **associative properties** hold for any matrices that can be added. The $m \times n$ **zero matrix**, denoted by O, is the matrix that contains all zeros. It is the **identity element for addition**. For example,

$$\begin{bmatrix} 2 & 3 & -1 & -5 \\ -7 & 6 & 2 & 8 \end{bmatrix} + \begin{bmatrix} 0 & 0 & 0 & 0 \\ 0 & 0 & 0 & 0 \end{bmatrix} = \begin{bmatrix} 2 & 3 & -1 & -5 \\ -7 & 6 & 2 & 8 \end{bmatrix}.$$

Every matrix A has an additive inverse, $-A$, that can be found by changing the sign of each element of A. For example, if

$$A = [2 \quad -3 \quad 0 \quad 4 \quad -7],$$

then

$$-A = [-2 \quad 3 \quad 0 \quad -4 \quad 7].$$

Furthermore, $A + (-A) = O$ for all matrices.

The definition we gave earlier for subtraction, $A - B = A + (-B)$, can be extended to any two matrices of the same dimension. For example,

$$\begin{aligned}[-4 \quad 3 \quad -5] - [7 \quad -4 \quad -1] &= [-4 \quad 3 \quad -5] + [-7 \quad 4 \quad 1]\\ &= [-11 \quad 7 \quad -4].\end{aligned}$$

The scalar product of any real number k and any $m \times n$ matrix $A = (a_{ij})$ is defined by

$$kA = (ka_{ij}).$$

In other words, to find kA, we simply multiply each element of A by k. For example,

$$(-4)\begin{bmatrix} 1 & -1 \\ -2 & 3 \\ 4 & 5 \\ 0 & -8 \end{bmatrix} = \begin{bmatrix} -4 & 4 \\ 8 & -12 \\ -16 & -20 \\ 0 & 32 \end{bmatrix}.$$

The properties $k(A + B) = kA + kB$, $(k + l)A = kA + lA$, and $(kl)A = k(lA)$ hold for all matrices. The matrices A and B must be of the same dimension to be added.

The *row-by-column* definition for multiplying two matrices can be extended, but a little care must be taken. In order to define the product AB of two matrices A and B, the number of columns of A must equal the number of rows of B. Therefore, suppose $A = (a_{ij})$ is $m \times n$ and $B = (b_{ij})$ is $n \times p$. Then

$$AB = \begin{bmatrix} a_{11} & a_{12} & \cdots & a_{1n} \\ \vdots & \vdots & & \vdots \\ a_{i1} & a_{i2} & \cdots & a_{in} \\ \vdots & \vdots & & \vdots \\ a_{m1} & a_{m2} & \cdots & a_{mn} \end{bmatrix} \begin{bmatrix} b_{11} & \cdots & b_{1j} & \cdots & b_{1p} \\ b_{21} & \cdots & b_{2j} & \cdots & b_{2p} \\ \vdots & & \vdots & & \vdots \\ b_{n1} & \cdots & b_{nj} & \cdots & b_{np} \end{bmatrix} = C.$$

The product matrix C is of dimension $m \times p$ and the general element, c_{ij}, is determined as follows.

$$c_{ij} = a_{i1}b_{1j} + a_{i2}b_{2j} + \cdots a_{in}b_{nj}$$

A specific element of the product matrix, such as c_{23}, is the result of multiplying the elements in row 2 of matrix A times the elements in column 3 of matrix B and adding the results. Therefore,

$$c_{23} = a_{21}b_{13} + a_{22}b_{23} + \cdots + a_{2n}b_{n3}.$$

The following example illustrates the product of a 2 × 3 matrix and a 3 × 2 matrix.

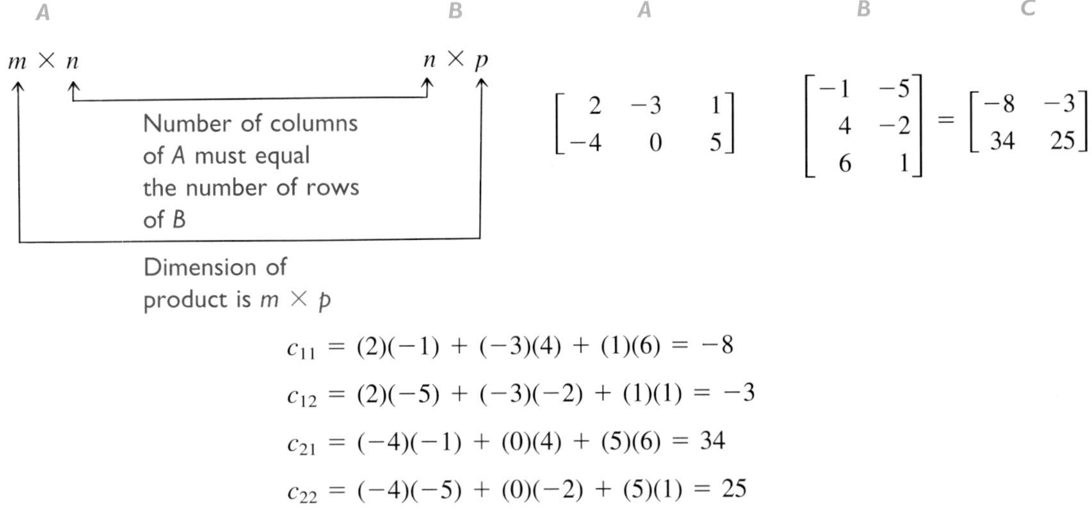

$$\begin{bmatrix} 2 & -3 & 1 \\ -4 & 0 & 5 \end{bmatrix} \begin{bmatrix} -1 & -5 \\ 4 & -2 \\ 6 & 1 \end{bmatrix} = \begin{bmatrix} -8 & -3 \\ 34 & 25 \end{bmatrix}$$

$$c_{11} = (2)(-1) + (-3)(4) + (1)(6) = -8$$

$$c_{12} = (2)(-5) + (-3)(-2) + (1)(1) = -3$$

$$c_{21} = (-4)(-1) + (0)(4) + (5)(6) = 34$$

$$c_{22} = (-4)(-5) + (0)(-2) + (5)(1) = 25$$

Recall that matrix multiplication is not commutative. In fact, it may be that AB is defined and BA is not defined. For example, if A is a 2 × 3 matrix and B is a 3 × 4 matrix, then the product AB is a 2 × 4 matrix but the product BA is not defined, because the number of columns of B does not equal the number of rows of A.

The associative property for multiplication and the two distributive properties hold if the matrices have the proper number of rows and columns for the operations to be defined. In that case, we have $(AB)C = A(BC)$, $A(B + C) = AB + AC$, and $(A + B)C = AC + BC$.

Square Matrices

Now let's extend some of the algebra of 2 × 2 matrices to all square matrices (where the number of rows equals the number of columns). For example, the general multiplicative identity element for square matrices contains 1s in the main diagonal from the upper left-hand corner to the lower right-hand corner and 0s elsewhere. Therefore, for 3 × 3 and 4 × 4 matrices, the multiplicative identity elements are as follows.

$$I_3 = \begin{bmatrix} 1 & 0 & 0 \\ 0 & 1 & 0 \\ 0 & 0 & 1 \end{bmatrix} \qquad I_4 = \begin{bmatrix} 1 & 0 & 0 & 0 \\ 0 & 1 & 0 & 0 \\ 0 & 0 & 1 & 0 \\ 0 & 0 & 0 & 1 \end{bmatrix}$$

We saw in Section 11.2 that some, but not all, 2×2 matrices have multiplicative inverses. In general, some, but not all, square matrices of a particular dimension have multiplicative inverses. If an $n \times n$ square matrix A does have a multiplicative inverse A^{-1}, then

$$AA^{-1} = A^{-1}A = I_n.$$

The technique used in Section 11.2 for finding multiplicative inverses of 2×2 matrices does generalize, but it becomes quite complicated. Therefore, we shall now describe another technique that works for all square matrices. Given an $n \times n$ matrix A, we begin by forming the $n \times 2n$ matrix

$$\left[\begin{array}{cccc|ccccc} a_{11} & a_{12} & \cdots & a_{1n} & 1 & 0 & 0 & \cdots & 0 \\ a_{21} & a_{22} & \cdots & a_{2n} & 0 & 1 & 0 & \cdots & 0 \\ \vdots & \vdots & & \vdots & \vdots & \vdots & \vdots & & \vdots \\ a_{n1} & a_{n2} & \cdots & a_{nn} & 0 & 0 & 0 & \cdots & 1 \end{array}\right],$$

where the identity matrix I_n appears to the right of A. Now we apply a succession of elementary row transformations to this double matrix until we obtain a matrix of the form

$$\left[\begin{array}{ccccc|cccc} 1 & 0 & 0 & \cdots & 0 & b_{11} & b_{12} & \cdots & b_{1n} \\ 0 & 1 & 0 & \cdots & 0 & b_{21} & b_{22} & \cdots & b_{2n} \\ \vdots & \vdots & \vdots & & \vdots & \vdots & \vdots & & \vdots \\ 0 & 0 & 0 & \cdots & 1 & b_{n1} & b_{n2} & \cdots & b_{nn} \end{array}\right].$$

The B matrix in this matrix is the desired inverse A^{-1}. If A does not have an inverse, then it is impossible to change the original matrix to this final form.

EXAMPLE 1

Find A^{-1} if $A = \begin{bmatrix} 2 & 4 \\ 3 & 5 \end{bmatrix}$.

Solution

First, we form the matrix

$$\left[\begin{array}{cc|cc} 2 & 4 & 1 & 0 \\ 3 & 5 & 0 & 1 \end{array}\right].$$

Now multiply row 1 by $\frac{1}{2}$.

$$\left[\begin{array}{cc|cc} 1 & 2 & \frac{1}{2} & 0 \\ 3 & 5 & 0 & 1 \end{array}\right]$$

Next, add -3 times row 1 to row 2 to form a new row 2.

$$\left[\begin{array}{cc|cc} 1 & 2 & \frac{1}{2} & 0 \\ 0 & -1 & -\frac{3}{2} & 1 \end{array}\right]$$

Next, multiply row 2 by -1.

$$\left[\begin{array}{cc|cc} 1 & 2 & \frac{1}{2} & 0 \\ 0 & 1 & \frac{3}{2} & -1 \end{array}\right]$$

Finally, add -2 times row 2 to row 1 to form a new row 1.

$$\left[\begin{array}{cc|cc} 1 & 0 & -\frac{5}{2} & 2 \\ 0 & 1 & \frac{3}{2} & -1 \end{array}\right]$$

The matrix inside the box is A^{-1}, that is,

$$A^{-1} = \begin{bmatrix} -\frac{5}{2} & 2 \\ \frac{3}{2} & -1 \end{bmatrix}.$$

This can be checked, as always, by showing that $AA^{-1} = A^{-1}A = I_2$.

EXAMPLE 2

Find A^{-1} if $A = \begin{bmatrix} 1 & 1 & 2 \\ 2 & 3 & -1 \\ -3 & 1 & -2 \end{bmatrix}$.

Solution

Form the matrix

$$\left[\begin{array}{ccc|ccc} 1 & 1 & 2 & 1 & 0 & 0 \\ 2 & 3 & -1 & 0 & 1 & 0 \\ -3 & 1 & -2 & 0 & 0 & 1 \end{array}\right].$$

Add -2 times row 1 to row 2, and add 3 times row 1 to row 3.

$$\left[\begin{array}{ccc|ccc} 1 & 1 & 2 & 1 & 0 & 0 \\ 0 & 1 & -5 & -2 & 1 & 0 \\ 0 & 4 & 4 & 3 & 0 & 1 \end{array}\right]$$

Add -1 times row 2 to row 1, and add -4 times row 2 to row 3.

$$\left[\begin{array}{ccc|ccc} 1 & 0 & 7 & 3 & -1 & 0 \\ 0 & 1 & -5 & -2 & 1 & 0 \\ 0 & 0 & 24 & 11 & -4 & 1 \end{array}\right]$$

Multiply row 3 by $\frac{1}{24}$.

$$\left[\begin{array}{ccc|ccc} 1 & 0 & 7 & 3 & -1 & 0 \\ 0 & 1 & -5 & -2 & 1 & 0 \\ 0 & 0 & 1 & \frac{11}{24} & -\frac{1}{6} & \frac{1}{24} \end{array}\right]$$

Add -7 times row 3 to row 1, and add 5 times row 3 to row 2.

$$\left[\begin{array}{ccc|ccc} 1 & 0 & 0 & -\frac{5}{24} & \frac{1}{6} & -\frac{7}{24} \\ 0 & 1 & 0 & \frac{7}{24} & \frac{1}{6} & \frac{5}{24} \\ 0 & 0 & 1 & \frac{11}{24} & -\frac{1}{6} & \frac{1}{24} \end{array}\right]$$

Therefore

$$A^{-1} = \begin{bmatrix} -\frac{5}{24} & \frac{1}{6} & -\frac{7}{24} \\ \frac{7}{24} & \frac{1}{6} & \frac{5}{24} \\ \frac{11}{24} & -\frac{1}{6} & \frac{1}{24} \end{bmatrix}.$$

Be sure to check this!

Systems of Equations

In Section 11.2 we used the concept of the multiplicative inverse to solve systems of two linear equations in two variables. This same technique can be applied to general systems of n linear equations in n variables. Let's consider one such example involving three equations in three variables.

EXAMPLE 3

Solve the system

$$\begin{pmatrix} x + y + 2z = -8 \\ 2x + 3y - z = 3 \\ -3x + y - 2z = 4 \end{pmatrix}.$$

Solution

By letting

$$A = \begin{bmatrix} 1 & 1 & 2 \\ 2 & 3 & -1 \\ -3 & 1 & -2 \end{bmatrix}, \quad X = \begin{bmatrix} x \\ y \\ z \end{bmatrix}, \quad \text{and} \quad B = \begin{bmatrix} -8 \\ 3 \\ 4 \end{bmatrix},$$

the given system can be represented by the matrix equation $AX = B$. Therefore we know that $X = A^{-1}B$, so we need to find A^{-1} and the product $A^{-1}B$. The matrix

A^{-1} was found in Example 2, so let's use that result and find $A^{-1}B$.

$$X = A^{-1}B = \begin{bmatrix} -\frac{5}{24} & \frac{1}{6} & -\frac{7}{24} \\ \frac{7}{24} & \frac{1}{6} & \frac{5}{24} \\ \frac{11}{24} & -\frac{1}{6} & \frac{1}{24} \end{bmatrix} \begin{bmatrix} -8 \\ 3 \\ 4 \end{bmatrix} = \begin{bmatrix} 1 \\ -1 \\ -4 \end{bmatrix}$$

The solution set of the given system is $\{(1, -1, -4)\}$.

PROBLEM SET 11.3

For Problems 1–8, find $A + B$, $A - B$, $2A + 3B$, and $4A - 2B$.

1. $A = \begin{bmatrix} 2 & -1 & 4 \\ -2 & 0 & 5 \end{bmatrix}$, $B = \begin{bmatrix} -1 & 4 & -7 \\ 5 & -6 & 2 \end{bmatrix}$

2. $A = \begin{bmatrix} 3 & -6 \\ 2 & -1 \\ -4 & 5 \end{bmatrix}$, $B = \begin{bmatrix} 1 & 0 \\ 5 & -7 \\ -6 & 9 \end{bmatrix}$

3. $A = [2 \quad -1 \quad 4 \quad 12]$, $B = [-3 \quad -6 \quad 9 \quad -5]$

4. $A = \begin{bmatrix} 3 \\ -9 \\ 7 \end{bmatrix}$, $B = \begin{bmatrix} -6 \\ 12 \\ 9 \end{bmatrix}$

5. $A = \begin{bmatrix} 3 & -2 & 1 \\ -1 & 4 & -7 \\ 0 & 5 & 9 \end{bmatrix}$, $B = \begin{bmatrix} 5 & -1 & -3 \\ 10 & -2 & 4 \\ 7 & 0 & 12 \end{bmatrix}$

6. $A = \begin{bmatrix} 7 & -4 \\ -5 & 9 \\ -1 & 2 \end{bmatrix}$, $B = \begin{bmatrix} 12 & 3 \\ -2 & -4 \\ -6 & 7 \end{bmatrix}$

7. $A = \begin{bmatrix} -1 & 0 \\ 2 & 3 \\ -5 & -4 \\ -7 & 11 \end{bmatrix}$, $B = \begin{bmatrix} 1 & 2 \\ -3 & 7 \\ 6 & -5 \\ 9 & -2 \end{bmatrix}$

8. $A = \begin{bmatrix} 0 & -1 & -2 \\ 3 & -4 & 6 \\ 5 & 4 & -9 \end{bmatrix}$, $B = \begin{bmatrix} 2 & 1 & -7 \\ -6 & 4 & 5 \\ 3 & -2 & -1 \end{bmatrix}$

For Problems 9–20, find AB and BA, whenever they exist.

9. $A = \begin{bmatrix} 2 & -1 \\ 0 & -4 \\ -5 & 3 \end{bmatrix}$, $B = \begin{bmatrix} 5 & -2 & 6 \\ -1 & 4 & -2 \end{bmatrix}$

10. $A = \begin{bmatrix} -2 & 3 & -1 \\ 7 & -4 & 5 \end{bmatrix}$, $B = \begin{bmatrix} 1 & -1 \\ -2 & 3 \\ -5 & -6 \end{bmatrix}$

11. $A = \begin{bmatrix} 2 & -1 & -3 \\ 0 & -4 & 7 \end{bmatrix}$, $B = \begin{bmatrix} 2 & 1 & -1 & 4 \\ 0 & -2 & 3 & 5 \\ -6 & 4 & -2 & 0 \end{bmatrix}$

12. $A = \begin{bmatrix} 3 & -1 & -4 \\ -5 & 2 & 2 \end{bmatrix}$, $B = \begin{bmatrix} 3 & -2 \\ -4 & -1 \end{bmatrix}$

13. $A = \begin{bmatrix} 1 & -1 & 2 \\ 0 & 1 & -2 \\ 3 & 1 & 4 \end{bmatrix}$, $B = \begin{bmatrix} 2 & 3 & -1 \\ 4 & 0 & 2 \\ -5 & 1 & -1 \end{bmatrix}$

14. $A = \begin{bmatrix} 1 & 0 & 1 \\ 0 & 1 & 1 \\ -1 & 2 & 3 \end{bmatrix}$, $B = \begin{bmatrix} -1 & -1 & 1 \\ 0 & 1 & 0 \\ 2 & -3 & 1 \end{bmatrix}$

15. $A = [2 \quad -1 \quad 3 \quad 4]$, $B = \begin{bmatrix} -1 \\ -3 \\ 2 \\ -4 \end{bmatrix}$

16. $A = \begin{bmatrix} -2 \\ 3 \\ -5 \end{bmatrix}$, $B = [3 \quad -4 \quad -5]$

17. $A = \begin{bmatrix} 2 \\ -7 \end{bmatrix}$, $B = \begin{bmatrix} 3 & -2 \\ 1 & 0 \\ -1 & 4 \end{bmatrix}$

18. $A = \begin{bmatrix} 3 & -2 & 2 & -4 \\ 1 & 0 & -1 & 2 \end{bmatrix}$, $B = \begin{bmatrix} 3 & -2 & 1 \\ -3 & 1 & 4 \\ 5 & 2 & 0 \\ -4 & -1 & -2 \end{bmatrix}$

19. $A = \begin{bmatrix} 3 \\ -4 \\ 2 \end{bmatrix}$, $B = [3 \quad -4]$

20. $A = [3 \quad -7]$, $B = \begin{bmatrix} 8 \\ -9 \end{bmatrix}$

For Problems 21–36, use the technique discussed in this section to find the multiplicative inverse (if it exists) of each of the following matrices.

21. $\begin{bmatrix} 1 & 3 \\ 4 & 2 \end{bmatrix}$

22. $\begin{bmatrix} 1 & 2 \\ 2 & -3 \end{bmatrix}$

23. $\begin{bmatrix} 2 & 1 \\ 7 & 4 \end{bmatrix}$

24. $\begin{bmatrix} 3 & 7 \\ 2 & 5 \end{bmatrix}$

25. $\begin{bmatrix} -2 & 1 \\ 3 & -4 \end{bmatrix}$

26. $\begin{bmatrix} -3 & 1 \\ 3 & -2 \end{bmatrix}$

27. $\begin{bmatrix} 1 & 2 & 3 \\ 1 & 3 & 4 \\ 1 & 4 & 3 \end{bmatrix}$

28. $\begin{bmatrix} 1 & 3 & -2 \\ 1 & 4 & -1 \\ -2 & -7 & 5 \end{bmatrix}$

29. $\begin{bmatrix} 1 & -2 & 1 \\ -2 & 5 & 3 \\ 3 & -5 & 7 \end{bmatrix}$

30. $\begin{bmatrix} 1 & 4 & -2 \\ -3 & -11 & 1 \\ 2 & 7 & 3 \end{bmatrix}$

31. $\begin{bmatrix} 2 & 3 & -4 \\ 3 & -1 & -2 \\ 1 & -4 & 2 \end{bmatrix}$

32. $\begin{bmatrix} -2 & 2 & 3 \\ 1 & -1 & 0 \\ 0 & 1 & 4 \end{bmatrix}$

33. $\begin{bmatrix} 1 & 2 & 3 \\ -3 & -4 & 3 \\ 2 & 4 & -1 \end{bmatrix}$

34. $\begin{bmatrix} 1 & -2 & 3 \\ -1 & 3 & -2 \\ -2 & 6 & 1 \end{bmatrix}$

35. $\begin{bmatrix} 2 & 0 & 0 \\ 0 & 4 & 0 \\ 0 & 0 & 10 \end{bmatrix}$

36. $\begin{bmatrix} 1 & -3 & 5 \\ 0 & 1 & 2 \\ 0 & 0 & 1 \end{bmatrix}$

Use the method of matrix inverses to solve each of the following systems. The required multiplicative inverses have been found in Problems 21–36.

37. $\begin{pmatrix} 2x + y = -4 \\ 7x + 4y = -13 \end{pmatrix}$

38. $\begin{pmatrix} 3x + 7y = -38 \\ 2x + 5y = -27 \end{pmatrix}$

39. $\begin{pmatrix} -2x + y = 1 \\ 3x - 4y = -14 \end{pmatrix}$

40. $\begin{pmatrix} -3x + y = -18 \\ 3x - 2y = 15 \end{pmatrix}$

41. $\begin{pmatrix} x + 2y + 3z = -2 \\ x + 3y + 4z = -3 \\ x + 4y + 3z = -6 \end{pmatrix}$

42. $\begin{pmatrix} x + 3y - 2z = 5 \\ x + 4y - z = 3 \\ -2x - 7y + 5z = -12 \end{pmatrix}$

43. $\begin{pmatrix} x - 2y + z = -3 \\ -2x + 5y + 3z = 34 \\ 3x - 5y + 7z = 14 \end{pmatrix}$

44. $\begin{pmatrix} x + 4y - 2z = 2 \\ -3x - 11y + z = -2 \\ 2x + 7y + 3z = -2 \end{pmatrix}$

45. $\begin{pmatrix} x + 2y + 3z = 2 \\ -3x - 4y + 3z = 0 \\ 2x + 4y - z = 4 \end{pmatrix}$

46. $\begin{pmatrix} x - 2y + 3z = -39 \\ -x + 3y - 2z = 40 \\ -2x + 6y + z = 45 \end{pmatrix}$

47. Five systems of linear equations can be generated from the system

$$\begin{pmatrix} x + y + 2z = a \\ 2x + 3y - z = b \\ -3x + y - 2z = c \end{pmatrix}$$

by letting a, b, and c assume five different sets of values. Solve the system for each set of values. The inverse of the coefficient matrix of these systems is given in Example 2 of this section.

a. $a = 7$, $b = 1$, and $c = -1$

b. $a = 7$, $b = 5$, and $c = 1$

c. $a = -9$, $b = -8$, and $c = 19$

d. $a = -1$, $b = -13$, and $c = -17$

e. $a = -2$, $b = 0$, and $c = -2$

MISCELLANEOUS PROBLEMS

48. Matrices can be used to code and decode messages. For example, suppose that we set up a one-to-one correspondence between the letters of the alphabet and the first 26 counting numbers, as follows.

A	B	C	...	Z
1	2	3		26

Now suppose that we want to code the message PLAY IT BY EAR. We can partition the letters of the message into groups of two. Since the last group will contain only one letter, let's arbitrarily stick in a Z to form a group of two. Let's also assign a number to each letter based on the letter/number association we exhibited.

P	L	A	Y	I	T	B	Y	E	A	R	Z
16	12	1	25	9	20	2	25	5	1	18	26

Each pair of numbers can be recorded as columns in a 2×6 matrix B.

$$B = \begin{bmatrix} 16 & 1 & 9 & 2 & 5 & 18 \\ 12 & 25 & 20 & 25 & 1 & 26 \end{bmatrix}$$

Now let's choose a 2×2 matrix that contains only integers and its inverse also contains only integers. For example, we can use

$$A = \begin{bmatrix} 3 & 1 \\ 5 & 2 \end{bmatrix}; \text{ then } A^{-1} = \begin{bmatrix} 2 & -1 \\ -5 & 3 \end{bmatrix}.$$

Next, let's find the product AB.

$$AB = \begin{bmatrix} 3 & 1 \\ 5 & 2 \end{bmatrix} \begin{bmatrix} 16 & 1 & 9 & 2 & 5 & 18 \\ 12 & 25 & 20 & 25 & 1 & 26 \end{bmatrix} = \begin{bmatrix} 60 & 28 & 47 & 31 & 16 & 80 \\ 104 & 55 & 85 & 60 & 27 & 142 \end{bmatrix}.$$

Now we have our **coded message**:

60 104 28 55 47 85 31 60 16 27 80 142.

A person decoding the message would put the numbers back into a 2×6 matrix, multiply it on the left by A^{-1}, and convert the numbers back to letters.

Each of the following coded messages was formed by using the matrix $A = \begin{bmatrix} 2 & 3 \\ 1 & 2 \end{bmatrix}$. Decode each of the messages.

a. 68 40 77 51 78 49 23 15 29 19 85 52 41 27

b. 62 40 78 47 64 36 19 11 93 57 93 56 88 57

c. 64 36 58 37 63 36 21 13 75 47 63 36 38 23 118 72

d. 61 38 115 69 93 57 36 20 78 49 68 40 77 51 60 37 47 26 84 51 21 11

49. Suppose that the ordered pair (x, y) of a rectangular coordinate system is recorded as a 2×1 matrix and then multiplied on the left by the matrix $\begin{bmatrix} 1 & 0 \\ 0 & -1 \end{bmatrix}$. We would obtain

$$\begin{bmatrix} 1 & 0 \\ 0 & -1 \end{bmatrix} \begin{bmatrix} x \\ y \end{bmatrix} = \begin{bmatrix} x \\ -y \end{bmatrix}.$$

The point $(x, -y)$ is an x-axis reflection of the point (x, y). Therefore, the matrix $\begin{bmatrix} 1 & 0 \\ 0 & -1 \end{bmatrix}$ performs an x-axis reflection. What type of geometric transformation is performed by each of the following matrices?

a. $\begin{bmatrix} -1 & 0 \\ 0 & 1 \end{bmatrix}$

b. $\begin{bmatrix} -1 & 0 \\ 0 & -1 \end{bmatrix}$

c. $\begin{bmatrix} 0 & -1 \\ 1 & 0 \end{bmatrix}$

[*Hint:* Check the slopes of lines through the origin.]

d. $\begin{bmatrix} 0 & 1 \\ -1 & 0 \end{bmatrix}$

GRAPHICS CALCULATOR ACTIVITIES

50. Use your calculator to check your answers for Problems 14, 18, 28, 30, 32, 34, 36, 42, 44, 46, and 47.

51. Use your calculator and the method of matrix inverses to solve each of the following systems. Be sure to check your solutions.

a. $\begin{pmatrix} 2x - 3y + 4z = 54 \\ 3x + y - z = 32 \\ 5x - 4y + 3z = 58 \end{pmatrix}$

b. $\begin{pmatrix} 17x + 15y - 19z = 10 \\ 18x - 14y + 16z = 94 \\ 13x + 19y - 14z = -23 \end{pmatrix}$

c. $\begin{pmatrix} 1.98x + 2.49y + 3.15z = 45.72 \\ 2.29x + 1.95y + 2.75z = 42.05 \\ 3.15x + 3.20y + 1.85z = 42 \end{pmatrix}$

d. $\begin{pmatrix} x_1 + 2x_2 - 4x_3 + 7x_4 = -23 \\ 2x_1 - 3x_2 + 5x_3 - x_4 = -22 \\ 5x_1 + 4x_2 - 2x_3 - 8x_4 = 59 \\ 3x_1 - 7x_2 + 8x_3 + 9x_4 = -103 \end{pmatrix}$

e. $\begin{pmatrix} 2x_1 - x_2 + 3x_3 - 4x_4 + 12x_5 = 98 \\ x_1 + 2x_2 - x_3 - 7x_4 + 5x_5 = 41 \\ 3x_1 + 4x_2 - 7x_3 + 6x_4 - 9x_5 = -41 \\ 4x_1 - 3x_2 + x_3 - x_4 + x_5 = 4 \\ 7x_1 + 8x_2 - 4x_3 - 6x_4 - 6x_5 = 12 \end{pmatrix}$

11.4 SYSTEMS INVOLVING LINEAR INEQUALITIES; LINEAR PROGRAMMING

Finding solution sets for systems of linear inequalities relies heavily on the graphing approach. (Recall that we discussed graphing of linear inequalities in Section 3.3.) The solution set of the system

$$\begin{pmatrix} x + y > 2 \\ x - y < 2 \end{pmatrix}$$

is the intersection of the solution sets of the individual inequalities. In Figure 11.1(a) we indicate the solution set for $x + y > 2$, and in Figure 11.1(b) we indicate the solution set for $x - y < 2$. Then, in Figure 11.1(c), the shaded region represents

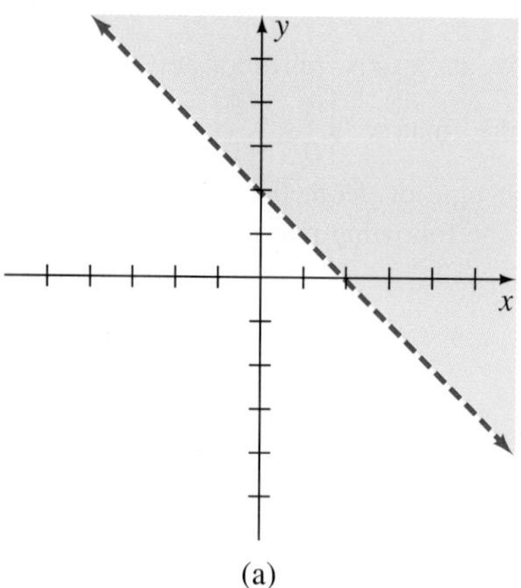

(a)

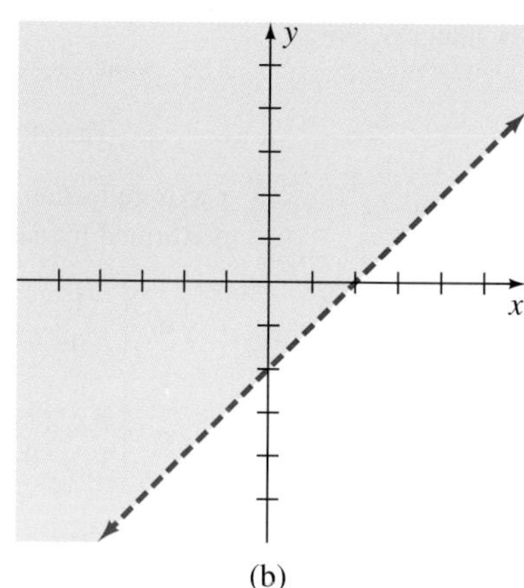

(b)

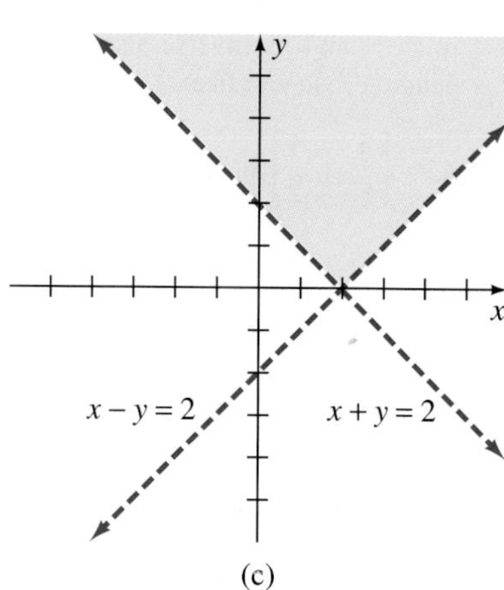

(c)

FIGURE 11.1

the intersection of the two solution sets; therefore it is the graph of the system. Remember that dashed lines are used to indicate that the points on the lines are not included in the solution set. In the following examples, we indicate only the final solution set for the system.

EXAMPLE 1

Solve the following system by graphing.

$$\begin{pmatrix} 2x - y \geq 4 \\ x + 2y < 2 \end{pmatrix}$$

Solution

The graph of $2x - y \geq 4$ consists of all points *on or below* the line $2x - y = 4$. The graph of $x + 2y < 2$ consists of all points *below* the line $x + 2y = 2$. The graph of the system is indicated by the shaded region in Figure 11.2. Notice that all points in the shaded region are on or below the line $2x - y = 4$ and below the line $x + 2y = 2$.

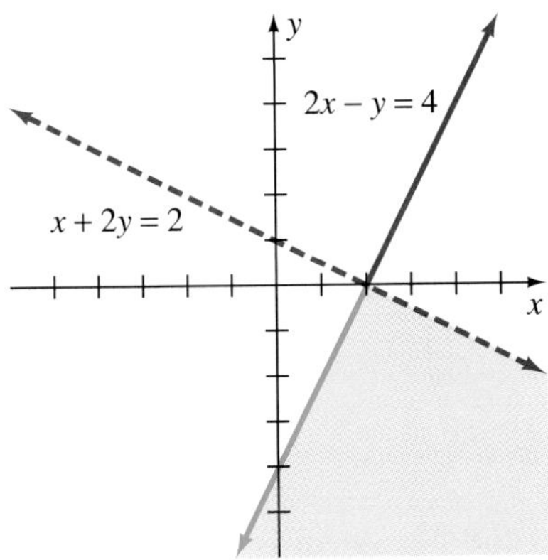

FIGURE 11.2

EXAMPLE 2

Solve the following system by graphing.

$$\begin{pmatrix} x \leq 2 \\ y \geq -1 \end{pmatrix}$$

Solution

Remember that even though each inequality contains only one variable, we are working in a rectangular coordinate system involving ordered pairs. That is to say, the system could also be written

$$\begin{pmatrix} x + 0(y) \leq 2 \\ 0(x) + y \geq -1 \end{pmatrix}.$$

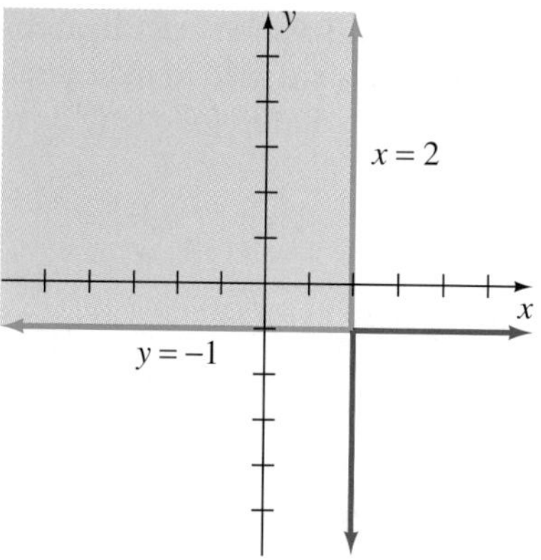

FIGURE 11.3

The graph of this system is the shaded region in Figure 11.3. Notice that all points in the shaded region are *on or to the left* of the line $x = 2$ and *on or above* the line $y = -1$.

A system may contain more than two inequalities, as the next example illustrates.

EXAMPLE 3

Solve the following system by graphing.

$$\begin{pmatrix} x \geq 0 \\ y \geq 0 \\ 2x + 3y \leq 12 \\ 3x + y \leq 6 \end{pmatrix}$$

Solution

The solution set for the system is the intersection of the solution sets of the four inequalities. The shaded region in Figure 11.4 indicates the solution set for the

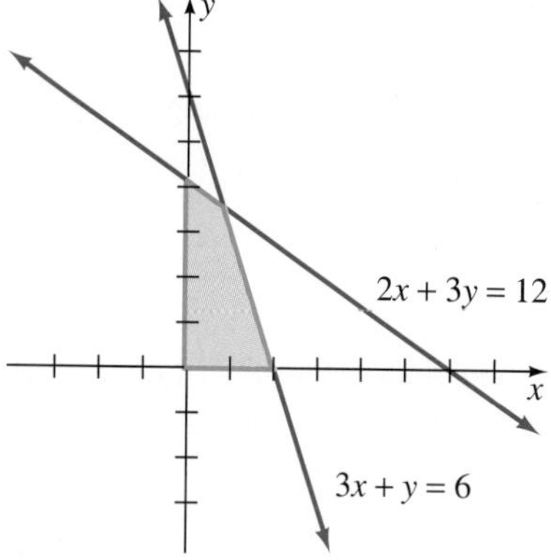

FIGURE 11.4

system. Notice that all points in the shaded region are *on or to the right* of the y-axis, *on or above* the x-axis, *on or below* the line $2x + 3y = 12$, and *on or below* the line $3x + y = 6$.

Linear Programming: Another Look at Problem Solving

Throughout this text, problem solving is a unifying theme. Therefore, it seems appropriate at this time to give you a brief glimpse of an area of mathematics that was developed in the 1940s specifically as a problem solving tool. Many applied problems involve the idea of *maximizing* or *minimizing* a certain function that is subject to various constraints; these can be expressed as linear inequalities. **Linear programming** was developed as one method for solving such problems.

> **REMARK** The term *programming* refers to the distribution of limited resources in order to maximize or minimize a certain function, such as cost, profit, distance, and so on. Thus, it is not synonymous with its meaning in computer programming. The constraints that govern the distribution of resources determine the linear inequalities and equations; thus, the term *linear programming* is used.

Before we introduce a linear programming type of problem, we need to extend one mathematical concept a bit. A **linear function in two variables** x and y is a function of the form $f(x, y) = ax + by + c$, where a, b, and c are real numbers. In other words, with each ordered pair (x, y) we associate a third number by the rule $ax + by + c$. For example, suppose the function f is described by $f(x, y) = 4x + 3y + 5$. Then $f(2, 1) = 4(2) + 3(1) + 5 = 16$.

First, let's take a look at some mathematical ideas that form the basis for solving a linear programming problem. Consider the shaded region in Figure 11.5 and the following linear functions in two variables.

$$f(x, y) = 4x + 3y + 5,$$

$$f(x, y) = 2x + 7y - 1,$$

$$f(x, y) = x - 2y$$

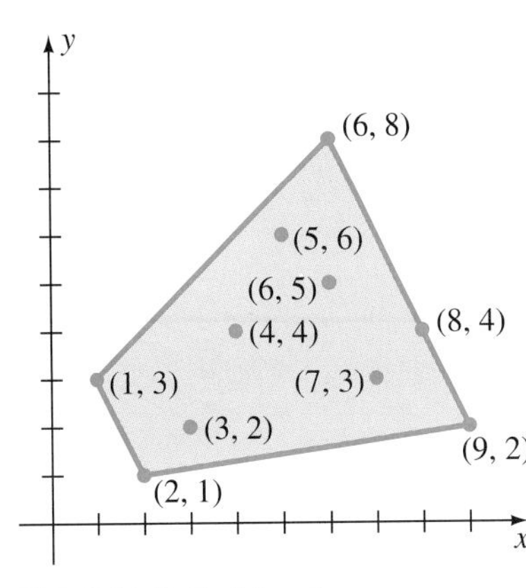

FIGURE 11.5

Suppose that we need to find the maximum and minimum value achieved by each of the functions in the indicated region. The following chart summarizes the values for the ordered pairs indicated in Figure 11.5.

	ORDERED PAIRS	VALUE OF $f(x, y) = 4x + 3y + 5$	VALUE OF $f(x, y) = 2x + 7y - 1$	VALUE OF $f(x, y) = x - 2y$
Vertex	(2, 1)	16 (*minimum*)	10 (*minimum*)	0
	(3, 2)	23	19	−1
Vertex	(9, 2)	47	31	5 (*maximum*)
Vertex	(1, 3)	18	22	−5
	(7, 3)	42	34	1
	(4, 4)	33	35	−4
	(8, 4)	49	43	0
	(6, 5)	44	46	−4
	(5, 6)	43	51	−7
Vertex	(6, 8)	53 (*maximum*)	67 (*maximum*)	−10 (*minimum*)

Notice that for each function, the maximum and minimum values are obtained at vertices of the region.

We claim that, for linear functions, maximum and minimum functional values are *always* obtained at vertices of the region. To substantiate this, let's consider the family of lines $x - 2y = k$, where k is an arbitrary constant. (We are now working only with the function $f(x, y) = x - 2y$.) In slope-intercept form, $x - 2y = k$ becomes $y = \frac{1}{2}x - \frac{1}{2}k$; so we have a family of parallel lines each having a slope of $\frac{1}{2}$. In Figure 11.6 we sketched some of these lines, so that each line has at least one point in common with the given region. Note that $x - 2y$ reaches a minimum value of -10 at the vertex (6, 8) and a maximum value of 5 at the vertex (9, 2).

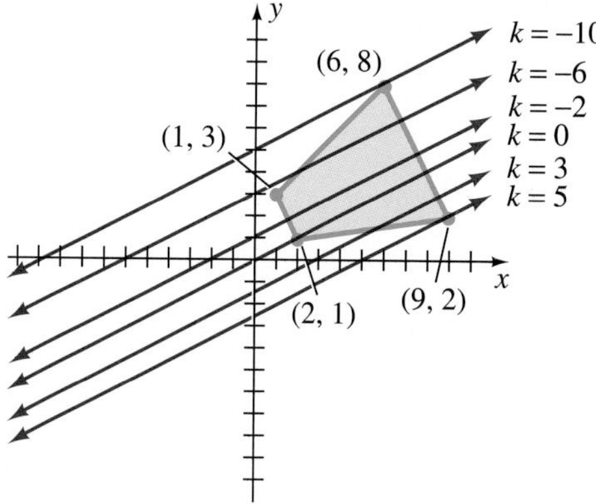

FIGURE 11.6

In general, **suppose that f is a linear function in two variables x and y and that S is a region of the xy-plane. If f attains a maximum (minimum) value in S, then that maximum (minimum) value is obtained at a vertex of S.**

REMARK A subset of the xy-plane is said to be **bounded** if there is a circle that contains all of its points; otherwise the subset if said to be **unbounded**. A bounded set will contain maximum and minimum values for a function, but an unbounded set may not contain such values.

Now we will consider two examples that illustrate a general graphing approach to solving a linear programming problem in two variables. The first example gives us the general makeup of such a problem; the second example will illustrate the type of setting from which the function and inequalities evolve.

EXAMPLE 4

Find the maximum value and the minimum value of the function $f(x, y) = 9x + 13y$ in the region determined by the following system of inequalities.

$$\begin{pmatrix} x \geq 0 \\ y \geq 0 \\ 2x + 3y \leq 18 \\ 2x + y \leq 10 \end{pmatrix}$$

Solution

First, let's graph the inequalities to determine the region, as indicated in Figure 11.7. (Such a region is called the **set of feasible solutions** and the inequalities are referred to as **constraints**.) The point (3, 4) is determined by solving the system

$$\begin{pmatrix} 2x + 3y = 18 \\ 2x + y = 10 \end{pmatrix}.$$

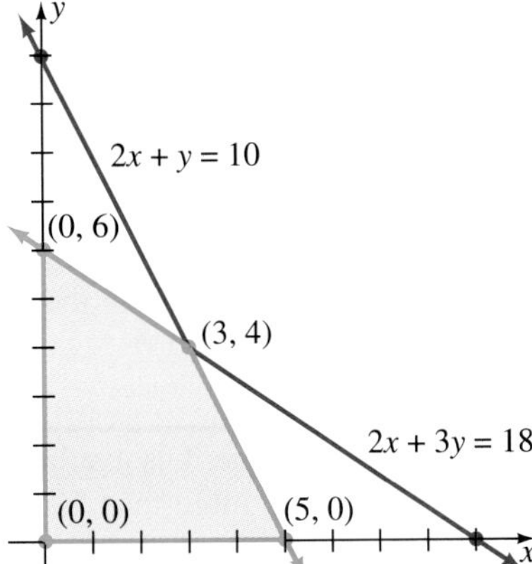

FIGURE 11.7

Next we can determine the values of the given function at the vertices of the region. (Such a function to be maximized or minimized is called the **objective function**.) A minimum value of 0 is obtained at (0, 0) and a maximum value of 79 is obtained at (3, 4).

VERTICES	VALUE OF $f(x, y) = 9x + 13y$
(0, 0)	0 (*minimum*)
(5, 0)	45
(3, 4)	79 (*maximum*)
(0, 6)	78

EXAMPLE 5

A company that manufactures gidgets and gadgets has the following production information available.

1. To produce a gidget requires 3 hours of working time on machine A and 1 hour on machine B.
2. To produce a gadget requires 2 hours on machine A and 1 hour on machine B.
3. Machine A is available for no more than 120 hours per week and machine B is available for no more than 50 hours per week.
4. Gidgets can be sold at a profit of \$3.75 each, while a profit of \$3 each can be realized on a gadget.

How many gidgets and how many gadgets should be produced each week to maximize profit? What would the maximum profit be?

Solution

Let x be the number of gidgets and y the number of gadgets. Thus, the profit function is $P(x, y) = 3.75x + 3y$. The constraints for the problem can be represented by the following inequalities.

$3x + 2y \leq 120$ Machine A is available for no more than 120 hours.

$x + y \leq 50$ Machine B is available for no more than 50 hours.

$x \geq 0$
$y \geq 0$ The number of gidgets and gadgets must be represented by a nonnegative number.

When we graph these inequalities, we obtain the set of feasible solutions indicated by the shaded region in Figure 11.8. Next we find the value of the profit function at the vertices; this produces the following chart.

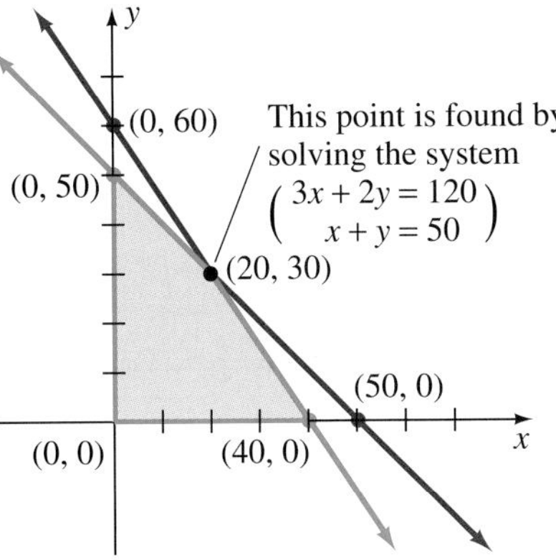

FIGURE 11.8

VERTICES	VALUE OF $P(x, y) = 3.75x + 3y$
(0, 0)	0
(40, 0)	150
(20, 30)	165 (*maximum*)
(0, 50)	150

Thus, a maximum profit of \$165 is realized by producing 20 gidgets and 30 gadgets.

PROBLEM SET 11.4

Indicate the solution set for each of the following systems of inequalities by graphing the system and shading the appropriate region.

1. $\begin{pmatrix} x + y > 3 \\ x - y > 1 \end{pmatrix}$

2. $\begin{pmatrix} x - y < 2 \\ x + y < 1 \end{pmatrix}$

3. $\begin{pmatrix} x - 2y \leq 4 \\ x + 2y > 4 \end{pmatrix}$

4. $\begin{pmatrix} 3x - y > 6 \\ 2x + y \leq 4 \end{pmatrix}$

5. $\begin{pmatrix} 2x + 3y \leq 6 \\ 3x - 2y \leq 6 \end{pmatrix}$

6. $\begin{pmatrix} 4x + 3y \geq 12 \\ 3x - 4y \geq 12 \end{pmatrix}$

7. $\begin{pmatrix} 2x - y \geq 4 \\ x + 3y < 3 \end{pmatrix}$

8. $\begin{pmatrix} 3x - y < 3 \\ x + y \geq 1 \end{pmatrix}$

9. $\begin{pmatrix} x + 2y > -2 \\ x - y < -3 \end{pmatrix}$

10. $\begin{pmatrix} x - 3y < -3 \\ 2x - 3y > -6 \end{pmatrix}$

11. $\begin{pmatrix} y > x - 4 \\ y < x \end{pmatrix}$

12. $\begin{pmatrix} y \leq x + 2 \\ y \geq x \end{pmatrix}$

13. $\begin{pmatrix} x - y > 2 \\ x - y > -1 \end{pmatrix}$

14. $\begin{pmatrix} x + y > 1 \\ x + y > 3 \end{pmatrix}$

15. $\begin{pmatrix} y \geq x \\ x > -1 \end{pmatrix}$

16. $\begin{pmatrix} y < x \\ y \leq 2 \end{pmatrix}$

17. $\begin{pmatrix} y < x \\ y > x + 3 \end{pmatrix}$

18. $\begin{pmatrix} x \leq 3 \\ y \leq -1 \end{pmatrix}$

19. $\begin{pmatrix} y > -2 \\ x > 1 \end{pmatrix}$

20. $\begin{pmatrix} x + 2y > 4 \\ x + 2y < 2 \end{pmatrix}$

21. $\begin{pmatrix} x \geq 0 \\ y \geq 0 \\ x + y \leq 4 \\ 2x + y \leq 6 \end{pmatrix}$

22. $\begin{pmatrix} x \geq 0 \\ y \geq 0 \\ x - y \leq 5 \\ 4x + 7y \leq 28 \end{pmatrix}$

23. $\begin{pmatrix} x \geq 0 \\ y \geq 0 \\ 2x + y \leq 4 \\ 2x - 3y \leq 6 \end{pmatrix}$

24. $\begin{pmatrix} x \geq 0 \\ y \geq 0 \\ 3x + 5y \geq 15 \\ 5x + 3y \geq 15 \end{pmatrix}$

For Problems 25–28 (Figures 11.9 through 11.12), find the maximum and minimum values of the given function in the indicated region.

25. $f(x, y) = 3x + 5y$

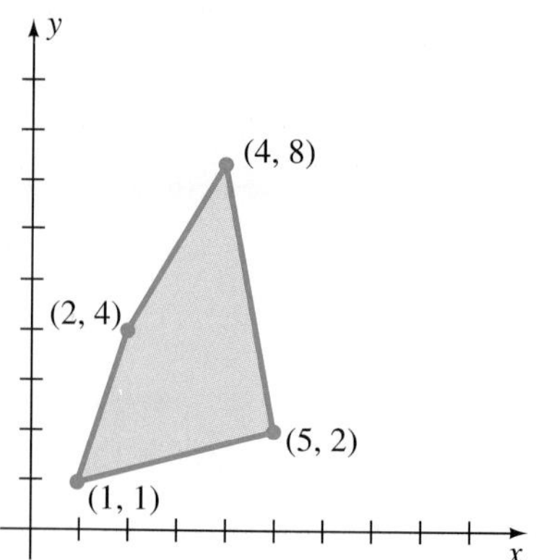

FIGURE 11.9

26. $f(x, y) = 8x + 3y$

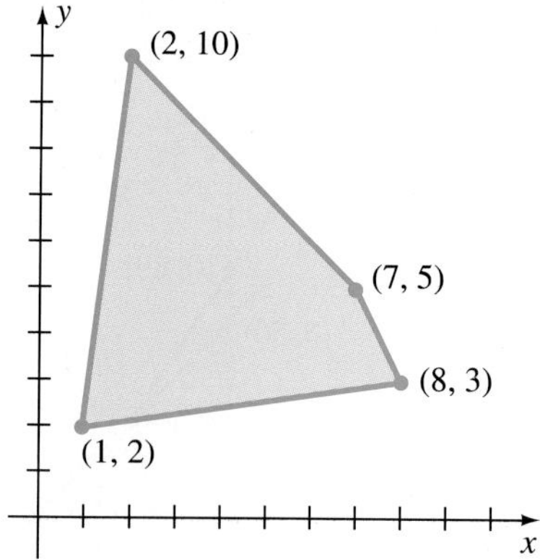

FIGURE 11.10

27. $f(x, y) = x + 4y$

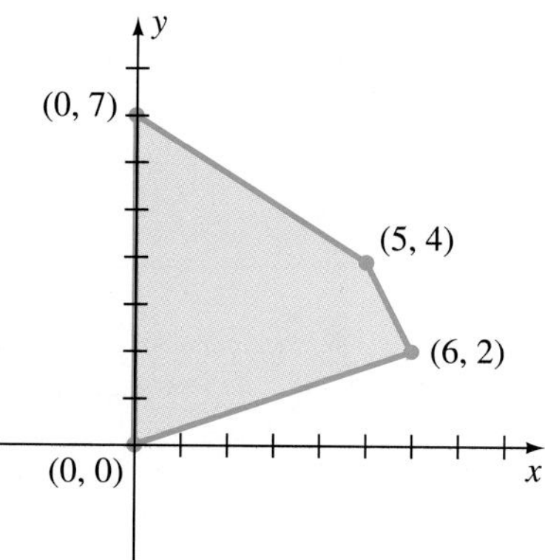

FIGURE 11.11

28. $f(x, y) = 2.5x + 3.5y$

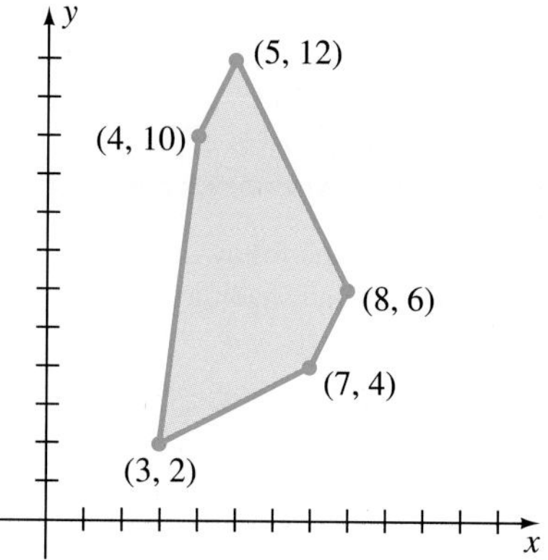

FIGURE 11.12

29. Maximize the function $f(x, y) = 3x + 7y$ in the region determined by the following constraints.

$$\begin{aligned} 3x + 2y &\leq 18 \\ 3x + 4y &\geq 12 \\ x &\geq 0 \\ y &\geq 0 \end{aligned}$$

30. Maximize the function $f(x, y) = 1.5x + 2y$ in the region determined by the following constraints.

$$\begin{aligned} 3x + 2y &\leq 36 \\ 3x + 10y &\leq 60 \\ x &\geq 0 \\ y &\geq 0 \end{aligned}$$

31. Maximize the function $f(x, y) = 40x + 55y$ in the region determined by the following constraints.

$$\begin{aligned} 2x + y &\leq 10 \\ x + y &\leq 7 \\ 2x + 3y &\leq 18 \\ x &\geq 0 \\ y &\geq 0 \end{aligned}$$

32. Maximize the function $f(x, y) = 0.08x + 0.09y$ in the region determined by the following constraints.

$$\begin{aligned} x + y &\leq 8000 \\ y &\leq \frac{1}{3}x \\ y &\geq 500 \\ x &\leq 7000 \\ x &\geq 0 \end{aligned}$$

33. Minimize the function $f(x, y) = 0.2x + 0.5y$ in the region determined by the following constraints.

$$\begin{aligned} 2x + y &\geq 12 \\ 2x + 5y &\geq 20 \\ x &\geq 0 \\ y &\geq 0 \end{aligned}$$

34. Minimize the function $f(x, y) = 3x + 7y$ in the region determined by the following constraints.

$$\begin{aligned} x + y &\geq 9 \\ 6x + 11y &\geq 84 \\ x &\geq 0 \\ y &\geq 0 \end{aligned}$$

35. Maximize the function $f(x, y) = 9x + 2y$ in the region determined by the following constraints.

$$\begin{aligned} 5y - 4x &\leq 20 \\ 4x + 5y &\leq 60 \\ x &\geq 0 \\ x &\leq 10 \\ y &\geq 0 \end{aligned}$$

36. Maximize the function $f(x, y) = 3x + 4y$ in the region determined by the following constraints.

$$\begin{aligned} 2y - x &\leq 6 \\ x + y &\leq 12 \\ x &\geq 2 \\ x &\leq 8 \\ y &\geq 0 \end{aligned}$$

Solve each of the following linear programming problems by using the graphing method illustrated in Example 5 on page 674.

37. Suppose that an investor wants to invest up to \$10,000. She plans to buy one speculative type of stock and one conservative type. The speculative stock is paying a 12% return and the conservative stock is paying a 9% return. She has decided to invest at least \$2000 in the conservative stock and no more than \$6000 in the speculative stock. Furthermore, she does not want the speculative investment to exceed the conservative one. How much should she invest at each rate to maximize her return?

38. A manufacturer of golf clubs makes a profit of \$50 per set on a model *A* set and \$45 per set on a model *B* set. Daily production of the model *A* clubs is between 30 and 50 sets, inclusive, and that of the model *B* clubs is between 10 and 20 sets, inclusive. The total daily production is not to exceed 50 sets. How many sets of each model should be manufactured per day to maximize the profit?

39. A company makes two types of calculators. Type *A* sells for \$12 and type *B* sells for \$10. It costs the company \$9 to produce one type *A* calculator and \$8 to produce one type *B* calculator. In one month, the company is equipped to produce between 200 and 300, inclusive, of the type *A* calculator and between 100 and 250, inclusive, of the type *B* calculator, but not more than 500 altogether. How many calculators of each type should be produced per month to maximize the difference between the total selling price and the total cost of production?

40. A manufacturer of small copiers makes a profit of \$200 on a deluxe model and \$250 on a standard model. The company wants to produce at least 50 deluxe models per week and at least 75 standard models per week. However, the weekly production is not to exceed 150 copiers. How many copiers of each kind should be produced in order to maximize the profit?

41. Products A and B are produced by a company according to the following production information.

a. To produce one unit of product A requires 1 hour of working time on machine I, 2 hours on machine II, and 1 hour on machine III.

b. To produce one unit of product B requires 1 hour of working time on machine I, one hour on machine II, and 3 hours on machine III.

c. Machine I is available for no more than 40 hours per week, machine II is available for no more than 40 hours per week, and machine III for no more than 60 hours per week.

d. Product A can be sold at a profit of \$2.75 per unit and product B at a profit of \$3.50 per unit.

How many units each of product A and product B should be produced per week to maximize profit?

42. Suppose that the company we refer to in Example 5 also manufactures widgets and wadgets and has the following production information available.

a. To produce a widget requires 4 hours of working time on machine A and 2 hours on machine B.

b. To produce a wadget requires 5 hours of working time on machine A and 5 hours on machine B.

c. Machine A is available for no more than 200 hours in a month and machine B is available for no more than 150 hours per month.

d. Widgets can be sold at a profit of \$7 each and wadgets at a profit of \$8 each.

How many widgets and wadgets should be produced per month in order to maximize profit?

CHAPTER 11 SUMMARY

Be sure that you understand the following ideas pertaining to the algebra of matrices.

1. Matrices of the same dimension are added by adding elements in corresponding positions.
2. Matrix addition is a commutative and an associative operation.
3. Matrices of any specific dimension have an additive identity element, namely, the matrix of that same dimension containing all zeros.
4. Every matrix A has an additive inverse, $-A$, which can be found by changing the sign of each element of A.
5. Matrices of the same dimension can be subtracted by the definition $A - B = A + (-B)$.
6. The scalar product of a real number k and a matrix A can be found by multiplying each element of A by k.
7. The following properties hold for scalar multiplication and matrix addition.

$$k(A + B) = kA + kB,$$
$$(k + l)A = kA + lA,$$
$$(kl)A = k(lA)$$

8. If A is an $m \times n$ matrix and B is an $n \times p$ matrix, then the product AB is an $m \times p$ matrix. The general term, c_{ij}, of the product matrix $C = AB$ is determined by the equation

$$c_{ij} = a_{i1}b_{1j} + a_{i2}b_{2j} + \cdots + a_{in}b_{nj}.$$

9. Matrix multiplication is not a commutative operation, but it is an associative operation.

10. Matrix multiplication has two distributive properties:

$$A(B + C) = AB + AC \qquad \text{and} \qquad (A + B)C = AC + BC.$$

11. The general multiplicative identity element, I_n, for square $n \times n$ matrices contains only 1s in the main diagonal and 0s elsewhere. For example,

$$I_2 = \begin{bmatrix} 1 & 0 \\ 0 & 1 \end{bmatrix} \qquad \text{and} \qquad I_3 = \begin{bmatrix} 1 & 0 & 0 \\ 0 & 1 & 0 \\ 0 & 0 & 1 \end{bmatrix}.$$

12. If a square matrix A has a multiplicative inverse A^{-1}, then $AA^{-1} = A^{-1}A = I_n$.

13. The multiplicative inverse of the 2×2 matrix

$$A = \begin{bmatrix} a_{11} & a_{12} \\ a_{21} & a_{22} \end{bmatrix}$$

is

$$A^{-1} = \frac{1}{|A|}\begin{bmatrix} a_{22} & -a_{12} \\ -a_{21} & a_{11} \end{bmatrix}$$

for $|A| \neq 0$. If $|A| = 0$, then the matrix A has no inverse.

14. When the inverse of a square matrix exists, a general technique for finding it is described on page 662.

15. The solution set of a system of n linear equations in n variables can be found by multiplying the inverse of the coefficient matrix times the column matrix consisting of the constant terms. For example, the solution set of the system

$$\begin{pmatrix} 2x + 3y - z = 4 \\ 3x - y + 2z = 5 \\ 5x - 7y - 4z = -1 \end{pmatrix}$$

can be found by the product

$$\begin{bmatrix} 2 & 3 & -1 \\ 3 & -1 & 2 \\ 5 & -7 & -4 \end{bmatrix}^{-1} \begin{bmatrix} 4 \\ 5 \\ -1 \end{bmatrix}.$$

The solution set of a system of linear inequalities is the intersection of the solution sets of the individual inequalities. Such solution sets are easily determined by the graphing approach.

Linear programming problems deal with the idea of maximizing or minimizing a certain linear function that is subject to various constraints. The constraints are expressed as linear inequalities. Look over Examples 4 and 5 of Section 11.4 to help summarize the general approach to linear programming problems in this chapter.

CHAPTER 11 REVIEW PROBLEM SET

For Problems 1–10, compute the indicated matrix, if it exists, using the following matrices.

$$A = \begin{bmatrix} 2 & -4 \\ -3 & 8 \end{bmatrix}, \quad B = \begin{bmatrix} 5 & -1 \\ 0 & 2 \end{bmatrix},$$

$$C = \begin{bmatrix} 3 & -1 \\ -2 & 4 \\ 5 & -6 \end{bmatrix} \quad D = \begin{bmatrix} -2 & -1 & 4 \\ 5 & 0 & -3 \end{bmatrix},$$

$$E = \begin{bmatrix} 1 \\ -3 \\ -7 \end{bmatrix}, \quad F = \begin{bmatrix} 1 & -2 \\ 4 & -4 \\ 7 & -8 \end{bmatrix}$$

1. $A + B$ **2.** $B - A$ **3.** $C - F$

4. $2A + 3B$ **5.** $3C - 2F$ **6.** CD

7. DC **8.** $DC + AB$ **9.** DE

10. EF

11. Use A and B from the preceding problems and show that $AB \neq BA$.

12. Use C, D, and F from the preceding problems and show that $D(C + F) = DC + DF$.

13. Use C, D, and F from the preceding problems and show that $(C + F)D = CD + FD$.

For each of the matrices in Problems 14–23, find the multiplicative inverse, if it exists.

14. $\begin{bmatrix} 9 & 5 \\ 7 & 4 \end{bmatrix}$ **15.** $\begin{bmatrix} 9 & 4 \\ 7 & 3 \end{bmatrix}$

16. $\begin{bmatrix} -2 & 1 \\ 2 & 3 \end{bmatrix}$ **17.** $\begin{bmatrix} 4 & -6 \\ 2 & -3 \end{bmatrix}$

18. $\begin{bmatrix} -1 & -3 \\ -4 & -5 \end{bmatrix}$ **19.** $\begin{bmatrix} 0 & -3 \\ 7 & 6 \end{bmatrix}$

20. $\begin{bmatrix} 1 & -2 & 1 \\ 2 & -5 & 2 \\ -3 & 7 & 5 \end{bmatrix}$ **21.** $\begin{bmatrix} 1 & 3 & -2 \\ 4 & 13 & -7 \\ 5 & 16 & -8 \end{bmatrix}$

22. $\begin{bmatrix} -2 & 4 & 7 \\ 1 & -3 & 5 \\ 1 & -5 & 22 \end{bmatrix}$ **23.** $\begin{bmatrix} -1 & 2 & 3 \\ 2 & -5 & -7 \\ -3 & 5 & 11 \end{bmatrix}$

For Problems 24–28, use the multiplicative inverse matrix approach to solve each of the systems. The required inverses have been found in Problems 14–23.

24. $\begin{pmatrix} 9x + 5y = 12 \\ 7x + 4y = 10 \end{pmatrix}$ **25.** $\begin{pmatrix} -2x + y = -9 \\ 2x + 3y = 5 \end{pmatrix}$

26. $\begin{pmatrix} x - 2y + z = 7 \\ 2x - 5y + 2z = 17 \\ -3x + 7y + 5z = -32 \end{pmatrix}$

27. $\begin{pmatrix} x + 3y - 2z = -7 \\ 4x + 13y - 7z = -21 \\ 5x + 16y - 8z = -23 \end{pmatrix}$

28. $\begin{pmatrix} -x + 2y + 3z = 22 \\ 2x - 5y - 7z = -51 \\ -3x + 5y + 11z = 71 \end{pmatrix}$

For Problems 29–32, indicate the solution set for each of the systems of linear inequalities by graphing the system and shading the appropriate region.

29. $\begin{pmatrix} 3x - 4y \geq 0 \\ 2x + 3y \leq 0 \end{pmatrix}$

30. $\begin{pmatrix} 3x - 2y < 6 \\ 2x - 3y < 6 \end{pmatrix}$

31. $\begin{pmatrix} x - 4y < 4 \\ 2x + y \geq 2 \end{pmatrix}$

32. $\begin{pmatrix} x \geq 0 \\ y \geq 0 \\ x + 2y \leq 4 \\ 2x - y \leq 4 \end{pmatrix}$

33. Maximize the function $f(x, y) = 8x + 5y$ in the region determined by the following constraints.

$$\begin{aligned} y &\leq 4x \\ x + y &\leq 5 \\ x &\geq 0 \\ y &\geq 0 \\ x &\leq 4 \end{aligned}$$

34. Maximize the function $f(x, y) = 2x + 7y$ in the region determined by the following constraints.

$$\begin{aligned} x &\geq 0 \\ y &\geq 0 \\ x + 2y &\leq 16 \\ x + y &\leq 9 \\ 3x + 2y &\leq 24 \end{aligned}$$

35. Maximize the function $f(x, y) = 7x + 5y$ in the region determined by the constraints of Problem 34.

36. Maximize the function $f(x, y) = 150x + 200y$ in the region determined by the constraints of Problem 34.

37. A manufacturer of electric ice cream freezers makes a profit of \$4.50 on a one-gallon freezer and a profit of \$5.25 on a two-gallon freezer. The company wants to produce at least 75 one-gallon and at least 100 two-gallon freezers per week. However, the weekly production is not to exceed a total of 250 freezers. How many freezers of each type should be produced per week in order to maximize the profit?

CONIC SECTIONS

Parabolas, circles, ellipses, and hyperbolas can be formed by intersecting a right circular conical surface with a plane, as shown in Figure 12.1; these figures are often referred to as **conic sections**. The conic sections are not new to you. You did some graphing of circles, parabolas, ellipses, and hyperbolas in Chapters 3 and 4. However, at that time, except for the circle, we did not present any formal definitions or standard forms of equations. In Chapter 3 we developed the standard form for the equation of a circle, $(x - h)^2 + (y - k)^2 = r^2$. We used this equation to solve a variety of problems that pertain to circles. It is now time to study the other conic sections in the same manner. We will define each conic section and derive the standard form of an equation. We will then use the standard forms to study specific conic sections.

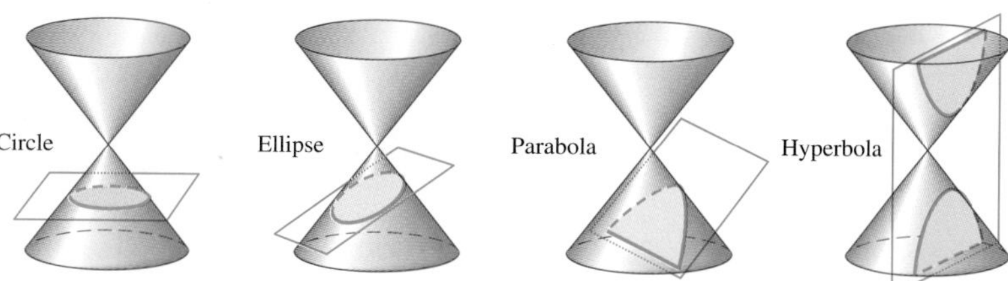

FIGURE 12.1

12

Parabolic surfaces are used in the construction of satellite dishes.

12.1 PARABOLAS

We discussed parabolas as the graphs of quadratic functions in Sections 4.2 and 4.3. All parabolas in those sections had vertical lines as axes of symmetry. Furthermore, we did not state the definition for a parabola at that time. We shall now define a parabola and derive standard forms of equations for those that have either vertical or horizontal axes of symmetry.

DEFINITION 12.1

A **parabola** is the set of all points in a plane such that the distance of each point from a fixed point F (the **focus**) is equal to its distance from a fixed line d (the **directrix**) in the plane.

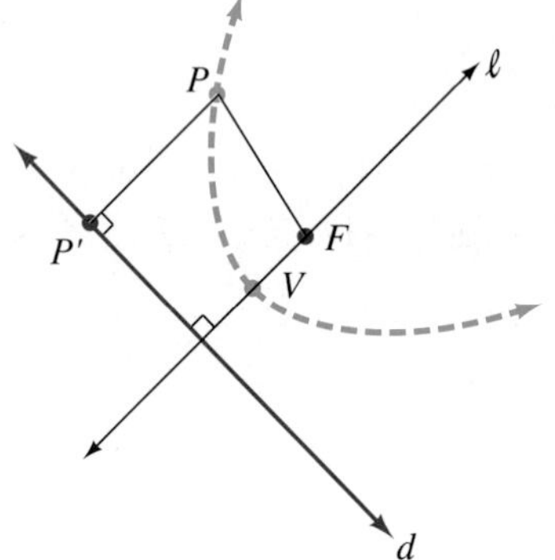

FIGURE 12.2

Using Definition 12.1, we can sketch a parabola by starting with a fixed line d and a fixed point F, not on d. Then a point P is on the parabola if and only if $PF = PP'$, where $\overline{PP'}$ is perpendicular to the directrix d (Figure 12.2). The dashed curved line in Figure 12.2 indicates the possible positions of P; it is the parabola. The line l, through F and perpendicular to the directrix d, is called the **axis of symmetry**. The point V, on the axis of symmetry halfway from F to the directrix d, is the **vertex** of the parabola.

We can derive a standard form for the equation of a parabola by coordinatizing the plane so that the origin is at the vertex of the parabola and the y-axis is the axis of symmetry (Figure 12.3). If the focus is at $(0, p)$, where $p \neq 0$, then the equation

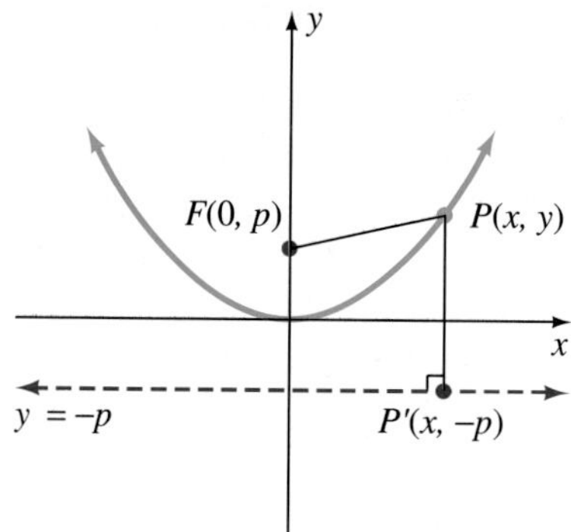

FIGURE 12.3

of the directrix is $y = -p$. Therefore, for any point P on the parabola, $PF = PP'$, and using the distance formula we obtain

$$\sqrt{(x-0)^2 + (y-p)^2} = \sqrt{(x-x)^2 + (y+p)^2}.$$

Squaring both sides and simplifying produces

$$\begin{aligned}(x - 0)^2 + (y - p)^2 &= (x - x)^2 + (y + p)^2 \\ x^2 + y^2 - 2py + p^2 &= y^2 + 2py + p^2 \\ x^2 &= 4py.\end{aligned}$$

Thus, the **standard form for the equation of a parabola**, with its vertex at the origin and y-axis as its axis of symmetry, is

$$x^2 = 4py.$$

If $p > 0$ the parabola opens upward, and if $p < 0$ the parabola opens downward.

In Figure 12.4 the line segment $\overline{QP}$ is called the **latus rectum**. It contains the focus and is parallel to the directrix. Since $FP = PP' = |2p|$, the entire length of the latus rectum is $|4p|$ units. You will see in a moment how we can use this fact when graphing parabolas.

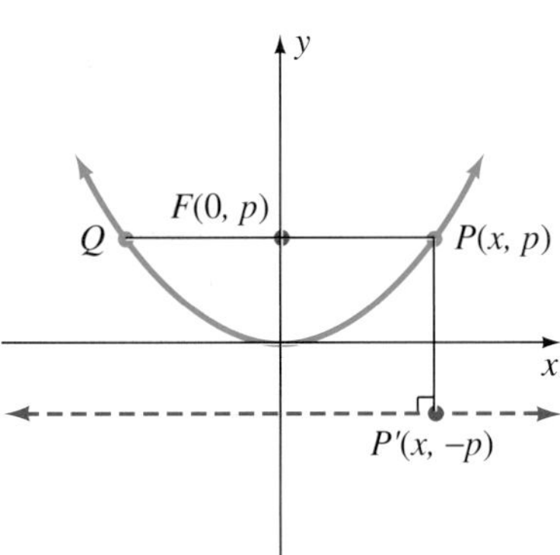

FIGURE 12.4

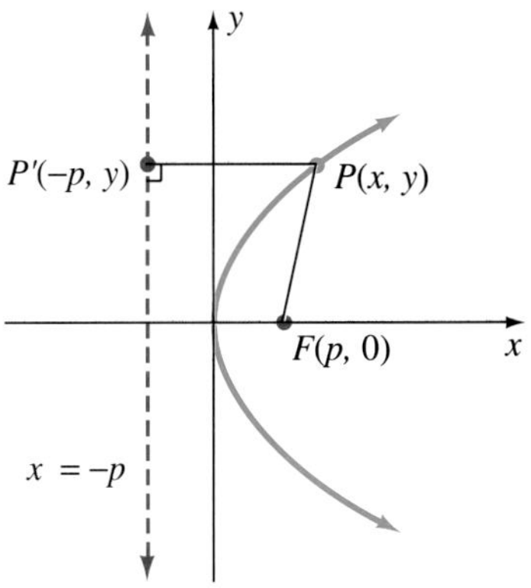

FIGURE 12.5

In a similar fashion, we can develop the standard form for the equation of a parabola with its vertex at the origin and x-axis as its axis of symmetry. By choosing a focus at $F(p, 0)$ and a directrix with an equation of $x = -p$ (see Figure 12.5), and by applying the definition of a parabola, we obtain the standard form for the equation:

$$y^2 = 4px.$$

If $p > 0$ the parabola opens to the right, as in Figure 12.5, and if $p < 0$ it opens to the left.

The concept of symmetry can be used to decide which of the two equations, $x^2 = 4py$ or $y^2 = 4px$, is to be used. The graph of $x^2 = 4py$ is symmetric with respect to the y-axis, since replacing x with $-x$ does not change the equation. Likewise, the graph of $y^2 = 4px$ is symmetric with respect to the x-axis, since replacing y with $-y$ leaves the equation unchanged.

The following property summarizes our previous discussion.

PROPERTY 12.1

The graph of each of the following equations is a parabola that has its vertex at the origin and has the indicated focus, directrix, and symmetry.

1. $x^2 = 4py$ focus $(0, p)$, directrix $y = -p$, y-axis symmetry
2. $y^2 = 4px$ focus $(p, 0)$, directrix $x = -p$, x-axis symmetry

Now let's illustrate some uses of the equations $x^2 = 4py$ and $y^2 = 4px$.

EXAMPLE 1

Find the focus and directrix of the parabola $x^2 = -8y$ and sketch its graph.

Solution

Compare $x^2 = -8y$ to the standard form $x^2 = 4py$ and we have $4p = -8$. Therefore $p = -2$ and the parabola opens downward. The focus is at $(0, -2)$ and the equation of the directrix is $y = -(-2) = 2$. The latus rectum is $|4p| = |-8| = 8$ units long. Therefore, the endpoints of the latus rectum are at $(4, -2)$ and $(-4, -2)$. The graph is sketched in Figure 12.6.

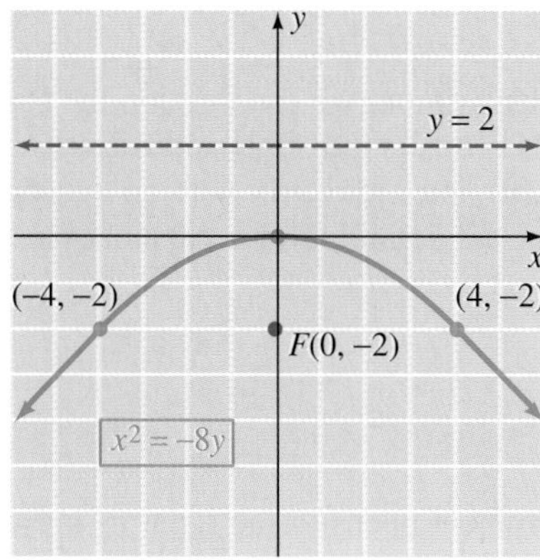

FIGURE 12.6

EXAMPLE 2

Write the equation of the parabola that is symmetric with respect to the y-axis, has its vertex at the origin, and contains the point $P(6, 3)$.

Solution

The standard form of the parabola is $x^2 = 4py$. Since P is on the parabola, the ordered pair (6, 3) must satisfy the equation. Therefore,

$$\begin{aligned} 6^2 &= 4p(3) \\ 36 &= 12p \\ 3 &= p. \end{aligned}$$

If $p = 3$, the equation becomes

$$\begin{aligned} x^2 &= 4(3)y \\ x^2 &= 12y. \end{aligned}$$

EXAMPLE 3

Find the focus and directrix of the parabola $y^2 = 6x$ and sketch its graph.

Solution

Compare $y^2 = 6x$ to the standard form $y^2 = 4px$; we see that $4p = 6$, and therefore $p = \frac{3}{2}$. So the focus is at $\left(\frac{3}{2}, 0\right)$ and the equation of the directrix is $x = -\frac{3}{2}$. The parabola opens to the right. The latus rectum is $|4p| = |6| = 6$ units long. Thus, the endpoints of the latus rectum are at $\left(\frac{3}{2}, 3\right)$ and $\left(\frac{3}{2}, -3\right)$. The graph is sketched in Figure 12.7.

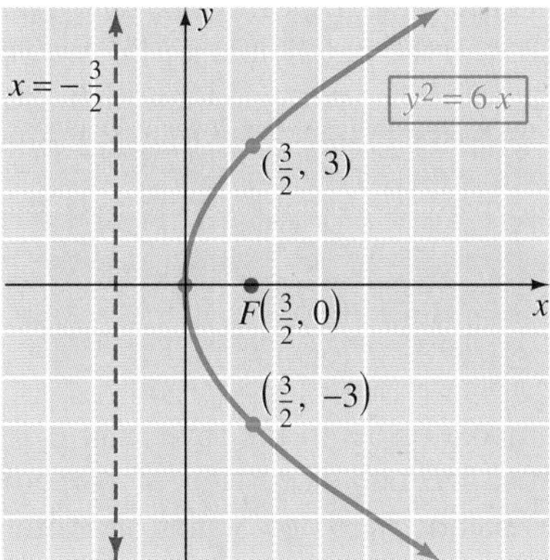

FIGURE 12.7

Other Parabolas

In much the same way we can develop the standard form for an equation of a parabola that is symmetric with respect to a line parallel to a coordinate axis. In

Figure 12.8 we have taken the vertex V at (h, k) and the focus F at $(h, k + p)$; the equation of the directrix is $y = k - p$. By the definition of a parabola, we know that $FP = PP'$. Therefore, applying the distance formula, we obtain

$$\sqrt{(x - h)^2 + (y - (k + p))^2} = \sqrt{(x - x)^2 + (y - (k - p))^2}.$$

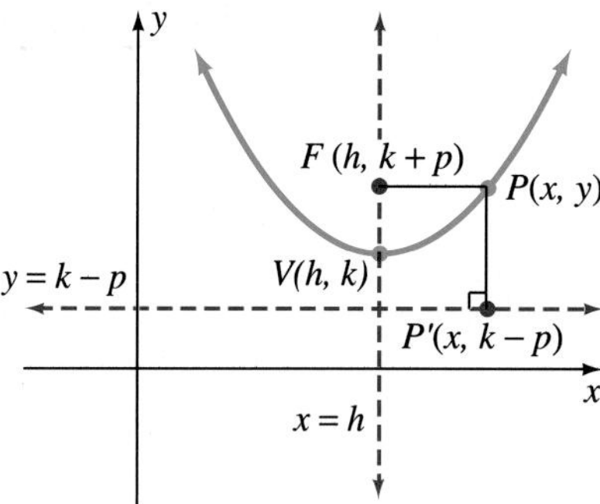

FIGURE 12.8

We leave it to the reader to show that this equation simplifies to

$$(x - h)^2 = 4p(y - k),$$

which is called the standard form for the equation of a parabola with its vertex at (h, k) and symmetric with respect to the line $x = h$. If $p > 0$ the parabola opens upward, and if $p < 0$ the parabola opens downward.

In a similar fashion, we can show that the standard form for the equation of a parabola, with its vertex at (h, k) and symmetric with respect to the line $y = k$, is

$$(y - k)^2 = 4p(x - h).$$

If $p > 0$ the parabola opens to the right, and if $p < 0$ it opens to the left.

Let's summarize our discussion of parabolas that have lines of symmetry parallel to the x-axis or y-axis by stating the following property.

PROPERTY 12.2

The graph of each of the following equations is a parabola that has its vertex at (h, k) and has the indicated focus, directrix, and symmetry:

1. $(x - h)^2 = 4p(y - k)$ focus $(h, k + p)$, directrix $y = k - p$, line of symmetry $x = h$;
2. $(y - k)^2 = 4p(x - h)$ focus $(h + p, k)$, directrix $x = h - p$, line of symmetry $y = k$.

EXAMPLE 4

Find the vertex, focus, and directrix of the parabola $y^2 + 4y - 4x + 16 = 0$, and sketch its graph.

Solution

Write the given equation as $y^2 + 4y = 4x - 16$ and we can complete the square on the left side by adding 4 to both sides.

$$y^2 + 4y + 4 = 4x - 16 + 4$$
$$(y + 2)^2 = 4x - 12$$
$$(y + 2)^2 = 4(x - 3)$$

Now let's compare this final equation to the form $(y - k)^2 = 4p(x - h)$.

$$(y - (-2))^2 = 4(x - 3).$$

$k = -2$ $4p = 4$ $h = 3$
$p = 1$

The vertex is at $(3, -2)$ and because $p > 0$, the parabola opens to the right and the focus is at $(4, -2)$. The equation of the directrix is $x = 2$. The latus rectum is $|4p| = |4| = 4$ units long and its endpoints are at $(4, 0)$ and $(4, -4)$. The graph is sketched in Figure 12.9.

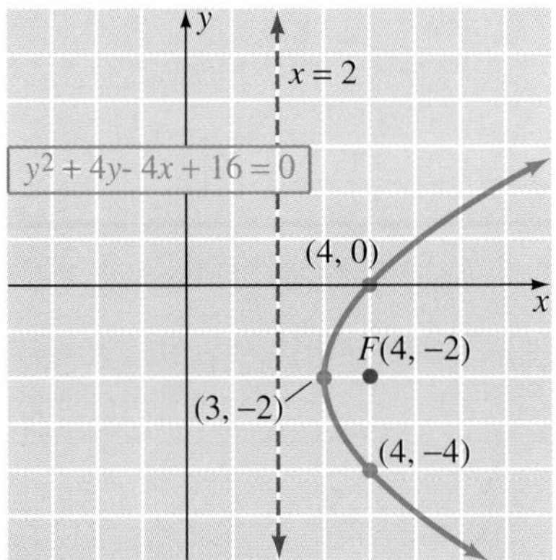

FIGURE 12.9

REMARK If we were using a graphing utility to graph the parabola in Example 4, then after the step $(y + 2)^2 = 4x - 12$ we would solve for y to obtain $y = -2 \pm \sqrt{4x - 12}$. Then we could enter $Y_1 = -2 + \sqrt{4x - 12}$ and $Y_2 = -2 - \sqrt{4x - 12}$ and obtain a figure that closely resembles Figure 12.9. (Perhaps you should do this if you have access to a graphing utility.)

EXAMPLE 5

Write the equation of the parabola if its focus is at $(-4, 1)$ and the equation of its directrix is $y = 5$.

Solution

Since the directrix is a horizontal line, we know that the equation of the parabola is of the form $(x - h)^2 = 4p(y - k)$. The vertex is halfway between the focus and directrix, so the vertex is at $(-4, 3)$. This means that $h = -4$ and $k = 3$. The parabola opens downward because the focus is below the directrix and the distance between the focus and the vertex is 2 units; so $p = -2$. Substitute -4 for h, 3 for k, and -2 for p in the equation $(x - h)^2 = 4p(y - k)$ to obtain

$$(x - (-4))^2 = 4(-2)(y - 3),$$

which simplifies to

$$\begin{aligned} (x + 4)^2 &= -8(y - 3) \\ x^2 + 8x + 16 &= -8y + 24 \\ x^2 + 8x + 8y - 8 &= 0. \end{aligned}$$

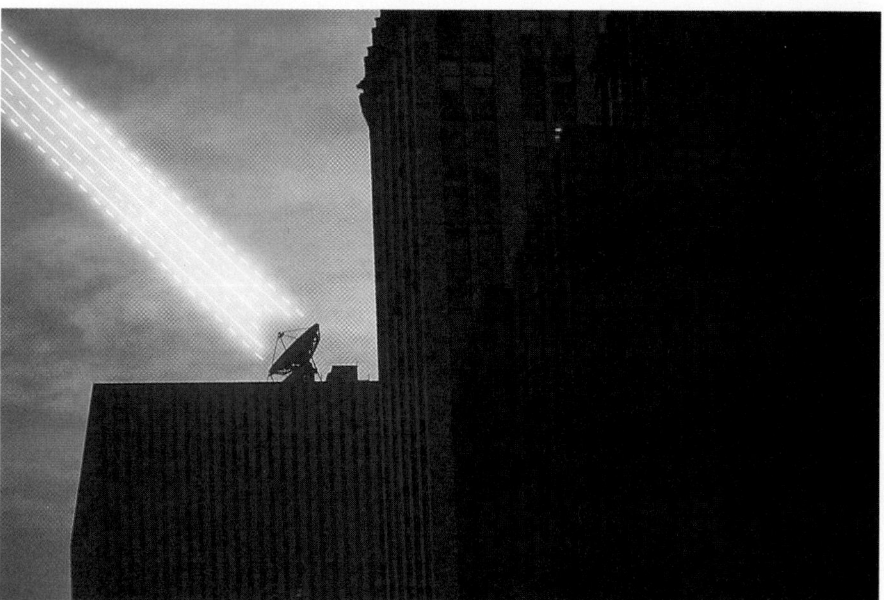

Parabolic reflectors

G. V. Faint/The Image Bank

REMARK For a problem such as Example 5, you may find it helpful to put the given information on a set of axes and draw a rough sketch of the parabola to help you with the analysis of the problem.

Parabolas possess various properties that make them very useful. For example, if a parabola is rotated about its axis, a parabolic surface is formed. The rays from a source of light placed at the focus of the surface will reflect from the surface parallel to the axis. It is for this reason that parabolic reflectors are used on searchlights as in Figure 12.10. Likewise, rays of light coming into a parabolic surface parallel to the axis will be reflected through the focus. This property of parabolas is useful in the design of mirrors for telescopes (see Figure 12.11) and in the construction of radar antennae.

A bullet fired into the air will follow the curvature of a parabola if air resistance and other outside factors are ignored, in other words, if only the force of gravity is considered. (See Figure 12.12.)

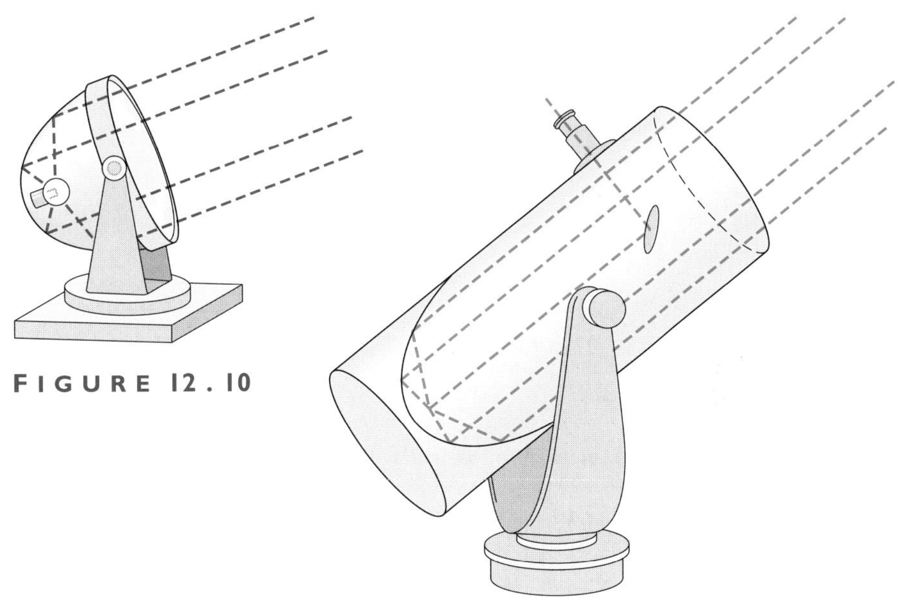

FIGURE 12.10

FIGURE 12.11

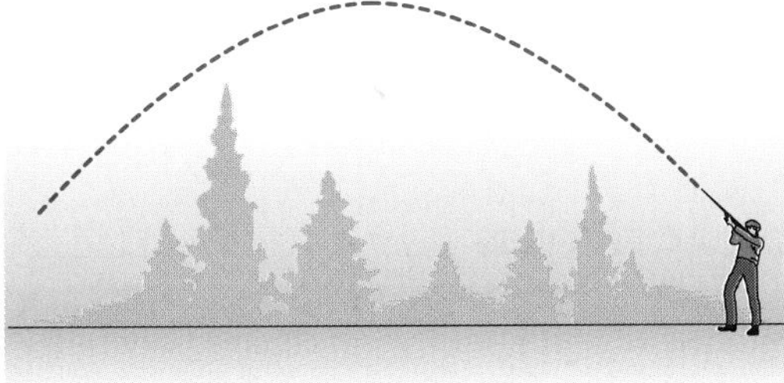

FIGURE 12.12

PROBLEM SET 12.1

For Problems 1–22, find the vertex, focus, and directrix of the given parabola and sketch its graph.

1. $y^2 = 8x$
2. $y^2 = -4x$
3. $x^2 = -12y$
4. $x^2 = 8y$
5. $y^2 = -2x$
6. $y^2 = 6x$
7. $x^2 = 6y$
8. $x^2 = -7y$
9. $x^2 - 4y + 8 = 0$
10. $x^2 - 8y - 24 = 0$
11. $x^2 + 8y + 16 = 0$
12. $x^2 + 4y - 4 = 0$
13. $y^2 - 12x + 24 = 0$
14. $y^2 + 8x - 24 = 0$
15. $x^2 - 2x - 4y + 9 = 0$
16. $x^2 + 4x - 8y - 4 = 0$
17. $x^2 + 6x + 8y + 1 = 0$
18. $x^2 - 4x + 4y - 4 = 0$
19. $y^2 - 2y + 12x - 35 = 0$
20. $y^2 + 4y + 8x - 4 = 0$
21. $y^2 + 6y - 4x + 1 = 0$
22. $y^2 - 6y - 12x + 21 = 0$

For Problems 23–42, find an equation of the parabola that satisfies the given conditions.

23. Focus $(0, 3)$, directrix $y = -3$
24. Focus $\left(0, -\frac{1}{2}\right)$, directrix $y = \frac{1}{2}$
25. Focus $(-1, 0)$, directrix $x = 1$
26. Focus $(5, 0)$, directrix $x = 1$
27. Focus $(0, 1)$, directrix $y = 7$
28. Focus $(0, -2)$, directrix $y = -10$
29. Focus $(3, 4)$, directrix $y = -2$
30. Focus $(-3, -1)$, directrix $y = 7$
31. Focus $(-4, 5)$, directrix $x = 0$
32. Focus $(5, -2)$, directrix $x = -1$
33. Vertex $(0, 0)$, symmetric with respect to the x-axis, and contains the point $(-3, 5)$
34. Vertex $(0, 0)$, symmetric with respect to the y-axis, and contains the point $(-2, -4)$
35. Vertex $(0, 0)$, focus $\left(\frac{5}{2}, 0\right)$
36. Vertex $(0, 0)$, focus $\left(0, -\frac{7}{2}\right)$
37. Vertex $(7, 3)$, focus $(7, 5)$, and symmetric with respect to the line $x = 7$
38. Vertex $(-4, -6)$, focus $(-7, -6)$, and symmetric with respect to the line $y = -6$
39. Vertex $(8, -3)$, focus $(11, -3)$, and symmetric with respect to the line $y = -3$
40. Vertex $(-2, 9)$, focus $(-2, 5)$, and symmetric with respect to the line $x = -2$
41. Vertex $(-9, 1)$, symmetric with respect to the line $x = -9$, and contains the point $(-8, 0)$
42. Vertex $(6, -4)$, symmetric with respect to the line $y = -4$, and contains the point $(8, -3)$

GRAPHICS CALCULATOR ACTIVITIES

43. The parabola determined by the equation $x^2 - 4x + 4y - 4 = 0$ (Problem 18) is easy to graph using a graphics calculator because it can be expressed as a function of x. That is, solving for y produces

$$y = \frac{-x^2 + 4x + 4}{4}.$$

However, an equation such as $y^2 + 4y + 8x - 4 = 0$ (Problem 20) requires a little more effort. Solving for y in terms of x produces the following.

$$y^2 + 4y = -8x + 4$$
$$y^2 + 4y + 4 = -8x + 4 + 4$$
$$(y + 2)^2 = -8x + 8$$
$$y + 2 = \pm\sqrt{-8x + 8}$$
$$y = -2 \pm\sqrt{-8x + 8}$$

Now we can graph the two functions, $y = -2 + \sqrt{-8x + 8}$ and $y = -2 - \sqrt{-8x + 8}$, on the same set of axes to produce the final parabola.

Use your graphics calculator to check your graphs for Problems 1–22.

12.2 ELLIPSES

Let's begin by defining the concept of an ellipse.

DEFINITION 12.2

An ellipse is the set of all points in a plane such that the sum of the distances of each point from two fixed points F and F' (the foci) in the plane is constant.

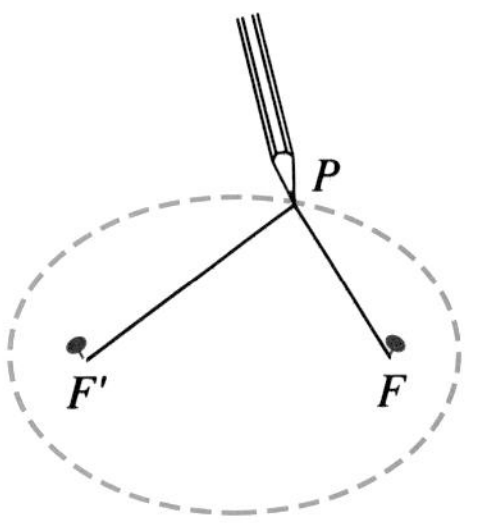

FIGURE 12.13

Using two thumbtacks, a piece of string, and a pencil, it is easy to draw an ellipse by satisfying the conditions of Definition 12.2. First, insert two thumbtacks in a piece of cardboard at points F and F' and fasten the ends of the piece of string to the thumbtacks, as in Figure 12.13. Then loop the string around the point of a pencil and hold the pencil so that the string is taut. Finally, move the pencil around the tacks, always keeping the string taut: You will draw an ellipse. The two points F and F' are the foci referred to in Definition 12.2, and the sum of the distances FP and $F'P$ is constant, since it represents the length of the piece of string. With the same piece of string, you can vary the shape of the ellipse by changing the positions of the foci. By moving F and F' farther apart, the ellipse will become flatter. Likewise, by moving F and F' closer together, the ellipse will resemble a circle. In fact, if $F = F'$, you will obtain a circle.

We can derive a standard form for the equation of an ellipse by coordinatizing the plane so that the foci are on the x-axis, equidistant from the origin (Figure 12.14). If F has coordinates $(c, 0)$, where $c > 0$, then F' has coordinates

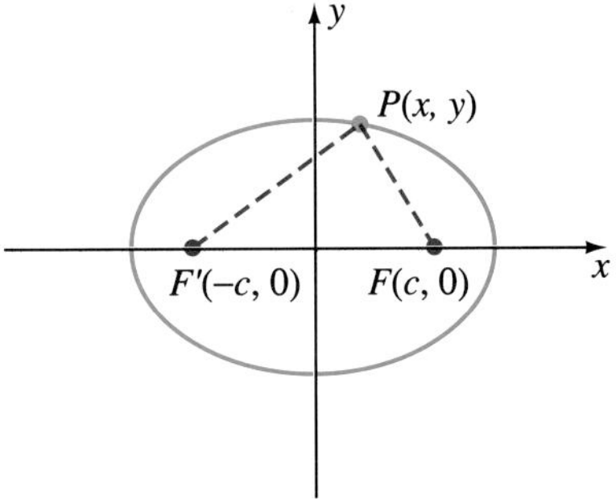

FIGURE 12.14

$(-c, 0)$, and the distance between F and F' is $2c$ units. We will let $2a$ represent the constant sum of $FP + F'P$. Note that $2a > 2c$ and therefore $a > c$. For any point P on the ellipse,

$$FP + F'P = 2a.$$

Use the distance formula to write this as

$$\sqrt{(x-c)^2 + (y-0)^2} + \sqrt{(x+c)^2 + (y-0)^2} = 2a.$$

Let's change the form of this equation to

$$\sqrt{(x-c)^2 + y^2} = 2a - \sqrt{(x+c)^2 + y^2}$$

and square both sides:

$$(x-c)^2 + y^2 = 4a^2 - 4a\sqrt{(x+c)^2 + y^2} + (x+c)^2 + y^2.$$

This can be simplified to

$$a^2 + cx = a\sqrt{(x+c)^2 + y^2}.$$

Again square both sides to produce

$$a^4 + 2a^2cx + c^2x^2 = a^2[(x+c)^2 + y^2],$$

which can be rewritten in the form

$$x^2(a^2 - c^2) + a^2y^2 = a^2(a^2 - c^2).$$

Divide both sides by $a^2(a^2 - c^2)$, which leads to the form

$$\frac{x^2}{a^2} + \frac{y^2}{a^2 - c^2} = 1.$$

Letting $b^2 = a^2 - c^2$, where $b > 0$, produces the equation

$$\frac{x^2}{a^2} + \frac{y^2}{b^2} = 1. \qquad \textbf{(1)}$$

Since $c > 0$, $a > c$, and $b^2 = a^2 - c^2$, it follows that $a^2 > b^2$ and hence $a > b$. This equation that we have derived is called the **standard form for the equation of an ellipse** with its foci on the x-axis and its center at the origin.

The x-intercepts of equation (1) can be found by letting $y = 0$. Doing this produces $x^2/a^2 = 1$ or $x^2 = a^2$; consequently, the x-intercepts are a and $-a$. The corresponding points on the graph (see Figure 12.15) are $A(a, 0)$ and $A'(-a, 0)$, and the line segment $\overline{A'A}$, which is of length $2a$, is called the **major axis** of the ellipse. The endpoints of the major axis are also referred to as the **vertices** of the ellipse. Similarly, letting $x = 0$ produces $y^2/b^2 = 1$ or $y^2 = b^2$; consequently the y-intercepts are b and $-b$. The corresponding points on the graph are $B(0, b)$ and $B'(0, -b)$, and the line segment $\overline{BB'}$, which is of length $2b$, is called the **minor axis**. Since $a > b$, **the major axis is always longer than the minor axis**. The point of intersection of the major and minor axis is called the **center** of the ellipse.

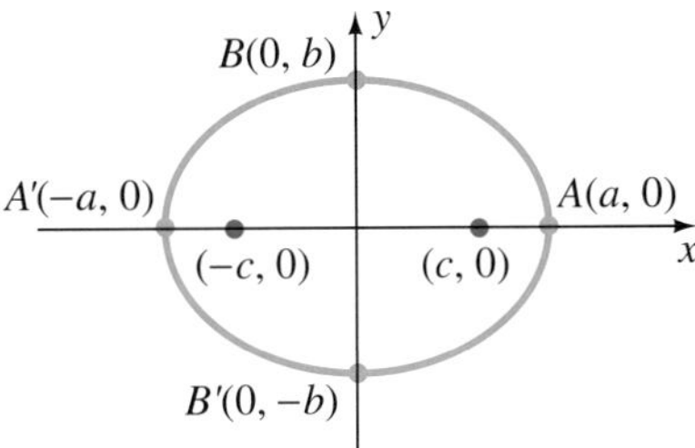

FIGURE 12.15

Let's summarize this discussion by stating the following property.

PROPERTY 12.3

The graph of the equation

$$\frac{x^2}{a^2} + \frac{y^2}{b^2} = 1,$$

for $a^2 > b^2$, is an ellipse with the endpoints of its major axis (the vertices) at $(a, 0)$ and $(-a, 0)$, and the endpoints of its minor axis at $(0, b)$ and $(0, -b)$. The foci are at $(c, 0)$ and $(-c, 0)$, where $c^2 = a^2 - b^2$.

Notice that replacing y with $-y$, or x with $-x$, or both x and y with $-x$ and $-y$, leaves the equation unchanged. Thus the graph of

$$\frac{x^2}{a^2} + \frac{y^2}{b^2} = 1$$

is symmetric with respect to the x-axis, the y-axis, and the origin.

EXAMPLE 1

Find the vertices, the endpoints of the minor axis, and the foci, and sketch the ellipse $4x^2 + 9y^2 = 36$.

Solution

The given equation can be changed to standard form by dividing both sides by 36.

$$\frac{4x^2}{36} + \frac{9y^2}{36} = \frac{36}{36}$$

$$\frac{x^2}{9} + \frac{y^2}{4} = 1$$

Therefore, $a^2 = 9$ and $b^2 = 4$; hence, the vertices are at $(3, 0)$ and $(-3, 0)$, and the ends of the minor axis are at (0, 2) and (0, −2). Since $c^2 = a^2 - b^2$, we have

$$c^2 = 9 - 4 = 5.$$

So the foci are at $(\sqrt{5}, 0)$ and $(-\sqrt{5}, 0)$. The ellipse is sketched in Figure 12.16.

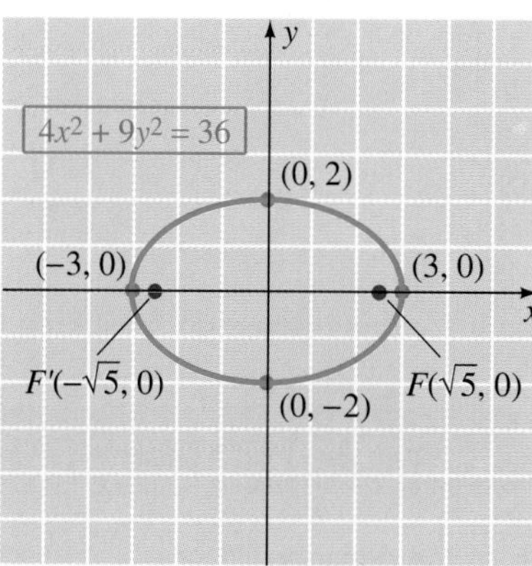

FIGURE 12.16

REMARK For a problem such as Example 1, it is not necessary to change to standard form to find the values for a and b. After all, $\pm a$ are the x-intercepts and $\pm b$ are the y-intercepts. These values can be found quite easily from the given form of the equation.

EXAMPLE 2

Find the equation of the ellipse with vertices at (±6, 0) and foci at (±4, 0).

Solution

From the given information we know that $a = 6$ and $c = 4$. Therefore,

$$b^2 = a^2 - c^2 = 36 - 16 = 20.$$

Substitute 36 for a^2 and 20 for b^2 in the standard form to produce

$$\frac{x^2}{36} + \frac{y^2}{20} = 1.$$

Multiply both sides by 180 to get

$$5x^2 + 9y^2 = 180.$$

Ellipses with Foci on y-Axis

We can develop a standard form for the equation of an ellipse with foci on the y-axis in a similar fashion. The following property summarizes the results of such a development with the foci at $(0, c)$ and $(0, -c)$, where $c > 0$.

PROPERTY 12.4

The graph of the equation

$$\frac{x^2}{b^2} + \frac{y^2}{a^2} = 1,$$

for $a^2 > b^2$, is an ellipse with the endpoints of its major axis (vertices) at $(0, a)$ and $(0, -a)$, and the endpoints of its minor axis at $(b, 0)$, and $(-b, 0)$. The foci are at $(0, c)$ and $(0, -c)$, where $c^2 = a^2 - b^2$.

From Properties 12.3 and 12.4 it is evident that an equation of an ellipse with its center at the origin and foci on a coordinate axis can be written in the form

$$\frac{x^2}{p} + \frac{y^2}{q} = 1 \qquad \text{or} \qquad qx^2 + py^2 = pq.$$

where p and q are positive. If $p > q$ the major axis lies on the x-axis, but if $q > p$ the major axis is on the y-axis. It is not necessary to memorize these facts, since for any specific problem the endpoints of the major and minor axes are determined by the x- and y-intercepts. However, it is necessary to remember the relationship $c^2 = a^2 - b^2$.

EXAMPLE 3

Find the vertices, the endpoints of the minor axis, and the foci, and sketch the ellipse $18x^2 + 4y^2 = 36$.

Solution

To find the x-intercepts, we let $y = 0$ and we obtain

$$18x^2 = 36$$
$$x^2 = 2$$
$$x = \pm\sqrt{2}.$$

Similarly, to find the y-intercepts we let $x = 0$ and we obtain

$$4y^2 = 36$$
$$y^2 = 9$$
$$y = \pm 3.$$

Since $3 > \sqrt{2}$, we know that $a = 3$ and $b = \sqrt{2}$. Therefore, the vertices are at $(0, 3)$ and $(0, -3)$, and the endpoints of the minor axis are at $(\sqrt{2}, 0)$ and $(-\sqrt{2}, 0)$. From the relationship $c^2 = a^2 - b^2$ we get

$$c^2 = 9 - 2 = 7.$$

So the foci are at $(0, \sqrt{7})$ and $(0, -\sqrt{7})$. The ellipse is sketched in Figure 12.17.

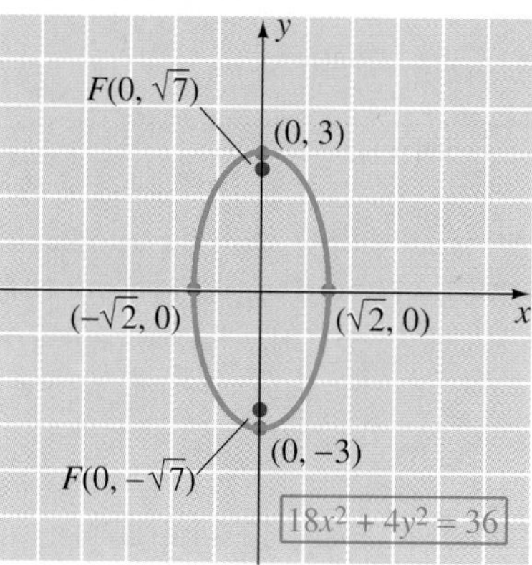

FIGURE 12.17

Other Ellipses

In the same way, we can develop the standard form for an equation of an ellipse that is symmetric with respect to a line parallel to a coordinate axis. We will not show such developments in this text, but Figures 12.18 and 12.19 indicate the basic facts we need to develop and use the resulting equations. Notice that in each case, the center of the ellipse is at a point (h, k). Furthermore, the physical significance of a, b, and c is the same as before. However, these values are used relative to the center (h, k) to find the endpoints of the major and minor axes, and the foci. Let's see how this works in a specific example.

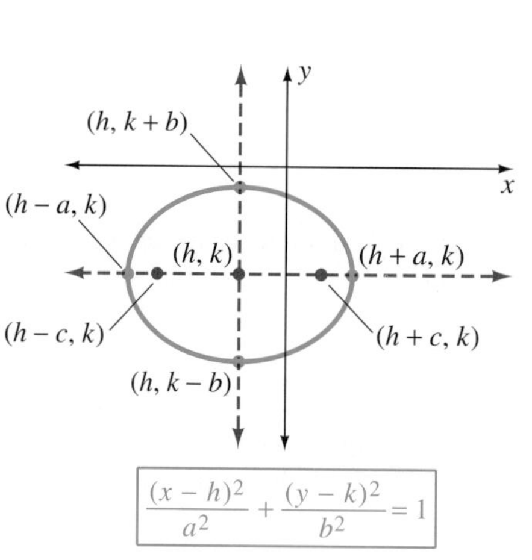

FIGURE 12.18

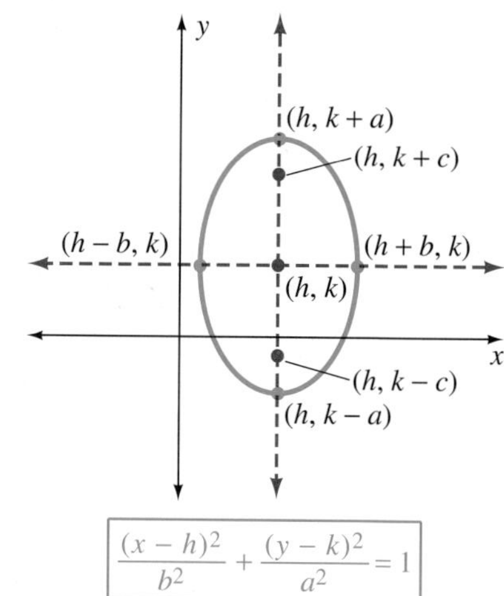

FIGURE 12.19

EXAMPLE 4 Find the vertices, the endpoints of the minor axis, and the foci, and sketch the ellipse $9x^2 + 54x + 4y^2 - 8y + 49 = 0$.

Solution

First, we need to change to a standard form by completing the square on both x and y.

$$9(x^2 + 6x + \_\_) + 4(y^2 - 2y + \_\_) = -49$$
$$9(x^2 + 6x + 9) + 4(y^2 - 2y + 1) = -49 + 81 + 4$$
$$9(x + 3)^2 + 4(y - 1)^2 = 36$$
$$\frac{(x + 3)^2}{4} + \frac{(y - 1)^2}{9} = 1$$

Since $a > b$, this last equation is of the form

$$\frac{(x - h)^2}{b^2} + \frac{(y - k)^2}{a^2} = 1,$$

where $h = -3$, $k = 1$, $a = 3$, and $b = 2$. Thus, the endpoints of the major axis (vertices) are up 3 units and down 3 units from the center, $(-3, 1)$, so they are at $(-3, 4)$ and $(-3, -2)$. Likewise, the endpoints of the minor axis are 2 units to the right and 2 units to the left of the center. Thus, they are at $(-1, 1)$ and $(-5, 1)$. From the relationship $c^2 = a^2 - b^2$, we get

$$c^2 = 9 - 4 = 5.$$

Thus, the foci are at $\left(-3, 1 + \sqrt{5}\right)$ and $\left(-3, 1 - \sqrt{5}\right)$. The ellipse is sketched in Figure 12.20.

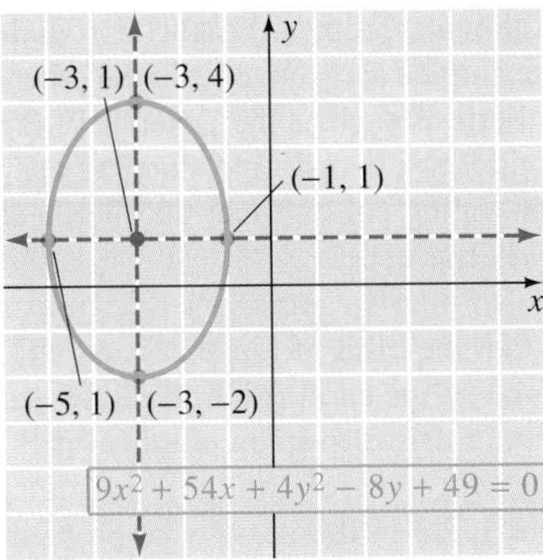

FIGURE 12.20

EXAMPLE 5

Write the equation of the ellipse that has vertices at $(-3, -5)$ and $(7, -5)$, and foci at $(-1, -5)$ and $(5, -5)$.

Solution

Since the vertices and foci are on the same horizontal line ($y = -5$), this ellipse has an equation of the form

$$\frac{(x-h)^2}{a^2} + \frac{(y-k)^2}{b^2} = 1.$$

The center of the ellipse is at the midpoint of the major axis. Therefore,

$$h = \frac{-3+7}{2} = 2 \quad \text{and} \quad k = \frac{-5+(-5)}{2} = -5.$$

The distance between the center $(2, -5)$ and a vertex $(7, -5)$ is 5 units; thus, $a = 5$. The distance between the center $(2, -5)$ and a focus $(5, -5)$ is 3 units; thus, $c = 3$. Using the relationship $c^2 = a^2 - b^2$, we obtain

$$b^2 = a^2 - c^2 = 25 - 9 = 16.$$

Now let's substitute 2 for h, -5 for k, 25 for a^2, and 16 for b^2 in the general form, and then we can simplify.

$$\begin{aligned} \frac{(x-2)^2}{25} + \frac{(y+5)^2}{16} &= 1 \\ 16(x-2)^2 + 25(y+5)^2 &= 400 \\ 16(x^2 - 4x + 4) + 25(y^2 + 10y + 25) &= 400 \\ 16x^2 - 64x + 64 + 25y^2 + 250y + 625 &= 400 \\ 16x^2 - 64x + 25y^2 + 250y + 289 &= 0 \end{aligned}$$

As with parabolas, ellipses also possess properties that make them very useful. For example, the elliptical surface formed by rotating an ellipse about its major axis has the following property: Light or sound waves emitted at one focus will reflect off of the surface and converge at the other focus. This is the principle behind "whispering galleries," such as the Rotunda of the Capitol Building in Washington, D.C. In such buildings, two people standing at two specific spots that are the foci of the elliptical ceiling can whisper and yet hear each other clearly, even though they may be quite far apart.

Ellipses also play an important role in astronomy. Johannes Kepler (1571–1630) showed that the orbit of a planet is an ellipse with the sun at one focus. For example, the orbit of the earth is elliptical but nearly circular; at the same time, the moon moves about the earth in an elliptical path (see Figure 12.21).

The arches for concrete bridges are sometimes elliptical. (One example is shown in Figure 12.23 in the next set of problems.) Also, elliptical gears are used in certain kinds of machinery that require a slow but powerful force at impact, such as a heavy-duty punch (see Figure 12.22).

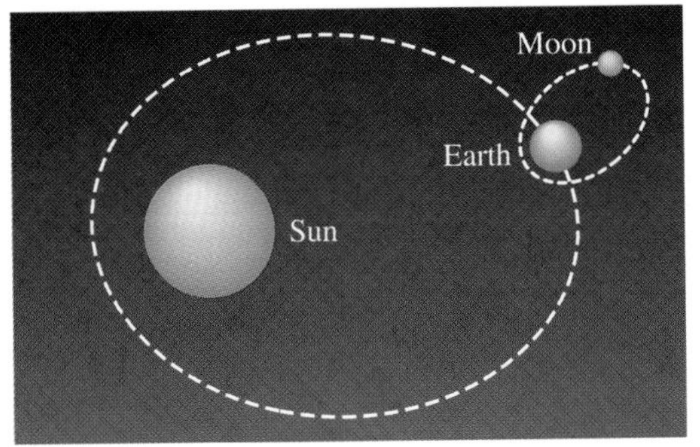

FIGURE 12.21

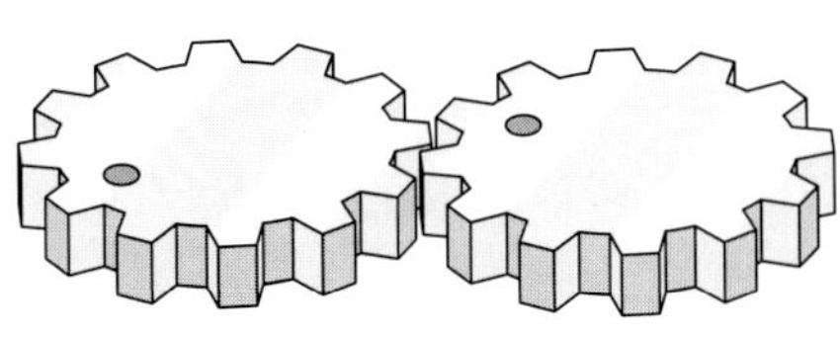

FIGURE 12.22

PROBLEM SET 12.2

For Problems 1–22, find the vertices, the endpoints of the minor axis, and the foci of the given ellipse, and sketch its graph.

1. $\frac{x^2}{4} + \frac{y^2}{1} = 1$

2. $\frac{x^2}{16} + \frac{y^2}{1} = 1$

3. $\frac{x^2}{4} + \frac{y^2}{9} = 1$

4. $\frac{x^2}{4} + \frac{y^2}{16} = 1$

5. $9x^2 + 3y^2 = 27$

6. $4x^2 + 3y^2 = 36$

7. $2x^2 + 5y^2 = 50$

8. $5x^2 + 36y^2 = 180$

9. $12x^2 + y^2 = 36$

10. $8x^2 + y^2 = 16$

11. $7x^2 + 11y^2 = 77$

12. $4x^2 + y^2 = 12$

13. $4x^2 - 8x + 9y^2 - 36y + 4 = 0$

14. $x^2 + 6x + 9y^2 - 36y + 36 = 0$

15. $4x^2 + 16x + y^2 + 2y + 1 = 0$

16. $9x^2 - 36x + 4y^2 + 16y + 16 = 0$

17. $x^2 - 6x + 4y^2 + 5 = 0$

18. $16x^2 + 9y^2 + 36y - 108 = 0$

19. $9x^2 - 72x + 2y^2 + 4y + 128 = 0$

20. $5x^2 + 10x + 16y^2 + 160y + 325 = 0$

21. $2x^2 + 12x + 11y^2 - 88y + 172 = 0$

22. $9x^2 + 72x + y^2 + 6y + 135 = 0$

For Problems 23–36, find an equation of the ellipse that satisfies the given conditions.

23. Vertices $(\pm 5, 0)$, foci $(\pm 3, 0)$

24. Vertices $(\pm 4, 0)$, foci $(\pm 2, 0)$

25. Vertices $(0, \pm 6)$, foci $(0, \pm 5)$

26. Vertices $(0, \pm 3)$, foci $(0, \pm 2)$

27. Vertices $(\pm 3, 0)$, length of minor axis is 2

28. Vertices $(0, \pm 5)$, length of minor axis is 4

29. Foci $(0, \pm 2)$, length of minor axis is 3

30. Foci $(\pm 1, 0)$, length of minor axis is 2

31. Vertices $(0, \pm 5)$, contains the point $(3, 2)$

32. Vertices $(\pm 6, 0)$, contains the point $(5, 1)$

33. Vertices $(5, 1)$ and $(-3, 1)$, foci $(3, 1)$ and $(-1, 1)$

34. Vertices $(2, 4)$ and $(2, -6)$, foci $(2, 3)$ and $(2, -5)$

35. Center $(0, 1)$, one focus at $(-4, 1)$, length of minor axis is 6

36. Center $(3, 0)$, one focus at $(3, 2)$, length of minor axis is 4

37. Find an equation of the set of points in a plane such that the sum of the distances between each point of the set and the points $(2, 0)$ and $(-2, 0)$ is 8 units.

38. Find an equation of the set of points in a plane such that the sum of the distances between each point of the set and the points $(0, 3)$ and $(0, -3)$ is 10 units.

MISCELLANEOUS PROBLEMS

39. An arch of a bridge is semielliptical and the major axis is horizontal (see Figure 12.23). The arch is 30 feet wide and 10 feet high. Find the height of the arch 10 feet from the center of the base.

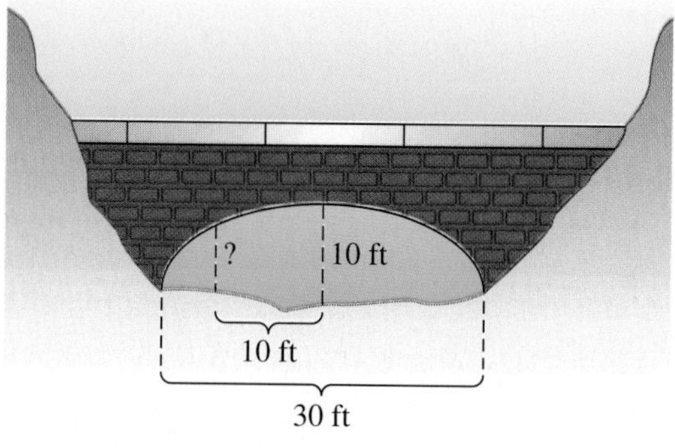

FIGURE 12.23

40. In Figure 12.23, how much clearance is there 10 feet from the bank?

GRAPHICS CALCULATOR ACTIVITIES

41. Use your graphics calculator to check your graphs for Problems 13–22.

42. Use your graphics calculator to graph each of the following ellipses.

a. $2x^2 - 40x + y^2 + 2y + 185 = 0$

b. $x^2 - 4x + 2y^2 - 48y + 272 = 0$

c. $4x^2 - 8x + y^2 - 4y - 136 = 0$

d. $x^2 + 6x + 2y^2 + 56y + 301 = 0$

12.3 HYPERBOLAS

A hyperbola and an ellipse are similar by definition; however an ellipse involves the *sum* of distances and the hyperbola involves the *difference* of distances.

DEFINITION 12.3

A **hyperbola** is the set of all points in a plane such that the difference of the distances of each point from two fixed points F and F' (the **foci**) in the plane is a positive constant.

Using Definition 12.3, we can sketch a hyperbola by starting with two fixed points F and F' as shown in Figure 12.24. Then we locate all points P such that $PF' - PF$ is a positive constant. Likewise, as shown in Figure 12.24, all points Q are located so that $QF - QF'$ is the same positive constant. The two dashed curved lines in Figure 12.24 make up the hyperbola. The two curves are sometimes referred to as the branches of the hyperbola.

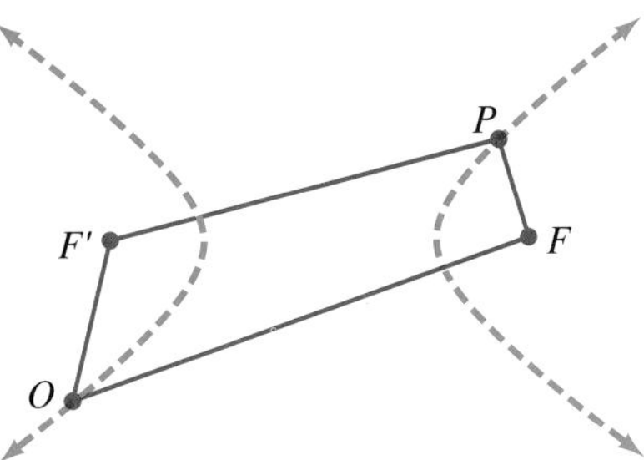

FIGURE 12.24

To develop a standard form for the equation of a hyperbola, let's coordinatize the plane so that the foci are located at $F(c, 0)$ and $F'(-c, 0)$, as indicated in Figure 12.25. Using the distance formula and setting $2a$ equal to the difference of the distances from any point P on the hyperbola to the foci, we have the following equation.

$$\left|\sqrt{(x - c)^2 + (y - 0)^2} - \sqrt{(x + c)^2 + (y - 0)^2}\right| = 2a$$

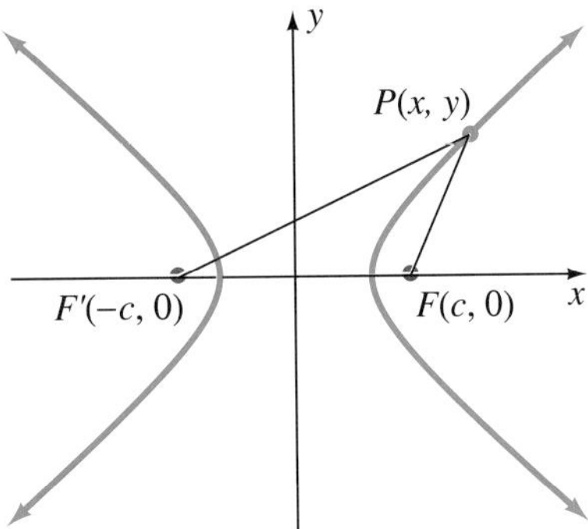

FIGURE 12.25

(The absolute value sign is used to allow the point P to be on either branch of the hyperbola.) Using the same type of simplification procedure that we used for deriving the standard form for the equation of an ellipse, this equation simplifies to

$$\frac{x^2}{a^2} - \frac{y^2}{c^2 - a^2} = 1.$$

Letting $b^2 = c^2 - a^2$, where $b > 0$, we obtain the standard form

$$\frac{x^2}{a^2} - \frac{y^2}{b^2} = 1. \quad \textbf{(1)}$$

Equation (1) indicates that this hyperbola is symmetric with respect to both axes and the origin. Furthermore, by letting $y = 0$, we obtain $x^2/a^2 = 1$ or $x^2 = a^2$, and therefore the x-intercepts are a and $-a$. The corresponding points $A(a, 0)$ and $A'(-a, 0)$ are the **vertices** of the hyperbola and the line segment AA' is called the **transverse axis**; it is of length $2a$ (see Figure 12.26). The midpoint of the transverse axis is called the **center** of the hyperbola; it is located at the origin. By letting $x = 0$ in equation (1), we obtain $-y^2/b^2 = 1$ or $y^2 = -b^2$. This implies that there are no y-intercepts, as indicated in Figure 12.26.

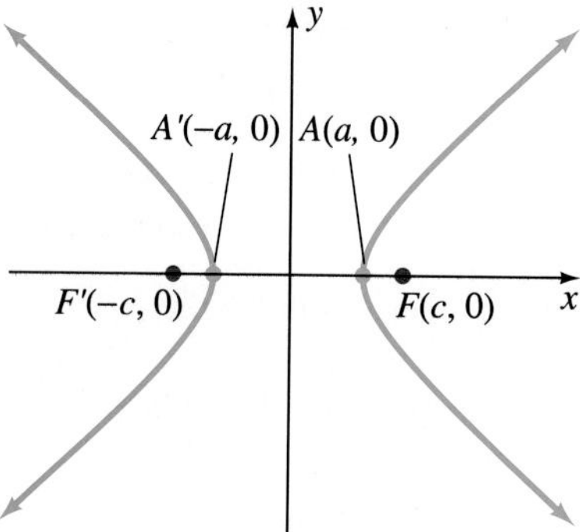

FIGURE 12.26

The following property summarizes the previous discussion.

PROPERTY 12.5

The graph of the equation

$$\frac{x^2}{a^2} - \frac{y^2}{b^2} = 1$$

is a hyperbola with vertices at $(a, 0)$ and $(-a, 0)$. The foci are at $(c, 0)$ and $(-c, 0)$, where $c^2 = a^2 + b^2$.

In conjunction with every hyperbola there are two intersecting lines that pass through the center of the hyperbola. These lines, referred to as *asymptotes*, are very helpful when sketching a hyperbola. Their equations are easily determined by using the following type of reasoning: Solving the equation

$$\frac{x^2}{a^2} - \frac{y^2}{b^2} = 1$$

for y produces $y = \pm\frac{b}{a}\sqrt{x^2 - a^2}$. From this form it is evident that there are no points on the graph for $x^2 - a^2 < 0$, that is, if $-a < x < a$. However, there are points on the graph if $x \geq a$ or $x \leq -a$. If $x \geq a$, then $y = \pm\frac{b}{a}\sqrt{x^2 - a^2}$ can be written

$$\begin{aligned} y &= \pm\frac{b}{a}\sqrt{x^2\left(1 - \frac{a^2}{x^2}\right)} \\ &= \pm\frac{b}{a}\sqrt{x^2}\sqrt{1 - \frac{a^2}{x^2}} \\ &= \pm\frac{b}{a}x\sqrt{1 - \frac{a^2}{x^2}}. \end{aligned}$$

Now suppose that we are going to determine some y-values for very large values of x. (Remember that a and b are arbitrary constants; they have specific values for a particular hyperbola.) When x is very large, a^2/x^2 will be close to zero, so the radicand will be close to 1. Therefore, the y-value will be close to either $(b/a)x$ or $-(b/a)x$. In other words, as x becomes larger and larger, the point $P(x, y)$ gets closer and closer to either the line $y = (b/a)x$ or the line $y = -(b/a)x$. A corresponding situation occurs when $x \leq a$. The lines with equations

$$y = \pm\frac{b}{a}x$$

are called the **asymptotes** of the hyperbola.

As we mentioned earlier, the asymptotes are very helpful for sketching hyperbolas. An easy way to sketch the asymptotes is first to plot the vertices $A(a, 0)$ and $A'(-a, 0)$, and the points $B(0, b)$ and $B'(0, -b)$, as in Figure 12.27. The line segment $\overline{BB'}$ is of length $2b$ and is called the **conjugate axis** of the hyperbola. The horizontal line segments drawn through B and B', together with the vertical line segments drawn through A and A', form a rectangle. The diagonals of this rectangle have slopes b/a and $-(b/a)$. Therefore, by extending the diagonals, we obtain the asymptotes $y = (b/a)x$ and $y = -(b/a)x$. The two branches of the hyperbola can be sketched using the asymptotes as guidelines, as shown in Figure 12.27.

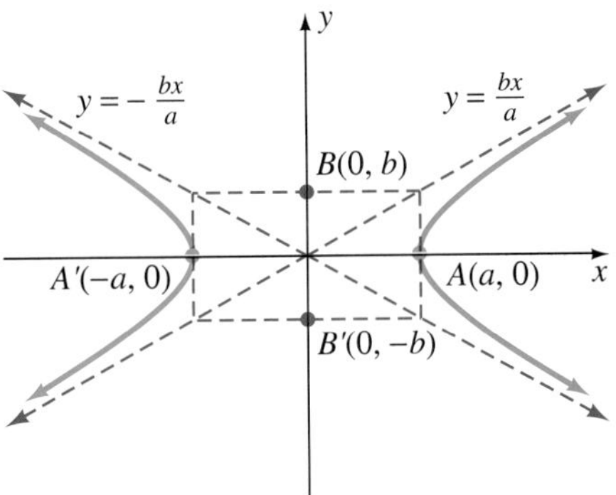

FIGURE 12.27

EXAMPLE 1

Find the vertices, the foci, and the equations of the asymptotes, and sketch the hyperbola $9x^2 - 4y^2 = 36$.

Solution

Dividing both sides of the given equation by 36 and simplifiying changes the equation to the standard form,

$$\frac{x^2}{4} - \frac{y^2}{9} = 1,$$

where $a^2 = 4$ and $b^2 = 9$. Hence, $a = 2$ and $b = 3$. The vertices are $(\pm 2, 0)$ and the endpoints of the conjugate axis are $(0, \pm 3)$; these points determine the rectangle whose diagonals extend to become the asymptotes. Using $a = 2$ and $b = 3$, the equations of the asymptotes are $y = \frac{3}{2}x$ and $y = -\frac{3}{2}x$. Then, using the relationship $c^2 = a^2 + b^2$, we obtain $c^2 = 4 + 9 = 13$. So the foci are at $(\sqrt{13}, 0)$ and $(-\sqrt{13}, 0)$. Using the vertices and the asymptotes, the hyperbola has been sketched in Figure 12.28.

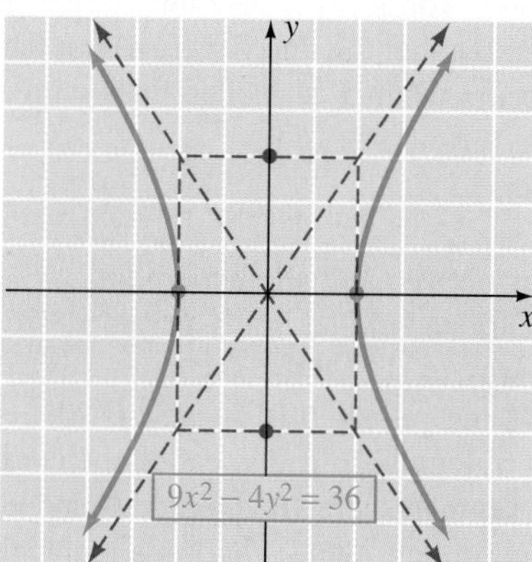

FIGURE 12.28

EXAMPLE 2

Find the equation of the hyperbola with vertices at $(\pm 4, 0)$ and foci at $(\pm 2\sqrt{5}, 0)$.

Solution

From the given information we know that $a = 4$ and $c = 2\sqrt{5}$. Then, using the relationship $b^2 = c^2 - a^2$, we obtain

$$b^2 = \left(2\sqrt{5}\right)^2 - 4^2 = 20 - 16 = 4.$$

Substituting 16 for a^2 and 4 for b^2 in the standard form produces

$$\frac{x^2}{16} - \frac{y^2}{4} = 1.$$

Multiplying both sides of this equation by 16 produces

$$x^2 - 4y^2 = 16.$$

Hyperbolas with Foci on the y-axis

In a similar fashion, we can develop a standard form for the equation of a hyperbola with foci on the y-axis. The following property summarizes the results of such a development, where the foci are at $(0, c)$ and $(0, -c)$.

PROPERTY 12.6

The graph of the equation

$$\frac{y^2}{a^2} - \frac{x^2}{b^2} = 1$$

is a hyperbola with vertices at $(0, a)$ and $(0, -a)$. The foci are at $(0, c)$ and $(0, -c)$, where $c^2 = a^2 + b^2$.

For this type of hyperbola, the endpoints of the conjugate axis are at $(b, 0)$ and $(-b, 0)$. In this case we can find the asymptotes by extending the diagonals of the rectangle determined by the horizontal lines through the vertices and the vertical lines through the endpoints of the conjugate axis. The slopes of these diagonals are a/b and $-a/b$; thus, the equations of these asymptotes are

$$y = \frac{a}{b}x \quad \text{and} \quad y = -\frac{a}{b}x.$$

EXAMPLE 3

Find the vertices, the foci, and the equations of the asymptotes, and sketch the hyperbola $4y^2 - x^2 = 12$.

Solution

Divide both sides of the given equation by 12 to change the equation to the standard form,

$$\frac{y^2}{3} - \frac{x^2}{12} = 1,$$

where $a^2 = 3$ and $b^2 = 12$. Hence, $a = \sqrt{3}$ and $b = 2\sqrt{3}$. The vertices, $(0, \pm\sqrt{3})$, and the endpoints of the conjugate axis, $(\pm 2\sqrt{3}, 0)$, determine the rectangle whose diagonals extend to become the asymptotes. Using $a = \sqrt{3}$ and $b = 2\sqrt{3}$, the equations of the asymptotes are $y = (\sqrt{3}/2\sqrt{3})x = \frac{1}{2}x$ and $y = -\frac{1}{2}x$. Then, using the relationship $c^2 = a^2 + b^2$, we obtain $c^2 = 3 + 12 = 15$, so the foci are at $(0, \sqrt{15})$ and $(0, -\sqrt{15})$. The hyperbola is sketched in Figure 12.29.

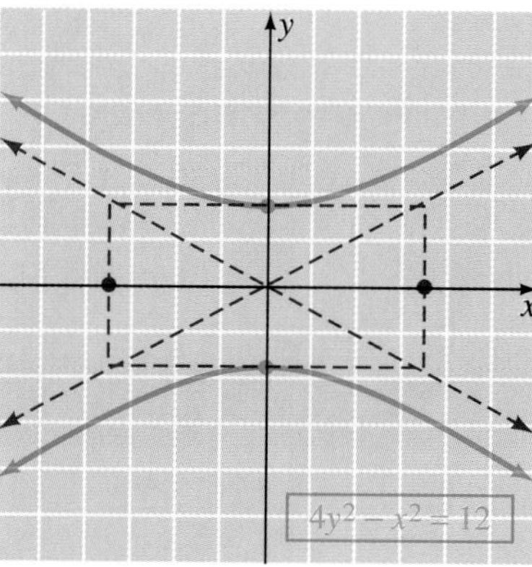

FIGURE 12.29

Other Hyperbolas

In the same way, we can develop the standard form for an equation of a hyperbola that is symmetric with respect to a line parallel to a coordinate axis. We will not show such developments in this text but we will simply state and use the results.

$$\frac{(x-h)^2}{a^2} - \frac{(y-k)^2}{b^2} = 1$$ A hyperbola with center at (h, k) and transverse axis on the horizontal line $y = k$

$$\frac{(y-k)^2}{a^2} - \frac{(x-h)^2}{b^2} = 1$$ A hyperbola with center at (h, k) and transverse axis on the vertical line $x = h$

The relationship $c^2 = a^2 + b^2$ still holds and the physical significance of a, b, and c remains the same. However, these values are used relative to the center (h, k) to find the endpoints of the transverse and conjugate axes, and the foci. Furthermore, the slopes of the asymptotes are as before, but these lines now contain the new center, (h, k). Let's see how all of this works in a specific example.

EXAMPLE 4 Find the vertices, the foci, the equations of the asymptotes, and sketch the hyperbola $9x^2 - 36x - 16y^2 + 96y - 252 = 0$.

Solution

First, we need to change to a standard form by completing the square on both x and y.

$$9(x^2 - 4x + \_\_) - 16(y^2 - 6y + \_\_) = 252$$
$$9(x^2 - 4x + 4) - 16(y^2 - 6y + 9) = 252 + 36 - 144$$

$$9(x-2)^2 - 16(y-3)^2 = 144$$
$$\frac{(x-2)^2}{16} - \frac{(y-3)^2}{9} = 1$$

The center is at (2, 3) and the tranverse axis is on the line $y = 3$. Since $a^2 = 16$, we know that $a = 4$. Therefore, the vertices are 4 units to the right and 4 units to the left of the center, (2, 3), so they are at (6, 3), and (−2, 3). Likewise, since $b^2 = 9$ or $b = 3$, the endpoints of the conjugate axis are 3 units up and 3 units down from the center, so they are at (2, 6) and (2, 0). Using $a = 4$ and $b = 3$, the slopes of the asymptotes are $\frac{3}{4}$ and $-\frac{3}{4}$. Then, using the slopes, the center (2, 3), and the point-slope form for writing the equation of a line, we can determine the equations of the asymptotes to be $3x - 4y = -6$ and $3x + 4y = 18$. From the relationship $c^2 = a^2 + b^2$ we obtain $c^2 = 16 + 9 = 25$. Thus, the foci are at (7, 3) and (−3, 3). The hyperbola is sketched in Figure 12.30.

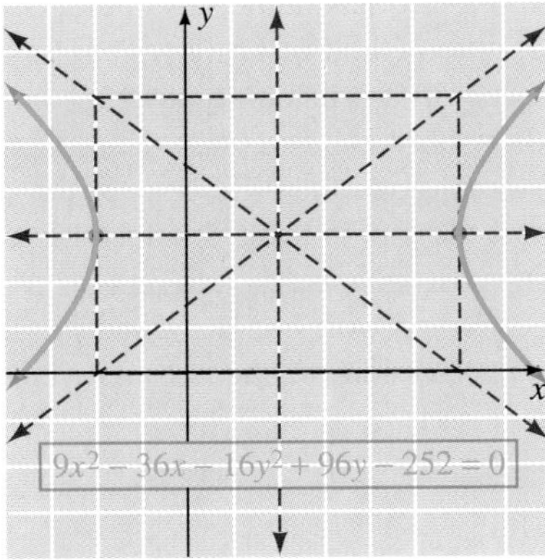

FIGURE 12.30

EXAMPLE 5

Find the equation of the hyperbola with vertices at (−4, 2) and (−4, −4), and with foci at (−4, 3) and (−4, −5).

Solution

Since the vertices and foci are on the same vertical line ($x = -4$), this hyperbola has an equation of the form

$$\frac{(y-k)^2}{a^2} - \frac{(x-h)^2}{b^2} = 1.$$

The center of the hyperbola is at the midpoint of the transverse axis. Therefore,

$$h = \frac{-4 + (-4)}{2} = -4 \qquad \text{and} \qquad k = \frac{2 + (-4)}{2} = -1.$$

The distance between the center, $(-4, -1)$, and a vertex, $(-4, 2)$, is 3 units, so $a = 3$. The distance between the center, $(-4, -1)$, and a focus, $(-4, 3)$, is 4 units, so $c = 4$. Then, using the relationship $c^2 = a^2 + b^2$, we obtain

$$b^2 = c^2 - a^2 = 16 - 9 = 7.$$

Now we can substitute -4 for h, -1 for k, 9 for a^2, and 7 for b^2 in the general form and simplify.

$$\begin{aligned} \frac{(y+1)^2}{9} - \frac{(x+4)^2}{7} &= 1 \\ 7(y+1)^2 - 9(x+4)^2 &= 63 \\ 7(y^2 + 2y + 1) - 9(x^2 + 8x + 16) &= 63 \\ 7y^2 + 14y + 7 - 9x^2 - 72x - 144 &= 63 \\ 7y^2 + 14y - 9x^2 - 72x - 200 &= 0 \end{aligned}$$

The hyperbola also has numerous applications, many you may not be aware of. For example, one method of artillery range-finding is based on the concept of a hyperbola. If each of two listening posts, P_1 and P_2 in Figure 12.31, records the time that an artillery blast is heard, then the difference between the times multiplied by the speed of sound gives the difference of the distances of the gun from the two fixed points. Thus, the gun is located somewhere on the hyperbola whose foci are the two listening posts. Now by bringing in a third listening post, P_3, another hyperbola can be formed with foci at P_2 and P_3. Then the location of the gun must be at one of the intersections of the two hyperbolas.

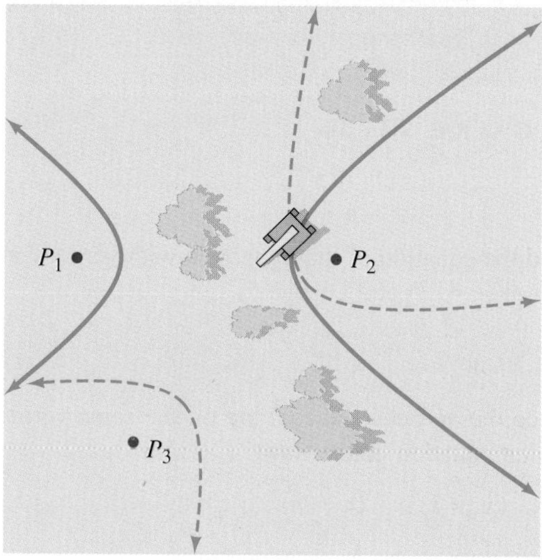

FIGURE 12.31

This same principle of intersecting hyperbolas is used in a long range navigation system known as LORAN. Radar stations serve as the foci of the hyperbolas and of course computers are used for the many calculations that are necessary to fix the location of a plane or a ship. At the present time, LORAN is probably used mostly for coastal navigation in connection with small pleasure boats.

Some rather unique architectural creations have used the concept of a hyperbolic paraboloid, pictured in Figure 12.32. For example, the TWA building at Kennedy Airport is so-designed. Some comets, upon entering the sun's gravitational field, will follow a hyperbolic path, with the sun as one of the foci (see Figure 12.33).

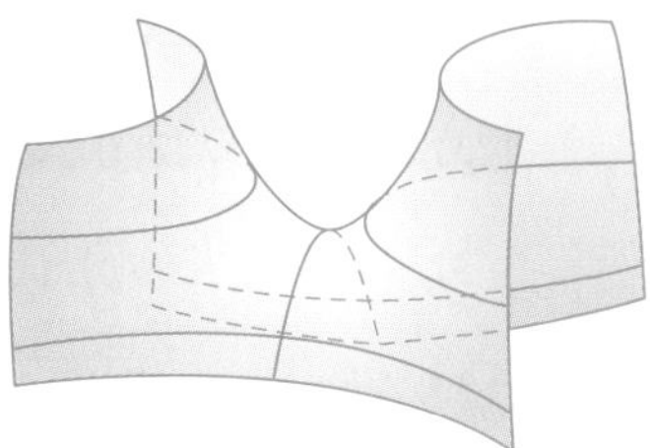

FIGURE 12.32

TWA building at Kennedy Airport
Max Hilaire/The Image Bank

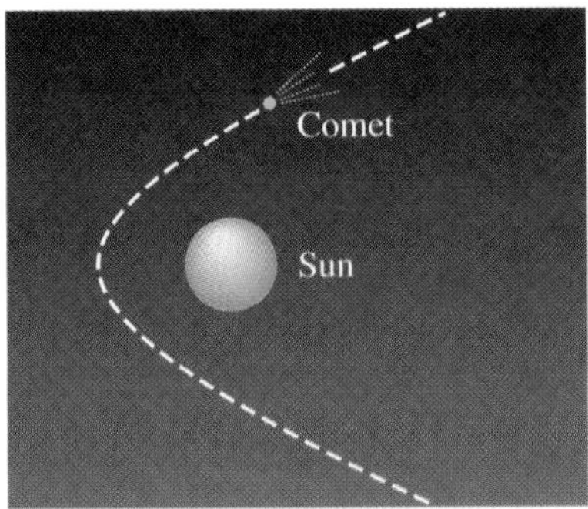

FIGURE 12.33

PROBLEM SET 12.3

For Problems 1–22, find the vertices, the foci, and the equations of the asymptotes, and sketch each of the hyperbolas.

1. $\frac{x^2}{9} - \frac{y^2}{4} = 1$
2. $\frac{x^2}{4} - \frac{y^2}{16} = 1$
3. $\frac{y^2}{4} - \frac{x^2}{9} = 1$
4. $\frac{y^2}{16} - \frac{x^2}{4} = 1$
5. $9y^2 - 16x^2 = 144$
6. $4y^2 - x^2 = 4$
7. $x^2 - y^2 = 9$
8. $x^2 - y^2 = 1$
9. $5y^2 - x^2 = 25$
10. $y^2 - 2x^2 = 8$
11. $y^2 - 9x^2 = -9$
12. $16y^2 - x^2 = -16$
13. $4x^2 - 24x - 9y^2 - 18y - 9 = 0$
14. $9x^2 + 72x - 4y^2 - 16y + 92 = 0$
15. $y^2 - 4y - 4x^2 - 24x - 36 = 0$
16. $9y^2 + 54y - x^2 + 6x + 63 = 0$
17. $2x^2 - 8x - y^2 + 4 = 0$
18. $x^2 + 6x - 3y^2 = 0$
19. $y^2 + 10y - 9x^2 + 16 = 0$
20. $4y^2 - 16y - x^2 + 12 = 0$
21. $x^2 + 4x - y^2 - 4y - 1 = 0$
22. $y^2 + 8y - x^2 + 2x + 14 = 0$

For Problems 23–38, find an equation of the hyperbola that satisfies the given conditions.

23. Vertices $(\pm 2, 0)$, foci $(\pm 3, 0)$
24. Vertices $(\pm 1, 0)$, foci $(\pm 4, 0)$
25. Vertices $(0, \pm 3)$, foci $(0, \pm 5)$
26. Vertices $(0, \pm 2)$, foci $(0, \pm 6)$
27. Vertices $(\pm 1, 0)$, contains the point $(2, 3)$
28. Vertices $(0, \pm 1)$, contains the point $(-3, 5)$
29. Vertices $\left(0, \pm\sqrt{3}\right)$, length of conjugate axis is 4
30. Vertices $\left(\pm\sqrt{5}, 0\right)$, length of conjugate axis is 6
31. Foci $\left(\pm\sqrt{23}, 0\right)$, length of transverse axis is 8
32. Foci $\left(0, \pm 3\sqrt{2}\right)$, length of conjugate axis is 4
33. Vertices $(6, -3)$ and $(2, -3)$, foci $(7, -3)$ and $(1, -3)$
34. Vertices $(-7, -4)$ and $(-5, -4)$, foci $(-8, -4)$ and $(-4, -4)$
35. Vertices $(-3, 7)$ and $(-3, 3)$, foci $(-3, 9)$ and $(-3, 1)$
36. Vertices $(7, 5)$ and $(7, -1)$, foci $(7, 7)$ and $(7, -3)$
37. Vertices $(0, 0)$ and $(4, 0)$, foci $(5, 0)$ and $(-1, 0)$
38. Vertices $(0, 0)$ and $(0, -6)$, foci $(0, 2)$ and $(0, -8)$

For problems 39–48, identify the graph of each of the equations as a straight line, a circle, a parabola, an ellipse, or a hyperbola. Do not sketch the graphs.

39. $x^2 - 7x + y^2 + 8y - 2 = 0$
40. $x^2 - 7x - y^2 + 8y - 2 = 0$
41. $5x - 7y = 9$
42. $4x^2 - x + y^2 + 2y - 3 = 0$
43. $10x^2 + y^2 = 8$
44. $-3x - 2y = 9$
45. $5x^2 + 3x - 2y^2 - 3y - 1 = 0$
46. $x^2 + y^2 - 3y - 6 = 0$
47. $x^2 - 3x + y - 4 = 0$
48. $5x + y^2 - 2y - 1 = 0$

GRAPHICS CALCULATOR ACTIVITIES

49. Use a graphics calculator to check your graphs for Problems 13–22. Be sure to graph the asymptotes for each hyperbola.

50. Use a graphics calculator to check your answers for Problems 39–48.

12.4 SYSTEMS INVOLVING NONLINEAR EQUATIONS

In Chapters 10 and 11, we used several techniques to solve systems of linear equations. We will use two of those techniques in this section to solve some systems that contain at least one nonlinear equation. Furthermore, we will use our knowledge of graphing lines, circles, parabolas, ellipses, and hyperbolas to get a pictorial view of the systems. That will give us a basis for predicting approximate real number solutions if there are any. In other words, we have once again arrived at a topic that vividly illustrates the merging of mathematical ideas. Let's begin by considering a system that contains one linear and one nonlinear equation.

EXAMPLE 1

Solve the system

$$\begin{pmatrix} x^2 + y^2 = 13 \\ 3x + 2y = 0 \end{pmatrix}.$$

Solution

From our previous graphing experiences, we should recognize that $x^2 + y^2 = 13$ is a circle and $3x + 2y = 0$ is a straight line. Thus, the system can be pictured as in Figure 12.34. The graph indicates that the solution set of this system should consist of two ordered pairs of real numbers, which represent the points of intersection in the second and fourth quadrants.

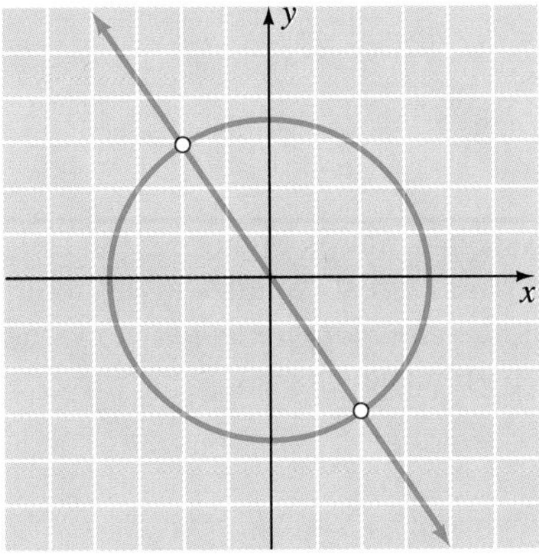

FIGURE 12.34

Now let's solve the system analytically by using the *substitution method.* Change the form of $3x + 2y = 0$ to $y = -3x/2$ and then substitute $-3x/2$ for y in the other equation to produce

$$x^2 + \left(-\frac{3x}{2}\right)^2 = 13.$$

This equation can now be solved for x.

$$\begin{aligned} x^2 + \frac{9x^2}{4} &= 13 \\ 4x^2 + 9x^2 &= 52 \\ 13x^2 &= 52 \\ x^2 &= 4 \\ x &= \pm 2 \end{aligned}$$

Substitute 2 for x and then -2 for x in the second equation of the system to produce two values for y.

$$\begin{aligned} 3x + 2y &= 0 \\ 3(2) + 2y &= 0 \\ 2y &= -6 \\ y &= -3 \end{aligned} \qquad \begin{aligned} 3x + 2y &= 0 \\ 3(-2) + 2y &= 0 \\ 2y &= 6 \\ y &= 3 \end{aligned}$$

Therefore, the solution set of the system is $\{(2, -3), (-2, 3)\}$.

REMARK Don't forget that, as always, the solutions can be checked by substituting them back into the original equations. Graphing the system only permits you to approximate any possible real number solutions before solving the system. Then, after solving the system, the graph can be used again to check that the answers are reasonable.

EXAMPLE 2

Solve the system

$$\begin{pmatrix} x^2 + y^2 = 16 \\ y^2 - x^2 = 4 \end{pmatrix}.$$

Solution

Graphing the system produces Figure 12.35. This figure indicates that there should be four ordered pairs of real numbers in the solution set of the system. Solving the system by using the *elimination method* works nicely. We can simply add the two equations, which eliminates the xs.

$$\begin{aligned} x^2 + y^2 &= 16 \\ -x^2 + y^2 &= 4 \\ \hline 2y^2 &= 20 \\ y^2 &= 10 \\ y &= \pm\sqrt{10} \end{aligned}$$

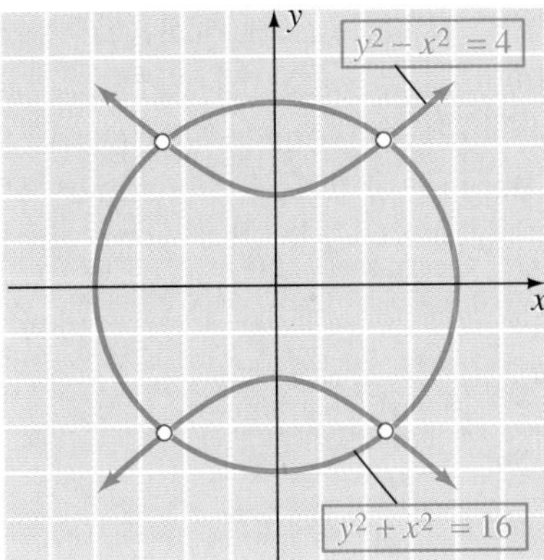

FIGURE 12.35

Substituting $\sqrt{10}$ for y in the first equation yields

$$\begin{aligned} x^2 + y^2 &= 16 \\ x^2 + \left(\sqrt{10}\right)^2 &= 16 \\ x^2 + 10 &= 16 \\ x^2 &= 6 \\ x &= \pm\sqrt{6}. \end{aligned}$$

Thus, $\left(\sqrt{6}, \sqrt{10}\right)$ and $\left(-\sqrt{6}, \sqrt{10}\right)$ are solutions. Substituting $-\sqrt{10}$ for y in the first equation yields

$$\begin{aligned} x^2 + y^2 &= 16 \\ x^2 + \left(-\sqrt{10}\right)^2 &= 16 \\ x^2 + 10 &= 16 \\ x^2 &= 6 \\ x &= \pm\sqrt{6}. \end{aligned}$$

Thus, $\left(\sqrt{6}, -\sqrt{10}\right)$ and $\left(-\sqrt{6}, -\sqrt{10}\right)$ are also solutions. The solution set is thus $\left\{\left(-\sqrt{6}, \sqrt{10}\right), \left(-\sqrt{6}, -\sqrt{10}\right), \left(\sqrt{6}, \sqrt{10}\right), \left(\sqrt{6}, -\sqrt{10}\right)\right\}$.

Sometimes a sketch of the graph of a system may not clearly indicate whether or not the system contains any real-number solutions. The next example illustrates such a situation.

EXAMPLE 3

Solve the system

$$\begin{pmatrix} y = x^2 + 2 \\ 6x - 4y = -5 \end{pmatrix}.$$

Solution

From previous graphing experiences, we recognize that $y = x^2 + 2$ is the basic parabola shifted upward two units and $6x - 4y = -5$ is a straight line (see Figure 12.36). Because of the close proximity of the curves, it is difficult to tell whether or not they intersect. In other words, the graph does not definitely indicate any real number solutions for the system.

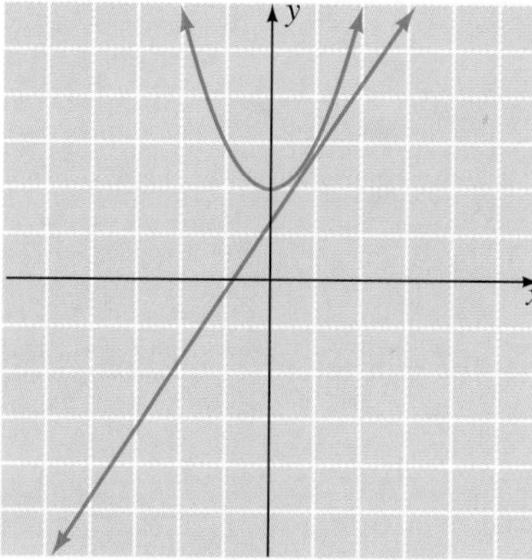

FIGURE 12.36

Let's solve the system using the substitution method. We can substitute $x^2 + 2$ for y in the second equation, which produces two values for x.

$$6x - 4(x^2 + 2) = -5$$
$$6x - 4x^2 - 8 = -5$$
$$-4x^2 + 6x - 3 = 0$$
$$4x^2 - 6x + 3 = 0$$
$$x = \frac{6 \pm \sqrt{36 - 48}}{8}$$
$$x = \frac{6 \pm \sqrt{-12}}{8}$$
$$x = \frac{6 \pm 2i\sqrt{3}}{8}$$
$$x = \frac{3 \pm i\sqrt{3}}{4}$$

It is now obvious that the system has no real number solutions. That is to say, the line and the parabola do not intersect in the real number plane. However, there will be two pairs of complex numbers in the solution set. We can substitute $(3 + i\sqrt{3})/4$ for x in the first equation.

$$y = \left(\frac{3 + i\sqrt{3}}{4}\right)^2 + 2$$

$$= \frac{6 + 6i\sqrt{3}}{16} + 2$$

$$= \frac{6 + 6i\sqrt{3} + 32}{16}$$

$$= \frac{38 + 6i\sqrt{3}}{16}$$

$$= \frac{19 + 3i\sqrt{3}}{8}$$

Likewise, we can substitute $(3 - i\sqrt{3})/4$ for x in the first equation.

$$y = \left(\frac{3 - i\sqrt{3}}{4}\right)^2 + 2$$

$$= \frac{6 - 6i\sqrt{3}}{16} + 2$$

$$= \frac{6 - 6i\sqrt{3} + 32}{16}$$

$$= \frac{38 - 6i\sqrt{3}}{16}$$

$$= \frac{19 - 3i\sqrt{3}}{8}$$

The solution set is $\left\{\left(\frac{3 + i\sqrt{3}}{4}, \frac{19 + 3i\sqrt{3}}{8}\right), \left(\frac{3 - i\sqrt{3}}{4}, \frac{19 - 3i\sqrt{3}}{8}\right)\right\}$.

In Example 3 the use of a graphing utility may not, at first, indicate whether or not the system has any real number solutions. Suppose that we graph the system using a viewing rectangle such that $-15 \le x \le 15$ and $-10 \le y \le 10$. As shown in the display in Figure 12.37 we cannot tell whether or not the line and parabola intersect. However, if we change the viewing rectangle so that $0 \le x \le 2$ and

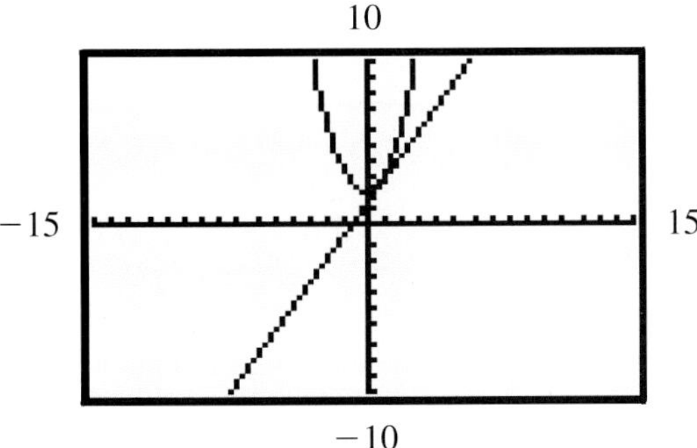

FIGURE 12.37

$0 \leq y \leq 4$ as shown in Figure 12.38, then it becomes apparent that the two graphs do not intersect.

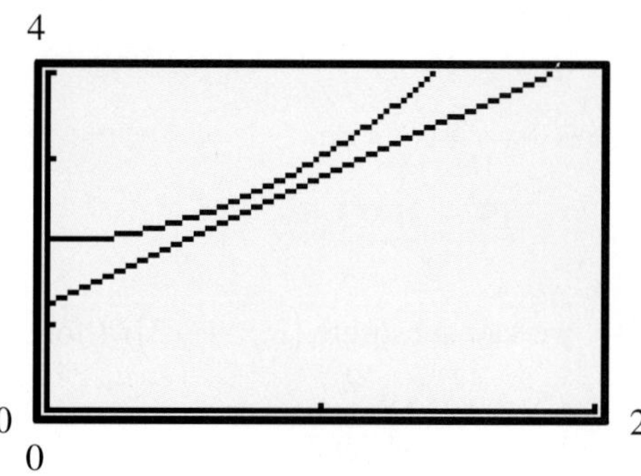

FIGURE 12.38

EXAMPLE 4

Find the real number solutions for the system $\begin{pmatrix} y = \log_2(x - 3) - 2 \\ y = -\log_2 x \end{pmatrix}$.

Solution

First, let's use a graphing utility to obtain a graph of the system (Figure 12.39). The two curves appear to intersect at approximately $x = 4$ and $y = -2$. To solve the system algebraically we can equate the two expressions for y and solve the resulting equation for x.

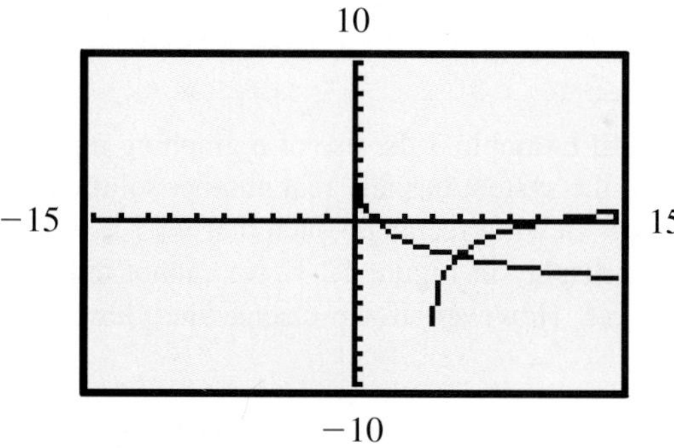

FIGURE 12.39

$$\begin{aligned} \log_2(x - 3) - 2 &= -\log_2 x \\ \log_2 x + \log_2(x - 3) &= 2 \\ \log_2 x(x - 3) &= 2 \end{aligned}$$

At this step we can either change to exponential form or rewrite 2 as $\log_2 4$.

$$\begin{aligned} \log_2 x(x - 3) &= \log_2 4 \\ x(x - 3) &= 4 \end{aligned}$$

$$x^2 - 3x - 4 = 0$$
$$(x - 4)(x + 1) = 0$$
$$x - 4 = 0 \quad \text{or} \quad x + 1 = 0$$
$$x = 4 \quad \text{or} \quad x = -1$$

Since logarithms are not defined for negative numbers, -1 is discarded. Therefore, if $x = 4$, then

$$y = -\log_2 x$$

becomes

$$y = -\log_2 4 = -2.$$

Therefore, the solution set is $\{(4, -2)\}$.

PROBLEM SET 12.4

For each of the following systems, (a) graph the system so that approximate real number solutions (if there are any) can be predicted, and (b) solve the system by the substitution or elimination method.

1. $\begin{pmatrix} x^2 + y^2 = 5 \\ x + 2y = 5 \end{pmatrix}$
2. $\begin{pmatrix} x^2 + y^2 = 13 \\ 2x + 3y = 13 \end{pmatrix}$
3. $\begin{pmatrix} x^2 + y^2 = 26 \\ x + y = -4 \end{pmatrix}$
4. $\begin{pmatrix} x^2 + y^2 = 10 \\ x + y = -2 \end{pmatrix}$
5. $\begin{pmatrix} x^2 + y^2 = 2 \\ x - y = 4 \end{pmatrix}$
6. $\begin{pmatrix} x^2 + y^2 = 3 \\ x - y = -5 \end{pmatrix}$
7. $\begin{pmatrix} y = x^2 + 6x + 7 \\ 2x + y = -5 \end{pmatrix}$
8. $\begin{pmatrix} y = x^2 - 4x + 5 \\ y - x = 1 \end{pmatrix}$
9. $\begin{pmatrix} 2x + y = -2 \\ y = x^2 + 4x + 7 \end{pmatrix}$
10. $\begin{pmatrix} 2x + y = 0 \\ y = -x^2 + 2x - 4 \end{pmatrix}$
11. $\begin{pmatrix} y = x^2 - 3 \\ x + y = -4 \end{pmatrix}$
12. $\begin{pmatrix} y = -x^2 + 1 \\ x + y = 2 \end{pmatrix}$
13. $\begin{pmatrix} x^2 + 2y^2 = 9 \\ x - 4y = -9 \end{pmatrix}$
14. $\begin{pmatrix} 2x - y = 7 \\ 3x^2 + y^2 = 21 \end{pmatrix}$
15. $\begin{pmatrix} x + y = -3 \\ x^2 + 2y^2 - 12y - 18 = 0 \end{pmatrix}$
16. $\begin{pmatrix} 4x^2 + 9y^2 = 25 \\ 2x + 3y = 7 \end{pmatrix}$
17. $\begin{pmatrix} x - y = 2 \\ x^2 - y^2 = 16 \end{pmatrix}$
18. $\begin{pmatrix} x^2 - 4y^2 = 16 \\ 2y - x = 2 \end{pmatrix}$
19. $\begin{pmatrix} y = -x^2 + 3 \\ y = x^2 + 1 \end{pmatrix}$
20. $\begin{pmatrix} y = x^2 \\ y = x^2 - 4x + 4 \end{pmatrix}$
21. $\begin{pmatrix} y = x^2 + 2x - 1 \\ y = x^2 + 4x + 5 \end{pmatrix}$
22. $\begin{pmatrix} y = -x^2 + 1 \\ y = x^2 - 2 \end{pmatrix}$
23. $\begin{pmatrix} x^2 - y^2 = 4 \\ x^2 + y^2 = 4 \end{pmatrix}$
24. $\begin{pmatrix} 2x^2 + y^2 = 8 \\ x^2 + y^2 = 4 \end{pmatrix}$
25. $\begin{pmatrix} 8y^2 - 9x^2 = 6 \\ 8x^2 - 3y^2 = 7 \end{pmatrix}$
26. $\begin{pmatrix} 2x^2 + y^2 = 11 \\ x^2 - y^2 = 4 \end{pmatrix}$
27. $\begin{pmatrix} 2x^2 - 3y^2 = -1 \\ 2x^2 + 3y^2 = 5 \end{pmatrix}$
28. $\begin{pmatrix} 4x^2 + 3y^2 = 9 \\ y^2 - 4x^2 = 7 \end{pmatrix}$
29. $\begin{pmatrix} xy = 3 \\ 2x + 2y = 7 \end{pmatrix}$
30. $\begin{pmatrix} x^2 + 4y^2 = 25 \\ xy = 6 \end{pmatrix}$

For Problems 31–36, solve each system for all real number solutions.

31. $\begin{pmatrix} y = \log_3(x - 6) - 3 \\ y = -\log_3 x \end{pmatrix}$
32. $\begin{pmatrix} y = \log_{10}(x - 9) - 1 \\ y = -\log_{10} x \end{pmatrix}$
33. $\begin{pmatrix} y = e^x - 1 \\ y = 2e^{-x} \end{pmatrix}$
34. $\begin{pmatrix} y = 28 - 11e^x \\ y = -e^{2x} \end{pmatrix}$
35. $\begin{pmatrix} y = x^3 \\ y = x^3 + 2x^2 + 5x - 3 \end{pmatrix}$
36. $\begin{pmatrix} y = 3(4^x) - 8 \\ y = 4^{2x} - 2(4^x) - 4 \end{pmatrix}$

THOUGHTS into WORDS

37. What happens if you try to graph the system $\begin{pmatrix} 7x^2 + 8y^2 = 36 \\ 11x^2 + 5y^2 = -4 \end{pmatrix}$?

38. For what value(s) of k will the line $x + y = k$ touch the ellipse $x^2 + 2y^2 = 6$ in one and only one point? Defend your answer.

39. The system

$$\begin{pmatrix} x^2 - 6x + y^2 - 4y + 4 = 0 \\ x^2 - 4x + y^2 + 8y - 5 = 0 \end{pmatrix}$$

represents two circles that intersect in two points. An equivalent system can be formed by replacing the second equation with the result of adding -1 times the first equation to the second equation. Thus, we obtain the system

$$\begin{pmatrix} x^2 - 6x + y^2 - 4y + 4 = 0 \\ 2x + 12y - 9 = 0 \end{pmatrix}.$$

Explain why the linear equation in this system is the equation of the common chord of the original two intersecting circles.

GRAPHICS CALCULATOR ACTIVITIES

40. Graph the system of equations $\begin{pmatrix} y = x^2 + 2 \\ 6x - 4y = -5 \end{pmatrix}$ and use the trace and zoom features of your calculator to show that this system has no real number solutions.

41. Use a graphics calculator to graph the systems in Problems 31–36 and check the reasonableness of your answers to those problems.

For Problems 42–47 use a graphics calculator to approximate to the nearest tenth the real number solutions for each system of equations.

42. $\begin{pmatrix} y = e^x + 1 \\ y = x^3 + x^2 - 2x - 1 \end{pmatrix}$

43. $\begin{pmatrix} y = x^3 + 2x^2 - 3x + 2 \\ y = -x^3 - x^2 + 1 \end{pmatrix}$

44. $\begin{pmatrix} y = 2^x + 1 \\ y = 2^{-x} + 2 \end{pmatrix}$

45. $\begin{pmatrix} y = \ln(x - 1) \\ y = x^2 - 16x + 64 \end{pmatrix}$

46. $\begin{pmatrix} x = y^2 - 2y + 3 \\ x^2 + y^2 = 25 \end{pmatrix}$

47. $\begin{pmatrix} y^2 - x^2 = 16 \\ 2y^2 - x^2 = 8 \end{pmatrix}$

CHAPTER 12 SUMMARY

The following standard forms for the equations of conic sections were developed in this chapter.

Parabolas

$x^2 = 4py$ — focus $(0, p)$, directrix $y = -p$, y-axis symmetry

$y^2 = 4px$ — focus $(p, 0)$, directrix $x = -p$, x-axis symmetry

$(x - h)^2 = 4p(y - k)$	focus $(h, k + p)$, directrix $y = k - p$, symmetric with respect to the line $x = h$
$(y - k)^2 = 4p(x - h)$	focus $(h + p, k)$, directrix $x = h - p$, symmetric with respect to the line $y = k$

Ellipses

$\dfrac{x^2}{a^2} + \dfrac{y^2}{b^2} = 1$	center $(0, 0)$, vertices $(\pm a, 0)$, endpoints of minor axis $(0, \pm b)$, foci $(\pm c, 0)$, $c^2 = a^2 - b^2$, $a^2 > b^2$
$\dfrac{x^2}{b^2} + \dfrac{y^2}{a^2} = 1$	center $(0, 0)$, vertices $(0, \pm a)$, endpoints of minor axis $(\pm b, 0)$, foci $(0, \pm c)$, $c^2 = a^2 - b^2$, $a^2 > b^2$
$\dfrac{(x - h)^2}{a^2} + \dfrac{(y - k)^2}{b^2} = 1$	center (h, k), vertices $(h \pm a, k)$, endpoints of minor axis $(h, k \pm b)$, foci $(h \pm c, k)$, $c^2 = a^2 - b^2$, $a^2 > b^2$
$\dfrac{(x - h)^2}{b^2} + \dfrac{(y - k)^2}{a^2} = 1$	center (h, k), vertices $(h, k \pm a)$, endpoints of minor axis $(h \pm b, k)$, foci $(h, k \pm c)$, $c^2 = a^2 - b^2$, $a^2 > b^2$

Hyperbolas

$\dfrac{x^2}{a^2} - \dfrac{y^2}{b^2} = 1$	center $(0, 0)$, vertices $(\pm a, 0)$, endpoints of conjugate axis $(0, \pm b)$, foci $(\pm c, 0)$, $c^2 = a^2 + b^2$, asymptotes $y = \pm \dfrac{b}{a}x$
$\dfrac{y^2}{a^2} - \dfrac{x^2}{b^2} = 1$	center $(0, 0)$, vertices $(0, \pm a)$, endpoints of conjugate axis $(\pm b, 0)$, foci $(0, \pm c)$, $c^2 = a^2 + b^2$, asymptotes $y = \pm \dfrac{a}{b}x$
$\dfrac{(x - h)^2}{a^2} - \dfrac{(y - k)^2}{b^2} = 1$	center (h, k), vertices $(h \pm a, k)$, endpoints of conjugate axis $(h, k \pm b)$, foci $(h \pm c, k)$, $c^2 = a^2 + b^2$, asymptotes $y - k = \pm \dfrac{b}{a}(x - h)$

$\dfrac{(y-k)^2}{a^2} - \dfrac{(x-h)^2}{b^2} = 1$ center (h, k), vertices $(h, k \pm a)$, endpoints of conjugate axis $(h \pm b, k)$, foci $(h, k \pm c)$, $c^2 = a^2 + b^2$, asymptotes $y - k = \pm \dfrac{a}{b}(x - h)$

Systems that contain at least one nonlinear equation can often be solved by substitution or by the elimination method. Graphing the system will often provide a basis for predicting approximate real number solutions, if there are any.

CHAPTER 12 REVIEW PROBLEM SET

For Problems 1–12, (a) identify the conic section as a parabola, an ellipse, or a hyperbola; (b) if it is a parabola, find its vertex, focus, and directrix; if it is an ellipse, find its vertices, endpoints of minor axis, and foci; if it is a hyperbola, find its vertices, endpoints of conjugate axis, foci, and asymptotes, and (c) sketch each of the curves.

1. $x^2 + 2y^2 = 32$

2. $y^2 = -12x$

3. $3y^2 - x^2 = 9$

4. $2x^2 - 3y^2 = 18$

5. $5x^2 + 2y^2 = 20$

6. $x^2 = 2y$

7. $x^2 - 8x - 2y^2 + 4y + 10 = 0$

8. $9x^2 - 54x + 2y^2 + 8y + 71 = 0$

9. $y^2 - 2y + 4x + 9 = 0$

10. $x^2 + 2x + 8y + 25 = 0$

11. $x^2 + 10x + 4y^2 - 16y + 25 = 0$

12. $3y^2 + 12y - 2x^2 - 8x - 8 = 0$

For Problems 13–24, find the equation of the indicated conic section that satisfies the given conditions.

13. Parabola with vertex $(0, 0)$, focus $(-5, 0)$, directrix $x = 5$

14. Ellipse with vertices $(0, \pm 4)$, foci $\left(0, \pm\sqrt{15}\right)$

15. Hyperbola with vertices $\left(\pm\sqrt{2}, 0\right)$, length of conjugate axis 10

16. Ellipse with vertices $(\pm 2, 0)$, contains the point $(1, -2)$

17. Parabola with vertex $(0, 0)$, symmetric with respect to the y-axis, contains the point $(2, 6)$

18. Hyperbola with vertices $(0, \pm 1)$, foci $\left(0, \pm\sqrt{10}\right)$

19. Ellipse with vertices $(6, 1)$ and $(6, 7)$, length of minor axis 2 units

20. Parabola with vertex $(4, -2)$, focus $(6, -2)$

21. Hyperbola with vertices $(-5, -3)$ and $(-5, -5)$, foci $(-5, -2)$ and $(-5, -6)$

22. Parabola with vertex $(-6, -3)$, symmetric with respect to the line $x = -6$, contains the point $(-5, -2)$

23. Ellipse with endpoints of minor axis $(-5, 2)$ and $(-5, -2)$, length of major axis 10 units

24. Hyperbola with vertices $(2, 0)$ and $(6, 0)$, length of conjugate axis 8 units

For Problems 25–30, (a) graph the system, and (b) solve the system by using the substitution or elimination method.

25. $\begin{pmatrix} x^2 + y^2 = 17 \\ x - 4y = -17 \end{pmatrix}$

26. $\begin{pmatrix} x^2 - y^2 = 8 \\ 3x - y = 8 \end{pmatrix}$

27. $\begin{pmatrix} x - y = 1 \\ y = x^2 + 4x + 1 \end{pmatrix}$

28. $\begin{pmatrix} 4x^2 - y^2 = 16 \\ 9x^2 + 9y^2 = 16 \end{pmatrix}$

29. $\begin{pmatrix} x^2 + 2y^2 = 8 \\ 2x^2 + 3y^2 = 12 \end{pmatrix}$

30. $\begin{pmatrix} y^2 - x^2 = 1 \\ 4x^2 + y^2 = 4 \end{pmatrix}$

SEQUENCES AND MATHEMATICAL INDUCTION

Suppose that an auditorium has 35 seats in the first row, 40 seats in the second row, 45 seats in the third row, and so on for ten rows. The numbers 35, 40, 45, 50, . . . , 80 represent the number of seats per row from row 1 through row 10. The list of numbers has a constant difference of 5 between any two successive numbers in the list; such a list is called an **arithmetic sequence**.

Suppose that a fungus culture growing under controlled conditions doubles in size each day. If today the size of the culture is 6 units, then the numbers 12, 24, 48, 96, 192 represent the size of the culture for the next five days. In this list of numbers, each number after the first is two times the previous number; such a list is called a **geometric sequence**. Arithmetic sequences and geometric sequences will be the center of our attention in this chapter.

13

If you could get a job that pays only a penny the first day of your employment, but then doubles each succeeding day, by the 31st working day your salary would be $10,737,418.24.

13.1 ARITHMETIC SEQUENCES

An **infinite sequence** is a function whose domain is the set of positive integers. For example, consider the function defined by the equation

$$f(n) = 5n + 1,$$

where the domain is the set of positive integers. If we substitute the numbers of the domain in order, starting with 1, we can list the resulting ordered pairs:

$$(1, 6), \quad (2, 11), \quad (3, 16), \quad (4, 21), \quad (5, 26),$$

and so on. However, since we know we are using the domain of positive integers in order, starting with 1, there is no need to use ordered pairs. We can simply express the infinite sequence as

$$6, 11, 16, 21, 26, \ldots .$$

Frequently the letter a is used to represent sequential functions and the functional value of a at n is written a_n (read a sub n) instead of $a(n)$. The sequence is then expressed

$$a_1, a_2, a_3, a_4, \ldots ,$$

where a_1 is the **first term**, a_2 is the **second term**, a_3 is the **third term**, and so on. The expression a_n, which defines the sequence, is called the **general term** of the sequence. Knowing the general term of a sequence allows us to find as many terms of the sequence as needed and also to find any specific terms. Consider the following example.

EXAMPLE 1

Find the first five terms of the sequence where $a_n = 2n^2 - 3$; find the twentieth term.

Solution

The first five terms are generated by replacing n with 1, 2, 3, 4, and 5.

$$a_1 = 2(1)^2 - 3 = -1 \qquad a_2 = 2(2)^2 - 3 = 5$$

$$a_3 = 2(3)^2 - 3 = 15 \qquad a_4 = 2(4)^2 - 3 = 29$$

$$a_5 = 2(5)^2 - 3 = 47$$

The first five terms are thus -1, 5, 15, 29, and 47. The twentieth term is

$$a_{20} = 2(20)^2 - 3 = 797.$$

Arithmetic Sequences

An **arithmetic sequence** (also called an arithmetic progression) is a sequence that has a common difference between successive terms. The following are examples of arithmetic sequences.

$1, 8, 15, 22, 29, \ldots$

$4, 7, 10, 13, 16, \ldots$

$4, 1, -2, -5, -8, \ldots$

$-1, -6, -11, -16, -21, \ldots$

The common difference in the first sequence is 7. That is to say, $8 - 1 = 7$, $15 - 8 = 7$, $22 - 15 = 7$, $29 - 22 = 7$, and so on. The common differences for the next three sequences are 3, -3, and -5, respectively.

In a more general setting we say that the sequence

$$a_1, a_2, a_3, a_4, \ldots, a_n, \ldots$$

is an arithmetic sequence if and only if there is a real number d such that

$$a_{k+1} - a_k = d$$

for every positive integer k. The number d is called the **common difference**.

From the definition we see that $a_{k+1} = a_k + d$. In other words, we can generate an arithmetic sequence that has a common difference of d by starting with a first term a_1 and then simply adding d to each successive term.

First term: a_1

Second term: $a_1 + d$

Third term: $a_1 + 2d$ $\quad (a_1 + d) + d = a_1 + 2d$

Fourth term: $a_1 + 3d$

$\vdots$

nth term: $a_1 + (n - 1)d$

Thus, the **general term** of an arithmetic sequence is given by

$$a_n = a_1 + (n - 1)d,$$

where a_1 is the first term and d is the common difference. This formula for the general term can be used to solve a variety of problems involving arithmetic sequences.

EXAMPLE 2

Find the general term expression for the arithmetic sequence $6, 2, -2, -6, \ldots$.

Solution

The common difference, d, is $2 - 6 = -4$ and the first term, a_1, is 6. Substitute these values into $a_n = a_1 + (n - 1)d$ and simplify to obtain

$$\begin{aligned} a_n &= a_1 + (n - 1)d \\ &= 6 + (n - 1)(-4) \\ &= 6 - 4n + 4 \\ &= -4n + 10. \end{aligned}$$

EXAMPLE 3

Find the fortieth term of the arithmetic sequence 1, 5, 9, 13, ...

Solution

Using $a_n = a_1 + (n - 1)d$, we obtain

$$\begin{aligned} a_{40} &= 1 + (40 - 1)4 \\ &= 1 + (39)(4) \\ &= 157. \end{aligned}$$

EXAMPLE 4

Find the first term of the arithmetic sequence where the fourth term is 26 and the ninth term is 61.

Solution

Using $a_n = a_1 + (n - 1)d$ with $a_4 = 26$ (the fourth term is 26) and $a_9 = 61$ (the ninth term is 61), we have

$$\begin{aligned} 26 &= a_1 + (4 - 1)d = a_1 + 3d, \\ 61 &= a_1 + (9 - 1)d = a_1 + 8d. \end{aligned}$$

Solving the system of equations

$$\begin{pmatrix} a_1 + 3d = 26 \\ a_1 + 8d = 61 \end{pmatrix}$$

yields $a_1 = 5$ and $d = 7$. Thus, the first term is 5.

Sums of Arithmetic Sequences

We often use sequences to solve problems, so we need to be able to find the sum of a certain number of terms of the sequence. Before we develop a general sum formula for arithmetic sequences, let's consider an approach to a specific problem that we can then use in a general setting.

EXAMPLE 5

Find the sum of the first 100 positive integers.

Solution

We are being asked to find the sum of $1 + 2 + 3 + 4 + \cdots + 100$. Rather than adding in the usual way, let's find the sum in the following manner.

$$\begin{array}{r} 1 + \quad 2 + \quad 3 + \quad 4 + \cdots + 100 \\ 100 + \quad 99 + \quad 98 + \quad 97 + \cdots + \quad 1 \\ \hline 101 + 101 + 101 + 101 + \cdots + 101 \end{array}$$

$$\frac{\overset{50}{\cancel{100}}(101)}{\cancel{2}} = 5050.$$

Note that we simply wrote the indicated sum forward and backward, and then we added the results. In so doing, we produced 100 sums of 101, but half of them are repeats. For example, $100 + 1$ and $1 + 100$ are both counted in this process. Thus, we divide the product (100)(101) by 2, which yields the final result of 5050.

The *forward-backward* approach we used in Example 5 can be used to develop a formula for finding the sum of the first n terms of any arithmetic sequence. Consider an arithmetic sequence $a_1, a_2, a_3, a_4, \ldots, a_n$ with a common difference of d. Use S_n to represent the sum of the first n terms and proceed as follows.

$$S_n = a_1 + (a_1 + d) + (a_1 + 2d) + \cdots + (a_n - 2d) + (a_n - d) + a_n$$

Now write this sum in reverse.

$$S_n = a_n + (a_n - d) + (a_n - 2d) + \cdots + (a_1 + 2d) + (a_1 + d) + a_1$$

Add the two equations to produce

$$2S_n = (a_1 + a_n) + (a_1 + a_n) + (a_1 + a_n) + \cdots + (a_1 + a_n) + (a_1 + a_n) + (a_1 + a_n).$$

That is, we have n sums $a_1 + a_n$, so

$$2S_n = n(a_1 + a_n),$$

from which we obtain a **sum formula**:

$$S_n = \frac{n(a_1 + a_n)}{2}.$$

Using the nth-term formula and/or the sum formula, we can solve a variety of problems involving arithmetic sequences.

EXAMPLE 6

Find the sum of the first 30 terms of the arithmetic sequence 3, 7, 11, 15, ...

Solution

Using $a_n = a_1 + (n - 1)d$, we can find the 30th term.

$$a_{30} = 3 + (30 - 1)4 = 3 + 29(4) = 119$$

Now we can use the sum formula.

$$S_{30} = \frac{30(3 + 119)}{2} = 1830$$

EXAMPLE 7

Find the sum $7 + 10 + 13 + \cdots + 157$.

Solution

To use the sum formula, we need to know the number of terms. The nth-term formula will do that for us.

$$\begin{aligned} a_n &= a_1 + (n - 1)d \\ 157 &= 7 + (n - 1)3 \\ 157 &= 7 + 3n - 3 \\ 157 &= 3n + 4 \\ 153 &= 3n \\ 51 &= n \end{aligned}$$

Now we can use the sum formula.

$$S_{51} = \frac{51(7 + 157)}{2} = 4182$$

Keep in mind that we developed the sum formula for an arithmetic sequence by using the forward-backward technique, which we had previously used on a specific problem. Now that we have the sum formula, we have two choices when solving problems: We can either memorize the formula and use it, or we can simply use the forward-backward technique. However, we should emphasize that even if you choose to use the formula and some day you forget it, don't panic, just use the forward-backward technique. In other words, understanding the development of a formula often allows you to do problems even when you forget the formula itself.

Summation Notation

Sometimes a special notation is used to indicate the sum of a certain number of terms of a sequence. The capital Greek letter, *sigma*, Σ, is used as a **summation symbol**. For example,

$$\sum_{i=1}^{5} a_i$$

represents the sum $a_1 + a_2 + a_3 + a_4 + a_5$. The letter i is frequently used as the **index of summation**; the letter i takes on all integer values from the lower limit to the upper limit, inclusive. Thus we can write the following examples.

$$\sum_{i=1}^{4} b_i = b_1 + b_2 + b_3 + b_4$$

$$\sum_{i=3}^{7} a_i = a_3 + a_4 + a_5 + a_6 + a_7$$

$$\sum_{i=1}^{15} i^2 = 1^2 + 2^2 + 3^2 + \cdots + 15^2$$

$$\sum_{i=1}^{n} a_i = a_1 + a_2 + a_3 + \cdots + a_n$$

If $a_1, a_2, a_3, \ldots$ represents an arithmetic sequence, we can now write the sum formula

$$\sum_{i=1}^{n} a_i = \frac{n}{2}(a_1 + a_n).$$

EXAMPLE 8

Find the sum $\sum_{i=1}^{50} (3i + 4)$.

Solution

This indicated sum means

$$\begin{aligned}\sum_{i=1}^{50} (3i + 4) &= [3(1) + 4] + [3(2) + 4] + [3(3) + 4] + \cdots + [3(50) + 4] \\ &= 7 + 10 + 13 + \cdots + 154.\end{aligned}$$

Since this is an indicated sum of an arithmetic sequence, we can use our sum formula.

$$S_{50} = \frac{50}{2}(7 + 154) = 4025$$

EXAMPLE 9

Find the sum $\sum_{i=2}^{7} 2i^2$.

Solution

This indicated sum means

$$\begin{aligned}\sum_{i=2}^{7} 2i^2 &= 2(2)^2 + 2(3)^2 + 2(4)^2 + 2(5)^2 + 2(6)^2 + 2(7)^2 \\ &= 8 + 18 + 32 + 50 + 72 + 98.\end{aligned}$$

This is not the indicated sum of an *arithmetic* sequence; therefore, let's simply add the numbers in the usual way. The sum is 278.

Example 9 should serve as a word of caution. Be sure that you analyze the sequence of numbers that is represented by the summation symbol. You may or may not be able to use a formula for adding the numbers.

PROBLEM SET 13.1

For Problems 1–10, write the first five terms of the sequence that has the indicated general term.

1. $a_n = 3n - 7$

2. $a_n = 5n - 2$

3. $a_n = -2n + 4$

4. $a_n = -4n + 7$

5. $a_n = 3n^2 - 1$

6. $a_n = 2n^2 - 6$

7. $a_n = n(n - 1)$

8. $a_n = (n + 1)(n + 2)$

9. $a_n = 2^{n+1}$

10. $a_n = 3^{n-1}$

11. Find the 15th and 30th terms of the sequence where $a_n = -5n - 4$.

12. Find the 20th and 50th terms of the sequence where $a_n = -n - 3$.

13. Find the 25th and 50th terms of the sequence where $a_n = (-1)^{n+1}$.

14. Find the 10th and 15th terms of the sequence where $a_n = -n^2 - 10$.

For Problems 15–24, find the general term (the nth term) for each of the arithmetic sequences.

15. 11, 13, 15, 17, 19, . . .

16. 7, 10, 13, 16, 19, . . .

17. 2, −1, −4, −7, −10, . . .

18. 4, 2, 0, −2, −4, . . .

19. $\frac{3}{2}, 2, \frac{5}{2}, 3, \frac{7}{2}, \ldots$

20. $0, \frac{1}{2}, 1, \frac{3}{2}, 2, \ldots$

21. 2, 6, 10, 14, 18, . . .

22. 2, 7, 12, 17, 22, . . .

23. −3, −6, −9, −12, −15, . . .

24. −4, −8, −12, −16, −20, . . .

For Problems 25–30, find the required term for each of the arithmetic sequences.

25. The 15th term of 3, 8, 13, 18, . . .

26. The 20th term of 4, 11, 18, 25, . . .

27. The 30th term of 15, 26, 37, 48, . . .

28. The 35th term of 9, 17, 25, 33, . . .

29. The 52nd term of $1, \frac{5}{3}, \frac{7}{3}, 3, \ldots$

30. The 47th term of $\frac{1}{2}, \frac{5}{4}, 2, \frac{11}{4}, \ldots$

31. If the 6th term of an arithmetic sequence is 12 and the 10th term is 16, find the first term.

32. If the 5th term of an arithmetic sequence is 14 and the 12th term is 42, find the first term.

33. If the 3rd term of an arithmetic sequence is 20 and the 7th term is 32, find the 25th term.

34. If the 5th term of an arithmetic sequence is −5 and the 15th term is −25, find the 50th term.

35. Find the sum of the first 50 terms of the arithmetic sequence 5, 7, 9, 11, 13, . . .

36. Find the sum of the first 30 terms of the arithmetic sequence 0, 2, 4, 6, 8, . . .

37. Find the sum of the first 40 terms of the arithmetic sequence 2, 6, 10, 14, 18, . . .

38. Find the sum of the first 60 terms of the arithmetic sequence −2, 3, 8, 13, 18, . . .

39. Find the sum of the first 75 terms of the arithmetic sequence 5, 2, −1, −4, −7, . . .

40. Find the sum of the first 80 terms of the arithmetic sequence 7, 3, −1, −5, −9, . . .

41. Find the sum of the first 50 terms of the arithmetic sequence $\frac{1}{2}, 1, \frac{3}{2}, 2, \frac{5}{2}, \ldots$

42. Find the sum of the first 100 terms of the arithmetic sequence $-\frac{1}{3}, \frac{1}{3}, 1, \frac{5}{3}, \frac{7}{3}, \ldots$

For Problems 43–50, find each of the indicated sums.

43. $1 + 5 + 9 + 13 + \cdots + 197$

44. $3 + 8 + 13 + 18 + \cdots + 398$

45. $2 + 8 + 14 + 20 + \cdots + 146$

46. $6 + 9 + 12 + 15 + \cdots + 93$

47. $(-7) + (-10) + (-13) + (-16) + \cdots + (-109)$

48. $(-5) + (-9) + (-13) + (-17) + \cdots + (-169)$

49. $(-5) + (-3) + (-1) + 1 + \cdots + 119$

50. $(-7) + (-4) + (-1) + 2 + \cdots + 131$

51. Find the sum of the first 200 odd whole numbers.

52. Find the sum of the first 175 positive even whole numbers.

53. Find the sum of all even numbers between 18 and 482, inclusive.

54. Find the sum of all odd numbers between 17 and 379, inclusive.

55. Find the sum of the first 30 terms of the arithmetic sequence with the general term $a_n = 5n - 4$.

56. Find the sum of the first 40 terms of the arithmetic sequence with the general term $a_n = 4n - 7$.

57. Find the sum of the first 25 terms of the arithmetic sequence with the general term $a_n = -4n - 1$.

58. Find the sum of the first 35 terms of the arithmetic sequence with the general term $a_n = -5n - 3$.

For Problems 59–70, find each of the following sums.

59. $\sum_{i=1}^{45} (5i + 2)$

60. $\sum_{i=1}^{38} (3i + 6)$

61. $\sum_{i=1}^{30} (-2i + 4)$

62. $\sum_{i=1}^{40} (-3i + 3)$

63. $\sum_{i=4}^{32} (3i - 10)$

64. $\sum_{i=6}^{47} (4i - 9)$

65. $\sum_{i=10}^{20} 4i$

66. $\sum_{i=15}^{30} (-5i)$

67. $\sum_{i=1}^{5} i^2$

68. $\sum_{i=1}^{6} (i^2 + 1)$

69. $\sum_{i=3}^{8} (2i^2 + i)$

70. $\sum_{i=4}^{7} (3i^2 - 2)$

MISCELLANEOUS PROBLEMS

The general term of a sequence can consist of one expression for certain values of n and another expression (or expressions) for other values of n. That is to say, a **multiple description** of the sequence can be given. For example,

$$a_n = \begin{cases} 2n + 3 & \text{for } n \text{ odd} \\ 3n - 2 & \text{for } n \text{ even} \end{cases}$$

means that we use $a_n = 2n + 3$ for $n = 1, 3, 5, 7, \ldots,$ and we use $a_n = 3n - 2$ for $n = 2, 4, 6, 8, \ldots.$ The first six terms of this sequence are 5, 4, 9, 10, 13, and 16.

For Problems 71–74, write the first six terms of each sequence.

71. $a_n = \begin{cases} 2n + 1 & \text{for } n \text{ odd} \\ 2n - 1 & \text{for } n \text{ even} \end{cases}$

72. $a_n = \begin{cases} \dfrac{1}{n} & \text{for } n \text{ odd} \\ n^2 & \text{for } n \text{ even} \end{cases}$

73. $a_n = \begin{cases} 3n + 1 & \text{for } n \leq 3 \\ 4n - 3 & \text{for } n > 3 \end{cases}$

74. $a_n = \begin{cases} 5n - 1 & \text{for } n \text{ a multiple of 3} \\ 2n & \text{otherwise} \end{cases}$

The multiple-description approach can also be used to give a **recursive description** for a sequence. A sequence is said to be **described recursively** if the first n terms are stated and then each succeeding term is defined as a function of one or more of the preceding terms. For example,

$$\begin{cases} a_1 = 2 \\ a_n = 2a_{n-1} & \text{for } n \geq 2 \end{cases}$$

means that the first term, a_1, is 2 and each succeeding term is 2 times the previous term. Thus, the first six terms are 2, 4, 8, 16, 32, and 64.

For Problems 75–80, write the first six terms of each sequence.

75. $\begin{cases} a_1 = 4 \\ a_n = 3a_{n-1} & \text{for } n \geq 2 \end{cases}$

76. $\begin{cases} a_1 = 3 \\ a_n = a_{n-1} + 2 & \text{for } n \geq 2 \end{cases}$

77. $\begin{cases} a_1 = 1 \\ a_2 = 1 \\ a_n = a_{n-2} + a_{n-1} & \text{for } n \geq 3 \end{cases}$

78. $\begin{cases} a_1 = 2 \\ a_2 = 3 \\ a_n = 2a_{n-2} + 3a_{n-1} & \text{for } n \geq 3 \end{cases}$

79. $\begin{cases} a_1 = 3 \\ a_2 = 1 \\ a_n = (a_{n-1} - a_{n-2})^2 & \text{for } n \geq 3 \end{cases}$

80. $\begin{cases} a_1 = 1 \\ a_2 = 2 \\ a_3 = 3 \\ a_n = a_{n-1} + a_{n-2} + a_{n-3} & \text{for } n \geq 4 \end{cases}$

13.2 GEOMETRIC SEQUENCES

A **geometric sequence** or **geometric progression** is a sequence in which we obtain each term after the first by multiplying the preceding term by a common multiplier, called the **common ratio** of the sequence. We can find the common ratio of a geometric sequence by dividing any term (other than the first) by the preceding term. The following geometric sequences have common ratios of 3, 2, $\frac{1}{2}$, and -4, respectively.

1, 3, 9, 27, 81, . . .

3, 6, 12, 24, 48, . . .

16, 8, 4, 2, 1, . . .

$-1, 4, -16, 64, -256, \ldots$

In a more general setting we say that the sequence $a_1, a_2, a_3, \ldots, a_n, \ldots$ is a geometric sequence if and only if there is a nonzero real number r such that

$$a_{k+1} = ra_k$$

for every positive integer k. The nonzero real number r is called the common ratio of the sequence.

The previous equation can be used to generate a general geometric sequence that has a_1 as a first term and r as a common ratio. We can proceed as follows.

First term: a_1

Second term: a_1r

Third term: a_1r^2 $\quad (a_1r)(r) = a_1r^2$

Fourth term: a_1r^3

$\vdots$

nth term: a_1r^{n-1}

Thus, the **general term** of a geometric sequence is given by

$$a_n = a_1r^{n-1},$$

where a_1 is the first term and r is the common ratio.

EXAMPLE 1

Find the general term for the geometric sequence 8, 16, 32, 64, ...

Solution

Using $a_n = a_1 r^{n-1}$, we obtain

$$a_n = 8(2)^{n-1} = (2^3)(2)^{n-1} = 2^{n+2}.$$

EXAMPLE 2

Find the ninth term of the geometric sequence 27, 9, 3, 1, ...

Solution

Using $a_n = a_1 r^{n-1}$, we can find the ninth term as follows.

$$a_9 = 27\left(\frac{1}{3}\right)^{9-1} = 27\left(\frac{1}{3}\right)^8 = \frac{3^3}{3^8} = \frac{1}{3^5} = \frac{1}{243}$$

Sums of Geometric Sequences

As with arithmetic sequences, we often need to find the sum of a certain number of terms of a geometric sequence. Before we develop a general sum formula for geometric sequences, let's consider an approach to a specific problem that we can then use in a general setting.

EXAMPLE 3

Find the sum $1 + 3 + 9 + 27 + \cdots + 6561$.

Solution

Let S represent the sum and proceed as follows.

$$S = 1 + 3 + 9 + 27 + \cdots + 6561 \qquad (1)$$

$$3S = 3 + 9 + 27 + \cdots + 6561 + 19683 \qquad (2)$$

Equation (2) is the result of multiplying equation (1) by the common ratio, 3. Subtracting equation (1) from equation (2) produces

$$2S = 19683 - 1 = 19682$$

$$S = 9841.$$

Now let's consider a general geometric sequence $a_1, a_1 r, a_1 r^2, \ldots, a_1 r^{n-1}$. By applying a procedure similar to the one we used in Example 3, we can develop a formula for finding the sum of the first n terms of any geometric sequence. We let S_n represent the sum of the first n terms:

$$S_n = a_1 + a_1 r + a_1 r^2 + \cdots + a_1 r^{n-1}. \qquad (3)$$

Next we multiply both sides of equation (3) by the common ratio r:

$$rS_n = a_1 r + a_1 r^2 + a_1 r^3 + \cdots + a_1 r^n. \qquad (4)$$

We then subtract equation (3) from equation (4):

$$rS_n - S_n = a_1r^n - a_1.$$

When we apply the distributive property to the left side and then solve for S_n, we obtain

$$S_n(r - 1) = a_1r^n - a_1$$
$$S_n = \frac{a_1r^n - a_1}{r - 1}, \qquad r \neq 1.$$

Therefore, the sum of the first n terms of a geometric sequence with a first term a_1 and a common ratio r is given by

$$S_n = \frac{a_1r^n - a_1}{r - 1}, \qquad r \neq 1.$$

EXAMPLE 4

Find the sum of the first eight terms of the geometric sequence 1, 2, 4, 8, ...

Solution

Use the sum formula to obtain

$$S_8 = \frac{1(2)^8 - 1}{2 - 1} = \frac{2^8 - 1}{1} = 255.$$

If the common ratio of a geometric sequence is less than 1, it may be more convenient to change the form of the sum formula. That is, the fraction

$$\frac{a_1r^n - a_1}{r - 1}$$

can be changed to

$$\frac{a_1 - a_1r^n}{1 - r}$$

by multiplying both the numerator and denominator by -1. Thus, using

$$S_n = \frac{a_1 - a_1r^n}{1 - r}$$

can sometimes avoid unnecessary work with negative numbers when $r < 1$, as the next example illustrates.

EXAMPLE 5

Find the sum $1 + \frac{1}{2} + \frac{1}{4} + \cdots + \frac{1}{256}$.

Solution A

To use the sum formula, we need to know the number of terms, which can be found by counting them or by applying the n-th term formula, as follows.

$$a_n = a_1 r^{n-1}$$
$$\frac{1}{256} = 1\left(\frac{1}{2}\right)^{n-1}$$
$$\left(\frac{1}{2}\right)^8 = \left(\frac{1}{2}\right)^{n-1}$$
$$8 = n - 1 \qquad \text{If } b^n = b^m\text{, then } n = m.$$
$$9 = n$$

Now we use $n = 9$, $a_1 = 1$, and $r = \frac{1}{2}$ in the sum formula of the form

$$S_n = \frac{a_1 - a_1 r^n}{1 - r}.$$

$$S_9 = \frac{1 - 1\left(\frac{1}{2}\right)^9}{1 - \frac{1}{2}} = \frac{1 - \frac{1}{512}}{\frac{1}{2}} = \frac{\frac{511}{512}}{\frac{1}{2}} = 1\frac{255}{256}$$

We can also do a problem like Example 5 without finding the number of terms; we use the general approach illustrated in Example 3. Solution B demonstrates this idea.

Solution B

Let S represent the desired sum.

$$S = 1 + \frac{1}{2} + \frac{1}{4} + \cdots + \frac{1}{256}$$

Multiply both sides by the common ratio, $\frac{1}{2}$.

$$\frac{1}{2}S = \frac{1}{2} + \frac{1}{4} + \frac{1}{8} + \cdots + \frac{1}{256} + \frac{1}{512}$$

Subtract the second equation from the first and solve for S.

$$\frac{1}{2}S = 1 - \frac{1}{512} = \frac{511}{512}$$
$$S = \frac{511}{256} = 1\frac{255}{256}.$$

Summation notation can also be used to indicate the sum of a certain number of terms of a geometric sequence.

EXAMPLE 6 Find the sum $\sum_{i=1}^{10} 2^i$.

Solution

This indicated sum means

$$\sum_{i=1}^{10} 2^i = 2^1 + 2^2 + 2^3 + \cdots + 2^{10}$$
$$= 2 + 4 + 8 + \cdots + 1024.$$

Since this is the indicated sum of a geometric sequence, we can use the sum formula, with $a_1 = 2$, $r = 2$, and $n = 10$.

$$S_{10} = \frac{2(2)^{10} - 2}{2 - 1} = \frac{2(2^{10} - 1)}{1} = 2046$$

The Sum of an Infinite Geometric Sequence

Let's take the formula

$$S_n = \frac{a_1 - a_1 r^n}{1 - r}$$

and rewrite the right side by applying the property

$$\frac{a - b}{c} = \frac{a}{c} - \frac{b}{c}.$$

Thus we obtain

$$S_n = \frac{a_1}{1 - r} - \frac{a_1 r^n}{1 - r}. \qquad \textbf{(1)}$$

Now let's examine the behavior of r^n for $|r| < 1$, that is, for $-1 < r < 1$. For example, suppose that $r = \frac{1}{2}$; then

$$r^2 = \left(\frac{1}{2}\right)^2 = \frac{1}{4} \qquad r^3 = \left(\frac{1}{2}\right)^3 = \frac{1}{8}$$
$$r^4 = \left(\frac{1}{2}\right)^4 = \frac{1}{16} \qquad r^5 = \left(\frac{1}{2}\right)^5 = \frac{1}{32}$$

and so on. We can make $\left(\frac{1}{2}\right)^n$ as close to zero as we please by choosing sufficiently large values for n. In general, for values of r such that $|r| < 1$, the expression r^n will approach zero as n gets larger and larger. Therefore, the fraction $a_1 r^n/(1 - r)$ in equation (1) will approach zero as n increases. We say that **the sum of the infinite geometric sequence** is given by

$$S_\infty = \frac{a_1}{1 - r}, \qquad |r| < 1.$$

EXAMPLE 7

Find the sum of the infinite geometric sequence

$$1, \frac{1}{2}, \frac{1}{4}, \frac{1}{8}, \ldots.$$

Solution

Since $a_1 = 1$ and $r = \frac{1}{2}$, we obtain

$$S_\infty = \frac{1}{1 - \frac{1}{2}} = \frac{1}{\frac{1}{2}} = 2.$$

When we state that $S_\infty = 2$ in Example 7, we mean that as we add more and more terms, the sum approaches 2. Observe what happens when we calculate the sum up to five terms.

First term:	1
Sum of first two terms:	$1 + \frac{1}{2} = 1\frac{1}{2}$
Sum of first three terms:	$1 + \frac{1}{2} + \frac{1}{4} = 1\frac{3}{4}$
Sum of first four terms:	$1 + \frac{1}{2} + \frac{1}{4} + \frac{1}{8} = 1\frac{7}{8}$
Sum of first five terms:	$1 + \frac{1}{2} + \frac{1}{4} + \frac{1}{8} + \frac{1}{16} = 1\frac{15}{16}.$

If $|r| > 1$, the absolute value of r^n increases without bound as n increases. Consider the following two examples and notice the unbounded growth of the absolute value of r^n.

Let $r = 3$	Let $r = -2$	
$r^2 = 3^2 = 9$	$r^2 = (-2)^2 = 4$	
$r^3 = 3^3 = 27$	$r^3 = (-2)^3 = -8$	$\lvert -8 \rvert = 8$
$r^4 = 3^4 = 81$	$r^4 = (-2)^4 = 16$	
$r^5 = 3^5 = 243$	$r^5 = (-2)^5 = -32$	$\lvert -32 \rvert = 32$

If $r = 1$, then $S_n = na_1$, and as n increases without bound, $|S_n|$ also increases without bound. If $r = -1$, then S_n will either be a_1 or 0. Therefore, we say that the sum of any infinite geometric sequence where $|r| \geq 1$ *does not exist*.

Repeating Decimals as Sums of Infinite Geometric Sequences

In Section 1.1, we defined rational numbers to be numbers that have either a terminating or repeating decimal representation. For example,

$$2.23, \quad 0.147, \quad 0.\overline{3}, \quad 0.\overline{14}, \quad \text{and} \quad 0.5\overline{6}$$

are examples of rational numbers. (Remember that $0.\overline{3}$ means 0.3333. . . .) Place value provides the basis for changing terminating decimals such as 2.23 and 0.147 to a/b form, where a and b are integers and $b \neq 0$:

$$2.23 = \frac{223}{100} \quad \text{and} \quad 0.147 = \frac{147}{1000}.$$

However, changing repeating decimals to a/b form requires a different technique, and our work with sums of infinite geometric sequences provides the basis for one such approach. Consider the following examples.

EXAMPLE 8

Change $0.\overline{14}$ to a/b form, where a and b are integers and $b \neq 0$.

Solution

The repeating decimal $0.\overline{14}$ can be written as the indicated sum of an infinite geometric sequence with first term 0.14 and common ratio 0.01.

$$0.14 + 0.0014 + 0.000014 + \cdots.$$

Using $S_\infty = a_1/(1 - r)$, we obtain

$$S_\infty = \frac{0.14}{1 - 0.01} = \frac{0.14}{0.99} = \frac{14}{99}.$$

Thus, $0.\overline{14} = \dfrac{14}{99}$.

If the repeating block of digits does not begin immediately after the decimal point, as in $0.5\overline{6}$, we can make a slight adjustment in the technique we used in Example 8.

EXAMPLE 9

Change $0.5\overline{6}$ to a/b form, where a and b are integers and $b \neq 0$.

Solution

The repeating decimal $0.5\overline{6}$ can be written

$$(0.5) + (0.06 + 0.006 + 0.0006 + \cdots),$$

where

$$0.06 + 0.006 + 0.0006 + \cdots$$

is the indicated sum of the infinite geometric sequence with $a_1 = 0.06$ and

$r = 0.1$. Therefore,

$$S_\infty = \frac{0.06}{1 - 0.1} = \frac{0.06}{0.9} = \frac{6}{90} = \frac{1}{15}.$$

Now we can add 0.5 and $\frac{1}{15}$.

$$0.5\overline{6} = 0.5 + \frac{1}{15} = \frac{1}{2} + \frac{1}{15} = \frac{15}{30} + \frac{2}{30} = \frac{17}{30}$$

PROBLEM SET 13.2

For Problems 1–12, find the general term (the nth term) for each of the geometric sequences.

1. 3, 6, 12, 24, . . .
2. 2, 6, 18, 54, . . .
3. 3, 9, 27, 81, . . .
4. 2, 4, 8, 16, . . .
5. $\frac{1}{4}, \frac{1}{8}, \frac{1}{16}, \frac{1}{32}, \ldots$
6. 8, 4, 2, 1, . . .
7. 4, 16, 64, 256, . . .
8. $6, 2, \frac{2}{3}, \frac{2}{9}, \ldots$
9. 1, 0.3, 0.09, 0.027, . . .
10. 0.2, 0.04, 0.008, 0.0016, . . .
11. $1, -2, 4, -8, \ldots$
12. $-3, 9, -27, 81, \ldots$

For Problems 13–20, find the required term for each of the given geometric sequences.

13. The 8th term of $\frac{1}{2}, 1, 2, 4, \ldots$
14. The 7th term of 2, 6, 18, 54, . . .
15. The 9th term of 729, 243, 81, 27, . . .
16. The 11th term of 768, 384, 192, 96, . . .
17. The 10th term of $1, -2, 4, -8, \ldots$
18. The 8th term of $-1, -\frac{3}{2}, -\frac{9}{4}, -\frac{27}{8}, \ldots$
19. The 8th term of $\frac{1}{2}, \frac{1}{6}, \frac{1}{18}, \frac{1}{54}, \ldots$
20. The 9th term of $\frac{16}{81}, \frac{8}{27}, \frac{4}{9}, \frac{2}{3}, \ldots$
21. Find the first term of the geometric sequence with 5th term $\frac{32}{3}$ and common ratio 2.
22. Find the first term of the geometric sequence with 4th term $\frac{27}{128}$ and common ratio $\frac{3}{4}$.
23. Find the common ratio of the geometric sequence with 3rd term 12 and 6th term 96.
24. Find the common ratio of the geometric sequence with 2nd term $\frac{8}{3}$ and 5th term $\frac{64}{81}$.
25. Find the sum of the first 10 terms of the geometric sequence 1, 2, 4, 8, . . .
26. Find the sum of the first 7 terms of the geometric sequence 3, 9, 27, 81, . . .
27. Find the sum of the first 9 terms of the geometric sequence 2, 6, 18, 54, . . .
28. Find the sum of the first 10 terms of the geometric sequence 5, 10, 20, 40, . . .
29. Find the sum of the first 8 terms of the geometric sequence 8, 12, 18, 27, . . .
30. Find the sum of the first 8 terms of the geometric sequence $9, 12, 16, \frac{64}{3}, \ldots$
31. Find the sum of the first 10 terms of the geometric sequence $-4, 8, -16, 32, \ldots$
32. Find the sum of the first 9 terms of the geometric sequence $-2, 6, -18, 54, \ldots$

For Problems 33–38, find each of the indicated sums.

33. $9 + 27 + 81 + \cdots + 729$
34. $2 + 8 + 32 + \cdots + 8192$
35. $4 + 2 + 1 + \cdots + \frac{1}{512}$
36. $1 + (-2) + 4 + \cdots + 256$

37. $(-1) + 3 + (-9) + \cdots + (-729)$

38. $16 + 8 + 4 + \cdots + \frac{1}{32}$

For Problems 39–44, find each of the indicated sums.

39. $\sum_{i=1}^{9} 2^{i-3}$ **40.** $\sum_{i=1}^{6} 3^i$ **41.** $\sum_{i=2}^{5} (-3)^{i+1}$

42. $\sum_{i=3}^{8} (-2)^{i-1}$ **43.** $\sum_{i=1}^{6} 3\left(\frac{1}{2}\right)^i$ **44.** $\sum_{i=1}^{5} 2\left(\frac{1}{3}\right)^i$

For Problems 45–56, find the sum of each infinite geometric sequence. If the sequence has no sum, so state.

45. $2, 1, \frac{1}{2}, \frac{1}{4}, \ldots$ **46.** $9, 3, 1, \frac{1}{3}, \ldots$

47. $1, \frac{2}{3}, \frac{4}{9}, \frac{8}{27}, \ldots$ **48.** $5, 3, \frac{9}{5}, \frac{27}{25}, \ldots$

49. $4, 8, 16, 32, \ldots$ **50.** $32, 16, 8, 4, \ldots$

51. $9, -3, 1, -\frac{1}{3}, \ldots$ **52.** $2, -6, 18, -54, \ldots$

53. $\frac{1}{2}, \frac{3}{8}, \frac{9}{32}, \frac{27}{128}, \ldots$ **54.** $4, -\frac{4}{3}, \frac{4}{9}, -\frac{4}{27}, \ldots$

55. $8, -4, 2, -1, \ldots$ **56.** $7, \frac{14}{5}, \frac{28}{25}, \frac{56}{125}, \ldots$

For Problems 57–68, change each repeating decimal to a/b form, where a and b are integers and $b \neq 0$. Express a/b in reduced form.

57. $0.\overline{3}$ **58.** $0.\overline{4}$ **59.** $0.\overline{26}$

60. $0.\overline{18}$ **61.** $0.\overline{123}$ **62.** $0.\overline{273}$

63. $0.2\overline{6}$ **64.** $0.4\overline{3}$ **65.** $0.2\overline{14}$

66. $0.3\overline{71}$ **67.** $2.\overline{3}$ **68.** $3.\overline{7}$

THOUGHTS into WORDS

69. Explain the difference between an arithmetic sequence and a geometric sequence.

70. What does it mean to say that the sum of the infinite geometric sequence $1 + \frac{1}{2} + \frac{1}{4} + \frac{1}{8} + \cdots$ is 2?

71. What do we mean when we state that the infinite geometric sequence $1 + 2 + 4 + 8 + \cdots$ has no sum?

13.3 ANOTHER LOOK AT PROBLEM SOLVING

In the previous two sections, many of the exercises fell into one of the following four categories.

1. Find the nth term of an arithmetic sequence

$$a_n = a_1 + (n - 1)d.$$

2. Find the sum of the first n terms of an arithmetic sequence

$$S_n = \frac{n(a_1 + a_n)}{2}.$$

3. Find the nth term of a geometric sequence

$$a_n = a_1 r^{n-1}.$$

4. Find the sum of the first n terms of a geometric sequence

$$S_n = \frac{a_1 r^n - a_1}{r - 1}.$$

In this section we want to use this knowledge of arithmetic sequences and geometric sequences to expand our problem solving capabilities. Let's begin by restating some old problem solving suggestions that continue to apply here; we will also consider some other suggestions that are directly related to problems that involve sequences of numbers. (We will indicate the new suggestions with an asterisk.)

Suggestions for Solving Word Problems

1. Read the problem carefully and make certain that you understand the meanings of all the words. Be especially alert for any technical terms used in the statement of the problem.

2. Read the problem a second time (perhaps even a third time) to get an overview of the situation being described and to determine the known facts, as well as what you are to find.

3. Sketch a figure, diagram, or chart that might be helpful in analyzing the problem.

***4.** Write down the first few terms of the sequence to describe what is taking place in the problem. Be sure that you understand, term by term, what the sequence represents in the problem.

***5.** Determine whether the sequence is arithmetic or geometric.

***6.** Determine whether the problem is asking for a specific term of the sequence or for the sum of a certain number of terms.

***7.** Carry out the necessary calculations and check your answer for reasonableness.

As we solve some problems, these suggestions will become more meaningful.

PROBLEM 1

Domenica started to work in 1975 at an annual salary of $14,500. She received a $1050 raise each year. What was her annual salary in 1984?

Solution

The following sequence represents her annual salary beginning in 1975.

$$14500, 15550, 16600, 17650, \ldots$$

This is an arithmetic sequence, with $a_1 = 14500$ and $d = 1050$. Since each term of the sequence represents her annual salary, we are looking for the the tenth term.

$$a_{10} = 14500 + (10 - 1)1050 = 14500 + 9(1050) = 23950$$

Her annual salary in 1984 was $23,950.

PROBLEM 2

An auditorium has 20 seats in the front row, 24 seats in the second row, 28 seats in the third row, and so on, for 15 rows. How many seats are there in the auditorium?

Solution

The following sequence represents the number of seats per row starting with the first row.

20, 24, 28, 32, . . .

This is an arithmetic sequence, with $a_1 = 20$ and $d = 4$. Therefore, the fifteenth term, which represents the number of seats in the fifteenth row, is given by

$$a_{15} = 20 + (15 - 1)4 = 20 + 14(4) = 76.$$

The total number of seats in the auditorium is represented by

$$20 + 24 + 28 + \cdots + 76.$$

Use the sum formula for an arithmetic sequence to obtain

$$S_{15} = \frac{15}{2}(20 + 76) = 720.$$

There are 720 seats in the auditorium.

PROBLEM 3

Suppose that you save 25 cents the first day of a week, 50 cents the second day, one dollar the third day, and continue to double your savings each day. How much will you save on the seventh day? What will be your total savings for the week?

Solution

The following sequence represents your savings per day, expressed in cents.

25, 50, 100, . . .

This is a geometric sequence, with $a_1 = 25$ and $r = 2$. Your savings on the seventh day is the seventh term of this sequence. Therefore, using $a_n = a_1 r^{n-1}$, we obtain

$$a_7 = 25(2)^6 = 1600.$$

So you will save \$16 on the seventh day. Your total savings for the 7 days is given by

$$25 + 50 + 100 + \cdots + 1600.$$

Use the sum formula for a geometric sequence to obtain

$$S_7 = \frac{25(2)^7 - 25}{2 - 1} = \frac{25(2^7 - 1)}{1} = 3175.$$

So your savings for the entire week is \$31.75.

PROBLEM 4

A pump is attached to a container for the purpose of creating a vacuum. For each stroke of the pump, $\frac{1}{4}$ of the air that remains in the container is removed. To the nearest tenth of a percent, how much of the air remains in the container after six strokes?

Solution

Let's draw a diagram to help with the analysis of this problem.

First stroke:	$\frac{1}{4}$ of the air is removed	$1 - \frac{1}{4} = \frac{3}{4}$ of the air remains
Second stroke:	$\frac{1}{4}\left(\frac{3}{4}\right) = \frac{3}{16}$ of the air is removed	$\frac{3}{4} - \frac{3}{16} = \frac{9}{16}$ of the air remains
Third stroke:	$\frac{1}{4}\left(\frac{9}{16}\right) = \frac{9}{64}$ of the air is removed	$\frac{9}{16} - \frac{9}{64} = \frac{27}{64}$ of the air remains

The diagram suggests two approaches to the problem.

Approach A The sequence $\frac{1}{4}, \frac{3}{16}, \frac{9}{64}, \ldots$ represents, term by term, the fractional amount of air that is removed with each successive stroke. Therefore, we can find the total amount removed and subtract it from 100%. The sequence is geometric with $a_1 = \frac{1}{4}$ and $r = \frac{3}{4}$.

$$S_6 = \frac{\frac{1}{4} - \frac{1}{4}\left(\frac{3}{4}\right)^6}{1 - \frac{3}{4}} = \frac{\frac{1}{4}\left[1 - \left(\frac{3}{4}\right)^6\right]}{\frac{1}{4}}$$

$$= 1 - \frac{729}{4096} = \frac{3367}{4096} = 82.2\%$$

Therefore, $100\% - 82.2\% = 17.8\%$ of the air remains after 6 strokes.

Approach B The sequence

$$\frac{3}{4}, \frac{9}{16}, \frac{27}{64}, \ldots$$

represents, term by term, the amount of air that remains in the container after each stroke. Therefore, when we find the sixth term of this geometric sequence, we will have the answer to the problem. Since $a_1 = \frac{3}{4}$ and $r = \frac{3}{4}$, we obtain

$$a_6 = \frac{3}{4}\left(\frac{3}{4}\right)^5 = \left(\frac{3}{4}\right)^6 = \frac{729}{4096} = 17.8\%.$$

Therefore, 17.8% of the air remains after 6 strokes.

It will be helpful for you to take another look at the two approaches we used to solve Problem 4. Notice in Approach B that finding the sixth term of the sequence produced the answer to the problem without any further calculations. In Approach A, we had to find the sum of 6 terms of the sequence and then subtract that amount from 100%. As we solve problems that involve sequences, it is necessary that we understand what a particular sequence represents on a term-by-term basis.

PROBLEM SET 13.3

Use your knowledge of arithmetic sequences and geometric sequences to help solve each of the following problems.

1. A man started to work in 1960 at an annual salary of \$9500. He received a \$700 raise each year. How much was his annual salary in 1981?

2. A woman started to work in 1970 at an annual salary of \$13,400. She received a \$900 raise per year. How much was her annual salary in 1985?

3. State University had an enrollment of 9600 students in 1960. Each year the enrollment increased by 150 students. What was the enrollment in 1973?

4. Math University had an enrollment of 12,800 students in 1977. Each year the enrollment decreased by 75 students. What was the enrollment in 1984?

5. The enrollment at University X is predicted to increase at the rate of 10% per year. If the enrollment for 1982 was 5000 students, find the predicted enrollment for 1986. Express your answer to the nearest whole number.

6. If you pay \$12,000 for a car and its value depreciates 20% per year, how much will it be worth in 5 years? Express your answer to the nearest dollar.

7. A tank contains 16,000 liters of water. Each day one-half of the water in the tank is removed and not replaced. How much water remains in the tank at the end of 7 days?

8. If the price of a pound of coffee is \$3.20 and the projected rate of inflation is 5% per year, how much per pound will coffee cost in 5 years? Express your answer to the nearest cent.

9. A tank contains 5832 gallons of water. Each day one-third of the water in the tank is removed and not replaced. How much water remains in the tank at the end of 6 days?

10. A fungus culture growing under controlled conditions doubles in size each day. How many units will the culture contain after 7 days if it originally contains 4 units?

11. Sue is saving quarters. She saves 1 quarter the first day, 2 quarters the second day, 3 quarters the third day, and so on for 30 days. How much money will she have saved in 30 days?

12. Suppose you save a penny the first day of a month, 2 cents the second day, 3 cents the third day, and so on for 31 days. What will be your total savings for the 31 days?

13. Suppose you save a penny the first day of a month, 2 cents the second day, 4 cents the third day, and continue to double your savings each day. How much will you save on the 15th day of the month? How much will your total savings be for the 15 days?

14. Eric saved a nickel the first day of a month, a dime the second day, 20 cents the third day, and continued to double this daily savings each day for 14 days. What was his daily savings on the 14th day? What was his total savings for the 14 days?

15. Ms. Bryan invested \$1500 at 12% simple interest at the beginning of each year for a period of 10 years. Find the total accumulated value of all the investments at the end of the 10-year period.

16. Mr. Woodley invested \$1200 at 11% simple interest at the beginning of each year for a period of 8 years. Find the total accumulated value of all the investments at the end of the 8-year period.

17. An object falling from rest in a vacuum falls approximately 16 feet the first second, 48 feet the second second, 80 feet the third second, 112 feet the fourth second, and so on. How far will it fall in 11 seconds?

18. A raffle is organized so that the amount paid for each ticket is determined by the number on the ticket. The tickets are numbered with the consecutive odd whole numbers 1, 3, 5, 7 . . . . Each contestant pays as many cents as the number on the ticket drawn. How much money will the raffle take in if 1000 tickets are sold?

19. Suppose an element has a half-life of 4 hours. This means that if n grams of it exist at a specific time, then only $\frac{1}{2}n$ grams remain 4 hours later. If at a particular moment we have 60 grams of the element, how much of it will remain 24 hours later?

20. Suppose an element has a half-life of 3 hours. (See Problem 19 for a definition of half-life.) If at a particular moment we have 768 grams of the element, how much of it will remain 24 hours later?

21. A rubber ball is dropped from a height of 1458 feet, and at each bounce it rebounds one-third of the height from which it last fell. How far has the ball traveled by the time it strikes the ground for the 6th time?

22. A rubber ball is dropped from a height of 100 feet, and at each bounce it rebounds one-half of the height from which it last fell. What distance has the ball traveled up to the instant it hits the ground for the 8th time?

23. A pile of logs has 25 logs in the bottom layer, 24 logs in the next layer, 23 logs in the next layer, and so on, until the top layer has 1 log. How many logs are in the pile?

24. A well-driller charges \$9.00 per foot for the first 10 feet, \$9.10 per foot for the next 10 feet, \$9.20 per foot for the next 10 feet, and so on, at a price increase of \$.10 per foot for succeeding intervals of 10 feet. How much does it cost to drill a well to a depth of 150 feet?

25. A pump is attached to a container for the purpose of creating a vacuum. For each stroke of the pump, $\frac{1}{3}$ of the air remaining in the container is removed. To the nearest tenth of a percent, how much of the air remains in the container after 7 strokes?

26. Suppose that in Problem 25, each stroke of the pump removes $\frac{1}{2}$ of the air remaining in the container. What fractional part of the air has been removed after 6 strokes?

27. A tank contains 20 gallons of water. One-half of the water is removed and replaced with antifreeze. Then one-half of this mixture is removed and replaced with antifreeze. This process is continued 8 times. How much water remains in the tank after the 8th replacement process?

28. The radiator of a truck contains 10 gallons of water. Suppose we remove 1 gallon of water and replace it with antifreeze. Then we remove 1 gallon of this mixture and replace it with antifreeze. This process is continued seven times. To the nearest tenth of a gallon, how much antifreeze is in the final mixture?

13.4 MATHEMATICAL INDUCTION

Is $2^n > n$ for all positive integer values of n? In an attempt to answer this question we might proceed as follows.

If $n = 1$, then $2^n > n$ becomes $2^1 > 1$, a true statement.

If $n = 2$, then $2^n > n$ becomes $2^2 > 2$, a true statement.

If $n = 3$, then $2^n > n$ becomes $2^3 > 3$, a true statement.

We can continue in this way as long as we want, but obviously we can never show that $2^n >$ n for *every* positive integer n in this manner. However, we do have a form of proof, called **proof by mathematical induction**, that can be used to verify the truth of many mathematical statements involving positive integers. This form of proof is based on the following principle.

Principle of Mathematical Induction

Let P_n be a statement in terms of n, where n is a positive integer. If

1. P_1 is true, and
2. the truth of P_k implies the truth of P_{k+1} for every positive integer k,

then P_n is true for every positive integer n.

The principle of mathematical induction, a proof that some statement is true for all positive integers, consists of two parts. First, we must show that the statement is true for the positive integer 1. Then we must show that if the statement is true for some positive integer, then it follows that it is also true for the next positive integer. Let's illustrate what this means.

EXAMPLE 1

Prove that $2^n > n$ for all positive integer values of n.

Proof

PART 1 If $n = 1$, then $2^n > n$ becomes $2^1 > 1$, which is a true statement.

PART 2 We must prove the statement, If $2^k > k$, then $2^{k+1} > k + 1$ for all positive integer values of k. In other words, we should be able to start with $2^k > k$ and from that deduce $2^{k+1} > k + 1$. This can be done as follows.

$$2^k > k$$
$$2(2^k) > 2(k) \qquad \text{Multiply both sides by 2.}$$
$$2^{k+1} > 2k$$

We know that $k \geq 1$, since we are working with positive integers. Therefore,

$$k + k \geq k + 1 \qquad \text{Add } k \text{ to both sides.}$$
$$2k \geq k + 1.$$

Since $2^{k+1} > 2k$ and $2k \geq k + 1$, by the transitive property we conclude that

$$2^{k+1} > k + 1.$$

Therefore, using part 1 and part 2 we have proven that $2^n > n$ for *all* positive integers.

It will be helpful for you to look back over the proof in Example 1. Notice that in part 1 we established that $2^n > n$ is true for $n = 1$. Then in part 2 we established that if $2^n > n$ is true for any positive integer, then it must be true for the next consecutive positive integer. Therefore, since $2^n > n$ is true for $n = 1$, then it must be true for $n = 2$. Likewise, if $2^n > n$ is true for $n = 2$, then it must be true for $n = 3$, and so on, for *all* positive integers.

We can depict proof by mathematical induction with dominoes. Suppose that in Figure 13.1 we have infinitely many dominoes lined up. If we can push the first

FIGURE 13.1

domino over (part 1 of a mathematical induction proof) and if the dominoes are spaced so that each time one falls over, it causes the next one to fall over (part 2 of a mathematical induction proof), then by pushing the first one over we will cause a chain reaction that will topple all of the dominoes (Figure 13.2).

FIGURE 13.2

Recall that in the first three sections of this chapter, we used a_n to represent the nth term of a sequence and S_n to represent the sum of the first n terms of a sequence. For example, if $a_n = 2n$, then the first three terms of the sequence are $a_1 = 2(1) = 2$, $a_2 = 2(2) = 4$, and $a_3 = 2(3) = 6$. Furthermore, the kth term is $a_k = 2(k) = 2k$ and the $(k + 1)$st term is $a_{k+1} = 2(k + 1) = 2k + 2$. Relative to this same sequence, we can state that $S_1 = 2$, $S_2 = 2 + 4 = 6$, and $S_3 = 2 + 4 + 6 = 12$.

There are numerous sum formulas for sequences that can be verified by mathematical induction. For such proofs, the following property of sequences is used.

$$S_{k+1} = S_k + a_{k+1}$$

This property states that **the sum of the first $k + 1$ terms is equal to the sum of the first k terms plus the $(k + 1)$st term**. Let's see how this can be used in a specific example.

EXAMPLE 2

Prove that $S_n = n(n + 1)$ for the sequence $a_n = 2n$, where n is any positive integer.

Proof

PART 1 If $n = 1$, then $1(1 + 1) = 2$ and 2 is the first term of the sequence $a_n = 2n$, so $S_1 = a_1 = 2$.

PART 2 We need to prove that if $S_k = k(k+1)$, then $S_{k+1} = (k+1) \cdot (k+2)$. Using the property $S_{k+1} = S_k + a_{k+1}$, we can proceed as follows.

$$\begin{aligned} S_{k+1} &= S_k + a_{k+1} \\ &= k(k+1) + 2(k+1) \\ &= (k+1)(k+2) \end{aligned}$$

Therefore, using part 1 and part 2 we have proved that $S_n = n(n+1)$ will yield the correct sum for any number of terms of the sequence $a_n = 2n$.

EXAMPLE 3

Prove that $S_n = 5n(n+1)/2$ for the sequence $a_n = 5n$, where n is any positive integer.

Proof

PART 1 Since $5(1)\,(1+1)/2 = 5$ and 5 is the first term of the sequence $a_n = 5n$, we have $S_1 = a_1 = 5$.

PART 2 We need to prove that if $S_k = 5k(k+1)/2$, then $S_{k+1} = \dfrac{5(k+1)(k+2)}{2}$.

$$\begin{aligned} S_{k+1} &= S_k + a_{k+1} \\ &= \frac{5k(k+1)}{2} + 5(k+1) \\ &= \frac{5k(k+1)}{2} + 5k + 5 \\ &= \frac{5k(k+1) + 2(5k+5)}{2} \\ &= \frac{5k^2 + 5k + 10k + 10}{2} \\ &= \frac{5k^2 + 15k + 10}{2} \\ &= \frac{5(k^2 + 3k + 2)}{2} \\ &= \frac{5(k+1)(k+2)}{2} \end{aligned}$$

Therefore, using part 1 and part 2 we have proved that $S_n = 5n(n+1)/2$ yields the correct sum for any number of terms of the sequence $a_n = 5n$.

EXAMPLE 4

Prove that $S_n = (4^n - 1)/3$ for the sequence $a_n = 4^{n-1}$, where n is any positive integer.

Proof

PART 1 Since $(4^1 - 1)/3 = 1$ and 1 is the first term of the sequence $a_n = 4^{n-1}$, we have $S_1 = a_1 = 1$.

PART 2 We need to prove that if $S_k = (4^k - 1)/3$, then $S_{k+1} = (4^{k+1} - 1)/3$.

$$\begin{aligned} S_{k+1} &= S_k + a_{k+1} \\ &= \frac{4^k - 1}{3} + 4^k \\ &= \frac{4_k - 1 + 3(4^k)}{3} \\ &= \frac{4^k + 3(4^k) - 1}{3} \\ &= \frac{4^k(1 + 3) - 1}{3} \\ &= \frac{4^k(4) - 1}{3} \\ &= \frac{4^{k+1} - 1}{3}. \end{aligned}$$

Therefore, using part 1 and part 2 we have proved that $S_n = (4^n - 1)/3$ yields the correct sum for any number of terms of the sequence $a_n = 4^{n-1}$.

As our final example of this section let's consider a proof by mathematical induction involving the concept of divisibility.

EXAMPLE 5

Prove that for all positive integers n, the number $3^{2n} - 1$ is divisible by 8.

Proof

PART 1 If $n = 1$, then $3^{2n} - 1$ becomes $3^{2(1)} - 1 = 3^2 - 1 = 8$, and of course 8 is divisible by 8.

PART 2 We need to prove the statement, If $3^{2k} - 1$ is divisible by 8, then $3^{2k+2} - 1$ is divisible by 8 for all integer values of k. This can be verified as follows. If $3^{2k} - 1$ is divisible by 8, then for some integer x, we have $3^{2k} - 1 = 8x$. Therefore,

$$\begin{aligned} 3^{2k} - 1 &= 8x \\ 3^{2k} &= 1 + 8x \\ 3^2(3^{2k}) &= 3^2(1 + 8x) && \text{Multiply both sides by } 3^2. \\ 3^{2k+2} &= 9(1 + 8x) \\ 3^{2k+2} &= 9 + 9(8x) \end{aligned}$$

$$3^{2k+2} = 1 + 8 + 9(8x) \qquad 9 = 1 + 8$$
$$3^{2k+2} = 1 + 8(1 + 9x) \qquad \text{Apply distributive property to } 8 + 9(8x)$$
$$3^{2k+2} - 1 = 8(1 + 9x).$$

Therefore, $3^{2k+2} - 1$ is divisible by 8. Thus, using part 1 and part 2 we have proven that $3^{2n} - 1$ is divisible by 8 for all positive integers n.

We conclude this section with a few final comments about proof by mathematical induction. Every mathematical induction proof is a two-part proof, and both parts are absolutely necessary. There can be mathematical statements that hold for one or the other of the two parts but not for both. For example, $(a + b)^n = a^n + b^n$ is true for $n = 1$, but it is false for every positive integer greater than 1. Therefore, if we were to attempt a mathematical induction proof for $(a + b)^n = a^n + b^n$, we could establish part 1 but part 2 would break down. Another example of this type is the statement, $n^2 - n + 41$ produces a prime number for all positive integer values of n. This statement is true for $n = 1, 2, 3, 4, \ldots, 40$, but it is false when $n = 41$ (since $41^2 - 41 + 41 = 41^2$, which is not a prime number).

It is also possible that part 2 of a mathematical induction proof can be established but part 1 breaks down. For example, consider the sequence $a_n = n$ and the sum formula $S_n = (n + 3)(n - 2)/2$. If $n = 1$, then $a_1 = 1$ but $S_1 = (4)(-1)/2 = -2$, so part 1 does not hold. However, it is possible to show that $S_k = (k + 3)(k - 2)/2$ implies $S_{k+1} = (k + 4)(k - 1)/2$. We will leave the details of this for you to do.

Finally, it is important to realize that some mathematical statements are true for all positive integers greater than some fixed positive integer other than 1. (Back in Figure 13.1, perhaps we cannot knock down the first four dominoes, whereas we can knock down the fifth domino and every one thereafter.) For example, we can prove by mathematical induction that $2^n > n^2$ for all positive integers $n > 4$. It requires a slight variation in the statement of the principle of mathematical induction. We will not concern ourselves with such problems in this text, but we do want you to be aware of their existence.

PROBLEM SET 13.4

In Problems 1–10, use mathematical induction to prove each of the sum formulas for the indicated sequences. They are to hold for all positive integers n.

1. $S_n = \dfrac{n(n + 1)}{2}$ for $a_n = n$

2. $S_n = n^2$ for $a_n = 2n - 1$

3. $S_n = \dfrac{n(3n + 1)}{2}$ for $a_n = 3n - 1$

4. $S_n = \dfrac{n(5n + 9)}{2}$ for $a_n = 5n + 2$

5. $S_n = 2(2^n - 1)$ for $a_n = 2^n$

6. $S_n = \dfrac{3(3^n - 1)}{2}$ for $a_n = 3^n$

7. $S_n = \dfrac{n(n + 1)(2n + 1)}{6}$ for $a_n = n^2$

8. $S_n = \dfrac{n^2(n + 1)^2}{4}$ for $a_n = n^3$

9. $S_n = \dfrac{n}{n + 1}$ for $a_n = \dfrac{1}{n(n + 1)}$

10. $S_n = \dfrac{n(n + 1)(n + 2)}{3}$ for $a_n = n(n + 1)$

In Problems 11–20, use mathematical induction to prove that each statement is true for all positive integers n.

11. $3^n \geq 2n + 1$

12. $4^n \geq 4n$

13. $n^2 \geq n$

14. $2^n \geq n + 1$

15. $4^n - 1$ is divisible by 3

16. $5^n - 1$ is divisible by 4

17. $6^n - 1$ is divisible by 5

18. $9^n - 1$ is divisible by 4

19. $n^2 + n$ is divisible by 2

20. $n^2 - n$ is divisible by 2

CHAPTER 13 SUMMARY

There are four main topics in this chapter: arithmetic sequences, geometric sequences, problem solving, and mathematical induction.

Arithmetic Sequences

The sequence $a_1, a_2, a_3, a_4, \ldots$ is called **arithmetic** if and only if

$$a_{k+1} - a_k = d$$

for every positive integer k. In other words, there is a **common difference**, d, between successive terms.

The **general term** of an arithmetic sequence is given by the formula

$$a_n = a_1 + (n - 1)d,$$

where a_1 is the first term, n is the number of terms, and d is the common difference.

The **sum** of the first n terms of an arithmetic sequence is given by the formula

$$S_n = \frac{n(a_1 + a_n)}{2}.$$

Summation notation can be used to indicate the sum of a certain number of terms of a sequence. For example,

$$\sum_{i=1}^{5} 4^i = 4^1 + 4^2 + 4^3 + 4^4 + 4^5.$$

Geometric Sequences

The sequence $a_1, a_2, a_3, a_4, \ldots$ is called **geometric** if and only if

$$a_{k+1} = ra_k$$

for every positive integer k. There is a **common ratio**, r, between successive terms.

The **general term** of a geometric sequence is given by the formula

$$a_n = a_1 r^{n-1},$$

where a_1 is the first term, n is the number of terms, and r is the common ratio.

The **sum** of the first n terms of a geometric sequence is given by the formula

$$S_n = \frac{a_1 r^n - a_1}{r - 1}, \qquad r \neq 1.$$

The **sum of an infinite geometric sequence** is given by the formula

$$S_\infty = \frac{a_1}{1 - r}, \qquad \text{for } |r| < 1.$$

If $|r| \geq 1$, the sequence has no sum.

Repeating decimals (such as $0.\overline{4}$) can be changed to a/b form, where a and b are integers and $b \neq 0$, by treating them as the sum of an infinite geometric sequence. For example, the repeating decimal $0.\overline{4}$ can be written $0.4 + 0.04 + 0.004 + 0.0004 + \cdots$.

Problem Solving

Many of the problem solving suggestions offered earlier in this text are still appropriate when solving problems dealing with sequences. However, there are also some special suggestions pertaining to sequence problems.

1. Write down the first few terms of the sequence to describe what is taking place in the problem. A picture or diagram may help with this step.
2. Be sure that you understand, term by term, what the sequence represents in the problem.
3. Determine whether the sequence is arithmetic or geometric. (Those are the only kinds of sequences we are working with in this text.)
4. Determine whether the problem is asking for a specific term or whether it is asking for the sum of a certain number of terms.

Mathematical Induction

Proof by mathematical induction relies on the following **principle of induction**.

Let P_n be a statement in terms of n, where n is a positive integer. If

1. P_1 is true, and
2. The truth of P_k implies the truth of P_{k+1}, for every positive integer k,

then P_n is true for every positive integer n.

CHAPTER 13 REVIEW PROBLEM SET

For Problems 1–10, find the general term (the nth term) for each of the following sequences. These problems contain a mixture of arithmetic sequences and geometric sequences.

1. 3, 9, 15, 21, . . .

2. $\frac{1}{3}$, 1, 3, 9, . . .

3. 10, 20, 40, 80, . . .

4. 5, 2, −1, −4, . . .

5. −5, −3, −1, 1, . . .

6. 9, 3, 1, $\frac{1}{3}$, . . .

7. −1, 2, −4, 8, . . .

8. 12, 15, 18, 21, . . .

9. $\frac{2}{3}$, 1, $\frac{4}{3}$, $\frac{5}{3}$, . . .

10. 1, 4, 16, 64, . . .

For Problems 11–16, find the required term of each of the sequences.

11. The 19th term of 1, 5, 9, 13, . . .

12. The 28th term of −2, 2, 6, 10, . . .

13. The 9th term of 8, 4, 2, 1, . . .

14. The 8th term of $\frac{243}{32}$, $\frac{81}{16}$, $\frac{27}{8}$, $\frac{9}{4}$, . . .

15. The 34th term of 7, 4, 1, −2, . . .

16. The 10th term of −32, 16, −8, 4, . . .

17. If the 5th term of an arithmetic sequence is −19 and the 8th term is −34, find the common difference of the sequence.

18. If the 8th term of an arithmetic sequence is 37 and the 13th term is 57, find the 20th term.

19. Find the first term of a geometric sequence if the 3rd term is 5 and the 6th term is 135.

20. Find the common ratio of a geometric sequence if the 2nd term is $\frac{1}{2}$ and the 6th term is 8.

21. Find the sum of the first 9 terms of the sequence 81, 27, 9, 3, . . .

22. Find the sum of the first 70 terms of the sequence −3, 0, 3, 6, . . .

23. Find the sum of the first 75 terms of the sequence 5, 1, −3, −7, . . .

24. Find the sum of the first 10 terms of the sequence where $a_n = 2^{5-n}$.

25. Find the sum of the first 95 terms of the sequence where $a_n = 7n + 1$.

26. Find the sum $5 + 7 + 9 + \cdots + 137$.

27. Find the sum $64 + 16 + 4 + \cdots + \frac{1}{64}$.

28. Find the sum of all even numbers between 8 and 384, inclusive.

29. Find the sum of all multiples of 3 between 27 and 276, inclusive.

For Problems 30–33, find each of the indicated sums.

30. $\sum_{i=1}^{45} (-2i + 5)$

31. $\sum_{i=1}^{5} i^3$

32. $\sum_{i=1}^{8} 2^{8-i}$

33. $\sum_{i=4}^{75} (3i - 4)$

34. Find the sum of the infinite geometric sequence 64, 16, 4, 1, . . .

35. Change $0.\overline{36}$ to reduced a/b, form, where a and b are integers and $b \neq 0$.

36. Change $0.4\overline{5}$ to reduced a/b form, where a and b are integers and $b \neq 0$.

Solve each of Problems 37–40 by using your knowledge of arithmetic sequences and geometric sequences.

37. Suppose that at the beginning of a year, your savings account contains \$3750. If you withdraw \$250 per month from the account, how much will it contain at the end of the year?

38. Sonya decides to start saving dimes. She plans to save 1 dime the first day of April, 2 dimes the second day, 3 dimes the third day, 4 dimes the fourth day, and so on for the 30 days of April. How much money will she save in April?

39. Nancy decides to start saving dimes. She plans to save 1 dime the first day of April, 2 dimes the second day, 4 dimes the third day, 8 dimes the fourth day, and so on for the first 15 days of April. How much will she save in 15 days?

40. A tank contains 61,440 gallons of water. Each day one-fourth of the water is drained out. How much remains in the tank at the end of 6 days?

For Problems 41–43, show a mathematical induction proof.

41. Prove that $5^n > 5n - 1$ for all positive integer values of n.

42. Prove that $n^3 - n + 3$ is divisible by 3 for all positive integer values of n.

43. Prove that

$$S_n = \frac{n(n+3)}{4(n+1)(n+2)}$$

is the sum formula for the sequence $a_n = \frac{1}{n(n+1)(n+2)}$, where n is any positive integer.

Counting Techniques, Probability, and the Binomial Theorem

Using an ordinary deck of 52 playing cards, there is 1 chance out of 54,145 that you will be dealt four aces in a five-card hand. The weatherman is predicting a 40% chance of locally severe thunderstorms by late afternoon. The odds in favor of the Cubs winning the pennant are 2 to 3. Suppose that in a box containing 50 light bulbs, 45 are good ones and 5 are burned out. If two bulbs are chosen at random, the probability of getting at least one good bulb is $\frac{243}{245}$. Historically, many basic probability concepts have been developed as a result of studying various games of chance. However, in recent years, probability applications have been surfacing at a phenomenal rate in a large variety of fields, such as physics, biology, psychology, economics, insurance, military science, manufacturing, and politics. It is our purpose in this chapter first to introduce some counting techniques and then to use those techniques to motivate some basic concepts of probability. The last section of the chapter will be devoted to the binomial theorem.

14

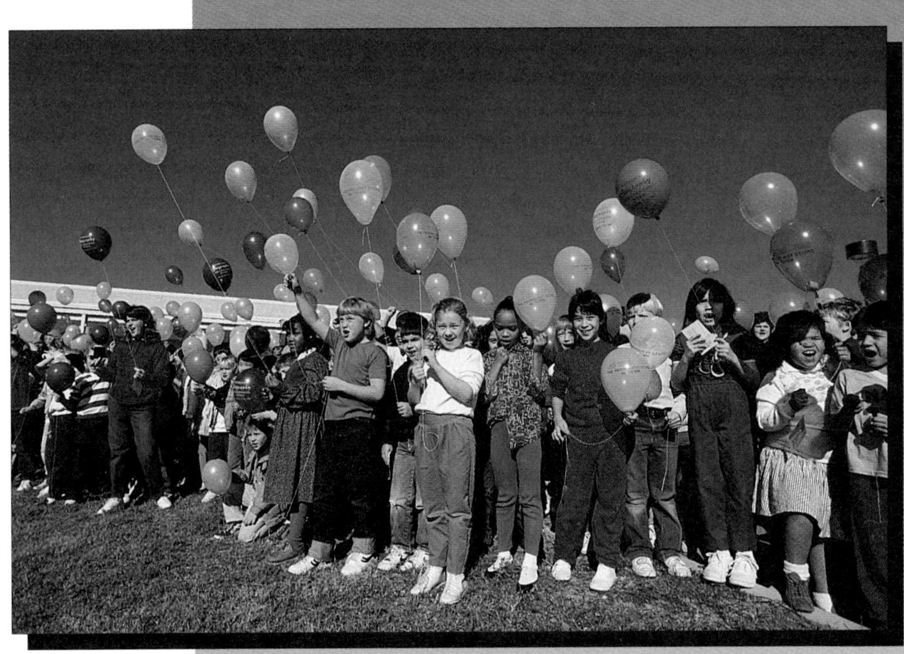

In a group of 30 people, there is approximately a 70% chance that at least two of them will have the same birthday (same month and same day of the month). In a group of 60 people, there is approximately a 99% chance that at least two of them will have the same birthday.

14.1 FUNDAMENTAL PRINCIPLE OF COUNTING

One very useful counting principle is referred to as the **fundamental principle of counting**. We will motivate this property with some examples, state the property, and then use it to solve a variety of counting-type problems. Let's consider two examples to lead up to the statement of the property.

EXAMPLE 1

A woman has 4 skirts and 5 blouses. Assuming that each blouse can be worn with each skirt, how many different skirt-blouse outfits does she have?

Solution

For *each* of the 4 skirts she has a choice of 5 blouses. Therefore, she has $4(5) = 20$ different skirt-blouse outfits from which to choose.

EXAMPLE 2

Eric is shopping for a new bicycle and has 2 different models (5-speed or 10-speed) and 4 different colors (red, white, blue, or silver) from which to choose. How many different choices does he have?

Solution

His different choices can be counted with the help of a **tree diagram** as follows.

Models	Colors	Choices
5-speed	red	5-speed red
	white	5-speed white
	blue	5-speed blue
	silver	5-speed silver
10-speed	red	10-speed red
	white	10-speed white
	blue	10-speed blue
	silver	10-speed silver

For each of the 2 model choices, there are 4 choices of color. So altogether Eric has $2 \cdot 4 = 8$ choices.

The two previous examples motivate the following general principle.

Fundamental Principle of Counting

If one task can be accomplished in x different ways and, following this task, a second task can be accomplished in y different ways, then the first task followed by the second task can be accomplished in $x \cdot y$ different ways. (This counting principle can be extended to any finite number of tasks.)

As you apply the fundamental principle of counting, it is often helpful to analyze a problem systematically in terms of the tasks to be accomplished. Let's consider some examples.

EXAMPLE 3

How many numbers of 3 different digits each can be formed by choosing from the digits 1, 2, 3, 4, 5, and 6?

Solution

Let's analyze this problem in terms of 3 tasks.

TASK 1 Choose the hundred's digit, for which there are 6 choices.

TASK 2 Now choose the ten's digit, for which there are only 5 choices, since one digit was used in the hundred's place.

TASK 3 Now choose the unit's digit, for which there are only 4 choices, since two digits have been used for the other places.

Therefore, task 1 followed by task 2 followed by task 3 can be accomplished in $(6)(5)(4) = 120$ ways. In other words, there are 120 numbers of 3 different digits that can be formed by choosing from the 6 given digits.

Now look back over the solution for Example 3 and think about each of the following questions.

1. Can we solve the problem by choosing the unit's digit first, then the ten's digit, and finally the hundred's digit?
2. How many 3-digit numbers can be formed from 1, 2, 3, 4, 5, and 6 if we do not require each number to have 3 *different* digits? (Your answer should be 216.)
3. Suppose that the digits from which to choose are 0, 1, 2, 3, 4, and 5. Now how many numbers of 3 different digits each can be formed, assuming that we do not want zero in the hundred's place? (Your answer should be 100.)
4. Suppose that we want to know the number of *even* numbers with 3 different digits each that can be formed by choosing from 1, 2, 3, 4, 5, and 6. How many are there? (Your answer should be 60.)

EXAMPLE 4

Employee ID numbers at a certain factory consist of one capital letter followed by a 3-digit number that contains no repeated digits. For example, A-014 is an ID number. How many such ID numbers can be formed? How many can be formed if repeated digits are allowed?

Solution

Again, let's analyze in terms of tasks to be completed.

TASK 1 Choose the letter part of the ID number: there are 26 choices.

TASK 2 Choose the first digit of the 3-digit number: there are 10 choices.

TASK 3 Choose the second digit: there are 9 choices.

TASK 4 Choose the third digit: there are 8 choices.

Therefore, applying the fundamental principle, we obtain $(26)(10)(9)(8) = 18{,}720$ possible ID numbers.

If repeat digits were allowed, then there would be $(26)(10)(10)(10) = 26{,}000$ possible ID numbers.

EXAMPLE 5

How many ways can Al, Barb, Chad, Dan, and Edna be seated in a row of 5 seats so that Al and Barb are seated side by side?

Solution

This problem can be analyzed in terms of three tasks.

TASK 1 Choose the 2 adjacent seats to be occupied by Al and Barb. An illustration such as Figure 14.1 helps us to see that there are 4 choices for the two adjacent seats.

FIGURE 14.1

TASK 2 Determine the number of ways that Al and Barb can be seated. Since Al can be seated on the left and Barb on the right, or vice versa, there are 2 ways to seat Al and Barb for each pair of adjacent seats.

TASK 3 The remaining 3 people must be seated in the remaining 3 seats. This can be done in $(3)(2)(1) = 6$ different ways.

Therefore, by the fundamental principle, task 1 followed by task 2 followed by task 3 can be done in $(4)(2)(6) = 48$ ways.

In Example 5, suppose instead that we wanted the number of ways the five people can sit so that Al and Barb are *not* side by side. We can determine this number by using two basically different techniques: (1) analyze and count the number of nonadjacent positions for Al and Barb, or (2) subtract the number of seating arrangements determined in Example 5 from the total number of ways that 5 people can be seated in 5 seats. Try doing this problem both ways and see if you agree with the answer of 72 ways.

As you apply the fundamental principle of counting, you may find that for certain problems, simply thinking about an appropriate tree diagram is helpful, even though the size of the problem may make it inappropriate to write out the diagram in detail. Consider the following example.

EXAMPLE 6

Suppose that the undergraduate students in three departments–geography, history, and psychology–are to be classified according to sex and year in school. How many categories are needed?

Solution

Let's represent the various classifications symbolically as follows.

M:	Male	1.	Freshman	G:	Geography
F:	Female	2.	Sophomore	H:	History
		3.	Junior	P:	Psychology
		4.	Senior		

We can mentally picture a tree diagram such that each of the two sex classifications branches into four school-year classifications, which in turn branch into three department classifications. Thus, we have $(2)(4)(3) = 24$ different categories.

Another technique that works on certain problems involves what some people call the *back door* approach. For example, suppose we know that the classroom contains 50 seats. On some days, it may be easier to determine the number of students present by counting the number of empty seats and subtracting from 50 than by counting the number of students in attendance. (We suggested this back door approach as one way to count the nonadjacent seating arrangements in the discussion following Example 5.) The next example further illustrates this approach.

EXAMPLE 7

When rolling a pair of dice, in how many ways can we obtain a sum greater than 4?

Solution

For clarification purposes, let's use a red die and a white die. (It is not necessary to use different colored dice, but it does help us analyze the different possible outcomes.) With a moment of thought you will see that there are more ways to get a sum greater than 4 than there are ways to get a sum of 4 or less. Therefore, let's determine the number of possibilities for getting a sum of 4 or less; then we'll subtract that number from the total number of possible outcomes when rolling a pair of dice.

First, we can simply list and count the ways of getting a sum of 4 or less.

Red die	White die
1	1
1	2
1	3
2	1
2	2
3	1

There are 6 ways of getting a sum of 4 or less.

Second, since there are 6 possible outcomes on the red die and 6 possible outcomes on the white die, there is a total of $(6)(6) = 36$ possible outcomes when rolling a pair of dice.

Therefore, subtracting the number of ways of getting 4 or less from the total number of possible outcomes, we obtain $36 - 6 = 30$ ways of getting a sum greater than 4.

PROBLEM SET 14.1

1. If a woman has two skirts and ten blouses, how many different skirt-blouse combinations does she have?

2. If a man has eight shirts, five pairs of slacks, and three pairs of shoes, how many different shirt-slack-shoe combinations does he have?

3. In how many ways can four people be seated in a row of four seats?

4. How many numbers of two different digits can be formed by choosing from the digits 1, 2, 3, 4, 5, 6, and 7?

5. How many *even* numbers of three different digits can be formed by choosing from the digits 2, 3, 4, 5, 6, 7, 8, and 9?

6. How many *odd* numbers of four different digits can be formed by choosing from the digits 1, 2, 3, 4, 5, 6, 7, and 8?

7. Suppose that the students at a certain university are to be classified according to their college (College of Applied Science, College of Arts and Sciences, College of Business, College of Education, College of Fine Arts, College of Health and Physical Education), sex (female, male), and year in school (1, 2, 3, 4). How many categories are possible?

8. A medical researcher classifies subjects according to sex (female, male), smoking habits (smoker, nonsmoker), and weight (below average, average, above average). How many different combined classifications are used?

9. A pollster classifies voters according to sex (female, male), party affiliation (Democrat, Republican, Independent), and family income (below \$10,000, \$10,000–\$19,999, \$20,000–\$29,999, \$30,000–\$39,999, \$40,000–\$49,999, \$50,000 and above). How many combined classifications does the pollster use?

10. A couple is planning to have four children. How many ways can this happen relative to a boy-girl classification? (For example, *BBBG* indicates that the first three children are boys and the last is a girl.)

11. In how many ways can three officers—president, secretary, and treasurer—be selected from a club that has twenty members?

12. In how many ways can three officers—president, secretary, and treasurer—be selected from a club with 15 female and 10 male members, so that the president is female and the secretary and treasurer are males?

13. A disc jockey wants to play six songs once each in a half-hour program. How many different ways can he order these songs?

14. A state has agreed to have its automobile license plates consist of two letters followed by four digits. They do not want to repeat any letters or digits in any license numbers. How many different license plates will be available?

15. In how many ways can six people be seated in a row of six seats?

16. In how many ways can Al, Bob, Carl, Don, Ed, and Fern be seated in a row of six seats if Al and Bob want to sit side by side?

17. In how many ways can Amy, Bob, Cindy, Dan, and Elmer be seated in a row of five seats so that neither Amy nor Bob occupies an end seat?

18. In how many ways can Al, Bob, Carl, Don, Ed, and Fern be seated in a row of six seats if Al and Bob are not to be seated side by side? [*Hint:* Either Al and Bob will be seated side by side or they will not be seated side by side.]

19. In how many ways can Al, Bob, Carol, Dawn, and Ed be seated in a row of five chairs if Al is to be seated in the middle chair?

20. In how many ways can three letters be dropped in five mailboxes?

21. In how many ways can five letters be dropped in three mailboxes?

22. In how many ways can four letters be dropped in six mailboxes so that no two letters go in the same box?

23. In how many ways can six letters be dropped in four mailboxes so that no two letters go in the same box?

24. If five coins are tossed, in how many ways can they fall?

25. If three dice are tossed, in how many ways can they fall?

26. In how many ways can a sum less than ten be obtained when tossing a pair of dice?

27. In how many ways can a sum greater than five be obtained when tossing a pair of dice?

28. In how many ways can a sum greater than four be obtained when tossing three dice?

29. If no number contains repeated digits, how many numbers greater than 400 can be formed by choosing from the digits 2, 3, 4, and 5? [*Hint:* Consider both three-digit and four-digit numbers.]

30. If no number contains repeated digits, how many numbers greater than 5000 can be formed by choosing from the digits 1, 2, 3, 4, 5, and 6?

31. In how many ways can four boys and three girls be seated in a row of seven seats so that boys and girls occupy alternate seats?

32. In how many ways can three different mathematics books and four different history books be exhibited on a shelf so that all of the books in a subject area are side-by-side?

33. In how many ways can a true-false test of ten questions be answered?

34. If no number contains repeated digits, how many even numbers greater than 3000 can be formed by choosing from the digits 1, 2, 3, and 4?

35. If no number contains repeated digits, how many odd numbers greater than 40,000 can be formed by choosing from the digits 1, 2, 3, 4, and 5?

36. In how many ways can Al, Bob, Carol, Don, Ed, Faye, and George be seated in a row of seven seats so that Al, Bob, and Carol occupy consecutive seats in some order?

37. The license plates for a certain state consist of two letters followed by a four-digit number such that the first digit of the number is not zero. An example would be PK-2446.

a. How many different license plates can be produced?

b. How many different plates do not have a repeated letter?

c. How many plates do not have any repeated digits in the number part of the plate?

d. How many plates do not have a repeated letter and also do not have any repeated digits?

14.2 PERMUTATIONS AND COMBINATIONS

As we develop the material in this section, **factorial notation** becomes very useful. The notation $n!$ (read *n factorial*) is used with positive integers as follows.

$$1! = 1$$

$$2! = 2 \cdot 1 = 2$$

$$3! = 3 \cdot 2 \cdot 1 = 6$$

$$4! = 4 \cdot 3 \cdot 2 \cdot 1 = 24$$

Notice that the factorial notation refers to an *indicated product*. In general, we write

$$n! = n(n-1)(n-2)\cdots 3 \cdot 2 \cdot 1.$$

We also define $0! = 1$ so that certain formulas will be true for all nonnegative integers.

Now, as an introduction to the first concept of this section, let's consider a counting problem that closely resembles problems from the previous section.

EXAMPLE 1

In how many ways can the three letters, A, B, and C, be arranged in a row?

Solution A

Certainly one approach to the problem is simply to list and count the arrangements.

ABC, ACB, BAC, BCA, CAB, CBA

There are six arrangements of the three letters.

Solution B

Another approach, one that can be generalized for more difficult problems, uses the fundamental principle of counting. Since there are three choices for the first letter of an arrangement, two choices for the second letter, and one choice for the third letter, there are $(3)(2)(1) = 6$ different arrangements.

Ordered arrangements are called **permutations**. In general, a permutation of a set of n elements is an ordered arrangement of the n elements; we will use the symbol $P(n, n)$ to denote the number of such permutations. For example, from Example 1 we know that $P(3, 3) = 6$. Furthermore, by using the same basic approach as in Solution B of Example 1, we can obtain the following results.

$$P(1, 1) = 1 = 1!$$

$$P(2, 2) = 2 \cdot 1 = 2!$$

$$P(4, 4) = 4 \cdot 3 \cdot 2 \cdot 1 = 4!$$

$$P(5, 5) = 5 \cdot 4 \cdot 3 \cdot 2 \cdot 1 = 5!$$

In general, the following formula becomes evident.

$$P(n, n) = n!$$

Now suppose that we are interested in the number of two-letter permutations that can be formed by choosing from the four letters A, B, C, and D. (Some examples of such permutations are AB, BA, AC, CA, BC, and CB.) In other words, we want to find the number of two-element permutations that can be formed from a set of four elements. We denote this number by $P(4, 2)$. To find $P(4, 2)$, we can reason as follows: First, we can choose any one of the four letters to occupy the first position in the permutation, and then we can choose any one of the three remaining letters for the second position. Therefore, by the fundamental principle of counting, we have $(4)(3) = 12$ different 2-letter permutations; that is, $P(4, 2) = 12$. By using a similar line of reasoning, the following numbers can be determined. (Make sure that you agree with each of these.)

$$P(4, 3) = 4 \cdot 3 \cdot 2 = 24$$

$$P(5, 2) = 5 \cdot 4 = 20$$

$$P(6, 4) = 6 \cdot 5 \cdot 4 \cdot 3 = 360$$

$$P(7, 3) = 7 \cdot 6 \cdot 5 = 210$$

In general, we say that **the number of r-element permutations that can be formed from a set of n elements is given by**

$$P(n, r) = \underbrace{n(n-1)(n-2) \cdots}_{r \text{ factors}} .$$

Notice that the indicated product for $P(n, \text{r})$ begins with n. Thereafter each factor is one less than the previous one and there is a total of r factors. For example,

$$P(6, 2) = 6 \cdot 5 = 30;$$

$$P(8, 3) = 8 \cdot 7 \cdot 6 = 336;$$

$$P(9, 4) = 9 \cdot 8 \cdot 7 \cdot 6 = 3024.$$

Let's consider two examples illustrating the use of $P(n, n)$ and $P(n, r)$.

EXAMPLE 2

In how many ways can five students be seated in a row of five seats?

Solution

The problem is asking for the number of five-element permutations that can be formed from a set of five elements. Thus, we can apply $P(n, n) = n!$.

$$P(5, 5) = 5! = 5 \cdot 4 \cdot 3 \cdot 2 \cdot 1 = 120$$

EXAMPLE 3

Suppose that seven people enter a swimming race. In how many ways can first, second, and third prizes be awarded?

Solution

This problem is asking for the number of three-element permutations that can be formed from a set of seven elements. Therefore, using the formula for $P(n, r)$, we obtain

$$P(7, 3) = 7 \cdot 6 \cdot 5 = 210.$$

It should be evident that both Example 2 and Example 3 could have been solved by applying the fundamental principle of counting. In fact, it should be noted that the formulas for $P(n, n)$ and $P(n, r)$ do not really give us much additional problem solving power. However, as we will see in a moment, they do provide the basis for developing a formula that is very useful as a problem solving tool.

Permutations Involving Nondistinguishable Objects

Suppose we have two identical Hs and one T in an arrangement such as HTH. If we switch the two identical Hs, the newly formed arrangement, HTH, will not be distinguishable from the original. In other words, there are fewer distinguishable permutations of n elements when some of those elements are identical than there are when the n elements are distinctly different.

To see the effect of identical elements on the number of distinguishable permutations, let's look at some specific examples.

2 identical Hs	1 permutation (HH)
2 different letters	2! permutations (HT, TH)

Therefore, having two different letters affects the number of permutations by a *factor of* 2!.

3 identical Hs	1 permutation (HHH)
3 different letters	3! permutations

Therefore, having three different letters affects the number of permutations by a *factor of* 3!.

4 identical Hs	1 permutation (HHHH)
4 different letters	4! permutations

Therefore, having 4 different letters affects the number of permutations by a *factor of* 4!.

Now let's solve a specific problem.

EXAMPLE 4

How many distinguishable permutations can be formed from three identical Hs and two identical Ts?

Solution

If we had five distinctly different letters, we could form 5! permutations. But the three identical Hs affect the number of distinguishable permutations by a factor of 3!, and the two identical Ts affect the number of permutations by a factor of 2!. Therefore, we must divide 5! by 3! and 2!. Thus, we obtain

$$\frac{5!}{(3!)(2!)} = \frac{5 \cdot \overset{2}{\cancel{4}} \cdot \cancel{3} \cdot \cancel{2} \cdot 1}{\cancel{3} \cdot \cancel{2} \cdot 1 \cdot \cancel{2} \cdot 1} = 10$$

distinguishable permutations of three Hs and two Ts.

The type of reasoning used in Example 4 leads us to the following general counting technique: If there are n elements to be arranged, where there are r_1 of one kind, r_2 of another kind, r_3 of another kind, . . . , r_k of a kth kind, then the total number of distinguishable permutations is given by the expression

$$\frac{n!}{(r_1!)(r_2!)(r_3!) \cdots (r_k!)}.$$

EXAMPLE 5

How many different 11-letter permutations can be formed from the 11 letters of the word MISSISSIPPI?

Solution

Since there are 4 Is, 4 Ss, and 2 Ps we can form

$$\frac{11!}{(4!)(4!)(2!)} = \frac{11 \cdot 10 \cdot 9 \cdot 8 \cdot 7 \cdot 6 \cdot 5 \cdot 4 \cdot 3 \cdot 2 \cdot 1}{4 \cdot 3 \cdot 2 \cdot 1 \cdot 4 \cdot 3 \cdot 2 \cdot 1 \cdot 2 \cdot 1} = 34{,}650$$

distinguishable permutations.

Combinations (Subsets)

Permutations are *ordered* arrangements; however, frequently *order* is not a consideration. For example, suppose that we want to determine the number of 3-person committees that can be formed from the five people: Al, Barb, Carol, Dawn, and Eric. Certainly the committee consisting of Al, Barb, and Eric is the same as the committee consisting of Barb, Eric, and Al. In other words, the order in which we choose or list the members is not important. Therefore, we are really dealing with subsets; that is, we are looking for the number of three-element subsets that can be formed from a set of five elements. Traditionally in this context, subsets have been called **combinations**. So, stated another way, we are looking for the number of combinations of five things taken three at a time. In general, r-element subsets taken from a set of n elements are called **combinations of n things taken r at a time**. The symbol $C(n, r)$ denotes the number of these combinations.

Now let's restate the previous committee problem and show a detailed solution that can be generalized to handle a variety of problems dealing with combinations.

EXAMPLE 6

How many three-person committees can be formed from the five people: Al, Barb, Carol, Dawn, and Eric?

Solution

Let's use the set $\{A, B, C, D, E\}$ to represent the five people. Consider one possible three-person committee (subset), such as $\{A, B, C\}$; there are 3! permutations of these three letters. Now take another committee, such as $\{A, B, D\}$; there are also 3! permutations of these three letters. If we were to continue this process with all of the three-letter subsets that can be formed from the five letters, we would be counting all possible three-letter permutations of the five letters. That is, we would obtain $P(5, 3)$. Therefore, if we let $C(5, 3)$ represent the number of three-element subsets, then

$$(3!) \cdot C(5, 3) = P(5, 3).$$

Solving this equation for $C(5, 3)$ yields

$$C(5, 3) = \frac{P(5, 3)}{3!} = \frac{5 \cdot 4 \cdot 3}{3 \cdot 2 \cdot 1} = 10.$$

So there are *ten* three-person committees that can be formed from the five people.

In general, $C(n, r)$ times $r!$ yields $P(n, r)$. Thus,

$$(r!) \cdot C(n, r) = P(n, r),$$

and solving this equation for $C(n, r)$ produces

$$C(n, r) = \frac{P(n, r)}{r!}.$$

In other words, we can find the number of *combinations* of n things taken r at a time by dividing by $r!$ the number of permutations of n things taken r at a time. The following examples illustrate this idea.

$$C(7, 3) = \frac{P(7, 3)}{3!} = \frac{7 \cdot 6 \cdot 5}{3 \cdot 2 \cdot 1} = 35$$

$$C(9, 2) = \frac{P(9, 2)}{2!} = \frac{9 \cdot 8}{2 \cdot 1} = 36$$

$$C(10, 4) = \frac{P(10, 4)}{4!} = \frac{10 \cdot 9 \cdot 8 \cdot 7}{4 \cdot 3 \cdot 2 \cdot 1} = 210$$

EXAMPLE 7

How many different five-card hands can be dealt from a deck of 52 playing cards?

Solution

Since the order in which the cards are dealt is not an issue, we are working with a combination (subset) problem. Thus, using the formula for $C(n, r)$, we obtain

$$C(52, 5) = \frac{P(52, 5)}{5!} = \frac{52 \cdot 51 \cdot 50 \cdot 49 \cdot 48}{5 \cdot 4 \cdot 3 \cdot 2 \cdot 1} = 2{,}598{,}960.$$

There are 2,598,960 different five-card hands that can be dealt from a deck of 52 playing cards.

Some counting problems can be solved by using the fundamental principle of counting along with the combination formula, as in the next example.

EXAMPLE 8

How many committees that consist of three women and two men can be formed from a group of five women and four men?

Solution

Let's think of this problem in terms of two tasks.

TASK 1 Choose a subset of three women from the five women. This can be done in

$$C(5, 3) = \frac{P(5, 3)}{3!} = \frac{5 \cdot 4 \cdot 3}{3 \cdot 2 \cdot 1} = 10 \text{ ways.}$$

TASK 2 Choose a subset of two men from the four men. This can be done in

$$C(4, 2) = \frac{P(4, 2)}{2!} = \frac{4 \cdot 3}{2 \cdot 1} = 6 \text{ ways.}$$

Task 1 followed by task 2 can be done in $(10)(6) = 60$ ways. Therefore, there are 60 committees consisting of three women and two men that can be formed.

Sometimes it takes a little thought to decide whether permutations or combinations should be used. Remember that **if order is to be considered, permutations should be used, but if order does not matter, then use combinations.** It is helpful to think of combinations as subsets.

EXAMPLE 9

A small accounting firm has 12 computer programmers. Three of these people are to be promoted to systems analysts. In how many ways can the firm select the three people to be promoted?

Solution

Let's call the people A, B, C, D, E, F, G, H, I, J, K, and L. Suppose A, B, and C are chosen for promotion. Is this any different than choosing B, C, and A? Obviously not, and therefore order does not matter and we are being asked a question about combinations. More specifically, we need to find the number of combinations of 12 people taken three at a time. Thus, there are

$$C(12, 3) = \frac{P(12, 3)}{3!} = \frac{12 \cdot 11 \cdot 10}{3 \cdot 2 \cdot 1} = 220$$

different ways to choose the three people to be promoted.

EXAMPLE 10

A club is to elect three officers–president, secretary, and treasurer–from a group of six people, all of whom are willing to serve in any office. How many different ways can the officers be chosen?

Solution

Let's call the candidates A, B, C, D, E, and F. Is electing A as president, B as secretary, and C as treasurer different than electing B as president, C as secretary, and A as treasurer? Obviously it is, and therefore we are working with permutations. Thus, there are

$$P(6, 3) = 6 \cdot 5 \cdot 4 = 120$$

different ways of filling the offices.

PROBLEM SET 14.2

In Problems 1–12, evaluate each of the following.

1. $P(5, 3)$ **2.** $P(8, 2)$ **3.** $P(6, 4)$

4. $P(9, 3)$ **5.** $C(7, 2)$ **6.** $C(8, 5)$

7. $C(10, 5)$ **8.** $C(12, 4)$ **9.** $C(15, 2)$

10. $P(5, 5)$ **11.** $C(5, 5)$ **12.** $C(11, 1)$

13. How many permutations of the four letters A, B, C, and D can be formed by using all the letters in each permutation?

14. In how many ways can six students be seated in a row of six seats?

15. How many three-person committees can be formed from a group of nine people?

16. How many two-card hands can be dealt from a deck of 52 playing cards?

17. How many three-letter permutations can be formed from the first eight letters of the alphabet if (a) repetitions are not allowed? (b) repetitions are allowed?

18. In a seven-team baseball league, in how many ways can the top three positions in the final standings be filled?

19. In how many ways can the manager of a baseball team arrange his batting order of nine starters if he wants his four best hitters in the top four positions?

20. In a baseball league of nine teams, how many games are needed to complete the schedule if each team plays 12 games with each other team?

21. How many committees consisting of four women and four men can be chosen from a group of seven women and eight men?

22. How many three-element subsets containing one vowel and two consonants can be formed from the set {a, b, c, d, e, f, g , h, i}?

23. Five associate professors are being considered for promotion to the rank of full professor, but only three will be promoted. How many different ways are there of selecting the three to be promoted?

24. How many numbers of four different digits can be formed from the digits 1, 2, 3, 4, 5, 6, 7, 8, and 9, if each number must consist of two odd and two even digits?

25. How many three-element subsets containing the letter A can be formed from the set {A, B, C, D, E, F}?

26. How many four-person committees can be chosen from five women and three men if each committee must contain at least one man?

27. How many different seven-letter permutations can be formed from four identical Hs and three identical Ts?

28. How many different eight-letter permutations can be formed from six identical Hs and two identical Ts?

29. How many different nine-letter permutations can be formed from three identical As, four identical Bs, and two identical Cs?

30. How many different ten-letter permutations can be formed from five identical As, four identical Bs, and one C?

31. How many different seven-letter permutations can be formed from the seven letters of the word ALGEBRA?

32. How many different 11-letter permutations can be formed from the 11 letters of the word MATHEMATICS?

33. In how many ways can x^4y^2 be written without using exponents? [*Hint:* One way is $xxxxyy$.]

34. In how many ways can $x^3y^4z^3$ be written without using exponents?

35. Ten basketball players are going to divide into two teams of five players each for a game. In how many ways can this be done?

36. Ten basketball players are going to divide into two teams of five in such a way that the two best players are on opposite teams. In how many ways can this be done?

37. A box contains nine good light bulbs and four defective bulbs. How many samples of three bulbs contain one defective bulb? How many samples of three bulbs contain *at least* one defective bulb?

38. How many five-person committees consisting of two juniors and three seniors can be formed from a group of six juniors and eight seniors?

39. In how many ways can six people be divided into two groups so that there are four in one group and two in the other? In how many ways can six people be divided into two groups of three each?

40. How many five-element subsets containing both A and B can be formed from the set {A, B, C, D, E, F, G, H}?

41. How many four-element subsets containing A or B but not both A and B can be formed from the set {A, B, C, D, E, F, G}?

42. How many different five-person committees can be selected from nine people if two of those people refuse to serve together on a committee?

43. How many different line segments are determined by five points? By six points? By seven points? By n points?

44. **a.** How many five-card hands consisting of two kings and three aces can be dealt from a deck of 52 playing cards?

    b. How many five-card hands consisting of three kings and two aces can be dealt from a deck of 52 playing cards?

    c. How many five-card hands consisting of three cards of one face value and two cards of another face value can be dealt from a deck of 52 playing cards?

MISCELLANEOUS PROBLEMS

45. In how many ways can six people be seated at a circular table? [*Hint:* Moving each person one place to the right (or left) does not create a new seating arrangement.]

46. The quantity $P(8, 3)$ can be expressed completely in factorial notation as follows

$$P(8, 3) = \frac{P(8, 3) \cdot 5!}{5!} = \frac{(8 \cdot 7 \cdot 6)(5 \cdot 4 \cdot 3 \cdot 2 \cdot 1)}{5!} = \frac{8!}{5!}$$

Express each of the following in terms of factorial notation.

a. $P(7, 3)$ **b.** $P(9, 2)$ **c.** $P(10, 7)$

d. $P(n, r)$, $r \leq n$ and $0!$ is defined to be 1

47. Sometimes the formula

$$C(n, r) = \frac{n!}{r!(n - r)!}$$

is used to find the number of combinations of n things taken r at a time. Use the result from Problem 46(d) and develop this formula.

48. Compute $C(7, 3)$ and $C(7, 4)$. Compute $C(8, 2)$ and $C(8, 6)$. Compute $C(9, 8)$ and $C(9, 1)$. Now argue that $C(n, r) = C(n, n - r)$ for $r \leq n$.

GRAPHICS CALCULATOR ACTIVITIES

Before doing Problems 49–54, be sure that you can use your calculator to compute the number of permutations and combinations. Your calculator may possess a special sequence of keys for such computations. You may need to refer to your calculator manual for this information.

49. Use your calculator to check your answers for Problems 1–12.

50. How many different five-card hands can be dealt from a deck of 52 playing cards?

51. How many different seven-card hands can be dealt from a deck of 52 playing cards?

52. How many different five-person committees can be formed from a group of 50 people?

53. How many different juries consisting of 11 people can be chosen from a group of 30 people?

54. How many seven-person committees consisting of three juniors and four seniors can be formed from 45 juniors and 53 seniors?

14.3 PROBABILITY

In order to introduce some terminology and notation, let's consider a simple experiment of tossing a regular six-sided die. There are six possible outcomes to this experiment: either the 1, the 2, the 3, the 4, the 5, or the 6 will land up. This set of possible outcomes is called a *sample space* and the individual elements of the sample space are called *sample points*. We will use S (sometimes with subscripts for identification purposes) to refer to a particular sample space of an experiment; then we will denote the number of sample points by $n(S)$. Thus, for the experiment of tossing a die, $S = \{1, 2, 3, 4, 5, 6\}$ and $n(S) = 6$.

In general, the set of all possible outcomes of a given experiment is called the **sample space** and the individual elements of the sample space are called **sample points**. (In this text we will be working only with sample spaces that are finite.)

Now suppose we are interested in some of the various possible outcomes in the die tossing experiment. For example, we might be interested in the event, *An even number comes up*. In this case we are satisfied if a 2, 4, or 6 appears on the top face of the die, and therefore the event *An even number comes up* is the subset $E = \{2, 4, 6\}$, where $n(E) = 3$. Perhaps, instead, we might be interested in the event, *A multiple of* 3 *comes up*. This event determines the subset $F = \{3, 6\}$, where $n(F) = 2$.

In general, any subset of a sample space is called an **event** or an **event space**. If the event consists of exactly one element of the sample space, then it is called a **simple event**. Any nonempty event that is not simple is called a **compound event**. A compound event can be represented as the union of simple events.

It is now possible to give a very simple definition for *probability* as we want to use it in this text.

DEFINITION 14.1

In an experiment where all possible outcomes in the sample space S are equally likely to occur, the **probability** of an event E is defined by

$$P(E) = \frac{n(E)}{n(S)},$$

where $n(E)$ denotes the number of elements in the event E and $n(S)$ denotes the number of elements in the sample space S.

Many probability problems can be solved by applying Definition 14.1. Such an approach requires that we be able to determine the number of elements in the sample space and the number of elements in the event space. For example, returning to the die tossing experiment, the probability of getting an even number with one toss of a die is given by

$$P(E) = \frac{n(E)}{n(S)} = \frac{3}{6} = \frac{1}{2}.$$

Let's consider two examples where the number of elements in both the sample space and the event space are quite easy to determine.

EXAMPLE 1

A coin is tossed. Find the probability that a head turns up.

Solution

Let the sample space be $S = \{H, T\}$; then $n(S) = 2$. The event of turning up a head is the subset $E = \{H\}$, so $n(E) = 1$. Therefore, the probability of getting a head with one flip of a coin is given by

$$P(E) = \frac{n(E)}{n(S)} = \frac{1}{2}.$$

EXAMPLE 2

Two coins are tossed. What is the probability that *at least* one head will turn up?

Solution

For clarification purposes, let the coins be a penny and a nickel. The possible outcomes of this experiment are (1) a head on both coins, (2) a head on the penny and a tail on the nickel, (3) a tail on the penny and a head on the nickel, or (4) a tail on both coins. Using ordered-pair notation, where the first entry of a pair represents the penny and the second entry the nickel, the sample space can be written

$$S = \{(H, H), (H, T), (T, H), (T, T)\}$$

and $n(S) = 4$.

Let E be the event of getting at least one head. Thus, $E = \{(H, H), (H, T), (T, H)\}$ and $n(E) = 3$. Therefore, the probability of getting at least one head with one toss of two coins is

$$P(E) = \frac{n(E)}{n(S)} = \frac{3}{4}.$$

As you might expect, the counting techniques discussed in the first two sections of this chapter can frequently be used to solve probability problems.

EXAMPLE 3

Four coins are tossed. Find the probability of getting three heads and one tail.

Solution

The sample space consists of the possible outcomes for tossing four coins. Since there are two things that can happen on each coin, by the fundamental principle of counting there are $2 \cdot 2 \cdot 2 \cdot 2 = 16$ possible outcomes for tossing four coins. So we know that $n(S) = 16$ without taking the time to list all of the elements. The event of getting three heads and one tail is the subset $E = \{(H, H, H, T), (H, H, T, H), (H, T, H, H), (T, H, H, H)\}$, where $n(E) = 4$. Therefore, the requested probability is

$$P(E) = \frac{n(E)}{n(S)} = \frac{4}{16} = \frac{1}{4}.$$

EXAMPLE 4

Al, Bob, Chad, Dawn, Eve, and Francis are randomly seated in a row of six chairs. What is the probability that Al and Bob are seated in the end seats?

Solution

The sample space consists of all possible ways of seating six people in six chairs, or in other words, the permutations of six things taken six at a time. Thus, $n(S) = P(6, 6) = 6! = 6 \cdot 5 \cdot 4 \cdot 3 \cdot 2 \cdot 1 = 720$.

The event space consists of all possible ways of seating the six people so that Al and Bob both occupy end seats. The number of these possibilities can be counted as follows.

TASK 1 Put Al and Bob in the end seats. This can be done in two ways since Al can be on the left end and Bob on the right end, or vice versa.

TASK 2 Put the other four people in the remaining four seats. This can be done in $4! = 4 \cdot 3 \cdot 2 \cdot 1 = 24$ different ways.

Therefore, task 1 followed by task 2 can be done in $(2)(24) = 48$ different ways; so $n(E) = 48$. Thus, the requested probability is

$$P(E) = \frac{n(E)}{n(S)} = \frac{48}{720} = \frac{1}{15}.$$

Notice that in Example 3, by using the fundamental principle of counting to determine the number of elements in the sample space, we did not actually have to list all of the elements. For the event space, we listed the elements and counted them in the usual way. In Example 4 we used the permutation formula $P(n, n) = n!$ to determine the number of elements in the sample space, and then we used the fundamental principle to determine the number of elements in the event space. There are no definite rules about when to list the elements and when to apply some sort of counting technique. In general, we suggest that if you do not immediately see a counting pattern for a particular problem, you should begin the listing process. If a counting pattern then emerges as you are listing the elements, use the pattern at that time.

The combination (subset) formula we developed in Section 14.2, $C(n, r) = P(n, r)/r!$, is also a very useful tool for solving certain kinds of probability problems. The next three examples illustrate some problems of this type.

EXAMPLE 5

A committee of three people is randomly selected from Alice, Barb, Chad, Dee, and Eric. What is the probability that Alice is on the committee?

Solution

The sample space, S, consists of all possible three-person committees that can be formed from the five people. Therefore,

$$n(S) = C(5, 3) = \frac{P(5, 3)}{3!} = \frac{5 \cdot 4 \cdot 3}{3 \cdot 2 \cdot 1} = 10.$$

The event space, E, consists of all the three-person committees that have Alice as a member. Each of these committees contains Alice and two other people chosen from the four remaining people. Thus, the number of such committees is $C(4, 2)$. So we obtain

$$n(E) = C(4, 2) = \frac{P(4, 2)}{2!} = \frac{4 \cdot 3}{2 \cdot 1} = 6.$$

The requested probability is

$$P(E) = \frac{n(E)}{n(S)} = \frac{6}{10} = \frac{3}{5}.$$

EXAMPLE 6

A committee of four is chosen at random from a group of five seniors and four juniors. Find the probability that the committee will contain two seniors and two juniors.

Solution

The sample space, S, consists of all possible four-person committees that can be formed from the nine people. Thus,

$$n(S) = C(9,4) = \frac{P(9, 4)}{4!} = \frac{9 \cdot 8 \cdot 7 \cdot 6}{4 \cdot 3 \cdot 2 \cdot 1} = 126.$$

The event space, E, consists of all four-person committees that contain two seniors and two juniors. They can be counted as follows.

TASK 1 Choose two seniors from the five available seniors in $C(5, 2) =$ ten ways.

TASK 2 Choose two juniors from the four available juniors in $C(4, 2) =$ six ways.

Therefore, there are $10 \cdot 6 = 60$ committees consisting of two seniors and two juniors. The requested probability is

$$P(E) = \frac{n(E)}{n(S)} = \frac{60}{126} = \frac{10}{21}.$$

EXAMPLE 7

Eight coins are tossed. Find the probability of getting two heads and six tails.

Solution

Since two things can happen on each coin, the total number of possible outcomes, $n(S)$, is $2^8 = 256$.

We can select two coins, which are to fall heads, in $C(8, 2) = 28$ ways. For each of these ways, there is only one way to select the other six coins that are to fall tails. Therefore, there are $28 \cdot 1 = 28$ ways of getting two heads and six tails; so $n(E) = 28$. The requested probability is

$$P(E) = \frac{n(E)}{n(S)} = \frac{28}{256} = \frac{7}{64}.$$

PROBLEM SET 14.3

For Problems 1–4, *two* coins are tossed. Find the probability of tossing each of the following events.

1. One head and one tail **2.** Two tails

3. At least one tail **4.** No tails

For Problems 5–8, *three* coins are tossed. Find the probability of tossing each of the following events.

5. Three heads **6.** Two heads and a tail

7. At least one head **8.** Exactly one tail

For Problems 9–12, *four* coins are tossed. Find the probability of tossing each of the following events.

9. Four heads **10.** Three heads and a tail

11. Two heads and two tails **12.** At least one head

For Problems 13–16, *one* die is tossed. Find the probability of rolling each of the following events.

13. A multiple of 3 **14.** A prime number

15. An even number **16.** A multiple of 7

For Problems 17–22, *two* dice are tossed. Find the probabiity of rolling each of the following events.

17. A sum of 6 **18.** A sum of 11

19. A sum less than 5 **20.** A 5 on exactly one die

21. A 4 on at least one die **22.** A sum greater than 4

For Problems 23–26, *one* card is drawn from a standard deck of 52 playing cards. Find the probability of each of the following events.

23. A heart is drawn.

24. A king is drawn.

25. A spade or a diamond is drawn.

26. A red jack is drawn.

For Problems 27–30, suppose that 25 slips of paper numbered 1 to 25, inclusive, are put in a hat and then one is drawn out at random. Find the probability of each of the following events.

27. The slip with the 5 on it is drawn.

28. A slip with an even number on it is drawn.

29. A slip with a prime number on it is drawn.

30. A slip with a multiple of 6 on it is drawn.

For Problems 31–34, suppose that a committee of two boys is to be chosen at random from the five boys, Al, Bill, Carl, Dan, and Elmer. Find the probability of each of the following events.

31. Dan is on the committee.

32. Dan and Elmer are both on the committee.

33. Bill and Carl are not both on the committee.

34. Dan or Elmer but not both of them are on the committee.

For Problems 35–38, suppose that a five-person committee is selected at random from the eight people, Al, Barb, Chad, Dawn, Eric, Fern, George, and Harriet. Find the probability of each of the following events.

35. Al and Barb are both on the committee.

36. George is not on the committee.

37. Either Chad or Dawn, but not both, is on the committee.

38. Neither Al nor Barb is on the committee.

For Problems 39–41, suppose that a box of 10 items from a manufacturing company is known to contain two defective and eight nondefective items. If a sample of three items is selected at random, find the probability of each of the following events.

39. The sample contains all nondefective items.

40. The sample contains one defective and two nondefective items.

41. The sample contains two defective and one nondefective item.

Solve the following problems.

42. A building has five doors. Find the probability that two people, entering the building at random, will choose the same door.

43. Bill, Carol, and Alice are seated at random in a row of three seats. Find the probability that Bill and Carol will be seated side by side.

44. April, Bill, Carl, and Denise are to be seated at random in a row of four chairs. What is the probability that April and Bill will occupy the end seats?

45. A committee of four girls is to be chosen at random from the five girls, Alice, Becky, Candy, Dee, and Elaine. Find the probability that Elaine is not on the committee.

46. Three boys and two girls are randomly seated in a row of five seats. What is the probability that the boys and girls will be in alternate seats?

47. Four different mathematics books and five different history books are randomly placed on a shelf. What is the probability that all of the books on a subject are side by side?

48. Each of three letters is to be mailed in any one of five different mailboxes. What is the probability that all are mailed in the same mailbox?

49. Randomly form a four-digit number by using the digits 2, 3, 4, and 6 once each. What is the probability that the number formed is greater than 4000?

50. Randomly select one of the 120 permutations of the letters a, b, c, d, and e. Find the probability that in the chosen permutation, the letter a precedes the b (the a is to the left of the b).

51. A committee of four is chosen at random from a group of six women and five men. Find the probability that the committee contains two women and two men.

52. A committee of three is chosen at random from a group of four women and five men. Find the probability that the committee will contain at least one man.

53. Al, Bob, Carl, Dan, Ed, Frank, Gino, Harry, Jerry, and Mike are randomly divided into two five-man teams for a basketball game. What is the probability that Al, Bob, and Carl are on the same team?

54. Seven coins are tossed. Find the probability of getting four heads and three tails.

55. Nine coins are tossed. Find the probability of getting three heads and six tails.

56. Six coins are tossed. Find the probability of getting at least four heads.

57. Five coins are tossed. Find the probability of getting no more than three heads.

58. Each arrangement of the 11 letters of the word MISSISSIPPI is put on a slip of paper and placed in a hat. One slip is drawn at random from the hat. Find the probability that the slip contains an arrangement of the letters with the four Ss at the beginning.

59. Each arrangement of the seven letters of the word OSMOSIS is put on a slip of paper and placed in a hat. One slip is drawn at random from the hat. Find the probability that the slip contains an arrangement of the letters with an O at the beginning and an O at the end.

60. Consider all possible arrangements of three identical Hs and three identical Ts. Suppose that one of these arrangements is selected at random. What is the probability that the selected arrangement has the three Hs in consecutive positions?

MISCELLANEOUS PROBLEMS

In Example 7 of Section 14.2, we found that there are 2,598,960 different five-card hands that can be dealt from a hand of 52 playing cards. Therefore, probabilities for certain kinds of five-card poker hands can be calculated using 2,598,960 as the number of elements in the sample space. For Problems 61–69, determine the number of different five-card poker hands of the indicated type that can be obtained.

61. A straight flush (five cards in sequence and of the same suit; aces are both low and high, so A2345 and 10JQKA are both acceptable)

62. Four of a kind (four of the same face value, such as four kings)

63. A full house (three cards of one face value and two cards of another face value)

64. A flush (five cards of the same suit but not in sequence)

65. A straight (five cards in sequence but not all of the same suit)

66. Three of a kind (three cards of one face value and two cards of two different face values)

67. Two pairs

68. Exactly one pair

69. No pairs

14.4 SOME PROPERTIES OF PROBABILITY; EXPECTED VALUES

There are several basic properties that are useful in the study of probability from both a theoretical and a computational viewpoint. We will discuss two of these properties at this time and some additional ones in the next section. The first property may seem to state the obvious, but it still needs to be mentioned.

PROPERTY 14.1

For all events E,

$$0 \leq P(E) \leq 1.$$

Property 14.1 simply states that probabilities must fall in the range from 0 to 1, inclusive. This should seem reasonable since $P(E) = n(E)/n(S)$, and E is a subset of S. The next two examples illustrate circumstances where $P(E) = 0$ and $P(E) = 1$.

EXAMPLE 1

Toss a regular six-sided die. What is the probability of getting a 7?

Solution

The sample space is $S = \{1, 2, 3, 4, 5, 6\}$, where $n(S) = 6$. The event space is $E = \varnothing$, so $n(E) = 0$. Therefore, the probability of getting a 7 is

$$P(E) = \frac{n(E)}{n(S)} = \frac{0}{6} = 0.$$

EXAMPLE 2

What is the probability of getting a head or a tail with one flip of a coin?

Solution

The sample space is $S = \{H, T\}$ and the event space is $E = \{H, T\}$. Therefore, $n(S) = n(E) = 2$ and

$$P(E) = \frac{n(E)}{n(S)} = \frac{2}{2} = 1.$$

An event that has a probability of 1 is sometimes called **certain success**, and an event with a probability of zero is called **certain failure**.

It should also be mentioned that Property 14.1 serves as a check for reasonableness of answers. In other words, when computing probabilities, we know that our answer must fall between 0 and 1, inclusive. Any other probability answer is simply not reasonable.

Complementary Events

Complementary events are complementary sets such that S, the sample space, serves as the universal set. The following examples illustrate this idea.

Sample space	Event space	Complement of event space
$S = \{1, 2, 3, 4, 5, 6\}$	$E = \{1, 2\}$	$E' = \{3, 4, 5, 6\}$
$S = \{H, T\}$	$E = \{T\}$	$E' = \{H\}$
$S = \{2, 3, 4, \ldots, 12\}$	$E = \{2, 3, 4\}$	$E' = \{5, 6, 7, \ldots, 12\}$
$S = \{1, 2, 3, \ldots, 25\}$	$E = \{3, 4, 5, \ldots, 25\}$	$E' = \{1, 2\}$

In each case, note that E' (the complement of E) consists of all elements of S that are *not* in E. Thus, E and E' are called *complementary events*. Also note that for each example, $P(E) + P(E') = 1$. We can state the following general property.

PROPERTY 14.2

If E is any event of a sample space S, and E' is the complementary event, then

$$P(E) + P(E') = 1.$$

From a computational viewpoint, Property 14.2 provides us with a double-barreled attack to some probability problems. That is to say, once we compute either $P(E)$ or $P(E')$, then the other one is determined simply by subtracting from 1. For example, suppose that for a particular problem we can determine that $P(E) = \frac{3}{13}$. Then we immediately know that $P(E') = 1 - P(E) = 1 - \frac{3}{13} = \frac{10}{13}$. The following examples further illustrate the usefulness of Property 14.2.

EXAMPLE 3

Two dice are tossed. Find the probability of getting a sum greater than 3.

Solution

Let S be the familiar sample space of ordered pairs for this problem, where $n(S) = 36$. Let E be the event of obtaining a sum greater than 3. Then E' is the event of obtaining a sum less than or equal to 3, that is, $E' = \{(1, 1), (1, 2), (2, 1)\}$. Thus,

$$P(E') = \frac{n(E')}{n(S)} = \frac{3}{36} = \frac{1}{12}.$$

From this we conclude that

$$P(E) = 1 - P(E') = 1 - \frac{1}{12} = \frac{11}{12}.$$

EXAMPLE 4

Toss three coins and find the probability of getting at least one head.

Solution

The sample space, S, consists of all possible outcomes for tossing three coins. Using the fundamental principle of counting, we know that there are $(2)(2)(2) = 8$ outcomes, so $n(S) = 8$. Let E be the event of getting at least one head. Then E' is the complementary event of not getting any heads. The set E' is easy to list, namely, $E' = \{(T, T, T)\}$. Thus, $n(E') = 1$ and $P(E') = \frac{1}{8}$. From this, $P(E)$ can be determined to be

$$P(E) = 1 - P(E') = 1 - \frac{1}{8} = \frac{7}{8}.$$

EXAMPLE 5

A three-person committee is chosen at random from a group of five women and four men. Find the probability that the committee contains at least one woman.

Solution

Let the sample space, S, be the set of all possible three-person committees that can be formed from nine people. There are $C(9, 3) = 84$ such committees; therefore, $n(S) = 84$.

Let E be the event, *The committee contains at least one woman*. Then E' is the complementary event, *The committee contains all men*. Thus, E' consists of all three-man committees that can be formed from four men. There are $C(4, 3) = 4$ such committees; thus $n(E') = 4$. So we have

$$P(E') = \frac{n(E')}{n(S)} = \frac{4}{84} = \frac{1}{21},$$

which determines $P(E)$ to be

$$P(E) = 1 - P(E') = 1 - \frac{1}{21} = \frac{20}{21}.$$

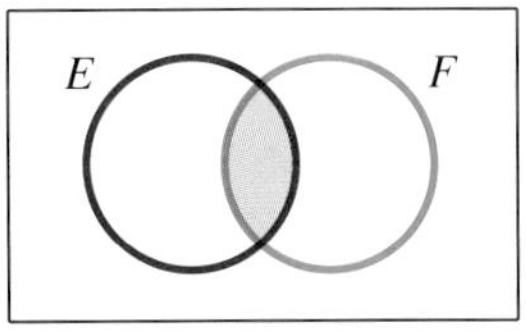

E F

$E \cap F$

FIGURE 14.2

The concepts of **set intersection** and **set union** play an important role in the study of probability. If E and F are two events in a sample space S,then $E \cap F$ is the event consisting of all sample points of S that are in both E and F as indicated in Figure 14.2. Likewise, $E \cup F$ is the event consisting of all sample points of S that are in E or F, or both, as shown in Figure 14.3.

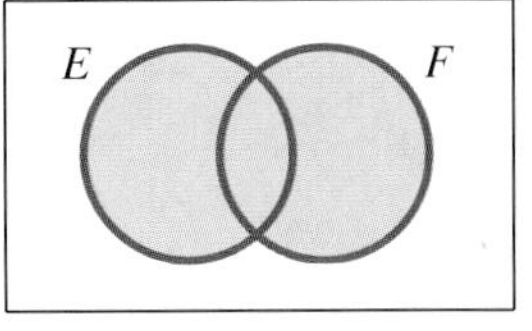

FIGURE 14.3

In Figure 14.4 there are 47 sample points in E, 38 sample points in F, and 15 sample points in $E \cap F$. How many sample points are there in $E \cup F$? Simply

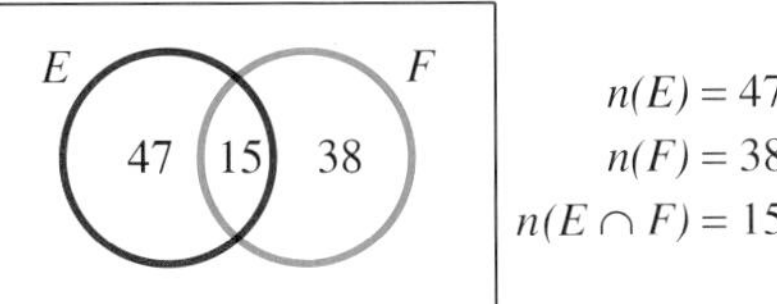

FIGURE 14.4

adding the number of points in E and F would result in counting the 15 points in $E \cap F$ twice. Therefore, 15 must be subtracted from the total number of points in E and F, yielding $47 + 38 - 15 = 70$ points in $E \cup F$. We can state the following general counting property:

$$n(E \cup F) = n(E) + n(F) - n(E \cap F).$$

If we divide both sides of this equation by $n(S)$, we obtain the following probability property.

PROPERTY 14.3

For events E and F of a sample space S,

$$P(E \cup F) = P(E) + P(F) - P(E \cap F).$$

EXAMPLE 6

What is the probability of getting an odd number or a prime number with one toss of a die?

Solution

Let $S = \{1, 2, 3, 4, 5, 6\}$ be the sample space, $E = \{1, 3, 5\}$ the event of getting an odd number, and $F = \{2, 3, 5\}$ the event of getting a prime number. Then $E \cap F = \{3, 5\}$ and, using Property 14.3, we obtain

$$P(E \cup F) = \frac{3}{6} + \frac{3}{6} - \frac{2}{6} = \frac{4}{6} = \frac{2}{3}.$$

EXAMPLE 7

Toss three coins. What is the probability of getting at least two heads or exactly one tail?

Solution

Using the fundamental principle of counting, we know that there are $2 \cdot 2 \cdot 2 = 8$ possible outcomes of tossing three coins; thus $n(S) = 8$. Let

$$E = \{(H, H, H), (H, H, T), (H, T, H), (T, H, H)\}$$

be the event of getting at least two heads and let

$$F = \{(H, H, T), (H, T, H), (T, H, H)\}$$

be the event of getting exactly one tail. Then

$$E \cap F = \{(H, H, T), (H, T, H), (T, H, H)\}$$

and we can compute $P(E \cup F)$ as follows.

$$\begin{aligned} P(E \cup F) &= P(E) + P(F) - P(E \cap F) \\ &= \frac{4}{8} + \frac{3}{8} - \frac{3}{8} \\ &= \frac{4}{8} = \frac{1}{2}. \end{aligned}$$

In Property 14.3, if $E \cap F = \varnothing$, then the events E and F are said to be **mutually exclusive**. In other words, mutually exclusive events are events that cannot occur at the same time. For example, when rolling a die, the event of getting a 4 is mutually exclusive of the event of getting a 5; they cannot both happen on the same roll. If $E \cap F = \varnothing$, then $P(E \cap F) = 0$ and Property 14.3 becomes $P(E \cup F) = P(E) + P(F)$ **for mutually exclusive events.**

EXAMPLE 8

Suppose we have a jar that contains five white, seven green, and nine red marbles. If one marble is drawn at random from the jar, find the probability that it is white or green.

Solution

The events of drawing a white marble and drawing a green marble are mutually exclusive. Therefore, the probability of drawing a white or green marble is

$$\frac{5}{21} + \frac{7}{21} = \frac{12}{21} = \frac{4}{7}.$$

Note in the solution for Example 8 that we did not explicitly name and list the elements of the sample space or event spaces. It was obvious that the sample space contained 21 elements (21 marbles in the jar) and the event spaces contained five elements (five white marbles) and seven elements (seven green marbles). Thus, it was unnecessary to name and list the sample space and event spaces.

EXAMPLE 9

Suppose that the data in the following table represent a survey of 1000 drivers after a holiday weekend.

	RAIN (R)	NO RAIN (R')	TOTAL
ACCIDENT (A)	35	10	45
NO ACCIDENT (A')	450	505	955
TOTAL	485	515	1000

If a person is selected at random, what is the probability that the person was in an accident or that it rained?

Solution

First, let's form a **probability table** by dividing each entry by 1000, the total number surveyed.

	RAIN (R)	NO RAIN (R')	TOTAL
ACCIDENT (A)	0.035	0.010	0.045
NO ACCIDENT (A')	0.450	0.505	0.955
TOTAL	0.485	0.515	1.000

Now we can use Property 14.3 and compute $P(A \cup R)$.

$$\begin{aligned} P(A \cup R) &= P(A) + P(R) - P(A \cap R) \\ &= 0.045 + 0.485 - 0.035 \\ &= 0.495 \end{aligned}$$

Expected Value

Suppose we toss a coin 500 times. We would expect to get approximately 250 heads. In other words, since the probability of getting a head with one toss of a coin is $\frac{1}{2}$, then in 500 tosses we should get approximately $500\left(\frac{1}{2}\right) = 250$ heads. The word *approximately* conveys a key idea. As we know from experience, it is possible to toss a coin several times and get all heads. However, with a large number of tosses, things should average out, so that we get about an equal number of heads and tails.

As another example, consider the fact that the probability of getting a sum of 6 with one toss of a pair of dice is $\frac{5}{36}$. Therefore, if a pair of dice is tossed 360 times, we should expect to get a sum of 6 approximately $360\left(\frac{5}{36}\right) = 50$ times.

Let us now define the concept of *expected value*.

DEFINITION 14.2

If the k possible outcomes of an experiment are assigned the values $x_1, x_2, x_3, \ldots, x_k$ and occur with probabilities of $p_1, p_2, p_3, \ldots, p_k$, respectively, then the **expected value** of the experiment (E_v) is given by

$$E_v = x_1p_1 + x_2p_2 + x_3p_3 + \cdots + x_kp_k.$$

The concept of expected value (also called **mathematical expectation**) is used in a variety of probability situations that deal with such things as fairness of games and decision making in business ventures. Let's consider some examples.

EXAMPLE 10

Suppose that you buy one ticket in a lottery where 1000 tickets are sold. Furthermore, suppose that three prizes are awarded: one of \$500, one of \$300, and one of \$100. What is your mathematical expectation?

Solution

Since you bought one ticket, your probability of winning \$500 is $\frac{1}{1000}$, that of winning \$300 is $\frac{1}{1000}$, and that of winning \$100 is $\frac{1}{1000}$. Multiplying each of these probabilities times the corresponding prize money and then adding the results yields your mathematical expectation.

$$E_v = \$500\left(\frac{1}{1000}\right) + \$300\left(\frac{1}{1000}\right) + \$100\left(\frac{1}{1000}\right)$$
$$= \$.50 + \$.30 + \$.10$$
$$= \$.90$$

In Example 10, if you pay more than \$.90 for a ticket, then it is not a **fair game** from your standpoint. If the price of the game is included in the calculation of the expected value, then a fair game is defined to be one where the expected value is zero.

EXAMPLE 11

A player pays \$5 to play a game where the probability of winning is $\frac{1}{5}$ and the probability of losing is $\frac{4}{5}$. If the player wins the game, he receives \$25. Is this a fair game for the player?

Solution

Using Definition 14.2, let $x_1 = \$20$, which represents the \$25 won minus the \$5 paid to play, and let $x_2 = -\$5$, the amount paid to play the game. We are also given that $p_1 = \frac{1}{5}$ and $p_2 = \frac{4}{5}$. Thus, the expected value is

$$E_v = \$20\left(\frac{1}{5}\right) + (-\$5)\left(\frac{4}{5}\right)$$
$$= \$4 - \$4$$
$$= 0.$$

Since the expected value is zero, it is a fair game.

EXAMPLE 12

Suppose you are interested in insuring a diamond ring for \$2000 against theft. An insurance company charges a premium of \$25 per year, claiming that there is a probability of 0.01 that the ring will be stolen during the year. What is your expected gain or loss if you take out the insurance?

Solution

Using Definition 14.2, let $x_1 = \$1975$, which represents the \$2000 minus the cost of the premium, \$25, and let $x_2 = -\$25$. We also are given that $p_1 = 0.1$ and so $p_2 = 1 - 0.01 = 0.99$. Thus, the expected value is

$$E_v = \$1975(0.01) + (-\$25)(0.99)$$
$$= \$19.75 - \$24.75$$
$$= -\$5.00.$$

This means that if you insure with this company over many years and the circumstances remain the same, you will have an average net loss of \$5 per year.

PROBLEM SET 14.4

For Problems 1–4, *two* dice are tossed. Find the probability of rolling each of the following events.

1. A sum of 6

2. A sum greater than 2

3. A sum less than 8

4. A sum greater than 1

For Problems 5–8, *three* dice are tossed. Find the probability of rolling each of the following events.

5. A sum of 3

6. A sum greater than 4

7. A sum less than 17

8. A sum greater than 18

For Problems 9–12, *four* coins are tossed. Find the probability of getting each of the following events.

9. Four heads

10. Three heads and a tail

11. At least one tail

12. At least one head

For Problems 13–16, *five* coins are tossed. Find the probability of each of the following events.

13. Five tails

14. Four heads and a tail

15. At least one tail

16. At least two heads

Solve the following problems.

17. Toss a pair of dice. What is the probability of not getting a double?

18. The probability that a certain horse will win the Kentucky Derby is $\frac{1}{20}$. What is the probability that it will lose the race?

19. One card is randomly drawn from a deck of 52 playing cards. What is the probability that it is not an ace?

20. Six coins are tossed. Find the probability of getting at least two heads.

21. A subset of two letters is chosen at random from the set {a, b, c, d, e, f, g, h, i}. Find the probability that the subset contains at least one vowel.

22. A two-person committee is chosen at random from a group of four men and three women. Find the probability that the committee contains at least one man.

23. A three-person committee is chosen at random from a group of seven women and five men. Find the probability that the committee contains at least one man.

For Problems 24–27, one die is tossed. Find the probability of rolling each of the following events.

24. A 3 or an odd number

25. A 2 or an odd number

26. An even number or a prime number

27. An odd number or a multiple of 3

For Problems 28–31, two dice are tossed. Find the probability of rolling each of the following events.

28. A double or a sum of 6

29. A sum of 10 or a sum greater than 8

30. A sum of 5 or a sum greater than 10

31. A double or a sum of 7

Solve the following problems.

32. Two coins are tossed. Find the probability of getting exactly one head or at least one tail.

33. Three coins are tossed. Find the probability of getting at least two heads or exactly one tail.

34. A jar contains 7 white, 6 blue, and 10 red marbles. If one marble is drawn at random from the jar, find the probability that (a) the marble is white or blue; (b) the marble is white or red; (c) the marble is blue or red.

35. A coin and a die are tossed. Find the probability of getting a head on the coin or a 2 on the die.

36. A card is randomly drawn from a deck of 52 playing cards. Find the probability that it is a red card or a face card. (Jacks, queens, and kings are the face cards.)

37. The data in the following table represent a survey of 1000 drivers after a holiday weekend.

	RAIN (R)	NO RAIN (R')	TOTAL
ACCIDENT (A)	45	15	60
NO ACCIDENT (A')	350	590	940
TOTAL	395	605	1000

If a person is selected at random from those surveyed, find the probability of each of the following events. (Express the probabilities in decimal form.)

a. The person was in an accident or it rained.

b. The person was not in an accident or it rained.

c. The person was not in an accident or it did not rain.

38. One hundred people were surveyed and one question pertained to their educational background. The results of this question are given in the following table.

	FEMALE (F)	MALE (F')	TOTAL
COLLEGE DEGREE (D)	30	20	50
NO COLLEGE DEGREE (D')	15	35	50
TOTAL	45	55	100

If a person is selected at random from those surveyed, find the probability of each of the following events. Express the probabilities in decimal form.

a. The person is female or has a college degree.

b. The person is male or does not have a college degree.

c. The person is female or does not have a college degree.

39. In a recent election there were 1000 eligible voters. They were asked to vote on two issues, A and B. The results were as follows: 300 people voted for A, 400 people voted for B, and 175 people voted for both A and B. If one person is chosen at random from the 1000 eligible voters, find the probability that the person voted for A or B.

40. A company has 500 employees of which 200 are females, 15 are high-level executives, and 7 of the high-level executives are females. If one of the 500 employees is chosen at random, find the probability that the person chosen is female or a high-level executive.

41. A die is tossed 360 times. How many times would you expect to get a 6?

42. Two dice are tossed 360 times. How many times would you expect to get a sum of 5?

43. Two dice are tossed 720 times. How many times would you expect to get a sum greater than 9?

44. Four coins are tossed 80 times. How many times would you expect to get one head and three tails?

45. Four coins are tossed 144 times. How many times would you expect to get four tails?

46. Two dice are tossed 300 times. How many times would you expect to get a double?

47. Three coins are tossed 448 times. How many times would you expect to get three heads?

48. Suppose 5000 tickets are sold in a lottery. There are three prizes: the first is \$1000, the second is \$500, and the third is \$100. What is the mathematical expectation of winning?

49. Your friend challenges you with the following game: You are to roll a pair of dice and he will give you \$5 if you roll a sum of 2 or 12, \$2 if you roll a sum of 3 or 11, \$1 if you roll a sum of 4 or 10; otherwise, you are to pay him \$1. Should you play the game?

50. A contractor bids on a building project. There is a probability of 0.8 that he can show a profit of \$30,000 and a probability of 0.2 that he will have to absorb a loss of \$10,000. What is his mathematical expectation?

51. Suppose a person tosses 2 coins and receives \$5 if 2 heads come up, \$2 if 1 head and 1 tail come up, and has to pay \$2 if 2 tails come up. Is it a fair game for him?

52. A "wheel of fortune" is divided into four colors: red, white, blue, and yellow. The probability of the spinner landing on each of the colors and the money received is given by the following chart. The price to spin the wheel is \$1.50. Is it a fair game?

COLOR	PROBABILITY OF LANDING ON THE COLOR	MONEY RECEIVED FOR LANDING ON THE COLOR
red	$\frac{4}{10}$	\$.50
white	$\frac{3}{10}$	1.00
blue	$\frac{2}{10}$	2.00
yellow	$\frac{1}{10}$	5.00

53. A contractor estimates a probability of 0.7 of making \$20,000 on a building project and a probability of 0.3 of losing \$10,000 on the project. What is his mathematical expectation?

54. A farmer estimates his corn crop at 30,000 bushels. Based on past experience, he also estimates a probability of $\frac{3}{5}$ that he will make a profit of \$.50 per bushel and a probability of $\frac{1}{5}$ of losing \$.30 per bushel. What is his expected income from the corn crop?

55. Bill finds that the annual premium for insuring a stereo system for \$2500 against theft is \$75. If the probability that the set will be stolen during the year is 0.02, what is Bill's expected gain or loss by taking out the insurance?

56. Sandra finds that the annual premium for a \$2000 insurance policy against the theft of a painting is \$100. If the probability that the painting will be stolen during the year is 0.01, what is Sandra's expected gain or loss by taking out the insurance?

MISCELLANEOUS PROBLEMS

The term **odds** is sometimes used to express a probability statement. For example, we might say, The odds in favor of the Cubs winning the pennant are 5 to 1, or, The odds against the Mets winning the pennant are 50 to 1. *Odds in favor* and *odds against* for equally likely outcomes can be defined as follows.

$$\text{odds in favor} = \frac{\text{number of favorable outcomes}}{\text{number of unfavorable outcomes}},$$

$$\text{odds against} = \frac{\text{number of unfavorable outcomes}}{\text{number of favorable outcomes}}$$

We have used the fractional form to define odds; however, in practice, the *to* vocabulary is commonly used. Thus, the odds in favor of rolling a 4 with one roll of a die are usually stated as *1 to 5* instead of $\frac{1}{5}$. The odds against rolling a 4 are stated as *5 to 1*.

The *odds in favor of* statement against the Cubs means that there are 5 favorable outcomes compared to 1 unfavorable, or a total of 6 possible outcomes. So the *5 to 1 in favor of* statement also means that the probability of the Cubs winning the pennant is $\frac{5}{6}$. Likewise, the *50 to 1 against* statement about the Mets means that the probability that the Mets will not win the pennant is $\frac{50}{51}$.

Odds are usually stated in reduced form. For example, odds of 6 to 4 are usually stated as 3 to 2. Likewise, a fraction representing probability is reduced before changing to a statement about odds.

57. What are the odds in favor of getting three heads with a toss of three coins?

58. What are the odds against getting four tails with a toss of four coins?

59. What are the odds against getting three heads and two tails with a toss of five coins?

60. What are the odds in favor of getting four heads and two tails with a toss of six coins?

61. What are the odds in favor of getting a sum of 5 with one toss of a pair of dice?

62. What are the odds against getting a sum greater than 5 with one toss of a pair of dice?

63. Suppose that one card is drawn at random from a deck of 52 playing cards. Find the odds against drawing a red card.

64. Suppose that one card is drawn at random from a deck of 52 playing cards. Find the odds in favor of drawing an ace or a king.

65. If $P(E) = \frac{4}{7}$ for some event E, find the odds in favor of E happening.

66. If $P(E) = \frac{5}{9}$ for some event E, find the odds against E happening.

67. Suppose that there is a predicted 40% chance of freezing rain. State the prediction in terms of the odds against getting freezing rain.

68. Suppose that there is a predicted 20% chance of thunderstorms. State the prediction in terms of the odds in favor of getting thunderstorms.

69. If the odds against an event happening are 5 to 2, find the probability that the event will occur.

70. The odds against Belly Dancer winning the fifth race are 20 to 9. What is the probability of Belly Dancer winning the fifth race?

71. The odds in favor of the Mets winning the pennant are stated as 7 to 5. What is the probability of the Mets winning the pennant?

72. The following chart contains some poker-hand probabilities. Complete the column *Odds against being dealt this hand*. Notice that fractions are reduced before being changed to odds.

5-CARD HAND	PROBABILITY OF BEING DEALT THIS HAND	ODDS AGAINST BEING DEALT THIS HAND
straight flush	$\frac{40}{2,598,960} = \frac{1}{64,974}$	64,973 to 1
four of a kind	$\frac{624}{2,598,960} =$	
full house	$\frac{3744}{2,598,960} =$	
flush	$\frac{5108}{2,598,960} =$	
straight	$\frac{10,200}{2,598,960} =$	
three of a kind	$\frac{54,912}{2,598,960} =$	
two pairs	$\frac{123,552}{2,598,960} =$	
one pair	$\frac{1,098,240}{2,598,960} =$	
no pairs	$\frac{1,302,540}{2,598,960} =$	

14.5 Conditional Probability; Dependent and Independent Events

Two events are often related in a way that the probability of one of them may vary depending upon whether or not the other event has occurred. For example, the probability of rain may change drastically if additional information is obtained indicating a front moving through the area. Mathematically, the additional information about the front changes the sample space for the probability of rain.

In general, the probability of the occurrence of an event E, given the occurrence of another event F, is called a **conditional probability** and is denoted $P(E|F)$. Let's look at a simple example and use it to motivate a definition for conditional probability.

What is the probability of rolling a prime number in one roll of a die? Let $S = \{1, 2, 3, 4, 5, 6\}$, so $n(S) = 6$; and let $E = \{2, 3, 5\}$, so $n(E) = 3$. Therefore,

$$P(E) = \frac{n(E)}{n(S)} = \frac{3}{6} = \frac{1}{2}.$$

Next, what is the probability of rolling a prime number in one roll of a die, *given that an odd number has turned up*? Let $F = \{1, 3, 5\}$ be the new sample space of odd numbers. Then $n(F) = 3$. We are now interested in only that part of E (rolling a prime number) that is also in F, in other words, $E \cap F$. Therefore, since $E \cap F = \{3, 5\}$, the probability of E given F is

$$P(E|F) = \frac{n(E \cap F)}{n(F)} = \frac{2}{3}.$$

If the numerator and denominator of $n(E \cap F)/n(F)$ are both divided by $n(S)$, we obtain

$$\frac{\dfrac{n(E \cap F)}{n(S)}}{\dfrac{n(F)}{n(S)}} = \frac{P(E \cap F)}{P(F)}.$$

Therefore, we can state the following general definition of the conditional probability of E given F for arbitrary events E and F.

DEFINITION 14.3 *Conditional Probability*

$$P(E|F) = \frac{P(E \cap F)}{P(F)}, \qquad P(F) \neq 0$$

In an example of the previous section, the following probability table was formed relative to car accidents and weather conditions on a holiday weekend.

	RAIN (R)	NO RAIN (R')	TOTAL
ACCIDENT (A)	0.035	0.010	0.045
NO ACCIDENT (A')	0.450	0.505	0.955
TOTAL	0.485	0.515	1.000

Some conditional probabilities that can be calculated from the table are as follows.

$$P(A|R) = \frac{P(A \cap R)}{P(R)} = \frac{0.035}{0.485} = \frac{35}{485} = \frac{7}{97}$$

$$P(A'|R) = \frac{P(A' \cap R)}{P(R)} = \frac{0.450}{0.485} = \frac{450}{485} = \frac{90}{97}$$

$$P(A|R') = \frac{P(A \cap R')}{P(R')} = \frac{0.010}{0.515} = \frac{10}{515} = \frac{2}{103}$$

Note that the probability of an accident given that it was raining ($P(A|R)$) is greater than the probability of an accident given that it was not raining ($P(A|R')$). This seems reasonable.

EXAMPLE 1

A die is tossed. Find the probability that a 4 came up if it is known that an even number turned up.

Solution

Let E be the event of rolling a 4 and let F be the event of rolling an even number. Therefore, $E = \{4\}$ and $F = \{2, 4, 6\}$, from which we obtain $E \cap F = \{4\}$. Using Definition 14.3, we obtain

$$P(E|F) = \frac{P(E \cap F)}{P(F)} = \frac{\frac{1}{6}}{\frac{3}{6}} = \frac{1}{3}.$$

EXAMPLE 2

Suppose the probability that a student will enroll in a mathematics course is 0.45, the probabilty that he (she) will enroll in a science course is 0.38, and the probability that he (she) will enroll in both courses is 0.26. Find the probability that a student will enroll in a mathematics course, given that he (she) is also enrolled in a science course. Also, find the probability that a student will enroll in a science course given that he (she) is enrolled in mathematics.

Solution

Let M be the event *Will enroll in mathematics* and S be the event *Will enroll in science*. Therefore, using Definition 14.3, we obtain

$$P(M|S) = \frac{P(M \cap S)}{P(S)} = \frac{0.26}{0.38} = \frac{26}{38} = \frac{13}{19}$$

and

$$P(S|M) = \frac{P(S \cap M)}{P(M)} = \frac{0.26}{0.45} = \frac{26}{45}.$$

Independent and Dependent Events

Suppose that, when computing a conditional probability, we find that

$$P(E|F) = P(E).$$

This means that the probability of E is not affected by the occurrence or nonoccurrence of F. In such a situation we say that event E is *independent* of event F. It can be shown that if event E is independent of event F, then F is also independent of E; thus, E and F are referred to as **independent events**. Furthermore, from the equations

$$P(E|F) = \frac{P(E \cap F)}{P(F)} \text{ and } P(E|F) = P(E),$$

we see that

$$\frac{P(E \cap F)}{P(F)} = P(E),$$

which can be written

$$P(E \cap F) = P(E)P(F).$$

Therefore, we state the following general definition.

DEFINITION 14.4

Two events E and F are said to be **independent** if and only if

$$P(E \cap F) = P(E)P(F).$$

Two events that are not independent are called **dependent events**.

In the probability table preceding Example 1, we see that $P(A) = 0.045$, $P(R) = 0.485$, and $P(A \cap R) = 0.035$. Since

$$P(A)P(R) = (0.045)(0.485) = 0.021825$$

and this does not equal $P(A \cap R)$, the events A (have a car accident) and R (rainy conditions) are not independent. This is not too surprising, since we would certainly expect rainy conditions and automobile accidents to be related.

EXAMPLE 3

Suppose we roll a white die and a red die. If we let E be the event *We roll a 4 on the white die* and F the event *We roll a 6 on the red die*, are E and F independent events?

Solution

The sample space for rolling a pair of dice has $(6)(6) = 36$ elements. Using ordered-pair notation, where the first entry represents the white die and the second entry the red die, we can list events E and F as follows.

$$E = \{(4, 1), (4, 2), (4, 3), (4, 4), (4, 5), (4, 6)\}$$

$$F = \{(1, 6), (2, 6), (3, 6), (4, 6), (5, 6), (6, 6)\}$$

Therefore, $E \cap F = \{4, 6\}$. Since $P(F) = \frac{1}{6}$, $P(E) = \frac{1}{6}$, and $P(E \cap F) = \frac{1}{36}$, we see that $P(E \cap F) = P(E)P(F)$, and the events E and F are independent.

EXAMPLE 4

Two coins are tossed. Let E be the event *Toss not more than one head* and F the event *Toss at least one of each face*. Are these independent events?

Solution

The sample space has $(2)(2) = 4$ elements. The events E and F can be listed as follows.

$$E = \{(H, T), (T, H), (T, T)\}$$

$$F = \{(H, T), (T, H)\}$$

Therefore, $E \cap F = \{(H, T), (T, H)\}$. Since $P(E) = \frac{3}{4}$, $P(F) = \frac{1}{2}$, and $P(E \cap F) = \frac{1}{2}$, we see that $P(E \cap F) \neq P(E)P(F)$, so the events E and F are dependent.

Sometimes the independence issue can be decided by the physical nature of the events in the problem. For instance, in Example 3 it should seem evident that rolling a 4 on the white die is not affected by rolling a 6 on the red die. However, as in Example 4, the description of the events may not clearly indicate whether the events are dependent.

From a problem solving viewpoint, the following two statements are very helpful.

1. If E and F are independent events, then

$$P(E \cap F) = P(E)P(F).$$

(This property generalizes to any finite number of independent events.)

2. If E and F are dependent events, then

$$P(E \cap F) = P(E)P(F \mid E).$$

Let's analyze some problems using these ideas.

EXAMPLE 5

A die is rolled three times. (This is equivalent to rolling three dice once each.) What is the probability of getting a 6 all three times?

Solution

The events of a 6 on the first roll, a 6 on the second roll, and a 6 on the third roll are independent events. Therefore, the probability of getting three 6s is

$$\left(\frac{1}{6}\right)\left(\frac{1}{6}\right)\left(\frac{1}{6}\right) = \frac{1}{216}.$$

EXAMPLE 6

A jar contains five white, seven green, and nine red marbles. If two marbles are drawn in succession *without replacement*, find the probability that both marbles are white.

Solution

Let E be the event of drawing a white marble on the first draw and F the event of drawing a white marble on the second draw. Since the marble drawn first is not to be replaced before drawing the second marble, we have dependent events. Therefore,

$$\begin{aligned} P(E \cap F) &= P(E)P(F|E) \\ &= \left(\frac{5}{21}\right)\left(\frac{4}{20}\right) = \frac{20}{420} = \frac{1}{21}. \end{aligned}$$

$P(F|E)$ means the probability of drawing a white marble on the second draw given that a white marble was obtained on the first draw.

The concept of *mutually exclusive events* may also enter the picture when working with independent or dependent events. Our final examples of this section illustrate this idea.

EXAMPLE 7

A coin is tossed three times. Find the probability of getting two heads and one tail.

Solution

Two heads and one tail can be obtained in three different ways: (1) *HHT* (head on first toss, head on second toss, and tail on third toss), (2) *HTH*, and (3) *THH*. Thus we have three *mutually exclusive* events, each of which can be broken into *independent* events: first toss, second toss, and third toss. Therefore, the probability can be computed as follows

$$\left(\frac{1}{2}\right)\left(\frac{1}{2}\right)\left(\frac{1}{2}\right) + \left(\frac{1}{2}\right)\left(\frac{1}{2}\right)\left(\frac{1}{2}\right) + \left(\frac{1}{2}\right)\left(\frac{1}{2}\right)\left(\frac{1}{2}\right) = \frac{3}{8}$$

EXAMPLE 8

A jar contains five white, seven green, and nine red marbles. If two marbles are drawn in succession *without replacement*, find the probability that one of them is white and the other is green.

Solution

The drawing of a white marble and a green marble can occur in two different ways: (1) by drawing a white first and then a green, or (2) by drawing a green first and then a white. Thus, we have two mutually exclusive events, each of which is broken into two *dependent* events: first draw and second draw. Therefore, the probability can be computed as follows.

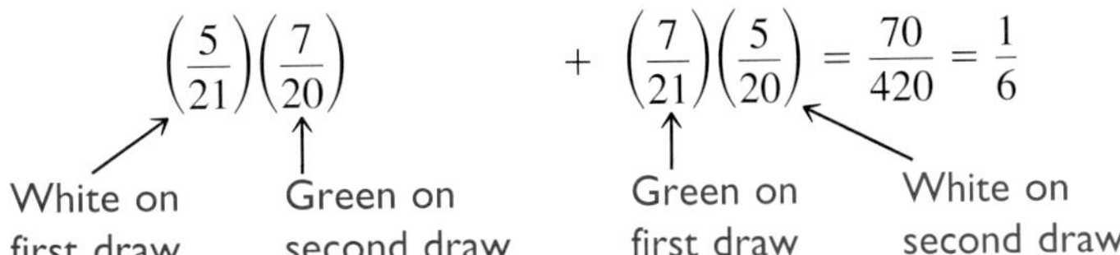

EXAMPLE 9 Two cards are drawn in succession *with replacement* from a deck of 52 playing cards. Find the probability of drawing a jack and a queen.

Solution

Drawing a jack and a queen can occur two different ways: (1) a jack on the first draw and a queen on the second, or (2) a queen on the first draw and a jack on the second. Thus (1) and (2) are mutually exclusive events and each is broken into the *independent* events of first draw and second draw with replacement. Therefore, the probability can be computed as follows.

$$\left(\frac{4}{52}\right)\left(\frac{4}{52}\right) + \left(\frac{4}{52}\right)\left(\frac{4}{52}\right) = \frac{32}{2704} = \frac{2}{169}$$

Jack on first draw — Queen on second draw — Queen on first draw — Jack on second draw

PROBLEM SET 14.5

1. A die is tossed. Find the probability that a 5 came up if it is known that an odd number came up.

2. A die is tossed. Find the probability that a prime number was obtained, given that an even number came up. Also find the probability that an even number came up, given that a prime number was obtained.

3. Two dice are rolled and someone indicates that the two numbers that come up are different. Find the probability that the sum of the two numbers is 6.

4. Two dice are rolled and someone indicates that the two numbers that come up are identical. Find the probability that the sum of the two numbers is 8.

5. One card is randomly drawn from a deck of 52 playing cards. Find the probability that it is a jack, given that the card is a face card. (We are considering jacks, queens, and kings as face cards.)

6. One card is randomly drawn from a deck of 52 playing cards. Find the probability that it is a spade, given the fact that it is a black card.

7. A coin and a die are tossed. Find the probability of getting a 5 on the die, given that a head comes up on the coin.

8. A family has three children. Assume that each child is as likely to be a boy as it is to be a girl. Find the probability that the family has three girls if it is known that they have at least one girl.

9. The probability that a student will enroll in a mathematics course is 0.7, the probability that he (she) will enroll in a history course is 0.3, and the probability that he (she) will enroll in both mathematics and history is 0.2. Find the probability that a student will enroll in mathematics, given that he (she) is also enrolled in history. Also find the probability that a student will enroll in history, given that he (she) is also enrolled in mathematics.

10. The following probability table contains data relative to car accidents and weather conditions on a holiday weekend.

	RAIN (R)	NO RAIN (R')	TOTAL
ACCIDENT (A)	0.025	0.015	0.040
NO ACCIDENT (A')	0.400	0.560	0.960
TOTAL	0.425	0.575	1.000

Find the probability that a person chosen at random from the survey was in an accident, given that it was raining. Also find the probability that a person was not in an accident, given that it was not raining.

11. One hundred people were surveyed and one question pertained to their educational background. The results of this question are given in the following table.

	FEMALE (F)	MALE (F')	TOTAL
COLLEGE DEGREE (D)	30	20	50
NO COLLEGE DEGREE (D')	15	35	50
TOTAL	45	55	100

Find the probability that a person chosen at random from the survey has a college degree, given that the person is female. Also find the probability that a person chosen is male, given that the person has a college degree.

12. In a recent election there were 1000 eligible voters. They were asked to vote on two issues, A and B. The results were as follows: 200 people voted for A, 400 people voted for B, and 50 people voted for both A and B. If one person is chosen at random from the 1000 eligible voters, find the probability that the person voted for A, given that he (she) voted for B. Also find the probability that the person voted for B, given that he (she) voted for A.

13. A small company has 100 employees of which 75 are males, 7 are administrators, and 5 of the administrators are males. If a person is chosen at random from the 100 employees, find the probability that the person is an administrator, given that he is a male. Also find the probability that the person chosen is female, given that she is an administrator.

14. A survey claims that 80% of the households in a certain town have a color TV, 10% have a microwave oven, and 2% have both a color TV and a microwave oven. Find the probability that a randomly selected household will have a microwave oven, given that it has a color TV.

15. Consider a family of three children. Let E be the event *The first child is a boy* and let F be the event *They have exactly one boy*. Are events E and F dependent or independent?

16. Roll a white die and a green die. Let E be the event *Roll a 2 on the white die* and let F be the event *Roll a 4 on the green die*. Are E and F dependent or independent events?

17. Toss 3 coins. Let E be the event *Toss not more than one head* and let F be the event *Toss at least one of each face*. Are E and F dependent or independent events?

18. A card is drawn at random from a standard deck of 52 playing cards. Let E be the event *The card is a 2* and let F be the event *The card is a 2 or 3*. Are the events E and F dependent or independent?

19. A coin is tossed four times. Find the probability of getting three heads and one tail.

20. A coin is tossed five times. Find the probability of getting four heads and one tail.

21. Toss a pair of dice three times. Find the probability that a double is obtained on all three tosses.

22. Toss a pair of dice three times. Find the probability that each toss will produce a sum of 4.

For Problems 23–26, suppose that two cards are drawn in succession *without replacement* from a deck of 52 playng cards. Find the probability of each of the following events.

23. Both cards are 4s.

24. One card is an ace and one card is a king.

25. One card is a spade and one card is a diamond.

26. Both cards are black.

For Problems 27–30, suppose that two cards are drawn in succession *with replacement* from a deck of 52 playing cards. Find the probability of each of the following events.

27. Both cards are spades.

28. One card is an ace and one card is a king.

29. One card is the ace of spades and one card is the king of spades.

30. Both cards are red.

31. A person holds three kings from a deck of 52 playing cards. If the person draws two cards without replacement from the 49 cards remaining in the deck, find the probability of drawing the fourth king.

32. A person removes two aces and a king from a deck of 52 playing cards and draws, without replacement, two more cards from the deck. Find the probability that the person will draw two aces, or two kings, or an ace and a king.

For Problems 33–36, a bag contains five red and four white marbles. Two marbles are drawn in succession *with replacement*. Find the probability of each of the following events.

33. Both marbles drawn are red.

34. Both marbles drawn are white.

35. The first marble is red and the second marble is white.

36. At least one marble is red.

For Problems 37–40, a bag contains five white, four red, and four blue marbles. Two marbles are drawn in succession *with replacement*. Find the probability of each of the following events.

37. Both marbles drawn are white.

38. Both marbles drawn are red.

39. One red and one blue marble are drawn.

40. One white and one blue marble are drawn.

For Problems 41–44, a bag contains one red and two white marbles. Two marbles are drawn in succession *without replacement*. Find the probability of each of the following events.

41. One marble drawn is red and one marble drawn is white.

42. The first marble is red and the second one is white.

43. Both marbles drawn are white.

44. Both marbles drawn are red.

For Problems 45–48, a bag contains five red and twelve white marbles. Two marbles are drawn in succession *without replacement*. Find the probability of each of the following events.

45. Both marbles drawn are red.

46. Both marbles drawn are white.

47. One red and one white marble are drawn.

48. At least one marble drawn is red.

For Problems 49–52, a bag contains two red, three white, and four blue marbles. Two marbles are drawn in succession *without replacement*. Find the probability of each of the following events.

49. Both marbles drawn are white.

50. One marble drawn is white and one is blue.

51. Both marbles drawn are blue.

52. At least one red marble is drawn.

For Problems 53–56, a bag contains five white, one blue, and three red marbles. *Three* marbles are drawn in succession *with replacement*. Find the probability of each of the following events.

53. All three marbles drawn are blue.

54. One marble of each color is drawn.

55. One white and two red marbles are drawn.

56. One blue and two white marbles are drawn.

For Problems 57–60, a bag contains four white, one red, and two blue marbles. *Three* marbles are drawn in succession *without replacement*. Find the probability of each of the following events.

57. All three marbles drawn are white.

58. One red and two blue marbles are drawn.

59. One marble of each color is drawn.

60. One white and two red marbles are drawn.

61. Two boxes with red and white marbles are shown here. A marble is drawn at random from Box 1 and then a second marble is drawn from Box 2. Find the probability that both marbles drawn are white. Find the probability that both marbles drawn are red. Find the probability that one red and one white marble are drawn.

Box 1	Box 2
3 red 4 white	2 red 1 white

62. Three boxes containing red and white marbles are shown here. Randomly draw a marble from Box 1 and put it in Box 2. Then draw a marble from Box 2 and put it in Box 3. Then draw a marble from Box 3. What is the probability that the last marble drawn, from Box 3, is red? What is the probability that it is white?

Box 1	Box 2	Box 3
2 red 2 white	3 red 1 white	3 white

14.6 BINOMIAL THEOREM

In Chapter 1 we developed a pattern for expanding binomials, using Pascal's triangle to determine the coefficients of each term. Now we will be more precise and develop a general formula, called the *binomial formula*. In other words, we want to develop a formula that will allow us to expand $(x + y)^n$, where n is any positive integer.

Let's begin, as we did in Chapter 1, by looking at some specific expansions, which can be verified by direct multiplication.

$$(x + y)^0 = 1$$
$$(x + y)^1 = x + y$$
$$(x + y)^2 = x^2 + 2xy + y^2$$
$$(x + y)^3 = x^3 + 3x^2y + 3xy^2 + y^3$$
$$(x + y)^4 = x^4 + 4x^3y + 6x^2y^2 + 4xy^3 + y^4$$
$$(x + y)^5 = x^5 + 5x^4y + 10x^3y^2 + 10x^2y^3 + 5xy^4 + y^5$$

First, note the pattern of the exponents for x and y on a term-by-term basis. The exponents of x begin with the exponent of the binomial and decrease by 1, term by term, until the last term has x^0, which is 1. The exponents of y begin with zero ($y^0 = 1$) and increase by 1, term by term, until the last term contains y to the power of the binomial. In other words, the variables in the expansion of $(x + y)^n$ have the following pattern.

$$x^n, \quad x^{n-1}y, \quad x^{n-2}y^2, \quad x^{n-3}y^3, \quad \ldots, \quad xy^{n-1}, \quad y^n$$

Notice that for each term, the sum of the exponents of x and y is n.

Now let's look for a pattern for the coefficients by looking specifically at the expansion of $(x + y)^5$.

$$(x + y)^5 = x^5 + 5x^4y^1 + 10x^3y^2 + 10x^2y^3 + 5x^1y^4 + 1y^5.$$

$$\uparrow\ C(5, 1) \quad \uparrow\ C(5, 2) \quad \uparrow\ C(5, 3) \quad \uparrow\ C(5, 4) \quad \uparrow\ C(5, 5)$$

As indicated by the arrows, the coefficients are numbers that arise as different-sized combinations of five things. To see why this happens, consider the coefficient for the term containing x^2y^2. The two ys (for y^2) come from two of the factors of $(x + y)$ and therefore the three xs (for x^3) must come from the other three factors of $(x + y)$. In other words, the coefficient is $C(5, 2)$.

We can now state a general expansion formula for $(x + y)^n$; this formula is often called the **binomial theorem**. But, before stating it, let's make a small switch in notation. Instead of $C(n, r)$ we shall write $\binom{n}{r}$, which will prove to be a little more convenient at this time. The symbol $\binom{n}{r}$ still refers to the number of combinations of n things taken r at a time, but in this context it is often called a **binomial coefficient**.

Binomial Theorem

For any binomial $(x + y)$ and any natural number n;

$$(x + y)^n = x^n + \binom{n}{1}x^{n-1}y + \binom{n}{2}x^{n-2}y^2 + \cdots + \binom{n}{n}y^n.$$

The binomial theorem can be proved by mathematical induction, but we will not do that in this text. Instead we'll consider a few examples that put the binomial theorem to work.

EXAMPLE 1

Expand $(x + y)^7$.

Solution

$$\begin{aligned}(x + y)^7 &= x^7 + \binom{7}{1}x^6y + \binom{7}{2}x^5y^2 + \binom{7}{3}x^4y^3 + \binom{7}{4}x^3y^4 + \binom{7}{5}x^2y^5 + \binom{7}{6}xy^6 + \binom{7}{7}y^7 \\ &= x^7 + 7x^6y + 21x^5y^2 + 35x^4y^3 + 35x^3y^4 + 21x^2y^5 + 7xy^6 + y^7\end{aligned}$$

EXAMPLE 2

Expand $(x - y)^5$.

Solution

We shall treat $(x - y)^5$ as $[x + (-y)]^5$.

$$\begin{aligned}[x + (-y)]^5 &= x^5 + \binom{5}{1}x^4(-y) + \binom{5}{2}x^3(-y)^2 + \binom{5}{3}x^2(-y)^3 + \binom{5}{4}x(-y)^4 + \binom{5}{5}(-y)^5 \\ &= x^5 - 5x^4y + 10x^3y^2 - 10x^2y^3 + 5xy^4 - y^5\end{aligned}$$

EXAMPLE 3

Expand $(2a + 3b)^4$.

Solution

Let $x = 2a$ and $y = 3b$ in the binomial theorem.

$$\begin{aligned}(2a + 3b)^4 &= (2a)^4 + \binom{4}{1}(2a)^3(3b) + \binom{4}{2}(2a)^2(3b)^2 + \binom{4}{3}(2a)(3b)^3 + \binom{4}{4}(3b)^4 \\ &= 16a^4 + 96a^3b + 216a^2b^2 + 216ab^3 + 81b^4\end{aligned}$$

EXAMPLE 4

Expand $\left(a + \frac{1}{n}\right)^5$.

Solution

$$\begin{aligned}\left(a + \frac{1}{n}\right)^5 &= a^5 + \binom{5}{1}a^4\left(\frac{1}{n}\right) + \binom{5}{2}a^3\left(\frac{1}{n}\right)^2 + \binom{5}{3}a^2\left(\frac{1}{n}\right)^3 + \binom{5}{4}a\left(\frac{1}{n}\right)^4 + \binom{5}{5}\left(\frac{1}{n}\right)^5 \\ &= a^5 + \frac{5a^4}{n} + \frac{10a^3}{n^2} + \frac{10a^2}{n^3} + \frac{5a}{n^4} + \frac{1}{n^5}\end{aligned}$$

EXAMPLE 5

Expand $(x^2 - 2y^3)^6$.

Solution

$$\begin{aligned}[x^2 + (-2y^3)]^6 &= (x^2)^6 + \binom{6}{1}(x^2)^5(-2y^3) + \binom{6}{2}(x^2)^4(-2y^3)^2 \\ &\quad + \binom{6}{3}(x^2)^3(-2y^3)^3 + \binom{6}{4}(x^2)^2(-2y^3)^4 \\ &\quad + \binom{6}{5}(x^2)(-2y^3)^5 + \binom{6}{6}(-2y^3)^6 \\ &= x^{12} - 12x^{10}y^3 + 60x^8y^6 - 160x^6y^9 + 240x^4y^{12} - 192x^2y^{15} + 64y^{18}\end{aligned}$$

Finding Specific Terms

Sometimes it is convenient to be able to write down the specific term of a binomial expansion without writing out the entire expansion. For example, suppose that we want the sixth term of the expansion $(x + y)^{12}$. We can proceed as follows: The sixth term will contain y^5. (Note in the binomial theorem that **the exponent of y is always one less than the number of the term.**) Since the sum of the exponents for x and y must be 12 (the exponent of the binomial), the sixth term will also contain x^7. The coefficient is $\binom{12}{5}$, where the 5 agreeswith the exponent of y^5.

Therefore, the sixth term of $(x + y)^{12}$ is

$$\binom{12}{5}x^7y^5 = 792x^7y^5.$$

EXAMPLE 6

Find the fourth term of $(3a + 2b)^7$.

Solution

The fourth term will contain $(2b)^3$ and therefore it will also contain $(3a)^4$. The coefficient is $\binom{7}{3}$. Thus, the fourth term is

$$\binom{7}{3}(3a)^4(2b)^3 = (35)(81a^4)(8b^3) = 22{,}680a^4b^3.$$

EXAMPLE 7

Find the sixth term of $(4x - y)^9$.

Solution

The sixth term will contain $(-y)^5$ and therefore it will also contain $(4x)^4$. The coefficient is $\binom{9}{5}$. Thus, the sixth term is

$$\binom{9}{5}(4x)^4(-y)^5 = (126)(256x^4)(-y^5)$$
$$= -32256x^4y^5.$$

PROBLEM SET 14.6

Expand and simplify each of the following binomials.

1. $(x + y)^8$ **2.** $(x + y)^9$

3. $(x - y)^6$ **4.** $(x - y)^4$

5. $(a + 2b)^4$ **6.** $(3a + b)^4$

7. $(x - 3y)^5$ **8.** $(2x - y)^6$

9. $(2a - 3b)^4$ **10.** $(3a - 2b)^5$

11. $(x^2 + y)^5$ **12.** $(x + y^3)^6$

13. $(2x^2 - y^2)^4$ **14.** $(3x^2 - 2y^2)^5$

15. $(x + 3)^6$ **16.** $(x + 2)^7$

17. $(x - 1)^9$ **18.** $(x - 3)^4$

19. $\left(1 + \frac{1}{n}\right)^4$ **20.** $\left(2 + \frac{1}{n}\right)^5$

21. $\left(a - \frac{1}{n}\right)^6$ **22.** $\left(2a - \frac{1}{n}\right)^5$

23. $\left(1 + \sqrt{2}\right)^4$ **24.** $\left(2 + \sqrt{3}\right)^3$

25. $\left(3 - \sqrt{2}\right)^5$ **26.** $\left(1 - \sqrt{3}\right)^4$

Write the first four terms of each of the following expansions.

27. $(x + y)^{12}$ **28.** $(x + y)^{15}$

29. $(x - y)^{20}$ **30.** $(a - 2b)^{13}$

31. $(x^2 - 2y^3)^{14}$ **32.** $(x^3 - 3y^2)^{11}$

33. $\left(a + \frac{1}{n}\right)^9$ **34.** $\left(2 - \frac{1}{n}\right)^6$

35. $(-x + 2y)^{10}$ **36.** $(-a - b)^{14}$

Find the specified term for each of the following binomial expansions.

37. The fourth term of $(x + y)^8$

38. The seventh term of $(x + y)^{11}$

39. The fifth term of $(x - y)^9$

40. The fourth term of $(x - 2y)^6$

41. The sixth term of $(3a + b)^7$

42. The third term of $(2x - 5y)^5$

43. The eighth term of $(x^2 + y^3)^{10}$

44. The ninth term of $(a + b^3)^{12}$

45. The seventh term of $\left(1 - \frac{1}{n}\right)^{15}$

46. The eighth term of $\left(1 - \frac{1}{n}\right)^{13}$

MISCELLANEOUS PROBLEMS

Expand and simplify each of the following complex numbers.

47. $(1 + 2i)^5$

48. $(2 + i)^6$

49. $(2 - i)^6$

50. $(3 - 2i)^5$

CHAPTER 14 SUMMARY

We can summarize this chapter with three main topics: counting techniques, probability, and the binomial theorem.

Counting Techniques

The **fundamental principle of counting** states that if a first task can be accomplished in x ways and, following this task, a second task can be accomplished in y ways, then task 1 followed by task 2 can be accomplished in $x \cdot y$ ways. The principle extends to any finite number of tasks. As you solve problems involving the fundamental principle of counting, it is often helpful to analyze the problem in terms of the tasks to be completed.

Ordered arrangements are called **permutations**. The number of permutations of n things taken n at a time is given by

$$P(n, n) = n!.$$

The number of r-element permutations that can be formed from a set of n elements is given by

$$P(n, r) = \underbrace{n(n-1)(n-2)\cdots}_{r \text{ factors}}.$$

If there are n elements to be arranged, where there are r_1 of one kind, r_2 of another kind, r_3 of another kind, . . . , r_k of a kth kind, then the number of distinguishable permutations is given by

$$\frac{n!}{(r_1!)(r_2!)(r_3!)\cdots(r_k!)}.$$

Combinations are subsets; the order in which the elements appear does not make a difference. The number of r-element combinations (subsets) that can be formed from a set of n elements is given by

$$C(n, r) = \frac{P(n, r)}{r!}.$$

Does the order in which the elements appear make any difference? This is a key question to consider when trying to decide whether a particular problem involves permutations or combinations. If the answer to the question is yes, then it is a permutation problem; if the answer is no, then it is a combination problem. Don't forget that combinations are subsets.

Probability

In an experiment where all possible outcomes in the sample space S are equally likely to occur, the **probability** of an event E is defined by

$$P(E) = \frac{n(E)}{n(S)},$$

where $n(E)$ denotes the number of elements in the event E and $n(S)$ denotes the number of elements in the sample space S. The numbers $n(E)$ and $n(S)$ can often be determined by using one or more of the previously listed counting techniques. For all events E, it is always true that $0 \leq P(E) \leq 1$. That is to say, all probabilities fall in the range from 0 to 1, inclusive.

If E and E' are **complementary events**, then $P(E) + P(E') = 1$. Therefore, if we can calculate either $P(E)$ or $P(E')$, then we can find the other one by subtracting from 1.

For two events E and F, the probability of E or F is given by

$$P(E \cup F) = P(E) + P(F) - P(E \cap F).$$

If $E \cap F = \varnothing$, then E and F are **mutually exclusive events**.

The probability that an event E occurs, given that another event F has already occurred, is called **conditional probability**, and it is given by the equation

$$P(E|F) = \frac{P(E \cap F)}{P(F)}.$$

Two events E and F are said to be **independent** if and only if

$$P(E \cap F) = P(E)P(F).$$

Two events that are not independent are called **dependent** events, and the probability of two dependent events is given by

$$P(E \cap F) = P(E)P(F|E).$$

The Binomial Theorem

For any binomial $(x + y)$ and any natural number n,

$$(x + y)^n = x^n + \binom{n}{1}x^{n-1}y + \binom{n}{2}x^{n-2}y^2 + \cdots + \binom{n}{n}y^n.$$

Note the following patterns in a binomial expansion.

1. In each term, the sum of the exponents of x and y is n.
2. The exponents of x begin with the exponent of the binomial and decrease by 1, term by term, until the last term has x^0, which is 1. The exponents of y begin with zero ($y^0 = 1$) and increase by 1, term by term, until the last term contains y to the power of the binomial.
3. The coefficient of any term is given by $\binom{n}{r}$, where the value of r agrees with the exponent of y for that term. For example, if the term contains y^3, then the coefficient of that term is $\binom{n}{3}$.
4. The expansion of $(x + y)^n$ contains $n + 1$ terms.

CHAPTER 14 REVIEW PROBLEM SET

Problems 1–14 are counting type problems.

1. How many different arrangements of the letters A, B, C, D, E, and F can be made?
2. How many different nine-letter arrangements can be formed from the nine letters of the word APPARATUS?
3. How many odd numbers of three different digits each can be formed by choosing from the digits 1, 2, 3, 5, 7, 8, and 9?
4. In how many ways can Arlene, Brent, Cindy, Dave, Ernie, Frank, and Gladys be seated in a row of seven seats so that Arlene and Cindy are side by side?
5. In how many ways can a committee of three people be chosen from six people?
6. How many committees consisting of three men and two women can be formed from seven men and six women?
7. How many different five-card hands consisting of all hearts can be formed from a deck of 52 playing cards?
8. If no number contains repeated digits, how many numbers greater than 500 can be formed by choosing from the digits 2, 3, 4, 5, and 6?
9. How many three-person committees can be formed from four men and five women so that each committee contains at least one man?

10. How many different four-person committees can be formed from eight people if two particular people refuse to serve together on a committee?

11. How many four-element subsets containing A or B but not both A and B can be formed from the set {A, B, C, D, E, F, G, H}?

12. How many different six-letter permutations can be formed from four identical Hs and two identical Ts?

13. How many four-person committees consisting of two seniors, one sophomore, and one junior can be formed from three seniors, four juniors, and five sophomores?

14. In a baseball league of six teams, how many games are needed to complete a schedule if each team plays eight games with each other team?

Problems 15–35 pose some probability questions.

15. If three coins are tossed, find the probability of getting two heads and one tail.

16. If five coins are tossed, find the probability of getting three heads and two tails.

17. What is the probability of getting a sum of eight with one roll of a pair of dice?

18. What is the probability of getting a sum more than 5 with one roll of a pair of dice?

19. Amy, Brenda, Chuck, Dave, and Elmer are randomly seated in a row of five seats. Find the probability that Amy and Chuck are not seated side by side.

20. Four girls and three boys are randomly seated in a row of seven seats. Find the probability that the girls and boys will be seated in alternate seats.

21. Six coins are tossed. Find the probability of getting at least two heads.

22. Two cards are randomly chosen from a deck of 52 playing cards. What is the probability that two jacks are drawn?

23. Each arrangement of the six letters of the word CYCLIC is put on a slip of paper and placed in a hat. One slip is drawn at random. Find the probability that the slip contains an arrangement with the Y at the beginning.

24. A committee of three is randomly chosen from one man and six women. What is the probability that the man is not on the committee?

25. A four-person committee is selected at random from the eight people, Alice, Bob, Carl, Dee, Edna, Fred, Gina, and Hilda. Find the probability that Alice or Bob, but not both, is on the committee.

26. A committee of three is chosen at random from a group of five men and four women. Find the probability that the committee contains two men and one woman.

27. A committee of four is chosen at random from a group of six men and seven women. Find the probability that the committee contains at least one woman.

28. A bag contains five red and eight white marbles. Two marbles are drawn in succession with replacement. What is the probability that at least one red marble is drawn?

29. A bag contains four red, five white, and three blue marbles. Two marbles are drawn in succession with replacement. Find the probability that one red and one blue marble are drawn.

30. A bag contains four red and seven blue marbles. Two marbles are drawn in succession without replacement. Find the probability of drawing one red and one blue marble.

31. A bag contains three red, two white, and two blue marbles. Two marbles are drawn in succession without replacement. Find the probability of drawing at least one red marble.

32. Each of three letters is to be mailed in any one of four different mailboxes. What is the probability that all three letters are mailed in the same mailbox?

33. The probability that a customer in a department store will buy a blouse is 0.15, the probability that she will buy a pair of shoes is 0.10, and the probability that she will buy both a blouse and a pair of shoes is 0.05. Find the probability that the customer will buy a blouse, given that she has already purchased a pair of shoes. Also find the probability that she will buy a pair of shoes, given that she has already purchased a blouse.

34. A survey of 500 employees of a company produced the following information.

EMPLOYMENT LEVEL	COLLEGE DEGREE	NO COLLEGE DEGREE
Managerial	45	5
Nonmanagerial	50	400

Find the probability that an employee chosen at random (a) is working in a managerial position, given that he (she) has a college degree; and (b) has a college degree, given that he (she) is working in a managerial position.

35. From a survey of 1000 college students, it was found that 450 of them owned cars, 700 of them owned stereos, and 200 of them owned both a car and a stereo. If a student is chosen at random from the 1000 students, find the probability that (a) he (she) owns a car, given the fact that he (she) owns a stereo; and (b) he (she) owns a stereo, given the fact that he (she) owns a car.

For Problems 36–41, expand each binomial and simplify.

36. $(x + 2y)^5$ **37.** $(x - y)^8$ **38.** $(a^2 - 3b^3)^4$

39. $\left(x + \frac{1}{n}\right)^6$ **40.** $\left(1 - \sqrt{2}\right)^5$ **41.** $(-a + b)^3$

42. Find the 4th term of the expansion of $(x - 2y)^{12}$.

43. Find the 10th term of the expansion of $(3a + b^2)^{13}$.

APPENDIX

COMMON LOGARITHMS

Table of Common Logarithms

N	0	1	2	3	4	5	6	7	8	9
1.0	.0000	.0043	.0086	.0128	.0170	.0212	.0253	.0294	.0334	.0374
1.1	.0414	.0453	.0492	.0531	.0569	.0607	.0645	.0682	.0719	.0755
1.2	.0792	.0828	.0864	.0899	.0934	.0969	.1004	.1038	.1072	.1106
1.3	.1139	.1173	.1206	.1239	.1271	.1303	.1335	.1367	.1399	.1430
1.4	.1461	.1492	.1523	.1553	.1584	.1614	.1644	.1673	.1703	.1732
1.5	.1761	.1790	.1818	.1847	.1875	.1903	.1931	.1959	.1987	.2014
1.6	.2041	.2068	.2095	.2122	.2148	.2175	.2201	.2227	.2253	.2279
1.7	.2304	.2330	.2355	.2380	.2405	.2430	.2455	.2480	.2504	.2529
1.8	.2553	.2577	.2601	.2625	.2648	.2672	.2695	.2718	.2742	.2765
1.9	.2788	.2810	.2833	.2856	.2878	.2900	.2923	.2945	.2967	.2989
2.0	.3010	.3032	.3054	.3075	.3096	.3118	.3139	.3160	.3181	.3201
2.1	.3222	.3243	.3263	.3284	.3304	.3324	.3345	.3365	.3385	.3404
2.2	.3424	.3444	.3464	.3483	.3502	.3522	.3541	.3560	.3579	.3598
2.3	.3617	.3636	.3655	.3674	.3692	.3711	.3729	.3747	.3766	.3784
2.4	.3802	.3820	.3838	.3856	.3874	.3892	.3909	.3927	.3945	.3962
2.5	.3979	.3997	.4014	.4031	.4048	.4065	.4082	.4099	.4116	.4133
2.6	.4150	.4166	.4183	.4200	.4216	.4232	.4249	.4265	.4281	.4298
2.7	.4314	.4330	.4346	.4362	.4378	.4393	.4409	.4425	.4440	.4456
2.8	.4472	.4487	.4502	.4518	.4533	.4548	.4564	.4579	.4594	.4609
2.9	.4624	.4639	.4654	.4669	.4683	.4698	.4713	.4728	.4742	.4757
3.0	.4771	.4786	.4800	.4814	.4829	.4843	.4857	.4871	.4886	.4900
3.1	.4914	.4928	.4942	.4955	.4969	.4983	.4997	.5011	.5024	.5038
3.2	.5051	.5065	.5079	.5092	.5105	.5119	.5132	.5145	.5159	.5172
3.3	.5185	.5198	.5211	.5224	.5237	.5250	.5263	.5276	.5289	.5302
3.4	.5315	.5328	.5340	.5353	.5366	.5378	.5391	.5403	.5416	.5428
3.5	.5441	.5453	.5465	.5478	.5490	.5502	.5514	.5527	.5539	.5551
3.6	.5563	.5575	.5587	.5599	.5611	.5623	.5635	.5647	.5658	.5670
3.7	.5682	.5694	.5705	.5717	.5729	.5740	.5752	.5763	.5775	.5786
3.8	.5798	.5809	.5821	.5832	.5843	.5855	.5866	.5877	.5888	.5899
3.9	.5911	.5922	.5933	.5944	.5955	.5966	.5977	.5988	.5999	.6010

Table of Common Logarithms (continued)

N	0	1	2	3	4	5	6	7	8	9
4.0	.6021	.6031	.6042	.6053	.6064	.6075	.6085	.6096	.6107	.6117
4.1	.6128	.6138	.6149	.6160	.6170	.6180	.6191	.6201	.6212	.6222
4.2	.6232	.6243	.6253	.6263	.6274	.6284	.6294	.6304	.6314	.6325
4.3	.6335	.6345	.6355	.6365	.6375	.6385	.6395	.6405	.6415	.6425
4.4	.6435	.6444	.6454	.6464	.6474	.6484	.6493	.6503	.6513	.6522
4.5	.6532	.6542	.6551	.6561	.6571	.6580	.6590	.6599	.6609	.6618
4.6	.6628	.6637	.6646	.6656	.6665	.6675	.6684	.6693	.6702	.6712
4.7	.6721	.6730	.6739	.6749	.6758	.6767	.6776	.6785	.6794	.6803
4.8	.6812	.6821	.6830	.6839	.6848	.6857	.6866	.6875	.6884	.6893
4.9	.6902	.6911	.6920	.6928	.6937	.6946	.6955	.6964	.6972	.6981
5.0	.6990	.6998	.7007	.7016	.7024	.7033	.7042	.7050	.7059	.7067
5.1	.7076	.7084	.7093	.7101	.7110	.7118	.7126	.7135	.7143	.7152
5.2	.7160	.7168	.7177	.7185	.7193	.7202	.7210	.7218	.7226	.7235
5.3	.7243	.7251	.7259	.7267	.7275	.7284	.7292	.7300	.7308	.7316
5.4	.7324	.7332	.7340	.7348	.7356	.7364	.7372	.7380	.7388	.7396
5.5	.7404	.7412	.7419	.7427	.7435	.7443	.7451	.7459	.7466	.7474
5.6	.7482	.7490	.7497	.7505	.7513	.7520	.7528	.7536	.7543	.7551
5.7	.7559	.7566	.7574	.7582	.7589	.7597	.7604	.7612	.7619	.7627
5.8	.7634	.7642	.7649	.7657	.7664	.7672	.7679	.7686	.7694	.7701
5.9	.7709	.7716	.7723	.7731	.7738	.7745	.7752	.7760	.7767	.7774
6.0	.7782	.7789	.7796	.7803	.7810	.7818	.7825	.7832	.7839	.7846
6.1	.7853	.7860	.7868	.7875	.7882	.7889	.7896	.7903	.7910	.7917
6.2	.7924	.7931	.7938	.7945	.7952	.7959	.7966	.7973	.7980	.7987
6.3	.7993	.8000	.8007	.8014	.8021	.8028	.8035	.8041	.8048	.8055
6.4	.8062	.8069	.8075	.8082	.8089	.8096	.8102	.8109	.8116	.8122
6.5	.8129	.8136	.8142	.8149	.8156	.8162	.8169	.8176	.8182	.8189
6.6	.8195	.8202	.8209	.8215	.8222	.8228	.8235	.8241	.8248	.8254
6.7	.8261	.8267	.8274	.8280	.8287	.8293	.8299	.8306	.8312	.8319
6.8	.8325	.8331	.8338	.8344	.8351	.8357	.8363	.8370	.8376	.8382
6.9	.8388	.8395	.8401	.8407	.8414	.8420	.8426	.8432	.8439	.8445
7.0	.8451	.8457	.8463	.8470	.8476	.8482	.8488	.8494	.8500	.8506
7.1	.8513	.8519	.8525	.8531	.8537	.8543	.8549	.8555	.8561	.8567
7.2	.8573	.8579	.8585	.8591	.8597	.8603	.8609	.8615	.8621	.8627
7.3	.8633	.8639	.8645	.8651	.8657	.8663	.8669	.8675	.8681	.8686
7.4	.8692	.8698	.8704	.8710	.8716	.8722	.8727	.8733	.8739	.8745
7.5	.8751	.8756	.8762	.8768	.8774	.8779	.8785	.8791	.8797	.8802
7.6	.8808	.8814	.8820	.8825	.8831	.8837	.8842	.8848	.8854	.8859
7.7	.8865	.8871	.8876	.8882	.8887	.8893	.8899	.8904	.8910	.8915
7.8	.8921	.8927	.8932	.8938	.8943	.8949	.8954	.8960	.8965	.8971
7.9	.8976	.8982	.8987	.8993	.8998	.9004	.9009	.9015	.9020	.9025
8.0	.9031	.9036	.9042	.9047	.9053	.9058	.9063	.9069	.9074	.9079
8.1	.9085	.9090	.9096	.9101	.9106	.9112	.9117	.9122	.9128	.9133
8.2	.9138	.9143	.9149	.9154	.9159	.9165	.9170	.9175	.9180	.9186
8.3	.9191	.9196	.9201	.9206	.9212	.9217	.9222	.9227	.9232	.9238
8.4	.9243	.9248	.9253	.9258	.9263	.9269	.9274	.9279	.9284	.9289

Table of Common Logarithms (continued)

N	0	1	2	3	4	5	6	7	8	9
8.5	.9294	.9299	.9304	.9309	.9315	.9320	.9325	.9330	.9335	.9340
8.6	.9345	.9350	.9355	.9360	.9365	.9370	.9375	.9380	.9385	.9390
8.7	.9395	.9400	.9405	.9410	.9415	.9420	.9425	.9430	.9435	.9440
8.8	.9445	.9450	.9455	.9460	.9465	.9469	.9474	.9479	.9484	.9489
8.9	.9494	.9499	.9504	.9509	.9513	.9518	.9523	.9528	.9533	.9538
9.0	.9542	.9547	.9552	.9557	.9562	.9566	.9571	.9576	.9581	.9586
9.1	.9590	.9595	.9600	.9605	.9609	.9614	.9619	.9624	.9628	.9633
9.2	.9638	.9643	.9647	.9652	.9657	.9661	.9666	.9671	.9675	.9680
9.3	.9685	.9689	.9694	.9699	.9703	.9708	.9713	.9717	.9722	.9727
9.4	.9731	.9736	.9741	.9745	.9750	.9754	.9759	.9763	.9768	.9773
9.5	.9777	.9782	.9786	.9791	.9795	.9800	.9805	.9809	.9814	.9818
9.6	.9823	.9827	.9832	.9836	.9841	.9845	.9850	.9854	.9859	.9863
9.7	.9868	.9872	.9877	.9881	.9886	.9890	.9894	.9899	.9903	.9908
9.8	.9912	.9917	.9921	.9926	.9930	.9934	.9939	.9943	.9948	.9952
9.9	.9956	.9961	.9965	.9969	.9974	.9978	.9983	.9987	.9991	.9996

Using a table to find a common logarithm is relatively easy but it does require a little more effort than pushing a key on a calculator. Each number in the column headed n represents the first two significant digits of a number between 1 and 10 and each of the column headings 0 through 9 represents the third significant digit. To find the logarithm of a number such as 1.75, we look at the intersection of the row that contains 1.7 and the column headed 5. Thus, we obtain

$$\log 1.75 = 0.2430.$$

Similarly, we can find that

$$\log 2.09 = 0.3201 \qquad \text{and} \qquad \log 2.40 = 0.3802;$$

keep in mind that these values are also rounded to four decimal places.

Now suppose that we want to use the table to find the logarithm of a positive number greater than 10 or less than 1. To accomplish this we represent the number in scientific notation and then apply the property $\log rs = \log r + \log s$. For example, to find log 134 we can proceed as follows.

$$\begin{aligned} \log 134 &= \log(1.34 \cdot 10^2) \\ &= \log 1.34 + \log 10^2 \\ &= 0.1271 + 2 = 2.1271 \end{aligned}$$

By inspection we know that the common logarithm of 10^2 is 2 (the exponent), and the common logarithm of 1.34 can be found in the table.

The decimal part (0.1271) of the logarithm 2.1271 is called the **mantissa**, and the integral part, (2), is called the **characteristic**. Thus, we can find the characteristic of a common logarithm by inspection (since it is the exponent of 10 when the number is written in scientific notation), and the mantissa we can get from a table. Let's consider two more examples.

$$\begin{aligned}\log 23.8 &= \log (2.38 \cdot 10^1)\\ &= \log 2.38 + \log 10^1\\ &= \underset{\substack{\uparrow\\ \text{From the table}}}{0.3766} + \underset{\substack{\nwarrow\\ \text{Exponent of 10}}}{1}\\ &= 1.3766\end{aligned}$$

$$\begin{aligned}\log 0.192 &= \log(1.92 \cdot 10^{-1})\\ &= \log 1.92 + \log 10^{-1}\\ &= \underset{\substack{\uparrow\\ \text{From the table}}}{0.2833} + \underset{\substack{\nwarrow\\ \text{Exponent of 10}}}{(-1)}\\ &= 0.2833 + (-1)\end{aligned}$$

Notice that in the last example we expressed the logarithm of 0.192 as $0.2833 + (-1)$; we did not add 0.2833 and -1. This is normal procedure when using a table of common logarithms because the mantissas given in the table are positive numbers. However, you should recognize that adding 0.2833 and -1 produces -0.7167, which agrees with the result obtained earlier with a calculator.

We can also use the table to find a number when given the common logarithm of the number. That is to say, given $\log x$ we can determine x from the table. Traditionally, x is referred to as the **antilogarithm** (abbreviated **antilog**) of $\log x$. Let's consider some examples.

EXAMPLE 1

Determine antilog 1.3365.

Solution

Finding an antilogarithm simply reverses the process used before for finding a logarithm. Thus, antilog 1.3365 means that 1 is the characteristic and 0.3365 the mantissa. We look for 0.3365 in the body of the common logarithm table and we find that it is located at the intersection of the 2.1-row and the 7-column. Therefore, the antilogarithm is

$$2.17 \cdot 10^1 = 21.7.$$

EXAMPLE 2

Determine antilog $(0.1523 + (-2))$.

Solution

The mantissa, 0.1523, is located at the intersection of the 1.4-row and 2-column. The characteristic is -2 and therefore the antilogarithm is

$$1.42 \cdot 10^{-2} = 0.0142.$$

EXAMPLE 3

Determine antilog -2.6038.

Solution

The mantissas given in a table are *positive* numbers. Thus, we need to express -2.6038 in terms of a positive mantissa and this can be done by adding and subtracting 3 as follows.

$$(-2.6038 + 3) - 3 = 0.3962 + (-3)$$

Now we can look for 0.3962 and find it at the intersection of the 2.4-row and 9-column. Therefore, the antilogarithm is

$$2.49 \cdot 10^{-3} = 0.00249.$$

Linear Interpolation

Now suppose that we want to determine log 2.774 from the table. Because the table contains only logarithms of numbers with, at most, three significant digits, we have a problem. However, by a process called **linear interpolation** we can extend the capabilities of the table to include numbers with four significant digits.

First, let's consider a geometric basis of linear interpolation and then we will use a systematic procedure for carrying out the necessary calculations. A portion of the graph of $y = \log x$, with the curvature exaggerated to help illustrate the principle involved, is shown in Figure A.1. The line segment that joins points P and Q is used to approximate the curve from P to Q. The actual value of log 2.744 is the ordinate of the point C, that is, the length of $\overline{AC}$. This cannot be determined from the table. Instead we will use the ordinate of point B (the length of $\overline{AB}$) as an approximation for log 2.744.

Consider Figure A.2 where line segments $\overline{DB}$ and $\overline{EQ}$ are drawn perpendicular to $\overline{PE}$. The right triangles formed, $\triangle PDB$ and $\triangle PEQ$, are similar and therefore the lengths of their corresponding sides are proportional. Thus, we can write

$$\frac{PD}{PE} = \frac{DB}{EQ}. \qquad (1)$$

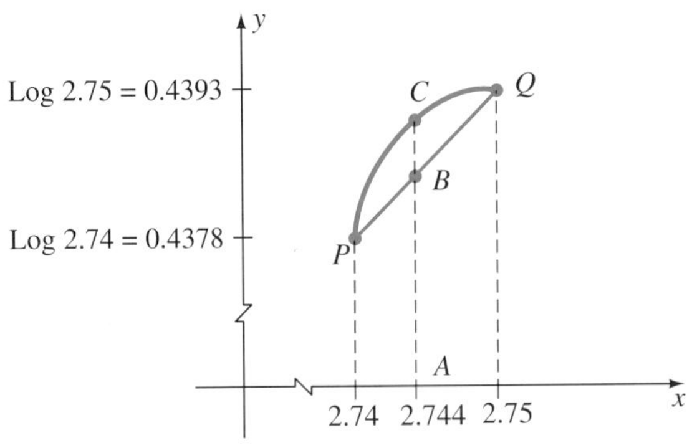

FIGURE A.1

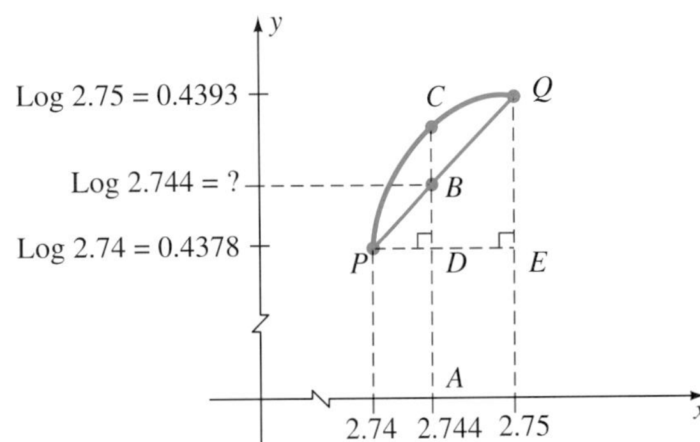

FIGURE A.2

From Figure A.2 we see that

$$PD = 2.744 - 2.74 = 0.004$$

$$PE = 2.75 - 2.74 = 0.01$$

$$EQ = 0.4393 - 0.4378 = 0.0015.$$

Therefore, the proportion (1) becomes

$$\frac{0.004}{0.01} = \frac{DB}{0.0015}.$$

Solving this proportion for DB yields

$$DB = 0.0006.$$

Since $AB = AD + DB$, we have

$$AB = 0.4378 + 0.0006 = 0.4384.$$

Thus, we obtain $\log 2.744 = 0.4384$.

Now let's suggest an abbreviated format for carrying out the calculations necessary to find log 2.744.

$$\begin{array}{cc} x & \log x \\ 4\left\{\begin{matrix}2.740\\2.744\end{matrix}\right. \left.\begin{matrix}\\ \\ \\ 2.750\end{matrix}\right\} 10 & k\left\{\begin{matrix}0.4378\\?\end{matrix}\right. \left.\begin{matrix}\\ \\ \\ 0.4393\end{matrix}\right\} 0.0015 \end{array}$$

Notice that we have used 4 and 10 for the differences for values of x instead of 0.004 and 0.01 because the ratio $\frac{0.004}{0.01}$ equals $\frac{4}{10}$. Setting up a proportion and solving for k yields

$$\frac{4}{10} = \frac{k}{0.0015}$$

$$10k = 4(0.0015) = 0.0060$$

$$k = 0.0006.$$

Thus, $\log 2.744 = 0.4378 + 0.0006 = 0.4384$.

Let's do another example to make sure of the process.

EXAMPLE 4

Find log 617.6.

Solution

$$\begin{aligned} \log 617.6 &= \log(6.176 \cdot 10^2) \\ &= \log 6.176 + \log 10^2 \end{aligned}$$

Thus, the characteristic is 2 and we can approximate the mantissa by using interpolation from the table as follows.

$$\begin{array}{cc} x & \log x \\ 6\left\{\begin{matrix} 6.170 \\ \\ 6.176 \end{matrix}\right. \left.\begin{matrix} \\ \\ \\ 6.180 \end{matrix}\right\} 10 & k\left\{\begin{matrix} 0.7903 \\ \\ ? \end{matrix}\right. \left.\begin{matrix} \\ \\ \\ 0.7910 \end{matrix}\right\} 0.007 \end{array}$$

$$\frac{6}{10} = \frac{k}{0.0007}$$

$$10k = 6(0.0007) = 0.0042$$

$$k = 0.00042 \approx 0.0004$$

Therefore, $\log 6.176 = 0.7903 + 0.0004 = 0.7907$ and we can complete the solution for $\log 617.6$ as follows.

$$\begin{aligned} \log 617.6 &= \log(6.176 \cdot 10^2) \\ &= \log 6.176 + \log 10^2 \\ &= 0.7907 + 2 \\ &= 2.7907 \end{aligned}$$

The process of linear interpolation can also be used to approximate an antilogarithm when the mantissa is in between two values in the table. The following example illustrates this procedure.

EXAMPLE 5

Find antilog 1.6157.

Solution

From the table we see that the mantissa, 0.6157, is between 0.6149 and 0.6160. We can carry out the interpolation as follows.

$$h\left\{\begin{matrix} 4.120 \\ \\ ? \end{matrix}\right. \left.\begin{matrix} \\ \\ \\ 4.130 \end{matrix}\right\} 0.0101 \qquad 8\left\{\begin{matrix} 0.6149 \\ \\ 0.6157 \end{matrix}\right. \left.\begin{matrix} \\ \\ \\ 0.6160 \end{matrix}\right\} 11 \qquad \frac{0.0008}{0.0011} = \frac{8}{11}$$

$$\frac{h}{0.010} = \frac{8}{11}$$

$$h = 8(0.010) = 0.080$$

$$h = \frac{1}{11}(0.080) = 0.007 \quad \text{to the nearest thousandth}$$

Thus, antilog 0.6157 = 4.120 + 0.007 = 4.127. Therefore,

$$\begin{aligned} \text{antilog } 1.6157 &= \text{antilog}(0.6157 + 1) \\ &= 4.127 \cdot 10^1 \\ &= 41.27. \end{aligned}$$

Computation with Common Logarithms

Let's first restate the basic properties of logarithms in terms of **common logarithms**. (Remember that we are writing $\log x$ instead of $\log_{10} x$.)

If x and y are positive real numbers, then

1. $\log xy = \log x + \log y$

2. $\log \frac{x}{y} = \log x - \log y$

3. $\log x^p = p \log x$. *p is any real number.*

The following two properties of equality that pertain to logarithms will also be used.

4. If $x = y$ (x and y are positive), then $\log x = \log y$.

5. If $\log x = \log y$, then $x = y$.

EXAMPLE 6

Find the product (49.1)(876).

Solution

Let $N = (49.1)(876)$; by Property 4,

$$\log N = \log(49.1)(876).$$

By Property 1,

$$\log N = \log 49.1 + \log 876.$$

From the table on page A2 we find that $\log 49.1 = 1.6911$ and that $\log 876 = 2.9425$. Thus,

$$\begin{aligned} \log N &= 1.6911 + 2.9425 \\ &= 4.6336. \end{aligned}$$

Therefore,

$$N = \text{antilog } 4.6336.$$

By using linear interpolation, we can determine antilog 0.6336 to four significant digits. Thus, we obtain

$$\begin{aligned} N &= \text{antilog } (0.6336 + 4) \\ &= 4.301 \cdot 10^4 \\ &= 43{,}010. \end{aligned}$$

Check By using a calculator we obtain

$$N = (49.1)(876) = 43011.6.$$

EXAMPLE 7

Find the quotient $\dfrac{942}{64.8}$.

Solution

Let $N = \dfrac{94.2}{64.8}$. Therefore,

$$\begin{aligned}\log N &= \log \frac{942}{64.8} \\ &= \log 942 - \log 64.8 && \log \frac{x}{y} = \log x - \log y \\ &= 2.9741 - 1.8116 && \text{from the table} \\ &= 1.1625.\end{aligned}$$

Therefore,

$$\begin{aligned}N &= \text{antilog } 1.1625 \\ &= \text{antilog}(0.01625 + 1) \\ &= 1.454 \cdot 10^1 \\ &= 14.54.\end{aligned}$$

Check By using a calculator we obtain

$$N = \frac{942}{64.8} = 14.537037.$$

EXAMPLE 8

Evaluate $\dfrac{(571.4)(8.236)}{71.68}$.

Solution

Let $N = \dfrac{(571.4)(8.236)}{71.68}$. Therefore,

$$\begin{aligned}\log N &= \log \frac{(571.4)(8.236)}{71.68} \\ &= \log 571.4 + \log 8.236 - \log 71.68 \\ &= 2.7569 + 0.9157 - 1.8554 = 1.8172.\end{aligned}$$

Therefore,

$$\begin{aligned}N &= \text{antilog } 1.8172 \\ &= \text{antilog}(0.8172 + 1) \\ &= 6.564 \cdot 10^1 = 65.64.\end{aligned}$$

Check By using a calculator we obtain

$$N = \frac{(571.4)(8.236)}{71.68} = 65.653605.$$

EXAMPLE 9

Evaluate $\sqrt[3]{3770}$.

Solution

Let $N = \sqrt[3]{3770} = (3770)^{1/3}$. Therefore,

$$\begin{aligned} \log N &= \log(3770)^{1/3} \\ &= \frac{1}{3}\log 3770 \qquad \log x^p = p \log x \\ &= \frac{1}{3}(3.5763) \\ &= 1.1921. \end{aligned}$$

Therefore,

$$\begin{aligned} N &= \text{antilog } 1.1921 \\ &= \text{antilog}(0.1921 + 1) \\ &= 1.556 \cdot 10^1 = 15.56. \end{aligned}$$

Check By using a calculator we obtain

$$N = \sqrt[3]{3370} = 15.563733.$$

When using tables of logarithms, it is sometimes necessary to change the form of writing a logarithm so that the decimal part (mantissa) is positive. The next example illustrates this idea.

EXAMPLE 10

Find the quotient $\frac{1.73}{5.08}$.

Solution

Let $N = \frac{1.73}{5.08}$. Therefore,

$$\begin{aligned} \log N &= \log \frac{1.73}{5.08} \\ &= \log 1.73 - \log 5.08 \\ &= 0.2380 - 0.7059 = -0.4679. \end{aligned}$$

Now by adding 1 and subtracting 1, which changes the form but not the value, we obtain

$$\begin{aligned} \log N &= -0.4679 + 1 - 1 \\ &= 0.5321 - 1 \\ &= 0.5321 + (-1). \end{aligned}$$

Therefore,

$$\begin{aligned} N &= \text{antilog}(0.5321 + (-1)) \\ &= 3.405 \cdot 10^{-1} = 0.3405. \end{aligned}$$

Check By using a calculator we obtain

$$N = \frac{1.73}{5.08} = 0.34055118.$$

Sometimes it is also necessary to change the form of a logarithm so that a subsequent calculation will produce an **integer for the characteristic part of the logarithm**. Let's consider an example to illustrate this idea.

EXAMPLE 11

Evaluate $\sqrt[4]{0.0767}$.

Solution

Let $N = \sqrt[4]{0.0767} = (0.0767)^{1/4}$. Therefore,

$$\begin{aligned}\log N &= \log(0.0767)^{1/4} = \frac{1}{4}\log 0.0767\\ &= \frac{1}{4}(0.8848 + (-2))\\ &= \frac{1}{4}(-2 + 0.8848).\end{aligned}$$

At this stage we recognize that applying the distributive property will produce a nonintegral characteristic, namely, $-\frac{1}{2}$. Therefore, let's add 4 and subtract 4 inside the parentheses, which will change the form as follows.

$$\begin{aligned}\log N &= \frac{1}{4}(-2 + 0.8848 + 4 - 4)\\ &= \frac{1}{4}(4 - 2 + 0.8848 - 4)\\ &= \frac{1}{4}(2.8848 - 4)\end{aligned}$$

Now applying the distributive property we obtain

$$\begin{aligned}\log N &= \frac{1}{4}(2.8848) - \frac{1}{4}(4)\\ &= 0.7212 - 1 = 0.7212 + (-1).\end{aligned}$$

Therefore,

$$\begin{aligned}N &= \text{antilog}(0.7212 + (-1))\\ &= 5.262 \cdot 10^{-1} = 0.5262.\end{aligned}$$

Check By using a calculator we obtain

$$N = \sqrt[4]{0.0767} = 0.5262816.$$

PRACTICE EXERCISES

Use the table of common logarithms and linear interpolation to find each of the following common logarithms.

1. log 4.327
2. log 27.43
3. log 128.9
4. log 3526
5. log 0.8761
6. log 0.07692
7. log 0.005186
8. log 0.0002558

Use the table of commmon logarithms and linear interpolation to find each of the following antilogarithms to four significant digits.

9. antilog 0.4690
10. antilog 1.7971
11. antilog 2.1925
12. antilog 3.7225
13. antilog(0.5026 + (−1))
14. antilog(0.9397 + (−2))

Use common logarithms and linear interpolation to help evaluate each of the following. Express your answers with four significant digits. Check your answers by using a calculator.

15. (294)(71.2)
16. (192.6)(4.017)
17. $\frac{23.4}{4.07}$
18. $\frac{718.5}{8.248}$
19. $(17.3)^5$
20. $(48.02)^3$
21. $\frac{(108)(76.2)}{13.4}$
22. $\frac{(126.3)(24.32)}{8.019}$
23. $\sqrt[5]{0.821}$
24. $\sqrt[4]{645.3}$
25. $(79.3)^{3/5}$
26. $(176.8)^{3/4}$
27. $\sqrt{\frac{(7.05)(18.7)}{0.521}}$
28. $\sqrt[3]{\frac{(41.3)(0.271)}{8.05}}$

APPENDIX

B

NATURAL LOGARITHMS

The following table contains the natural logarithms for numbers between 0.1 and 10, inclusive, at intervals of 0.1. Be sure that you agree with the following values taken directly from the table.

$$\ln 1.6 = 0.4700$$

$$\ln 0.5 = -0.6931$$

$$\ln 4.8 = 1.5686$$

$$\ln 9.2 = 2.2192$$

Table of Natural Logarithms

n	$\ln n$	n	$\ln n$	n	$\ln n$	n	$\ln n$
0.1	−2.3026	2.6	0.9555	5.1	1.6292	7.6	2.0281
0.2	−1.6094	2.7	0.9933	5.2	1.6487	7.7	2.0412
0.3	−1.2040	2.8	1.0296	5.3	1.6677	7.8	2.0541
0.4	−0.9163	2.9	1.0647	5.4	1.6864	7.9	2.0669
0.5	−0.6931	3.0	1.0986	5.5	1.7047	8.0	2.0794
0.6	−0.5108	3.1	1.1314	5.6	1.7228	8.1	2.0919
0.7	−0.3567	3.2	1.1632	5.7	1.7405	8.2	2.1041
0.8	−0.2231	3.3	1.1939	5.8	1.7579	8.3	2.1163
0.9	−0.1054	3.4	1.2238	5.9	1.7750	8.4	2.1282
1.0	0.0000	3.5	1.2528	6.0	1.7918	8.5	2.1401
1.1	0.0953	3.6	1.2809	6.1	1.8083	8.6	2.1518
1.2	0.1823	3.7	1.3083	6.2	1.8245	8.7	2.1633
1.3	0.2624	3.8	1.3350	6.3	1.8405	8.8	2.1748
1.4	0.3365	3.9	1.3610	6.4	1.8563	8.9	2.1861
1.5	0.4055	4.0	1.3863	6.5	1.8718	9.0	2.1972

Table of Natural Logarithms (continued)

n	ln n	n	ln n	n	ln n	n	ln n
1.6	0.4700	4.1	1.4110	6.6	1.8871	9.1	2.2083
1.7	0.5306	4.2	1.4351	6.7	1.9021	9.2	2.2192
1.8	0.5878	4.3	1.4586	6.8	1.9169	9.3	2.2300
1.9	0.6419	4.4	1.4816	6.9	1.9315	9.4	2.2407
2.0	0.6931	4.5	1.5041	7.0	1.9459	9.5	2.2513
2.1	0.7419	4.6	1.5261	7.1	1.9601	9.6	2.2618
2.2	0.7885	4.7	1.5476	7.2	1.9741	9.7	2.2721
2.3	0.8329	4.8	1.5686	7.3	1.9879	9.8	2.2824
2.4	0.8755	4.9	1.5892	7.4	2.0015	9.9	2.2925
2.5	0.9163	5.0	1.6094	7.5	2.0149	10	2.3026

When using a table, the natural logarithm of a positive number less than 0.1 or greater than 10 can be approximated by using the property $\ln rs = \ln r + \ln s$ as follows.

$$\begin{aligned}
\ln 190 &= \ln(1.9 \cdot 10^2) \\
&= \ln 1.9 + \ln 10^2 \\
&= \ln 1.9 + 2 \ln 10 \\
&= \underset{\substack{\uparrow \\ \text{From the} \\ \text{table}}}{0.6419} + 2\underset{\substack{\uparrow \\ \text{From the} \\ \text{table}}}{(2.3026)} \\
&= 5.2471
\end{aligned}$$

$$\begin{aligned}
\ln 0.0084 &= \ln(8.4 \cdot 10^{-3}) \\
&= \ln 8.4 + \ln 10^{-3} \\
&= \ln 8.4 + (-3)(\ln 10) \\
&= \underset{\substack{\uparrow \\ \text{From the} \\ \text{table}}}{2.1282} - 3\underset{\substack{\uparrow \\ \text{From the} \\ \text{table}}}{(2.3026)} \\
&= 2.1282 - 6.9078 = -4.7796
\end{aligned}$$

APPENDIX

C

TABLE OF TRIGONOMETRIC VALUES

The table on page A17 contains trigonometric values for angles measured in degrees or radians. Degree measures are given in 0.1° intervals from 0.0° to 45.0° in the second column from the left, and from 45.0° to 90.0° in the second column from the right. Since $0.1° \approx 0.0017$ radians, the radian measures are given in intervals of approximately 0.0017. The following examples illustrate the use of the table.

Degree Measure

EXAMPLE 1

Find cos 32.4°.

Solution

We locate 32.4° by reading down in the second column from the left. Then we read across to the column labeled COS on *top* and we obtain

$$\cos 32.4° = 0.8443.$$

EXAMPLE 2

Find sin 73.8°.

Solution

We locate 73.8° by reading up in the second column from the right. Then we read across to the column labeled SIN at the *bottom* to obtain

$$\sin 73.8° = 0.9603.$$

EXAMPLE 3

Find θ if $\tan \theta = 1.076$.

Solution

We locate 1.076 in the column labeled TAN at the *bottom*. (If $0 < \tan \theta < 1$, then it would be located in the column labeled TAN at the *top*.) Reading across to the *right* in the degree column, we obtain

$$\theta = 47.1.$$

EXAMPLE 4

Find θ, to the nearest tenth of a degree, if $\sin \theta = 0.0562$.

Solution

Reading down in the column labeled SIN at the *top*, we see that 0.0562 falls between 0.0558 and 0.0576. Since 0.0562 is closer to 0.0558, we obtain

$$\theta = 3.2°$$

Radian Measure

EXAMPLE 5

Find cos 1.4556.

Solution

We locate 1.4556 in the column on the far right labeled RADIANS at the *bottom*. Then we read across to the column labeled COS at the *bottom* to obtain

$$\cos 1.4556 = 0.1149.$$

EXAMPLE 6

Find x to the nearest hundredth of a radian, if $\cos \theta = 0.0650$.

Solution

In the column labeled COS at the *bottom*, we find that 0.0650 falls between 0.0645 and 0.0663. Since 0.0650 is closer to 0.0645, we obtain $x = 1.5062$ by reading across in the far right column. Therefore, to the nearest hundredth of a radian, we obtain

$$x = 1.51.$$

Values of the Trigonometric Functions

RADIANS	DEGREES	SIN	COS	TAN	COT		
0.0000	0.0°	0.0000	1.0000	0.0000	—	90.0°	1.5708
0.0017	0.1°	0.0017	1.0000	0.0017	573.0	89.9°	1.5691
0.0035	0.2°	0.0035	1.0000	0.0035	286.5	89.8°	1.5673
0.0052	0.3°	0.0052	1.0000	0.0052	191.0	89.7°	1.5656
0.0070	0.4°	0.0070	1.0000	0.0070	143.2	89.6°	1.5638
0.0087	0.5°	0.0087	1.0000	0.0087	114.6	89.5°	1.5621
0.0105	0.6°	0.0105	0.9999	0.0105	95.49	89.4°	1.5603
0.0122	0.7°	0.0122	0.9999	0.0122	81.85	89.3°	1.5586
0.0140	0.8°	0.0140	0.9999	0.0140	71.62	89.2°	1.5568
0.0157	0.9°	0.0157	0.9999	0.0157	63.66	89.1°	1.5551
0.0175	1.0°	0.0175	0.9998	0.0175	57.29	89.0°	1.5533
0.0192	1.1°	0.0192	0.9998	0.0192	52.08	88.9°	1.5516
0.0209	1.2°	0.0209	0.9998	0.0209	47.74	88.8°	1.5499
0.0227	1.3°	0.0227	0.9997	0.0227	44.07	88.7°	1.5481
0.0244	1.4°	0.0244	0.9997	0.0244	40.92	88.6°	1.5464
0.0262	1.5°	0.0262	0.9997	0.0262	38.19	88.5°	1.5446
0.0279	1.6°	0.0279	0.9996	0.0279	35.80	88.4°	1.5429
0.0297	1.7°	0.0297	0.9996	0.0297	33.69	88.3°	1.5411
0.0314	1.8°	0.0314	0.9995	0.0314	31.82	88.2°	1.5394
0.0332	1.9°	0.0332	0.9995	0.0332	30.14	88.1°	1.5376
0.0349	2.0°	0.0349	0.9994	0.0349	28.64	88.0°	1.5359
0.0367	2.1°	0.0366	0.9993	0.0367	27.27	87.9°	1.5341
0.0384	2.2°	0.0384	0.9993	0.0384	26.03	87.8°	1.5324
0.0401	2.3°	0.0401	0.9992	0.0402	24.90	87.7°	1.5307
0.0419	2.4°	0.0419	0.9991	0.0419	23.86	87.6°	1.5289
0.0436	2.5°	0.0436	0.9990	0.0437	22.90	87.5°	1.5272
0.0454	2.6°	0.0454	0.9990	0.0454	22.02	87.4°	1.5254
0.0471	2.7°	0.0471	0.9989	0.0472	21.20	87.3°	1.5237
0.0489	2.8°	0.0488	0.9988	0.0489	20.45	87.2°	1.5219
0.0506	2.9°	0.0506	0.9987	0.0507	19.74	87.1°	1.5202
0.0524	3.0°	0.0523	0.9986	0.0524	19.08	87.0°	1.5184
0.0541	3.1°	0.0541	0.9985	0.0542	18.46	86.9°	1.5167
0.0559	3.2°	0.0558	0.9984	0.0559	17.89	86.8°	1.5149
0.0576	3.3°	0.0576	0.9983	0.0577	17.34	86.7°	1.5132
0.0593	3.4°	0.0593	0.9982	0.0594	16.83	86.6°	1.5115
0.0611	3.5°	0.0610	0.9981	0.0612	16.35	86.5°	1.5097
0.0628	3.6°	0.0628	0.9980	0.0629	15.89	86.4°	1.5080
0.0646	3.7°	0.0645	0.9979	0.0647	15.46	86.3°	1.5062
0.0663	3.8°	0.0663	0.9978	0.0664	15.06	86.2°	1.5045
0.0681	3.9°	0.0680	0.9977	0.0682	14.67	86.1°	1.5027
		COS	SIN	COT	TAN	DEGREES	RADIANS

Values of the Trigonometric Functions (continued)

RADIANS	DEGREES	SIN	COS	TAN	COT		
0.0698	4.0°	0.0698	0.9976	0.0699	14.30	86.0°	1.5010
0.0716	4.1°	0.0715	0.9974	0.0717	13.95	85.9°	1.4992
0.0733	4.2°	0.0732	0.9973	0.0734	13.62	85.8°	1.4975
0.0750	4.3°	0.0750	0.9972	0.0752	13.30	85.7°	1.4957
0.0768	4.4°	0.0767	0.9971	0.0769	13.00	85.6°	1.4940
0.0785	4.5°	0.0785	0.9969	0.0787	12.71	85.5°	1.4923
0.0803	4.6°	0.0802	0.9968	0.0805	12.43	85.4°	1.4905
0.0820	4.7°	0.0819	0.9966	0.0822	12.16	85.3°	1.4888
0.0838	4.8°	0.0837	0.9965	0.0840	11.91	85.2°	1.4870
0.0855	4.9°	0.0854	0.9963	0.0857	11.66	85.1°	1.4853
0.0873	5.0°	0.0872	0.9962	0.0875	11.43	85.0°	1.4835
0.0890	5.1°	0.0889	0.9960	0.0892	11.20	84.9°	1.4818
0.0908	5.2°	0.0906	0.9959	0.0910	10.99	84.8°	1.4800
0.0925	5.3°	0.0924	0.9957	0.0928	10.78	84.7°	1.4783
0.0942	5.4°	0.0941	0.9956	0.0945	10.58	84.6°	1.4765
0.0960	5.5°	0.0958	0.9954	0.0963	10.39	84.5°	1.4748
0.0977	5.6°	0.0976	0.9952	0.0981	10.20	84.4°	1.4731
0.0995	5.7°	0.0993	0.9951	0.0998	10.02	84.3°	1.4713
0.1012	5.8°	0.1011	0.9949	0.1016	9.845	84.2°	1.4696
0.1030	5.9°	0.1028	0.9947	0.1033	9.677	84.1°	1.4678
0.1047	6.0°	0.1045	0.9945	0.1051	9.514	84.0°	1.4661
0.1065	6.1°	0.1063	0.9943	0.1069	9.357	83.9°	1.4643
0.1082	6.2°	0.1080	0.9942	0.1086	9.205	83.8°	1.4626
0.1100	6.3°	0.1097	0.9940	0.1104	9.058	83.7°	1.4608
0.1117	6.4°	0.1115	0.9938	0.1122	8.915	83.6°	1.4591
0.1134	6.5°	0.1132	0.9936	0.1139	8.777	83.5°	1.4573
0.1152	6.6°	0.1149	0.9934	0.1157	8.643	83.4°	1.4556
0.1169	6.7°	0.1167	0.9932	0.1175	8.513	83.3°	1.4539
0.1187	6.8°	0.1184	0.9930	0.1192	8.386	83.2°	1.4521
0.1204	6.9°	0.1201	0.9928	0.1210	8.264	83.1	1.4504
0.1222	7.0°	0.1219	0.9925	0.1228	8.144	83.0°	1.4486
0.1239	7.1°	0.1236	0.9923	0.1246	8.028	82.9°	1.4469
0.1257	7.2°	0.1253	0.9921	0.1263	7.916	82.8°	1.4451
0.1274	7.3°	0.1271	0.9919	0.1281	7.806	82.7°	1.4434
0.1292	7.4°	0.1288	0.9917	0.1299	7.700	82.6°	1.4416
0.1309	7.5°	0.1305	0.9914	0.1317	7.596	82.5°	1.4399
0.1326	7.6°	0.1323	0.9912	0.1334	7.495	82.4°	1.4382
0.1344	7.7°	0.1340	0.9910	0.1352	7.396	82.3°	1.4364
0.1361	7.8°	0.1357	0.9907	0.1370	7.300	82.2°	1.4347
0.1379	7.9°	0.1374	0.9905	0.1388	7.207	82.1°	1.4329
		COS	SIN	COT	TAN	DEGREES	RADIANS

Values of the Trigonometric Functions (continued)

RADIANS	DEGREES	SIN	COS	TAN	COT		
0.1396	8.0°	0.1392	0.9903	0.1405	7.115	82.0°	1.4312
0.1414	8.1°	0.1409	0.9900	0.1423	7.026	81.9°	1.4294
0.1431	8.2°	0.1426	0.9898	0.1441	6.940	81.8°	1.4277
0.1449	8.3°	0.1444	0.9895	0.1459	6.855	81.7°	1.4259
0.1466	8.4°	0.1461	0.9893	0.1477	6.772	81.6°	1.4242
0.1484	8.5°	0.1478	0.9890	0.1495	6.691	81.5°	1.4224
0.1501	8.6°	0.1495	0.9888	0.1512	6.612	81.4°	1.4207
0.1518	8.7°	0.1513	0.9885	0.1530	6.535	81.3°	1.4190
0.1536	8.8°	0.1530	0.9882	0.1548	6.460	81.2°	1.4172
0.1553	8.9°	0.1547	0.9880	0.1566	6.386	81.1°	1.4155
0.1571	9.0°	0.1564	0.9877	0.1584	6.314	81.0°	1.4137
0.1588	9.1°	0.1582	0.9874	0.1602	6.243	80.9°	1.4120
0.1606	9.2°	0.1599	0.9871	0.1620	6.174	80.8°	1.4102
0.1623	9.3°	0.1616	0.9869	0.1638	6.107	80.7°	1.4085
0.1641	9.4°	0.1633	0.9866	0.1655	6.041	80.6°	1.4067
0.1658	9.5°	0.1650	0.9863	0.1673	5.976	80.5°	1.4050
0.1676	9.6°	0.1668	0.9860	0.1691	5.912	80.4°	1.4032
0.1693	9.7°	0.1685	0.9857	0.1709	5.850	80.3°	1.4015
0.1710	9.8°	0.1702	0.9854	0.1727	5.789	80.2°	1.3998
0.1728	9.9°	0.1719	0.9851	0.1745	5.730	80.1°	1.3980
0.1745	10.0°	0.1736	0.9848	0.1763	5.671	80.0°	1.3963
0.1763	10.1°	0.1754	0.9845	0.1781	5.614	79.9°	1.3945
0.1780	10.2°	0.1771	0.9842	0.1799	5.558	79.8°	1.3928
0.1798	10.3°	0.1788	0.9839	0.1817	5.503	79.7°	1.3910
0.1815	10.4°	0.1805	0.9836	0.1835	5.449	79.6°	1.3893
0.1833	10.5°	0.1822	0.9833	0.1853	5.396	79.5°	1.3875
0.1850	10.6°	0.1840	0.9829	0.1871	5.343	79.4°	1.3858
0.1868	10.7°	0.1857	0.9826	0.1890	5.292	79.3°	1.3840
0.1885	10.8°	0.1874	0.9823	0.1908	5.242	79.2°	1.3823
0.1902	10.9°	0.1891	0.9820	0.1926	5.193	79.1°	1.3806
0.1920	11.0°	0.1908	0.9816	0.1944	5.145	79.0°	1.3788
0.1937	11.1°	0.1925	0.9813	0.1962	5.097	78.9°	1.3771
0.1955	11.2°	0.1942	0.9810	0.1980	5.050	78.8°	1.3753
0.1972	11.3°	0.1959	0.9806	0.1998	5.005	78.7°	1.3736
0.1990	11.4°	0.1977	0.9803	0.2016	4.959	78.6°	1.3718
0.2007	11.5°	0.1994	0.9799	0.2035	4.915	78.5°	1.3701
0.2025	11.6°	0.2011	0.9796	0.2053	4.872	78.4°	1.3683
0.2042	11.7°	0.2028	0.9792	0.2071	4.829	78.3°	1.3666
0.2059	11.8°	0.2045	0.9789	0.2089	4.787	78.2°	1.3648
0.2077	11.9°	0.2062	0.9785	0.2107	4.745	78.1°	1.3631
		COS	SIN	COT	TAN	DEGREES	RADIANS

Values of the Trigonometric Functions (continued)

RADIANS	DEGREES	SIN	COS	TAN	COT		
0.2094	12.0°	0.2079	0.9781	0.2126	4.705	78.0°	1.3614
0.2112	12.1°	0.2096	0.9778	0.2144	4.665	77.9°	1.3596
0.2129	12.2°	0.2113	0.9774	0.2162	4.625	77.8°	1.3579
0.2147	12.3°	0.2130	0.9770	0.2180	4.586	77.7°	1.3561
0.2164	12.4°	0.2147	0.9767	0.2199	4.548	77.6°	1.3544
0.2182	12.5°	0.2164	0.9763	0.2217	4.511	77.5°	1.3526
0.2199	12.6°	0.2181	0.9759	0.2235	4.474	77.4°	1.3509
0.2217	12.7°	0.2198	0.9755	0.2254	4.437	77.3°	1.3491
0.2234	12.8°	0.2215	0.9751	0.2272	4.402	77.2°	1.3474
0.2251	12.9°	0.2233	0.9748	0.2290	4.366	77.1°	1.3456
0.2269	13.0°	0.2250	0.9744	0.2309	4.331	77.0°	1.3439
0.2286	13.1°	0.2267	0.9740	0.2327	4.297	76.9°	1.3422
0.2304	13.2°	0.2284	0.9736	0.2345	4.264	76.8°	1.3404
0.2321	13.3°	0.2300	0.9732	0.2364	4.230	76.7°	1.3387
0.2339	13.4°	0.2317	0.9728	0.2382	4.198	76.6°	1.3369
0.2356	13.5°	0.2334	0.9724	0.2401	4.165	76.5°	1.3352
0.2374	13.6°	0.2351	0.9720	0.2419	4.134	76.4°	1.3334
0.2391	13.7°	0.2368	0.9715	0.2438	4.102	76.3°	1.3317
0.2409	13.8°	0.2385	0.9711	0.2456	4.071	76.2°	1.3299
0.2426	13.9°	0.2402	0.9707	0.2475	4.041	76.1°	1.3282
0.2443	14.0°	0.2419	0.9703	0.2493	4.011	76.0°	1.3265
0.2461	14.1°	0.2436	0.9699	0.2512	3.981	75.9°	1.3247
0.2478	14.2°	0.2453	0.9694	0.2530	3.952	75.8°	1.3230
0.2496	14.3°	0.2470	0.9690	0.2549	3.923	75.7°	1.3212
0.2513	14.4°	0.2487	0.9686	0.2568	3.895	75.6°	1.3195
0.2531	14.5°	0.2504	0.9681	0.2586	3.867	75.5°	1.3177
0.2548	14.6°	0.2521	0.9677	0.2605	3.839	75.4°	1.3160
0.2566	14.7°	0.2538	0.9673	0.2623	3.812	75.3°	1.3142
0.2583	14.8°	0.2554	0.9668	0.2642	3.785	75.2°	1.3125
0.2601	14.9°	0.2571	0.9664	0.2661	3.758	75.1°	1.3107
0.2618	15.0°	0.2588	0.9659	0.2679	3.732	75.0°	1.3090
0.2635	15.1°	0.2605	0.9655	0.2698	3.706	74.9°	1.3073
0.2653	15.2°	0.2622	0.9650	0.2717	3.681	74.8°	1.3055
0.2670	15.3°	0.2639	0.9646	0.2736	3.655	74.7°	1.3038
0.2688	15.4°	0.2656	0.9641	0.2754	3.630	74.6°	1.3020
0.2705	15.5°	0.2672	0.9636	0.2773	3.606	74.5°	1.3003
0.2723	15.6°	0.2689	0.9632	0.2792	3.582	74.4°	1.2985
0.2740	15.7°	0.2706	0.9627	0.2811	3.558	74.3°	1.2968
0.2758	15.8°	0.2723	0.9622	0.2830	3.534	74.2°	1.2950
0.2775	15.9°	0.2740	0.9617	0.2849	3.511	74.1°	1.2933
		COS	SIN	COT	TAN	DEGREES	RADIANS

Values of the Trigonometric Functions (continued)

RADIANS	DEGREES	SIN	COS	TAN	COT		
0.2793	16.0°	0.2756	0.9613	0.2867	3.487	74.0°	1.2915
0.2810	16.1°	0.2773	0.9608	0.2886	3.465	73.9°	1.2898
0.2827	16.2°	0.2790	0.9603	0.2905	3.442	73.8°	1.2881
0.2845	16.3°	0.2807	0.9598	0.2924	3.420	73.7°	1.2863
0.2862	16.4°	0.2823	0.9593	0.2943	3.398	73.6°	1.2846
0.2880	16.5°	0.2840	0.9588	0.2962	3.376	73.5°	1.2828
0.2897	16.6°	0.2857	0.9583	0.2981	3.354	73.4°	1.2811
0.2915	16.7°	0.2874	0.9578	0.3000	3.333	73.3°	1.2793
0.2932	16.8°	0.2890	0.9673	0.3019	3.312	73.2°	1.2776
0.2950	16.9°	0.2907	0.9568	0.3038	3.291	73.1°	1.2758
0.2967	17.0°	0.2924	0.9563	0.3057	3.271	73.0°	1.2741
0.2985	17.1°	0.2940	0.9558	0.3076	3.251	72.9°	1.2723
0.3002	17.2°	0.2957	0.9553	0.3096	3.230	72.8°	1.2706
0.3019	17.3°	0.2974	0.9548	0.3115	3.211	72.7°	1.2689
0.3037	17.4°	0.2990	0.9542	0.3134	3.191	72.6°	1.2671
0.3054	17.5°	0.3007	0.9537	0.3153	3.172	72.5°	1.2654
0.3072	17.6°	0.3024	0.9532	0.3172	3.152	72.4°	1.2636
0.3089	17.7°	0.3040	0.9527	0.3191	3.133	72.3°	1.2619
0.3107	17.8°	0.3057	0.9521	0.3211	3.115	72.2°	1.2601
0.3124	17.9°	0.3074	0.9516	0.3230	3.096	72.1°	1.2584
0.3142	18.0°	0.3090	0.9511	0.3249	3.078	72.0°	1.2566
0.3159	18.1°	0.3107	0.9505	0.3269	3.060	71.9°	1.2549
0.3176	18.2°	0.3123	0.9500	0.3288	3.042	71.8°	1.2531
0.3194	18.3°	0.3140	0.9494	0.3307	3.024	71.7°	1.2514
0.3211	18.4°	0.3156	0.9489	0.3327	3.006	71.6°	1.2497
0.3229	18.5°	0.3173	0.9483	0.3346	2.989	71.5°	1.2479
0.3246	18.6°	0.3190	0.9478	0.3365	2.971	71.4°	1.2462
0.3264	18.7°	0.3206	0.9472	0.3385	2.954	71.3°	1.2444
0.3281	18.8°	0.3223	0.9466	0.3404	2.937	71.2°	1.2427
0.3299	18.9°	0.3239	0.9461	0.3424	2.921	71.1°	1.2409
0.3316	19.0°	0.3256	0.9455	0.3443	2.904	71.0°	1.2392
0.3334	19.1°	0.3272	0.9449	0.3463	2.888	70.9°	1.2374
0.3351	19.2°	0.3289	0.9444	0.3482	2.872	70.8°	1.2357
0.3368	19.3°	0.3305	0.9438	0.3502	2.856	70.7°	1.2339
0.3386	19.4°	0.3322	0.9432	0.3522	2.840	70.6°	1.2322
0.3403	19.5°	0.3338	0.9426	0.3541	2.824	70.5°	1.2305
0.3421	19.6°	0.3355	0.9421	0.3561	2.808	70.4°	1.2287
0.3438	19.7°	0.3371	0.9415	0.3581	2.793	70.3°	1.2270
0.3456	19.8°	0.3387	0.9409	0.3600	2.778	70.2°	1.2252
0.3473	19.9°	0.3404	0.9403	0.3620	2.762	70.1°	1.2235
		COS	SIN	COT	TAN	DEGREES	RADIANS

Values of the Trigonometric Functions (continued)

RADIANS	DEGREES	SIN	COS	TAN	COT		
0.3491	20.0°	0.3420	0.9397	0.3640	2.747	70.0°	1.2217
0.3508	20.1°	0.3437	0.9391	0.3659	2.733	69.9°	1.2200
0.3526	20.2°	0.3453	0.9385	0.3679	2.718	69.8°	1.2182
0.3543	20.3°	0.3469	0.9379	0.3699	2.703	69.7°	1.2165
0.3560	20.4°	0.3486	0.9373	0.3719	2.689	69.6°	1.2147
0.3578	20.5°	0.3502	0.9367	0.3739	2.675	69.5°	1.2130
0.3595	20.6°	0.3518	0.9361	0.3759	2.660	69.4°	1.2113
0.3613	20.7°	0.3535	0.9354	0.3779	2.646	69.3°	1.2095
0.3630	20.8°	0.3551	0.9348	0.3799	2.633	69.2°	1.2078
0.3648	20.9°	0.3567	0.9342	0.3819	2.619	69.1°	1.2060
0.3665	21.0°	0.3584	0.9336	0.3839	2.605	69.0°	1.2043
0.3683	21.1°	0.3600	0.9330	0.3859	2.592	68.9°	1.2025
0.3700	21.2°	0.3616	0.9323	0.3879	2.578	68.8°	1.2008
0.3718	21.3°	0.3633	0.9317	0.3899	2.565	68.7°	1.1990
0.3735	21.4°	0.3649	0.9311	0.3919	2.552	68.6°	1.1973
0.3752	21.5°	0.3665	0.9304	0.3939	2.539	68.5°	1.1956
0.3770	21.6°	0.3681	0.9298	0.3959	2.526	68.4°	1.1938
0.3787	21.7°	0.3697	0.9291	0.3979	2.513	68.3°	1.1921
0.3805	21.8°	0.3714	0.9285	0.4000	2.500	68.2°	1.1903
0.3822	21.9°	0.3730	0.9278	0.4020	2.488	68.1°	1.1886
0.3840	22.0°	0.3746	0.9272	0.4040	2.475	68.0°	1.1868
0.3857	22.1°	0.3762	0.9265	0.4061	2.463	67.9°	1.1851
0.3875	22.2°	0.3778	0.9259	0.4081	2.450	67.8°	1.1833
0.3892	22.3°	0.3795	0.9252	0.4101	2.438	67.7°	1.1816
0.3910	22.4°	0.3811	0.9245	0.4122	2.426	67.6°	1.1798
0.3927	22.5°	0.3827	0.9239	0.4142	2.414	67.5°	1.1781
0.3944	22.6°	0.3843	0.9232	0.4163	2.402	67.4°	1.1764
0.3962	22.7°	0.3859	0.9225	0.4183	2.391	67.3°	1.1746
0.3979	22.8°	0.3875	0.9219	0.4204	2.379	67.2°	1.1729
0.3997	22.9°	0.3891	0.9212	0.4224	2.367	67.1°	1.1711
0.4014	23.0°	0.3907	0.9205	0.4245	2.356	67.0°	1.1694
0.4032	23.1°	0.3923	0.9198	0.4265	2.344	66.9°	1.1676
0.4049	23.2°	0.3939	0.9191	0.4286	2.333	66.8°	1.1659
0.4067	23.3°	0.3955	0.9184	0.4307	2.322	66.7°	1.1641
0.4084	23.4°	0.3971	0.9178	0.4327	2.311	66.6°	1.1624
0.4102	23.5°	0.3987	0.9171	0.4348	2.300	66.5°	1.1606
0.4119	23.6°	0.4003	0.9164	0.4369	2.289	66.4°	1.1589
0.4136	23.7°	0.4019	0.9157	0.4390	2.278	66.3°	1.1572
0.4154	23.8°	0.4035	0.9150	0.4411	2.267	66.2°	1.1554
0.4171	23.9°	0.4051	0.9143	0.4431	2.257	66.1°	1.1537
		COS	SIN	COT	TAN	DEGREES	RADIANS

Values of the Trigonometric Functions (continued)

RADIANS	DEGREES	SIN	COS	TAN	COT		
0.4189	24.0°	0.4067	0.9135	0.4452	2.246	66.0°	1.1519
0.4206	24.1°	0.4083	0.9128	0.4473	2.236	65.9°	1.1502
0.4224	24.2°	0.4099	0.9121	0.4494	2.225	65.8°	1.1484
0.4241	24.3°	0.4115	0.9114	0.4515	2.215	65.7°	1.1467
0.4259	24.4°	0.4131	0.9107	0.4536	2.204	65.6°	1.1449
0.4276	24.5°	0.4147	0.9100	0.4557	2.194	65.5°	1.1432
0.4294	24.6°	0.4163	0.9092	0.4578	2.184	65.4°	1.1414
0.4311	24.7°	0.4179	0.9085	0.4599	2.174	65.3°	1.1397
0.4328	24.8°	0.4195	0.9078	0.4621	2.164	65.2°	1.1380
0.4346	24.9°	0.4210	0.9070	0.4642	2.154	65.1°	1.1362
0.4363	25.0°	0.4226	0.9063	0.4663	2.145	65.0°	1.1345
0.4381	25.1°	0.4242	0.9056	0.4684	2.135	64.9°	1.1327
0.4398	25.2°	0.4258	0.9048	0.4706	2.125	64.8°	1.1310
0.4416	25.3°	0.4274	0.9041	0.4727	2.116	64.7°	1.1292
0.4433	25.4°	0.4289	0.9033	0.4748	2.106	64.6°	1.1275
0.4451	25.5°	0.4305	0.9026	0.4770	2.097	64.5°	1.1257
0.4468	25.6°	0.4321	0.9018	0.4791	2.087	64.4°	1.1240
0.4485	25.7°	0.4337	0.9011	0.4813	2.078	64.3°	1.1222
0.4503	25.8°	0.4352	0.9003	0.4834	2.069	64.2°	1.1205
0.4520	25.9°	0.4368	0.8996	0.4856	2.059	64.1°	1.1188
0.4538	26.0°	04384	0.8988	0.4877	2.050	64.0°	1.1170
0.4555	26.1°	0.4399	0.8980	0.4899	2.041	63.9°	1.1153
0.4573	26.2°	0.4415	0.8973	0.4921	2.032	63.8°	1.1135
0.4590	26.3°	0.4431	0.8965	0.4942	2.023	63.7°	1.1118
0.4608	26.4°	0.4446	0.8957	0.4964	2.014	63.6°	1.1100
0.4625	26.5°	0.4462	0.8949	0.4986	2.006	63.5°	1.1083
0.4643	26.6°	0.4478	0.8942	0.5008	1.997	63.4°	1.1065
0.4660	26.7°	0.4493	0.8934	0.5029	1.988	63.3°	1.1048
0.4677	26.8°	0.4509	0.8926	0.5051	1.980	63.2°	1.1030
0.4695	26.9°	0.4524	0.8919	0.5073	1.971	63.1°	1.1013
0.4712	27.0°	0.4540	0.8910	0.5095	1.963	63.0°	1.0996
0.4730	27.1°	0.4555	0.8902	0.5117	1.954	62.9°	1.0978
0.4747	27.2°	0.4571	0.8894	0.5139	1.946	62.8°	1.0961
0.4765	27.3°	0.4586	0.8886	0.5161	1.937	62.7°	1.0943
0.4782	27.4°	0.4602	0.8878	0.5184	1.929	62.6°	1.0926
0.4800	27.5°	0.4617	0.8870	0.5206	1.921	62.5°	1.0908
0.4817	27.6°	0.4633	0.8862	0.5228	1.913	62.4°	1.0891
0.4835	27.7°	0.4648	0.8854	0.5250	1.905	62.3°	1.0873
0.4852	27.8°	0.4664	0.8846	0.5272	1.897	62.2°	1.0856
0.4869	27.9°	0.4679	0.8838	0.5295	1.889	62.1°	1.0838
		COS	SIN	COT	TAN	DEGREES	RADIANS

Values of the Trigonometric Functions (continued)

RADIANS	DEGREES	SIN	COS	TAN	COT		
0.4887	28.0°	0.4695	0.8829	0.5317	1.881	62.0°	1.0821
0.4904	28.1°	0.4710	0.8821	0.5340	1.873	61.9°	1.0804
0.4922	28.2°	04726	0.8813	0.5362	1.865	61.8°	1.0786
0.4939	28.3°	0.4741	0.8805	0.5384	1.857	61.7°	1.0769
0.4957	28.4°	0.4756	0.8796	0.5407	1.849	61.6°	1.0751
0.4974	28.5°	0.4772	0.8788	0.5430	1.842	61.5°	1.0734
0.4992	28.6°	0.4787	0.8780	0.5452	1.834	61.4°	1.0716
0.5009	28.7°	0.4802	0.8771	0.5475	1.827	61.3°	1.0699
0.5027	28.8°	0.4818	0.8763	0.5498	1.819	61.2°	1.0681
0.5044	28.9°	0.4833	0.8755	0.5520	1.811	61.1°	1.0664
0.5061	29.0°	0.4848	0.8746	0.5543	1.804	61.0°	1.0647
0.5079	29.1°	0.4863	0.8738	0.5566	1.797	60.9°	1.0629
0.5096	29.2°	0.4879	0.8729	0.5589	1.789	60.8°	1.0612
0.5114	29.3°	0.4894	0.8721	0.5612	1.782	60.7°	1.0594
0.5131	29.4°	0.4909	0.8712	0.5635	1.775	60.6°	1.0577
0.5149	29.5°	0.4924	0.8704	0.5658	1.767	60.5°	1.0559
0.5166	29.6°	0.4939	0.8695	0.5681	1.760	60.4°	1.0542
0.5184	29.7°	0.4955	0.8686	0.5704	1.753	60.3°	1.0524
0.5201	29.8°	0.4970	0.8678	0.5727	1.746	60.2°	1.0507
0.5219	29.9°	0.4985	0.8669	0.5750	1.739	60.1°	1.0489
0.5236	30.0°	0.5000	0.8660	0.5774	1.732	60.0°	1.0472
0.5253	30.1°	0.5015	0.8652	0.5797	1.725	59.9°	1.0455
0.5271	30.2°	0.5030	0.8643	0.5820	1.718	59.8°	1.0437
0.5288	30.3°	0.5045	0.8634	0.5844	1.711	59.7°	1.0420
0.5306	30.4°	0.5060	0.8625	0.5867	1.704	59.6°	1.0402
0.5323	30.5°	0.5075	0.8616	0.5890	1.698	59.5°	1.0385
0.5341	30.6°	0.5090	0.8607	0.5914	1.691	59.4°	1.0367
0.5358	30.7°	0.5105	0.8599	0.5938	1.684	59.3°	1.0350
0.5376	30.8°	0.5120	0.8590	0.5961	1.678	59.2°	1.0332
0.5393	30.9°	0.5135	0.8581	0.5985	1.671	59.1°	1.0315
0.5411	31.0°	0.5150	0.8572	0.6009	1.664	59.0°	1.0297
0.5428	31.1°	0.5165	0.8563	0.6032	1.658	58.9°	1.0280
0.5445	31.2°	0.5180	0.8554	0.6056	1.651	58.8°	1.0263
0.5463	31.3°	0.5195	0.8545	0.6080	1.645	58.7°	1.0245
0.5480	31.4°	0.5210	0.8536	0.6104	1.638	58.6°	1.0228
0.5498	31.5°	0.5225	0.8526	0.6128	1.632	58.5°	1.0210
0.5515	31.6°	0.5240	0.8517	0.6152	1.625	58.4°	1.0193
0.5533	31.7°	0.5255	0.8508	0.6176	1.619	58.3°	1.0175
0.5550	31.8°	0.5270	0.8499	0.6200	1.613	58.2°	1.0158
0.5568	31.9°	0.5284	0.8490	0.6224	1.607	58.1°	1.0140
		COS	SIN	COT	TAN	DEGREES	RADIANS

Values of the Trigonometric Functions (continued)

RADIANS	DEGREES	SIN	COS	TAN	COT		
0.5585	32.0°	0.5299	0.8480	0.6249	1.600	58.0°	1.0123
0.5603	32.1°	0.5314	0.8471	0.6273	1.594	57.9°	1.0105
0.5620	32.2°	0.5329	0.8462	0.6297	1.588	57.8°	1.0088
0.5637	32.3°	0.5344	0.8453	0.6322	1.582	57.7°	1.0071
0.5655	32.4°	0.5358	0.8443	0.6346	1.576	57.6°	1.0053
0.5672	32.5°	0.5373	0.8434	0.6371	1.570	57.5°	1.0036
0.5690	32.6°	0.5388	0.8425	0.6395	1.564	57.4°	1.0018
0.5707	32.7°	0.5402	0.8415	0.6420	1.558	57.3°	1.0001
0.5725	32.8°	0.5417	0.8406	0.6445	1.552	57.2°	0.9983
0.5742	32.9°	0.5432	0.8396	0.6469	1.546	57.1°	0.9966
0.5760	33.0°	0.5446	0.8387	0.6494	1.540	57.0°	0.9948
0.5777	33.1°	0.5461	0.8377	0.6519	1.534	56.9°	0.9931
0.5794	33.2°	0.5476	0.8368	0.6544	1.528	56.8°	0.9913
0.5812	33.3°	0.5490	0.8358	0.6569	1.522	56.7°	0.9896
0.5829	33.4°	0.5505	0.8348	0.6494	1.517	56.6°	0.9879
0.5847	33.5°	0.5519	0.8339	0.6619	1.511	56.5°	0.9861
0.5864	33.6°	0.5534	0.8329	0.6644	1.505	56.4°	0.9844
0.5882	33.7°	0.5548	0.8320	0.6669	1.499	56.3°	0.9826
0.5899	33.8°	0.5563	0.8310	0.6694	1.494	56.2°	0.9809
0.5917	33.9°	0.5577	0.8300	0.6720	1.488	56.1°	0.9791
0.5934	34.0°	0.5592	0.8290	0.6745	1.483	56.0°	0.9774
0.5952	34.1°	0.5606	0.8281	0.6771	1.477	55.9°	0.9756
0.5969	34.2°	0.5621	0.8271	0.6796	1.471	55.8°	0.9739
0.5986	34.3°	0.5635	0.8261	0.6822	1.466	55.7°	0.9721
0.6004	34.4°	0.5650	0.8251	0.6847	1.460	55.6°	0.9704
0.6021	34.5°	0.5664	0.8241	0.6873	1.455	55.5°	0.9687
0.6039	34.6°	0.5678	0.8231	0.6899	1.450	55.4°	0.9669
0.6056	34.7°	0.5693	0.8221	0.6924	1.444	55.3°	0.9652
0.6074	34.8°	0.5707	0.8211	0.6950	1.439	55.2°	0.9634
0.6091	34.9°	0.5721	0.8202	0.6976	1.433	55.1°	0.9617
0.6109	35.0°	0.5736	0.8192	0.7002	1.428	55.0°	0.9599
0.6126	35.1°	0.5750	0.8181	0.7028	1.423	54.9°	0.9582
0.6144	35.2°	0.5764	0.8171	0.7054	1.418	54.8°	0.9564
0.6161	35.3°	0.5779	0.8161	0.7080	1.412	54.7°	0.9547
0.6178	35.4°	0.5793	0.8151	0.7107	1.407	54.6°	0.9530
0.6196	35.5°	0.5807	0.8141	0.7133	1.402	54.5°	0.9512
0.6213	35.6°	0.5821	0.8131	0.7159	1.397	54.4°	0.9495
0.6231	35.7°	0.5835	0.8121	0.7186	1.392	54.3°	0.9477
0.6248	35.8°	0.5850	0.8111	0.7212	1.387	54.2°	0.9460
0.6266	35.9°	0.5864	0.8100	0.7239	1.381	54.1°	0.9442
		COS	SIN	COT	TAN	DEGREES	RADIANS

Values of the Trigonometric Functions (continued)

RADIANS	DEGREES	SIN	COS	TAN	COT		
0.6283	36.0°	0.5878	0.8090	0.7265	1.376	54.0°	0.9425
0.6301	36.1°	0.5892	0.8080	0.7292	1.371	53.9°	0.9407
0.6318	36.2°	0.5906	0.8070	0.7319	1.366	53.8°	0.9390
0.6336	36.3°	0.5920	0.8059	0.7346	1.361	53.7°	0.9372
0.6353	36.4°	0.5934	0.8049	0.7373	1.356	53.6°	0.9355
0.6370	36.5°	0.5948	0.8039	0.7400	1.351	53.5°	0.9338
0.6388	36.6°	0.5962	0.8028	0.7427	1.347	53.4°	0.9320
0.6405	36.7°	0.5976	0.8018	0.7454	1.342	53.3°	0.9303
0.6423	36.8°	0.5990	0.8007	0.7481	1.337	53.2°	0.9285
0.6440	36.9°	0.6004	0.7997	0.7508	1.332	53.1°	0.9268
0.6458	37.0°	0.6018	0.7986	0.7536	1.327	53.0°	0.9250
0.6475	37.1°	0.6032	0.7976	0.7563	1.322	52.9°	0.9233
0.6493	37.2°	0.6046	0.7965	0.7590	1.317	52.8°	0.9215
0.6510	37.3°	0.6060	0.7955	0.7618	1.313	52.7°	0.9198
0.6528	37.4°	0.6074	0.7944	0.7646	1.308	52.6°	0.9180
0.6545	37.5°	0.6088	0.7934	0.7673	1.303	52.5°	0.9163
0.6562	37.6°	0.6101	0.7923	0.7701	1.299	52.4°	0.9146
0.6580	37.7°	0.6115	0.7912	0.7729	1.294	52.3°	0.9128
0.6597	37.8°	0.6129	0.7902	0.7757	1.289	52.2°	0.9111
0.6615	37.9°	0.6143	0.7891	0.7785	1.285	52.1°	0.9093
0.6632	38.0°	0.6157	0.7880	0.7813	1.280	52.0°	0.9076
0.6650	38.1°	0.6170	0.7869	0.7841	1.275	51.9°	0.9058
0.6667	38.2°	0.6184	0.7859	0.7869	1.271	51.8°	0.9041
0.6685	38.3°	0.6198	0.7848	0.7898	1.266	51.7°	0.9023
0.6702	38.4°	0.6211	0.7837	0.7926	1.262	51.6°	0.9006
0.6720	38.5°	0.6225	0.7826	0.7954	1.257	51.5°	0.8988
0.6737	38.6°	0.6239	0.7815	0.7983	1.253	51.4°	0.8971
0.6754	38.7°	0.6252	0.7804	0.8012	1.248	51.3°	0.8954
0.6772	38.8°	0.6266	0.7793	0.8040	1.244	51.2°	0.8936
0.6790	38.9°	0.6280	0.7782	0.8069	1.239	51.1°	0.8919
0.6807	39.0°	0.6293	0.7771	0.8098	1.235	51.0°	0.8901
0.6824	39.1°	0.6307	0.7760	0.8127	1.230	50.9°	0.8884
0.6842	39.2°	0.6320	0.7749	0.8156	1.226	50.8°	0.8866
0.6859	39.3°	0.6334	0.7738	0.8185	1.222	50.7°	0.8849
0.6877	39.4°	0.6347	0.7727	0.8214	1.217	50.6°	0.8831
0.6894	39.5°	0.6361	0.7716	0.8243	1.213	50.5°	0.8814
0.6912	39.6°	0.6374	0.7705	0.8273	1.209	50.4°	0.8796
0.6929	39.7°	0.6388	0.7694	0.8302	1.205	50.3°	0.8779
0.6946	39.8°	0.6401	0.7683	0.8332	1.200	50.2°	0.8762
0.6964	39.9°	0.6414	0.7672	0.8361	1.196	50.1°	0.8744
		COS	SIN	COT	TAN	DEGREES	RADIANS

Values of the Trigonometric Functions (continued)

RADIANS	DEGREES	SIN	COS	TAN	COT		
0.6981	40.0°	0.6428	0.7660	0.8391	1.192	50.0°	0.8727
0.6999	40.1°	0.6441	0.7649	0.8421	1.188	49.9°	0.8709
0.7016	40.2°	0.6455	0.7638	0.8451	1.183	49.8°	0.8692
0.7034	40.3°	0.6468	0.7627	0.8481	1.179	49.7°	0.8674
0.7051	40.4°	0.6481	0.7615	0.8511	1.175	49.6°	0.8657
0.7069	40.5°	0.6494	0.7604	0.8541	1.171	49.5°	0.8639
0.7086	40.6°	0.6508	0.7593	0.8571	1.167	49.4°	0.8622
0.7103	40.7°	0.6521	0.7581	0.8601	1.163	49.3°	0.8604
0.7121	40.8°	0.6534	0.7570	0.8632	1.159	49.2°	0.8587
0.7138	40.9°	0.6547	0.7559	0.8662	1.154	49.1°	0.8570
0.7156	41.0°	0.6561	0.7547	0.8693	1.150	49.0°	0.8552
0.7173	41.1°	0.6574	0.7536	0.8724	1.146	48.9°	0.8535
0.7191	41.2°	0.6587	0.7524	0.8754	1.142	48.8°	0.8517
0.7208	41.3°	0.6600	0.7513	0.8785	1.138	48.7°	0.8500
0.7226	41.4°	0.6613	0.7501	0.8816	1.134	48.6°	0.8482
0.7243	41.5°	0.6626	0.7490	0.8847	1.130	48.5°	0.8465
0.7261	41.6°	0.6639	0.7478	0.8878	1.126	48.4°	0.8447
0.7278	41.7°	0.6652	0.7466	0.8910	1.122	48.3°	0.8430
0.7295	41.8°	0.6665	0.7455	0.8941	1.118	48.2°	0.8412
0.7313	41.9°	0.6678	0.7443	0.8972	1.115	48.1°	0.8395
0.7330	42.0°	0.6691	0.7431	0.9004	1.111	48.0°	0.8378
0.7348	42.1°	0.6704	0.7420	0.9036	1.107	47.9°	0.8360
0.7365	42.2°	0.6717	0.7408	0.9067	1.103	47.8°	0.8343
0.7383	42.3°	0.6730	0.7396	0.9099	1.099	47.7°	0.8325
0.7400	42.4°	0.6743	0.7385	0.9131	1.095	47.6°	0.8308
0.7418	42.5°	0.6756	0.7373	0.9163	1.091	47.5°	0.8290
0.7435	42.6°	0.6769	0.7361	0.9195	1.087	47.4°	0.8273
0.7453	42.7°	0.6782	0.7349	0.9228	1.084	47.3°	0.8255
0.7470	42.8°	0.6794	0.7337	0.9260	1.080	47.2°	0.8238
0.7487	42.9°	0.6807	0.7325	0.9293	1.076	47.1°	0.8221
0.7505	43.0°	0.6820	0.7314	0.9325	1.072	47.0°	0.8203
0.7522	43.1°	0.6833	0.7302	0.9358	1.069	46.9°	0.8186
0.7540	43.2°	0.6845	0.7290	0.9391	1.065	46.8°	0.8168
0.7557	43.3°	0.6858	0.7278	0.9424	1.061	46.7°	0.8151
0.7575	43.4°	0.6871	0.7266	0.9457	1.057	46.6°	0.8133
0.7592	43.5°	0.6884	0.7254	0.9490	1.054	46.5°	0.8116
0.7610	43.6°	0.6896	0.7242	0.9523	1.050	46.4°	0.8098
0.7627	43.7°	0.6909	0.7230	0.9556	1.046	46.3°	0.8081
0.7645	43.8°	0.6921	0.7218	0.9590	1.043	46.2°	0.8063
0.7662	43.9°	0.6934	0.7206	0.9623	1.039	46.1°	0.8046
		COS	SIN	COT	TAN	DEGREES	RADIANS

Values of the Trigonometric Functions (continued)

RADIANS	DEGREES	SIN	COS	TAN	COT		
0.7679	44.0°	0.6947	0.7193	0.9657	1.036	46.0°	0.8029
0.7697	44.1°	0.6959	0.7181	0.9691	1.032	45.9°	0.8011
0.7714	44.2°	0.6972	0.7169	0.9725	1.028	45.8°	0.7994
0.7732	44.3°	0.6984	0.7157	0.9759	1.025	45.7°	0.7976
0.7749	44.4°	0.6997	0.7145	0.9793	1.021	45.6°	0.7959
0.7767	44.5°	0.7009	0.7133	0.9827	1.018	45.5°	0.7941
0.7784	44.6°	0.7022	0.7120	0.9861	1.014	45.4°	0.7924
0.7802	44.7°	0.7034	0.7108	0.9896	1.011	45.3°	0.7906
0.7819	44.8°	0.7046	0.7096	0.9930	1.007	45.2°	0.7889
0.7837	44.9°	0.7059	0.7083	0.9965	1.003	45.1°	0.7871
0.7854	45.0°	0.7071	0.7071	1.0000	1.000	45.0°	0.7854
		COS	SIN	COT	TAN	DEGREES	RADIANS

ANSWERS TO ALL PROBLEMS FOR THE INSTRUCTOR

CHAPTER 1

Problem Set 1.1 (page 12)

1. True **2.** True **3.** False **4.** False
5. False **6.** True **7.** True **8.** False
9. False **10.** True **11.** $\{46\}$ **12.** $\{0, 46\}$
13. $\{0, -14, 46\}$
14. $\left\{0, \frac{7}{8}, -\frac{10}{13}, 7\frac{1}{8}, 0.279, 0.4\overline{67}, -14, 46, 6.75\right\}$
15. $\{\sqrt{5}, -\sqrt{2}, -\pi\}$ **16.** $\{0, 46\}$ **17.** $\{0, -14\}$
18. All of them **19.** $\subseteq$ **20.** $\not\subseteq$ **21.** $\subseteq$
22. $\subseteq$ **23.** $\not\subseteq$ **24.** $\not\subseteq$ **25.** $\subseteq$ **26.** $\subseteq$
27. $\not\subseteq$ **28.** $\not\subseteq$ **29.** $\subseteq$ **30.** $\subseteq$ **31.** $\not\subseteq$
32. $\not\subseteq$ **33.** $\{1\}$ **34.** $\{6, 7, 8, \ldots\}$
35. $\{0, 1, 2, 3\}$ **36.** $\{-2, -1, 0, 1, \ldots\}$
37. $\{\ldots -2, -1, 0, 1\}$ **38.** $\{1, 2, 3, 4, \ldots\}$
39. $\varnothing$ **40.** $\{-4, -3, -2, -1\}$ **41.** $\{0, 1, 2\}$
42. $\varnothing$ **43.** **(a)** 18 **(b)** 26 **(c)** 39
(d) 25 **(e)** 35 **(f)** 37
44. **(a)** 1 if $x > 0$ and -1 if $x < 0$
(b) 1 if $x > 0$ and -1 if $x < 0$
(c) -1 if $x > 0$ and 1 if $x < 0$ **(d)** 0
45. Commutative property of multiplication
46. Associative property of addition
47. Identity property of multiplication
48. Commutative property of addition
49. Multiplication property of negative one
50. Additive inverse property
51. Distributive property
52. Multiplication property of negative one
53. Commutative property of multiplication
54. Associative property of addition
55. Distributive property
56. Multiplicative inverse property
57. Associative property of multiplication
58. Associative property of multiplication **59.** -22
60. -31 **61.** 100 **62.** -22 **63.** -21
64. -31 **65.** 8 **66.** 47 **67.** 19
68. 495 **69.** 66 **70.** 36 **71.** -75
72. -8 **73.** 34 **74.** 1 **75.** 1 **76.** 59
77. 11 **78.** -8 **79.** 4

Problem Set 1.2 (page 22)

1. $\frac{1}{8}$ **2.** $\frac{1}{9}$ **3.** $-\frac{1}{1000}$ **4.** $\frac{1}{10000}$
5. 27 **6.** 32 **7.** 4 **8.** -9 **9.** $-\frac{27}{8}$
10. $\frac{36}{25}$ **11.** 1 **12.** $\frac{9}{25}$ **13.** $\frac{16}{25}$ **14.** 1
15. 4 **16.** 27 **17.** $\frac{1}{100}$ or 0.01
18. $\frac{1}{1000}$ or 0.001 **19.** $\frac{1}{100{,}000}$ or 0.00001
20. $\frac{1}{1{,}000{,}000}$ or 0.000001 **21.** 81 **22.** -8
23. $\frac{1}{16}$ **24.** $\frac{1}{27}$ **25.** $\frac{3}{4}$ **26.** $\frac{81}{64}$
27. $\frac{256}{25}$ **28.** $\frac{1}{4096}$ **29.** $\frac{16}{25}$ **30.** $\frac{9}{64}$
31. $\frac{64}{81}$ **32.** $\frac{1}{80}$ **33.** 64 **34.** $\frac{1}{64}$
35. $\frac{1}{100{,}000}$ or 0.00001 **36.** 10,000 **37.** $\frac{17}{72}$
38. $\frac{13}{40}$ **39.** $\frac{1}{6}$ **40.** $-\frac{71}{9}$ **41.** $\frac{48}{19}$
42. $-\frac{45}{4}$ **43.** $\frac{1}{x^4}$ **44.** $\frac{1}{x^5}$ **45.** $\frac{1}{a^2}$

46. $\frac{1}{b^2}$ **47.** $\frac{1}{a^6}$ **48.** $\frac{1}{b^{10}}$ **49.** $\frac{y^4}{x^3}$ **50.** $\frac{y^4}{x^8}$
51. $\frac{c^3}{a^3b^6}$ **52.** $\frac{b^4c^8}{a^8}$ **53.** $\frac{y^2}{4x^4}$ **54.** $\frac{y^2}{3x^4}$
55. $\frac{x^4}{y^6}$ **56.** $\frac{1}{x^3y^{12}}$ **57.** $\frac{9a^2}{4b^4}$ **58.** $\frac{4}{3ab^3x^2y}$
59. $\frac{1}{x^3}$ **60.** $\frac{1}{a^8}$ **61.** $\frac{a^3}{b}$ **62.** $\frac{1}{x^4y}$
63. $-20x^4y^5$ **64.** $-18x^3y^5$ **65.** $-27x^3y^9$
66. $16x^8y^{16}$ **67.** $\frac{8x^6}{27y^9}$ **68.** $\frac{64x^3}{125y^6}$
69. $-8x^6$ **70.** $-9x^4$ **71.** $\frac{6}{x^3y}$ **72.** $-\frac{20x}{y}$
73. $\frac{6}{a^2y^3}$ **74.** $\frac{48b^3}{a^5}$ **75.** $\frac{4x^3}{y^5}$ **76.** $\frac{7}{xy^5}$
77. $-\frac{5}{a^2b}$ **78.** $-\frac{9}{a^2b}$ **79.** $\frac{1}{4x^2y^4}$
80. $-\frac{x^3y^6}{27}$ **81.** $\frac{x+1}{x^2}$ **82.** $\frac{x^2+1}{x^4}$
83. $\frac{y-x^2}{x^2y}$ **84.** $\frac{2y^3-3x}{xy^3}$ **85.** $\frac{3b^3+2a^2}{a^2b^3}$
86. $\frac{b^2+a}{a^2b^2}$ **87.** $\frac{y^2-x^2}{xy}$ **88.** $\frac{x^5-y^3}{x^3y}$
89. $12x^{3a+1}$ **90.** $-30x^{2a-1}$ **91.** 1
92. $8y^{4b+1}$ **93.** x^{2a} **94.** $2x^{a+3}$
95. $-4y^{6b+2}$ **96.** x^{3ab} **97.** x^b **98.** $-x^{4b-1}$
99. $(6.2)(10)^7$ **100.** $(1.7)(10)^{10}$
101. $(4.12)(10)^{-4}$ **102.** $(7.8)(10)^{-8}$
103. 180,000 **104.** 54,100,000 **105.** 0.0000023
106. 0.00000000413 **107.** 0.04 **108.** 20
109. 30,000 **110.** 0.002 **111.** 0.03
112. 60

Problem Set 1.3 (page 29)

1. $14x^2 + x - 6$ **2.** $-2x^2 + 3x + 1$
3. $-x^2 - 4x - 9$ **4.** $-7x^2 - 4x + 9$
5. $6x - 11$ **6.** $-11a + 1$ **7.** $6x^2 - 5x - 7$
8. $5x^2 - 4x - 9$ **9.** $-x - 34$ **10.** $-20x - 7$
11. $12x^3y^2 + 15x^2y^3$ **12.** $-6a^3b^3 + 8a^2b^5$
13. $30a^4b^3 - 24a^5b^3 + 18a^4b^4$
14. $-5x^3y^5 + 4x^2y^6 - 3x^3y^6$
15. $x^2 + 20x + 96$
16. $x^2 - 3x - 54$ **17.** $n^2 - 16n + 48$
18. $n^2 - 4n - 60$ **19.** $sx + sy - tx - ty$
20. $ac + ad + bc + bd$ **21.** $6x^2 + 7x - 3$
22. $15x^2 + 26x + 8$ **23.** $12x^2 - 37x + 21$
24. $24n^2 + 14n - 3$ **25.** $x^2 + 8x + 16$
26. $x^2 - 12x + 36$ **27.** $4n^2 + 12n + 9$
28. $9n^2 - 30n + 25$ **29.** $x^3 + x^2 - 14x - 24$
30. $x^3 - 31x + 30$ **31.** $6x^3 - x^2 - 11x + 6$
32. $6x^3 - 7x^2 - 63x - 20$ **33.** $x^3 + 2x^2 - 7x + 4$
34. $t^3 - t^2 - 6t - 4$ **35.** $t^3 - 1$
36. $2x^3 + 7x^2 + 2x - 3$ **37.** $6x^3 + x^2 - 5x - 2$
38. $6x^3 + 5x^2 + 6x - 8$
39. $x^4 + 8x^3 + 15x^2 + 2x - 4$
40. $2x^4 - 5x^3 + 10x^2 - 11x - 4$ **41.** $25x^2 - 4$
42. $9x^2 - 16$ **43.** $x^4 - 10x^3 + 21x^2 + 20x + 4$
44. $x^4 - 2x^3 + 3x^2 - 2x + 1$ **45.** $4x^2 - 9y^2$
46. $81x^2 - y^2$ **47.** $x^3 + 15x^2 + 75x + 125$
48. $x^3 - 18x^2 + 108x - 216$
49. $8x^3 + 12x^2 + 6x + 1$
50. $27x^3 + 108x^2 + 144x + 64$
51. $64x^3 - 144x^2 + 108x - 27$
52. $8x^3 - 60x^2 + 150x - 125$
53. $125x^3 - 150x^2y + 60xy^2 - 8y^3$
54. $x^3 + 9x^2y + 27xy^2 + 27y^3$
55. $a^7 + 7a^6b + 21a^5b^2 + 35a^4b^3 + 35a^3b^4 + 21a^2b^5 + 7ab^6 + b^7$
56. $a^8 + 8a^7b + 28a^6b^2 + 56a^5b^3 + 70a^4b^4 + 56a^3b^5 + 28a^2b^6 + 8ab^7 + b^8$
57. $x^5 - 5x^4y + 10x^3y^2 - 10x^2y^3 + 5xy^4 - y^5$
58. $x^6 - 6x^5y + 15x^4y^2 - 20x^3y^3 + 15x^2y^4 - 6xy^5 + y^6$
59. $x^4 + 8x^3y + 24x^2y^2 + 32xy^3 + 16y^4$
60. $32x^5 + 80x^4y + 80x^3y^2 + 40x^2y^3 + 10xy^4 + y^5$
61. $64a^6 - 192a^5b + 240a^4b^2 - 160a^3b^3 + 60a^2b^4 - 12ab^5 + b^6$
62. $81a^4 - 108a^3b + 54a^2b^2 - 12ab^3 + b^4$
63. $x^{14} + 7x^{12}y + 21x^{10}y^2 + 35x^8y^3 + 35x^6y^4 + 21x^4y^5 + 7x^2y^6 + y^7$
64. $x^7 + 14x^6y^2 + 84x^5y^4 + 280x^4y^6 + 560x^3y^8 + 672x^2y^{10} + 448xy^{12} + 128y^{14}$
65. $32a^5 - 240a^4b + 720a^3b^2 - 1080a^2b^3 + 810ab^4 - 243b^5$ **66.** $64a^3 - 144a^2b + 108ab^2 - 27b^3$
67. $3x^2 - 5x$ **68.** $6x^4 + 9x^2$
69. $-5a^4 + 4a^2 - 9a$ **70.** $6x^2y + 9xy^2$
71. $5ab + 11a^2b^4$ **72.** $3x^3y^3 + 4x^2 - 5x^3y$
73. $x^{2a} - y^{2b}$ **74.** $x^{4a} - 2x^{2a} - 3$
75. $x^{2b} - 3x^b - 28$ **76.** $3x^{2a} + 13x^a - 10$
77. $6x^{2b} + x^b - 2$ **78.** $4x^{2a} - 9$
79. $x^{4a} - 2x^{2a} + 1$
80. $x^{6b} + 4x^{3b} + 4$ **81.** $x^{3a} - 6x^{2a} + 12x^a - 8$
82. $x^{3b} + 9x^{2b} + 27x^b + 27$

Problem Set 1.4 (page 39)

1. $2xy(3 - 4y)$ **2.** $4ab^2(a + 3b)$
3. $(z + 3)(x + y)$ **4.** $(x + y)(5 + a)$

5. $(x + y)(3 + a)$ **6.** $(a + b)(c + 1)$
7. $(x - y)(a - b)$ **8.** $(a - b)(2a + 3c)$
9. $(3x + 5)(3x - 5)$ **10.** Not factorable
11. $(1 + 9n)(1 - 9n)$ **12.** $(3xy + 8)(3xy - 8)$
13. $(x + 4 + y)(x + 4 - y)$
14. $(x + y - 1)(x - y + 1)$
15. $(3s + 2t - 1)(3s - 2t + 1)$
16. $(2a + 3b + 1)(2a - 3b - 1)$
17. $(x - 7)(x + 2)$ **18.** $(a + 8)(a - 3)$
19. $(5 + x)(3 - x)$ **20.** $(10 + x)(4 - x)$
21. Not factorable **22.** $(x - 5y)(x + y)$
23. $(3x - 5)(x - 2)$ **24.** $(2x + 5)(x - 6)$
25. $(5x + 1)(2x - 7)$ **26.** $(4y - 3)(2y + 7)$
27. $(x - 2)(x^2 + 2x + 4)$
28. $(x + 4)(x^2 - 4x + 16)$
29. $(4x + 3y)(16x^2 - 12xy + 9y^2)$
30. $(3x - 2y)(9x^2 + 6xy + 4y^2)$ **31.** $4(x^2 + 4)$
32. $n(n + 7)(n - 7)$ **33.** $x(x + 3)(x - 3)$
34. $(4n + 9)(3n + 8)$ **35.** $(3a - 7)^2$
36. $(1 + 2x)(1 - 2x)(1 + 4x^2)$
37. $2n(n^2 + 3n + 5)$ **38.** $(x + y - 7)(x - y + 7)$
39. $(5x - 3)(2x + 9)$ **40.** Not factorable
41. $(6a - 1)^2$ **42.** $3n(2n + 5)(3n - 1)$
43. $(4x - y)(2x + y)$ **44.** $(4x + 5y)(3x - 2y)$
45. Not factorable **46.** $25(t + 2)(t - 2)$
47. $2n(n^2 + 7n - 10)$ **48.** Not factorable
49. $4(x + 2)(x^2 - 2x + 4)$
50. $2(x - 3)(x^2 + 3x + 9)$
51. $(x + 3)(x - 3)(x^2 + 5)$
52. $(x + 2)(x - 2)(x^2 + 3)$
53. $2y(x + 4)(x - 4)(x^2 + 3)$
54. $3y(x + 3)(x - 3)(x^2 + 4)$
55. $(a + b + c + d)(a + b - c - d)$
56. $(a - b + c - d)(a - b - c + d)$
57. $(x + 4 + y)(x + 4 - y)$
58. $(2x + 3 + y)(2x + 3 - y)$
59. $(x + y + 5)(x - y - 5)$
60. $(y + x - 8)(y - x + 8)$
61. $(10x + 3)(6x - 5)$
62. $(8x - 7)(5x + 9)$ **63.** $3x(7x - 4)(4x + 5)$
64. $6x(5x - 6)(7x + 5)$ **65.** $(x^a + 4)(x^a - 4)$
66. $(x^{2n} + 3)(x^{2n} - 3)$
67. $(x^n - y^n)(x^{2n} + x^ny^n + y^{2n})$
68. $(x^a + y^{2a})(x^{2a} - x^ay^{2a} + y^{4a})$
69. $(x^a + 4)(x^a - 7)$ **70.** $(x^a + 3)(x^a + 7)$
71. $(2x^n - 5)(x^n + 6)$ **72.** $(3x^n + 2)(x^n - 6)$
73. $(x^{2n} + y^{2n})(x^n + y^n)(x^n - y^n)$ **74.** $(4x^a + 3)^2$
75. **(a)** $(x + 32)(x + 3)$ **(b)** $(x + 11)(x + 16)$
(c) $(x - 21)(x - 24)$ **(d)** $(x - 14)(x - 12)$
(e) $(x + 28)(x + 32)$ **(f)** $(x - 36)(x - 48)$

Problem Set 1.5 (page 49)

1. $\frac{2x}{3}$ **2.** $-\frac{2xy}{5}$ **3.** $\frac{7y^3}{9x}$ **4.** $\frac{x - y}{x}$
5. $\frac{a + 4}{a - 9}$ **6.** $\frac{3x + 5}{4x + 1}$ **7.** $\frac{x(2x + 7)}{y(x + 9)}$
8. $-\frac{x}{x + 3}$ **9.** $\frac{x^2 + xy + y^2}{x + 2y}$ **10.** $\frac{x + 2y}{2x + y}$
11. $-\frac{2}{x + 1}$ **12.** $\frac{2x(x + 2y)}{3(x + y)}$ **13.** $\frac{x}{2y^3}$
14. $\frac{3x^3}{4}$ **15.** $-\frac{8x^3y^3}{15}$ **16.** $-\frac{5x^3}{12y^2}$ **17.** $\frac{14}{27a}$
18. $\frac{ac^2}{2b^2}$ **19.** $5y$ **20.** $\frac{a(a^2 + 3)}{2(2a - 1)}$
21. $\frac{5(a + 3)}{a(a - 2)}$ **22.** $\frac{(t^2 + 9)(6t + 7)}{5t - 7}$
23. $\frac{(x + 6y)^2(2x + 3y)}{y^3(x + 4y)}$ **24.** 2 **25.** $\frac{3xy}{4(x + 6)}$
26. $\frac{5(x - 2y)}{7y}$ **27.** $\frac{x - 9}{42x^2}$ **28.** $\frac{2(a - 2b)}{a(3a - 2b)}$
29. $\frac{8x + 5}{12}$ **30.** $\frac{9n - 10}{36}$ **31.** $\frac{7x}{24}$
32. $\frac{3x - 25}{30}$ **33.** $\frac{35b + 12a^3}{80a^2b^2}$ **34.** $\frac{20b^2 - 33a^3}{96a^2b}$
35. $\frac{12 + 9n - 10n^2}{12n^2}$ **36.** $\frac{45 - 6n + 20n^2}{15n^2}$
37. $\frac{9y + 8x - 12xy}{12xy}$ **38.** $\frac{10y - 9x + 24xy}{12xy}$
39. $\frac{13x + 14}{(2x + 1)(3x + 4)}$ **40.** $\frac{7x - 12}{(x - 1)(2x - 3)}$
41. $\frac{7x + 21}{x(x + 7)}$ **42.** $\frac{-3x - 18}{x(x + 8)}$ **43.** $\frac{1}{a - 2}$
44. $\frac{1}{a + 1}$ **45.** $\frac{1}{x + 1}$ **46.** $\frac{x^2 + 60}{x(x + 6)}$
47. $\frac{9x + 73}{(x + 3)(x + 7)(x + 9)}$
48. $\frac{-2a - 20}{(a - 6)(a + 3)(a - 10)}$
49. $\frac{3x^2 + 30x - 78}{(x + 1)(x - 1)(x + 8)(x - 2)}$
50. $\frac{-3x^2 + 4x - 62}{(x^2 + 2)(x + 4)(x - 3)}$ **51.** $\frac{-x^2 - x + 1}{(x + 1)(x - 1)}$
52. $\frac{7x^2 - 50x}{(x + 7)(x - 7)}$ **53.** $\frac{-8}{(n^2 + 4)(n + 2)(n - 2)}$
54. $\frac{1}{(n^2 + 1)(n + 1)}$ **55.** $\frac{5x^2 + 16x + 5}{(x + 1)(x - 4)(x + 7)}$
56. $\frac{5x^2 - 15}{(2x + 1)(x - 5)(3x + 4)}$ **57.** **(a)** $\frac{5}{x - 1}$

(b) $-\frac{3}{2x-1}$ (c) $\frac{5}{a-3}$ (d) $\frac{15}{a-9}$

(e) $x+3$ (f) $x+7$ **58.** $\frac{2y+7x}{3y-10x}$

59. $\frac{5y^2-3xy^2}{x^2y+2x^2}$ **60.** $\frac{y+3xy}{2x+4xy}$ **61.** $\frac{x+1}{x-1}$

62. $\frac{3n-14}{5n-16}$ **63.** $\frac{n-1}{n+1}$ **64.** $\frac{-x+15}{-2x-1}$

65. $\frac{-6x-4}{3x+9}$ **66.** $\frac{-x+5y-10}{3y-10}$

67. $\frac{x^2+x+1}{x+1}$ **68.** $\frac{-x^2+6x-4}{3x-2}$

69. $\frac{a^2+4a+1}{4a+1}$ **70.** $\frac{3a^2-2a+1}{2a-1}$

71. $-\frac{2x+h}{x^2(x+h)^2}$ **72.** $-\frac{3x^2+3xh+h^2}{x^3(x+h)^3}$

73. $-\frac{1}{(x+1)(x+h+1)}$ **74.** $-\frac{3}{x(x+h)}$

75. $-\frac{4}{(2x-1)(2x+2h-1)}$

76. $-\frac{12}{(4x+5)(4x+4h+5)}$ **77.** $\frac{y+x}{x^2y-xy^2}$

78. xy **79.** $\frac{x^2y^2+1}{y^2-1x}$ **80.** $\frac{y^2-2x^2y}{3xy^2+x^2}$

Problem Set 1.6 (page 60)

1. 9 **2.** −7 **3.** 5 **4.** 3 **5.** $\frac{6}{7}$

6. 2 **7.** $-\frac{3}{2}$ **8.** $\frac{4}{3}$ **9.** $2\sqrt{6}$

10. $3\sqrt{6}$ **11.** $4\sqrt{7}$ **12.** $12\sqrt{7}$

13. $-6\sqrt{11}$ **14.** $-10\sqrt{17}$ **15.** $\frac{3\sqrt{5}}{2}$

16. $\frac{9\sqrt{2}}{4}$ **17.** $2x\sqrt{3}$ **18.** $3y\sqrt{5x}$

19. $8x^2y^3\sqrt{y}$ **20.** $12a\sqrt{2a}$ **21.** $\frac{9y^3\sqrt{5x}}{7}$

22. $2\sqrt[3]{4}$ **23.** $4\sqrt[3]{2}$ **24.** $3x\sqrt[3]{2}$

25. $2x\sqrt[3]{2x}$ **26.** $3xy^2\sqrt[3]{3x^2}$ **27.** $2x\sqrt[4]{3x}$

28. $3xy\sqrt[4]{2x^2y^3}$ **29.** $\frac{2\sqrt{3}}{5}$ **30.** $\frac{5\sqrt{3}}{9}$

31. $\frac{\sqrt{14}}{4}$ **32.** $\sqrt{5}$ **33.** $\frac{4\sqrt{15}}{5}$ **34.** $\frac{\sqrt{6}}{2}$

35. $\frac{3\sqrt{2}}{7}$ **36.** $\frac{\sqrt{6xy}}{2y}$ **37.** $\frac{\sqrt{15}}{6x^2}$ **38.** $\frac{\sqrt{10xy}}{6x^2}$

39. $\frac{2\sqrt{15a}}{5ab}$ **40.** $\frac{5\sqrt[3]{9}}{3}$ **41.** $\frac{3\sqrt[3]{2}}{2}$ **42.** $\frac{\sqrt[3]{20x^2}}{2x}$

43. $\frac{\sqrt[3]{18x^2y}}{3x}$ **44.** $\frac{\sqrt[3]{4x^2y^2}}{xy^2}$ **45.** $12\sqrt{3}$

46. $-16\sqrt{2}$ **47.** $3\sqrt{7}$ **48.** $5\sqrt[3]{2}$

49. $\frac{11\sqrt{3}}{6}$ **50.** $\frac{13\sqrt{10}}{10}$ **51.** $-\frac{89\sqrt{2}}{30}$

52. $\frac{47\sqrt[3]{2}}{6}$ **53.** $48\sqrt{6}$ **54.** $30\sqrt{14}$

55. $10\sqrt{6}+8\sqrt{30}$ **56.** $24\sqrt{3}-54\sqrt{2}$

57. $3x\sqrt{6y}-6\sqrt{2xy}$ **58.** $4\sqrt{3xy}+2y\sqrt{15y}$

59. $13+7\sqrt{3}$ **60.** $-10+\sqrt{2}$

61. $30+11\sqrt{6}$ **62.** $96+17\sqrt{30}$ **63.** 16

64. 31 **65.** $x+2\sqrt{xy}+y$

66. $4x-12\sqrt{xy}+9y$ **67.** $a-b$

68. $9x-25y$ **69.** $3\sqrt{5}-6$ **70.** $7\sqrt{10}+21$

71. $\sqrt{7}+\sqrt{3}$ **72.** $\sqrt{5}-\sqrt{3}$

73. $\frac{-2\sqrt{10}+3\sqrt{14}}{43}$ **74.** $5\sqrt{2}+3\sqrt{5}$

75. $\frac{x+\sqrt{x}}{x-1}$ **76.** $\frac{x-2\sqrt{x}}{x-4}$ **77.** $\frac{x-\sqrt{xy}}{x-y}$

78. $\frac{2x+2\sqrt{xy}}{x-y}$ **79.** $\frac{6x+7\sqrt{xy}+2y}{9x-4y}$

80. $\frac{6x-19\sqrt{xy}-10y}{4x-25y}$ **81.** $\frac{2}{\sqrt{2x+2h}+\sqrt{2x}}$

82. $\frac{1}{\sqrt{x+h+1}+\sqrt{x+1}}$

83. $\frac{1}{\sqrt{x+h-3}+\sqrt{x-3}}$ **84.** $\frac{2}{\sqrt{x+h}+\sqrt{x}}$

Problem Set 1.7 (page 65)

1. 7 **2.** 4 **3.** 8 **4.** −2 **5.** −4

6. $\frac{1}{8}$ **7.** 2 **8.** $-\frac{2}{3}$ **9.** 64 **10.** 0.2

11. 0.001 **12.** 9 **13.** $\frac{1}{32}$ **14.** −32

15. 2 **16.** 4 **17.** $15x^{7/12}$ **18.** $12x^{13/20}$

19. $y^{5/12}$ **20.** $\frac{2}{x^{1/6}}$ **21.** $64x^{3/4}y^{3/2}$

22. $25xy^2$ **23.** $4x^{4/15}$ **24.** $2x^{1/6}$ **25.** $\frac{7}{a^{1/12}}$

26. $\frac{4}{b^{5/12}}$ **27.** $\frac{16x^{4/3}}{81y}$ **28.** $\frac{36x^{4/5}}{49y^{4/3}}$ **29.** $\frac{y^{3/2}}{x}$

30. $\frac{1}{ab^{2/3}}$ **31.** $8a^{9/2}x^2$ **32.** $9ax^2$ **33.** $\sqrt[4]{8}$

34. $\sqrt[6]{243}$ **35.** $\sqrt[12]{x^7}$ **36.** $x\sqrt[15]{x^4}$

37. $xy\sqrt[4]{xy^3}$ **38.** $xy\sqrt[12]{x^5y^7}$ **39.** $a\sqrt[12]{a^5b^{11}}$

40. $ab^2\sqrt[6]{a^5b}$ **41.** $4\sqrt[6]{2}$ **42.** $9\sqrt[6]{3}$ **43.** $\sqrt[6]{2}$

44. $\sqrt[3]{9}$ **45.** $\sqrt{2}$ **46.** 2 **47.** $x\sqrt[12]{x^7}$
48. $x\sqrt[15]{x}$ **49.** $\frac{5\sqrt[3]{x^2}}{x}$ **50.** $\frac{3\sqrt[3]{x}}{x}$ **51.** $\frac{\sqrt[6]{x^3y^4}}{y}$
52. $\frac{\sqrt[4]{xy^2}}{y}$ **53.** $\frac{\sqrt[20]{x^{15}y^8}}{y}$ **54.** $\frac{2\sqrt[6]{x^3y^4}}{3y}$
55. $\frac{5\sqrt[12]{x^9y^8}}{4x}$ **56.** $\frac{\sqrt[6]{x^3y^3a^2b^4}}{ab}$ **57.** **(a)** $\sqrt[6]{2}$
(b) $\sqrt[12]{3}$ **(c)** $\sqrt{x}$ **(d)** $\sqrt[3]{x^2}$ **58.** **(a)** 12
(b) 18 **(c)** 7 **(d)** 16 **(e)** 11 **(f)** 23
60. **(a)** 1024 **(b)** 78,125 **(c)** 512 **(d)** 243
(e) 49 **(f)** 4096 **61.** **(a)** 13.391 **(b)** 6.310
(c) 2.702 **(d)** 3.247 **(e)** 4.304 **(f)** 17.783
62. $\frac{x+2}{(x+1)^{3/2}}$ **63.** $\frac{2x-2}{(2x-1)^{3/2}}$ **64.** $\frac{6x^2+2x}{(4x+1)^{3/2}}$
65. $\frac{x}{(x^2+2x)^{3/2}}$ **66.** $\frac{2x}{(3x)^{4/3}}$ **67.** $\frac{4x}{(2x)^{4/3}}$

Problem Set 1.8 (page 72)

1. $13+8i$ **2.** $-5+8i$ **3.** $3+4i$
4. $-10-2i$ **5.** $-11+i$ **6.** $-1-i$
7. $-1-2i$ **8.** $\frac{5}{6}+\frac{13}{20}i$ **9.** $-\frac{3}{20}+\frac{5}{12}i$
10. $-\frac{1}{4}+\frac{3}{10}i$ **11.** $\frac{7}{10}-\frac{11}{12}i$ **12.** $-2-i\sqrt{3}$
13. $4+0i$ **14.** $-3i$ **15.** $3i$ **16.** $7i$
17. $i\sqrt{19}$ **18.** $i\sqrt{31}$ **19.** $\frac{2}{3}i$ **20.** $\frac{5}{6}i$
21. $2i\sqrt{2}$ **22.** $3i\sqrt{2}$ **23.** $3i\sqrt{3}$
24. $4i\sqrt{2}$ **25.** $3i\sqrt{6}$ **26.** $2i\sqrt{10}$ **27.** $18i$
28. $40i$ **29.** $12i\sqrt{2}$ **30.** $12i\sqrt{2}$ **31.** -8
32. -15 **33.** $-\sqrt{6}$ **34.** $-\sqrt{21}$
35. $-2\sqrt{5}$ **36.** $-3\sqrt{7}$ **37.** $-2\sqrt{15}$
38. $-2\sqrt{6}$ **39.** $-2\sqrt{14}$ **40.** $-2\sqrt{15}$
41. 3 **42.** 2 **43.** $\sqrt{6}$ **44.** $\sqrt{6}$
45. -21 **46.** 40 **47.** $8+12i$
48. $-30+10i$ **49.** $0+26i$ **50.** $44+52i$
51. $53-26i$ **52.** $3+i$ **53.** $10-24i$
54. $64-44i$ **55.** $-14-8i$ **56.** $-38-50i$
57. $-7+24i$ **58.** $12-16i$ **59.** $-3+4i$
60. $-21-20i$ **61.** $113+0i$ **62.** $34+0i$
63. $13+0i$ **64.** $85+0i$ **65.** $-\frac{8}{13}+\frac{12}{13}i$
66. $\frac{3}{20}+\frac{9}{20}i$ **67.** $1-\frac{2}{3}i$ **68.** $-\frac{5}{4}-\frac{3}{4}i$
69. $0-\frac{3}{2}i$ **70.** $0-\frac{7}{4}i$ **71.** $\frac{22}{41}-\frac{7}{41}i$
72. $\frac{41}{58}+\frac{1}{58}i$ **73.** $-1+2i$ **74.** $\frac{3}{17}+\frac{39}{17}i$
75. $-\frac{17}{10}+\frac{1}{10}i$ **76.** $-\frac{41}{29}+\frac{1}{29}i$ **77.** $\frac{5}{13}-\frac{1}{13}i$
78. $-\frac{14}{15}-\frac{17}{15}i$ **82.** **(a)** i **(b)** -1 **(c)** $-i$
(d) 1 **(e)** 1 **(f)** -1 **(g)** i **(h)** $-i$
83. **(a)** $2+11i$ **(b)** $-2-2i$ **(c)** $-11+2i$
(d) $-4+0i$ **(e)** $-7-24i$ **(f)** $24-4i$

Chapter 1 Review Problem Set (page 77)

1. $\frac{1}{125}$ **2.** $-\frac{1}{81}$ **3.** $\frac{16}{9}$ **4.** $\frac{1}{9}$ **5.** -8
6. $\frac{3}{2}$ **7.** $-\frac{1}{2}$ **8.** $\frac{1}{6}$ **9.** 4 **10.** -8
11. $12x^2y$ **12.** $-30x^{7/6}$ **13.** $\frac{48}{a^{1/6}}$
14. $\frac{27y^{3/5}}{x^2}$ **15.** $\frac{4y^5}{x^5}$ **16.** $\frac{8y}{x^{7/12}}$ **17.** $\frac{16x^6}{y^6}$
18. $-\frac{a^3b^1}{3}$ **19.** $4x-1$ **20.** $-3x+8$
21. $12a-19$ **22.** $20x^2-11x-42$
23. $-12x^2+17x-6$ **24.** $-35x^2+22x-3$
25. $x^3+x^2-19x-28$ **26.** $2x^3-x^2+10x+6$
27. $25x^2-30x+9$ **28.** $9x^2+42x+49$
29. $8x^3-12x^2+6x-1$
30. $27x^3+135x^2+225x+125$
31. $x^4+2x^3-6x^2-22x-15$
32. $2x^4+11x^3-16x^2-8x+8$
33. $-4x^2y^3+8xy^2$ **34.** $-7y+9xy^2$
35. $(3x+2y)(3x-2y)$ **36.** $3x(x+5)(x-8)$
37. $(2x+5)^2$ **38.** $(x-y+3)(x-y-3)$
39. $(x-2)(x-y)$
40. $(4x-3y)(16x^2+12xy+9y^2)$
41. $(3x-4)(5x+2)$ **42.** $3(x^3+12)$
43. Not factorable **44.** $3(x+2)(x^2-2x+4)$
45. $(x+3)(x-3)(x+2)(x-2)$
46. $(2x-1-y)(2x-1+y)$ **47.** $\frac{2}{3y}$
48. $\frac{-5a^2}{3}$ **49.** $\frac{3x+5}{x}$ **50.** $\frac{2(3x-1)}{x^2+4}$
51. $\frac{29x-10}{12}$ **52.** $\frac{x-38}{15}$ **53.** $\frac{-6n+15}{5n^2}$
54. $\frac{-3x-16}{x(x+7)}$ **55.** $\frac{3x^2-8x-40}{(x+4)(x-4)(x-10)}$
56. $\frac{8x-4}{x(x+2)(x-2)}$ **57.** $\frac{3xy-2x^2}{5y+7x^2}$
58. $\frac{3x-2}{4x+3}$ **59.** $-\frac{6x+3h}{x^2(x+h)^2}$ **60.** $\frac{12}{(x^2+2)^{3/2}}$
61. $20\sqrt{3}$ **62.** $6x\sqrt{6x}$ **63.** $2xy\sqrt[3]{4xy^2}$

64. $\sqrt{3}$ **65.** $\frac{\sqrt{10x}}{2y}$ **66.** $\frac{15 - 3\sqrt{2}}{23}$

67. $\frac{24 - 4\sqrt{6}}{15}$ **68.** $\frac{3x + 6\sqrt{xy}}{x - 4y}$ **69.** $\sqrt[6]{5^5}$

70. $\sqrt[12]{x^{11}}$ **71.** $x^2\sqrt[6]{x^5}$ **72.** $x\sqrt[10]{xy^9}$ **73.** $\sqrt[6]{5}$

74. $\frac{\sqrt[12]{x^{11}}}{x}$ **75.** $-11 - 6i$ **76.** $-1 - 2i$

77. $1 - 2i$ **78.** $21 + 0i$ **79.** $26 - 7i$

80. $-25 + 15i$ **81.** $-14 - 12i$ **82.** $29 + 0i$

83. $0 - \frac{5}{3}i$ **84.** $-\frac{6}{25} + \frac{17}{25}i$ **85.** $0 + i$

86. $-\frac{12}{29} - \frac{30}{29}i$ **87.** $10i$ **88.** $2i\sqrt{10}$

89. $16i\sqrt{5}$ **90.** -12 **91.** $-4\sqrt{3}$

92. $2\sqrt{2}$ **93.** 600,000,000 **94.** 800,000

CHAPTER 2

Problem Set 2.1 (page 88)

1. $\{-2\}$ **2.** $\{3\}$ **3.** $\left\{-\frac{1}{2}\right\}$ **4.** $\{2\}$

5. $\{7\}$ **6.** $\{12\}$ **7.** $\left\{-\frac{3}{2}\right\}$ **8.** $\{0\}$

9. $\left\{-\frac{10}{3}\right\}$ **10.** $\left\{\frac{9}{5}\right\}$ **11.** $\{-10\}$ **12.** $\{-35\}$

13. $\{-17\}$ **14.** $\left\{\frac{1}{14}\right\}$ **15.** $\left\{-\frac{21}{16}\right\}$

16. $\left\{\frac{11}{4}\right\}$ **17.** $\left\{\frac{3}{5}\right\}$ **18.** $\{-6\}$ **19.** $\{-14\}$

20. $\{5\}$ **21.** $\{9\}$ **22.** $\{15\}$ **23.** $\left\{\frac{10}{7}\right\}$

24. $\{10\}$ **25.** $\{-10\}$ **26.** $\{7\}$ **27.** $\{1\}$

28. $\{-8\}$ **29.** $\{-12\}$ **30.** $\{1\}$ **31.** $\{27\}$

32. $\{19\}$ **33.** $\left\{\frac{159}{5}\right\}$ **34.** $\left\{\frac{15}{17}\right\}$ **35.** $\{3\}$

36. $\left\{\frac{25}{9}\right\}$ **37.** $\{0\}$ **38.** $\left\{\frac{24}{5}\right\}$ **39.** $\left\{-\frac{2}{3}\right\}$

40. $\{-1\}$ **41.** $\left\{\frac{1}{2}\right\}$ **42.** $\{-2\}$ **43.** 41

44. -12 **45.** 7 and 23 **46.** 9 and 14

47. 18, 19, and 20 **48.** 14, 16, and 18

49. 31, 33, and 35 **50.** 18 and 19

51. 6, 7, and 8 **52.** 10, 11, 12, and 13

53. \$24,000 for Renee, \$20,000 for Kelly, \$16,000 for Nina **54.** \$7 per hour

55. 48 pennies, 21 nickels, and 11 dimes

56. 45 nickels, 20 dimes, and 40 quarters

57. 17 females and 26 males

58. 130 Republicans and 186 Democrats

59. 13 years old **60.** Eric is 16 and his father is 42.

61. Brad is 29 and Pedro is 23.

62. Tina is 21 and Sherry is 17.

Problem Set 2.2 (page 99)

1. $\{1\}$ **2.** $\left\{\frac{11}{21}\right\}$ **3.** $\{9\}$ **4.** $\{6\}$

5. $\left\{\frac{10}{3}\right\}$ **6.** $\left\{-\frac{5}{2}\right\}$ **7.** $\{4\}$ **8.** $\{39\}$

9. $\{14\}$ **10.** $\{17\}$ **11.** $\{9\}$ **12.** $\left\{\frac{1}{2}\right\}$

13. $\left\{\frac{1}{2}\right\}$ **14.** $\left\{\frac{7}{19}\right\}$ **15.** $\left\{\frac{1}{4}\right\}$ **16.** $\left\{-\frac{5}{2}\right\}$

17. $\left\{\frac{2}{3}\right\}$ **18.** $\varnothing$ **19.** $\{-8\}$ **20.** $\{7\}$

21. $\varnothing$ **22.** $\{-3\}$ **23.** $\{12\}$ **24.** $\{3\}$

25. $\{300\}$ **26.** $\{450\}$ **27.** $\{275\}$ **28.** $\{400\}$

29. $\left\{-\frac{66}{37}\right\}$ **30.** $\{10\}$ **31.** $\{6\}$ **32.** $\{13.5\}$

33. $w = \frac{P - 2l}{2}$ **34.** $B = \frac{3v}{h}$

35. $h = \frac{A - 2\pi r^2}{2\pi r}$ **36.** $h = \frac{2A}{b_1 + b_2}$

37. $\text{F} = \frac{9\text{C} + 160}{5}$ or $\text{F} = \frac{9}{5}\text{C} + 32$

38. $\text{C} = \frac{5}{9}(\text{F} - 32)$ or $\text{C} = \frac{5\text{F} - 160}{9}$

39. $T = \frac{NC - NV}{C}$ **40.** $N = \frac{CT}{C - V}$

41. $T = \frac{I + klt}{kl}$ **42.** $d = \frac{CRD}{12S}$

43. $R_n = \frac{R_1R_2}{R_1 + R_2}$ **44.** $b = \frac{af}{a - f}$

45. 17 and 81 **46.** 7 and 107 **47.** 9 **48.** $\frac{5}{9}$

49. \$900 and \$1350

50. 1.25 liters of oil and 18.75 liters of gasoline

51. 37 teachers and 740 students

52. 125 pounds of sodium and 75 pounds of chlorine

53. \$65 **54.** \$48 **55.** \$950 per month

56. \$10,500 **57.** \$32.20 **58.** \$.58

59. \$75 **60.** \$625 **61.** \$30 **62.** \$25

63. 14 nickels and 29 dimes

64. 50 nickels, 60 dimes, and 120 quarters

65. 15 dimes, 45 quarters, and 10 half-dollars
66. \$2500 at 8% and \$4000 at 9%
67. \$2000 at 9% and \$3500 at 10%
68. \$1500 at 9% and \$2000 at 11%
69. \$3500
70. 14 inches by 40 inches
71. 6 centimeters by 10 centimeters
72. 12 inches by 17 inches

Problem Set 2.3 (page 110)

1. $\{-4, 7\}$ **2.** $\{-2, 6\}$ **3.** $\left\{-3, \frac{4}{3}\right\}$
4. $\left\{\frac{1}{2}, 6\right\}$ **5.** $\left\{0, \frac{3}{2}\right\}$ **6.** $\{0, 1\}$
7. $\left\{\pm\frac{2\sqrt{3}}{3}\right\}$ **8.** $\left\{-\frac{3}{4}, \frac{5}{4}\right\}$ **9.** $\left\{\frac{-1 \pm 2\sqrt{5}}{2}\right\}$
10. $\left\{\frac{1 \pm \sqrt{5}}{4}\right\}$ **11.** $\left\{-\frac{5}{3}, \frac{2}{5}\right\}$ **12.** $\left\{-4, \frac{1}{6}\right\}$
13. $\{2 \pm 2i\}$ **14.** $\left\{-\frac{4}{3}, \frac{3}{8}\right\}$ **15.** $\left\{-\frac{7}{2}, \frac{1}{5}\right\}$
16. $\{3 \pm 3i\}$ **17.** $\{4, 6\}$ **18.** $\{-5, 4\}$
19. $\{-5 \pm 3\sqrt{3}\}$ **20.** $\{-3 \pm \sqrt{10}\}$
21. $\left\{\frac{3 \pm \sqrt{5}}{2}\right\}$ **22.** $\left\{\frac{-5 \pm \sqrt{17}}{2}\right\}$
23. $\{-2 \pm i\sqrt{2}\}$ **24.** $\{3 \pm 2i\sqrt{3}\}$
25. $\left\{\frac{-6 \pm \sqrt{46}}{2}\right\}$ **26.** $\left\{\frac{-6 \pm \sqrt{42}}{3}\right\}$
27. $\{-16, 18\}$ **28.** $\{-17, 13\}$
29. $\left\{\frac{-5 \pm \sqrt{37}}{6}\right\}$ **30.** $\left\{\frac{-1 \pm \sqrt{33}}{4}\right\}$
31. $\{-6, 9\}$ **32.** $\{-11, -2\}$ **33.** $\left\{-5, -\frac{1}{3}\right\}$
34. $\left\{-\frac{3}{5}, \frac{7}{2}\right\}$ **35.** $\{1 \pm \sqrt{5}\}$ **36.** $\{3 \pm 2\sqrt{3}\}$
37. $\left\{\frac{3 \pm \sqrt{7}}{2}\right\}$ **38.** $\left\{\frac{-3 \pm \sqrt{17}}{4}\right\}$
39. $\left\{\frac{3 \pm i\sqrt{19}}{2}\right\}$ **40.** $\left\{\frac{5 \pm i\sqrt{7}}{2}\right\}$
41. $\{4 \pm 2\sqrt{3}\}$ **42.** $\{-7 \pm 3\sqrt{2}\}$ **43.** $\left\{\frac{1}{2}\right\}$
44. $\{\pm 2i\sqrt{6}\}$ **45.** $\left\{-\frac{3}{2}, \frac{1}{4}\right\}$ **46.** $\left\{\frac{5}{6}, \frac{4}{3}\right\}$
47. $\{-14, 12\}$ **48.** $\{-17, -11\}$
49. $\left\{\frac{3 \pm i\sqrt{47}}{4}\right\}$ **50.** $\left\{\frac{1 \pm i\sqrt{14}}{3}\right\}$
51. $\left\{-1, \frac{5}{3}\right\}$ **52.** $\left\{-\frac{5}{4}, \frac{2}{5}\right\}$ **53.** $\left\{\frac{-1 \pm \sqrt{2}}{2}\right\}$
54. $\left\{\frac{-2 \pm 2i\sqrt{7}}{5}\right\}$ **55.** $\{8 \pm 5\sqrt{2}\}$
56. $\{9 \pm \sqrt{66}\}$ **57.** $\{-10 \pm 5\sqrt{5}\}$
58. $\{9 \pm 3\sqrt{10}\}$ **59.** $\left\{\frac{1 \pm \sqrt{6}}{5}\right\}$ **60.** $\left\{\frac{3}{7}, 2\right\}$
61. **(a)** One real solution
(b) Two complex but nonreal solutions
(c) One real solution **(d)** Two unequal real solutions
(e) Two complex but nonreal solutions
(f) One real solution **(g)** Two unequal real solutions
(h) Two unequal real solutions **62.** 22 and 24
63. 11 and 12 **64.** 10 and −6 **65.** 8 and 14
66. 12 inches and 16 inches
67. 10 meters and 24 meters **68.** 6, 8, and 10 units
69. 8 inches by 14 inches **70.** 9 inches by 18 inches
71. 7 meters wide and 18 meters long
72. A side of 13 centimeters and an altitude of 8 centimeters
73. 1 meter **74.** 24 inches and 36 inches
75. 7 inches by 3 inches
76. 8 feet by 6.5 feet or 13 feet by 4 feet **77.** 8 units
81. **(a)** $\{0, 7k\}$ **(b)** $\{0, 25k\}$ **(c)** $\{-2k, 5k\}$
(d) $\left\{\frac{k}{2}, -\frac{2k}{3}\right\}$ **(e)** $\left\{\frac{k}{3}\right\}$ **(f)** $\left\{-\frac{2}{k}, \frac{3}{k}\right\}$
(g) $\left\{\frac{-\sqrt{2} \pm \sqrt{14}}{2}\right\}$ **(h)** $\left\{\frac{\sqrt{3} \pm i\sqrt{17}}{2}\right\}$
82. **(a)** $r = \frac{\sqrt{A\pi}}{\pi}$ **(b)** $c = \frac{\sqrt{E(m - m_0)}}{m - m_0}$
(c) $t = \frac{\sqrt{2gs}}{g}$ **(d)** $x = \frac{a\sqrt{b^2 - y^2}}{b}$
(e) $y = \frac{b\sqrt{x^2 - a^2}}{a}$ **(f)** $t = \frac{-V_0 + \sqrt{V_0^2 + 2gs}}{g}$
83. $k > 1$ **84.** $k = 4$ or $k = -4$
85. Any real number value for k
86. **(a)** $\{-1.359, 7.359\}$ **(b)** $\{-1.381, 17.381\}$
(c) $\{-10.280, 4.280\}$ **(d)** $\{-13.426, 3.426\}$
(e) $\{-0.259, -7.742\}$ **(f)** $\{-0.347, -8.653\}$
(g) $\{0.191, 1.309\}$ **(h)** $\{0.119, 1.681\}$
(i) $\{-0.422, 5.922\}$ **(j)** $\{-0.708, 4.708\}$

Problem Set 2.4 (page 121)

1. $\{-12\}$ **2.** $\left\{-\frac{10}{3}\right\}$ **3** $\left\{\frac{37}{15}\right\}$ **4.** $\{-21\}$
5. $\{-2\}$ **6.** $\{-1, 2\}$ **7.** $\{-8, 1\}$

8. $\left\{-\frac{5}{2}, 1\right\}$ **9.** $\left\{\frac{6}{29}\right\}$ **10.** $\left\{-\frac{1}{5}\right\}$

11. $\left\{n \mid n \neq \frac{3}{2} \text{ and } n \neq 3\right\}$ **12.** $\{-2, 1\}$

13. $\left\{-4, \frac{4}{3}\right\}$ **14.** $\left\{-3, \frac{3}{2}\right\}$ **15.** $\left\{-\frac{1}{4}\right\}$

16. $\{x \mid x \neq 2 \text{ and } x \neq -3\}$ **17.** $\varnothing$ **18.** $\left\{\frac{7}{2}\right\}$

19. $\{-1\}$ **20.** $\left\{\frac{1}{15}\right\}$

21. 9 rows and 14 trees per row **22.** $\frac{1}{3}$ or 3

23. $4\frac{1}{2}$ hours **24.** 48 mph and 54 mph

25. 50 miles **26.** 12 minutes

27. 50 mph for the freight and 70 mph for the express

28. 410 mph **29.** 3 liters **30.** 6 liters

31. 3.5 liters of the 50% solution and 7 liters of the 80% solution **32.** 12 pounds **33.** 5 quarts

34. 10 gallons **35.** $2\frac{2}{5}$ hours **36.** 5 hours

37. 60 minutes **38.** 9 hours

39. 9 minutes for Pat and 18 minutes for Mike

40. 750 cards per minute for A and 1000 cards per minute for B

41. 60 words per minute for Amelia and 40 words per minute for Paul **42.** 32 miles

43. 7 golf balls **44.** 40-hour week **45.** 60 hours

Problem Set 2.5 (page 129)

1. $\{-2, -1, 2\}$ **2.** $\{-1, 1, 5\}$ **3.** $\left\{\pm i, \frac{3}{2}\right\}$

4. $\left\{\pm 2i, -\frac{5}{3}\right\}$ **5.** $\left\{-\frac{5}{4}, 0, \pm\frac{\sqrt{2}}{2}\right\}$

6. $\left\{-\frac{5}{2}, 0, \pm\frac{\sqrt{5}}{5}\right\}$ **7.** $\{0, 16\}$ **8.** $\left\{0, \frac{6}{5}\right\}$

9. $\{1\}$ **10.** $\{0, 64\}$ **11.** $\{6\}$ **12.** $\varnothing$

13. $\{3\}$ **14.** $\{2\}$ **15.** $\varnothing$ **16.** $\{3\}$

17. $\{-15\}$ **18.** $\{0\}$ **19.** $\{9\}$ **20.** $\{12\}$

21. $\left\{\frac{2}{3}, 1\right\}$ **22.** $\left\{\frac{4}{5}, 1\right\}$ **23.** $\{5\}$ **24.** $\{4\}$

25. $\{7\}$ **26.** $\{11\}$ **27.** $\{-2, -1\}$

28. $\{-3, 1\}$ **29.** $\{0\}$ **30.** $\{13\}$ **31.** $\{6\}$

32. $\left\{-8, -\frac{9}{2}\right\}$ **33.** $\{0, 4\}$ **34.** $\{9\}$

35. $\{\pm 1, \pm 2\}$ **36.** $\{\pm 3, \pm 4\}$ **37.** $\left\{\pm\frac{\sqrt{2}}{2}, \pm 2\right\}$

38. $\left\{\pm\frac{\sqrt{3}}{3}, \pm 1\right\}$ **39.** $\{\pm i\sqrt{5}, \pm\sqrt{7}\}$

40. $\left\{\pm 2i, \pm\frac{\sqrt{6}}{2}\right\}$

41. $\{\pm\sqrt{2+\sqrt{3}}, \pm\sqrt{2-\sqrt{3}}\}$

42. $\{\pm\sqrt{4+\sqrt{5}}, \pm\sqrt{4-\sqrt{5}}\}$ **43.** $\{-125, 8\}$

44. $\{-8, 1\}$ **45.** $\left\{-\frac{8}{27}, \frac{27}{8}\right\}$ **46.** $\left\{-\frac{1}{27}, 64\right\}$

47. $\left\{-\frac{1}{6}, \frac{1}{2}\right\}$ **48.** $\left\{-4, \frac{3}{5}\right\}$ **49.** $\{25, 36\}$

50. $\left\{\frac{9}{4}, 16\right\}$ **51.** $\{4\}$ **52.** $\left\{\frac{49}{4}\right\}$

54. **(a)** $\{\pm 1.62, \pm 0.62\}$ **(b)** $\{\pm 2.14, \pm 0.66\}$
(c) $\{\pm 1.78, \pm 0.56\}$ **(d)** $\{\pm 1.70, \pm 0.34\}$
(e) $\{\pm 8.00, \pm 6.00\}$ **(f)** $\{\pm 9.00, \pm 3.50\}$

Problem Set 2.6 (page 137)

1. $(-\infty, -2]$

2. $(-1, \infty)$

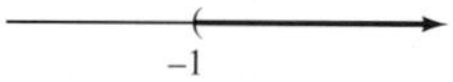

3. $(1, 4)$

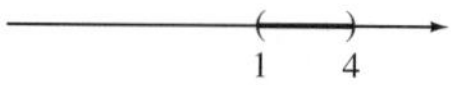

4. $(-1, 2]$

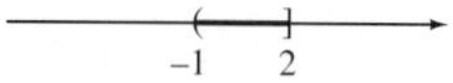

5. $(0, 2)$

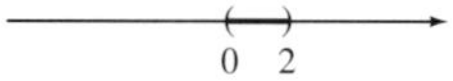

6. $(-\infty, -3]$

7. $[-2, -1]$

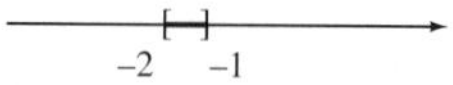

8. $[1, \infty)$

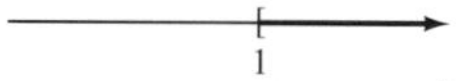

9. $(-\infty, -2)$ **10.** $(-2, \infty)$ **11.** $\left[-\frac{5}{3}, \infty\right)$

12. $(-\infty, -2)$ **13.** $[7, \infty)$ **14.** $\left(\frac{3}{13}, \infty\right)$

15. $\left(-\infty, \frac{17}{5}\right]$ **16.** $\left(-\infty, \frac{1}{7}\right]$ **17.** $\left(-\infty, \frac{7}{3}\right)$

18. $\left(\frac{5}{28}, \infty\right)$ **19.** $[-20, \infty)$ **20.** $\left(-\infty, -\frac{19}{6}\right)$

21. $(300, \infty)$ **22.** $(-\infty, 50]$ **23.** $\left(\frac{1}{5}, \frac{7}{5}\right)$

24. $\left[-\frac{9}{4}, -\frac{1}{4}\right]$ **25.** $[1, 5]$ **26.** $\left[1, \frac{13}{3}\right]$

27. $(-4, 1)$ **28.** $(-2, 2)$

29. $(-\infty, -3) \cup (5, \infty)$ **30.** $(-\infty, 4] \cup [8, \infty)$
31. $[-1, 2]$ **32.** $[-6, 1]$
33. $(-\infty, -4) \cup \left(\frac{1}{3}, \infty\right)$ **34.** $\left(-\infty, -\frac{1}{2}\right) \cup (5, \infty)$
35. $\left[\frac{2}{5}, \frac{4}{3}\right]$ **36.** $\left[-\frac{7}{3}, -\frac{2}{3}\right]$
37. $\left(-\infty, \frac{1}{2}\right) \cup \left(\frac{1}{2}, \infty\right)$ **38.** $\left\{-\frac{1}{3}\right\}$
39. $(-2, -1)$ **40.** $(-\infty, -8) \cup (2, \infty)$
41. $\left(-\infty, \frac{22}{3}\right]$ **42.** $\left[\frac{1}{12}, \infty\right)$
43. $(-4, 1) \cup (2, \infty)$ **44.** $[-7, -1] \cup [3, \infty)$
45. $(-\infty, -2] \cup \left[\frac{1}{2}, 5\right]$ **46.** $(-\infty, -4) \cup \left(-\frac{2}{3}, 3\right)$
47. $[-4, 0] \cup [6, \infty)$ **48.** $(-3, 0) \cup (1, \infty)$
49. $(-3, 2) \cup (2, \infty)$ **50.** $(-5, -4) \cup (-4, \infty)$
51. Greater than 12% **52.** More than \$900
53. 98 or better **54.** 5 feet and 10 inches or better
55. Between 5°C and 15°C, inclusive
56. Between −4°F and 23°F, inclusive
57. $8.8 \le M \le 15.4$ **58.** $6.3 \le M \le 11.25$
59. More than 250 miles
60. Between $t = \frac{7}{8}$ and $t = 1$, inclusive
63. **(a)** $(-\infty, -5) \cup (1, \infty)$ **(b)** $(-\infty, -2] \cup [4, \infty)$
(c) $(-4, 3)$ **(d)** $\left[-5, \frac{1}{2}\right]$
(e) $(-4, -1) \cup (2, \infty)$ **(f)** $(-\infty, -2) \cup (1, 3)$

Problem Set 2.7 (page 146)

1. $(-\infty, 1) \cup (5, \infty)$ **2.** $(-4, -2]$
3. $\left(-2, \frac{1}{2}\right)$ **4.** $\left(-\infty, -\frac{2}{3}\right) \cup (1, \infty)$
5. $\left(\frac{1}{3}, 3\right]$ **6.** $(-\infty, -4) \cup (-2, \infty)$
7. $[-3, -2)$ **8.** $(1, 2)$
9. $(-\infty, -5) \cup (-2, \infty)$ **10.** $(-\infty, 5) \cup [9, \infty)$
11. $(-\infty, -5)$ **12.** $(-7, \infty)$ **13.** $(-3, 2)$
14. $(-\infty, -11) \cup (-1, 4)$ **15.** $\{-4, 8\}$
16. $\{-7, 1\}$ **17.** $\left\{-\frac{13}{20}, \frac{3}{20}\right\}$ **18.** $\left\{-\frac{1}{12}, \frac{17}{12}\right\}$
19. $\{-3, 4\}$ **20.** $\{-6, 5\}$ **21.** $\left\{-3, \frac{1}{3}\right\}$
22. $\left\{-\frac{7}{5}, \frac{13}{5}\right\}$ **23.** $\varnothing$ **24.** $\left\{-\frac{7}{2}, \frac{5}{2}\right\}$
25. $\left\{-\frac{10}{3}, 2\right\}$ **26.** $\varnothing$ **27.** $\left\{\frac{1}{4}, \frac{7}{4}\right\}$
28. $\left\{-\frac{17}{5}, -\frac{13}{5}\right\}$ **29.** $\left\{-\frac{2}{5}, 4\right\}$ **30.** $\left\{\frac{1}{3}, 2\right\}$
31. $\{-2, 0\}$ **32.** $\{0, 10\}$ **33.** $\{-1\}$
34. $\left\{-\frac{1}{2}\right\}$ **35.** $(-6, 6)$
36. $(-\infty, -4] \cup [4, \infty)$ **37.** $(-\infty, -8) \cup (8, \infty)$
38. $[-1, 1]$ **39.** $(-\infty, \infty)$ **40.** $\varnothing$
41. $(-\infty, -2) \cup (8, \infty)$ **42.** $(-3, -1)$
43. $[-3, 4]$ **44.** $(-\infty, -2] \cup [1, \infty)$
45. $\left(-\infty, -\frac{11}{3}\right) \cup \left(\frac{7}{3}, \infty\right)$ **46.** $\left(0, \frac{4}{5}\right)$ **47.** $\varnothing$
48. $(-\infty, 1) \cup (3, \infty)$ **49.** $\left(-\frac{1}{2}, \frac{7}{2}\right)$
50. $(-\infty, \infty)$ **51.** $(-\infty, \infty)$
52. $(-\infty, 9] \cup [7, \infty)$ **53.** $[-7, 3]$ **54.** $(-1, 3)$
55. $(-6, 0)$ **56.** $\varnothing$ **57.** $(-\infty, -6) \cup (-2, \infty)$
58. $(-\infty, -1) \cup (5, \infty)$ **59.** $(-\infty, 0] \cup [4, \infty)$
60. $(-5, 3)$ **61.** $(-6, 4)$ **62.** $(-\infty, 0] \cup [8, \infty)$
63. $\left(-\infty, \frac{5}{4}\right) \cup \left(\frac{7}{2}, \infty\right)$ **64.** $(-\infty, 3) \cup (7, \infty)$
65. $(-\infty, -3) \cup (-3, -1)$ **66.** $\left[\frac{11}{4}, 5\right) \cup \left(5, \frac{19}{2}\right]$
67. $\left[-\frac{2}{5}, 0\right) \cup \left(0, \frac{2}{3}\right]$ **68.** $(-\infty, -2)$
69. $\left(-\infty, \frac{2}{5}\right] \cup \left[\frac{2}{3}, \infty\right)$
70. $\left(-\frac{8}{3}, -2\right) \cup \left(-2, -\frac{8}{5}\right)$ **77.** **(a)** $(-16, 6)$
(b) $[-6, 14]$ **(c)** $(-\infty, -3) \cup (4, \infty)$
(d) $(-\infty, -1] \cup \left[-\frac{1}{3}, \infty\right)$ **(e)** $(-3, 7)$
(f) $(-\infty, -3) \cup (9, \infty)$

Chapter 2 Review Problem Set (page 149)

1. $\{-14\}$ **2.** $\{-19\}$ **3.** $\left\{\frac{10}{7}\right\}$ **4.** $\{200\}$
5. $\left\{-1, \frac{5}{3}\right\}$ **6.** $\left\{\frac{5}{4}, 6\right\}$ **7.** $\{3 \pm i\}$
8. $\{-22, 18\}$ **9.** $\left\{-\frac{2}{5}, 0, \frac{1}{3}\right\}$ **10.** $\{-5\}$
11. $\left\{\frac{1}{2}, 6\right\}$ **12.** $\{\pm 3i, \pm\sqrt{5}\}$
13. $\left\{\pm\frac{\sqrt{5}}{5}, \pm\sqrt{2}\right\}$ **14.** $\left\{-1, 2, \frac{-5 \pm \sqrt{33}}{2}\right\}$
15. $\{2\}$ **16.** $\left\{-1, \frac{1}{2}\right\}$ **17.** $\{0\}$

18. $\left\{-\frac{6}{5}, \frac{8}{5}\right\}$ **19.** $\left\{\frac{2}{5}, 12\right\}$ **20.** $\left\{\frac{1}{4}, \frac{7}{4}\right\}$
21. $\{-\sqrt{2}, -1, \sqrt{2}\}$ **22.** $\left\{-64, \frac{27}{8}\right\}$
23. $(-8, \infty)$ **24.** $\left[-\frac{65}{4}, \infty\right)$ **25.** $\left(-\infty, -\frac{9}{2}\right)$
26. $(-\infty, 400]$ **27.** $[-2, 1]$ **28.** $\left(-\frac{2}{3}, 2\right)$
29. $(-3, 6)$ **30.** $(-\infty, -2] \cup [7, \infty)$
31. $(-\infty, -2) \cup (1, 4)$ **32.** $\left[-4, \frac{3}{2}\right)$
33. $\left(-\infty, \frac{1}{5}\right) \cup (2, \infty)$ **34.** $[-7, -3)$
35. $(-\infty, 4)$ **36.** $\left(-\infty, -\frac{1}{2}\right) \cup (2, \infty)$
37. $\left[-\frac{19}{3}, 3\right]$ **38.** $\left(-\frac{9}{2}, \frac{3}{2}\right)$
39. $(-1, 0) \cup \left(0, \frac{1}{3}\right)$ **40.** $\left(-\frac{3}{2}, \infty\right)$
41. 21, 23, and 25 **42.** 9 and 65
43. 7 centimeters by 12 centimeters
44. 13 nickels, 39 dimes, and 36 quarters **45.** \$20
46. 20 gallons
47. Rosie is 14 years old and her mother is 33 years old.
48. \$350 at 9% and \$450 at 12% **49.** 95 or better
50. $10\frac{10}{11}$ minutes **51.** $26\frac{2}{3}$ minutes
52. 40 shares at \$15/share
53. 54 mph for Mike and 52 mph for Larry
54. Cindy 4 hours and Bill 6 hours
55. 15 centimeters and 20 centimeters
56. 5 inches by 7 inches

CHAPTER 3

Problem Set 3.1 (page 163)

1. 10 **2.** −19 **3.** −5 **4.** 17 **5.** 6
6. 8 **7.** 15 **8.** 15 **9.** 7 **10.** 10
11. $\frac{1}{3}$ **12.** $-\frac{3}{5}$ **13.** −7 **14.** $-\frac{16}{3}$
15. 10; (6, 4) **16.** 15; $\left(\frac{5}{2}, 5\right)$
17. $\sqrt{13}$; $\left(2, -\frac{5}{2}\right)$ **18.** $4\sqrt{2}$; $(-3, 4)$
19. $3\sqrt{2}$; $\left(\frac{15}{2}, -\frac{11}{2}\right)$ **20.** $3\sqrt{5}$; $\left(-\frac{3}{2}, 0\right)$
21. $\frac{\sqrt{74}}{6}$; $\left(\frac{1}{12}, \frac{11}{12}\right)$ **22.** $\frac{\sqrt{170}}{4}$; $\left(-\frac{7}{8}, \frac{3}{8}\right)$
23. (3, 5) **24.** (5, 10) **25.** (2, 5)
26. $\left(-1, \frac{18}{5}\right)$ **27.** $\left(\frac{17}{8}, -7\right)$ **28.** $\left(-\frac{9}{8}, -\frac{15}{2}\right)$
29. $\left(4, \frac{25}{4}\right)$ **30.** (2, 10) **35.** $15 + 9\sqrt{5}$
38. (8, 9), (−12, 1), or (4, −3) **39.** 3 or −7
41. (3, 8) **42.** All three distances equal 13 units.
43. Both midpoints are at $\left(\frac{7}{2}, \frac{5}{2}\right)$.
44. Both midpoints are at $\left(\frac{5}{4}, 0\right)$.

Problem Set 3.2 (page 173)

1.
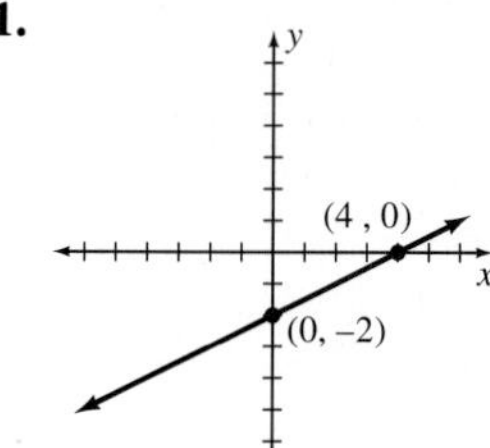

2.
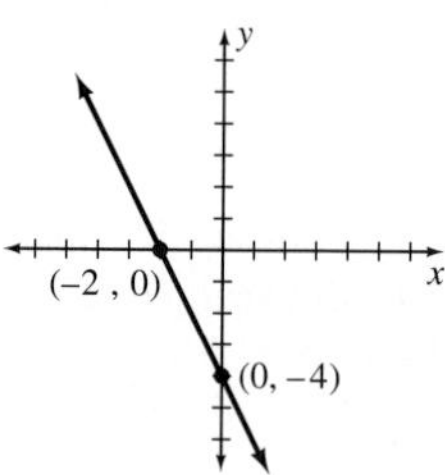

3.
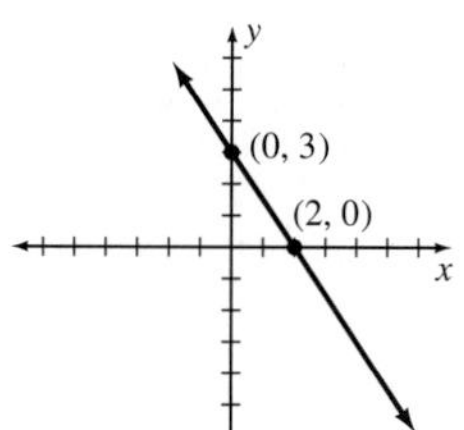

4.
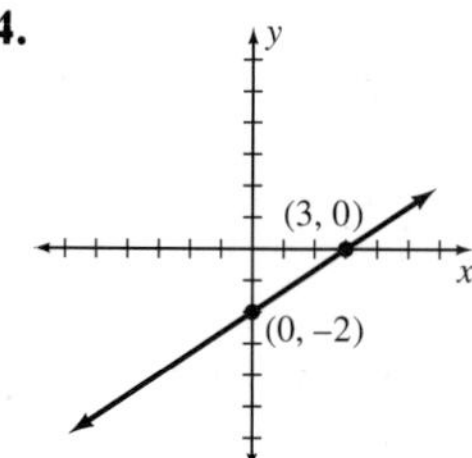

5.
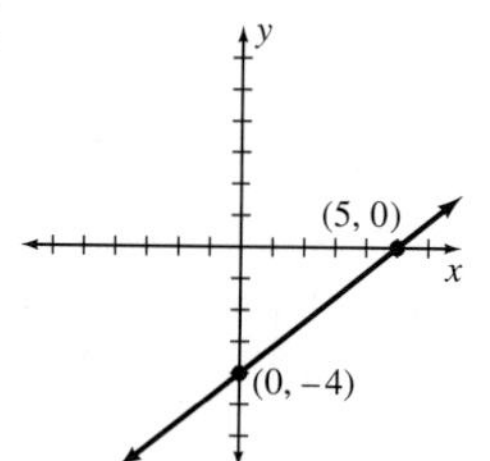

6.
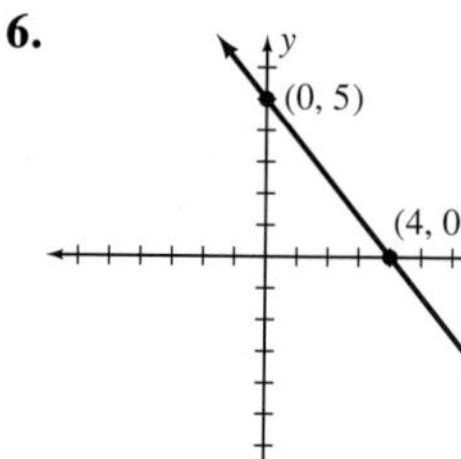

7.
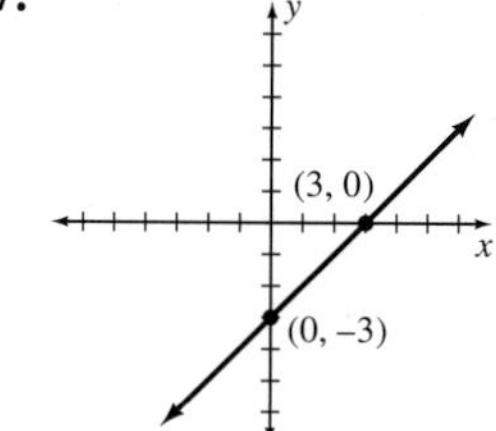

8.
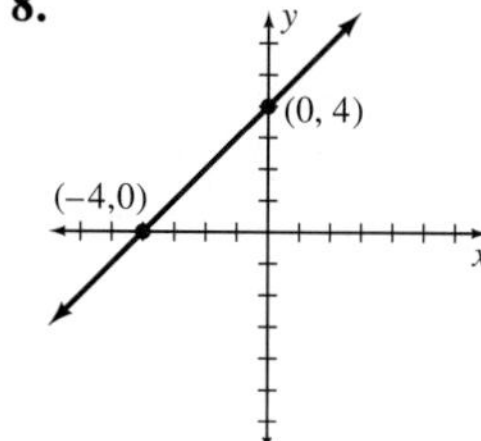

9.

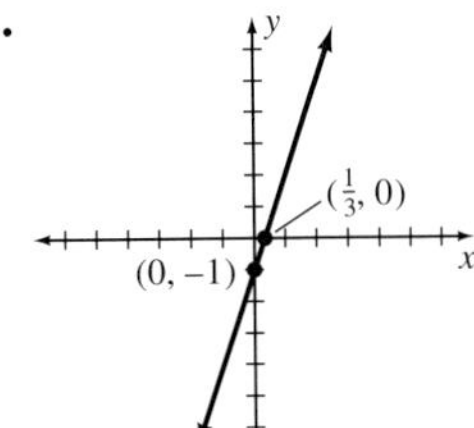

10.

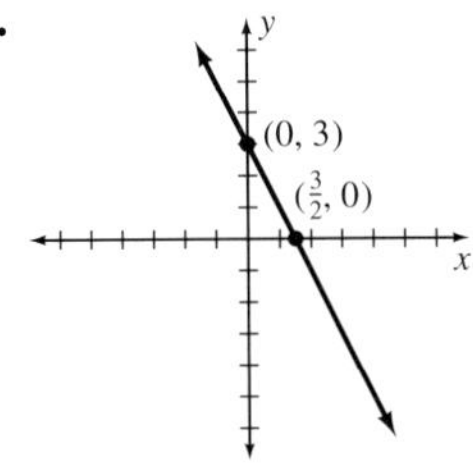

11.

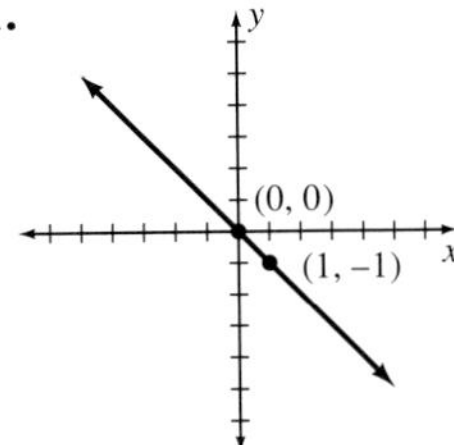

12.

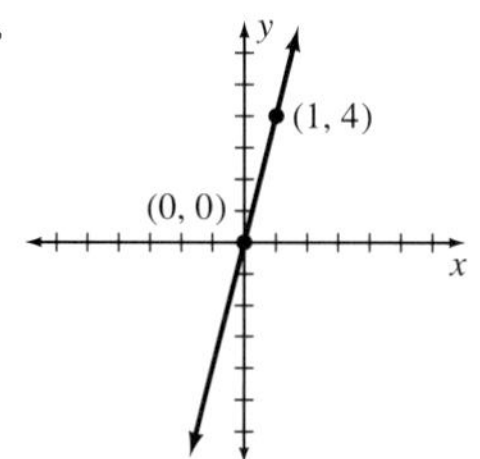

13.

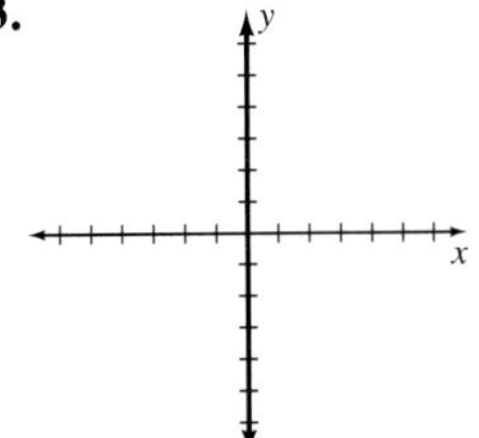

The graph is the y-axis.

14.

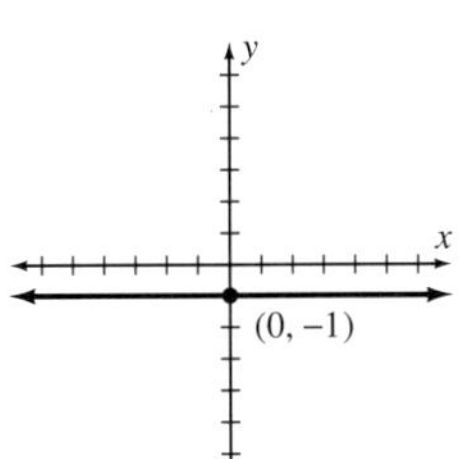

15.

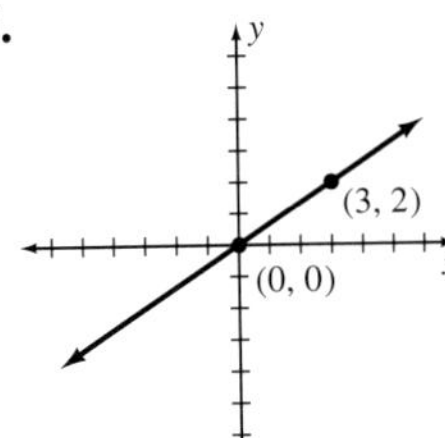

16.

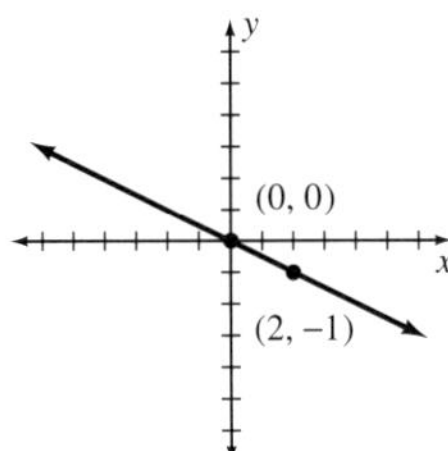

17.

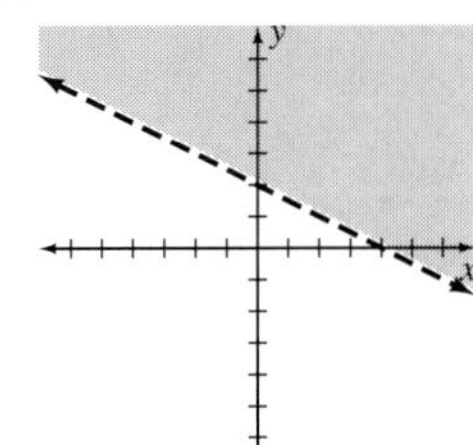

18.

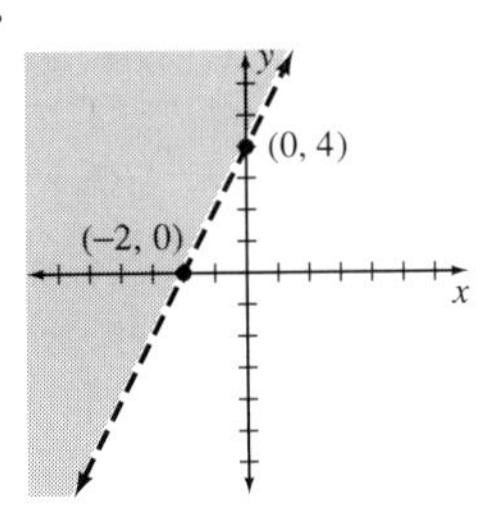

19.

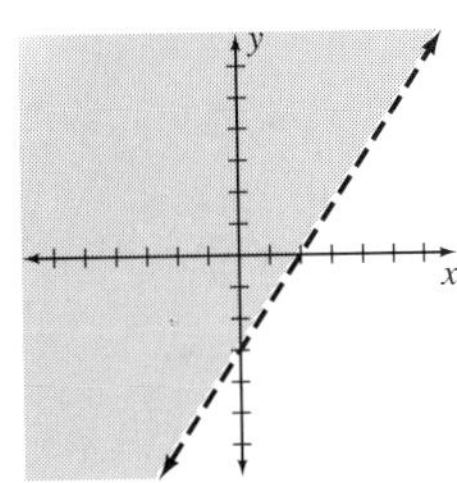

20.

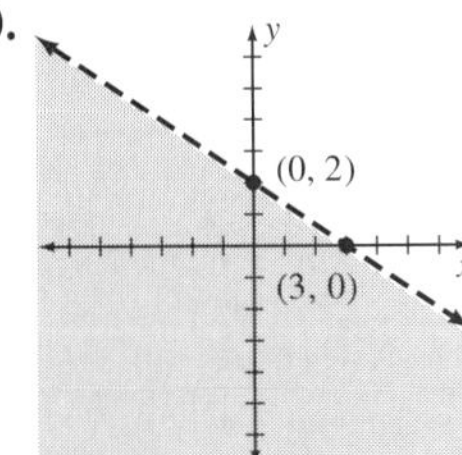

21.

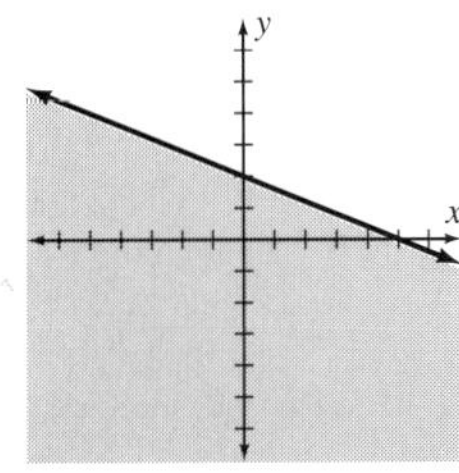

22.

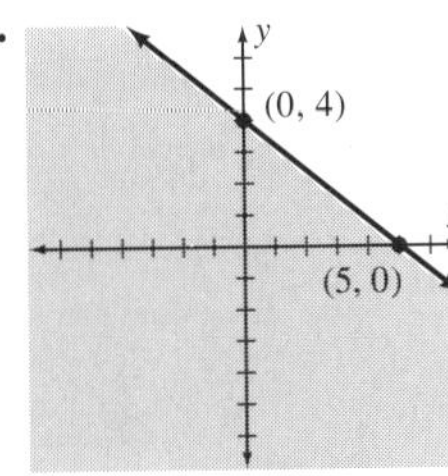

23.

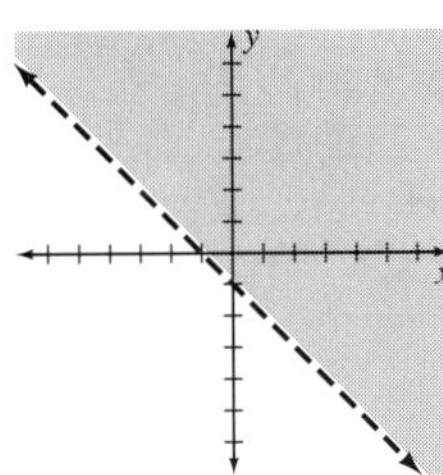

24.

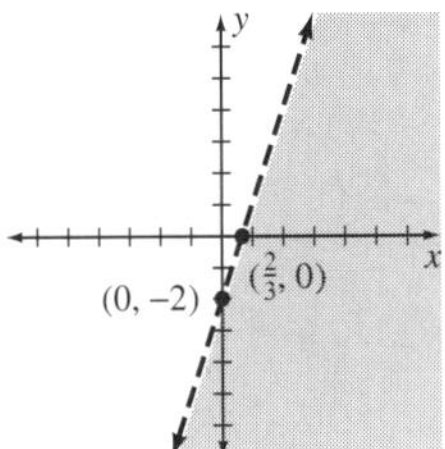

25.

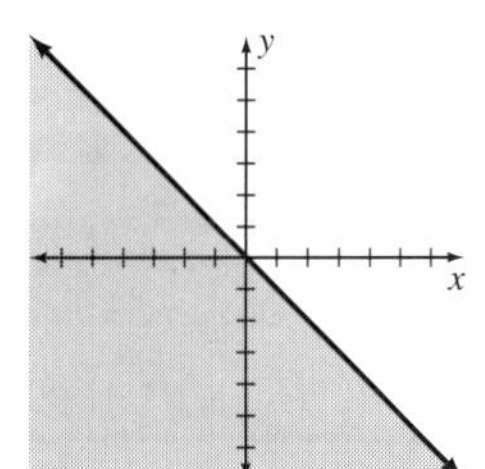

26.

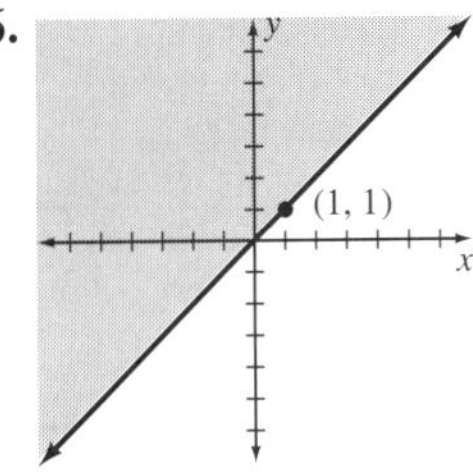

27.

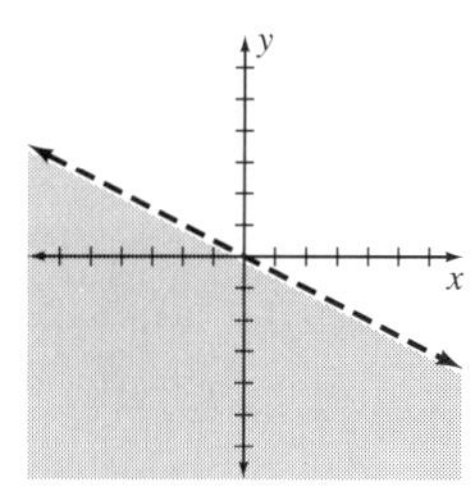

28.

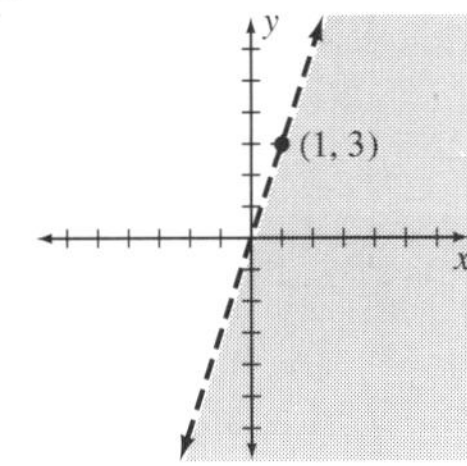

29.

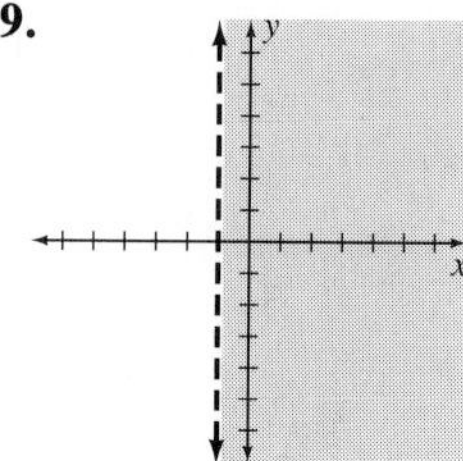

30.

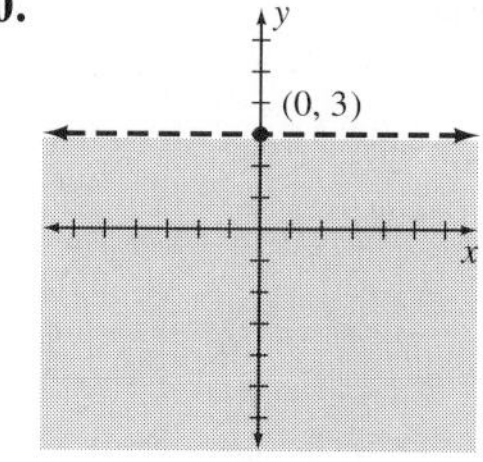

33.

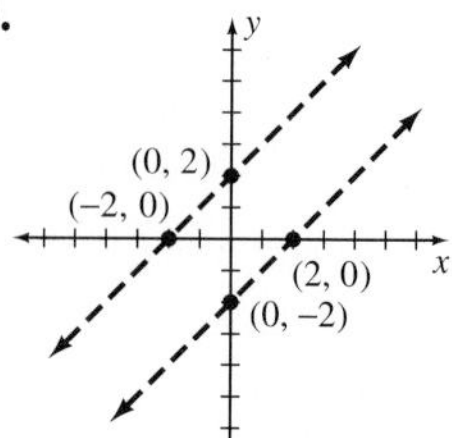

34.

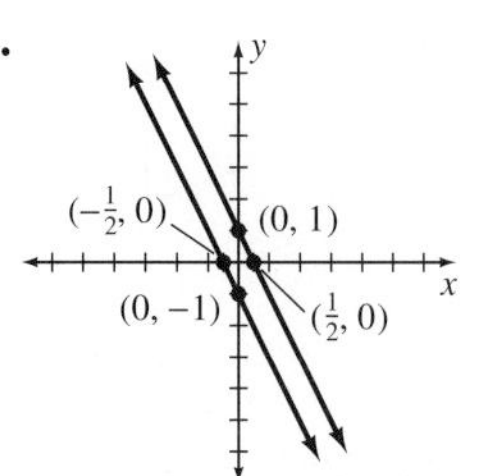

35.

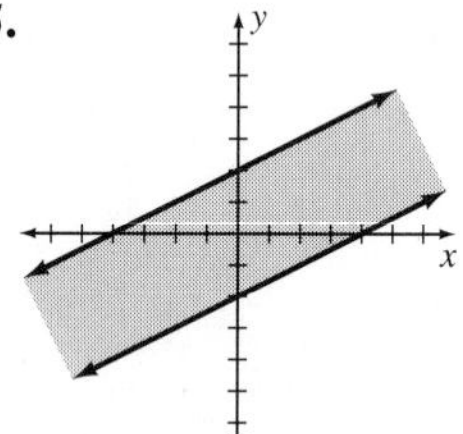

36.

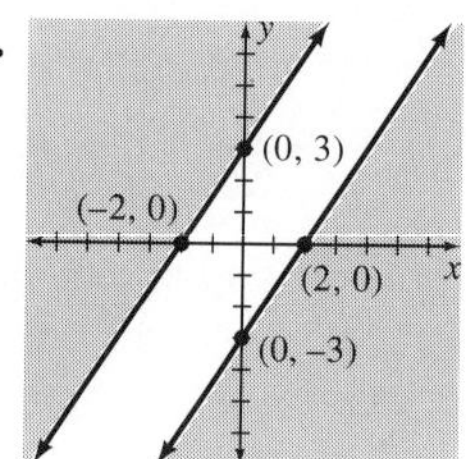
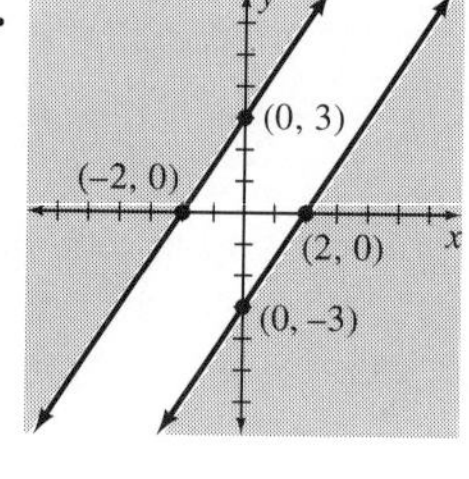

37.

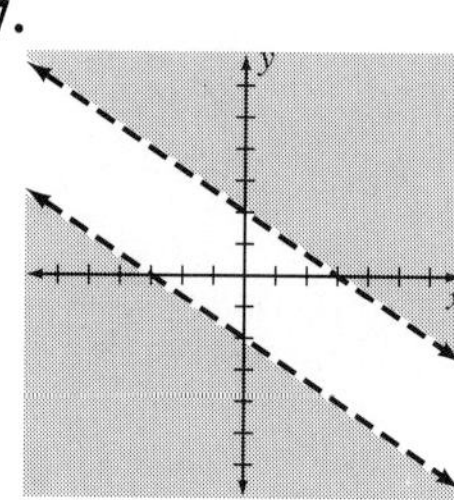

38.

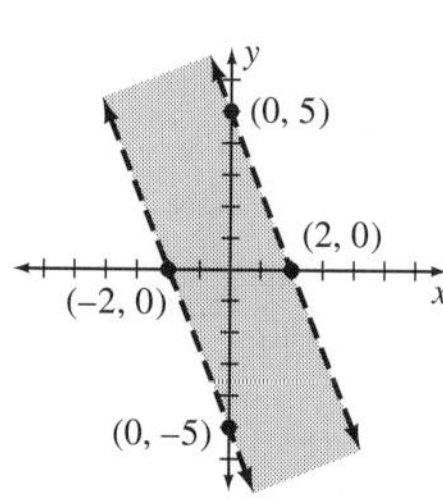

39.

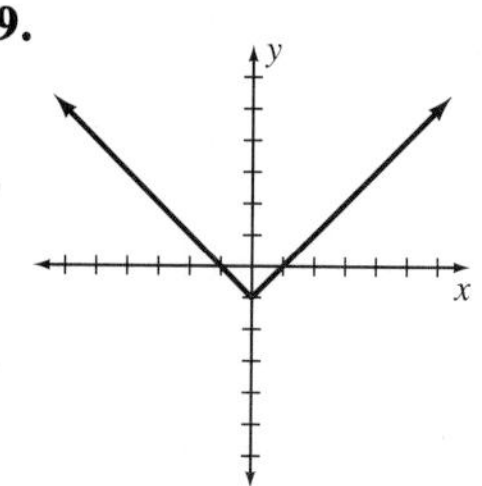

40.

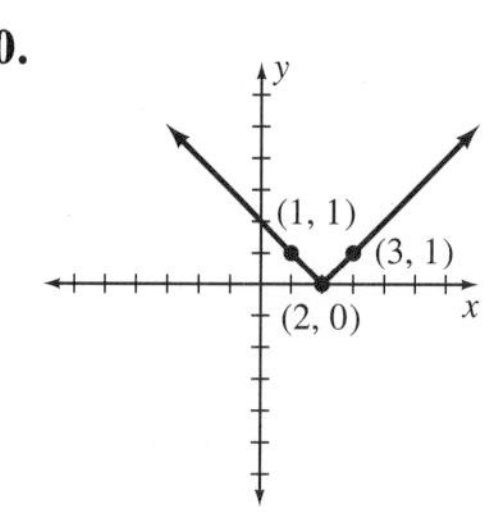

41.

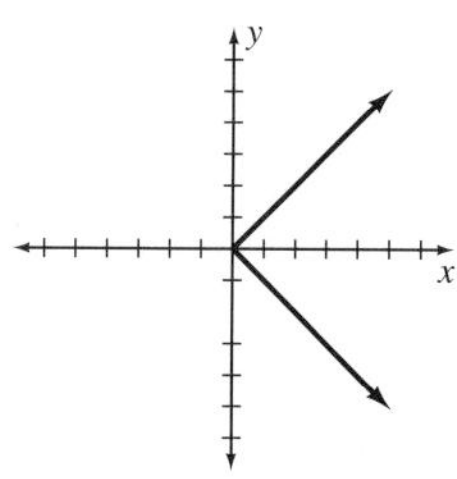

42.

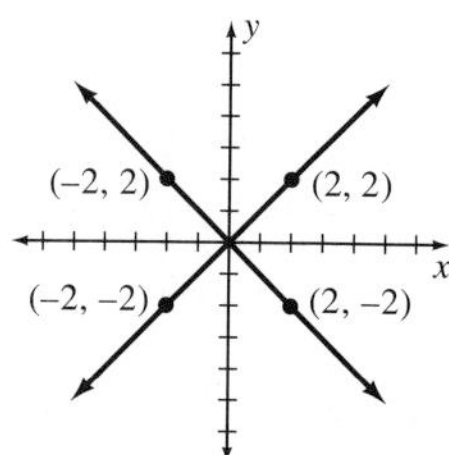

43.

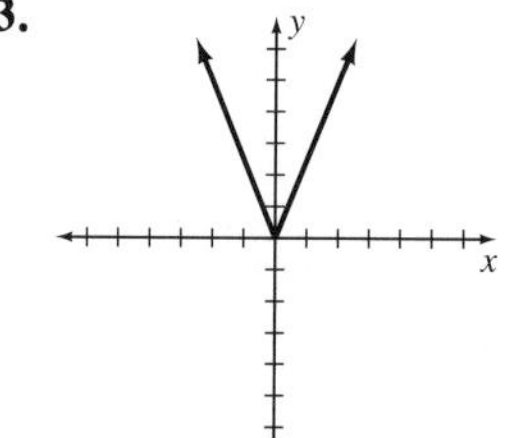

44.

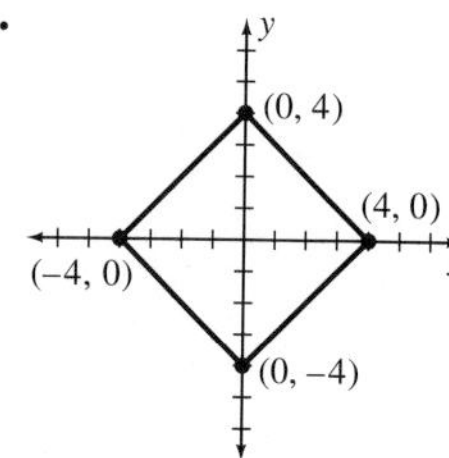

Problem Set 3.3 (page 187)

1. $\frac{3}{4}$ **2.** $-\frac{5}{6}$ **3.** -5 **4.** $\frac{11}{5}$ **5.** 0
6. 0 **7.** $-\frac{b}{a}$ **8.** $\frac{d-b}{c-a}$ **9.** $x = 7$
10. $y = -3$ **11.** $x = -2$ **12.** $y = 9$
13–18. Answers will vary. **19.** $x - 3y = -10$
20. $3x - 5y = -23$ **21.** $2x - y = 0$
22. $3x + y = 11$ **23.** $2x + 3y = -1$
24. $x + 5y = 32$ **25.** $y = -2$
26. $4x - 3y = -1$ **27.** $5x - 7y = -11$
28. $8x - 3y = 20$ **29.** $5x + 6y = 37$
30. $x + 3y = 0$ **31.** $x + 5y = 14$ **32.** $x = 2$
33. $y = -3$ **34.** $5x + 6y = -8$
35. $y = \frac{1}{2}x + 3$ **36.** $y = \frac{5}{3}x - 1$
37. $y = -\frac{3}{7}x + 2$ **38.** $y = -3x - 4$
39. $y = 4x + \frac{3}{2}$ **40.** $y = \frac{2}{3}x + \frac{3}{5}$
41. $y = -\frac{5}{6}x + \frac{1}{4}$ **42.** $y = -\frac{4}{5}x + 0$
43. $5x - 4y = 20$ **44.** $y = -1$ **45.** $x = -4$
46. $3x - y = 1$ **47.** $5x + 2y = 14$
48. $5x - 2y = 0$ **49.** $4x + y = -2$
50. $7x - 3y = -6$ **51.** Parallel
52. Perpendicular **53.** Perpendicular
54. Parallel

55. Intersecting lines that are not perpendicular
56. Intersecting lines that are not perpendicular
57. Perpendicular
58. Intersecting lines that are not perpendicular
59. $m = \frac{2}{3}, b = -\frac{4}{3}$ **60.** $m = -\frac{3}{4}, b = \frac{7}{4}$
61. $m = \frac{1}{2}, b = -\frac{7}{2}$ **62.** $m = -2, b = 9$
63. $m = -3, b = 0$ **64.** $m = \frac{1}{5}, b = 0$
65. $m = \frac{7}{5}, b = -\frac{12}{5}$ **66.** $m = \frac{5}{6}, b = \frac{13}{6}$
67. 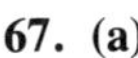 **(a)**

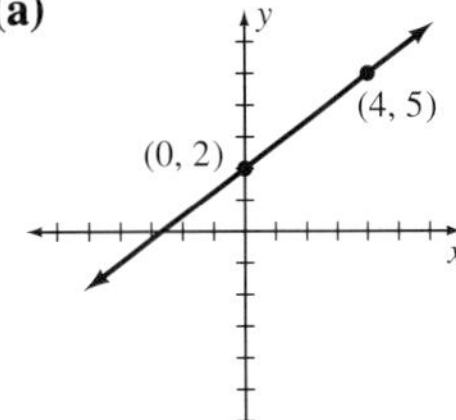

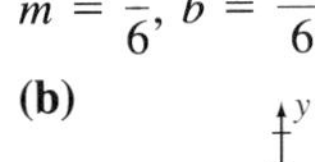 **(b)**

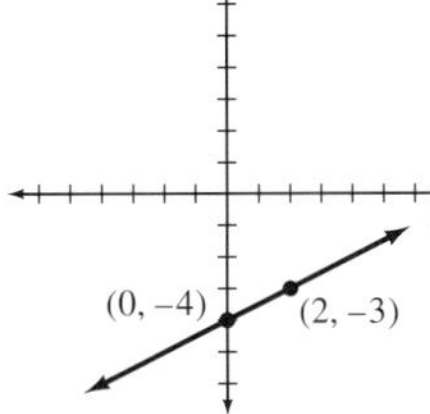

(c)

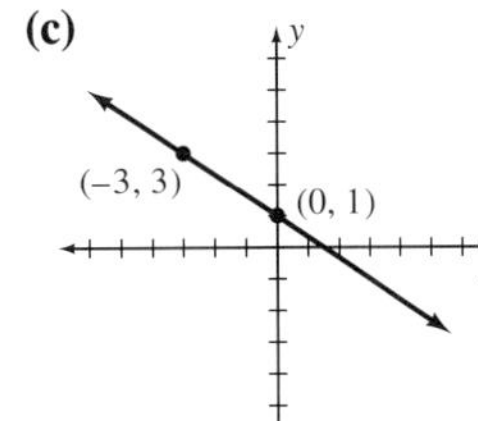

(d)

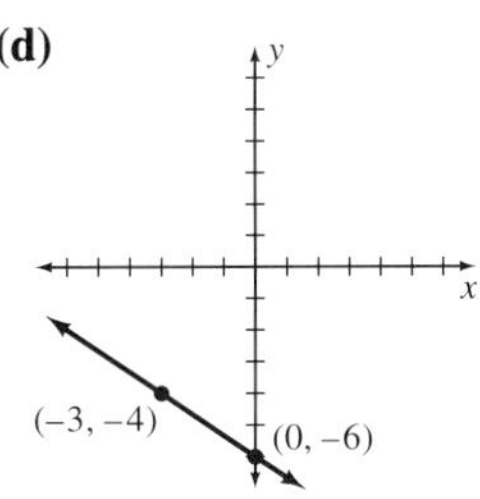

(e)

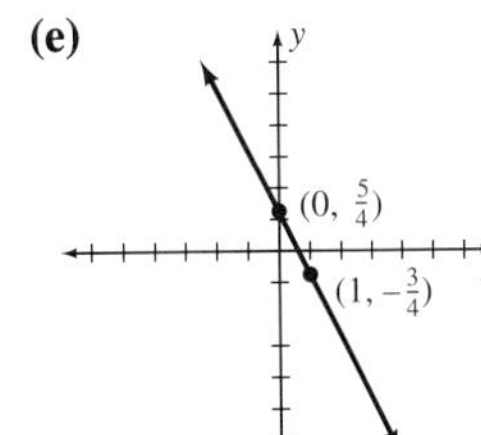

(f)

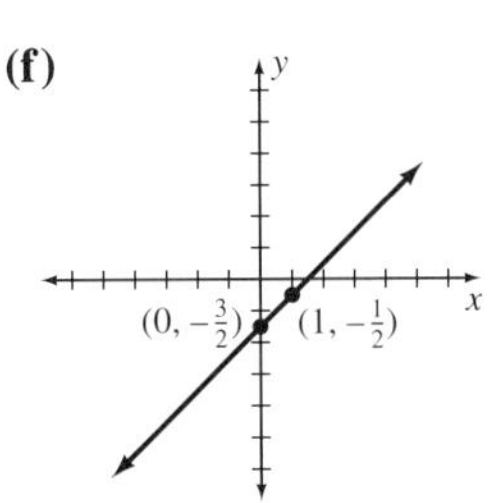

73. $x - y = -7, x + 5y = -19, 2x + y = 16$
74. $4x + 5y = 38, 3x - 5y = 23, x + 10y = 15$
75. $9x + 8y = -2, 6x - 7y = 11, 15x + y = 9$
76. **(a)** 105.6 feet **(b)** 250 feet
77. **(a)** 32 centimeters **(b)** 19 centimeters
78. 1.0 feet **79.** **(a)** $3x - y = 9$
(b) $6x + 5y = 7$ **(c)** $2x + 7y = 0$
(d) $3x - 8y = 23$ **80.** **(a)** $5x + 2y = 10$
(b) $x - y = -3$ **(c)** $2x - 3y = 12$
(d) $2x + y = -2$ **83.** **(a)** $2x - y = 4$
(b) $3x + 7y = 19$ **(c)** $5x + 2y = 2$
(d) $3x - 2y = 1$ **84.** **(a)** $200x - y = -15$
(b) $5x - 4y = -2$ **(c)** $9x - 5y = -160$

Problem Set 3.4 (page 198)

1. (4, −3); (−4, 3); (−4, −3)
2. (−2, −5); (2, 5); (2, −5)
3. (−6, 1); (6, −1); (6, 1)
4. (3, 7); (−3, −7); (−3, 7)
5. (0, −4); (0, 4); (0, −4)
6. (−5, 0); (5, 0); (5, 0) **7.** y-axis **8.** x-axis
9. x-axis **10.** x-axis, y-axis, and origin
11. x-axis, y-axis, and origin **12.** x-axis
13. None **14.** Origin **15.** Origin **16.** None
17. None **18.** y-axis **19.** y-axis **20.** y-axis

21.

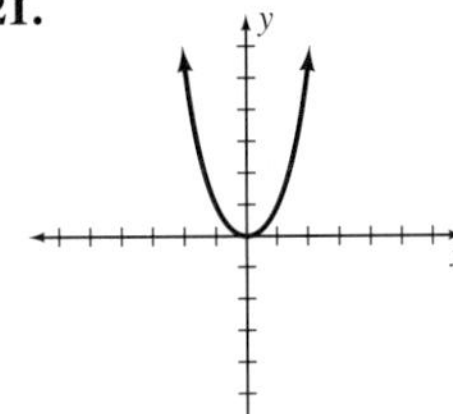

22.

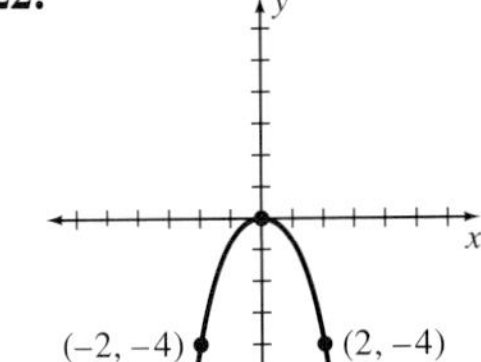

23.

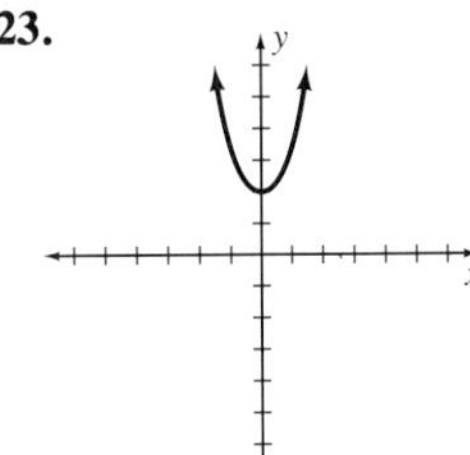

24.

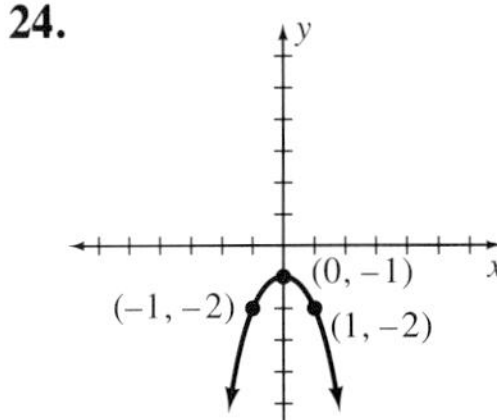

25.

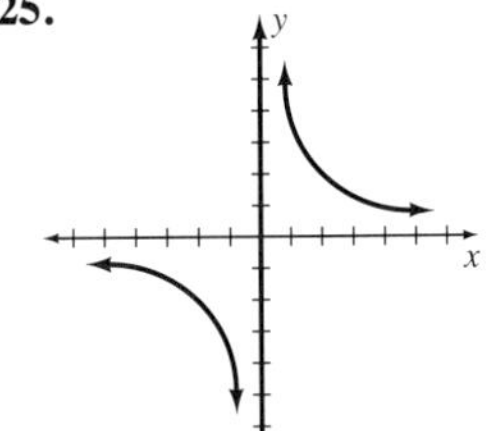

26.

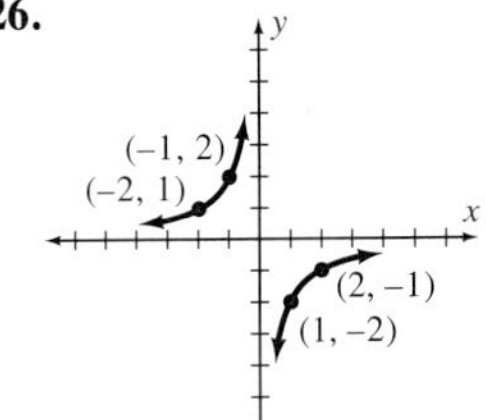

27.

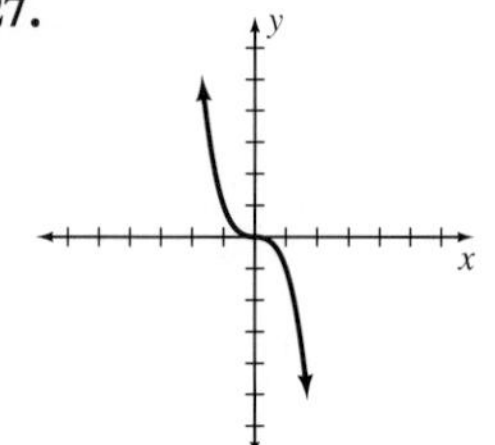

28.

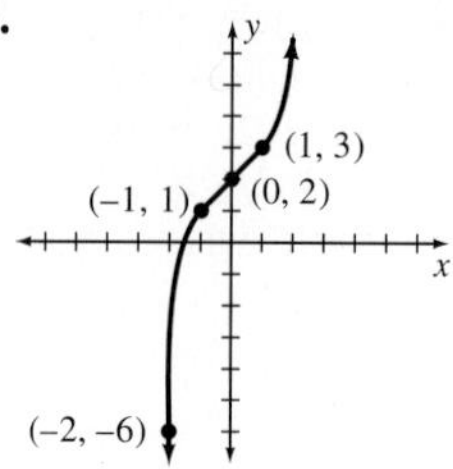

37.

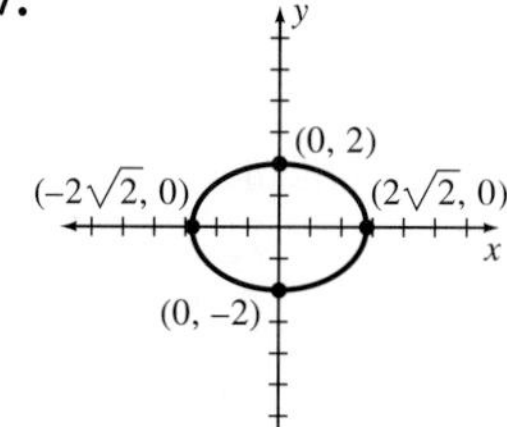

38.

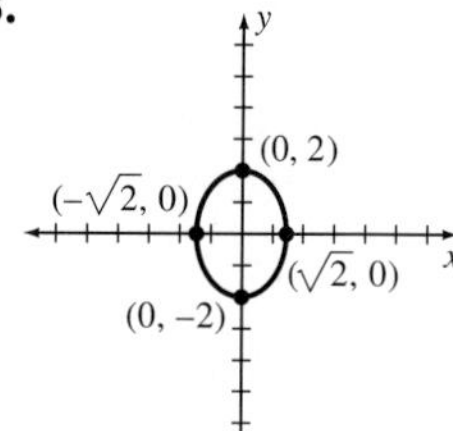

29.

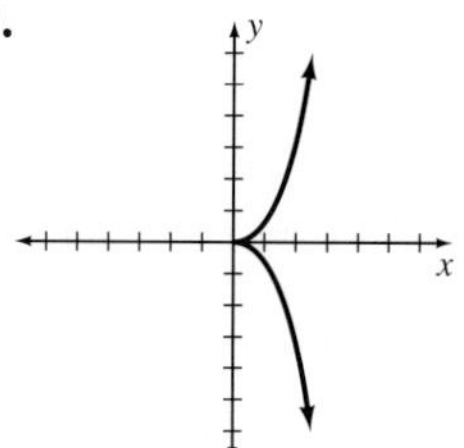

30.

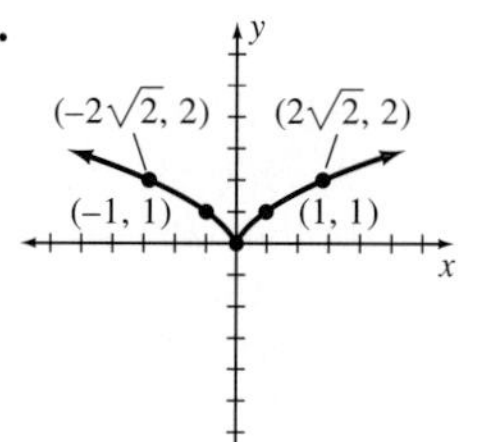

39.

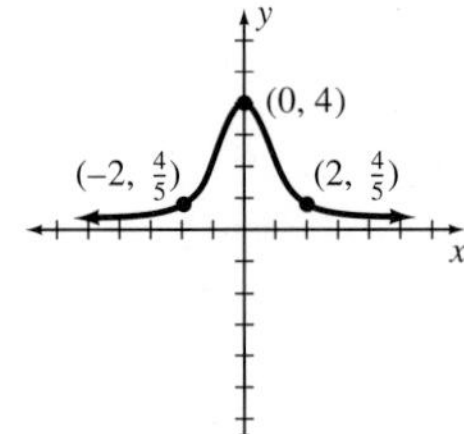

40.

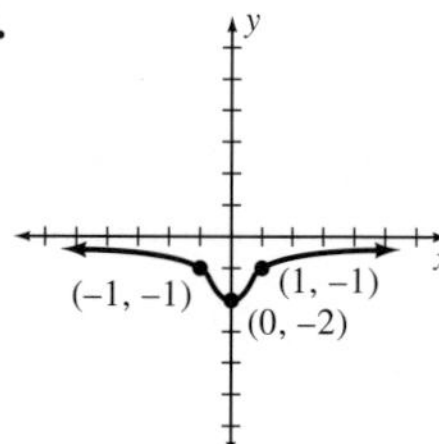

31.

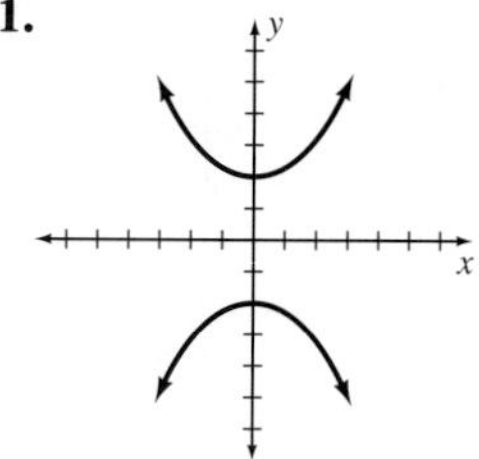

32.

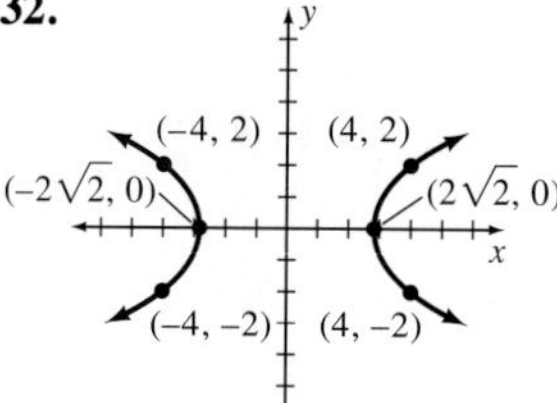

41.

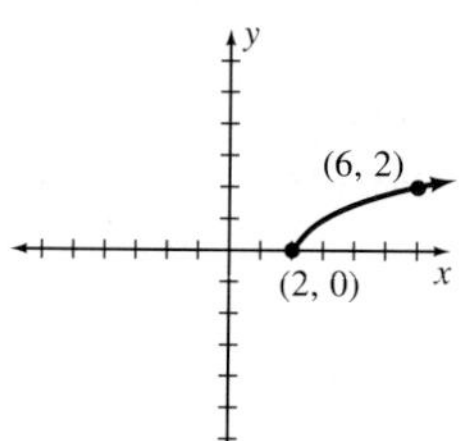

42.

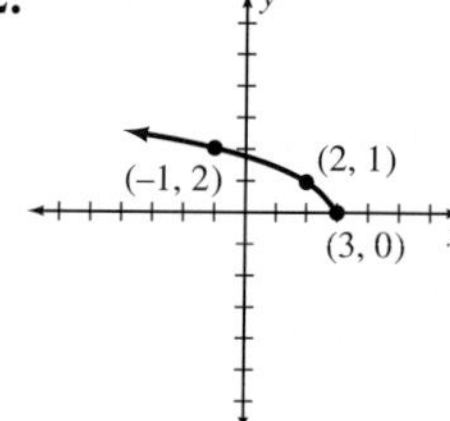

33.

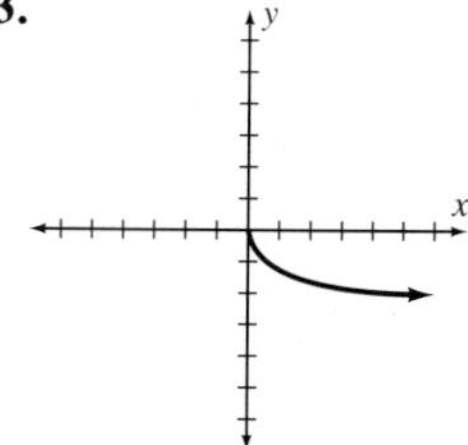

34.

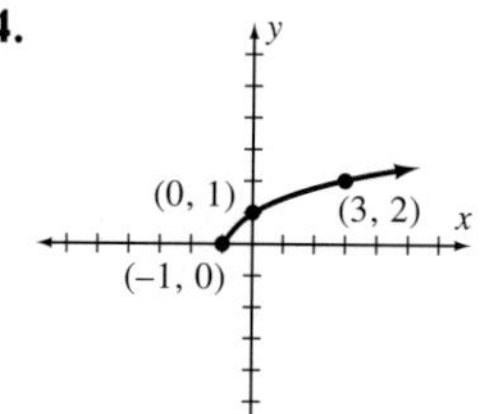

43.

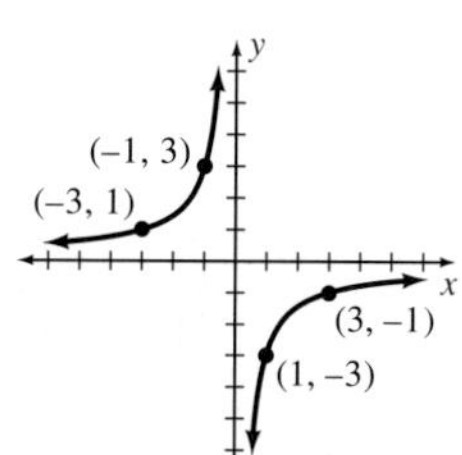

44.

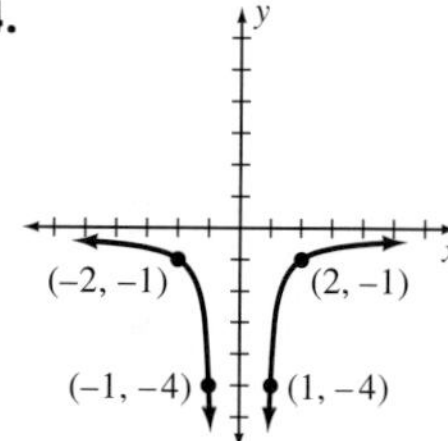

35.

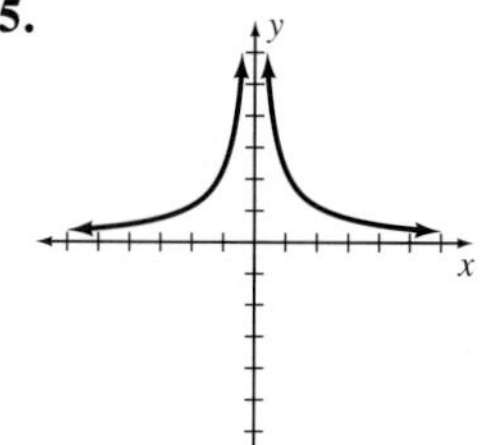

36.

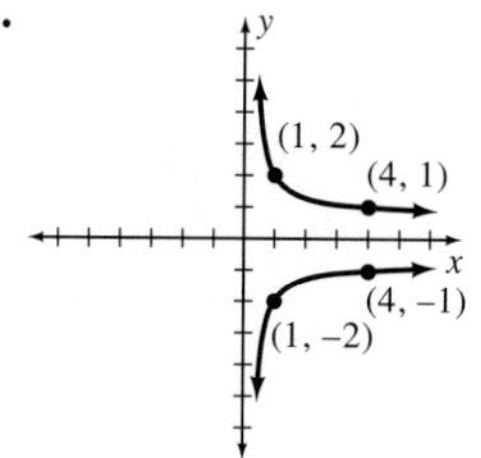

45.

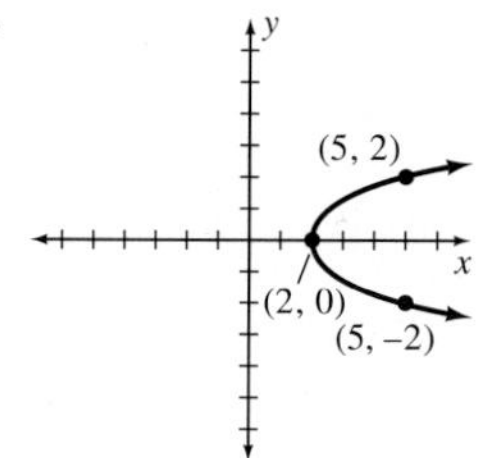

46.

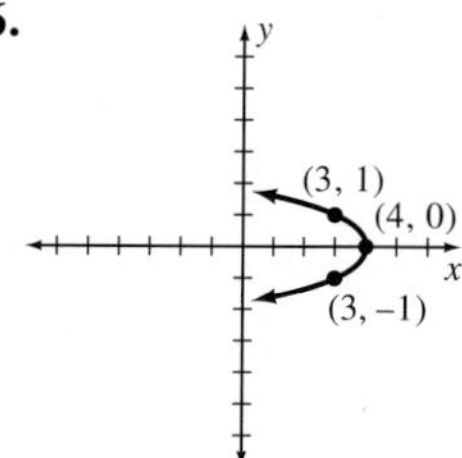

47.

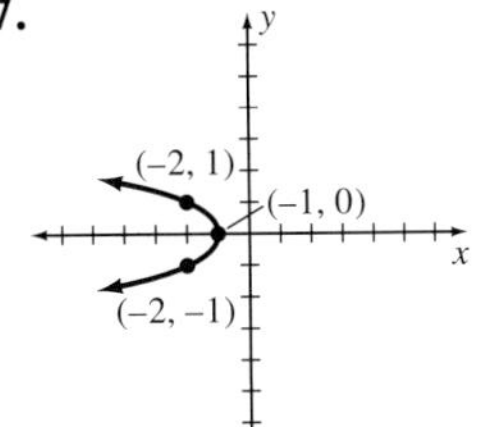

48.

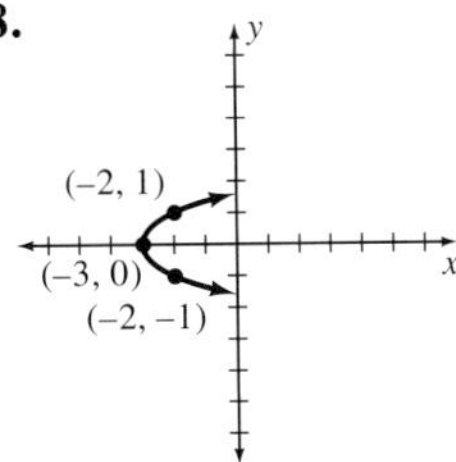

Problem Set 3.5 (page 210)

1. $x^2 + y^2 - 4x - 6y - 12 = 0$
2. $x^2 + y^2 + 6x - 8y + 21 = 0$
3. $x^2 + y^2 + 2x + 10y + 17 = 0$
4. $x^2 + y^2 - 8x + 4y + 19 = 0$
5. $x^2 + y^2 - 6x = 0$ **6.** $x^2 + y^2 + 8y - 20 = 0$
7. $x^2 + y^2 = 49$ **8.** $x^2 + y^2 = 1$
9. $(3, 5), r = 2$ **10.** $(-4, 6), r = 3$
11. $(-5, -7), r = 1$ **12.** $(0, -3), r = 4$
13. $(5, 0), r = 5$ **14.** $(2, -1), r = \sqrt{5}$
15. $(0, 0), r = 2\sqrt{2}$ **16.** $(0, 0), r = \frac{1}{2}$
17. $\left(\frac{1}{2}, 1\right), r = 2$ **18.** $\left(-\frac{2}{3}, \frac{1}{2}\right), r = 1$
19. $x^2 + y^2 - 6x - 6y - 67 = 0$
20. $x^2 + y^2 + 6x + 8y = 0$
21. $x^2 + y^2 - 14x + 14y + 49 = 0$
22. $x^2 + y^2 + 6x - 12y = 0$
23. $x^2 + y^2 + 6x - 10y + 9 = 0$ and $x^2 + y^2 + 6x + 10y + 9 = 0$

24.

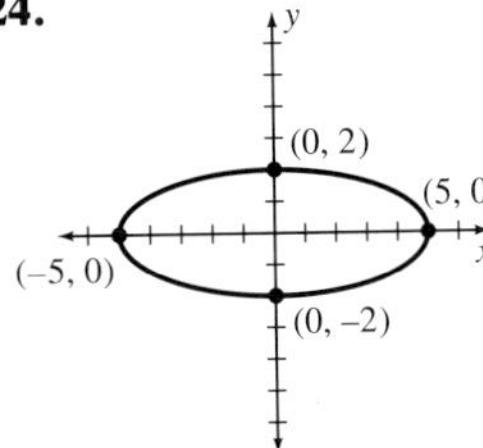

25.

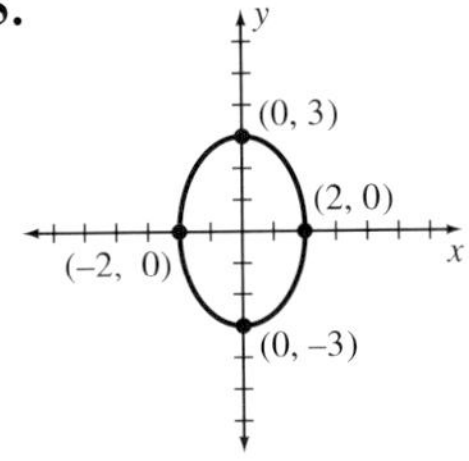

26.

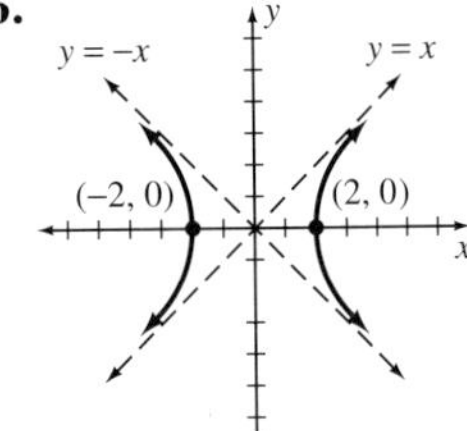

27.

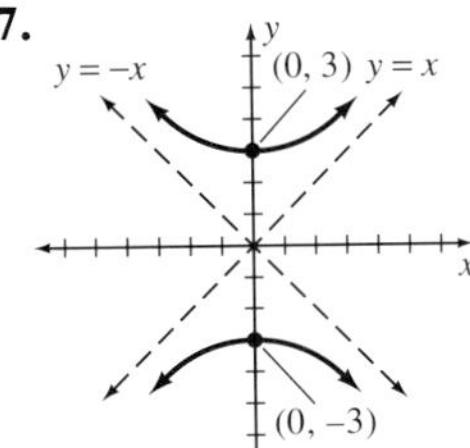

28.

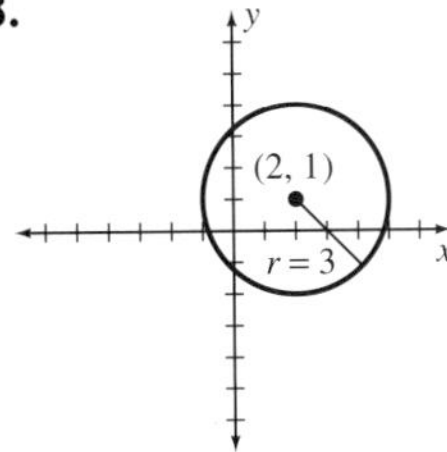

29.

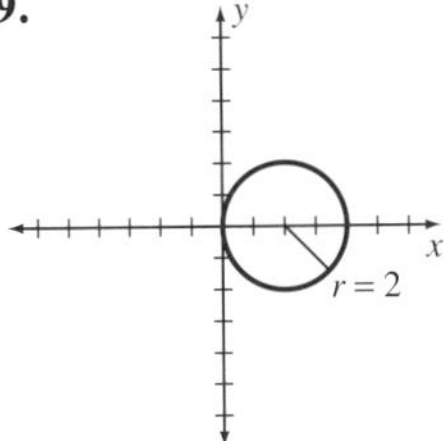
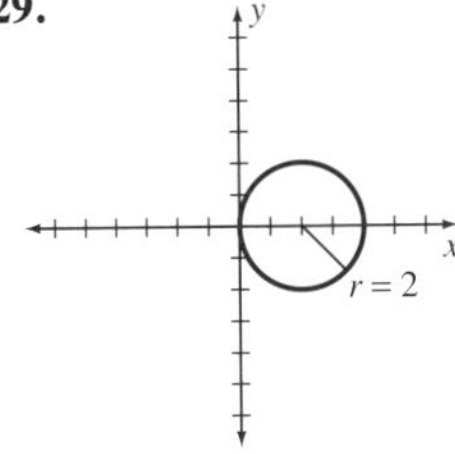

30.

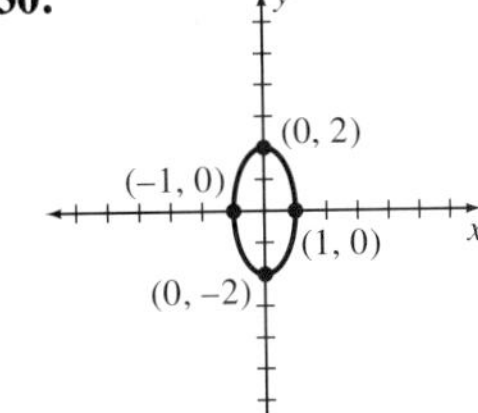

31.

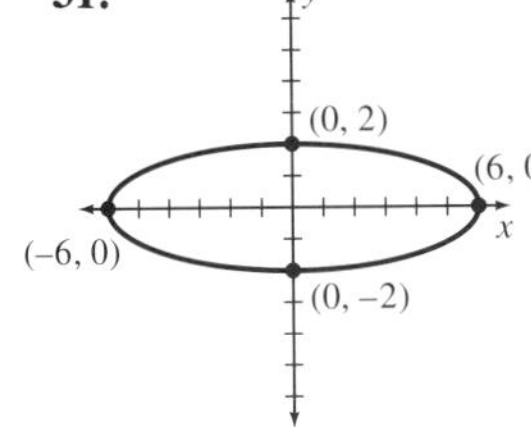

32.

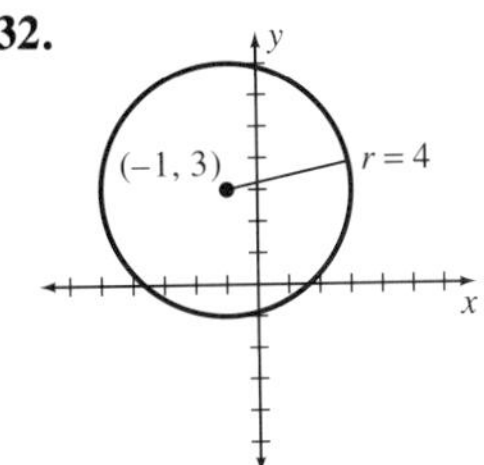

33.

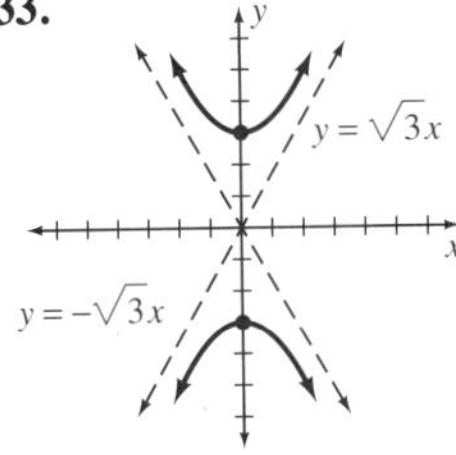

34.

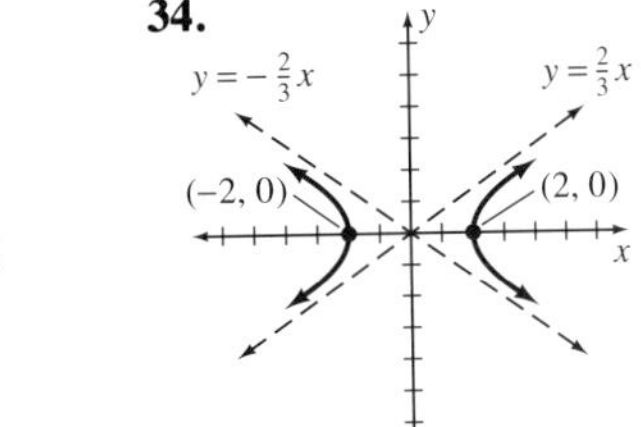

35.

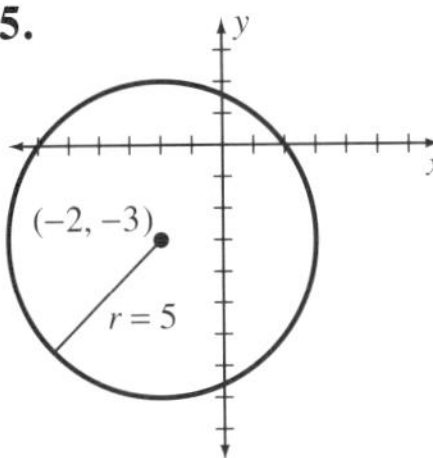

36.

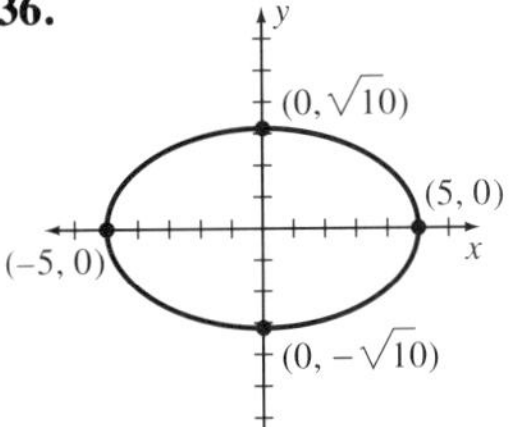

37.

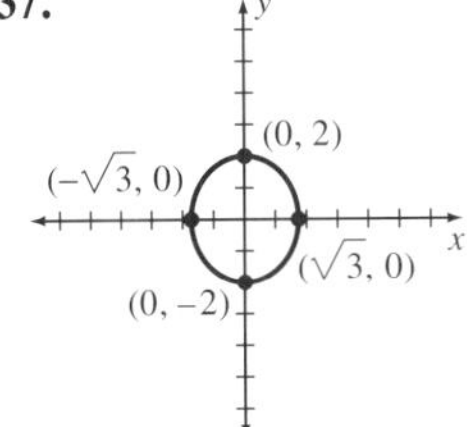

38.

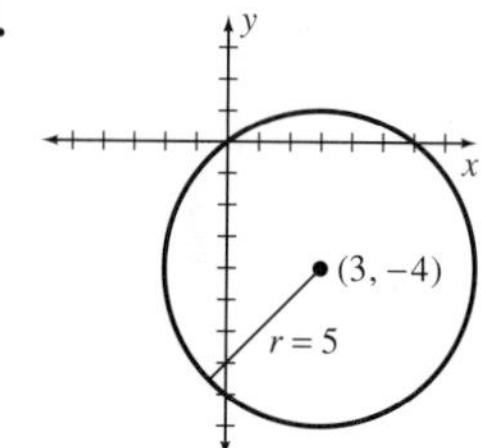

39.

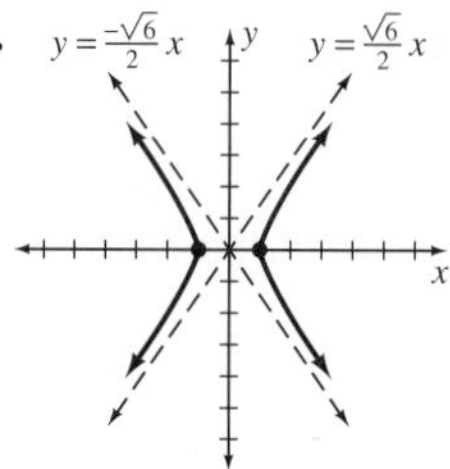

40.

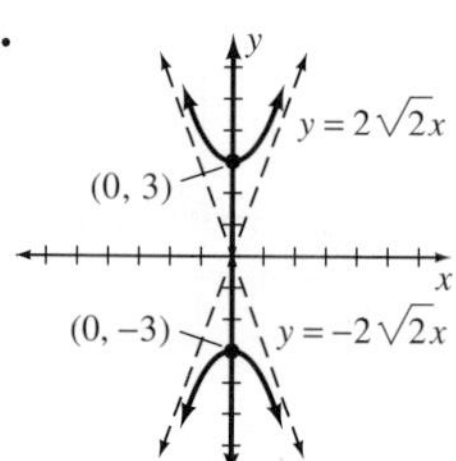

41.

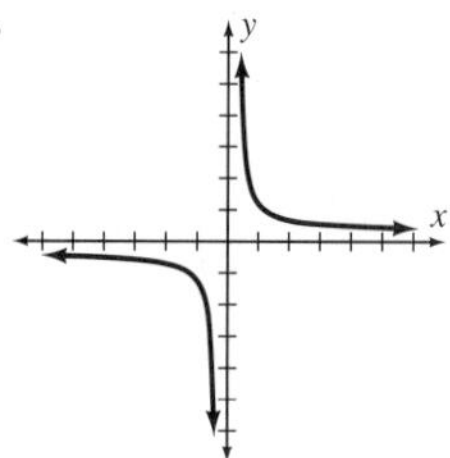

42.

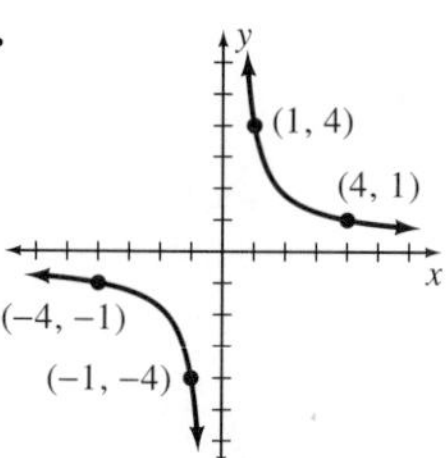

43.

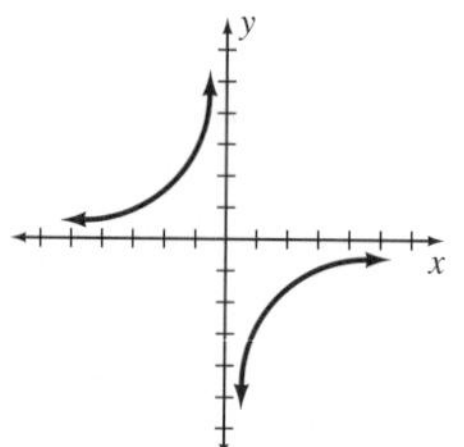

44.

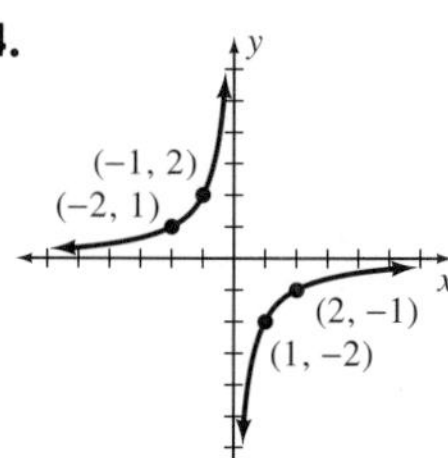

45. The x- and y-axes

46. **(a)** Origin **(b)** Origin **(c)** The lines $y = \pm x$ **(d)** The lines $y = \pm\frac{2}{3}x$

47. **(a)** $(1, 4)$, $r = 3$ **(b)** $(-2, 7)$, $r = 2$ **(c)** $(-6, -4)$, $r = 8$ **(d)** $(8, -10)$, $r = 7$ **(e)** $(0, 6)$, $r = 9$ **(f)** $(-7, 0)$, $r = 7$

Chapter 3 Review Problem Set (page 213)

1. 5 **2.** −5 **3.** $\left(9, \frac{1}{3}\right)$ **4.** $(-2, 6)$

7. x-axis **8.** None **9.** x-axis, y-axis, and origin

10. y-axis **11.** Origin **12.** y-axis

13.

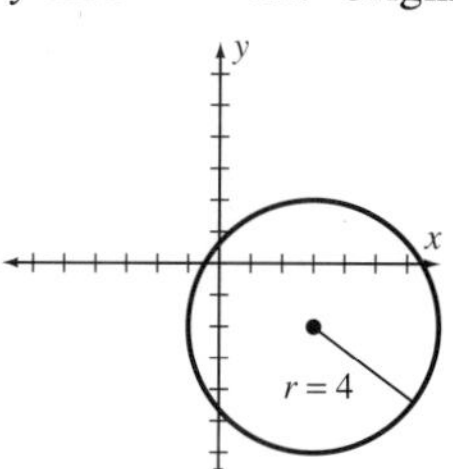

14.

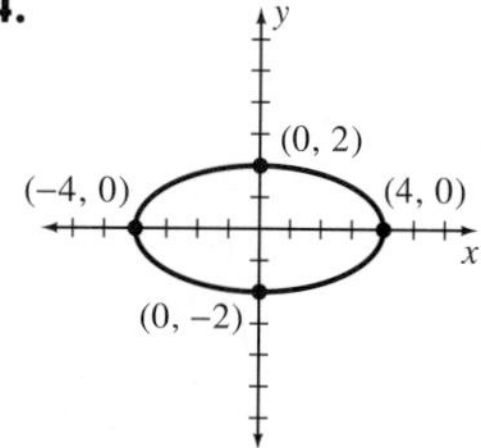

15.

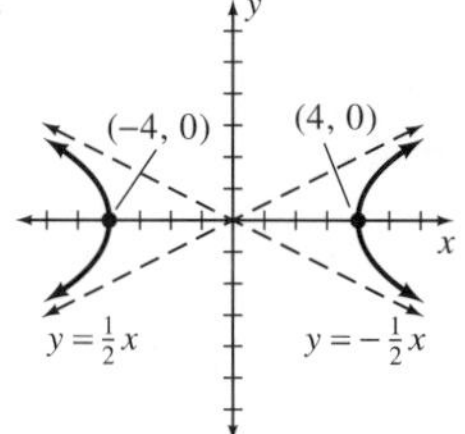

16.

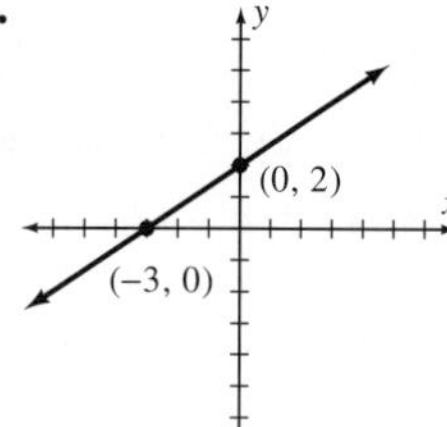

17.

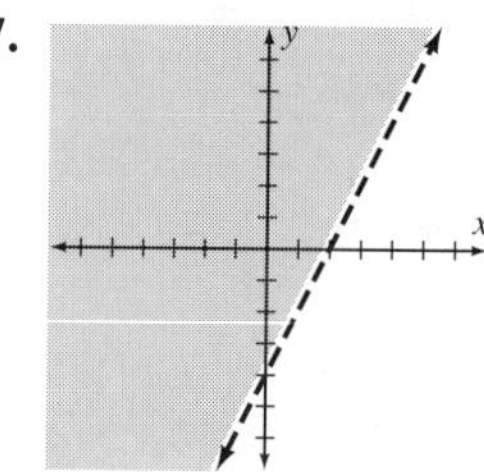

18.

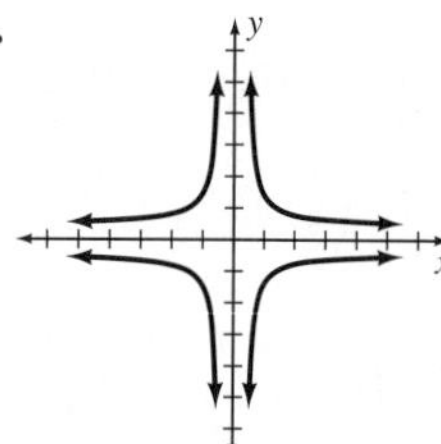

19.

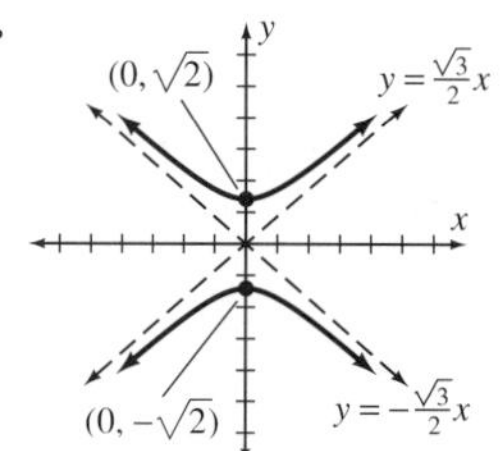

20.

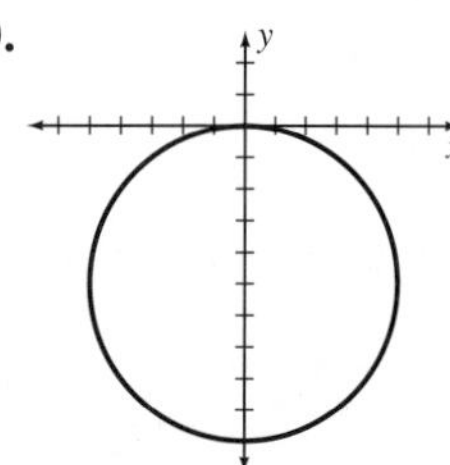

21.

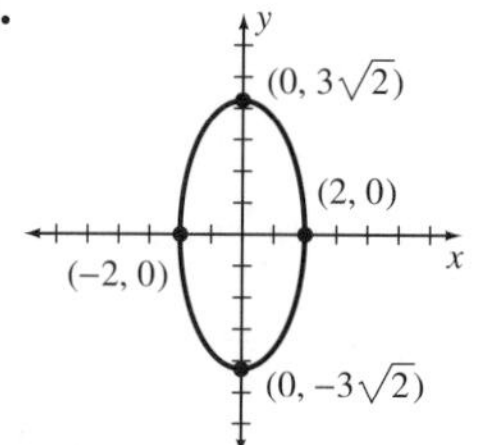

22.

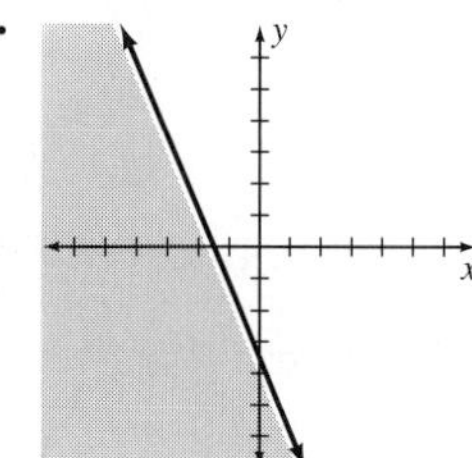

23. −5 **24.** $\frac{5}{7}$ **25.** $3x + 4y = 29$

26. $2x - y = -4$ **27.** $4x + 3y = -4$
28. $x + 2y = 3$
29. $x^2 + y^2 - 10x + 12y + 60 = 0$
30. $x^2 + y^2 - 4x - 6y - 4 = 0$
31. $x^2 + y^2 + 10x - 24y = 0$
32. $x^2 + y^2 + 8x + 8y + 16 = 0$

Chapters 1, 2, and 3 Cumulative Review Problem Set (page 214)

1. $\frac{1}{27}$ **2.** $-\frac{1}{16}$ **3.** $\frac{9}{4}$ **4.** $-\frac{2}{3}$ **5.** 9
6. $\frac{9}{16}$ **7.** $\frac{20}{x^2y^3}$ **8.** $-\frac{56a}{b}$ **9.** $4x^4y^2$
10. $\frac{5y^2}{x^4}$ **11.** $\frac{x^{1/3}}{17y^{7/4}}$ **12.** $\frac{4a^8}{b^{14}}$ **13.** $-30\sqrt{2}$
14. $6xy\sqrt{3x}$ **15.** $2xy^2\sqrt[3]{7xy}$ **16.** $\frac{3\sqrt{6}}{10}$
17. $\frac{\sqrt{21xy}}{7y}$ **18.** $-\frac{5(\sqrt{2} + 3)}{7}$
19. $\frac{6\sqrt{14} + 3\sqrt{42}}{2}$ **20.** $\frac{4x - 12\sqrt{xy}}{x - 9y}$
21. $\frac{3x^3y^2}{8}$ **22.** $-\frac{3b^3}{8a^3}$ **23.** $\frac{5x + 1}{x}$
24. $\frac{21x + 5}{24}$ **25.** $\frac{10 - 3n}{6n^2}$
26. $\frac{5x^2 + 18x + 27}{(x + 9)(x - 3)(x + 3)}$ **27.** $\left\{-\frac{23}{4}\right\}$ **28.** $\{3\}$
29. $\left\{\frac{3}{7}\right\}$ **30.** $\left\{\pm\frac{2}{3}\right\}$ **31.** $\{-4, 0, 2\}$
32. $\left\{\frac{3}{7}, 4\right\}$ **33.** $\{\pm 4i, \pm 1\}$ **34.** $\left\{-\frac{1}{5}, 1\right\}$
35. $\left\{\frac{3 \pm \sqrt{17}}{4}\right\}$ **36.** $\left\{\frac{-13 \pm \sqrt{205}}{2}\right\}$ **37.** $\{1\}$
38. $\left\{\frac{1 \pm 2i}{2}\right\}$ **39.** $\left(-\infty, \frac{1}{8}\right)$ **40.** $\left[-\frac{5}{9}, \infty\right)$
41. $(-\infty, 250]$ **42.** $(-\infty, -8,) \cup (3, \infty)$
43. $\left(-\frac{3}{2}, \frac{1}{3}\right)$ **44.** $\left(-3, \frac{1}{2}\right) \cup (4, \infty)$
45. $\left(-1, \frac{2}{3}\right]$ **46.** $(1, 7]$
47. $\left(-\infty, -\frac{4}{3}\right) \cup (2, \infty)$ **48.** $\left(-\frac{9}{5}, 3\right)$

49.

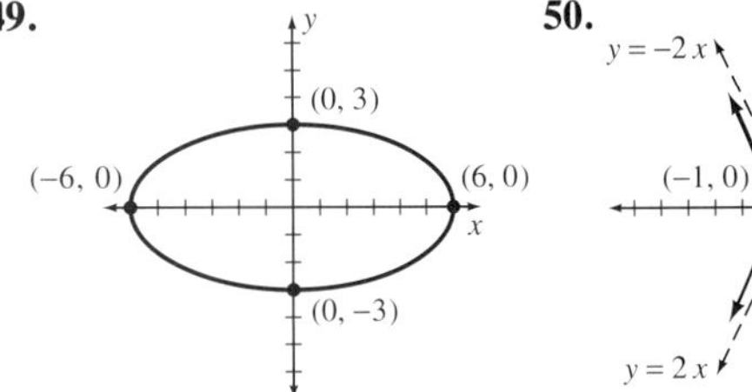

51.

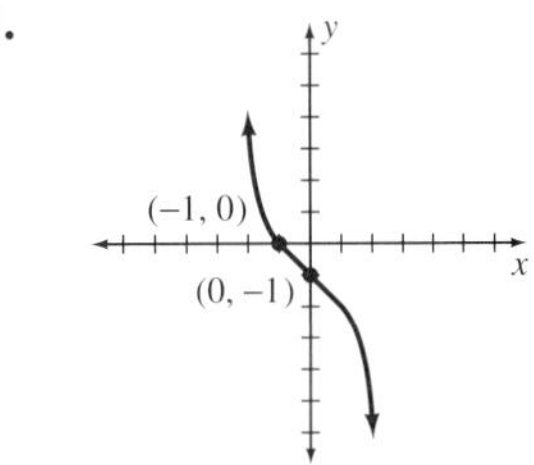

52.

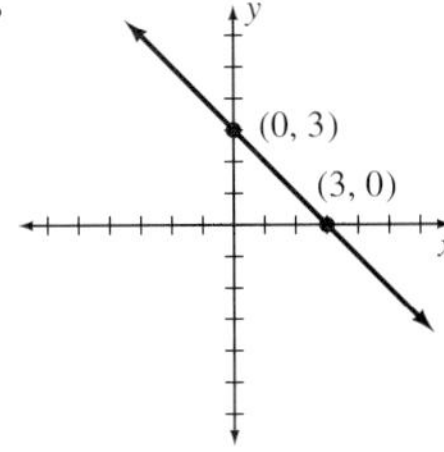

53.

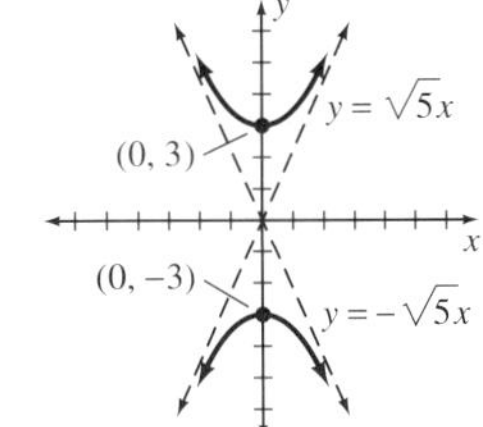

54.

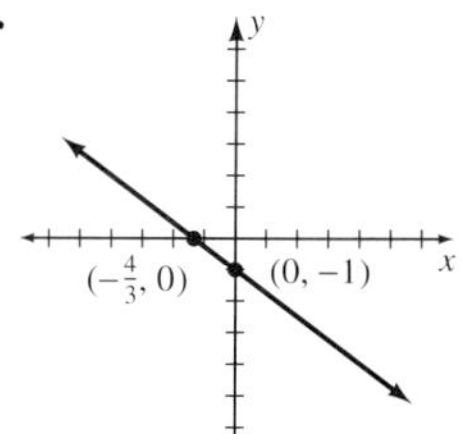

55. $(-7, 4)$; $r = 3$ **56.** $3x - 4y = -26$
57. $(-2, 6)$ **58.** $4x + 3y = 25$
59. \$28.60; \$31.43
60. \$3000 at 5% and \$4500 at 6%
61. Length of 8 inches and width of 4.5 inches
62. The side is 20 centimeters long and altitude is 8 centimeters long. **63.** 16 milliliters
64. 30 miles **65.** 3 hours

CHAPTER 4

Problem Set 4.1 (page 225)

1. $f(3) = -1, f(5) = -5; f(-2) = 9$
2. $f(2) = -6; f(4) = 0; f(-3) = 14$
3. $g(3) = -20; g(-1) = -8; g(-4) = -41$
4. $g(0) = 6; g(5) = -39; g(-5) = 1$

5. $h(3) = \frac{5}{4}$; $h(4) = \frac{23}{12}$; $h\left(-\frac{1}{2}\right) = -\frac{13}{12}$

6. $h(-2) = \frac{5}{3}$; $h(6) = -\frac{7}{3}$; $h\left(-\frac{2}{3}\right) = 1$

7. $f(5) = 3$; $f\left(\frac{1}{2}\right) = 0$; $f(23) = 3\sqrt{5}$

8. $f\left(\frac{14}{3}\right) = 4$; $f(10) = 4\sqrt{2}$; $f\left(-\frac{1}{3}\right) = 1$

9. $f(4) = 4$; $f(10) = 10$; $f(-3) = 9$; $f(-5) = 25$

10. $f(2) = 8$; $f(6) = 20$; $f(-1) = -6$; $f(-4) = -21$

11. $f(3) = 6$; $f(5) = 10$; $f(-3) = 6$; $f(-5) = 10$

12. $f(3) = 10$; $f(6) = -1$; $f(0) = 1$; $f(-3) = 2$

13. $f(2) = 1$; $f(0) = 0$; $f\left(-\frac{1}{2}\right) = 0$; $f(-4) = -1$

14. 4 **15.** -7 **16.** $2a + h - 3$

17. $-2a - h + 4$ **18.** $4a + 2h + 7$

19. $6a + 3h - 1$ **20.** $3a^2 + 3ah + h^2$

21. $3a^2 + 3ah + h^2 - 2a - h + 2$

22. $-\frac{1}{(a + 1)(a + h + 1)}$

23. $-\frac{2}{(a - 1)(a + h - 1)}$ **24.** $\frac{1}{(a + 1)(a + h + 1)}$

25. $-\frac{2a + h}{a^2(a + h)^2}$ **26.** Yes **27.** Yes

28. No **29.** No **30.** Yes **31.** Yes

32. No **33.** Yes

34. $D = \{x \mid x \geq 0\}$
$R = \{f(x) \mid f(x) \geq 0\}$

35. $D = \left\{x \mid x \geq \frac{4}{3}\right\}$
$R = \{f(x) \mid f(x) \geq 0\}$

36. $D = \{x \mid x$ is any real number$\}$
$R = \{f(x) \mid f(x) \geq 1\}$

37. $D = \{x \mid x$ is any real number$\}$
$R = \{f(x) \mid f(x) \geq -2\}$

38. The domain and the range each consist of the set of all real numbers.

39. $D = \{x \mid x$ is any real number$\}$
$R = \{f(x) \mid f(x)$ is any nonnegative real number$\}$

40. $D = \{x \mid x$ is any real number$\}$
$R = \{f(x) \mid f(x)$ is any nonnegative real number$\}$

41. $D = \{x \mid x$ is any nonnegative real number$\}$
$R = \{f(x) \mid f(x)$ is any nonpositive real number$\}$

42. $D = \{x \mid x \neq 4\}$ **43.** $D = \{x \mid x \neq -2\}$

44. $D = \{x \mid x \neq 2$ and $x \neq -3\}$

45. $D = \left\{x \mid x \neq \frac{1}{2} \text{ and } x \neq -4\right\}$

46. $D = \left\{x \mid x \geq -\frac{1}{5}\right\}$

47. $D = \{x \mid x \neq 2$ and $x \neq -2\}$

48. $D = \{x \mid x \neq -2$ and $x \neq -3\}$

49. $D = \{x \mid x \neq -3$ and $x \neq 4\}$

50. $D = \{x \mid x \neq 0$ and $x \neq -4\}$

51. $D = \left\{x \mid x \neq -\frac{5}{2} \text{ and } x \neq \frac{1}{3}\right\}$

52. $(-\infty, -1] \cup [1, \infty)$ **53.** $(-\infty, -4] \cup [4, \infty)$

54. $(-\infty, \infty)$ **55.** $(-\infty, \infty)$

56. $(-\infty, -4] \cup [6, \infty)$ **57.** $(-\infty, -5] \cup [8, \infty)$

58. $\left(-\infty, -\frac{3}{4}\right] \cup \left[\frac{2}{3}, \infty\right)$

59. $\left(-\infty, -\frac{5}{2}\right] \cup \left[\frac{7}{4}, \infty\right)$

60. -1500; 600; 1000; 900

61. 12.57; 28.27; 452.39; 907.92

62. 97.3; 93.9; 55.8 **63.** 48; 64; 48; 0

64. \$74; \$98; \$122; \$258 **65.** \$55; \$60; \$67.50; \$75

66. 8; 4; 3; 0 **67.** 125.66; 301.59; 804.25

68. Even **69.** Odd **70.** Even **71.** Neither

72. Neither **73.** Neither **74.** Odd

75. Even **76.** Odd **77.** Odd

Problem Set 4.2 (page 238)

1.

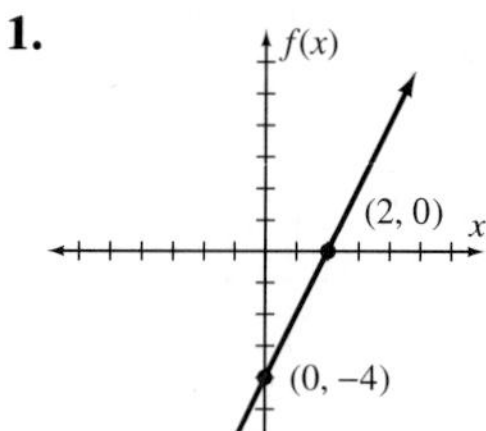

2.

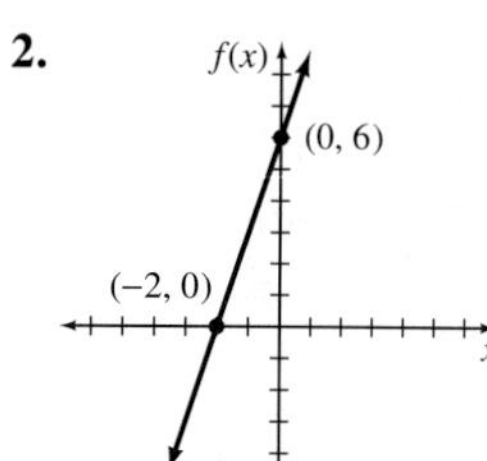

3.

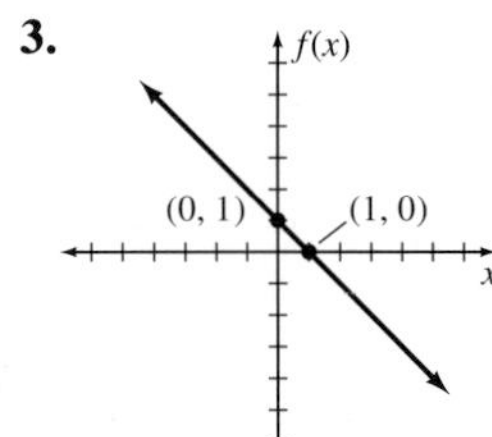

4.

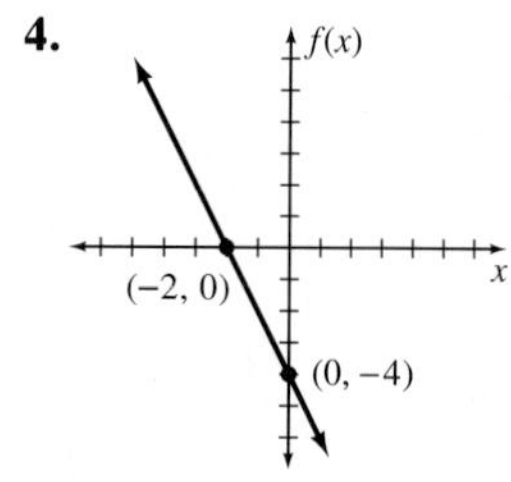

5.

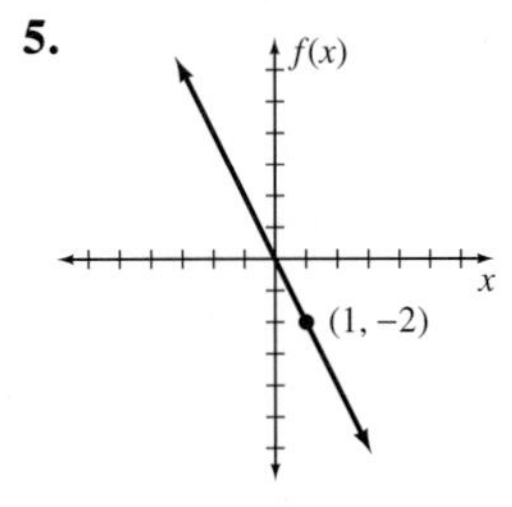

6.

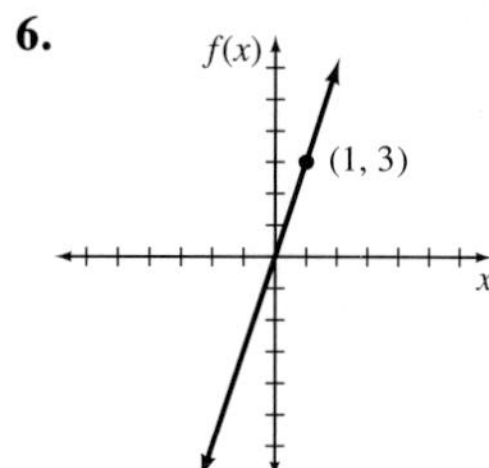

7.

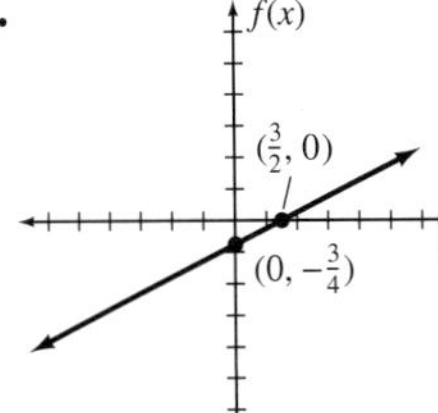

8.

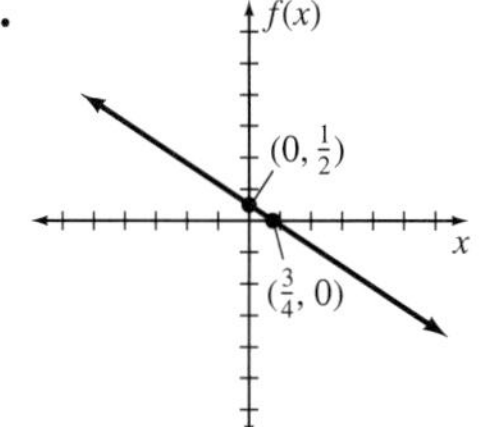

17.

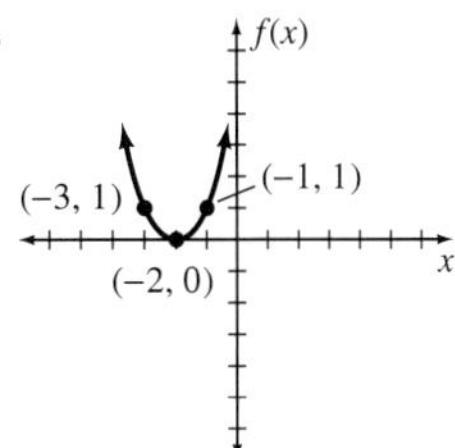

18.

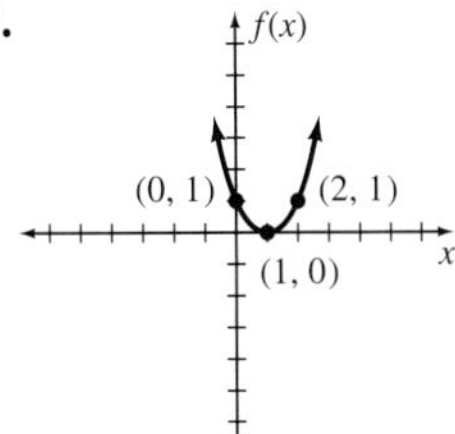

9.

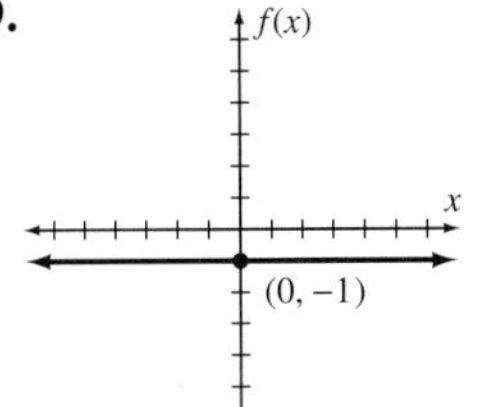

10.

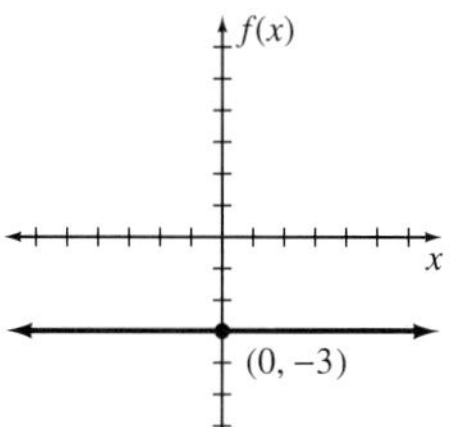

19.

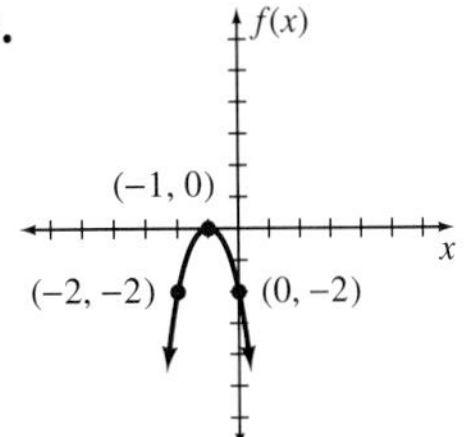

20.

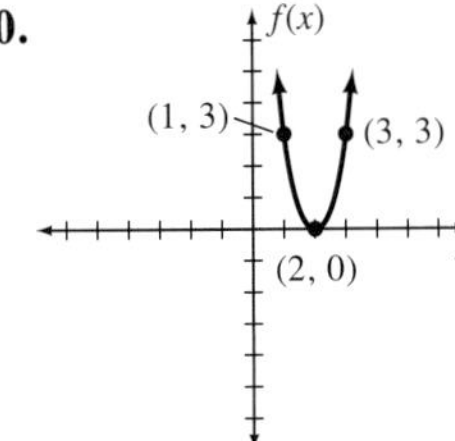

11.

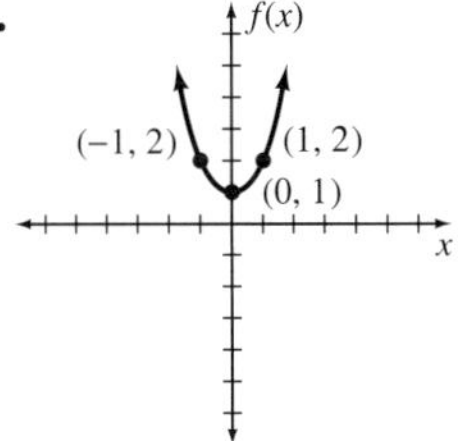

12.

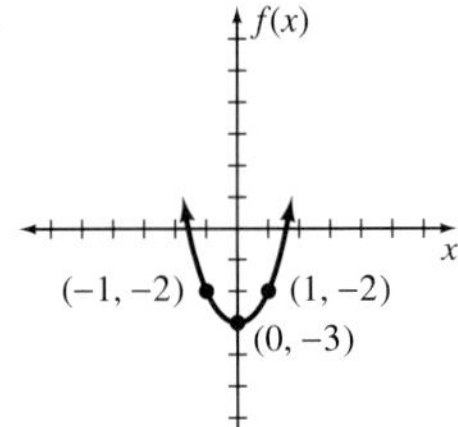

21.

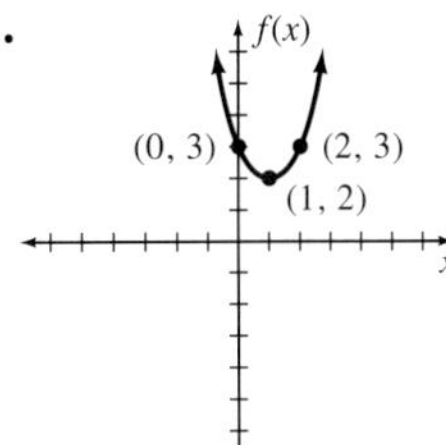

22.

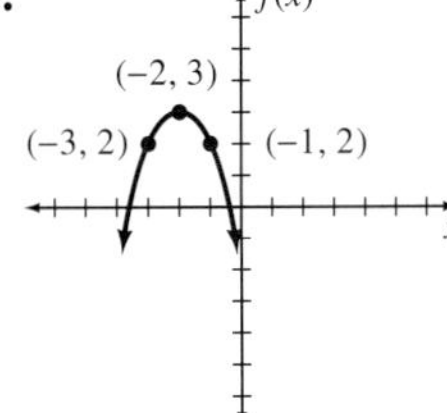

13.

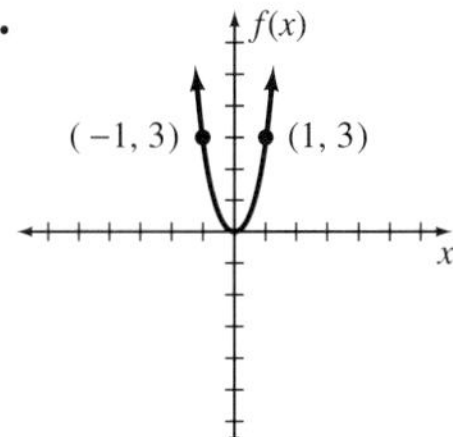

14.

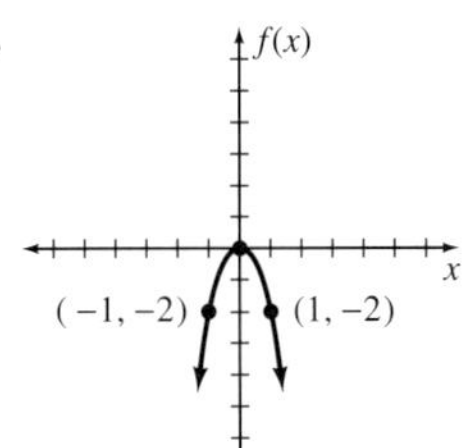

23.

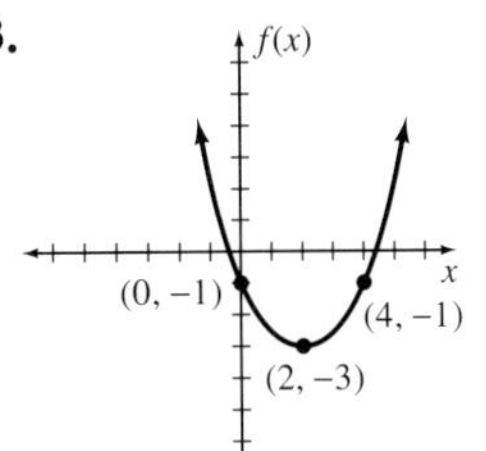

24.

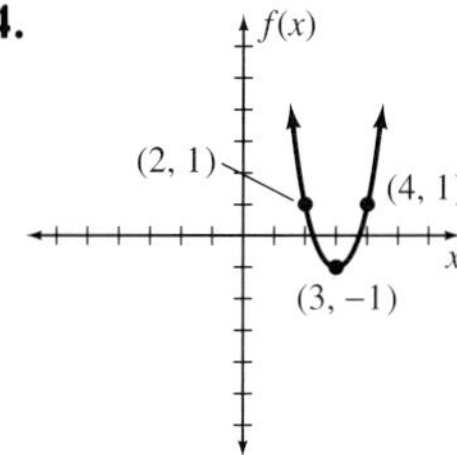

15.

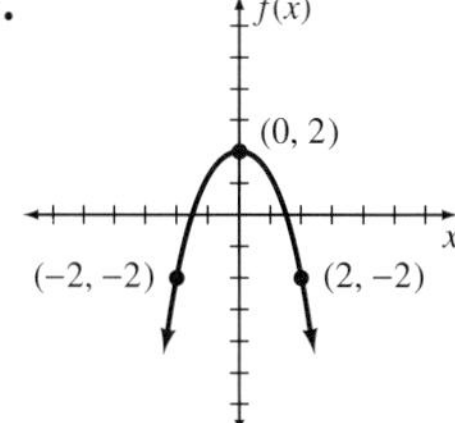

16.

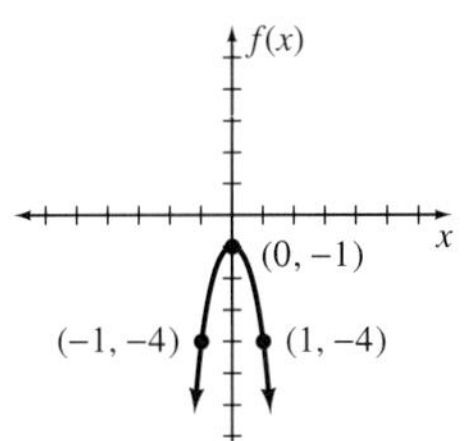

25.

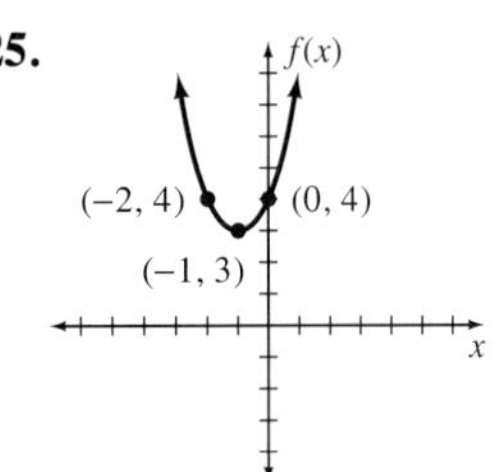

26.

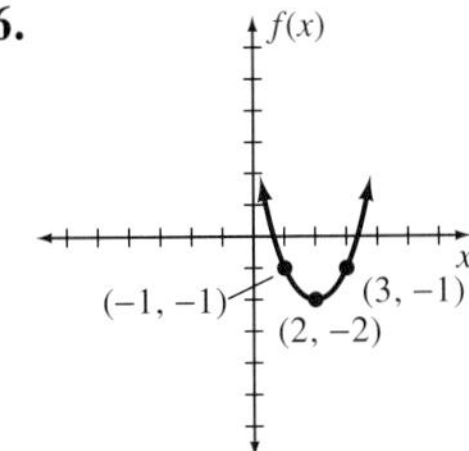

27.

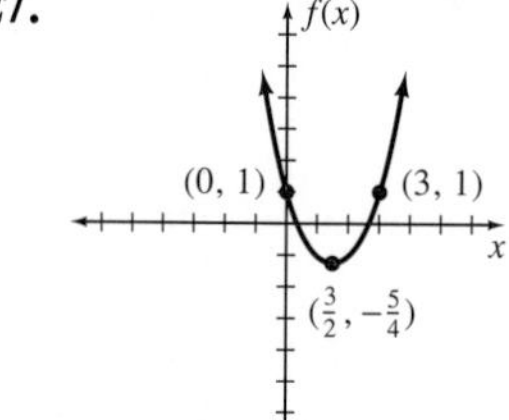

28.

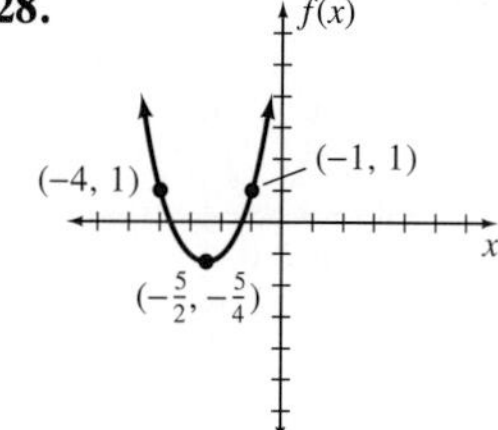

37.

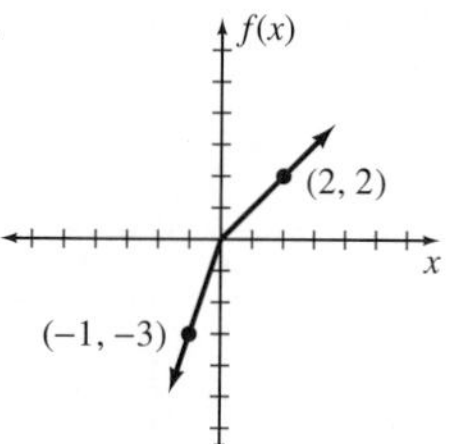

38.

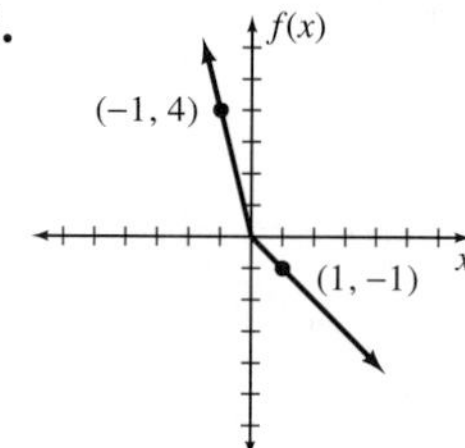

29.

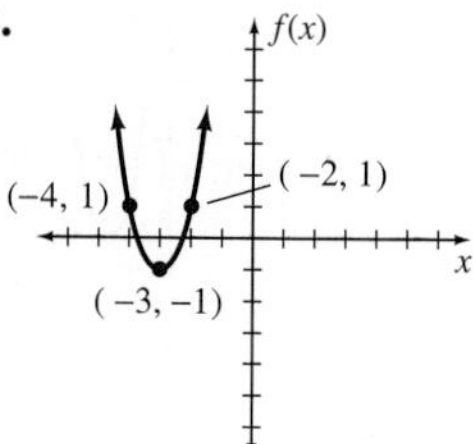

30.

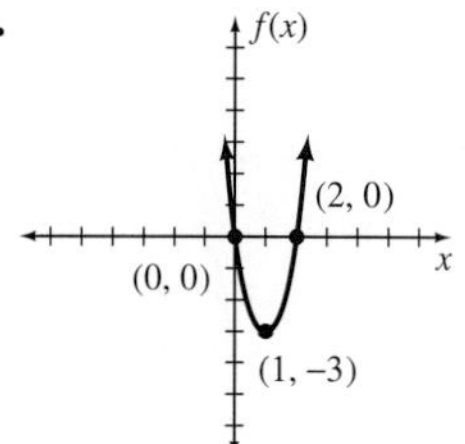

39.

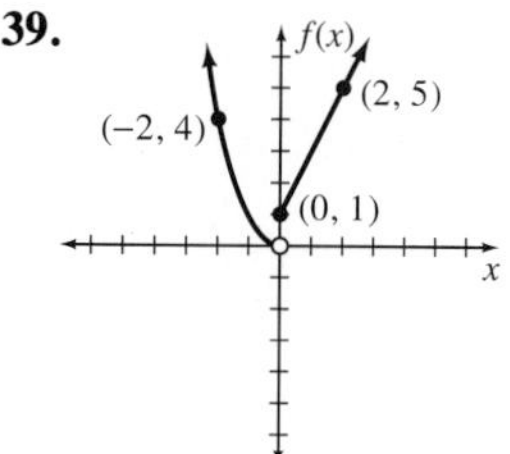

40.

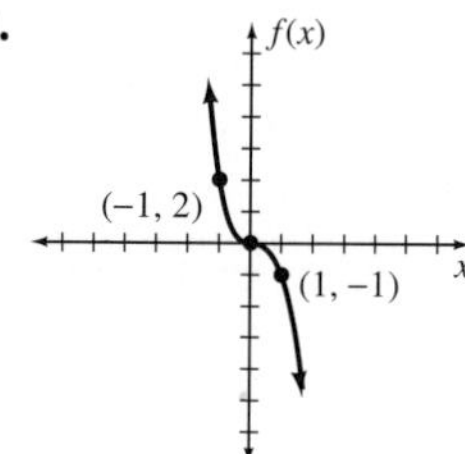

31.

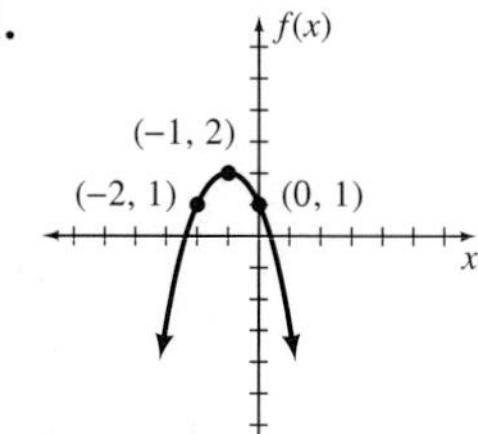

32.

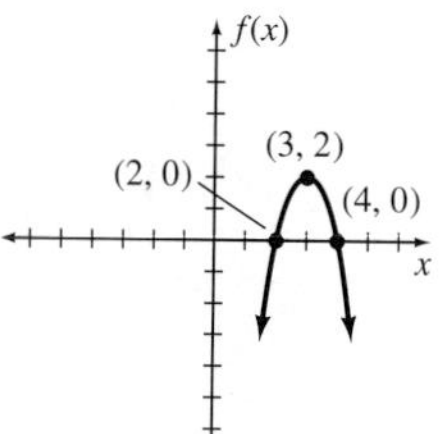

41.

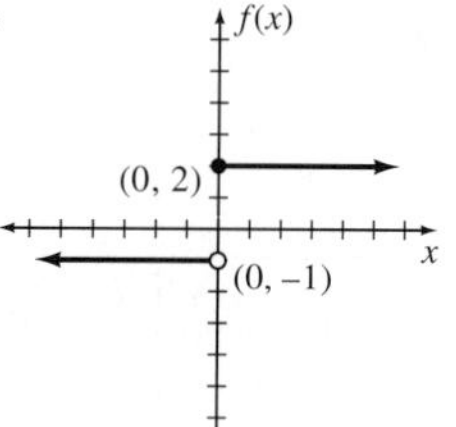

42.

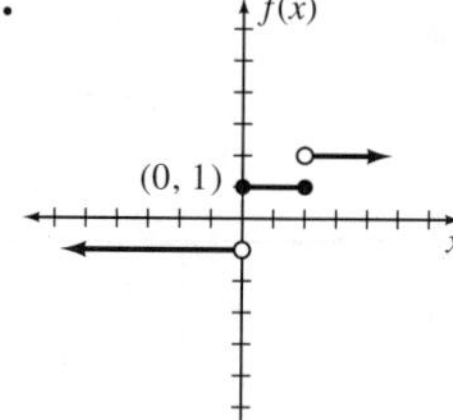

33.

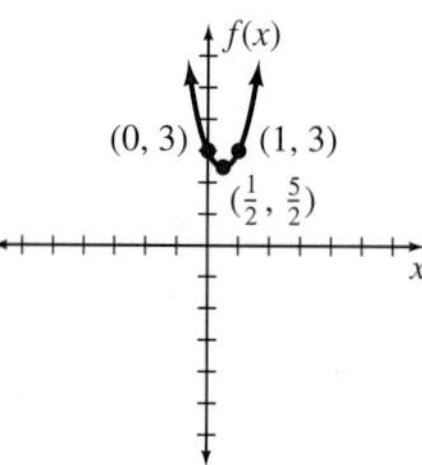

34.

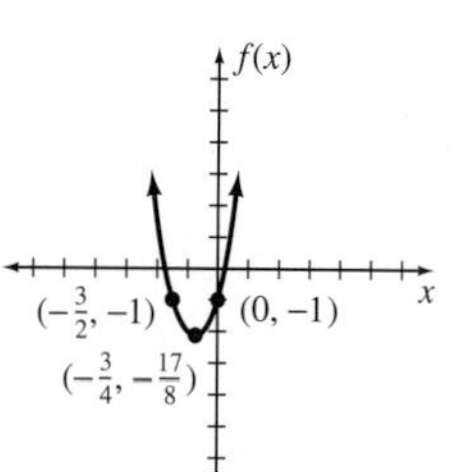

43.

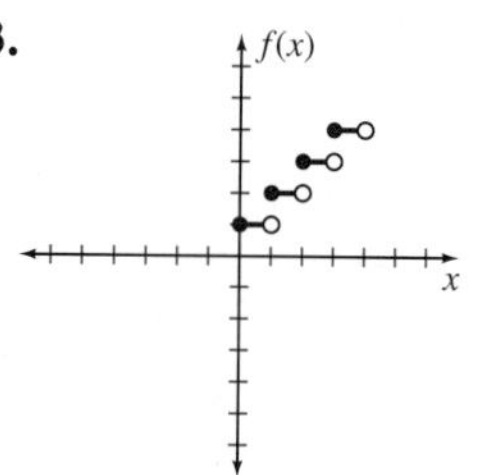

44.

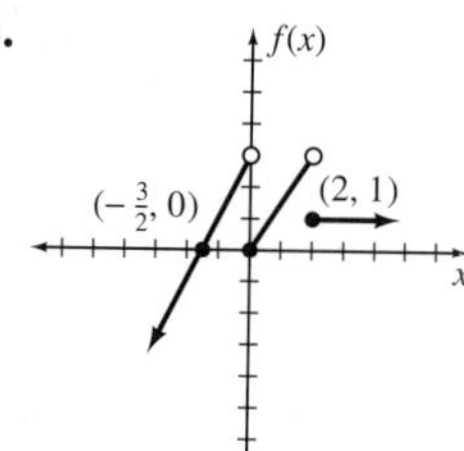

35.

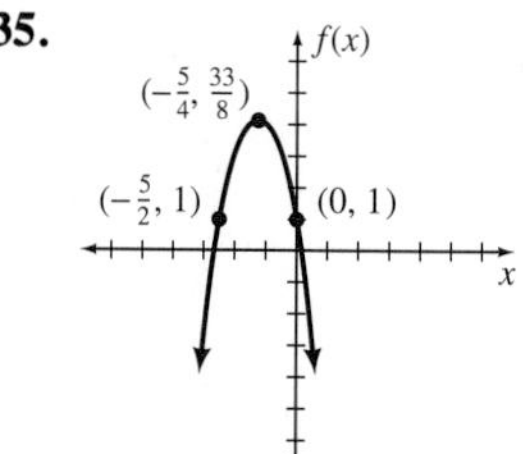

36.

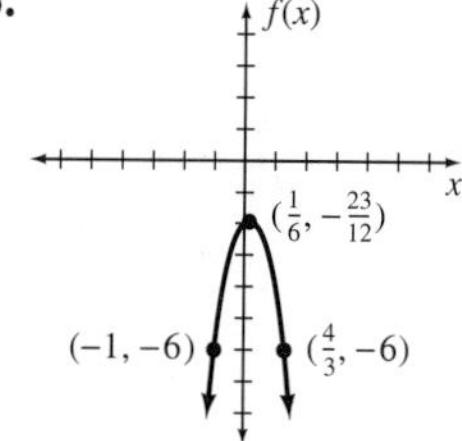

45.

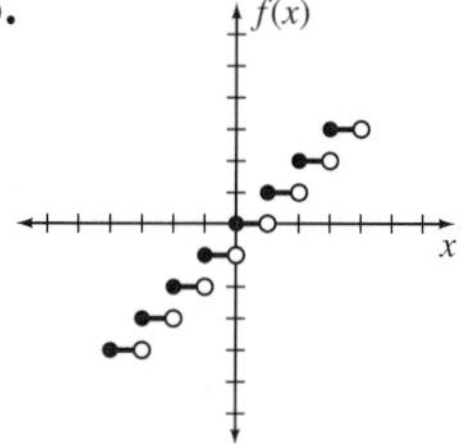

Problem Set 4.3 (page 248)

1.

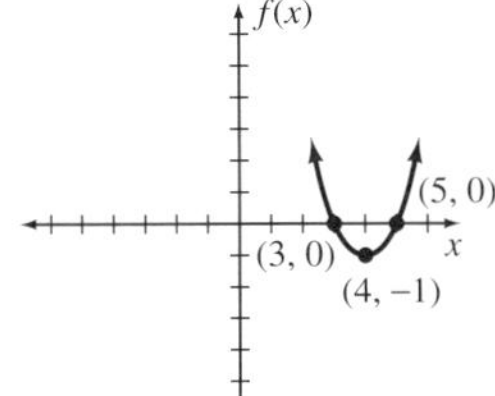

2.

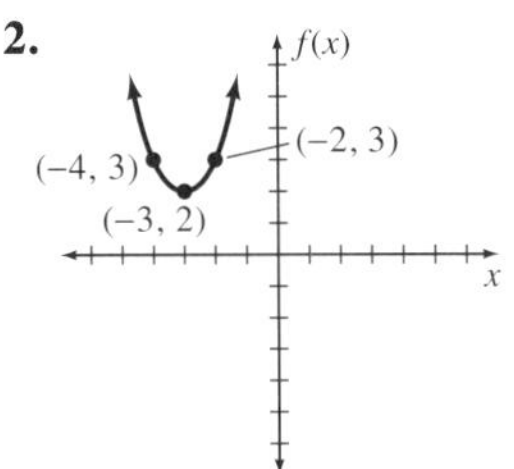

3.

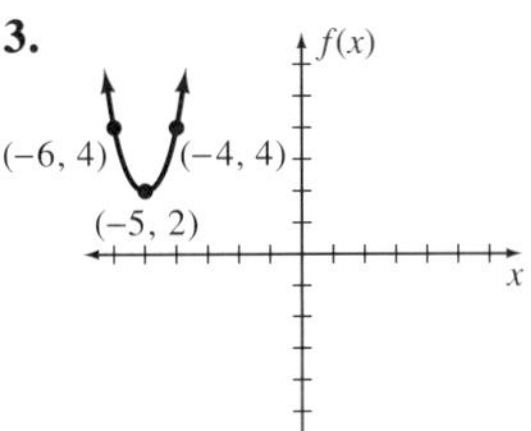

4.

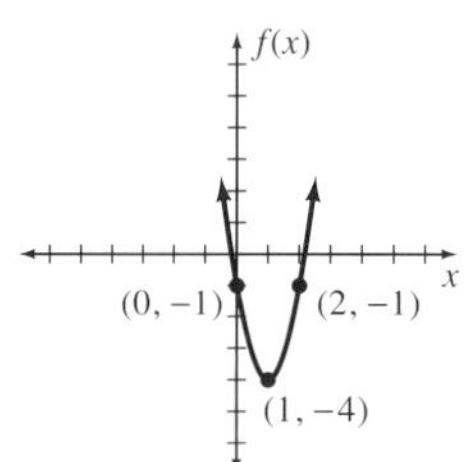

5.

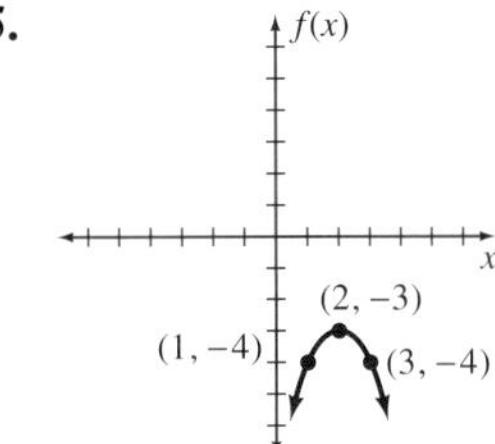

6.

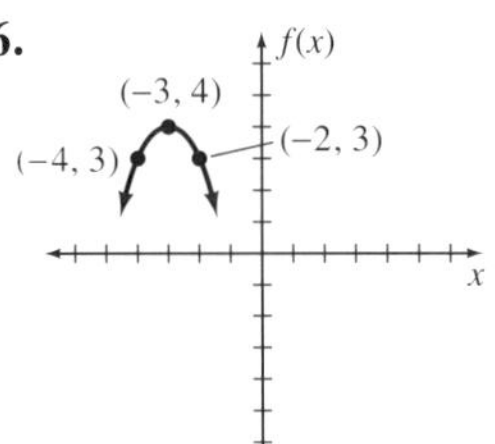

7.

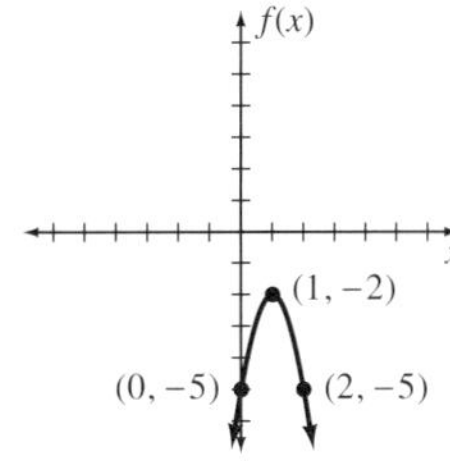

8.

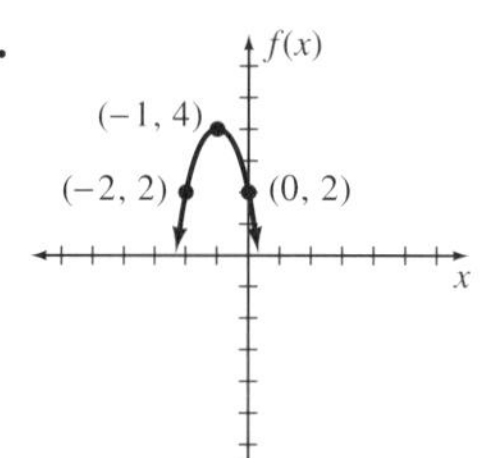

9.

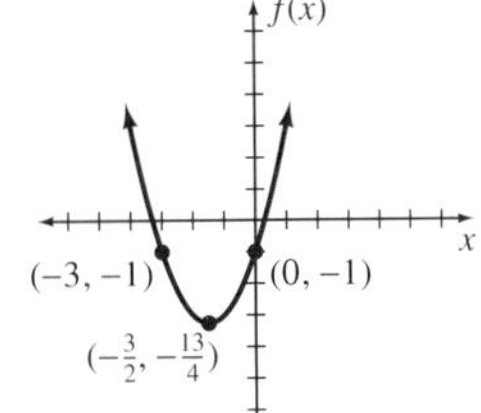

10.

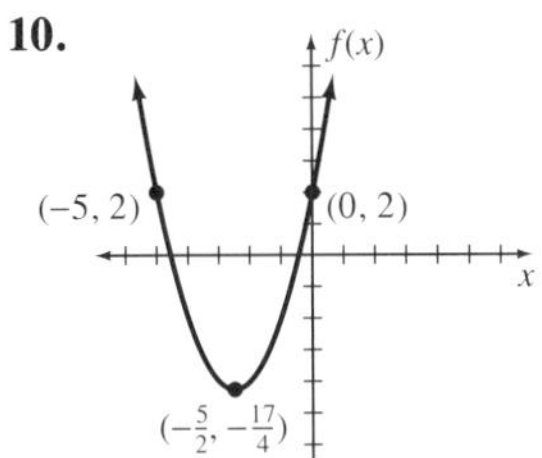

11.

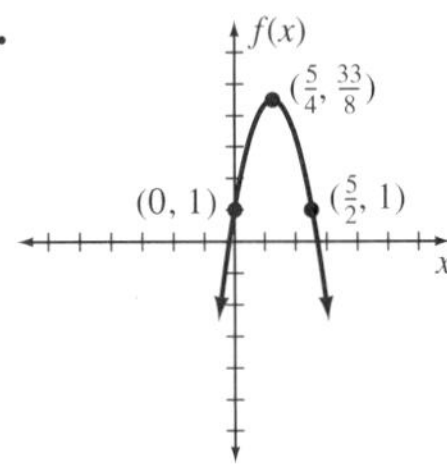

12.

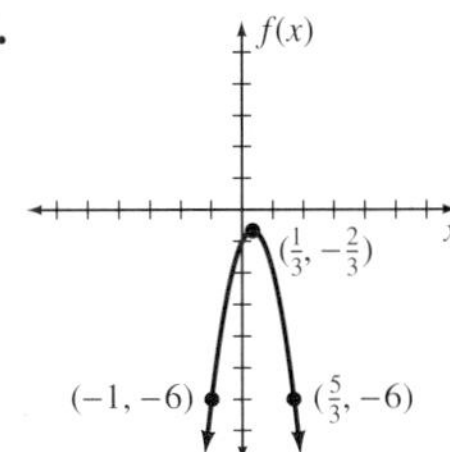

13.

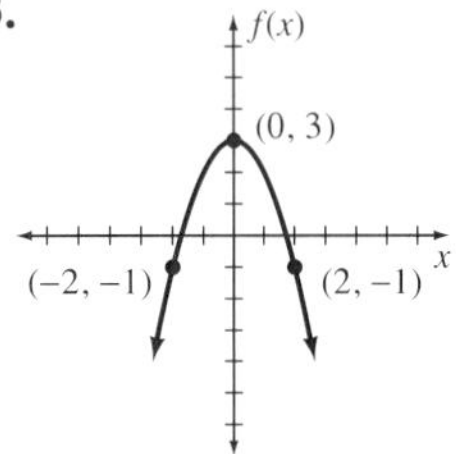

14.

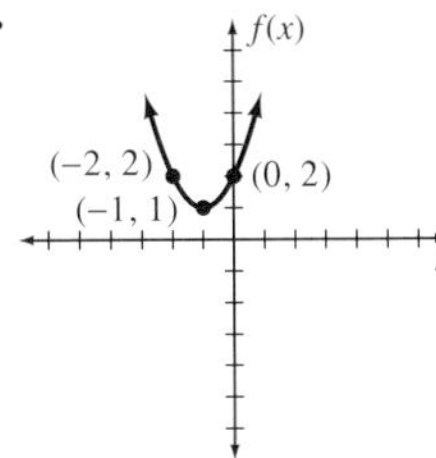

15.

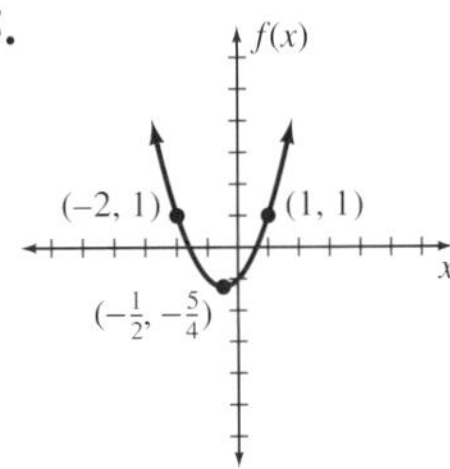

16.

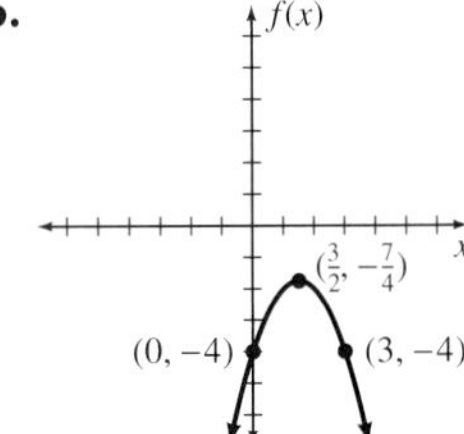

17.

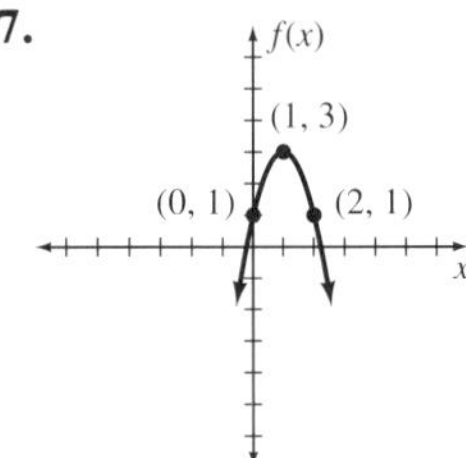

18.

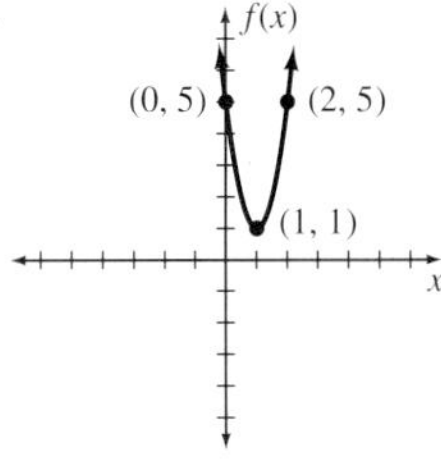

19.

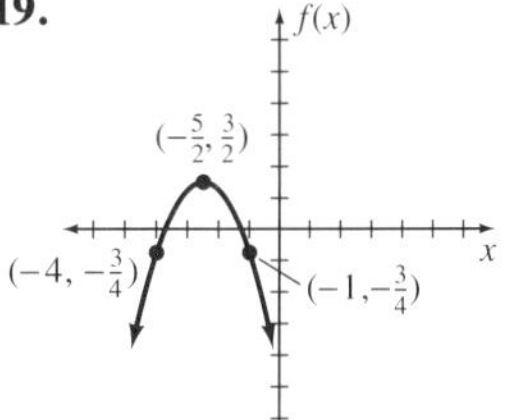

20.

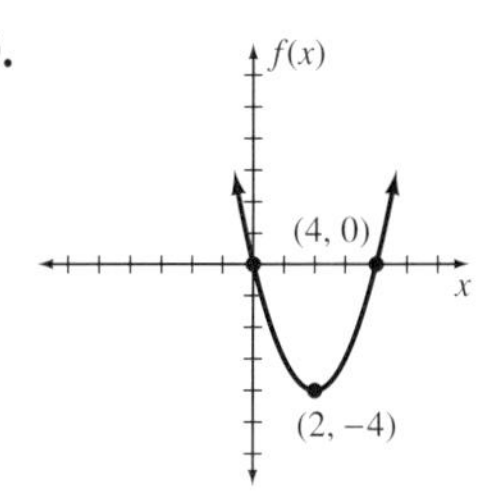

21. 3 and 5; $(4, -1)$ **22.** 7 and 9; $(8, -1)$
23. 6 and 8; $(7, -2)$ **24.** 9 and 11; $(10, -3)$
25. 4 and 6; $(5, 1)$ **26.** 8 and 10; $(9, 2)$
27. $7 + \sqrt{5}$ and $7 - \sqrt{5}$; $(7, -5)$
28. $9 + \sqrt{13}$ and $9 - \sqrt{13}$; $(9, -13)$
29. No x-intercepts; $\left(\frac{9}{2}, -\frac{3}{4}\right)$
30. No x-intercepts; $\left(-\frac{3}{4}, \frac{15}{8}\right)$
31. $\frac{1 + \sqrt{5}}{2}$ and $\frac{1 - \sqrt{5}}{2}$; $\left(\frac{1}{2}, 5\right)$
32. $\frac{3 + \sqrt{65}}{4}$ and $\frac{3 - \sqrt{65}}{4}$; $\left(\frac{3}{4}, \frac{65}{8}\right)$ **33.** 70
34. 80 **35.** 144 **36.** 5 and 25
37. 25 and 25 **38.** -20 and 20
39. 60 meters by 60 meters **40.** 65 couples
41. 1100 subscribers at \$13.75 per month
42. 750 units **43.** $10\sqrt{2}$ feet **44.** 75 feet
45. 62.5 feet **47. (a)** $3 + \sqrt{6}$ and $3 - \sqrt{6}$; $(3, -6)$
(b) $9 + \sqrt{15}$ and $9 - \sqrt{15}$; $(9, -15)$
(c) $4 + \sqrt{13}$ and $4 - \sqrt{13}$; $(4, 13)$
(d) $12 + \sqrt{15}$ and $12 - \sqrt{15}$; $(12, 15)$
(e) No x-intercept; $\left(\frac{1}{4}, \frac{1}{8}\right)$
(f) $5 + 2\sqrt{2}$ and $5 - 2\sqrt{2}$; $(5, 4)$

5.
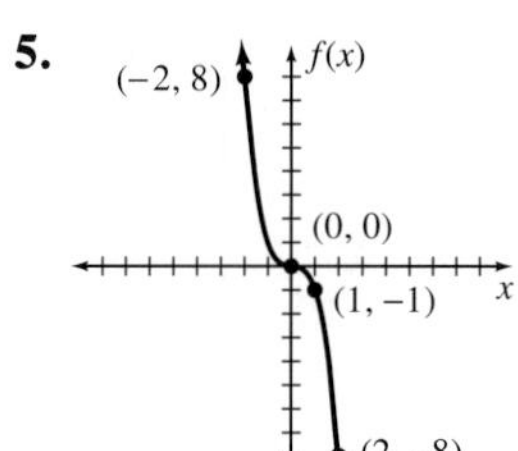

6.
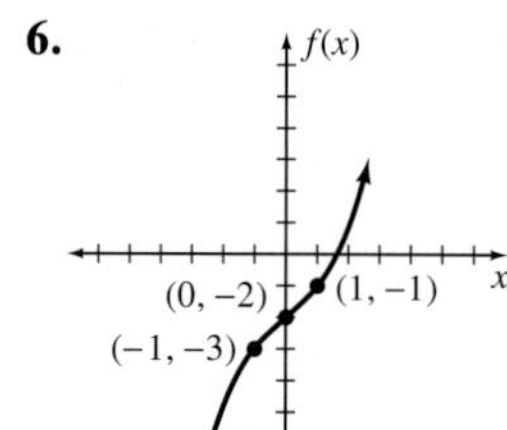

7.
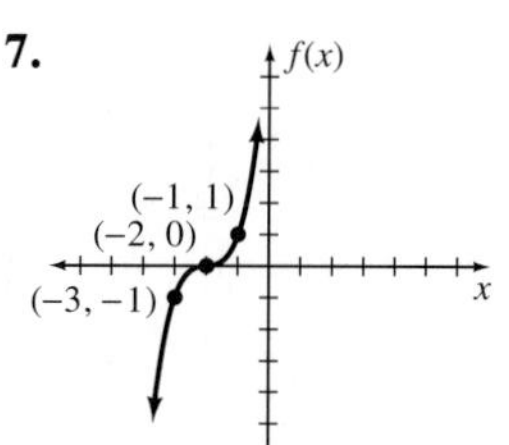

8.
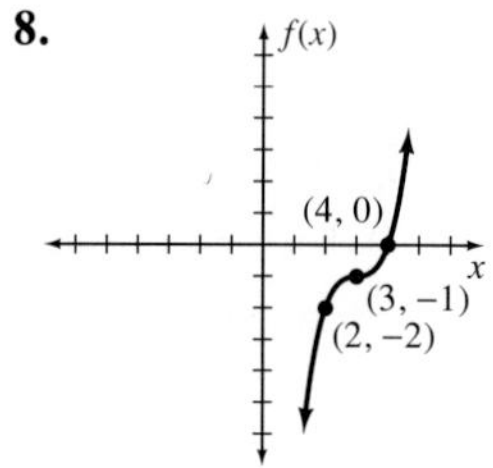

9.
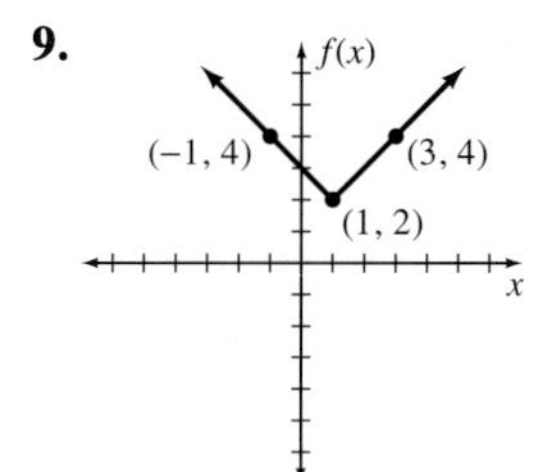

10.
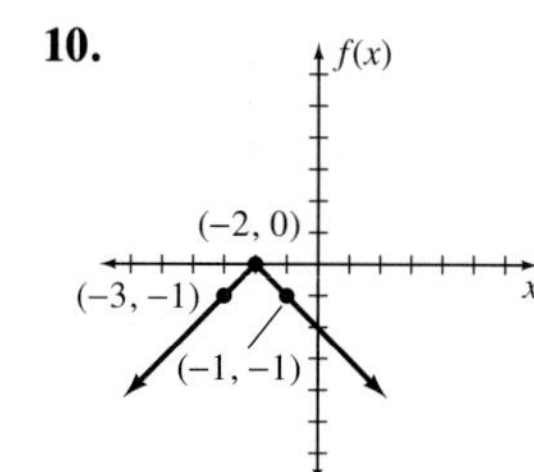

Problem Set 4.4 (page 260)

1.
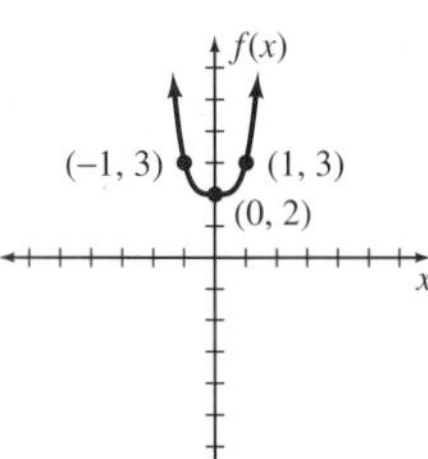

2.
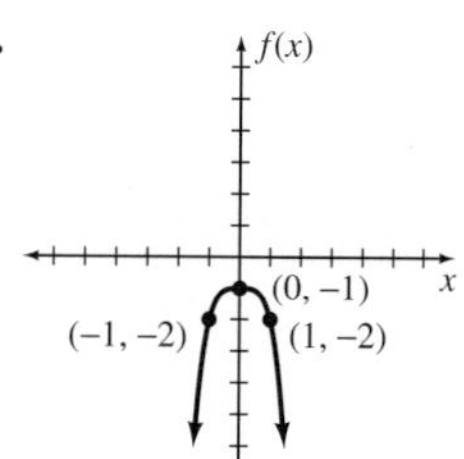

11.
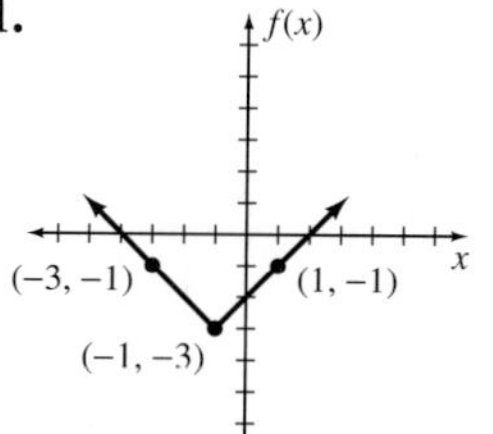

12.
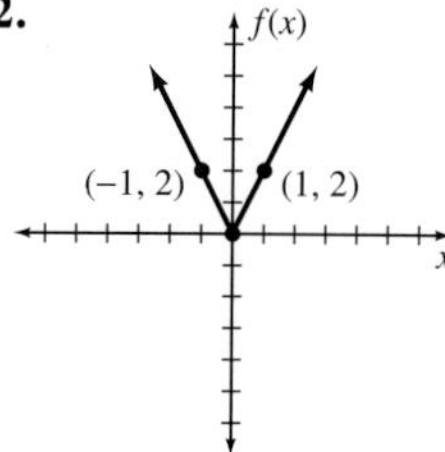

3.
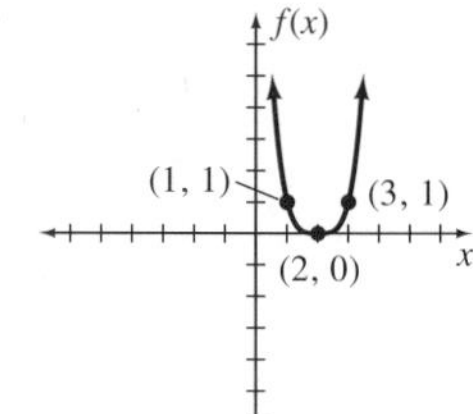

4.
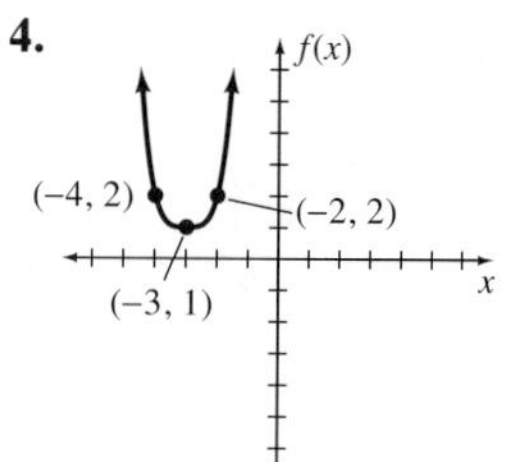

13.
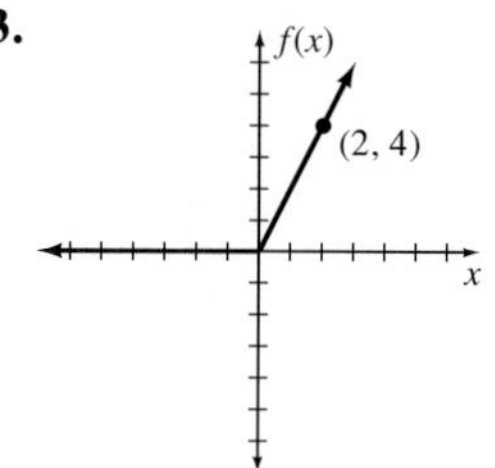

14.
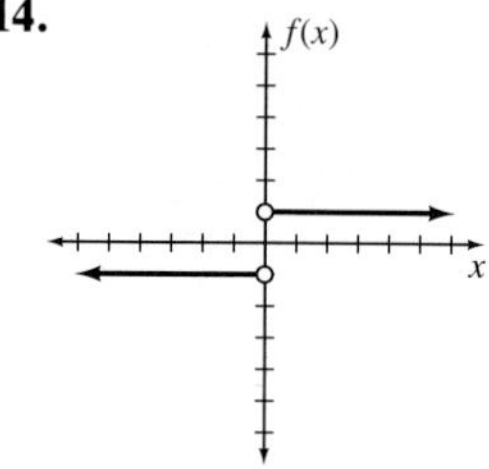

15.
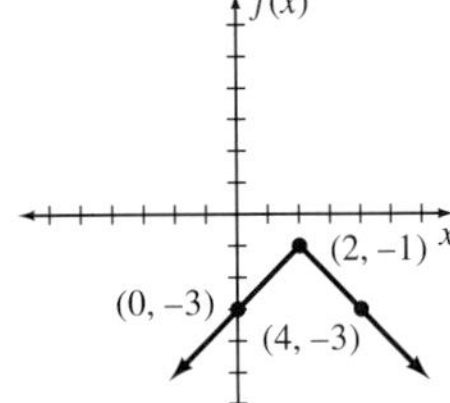

16.
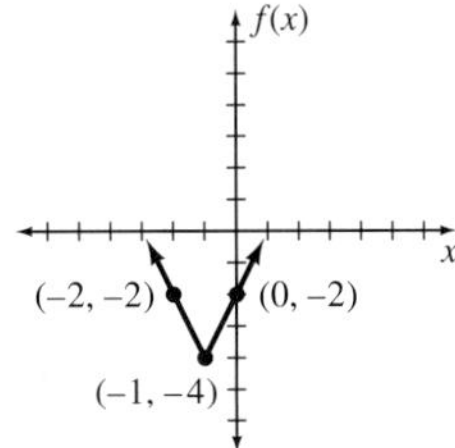

25.
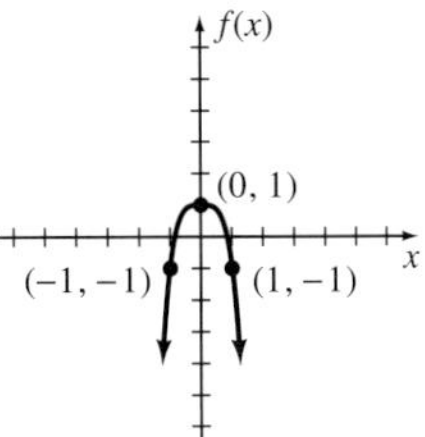

26.
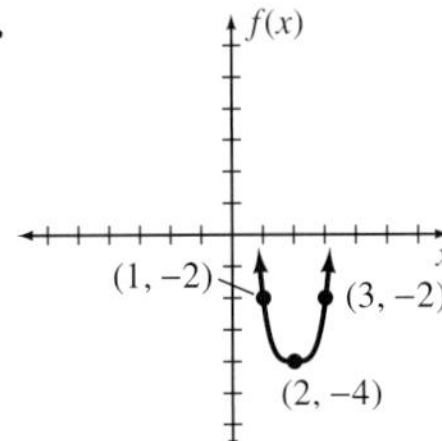

17.
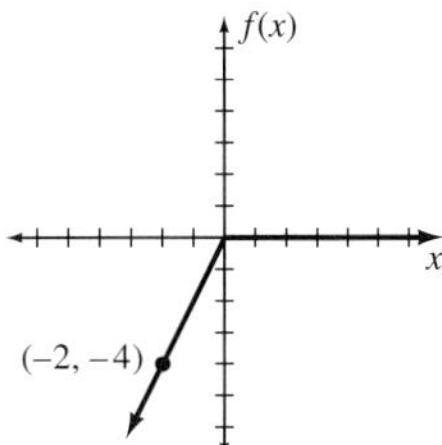

18.
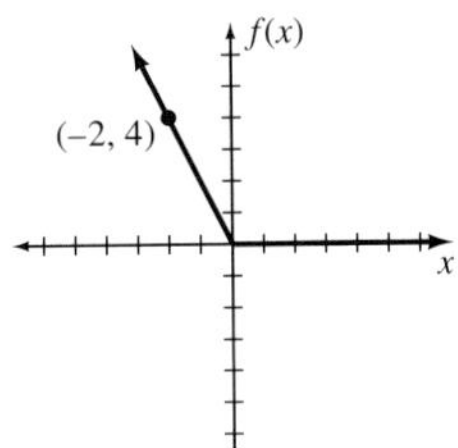

27.
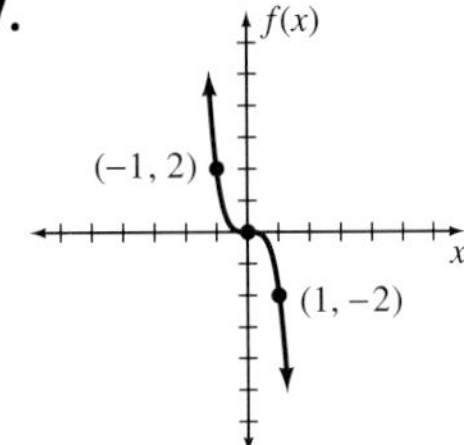

28.
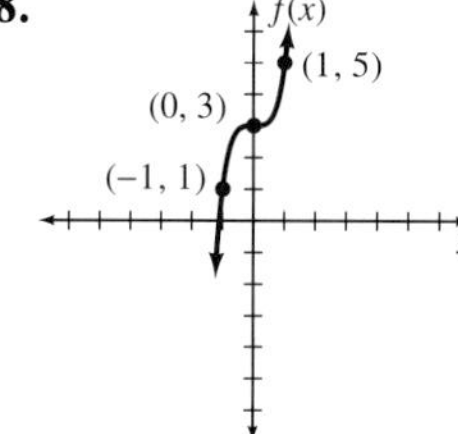

19.
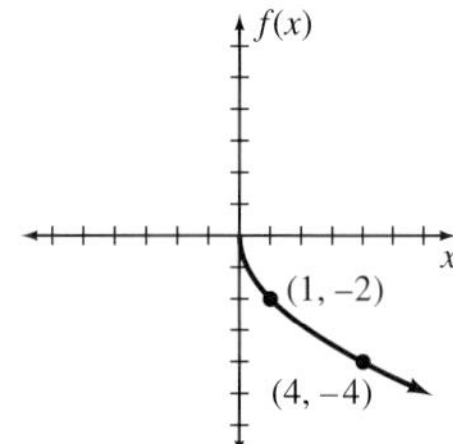

20.
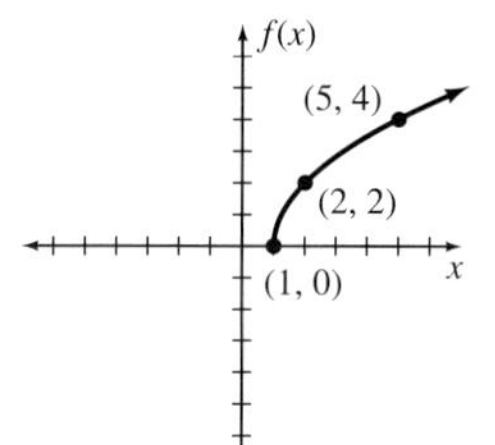

29.
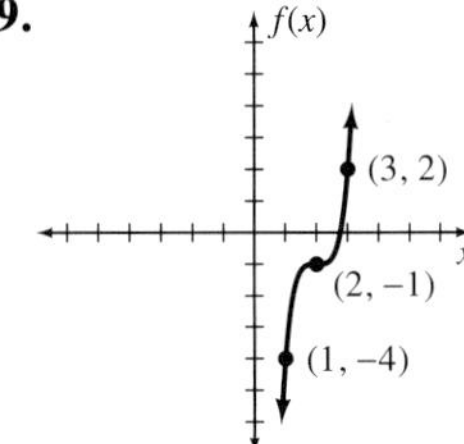

30.
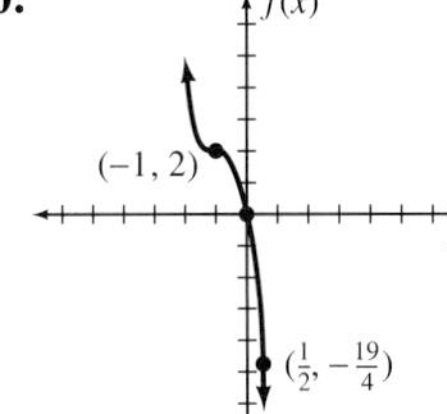

21.
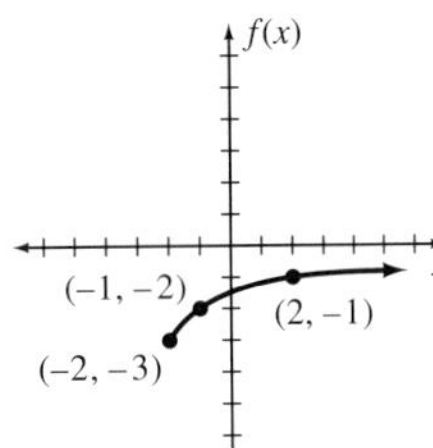

22.
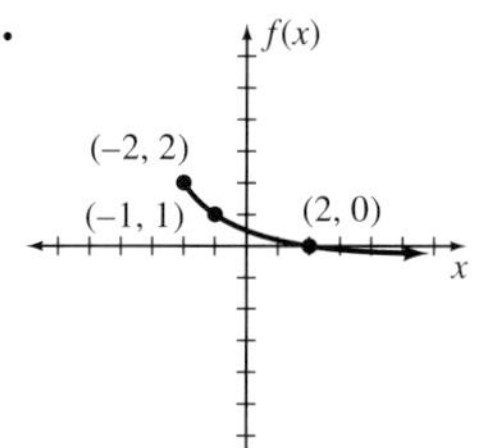

31. (a)
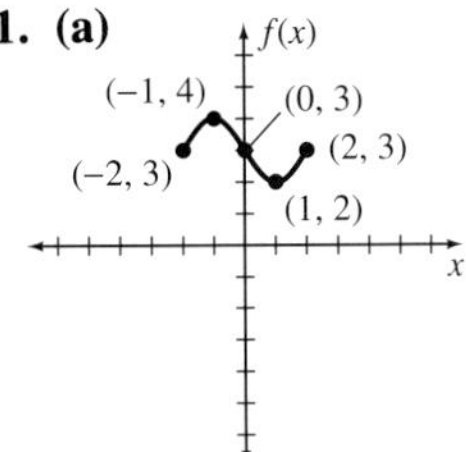

(b)
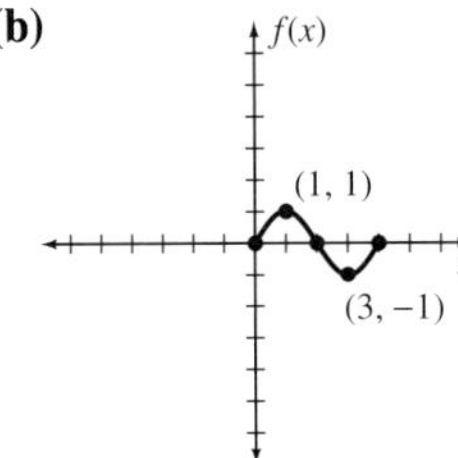

23.
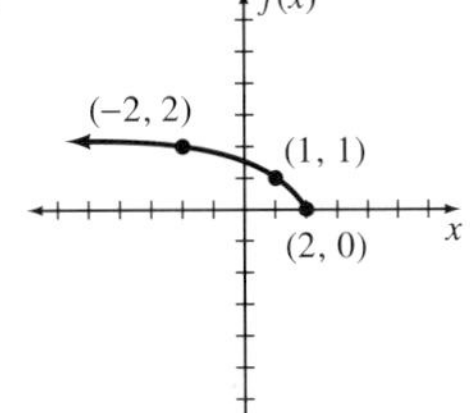

24.
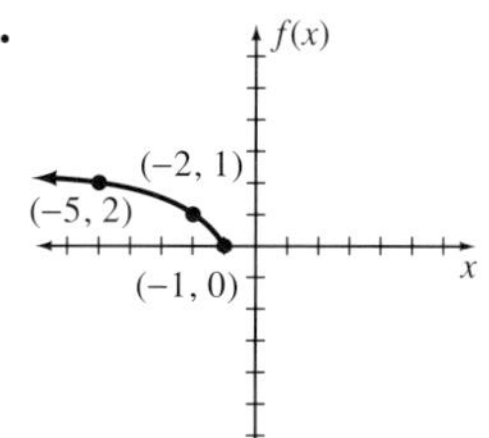

(c)
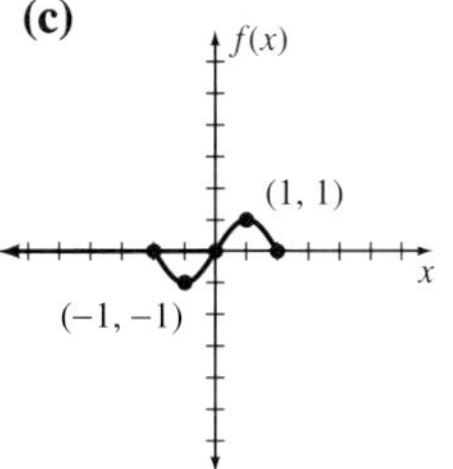

(d)
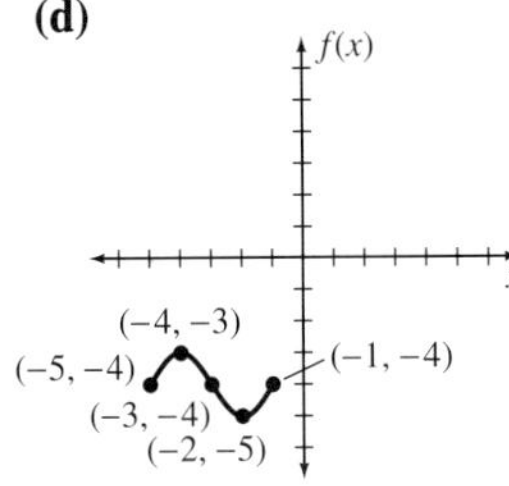

Problem Set 4.5 (page 267)

1. $8x - 2$; $-2x - 6$; $15x^2 - 14x - 8$; $\dfrac{3x - 4}{5x + 2}$

2. $-14x + 6$; $2x - 8$; $48x^2 - 34x - 7$; $\dfrac{-6x - 1}{-8x + 7}$

3. $x^2 - 7x + 3$; $x^2 - 5x + 5$; $-x^3 + 5x^2 + 2x - 4$; $\dfrac{x^2 - 6x + 4}{-x - 1}$

4. $3x^2 - 3x + 1$; $x^2 - 3x + 9$; $2x^4 - 3x^3 - 3x^2 + 12x - 20$; $\dfrac{2x^2 - 3x + 5}{x^2 - 4}$

5. $2x^2 + 3x - 6$; $-5x + 4$; $x^4 + 3x^3 - 10x^2 + x + 5$; $\dfrac{x^2 - x - 1}{x^2 + 4x - 5}$

6. $2x^2 - 3x + 54$; $-x + 6$; $x^4 - 3x^3 - 52x^2 + 84x + 720$; $\dfrac{x + 6}{x + 5}$

7. $\sqrt{x - 1} + \sqrt{x}$; $\sqrt{x - 1} - \sqrt{x}$; $\sqrt{x^2 - x}$; $\dfrac{\sqrt{x(x - 1)}}{x}$

8. $\sqrt{x + 2} + \sqrt{3x - 1}$; $\sqrt{x + 2} - \sqrt{3x - 1}$; $\sqrt{3x^2 + 4x - 2}$; $\dfrac{\sqrt{3x^2 + 4x - 2}}{3x - 1}$

9. $(f \circ g)(x) = 6x - 2$, $D = \{\text{all reals}\}$
$(g \circ f)(x) = 6x - 1$, $D = \{\text{all reals}\}$

10. $(f \circ g)(x) = 12x + 1$, $D = \{\text{all reals}\}$
$(g \circ f)(x) = 12x + 3$, $D = \{\text{all reals}\}$

11. $(f \circ g)(x) = 10x + 2$, $D = \{\text{all reals}\}$
$(g \circ f)(x) = 10x - 5$, $D = \{\text{all reals}\}$

12. $(f \circ g)(x) = 8x + 3$, $D = \{\text{all reals}\}$
$(g \circ f)(x) = 8x - 12$, $D = \{\text{all reals}\}$

13. $(f \circ g)(x) = 3x^2 + 7$, $D = \{\text{all reals}\}$
$(g \circ f)(x) = 9x^2 + 24x + 17$, $D = \{\text{all reals}\}$

14. $(f \circ g)(x) = 3$, $D = \{\text{all reals}\}$
$(g \circ f)(x) = -28$, $D = \{\text{all reals}\}$

15. $(f \circ g)(x) = 3x^2 + 9x - 16$, $D = \{\text{all reals}\}$
$(g \circ f)(x) = 9x^2 - 15x$, $D = \{\text{all reals}\}$

16. $(f \circ g)(x) = 2x^2 + 15x + 27$, $D = \{\text{all reals}\}$
$(g \circ f)(x) = 2x^2 - x + 3$, $D = \{\text{all reals}\}$

17. $(f \circ g)(x) = \dfrac{1}{2x + 7}$, $D = \left\{x \mid x \neq -\dfrac{7}{2}\right\}$
$(g \circ f)(x) = \dfrac{7x + 2}{x}$, $D = \{x \mid x \neq 0\}$

18. $(f \circ g)(x) = \dfrac{1}{x^2}$, $D = \{x \mid x \neq 0\}$
$(g \circ f)(x) = \dfrac{1}{x^2}$, $D = \{x \mid x \neq 0\}$

19. $(f \circ g)(x) = \sqrt{3x - 3}$, $D = \{x \mid x \geq 1\}$
$(g \circ f)(x) = 3\sqrt{x - 2} - 1$, $D = \{x \mid x \geq 2\}$

20. $(f \circ g)(x) = x^2$, $D = \{x \mid x \neq 0\}$
$(g \circ f)(x) = x^2$, $D = \{x \mid x \neq 0\}$

21. $(f \circ g)(x) = \dfrac{x}{2 - x}$, $D = \{x \mid x \neq 0 \text{ and } x \neq 2\}$
$(g \circ f)(x) = 2x - 2$, $D = \{x \mid x \neq 1\}$

22. $(f \circ g)(x) = \dfrac{8x}{4x + 3}$, $D = \left\{x \mid x \neq 0 \text{ and } x \neq -\dfrac{3}{4}\right\}$
$(g \circ f)(x) = \dfrac{3x + 6}{8}$, $D = \{x \mid x \neq -2\}$

23. $(f \circ g)(x) = 2\sqrt{x - 1} + 1$, $D = \{x \mid x \geq 1\}$
$(g \circ f)(x) = \sqrt{2x}$, $D = \{x \mid x \geq 0\}$

24. $(f \circ g)(x) = \sqrt{5x - 1}$, $D = \left\{x \mid x \geq \dfrac{1}{5}\right\}$
$(g \circ f)(x) = 5\sqrt{x + 1} - 2$, $D = \{x \mid x \geq -1\}$

25. $(f \circ g)(x) = x$, $D = \{x \mid x \neq 0\}$
$(g \circ f)(x) = x$, $D = \{x \mid x \neq 1\}$

26. $(f \circ g)(x) = \dfrac{1 - x}{1 + 2x}$, $D = \left\{x \mid x \neq 0 \text{ and } x \neq -\dfrac{1}{2}\right\}$
$(g \circ f)(x) = \dfrac{x + 2}{x - 1}$, $D = \{x \mid x \neq -2 \text{ and } x \neq 1\}$

27. 4; 50 **28.** 34; 18 **29.** 9; 0 **30.** $\dfrac{1}{3}$; 2

31. $\sqrt{11}$; 5 **32.** 9; 1 **41.** $b + 2a = 2$

46. **(a)** $f(x) = x^2$ and $g(x) = x + 5$
(b) $f(x) = x^3$ and $g(x) = x + 3$
(c) $f(x) = x - 6$ and $g(x) = -x^3$
(d)

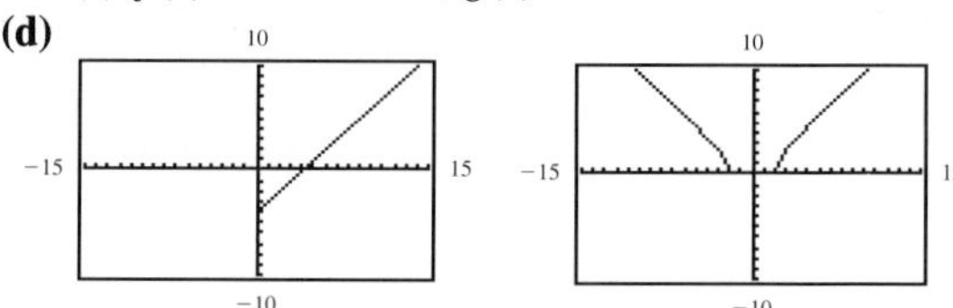

(e)

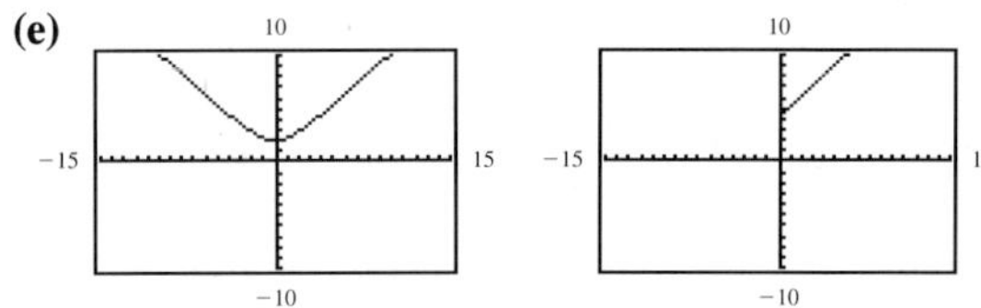

(f)

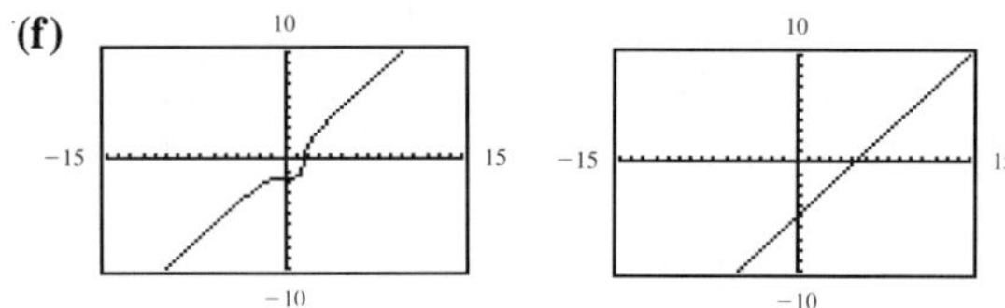

Problem Set 4.6 (page 277)

1. Yes **2.** Yes **3.** No **4.** Yes
5. Yes **6.** No **7.** Yes **8.** Yes
9. Yes **10.** Yes **11.** No **12.** No
13. No **14.** No

15. Domain of f: $\{1, 2, 5\}$
Range of f: $\{5, 9, 21\}$
$f^{-1} = \{(5, 1), (9, 2), (21, 5)\}$
Domain of f^{-1}: $\{5, 9, 21\}$
Range of f^{-1}: $\{1, 2, 5\}$

16. Domain of f: $\{1, 4, 9, 16\}$
Range of f: $\{1, 2, 3, 4\}$
$f^{-1} = \{(1, 1), (2, 4), (3, 9), (4, 16)\}$
Domain of f^{-1}: $\{1, 2, 3, 4\}$
Range of f^{-1}: $\{1, 4, 9, 16\}$

17. Domain of f: $\{0, 2, -1, -2\}$
Range of f: $\{0, 8, -1, -8\}$
f^{-1}: $\{(0, 0), (8, 2), (-1, -1), (-8, -2)\}$
Domain of f^{-1}: $\{0, 8, -1, -8\}$
Range of f^{-1}: $\{0, 2, -1, -2\}$

18. Domain of f: $\{-1, -2, -3, -4\}$
Range of f: $\{1, 4, 9, 16\}$
f^{-1}: $\{(1, -1), (4, -2), (9, -3), (16, -4)\}$
Domain of f^{-1}: $\{1, 4, 9, 16\}$
Range of f^{-1}: $\{-1, -2, -3, -4\}$

27. No **28.** Yes **29.** Yes **30.** Yes
31. No **32.** No **33.** Yes **34.** Yes
35. Yes **36.** No **37.** $f^{-1}(x) = x + 4$

38. $f^{-1}(x) = \dfrac{x + 1}{2}$

39. $f^{-1}(x) = \dfrac{-x - 4}{3}$

40. $f^{-1}(x) = \dfrac{-x + 6}{5}$

41. $f^{-1}(x) = \dfrac{12x + 10}{9}$

42. $f^{-1}(x) = \dfrac{12x + 3}{8}$ **43.** $f^{-1}(x) = -\dfrac{3}{2}x$

44. $f^{-1}(x) = \dfrac{3}{4}x$ **45.** $f^{-1}(x) = x^2$ for $x \geq 0$

46. $f^{-1}(x) = \dfrac{1}{x}$ **47.** $f^{-1}(x) = \sqrt{x - 4}$ for $x \geq 4$

48. $f^{-1}(x) = -\sqrt{x - 1}$ for $x \geq 1$

49. $f^{-1}(x) = \dfrac{1}{x - 1}$ for $x > 1$

50. $f^{-1}(x) = \dfrac{x}{1 - x}$ for $x < 1$

51. $f^{-1}(x) = \dfrac{1}{3}x$

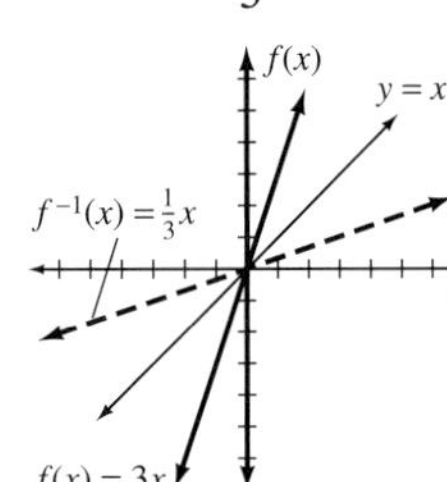

52. $f^{-1}(x) = -x$

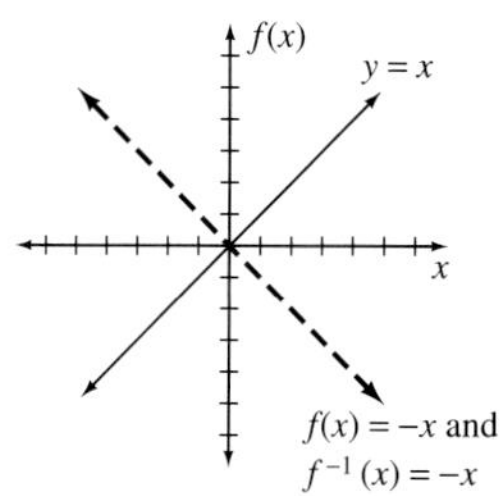

53. $f^{-1}(x) = \dfrac{x - 1}{2}$

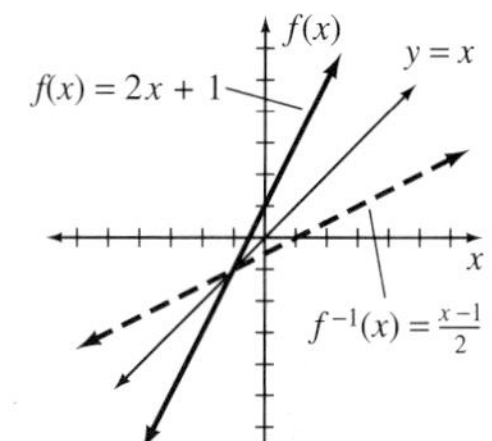

54. $f^{-1}(x) = \dfrac{-x - 3}{3}$

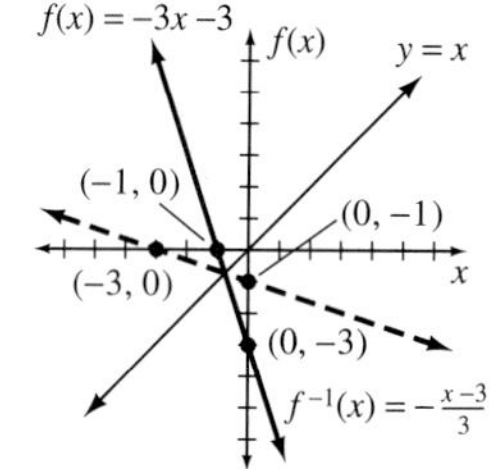

55. $f^{-1}(x) = \dfrac{x + 2}{x}$ for $x > 0$

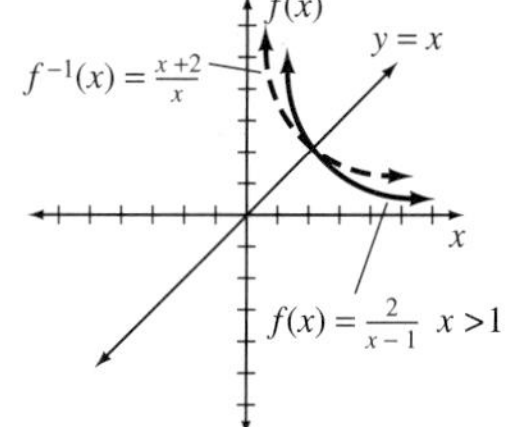

56. $f^{-1}(x) = \dfrac{2x - 1}{x}$ for $x < 0$

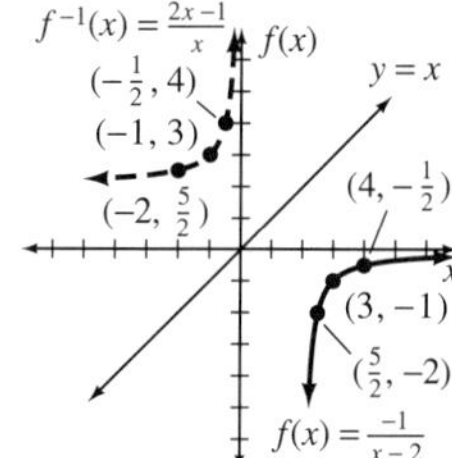

57. $f^{-1}(x) = \sqrt{x + 4}$ for $x \geq -4$

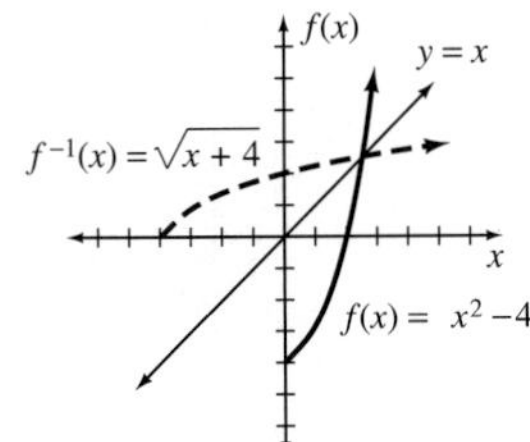

58. $f^{-1}(x) = x^2 + 3$ for $x \geq 0$

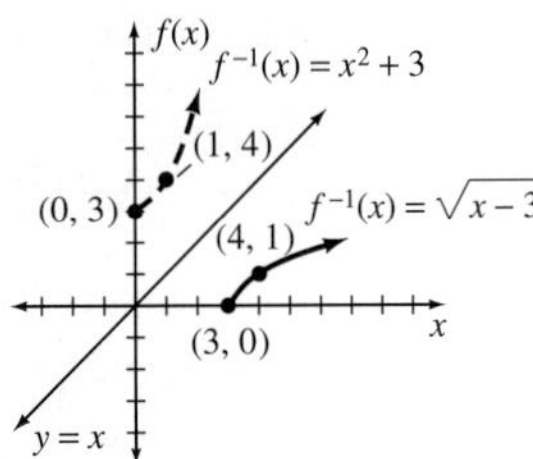

59. Increasing on $[0, \infty)$ and decreasing on $(-\infty, 0]$
60. Increasing on $(-\infty, \infty)$
61. Decreasing on $(-\infty, \infty)$
62. Increasing on $[3, \infty)$ and decreasing on $(-\infty, 3]$
63. Increasing on $(-\infty, -2]$ and decreasing on $[-2, \infty)$
64. Increasing on $[1, \infty)$ and decreasing on $(-\infty, 1]$
65. Increasing on $(-\infty, -4]$ and decreasing on $[-4, \infty)$
66. Increasing on $\left[-\frac{3}{2}, \infty\right)$ and decreasing on $\left(-\infty, -\frac{3}{2}\right]$
68. **(a)** $f^{-1}(x) = \frac{x + 9}{3}$ **(b)** $f^{-1}(x) = \frac{-x + 6}{2}$
(c) $f^{-1}(x) = -x + 1$ **(d)** $f^{-1}(x) = \frac{1}{2}x$
(e) $f^{-1}(x) = -\frac{1}{5}x$ **(f)** $f^{-1}(x) = \sqrt{x - 6}$ for $x \geq 6$
69. **(a)** $\frac{x + 7}{6}$ **(b)** $\frac{x - 4}{6}$ **(c)** $\frac{x + 7}{6}$

Problem Set 4.7 (page 285)

1. $y = kx^3$ **2.** $a = \frac{k}{b^2}$ **3.** $A = klw$
4. $s = kgt^2$ **5.** $V = \frac{k}{p}$ **6.** $y = \frac{kx^2}{w^3}$
7. $V = khr^2$ **8.** $I = krt$ **9.** 24 **10.** 16
11. $\frac{22}{7}$ **12.** $\frac{1}{3}$ **13.** $\frac{1}{2}$ **14.** $-\frac{1}{2}$ **15.** 7
16. $\frac{22}{7}$ **17.** 6 **18.** $\frac{\sqrt{10}}{45}$ **19.** 8
20. $\frac{1}{32}$ **21.** 96 **22.** 600 **23.** 5 hours
24. 400 feet **25.** 2 seconds **26.** $5\frac{3}{5}$ days
27. 24 days **28.** 12 cubic centimeters **29.** 28
30. 3080 cm^3 **31.** \$2400 **32.** \$128
33. **(a)** \$210 **(b)** \$637 **(c)** \$1050
34. 2.8 seconds **35.** 3560.76 m^3 **36.** 1.4
37. 0.048

Chapter 4 Review Problem Set (page 290)

1. 7; 4; 32 **2.** **(a)** -5 **(b)** $4a + 2h - 1$
(c) $-6a - 3h + 2$
3. The domain is the set of all real numbers and the range is the set of all real numbers greater than or equal to 5.
4. The domain is the set of all real numbers except $\frac{1}{2}$ and -4.
5. $(-\infty, 2] \cup [5, \infty)$
6.

7.

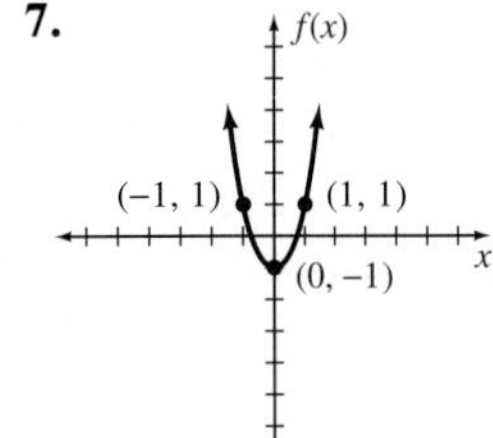

8.

9.

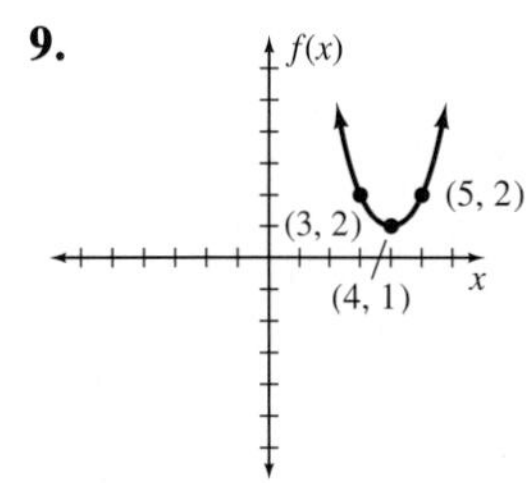

10.

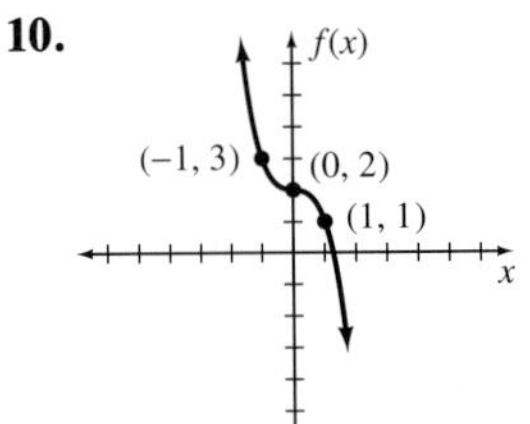

11.

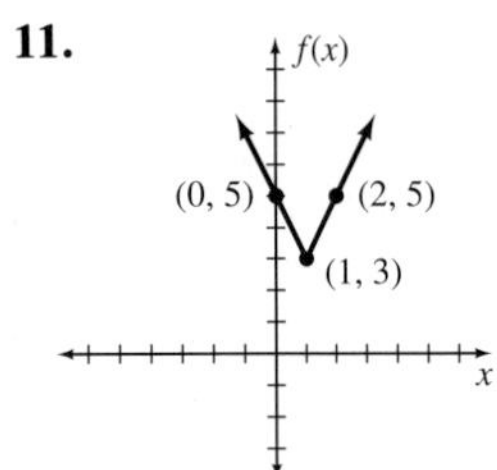

12.
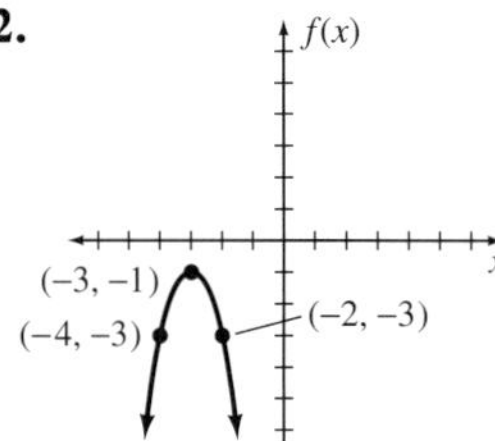

13.
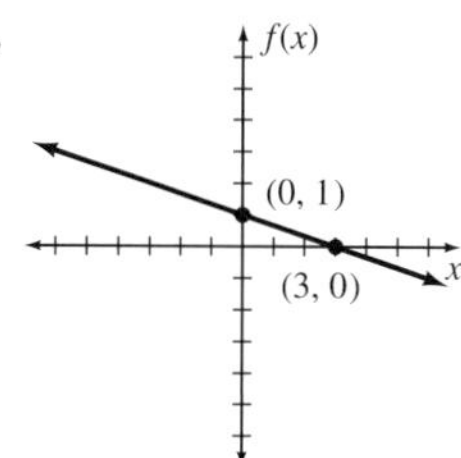

14.
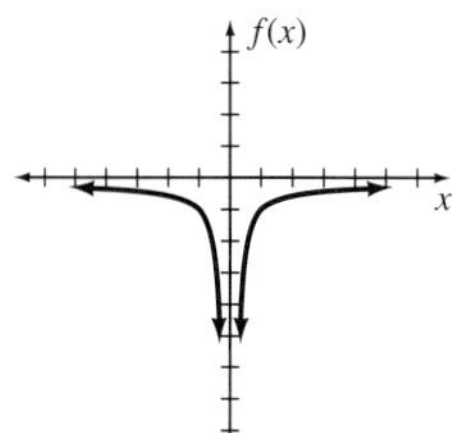

15.
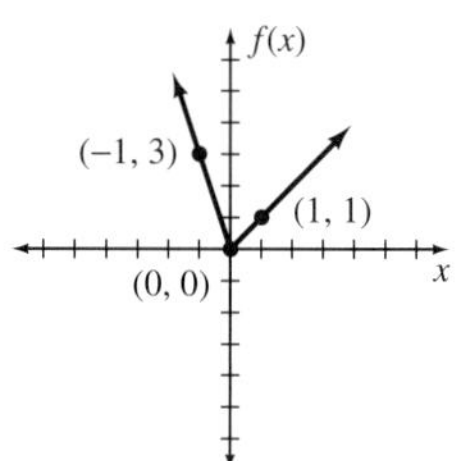

16. $x^2 - 2x$; $-x^2 + 6x + 6$; $2x^3 - 5x^2 - 18x - 9$; $\dfrac{2x + 3}{x^2 - 4x - 3}$

17. $(f \circ g)(x) = -6x + 12$; $D = \{\text{all reals}\}$
$(g \circ f)(x) = -6x + 25$; $D = \{\text{all reals}\}$

18. $(f \circ g)(x) = 25x^2 - 40x + 11$; $D = \{\text{all reals}\}$
$(g \circ f)(x) = 5x^2 - 29$; $D = \{\text{all reals}\}$

19. $(f \circ g)(x) = \sqrt{x - 3}$; $D = \{x|x \geq 3\}$
$(g \circ f)(x) = \sqrt{x - 5} + 2$; $D = \{x|x \geq 5\}$

20. $(f \circ g)(x) = \dfrac{x + 2}{-3x - 5}$; $D = \left\{x|x \neq -2 \text{ and } x \neq -\dfrac{5}{3}\right\}$
$(g \circ f)(x) = \dfrac{x - 3}{2x - 5}$; $D = \left\{x|x \neq 3 \text{ and } x \neq \dfrac{5}{2}\right\}$

21. 1; 5

22. The domain of f equals the range of g, and the range of f equals the domain of g. Furthermore, $(f \circ g)(x) = x$ and $(g \circ f)(x) = x$.

23. Yes **24.** No **25.** Yes **26.** Yes

27. $f^{-1}(x) = \dfrac{x - 5}{4}$

28. $f^{-1}(x) = \dfrac{-x - 7}{3}$

29. $f^{-1}(x) = \dfrac{6x + 2}{5}$

30. $f^{-1}(x) = \sqrt{-2 - x}$

31. Increasing on $(-\infty, 4]$ and decreasing on $[4, \infty)$

32. Increasing on $[3, \infty)$ **33.** 112 students

34. $k = 9$ **35.** 441 **36.** 128 pounds

CHAPTER 5

Problem Set 5.1 (page 301)

1. $\{3\}$ **2.** $\{6\}$ **3.** $\{3\}$ **4.** $\{-2\}$ **5.** $\{4\}$
6. $\{1\}$ **7.** $\{2\}$ **8.** $\{0\}$ **9.** $\{-2\}$
10. $\{-3\}$ **11.** $\left\{\dfrac{5}{3}\right\}$ **12.** $\left\{\dfrac{3}{2}\right\}$ **13.** $\left\{\dfrac{3}{2}\right\}$
14. $\left\{\dfrac{1}{5}\right\}$ **15.** $\left\{\dfrac{4}{9}\right\}$ **16.** $\left\{\dfrac{3}{5}\right\}$ **17.** $\left\{\dfrac{4}{3}\right\}$
18. $\{3\}$ **19.** $\left\{\dfrac{2}{3}\right\}$ **20.** $\left\{\dfrac{1}{2}\right\}$

21.
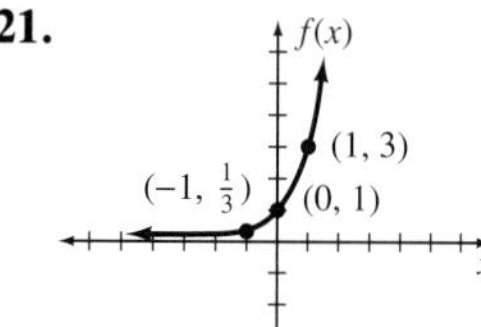

22.
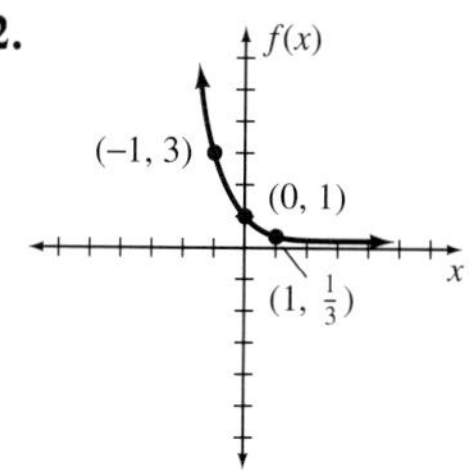

23.
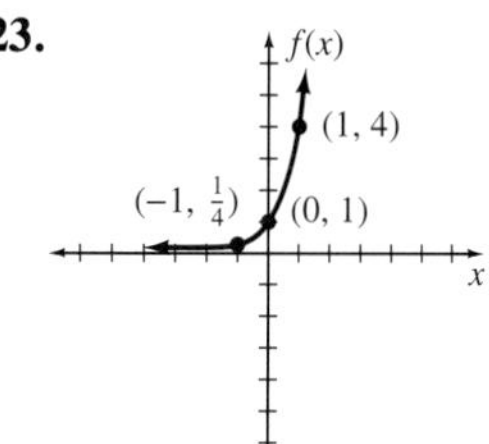

24.
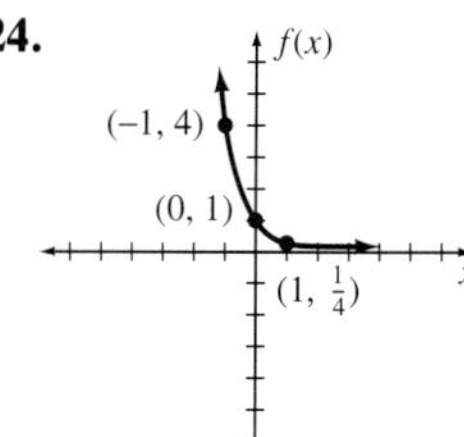

25.
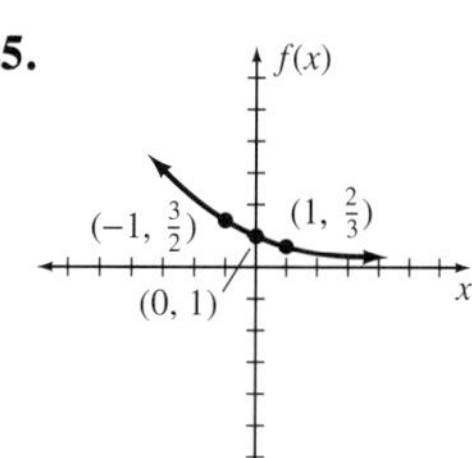

26.
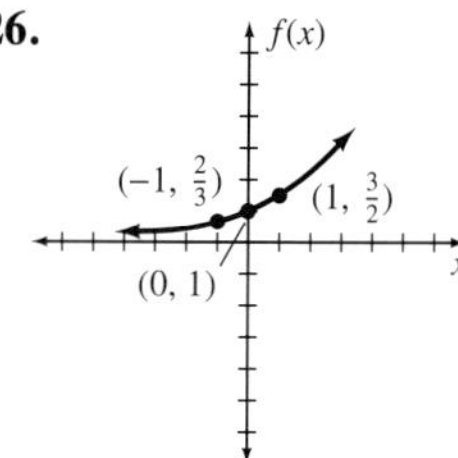

27.
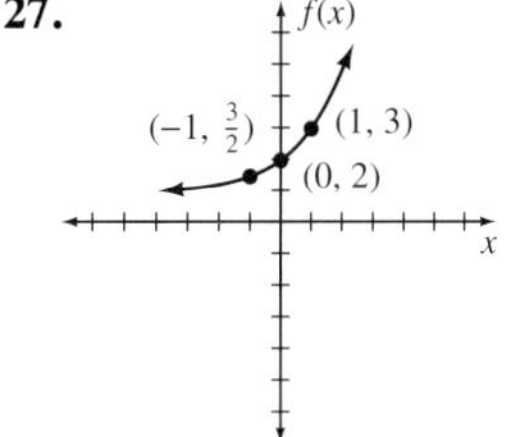

28.
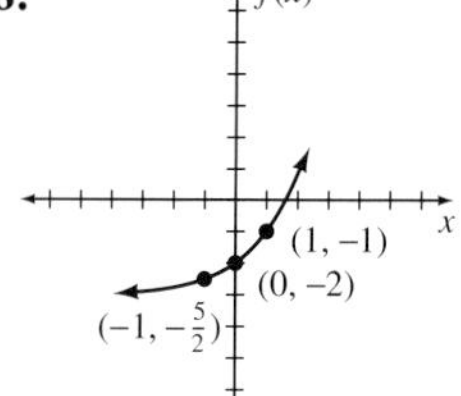

29.

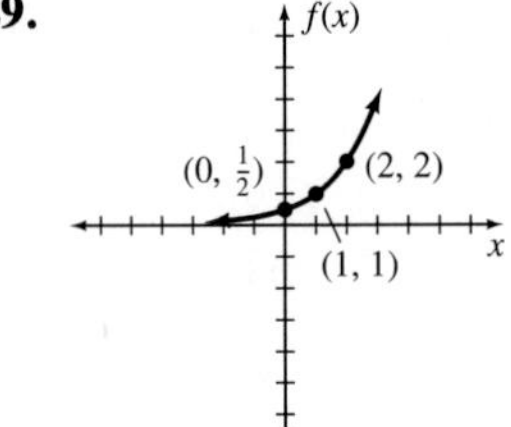

30.

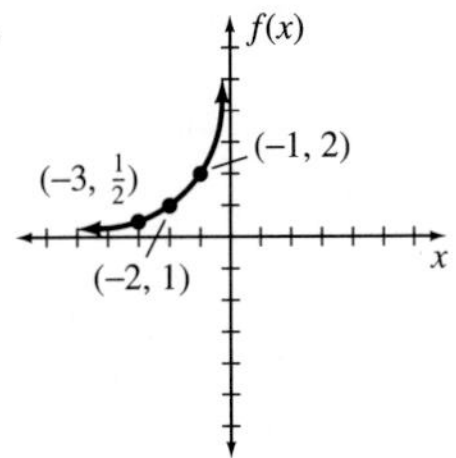

31.

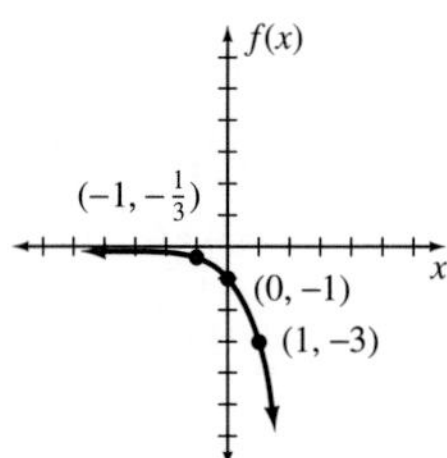

32.

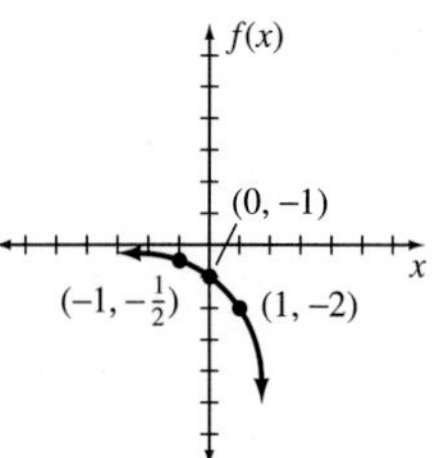

33.

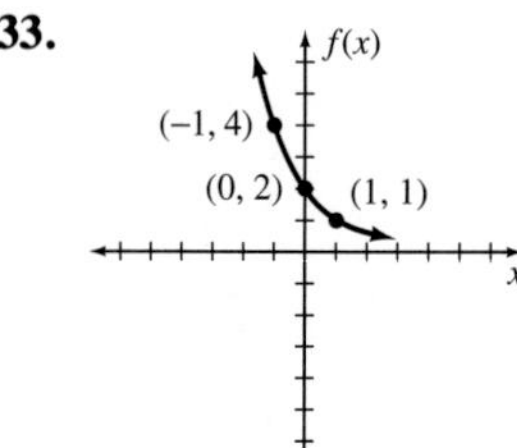

34.

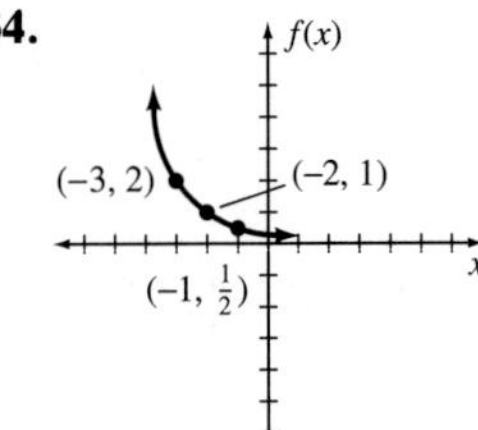

35.

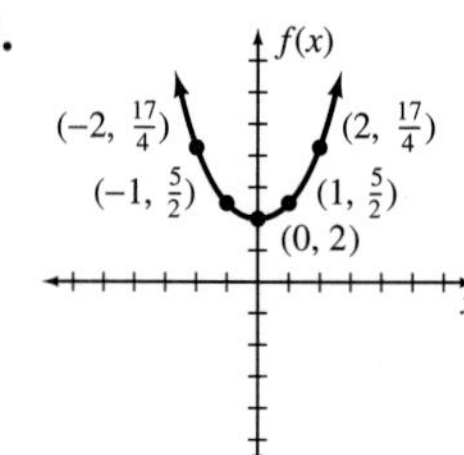

36.

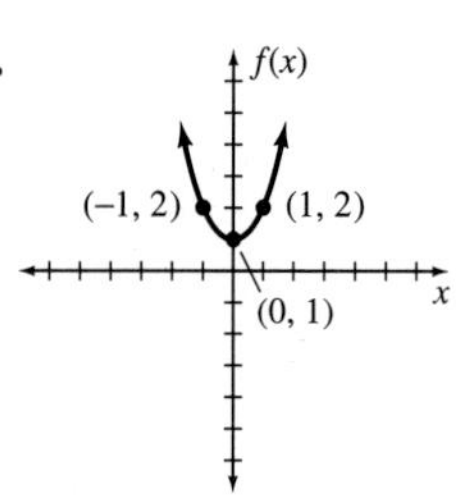

37.

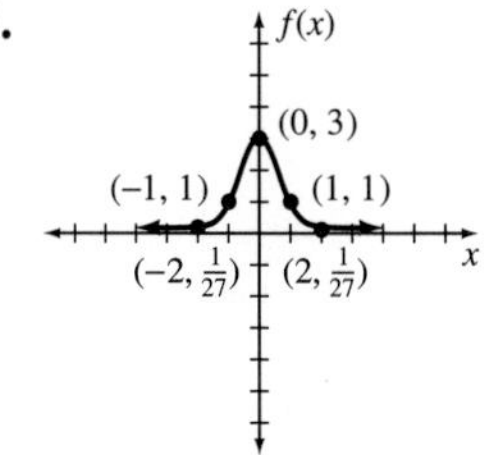

38.

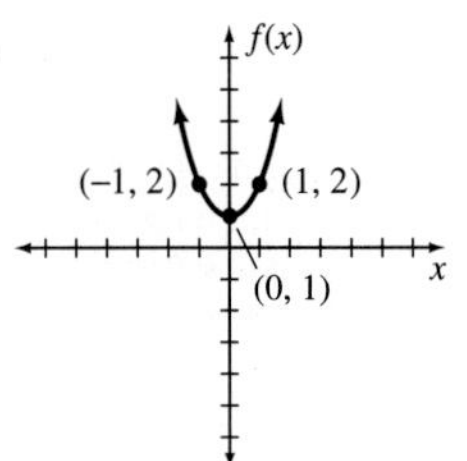

39.

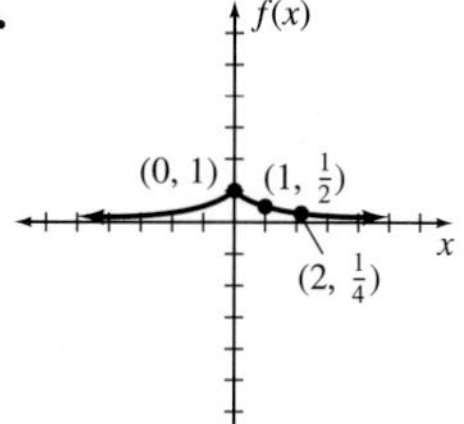

40.

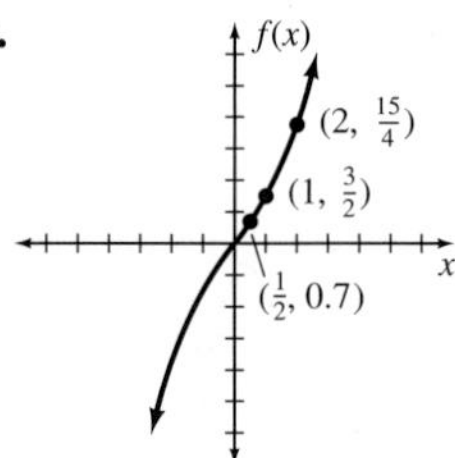

41.

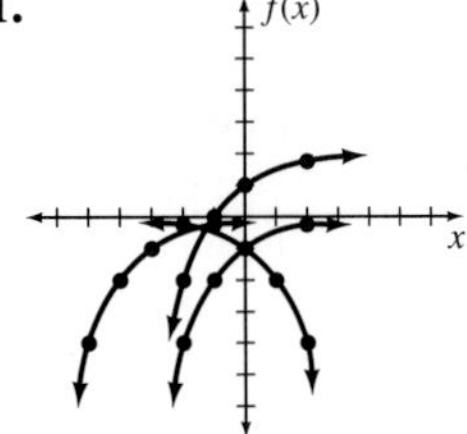

50. **(a)** {4.25} **(b)** {3.56} **(c)** {2.78} **(d)** {2.97} **(e)** {10.55} **(f)** {4.21}

Problem Set 5.2 (page 310)

1. **(a)** \$.67 **(b)** \$4.81 **(c)** \$2.31 **(d)** \$.87 **(e)** \$12,623 **(f)** \$85,909 **(g)** \$803 **2.** **(a)** \$3686 **(b)** \$3389 **(c)** \$2096 **(d)** \$7985 **3.** \$384.66 **4.** \$726.66 **5.** \$480.31 **6.** \$1193.98 **7.** \$2479.35 **8.** \$2850.10 **9.** \$1816.70 **10.** \$2561.15 **11.** \$1356.59 **12.** \$5823.12 **13.** \$22,553.65 **14.** \$100,104.83 **15.** \$567.63 **16.** \$760.98 **17.** \$1422.36 **18.** \$2459.60 **19.** \$8963.38 **20.** \$45,125.07 **21.** \$17,547.35 **22.** \$100,996.42 **23.** \$32,558.88 **24.** 11.8% **25.** 5.9% **26.** 7.8% **27.** 8.06% **28.** 7% compounded monthly **29.** 8.25% compounded quarterly **30.** 625 milligrams; 442 milligrams **31.** 50 grams; 37 grams **32.** 354 grams **33.** 2226; 3320; 7389 **34.** 8244; 22,408; 100,428 **35.** 2000 **36.** 203; 27; 1 **37.** **(a)** 6.5 pounds per square inch **(b)** 11.9 pounds per square inch **(c)** 13.6 pounds per square inch **(d)** 14.1 pounds per square inch **38.** **(a)** 82,888 **(b)** 87,138 **(c)** 96,302

39.

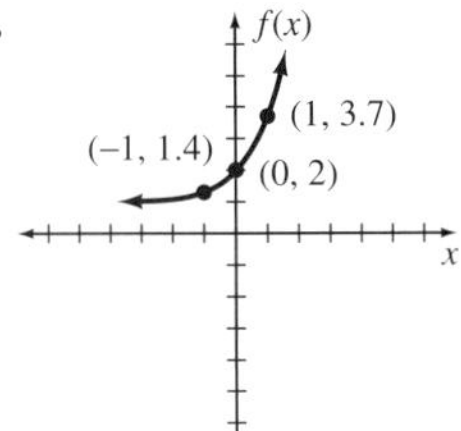

40.

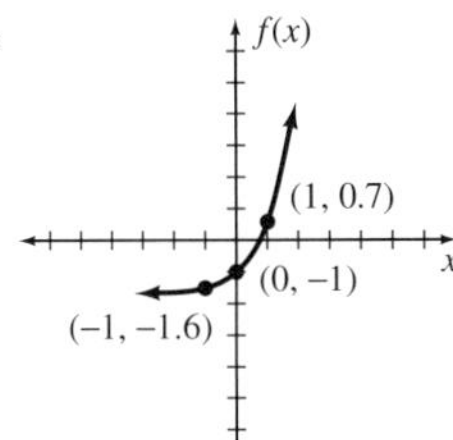

41.

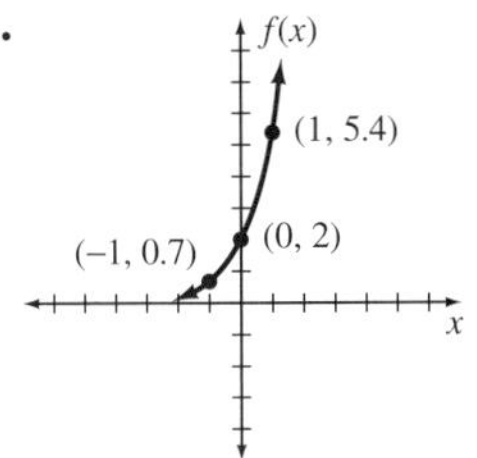

42.

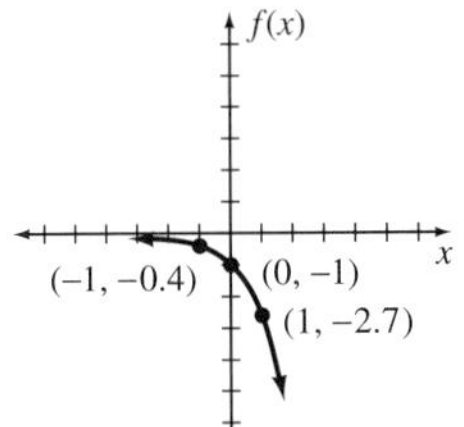

43.

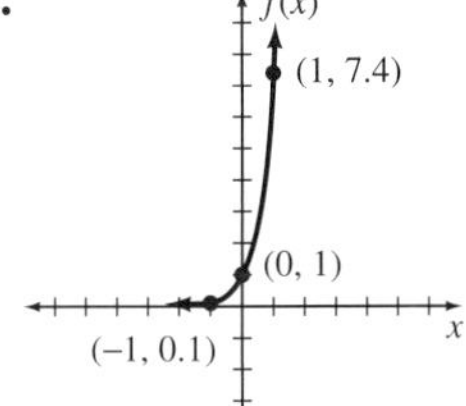

44.

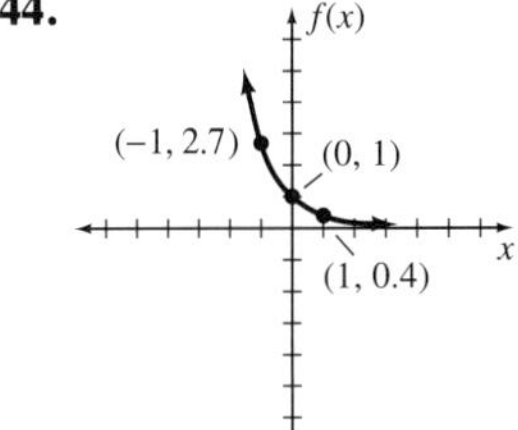

45.

	8%	10%	12%	14%
5 years	$1492	1649	1822	2014
10 years	2226	2718	3320	4055
15 years	3320	4482	6050	8166
20 years	4953	7389	11,023	16,445
25 years	7389	12,182	20,086	33,115

46.

	1 yr	5 yrs	10 yrs	20 yrs
Compounded annually	$1120	1762	3106	9646
Compounded semiannually	1124	1791	3207	10286
Compounded quarterly	1126	1806	3262	10641
Compounded monthly	1127	1817	3300	10893
Compounded continuously	1127	1822	3320	11023

47.

	8%	10%	12%	14%
Compounded annually	$2159	2594	3106	3707
Compounded semiannually	2191	2653	3207	3870
Compounded quarterly	2208	2685	3262	3959
Compounded monthly	2220	2707	3300	4022
Compounded continuously	2226	2718	3320	4055

48.

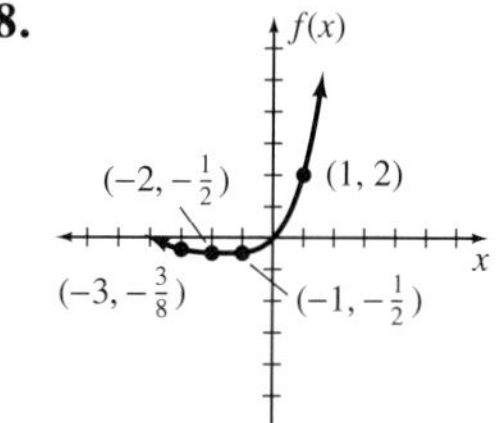

49.

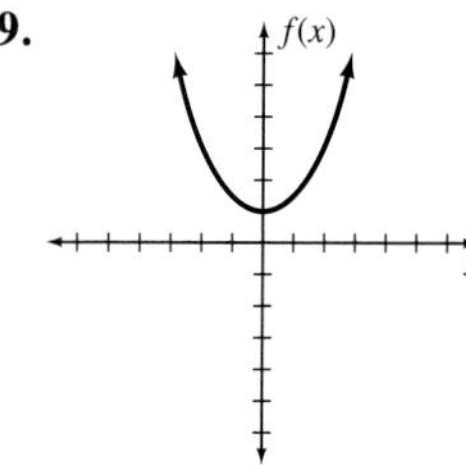

50.

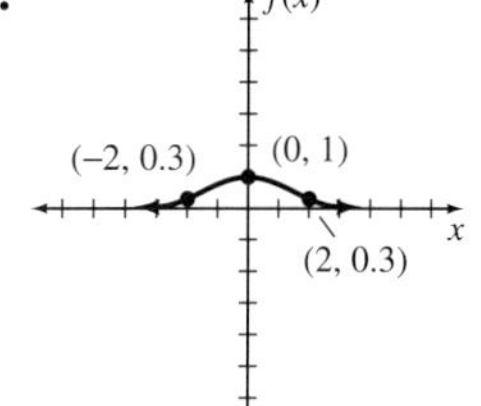

51.

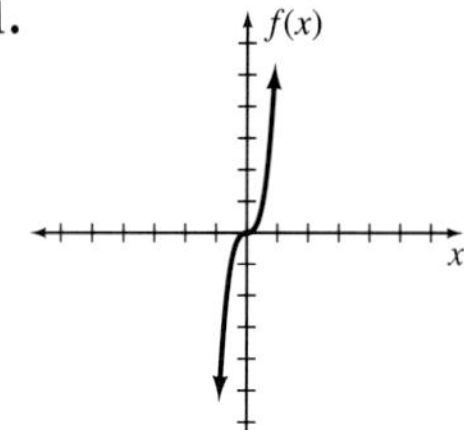

52.

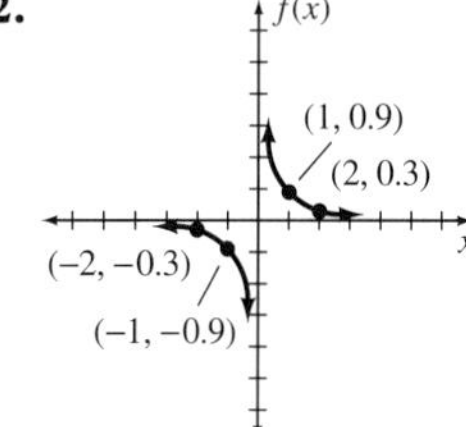

58. **(a)** {1.95} **(b)** {3.04} **(c)** {3.97}
(d) {3.40} **(e)** {4.01} **(f)** {7.70}

Problem Set 5.3 (page 321)

1. $\log_3 9 = 2$ **2.** $\log_2 32 = 5$
3. $\log_5 125 = 3$ **4.** $\log_{10} 10 = 1$
5. $\log_2\left(\frac{1}{16}\right) = -4$ **6.** $\log_{2/3}\left(\frac{27}{8}\right) = -3$

7. $\log_{10} 0.01 = -2$ **8.** $\log_{10} 100{,}000 = 5$
9. $2^6 = 64$ **10.** $3^3 = 27$ **11.** $10^{-1} = 0.1$
12. $5^{-2} = \frac{1}{25}$ **13.** $2^{-4} = \frac{1}{16}$
14. $10^{-5} = 0.00001$ **15.** 2 **16.** 5 **17.** -1
18. -3 **19.** 1 **20.** 0 **21.** $\frac{1}{2}$ **22.** $\frac{2}{3}$
23. $\frac{1}{2}$ **24.** $\frac{1}{4}$ **25.** $-\frac{1}{8}$ **26.** $-\frac{2}{3}$ **27.** 7
28. 13 **29.** 0 **30.** 1 **31.** {25}
32. {1000} **33.** {32} **34.** {8} **35.** {9}
36. {4} **37.** {1} **38.** {10} **39.** 5.1293
40. 0.4855 **41.** 6.9657 **42.** 5.6148
43. 1.4037 **44.** 0.7740 **45.** 7.4512
46. 5.8074 **47.** 6.3219 **48.** 1.9271
49. -0.3791 **50.** 1.5480 **51.** 0.5766
52. 0.5160 **53.** 2.1531 **54.** 2.7740
55. 0.3949 **56.** 0.7582
57. $\log_b x + \log_b y + \log_b z$ **58.** $2 \log_b x - \log_b y$
59. $2 \log_b x + 3 \log_b y$ **60.** $\frac{2}{3} \log_b x + \frac{3}{4} \log_b y$
61. $\frac{1}{2} \log_b x + \frac{1}{2} \log_b y$ **62.** $\frac{2}{3} \log_b x + \frac{1}{3} \log_b z$
63. $\frac{1}{2} \log_b x - \frac{1}{2} \log_b y$ **64.** $\frac{3}{2} \log_b x - \frac{1}{2} \log_b y$
65. $\log_b\left(\frac{xy}{z}\right)$ **66.** $\log_b\left(\frac{x^2}{y^4}\right)$ **67.** $\log_b\left(\frac{x}{yz}\right)$
68. $\log_b\left(\frac{xz}{y}\right)$ **69.** $\log_b(x\sqrt{y})$ **70.** $\log_b\left(\frac{x^2y^4}{z^3}\right)$
71. $\log_b\left(\frac{x^2\sqrt{x-1}}{(2x+5)^4}\right)$ **72.** $\log_b\left(\frac{y^4\sqrt{x}}{x^3}\right)$
73. $\left\{\frac{9}{4}\right\}$ **74.** $\left\{\frac{7}{5}\right\}$ **75.** {25} **76.** {5}
77. {4} **78.** {3} **79.** $\left\{\frac{19}{8}\right\}$ **80.** {38}
81. {9} **82.** {4} **83.** {1} **84.** {8}

Problem Set 5.4 (page 329)

1. 0.8597 **2.** 0.3118 **3.** 1.7179
4. 2.9169 **5.** 3.5071 **6.** 4.1520
7. -0.1373 **8.** -1.3589 **9.** -3.4685
10. -4.1612 **11.** 411.43 **12.** 33.597
13. 90095 **14.** 8580.3 **15.** 79.543
16. 3.5620 **17.** 0.048440 **18.** 0.71581
19. 0.0064150 **20.** 0.0022172 **21.** 1.6094
22. 2.8904 **23.** 3.4843 **24.** 4.3758
25. 6.0638 **26.** 5.9184 **27.** -0.7765
28. -0.6463 **29.** -3.4609 **30.** -4.8107
31. 1.6034 **32.** 2.5633 **33.** 3.1346
34. 15.830 **35.** 108.56 **36.** 20.613
37. 0.48268 **38.** 0.19699 **39.** 0.035994
40. 0.093061

41.

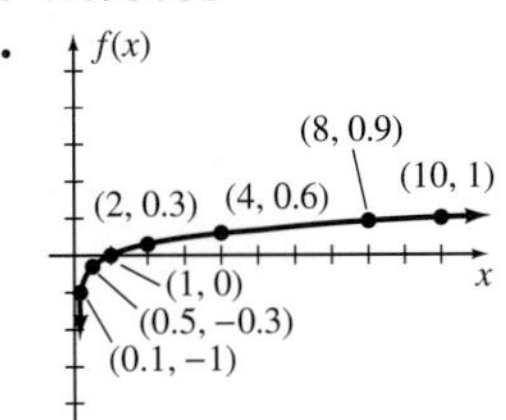

42.

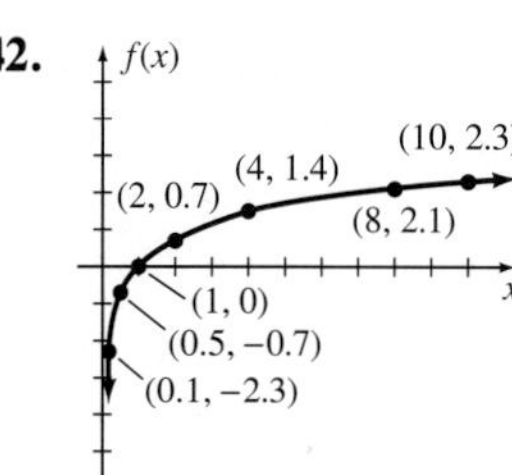

43.

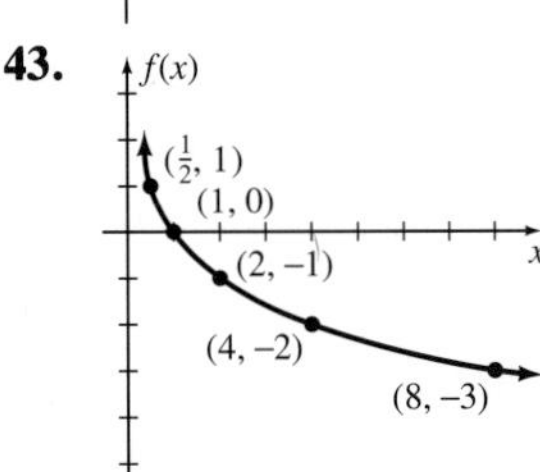

44. This graph is the same as Figure 5.6 in the text.

45. f(x); (3, 1); (9, 2); (1, 0); (⅓, −1); x

46.

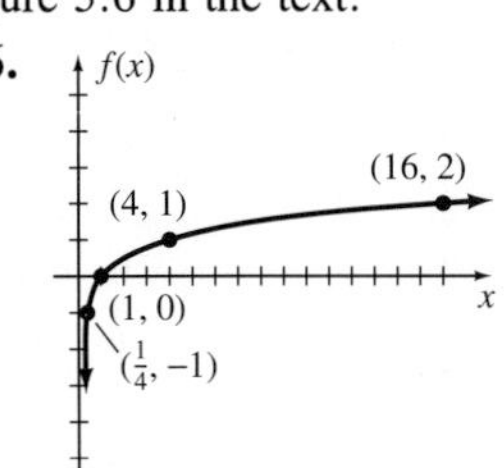

47. f(x); (8, 6); (2, 4); (4, 5); (1, 3); (½, 2); x

48.

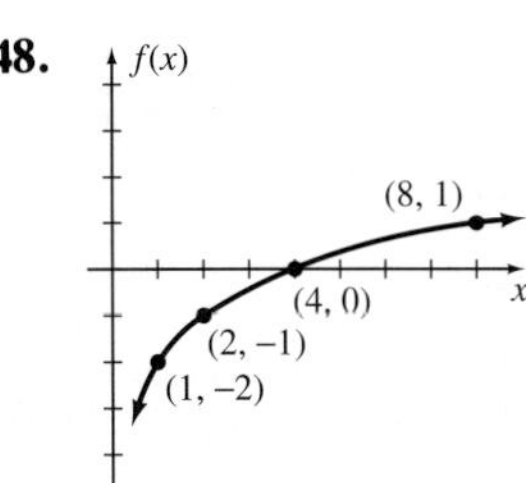

49. f(x); (5, 3); (−1, 1); (1, 2); (−2, 0); x

50.

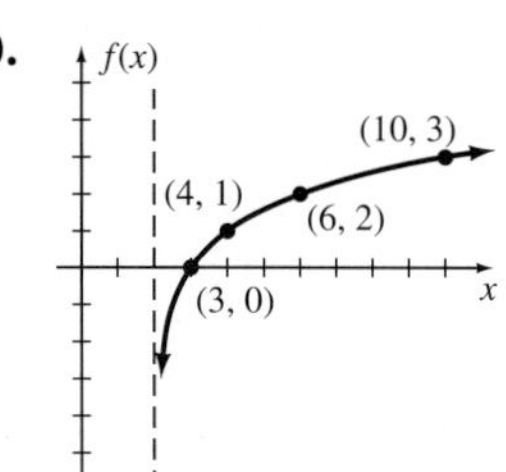

51.

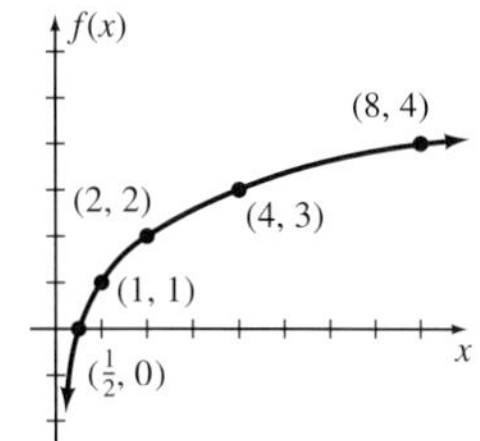

52.

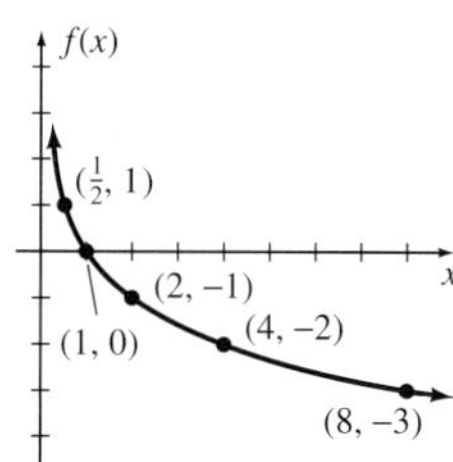

53.

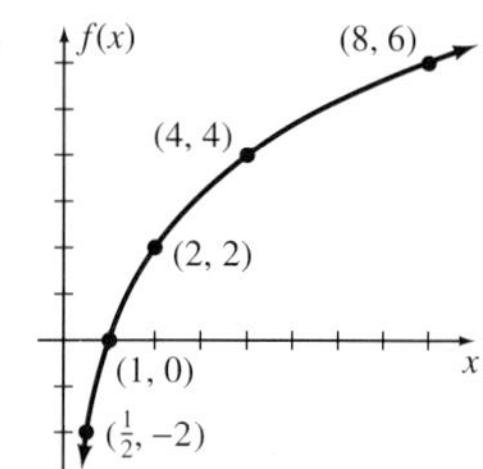

Problem Set 5.5 (page 339)

1. {3.17} **2.** {2.73} **3.** {2.99} **4.** {1.79}
5. {1.81} **6.** {4.18} **7.** {1.41} **8.** {0.12}
9. {1.41} **10.** {3.40} **11.** {3.10}
12. {4.57} **13.** {1.82} **14.** {1.79}
15. {7.84} **16.** {3.32} **17.** {10.32}
18. {2.44} **19.** {2} **20.** {4} **21.** $\left\{\frac{29}{8}\right\}$
22. $\left\{\frac{19}{47}\right\}$ **23.** $\left\{\frac{-1+\sqrt{65}}{2}\right\}$
24. $\left\{\frac{-1+\sqrt{33}}{4}\right\}$ **25.** $\{\sqrt{2}\}$ **26.** {1}
27. {6} **28.** {8} **29.** {1, 100}
30. {1, 10000} **31.** 2.402 **32.** 3.277
33. 0.461 **34.** −0.594 **35.** 2.657
36. 2.736 **37.** 1.211 **38.** −1.917
39. 7.9 years **40.** 2.4 years **41.** 12.2 years
42. 5.3 years **43.** 11.8% **44.** 5.9%
45. 6.6 years **46.** 6.8 hours **47.** 1.5 hours
48. 6100 feet **49.** 34.7 years **50.** 3.5 hours
51. 6.7
52. Approximately 200 million times the reference intensity
53. Approximately 8 times
54. Approximately 500 times **61.** {1.13}
62. $x = \left(y \pm \sqrt{y^2-1}\right)$ **63.** $x = \ln\left(y + \sqrt{y^2+1}\right)$

Chapter 5 Review Problem Set (page 344)

1. 32 **2.** −125 **3.** 81 **4.** 3 **5.** −2
6. $\frac{1}{3}$ **7.** $\frac{1}{4}$ **8.** −5 **9.** 1 **10.** 12
11. {5} **12.** $\left\{\frac{1}{9}\right\}$ **13.** $\left\{\frac{7}{2}\right\}$ **14.** {3.40}
15. {8} **16.** $\left\{\frac{1}{11}\right\}$ **17.** {1.95} **18.** {1.41}
19. {1.56} **20.** {20} **21.** $\{10^{100}\}$ **22.** {2}
23. $\left\{\frac{11}{2}\right\}$ **24.** {0} **25.** 0.3680 **26.** 1.3222
27. 1.4313 **28.** 0.5634
29. **(a)** $\log_b x - 2\log_b y$ **(b)** $\frac{1}{4}\log_b x + \frac{1}{2}\log_b y$
(c) $\frac{1}{2}\log_b x - 3\log_b y$ **30.** **(a)** $\log_b x^3y^2$
(b) $\log_b\left(\frac{\sqrt{y}}{x^4}\right)$ **(c)** $\log_b\left(\frac{\sqrt{xy}}{z^2}\right)$ **31.** 1.58
32. 0.63 **33.** 3.79 **34.** −2.12

35.

36.

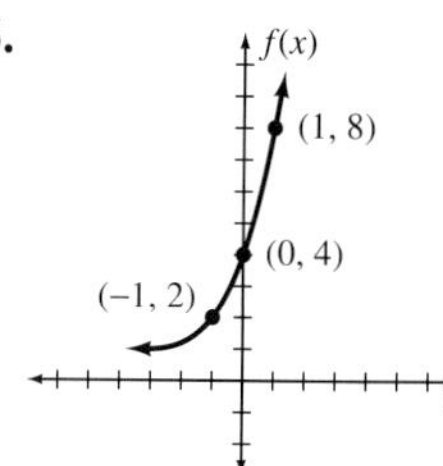

37.

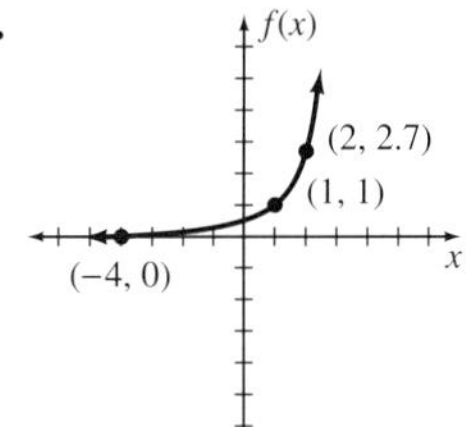

38.

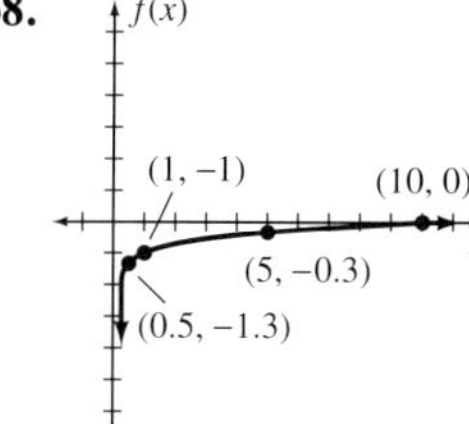

39.

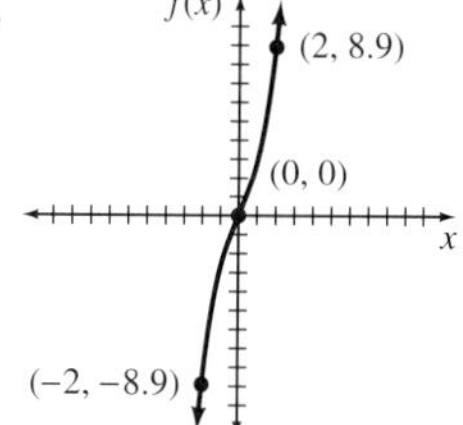

40.

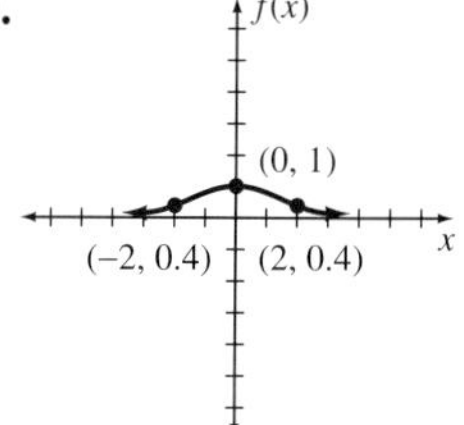

41. 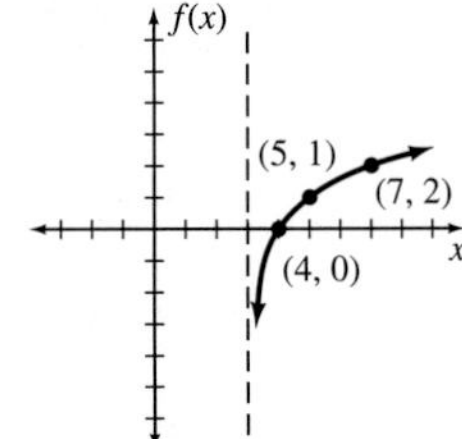

42. 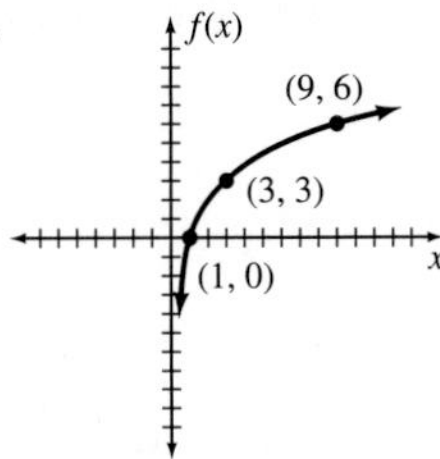

43. \$2219.91 **44.** \$4797.55 **45.** \$15,999.31
46. Approximately 5.3 years
47. Approximately 12.1 years
48. Approximately 8.7% **49.** 61,070; 67,493; 74,591
50. Approximately 4.8 hours **51.** 133 grams
52. 8.1

CHAPTER 6

Problem Set 6.1 (page 353)

1. Q: $4x + 5$
R: 0
2. Q: $4x - 3$
R: 0
3. Q: $t^2 + 2t - 4$
R: 0
4. Q: $t^2 - 3t - 2$
R: 0
5. Q: $2x + 5$
R: 1
6. Q: $4x - 1$
R: 1
7. Q: $3x - 4$
R: $3x - 1$
8. Q: $4x + 7$
R: $23x - 6$
9. Q: $5y - 1$
R: $-8y - 2$
10. Q: $8y - 9$
R: $8y + 5$
11. Q: $4a + 6$
R: $7a - 19$
12. Q: $5a - 8$
R: $42a - 41$
13. Q: $3x + 4y$
R: 0
14. Q: $4a - 4b$
R: 0
15. Q: $3x + 4$
R: 0
16. Q: $2x + 1$
R: 0
17. Q: $x + 6$
R: 14
18. Q: $x - 2$
R: -1
19. Q: $4x - 3$
R: 2
20. Q: $5x - 2$
R: 0
21. Q: $x^2 - 1$
R: 0
22. Q: $x^2 - 6x + 8$
R: 0
23. Q: $3x^3 - 4x^2 + 6x - 13$
R: 12
24. Q: $2x^2 - x - 6$
R: -6
25. Q: $x^2 - 2x - 3$
R: 0
26. Q: $x^2 + 7x + 2$
R: 1
27. Q: $x^3 + 7x^2 + 21x + 56$
R: 167
28. Q: $2x^3 - 4x^2 + 11x - 22$
R: 47
29. Q: $x^2 + 3x + 2$
R: 0
30. Q: $x^2 + x - 6$
R: 0
31. Q: $x^4 + x^3 + x^2 + x + 1$
R: 0
32. Q: $x^4 - x^3 + x^2 - x + 1$
R: -2
33. Q: $x^4 + x^3 + x^2 + x + 1$
R: 2
34. Q: $x^4 - x^3 + x^2 - x + 1$
R: 0
35. Q: $2x^2 + 2x - 3$
R: $\frac{9}{2}$
36. Q: $9x^2 - 3x + 2$
R: $-\frac{10}{3}$
37. Q: $4x^3 + 2x^2 - 4x - 2$
R: 0
38. Q: $3x^3 - 3x^2 + 6x - 3$
R: 0

Problem Set 6.2 (page 358)

1. $f(2) = -2$ **2.** $f(-1) = -2$
3. $f(-4) = -105$ **4.** $f(1) = 9$ **5.** $f(-2) = 9$
6. $f(2) = 23$ **7.** $f(6) = 74$ **8.** $f(-2) = -65$
9. $f(3) = 200$ **10.** $f(-3) = -179$
11. $f(-1) = 5$ **12.** $f(4) = 64$ **13.** $f(7) = -5$
14. $f(-5) = -874$ **15.** $f(-2) = -27$
16. $f(3) = 8751$ **17.** $f\left(\frac{1}{2}\right) = -2$
18. $f\left(-\frac{1}{3}\right) = -5$ **19.** Yes **20.** Yes
21. Yes **22.** Yes **23.** No **24.** No
25. Yes **26.** Yes **27.** Yes **28.** Yes
29. $(x + 2)(x + 6)(x - 1)$
30. $(x - 1)(3x - 2)(x + 8)$
31. $(x - 3)(2x - 1)(3x + 2)$
32. $(x + 2)(4x - 1)(3x + 2)$ **33.** $(x + 1)^2(x - 4)$
34. $(x - 5)(2x - 1)(x + 6)$ **35.** $k = 6$
36. $k = 1$ or $k = -4$ **37.** $k = -30$ **38.** $k = 6$
39. Let $f(x) = x^{12} - 4096$; then $f(-2) = 0$; therefore, $x + 2$ is a factor of $f(x)$.
40. $f(c) > 0$ for all values of c.
41. Let $f(x) = x^n - 1$. Since $1^n = 1$ for all positive integral values of n, then $f(1) = 0$ and $x - 1$ is a factor.
42. Let $f(x) = x^n - 1$. Since $(-1)^n = 1$ for all even positive integral values of n, then $f(-1) = 0$ and $x - (-1) = x + 1$ is a factor.

43. **(a)** Let $f(x) = x^n - y^n$. Therefore, $f(y) = y^n - y^n = 0$ and $x - y$ is a factor of $f(x)$.
(b) Let $f(x) = x^n - y^n$. Therefore, $f(-y) = (-y)^n - y^n = y^n - y^n = 0$ when n is even, and $x - (-y) = x + y$ is a factor of $f(x)$.
(c) Let $f(x) = x^n + y^n$. Therefore, $f(-y) = (-y)^n + y^n = -y^n + y^n = 0$ when n is odd, and $x - (-y) = x + y$ is a factor of $f(x)$.

44. $f(c) = 6 + i$ **45.** $f(1 + i) = 2 + 6i$

46. $f(2 - 3i) = -56 - 36i$

49. **(a)** $f(4) = 137; f(-5) = 11; f(7) = 575$
(b) $f(3) = 11; f(6) = 272; f(-7) = -859$
(c) $f(4) = -79; f(5) = -162; f(-3) = 110$
(d) $f(5) = 975; f(6) = 1901; f(-3) = -34$

Problem Set 6.3 (page 369)

1. $\{-2, -1, 2\}$ **2.** $\{-3, 1, 4\}$ **3.** $\left\{-\frac{3}{2}, \frac{1}{3}, 1\right\}$

4. $\left\{-2, -\frac{1}{4}, \frac{5}{2}\right\}$ **5.** $\left\{-7, \frac{2}{3}, 2\right\}$

6. $\left\{-1, -\frac{1}{3}, \frac{2}{5}\right\}$ **7.** $\{-1, 4\}$ **8.** $\{-3, 2\}$

9. $\{-3, 1, 2, 4\}$ **10.** $\{-3, -2, -1, 2\}$

11. $\{-2, 1 \pm \sqrt{7}\}$ **12.** $\{2, 1 \pm \sqrt{5}\}$

13. $\left\{-\frac{2}{3}, 1, \pm\sqrt{2}\right\}$ **14.** $\left\{-\frac{5}{2}, 1, \pm\sqrt{3}\right\}$

15. $\left\{-\frac{4}{3}, 0, \frac{1}{2}, 3\right\}$ **16.** $\{1, \pm i\}$

17. $\{-1, 2, 1 \pm i\}$ **18.** $\{-2, 3, -1 \pm 2i\}$

19. $\left\{-1, \frac{3}{2}, 2, \pm i\right\}$ **20.** $\left\{-2, \frac{1}{2}\right\}$

27. **(a)** $\{-4, -2, 1\}$ **(b)** $\{-3, -1, 2\}$
(c) $\left\{-4, -2, \frac{3}{2}\right\}$ **(d)** $\left\{-\frac{5}{2}, \frac{1}{3}, 3\right\}$

28. 1 positive and 1 negative solution

29. 2 positives or 2 nonreal complex solutions

30. 1 positive and 2 nonreal complex solutions

31. 1 negative and 2 nonreal complex solutions

32. 1 negative and 2 positive *or* 1 negative and 2 nonreal complex solutions

33. 1 positive and 2 negative *or* 1 positive and 2 nonreal complex solutions

34. 5 positive *or* 3 positive and 2 nonreal complex solutions *or* 1 positive and 4 nonreal complex solutions

35. 1 negative and 2 positive and 2 nonreal complex solutions *or* 1 negative and 4 nonreal complex solutions

36. 1 negative and 4 nonreal complex solutions

37. 1 positive and 1 negative and 4 nonreal complex solutions

38. $x^3 - 3x^2 - 10x + 24 = 0$

39. $x^4 + 2x^3 - 9x^2 - 2x + 8 = 0$

40. $6x^3 + 5x^2 - 12x + 4 = 0$

41. $12x^3 - 37x^2 - 3x + 18 = 0$

42. $x^5 - 5x^4 + 10x^3 - 10x^2 + 5x - 1 = 0$

43. $x^4 + 12x^3 + 54x^2 + 108x + 81 = 0$

44. $x^3 - 7x^2 + 25x - 39 = 0$

45. $x^3 + 13x + 34 = 0$

46. $x^4 - 2x^3 + 6x^2 - 8x + 8 = 0$

47. $x^4 + 4x^3 + 14x^2 + 4x + 13 = 0$

51. **(a)** Upper bound of 3 and lower bound of -1
(b) Upper bound of 2 and lower bound of -3
(c) Upper bound of 3 and lower bound of -6
(d) Upper bound of 2 and lower bound of -5
(e) Upper bound of 5 and lower bound of -3

52. **(a)** $\{4, -3 \pm i\}$ **(b)** $\{-5, 2 \pm 3i\}$
(c) $\{-2, 6, 1 \pm \sqrt{3}\}$ **(d)** $\{-8, 1, 2 \pm \sqrt{7}\}$
(e) $\{12, 1 \pm i\}$ **(f)** $\{-2, 1\}$

53. **(a)** 1 negative and 2 positive solutions
(b) 1 negative and 2 nonreal complex solutions
(c) 4 nonreal complex solutions
(d) 2 negative and 3 positive solutions
(e) 1 negative, 1 positive, and 2 nonreal complex solutions
(f) 1 positive and 4 nonreal complex solutions
(g) 1 negative and 3 positive solutions
(h) 1 negative and 2 nonreal complex solutions

54. **(a)** 0.28 and 3.73 **(b)** 0.67
(c) 2.27 and 5.76 **(d)** -3.59 and -6.44
(e) 0.71

Problem Set 6.4 (page 381)

1.

f(x)
(2, 5)
x
(0, −3)
(−1, −4)

2.

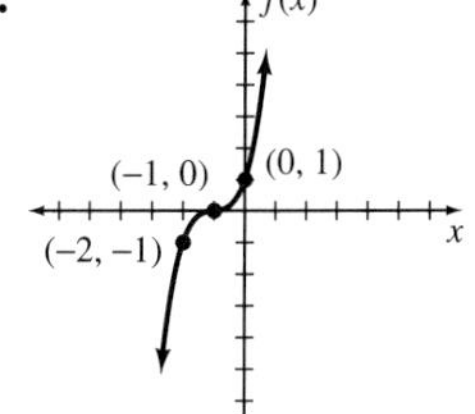

3.

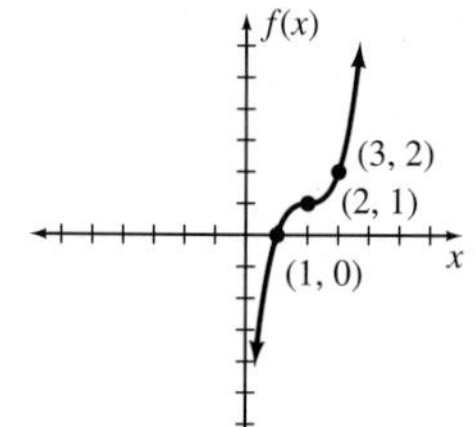

4.

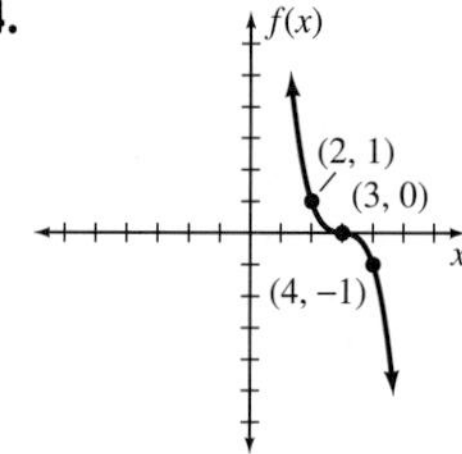

13.

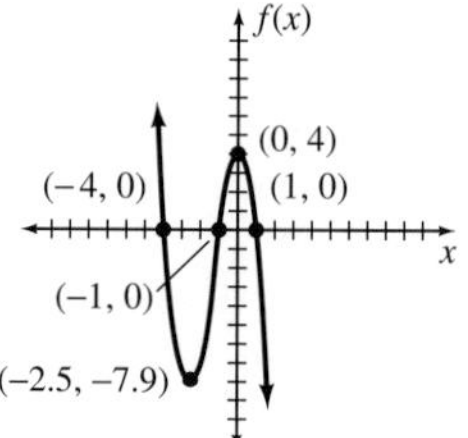

14.

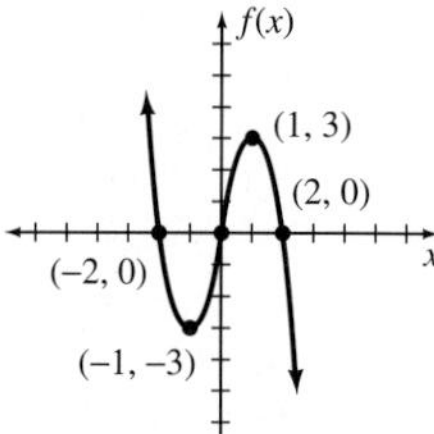

5.

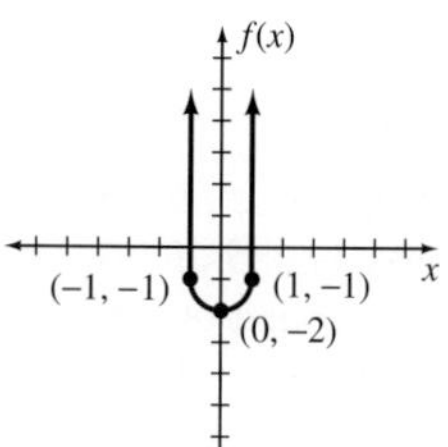

6.

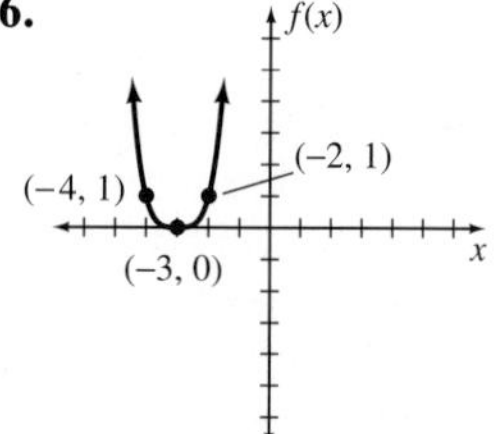

15.

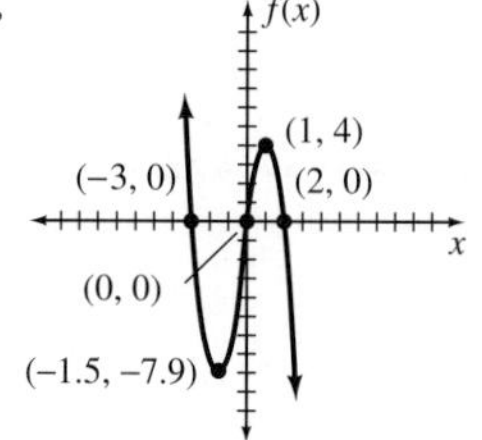

16.

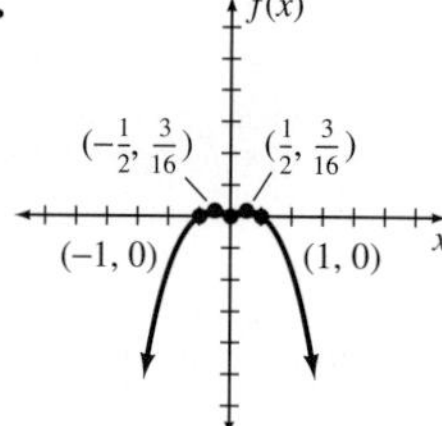

7.

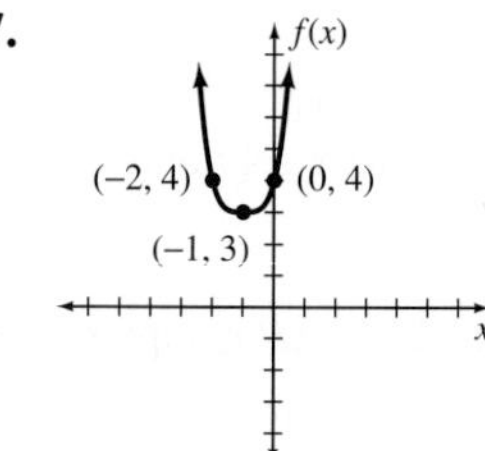

8.

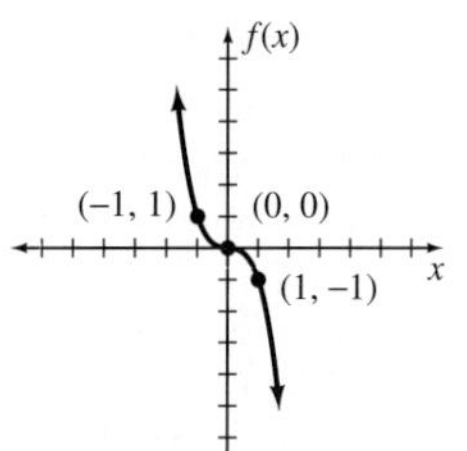

17.

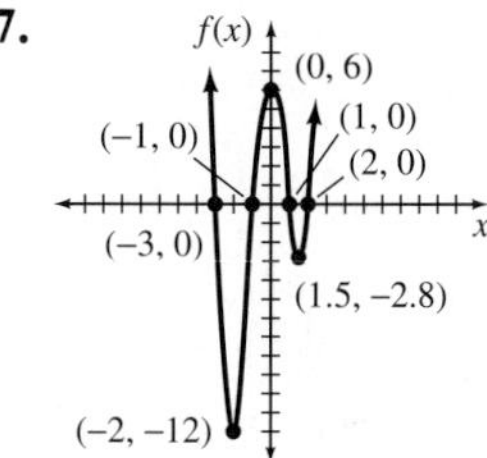

18.

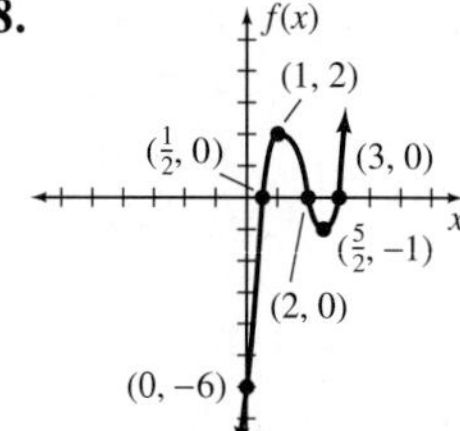

9.

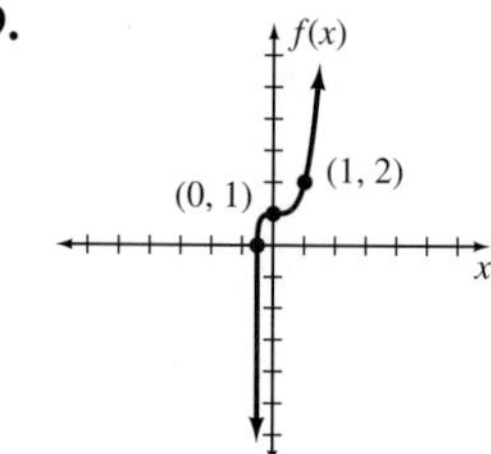

10.

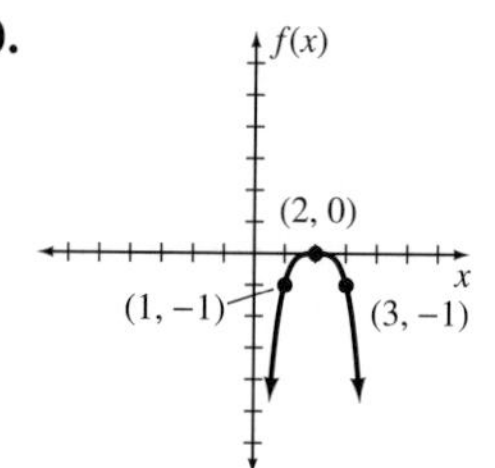

19.

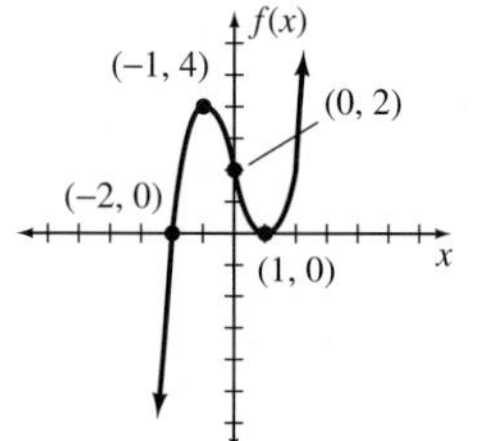

20.

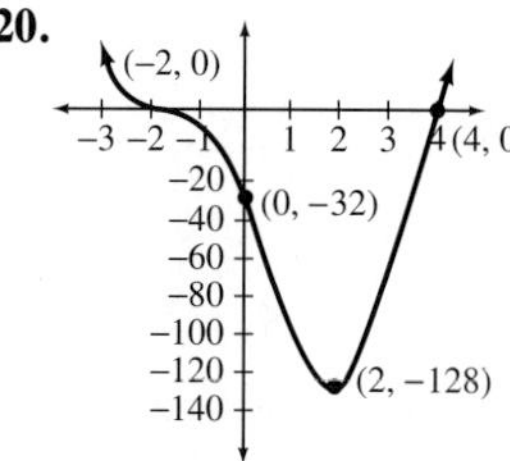

11.

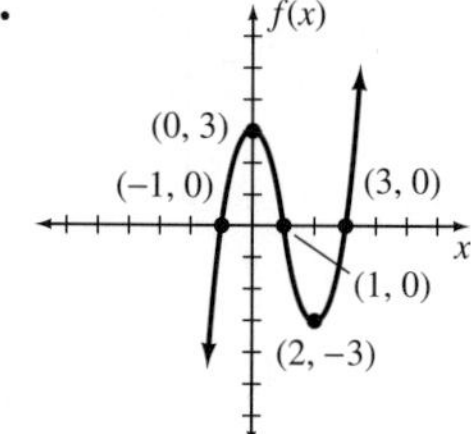

12.

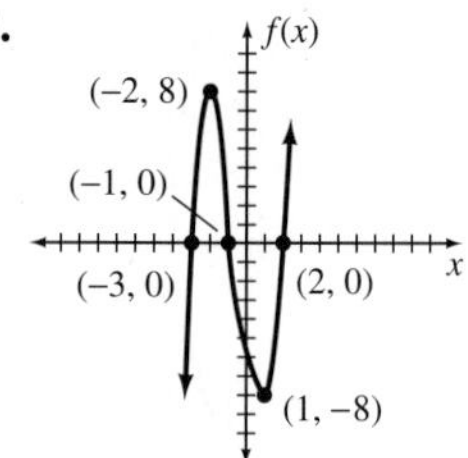

21.

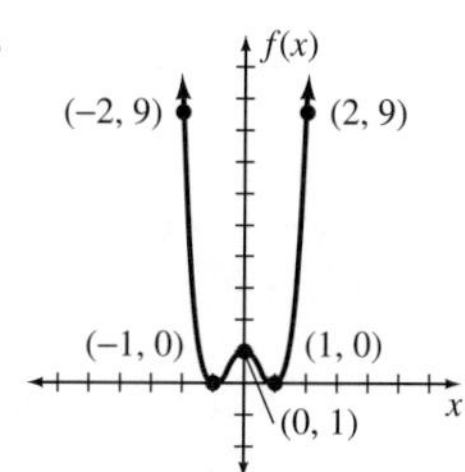

22.

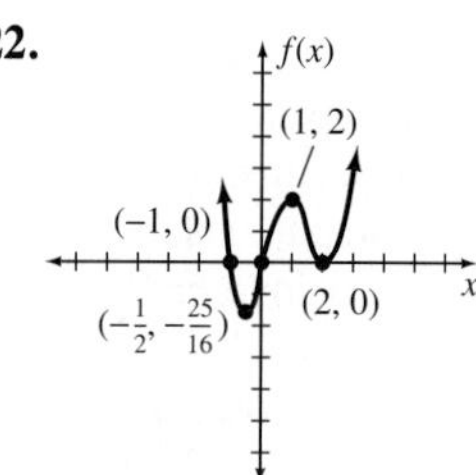

23.

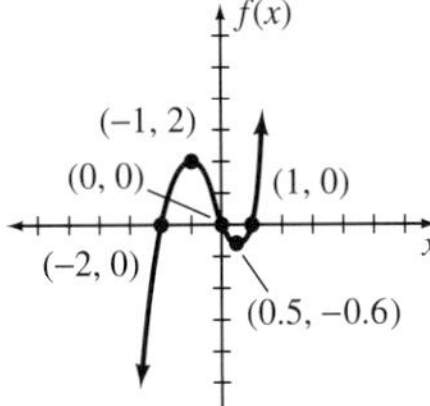

24.

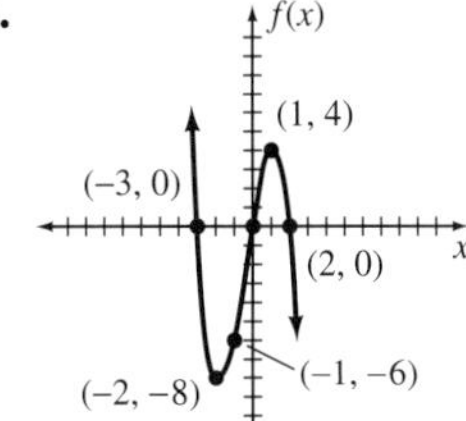

25.

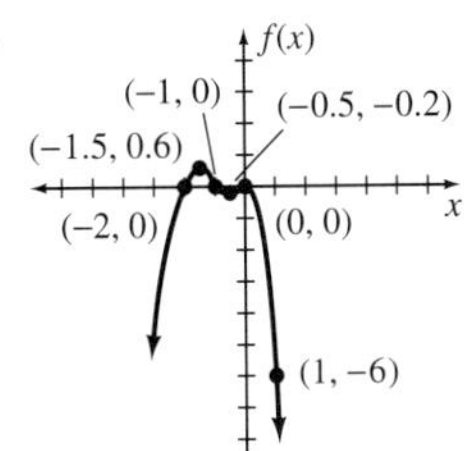

26.

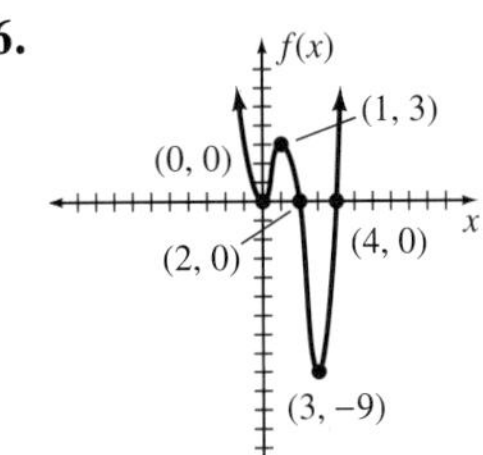

27.

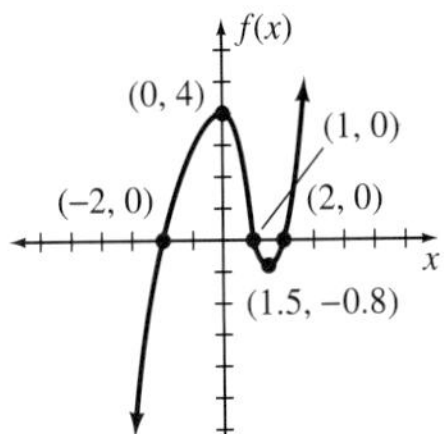

28.

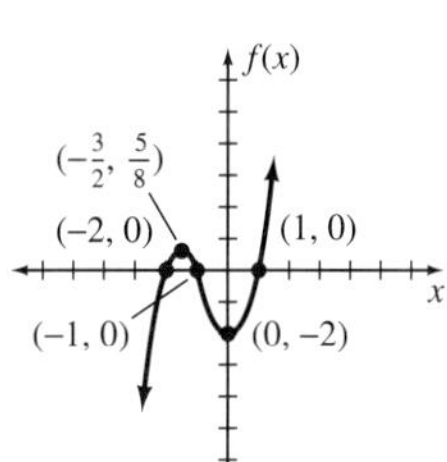

29.

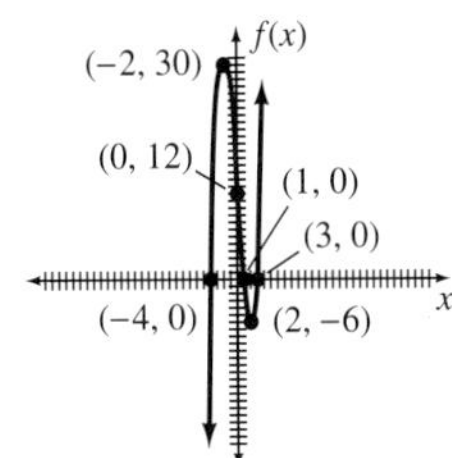

30.

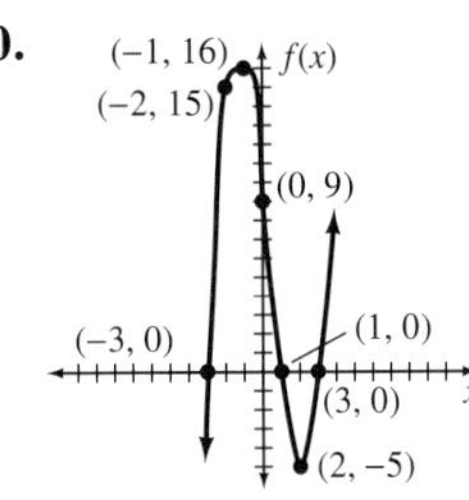

31.

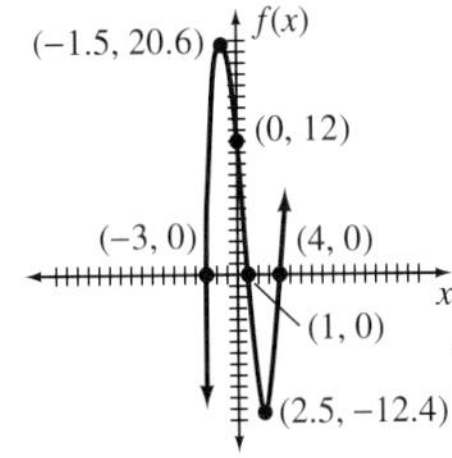

32.

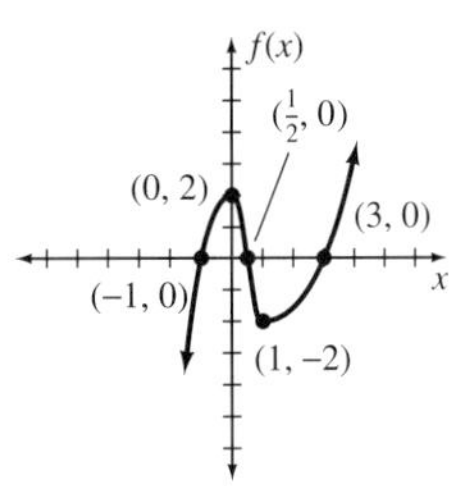

33.

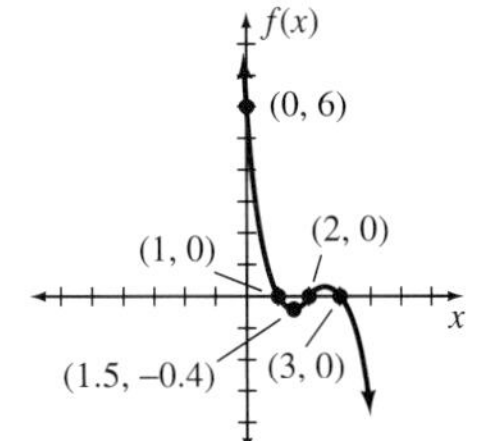

34.

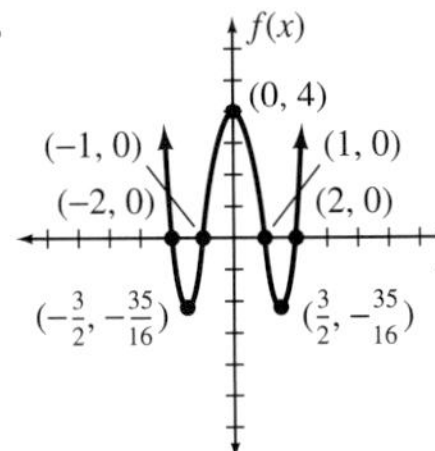

35. **(a)** 60 **(b)** −4, 3, and 5
(c) $f(x) > 0$ for $(-4, 3) \cup (5, \infty)$
$f(x) < 0$ for $(-\infty, -4) \cup (3, 5)$

36. **(a)** −144 **(b)** −3, 6, and 8
(c) $f(x) > 0$ for $(-\infty, -3) \cup (6, 8)$
$f(x) < 0$ for $(-3, 6) \cup (8, \infty)$

37. **(a)** 432 **(b)** −3 and 4
(c) $f(x) > 0$ for $(-3, 4) \cup (4, \infty)$
$f(x) < 0$ for $(-\infty, -3)$

38. **(a)** −81 **(b)** −3 and 1
(c) $f(x) > 0$ for $(1, \infty)$
$f(x) < 0$ for $(-\infty, -3) \cup (-3, 1)$

39. **(a)** 8 **(b)** −2, 1, and 2
(c) $f(x) > 0$ for $(-\infty, -2) \cup (-2, 1) \cup (2, \infty)$
$f(x) < 0$ for $(1, 2)$

40. **(a)** 0 **(b)** −4, 0, and 6
(c) $f(x) > 0$ for $(-\infty, -4) \cup (0, 6) \cup (6, \infty)$
$f(x) < 0$ for $(-4, 0)$

41. **(a)** 512 **(b)** −2 and 4
(c) $f(x) > 0$ for $(-2, 4) \cup (4, \infty)$
$f(x) < 0$ for $(-\infty, -2)$

45. **(a)** 1.6 **(b)** −0.7 **(c)** 4.4 **(d)** 1.8
(e) −1.4 **(f)** 4.9

51. **(a)** −2, 1, and 4; $f(x) > 0$ for $(-2, 1) \cup (4, \infty)$ and $f(x) < 0$ for $(-\infty, -2) \cup (1, 4)$
(b) −1, 1, and 8; $f(x) > 0$ for $(-1, 1) \cup (8, \infty)$ and $f(x) < 0$ for $(-\infty, -1) \cup (1, 8)$
(c) 2 and 3; $f(x) > 0$ for $(3, \infty)$ and $f(x) < 0$ for $(2, 3) \cup (-\infty, 2)$
(d) 1, 6, and 12; $f(x) > 0$ for $(1, 6) \cup (12, \infty)$ and $f(x) < 0$ for $(-\infty, 1) \cup (6, 12)$
(e) −3, −1, and 2; $f(x) > 0$ for $(-\infty, -3) \cup (2, \infty)$ and $f(x) < 0$ for $(-3, -1) \cup (-1, 2)$
(f) −2 and 2; $f(x) > 0$ for $(-\infty, -2) \cup (2, \infty)$ and $f(x) < 0$ for $(-2, 2)$

52. **(a)** (−1, 47) and (2, 20) **(b)** (1, 1079) and (10, 350) **(c)** (−4, −12) and (1, 113)
(d) (−1, −6), (1, 10) and (3, −6)
(e) (8, −4) and (12, −36)
(f) (0, −1) and (2, −38)

53. **(a)** -3, 3; (0.5, 3.1), (-1.8, 10.1)
(b) 2.5, $-.1$, -2.3; (-1.4, 192.2), (1.4, -266.1)
(c) -2.2, 2.2; (-1.4, -8.0), (0.0, -4.0), (1.4, 8.0)

54. 1.7 inches

55. 32 units

Problem Set 6.5 (page 392)

1.

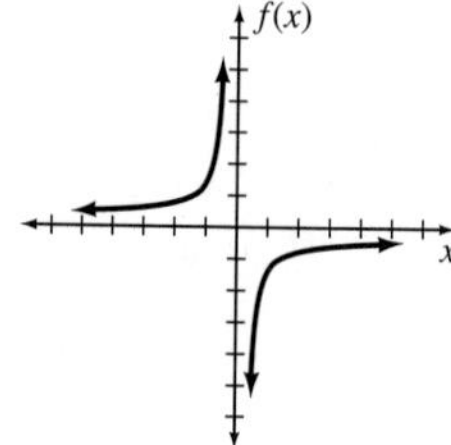

2.

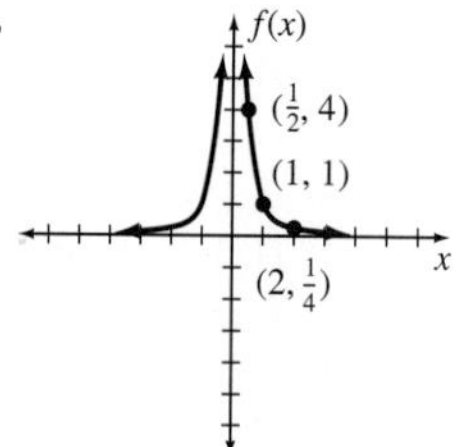

3.

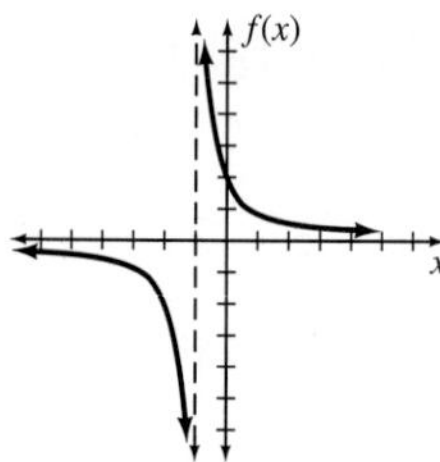

4.

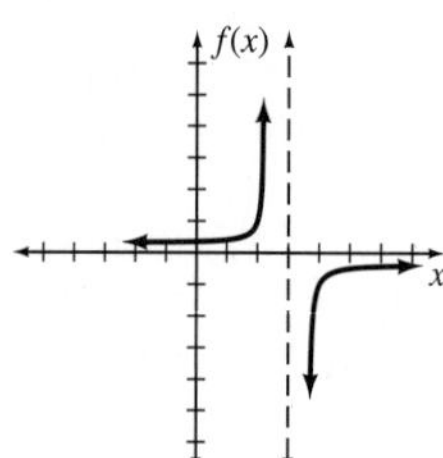

5.

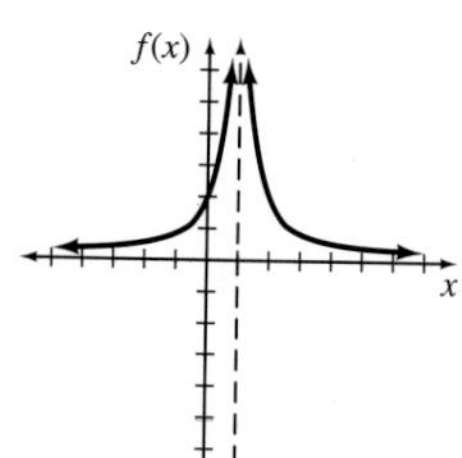

6.

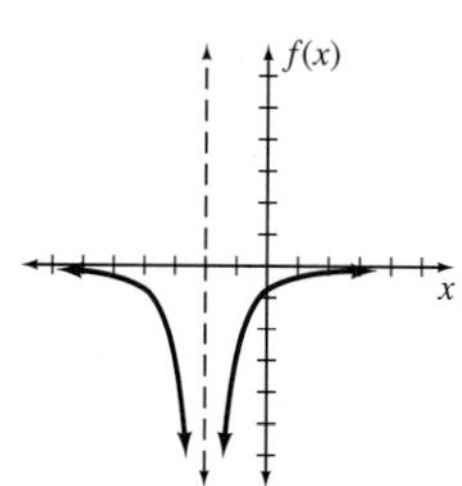

7.

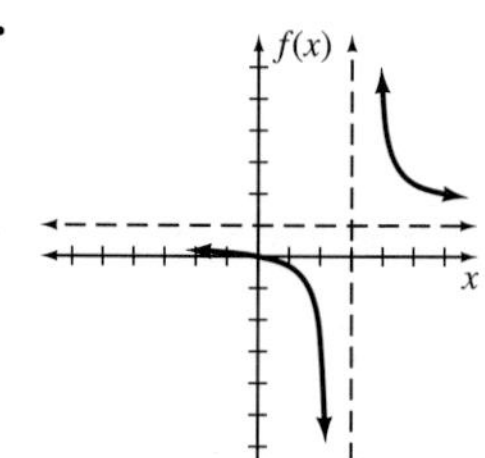

8.

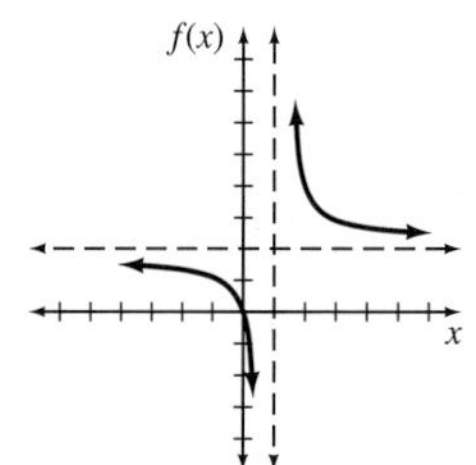

9.

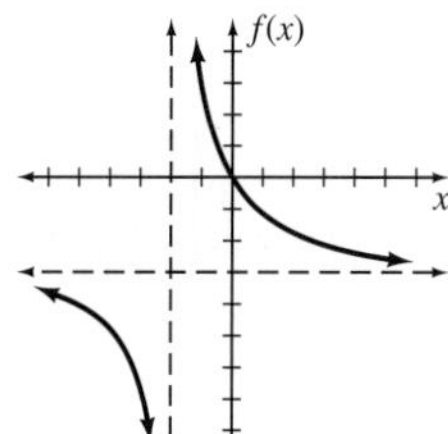

10.

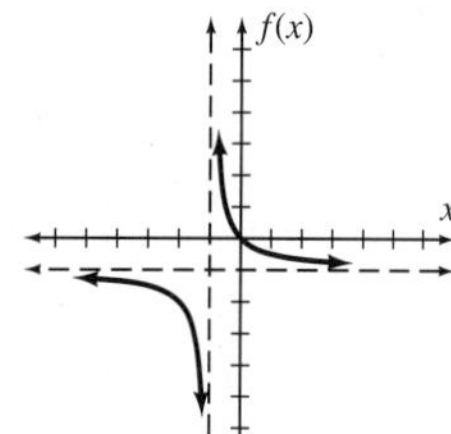

11.

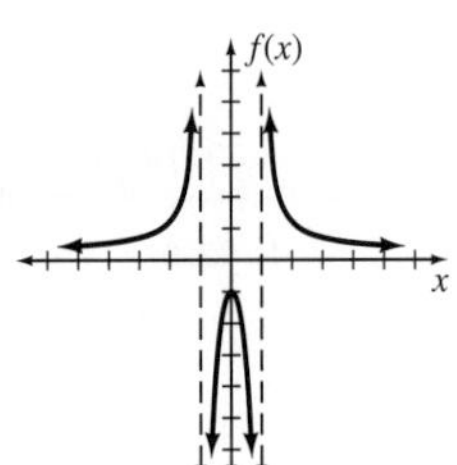

12.

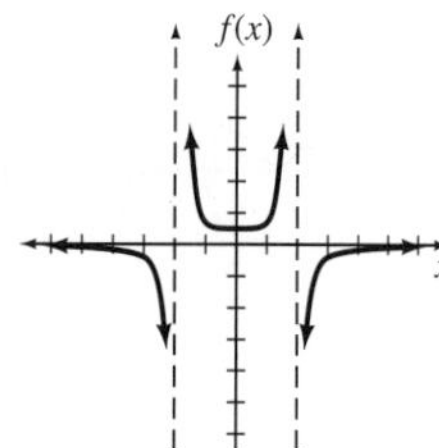

13.

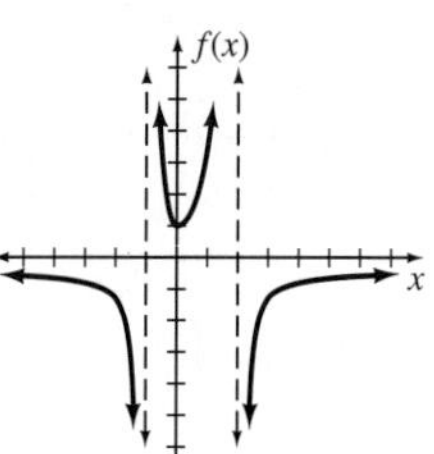

14.

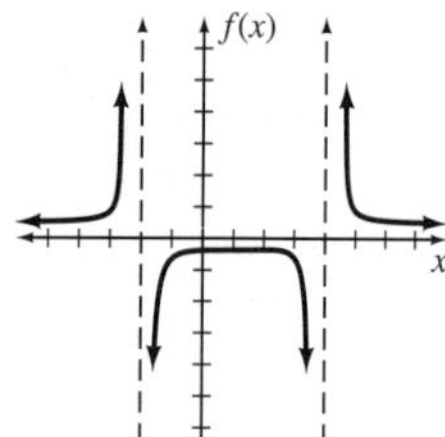

15.

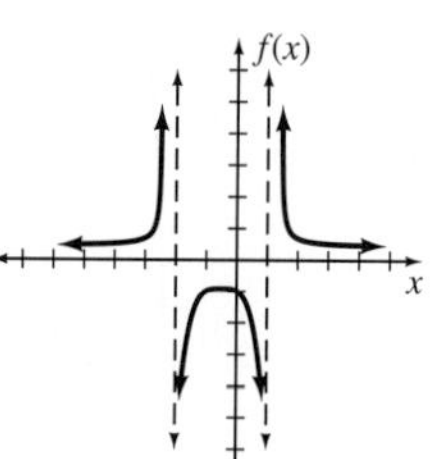

16.

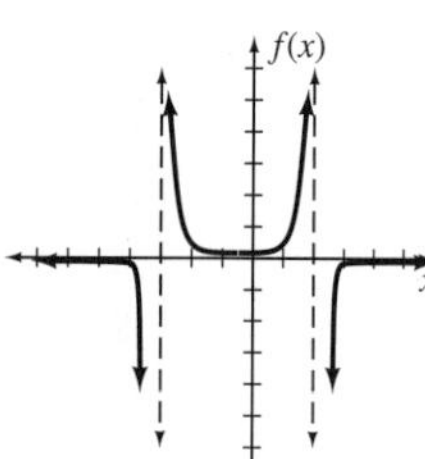

17.

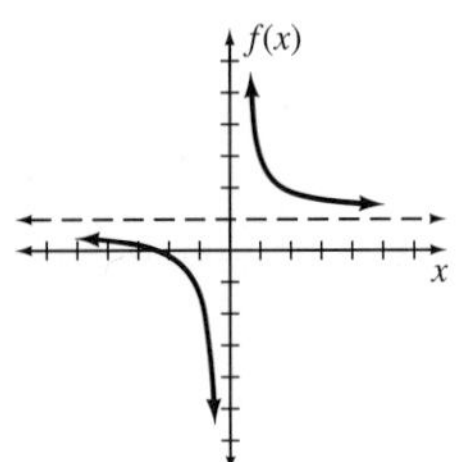

18.

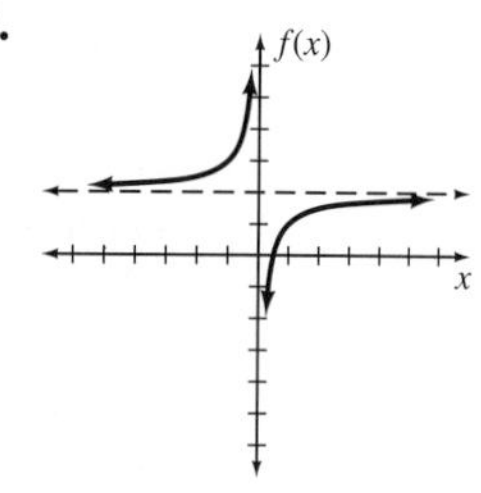

19.
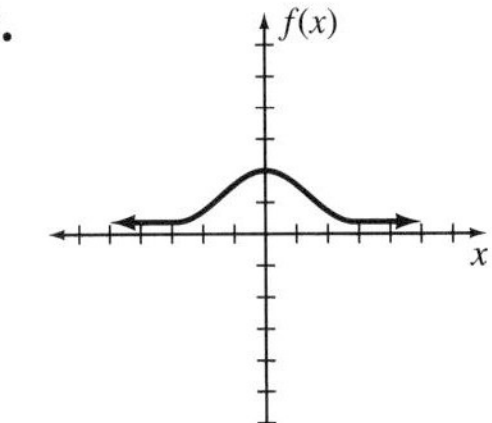

20.
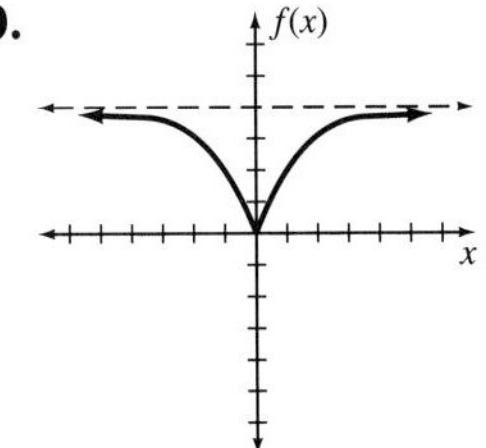

21.
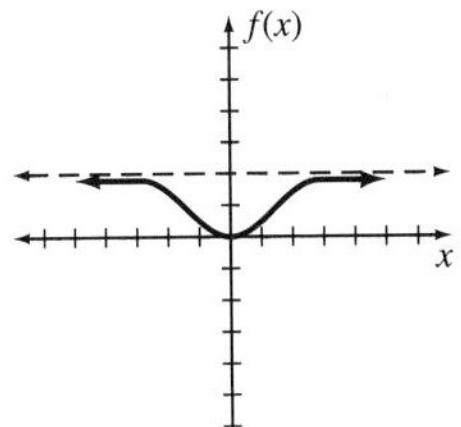

22.
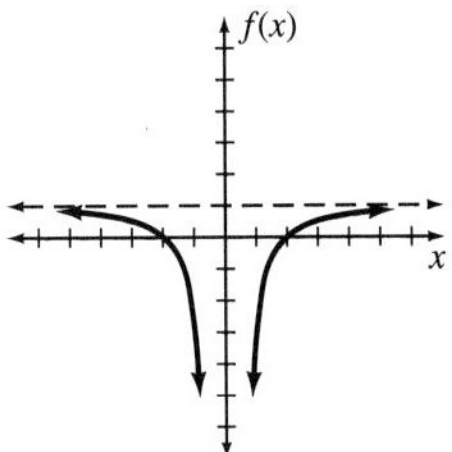

23. (a)
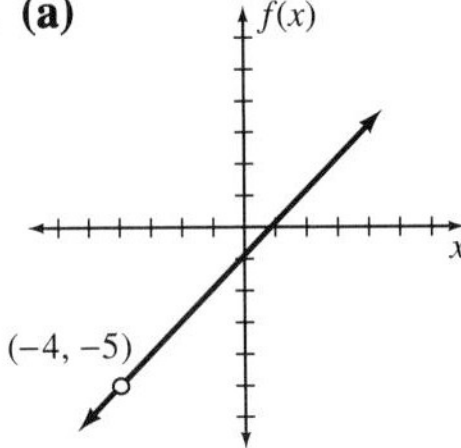

(b)
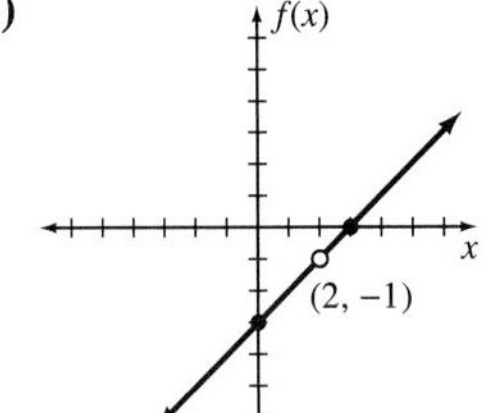

(c)
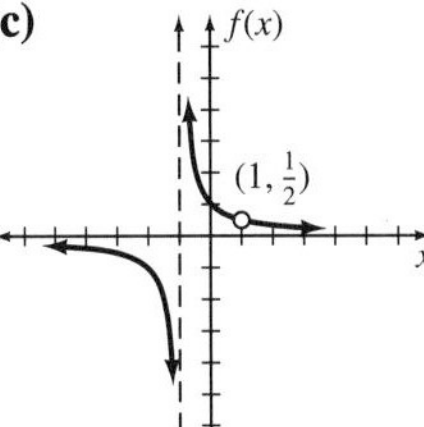

(d)
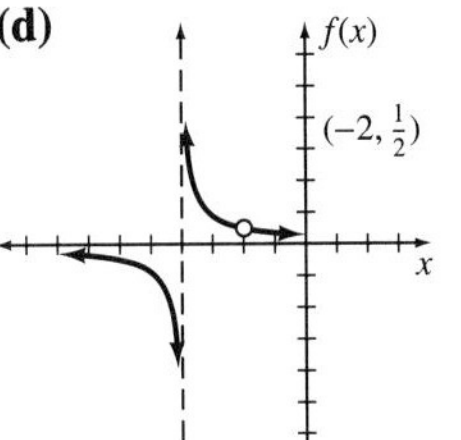

3.
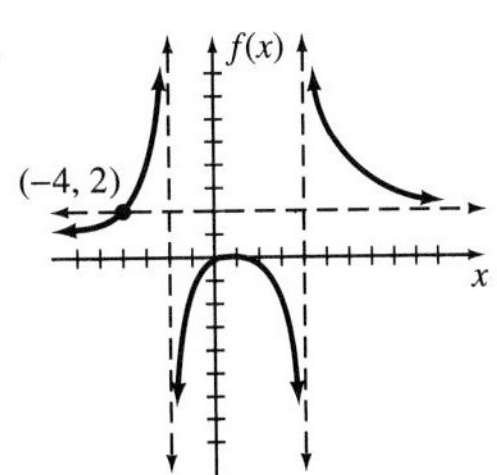

4.
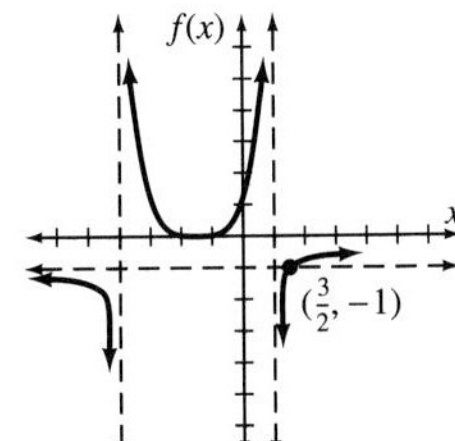

5.
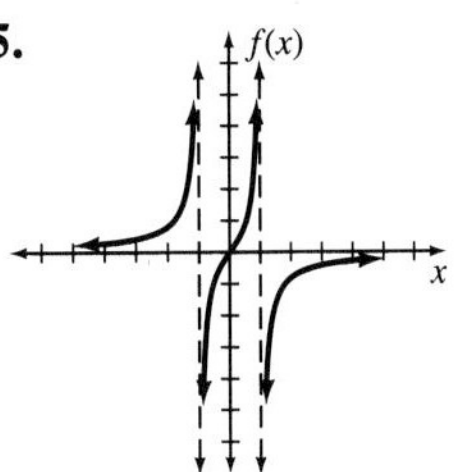

6.
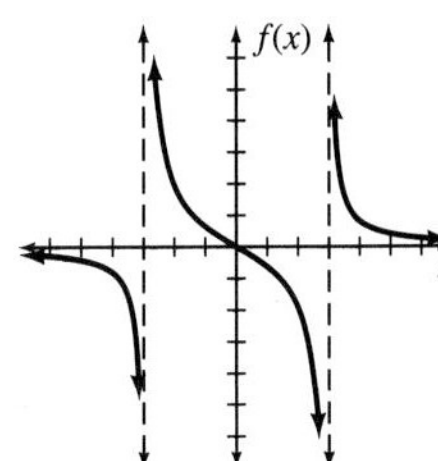

7.
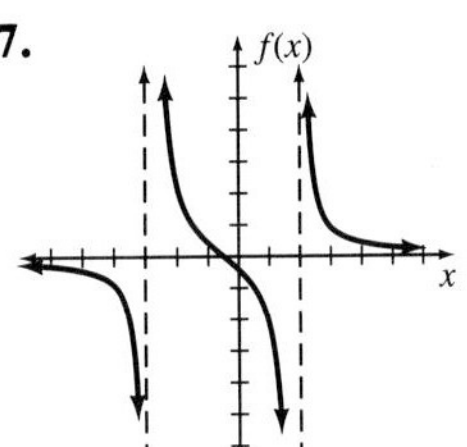

8.
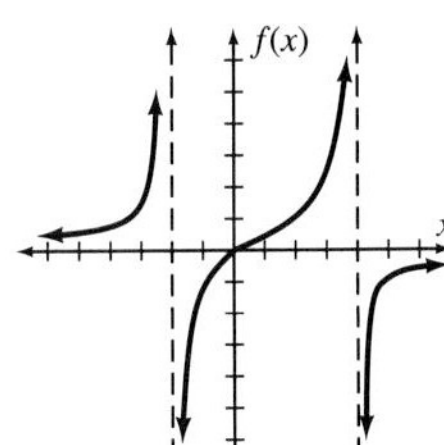

9.
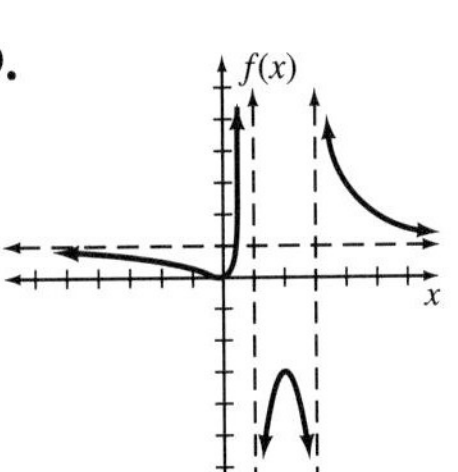

10.
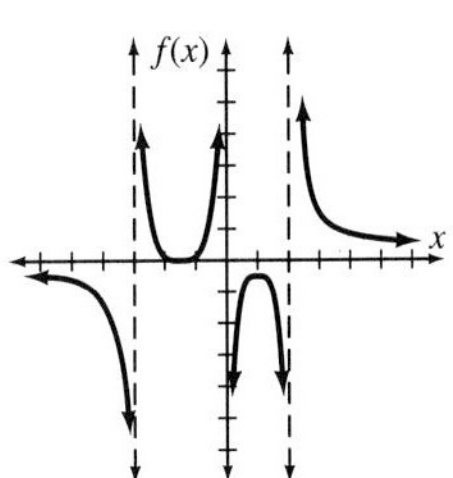

Problem Set 6.6 (page 400)

1.
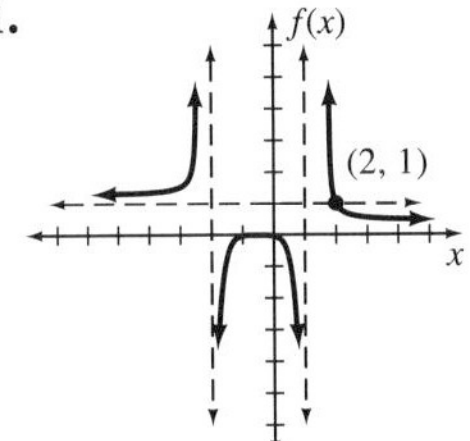

2.
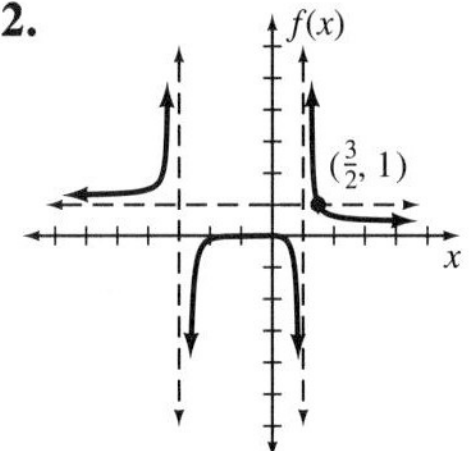

11.
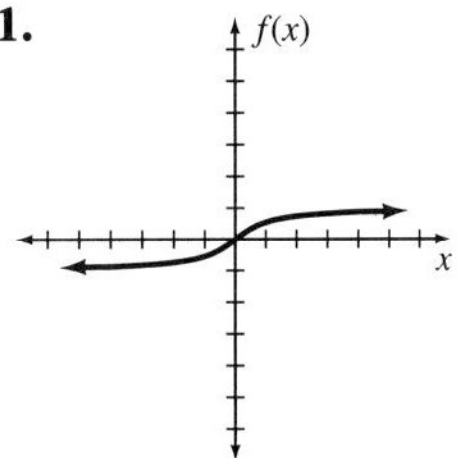

12.
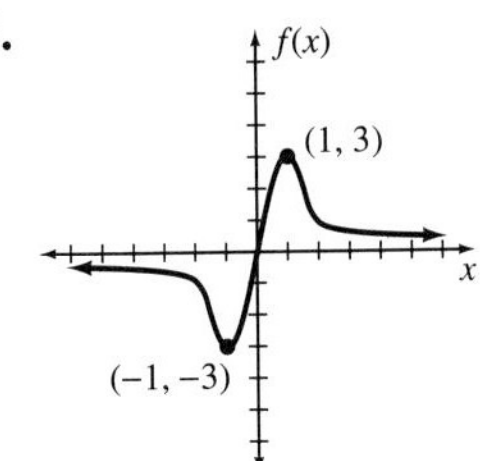

13. 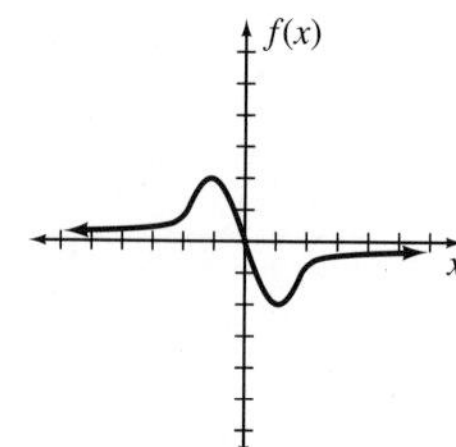

14. 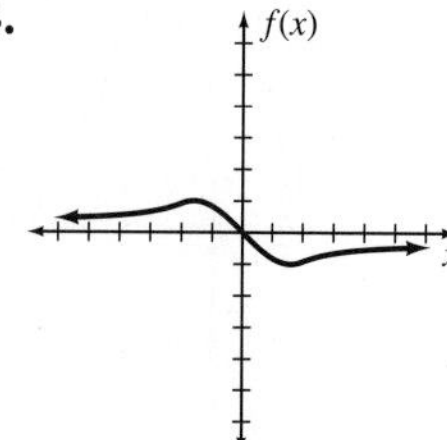

15. 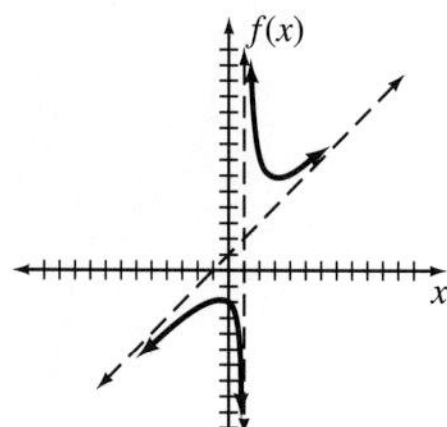

16. 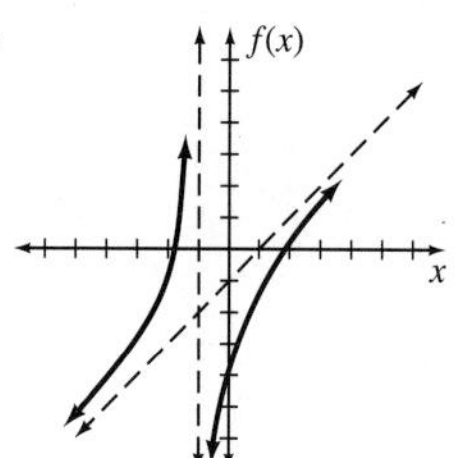

17. 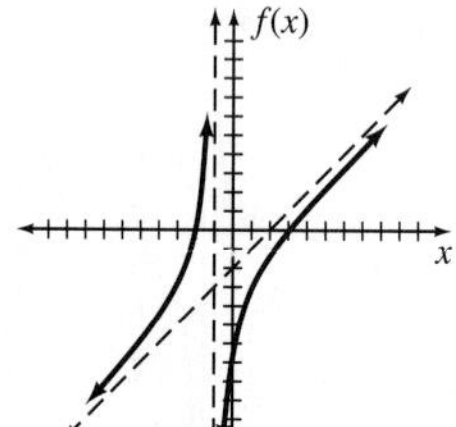

18. 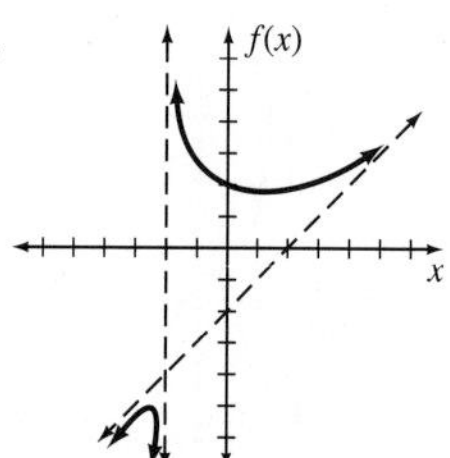

19. 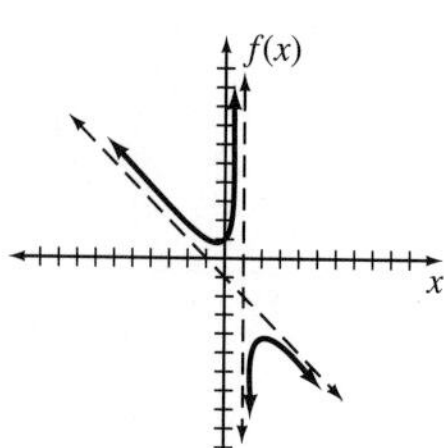

20. 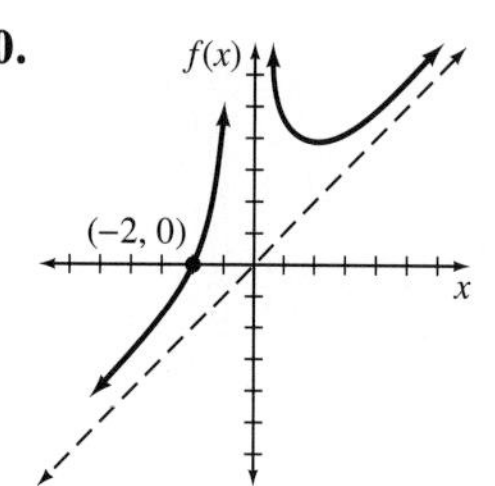

Problem Set 6.7 (page 406)

1. $\frac{4}{x-2}+\frac{7}{x+1}$ 2. $\frac{5}{x+3}+\frac{6}{x-4}$

3. $\frac{3}{x+1}-\frac{5}{x-1}$ 4. $\frac{7}{x-2}-\frac{9}{x+2}$

5. $\frac{1}{3x-1}+\frac{6}{2x+3}$ 6. $\frac{-2}{2x-1}+\frac{4}{5x+2}$

7. $\frac{2}{x-1}+\frac{3}{x+2}-\frac{4}{x-3}$ 8. $\frac{1}{x}-\frac{4}{x+1}-\frac{6}{x-4}$

9. $\frac{-1}{x}+\frac{2}{2x-1}-\frac{3}{4x+1}$

10. $\frac{5}{x+3}+\frac{6}{3x+2}-\frac{9}{2x+3}$

11. $\frac{2}{x-2}+\frac{5}{(x-2)^2}$ 12. $\frac{-3}{x+1}+\frac{4}{(x+1)^2}$

13. $\frac{4}{x}+\frac{7}{x^2}-\frac{10}{x+3}$ 14. $\frac{9}{x}+\frac{1}{x-4}+\frac{3}{(x-4)^2}$

15. $\frac{-3}{x^2+1}-\frac{2}{x-4}$ 16. $\frac{5}{x^2+4}+\frac{8}{3x-4}$

17. $\frac{3}{x+2}-\frac{2}{(x+2)^2}+\frac{1}{(x+2)^3}$

18. $\frac{-1}{x+1}+\frac{2}{(x+1)^2}+\frac{3x+1}{x^2+3}$

19. $\frac{2}{x}+\frac{3x+5}{x^2-x+3}$ 20. $\frac{x+1}{x^2+2}-\frac{2x}{(x^2+2)^2}$

21. $\frac{2x}{x^2+1}+\frac{3-x}{(x^2+1)^2}$ 22. $\frac{3}{x-2}+\frac{x-1}{x^2+2x+4}$

Chapter 6 Review Problem Set (page 408)

1. Q: $3x^2-5x+4$
R: 4

2. Q: $2a-1$
R: 5

3. Q: $3x^2-x+5$
R: 3

4. Q: $5x^2-3x-3$
R: 16

5. Q: $-2x^3+9x^2-38x+151$
R: -605

6. Q: $-3x^3+9x^2-32x+96$
R: -279

7. $f(1)=1$

8. $f(-3)=-197$ 9. $f(-2)=20$

10. $f(8)=0$ 11. Yes 12. No 13. Yes

14. Yes 15. $\{-3, 1, 5\}$ 16. $\left\{-\frac{7}{2}, -1, \frac{5}{4}\right\}$

17. $\{1, 2, 1 \pm 5i\}$ 18. $\{-2, 3 \pm \sqrt{7}\}$

19. 2 positive and 2 negative solutions *or* 2 positive and 2 nonreal complex solutions *or* 2 negative and 2 nonreal complex solutions *or* 4 nonreal complex solutions

20. 1 negative and 4 nonreal complex solutions

21. 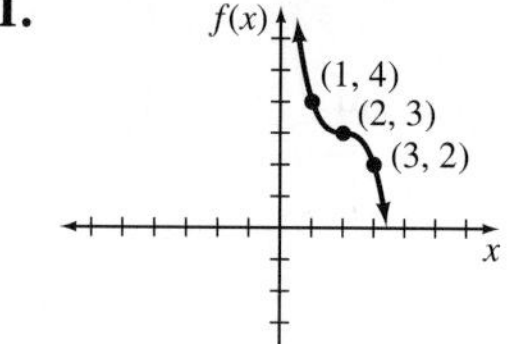

22. 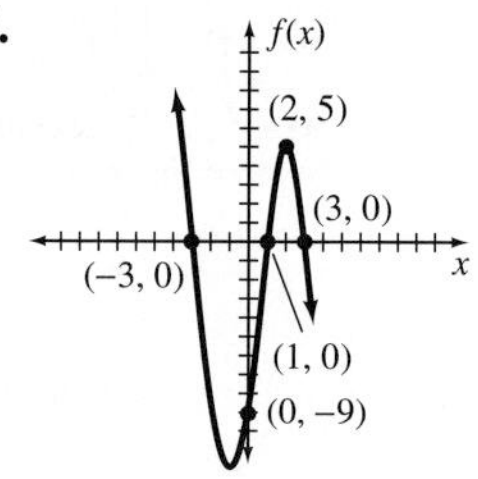

23.

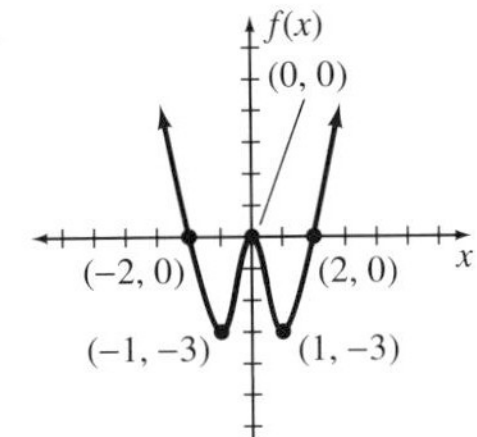

24.

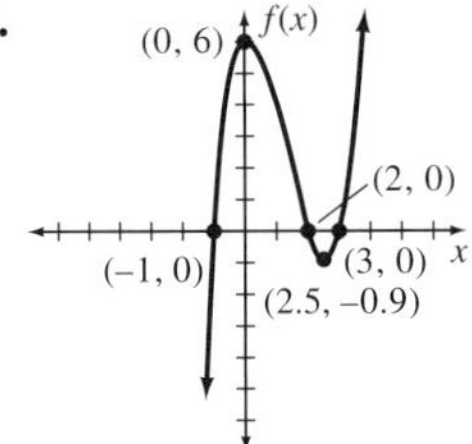

25.

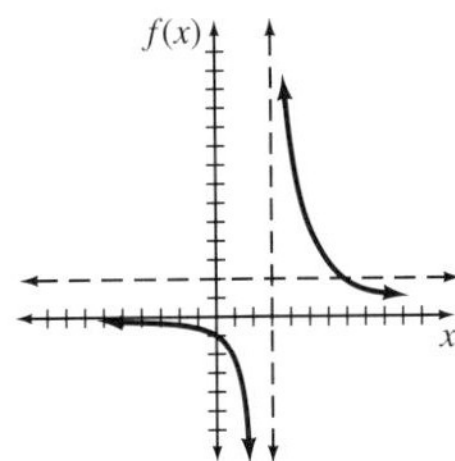

26.

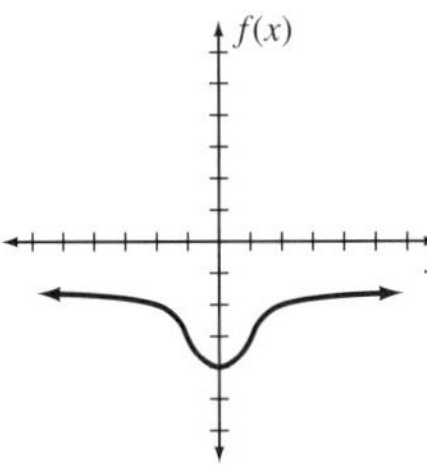

27.

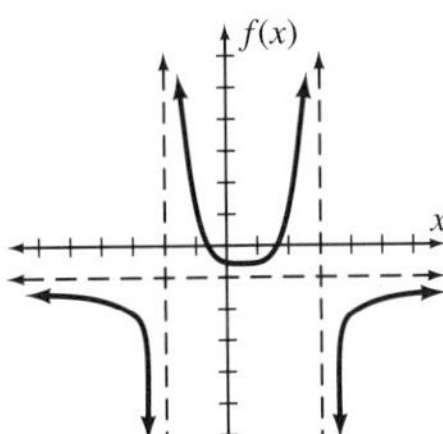

28.

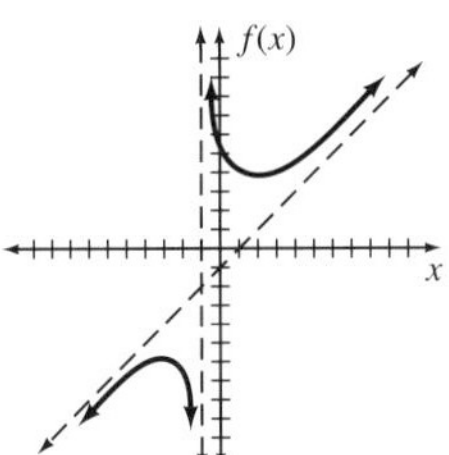

29. $\frac{1}{x} - \frac{2}{x^2} + \frac{4}{x + 2}$ **30.** $\frac{3x + 1}{x^2 + 4} - \frac{5}{2x - 1}$

Chapters 1–6 Cumulative Review Problem Set (page 409)

1. $\frac{64}{27}$ **2.** $-\frac{2}{3}$ **3.** $-\frac{1}{25}$ **4.** 16 **5.** $\frac{1}{27}$
6. 3 **7.** −4 **8.** −5 **9.** 16 **10.** 3
11. $(-\infty, -6] \cup \left[\frac{1}{2}, \infty\right)$
12. $(f \circ g)(-2) = 26$ and $(g \circ f)(3) = 59$
13. $(f \circ g)(x) = -2x + 8$ and $D = \{x|x \neq 4\}$

$(g \circ f)(x) = -\frac{x}{4x + 2}$ and

$D = \left\{x|x \neq 0 \text{ and } x \neq -\frac{1}{2}\right\}$

14. $f^{-1}(x) = \frac{-x + 7}{2}$ **15.** $2a + h + 7$
16. $f(9) = 33$ **17.** $3x^4 + 9x^3 + 2x^2 - x - 2$
18. No **19.** 5.64 **20.** $(-3, 2)$ and $r = 3$
21. $x + 3y = 2$ **22.** $4x + 3y = 5$ **23.** 16 units
24. $y = \pm\frac{1}{3}x$ **25.** 12 **26.** $\frac{2}{7}$ **27.** \$784
28. 8.7 years
29. 10 nickels, 15 dimes, and 32 quarters **30.** \$125
31. $1\frac{1}{3}$ quarts **32.** 45 miles **33.** 4 hours
34. $\left\{\frac{3}{5}\right\}$ **35.** $\left\{\frac{-13 \pm \sqrt{193}}{2}\right\}$ **36.** $\{-7, 0, 2\}$
37. $\left\{-\frac{5}{2}, -1, \frac{2}{3}\right\}$ **38.** $\left\{-1, \frac{5}{2}\right\}$ **39.** $\{0\}$
40. $\{-1\}$ **41.** $\left\{\frac{2}{3}\right\}$ **42.** $\{3\}$
43. $\{\pm 3i, \pm\sqrt{6}\}$ **44.** $\left\{-\frac{13}{2}, 4\right\}$
45. $\left\{1, 2, \frac{-1 \pm i\sqrt{11}}{2}\right\}$ **46.** $(-\infty, -5)$
47. $\left[-\frac{11}{17}, \infty\right)$ **48.** $(-3, 6)$
49. $[-3, 1] \cup [2, \infty)$ **50.** $\left(-\infty, -\frac{5}{2}\right) \cup \left(\frac{7}{2}, \infty\right)$
51. $\left[-\frac{10}{3}, 2\right]$ **52.** $\left(-\infty, \frac{3}{4}\right] \cup (2, \infty)$
53. $(-\infty, 4) \cup \left(\frac{15}{2}, \infty\right)$

54.

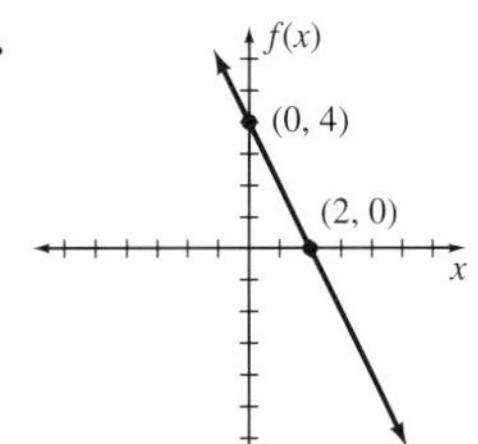

55.

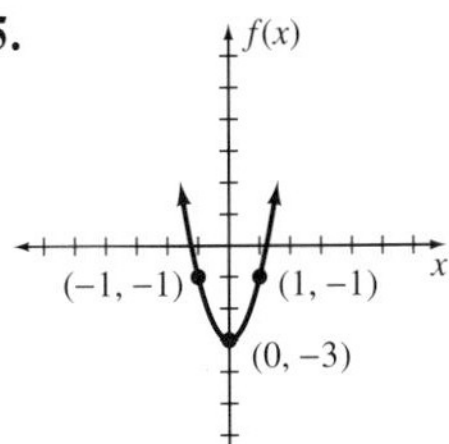

56.

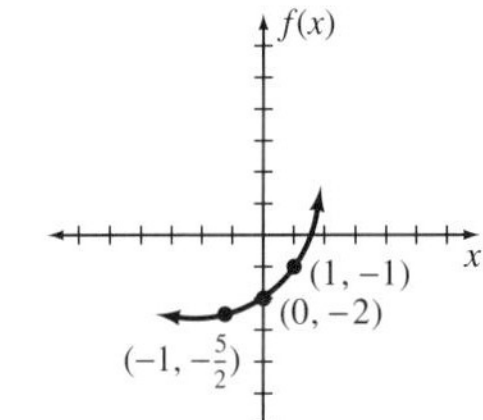

57.

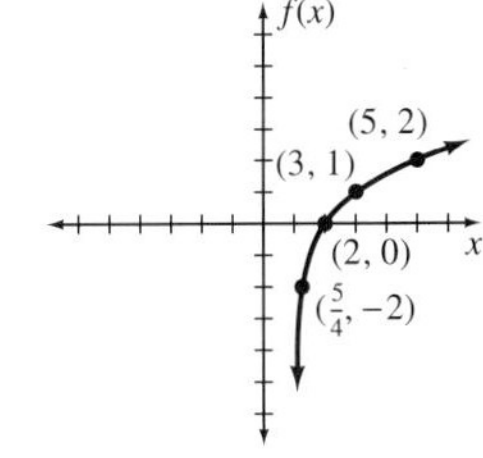

58.

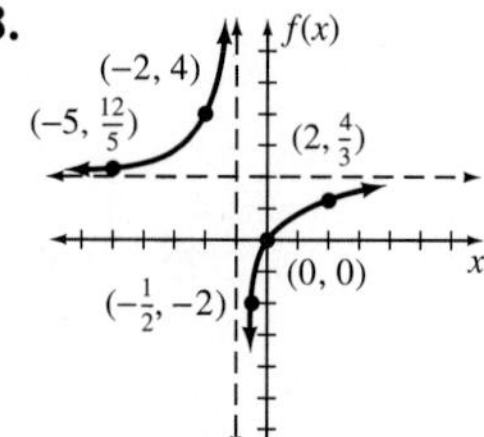

59.

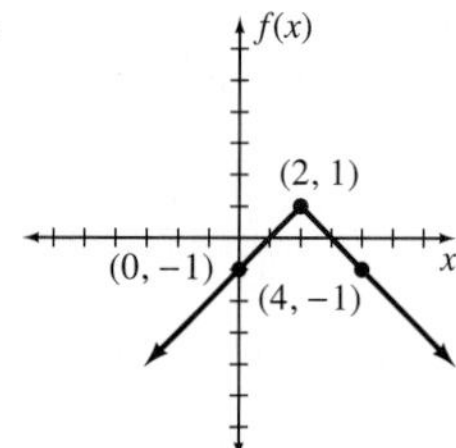

60.

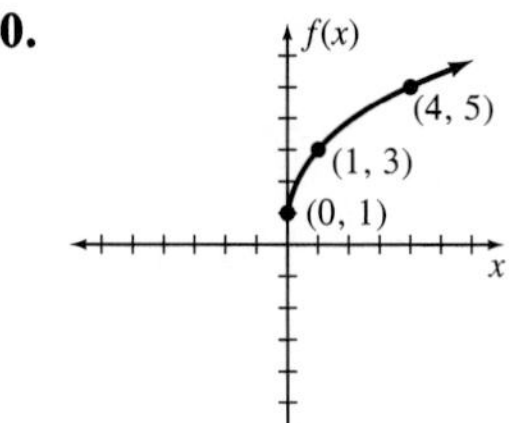

61.

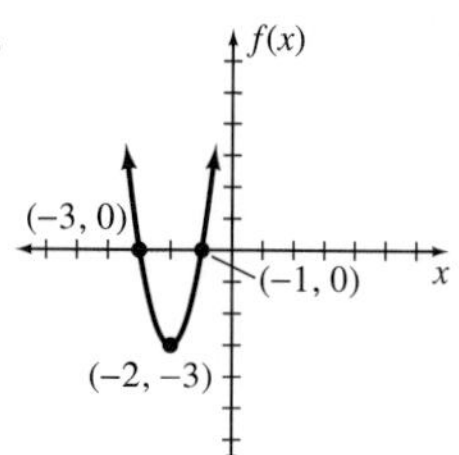

62.

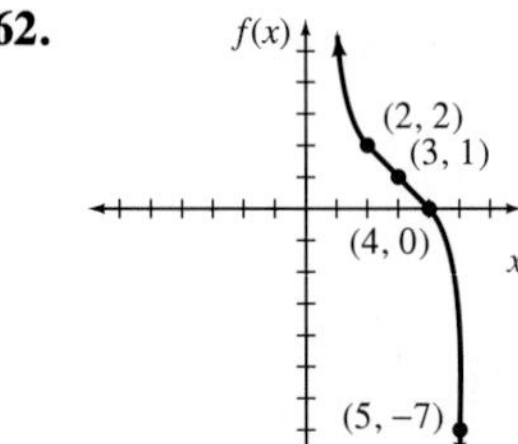

63.

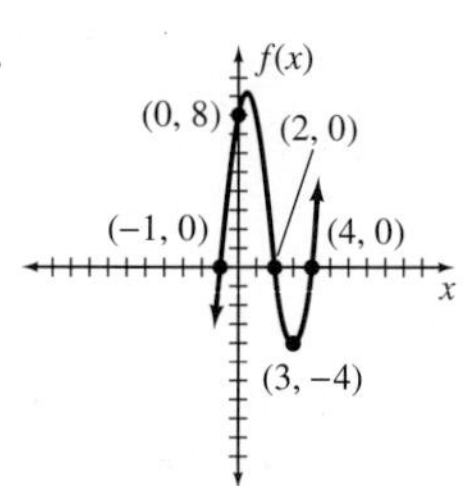

64.

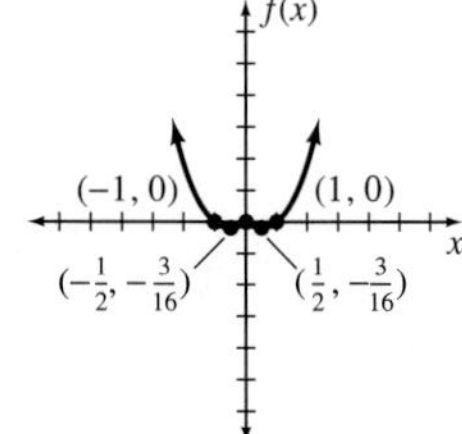

CHAPTER 7

Problem Set 7.1 (page 422)

1. 52° **2.** 108° **3.** 25° and 65° **4.** 40° **5.** 20° and 160° **6.** 40° and 140° **7.** 14.50° **8.** 62.25° **9.** 22°18′ **10.** 114°36′ **11.** 8.76° **12.** 34.84° **13.** 45°19′12″ **14.** 132°9′ **15.** 150.17° **16.** 94.75° **17.** 9°7′48″ **18.** 73°28′12″ **19.** $\frac{\pi}{18}$ **20.** $\frac{\pi}{12}$ **21.** $\frac{4\pi}{9}$ **22.** $\frac{2\pi}{3}$ **23.** $\frac{5\pi}{6}$ **24.** $\frac{7\pi}{6}$ **25.** $\frac{5\pi}{4}$ **26.** $\frac{5\pi}{3}$ **27.** $-\frac{\pi}{6}$ **28.** $-\frac{11\pi}{6}$ **29.** $-\frac{19\pi}{6}$ **30.** $\frac{8\pi}{3}$ **31.** 20° **32.** 50° **33.** 130° **34.** 105° **35.** 240° **36.** 315° **37.** 390° **38.** 510° **39.** −45° **40.** −100° **41.** −210° **42.** −420° **43.** 114.6° **44.** 171.9° **45.** 401.1° **46.** 234.9° **47.** −229.2° **48.** −355.2° **49.** 0.5 **50.** 3.7 **51.** 0.3 **52.** 2.5 **53.** −4.4 **54.** −6.5 **55.** 46.1 inches **56.** 18.2 meters **57.** 17.9 centimeters **58.** 6.4 feet **59.** 127.2° **60.** 26.2 inches **61.** 412.5° **62.** 205.8° **63.** 630° **64.** 133.7° **65.** 3.8 inches **66.** 218 inches **67.** 2.8 revolutions **68.** 13 inches **69.** 7 feet **70.** $6\sqrt{3}$ meters **71.** $4\sqrt{3}$ centimeters **72.** $b = 3\sqrt{3}$ yards and $c = 6$ yards **73.** $a = 6$ yards and $b = 6\sqrt{3}$ yards **74.** $4\sqrt{2}$ meters **75.** $a = b = 5\sqrt{2}$ meters **76.** 5.7 meters **77.** 7.8 centimeters **78.** 8.2 feet **79.** 36.4 feet **80.** 30 meters **81.** 127.3 feet **82.** 35.4 meters **83.** $3\sqrt{3}$ centimeters **84.** 20.8 centimeters **85.** 10.8 centimeters **88.** **(a)** Yes **(b)** Yes **(c)** No **(d)** No **(e)** Yes **(f)** Yes **89.** **(a)** 3000π radians per minute; 9000π inches per minute **(b)** 1700π radians per minute; 425π feet per minute **(c)** Approximately 616 revolutions per minute; approximately 3870 radians per minute

Problem Set 7.2 (page 431)

For Problems 1–16, the answers are given in the order of sin θ, cos θ, tan θ, csc θ, sec θ, and cot θ.

1. $-\frac{4}{5}, \frac{3}{5}, -\frac{4}{3}, -\frac{5}{4}, \frac{5}{3}, -\frac{3}{4}$

2. $-\frac{4}{5}, -\frac{3}{5}, \frac{4}{3}, -\frac{5}{4}, -\frac{5}{3}, \frac{3}{4}$

3. $\frac{12}{13}, -\frac{5}{13}, -\frac{12}{5}, \frac{13}{12}, -\frac{13}{5}, -\frac{5}{12}$

4. $\frac{5}{13}, \frac{12}{13}, \frac{5}{12}, \frac{13}{5}, \frac{13}{12}, \frac{12}{5}$

5. $-\frac{\sqrt{2}}{2}, \frac{\sqrt{2}}{2}, -1, -\sqrt{2}, \sqrt{2}, -1$

6. $-\frac{\sqrt{2}}{2}, -\frac{\sqrt{2}}{2}, -1, -\sqrt{2}, -\sqrt{2}, 1$

7. $-\frac{3\sqrt{13}}{13}, -\frac{2\sqrt{13}}{13}, \frac{3}{2}, -\frac{\sqrt{13}}{3}, -\frac{\sqrt{13}}{2}, \frac{2}{3}$

8. $-\frac{2\sqrt{13}}{13}, \frac{3\sqrt{13}}{13}, -\frac{2}{3}, -\frac{\sqrt{13}}{2}, \frac{\sqrt{13}}{3}, -\frac{3}{2}$

9. $\frac{2\sqrt{5}}{5}, \frac{\sqrt{5}}{5}, 2, \frac{\sqrt{5}}{2}, \sqrt{5}, \frac{1}{2}$

10. $-\frac{3\sqrt{10}}{10}, \frac{\sqrt{10}}{10}, -3, -\frac{\sqrt{10}}{3}, \sqrt{10}, -\frac{1}{3}$

11. $-\frac{\sqrt{10}}{10}, \frac{3\sqrt{10}}{10}, -\frac{1}{3}, -\sqrt{10}, \frac{\sqrt{10}}{3}, -3$

12. $\frac{\sqrt{2}}{2}, -\frac{\sqrt{2}}{2}, -1, \sqrt{2}, -\sqrt{2}, -1$

13. 1, 0, undefined, 1, undefined, 0

14. 0, −1, 0, undefined, −1, undefined

15. −1, 0, undefined, −1, undefined, 0

16. $\frac{\sqrt{2}}{2}, \frac{\sqrt{2}}{2}, 1, \sqrt{2}, \sqrt{2}, 1$

For Problems 17–34, the answers are given in the order of $\sin\theta$, $\cos\theta$, and $\tan\theta$.

17. $\frac{\sqrt{3}}{2}, \frac{1}{2}, \sqrt{3}$ **18.** $\frac{1}{2}, -\frac{\sqrt{3}}{2}, -\frac{\sqrt{3}}{3}$

19. $\frac{\sqrt{2}}{2}, -\frac{\sqrt{2}}{2}, -1$ **20.** $-\frac{1}{2}, -\frac{\sqrt{3}}{2}, \frac{\sqrt{3}}{3}$

21. $-\frac{\sqrt{3}}{2}, \frac{1}{2}, -\sqrt{3}$ **22.** $-\frac{1}{2}, \frac{\sqrt{3}}{2}, -\frac{\sqrt{3}}{3}$

23. $-\frac{\sqrt{2}}{2}, \frac{\sqrt{2}}{2}, -1$ **24.** $-\frac{\sqrt{3}}{2}, \frac{1}{2}, -\sqrt{3}$

25. $-\frac{1}{2}, \frac{\sqrt{3}}{2}, -\frac{\sqrt{3}}{3}$ **26.** $\frac{1}{2}, -\frac{\sqrt{3}}{2}, -\frac{\sqrt{3}}{3}$

27. $-\frac{\sqrt{2}}{2}, -\frac{\sqrt{2}}{2}, 1$ **28.** $-\frac{\sqrt{2}}{2}, \frac{\sqrt{2}}{2}, -1$

29. $\frac{1}{2}, \frac{\sqrt{3}}{2}, \frac{\sqrt{3}}{3}$ **30.** $\frac{\sqrt{3}}{2}, -\frac{1}{2}, -\sqrt{3}$

31. $-\frac{\sqrt{2}}{2}, -\frac{\sqrt{2}}{2}, 1$ **32.** $-\frac{\sqrt{3}}{2}, \frac{1}{2}, -\sqrt{3}$

33. $-\frac{1}{2}, \frac{\sqrt{3}}{2}, -\frac{\sqrt{3}}{3}$ **34.** $\frac{\sqrt{2}}{2}, -\frac{\sqrt{2}}{2}, -1$

35.

θ	radians	$\sin\theta$	$\cos\theta$	$\tan\theta$	$\csc\theta$	$\sec\theta$	$\cot\theta$
0°	0	0	1	0	U*	1	U*
30°	$\frac{\pi}{6}$	$\frac{1}{2}$	$\frac{\sqrt{3}}{2}$	$\frac{\sqrt{3}}{3}$	2	$\frac{2\sqrt{3}}{3}$	$\sqrt{3}$
45°	$\frac{\pi}{4}$	$\frac{\sqrt{2}}{2}$	$\frac{\sqrt{2}}{2}$	1	$\sqrt{2}$	$\sqrt{2}$	1
60°	$\frac{\pi}{3}$	$\frac{\sqrt{3}}{2}$	$\frac{1}{2}$	$\sqrt{3}$	$\frac{2\sqrt{3}}{3}$	2	$\frac{\sqrt{3}}{3}$
90°	$\frac{\pi}{2}$	1	0	U*	1	U*	0
180°	π	0	−1	0	U*	−1	U*
270°	$\frac{3}{2}\pi$	−1	0	U*	−1	U*	0

U* = Undefined

36. $-\frac{\sqrt{2}}{2}$ **37.** $-\frac{\sqrt{2}}{2}$ **38.** −2 **39.** $\frac{3\sqrt{10}}{10}$

40. $\cos\theta = \frac{3}{5}$ and $\tan\theta = -\frac{4}{3}$

41. $\sin\theta = -\frac{3}{5}$ and $\cot\theta = \frac{4}{3}$

42. $\sin\theta = \frac{5}{13}$ and $\cos\theta = -\frac{12}{13}$

43. $\sin\theta = \frac{7}{25}$ and $\sec\theta = \frac{25}{24}$ **44.** I or IV

45. IV **46.** I **47.** II or IV **48.** $\theta = 225°$

49. $\theta = 60°$ **50.** $\theta = 120°$ **51.** $\theta = 210°$

52. $\theta = 240°$ **53.** $\theta = 270°$

Problem Set 7.3 (page 438)

1. I **2.** III **3.** IV **4.** IV **5.** III

6. II **7.** II **8.** IV **9.** 150° **10.** 210°

11. 240° **12.** 30° **13.** 300° **14.** 150°

15. 240° **16.** 60° **17.** $\frac{3\pi}{2}$ **18.** $\frac{3\pi}{4}$

19. $\frac{7\pi}{6}$ **20.** $\frac{5\pi}{3}$ **21.** $\frac{3\pi}{4}$ **22.** $\frac{4\pi}{3}$

23. 85° **24.** 74.7° **25.** 71.8° **26.** 30°

27. 73° **28.** 10° **29.** $\frac{\pi}{4}$ **30.** $\frac{\pi}{6}$ **31.** $\frac{\pi}{3}$

32. $\frac{\pi}{4}$ **33.** $\frac{\pi}{3}$ **34.** $\frac{\pi}{6}$ **35.** $\frac{\sqrt{3}}{2}$

36. $-\frac{\sqrt{3}}{2}$ **37.** $-\frac{\sqrt{3}}{2}$ **38.** $-\frac{1}{2}$ **39.** $-\sqrt{3}$

40. −1 **41.** $\sqrt{2}$ **42.** −2 **43.** 2

44. $\frac{2\sqrt{3}}{3}$ **45.** $-\frac{1}{2}$ **46.** $-\frac{\sqrt{3}}{2}$ **47.** $\frac{1}{2}$
48. $-\frac{1}{2}$ **49.** $-\sqrt{3}$ **50.** $\frac{\sqrt{3}}{3}$ **51.** -1
52. -1 **53.** $\frac{\sqrt{2}}{2}$ **54.** $\frac{\sqrt{2}}{2}$ **55.** $\frac{\sqrt{3}}{2}$
56. $-\frac{\sqrt{2}}{2}$ **57.** $\sqrt{3}$ **58.** $-\frac{\sqrt{3}}{3}$ **59.** $-\frac{\sqrt{2}}{2}$
60. $-\frac{\sqrt{2}}{2}$ **61.** $\frac{\sqrt{3}}{3}$ **62.** $\frac{\sqrt{3}}{3}$ **63.** $\frac{1}{2}$
64. $\frac{1}{2}$ **65.** Undefined **66.** 0 **67.** 0.9659
68. 0.1736 **69.** 4.0108 **70.** −0.1512
71. 0.8607 **72.** −0.4633 **73.** −0.9135
74. −0.6909 **75.** 1.0358 **76.** 1.1025
77. −3.9408 **78.** 1.4659 **79.** −1.6003
80. 1.6353 **81.** $-\frac{1}{2}$ **82.** $-\frac{\sqrt{3}}{2}$
83. $-\frac{2\sqrt{3}}{3}$ **84.** −2 **85.** −0.1080
86. 0.2062 **87.** −1.897 **88.** $-\frac{3\sqrt{7}}{7}$
91. 0.8902 **92.** −0.5850 **93.** 2.1642
94. −1.3373 **95.** −1.4916 **96.** 4.1976

Problem Set 7.4 (page 447)

1. 12.9° **2.** 37.8° **3.** 66.1° **4.** 55.9°
5. 18.6° **6.** 51.0° **7.** 60.3° **8.** 8.2°
9. 31.2° **10.** 39.0° **11.** 81.1° **12.** 26.3°
13. 71.2° **14.** 47.6°
15. $B = 53°$, $a = 11$, and $c = 18$
16. $B = 32°$, $b = 12$, and $c = 22$
17. $A = 67°$, $a = 28$, and $c = 31$
18. $A = 48°$, $b = 8$, and $c = 12$
19. $B = 23°$, $a = 24$, and $b = 10$
20. $A = 71°$, $a = 32$, and $b = 11$
21. $c = 13$, $A = 23°$, and $B = 67°$
22. $b = 7$, $A = 74°$, and $B = 16°$
23. $a = 26$, $A = 66°$, and $B = 24°$
24. $c = 23$, $A = 52°$, and $B = 38°$ **25.** 23.0 feet
26. 22.6 feet and 8.2 feet **27.** 92.1 meters
28. 316 feet **29.** 1454 feet
30. The altitude is 14.1 kilometers and the horizontal range is 34.0 kilometers.
31. 42 yards **32.** 15 centimeters and 8 centimeters
33. 630 feet **34.** 250 feet **35.** 19.6 feet
36. 70.5°, 70.5°, and 39.0° **37.** 222 feet
38. 81.6 meters **39.** 44.0°
40. 100 feet and 300 feet **41.** 74 feet
42. 100 feet **43.** 20 feet **44.** 27.3 meters
45. 29.5° **46.** 4043 miles **47.** 71 feet

Problem Set 7.5 (page 454)

1. 9.6 centimeters **2.** 5.3 meters **3.** 12.3 feet
4. 22.5 yards **5.** 39.9 centimeters **6.** 51.0°
7. 94.7° **8.** 10.3° **9.** 28.8° **10.** 135.2°
11. $A = 33.1°$, $B = 128.7°$, and $C = 18.2°$
12. $A = 18.2°$, $B = 128.4°$, and $C = 33.4°$
13. $b = 56.2$ feet, $A = 45.9°$, and $C = 34.1°$
14. $c = 6.6$ inches, $A = 14.9°$, and $B = 153.1°$
15. $C = 90°$
16. Two sides of length 14.7 centimeters and two sides of length 21.8 centimeters
17. 56.2 centimeters and 35.9 centimeters **18.** 87.1°
19. 19 miles **20.** 421.5 miles **21.** 46 meters
22. 63.7 feet **23.** 318 feet **24.** 204 feet
25. 77.4° **26.** 24.9 feet **27.** 1479 miles
28. 308 feet **29.** 66.7°

Problem Set 7.6 (page 463)

1. $c = 13.8$ centimeters **2.** $c = 11.2$ feet
3. $a = 16.8$ meters **4.** $b = 10.4$ yards
5. $B = 102°$ and $b = 20.9$ centimeters
6. $C = 59.2°$ and $c = 17.8$ feet
7. $B = 31°$ and $b = 52.0$ miles
8. $B = 121.4°$ and $C = 24.5°$
9. $A = 56.9°$ and $C = 51.4°$
10. $A = 18.7°$ and $B = 25.3°$
11. $b = 34.6$ feet and $c = 18.8$ feet
12. $a = 21.7$ yards and $c = 10.8$ yards
13. One triangle; $C = 33.4°$
14. Two triangles; $C = 66.8°$ or $C = 113.2°$
15. No triangle determined
16. One triangle; $C = 24.8°$
17. Two triangles; $C = 52.2°$, $B = 99.8°$, and $b = 39.9$ miles or $C = 127.8°$, $B = 24.2°$, and $b = 16.6$ miles
18. Two triangles; $C = 64.0°$ and $B = 61.0°$, or $C = 116.0°$ and $B = 9.0°$
19. One triangle; $C = 90°$ and $b = 36.4$ centimeters
20. No triangle determined **22.** 222 meters
23. 48.7 feet **24.** 13.3 feet **25.** 79 feet
26. 104.6 feet **27.** 229.6 feet **28.** 178.0 feet
29. 168 meters or 43 meters **30.** 880 yards
31. 3.2 feet **32.** 163.1 feet **33.** 500 yards
34. **(a)** 158 square meters **(b)** 3059 square yards **(c)** 655 square feet
35. **(a)** 60 square centimeters **(b)** 1907 square feet **(c)** 31 square inches

36. **(a)** 107 square centimeters
(b) 1350 square yards **(c)** 140 square inches

Problem Set 7.7 (page 473)

1.
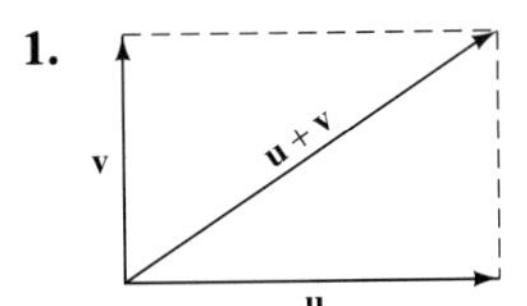

2.
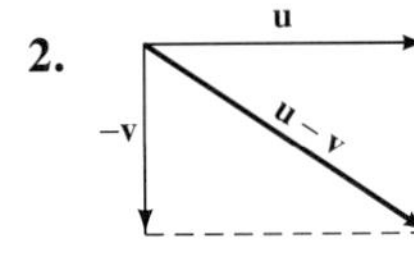

3.
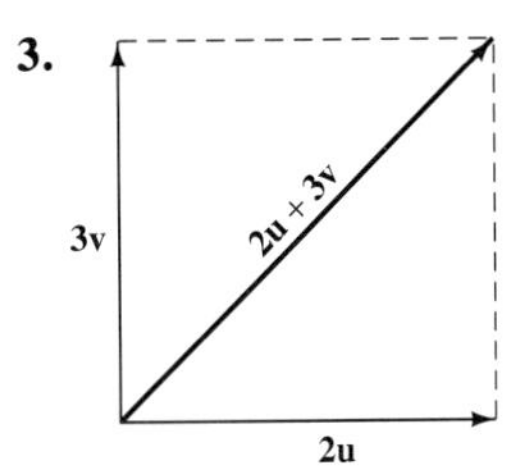

4.
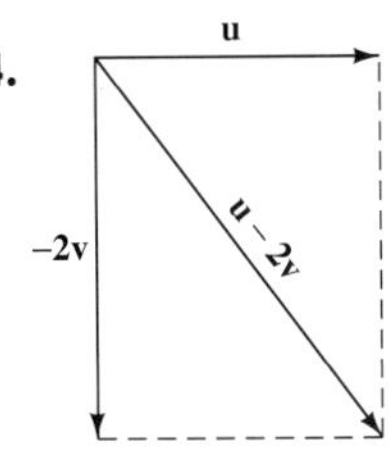

5.
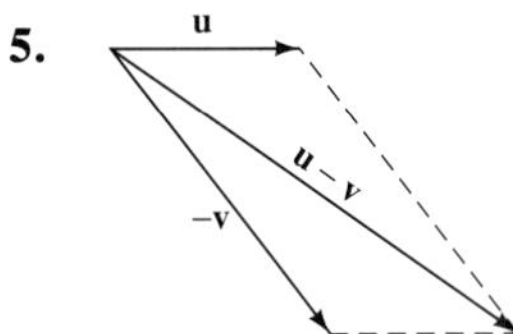

6.
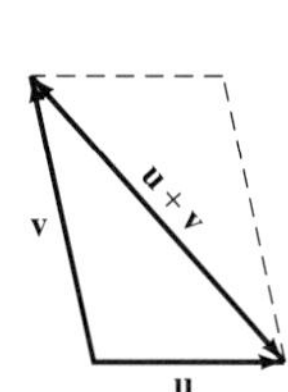

7.
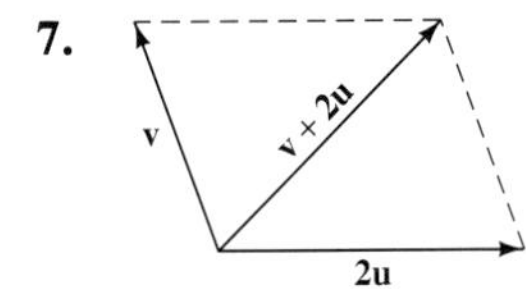

8.
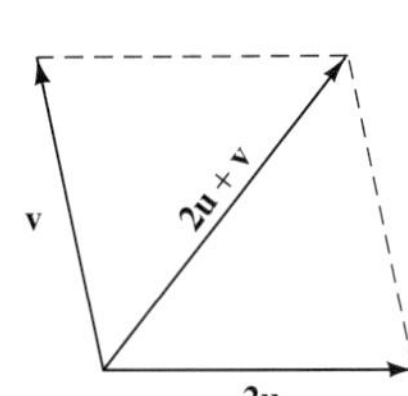

9.
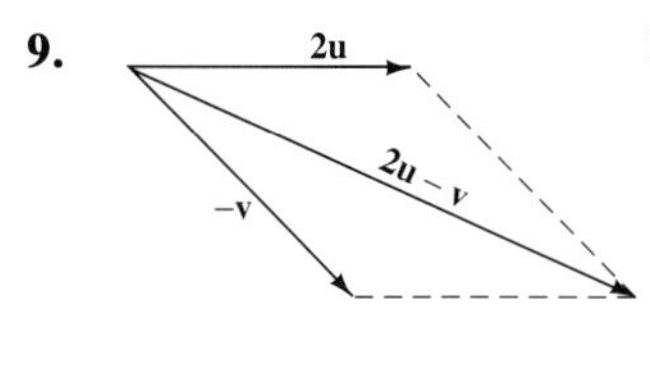

10.
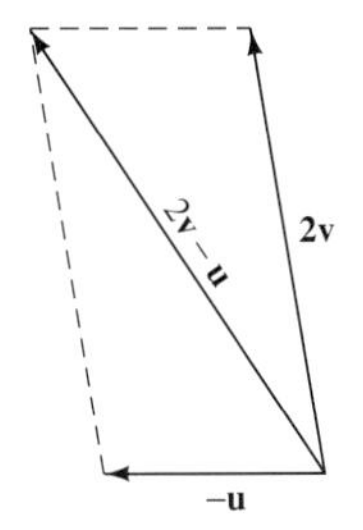

11. 9.1° and 253.2 miles per hour **12.** 15.0°
13. N75.5°E **14.** 15.8 feet per second and 18.4°
15. 90.1 pounds and 33.7°
16. 9.8 kilograms and 66.0°
17. 35.3 kilometers per hour **18.** 28.9 miles per hour
19. 42.3 pounds and 15.4 pounds
20. 225.3 miles per hour and 157.8 miles per hour
21. 16.4 pounds and 11.5 pounds
22. 828.2 pounds and 3091.0 pounds
23. 794.5 pounds **24.** 56.6 pounds **25.** 44.0°
26. 48.0 pounds **27.** 334.0 kilograms and 28.4°
28. 177.5 pounds and 23.3°
29. 93.4 miles and N71.5°W
30. 121.5 miles and S73.7°E
31. 131.0 miles and N86.0°E
32. 80.2 miles and S63.6°E

Chapter 7 Review Problem Set (page 477)

1. 35° and 55° **2.** 35.28° and 82.26°
3. 93°21′ and 163°16′12″
4. **(a)** $\frac{7\pi}{3}$ **(b)** $\frac{19\pi}{6}$ **(c)** $-\frac{\pi}{4}$
5. **(a)** 210° **(b)** −240° **(c)** 765°
6. 71.2 centimeters **7.** 31.8 inches
8. 25.6 feet **9.** 32.5 meters **10.** 43.3 feet
11. $\frac{4}{5}, -\frac{3}{5}, -\frac{4}{3}$ **12.** $-\frac{3\sqrt{10}}{10}, \frac{\sqrt{10}}{10}, -3$
13. $-\frac{2\sqrt{5}}{5}, -\frac{\sqrt{5}}{5}, 2$ **14.** $-\frac{12}{13}, -\frac{5}{13}, \frac{12}{5}$
15. $-\frac{\sqrt{3}}{2}$ **16.** $\frac{1}{2}$ **17.** $\sqrt{2}$ **18.** −2
19. −1 **20.** $-\frac{\sqrt{3}}{3}$ **21.** $\frac{\sqrt{3}}{2}$ **22.** $\frac{1}{2}$
23. Undefined **24.** Undefined **25.** −1
26. 1 **27.** II **28.** IV **29.** −5 **30.** $\frac{12}{13}$
31. −0.7986 **32.** −0.4067 **33.** 49.6°
34. 68.3° **35.** 85.8° **36.** 61.9°
37. 42.5 feet **38.** 135 feet **39.** 28.4 miles
40. 132.0 yards **41.** Two **42.** 84.5 feet
43. 118.2° **44.** 518 miles **45.** 257 yards
46. 83.3 feet **47.** 34.5 pounds and 28.9 pounds
48. 1457.3 pounds **49.** 286.4 pounds and 12.3°
50. 149.2 miles and N46.1°W

CHAPTER 8

Problem Set 8.1 (page 489)

The answers for Problems 1–12 are given in the order sin *s*, cos *s*, tan *s*, csc *s*, sec *s*, and cot *s*.

1. $\frac{\sqrt{2}}{2}, -\frac{\sqrt{2}}{2}, -1, \sqrt{2}, -\sqrt{2}, -1$
2. $-\frac{\sqrt{2}}{2}, \frac{\sqrt{2}}{2}, -1, -\sqrt{2}, \sqrt{2}, -1$
3. −1, 0, undefined, −1, undefined, 0
4. 0, 1, 0, undefined, 1, undefined

5. $-\frac{\sqrt{2}}{2}, \frac{\sqrt{2}}{2}, -1, -\sqrt{2}, \sqrt{2}, -1$

6. 0, −1, 0, undefined, −1, undefined

7. 0, −1, 0, undefined, −1, undefined

8. $\frac{\sqrt{2}}{2}, \frac{\sqrt{2}}{2}, 1, \sqrt{2}, \sqrt{2}, 1$

9. 1, 0, undefined, 1, undefined, 0

10. $\frac{\sqrt{2}}{2}, -\frac{\sqrt{2}}{2}, -1, \sqrt{2}, -\sqrt{2}, -1$

11. $-\frac{\sqrt{2}}{2}, -\frac{\sqrt{2}}{2}, 1, -\sqrt{2}, -\sqrt{2}, 1$

12. $\frac{\sqrt{2}}{2}, \frac{\sqrt{2}}{2}, 1, \sqrt{2}, \sqrt{2}, 1$

13.

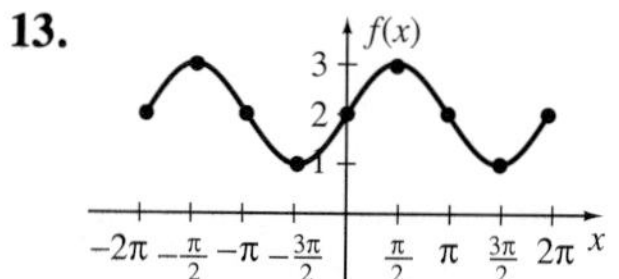

14.

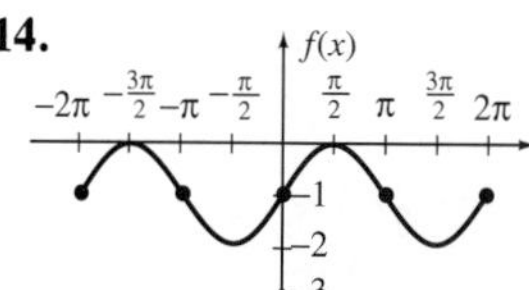

15.

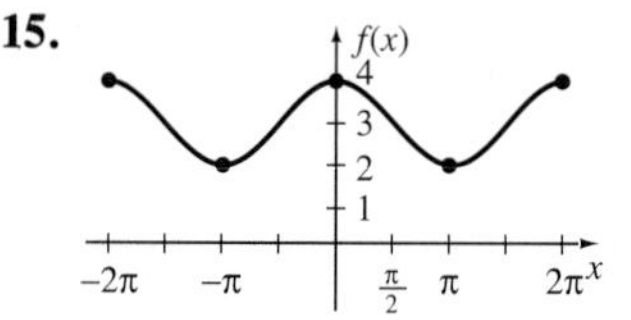

16.

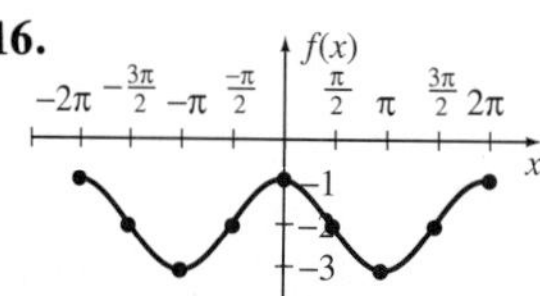

17.

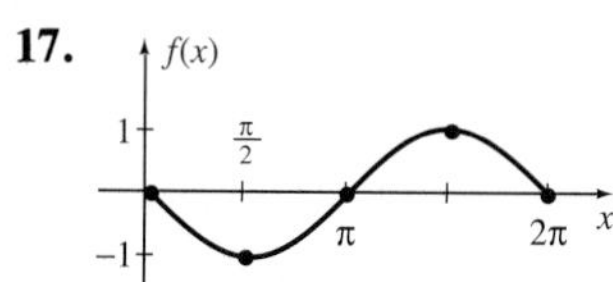

18.

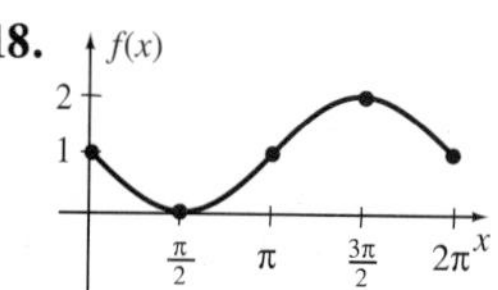

19.

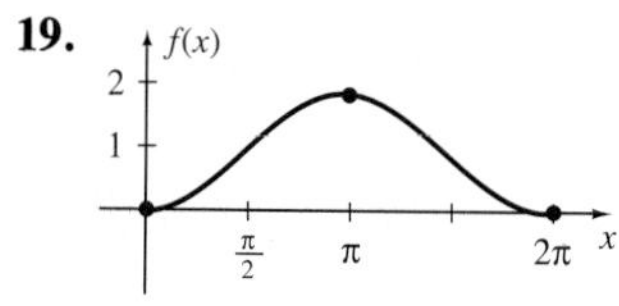

20.

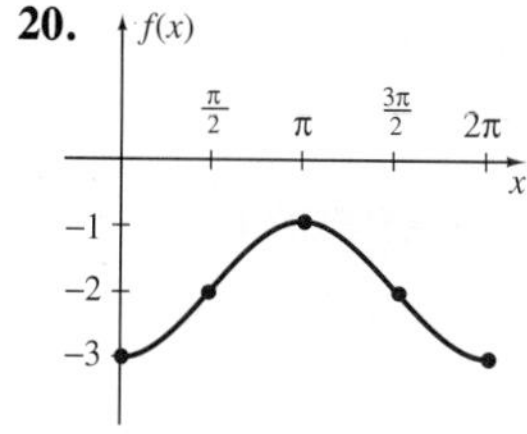

21.

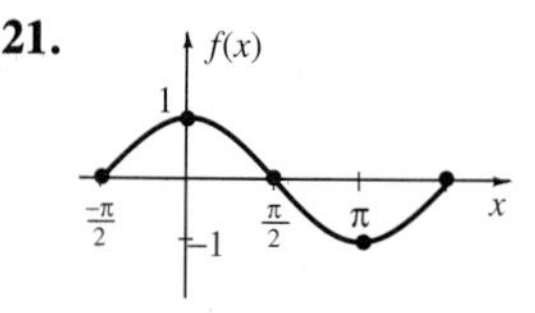

22.

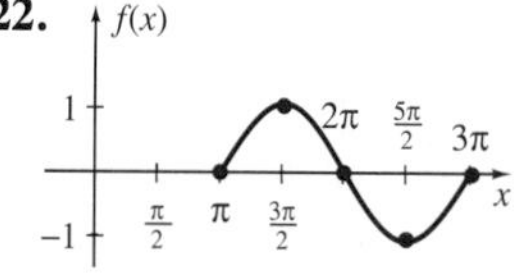

23.

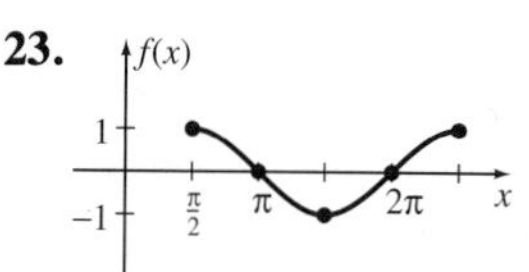

24.

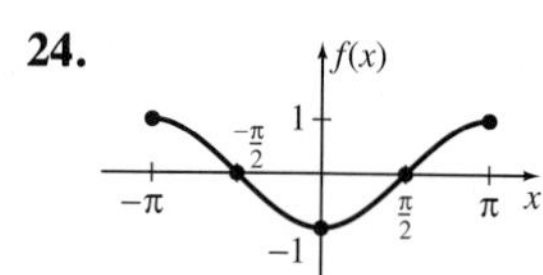

25.

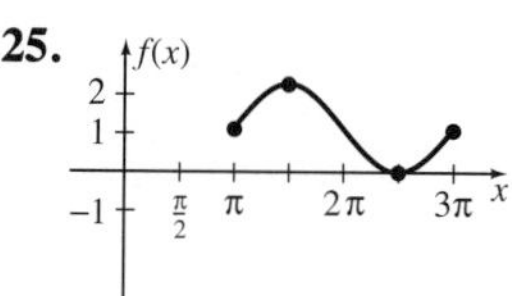

26.

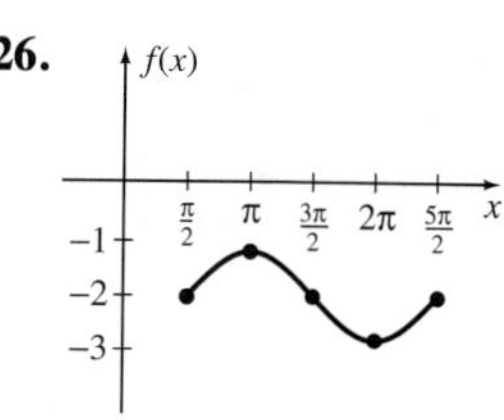

27.

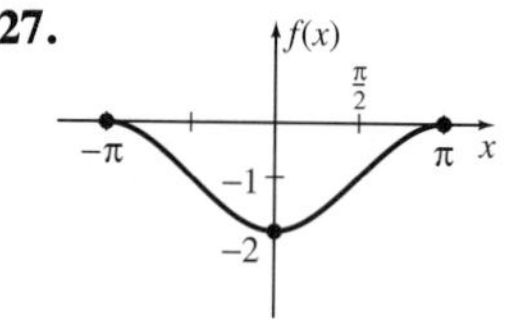

28.

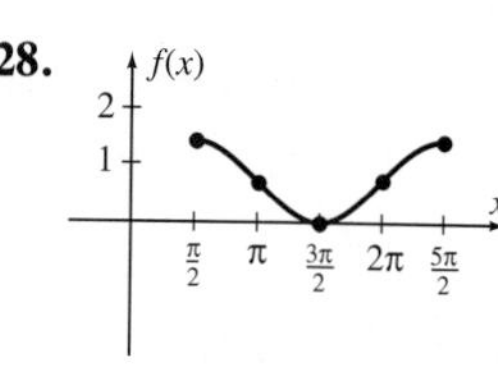

29.

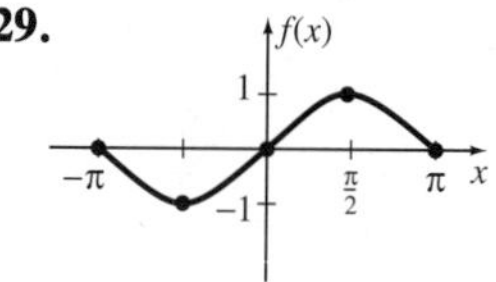

30.

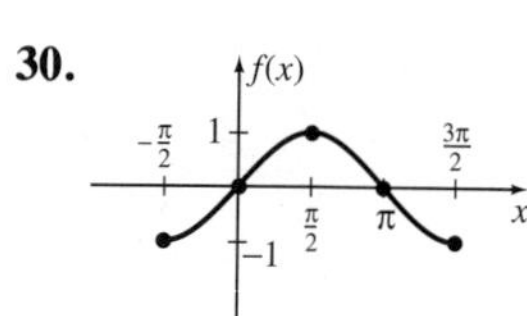

31.

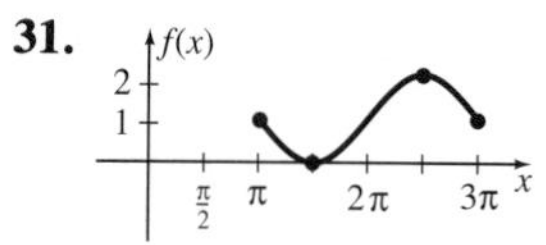

32.

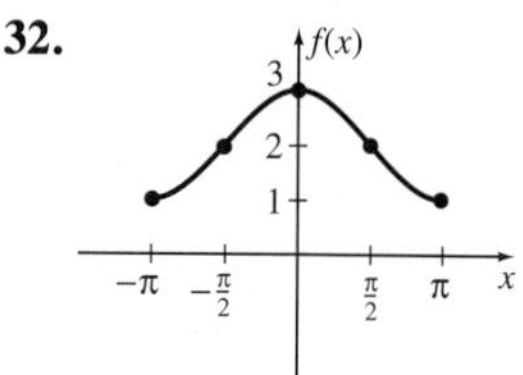

33.

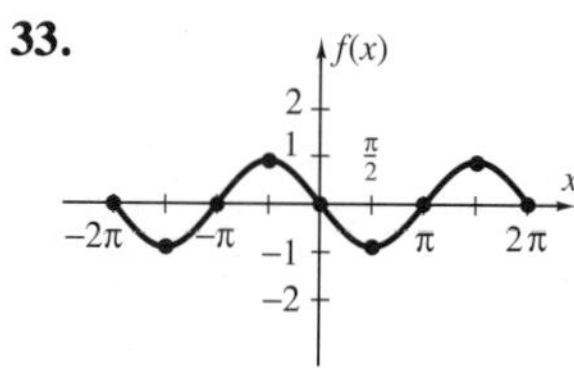

34.

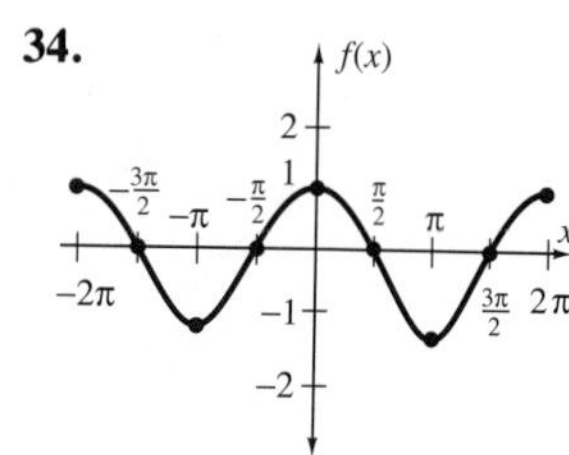

37.

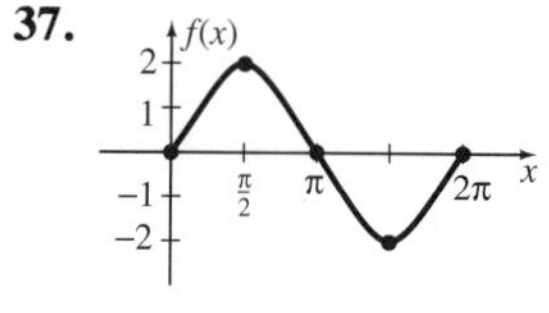

38.

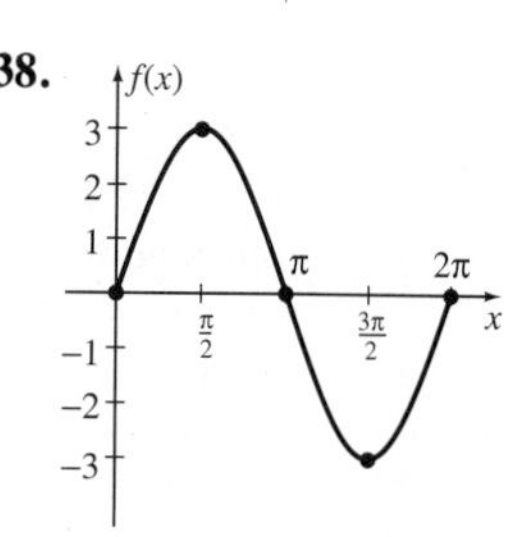

39. 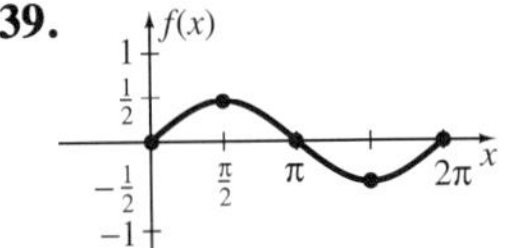

40. 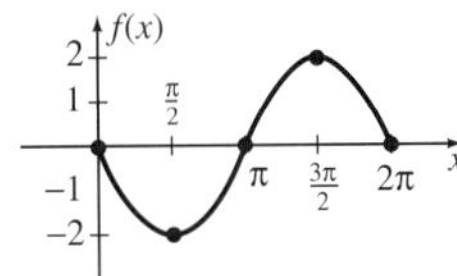

41. 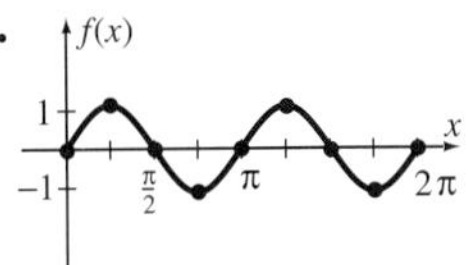

42. 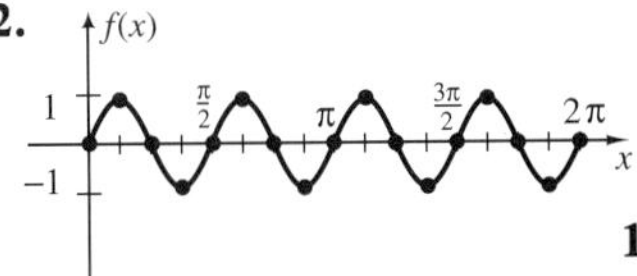

43. 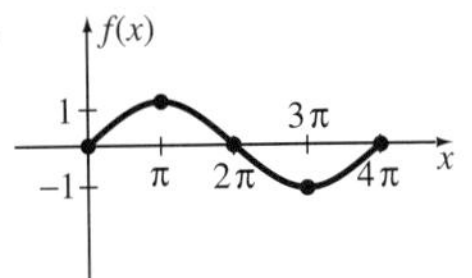

Problem Set 8.2 (page 498)

1. 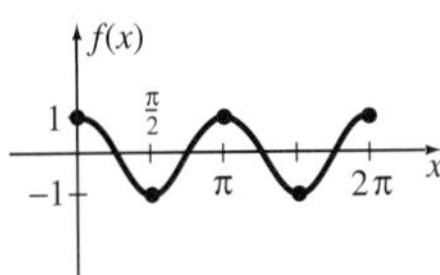

2. 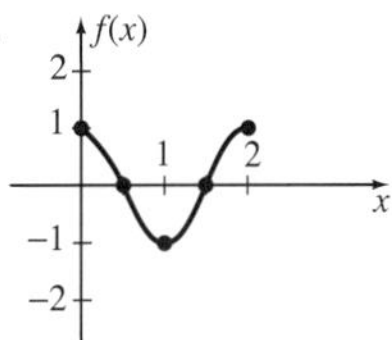

3. 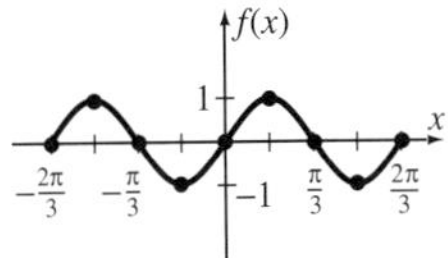

4. 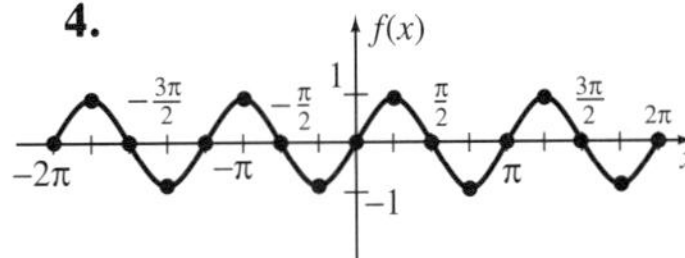

5. 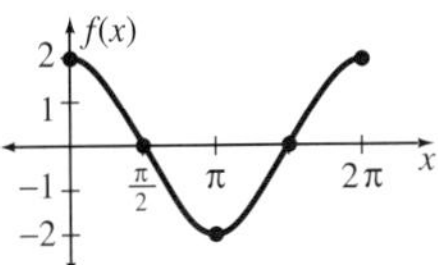

6. 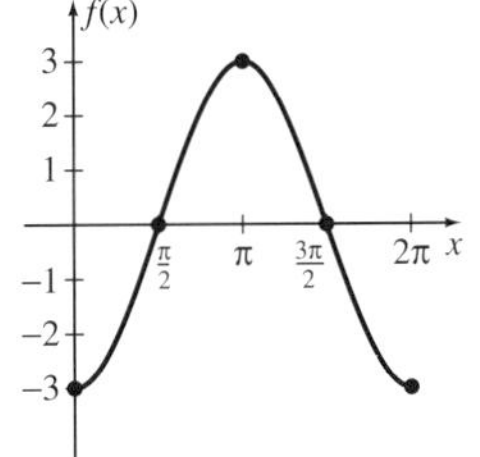

7. 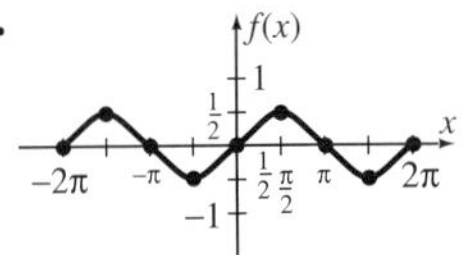

8. 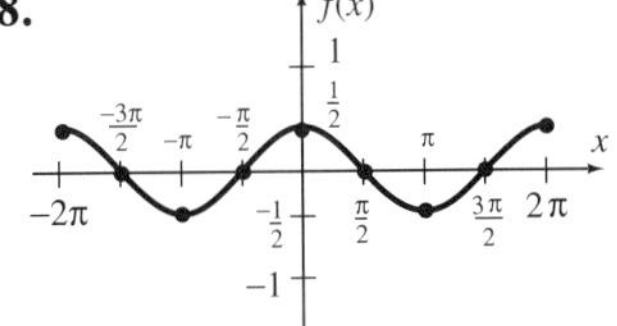

9. 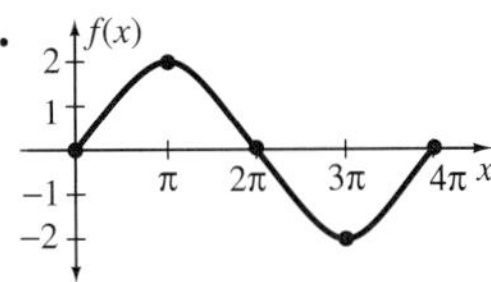

10. 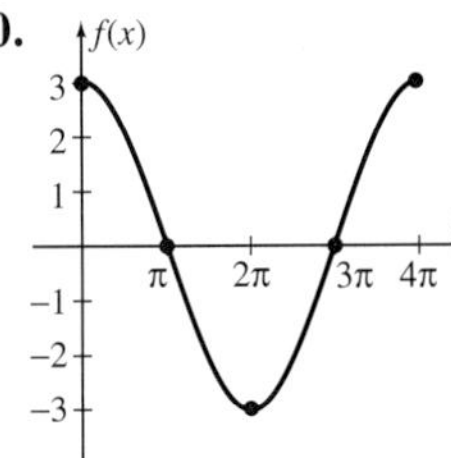

11. 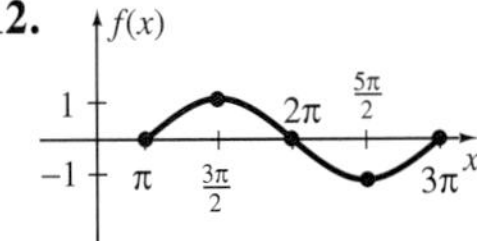

12. 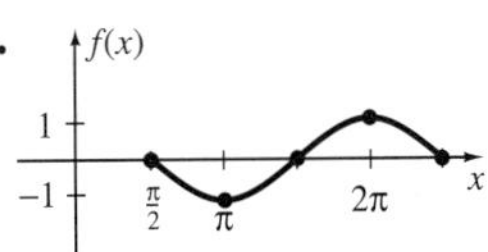

13. 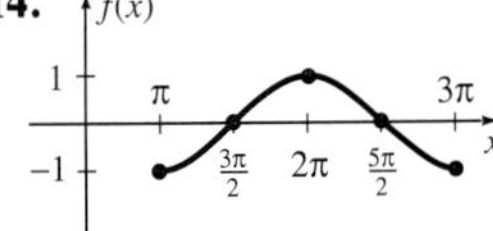

14. 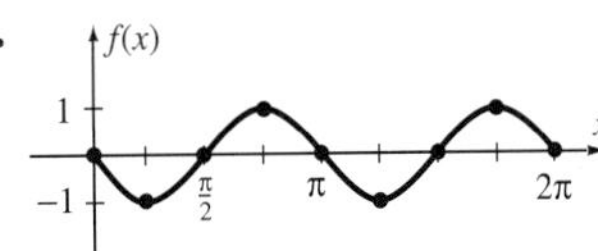

15. 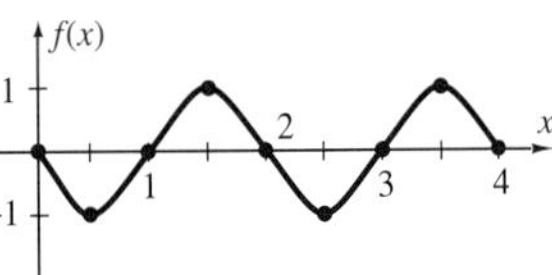

16. 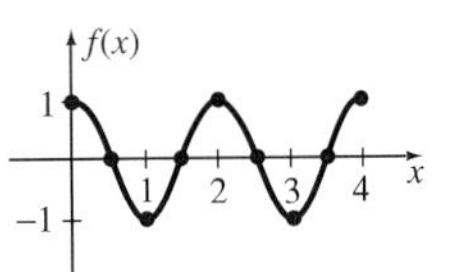

17. 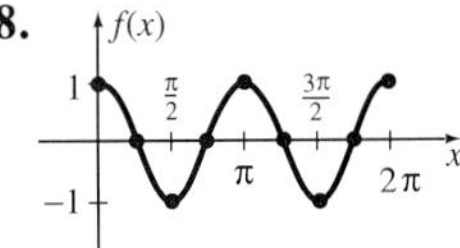

18.

19. period: 2π
amplitude: 3
phase shift: $\frac{\pi}{2}$ to the left

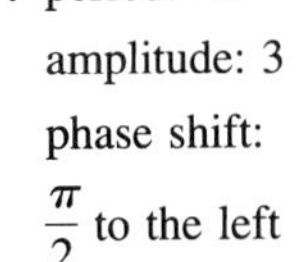

20. period: 2π

amplitude: $\frac{1}{2}$

phase shift: $\frac{\pi}{2}$ to the right

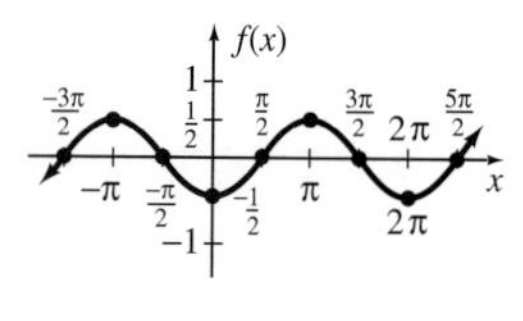

21. period: 2π

amplitude: 2

phase shift: π to the right

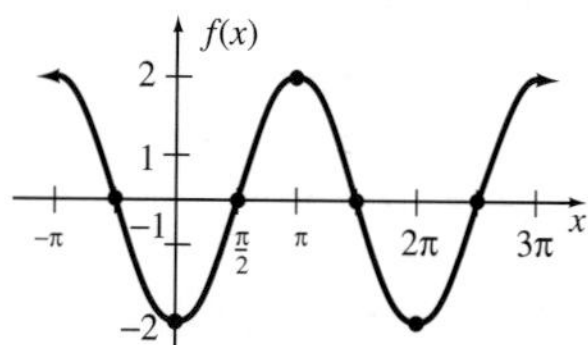

22. period: 2π

amplitude: 3

phase shift: $\frac{\pi}{2}$ to the left

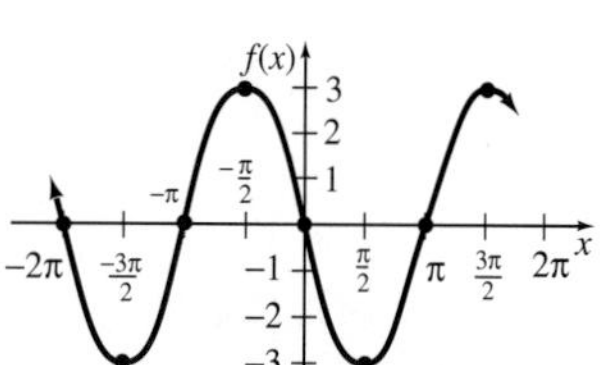

23. period: 2π

amplitude: $\frac{1}{2}$

phase shift: $\frac{\pi}{4}$ to the left

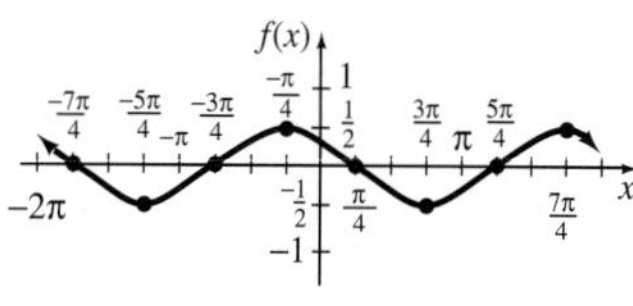

24. period: 2π

amplitude: 2

phase shift: $\frac{\pi}{3}$ to the right

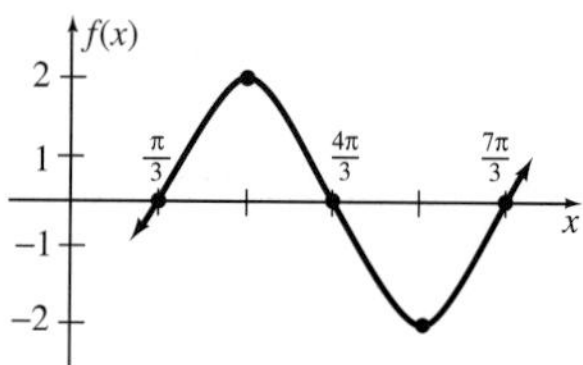

25. period: 2π

amplitude: 2

phase shift: π to the left

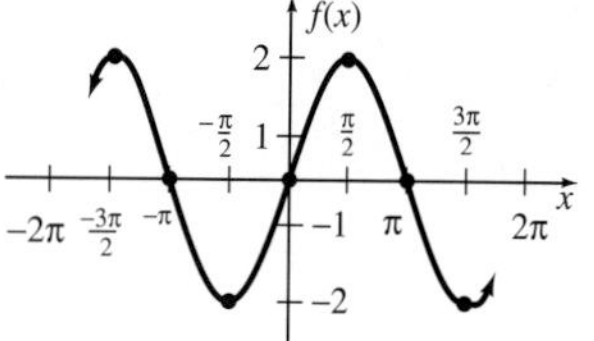

26. period: 2π

amplitude: 3

phase shift: π to the left

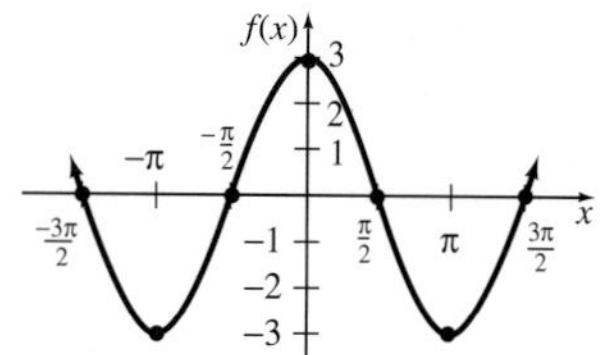

27. period: π

amplitude: 2

phase shift: π to the right

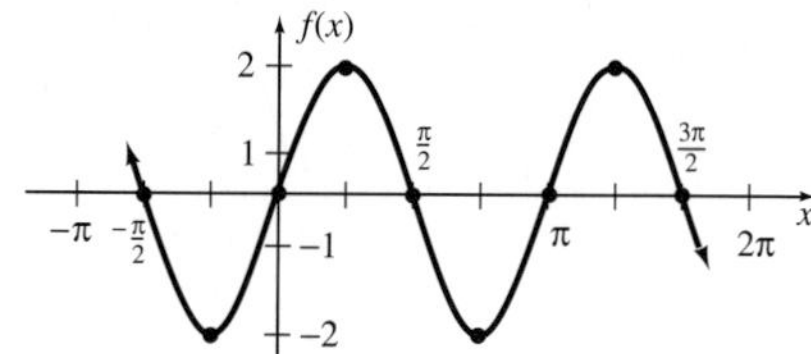

28. period: π

amplitude: 3

phase shift: $\frac{\pi}{2}$ to the right

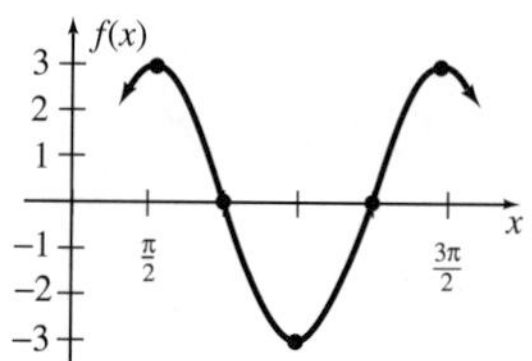

29. period: $\frac{2\pi}{3}$

amplitude: $\frac{1}{2}$

phase shift: π to the left

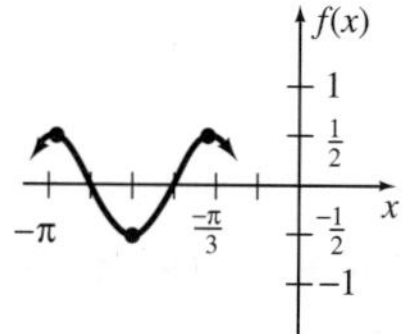

30. period: $\frac{2\pi}{3}$

amplitude: 2

phase shift: π to the right

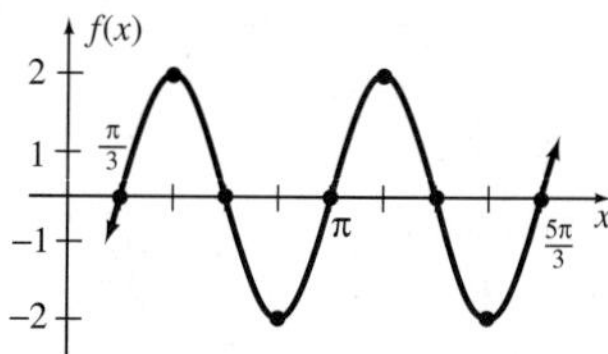

31. period: π

amplitude: 1

phase shift: $\frac{\pi}{2}$ to the right

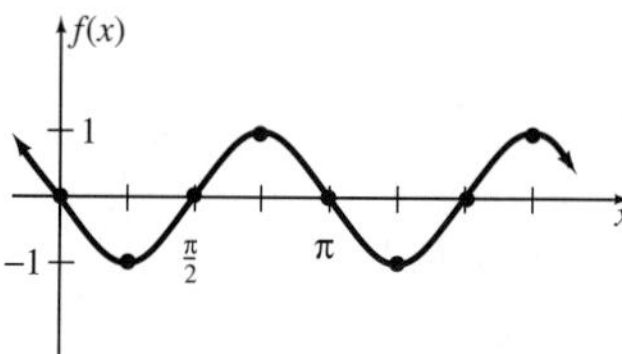

32. period: π

amplitude: 1

phase shift: $\frac{\pi}{4}$ to the left

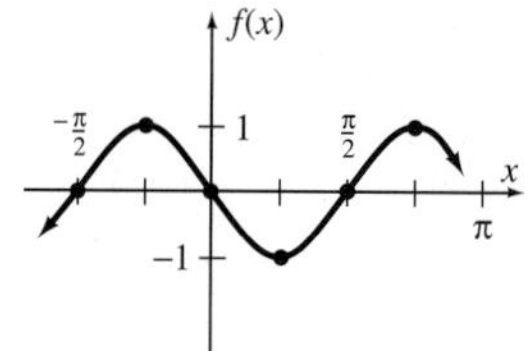

33. period: $\frac{2\pi}{3}$

amplitude: 2

phase shift: $\frac{\pi}{3}$ to the right

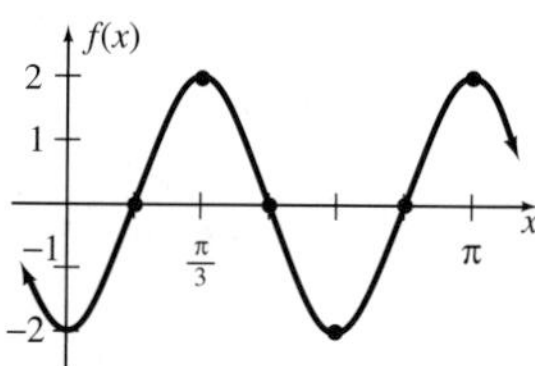

34. period: π
amplitude: 4
phase shift:
$\frac{\pi}{2}$ to the left

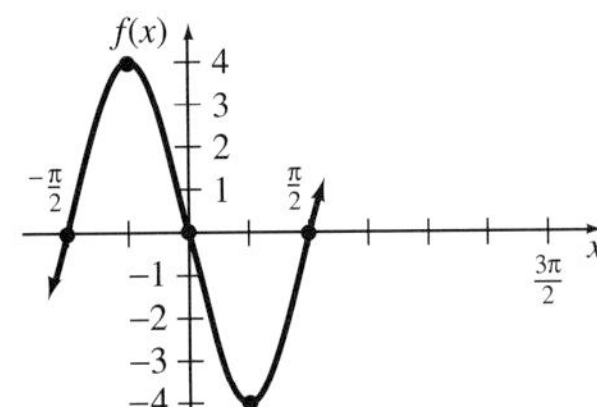

35. period: $\frac{2\pi}{3}$
amplitude: $\frac{1}{2}$
phase shift:
$\frac{\pi}{3}$ to the right

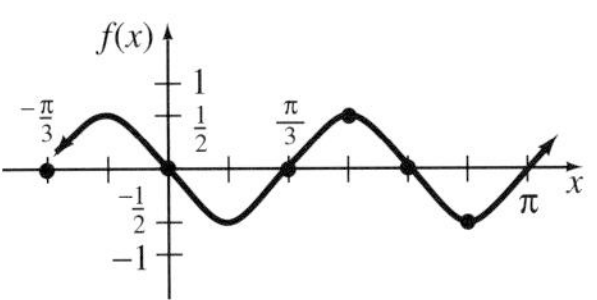

36. period: 4π
amplitude: 2
phase shift:
π to the right

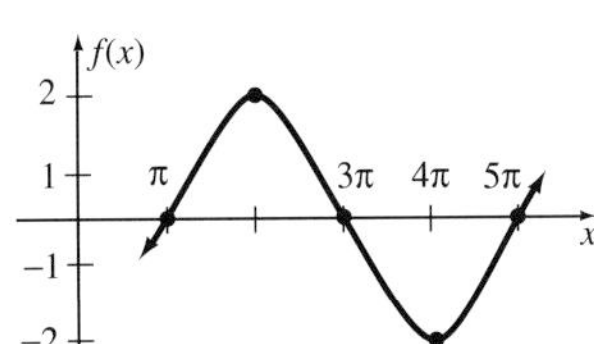

37. Period: 4π
amplitude: 2
phase shift:
π to the left

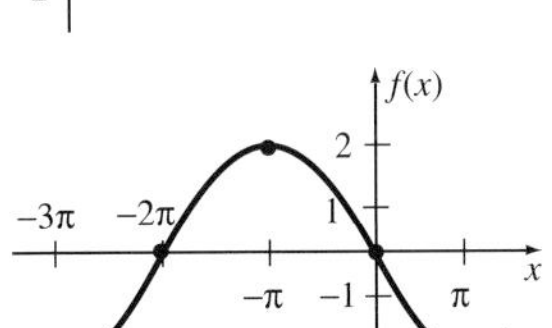

38. period: π
amplitude: 2
phase shift:
$\frac{\pi}{2}$ to the right

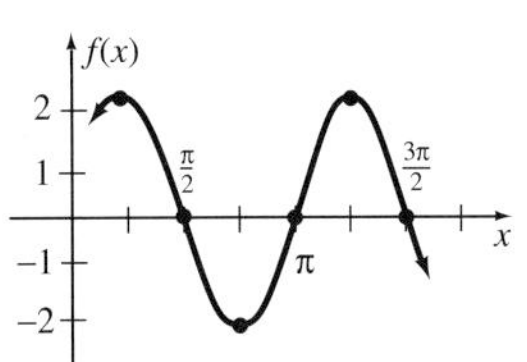

39. period: π
amplitude: 3
phase shift:
$\frac{\pi}{2}$ to the left

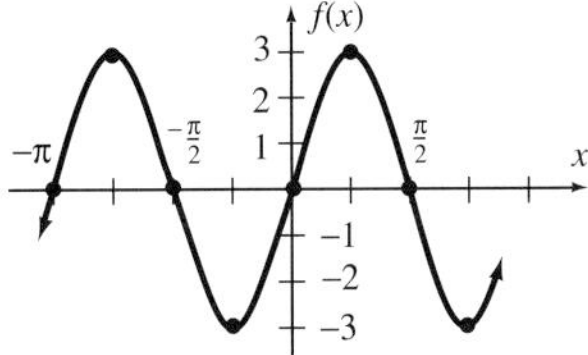

40. period: π
amplitude: $\frac{1}{2}$
phase shift:
$\frac{\pi}{4}$ to the right

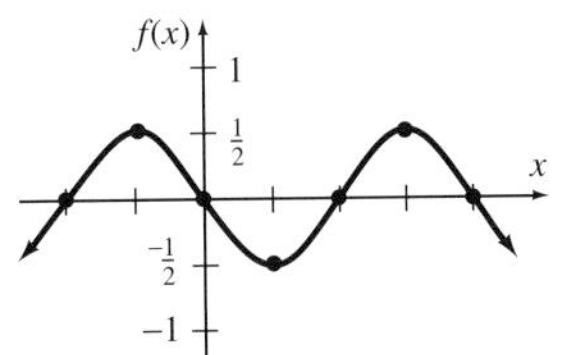

41.

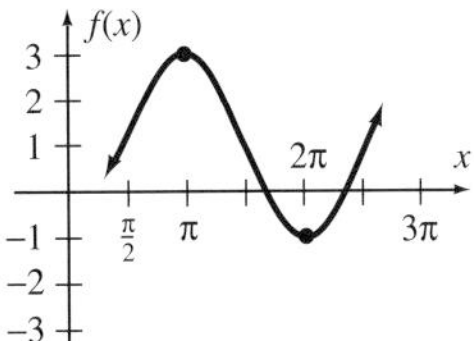

42.

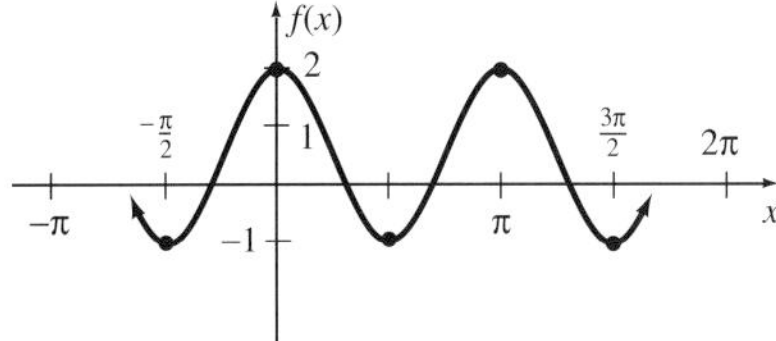

43.

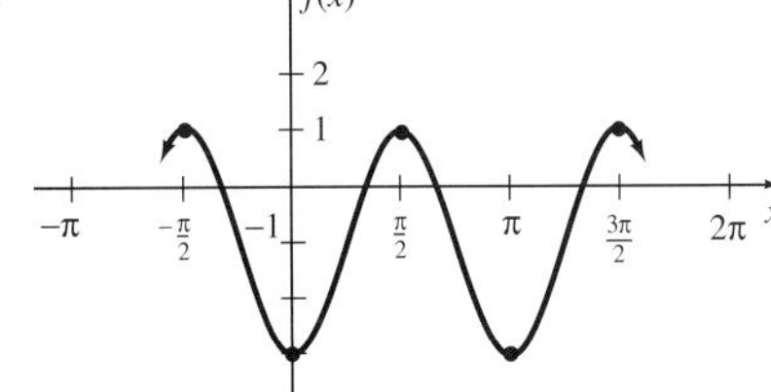

44.

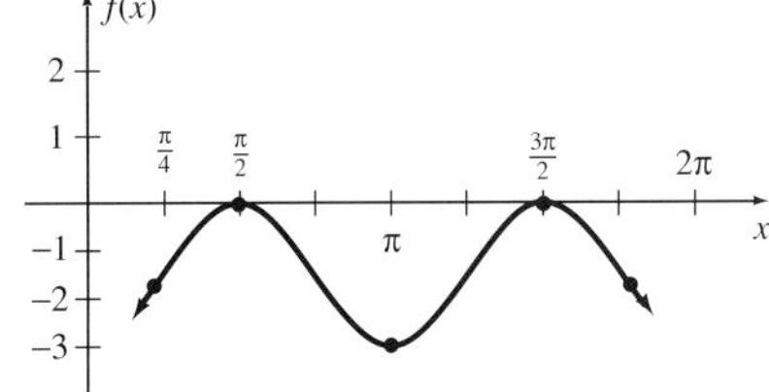

Problem Set 8.3 (page 506)

1.

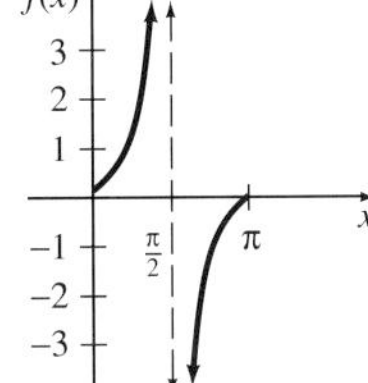

2.

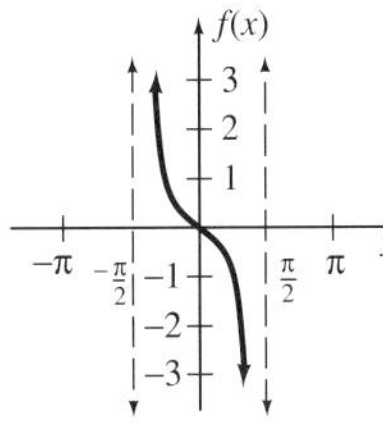

3.

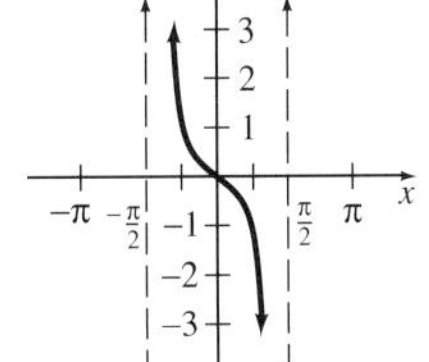

4.

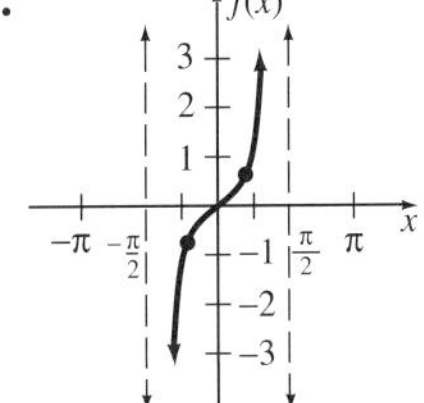

5. 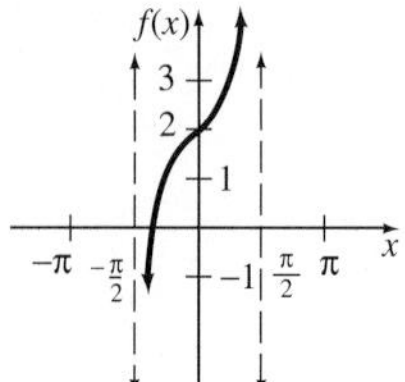

6. 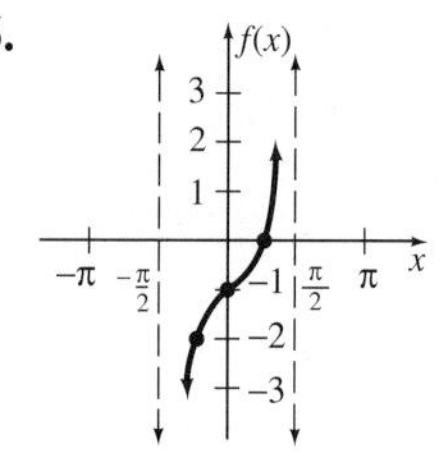

7. 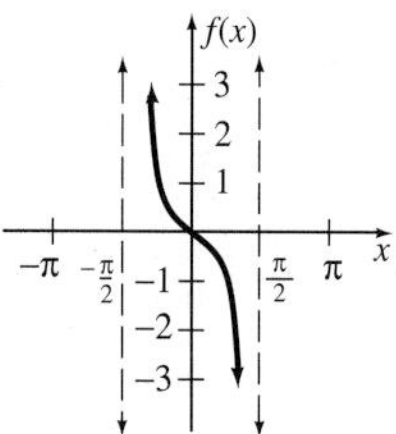

8. 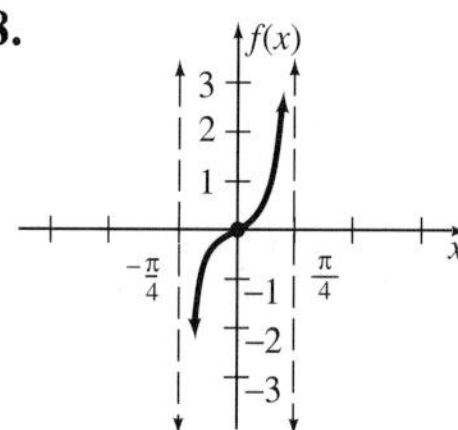

9. 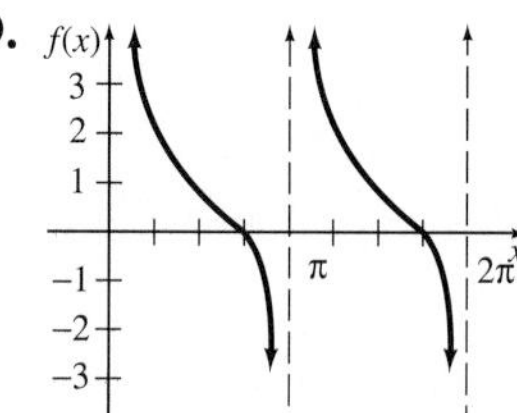

10. 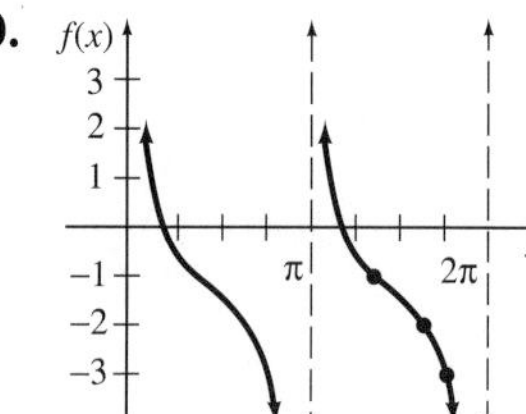

11. 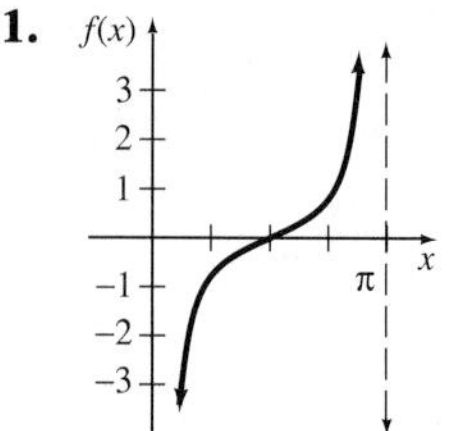

12. 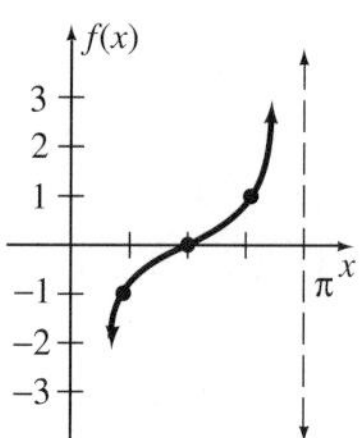

13. 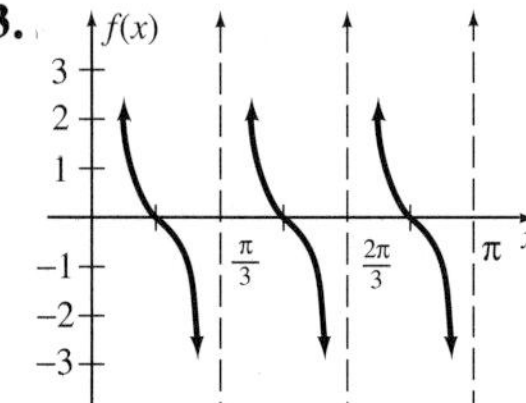

14. 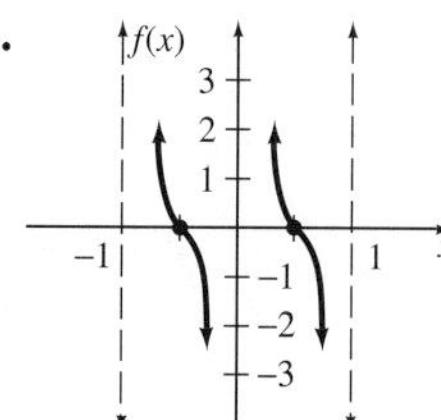

15. 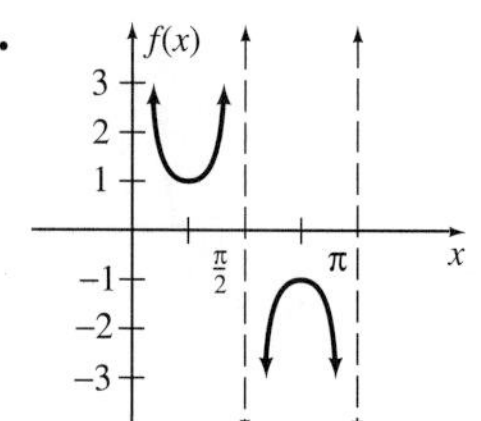

16. 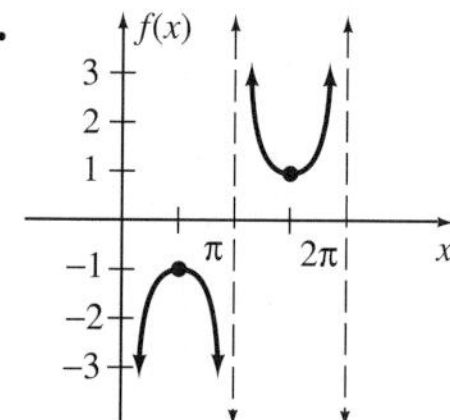

17. 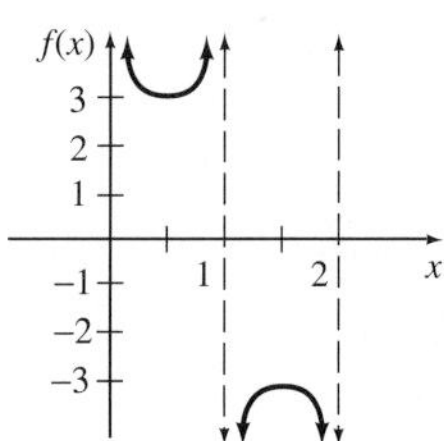

18. 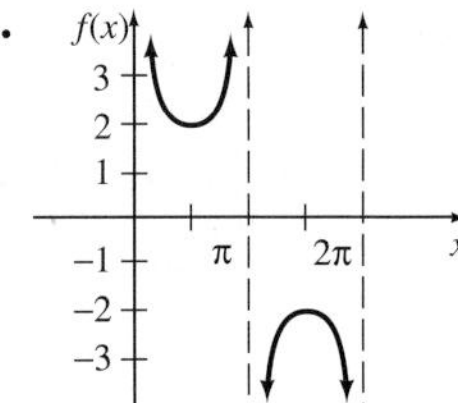

19. 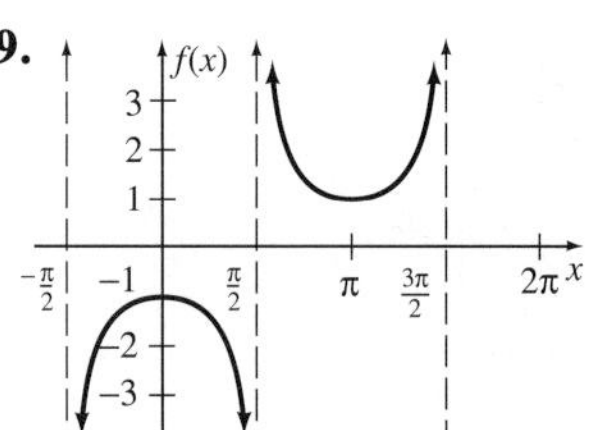

20. 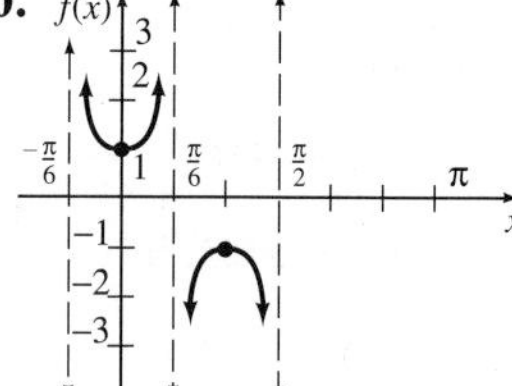

21. 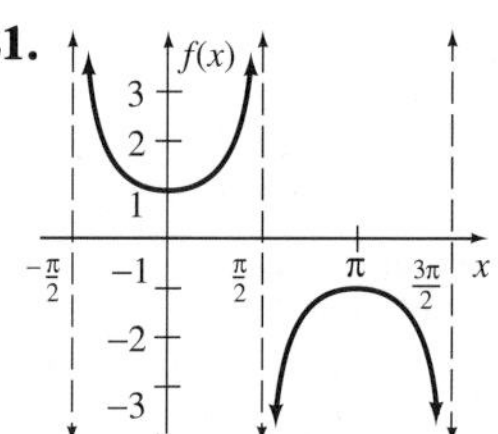

22. 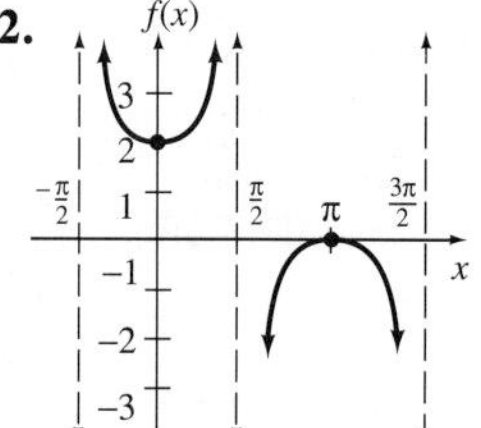

23. period: π

phase shift:

$\dfrac{\pi}{2}$ to the left

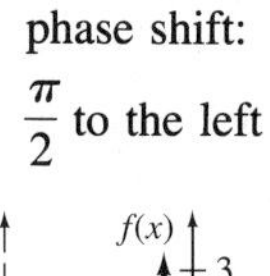

24. period: π

phase shift:

π to the right

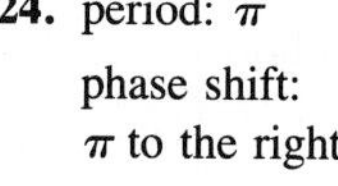

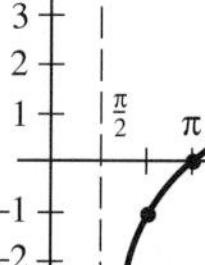

25. period: π

phase shift:

$\dfrac{\pi}{4}$ to the right

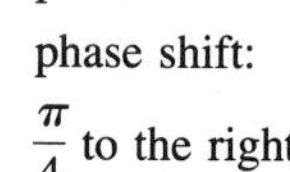

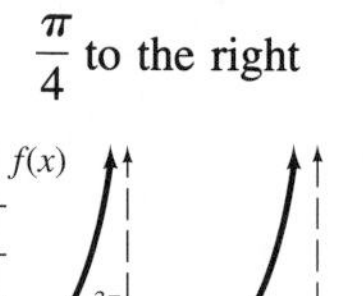

26. period: π

phase shift:

$\dfrac{\pi}{4}$ to the left

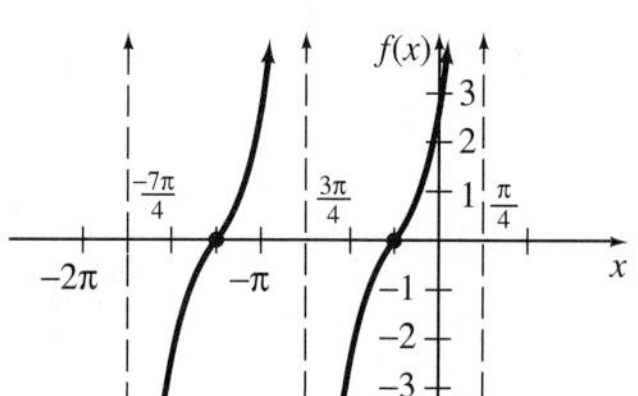

27. period: $\frac{\pi}{2}$

phase shift:

$\frac{\pi}{2}$ to the left

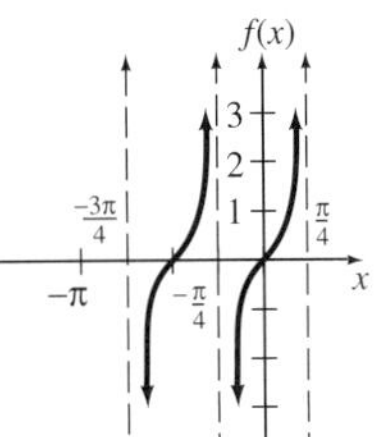

28. period: $\frac{\pi}{3}$

phase shift:

$\frac{\pi}{3}$ to the right

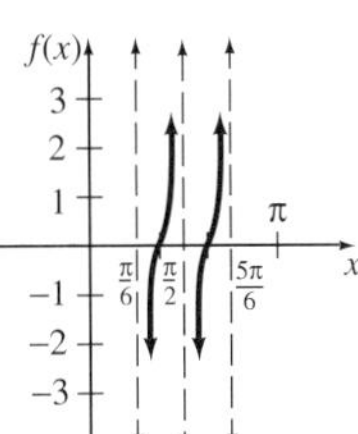

29. period: π

phase shift:

$\frac{\pi}{2}$ to the right

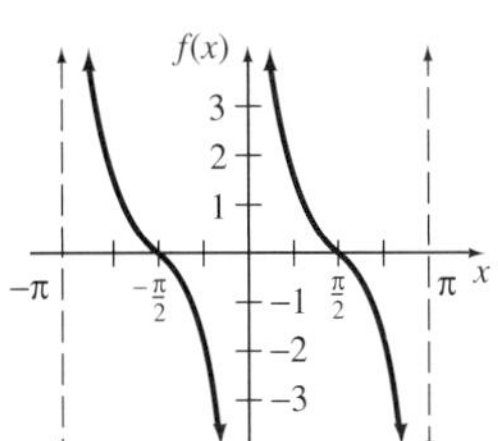

30. period: π

phase shift:

π to the left

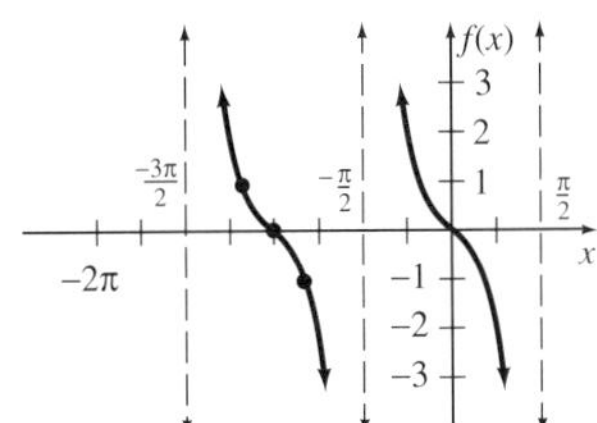

31. period: $\frac{\pi}{3}$

phase shift:

$\frac{2\pi}{3}$ to the right

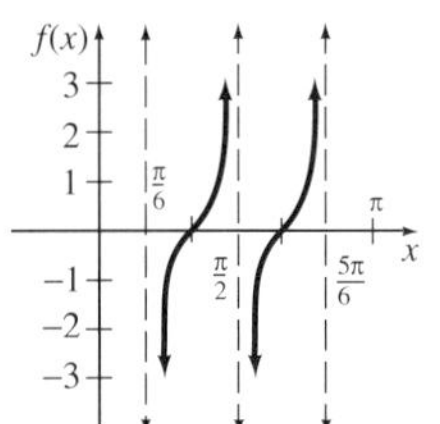

32. period: $\frac{\pi}{2}$

phase shift:

$\frac{3\pi}{2}$ to the left

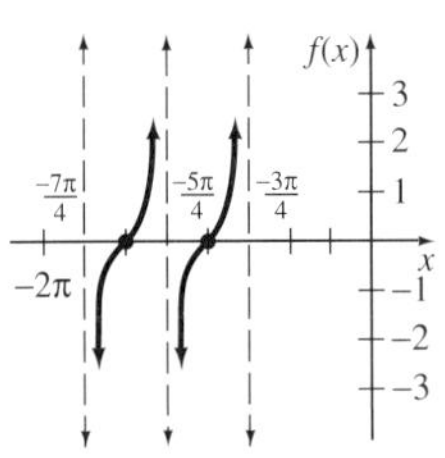

33. period: π

phase shift:

$\frac{\pi}{4}$ to the right

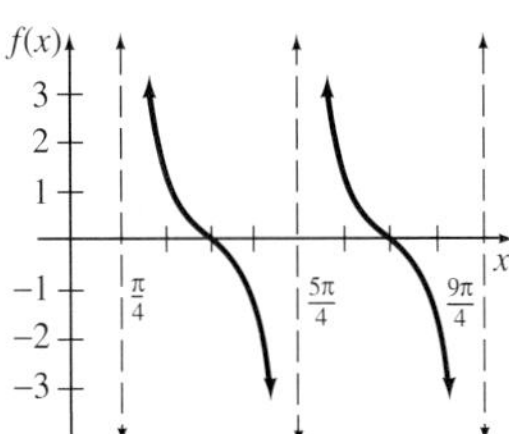

34. period: π

phase shift:

$\frac{\pi}{2}$ to the left

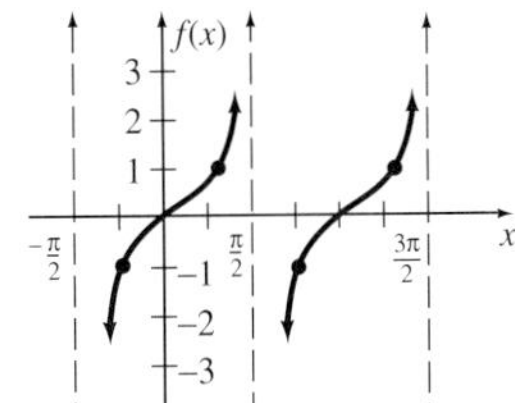

35. period: 2π

phase shift:

$\frac{\pi}{2}$ to the left

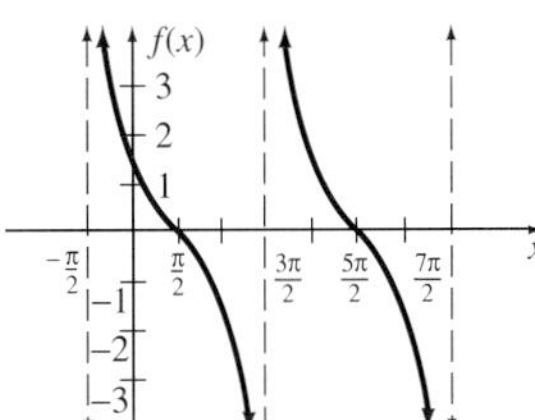

36. period: $\frac{\pi}{2}$

phase shift:

π to the right

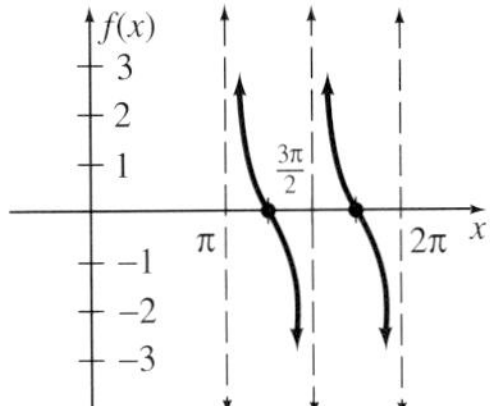

37. period: 2π

phase shift:

$\frac{\pi}{2}$ to the right

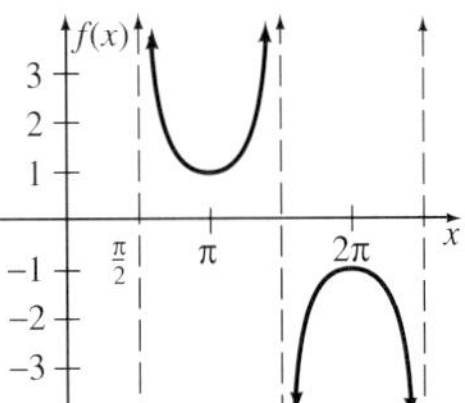

38. period: 2π

phase shift:

π to the left

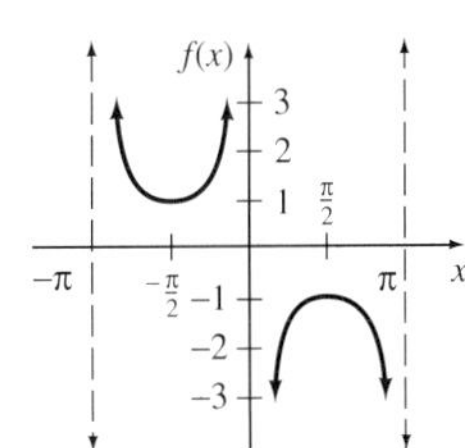

39. period: 2π
phase shift:
$\frac{\pi}{4}$ to the left

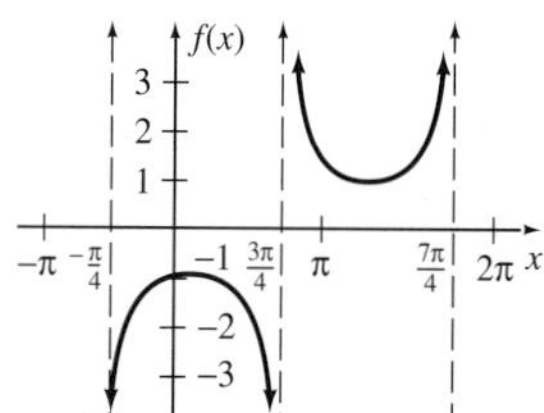

40. period: 2π
phase shift:
$\frac{\pi}{2}$ to the right

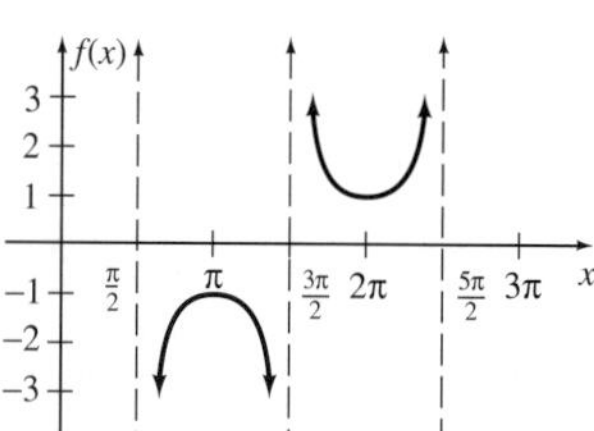

41. period: π
phase shift:
$\frac{\pi}{2}$ to the left

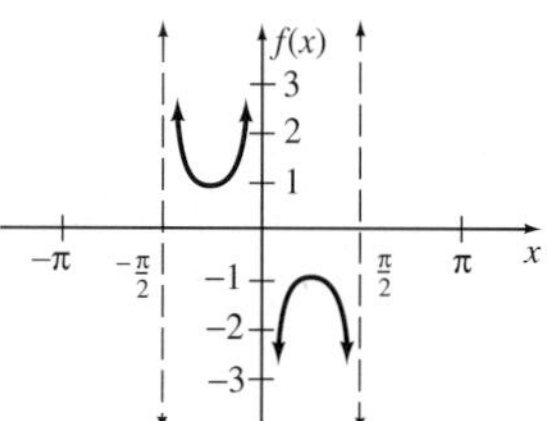

42. period: 4π
phase shift:
$\frac{\pi}{2}$ to the left

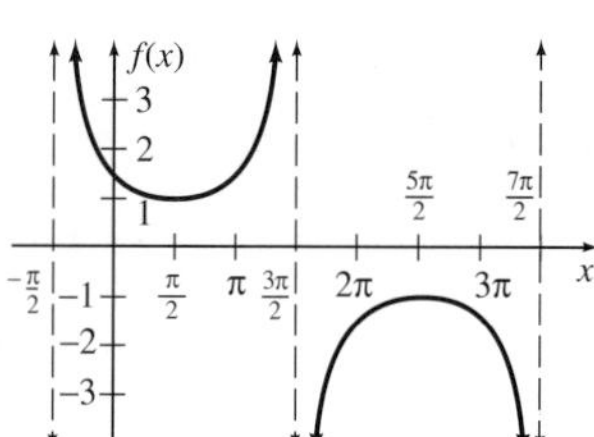

43. period: 2π
phase shift:
π to the left

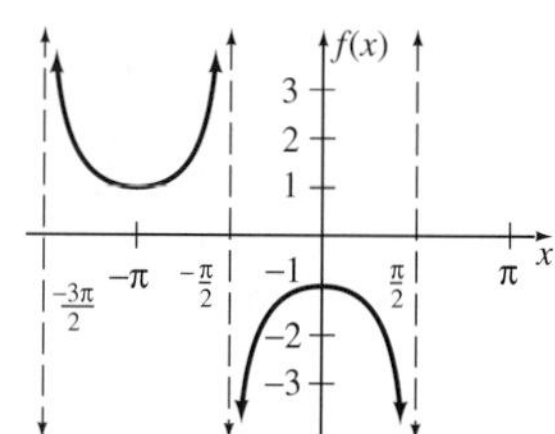

44. period: 2π
phase shift:
$\frac{\pi}{2}$ to the right

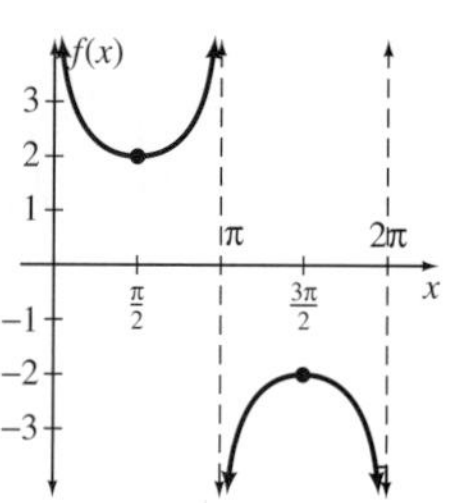

45. period: 2π
phase shift:
$\frac{\pi}{4}$ to the right

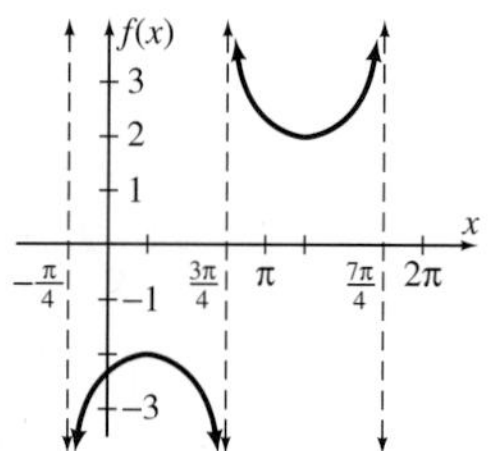

46. period: $\frac{2\pi}{3}$
phase shift:
π to the right

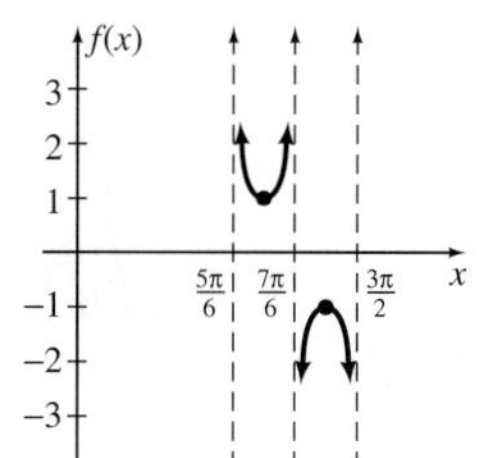

Problem Set 8.4 (page 514)

1. $\frac{\pi}{4}$ **2.** $\frac{\pi}{3}$ **3.** $-\frac{\pi}{3}$ **4.** $\frac{\pi}{2}$ **5.** $\frac{\pi}{3}$
6. $\frac{2\pi}{3}$ **7.** $\frac{\pi}{6}$ **8.** $\frac{\pi}{2}$ **9.** $\frac{\pi}{4}$ **10.** $-\frac{\pi}{4}$
11. $\frac{\pi}{3}$ **12.** 0 **13.** $-\frac{\pi}{2}$ **14.** π **15.** 30°
16. −30° **17.** 135° **18.** 150° **19.** 0°
20. −30° **21.** −45° **22.** −60° **23.** 0.366
24. 1.041 **25.** −1.154 **26.** −0.166
27. 0.940 **28.** 1.528 **29.** 2.000 **30.** 2.667
31. 1.455 **32.** 0.741 **33.** −0.196
34. −1.553 **35.** 25.5° **36.** 46.2°
37. −55.2° **38.** −13.4° **39.** 74.7°
40. 32.7° **41.** 99.3° **42.** 128.4° **43.** 86.0°
44. 44.6° **45.** −83.1° **46.** 87.3° **47.** $\frac{\sqrt{3}}{2}$
48. $\frac{\sqrt{3}}{2}$ **49.** 0 **50.** 0 **51.** 1
52. $-\frac{\sqrt{3}}{3}$ **53.** $\frac{\sqrt{3}}{2}$ **54.** $\frac{\sqrt{3}}{2}$ **55.** $\sqrt{2}$
56. 0 **57.** $\frac{3}{5}$ **58.** $\frac{12}{13}$ **59.** $\frac{3}{5}$ **60.** $\frac{3}{5}$
61. $-\frac{4}{3}$ **62.** $-\frac{12}{5}$ **63.** $\frac{\sqrt{5}}{3}$ **64.** $-\frac{2\sqrt{13}}{13}$
65. $-2\sqrt{2}$ **66.** $\frac{2\sqrt{21}}{21}$ **67.** $\frac{4\sqrt{7}}{7}$
68. $\frac{3\sqrt{5}}{5}$ **69.** $-\frac{3\sqrt{10}}{20}$ **70.** $\frac{5\sqrt{6}}{12}$

71.

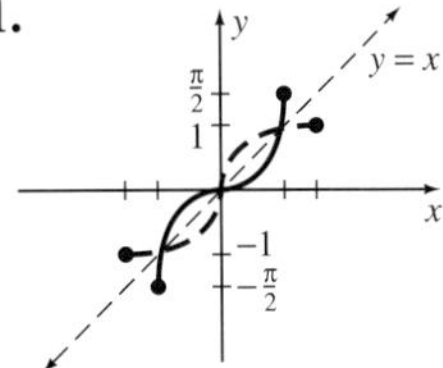

72.

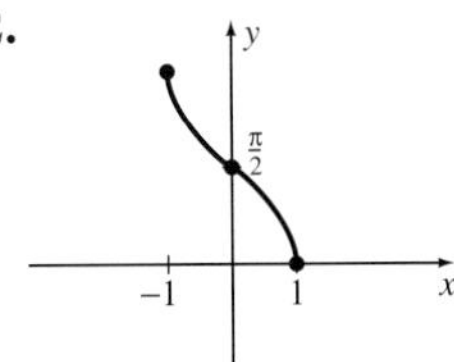

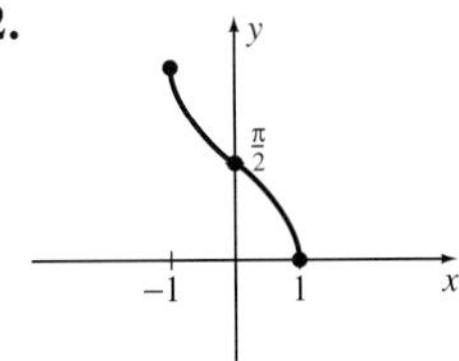

73.

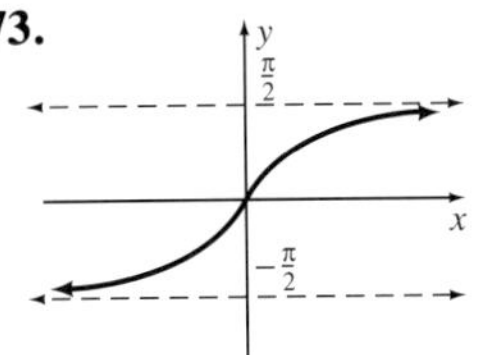

74.

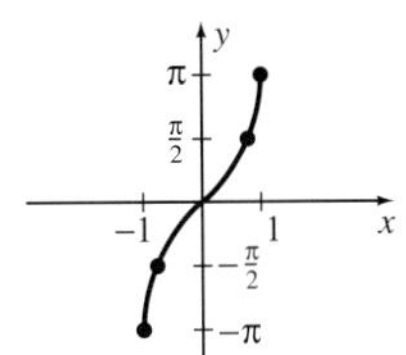

75.

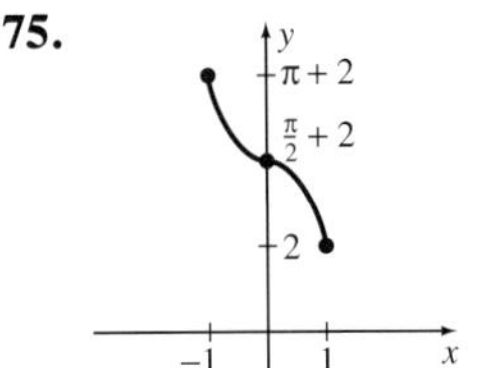

76.

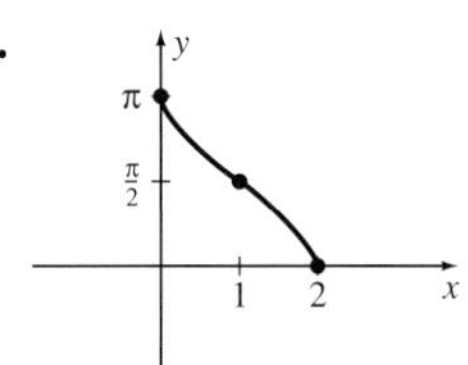

77.

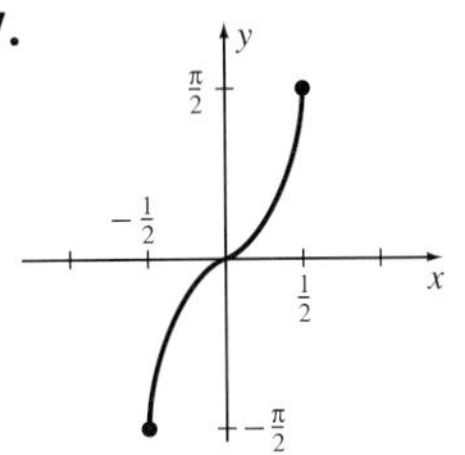

78.

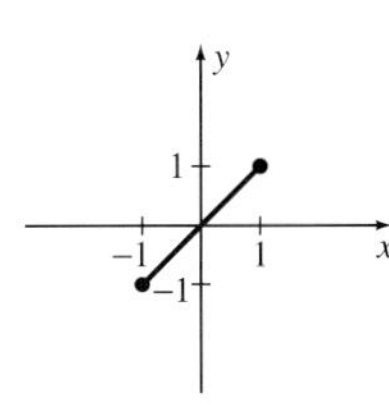

79.

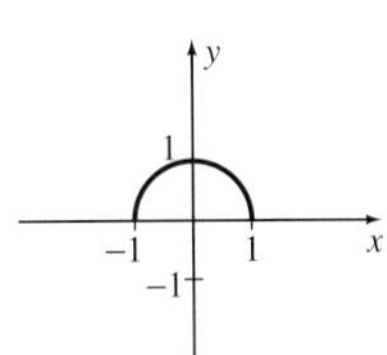

80.

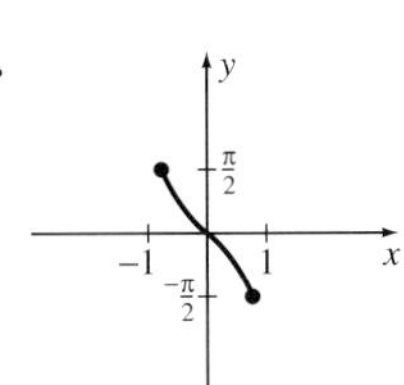

81.

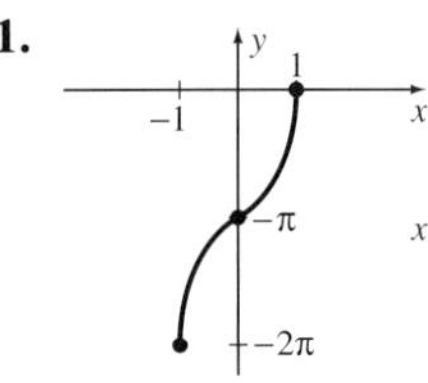

Problem Set 8.5 (page 522)

The points in Problems 1, 3, 5, 7, 9, and 11 are located on the following figure.

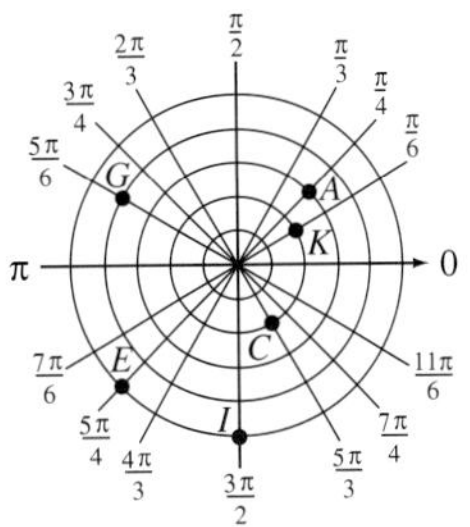

The points in Problems 2, 4, 6, 8, 10, and 12 are located on the following figure.

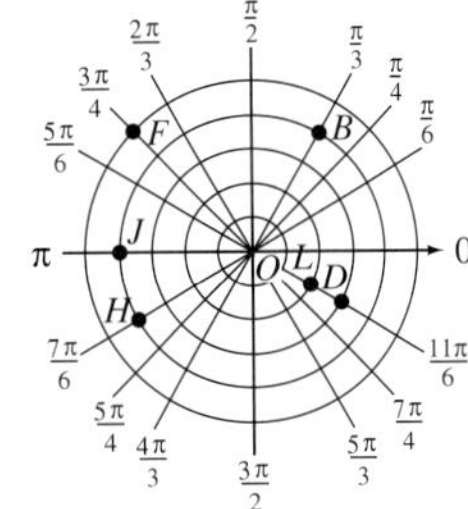

13. $\left(\frac{3\sqrt{3}}{2}, \frac{3}{2}\right)$ **14.** $\left(-3\sqrt{3}, 3\right)$

15. $\left(2\sqrt{2}, 2\sqrt{2}\right)$ **16.** $\left(-\sqrt{2}, \sqrt{2}\right)$

17. $\left(1, \sqrt{3}\right)$ **18.** $\left(-\frac{5\sqrt{3}}{2}, -\frac{5}{2}\right)$

19. $\left(-\frac{3}{2}, -\frac{3\sqrt{3}}{2}\right)$ **20.** $\left(\frac{7\sqrt{3}}{2}, -\frac{7}{2}\right)$

21. $\left(-2, -2\sqrt{3}\right)$ **22.** $\left(3, -3\sqrt{3}\right)$

23. $\left(-\frac{1}{2}, -\frac{\sqrt{3}}{2}\right)$ **24.** $\left(\frac{3\sqrt{3}}{2}, \frac{3}{2}\right)$

25. $\left(-\frac{7\sqrt{2}}{2}, -\frac{7\sqrt{2}}{2}\right)$ **26.** $\left(2\sqrt{2}, -2\sqrt{2}\right)$

27. $\left(\sqrt{3}, 1\right)$ **28.** $\left(-\frac{1}{2}, -\frac{\sqrt{3}}{2}\right)$ **29.** $(0, -3)$

30. $(0, -8)$ **31.** $\left(2, \frac{3\pi}{4}\right)$ **32.** $\left(4, \frac{7\pi}{4}\right)$

33. $\left(5, \frac{7\pi}{6}\right)$ **34.** $\left(6, \frac{5\pi}{6}\right)$ **35.** $\left(6, \frac{5\pi}{3}\right)$

36. $\left(1, \frac{4\pi}{3}\right)$ **37.** $(4, \pi)$ **38.** $\left(3, \frac{3\pi}{2}\right)$

39. $\left(3, \frac{\pi}{6}\right)$ **40.** $\left(2, \frac{\pi}{6}\right)$ **41.** $\left(-2, \frac{5\pi}{4}\right)$

42. $\left(-3, \frac{7\pi}{4}\right)$ **43.** $\left(-4, \frac{2\pi}{3}\right)$ **44.** $\left(-1, \frac{4\pi}{3}\right)$
45. $\left(-5, \frac{11\pi}{6}\right)$ **46.** $\left(-2, \frac{\pi}{6}\right)$
47. $(\sqrt{13}, 33.7°)$ **48.** $(\sqrt{29}, 68.2°)$
49. $(5, 143.1°)$ **50.** $(2\sqrt{10}, 341.6°)$
51. $(\sqrt{17}, 194.0°)$ **52.** $(5, 233.1°)$
53. $r \sin\theta = 2$ **54.** $r\cos\theta = 7$
55. $r(3\cos\theta - 2\sin\theta) = 4$
56. $r(5\cos\theta + 4\sin\theta) = 10$ **57.** $\tan\theta = 1$
58. $\tan\theta = -2$ **59.** $r = 8\cos\theta$
60. $r = -6\sin\theta$ **61.** $r = 1 - \cos\theta$
62. $r = 2 + 2\sin\theta$ **63.** $r = 4\tan\theta\sec\theta$
64. $r = \cot\theta\csc\theta$ **65.** $y = -4$ **66.** $x = 6$
67. $x^2 + y^2 - 3x = 0$ **68.** $x^2 + y^2 - 2y = 0$
69. $x^2 + y^2 - 2x - 3y = 0$
70. $x^2 + y^2 - 3x + 4y = 0$ **71.** $y + 4x = 5$
72. $2y - 3x = -4$ **73.** $3x^2 + 4y^2 + 8x - 16 = 0$
74. $4x^2 - 5y^2 - 30y - 25 = 0$
75. $x^2 + y^2 - 3y = 2\sqrt{x^2 + y^2}$
76. $x^2 + y^2 + 2x = -3\sqrt{x^2 + y^2}$

77.
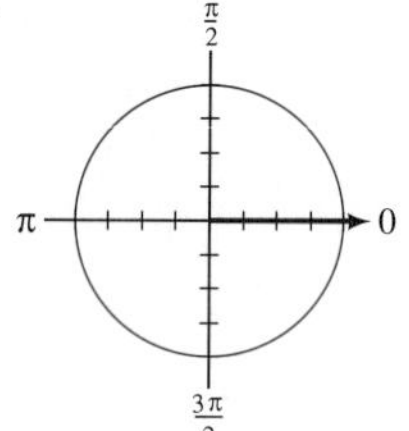

78.
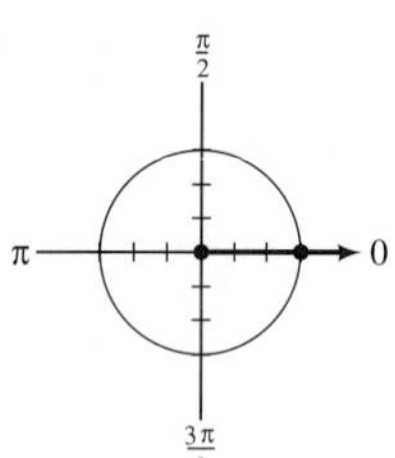

79.
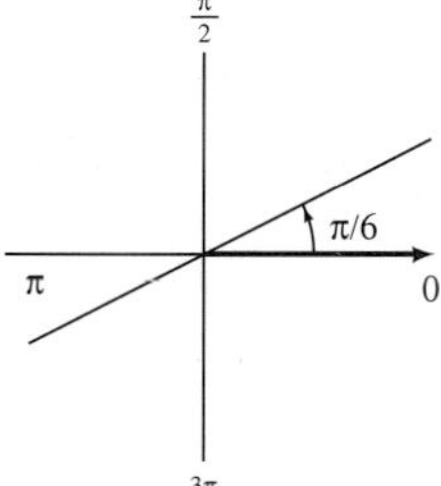

80.
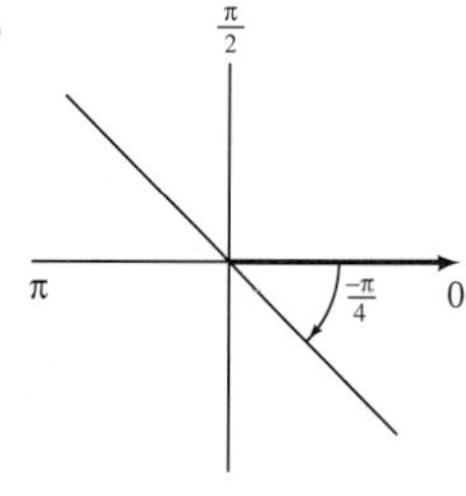

81.
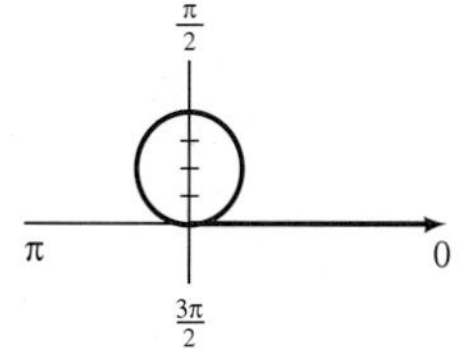

82.
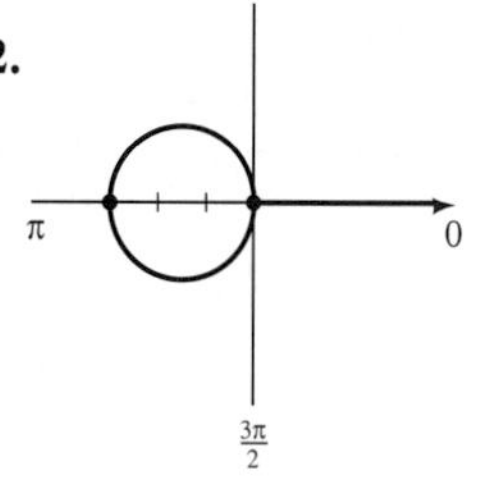

83.
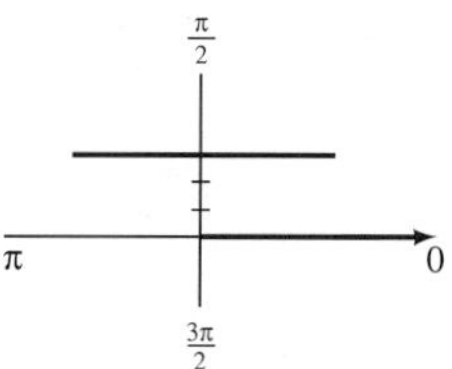

84.
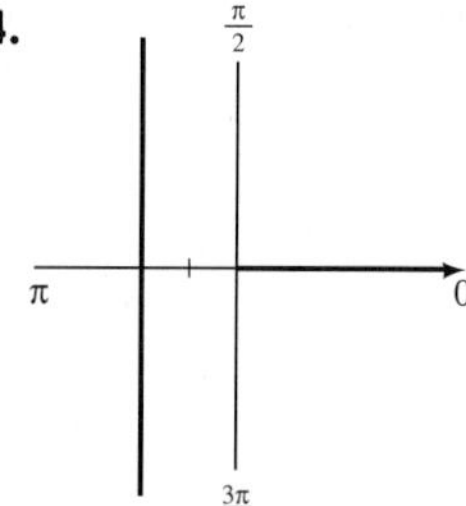

85.
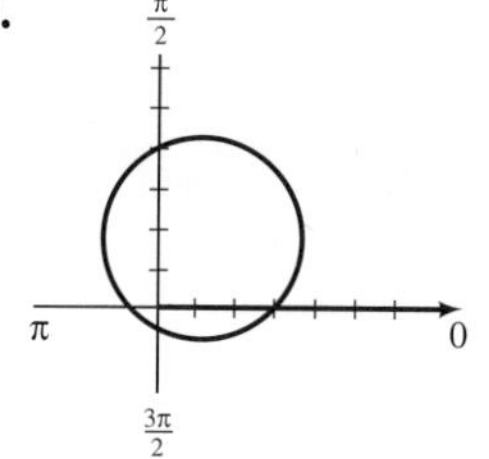

86.
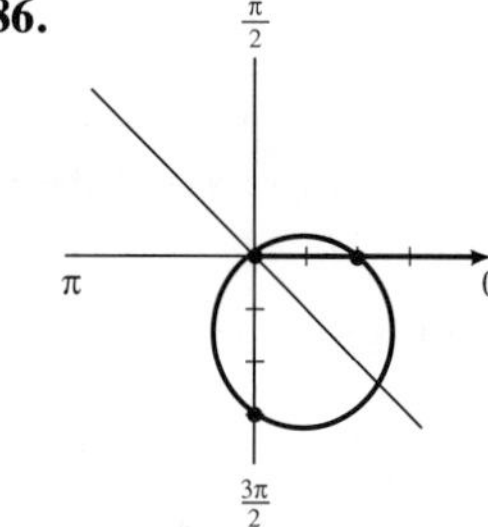

87.
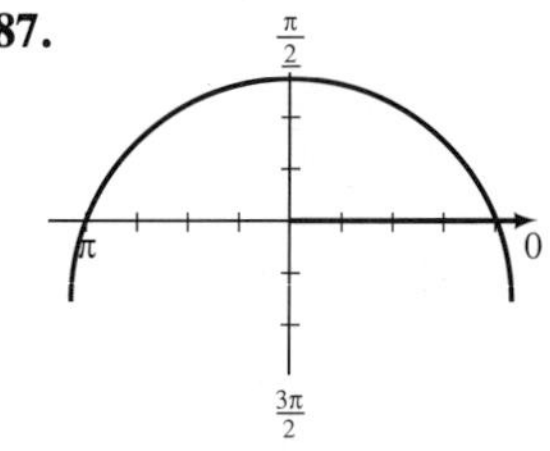

88.
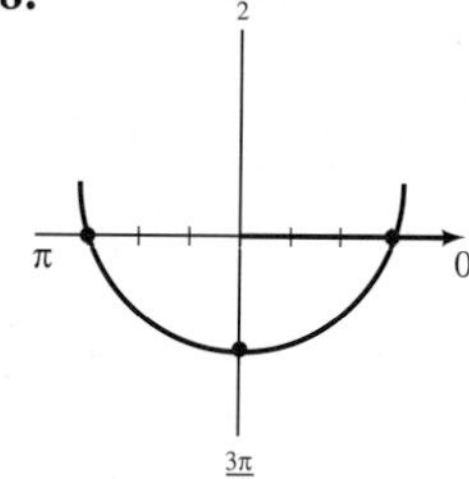

89.
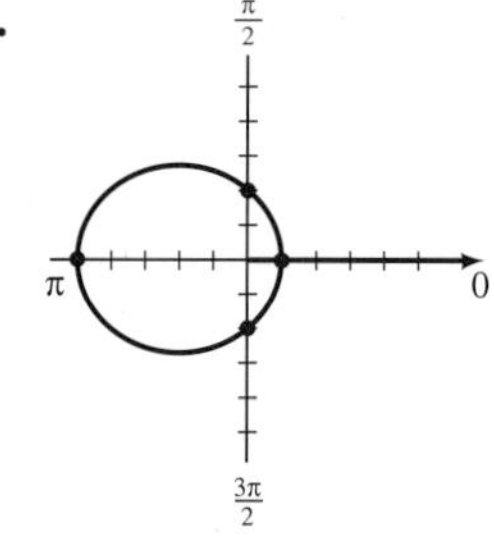

90.
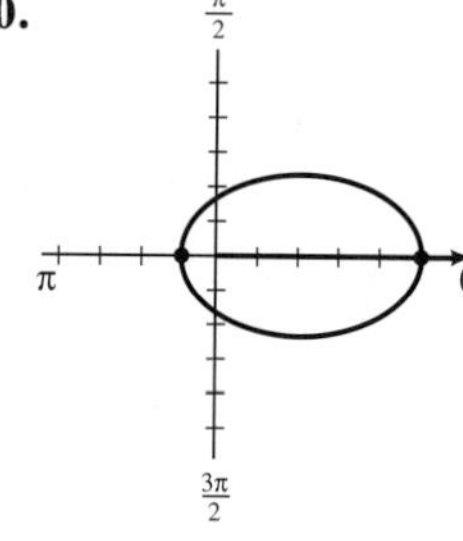

91.
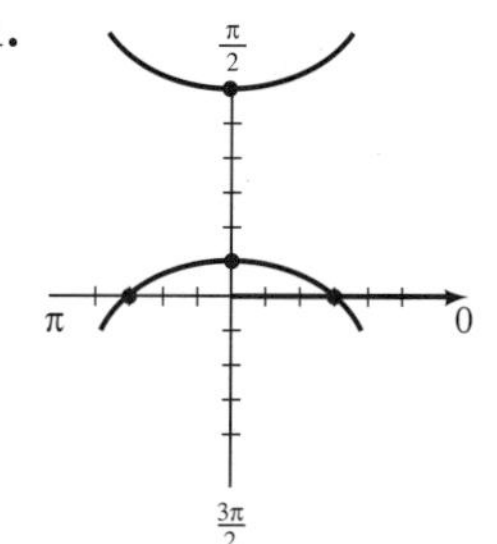

92.
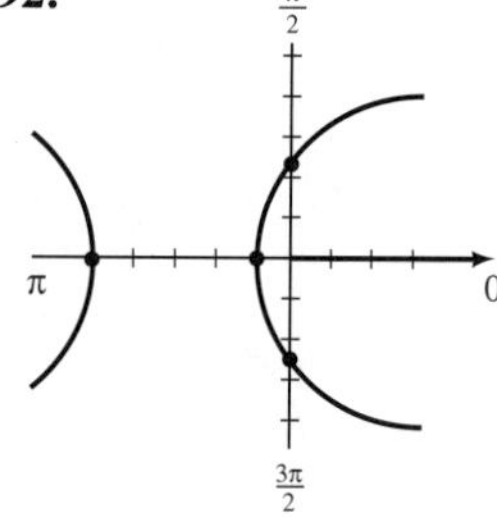

94. **(a)** $\sqrt{13}$ **(b)** 10 **(c)** $2\sqrt{26}$ **(d)** $3\sqrt{7}$

95. **(a)** The line segment $3x - y = -7$ with endpoints at $(-2, 1)$ and $(2, 13)$

(b) The line segment $2x + y = 4$ with endpoints at $(-1, 6)$ and $(4, -4)$

(c) The portion of the parabola $y = x^2 - 4x + 8$ between and including the points $(-1, 13)$ and $(4, 8)$

(d) The portion of the parabola $y = -x^2 - 2x + 1$ between and including the points $(-3, -2)$ and $(2, -7)$

(e) The circle $x^2 + y^2 = 4$

(f) The ellipse $x^2 + 4y^2 = 16$

(g) The hyperbola $x^2 - 4y^2 = 4$

(h) A portion of the right-hand branch of the curve $x^2y = 1$

Problem Set 8.6 (page 531)

1. Polar axis **2.** $\frac{\pi}{2}$-axis **3.** $\frac{\pi}{2}$-axis

4. Polar axis **5.** Polar axis **6.** $\frac{\pi}{2}$-axis

7. $\frac{\pi}{2}$-axis **8.** Polar axis **9.** None

10. None **11.** Polar axis **12.** $\frac{\pi}{2}$-axis

13. Polar axis **14.** $\frac{\pi}{2}$-axis **15.** Pole

16. Polar axis, $\frac{\pi}{2}$-axis, and pole

17.

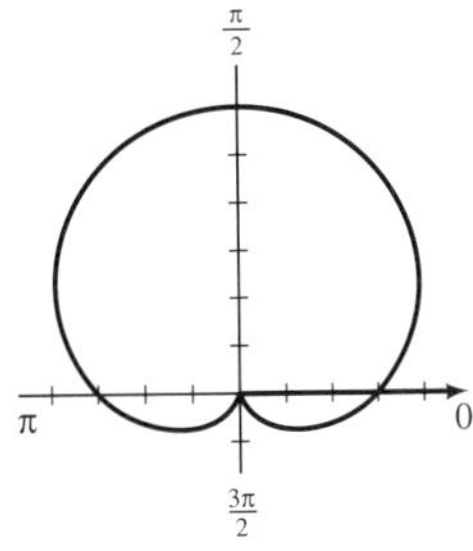

18.

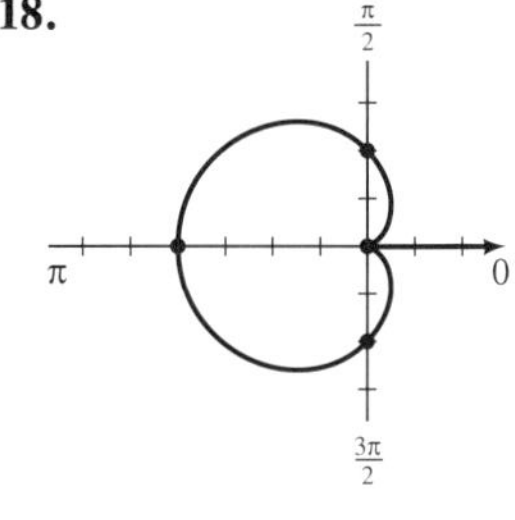

19.

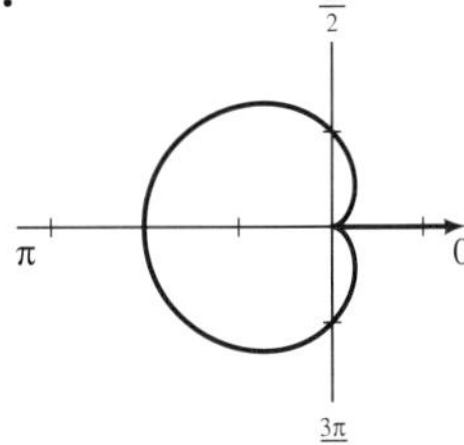

20.

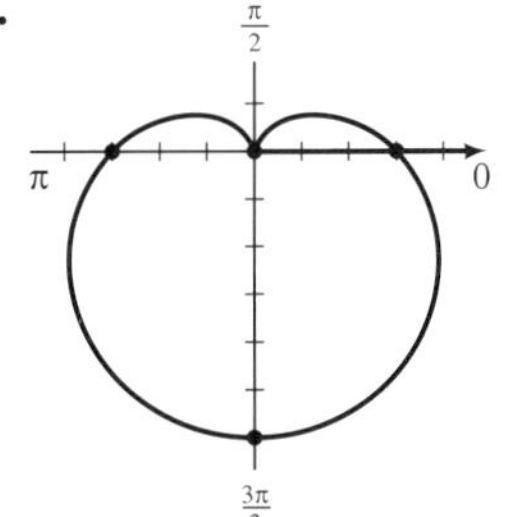
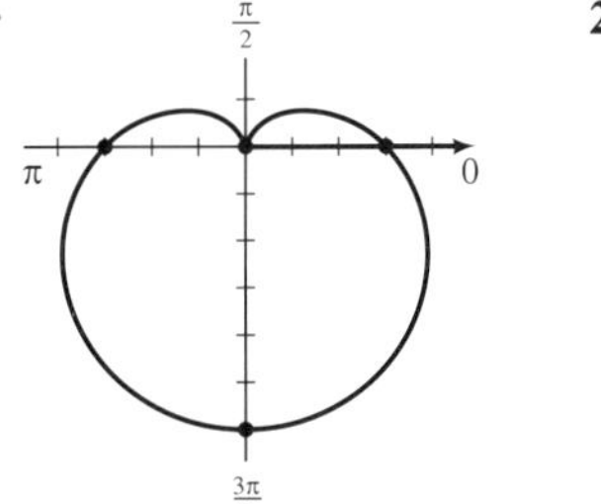

21.

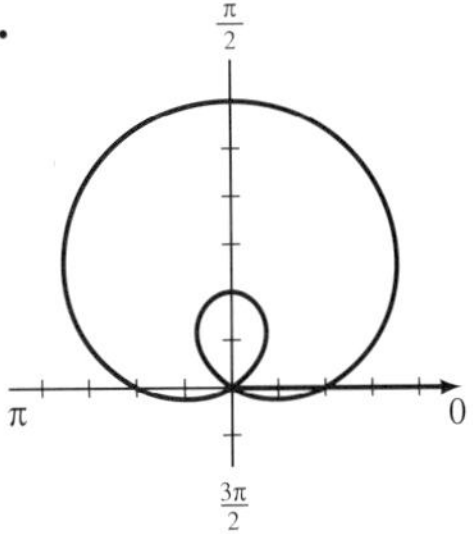

22.

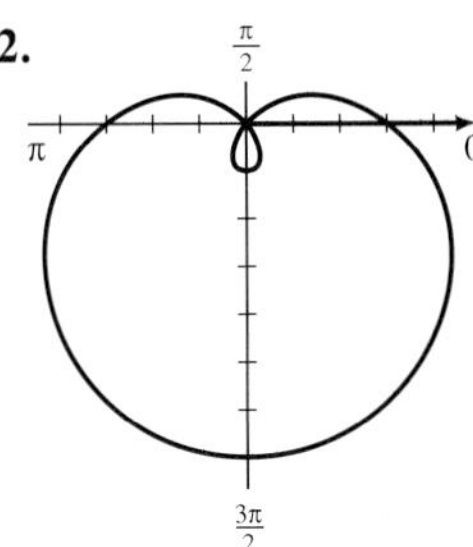

23.

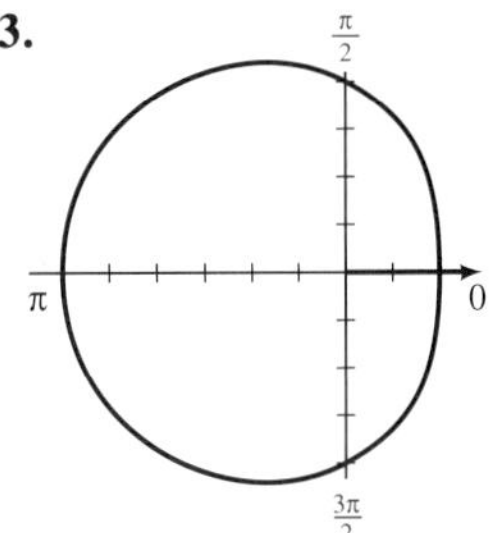

24.

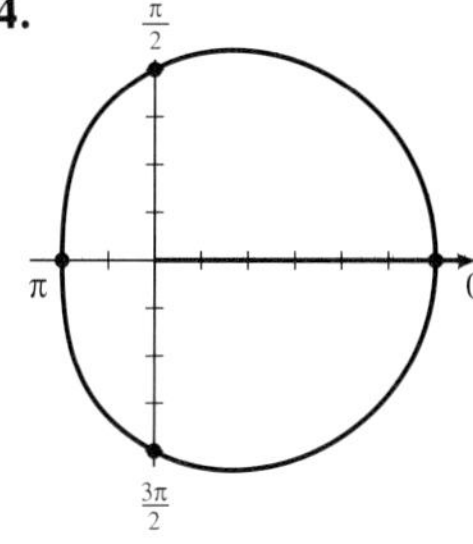

25.

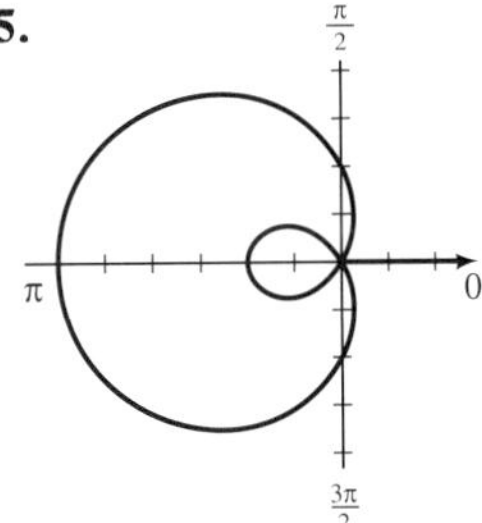

26.

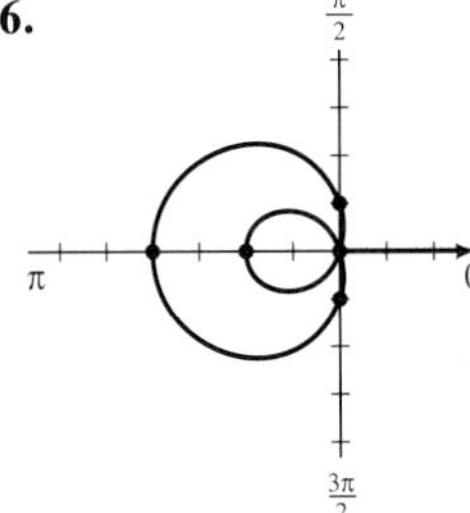

27.

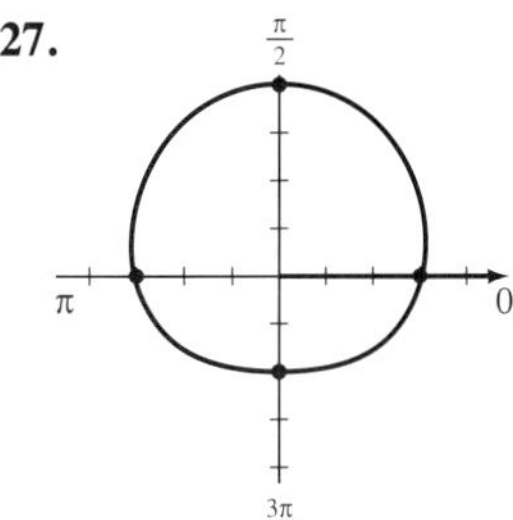

28.

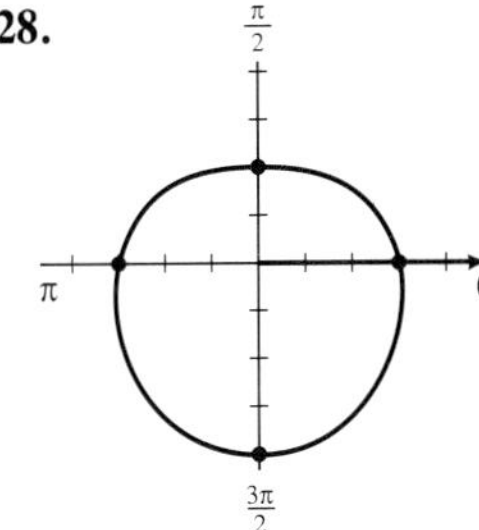

29.

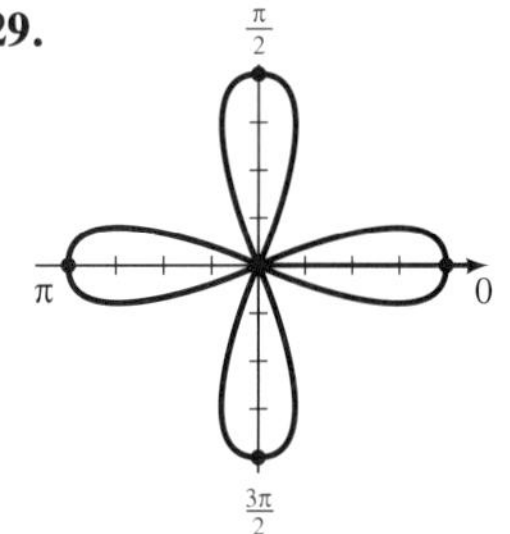

30.

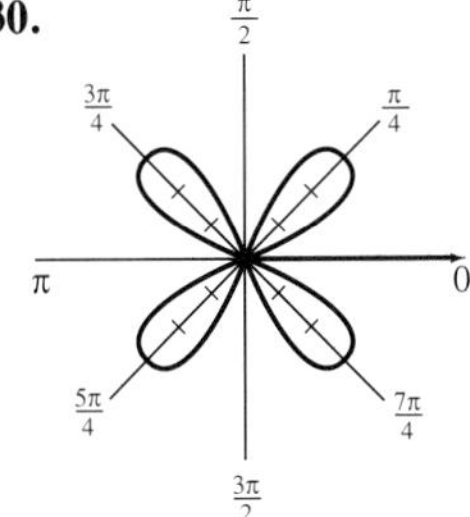

31.

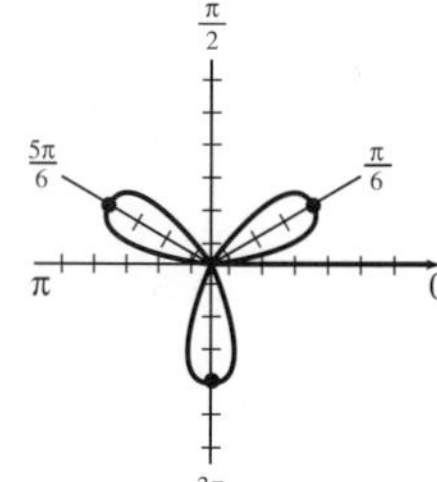

32.

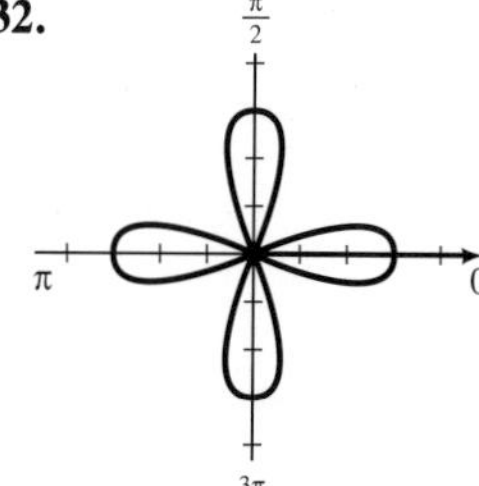

33.

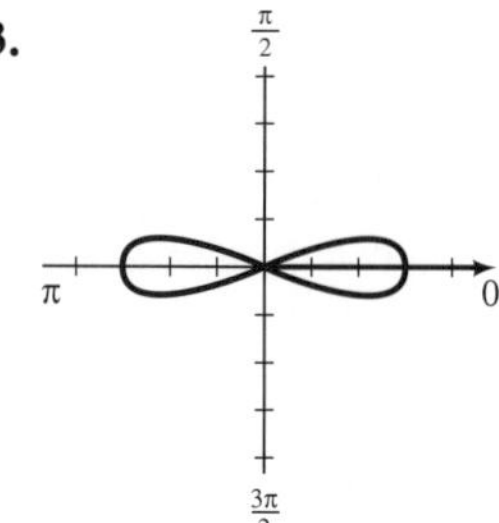

34.

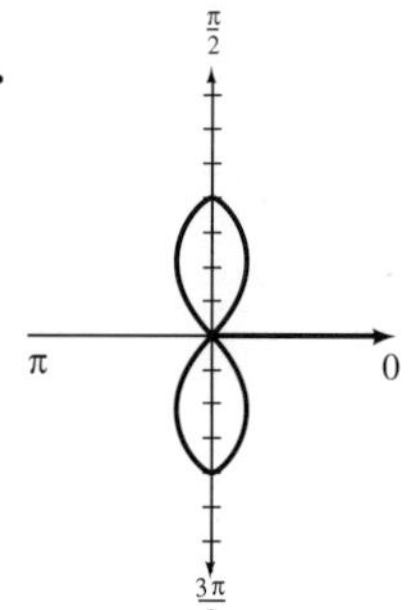

35.

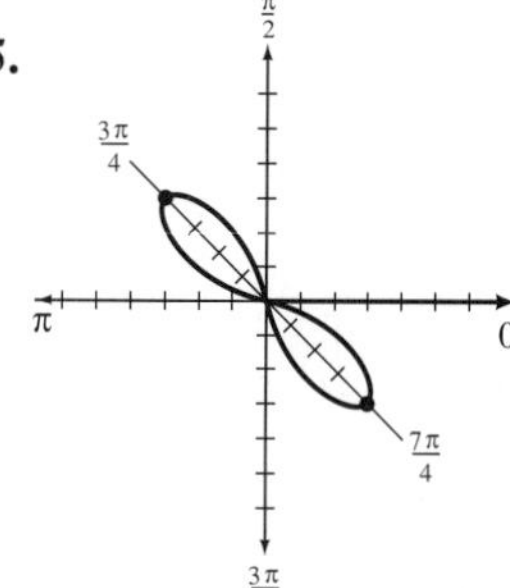

36.

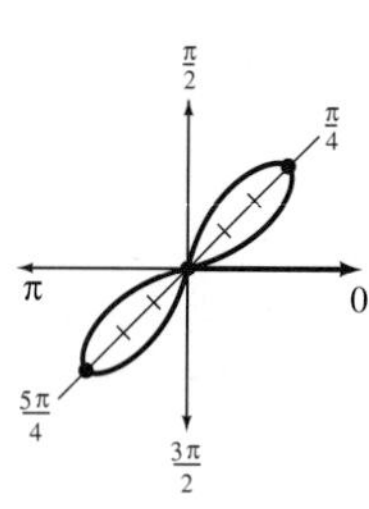

37.

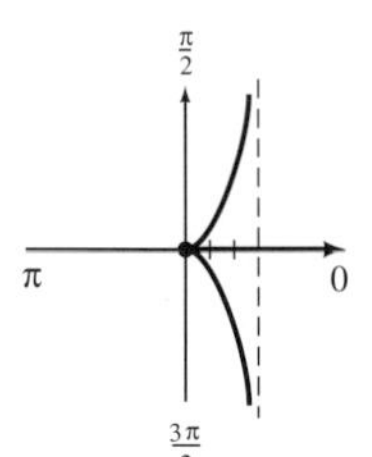

38.

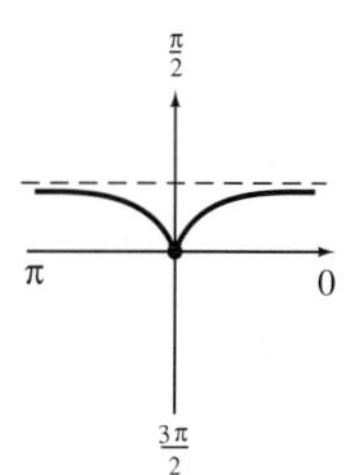

39.

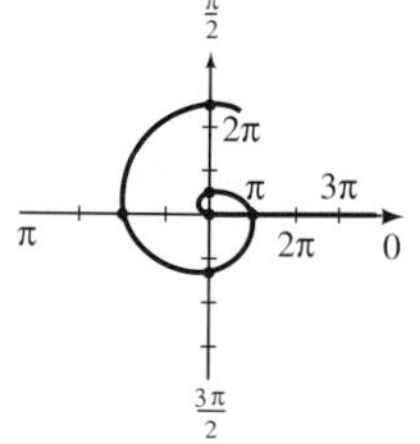

40.

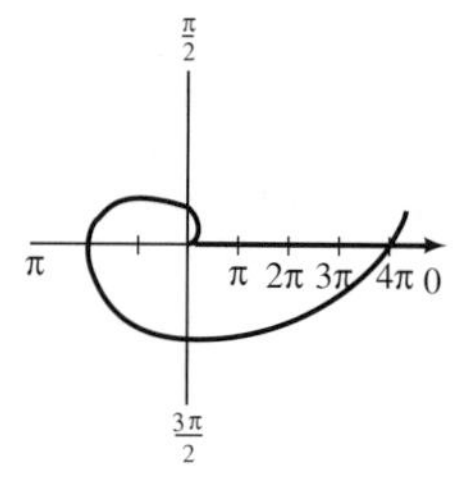

Chapter 8 Review Problem Set (page 534)

1. $-\frac{1}{2}$ **2.** 1 **3.** $\frac{\sqrt{2}}{2}$ **4.** $\frac{5}{13}$ **5.** $\frac{\sqrt{3}}{3}$

6. $-\frac{2\sqrt{13}}{13}$ **7.** $-\frac{\pi}{6}$ **8.** $\frac{5\pi}{6}$ **9.** $-\frac{\pi}{6}$

10. $-\frac{\pi}{3}$ **11.** 60° **12.** 120° **13.** −45°

14. −30° **15.** 2.18° **16.** −0.807

17. −1.557 **18.** 1.228 **19.** 75.6°

20. −24.2° **21.** −83.8° **22.** 124.2°

23. Period 2π, amplitude 4, phase shift $\pi/4$ units to the left

24. Period π, amplitude 3, phase shift $\pi/3$ units to the right

25. Period $\pi/2$, no amplitude, phase shift π units to the left

26. Period $2\pi/3$, amplitude 1, phase shift $\pi/3$ units to the right

27. Period 2, amplitude 2, phase shift 1 unit to the left

28. Period 2, amplitude 2, phase shift $\frac{1}{2}$ unit to the right

29. Period $2\pi/3$, amplitude 4, no phase shift

30. Period π, amplitude 5, no phase shift

31. Period $\pi/3$, no amplitude, phase shift $\pi/3$ to the right

32. Period $\pi/4$, no amplitude, phase shift $\pi/4$ to the left

33. Period $\pi/2$, no amplitude, phase shift $\pi/8$ to the left

34. Period $2\pi/3$, no amplitude, phase shift $\pi/12$ to the right

35. Period $2\pi/3$, no amplitude, phase shift 2 to the right

36. Period π, no amplitude, phase shift $\pi/2$ to the right

37.

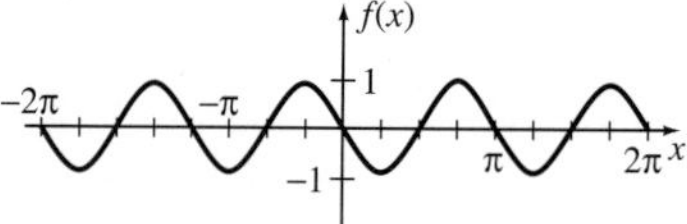

38.

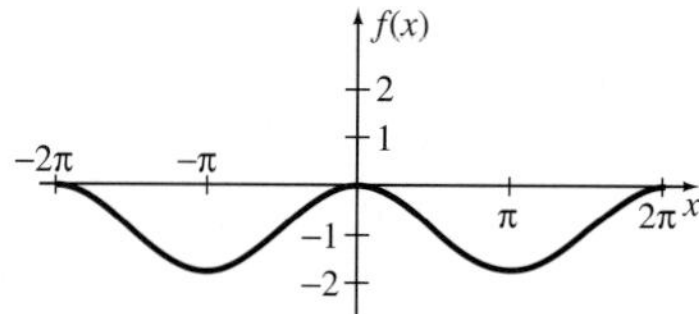

39.

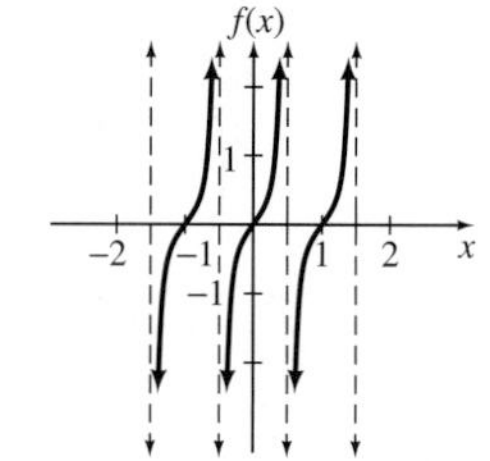

40.

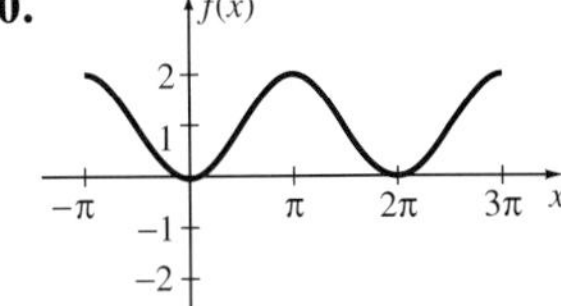

41.

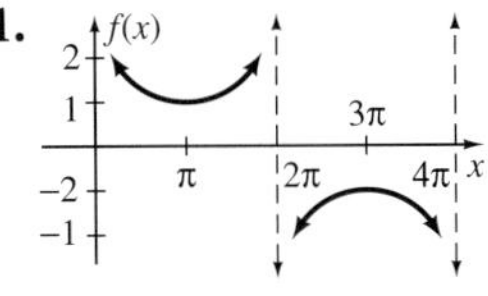

42.

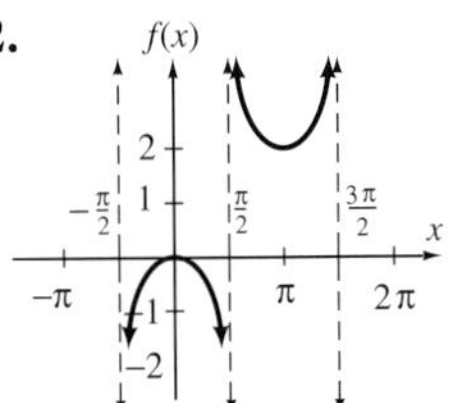

43.

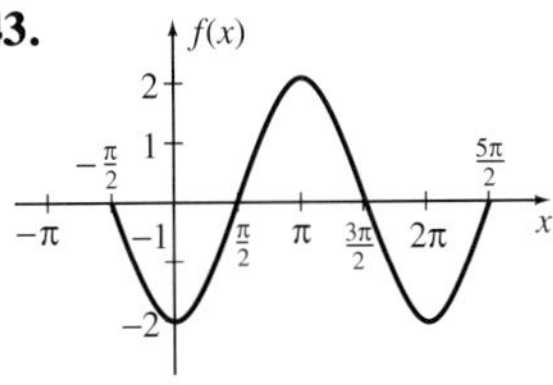

44.

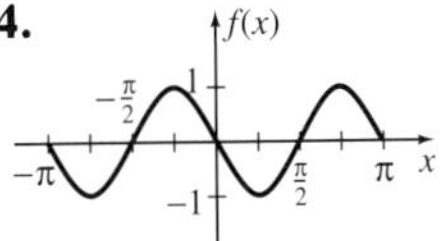

45.

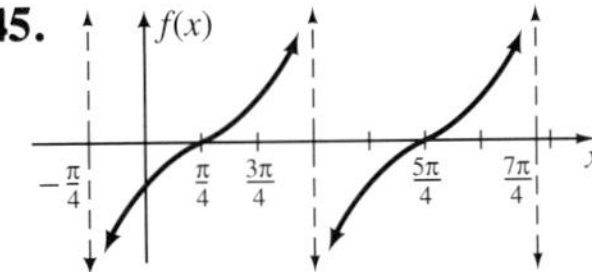

46.

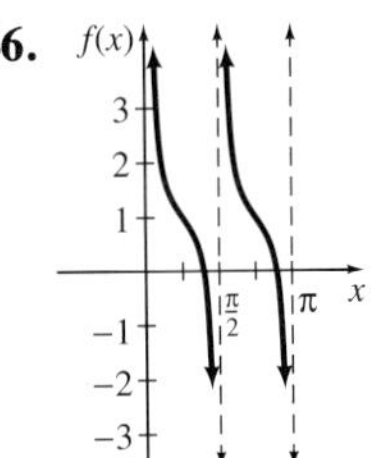

47.

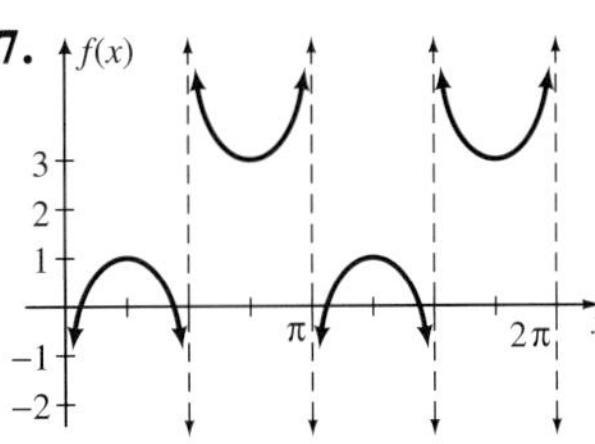

48.

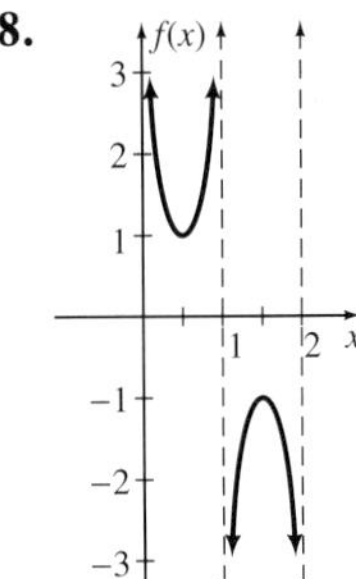

49.

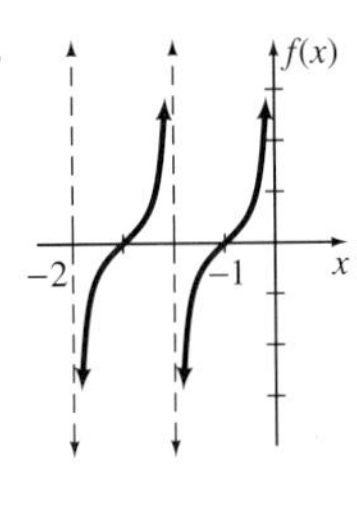

50.

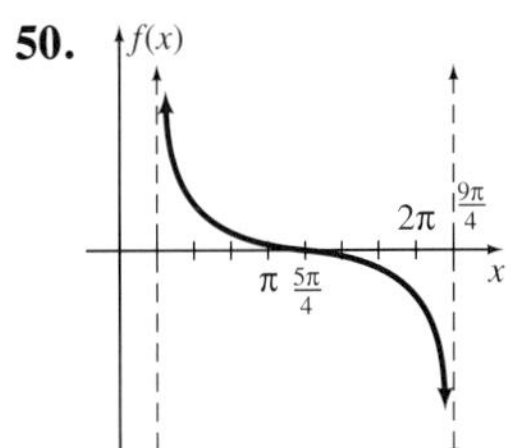

51.

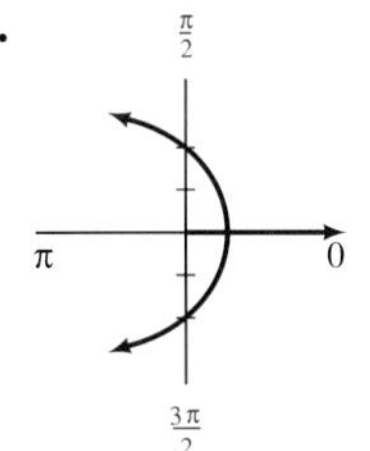

52.

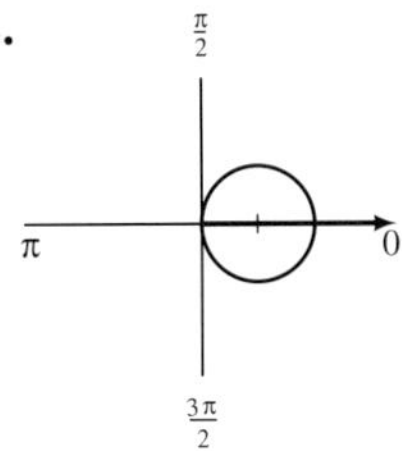

53.

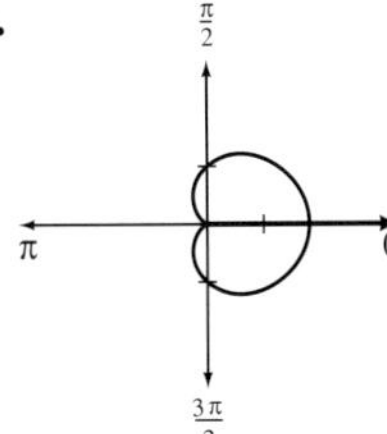

54.

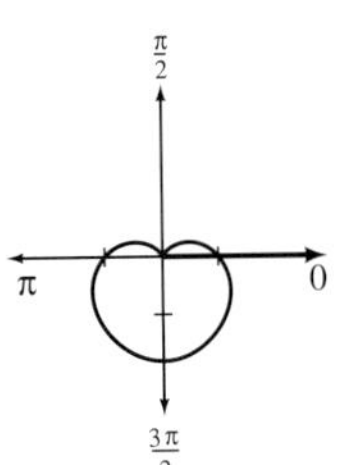

55.

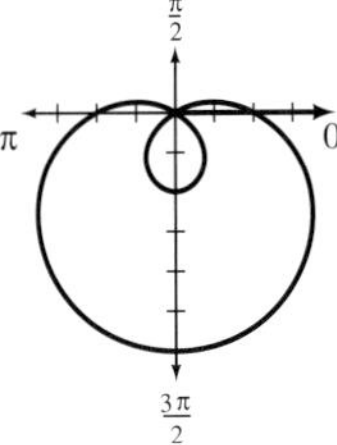

56.

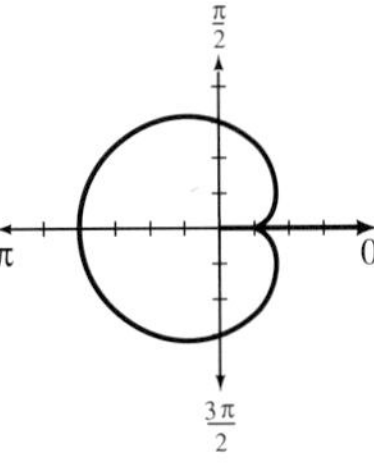

57. $x^2 + y^2 + x = \sqrt{x^2 + y^2}$, cardioid

58. $r = -3 \sec\theta \tan\theta$, parabola

59. $x^2 + y^2 - 3y = 0$, circle

60. $r = 2 + 3\sin\theta$, limacon

CHAPTER 9

Problem Set 9.1 (page 545)

1. $\sec\theta = \dfrac{r}{x} = \dfrac{1}{\dfrac{x}{r}} = \cos\theta$

$\cot\theta = \dfrac{x}{y} = \dfrac{1}{\dfrac{y}{x}} = \dfrac{1}{\tan\theta}$

2. $\dfrac{\cos\theta}{\sin\theta} = \dfrac{\dfrac{x}{r}}{\dfrac{y}{r}} = \dfrac{x}{r} \cdot \dfrac{r}{y} = \dfrac{x}{y} = \cot\theta$

For Problems 3–12, the answers are arranged in the order $\sin\theta$, $\cos\theta$, $\tan\theta$, $\csc\theta$, $\sec\theta$, and $\cot\theta$ omitting the value given in the problem.

3. $\dfrac{3}{5}, \dfrac{4}{3}, \dfrac{5}{4}, \dfrac{5}{3}, \dfrac{3}{4}$ **4.** $\dfrac{12}{13}, -\dfrac{12}{5}, \dfrac{13}{12}, -\dfrac{13}{5}, -\dfrac{5}{12}$

5. $-\frac{12}{13}, -\frac{5}{13}, -\frac{13}{12}, -\frac{13}{5}, \frac{5}{12}$
6. $\frac{15}{17}, -\frac{8}{15}, -\frac{17}{8}, \frac{17}{15}, -\frac{15}{8}$
7. $-\frac{3}{5}, -\frac{4}{3}, \frac{5}{4}, -\frac{5}{3}, -\frac{3}{4}$ 8. $-\frac{4}{5}, -\frac{3}{5}, \frac{4}{3}, -\frac{5}{3}, \frac{3}{4}$
9. $-\frac{\sqrt{5}}{5}, \frac{2\sqrt{5}}{5}, -\sqrt{5}, \frac{\sqrt{5}}{2}, -2$
10. $-\frac{2\sqrt{2}}{3}, \frac{1}{3}, -2\sqrt{2}, -\frac{3\sqrt{2}}{4}, -\frac{\sqrt{2}}{4}$
11. $\frac{2}{3}, -\frac{\sqrt{5}}{3}, -\frac{2\sqrt{5}}{5}, -\frac{3\sqrt{5}}{5}, -\frac{\sqrt{5}}{2}$
12. $-\frac{3\sqrt{10}}{10}, -\frac{\sqrt{10}}{10}, 3, -\frac{\sqrt{10}}{3}, -\sqrt{10}$ 13. $\sin\theta$
14. $\sec\theta$ 15. $\cot x$ 16. $\cos x$
17. $-\tan^2 x$ 18. $\csc x$ 19. 1 20. 1
21. $\sec\theta$ 22. $\sin\theta$ 23. $\cos x$ 24. $\sec x$

Problem Set 9.2 (page 552)

1. 60° and 120° 2. 120° and 240° 3. 180°
4. 225° and 315° 5. 120° and 300°
6. 60°, 120°, 240°, 300° 7. 270° 8. 0°
9. $\frac{4\pi}{3}$ and $\frac{5\pi}{3}$ 10. $\frac{3\pi}{4}$ and $\frac{5\pi}{4}$ 11. 0
12. $\frac{\pi}{2}$ and $\frac{3\pi}{2}$ 13. $\frac{\pi}{4}, \frac{7\pi}{6}, \frac{5\pi}{4}$, and $\frac{11\pi}{6}$
14. 0 and π 15. No solutions 16. 0 and π
17. $0, \pi, \frac{\pi}{6}, \frac{5\pi}{6}$ 18. $0, \pi, \frac{\pi}{4}, \frac{3\pi}{4}, \frac{5\pi}{4}$, and $\frac{7\pi}{4}$
19. 0°, 45°, 180°, 225° 20. 0° and 180°
21. 45°, 90°, 135°, 225°, 270°, and 315°
22. $\frac{\pi}{6}, \frac{\pi}{2}, \frac{5\pi}{6}$, and $\frac{3\pi}{2}$ 23. $\frac{2\pi}{3}, \pi$, and $\frac{4\pi}{3}$
24. $\frac{\pi}{2}, \frac{7\pi}{6}$, and $\frac{11\pi}{6}$ 25. 60°, 180°, and 300°
26. 90° and 180° 27. $\frac{\pi}{3}, \pi$, and $\frac{5\pi}{3}$
28. $\frac{\pi}{6}, \frac{5\pi}{6}$, and $\frac{3\pi}{2}$ 29. $0, \frac{\pi}{2}$, and $\frac{3\pi}{2}$
30. $0, \frac{\pi}{6}, \frac{5\pi}{6}$, and π 31. 0 and $\frac{\pi}{2}$ 32. 0
33. $\frac{\pi}{4}, \frac{3\pi}{4}, \frac{5\pi}{4}$, and $\frac{7\pi}{4}$ 34. $\frac{\pi}{2}$ and π

For Problems 35–46, n is an integer.

35. $30° + n \cdot 360°$ and $330° + n \cdot 360°$
36. $270° + n \cdot 360°$ 37. $\frac{4\pi}{3} + 2\pi n$ and $\frac{5\pi}{3} + 2\pi n$
38. $\frac{\pi}{3} + 2\pi n$ and $\frac{5\pi}{3} + 2\pi n$ 39. $\frac{3\pi}{4} + \pi n$
40. $\frac{\pi}{4} + \pi n$ 41. $60° + n \cdot 180°$ and $120° + n \cdot 180°$
42. $30° + n \cdot 180°$ and $150° + n \cdot 180°$
43. $\frac{\pi}{4} + \pi n$ and $\frac{\pi}{2} + \pi n$ 44. $2\pi n$
45. $\frac{\pi}{6} + 2\pi n$ and $\frac{5\pi}{6} + 2\pi n$ and $\frac{3\pi}{2} + 2\pi n$
(All of these can be represented by the one expression $\frac{\pi}{6} + \frac{2\pi n}{3}$.)
46. $\frac{\pi}{4} + \pi n$ and $\frac{\pi}{3} + \pi n$ 47. 347.5° and 192.5°
48. 55.3° and 124.7° 49. 130.0° and 230.0°
50. 102.6° and 257.4° 51. 287.7° and 107.7°
52. 86.0° and 266.0° 53. 19.5° and 160.5°
54. 14.5° and 165.5° 55. 48.2° and 311.8°
56. 41.4°, 70.5°, 289.5°, and 318.6°
57. 15.5° and 164.5° 58. 67.5° and 292.5°
59. 3.93 and 5.49 60. 0.23 and 2.91
61. 2.53 and 3.75 62. 1.71 and 4.57
63. 2.23 and 5.37 64. 1.46 and 4.60
65. 0.25 and 2.89 66. 1.23, 2.42, 3.86, and 5.05
67. 3.74 and 5.68 68. 3.81 and 5.61
69. 0.90 and 5.38 70. 1.86 and 4.42
74. (a) 0.7 (b) −1.9, 0, and 1.9 (c) −0.5
75. 7

Problem Set 9.3 (page 561)

1. $\frac{\sqrt{6}-\sqrt{2}}{4}$ 2. $2-\sqrt{3}$ 3. $2+\sqrt{3}$
4. $\frac{\sqrt{6}+\sqrt{2}}{4}$ 5. $\frac{\sqrt{6}+\sqrt{2}}{4}$ 6. $\frac{\sqrt{2}-\sqrt{6}}{4}$
7. $\frac{-\sqrt{6}-\sqrt{2}}{4}$ 8. $\frac{\sqrt{2}-\sqrt{6}}{4}$ 9. $2+\sqrt{3}$
10. $\frac{\sqrt{2}+\sqrt{6}}{4}$ 11. $\frac{\sqrt{6}+\sqrt{2}}{4}$ 12. $\frac{\sqrt{6}-\sqrt{2}}{4}$
13. $\frac{\sqrt{6}+\sqrt{2}}{4}$ 14. $\frac{\sqrt{6}-\sqrt{2}}{4}$ 15. $-\frac{77}{85}; -\frac{13}{84}$
16. $\frac{36}{85}; \frac{77}{85}$ 17. $-\frac{36}{85}; \frac{13}{85}$ 18. $\frac{7}{25}$; undefined
19. $\frac{416}{425}; -\frac{87}{425}$ 20. $-\frac{1}{21}; -\frac{19}{9}$ 21. $\frac{4}{5}$
22. $\frac{4}{5}$ 23. $-\frac{84}{13}$ 24. $-\frac{7}{24}$ 25. $\frac{9\sqrt{10}}{170}$
26. $-\frac{2\sqrt{5}}{25}$

45. **(a)** $\frac{1}{2}\cos 5\theta + \frac{1}{2}\cos\theta$ **(b)** $\frac{1}{2}\cos 6\theta + \frac{1}{2}\cos 2\theta$
(c) $\frac{1}{2}\cos 2\theta - \frac{1}{2}\cos 4\theta$ **(d)** $\frac{1}{2}\sin 7\theta + \frac{1}{2}\sin\theta$
(e) $\sin 12\theta + \sin 6\theta$ **(f)** $\frac{3}{2}\sin 3\theta + \frac{3}{2}\sin\theta$

48. **(a)** $2\cos 3\theta\cos\theta$ **(b)** $-2\sin 4\theta\sin 2\theta$
(c) $2\sin 3\theta\sin 2\theta$ **(d)** $2\sin 4\theta\cos 2\theta$
(e) $2\cos 2\theta\sin\theta$ **(f)** $-2\cos 3\theta\sin\theta$

49. **(a)** Conditional equation **(b)** Identity
(c) Identity **(d)** Identity
(e) Conditional equation **(f)** Identity
(g) Conditional equation **(h)** Identity

Problem Set 9.4 (page 571)

1. $\frac{24}{25}, \frac{7}{25}, \frac{24}{7}$ **2.** $\frac{24}{25}, -\frac{7}{25}, -\frac{24}{7}$
3. $-\frac{120}{169}, -\frac{119}{169}, \frac{120}{119}$ **4.** $\frac{120}{169}, \frac{119}{169}, \frac{120}{119}$
5. $-\frac{336}{625}, \frac{527}{625}, -\frac{336}{527}$ **6.** $-\frac{240}{289}, \frac{161}{289}, -\frac{240}{161}$
7. $\frac{4}{5}, \frac{3}{5}, \frac{4}{3}$ **8.** $-\frac{12}{13}, -\frac{5}{13}, \frac{12}{5}$ **9.** $\frac{\sqrt{2-\sqrt{3}}}{2}$
10. $\frac{\sqrt{2+\sqrt{3}}}{2}$ **11.** $\sqrt{7-4\sqrt{3}}$ or $2-\sqrt{3}$
12. $\frac{\sqrt{2+\sqrt{2}}}{2}$ **13.** $1-\sqrt{2}$ **14.** $\sqrt{2}-1$
15. $\frac{\sqrt{2-\sqrt{2}}}{2}$ **16.** $\frac{\sqrt{2+\sqrt{3}}}{2}$ **17.** $2+\sqrt{3}$
18. $-2-\sqrt{3}$ **19.** $-\frac{\sqrt{2-\sqrt{3}}}{2}$
20. $-\frac{\sqrt{2-\sqrt{2}}}{2}$ **21.** $\frac{\sqrt{10}}{10}, \frac{3\sqrt{10}}{10}, \frac{1}{3}$
22. $\frac{2\sqrt{5}}{5}, \frac{\sqrt{5}}{5}, 2$ **23.** $\frac{2\sqrt{5}}{5}, -\frac{\sqrt{5}}{5}, -2$
24. $\frac{\sqrt{26}}{26}, -\frac{5\sqrt{26}}{26}, -\frac{1}{5}$ **25.** $\frac{3\sqrt{13}}{13}, \frac{2\sqrt{13}}{13}, \frac{3}{2}$
26. $\frac{\sqrt{3}}{3}, \frac{\sqrt{6}}{3}, \frac{\sqrt{2}}{2}$ **27.** $\frac{\sqrt{6}}{6}, -\frac{\sqrt{30}}{6}, -\frac{\sqrt{5}}{5}$
28. $\frac{\sqrt{14}}{4}, -\frac{\sqrt{2}}{4}, -\sqrt{7}$ **29.** 0, 60°, 180°, and 300°
30. 60°, 180°, and 300° **31.** 30°, 90°, and 150°
32. 0° and 180° **33.** 90° and 270°
34. 90° and 270° **35.** 0° and 240°
36. 0°, 60°, and 300°
37. 0°, 30°, 90°, 150°, 180°, 210°, 270°, and 330°
38. 0°, 60°, 120°, 180°, 240°, and 300°
39. 0°, 90°, and 270° **40.** 270°
41. $\frac{\pi}{6}, \frac{\pi}{2}, \frac{5\pi}{6}$, and $\frac{3\pi}{2}$ **42.** $\frac{\pi}{2}, \frac{5\pi}{4}, \frac{3\pi}{2}$, and $\frac{7\pi}{4}$
43. $\frac{7\pi}{6}, \frac{3\pi}{2}$, and $\frac{11\pi}{6}$ **44.** $\frac{\pi}{2}, \frac{2\pi}{3}$, and $\frac{4\pi}{3}$
45. 0, $\frac{\pi}{3}$, and $\frac{5\pi}{3}$ **46.** 0, $\frac{\pi}{6}, \frac{5\pi}{6}$, and π
47. $\frac{\pi}{6}, \frac{\pi}{2}, \frac{5\pi}{6}$, and $\frac{3\pi}{2}$ **48.** 0 and π
49. 0, $\frac{\pi}{2}$, and $\frac{3\pi}{2}$ **50.** $\frac{2\pi}{3}$ and $\frac{4\pi}{3}$
71. $\cos 3\theta = 4\cos^3\theta - 3\cos\theta$
72. $\cos 4\theta = 8\cos^4\theta - 8\cos^2\theta + 1$
73. $\cos 6\theta = 32\cos^6\theta - 48\cos^4\theta + 18\cos^2\theta - 1$
76. −1.8 **77.** 0.9 **78.** 1.4 **79.** −0.5

Problem Set 9.5 (page 578)

1. 5 **2.** 5 **3.** 13 **4.** 13 **5.** 5
6. 4 **7.** $\sqrt{5}$ **8.** $\sqrt{2}$ **9.** $\sqrt{13}$
10. $\sqrt{13}$ **11.** 1 **12.** 1
13. $2\sqrt{2}\left(\cos\frac{3\pi}{4} + i\sin\frac{3\pi}{4}\right)$
14. $4\sqrt{2}\left(\cos\frac{5\pi}{4} + i\sin\frac{5\pi}{4}\right)$
15. $3\left(\cos\frac{3\pi}{2} + i\sin\frac{3\pi}{2}\right)$ **16.** $4(\cos\pi + i\sin\pi)$
17. $4\left(\cos\frac{11\pi}{6} + i\sin\frac{11\pi}{6}\right)$
18. $6\left(\cos\frac{5\pi}{6} + i\sin\frac{5\pi}{6}\right)$
19. $\sqrt{2}\left(\cos\frac{5\pi}{4} + i\sin\frac{5\pi}{4}\right)$
20. $\sqrt{2}\left(\cos\frac{7\pi}{4} + i\sin\frac{7\pi}{4}\right)$
21. $2\left(\cos\frac{2\pi}{3} + i\sin\frac{2\pi}{3}\right)$
22. $4\left(\cos\frac{5\pi}{3} + i\sin\frac{5\pi}{3}\right)$
23. $5\sqrt{2}(\cos 315° + i\sin 315°)$
24. $6\sqrt{2}(\cos 45° + i\sin 45°)$
25. $2(\cos 180° + i\sin 180°)$
26. $7(\cos 90° + i\sin 90°)$ **27.** $2(\cos 30° + i\sin 30°)$
28. $2(\cos 210° + i\sin 210°)$
29. $8(\cos 240° + i\sin 240°)$
30. $10(\cos 300° + i\sin 300°)$
31. $12(\cos 330° + i\sin 330°)$
32. $14(\cos 150° + i\sin 150°)$
33. $\sqrt{13}(\cos 56.3° + i\sin 56.3°)$

34. $5(\cos 323.1° + i \sin 323.1°)$
35. $\sqrt{5}(\cos 206.6° + i \sin 206.6°)$
36. $\sqrt{34}(\cos 149.0° + i \sin 149.0°)$
37. $\sqrt{17}(\cos 346.0° + i \sin 346.0°)$
38. $\sqrt{5}(\cos 26.6° + i \sin 26.6°)$
39. $2\sqrt{5}(\cos 116.6° + i \sin 116.6°)$
40. $3\sqrt{5}(\cos 206.6° + i \sin 206.6°)$ 41. $2\sqrt{3} + 2i$
42. $-\frac{5}{2} + \frac{5\sqrt{3}}{2}i$ 43. $-\frac{3\sqrt{2}}{2} - \frac{3\sqrt{2}}{2}i$
44. $\frac{\sqrt{3}}{2} - \frac{1}{2}i$ 45. $-1 - \sqrt{3}i$ 46. $-\frac{\sqrt{3}}{4} + \frac{1}{4}i$
47. $\frac{1}{3} - \frac{\sqrt{3}}{3}i$ 48. $-7 + 0i$ 49. $6 + 0i$
50. $4 + 4\sqrt{3}i$ 51. $-4 - 4\sqrt{3}i$
52. $-12 - 12\sqrt{3}i$ 53. $60 + 60\sqrt{3}i$
54. $1 - \sqrt{3}i$ 55. $\frac{3}{2} - \frac{3\sqrt{3}}{2}i$ 56. $-2 + 0i$
57. $10\sqrt{3} - 10i$ 58. $28 + 0i$ 59. $0 - 24i$
60. $\frac{15\sqrt{2}}{2} - \frac{15\sqrt{2}}{2}i$
61. $20(\cos 75° + i \sin 75°)$
62. $3(\cos 138° + i \sin 138°)$
63. $6(\cos 70° + i \sin 70°)$
64. $24(\cos 220° + i \sin 220°)$
65. $35\left(\cos \frac{11\pi}{10} + i \sin \frac{11\pi}{10}\right)$
66. $24\left(\cos \frac{17\pi}{12} + i \sin \frac{17\pi}{12}\right)$ 68. $1 - i$
69. $-\frac{2}{3} - \frac{2}{3}i$ 70. $0 + i$ 71. $0 + i$
72. $-\frac{1}{2} - \frac{\sqrt{3}}{2}i$ 73. $-\sqrt{3} - i$

Problem Set 9.6 (page 585)

1. $-1024 + 0i$ 2. $-64 + 0i$ 3. $0 - 32i$
4. $-64 + 0i$ 5. $-972 - 972i$
6. $-64\sqrt{3} - 64i$ 7. $-8 + 8\sqrt{3}i$
8. $512\sqrt{3} + 512i$ 9. $\frac{1}{2} - \frac{\sqrt{3}}{2}i$ 10. $\frac{1}{2} + \frac{\sqrt{3}}{2}i$
11. $-\frac{\sqrt{2}}{2} - \frac{\sqrt{2}}{2}i$ 12. $\frac{\sqrt{2}}{2} + \frac{\sqrt{2}}{2}i$
13. $8 + 8\sqrt{3}i$ 14. $32 - 32\sqrt{3}i$
15. $-\frac{\sqrt{2}}{2} - \frac{\sqrt{2}}{2}i$ 16. $-\frac{1}{2} + \frac{\sqrt{3}}{2}i$
17. $2 + 0i$, $-1 + \sqrt{3}i$, and $-1 - \sqrt{3}i$
18. $\frac{3}{2} + \frac{3\sqrt{3}}{2}i$, $-3 + 0i$, $\frac{3}{2} - \frac{3\sqrt{3}}{2}i$
19. $-2\sqrt{2} + 2\sqrt{2}i$ and $2\sqrt{2} - 2\sqrt{2}i$
20. $\frac{3\sqrt{2}}{2} + \frac{3\sqrt{2}}{2}i$ and $-\frac{3\sqrt{2}}{2} - \frac{3\sqrt{2}}{2}i$
21. $\sqrt{3} + i$, $-1 + \sqrt{3}i$, $-\sqrt{3} - i$, and $1 - \sqrt{3}i$
22. $1 + \sqrt{3}i$, $-\sqrt{3} + i$, $-1 - \sqrt{3}i$, and $\sqrt{3} - i$
23. $\sqrt[10]{2}(\cos \theta + i \sin \theta)$ where $\theta = 9°, 81°, 153°, 225°$, and $297°$
24. $\sqrt[10]{2}(\cos \theta + i \sin \theta)$ where $\theta = 63°, 135°, 207°, 279°$, and $351°$
25. $1 + 0i$, $\frac{1}{2} + \frac{\sqrt{3}}{2}i$, $-\frac{1}{2} + \frac{\sqrt{3}}{2}i$, $-1 + 0i$, $-\frac{1}{2} - \frac{\sqrt{3}}{2}i$, and $\frac{1}{2} - \frac{\sqrt{3}}{2}i$
26. $1 + 0i$, $\frac{\sqrt{2}}{2} + \frac{\sqrt{2}}{2}i$, $0 + i$, $-\frac{\sqrt{2}}{2} + \frac{\sqrt{2}}{2}i$, $-1 + 0i$, $-\frac{\sqrt{2}}{2} - \frac{\sqrt{2}}{2}i$, $0 - i$, and $\frac{\sqrt{2}}{2} - \frac{\sqrt{2}}{2}i$
27. $\sqrt[3]{2}(\cos \theta + i \sin \theta)$ where $\theta = 40°, 160°$, and $280°$
28. $\sqrt[3]{2}(\cos \theta + i \sin \theta)$ where $\theta = 100°, 220°$, and $340°$
29. $\sqrt[5]{2}(\cos \theta + i \sin \theta)$ where $\theta = 27°, 99°, 171°, 243°$, and $315°$
30. $\sqrt[5]{2}(\cos \theta + i \sin \theta)$ where $\theta = 63°, 135°, 207°, 279°$, and $351°$
31. $\frac{3\sqrt{3}}{2} + \frac{3}{2}i$ and $-\frac{3\sqrt{3}}{2} - \frac{3}{2}i$
32. $-1 + \sqrt{3}i$ and $1 - \sqrt{3}i$ 33. $\{\pm 2, \pm 2i\}$
34. $\left\{3, \frac{-3 \pm 3\sqrt{3}i}{2}\right\}$
35. $\{\sqrt{2} \pm \sqrt{2}i, -\sqrt{2} \pm \sqrt{2}i\}$
36. $\left\{-3, \frac{3 \pm 3\sqrt{3}i}{2}\right\}$
37. $(\cos \theta + i \sin \theta)$ where $\theta = 0°, 72°, 144°, 216°$, and $288°$
38. $(\cos \theta + i \sin \theta)$ where $\theta = 36°, 108°, 252°$, and $324°$
39. $\left\{\pm i, \frac{\sqrt{3}}{2} \pm \frac{1}{2}i, -\frac{\sqrt{3}}{2} \pm \frac{1}{2}i\right\}$
40. $\left\{\pm 1, \frac{1}{2} \pm \frac{\sqrt{3}}{2}i, -\frac{1}{2} \pm \frac{\sqrt{3}}{2}i\right\}$

Chapter 9 Review Problem Set (page 588)

1. $-\frac{5}{13}, \frac{13}{12}, -\frac{12}{5}$ 2. $-\frac{63}{65}; -\frac{56}{65}; \frac{63}{16}$
3. $-\frac{120}{169}; \frac{119}{169}; -\frac{120}{119}$ 4. $\frac{\sqrt{10}}{10}; -\frac{3\sqrt{10}}{10}; -\frac{1}{3}$
5. $\frac{\sqrt{6} - \sqrt{2}}{4}$ or $\frac{\sqrt{2 - \sqrt{3}}}{2}$

6. $\frac{\sqrt{6}-\sqrt{2}}{4}$ or $\frac{\sqrt{2-\sqrt{3}}}{2}$
7. $\frac{\sqrt{6}+\sqrt{2}}{4}$ or $\frac{\sqrt{2+\sqrt{3}}}{2}$ **8.** $\sqrt{2}-1$
9. $\frac{24}{25}$ **10.** $\frac{33}{56}$ **11.** 0°, 90°, and 180°
12. 30° and 150° **13.** 30° and 150°
14. 0°, 60°, 180°, and 300° **15.** 120°, 180°, and 240°
16. 60° and 300° **17.** 0° and 240°
18. 0°, 60°, 180°, and 300° **19.** $\frac{7\pi}{6}$ and $\frac{11\pi}{6}$
20. $\frac{\pi}{4}, \frac{\pi}{2}, \frac{3\pi}{4}, \frac{5\pi}{4}$, and $\frac{7\pi}{4}$ **21.** $\frac{\pi}{6}, \frac{\pi}{3}, \frac{5\pi}{6}$, and $\frac{5\pi}{3}$
22. $\frac{\pi}{8}, \frac{5\pi}{8}, \frac{9\pi}{8}$, and $\frac{13\pi}{8}$ **23.** $\frac{\pi}{3}$
24. $0, \frac{\pi}{2}, \pi$, and $\frac{3\pi}{2}$ **25.** $2\pi n$ where n is an integer
26. $45° + n \cdot 180°$, $210° + n \cdot 360°$, and $330° + n \cdot 360°$ where n is an integer
27. 220° and 158.0° **28.** 196.3° and 343.7°
29. 0.98, 1.81, 4.12, and 4.95 **30.** 2.24 and 4.04
45. $2\sqrt{5}$ **46.** $2\left(\cos\frac{11\pi}{6} + i\sin\frac{11\pi}{6}\right)$
47. $6(\cos 225° + i \sin 225°)$ **48.** $0 - 5i$
49. $4 - 4\sqrt{3}i$ **50.** $16 + 0i$
51. $-512 + 512\sqrt{3}i$ **52.** $\frac{\sqrt{2}}{2} + \frac{\sqrt{2}}{2}i$
53. $0 + 3i$, $-\frac{3\sqrt{3}}{2} - \frac{3}{2}i$, and $\frac{3\sqrt{3}}{2} - \frac{3}{2}i$
54. $\frac{\sqrt{6}}{2} + \frac{\sqrt{2}}{2}i$, $-\frac{\sqrt{2}}{2} + \frac{\sqrt{6}}{2}i$, $-\frac{\sqrt{6}}{2} - \frac{\sqrt{2}}{2}i$, and $\frac{\sqrt{2}}{2} - \frac{\sqrt{6}}{2}i$
55. $-2\sqrt{3} + 2i$ and $2\sqrt{3} - 2i$

Chapters 7–9 Cumulative Review Problem Set (page 590)

1. $\frac{\sqrt{3}}{2}$ **2.** $\frac{\sqrt{3}}{2}$ **3.** 1 **4.** $-\frac{2\sqrt{3}}{3}$
5 $-\frac{2\sqrt{3}}{3}$ **6.** $\frac{\sqrt{3}}{3}$ **7.** $-\frac{\sqrt{3}}{2}$ **8.** $\frac{1}{2}$
9. $\frac{\sqrt{3}}{2}$ **10.** $\frac{1}{2}$ **11.** $\frac{4}{5}$ **12.** $-\frac{4}{3}$
13. $\frac{\sqrt{6}+\sqrt{2}}{4}$ **14.** $\frac{\sqrt{2+\sqrt{2}}}{2}$ **15.** $\frac{56}{33}$
16. $\frac{33}{65}$ **17.** $-\frac{25}{24}$ **18.** $\sqrt{5}$
19. 30° and 150° **20.** $\frac{\pi}{4}$ and $\frac{5\pi}{4}$
21. $0, \frac{\pi}{6}, \frac{5\pi}{6}$, and π **22.** 0° **23.** $\frac{\pi}{3}, \pi$, and $\frac{5\pi}{3}$
24. $\frac{2\pi}{3}$ and $\frac{4\pi}{3}$ **25.** 30°, 120°, 210°, and 300°
26. 0°, 60°, and 300°
27. 41.8°, 138.2°, 228.6°, and 311.4°
28. 0.66 and 5.62 **29.** 80 feet **30.** 33.7°
31. 125.6° **32.** 6.3 inches **33.** 9.1 meters
34. 34 feet **35.** 699 miles **36.** 1352 pounds
37. $x^2 + y^2 - 2x + 3y = 0$ **38.** $r = 4\cos\theta$
39. $2\sqrt{10}$ **40.** $6\left(\cos\frac{5\pi}{3} + i\sin\frac{5\pi}{3}\right)$
41. $-6\sqrt{2} - 6\sqrt{2}i$ **42.** $0 + 8i$
43. $1 + \sqrt{3}i$, $-\sqrt{3} + i$, $-1 - \sqrt{3}i$, and $\sqrt{3} - i$
44. period of $\frac{2\pi}{3}$, amplitude of 2, and phase shift of $\frac{2\pi}{3}$ units to the right
45. period of 2, amplitude of 1, and phase shift of 1 unit to the left
46. period of π, phase shift of $\frac{\pi}{4}$ units to the left

47.

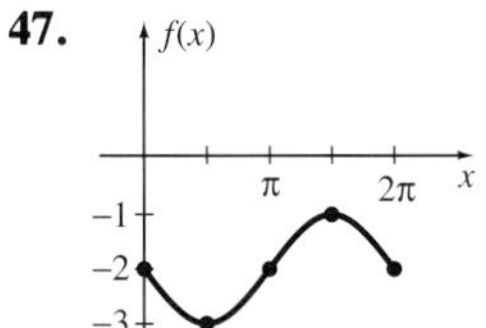

48.

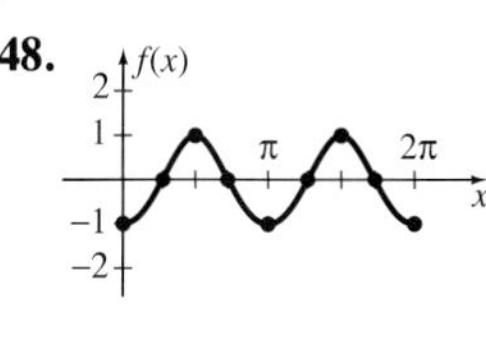

49.

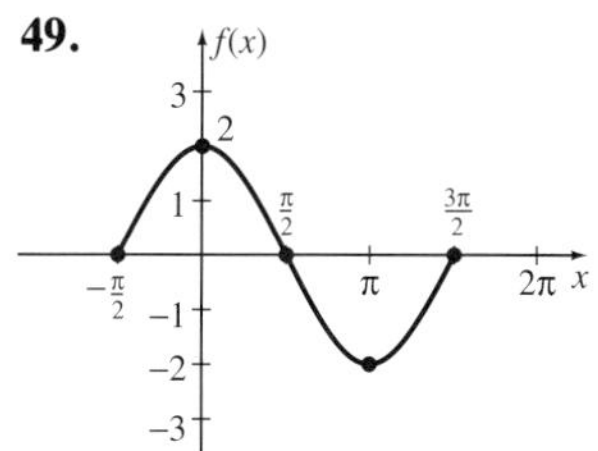

50.

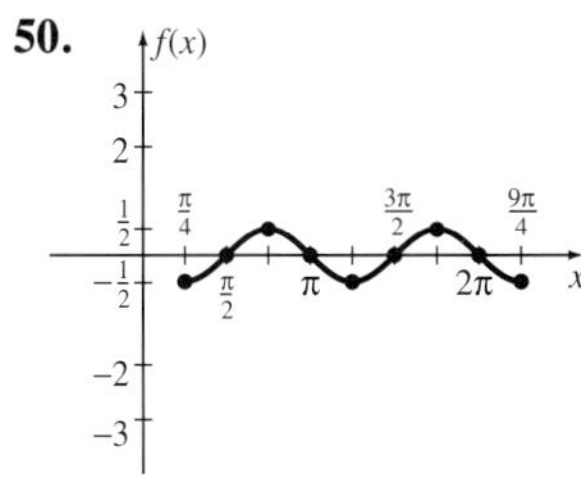

51.

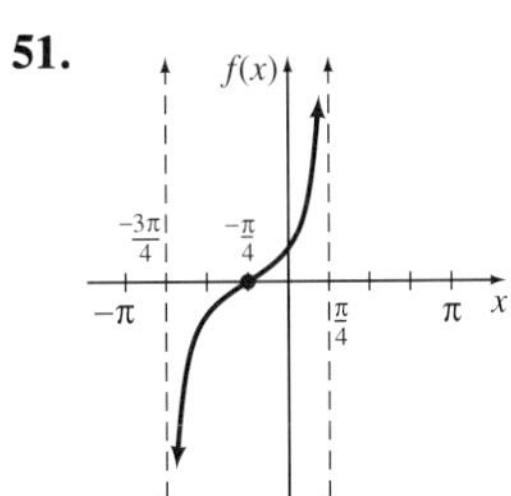

52.

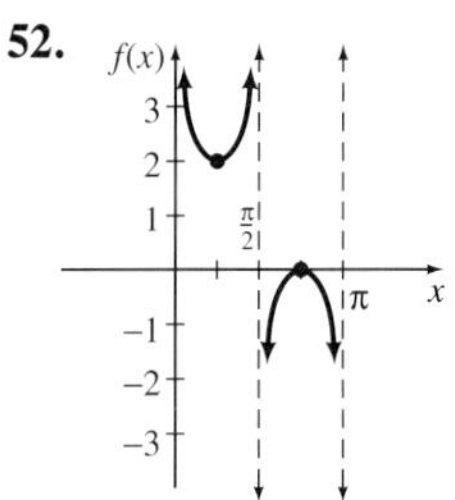

53.

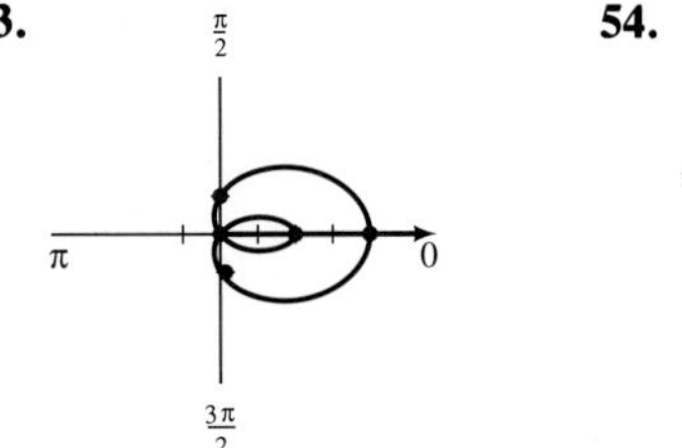

54.

CHAPTER 10

Problem Set 10.1 (page 602)

1. $\{(7, 9)\}$ **2.** $\{(-4, 1)\}$ **3.** $\{(-4, 7)\}$
4. $\{(5, -2)\}$ **5.** $\{(6, 3)\}$ **6.** $\{(8, 11)\}$
7. $a = -3$ and $b = -4$ **8.** $a = 2$ and $b = -5$
9. $\left\{\left(k, \frac{2}{3}k - \frac{4}{3}\right)\right\}$, a dependent system
10. $u = 2$ and $t = 9$ **11.** $u = 5$ and $t = 7$
12. $\{(k, 5k - 9)\}$, a dependent system **13.** $\{(2, -5)\}$
14. $\{(-8, -2)\}$ **15.** $\varnothing$, an inconsistent system
16. $\left\{\left(\frac{1}{2}, \frac{2}{3}\right)\right\}$ **17.** $\left\{\left(-\frac{3}{4}, -\frac{6}{5}\right)\right\}$
18. $\varnothing$, an inconsistent system **19.** $\{(3, -4)\}$
20. $\{(-1, -6)\}$ **21.** $\{(2, 8)\}$ **22.** $\{(0, -3)\}$
23. $\{(-1, -5)\}$ **24.** $\{(-8, -2)\}$
25. $\varnothing$, an inconsistent system
26. $a = \frac{53}{25}$ and $b = \frac{17}{25}$ **27.** $a = \frac{5}{27}$ and $b = -\frac{26}{27}$
28. $\varnothing$, an inconsistent system
29. $s = -6$ and $t = 12$ **30.** $s = 12$ and $t = 9$
31. $\left\{\left(-\frac{1}{2}, \frac{1}{3}\right)\right\}$ **32.** $\left\{\left(\frac{3}{4}, -\frac{1}{5}\right)\right\}$
33. $\left\{\left(\frac{13}{22}, \frac{2}{11}\right)\right\}$ **34.** $\left\{\left(-\frac{6}{37}, \frac{98}{37}\right)\right\}$
35. $\{(-4, 2)\}$ **36.** $\{(1, -9)\}$ **37.** $\{(5, 5)\}$
38. $\{(6, 0)\}$ **39.** $\varnothing$, an inconsistent system
40. $\{(10, 1)\}$ **41.** $\{(12, -24)\}$ **42.** $\{(-20, 3)\}$
43. $t = 8$ and $u = 3$ **44.** $t = 3$ and $u = 7$
45. $\{(200, 800)\}$ **46.** $\{(7.5, 2.5)\}$
47. $\{(400, 800)\}$ **48.** $\{(150, 450)\}$
49. $\{(3, 5, 7)\}$ **50.** $\{(3, 1.75)\}$ **51.** 17 and 36
52. -14 and 11 **53.** 5 and 15 **54.** 7 and 19
55. 72 **56.** 35 **57.** 34 **58.** 47
59. 8 single rooms and 15 double rooms
60. 24 one-bedroom and 12 two-bedroom
61. 2500 student tickets and 500 nonstudent tickets
62. 20 20-cent stamps and 30 22-cent stamps
63. \$500 at 9% and \$1500 at 11%
64. \$750 at 10% and \$1200 at 12%
65. 8 liters of the 40% solution and 12 liters of the 60% solution
66. 7.5 liters of the 30% solution and 2.5 liters of the 70% solution
67. \$1.25 per tennis ball and \$1.75 per golf ball
68. \$.42 per can for the pop and \$1.30 per bag for the potato chips
69. 30 five-dollar bills and 18 ten-dollar bills
70. 75 dimes and 160 quarters **71.** $\{(4, 6)\}$
72. $\{(2, 4)\}$ **73.** $\{(2, -3)\}$ **74.** $\left\{\left(\frac{1}{2}, \frac{1}{3}\right)\right\}$
75. $\left\{\left(\frac{1}{4}, -\frac{2}{3}\right)\right\}$ **76.** $\{(-4, -5)\}$
78. **(a)** Consistent **(b)** Consistent
(c) Inconsistent **(d)** Inconsistent **(e)** Dependent
(f) Consistent **(g)** Consistent **(h)** Dependent

Problem Set 10.2 (page 611)

1. $\{(-4, -2, 3)\}$ **2.** $\{(3, 1, -2)\}$
3. $\{(-2, 5, 2)\}$ **4.** $\{(0, 2, 4)\}$
5. $\{(4, -1, -2)\}$ **6.** $\{(-2, -1, 3)\}$
7. $\{(3, 1, 2)\}$ **8.** $\left\{\left(-2, \frac{3}{2}, 1\right)\right\}$
9. $\{(-1, 3, 5)\}$ **10.** $\left\{\left(\frac{1}{2}, 3, -4\right)\right\}$
11. $\{(-2, -1, 3)\}$ **12.** $\left\{\left(\frac{1}{3}, -\frac{1}{2}, 1\right)\right\}$
13. $\{(0, 2, 4)\}$ **14.** $\{(-2, 5, 2)\}$
15. $\{(4, -1, -2)\}$ **16.** $\{(-1, -1, 2)\}$
17. $\{(-4, 0, -1)\}$ **18.** $\{(-1, -2, 7)\}$
19. $\{(2, 2, -3)\}$ **20.** $\{(-6, 0, 1)\}$
21. 8, 15, and 20 **22.** -2, 6, and 16
23. 7 nickels, 13 dimes, and 22 quarters
24. 12 pennies, 20 nickels, and 33 dimes
25. 40°, 60°, and 80°
26. 12 centimeters, 13 centimeters, and 20 centimeters
27. \$500 at 12%, \$1000 at 13%, and \$1500 at 14%
28. \$700 at 10%, \$1000 at 11%, and \$1200 at 12%
29. 50 of type A, 75 of type B, and 150 of type C
30. 3 units of dish A, 2 units of dish B, and 4 units of dish C
31. $x^2 + y^2 - 4x - 20y - 2 = 0$

Problem Set 10.3 (page 620)

1. Yes **2.** Yes **3.** Yes **4.** No **5.** No
6. Yes **7.** No **8.** Yes **9.** Yes
10. No **11.** $\{(-1, -5)\}$ **12.** $\{(2, -4)\}$

13. $\{(3, -6)\}$ **14.** $\{(-3, -7)\}$ **15.** $\varnothing$
16. $\{(0, 4)\}$ **17.** $\{(-2, -9)\}$ **18.** $\varnothing$
19. $\{(-1, -2, 3)\}$ **20.** $\{(1, 0, -3)\}$
21. $\{(3, -1, 4)\}$ **22.** $\{(-2, 5, 1)\}$
23. $\{(0, -2, 4)\}$ **24.** $\{(-3, -4, 1)\}$
25. $\{(-7k + 8, -5k + 7, k)\}$ **26.** $\varnothing$
27. $\{(-4, -3, -2)\}$ **28.** $\{(-4, 0, -1)\}$
29. $\{(4, -1, -2)\}$ **30.** $\{(1, -1, 1)\}$
31. $\{(1, -1, 2, -3)\}$ **32.** $\{(1, 0, -2, -1)\}$
33. $\{(2, 1, 3, -2)\}$ **34.** $\{(-1, -2, 0, 3)\}$
35. $\{(-2, 4, -3, 0)\}$ **36.** $\{(0, -5, 0, 4)\}$
37. $\varnothing$ **38.** $\{(2, -2k - 3, -3k + 4, k)\}$
39. $\{(-3k + 5, -1, -4k + 2, k)\}$ **40.** $\varnothing$
41. $\{(-3k + 9, k, 2, -3)\}$ **42.** $\{(7, -3, 2k + 5, k)\}$
43. $\{(17k - 6, 10k - 5, k)\}$
44. $\{(11k - 7, -3k + 2, k)\}$
45. $\left\{\left(-\frac{1}{2}k + \frac{34}{11}, \frac{1}{2}k - \frac{5}{11}, k\right)\right\}$
46. $\left\{\left(-k + \frac{11}{7}, \frac{2}{3}k + \frac{5}{7}, k\right)\right\}$ **47.** $\varnothing$
48. All ordered triples that satisfy $x + y - 2z = -1$

Problem Set 10.4 (page 630)

1. 22 **2.** -18 **3.** -29 **4.** -23
5. 20 **6.** 20 **7.** 5 **8.** 13 **9.** -2
10. -2 **11.** $-\frac{2}{3}$ **12.** $\frac{21}{20}$ **13.** -25
14. 29 **15.** 58 **16.** 14 **17.** 39
18. -285 **19.** -12 **20.** -36 **21.** -41
22. 83 **23.** -8 **24.** -90 **25.** 1088
26. -48 **27.** -140 **28.** 56 **29.** 81
30. 60 **31.** 146 **32.** -142
33. Property 7.3 **34.** Property 7.3
35. Property 7.2 **36.** Property 7.5
37. Property 7.4 **38.** Property 7.4
39. Property 7.3 **40.** Property 7.2
41. Property 7.5 **42.** Property 7.4

Problem Set 10.5 (page 638)

1. $\{(1, 4)\}$ **2.** $\{(-2, -3)\}$ **3.** $\{(3, -5)\}$
4. $\{(-4, 1)\}$ **5.** $\{(2, -1)\}$ **6.** $\{(-2, 4)\}$
7. $\varnothing$ **8.** $\{(2, -5)\}$ **9.** $\left\{\left(-\frac{1}{4}, \frac{2}{3}\right)\right\}$
10. Infinitely many solutions **11.** $\left\{\left(\frac{2}{17}, \frac{52}{17}\right)\right\}$
12. $\left\{\left(\frac{17}{62}, \frac{4}{31}\right)\right\}$ **13.** $\{(9, -2)\}$ **14.** $\{(-8, -3)\}$
15. $\left\{\left(2, -\frac{5}{7}\right)\right\}$ **16.** $\left\{\left(\frac{14}{5}, 4\right)\right\}$
17. $\{(0, 2, -3)\}$ **18.** $\{(-1, -1, 2)\}$
19. $\{(2, 6, 7)\}$ **20.** $\{(3, -4, -5)\}$
21. $\{(4, -4, 5)\}$ **22.** $\{(0, 0, -1)\}$
23. $\{(-1, 3, -4)\}$ **24.** $\{(-5, -2, 1)\}$
25. Infinitely many solutions **26.** $\varnothing$
27. $\left\{\left(-2, \frac{1}{2}, -\frac{2}{3}\right)\right\}$ **28.** $\left\{\left(\frac{15}{4}, -\frac{7}{12}, -\frac{17}{12}\right)\right\}$
29. $\left\{\left(3, \frac{1}{2}, -\frac{1}{3}\right)\right\}$ **30.** $\left\{\left(-\frac{1}{4}, 2, \frac{3}{4}\right)\right\}$
31. $\{(-4, 6, 0)\}$ **32.** $\{(5, 0, -2)\}$
37. Infinitely many solutions **38.** $\{(0, 0, 0)\}$
39. $\{(0, 0, 0)\}$ **40.** Infinitely many solutions

Chapter 10 Review Problem Set (page 642)

1. $\{(3, -7)\}$ **2.** $\{(-1, -3)\}$ **3.** $\{(0, -4)\}$
4. $\left\{\left(\frac{23}{3}, -\frac{14}{3}\right)\right\}$ **5.** $\{(4, -6)\}$
6. $\left\{\left(-\frac{6}{7}, -\frac{15}{7}\right)\right\}$ **7.** $\{(-1, 2, -5)\}$
8. $\{(2, -3, -1)\}$ **9.** $\{(5, -4)\}$ **10.** $\{(2, 7)\}$
11. $\{(-2, 2, -1)\}$ **12.** $\{(0, -1, 2)\}$
13. $\{(-3, -1)\}$ **14.** $\{(4, 6)\}$ **15.** $\{(2, -3, -4)\}$
16. $\{(-1, 2, -5)\}$ **17.** $\{(5, -5)\}$
18. $\{(-12, 12)\}$ **19.** $\left\{\left(\frac{5}{7}, \frac{4}{7}\right)\right\}$
20. $\{(-10, -7)\}$ **21.** $\{(1, 1, -4)\}$
22. $\{(-4, 0, 1)\}$ **23.** $\varnothing$ **24.** $\{(-2, -4, 6)\}$
25. -34 **26.** 13 **27.** -40 **28.** 16
29. 51 **30.** 125 **31.** 72
32. \$900 at 10% and \$1600 at 12%
33. 20 nickels, 32 dimes, and 54 quarters
34. 25°, 45°, and 110°

Problem Set 11.1 (page 650)

1. $\begin{bmatrix} 3 & -5 \\ 8 & 3 \end{bmatrix}$ **2.** $\begin{bmatrix} 2 & -9 \\ 9 & -3 \end{bmatrix}$ **3.** $\begin{bmatrix} -2 & 21 \\ -7 & 2 \end{bmatrix}$
4. $\begin{bmatrix} -6 & 1 \\ 3 & -11 \end{bmatrix}$ **5.** $\begin{bmatrix} -2 & 1 \\ -3 & 19 \end{bmatrix}$
6. $\begin{bmatrix} -2 & 3 \\ 25 & -14 \end{bmatrix}$ **7.** $\begin{bmatrix} -1 & -5 \\ 2 & 3 \end{bmatrix}$ **8.** $\begin{bmatrix} 6 & -1 \\ 7 & 6 \end{bmatrix}$
9. $\begin{bmatrix} -12 & -14 \\ -18 & -20 \end{bmatrix}$ **10.** $\begin{bmatrix} -5 & -26 \\ -19 & 0 \end{bmatrix}$
11. $\begin{bmatrix} 2 & -11 \\ -7 & 0 \end{bmatrix}$ **12.** $\begin{bmatrix} -1 & -5 \\ 2 & 3 \end{bmatrix}$

13. $AB = \begin{bmatrix} 4 & -6 \\ 8 & -12 \end{bmatrix}$, $BA = \begin{bmatrix} -5 & 5 \\ 3 & -3 \end{bmatrix}$

14. $AB = \begin{bmatrix} 30 & -19 \\ 2 & 9 \end{bmatrix}$, $BA = \begin{bmatrix} 16 & -3 \\ -20 & 23 \end{bmatrix}$

15. $AB = \begin{bmatrix} -5 & -18 \\ -4 & 42 \end{bmatrix}$, $BA = \begin{bmatrix} 19 & -39 \\ -16 & 18 \end{bmatrix}$

16. $AB = \begin{bmatrix} -15 & 30 \\ 18 & -9 \end{bmatrix}$, $BA = \begin{bmatrix} -27 & 18 \\ 18 & 3 \end{bmatrix}$

17. $AB = \begin{bmatrix} 14 & -28 \\ 7 & -14 \end{bmatrix}$, $BA = \begin{bmatrix} 0 & 0 \\ 0 & 0 \end{bmatrix}$

18. $AB = \begin{bmatrix} 0 & 0 \\ 0 & 0 \end{bmatrix}$, $BA = \begin{bmatrix} 4 & 8 \\ -2 & -4 \end{bmatrix}$

19. $AB = \begin{bmatrix} -14 & -7 \\ -12 & -1 \end{bmatrix}$, $BA = \begin{bmatrix} -2 & -3 \\ -32 & -13 \end{bmatrix}$

20. $AB = \begin{bmatrix} -13 & -15 \\ -34 & -46 \end{bmatrix}$, $BA = \begin{bmatrix} 5 & -24 \\ 17 & -64 \end{bmatrix}$

21. $AB = \begin{bmatrix} 1 & 0 \\ 0 & 1 \end{bmatrix}$, $BA = \begin{bmatrix} 1 & 0 \\ 0 & 1 \end{bmatrix}$

22. $AB = \begin{bmatrix} 1 & 0 \\ 0 & 1 \end{bmatrix}$, $BA = \begin{bmatrix} 1 & 0 \\ 0 & 1 \end{bmatrix}$

23. $AB = \begin{bmatrix} 0 & -\frac{5}{3} \\ \frac{17}{6} & -3 \end{bmatrix}$, $BA = \begin{bmatrix} 0 & -\frac{17}{6} \\ \frac{5}{3} & -3 \end{bmatrix}$

24. $AB = \begin{bmatrix} -8 & 0 \\ -17 & -19 \end{bmatrix}$, $BA = \begin{bmatrix} -29 & 15 \\ -14 & 2 \end{bmatrix}$

25. $AB = \begin{bmatrix} 1 & 0 \\ 0 & 1 \end{bmatrix}$, $BA = \begin{bmatrix} 1 & 0 \\ 0 & 1 \end{bmatrix}$

26. $AB = \begin{bmatrix} 1 & 0 \\ 0 & 1 \end{bmatrix}$, $BA = \begin{bmatrix} 1 & 0 \\ 0 & 1 \end{bmatrix}$

27. $AB = \begin{bmatrix} 3 & -2 \\ 4 & 5 \end{bmatrix}$, $BA = \begin{bmatrix} 5 & 4 \\ -2 & 3 \end{bmatrix}$

28. $AC = \begin{bmatrix} 1 & 0 \\ 9 & 0 \end{bmatrix}$, $CA = \begin{bmatrix} -2 & 3 \\ -2 & 3 \end{bmatrix}$

29. $AD = \begin{bmatrix} 1 & 1 \\ 9 & 9 \end{bmatrix}$, $DA = \begin{bmatrix} 3 & 7 \\ 3 & 7 \end{bmatrix}$

30. $AI = \begin{bmatrix} -2 & 3 \\ 5 & 4 \end{bmatrix}$, $IA = \begin{bmatrix} -2 & 3 \\ 5 & 4 \end{bmatrix}$

44. $A^2 = \begin{bmatrix} 4 & 0 \\ 0 & 9 \end{bmatrix}$, $A^3 = \begin{bmatrix} 8 & 0 \\ 0 & 27 \end{bmatrix}$

45. $A^2 = \begin{bmatrix} -1 & -4 \\ 8 & 7 \end{bmatrix}$, $A^3 = \begin{bmatrix} -9 & -11 \\ 22 & 13 \end{bmatrix}$

46. No

Problem Set 11.2 (page 657)

1. $\begin{bmatrix} 3 & -7 \\ -2 & 5 \end{bmatrix}$ **2.** $\begin{bmatrix} 3 & -4 \\ -2 & 3 \end{bmatrix}$ **3.** $\begin{bmatrix} -5 & 8 \\ 2 & -3 \end{bmatrix}$

4. $\begin{bmatrix} -13 & 9 \\ 3 & -2 \end{bmatrix}$ **5.** $\begin{bmatrix} -\frac{2}{5} & \frac{1}{5} \\ \frac{3}{10} & \frac{1}{10} \end{bmatrix}$

6. $\begin{bmatrix} -\frac{3}{5} & \frac{2}{5} \\ -\frac{4}{5} & \frac{1}{5} \end{bmatrix}$ **7.** Does not exist

8. $\begin{bmatrix} \frac{4}{23} & \frac{1}{23} \\ -\frac{3}{23} & \frac{5}{23} \end{bmatrix}$ **9.** $\begin{bmatrix} -\frac{5}{7} & \frac{2}{7} \\ -\frac{4}{7} & \frac{3}{7} \end{bmatrix}$

10. Does not exist **11.** $\begin{bmatrix} -\frac{3}{5} & \frac{1}{5} \\ 1 & 0 \end{bmatrix}$

12. $\begin{bmatrix} -\frac{1}{2} & 0 \\ -\frac{3}{10} & \frac{1}{5} \end{bmatrix}$ **13.** $\begin{bmatrix} -\frac{4}{5} & \frac{3}{5} \\ \frac{1}{5} & -\frac{2}{5} \end{bmatrix}$

14. $\begin{bmatrix} 2 & -\frac{5}{3} \\ -1 & \frac{2}{3} \end{bmatrix}$ **15.** $\begin{bmatrix} 2 & -\frac{5}{3} \\ 1 & -\frac{2}{3} \end{bmatrix}$ **16.** $\begin{bmatrix} -1 & -2 \\ -\frac{1}{2} & -\frac{3}{2} \end{bmatrix}$

17. $\begin{bmatrix} \frac{1}{2} & \frac{1}{2} \\ \frac{1}{2} & -\frac{1}{2} \end{bmatrix}$ **18.** $\begin{bmatrix} \frac{1}{2} & \frac{1}{2} \\ -\frac{1}{2} & \frac{1}{2} \end{bmatrix}$ **19.** $\begin{bmatrix} 30 \\ 36 \end{bmatrix}$

20. $\begin{bmatrix} 9 \\ 23 \end{bmatrix}$ **21.** $\begin{bmatrix} 0 \\ 5 \end{bmatrix}$ **22.** $\begin{bmatrix} 5 \\ 12 \end{bmatrix}$ **23.** $\begin{bmatrix} -4 \\ 13 \end{bmatrix}$

24. $\begin{bmatrix} 18 \\ -60 \end{bmatrix}$ **25.** $\begin{bmatrix} -4 \\ -13 \end{bmatrix}$ **26.** $\begin{bmatrix} 59 \\ 58 \end{bmatrix}$

27. $\{(2, 3)\}$ **28.** $\{(4, -1)\}$ **29.** $\{(-2, 5)\}$

30. $\{(-3, -4)\}$ **31.** $\{(0, -1)\}$ **32.** $\{(-5, 0)\}$

33. $\{(-1, -1)\}$ **34.** $\{(2, 2)\}$ **35.** $\{(4, 7)\}$

36. $\{(7, 1)\}$ **37.** $\left\{\left(-\frac{1}{3}, \frac{1}{2}\right)\right\}$ **38.** $\left\{\left(\frac{1}{4}, \frac{2}{3}\right)\right\}$

39. $\{(-9, 20)\}$

40. $\{(6, 12)\}$

Problem Set 11.3 (page 665)

1. $\begin{bmatrix} 1 & 3 & -3 \\ 3 & -6 & 7 \end{bmatrix}$; $\begin{bmatrix} 3 & -5 & 11 \\ -7 & 6 & 3 \end{bmatrix}$; $\begin{bmatrix} 1 & 10 & -13 \\ 11 & -18 & 16 \end{bmatrix}$; $\begin{bmatrix} 10 & -12 & 30 \\ -18 & 12 & 16 \end{bmatrix}$

2. $\begin{bmatrix} 4 & -6 \\ 7 & -8 \\ -10 & 14 \end{bmatrix}$; $\begin{bmatrix} 2 & -6 \\ -3 & 6 \\ 2 & -4 \end{bmatrix}$; $\begin{bmatrix} 9 & -12 \\ 19 & -23 \\ -26 & 37 \end{bmatrix}$; $\begin{bmatrix} 10 & -24 \\ -2 & 10 \\ -4 & 2 \end{bmatrix}$

3. $[-1 \quad -7 \quad 13 \quad 7]$; $[5 \quad 5 \quad -5 \quad 17]$; $[-5 \quad -20 \quad 35 \quad 9]$; $[14 \quad 8 \quad -2 \quad 58]$

4. $\begin{bmatrix} -3 \\ 3 \\ 16 \end{bmatrix}$; $\begin{bmatrix} 9 \\ -21 \\ -2 \end{bmatrix}$; $\begin{bmatrix} -12 \\ 18 \\ 41 \end{bmatrix}$; $\begin{bmatrix} 24 \\ -60 \\ 10 \end{bmatrix}$

5. $\begin{bmatrix} 8 & -3 & -2 \\ 9 & 2 & -3 \\ 7 & 5 & 21 \end{bmatrix}$; $\begin{bmatrix} -2 & -1 & 4 \\ -11 & 6 & -11 \\ -7 & 5 & -3 \end{bmatrix}$; $\begin{bmatrix} 21 & -7 & -7 \\ 28 & 2 & -2 \\ 21 & 10 & 54 \end{bmatrix}$; $\begin{bmatrix} 2 & -6 & 10 \\ -24 & 20 & -36 \\ -14 & 20 & 12 \end{bmatrix}$

6. $\begin{bmatrix} 19 & -1 \\ -7 & 5 \\ -7 & 9 \end{bmatrix}$; $\begin{bmatrix} -5 & -7 \\ -3 & 13 \\ 5 & -5 \end{bmatrix}$; $\begin{bmatrix} 50 & 1 \\ -16 & 6 \\ -20 & 25 \end{bmatrix}$; $\begin{bmatrix} 4 & -22 \\ -16 & 44 \\ 8 & -6 \end{bmatrix}$

7. $\begin{bmatrix} 0 & 2 \\ -1 & 10 \\ 1 & -9 \\ 2 & 9 \end{bmatrix}$; $\begin{bmatrix} -2 & -2 \\ 5 & -4 \\ -11 & 1 \\ -16 & 13 \end{bmatrix}$; $\begin{bmatrix} 1 & 6 \\ -5 & 27 \\ 8 & -23 \\ 13 & 16 \end{bmatrix}$; $\begin{bmatrix} -6 & -4 \\ 14 & -2 \\ -32 & -6 \\ -46 & 48 \end{bmatrix}$

8. $\begin{bmatrix} 2 & 0 & -9 \\ -3 & 0 & 11 \\ 8 & 2 & -10 \end{bmatrix}$; $\begin{bmatrix} -2 & -2 & 5 \\ 9 & -8 & 1 \\ 2 & 6 & -8 \end{bmatrix}$; $\begin{bmatrix} 6 & 1 & -25 \\ -12 & 4 & 27 \\ 19 & 2 & -21 \end{bmatrix}$; $\begin{bmatrix} -4 & -6 & 6 \\ 24 & -24 & 14 \\ 14 & 20 & -34 \end{bmatrix}$

9. $AB = \begin{bmatrix} 11 & -8 & 14 \\ 4 & -16 & 8 \\ -28 & 22 & -36 \end{bmatrix}$; $BA = \begin{bmatrix} -20 & 21 \\ 8 & -21 \end{bmatrix}$

10. $AB = \begin{bmatrix} -3 & 17 \\ -10 & -49 \end{bmatrix}$; $BA = \begin{bmatrix} -9 & 7 & -6 \\ 25 & -18 & 17 \\ -32 & 9 & -25 \end{bmatrix}$

11. $AB = \begin{bmatrix} 22 & -8 & 1 & 3 \\ -42 & 36 & -26 & -20 \end{bmatrix}$; BA does not exist.

12. AB does not exist; $BA = \begin{bmatrix} 19 & -7 & -16 \\ -17 & 6 & 18 \end{bmatrix}$

13. $AB = \begin{bmatrix} -12 & 5 & -5 \\ 14 & -2 & 4 \\ -10 & 13 & -5 \end{bmatrix}$; $BA = \begin{bmatrix} -1 & 0 & -6 \\ 10 & -2 & 16 \\ -8 & 5 & -16 \end{bmatrix}$

14. $AB = \begin{bmatrix} 1 & -4 & 2 \\ 2 & -2 & 1 \\ 7 & -6 & 2 \end{bmatrix}$; $BA = \begin{bmatrix} -2 & 1 & 1 \\ 0 & 1 & 1 \\ 1 & -1 & 2 \end{bmatrix}$

15. $AB = [-9]$; $BA = \begin{bmatrix} -2 & 1 & -3 & -4 \\ -6 & 3 & -9 & -12 \\ 4 & -2 & 6 & 8 \\ -8 & 4 & -12 & -16 \end{bmatrix}$

16. $AB = \begin{bmatrix} -6 & 8 & 10 \\ 9 & -12 & -15 \\ -15 & 20 & 25 \end{bmatrix}$; $BA = [7]$

17. AB does not exist; $BA = \begin{bmatrix} 20 \\ 2 \\ -30 \end{bmatrix}$

18. $AB = \begin{bmatrix} 41 & 0 & 3 \\ -10 & -6 & -3 \end{bmatrix}$; BA does not exist.

19. $AB = \begin{bmatrix} 9 & -12 \\ -12 & 16 \\ 6 & -8 \end{bmatrix}$; BA does not exist.

20. $AB = [87]$; $BA = \begin{bmatrix} 24 & -56 \\ -27 & 63 \end{bmatrix}$

21. $\begin{bmatrix} -\frac{1}{5} & \frac{3}{10} \\ \frac{2}{5} & -\frac{1}{10} \end{bmatrix}$

22. $\begin{bmatrix} \frac{3}{7} & \frac{2}{7} \\ \frac{2}{7} & -\frac{1}{7} \end{bmatrix}$

23. $\begin{bmatrix} 4 & -1 \\ -7 & 2 \end{bmatrix}$

24. $\begin{bmatrix} 5 & -7 \\ -2 & 3 \end{bmatrix}$

25. $\begin{bmatrix} -\frac{4}{5} & -\frac{1}{5} \\ -\frac{3}{5} & -\frac{2}{5} \end{bmatrix}$

26. $\begin{bmatrix} -\frac{2}{3} & -\frac{1}{3} \\ -1 & -1 \end{bmatrix}$

27. $\begin{bmatrix} \frac{7}{2} & -3 & \frac{1}{2} \\ -\frac{1}{2} & 0 & \frac{1}{2} \\ -\frac{1}{2} & 1 & -\frac{1}{2} \end{bmatrix}$

28. $\begin{bmatrix} \frac{13}{2} & -\frac{1}{2} & \frac{5}{2} \\ -\frac{3}{2} & \frac{1}{2} & -\frac{1}{2} \\ \frac{1}{2} & \frac{1}{2} & \frac{1}{2} \end{bmatrix}$

29. $\begin{bmatrix} -50 & -9 & 11 \\ -23 & -4 & 5 \\ 5 & 1 & -1 \end{bmatrix}$

30. $\begin{bmatrix} -20 & -13 & -9 \\ \frac{11}{2} & \frac{7}{2} & \frac{5}{2} \\ \frac{1}{2} & \frac{1}{2} & \frac{1}{2} \end{bmatrix}$ **31.** Does not exist

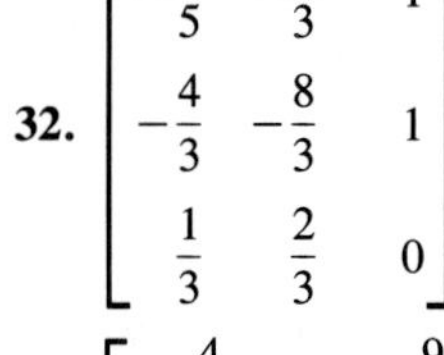

32. $\begin{bmatrix} -\frac{4}{5} & -\frac{5}{3} & 1 \\ -\frac{4}{3} & -\frac{8}{3} & 1 \\ \frac{1}{3} & \frac{2}{3} & 0 \end{bmatrix}$

33. $\begin{bmatrix} \frac{4}{7} & -1 & -\frac{9}{7} \\ -\frac{3}{14} & \frac{1}{2} & \frac{6}{7} \\ \frac{2}{7} & 0 & -\frac{1}{7} \end{bmatrix}$ **34.** $\begin{bmatrix} 3 & 4 & -1 \\ 1 & \frac{7}{5} & -\frac{1}{5} \\ 0 & -\frac{2}{5} & \frac{1}{5} \end{bmatrix}$

35. $\begin{bmatrix} \frac{1}{2} & 0 & 0 \\ 0 & \frac{1}{4} & 0 \\ 0 & 0 & \frac{1}{10} \end{bmatrix}$ **36.** $\begin{bmatrix} 1 & 3 & -11 \\ 0 & 1 & -2 \\ 0 & 0 & 1 \end{bmatrix}$

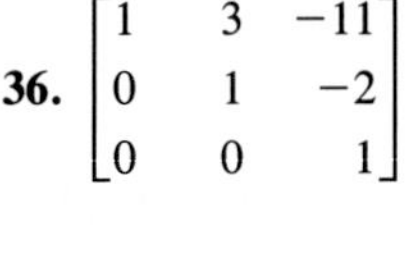

37. $\{(-3, 2)\}$ **38.** $\{(-1, -5)\}$ **39.** $\{(2, 5)\}$
40. $\{(7, 3)\}$ **41.** $\{(-1, -2, 1)\}$ **42.** $\{(1, 0, -2)\}$
43. $\{(-2, 3, 5)\}$ **44.** $\{(4, -1, -1)\}$
45. $\{(-4, 3, 0)\}$ **46.** $\{(-2, 8, -7)\}$
47. **(a)** $\{(-1, 2, 3)\}$ **(b)** $\{(2, -1, -4)\}$
(c) $\{(-5, 0, -2)\}$ **(d)** $\{(3, -6, 1)\}$
(e) $\{(1, -1, -1)\}$
48. **(a)** Play It Again Sam **(b)** Drop The Course
(c) The Price Is Right
(d) How Would It Play In Peoria
49. **(a)** y-axis reflection **(b)** Origin rotation
(c) 90° counterclockwise rotation
(d) 90° clockwise rotation

Problem Set 11.4 (page 675)

1.

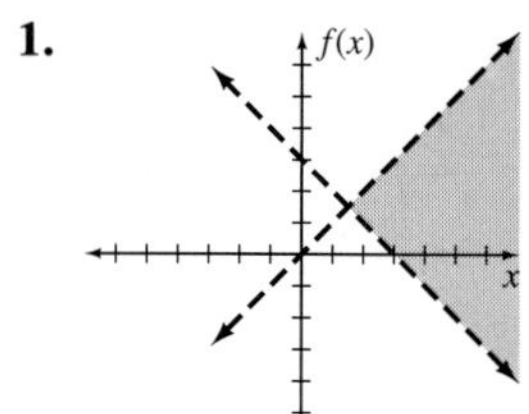

2.

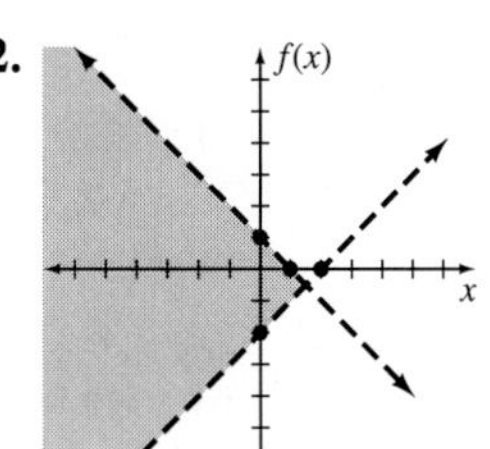

3.

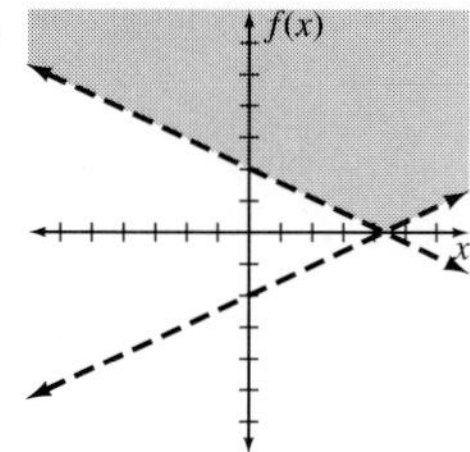

4.

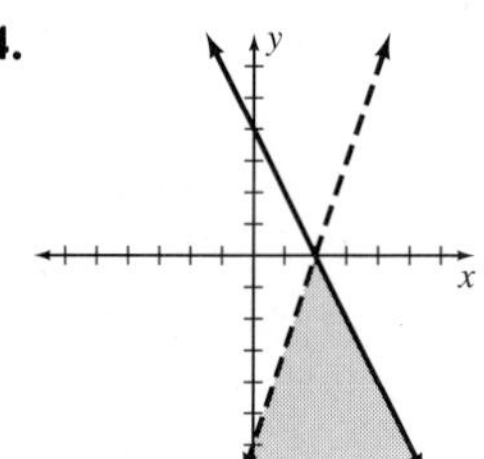

5.

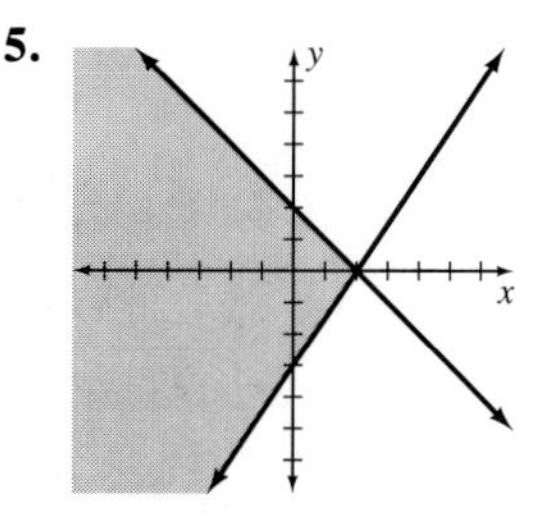

6.

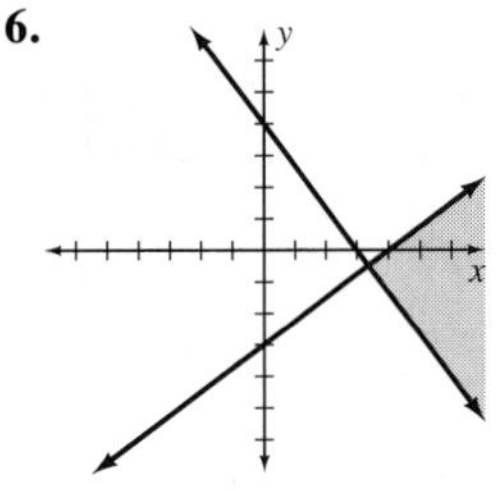

7.

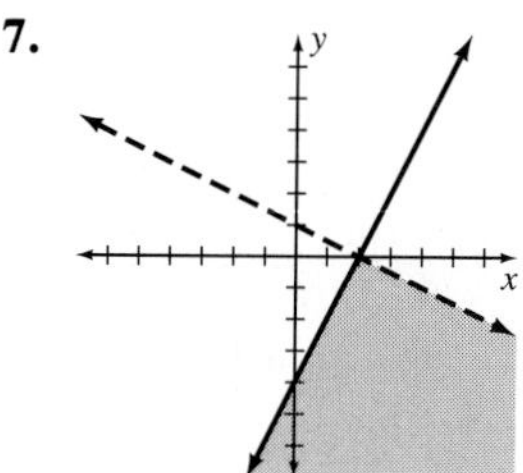

8.

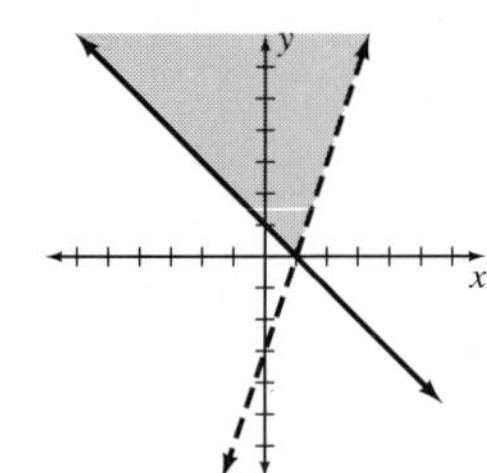

9.

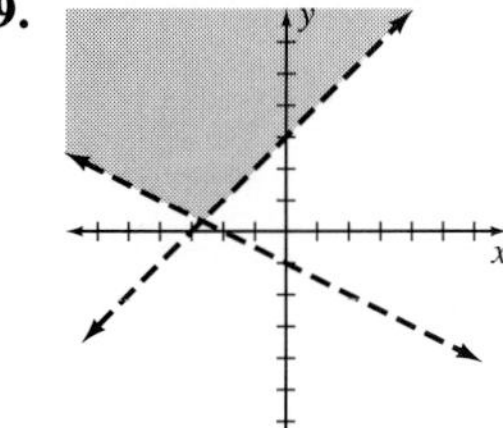

10.

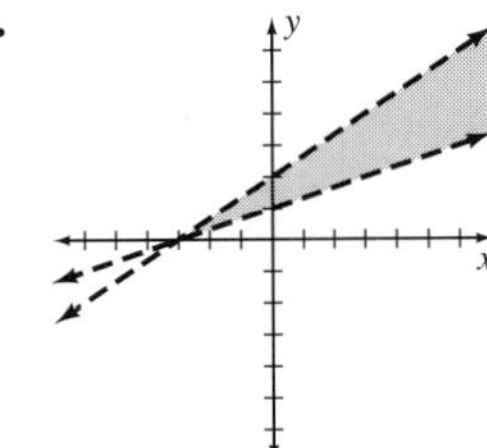

11.

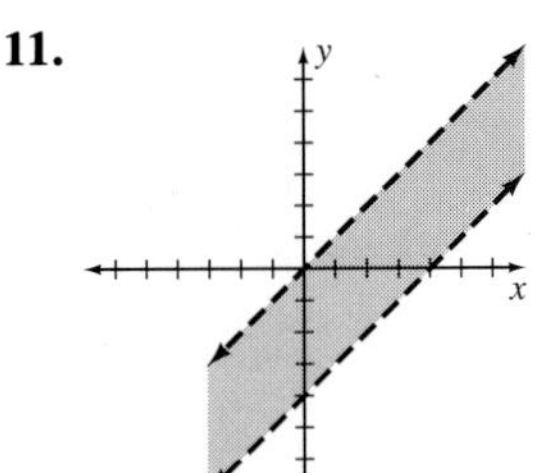

12.

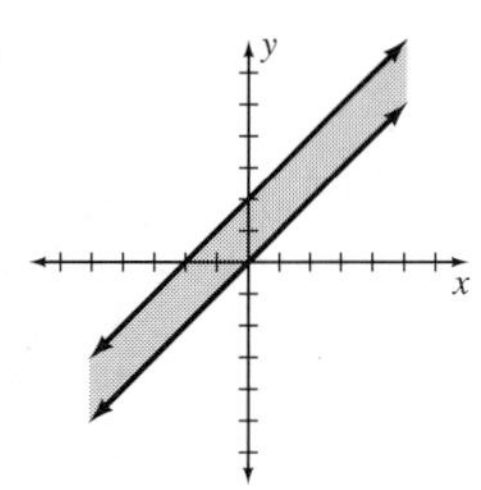

13. 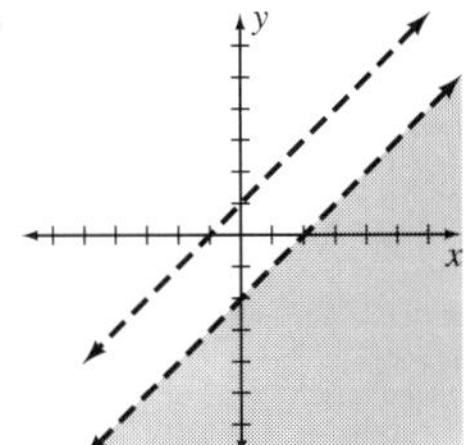

14. 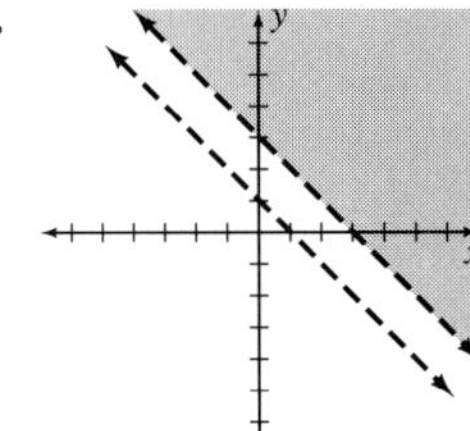

23. 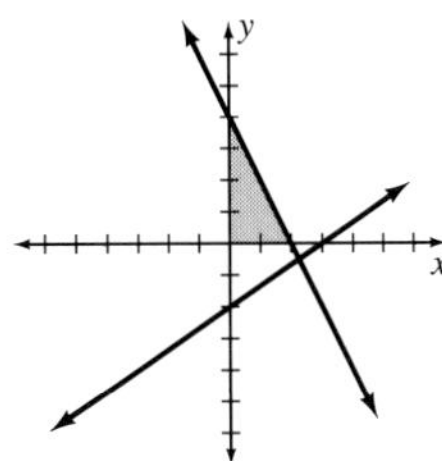

24. 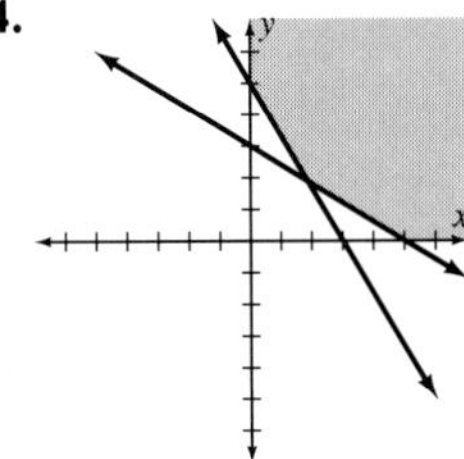

15. 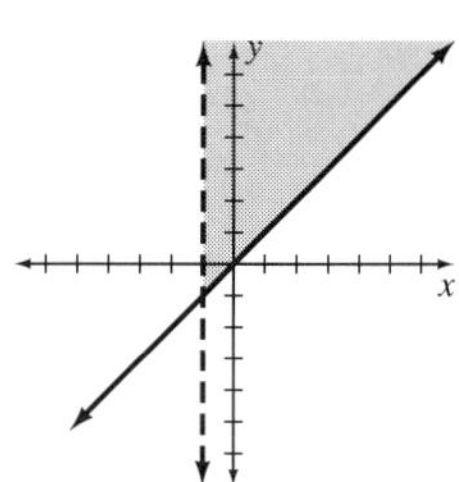

16. 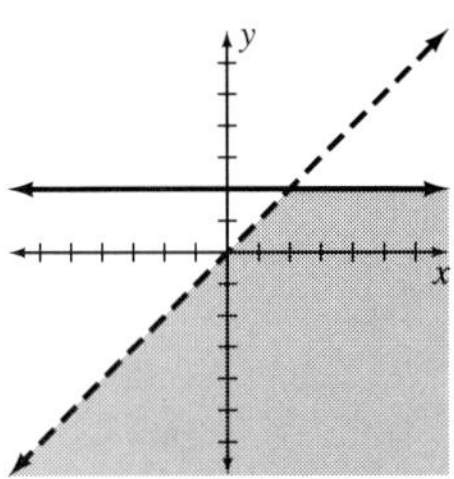

17. ∅ 

18. 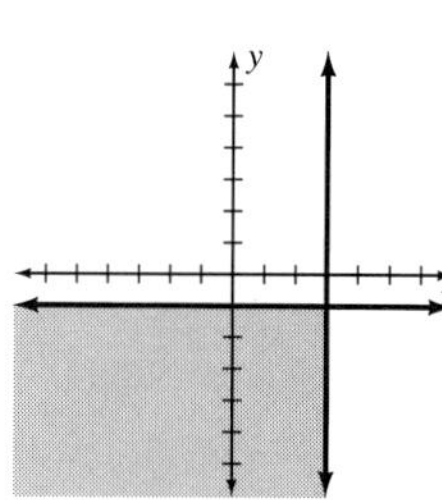

19. 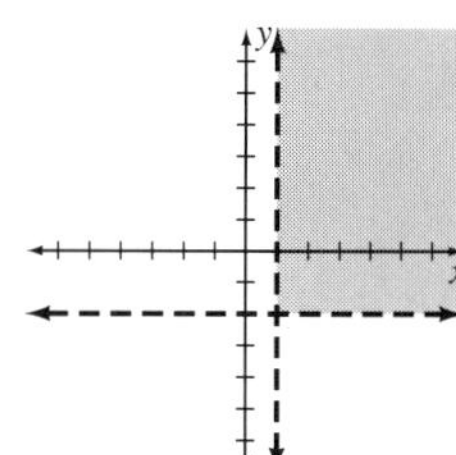

20. ∅

21. 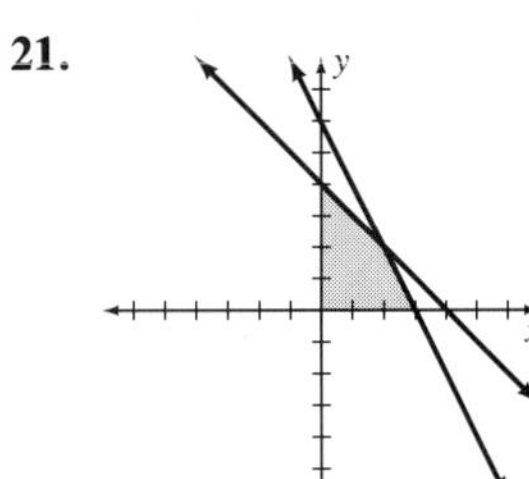

22. 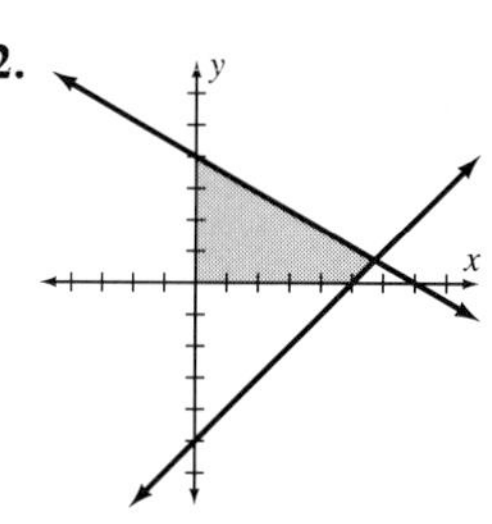

25. Minimum of 8 and maximum of 52
26. Minimum of 14 and maximum of 73
27. Minimum of 0 and maximum of 28
28. Minimum of 14.5 and maximum of 54.5 **29.** 63
30. 21 **31.** 340 **32.** 660 **33.** 2
34. 42 **35.** 98 **36.** 42
37. \$5000 at 9% and \$5000 at 12%
38. 40 sets of model A and 10 sets of model B
39. 300 of type A and 200 of type B
40. 50 deluxe and 100 standard models
41. 12 units of A and 16 units of B
42. 50 widgets and 0 wadgets

Chapter 11 Review Problem Set (page 680)

1. $\begin{bmatrix} 7 & -5 \\ -3 & 10 \end{bmatrix}$ **2.** $\begin{bmatrix} 3 & 3 \\ 3 & -6 \end{bmatrix}$ **3.** $\begin{bmatrix} 2 & 1 \\ -6 & 8 \\ -2 & 2 \end{bmatrix}$

4. $\begin{bmatrix} 19 & -11 \\ -6 & 22 \end{bmatrix}$ **5.** $\begin{bmatrix} 7 & 1 \\ -14 & 20 \\ 1 & -2 \end{bmatrix}$

6. $\begin{bmatrix} -11 & -3 & 15 \\ 24 & 2 & -20 \\ -40 & -5 & 38 \end{bmatrix}$ **7.** $\begin{bmatrix} 16 & -26 \\ 0 & 13 \end{bmatrix}$

8. $\begin{bmatrix} 26 & -36 \\ -15 & 32 \end{bmatrix}$ **9.** $\begin{bmatrix} -27 \\ 26 \end{bmatrix}$

10. EF does not exist. **14.** $\begin{bmatrix} 4 & -5 \\ -7 & 9 \end{bmatrix}$

15. $\begin{bmatrix} -3 & 4 \\ 7 & -9 \end{bmatrix}$ **16.** $\begin{bmatrix} -\frac{3}{8} & \frac{1}{8} \\ \frac{1}{4} & \frac{1}{4} \end{bmatrix}$

17. Inverse does not exist. **18.** $\begin{bmatrix} \frac{5}{7} & -\frac{3}{7} \\ -\frac{4}{7} & \frac{1}{7} \end{bmatrix}$

19. $\begin{bmatrix} \frac{2}{7} & \frac{1}{7} \\ -\frac{1}{3} & 0 \end{bmatrix}$ **20.** $\begin{bmatrix} \frac{39}{8} & -\frac{17}{8} & -\frac{1}{8} \\ 2 & -1 & 0 \\ \frac{1}{8} & \frac{1}{8} & \frac{1}{8} \end{bmatrix}$

21. $\begin{bmatrix} 8 & -8 & 5 \\ -3 & 2 & -1 \\ -1 & -1 & 1 \end{bmatrix}$ **22.** Inverse does not exist.

23. $\begin{bmatrix} -\frac{20}{3} & -\frac{7}{3} & \frac{1}{3} \\ -\frac{1}{3} & -\frac{2}{3} & -\frac{1}{3} \\ -\frac{5}{3} & -\frac{1}{3} & \frac{1}{3} \end{bmatrix}$

24. $\{(-2, 6)\}$ **25.** $\{(4, -1)\}$ **26.** $\{(2, -3, -1)\}$

27. $\{(-3, 2, 5)\}$ **28.** $\{(-4, 3, 4)\}$

29.

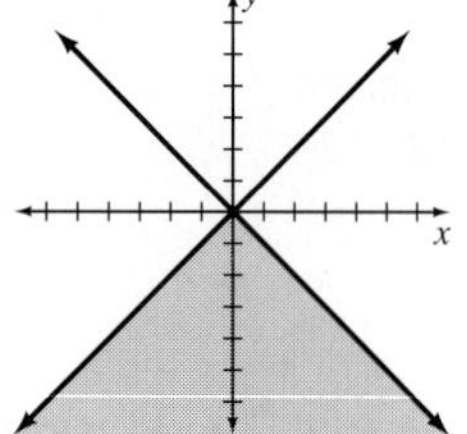

30.

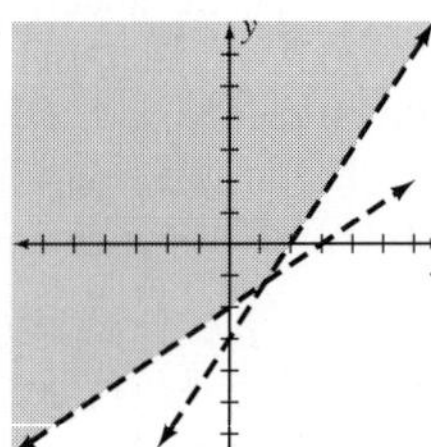

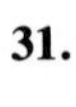

31.

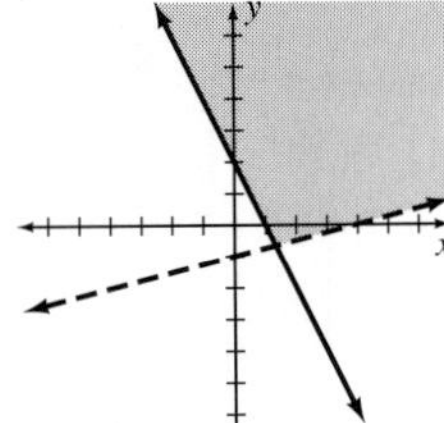

32.

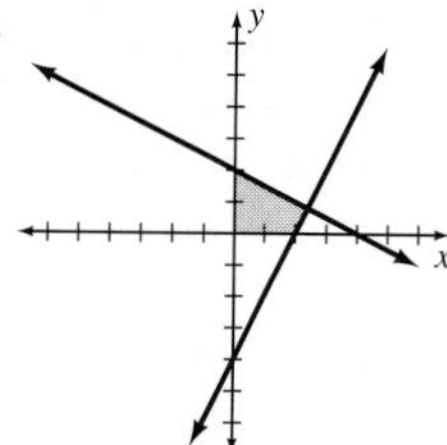

33. 37 **34.** 56 **35.** 57 **36.** 1700

35. 75 one-gallon and 175 two-gallon freezers

CHAPTER 12

Problem Set 12.1 (page 692)

1. $V(0, 0)$, $F(2, 0)$, $x = -2$

2. $V(0, 0)$, $F(-1, 0)$, $x = 1$

3. $V(0, 0)$, $F(0, -3)$, $y = 3$

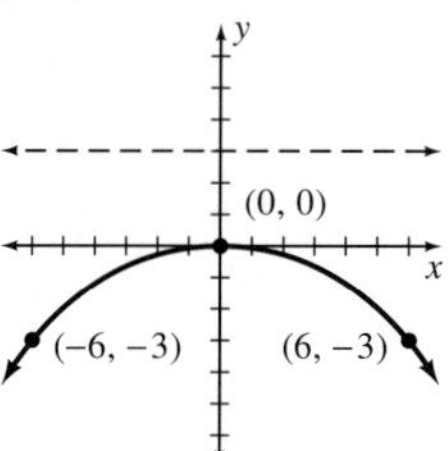

4. $V(0, 0)$, $F(0, 2)$, $y = -2$

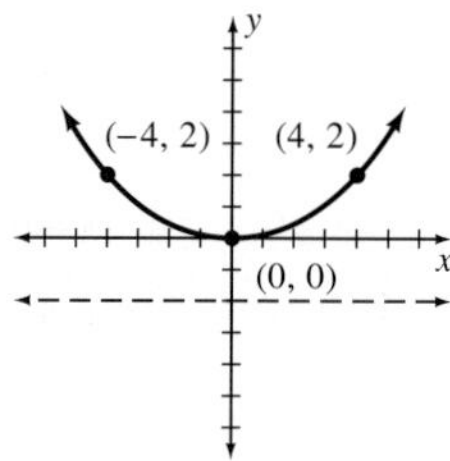

5. $V(0, 0)$, $F\left(-\frac{1}{2}, 0\right)$, $x = \frac{1}{2}$

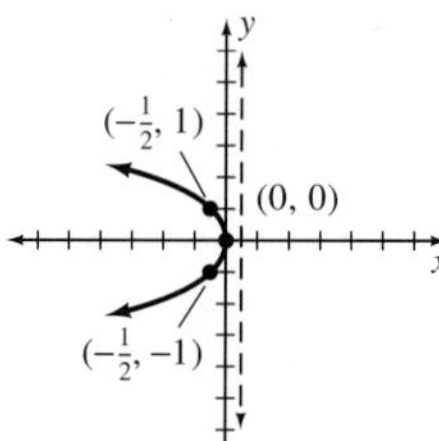

6. $V(0, 0)$, $F\left(\frac{3}{2}, 0\right)$, $x = -\frac{3}{2}$

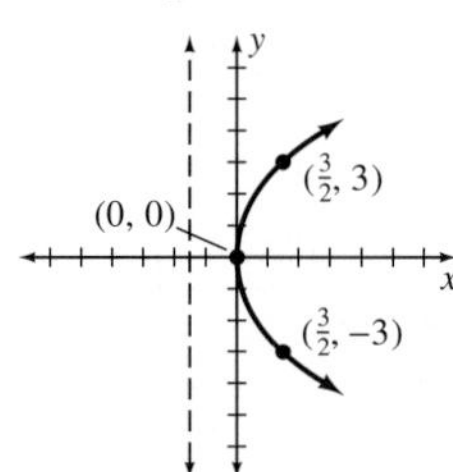

7. $V(0, 0)$, $F\left(0, \frac{3}{2}\right)$, $y = -\frac{3}{2}$

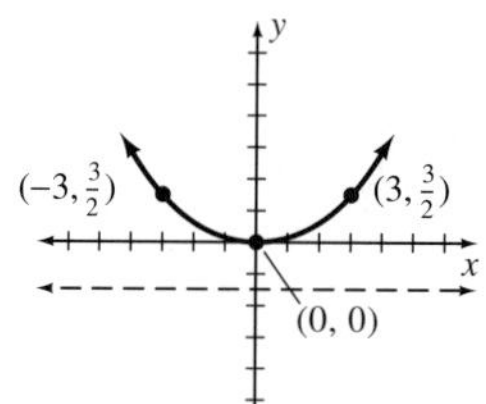

8. $V(0, 0)$, $F\left(0, -\frac{7}{4}\right)$, $y = \frac{7}{4}$

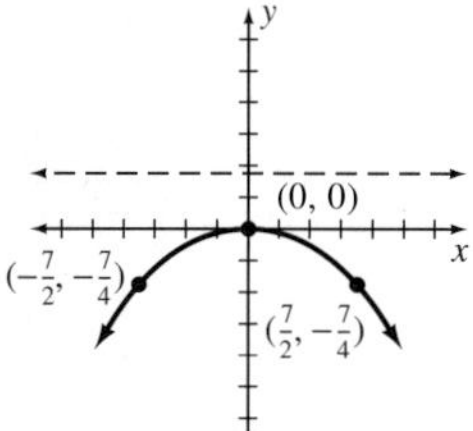

9. $V(0, 2)$, $F(0, 3)$, $y = 1$

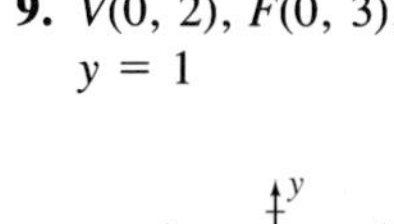

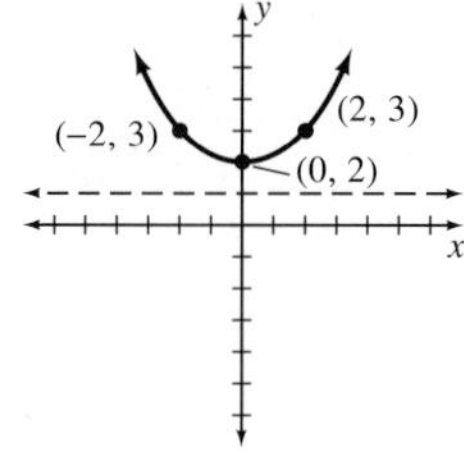

10. $V(0, -3)$, $F(0, -1)$, $y = -5$

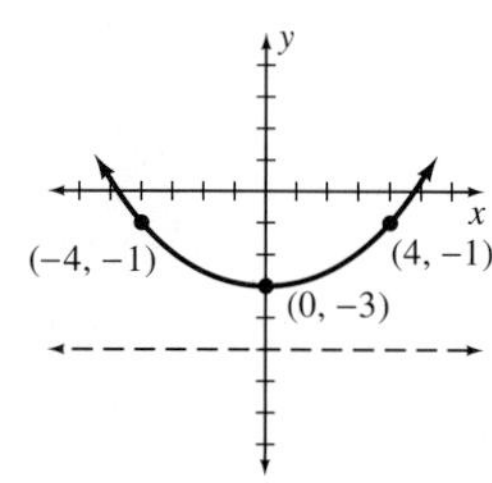

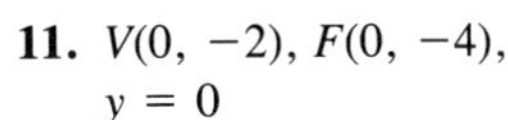

11. $V(0, -2)$, $F(0, -4)$, $y = 0$

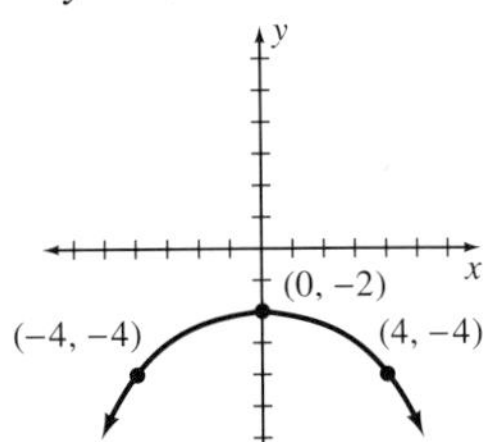

12. $V(0, 1)$, $F(0, 0)$, $y = 2$

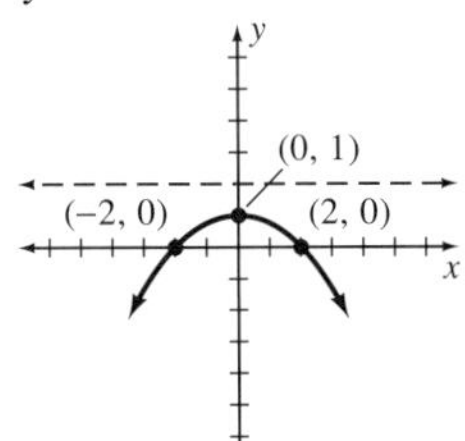

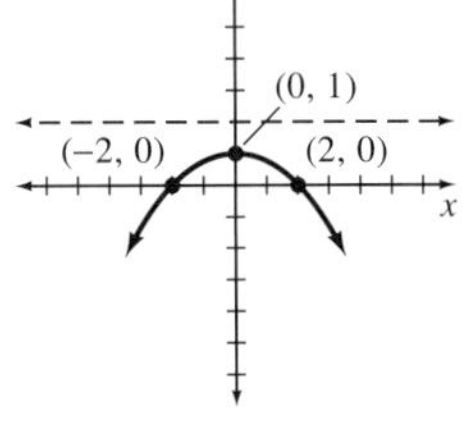

13. $V(2, 0)$, $F(5, 0)$, $x = -1$

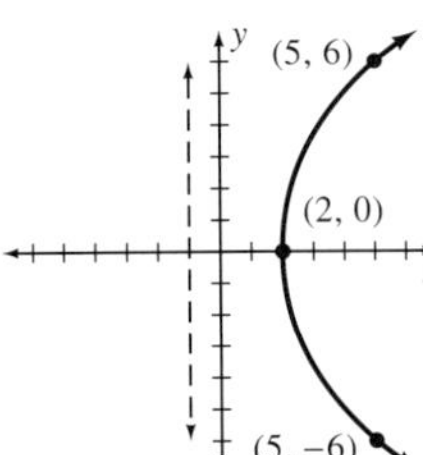

14. $V(3, 0)$, $F(1, 0)$, $x = 5$

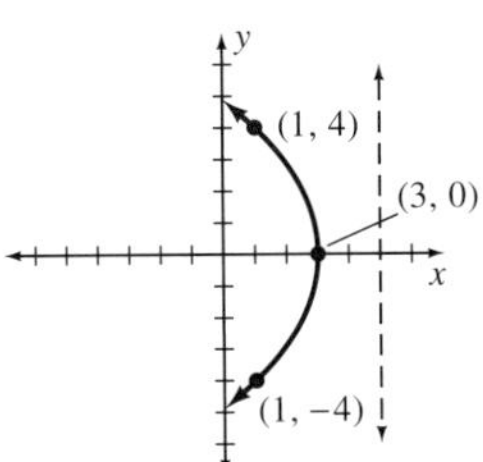

15. $V(1, 2)$, $F(1, 3)$, $y = 1$

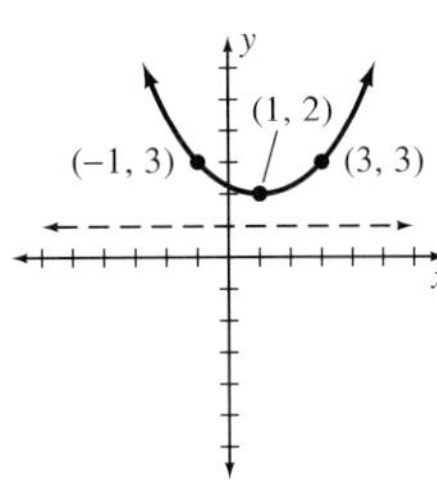

16. $V(-2, -1)$, $F(-2, 1)$, $y = -3$

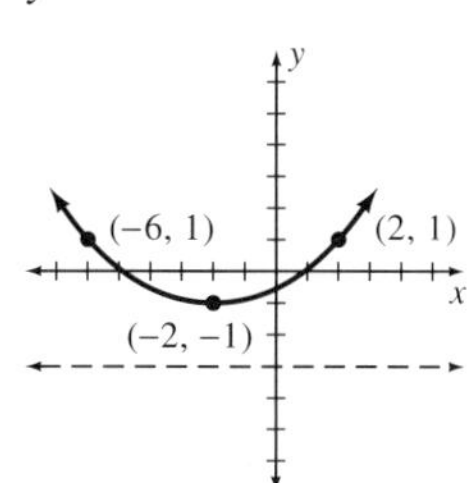

17. $V(-3, 1)$, $F(-3, -1)$, $y = 3$

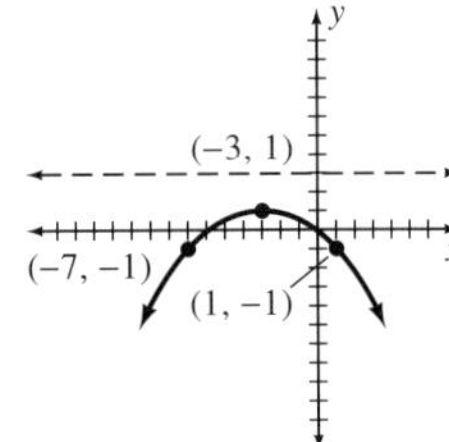

18. $V(2, 2)$, $F(2, 1)$, $y = 3$

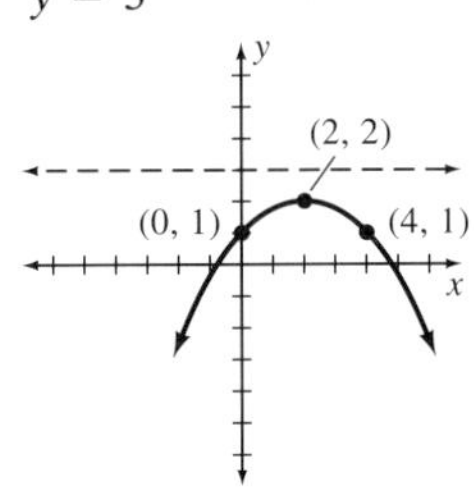

19. $V(3, 1)$, $F(0, 1)$, $x = 6$

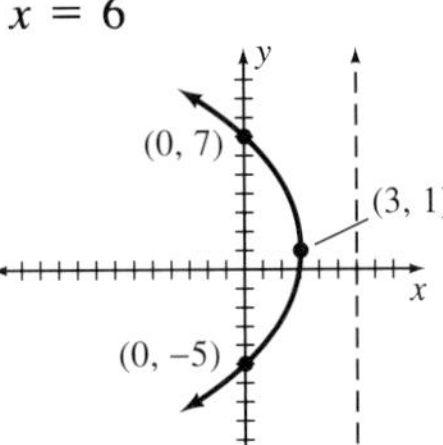

20. $V(1, -2)$, $F(-1, -2)$, $x = 3$

21. $V(-2, -3)$, $F(-1, -3)$, $x = -3$

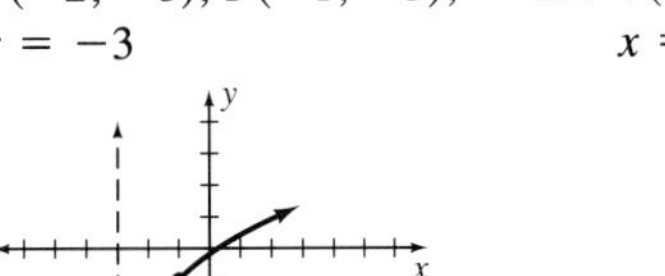

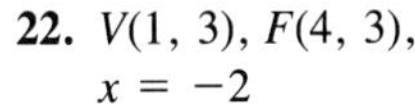

22. $V(1, 3)$, $F(4, 3)$, $x = -2$

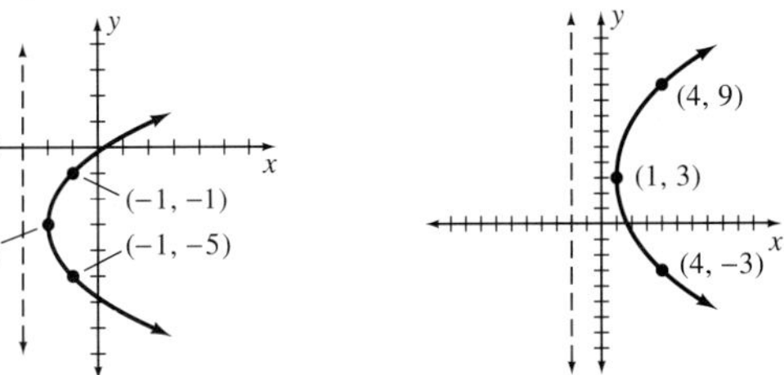

23. $x^2 = 12y$ **24.** $x^2 = -2y$ **25.** $y^2 = -4x$
26. $y^2 - 8x + 24 = 0$ **27.** $x^2 + 12y - 48 = 0$
28. $x^2 - 16y - 96 = 0$
29. $x^2 - 6x - 12y + 21 = 0$
30. $x^2 + 6x + 16y - 39 = 0$
31. $y^2 - 10y + 8x + 41 = 0$
32. $y^2 + 4y - 12x + 28 = 0$ **33.** $3y^2 = -25x$
34. $x^2 = -y$ **35.** $y^2 = 10x$ **36.** $x^2 = -14y$
37. $x^2 - 14x - 8y + 73 = 0$
38. $y^2 + 12y + 12x + 84 = 0$
39. $y^2 + 6y - 12x + 105 = 0$
40. $x^2 + 4x + 16y - 140 = 0$
41. $x^2 + 18x + y + 80 = 0$
42. $2y^2 + 16y - x + 38 = 0$

Problem Set 12.2 (page 701)

For Problems 1–22, the foci are indicated above the graph, and the vertices and endpoints of the minor axes are indicated on the graph.

1. $F(\sqrt{3}, 0)$, $F'(-\sqrt{3}, 0)$

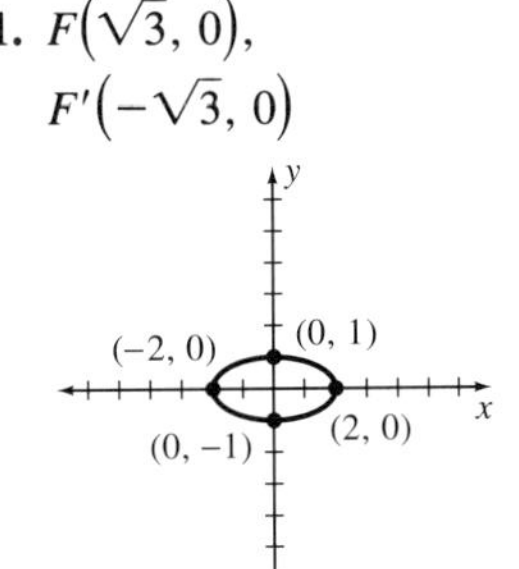

2. $F(\sqrt{15}, 0)$, $F'(-\sqrt{15}, 0)$

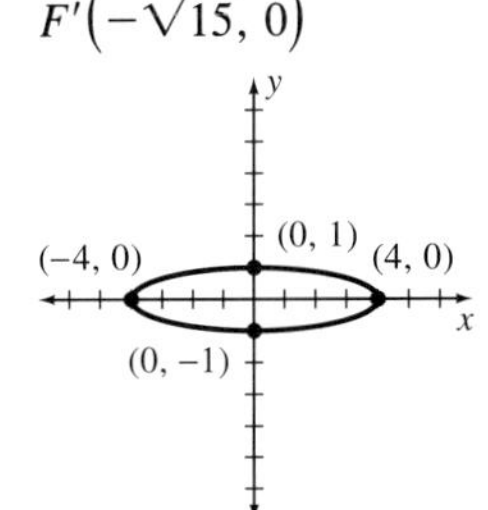

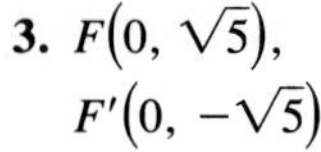

3. $F(0, \sqrt{5})$, $F'(0, -\sqrt{5})$

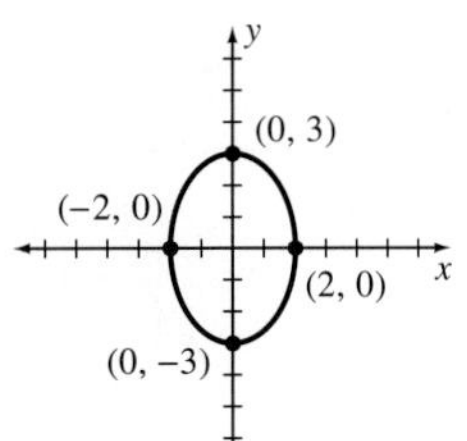

4. $F(0, 2\sqrt{3})$, $F'(0, -2\sqrt{3})$

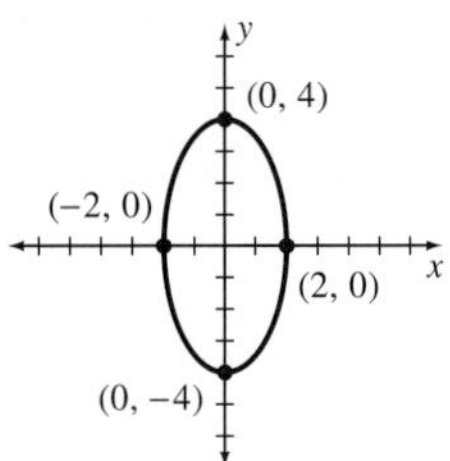

5. $F(0, \sqrt{6})$, $F'(0, -\sqrt{6})$

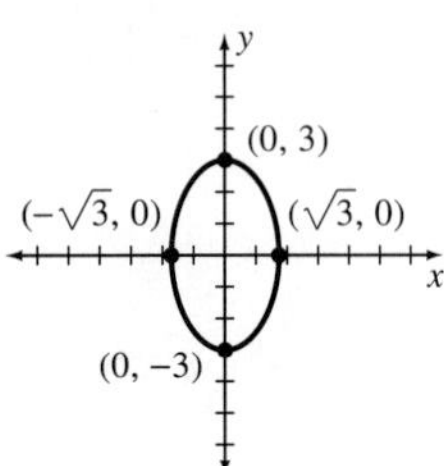

6. $F(0, \sqrt{3})$, $F'(0, -\sqrt{3})$

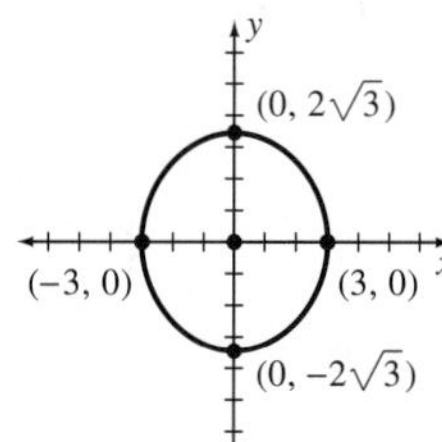

7. $F(\sqrt{15}, 0)$, $F'(-\sqrt{15}, 0)$

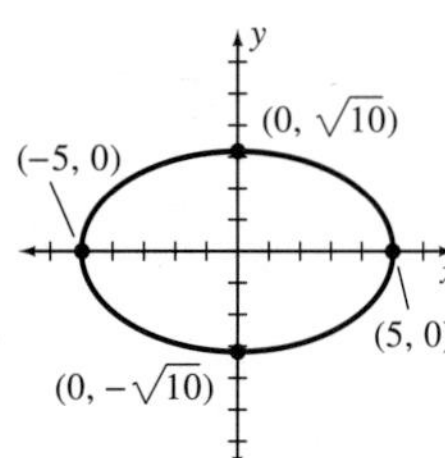

8. $F(\sqrt{31}, 0)$, $F'(-\sqrt{31}, 0)$

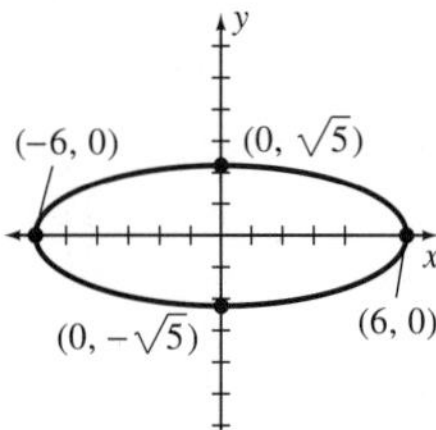

9. $F(0, \sqrt{33})$, $F'(0, -\sqrt{33})$

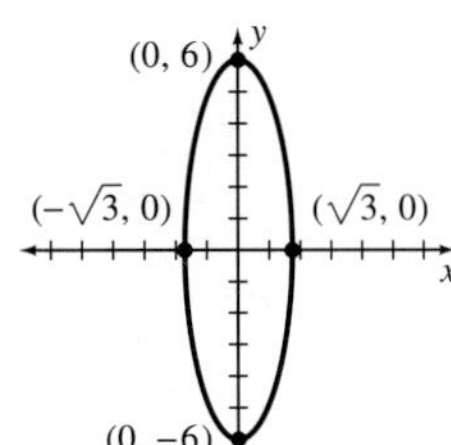

10. $F(0, \sqrt{14})$, $F'(0, -\sqrt{14})$

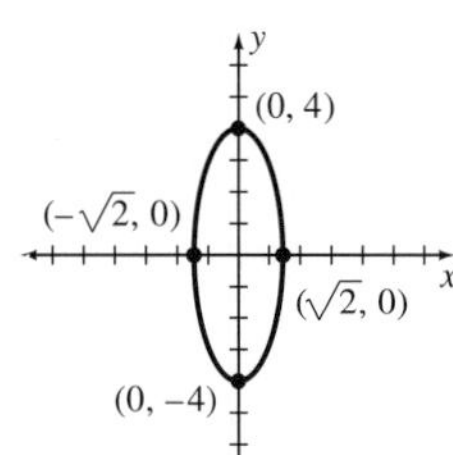

11. $F(2, 0)$, $F'(-2, 0)$

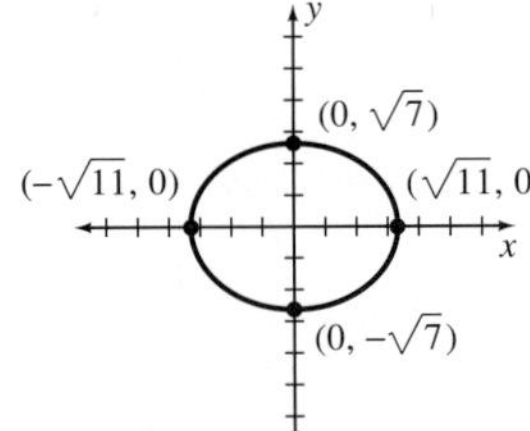

12. $F(0, 3)$, $F'(0, -3)$

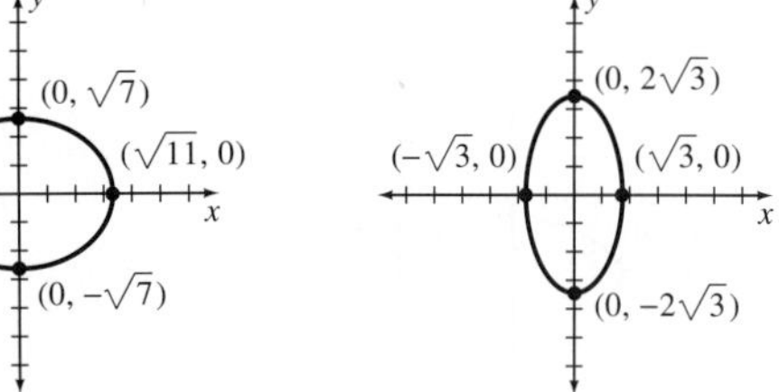

13. $F(1 + \sqrt{5}, 2)$, $F'(1 - \sqrt{5}, 2)$

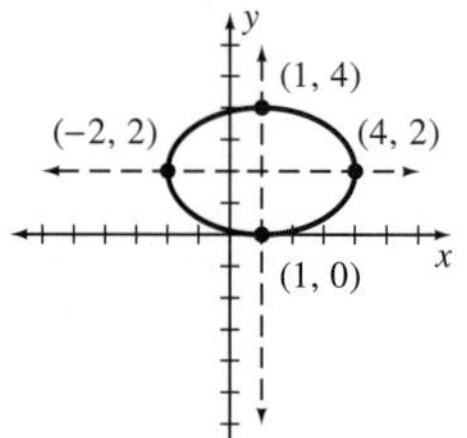

14. $F(-3 + 2\sqrt{2}, 2)$, $F'(-3 - 2\sqrt{2}, 2)$

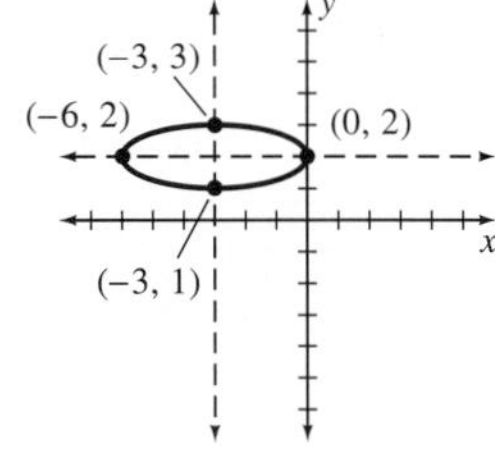

15. $F(-2, -1 + 2\sqrt{3})$, $F'(-2, -1 - 2\sqrt{3})$

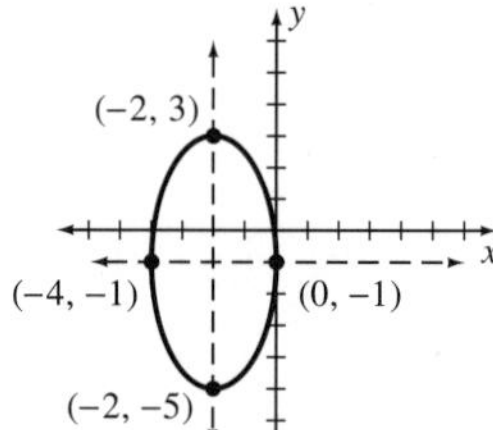

16. $F(2, -2 + \sqrt{5})$, $F'(2, -2 - \sqrt{5})$

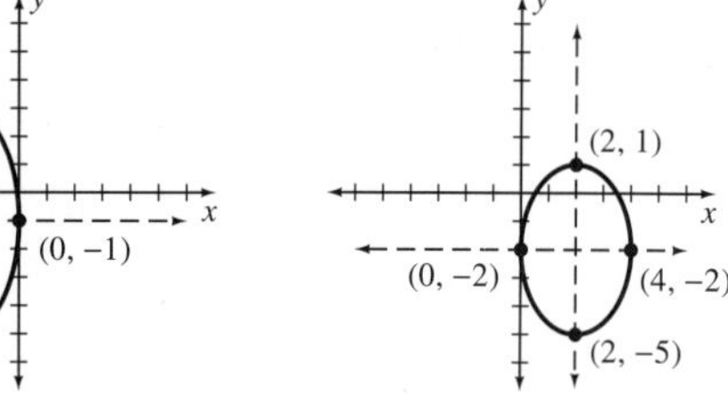

17. $F(3 + \sqrt{3}, 0)$, $F'(3 - \sqrt{3}, 0)$

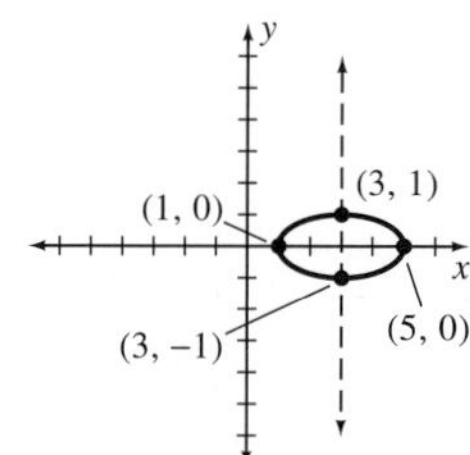

18. $F(0, -2 + \sqrt{7})$, $F'(0, -2 - \sqrt{7})$

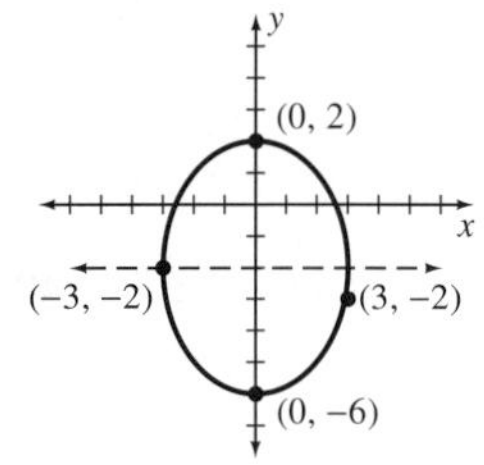

19. $F\left(4, -1 + \sqrt{7}\right)$,
$F'\left(4, -1 - \sqrt{7}\right)$

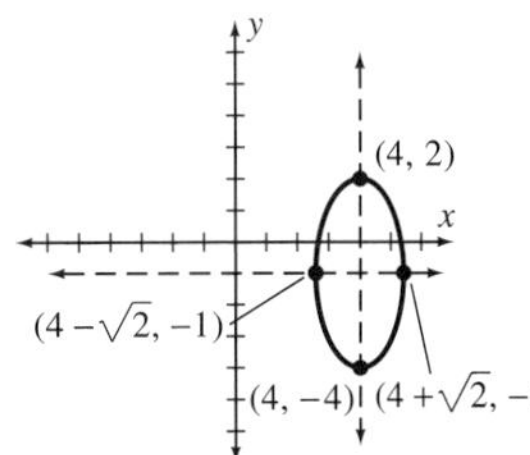

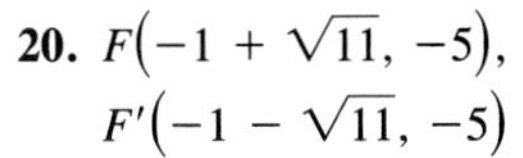

20. $F\left(-1 + \sqrt{11}, -5\right)$,
$F'\left(-1 - \sqrt{11}, -5\right)$

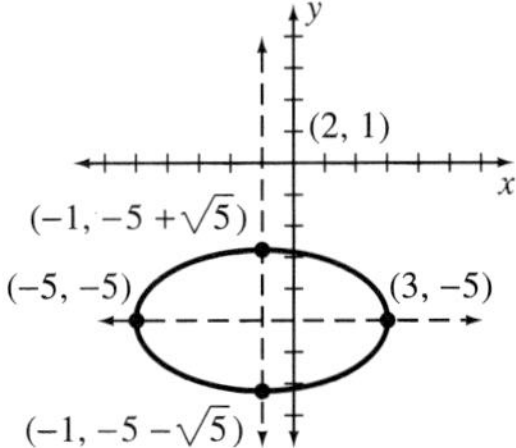

21. $F(0, 4)$, $F'(-6, 4)$

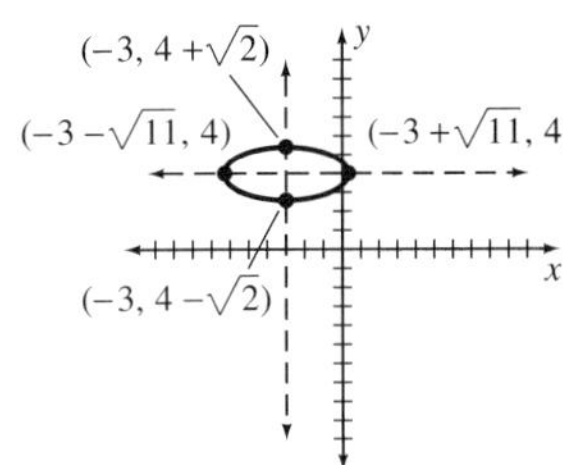

22. $F(-4, 1)$, $F'(-4, -7)$

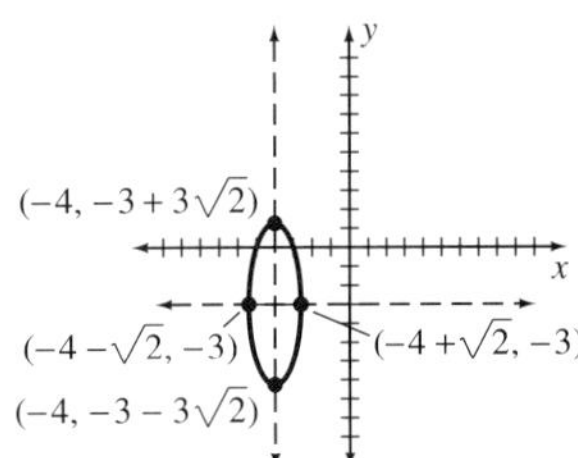

23. $16x^2 + 25y^2 = 400$ **24.** $3x^2 + 4y^2 = 48$
25. $36x^2 + 11y^2 = 396$ **26.** $9x^2 + 5y^2 = 45$
27. $x^2 + 9y^2 = 9$ **28.** $25x^2 + 4y^2 = 100$
29. $100x^2 + 36y^2 = 225$ **30.** $x^2 + 2y^2 = 2$
31. $7x^2 + 3y^2 = 75$ **32.** $x^2 + 11y^2 = 36$
33. $3x^2 - 6x + 4y^2 - 8y - 41 = 0$
34. $25x^2 - 100x + 9y^2 + 18y - 116 = 0$
35. $9x^2 + 25y^2 - 50y - 200 = 0$
36. $2x^2 - 12x + y^2 + 10 = 0$ **37.** $3x^2 + 4y^2 = 48$
38. $25x^2 + 16y^2 = 400$ **39.** $\dfrac{10\sqrt{5}}{3}$ feet
40. $\dfrac{20\sqrt{2}}{3}$ feet

Problem Set 12.3 (page 712)

For Problems 1–22, the foci and equations of the asymptotes are indicated above the graphs. The vertices are given on the graphs.

1. $F\left(\sqrt{13}, 0\right)$,
$F'\left(-\sqrt{13}, 0\right)$,
$y = \pm\frac{2}{3}x$

2. $F\left(2\sqrt{5}, 0\right)$,
$F'\left(-2\sqrt{5}, 0\right)$,
$y = \pm 2x$

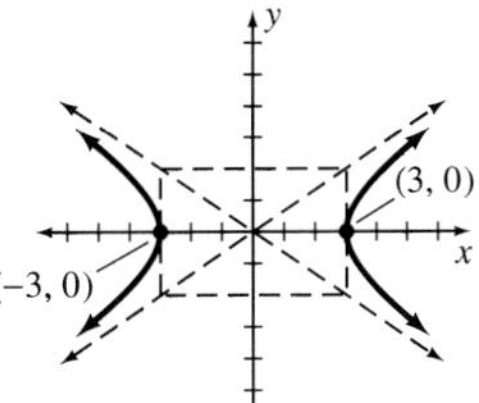

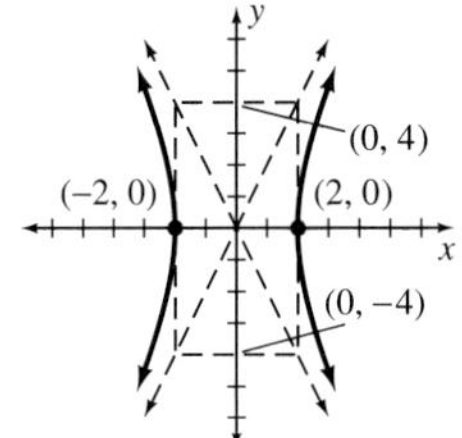

3. $F\left(0, \sqrt{13}\right)$,
$F'\left(0, -\sqrt{13}\right)$,
$y = \pm\frac{2}{3}x$

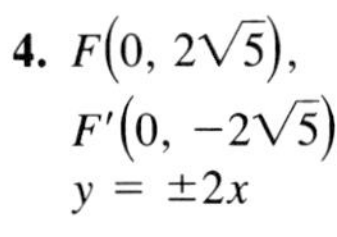

4. $F\left(0, 2\sqrt{5}\right)$,
$F'\left(0, -2\sqrt{5}\right)$,
$y = \pm 2x$

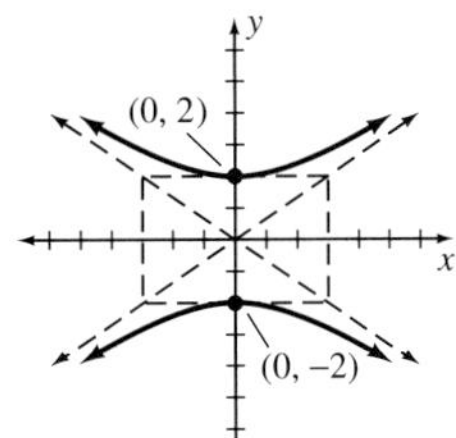

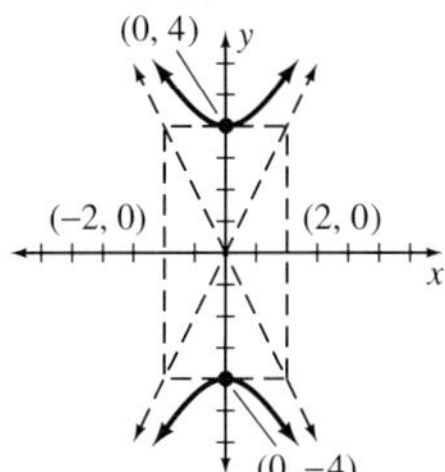

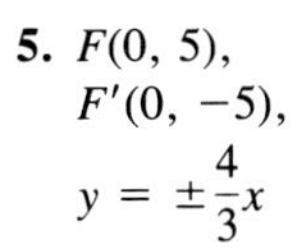

5. $F(0, 5)$,
$F'(0, -5)$,
$y = \pm\frac{4}{3}x$

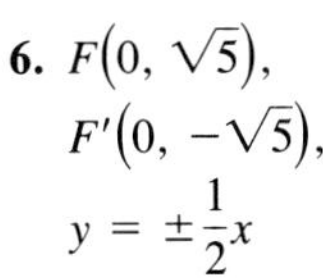

6. $F\left(0, \sqrt{5}\right)$,
$F'\left(0, -\sqrt{5}\right)$,
$y = \pm\frac{1}{2}x$

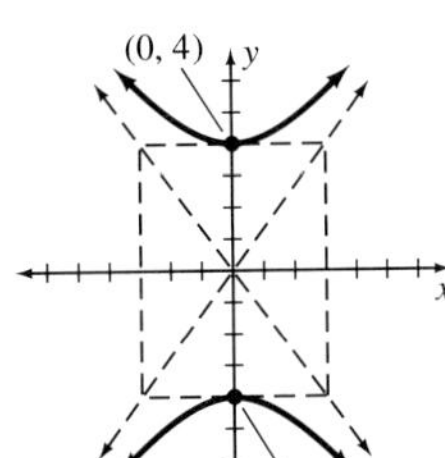

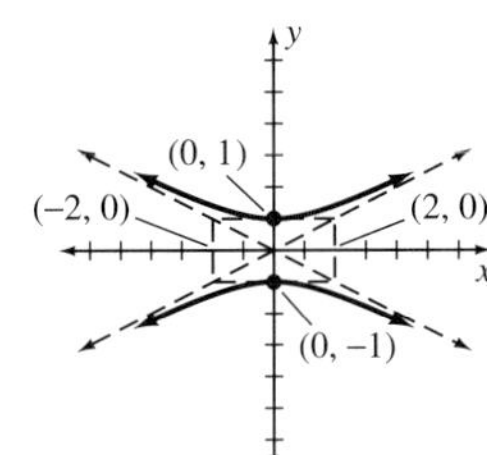

7. $F(3\sqrt{2}, 0)$, $F'(-3\sqrt{2}, 0)$, $y = \pm x$

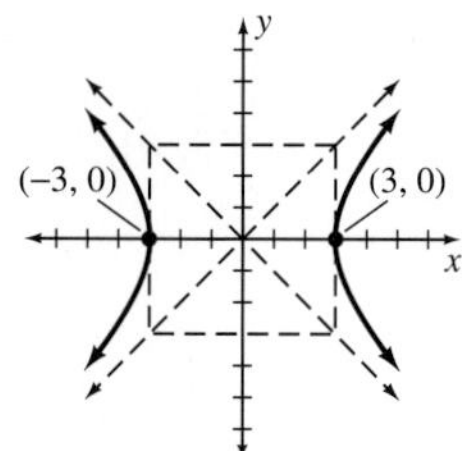

8. $F(\sqrt{2}, 0)$, $F'(-\sqrt{2}, 0)$, $y = \pm x$

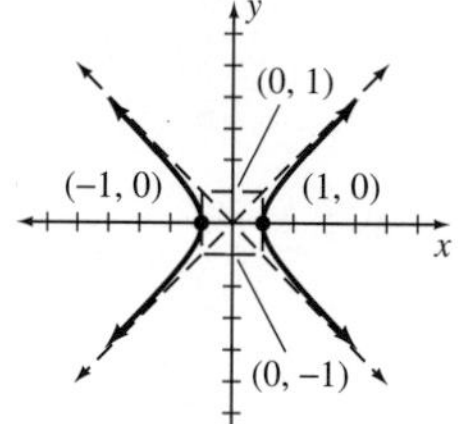

9. $F(0, \sqrt{30})$, $F'(0, -\sqrt{30})$, $y = \pm\frac{\sqrt{5}}{5}x$

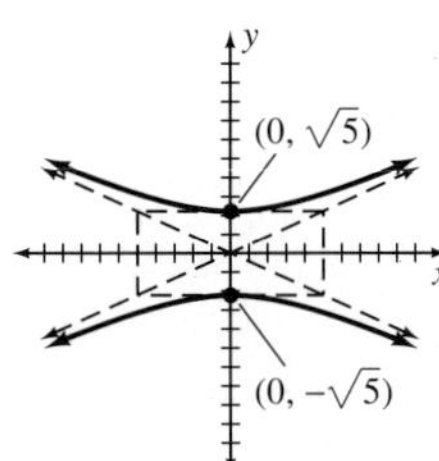

10. $F(0, 2\sqrt{3})$, $F'(0, -2\sqrt{3})$, $y = \pm\sqrt{2}x$

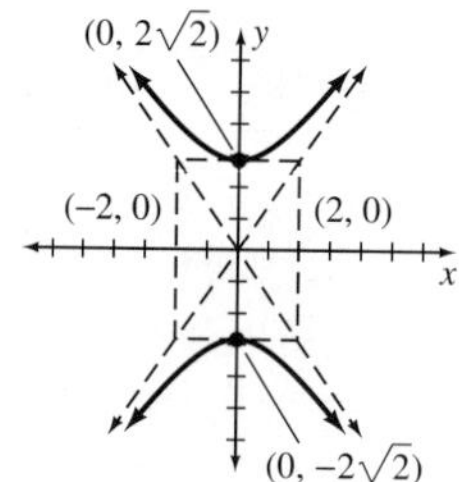

11. $F(\sqrt{10}, 0)$, $F'(-\sqrt{10}, 0)$, $y = \pm 3x$

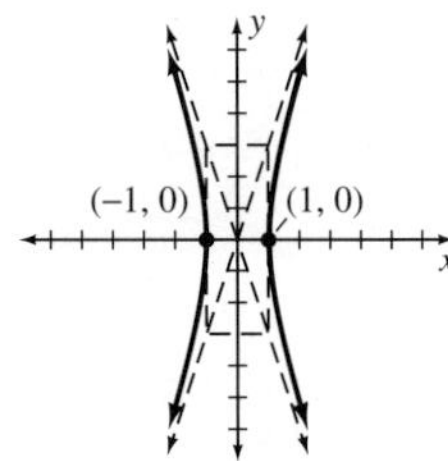

12. $F(\sqrt{17}, 0)$, $F'(-\sqrt{17}, 0)$, $y = \pm\frac{1}{4}x$

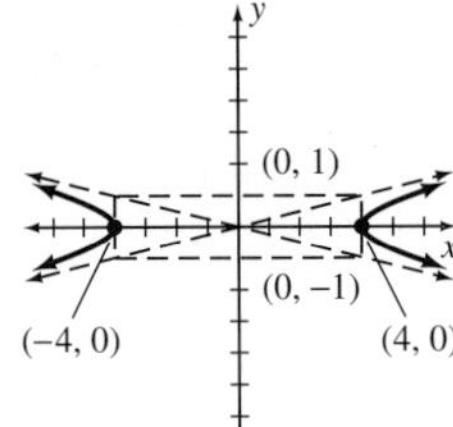

13. $F(3 + \sqrt{13}, -1)$, $F'(3 - \sqrt{13}, -1)$ $2x - 3y = 9$ and $2x + 3y = 3$

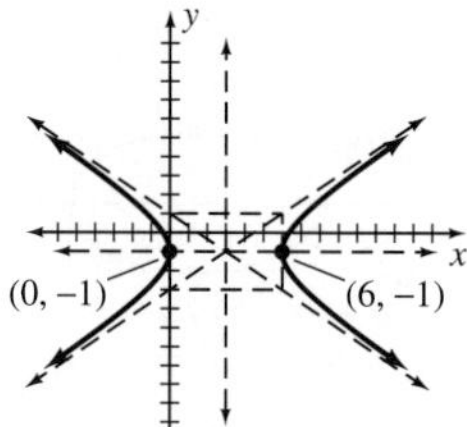

14. $F(-4 + \sqrt{13}, -2)$, $F'(-4 - \sqrt{13}, -2)$ $3x - 2y = -8$ and $3x + 2y = -16$

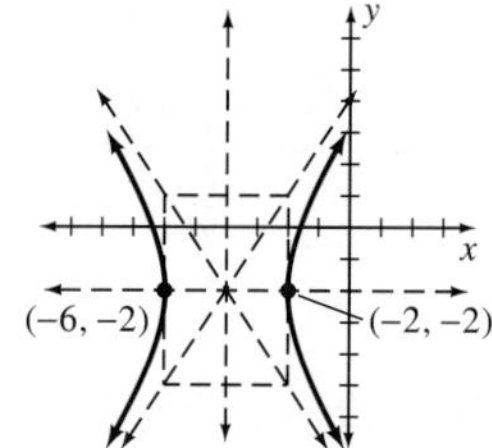

15. $F(-3, 2 + \sqrt{5})$, $F'(-3, 2 - \sqrt{5})$ $2x - y = -8$ and $2x + y = -4$

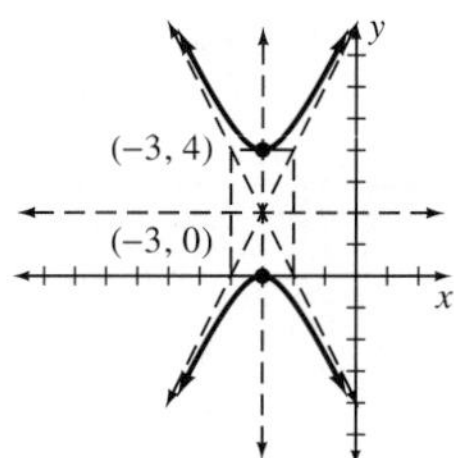

16. $F((3, -3 + \sqrt{10})$, $F'(3, -3 - \sqrt{10})$ $x - 3y = 12$ and $x + 3y = -6$

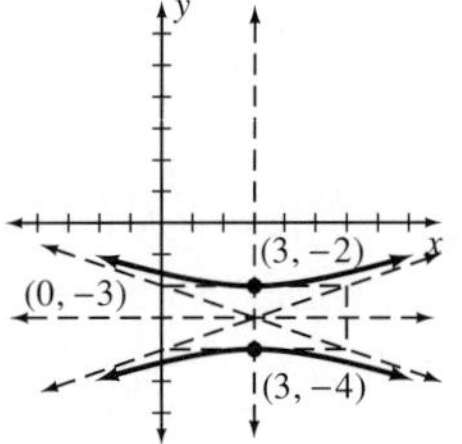

17. $F(2 + \sqrt{6}, 0)$, $F'(2 - \sqrt{6}, 0)$ $\sqrt{2}x - y = 2\sqrt{2}$ and $\sqrt{2}x + y = 2\sqrt{2}$

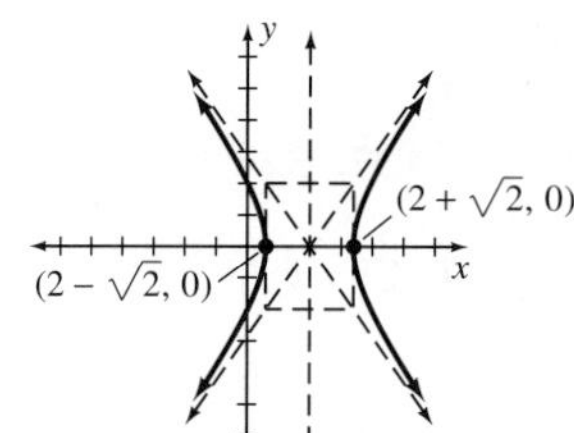

18. $F(-3 + 2\sqrt{3}, 0)$, $F'(-3 - 2\sqrt{3}, 0)$ $\sqrt{3}x - 3y = -3\sqrt{3}$ and $\sqrt{3}x + 3y = -3\sqrt{3}$

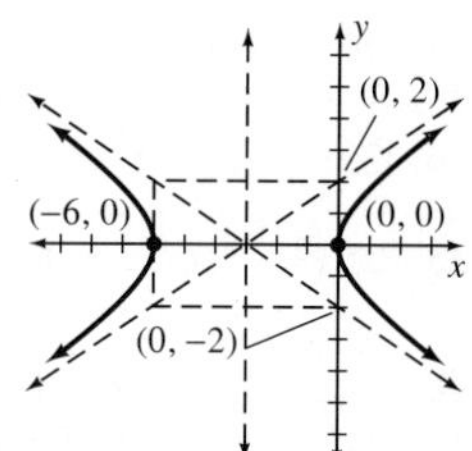

19. $F(0, -5 + \sqrt{10})$,
$F'(0, -5 - \sqrt{10})$
$3x - y = 5$ and
$3x + y = -5$

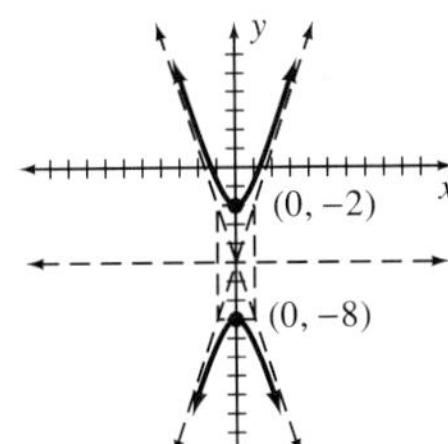

20. $F(0, 2 + \sqrt{5})$,
$F'(0, 2 - \sqrt{5})$
$x - 2y = -4$ and
$x + 2y = 4$

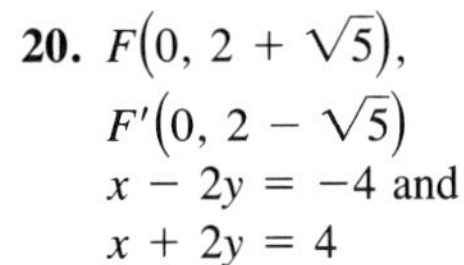

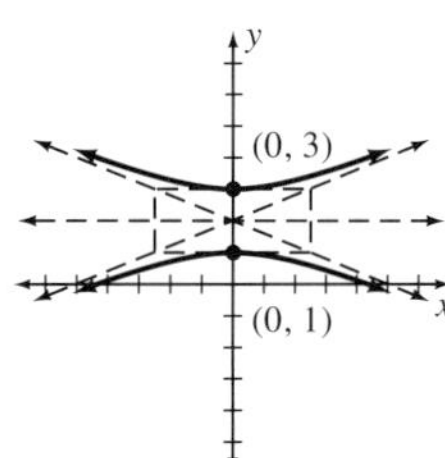

21. $F(-2 + \sqrt{2}, -2)$,
$F'(-2 - \sqrt{2}, -2)$
$x - y = 0$ and
$x + y = -4$

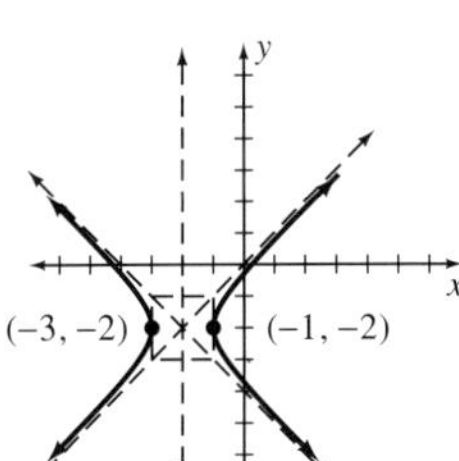

22. $F(1, -4 + \sqrt{2})$,
$F'(1, -4 - \sqrt{2})$
$x - y = 5$ and
$x + y = -3$

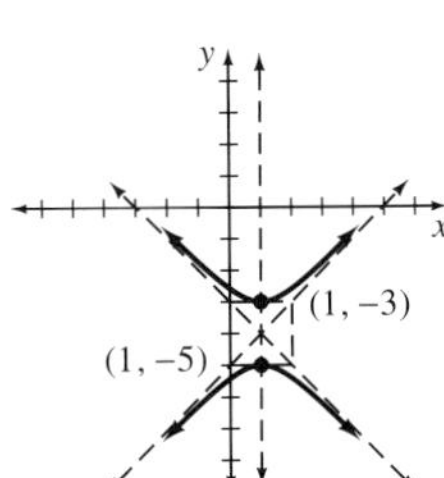

23. $5x^2 - 4y^2 = 20$ **24.** $15x^2 - y^2 = 15$
25. $16y^2 - 9x^2 = 144$ **26.** $8y^2 - x^2 = 32$
27. $3x^2 - y^2 = 3$ **28.** $9y^2 - 24x^2 = 9$
29. $4y^2 - 3x^2 = 12$ **30.** $9x^2 - 5y^2 = 45$
31. $7x^2 - 16y^2 = 112$ **32.** $2y^2 - 7x^2 = 28$
33. $5x^2 - 40x - 4y^2 - 24y + 24 = 0$
34. $3x^2 + 36x - y^2 - 8y + 89 = 0$
35. $3y^2 - 30y - x^2 - 6x + 54 = 0$
36. $16y^2 - 64y - 9x^2 + 126x - 521 = 0$
37. $5x^2 - 20x - 4y^2 = 0$
38. $16y^2 + 96y - 9x^2 = 0$ **39.** Circle
40. Hyperbola **41.** Straight line **42.** Ellipse
43. Ellipse **44.** Straight line **45.** Hyperbola
46. Circle **47.** Parabola **48.** Parabola

Problem Set 12.4 (page 719)

1. $\{(1, 2)\}$ **2.** $\{(2, 3)\}$ **3.** $\{(1, -5), (-5, 1)\}$
4. $\{(-3, 1), (1, -3)\}$
5. $\{(2 + i\sqrt{3}, -2 + i\sqrt{3}), (2 - i\sqrt{3}, -2 - i\sqrt{3})\}$
6. $\left\{\left(\frac{-5 + i\sqrt{19}}{2}, \frac{5 + i\sqrt{19}}{2}\right), \left(\frac{-5 - i\sqrt{19}}{2}, \frac{5 - i\sqrt{19}}{2}\right)\right\}$
7. $\{(-6, 7), (-2, -1)\}$ **8.** $\{(1, 2), (4, 5)\}$
9. $\{(-3, 4)\}$ **10.** $\{(2, -4)\}$
11. $\left\{\left(\frac{-1 + i\sqrt{3}}{2}, \frac{-7 - i\sqrt{3}}{2}\right), \left(\frac{-1 - i\sqrt{3}}{2}, \frac{-7 + i\sqrt{3}}{2}\right)\right\}$
12. $\left\{\left(\frac{1 + i\sqrt{3}}{2}, \frac{3 - i\sqrt{3}}{2}\right), \left(\frac{1 - i\sqrt{3}}{2}, \frac{3 + i\sqrt{3}}{2}\right)\right\}$
13. $\{(-1, 2)\}$ **14.** $\{(2, -3)\}$
15. $\{(-6, 3), (-2, -1)\}$ **16.** $\left\{(2, 1), \left(\frac{3}{2}, \frac{4}{3}\right)\right\}$
17. $\{(5, 3)\}$ **18.** $\left\{\left(-5, -\frac{3}{2}\right)\right\}$
19. $\{(1, 2), (-1, 2)\}$ **20.** $\{(1, 1)\}$ **21.** $\{(-3, 2)\}$
22. $\left\{\left(-\frac{\sqrt{6}}{2}, -\frac{1}{2}\right), \left(\frac{\sqrt{6}}{2}, -\frac{1}{2}\right)\right\}$
23. $\{(2, 0), (-2, 0)\}$ **24.** $\{(2, 0), (-2, 0)\}$
25. $\{(\sqrt{2}, \sqrt{3}), (\sqrt{2}, -\sqrt{3}), (-\sqrt{2}, \sqrt{3}), (-\sqrt{2}, -\sqrt{3})\}$
26. $\{(\sqrt{5}, 1), (\sqrt{5}, -1), (-\sqrt{5}, 1), (-\sqrt{5}, -1)\}$
27. $\{(1, 1), (1, -1), (-1, 1), (-1, -1)\}$
28. $\left\{\left(\frac{i\sqrt{3}}{2}, 2\right), \left(\frac{i\sqrt{3}}{2}, -2\right), \left(-\frac{i\sqrt{3}}{2}, 2\right), \left(-\frac{i\sqrt{3}}{2}, -2\right)\right\}$ **29.** $\left\{\left(2, \frac{3}{2}\right), \left(\frac{3}{2}, 2\right)\right\}$
30. $\left\{(3, 2), (-3, -2), \left(4, \frac{3}{2}\right), \left(-4, -\frac{3}{2}\right)\right\}$
31. $\{(9, -2)\}$ **32.** $\{(10, -1)\}$ **33.** $\{(\ln 2, 1)\}$
34. $\{(\ln 4, -16), (\ln 7, -49)\}$
35. $\left\{\left(\frac{1}{2}, \frac{1}{8}\right), (-3, -27)\right\}$ **36.** $\{(1, 4), (0, -5)\}$
42. $\{(2.2, 10.0)\}$ **43.** $\{(-2.3, 7.4)\}$
44. $\{(0.7, 2.6)\}$ **45.** $\{(6.7, 1.7), (9.5, 2.1)\}$
46. $\{(4.3, 2.5), (4.9, -0.7)\}$ **47.** None

Chapter 12 Review Problem Set (page 722)

1. $F(4, 0)$, $F'(-4, 0)$

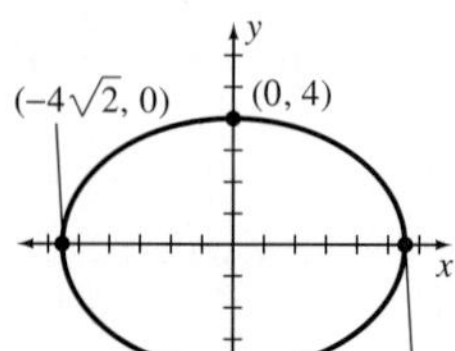

2. $F(-3, 0)$

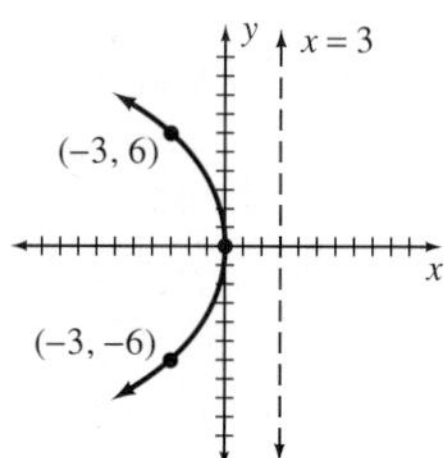

3. $F(0, 2\sqrt{3})$, $F'(0, -2\sqrt{3})$ $y = \pm\frac{\sqrt{3}}{3}x$

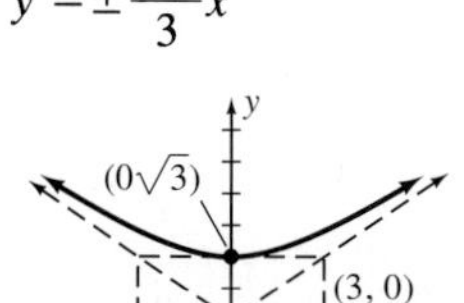

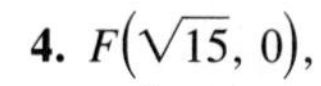

4. $F(\sqrt{15}, 0)$, $F'(-\sqrt{15}, 0)$ $y = \pm\frac{\sqrt{6}}{3}x$

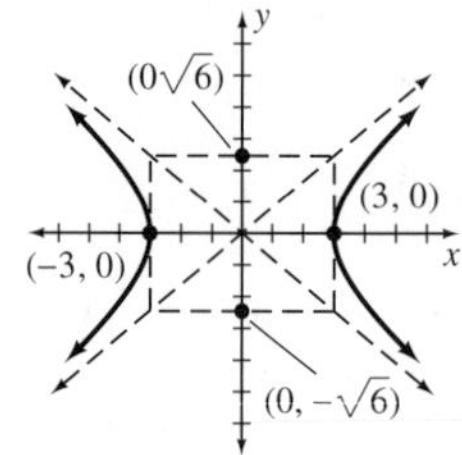

5. $F(0, \sqrt{6})$, $F'(0, -\sqrt{6})$

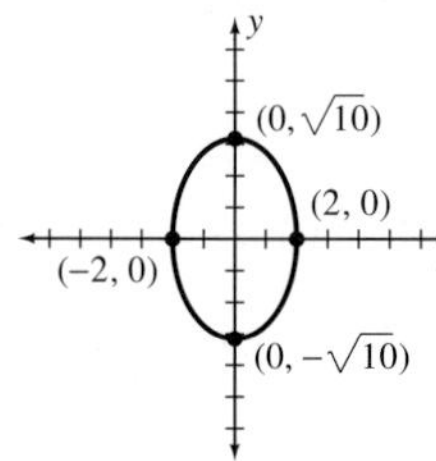

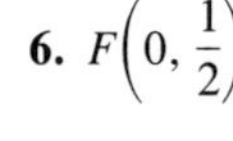

6. $F\left(0, \frac{1}{2}\right)$

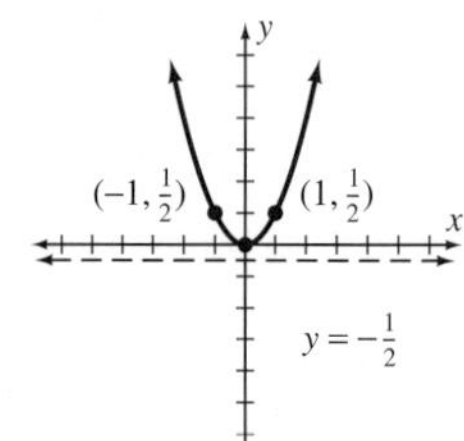

7. $F(4 + \sqrt{6}, 1)$, $F'(4 - \sqrt{6}, 1)$
$\sqrt{2}x - 2y = 4\sqrt{2} - 2$ and $\sqrt{2}x + 2y = 4\sqrt{2} + 2$

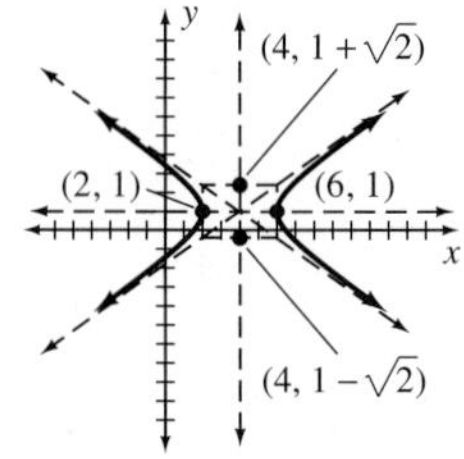

8. $F(3, -2 + \sqrt{7})$, $F'(3, -2 - \sqrt{7})$

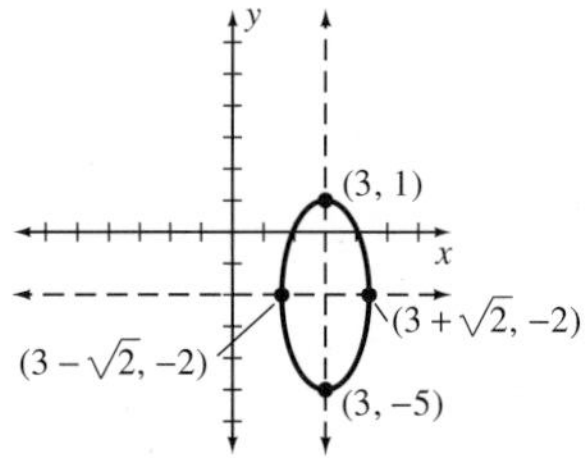

9. $F(-3, 1)$, $x = -1$

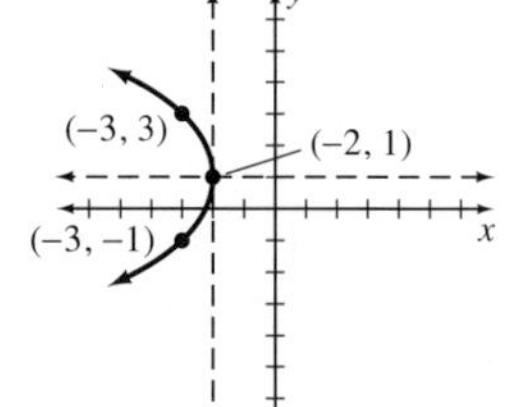

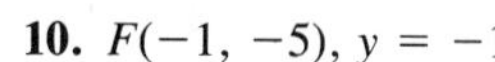

10. $F(-1, -5)$, $y = -1$

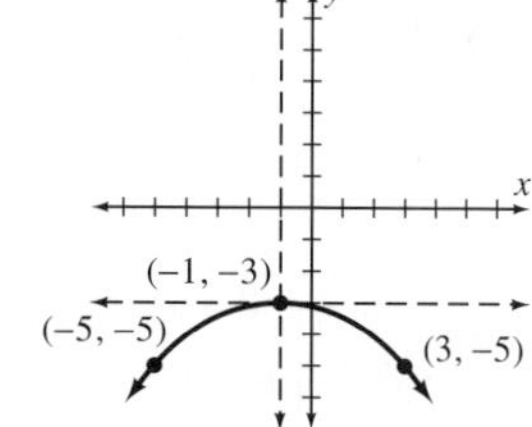

11. $F(-5 + 2\sqrt{3}, 2)$, $F'(-5 - 2\sqrt{3}, 2)$

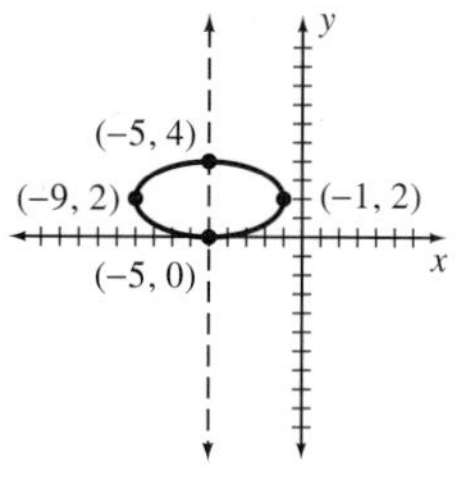

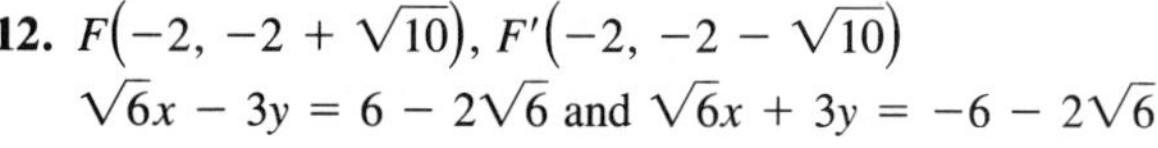

12. $F(-2, -2 + \sqrt{10})$, $F'(-2, -2 - \sqrt{10})$
$\sqrt{6}x - 3y = 6 - 2\sqrt{6}$ and $\sqrt{6}x + 3y = -6 - 2\sqrt{6}$

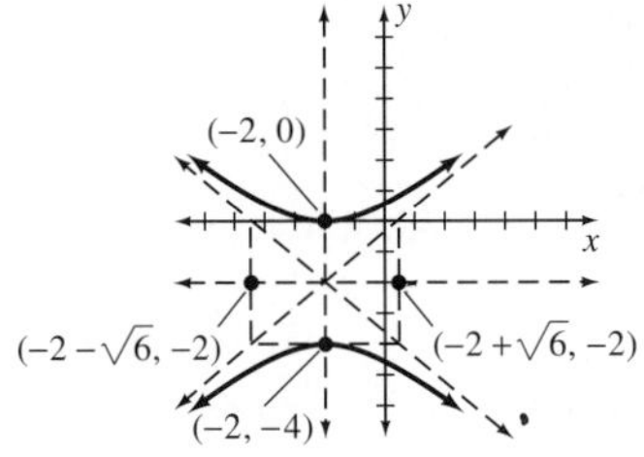

13. $y^2 = -20x$ **14.** $y^2 + 16x^2 = 16$
15. $25x^2 - 2y^2 = 50$ **16.** $4x^2 + 3y^2 = 16$
17. $3x^2 = 2y$ **18.** $9y^2 - x^2 = 9$
19. $9x^2 - 108x + y^2 - 8y + 331 = 0$
20. $y^2 + 4y - 8x + 36 = 0$
21. $3y^2 + 24y - x^2 - 10x + 20 = 0$
22. $x^2 + 12x - y + 33 = 0$
23. $4x^2 + 40x + 25y^2 = 0$
24. $4x^2 - 32x - y^2 + 48 = 0$ **25.** $\{(-1, 4)\}$
26. $\{(3, 1)\}$ **27.** $\{(-1, -2), (-2, -3)\}$
28. $\left\{\left(\frac{4\sqrt{2}}{3}, \frac{4}{3}i\right), \left(\frac{4\sqrt{2}}{3}, -\frac{4}{3}i\right), \left(-\frac{4\sqrt{2}}{3}, \frac{4}{3}i\right), \left(-\frac{4\sqrt{2}}{3}, -\frac{4}{3}i\right)\right\}$ **29.** $\{(0, 2), (0, -2)\}$
30. $\left\{\left(\frac{\sqrt{15}}{5}, \frac{2\sqrt{10}}{5}\right), \left(\frac{\sqrt{15}}{5}, -\frac{2\sqrt{10}}{5}\right), \left(-\frac{\sqrt{15}}{5}, \frac{2\sqrt{10}}{5}\right), \left(-\frac{\sqrt{15}}{5}, -\frac{2\sqrt{10}}{5}\right)\right\}$

CHAPTER 13

Problem Set 13.1 (page 732)

1. −4, −1, 2, 5, 8 **2.** 3, 8, 13, 18, 23
3. 2, 0, −2, −4, −6 **4.** 3, −1, −5, −9, −13
5. 2, 11, 26, 47, 74 **6.** −4, 2, 12, 26, 44
7. 0, 2, 6, 12, 20 **8.** 6, 12, 20, 30, 42
9. 4, 8, 16, 32, 64 **10.** 1, 3, 9, 27, 81
11. $a_{15} = -79$; $a_{30} = -154$
12. $a_{20} = -23$; $a_{50} = -53$ **13.** $a_{25} = 1$; $a_{50} = -1$
14. $a_{10} = -110$; $a_{15} = -235$ **15.** $a_n = 2n + 9$
16. $a_n = 3n + 4$ **17.** $a_n = -3n + 5$
18. $a_n = -2n + 6$ **19.** $a_n = \frac{n+2}{2}$
20. $a_n = \frac{n-1}{2}$ **21.** $a_n = 4n - 2$
22. $a_n = 5n - 3$ **23.** $a_n = -3n$ **24.** $a_n = -4n$
25. 73 **26.** 137 **27.** 334 **28.** 281
29. 35 **30.** 35 **31.** 7 **32.** −2 **33.** 86
34. −95 **35.** 2700 **36.** 870 **37.** 3200
38. 8730 **39.** −7950 **40.** −12,080
41. 637.5 **42.** $32.66.\overline{6}$ **43.** 4950
44. 16,040 **45.** 1850 **46.** 1485
47. −2030 **48.** −3654 **49.** 3591
50. 2914 **51.** 40,000 **52.** 30,800
53. 58,250 **54.** 36,036 **55.** 2205
56. 3000 **57.** −1325 **58.** −3255
59. 5265 **60.** 2451 **61.** −810 **62.** −2340
63. 1276 **64.** 4074 **65.** 660 **66.** −1800
67. 55 **68.** 97 **69.** 431 **70.** 370
71. 3, 3, 7, 7, 11, 11 **72.** 1, 4, $\frac{1}{3}$, 16, $\frac{1}{5}$, 36
73. 4, 7, 10, 13, 17, 21 **74.** 2, 4, 14, 8, 10, 29
75. 4, 12, 36, 108, 324, 972 **76.** 3, 5, 7, 9, 11, 13
77. 1, 1, 2, 3, 5, 8 **78.** 2, 3, 13, 45, 161, 573
79. 3, 1, 4, 9, 25, 256 **80.** 1, 2, 3, 6, 11, 20

Problem Set 13.2 (page 741)

1. $a_n = 3(2)^{n-1}$ **2.** $a_n = 2(3)^{n-1}$ **3.** $a_n = 3^n$
4. $a_n = 2^n$ **5.** $a_n = \left(\frac{1}{2}\right)^{n+1}$ **6.** $a_n = 2^{4-n}$
7. $a_n = 4^n$ **8.** $a_n = 6\left(\frac{1}{3}\right)^{n-1}$ **9.** $a_n = (0.3)^{n-1}$
10. $a_n = (0.2)^n$ **11.** $a_n = (-2)^{n-1}$ **12.** $(-3)^n$
13. 64 **14.** 1458 **15.** $\frac{1}{9}$ **16.** $\frac{3}{4}$
17. −512 **18.** $-\frac{2187}{128}$ **19.** $\frac{1}{4374}$ **20.** $\frac{81}{16}$
21. $\frac{2}{3}$ **22.** $\frac{1}{2}$ **23.** 2 **24.** $\frac{2}{3}$ **25.** 1023
26. 3279 **27.** 19,682 **28.** 5115
29. $394\frac{1}{16}$ **30.** $242\frac{69}{243}$ **31.** 1364
32. −9842 **33.** 1089 **34.** 10,922
35. $7\frac{511}{512}$ **36.** 171 **37.** −547 **38.** $31\frac{31}{32}$
39. $127\frac{3}{4}$ **40.** 1092 **41.** 540 **42.** −84
43. $2\frac{61}{64}$ **44.** $\frac{242}{243}$ **45.** 4 **46.** $\frac{27}{2}$
47. 3 **48.** $\frac{25}{2}$ **49.** No sum **50.** 64
51. $\frac{27}{4}$ **52.** No sum **53.** 2 **54.** 3
55. $\frac{16}{3}$ **56.** $\frac{35}{3}$ **57.** $\frac{1}{3}$ **58.** $\frac{4}{9}$ **59.** $\frac{26}{99}$
60. $\frac{18}{99}$ **61.** $\frac{41}{333}$ **62.** $\frac{91}{333}$ **63.** $\frac{4}{15}$
64. $\frac{13}{30}$ **65.** $\frac{106}{495}$ **66.** $\frac{184}{495}$ **67.** $\frac{7}{3}$
68. $\frac{34}{9}$

Problem Set 13.3 (page 746)

1. \$24,200 **2.** \$26,900 **3.** 11,550
4. 12,275 **5.** 7320 **6.** \$3932
7. 125 liters **8.** \$4.08 **9.** 512 gallons
10. 512 units **11.** \$116.25 **12.** \$4.96
13. \$163.84; \$327.67 **14.** \$409.60; \$819.15
15. \$24,900 **16.** \$14,352 **17.** 1936 feet
18. \$10,000 **19.** $\frac{15}{16}$ of a gram **20.** 3 grams
21. 2910 feet **22.** $298\frac{7}{16}$ feet **23.** 325 logs
24. \$1455 **25.** 5.9% **26.** $\frac{63}{64}$
27. $\frac{5}{64}$ of a gallon **28.** 5.2 gallons

Problem Set 13.4 (page 752)

These problems are proofs by mathematical induction and require class discussion.

Chapter 13 Review Problem Set (page 755)

1. $a_n = 6n - 3$ **2.** $a_n = 3^{n-2}$ **3.** $a_n = 5 \cdot 2^n$
4. $a_n = -3n + 8$ **5.** $a_n = 2n - 7$
6. $a_n = 3^{3-n}$ **7.** $a_n = -(-2)^{n-1}$
8. $a_n = 3n + 9$ **9.** $a_n = \frac{n+1}{3}$
10. $a_n = 4^{n-1}$ **11.** 73 **12.** 106 **13.** $\frac{1}{32}$
14. $\frac{4}{9}$ **15.** -92 **16.** $\frac{1}{16}$ **17.** -5
18. 85 **19.** $\frac{5}{9}$ **20.** 2 or -2 **21.** $121\frac{40}{81}$
22. 7035 **23.** $-10{,}725$ **24.** $31\frac{31}{32}$
25. 32,015 **26.** 4757 **27.** $85\frac{21}{64}$
28. 37,044 **29.** 12,726 **30.** -1845
31. 225 **32.** 255 **33.** 8244 **34.** $85\frac{1}{3}$
35. $\frac{4}{11}$ **36.** $\frac{41}{90}$ **37.** \$750 **38.** \$46.50
39. \$3276.70 **40.** 10,935 gallons

CHAPTER 14

Problem Set 14.1 (page 764)

1. 20 **2.** 120 **3.** 24 **4.** 42 **5.** 168
6. 840 **7.** 48 **8.** 12 **9.** 36 **10.** 16
11. 6840 **12.** 1350 **13.** 720
14. 3,276,000 **15.** 720 **16.** 240 **17.** 36
18. 480 **19.** 24 **20.** 125 **21.** 243
22. 360 **23.** Impossible **24.** 32 **25.** 216
26. 30 **27.** 26 **28.** 212 **29.** 36
30. 1560 **31.** 144 **32.** 288 **33.** 1024
34. 6 **35.** 30 **36.** 720 **37. (a)** 6,084,000
(b) 5,850,000 **(c)** 3,066,336 **(d)** 2,948,400

Problem Set 14.2 (page 772)

1. 60 **2.** 56 **3.** 360 **4.** 504 **5.** 21
6. 56 **7.** 252 **8.** 495 **9.** 105
10. 120 **11.** 1 **12.** 11 **13.** 24
14. 720 **15.** 84 **16.** 1326 **17. (a)** 336
(b) 512 **18.** 210 **19.** 2880 **20.** 432
21. 2450 **22.** 45 **23.** 10 **24.** 1440
25. 10 **26.** 65 **27.** 35 **28.** 28
29. 1260 **30.** 1260 **31.** 2520
32. 4,989,600 **33.** 15 **34.** 4200 **35.** 126
36. 70 **37.** 144; 202 **38.** 840 **39.** 15; 10
40. 20 **41.** 20 **42.** 91
43. 10; 15; 21; $\frac{n(n-1)}{2}$ **44. (a)** 24 **(b)** 24
(c) 3744 **45.** 120 **46. (a)** $\frac{7!}{4!}$ **(b)** $\frac{9!}{7!}$
(c) $\frac{10!}{3!}$ **(d)** $\frac{n!}{(n-r)!}$ **48.** 35; 28; 9
50. 2,598,960 **51.** 133,784,560 **52.** 2,118,760
53. 54,627,300 **54.** 4,155,186,750

Problem Set 14.3 (page 778)

1. $\frac{1}{2}$ **2.** $\frac{1}{4}$ **3.** $\frac{3}{4}$ **4.** $\frac{1}{4}$ **5.** $\frac{1}{8}$
6. $\frac{3}{8}$ **7.** $\frac{7}{8}$ **8.** $\frac{3}{8}$ **9.** $\frac{1}{16}$ **10.** $\frac{1}{4}$
11. $\frac{3}{8}$ **12.** $\frac{15}{16}$ **13.** $\frac{1}{3}$ **14.** $\frac{1}{2}$ **15.** $\frac{1}{2}$
16. 0 **17.** $\frac{5}{36}$ **18.** $\frac{1}{18}$ **19.** $\frac{1}{6}$ **20.** $\frac{5}{18}$
21. $\frac{11}{36}$ **22.** $\frac{5}{6}$ **23.** $\frac{1}{4}$ **24.** $\frac{1}{13}$ **25.** $\frac{1}{2}$

26. $\frac{1}{26}$ **27.** $\frac{1}{25}$ **28.** $\frac{12}{25}$ **29.** $\frac{9}{25}$
30. $\frac{4}{25}$ **31.** $\frac{2}{5}$ **32.** $\frac{1}{10}$ **33.** $\frac{9}{10}$ **34.** $\frac{3}{5}$
35. $\frac{5}{14}$ **36.** $\frac{3}{8}$ **37.** $\frac{15}{28}$ **38.** $\frac{3}{28}$ **39.** $\frac{7}{15}$
40. $\frac{7}{15}$ **41.** $\frac{1}{15}$ **42.** $\frac{1}{5}$ **43.** $\frac{2}{3}$ **44.** $\frac{1}{6}$
45. $\frac{1}{5}$ **46.** $\frac{1}{10}$ **47.** $\frac{1}{63}$ **48.** $\frac{1}{25}$ **49.** $\frac{1}{2}$
50. $\frac{1}{2}$ **51.** $\frac{5}{11}$ **52.** $\frac{20}{21}$ **53.** $\frac{1}{6}$ **54.** $\frac{35}{128}$
55. $\frac{21}{128}$ **56.** $\frac{11}{32}$ **57.** $\frac{13}{16}$ **58.** $\frac{1}{330}$
59. $\frac{1}{21}$ **60.** $\frac{1}{5}$ **61.** 40 **62.** 624
63. 3744 **64.** 5108 **65.** 10,200
66. 54,912 **67.** 123,552 **68.** 1,098,240
69. 1,302,540

Problem Set 14.4 (page 788)

1. $\frac{5}{36}$ **2.** $\frac{35}{36}$ **3.** $\frac{7}{12}$ **4.** 1 **5.** $\frac{1}{216}$
6. $\frac{53}{54}$ **7.** $\frac{53}{54}$ **8.** 0 **9.** $\frac{1}{16}$ **10.** $\frac{1}{4}$
11. $\frac{15}{16}$ **12.** $\frac{15}{16}$ **13.** $\frac{1}{32}$ **14.** $\frac{5}{32}$
15. $\frac{31}{32}$ **16.** $\frac{13}{16}$ **17.** $\frac{5}{6}$ **18.** $\frac{19}{20}$ **19.** $\frac{12}{13}$
20. $\frac{57}{64}$ **21.** $\frac{7}{12}$ **22.** $\frac{6}{7}$ **23.** $\frac{37}{44}$ **24.** $\frac{1}{2}$
25. $\frac{2}{3}$ **26.** $\frac{5}{6}$ **27.** $\frac{2}{3}$ **28.** $\frac{5}{18}$ **29.** $\frac{5}{18}$
30. $\frac{7}{36}$ **31.** $\frac{1}{3}$ **32.** $\frac{3}{4}$ **33.** $\frac{1}{2}$
34. **(a)** $\frac{13}{23}$ **(b)** $\frac{17}{23}$ **(c)** $\frac{16}{23}$ **35.** $\frac{7}{12}$
36. $\frac{8}{13}$ **37.** **(a)** 0.410 **(b)** 0.985
(c) 0.955 **38.** **(a)** 0.65 **(b)** 0.70 **(c)** 0.80
39. 0.525 **40.** 0.416 **41.** 60 **42.** 40
43. 120 **44.** 20 **45.** 9 **46.** 50
47. 56 **48.** \$.32 **49.** It is a fair game.
50. \$22,000 **51.** Yes **52.** No **53.** \$11,000
54. \$7200 **55.** −\$25 **56.** −\$80
57. 1 to 7 **58.** 15 to 1 **59.** 11 to 5
60. 15 to 49 **61.** 1 to 8 **62.** 5 to 13
63. 1 to 1 **64.** 2 to 11 **65.** 4 to 3
66. 4 to 5 **67.** 3 to 2 **68.** 1 to 4 **69.** $\frac{2}{7}$
70. $\frac{9}{29}$ **71.** $\frac{7}{12}$
72. 4164 to 1; 4159 to 6; 648,463 to 1277; 1269 to 5; 4077 to 88; 3967 to 198; 481 to 352; 1271 to 1277

Problem Set 14.5 (page 797)

1. $\frac{1}{3}$ **2.** $\frac{1}{3}; \frac{1}{3}$ **3.** $\frac{2}{15}$ **4.** $\frac{1}{6}$ **5.** $\frac{1}{3}$
6. $\frac{1}{2}$ **7.** $\frac{1}{6}$ **8.** $\frac{1}{7}$ **9.** $\frac{2}{3}; \frac{2}{7}$
10. $\frac{1}{17}; \frac{112}{115}$ **11.** $\frac{2}{3}; \frac{2}{5}$ **12.** $\frac{1}{8}; \frac{1}{4}$ **13.** $\frac{1}{15}; \frac{2}{7}$
14. $\frac{1}{40}$ **15.** Dependent **16.** Independent
17. Independent **18.** Dependent **19.** $\frac{1}{4}$
20. $\frac{5}{32}$ **21.** $\frac{1}{216}$ **22.** $\frac{1}{1728}$ **23.** $\frac{1}{221}$
24. $\frac{8}{663}$ **25.** $\frac{13}{102}$ **26.** $\frac{25}{102}$ **27.** $\frac{1}{16}$
28. $\frac{2}{169}$ **29.** $\frac{1}{1352}$ **30.** $\frac{1}{4}$ **31.** $\frac{2}{49}$
32. $\frac{5}{588}$ **33.** $\frac{25}{81}$ **34.** $\frac{16}{81}$ **35.** $\frac{20}{81}$
36. $\frac{65}{81}$ **37.** $\frac{25}{169}$ **38.** $\frac{16}{169}$ **39.** $\frac{32}{169}$
40. $\frac{40}{169}$ **41.** $\frac{2}{3}$ **42.** $\frac{1}{3}$ **43.** $\frac{1}{3}$ **44.** 0
45. $\frac{5}{68}$ **46.** $\frac{33}{68}$ **47.** $\frac{15}{34}$ **48.** $\frac{35}{68}$
49. $\frac{1}{12}$ **50.** $\frac{1}{3}$ **51.** $\frac{1}{6}$ **52.** $\frac{5}{12}$ **53.** $\frac{1}{729}$
54. $\frac{10}{81}$ **55.** $\frac{5}{27}$ **56.** $\frac{25}{243}$ **57.** $\frac{4}{35}$
58. $\frac{1}{35}$ **59.** $\frac{8}{35}$ **60.** 0 **61.** $\frac{4}{21}; \frac{2}{7}; \frac{11}{21}$
62. $\frac{7}{40}; \frac{33}{40}$

Problem Set 14.6 (page 803)

1. $x^8 + 8x^7y + 28x^6y^2 + 56x^5y^3 + 70x^4y^4 + 56x^3y^5 + 28x^2y^6 + 8xy^7 + y^8$

2. $x^9 + 9x^8y + 36x^7y^2 + 84x^6y^3 + 126x^5y^4 + 126x^4y^5 + 84x^3y^6 + 36x^2y^7 + 9xy^8 + y^9$
3. $x^6 - 6x^5y + 15x^4y^2 - 20x^3y^3 + 15x^2y^4 - 6xy^5 + y^6$
4. $x^4 - 4x^3y + 6x^2y^2 - 4xy^3 + y^4$
5. $a^4 + 8a^3b + 24a^2b^2 + 32ab^3 + 16b^4$
6. $81a^4 + 108a^3b + 54a^2b^2 + 12ab^3 + b^4$
7. $x^5 - 15x^4y + 90x^3y^2 - 270x^2y^3 + 405xy^4 - 243y^5$
8. $64x^6 - 192x^5y + 240x^4y^2 - 160x^3y^3 + 60x^2y^4 - 12xy^5 + y^6$
9. $16a^4 - 96a^3b + 216a^2b^2 - 216ab^3 + 81b^4$
10. $243a^5 - 810a^4b + 1080a^3b^2 - 720a^2b^3 + 240ab^4 - 32b^5$
11. $x^{10} + 5x^8y + 10x^6y^2 + 10x^4y^3 + 5x^2y^4 + y^5$
12. $x^6 + 6x^5y^3 + 15x^4y^6 + 20x^3y^9 + 15x^2y^{12} + 6xy^{15} + y^{18}$
13. $16x^8 - 32x^6y^2 + 24x^4y^4 - 8x^2y^6 + y^8$
14. $243x^{10} - 810x^8y^2 + 1080x^6y^4 - 720x^4y^6 + 240x^2y^8 - 32y^{10}$
15. $x^6 + 18x^5 + 135x^4 + 540x^3 + 1215x^2 + 1458x + 729$
16. $x^7 + 14x^6 + 84x^5 + 280x^4 + 560x^3 + 672x^2 + 448x + 128$
17. $x^9 - 9x^8 + 36x^7 - 84x^6 + 126x^5 - 126x^4 + 84x^3 - 36x^2 + 9x - 1$
18. $x^4 - 12x^3 + 54x^2 - 108x + 81$
19. $1 + \frac{4}{n} + \frac{6}{n^2} + \frac{4}{n^3} + \frac{1}{n^4}$
20. $32 + \frac{80}{n} + \frac{80}{n^2} + \frac{40}{n^3} + \frac{10}{n^4} + \frac{1}{n^5}$
21. $a^6 - \frac{6a^5}{n} + \frac{15a^4}{n^2} - \frac{20a^3}{n^3} + \frac{15a^2}{n^4} - \frac{6a}{n^5} + \frac{1}{n^6}$
22. $32a^5 - \frac{80a^4}{n} + \frac{80a^3}{n^2} - \frac{40a^2}{n^3} + \frac{10a}{n^4} - \frac{1}{n^5}$
23. $17 + 12\sqrt{2}$ 24. $26 + 15\sqrt{3}$
25. $843 - 589\sqrt{2}$ 26. $28 - 16\sqrt{3}$
27. $x^{12} + 12x^{11}y + 66x^{10}y^2 + 220x^9y^3$
28. $x^{15} + 15x^{14}y + 105x^{13}y^2 + 455x^{12}y^3$
29. $x^{20} - 20x^{19}y + 190x^{18}y^2 - 1140x^{17}y^3$
30. $a^{13} - 26a^{12}b + 312a^{11}b^2 - 2288a^{10}b^3$
31. $x^{28} - 28x^{26}y^3 + 364x^{24}y^6 - 2912x^{22}y^9$
32. $x^{33} - 33x^{30}y^2 + 495x^{27}y^4 - 4455x^{24}y^6$
33. $a^9 + \frac{9a^8}{n} + \frac{36a^7}{n^2} + \frac{84a^6}{n^3}$
34. $64 - \frac{192}{n} + \frac{240}{n^2} - \frac{160}{n^3}$
35. $x^{10} - 20x^9y + 180x^8y^2 - 960x^7y^3$
36. $a^{14} + 14a^{13}b + 91a^{12}b^2 + 546a^{11}b^3$ 37. $56x^5y^3$
38. $462x^5y^6$ 39. $126x^5y^4$ 40. $-160x^3y^3$
41. $189a^2b^5$ 42. $2000x^3y^2$ 43. $120x^6y^{21}$
44. $495a^4b^{24}$ 45. $\frac{5005}{n^6}$ 46. $-\frac{1716}{n^7}$
47. $41 - 38i$ 48. $-117 + 44i$
49. $-117 - 44i$ 50. $-597 - 122i$

Chapter 14 Review Problem Set (page 806)

1. 720 2. 30,240 3. 150 4. 1440
5. 20 6. 525 7. 1287 8. 264
9. 74 10. 55 11. 40 12. 15
13. 60 14. 120 15. $\frac{3}{8}$ 16. $\frac{5}{16}$
17. $\frac{5}{36}$ 18. $\frac{13}{18}$ 19. $\frac{3}{5}$ 20. $\frac{1}{35}$ 21. $\frac{57}{64}$
22. $\frac{1}{221}$ 23. $\frac{1}{6}$ 24. $\frac{4}{7}$ 25. $\frac{4}{7}$ 26. $\frac{10}{21}$
27. $\frac{140}{143}$ 28. $\frac{105}{169}$ 29. $\frac{1}{6}$ 30. $\frac{28}{55}$
31. $\frac{5}{7}$ 32. $\frac{1}{16}$ 33. $\frac{1}{2}; \frac{1}{3}$ 34. (a) $\frac{9}{19}$
(b) $\frac{9}{10}$ 35. (a) $\frac{2}{7}$ (b) $\frac{4}{9}$
36. $x^5 + 10x^4y + 40x^3y^2 + 80x^2y^3 + 80xy^4 + 32y^5$
37. $x^8 - 8x^7y + 28x^6y^2 - 56x^5y^3 + 70x^4y^4 - 56x^3y^5 + 28x^2y^6 - 8xy^7 + y^8$
38. $a^8 - 12a^6b^3 + 54a^4b^6 - 108a^2b^9 + 81b^{12}$
39. $x^6 + \frac{6x^5}{n} + \frac{15x^4}{n^2} + \frac{20x^3}{n^3} + \frac{15x^2}{n^4} + \frac{6x}{n^5} + \frac{1}{n^6}$
40. $41 - 29\sqrt{2}$ 41. $-a^3 + 3a^2b - 3ab^2 + b^3$
42. $-1760x^9y^3$ 43. $57915a^4b^{18}$

INDEX